Geology of México: Celebrating the Centenary of the Geological Society of México

edited by
Susana A. Alaniz-Álvarez
Centro de Geociencias
Universidad Nacional Autónoma de México
Apartado postal 1-742
Querétaro, Qro. 76230
México

and

Ángel F. Nieto-Samaniego
Centro de Geociencias
Universidad Nacional Autónoma de México
Apartado postal 1-742
Querétaro, Qro. 76230
México

THE
GEOLOGICAL
SOCIETY
OF AMERICA®

Special Paper 422

3300 Penrose Place, P.O. Box 9140 ▪ Boulder, Colorado 80301-9140 USA

2007

Published by The Geological Society of America, Inc.
3300 Penrose Place, P.O. Box 9140, Boulder, Colorado 80301-9140, USA
www.geosociety.org

Printed in U.S.A.

GSA Books Science Editors: Marion E. Bickford and Abhijit Basu

Papers in this volume were published (in Spanish) by the Sociedad Geológica Mexicana in the following issues of the *Boletín de la Sociedad Geológica Mexicana, Volumen Conmemorativo del Centenario* (ISSN 1405-3322).

Volume 57 (2005), no. 1, *Grandes Fronteras Tectónicas de México*, and no. 2, *Aspectos Históricos de la Geología en México*, edited by Susana A. Álaniz-Alvarez and Ángel F. Nieto-Samaniego; and no. 3, *Temas Selectos de la Geología Mexicana*, edited by Ángel F. Nieto-Samaniego and Susana A. Álaniz-Alvarez

Volume 58 (2006), no. 1, *Revisión de algunas tipologías de depósitos minerales de México*, edited by Jordi Tritlla and Antoni Camprubí

Library of Congress Cataloging-in-Publication Data

Geology of México : celebrating the centenary of the Geological Society of México / edited by Susana A. Alaniz-Álvarez and Ángel F. Nieto-Samaniego.
p. cm. — (Special paper ; 422)
Includes bibliographical references and index.
ISBN-13 978-0-8137-2422-5 (pbk.)
1. Geology—Mexico. 2. Geology, Structural—Mexico. 3. Faults (Geology)—Mexico. 4. Ore deposits—Mexico. I. Alaniz-Álvarez, Susana A., 1958–. II. Nieto-Samaniego, Ángel F., 1961–. III. Sociedad Geológica Mexicana.

QE201 .G44 2007
557.2—dc22

20060525972

Cover: Panoramic view of the southern flank of Popocatépetl, the second highest peak in México. The snow cover reaches up to ~4000 meters above sea level. In the foreground is the surface of a debris avalanche deposit with hummocky topography. Photograph by José Luis Macías taken in January 2004.

10 9 8 7 6 5 4 3 2 1

Contents

Preface

In 2004, the Sociedad Geológica Mexicana (Geological Society of México) celebrated its first centenary. The Society's *Boletín*, despite two interruptions during politically troubled times, continues to be active. To celebrate the centenary, the Society published three special volumes whose main purpose was to bring together a collection of papers that present the state of the art in some areas of Mexican geology. Considering that a broader audience would be interested in the papers, Gustavo Tolson, then president of the society, proposed to the Geological Society of America that an English-language version of the centenary volumes be published as a Special Paper. The wide variety of geological settings preserved in the rocks of México, representing 1.8 billion years of geologic history, provides us with unique opportunities to study a variety of geologic processes. The focus of this Special Paper is three subjects of Mexican geology: reviews of some of the geological provinces, major faults that constitute tectonic borders, and ore deposits.

The first group of papers is a collection of articles reviewing different aspects of the geology of México that presents a thorough reflection of the knowledge gathered so far on some of the important geological provinces. The authors were asked to analyze and confront data from the literature in order to assess the state of the current hypotheses. In each case, however, the authors were given free rein to include the information and ideas they considered pertinent.

One of the most extensive geologic provinces is the Sierra Madre Occidental, which consists of volcanic complexes that resulted from tectono-magmatic events during the Cretaceous and Cenozoic along the western North American plate margin in México. It is this province that is the subject of the paper by Ferrari et al. This article presents, for the first time, a comprehensive view of the province, including its structural and stratigraphic characteristics, which are presented by means of up-to-date structural and geologic maps; much of the information contained in the maps was compiled from previous literature. An important portion of the paper is devoted to an analysis of the magmatic and tectonic evolution of the province as well as a discussion of the age-old debate on the origin of the magmas, contrasting two extremes: differentiation of mantle-derived magmas versus partial fusion of the continental crust. Finally, Ferrari et al. put forth their hypothesis to explain the magmatic pulses that characterize the Sierra Madre Occidental and that make it one of the largest silicic volcanic provinces of the world.

The Mesa Central (Mexican Plateau) is a geologic province whose stratigraphy and structure are very different in the Cenozoic from the Mesozoic. The Mesozoic geologic record has been the subject of much research during the last century, while its Cenozoic counterpart has not. The Cenozoic stratigraphic framework, structural geology, and petrology of the plateau have been studied formally only since 1980. It is precisely the stratigraphy and structural evolution of the Cenozoic in Mesa Central that is the focus of the Nieto-Samaniego et al. paper, in which they review the available information and propose the crustal structure of this geologic province, taking into account the tectonics of the oceanic margin of western México. The authors also consider the mineral deposits of the region, indicating their spatial and temporal association with two major structures which bound the plateau: the San Luis–Tepehuanes fault and the Taxco–San Miguel de Allende fault system. The paper by Nieto-Samaniego et al. fills an important, and until now empty, niche in the literature of Mexican geology.

Morán-Zenteno et al. address the tectonic and magmatic evolution of southwestern México, topics which are currently the subject of much debate and lively scientific research. The authors establish quite clearly that southern México is a complex region with a stratigraphic record that begins in the Proterozoic and whose Cenozoic evolution, in addition to having its own degree of complexity, is strongly influenced by the marked heterogeneity of the preexisting rocks. Morán-Zenteno et al. argue that Laramide deformation toward the end of the Mesozoic could not have been the result of a shallow subduction angle or the product of an island arc collision along the margin of southwest México. They conclude that an explanation for the time-space distribution of magmatic products since the Cretaceous is still pending. The fundamental questions posed by this paper and the thorough assessment of the different tectonic models proposed to date let the reader feel the intensity of the ongoing debate about the geology of this part of México.

The paper by Aranda-Gómez et al. discusses intraplate volcanism during the Cenozoic, represented by rocks amply distributed around the country and, paradoxically, one of the least studied rock types. The authors show that the relationship between these rocks and faults bounding tectonostratigraphic units or geological provinces is unclear. They observe systematic variations in chemistry with age, since the older rocks (late Oligocene to Miocene) exhibit greater degrees of differentiation and contamination while the younger samples (of Pliocene or Quaternary age) are more primitive with little or no evidence of assimilation. Aranda-Gómez et al. attribute this chemical variation to the thermal evolution of the crust, which cooled steadily between the Oligocene and the Quaternary, affecting the depth of the brittle-ductile transition. They propose that the depth to which brittle structures extend below the surface in an extensional setting is a factor that significantly influences the rate of magma ascent. This paper by Aranda-Gómez et al. is, in fact, the first comprehensive study of this type of magmatism in México and represents without a doubt the state-of-the-art research on the subject. It will surely spark the development of research initiatives not only on the petrogenetic aspects of these rocks, but also on their relationship with tectonics.

The Trans-Mexican Volcanic Belt is the most studied geologic province of México and the paper by Gómez-Tuena et al. addresses one of its most controversial issues: its petrogenesis. The authors present an extensive treatise in which they succinctly review the copious literature and discuss the different petrogenetic models that have been proposed. They do not avoid the numerous aspects related to the evolution of this volcanic arc, taking into account its structural, stratigraphic, and tectonic peculiarities. The paper presents a clear picture of this geologic province; its authors consider that the petrological diversity of the arc reflects the complexity of the different factors that intervene in its evolution. An additional noteworthy contribution of this paper is a geologic map of the Trans-Mexican Volcanic Belt which is accompanied by an online database that includes major-element, trace-element, and isotopic geochemical data for 2600 samples. The contribution by Gómez-Tuena et al. leaves the reader with a clear impression of the enormous amount of work that has been invested in understanding this volcanic arc, of the great advances in our understanding of it, and of the still considerable number of questions yet to be answered. This paper will be of interest to all scientists interested in this and other volcanic arcs.

The last of this group of papers, by Macías, reviews the geology and eruptive history of several of the great volcanoes of México. By its very nature, the paper contains a great deal of up-to-date historical and geological information. Data on Colima, Nevado de Toluca, Popocatépetl, Citlaltepetl, Tacaná, and Chichón volcanoes are presented in an orderly and systematic manner, setting a benchmark by way of analysis, discussion, illustrations, and bibliography, which will be difficult to supersede in the near future. Since all the volcanoes discussed represent a vested interest to the population at large, owing to their activity, we have no doubt that this article will be of great interest to geologists, volcanologists, engineers, and other earth scientists, as well as nonspecialists and those in the general public interested in knowing more about the volcanoes of México.

The second part of the book consists of a series of papers related to the development of "tectonic boundaries": regional-scale shear zones with distinct periods of activity covering an important span of geological time that separate crustal blocks with different geologic histories. This has been an important topic in Mexican geology since several proposals of major tectonic structures (*megashears*) were put forth in the 1980s to explain the opening of the Gulf of México, the southward displacement of the Yucatan peninsula, and the displacement of crustal blocks to accommodate southern México in its present position (e.g., Anderson and Schmidt, 1983). On the other hand, during the same period of time, the concept of *tectonostratigraphic terranes* was introduced, which required faults or shear zones to bound them; the broad application of this methodology in México also led to the definition of major tectonic boundaries, some mapped and some hypothetical (Campa and Coney, 1983; Sedlock et al., 1993). The systematic study of these tectonic boundaries from a geological point of view is still incipient in México. For the centenary volume, contributors were invited who have carried out detailed research on México's major faults. Included are papers about the Mojave-Sonora megashear (Molina-Garza and Iriondo), the San Marcos fault (Chávez-Cabello et al.), and the Trans-Mexican Volcanic Belt (Alaniz-Álvarez and Nieto-Samaniego). This volume includes articles on tectonostratigraphic terrane boundaries on the Chacalapa fault as the northern limit of the Xolapa terrane (Tolson), the Caltepec fault (Elías-Herrera et al.), between the Mixteco and Zapoteco terranes (using the nomenclature of Sedlock et al., 1993) as well as the Taxco–San Miguel Allende fault system (Alaniz-Álvarez and Nieto-Samaniego), which defines part of the eastern limit of the Guerrero terrane. Also included is a paper about the Tosco-Abreojos fault (Michaud et al.), a structure in the transition zone between the North American plate and the Pacific plate.

The examples in this volume represent a spectrum of structures sampled at different depths. Thus, in the Caltepec and Chacalapa shear zones are exposed rocks formed at 15 km depth by crystal-plastic recrystallization processes (mylonites). Toward the edges of these zones, brittle-ductile and strictly brittle structures are found, which allowed the deeper (inner) portions of the shear systems to be exposed. In the other cases, the strictly brittle, upper crustal regime is represented. The study of tectonic boundaries also allows inferences to be made concerning the relationship between faulting and magmatism at different depths. According to the history reported for the Chacalapa, Caltepec, and San Marcos faults and the Taxco–San Miguel fault system, the main activity of these shear zones is related to magmatism and, in the case of the Caltepec shear zone, the granitic intrusives are shown to be synkinematic and contemporaneous with migmatization of the host rocks.

The origin of a tectonic boundary is difficult to establish; however, the data garnered from the Mexican examples presented in this volume provide us with insight into this problem. The displacement of blocks of crustal dimensions, or with contrasting bathymetries, along some of the faults discussed in this volume, suggests a first-order control of paleogeographic domain limits on tectonic boundaries. See, for example, the case studies of the Taxco–San Miguel Allende system or the Mojave-Sonora megashear.

The papers of this *Centenary* volume document the age and kinematics of the different periods of activity of the major faults analyzed, and, in the majority of cases, the reactivation of the structures occurred under stress regimes quite different from the original stress configuration. Based on detailed knowledge of the periods of activity of the structures, the tectonic models associated with each structure must be reassessed, paying particular attention to those cases in which the proposed model is incompatible with the documented timing and kinematics of displacement along the faults or shear zones in question.

The last group of articles deals with the characterization of some of the ore deposits of México. México is host to a plethora of world-class mineral deposits formed in different periods of geologic time and a broad variety of tectonic settings. They constitute great mineral-deposit belts which extend over the entire country. Given their diversity and their ample distribution in space and time, the ore deposits of México present fascinating case studies.

Canet and Prol-Ledesma describe mineral deposits and shallow (<200 m) submarine hot springs in island arc and continental margin settings. Deep hot-spring and submarine hydrothermal deposits comprise one of the most studied types of mineral deposits, while their shallow counterparts are little known. This alone sets the paper by Canet and Prol-Ledesma apart.

Camprubí and Albinson present a thorough review of the epithermal deposits of México, examples of which have a mining history going back to pre-Columbian times. The authors define three main epithermal deposit types: low sulfuration, intermediate sulfuration, and mixed low/intermediate sulfuration. Thus, they confront the status quo which states that low and intermediate sulfuration hydrothermal systems are mutually exclusive. They also describe details of the argillic alteration haloes typical of high sulfuration systems at depth within neutral or moderately alkaline epithermal deposits without a connection with similar shallow-level alteration zones; this the authors interpret as early incursions of high sulfuration fluids in systems of neutral or alkaline pH.

Tritlla et al. present a review of epigenetic stratabound deposits of the Mississippi Valley type and related deposits in México. Previous authors have generally assigned these deposits to different settings without proper attention to the distinctive geological and mineralogical features that characterize them correctly, as Tritlla et al. point out in their paper. This paper presents the first thorough description of Mississippi Valley–type deposits in México.

Valencia-Moreno et al. present a review of Mexican Cu-Au-Mo porphyry deposits, which define a broad mineral belt extending from the southwest United States to northwest México. The spatial distribution of Cu, Cu-Au, Cu-Mo, Mo, etc., deposits appears to be related to the distribution of the regional basement, inasmuch as three domains can be distinguished, at least in México: (1) the northern domain with a Proterozoic basement, (2) a central domain with a deep marine Paleozoic basin underlain by Proterozoic basement, and (3) a southern domain with Mesozoic island arcs.

Finally, we would like to use these last few lines to thank all those people who were involved in the realization of this *Centenary* volume of the Sociedad Geológica Mexicana: the authors, reviewers, and technical editors, without whose selfless dedication the publication of this work could not have been possible.

Susana A. Alaniz-Álvarez
Ángel F. Nieto-Samaniego

REFERENCES CITED

Anderson, T.H., and Schmidt, V.A., 1983, The evolution of Middle America and the Gulf of Mexico-Caribbean Sea region during Mesozoic time: Geological Society of America Bulletin, v. 94, p. 941–966.
Campa, M.F., and Coney, P.J., 1983, Tectono-stratigraphic terranes and mineral resource distributions of México: Canadian Journal of Earth Sciences, v. 20, p. 1040–1051.
Sedlock, R.L., Ortega-Gutiérrez, F., and Speed, R.C., 1993, Tectonostratigraphic Terranes and Tectonic Evolution of Mexico: Geological Society of America Special Paper 278, 153 p.

Geological Society of America
Special Paper 422
2007

Magmatism and tectonics of the Sierra Madre Occidental and its relation with the evolution of the western margin of North America

Luca Ferrari*
Centro de Geociencias, Universidad Nacional Autónoma de México, Campus Juriquilla, Querétaro, Qro., 76230 México

Martín Valencia-Moreno*
Estación Regional del Noroeste, Instituto de Geología, Universidad Nacional Autónoma de México, Hermosillo, Son. México

Scott Bryan*
Department of Geology and Geophysics, Yale University, New Haven, Connecticut 06520-8109, USA

ABSTRACT

The Sierra Madre Occidental is the result of Cretaceous-Cenozoic magmatic and tectonic episodes related to the subduction of the Farallon plate beneath North America and to the opening of the Gulf of California. The stratigraphy of the Sierra Madre Occidental consists of five main igneous complexes: (1) Late Cretaceous to Paleocene plutonic and volcanic rocks; (2) Eocene andesites and lesser rhyolites, traditionally grouped into the so-called Lower Volcanic Complex; (3) silicic ignimbrites mainly emplaced during two pulses in the Oligocene (ca. 32–28 Ma) and Early Miocene (ca. 24–20 Ma), and grouped into the "Upper Volcanic Supergroup"; (4) transitional basaltic-andesitic lavas that erupted toward the end of, and after, each ignimbrite pulse, which have been correlated with the Southern Cordillera Basaltic Andesite Province of the southwestern United States; and (5) postsubduction volcanism consisting of alkaline basalts and ignimbrites emplaced in the Late Miocene, Pliocene, and Pleistocene, directly related to the separation of Baja California from the Mexican mainland. The products of all these magmatic episodes, partially overlapping in space and time, cover a poorly exposed, heterogeneous basement with Precambrian to Paleozoic ages in the northern part (Sonora and Chihuahua) and Mesozoic ages beneath the rest of the Sierra Madre Occidental.

The oldest intrusive rocks of the Lower Volcanic Complex (ca. 101 to ca. 89 Ma) in Sinaloa, and Maastrichtian volcanics of the Lower Volcanic Complex in central Chihuahua, were affected by moderate contractile deformation during the Laramide orogeny. In the final stages of this deformation cycle, during the Paleocene and Early Eocene, ~E-W to ENE-WSW–trending extensional structures formed within the

*E-mails, Ferrari and Valencia-Moreno: luca@geociencias.unam.mx; valencia@geologia.unam.mx. Current address, Bryan: School of Earth Sciences & Geography, Kingston University, Kingston Upon Thames, Surrey KT1 2EE, UK; s.bryan@kingston.ac.uk.

Ferrari, L., Valencia-Moreno, M., and Bryan, S., 2007, Magmatism and tectonics of the Sierra Madre Occidental and its relation with the evolution of the western margin of North America, *in* Alaniz-Álvarez, S.A., and Nieto-Samaniego, Á.F., eds., Geology of México: Celebrating the Centenary of the Geological Society of México: Geological Society of America Special Paper 422, p. 1–39, doi: 10.1130/2007.2422(01). For permission to copy, contact editing@geosociety.org.

Lower Volcanic Complex, along which the world-class porphyry copper deposits of the Sierra Madre Occidental were emplaced. Extensional tectonics began as early as the Oligocene along the entire eastern half of the Sierra Madre Occidental, forming grabens bounded by high-angle normal faults, which have traditionally been referred to as the southern (or Mexican) Basin and Range Province. In the Early to Middle Miocene, extension migrated westward. In northern Sonora, the deformation was sufficiently intense to exhume lower crustal rocks, whereas in the rest of the Sierra Madre Occidental, crustal extension did not exceed 20%. By the Late Miocene, extension became focused in the westernmost part of the Sierra Madre Occidental, adjacent to the Gulf of California, where NNW-striking normal fault systems produced both ENE and WSW tilt domains separated by transverse accommodation zones. It is worth noting that most of the extension occurred when subduction of the Farallon plate was still active off Baja California.

Geochemical data show that the Sierra Madre Occidental rocks form a typical calc-alkaline rhyolite suite with intermediate to high K and relatively low Fe contents. Late Eocene to Miocene volcanism is clearly bimodal, but silicic compositions are volumetrically dominant. Initial $^{87}Sr/^{86}Sr$ ratios mostly range between 0.7041 and 0.7070, and initial εNd values are generally intermediate between crust and mantle values (+2.3 and −3.2). Based on isotopic data of volcanic rocks and crustal xenoliths from a few sites in the Sierra Madre Occidental, contrasting models for the genesis of the silicic volcanism have been proposed. A considerable body of work led by Ken Cameron and others considered the mid-Tertiary Sierra Madre Occidental silicic magmas to have formed by fractional crystallization of mantle-derived mafic magmas with little (<15%) or no crustal involvement. In contrast, other workers have suggested the rhyolites, taken to the extreme case, could be entirely the result of partial melting of the crust in response to thermal and material input from basaltic underplating. Several lines of evidence suggest that Sierra Madre Occidental ignimbrite petrogenesis involved large-scale mixing and assimilation-fractional crystallization processes of crustal and mantle-derived melts.

Geophysical data indicate that the crust in the unextended core of the northern Sierra Madre Occidental is ~55 km thick, but thins to ~40 km to the east. The anomalous thickness in the core of the Sierra Madre Occidental suggests that the lower crust was largely intruded by mafic magmas. In the westernmost Sierra Madre Occidental adjacent to the Gulf of California, crustal thickness is ~25 km, implying over 100% of extension. However, structures at the surface indicate no more than ~50% extension. The upper mantle beneath the Sierra Madre Occidental is characterized by a low-velocity anomaly, typical of the asthenosphere, which also occurs beneath the Basin and Range Province of the western United States.

The review of the magmatic and tectonic history presented in this work suggests that the Sierra Madre Occidental has been strongly influenced by the Cretaceous-Cenozoic evolution of the western North America subduction system. In particular, the Oligo-Miocene Sierra Madre Occidental is viewed as a silicic large igneous province formed as the precursor to the opening of the Gulf of California during and immediately following the final stages of the subduction of the Farallon plate. The mechanism responsible for the generation of the ignimbrite pulses seems related to the removal of the Farallon plate from the base of the North American plate after the end of the Laramide orogeny. The rapid increase in the subduction angle due to slab roll-back and, possibly, the detachment of the deeper part of the subducted slab as younger and buoyant oceanic lithosphere arrived at the paleotrench, resulted in extension of the continental margin, eventually leading to direct interaction between the Pacific and North American plates.

Keywords: Sierra Madre Occidental, Gulf of California, continental magmatism, silicic large igneous province, extensional tectonics, subduction dynamics.

1. INTRODUCTION

Large volcano-plutonic belts dominated by rhyolitic and/or granitic rocks are common features of most continents. They typically display rock volumes between 10^5 to $>10^6$ km^3, and were emplaced in time periods of 10–40 millions of years. These belts have been referred to as "silicic large igneous provinces" by Bryan et al. (2002) (Table 1). The silicic magmatic event that produced these large provinces is not a common phenomenon in the geologic record, and must be associated with global tectonic processes. This type of magmatism strongly contributes to the modification of the rheologic structure and the composition of the continental lithosphere, as well as to the generation of a variety of ore deposits. Additionally, the concentration of large explosive eruptions over a relatively short time span may have significant impact on the global climate. The main pulses of ignimbritic volcanism of the Sierra Madre Occidental overlap with a global low-temperature paleoclimatic event at the Eocene-Oligocene boundary, as well as with a cooling event of shorter duration in the Early Miocene (Cather et al., 2003).

The Sierra Madre Occidental (Fig. 1) is one of the larger silicic igneous provinces on Earth, and is the largest of the Cenozoic era (Table 1). The ignimbritic carpet of the Sierra Madre Occidental covers an area of ~300,000 km^2 (McDowell and Keizer, 1977; McDowell and Clabaugh, 1979; Ward, 1995) with an upper estimate of ~393,000 km^2 (Aguirre-Díaz and Labarthe, 2003). However, intracontinental Basin and Range–type extension and the opening of the Gulf of California have obscured a significant part of the original size of this province, which may have been significantly larger. Although the ignimbritic cover is the most obvious feature of the Sierra Madre Occidental, the "Lower Volcanic Complex" (McDowell and Keizer, 1977) that underlies it is also important. The Lower Volcanic Complex comprises Cretaceous to Paleocene plutonic and volcanic rocks that are similar in age and composition to the Peninsular Ranges batholith and the Jalisco block, and is the main host to world-class silver and copper ore deposits (e.g., Fresnillo and Juanicipio, Zacatecas; Cananea and Nacozari, Sonora; see also Staude and Barton, 2001; Damon et al., 1983). The location and age of the Lower Volcanic Complex, interpreted to be the remnants of the suprasubduction zone magmatic arc, are critical to understand the Laramide orogeny that affected the western margin of the North American continent.

Given the scientific and economic importance of the Sierra Madre Occidental, the available literature concerning this major geologic feature of México is still relatively limited. Despite the passage of ~30 years since the first formal publications on the Sierra Madre Occidental, detailed geological knowledge is mostly restricted to relatively small areas accessible by roads, or has been based largely on remote sensing studies. Geochemical studies have been from small areas, often with results extrapolated to encompass the entire province. In contrast, geophysical studies are essentially at regional or continental scale. Consequently, many problems on the origin and evolution of the Sierra Madre Occidental remain, and are open to scientific discussion. There is no consensus, for example, regarding the mechanism(s) that produced this huge magmatic pulse and, in particular, how much the continental crust contributed to the generation of the silicic magmas. In addition, the causes of Cenozoic extension and its relationship to the final stages of subduction are little understood. Thus, the Sierra Madre Occidental in many ways is an important frontier in the geologic knowledge of México.

This work intends to summarize the current state of geologic knowledge of the Sierra Madre Occidental, particularly focusing on the evolution of the Cretaceous-Cenozoic magmatism and associated tectonics. In the first part, we present a synthesis of the stratigraphy based on the available geochronologic data, as well as a summary of the geometry, kinematics, and age of the faulting that have affected the Sierra Madre Occidental. We then summarize the available geophysical data, and geochemical and petrological studies. Finally, we discuss the time-space evolution of the magmatism and tectonics, the models to explain the origin of the silicic volcanism, and the problems that still remain unsolved to better understand this volcanic province. To support the present work, we

TABLE 1. CATALOGUE OF LARGE SILICIC IGNEOUS PROVINCES ORDERED IN TERMS OF MAXIMUM EXTRUSIVE VOLUMES

Province	Age (Ma)	Volume (km^3)	Size (km)	Magma flux (km^3 kyr^{-1})*	Reference
Whitsunday (Eastern Australia)	ca. 132–95	>2.2 × 10^6	>2500 × 200	>55	Bryan et al. (1997, 2000); Bryan (2005b)
Kennedy-Connors-Auburn (Northeast Australia)	ca. 320–280	>5 × 10^5	>1900 × 300	>12.5	Bain and Draper (1997); Bryan et al. (2003)
Sierra Madre Occidental	ca. 38–20	>3.9 × 10^5	>2000 × 200 to 500	>22	McDowell and Clabaugh (1979); Ferrari et al. (2002)
Chon Aike (South America–Antarctica)	188–153	>2.3 × 10^5	>3000 × 1000	>7.1	Pankhurst et al. (1998, 2000)
Altiplano–Puna (Central Andes)	ca. 10–3	>3 × 10^4	~300 × 200	>4.3	De Silva (1989a, 1989b)
Taupo Volcanic Zone (New Zealand)	1.6–0	~2 × 10^4	300 × 60	~9.4–13	Wilson et al. (1995); Houghton et al. (1995)

Note: From Bryan (2005a). All provinces are dominated by rhyolitic igneous compositions and ignimbrite.
*Magma flux rate is averaged eruptive flux, based on known extrusive volumes for the provinces.

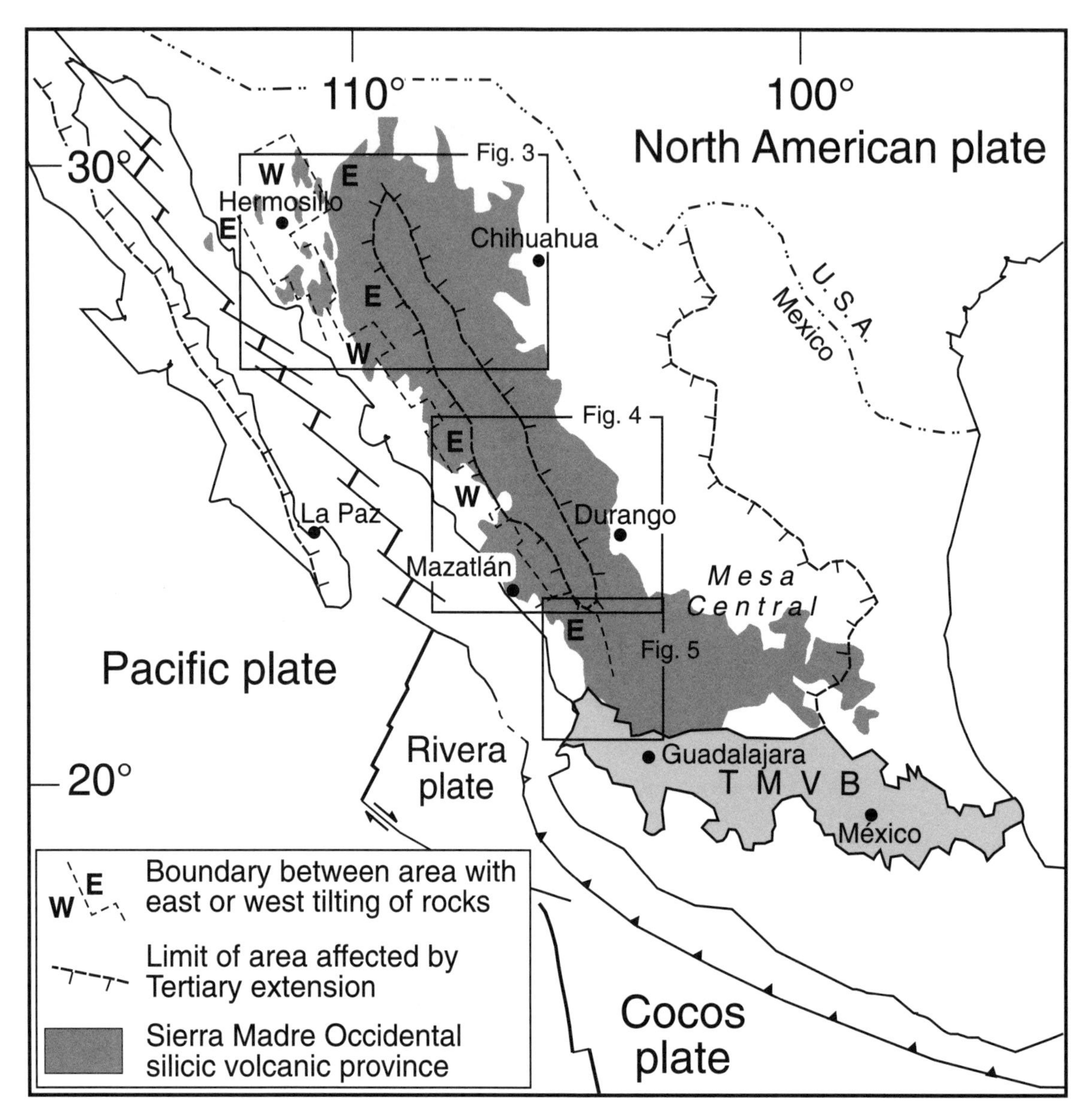

Figure 1. Tectonic sketch map of México showing Tertiary extension (after Henry and Aranda-Gómez, 2000, and Ferrari et al., 2002), the Sierra Madre Occidental (SMO) volcanic province, and the present plate configuration. TMVB—Trans-Mexican Volcanic Belt. The regional tilt domains are according to Stewart et al. (1998).

offer a series of regional synthetic geological maps of the magmatic episodes (Figs. 2–5), and tectonic maps based on interpretation of high-resolution satellite images, plus literature data (Figs. 6–9).

Several summary and review works previously published on the Sierra Madre Occidental are relevant to this paper. McDowell and Clabaugh (1979) summarized the pioneering studies in the 1970s, and provided the first general stratigraphic framework of the Sierra Madre Occidental. Most of the geologic and geochronologic knowledge of the northern and central parts of the Sierra Madre Occidental results from work done by researchers and students of the University of Texas at Austin and the Hermosillo regional station of the *Instituto de Geología* of *Universidad Nacional Autónoma de México* (UNAM), studies which were largely coordinated by Fred W. McDowell for about three decades, and summarized in various papers mainly published in the *GSA Bulletin* between 1994 and 2001. Other important contributions to the geology and tectonics of the central part of the Sierra Madre Occidental are the works by Chris Henry of the University of Nevada, and Jorge Aranda-Gómez and Gerardo Aguirre-Díaz of *Centro de Geociencias* of UNAM. An updated synthesis of these works was provided by Aranda-Gómez et al. (2003) and Henry et al. (2003). Until a few years ago, the southern part of the Sierra Madre Occidental remained little studied compared to the rest of the province. Nieto-Samaniego et al. (1999) provided

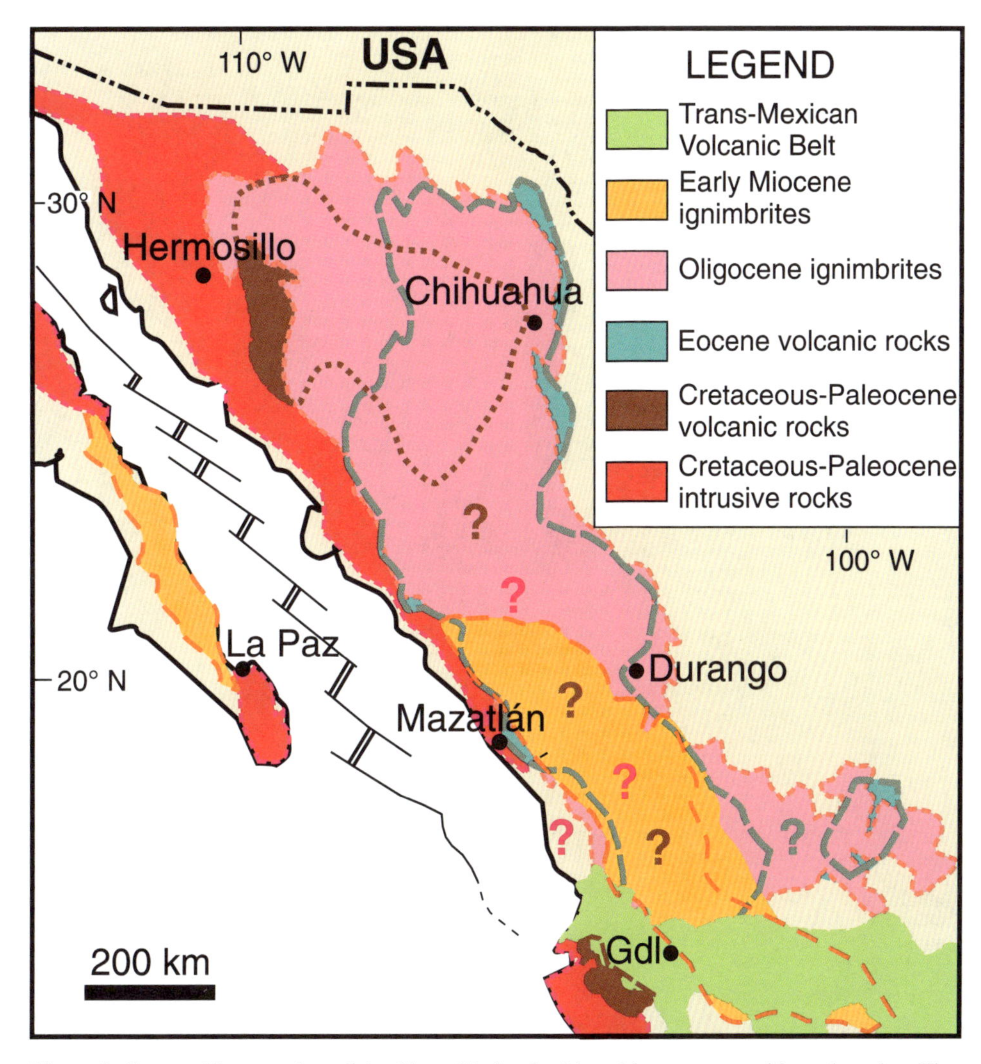

Figure 2. Geographic extension of the Sierra Madre Occidental igneous assemblages based on Figures 3, 4, and 5. The extension of the Cretaceous to Eocene assemblages is partly inferred from the extensive cover by the Oligocene and Early Miocene ignimbrites. Scarce Oligocene volcanic rocks in Baja California are not shown because of the scale of the figure.

the first review of this region along with the magmatic evolution of the Mesa Central. More recently, Ferrari et al. (2002) provided a geologic and tectonic framework for this southern section, and proposed a general model for the occurrence of the ignimbritic pulses of the Sierra Madre Occidental. Review papers on the age, geology, and tectonics of ore deposits within the Sierra Madre Occidental have been published by Damon et al. (1983), Staude and Barton (2001), and Camprubí et al. (2003).

2. REGIONAL STRATIGRAPHY

Introduction

The term "Sierra Madre Occidental" has traditionally been used to describe a large physiographic province of western México, characterized by a high plateau with an average elevation exceeding 2000 m asl, and covering an area ~1200 km long and 200–400 km wide. It extends south from the border with the United States to the Trans-Mexican Volcanic Belt, and is bounded to the west by the Gulf of California, and by the Mesa Central (Mexican Central Plateau) to the east (Fig. 1). The opening of the Gulf of California promoted the formation of deep canyons along the west flank of the Sierra Madre Occidental, whereas *Basin and Range* extensional tectonics produced wide tectonic depressions along the eastern flank of the province. The term "Sierra Madre Occidental" is also used to describe the Tertiary volcanic province characterized by large volumes of silicic ignimbrites (Fig. 1). The Tertiary volcanic province extends beyond the physiographic province to include the Mesa Central and part of eastern Chihuahua, although less abundant Late Eocene to Oligocene silicic ignimbrites are also present to the south of the Trans-Mexican Volcanic Belt in the Sierra Madre del Sur (Michoacan, Guerrero, México, and Oaxaca

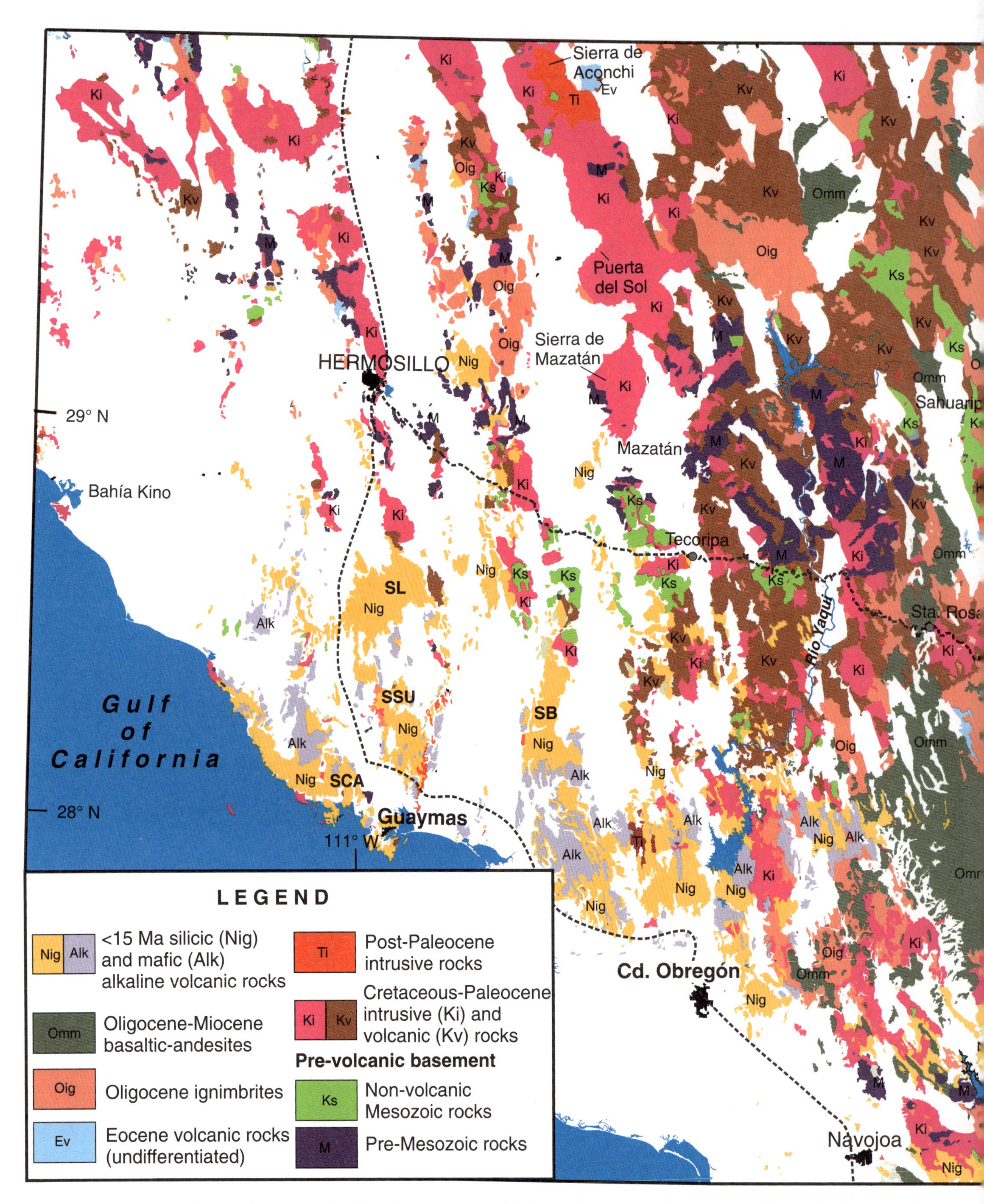

Figure 3. Geologic map of the northern part of the Sierra Madre Occidental elaborated based on a reinterpretation of the geologic cartography at 1:250,000 scale of Servicio Geológico Mexicano (sheets Hermosillo, Madera, Buenaventura, Sierra Libre, Tecoripa, Chihuahua, Guaymas, Ciudad Obregón, and San Juanito). SLF—Sierra de Los Filtros; SSU—Sierra Santa Úrsula; SB—Sierra del Bacatete; SL—Sierra Libre.

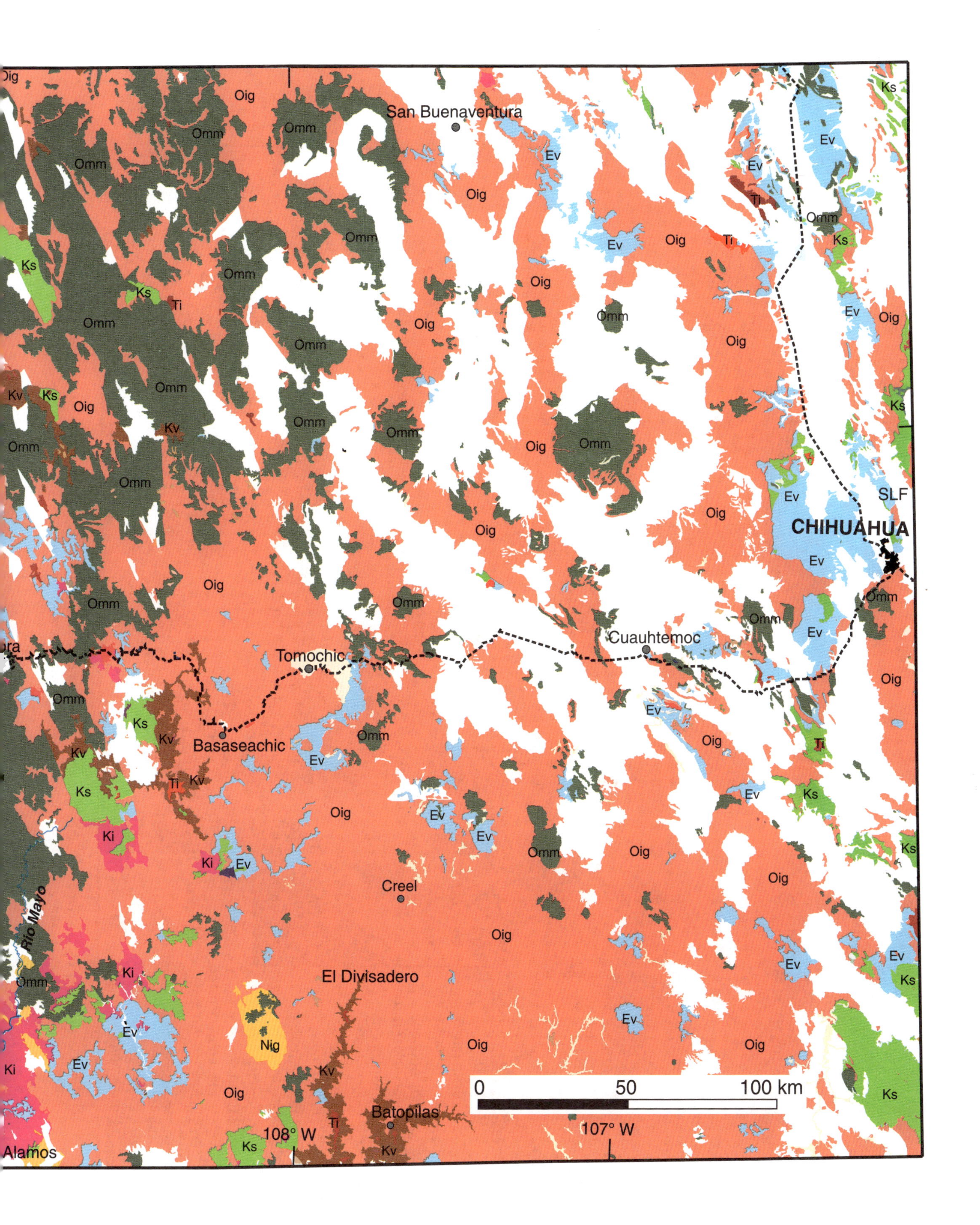
San Buenaventura
CHIHUAHUA
SLF
Cuauhtemoc
Tomochic
Basaseachic
Creel
El Divisadero
Batopilas
Alamos
Rio Mayo
0
50
100 km
108° W
107° W

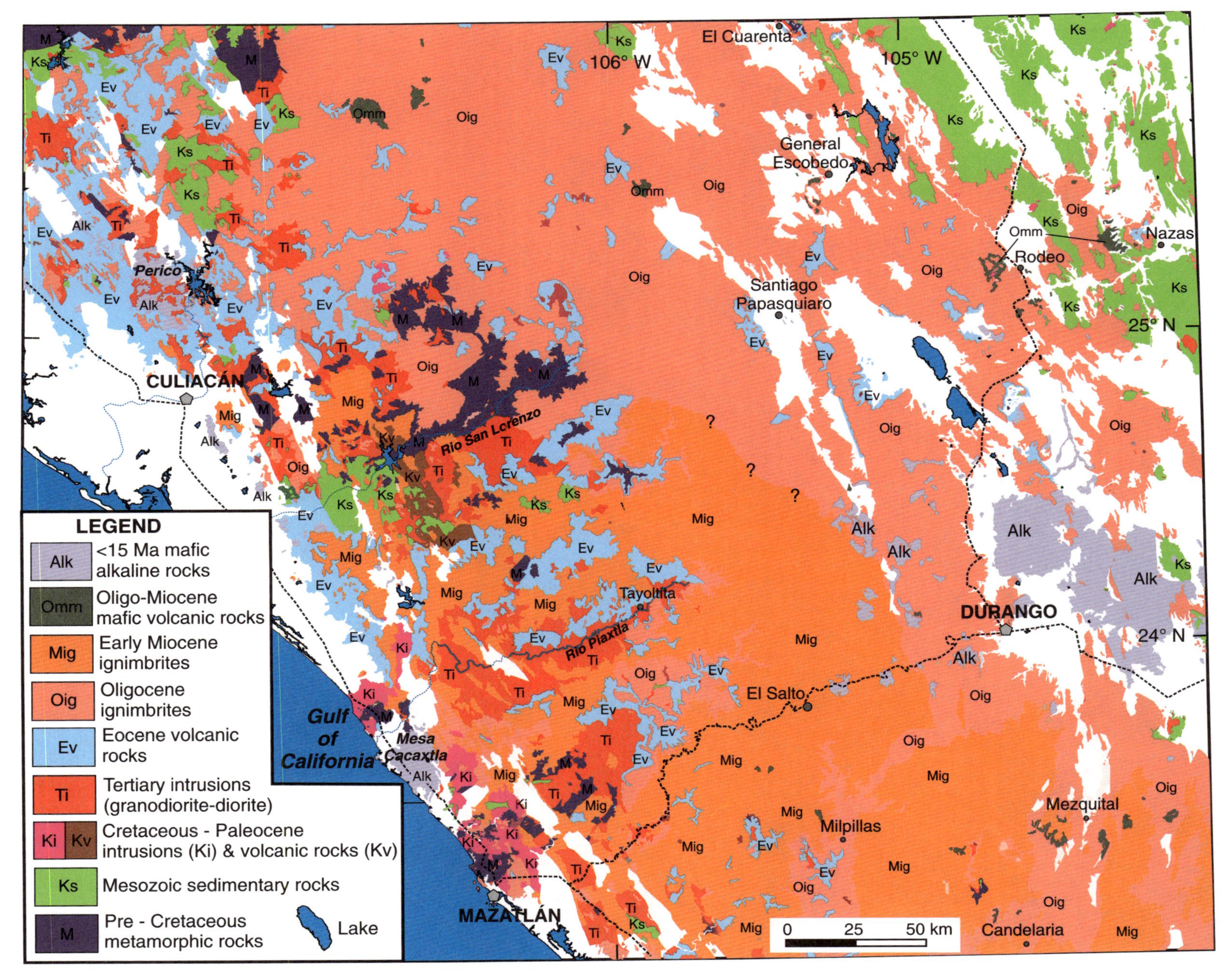

Figure 4. Geologic map of the central part of the Sierra Madre Occidental elaborated based on a reinterpretation of the geologic cartography at 1:250,000 scale of Servicio Geológico Mexicano (sheets Pericos, Santiago Papasquiaro, Culiacán, Durango, Mazatlán, and El Salto).

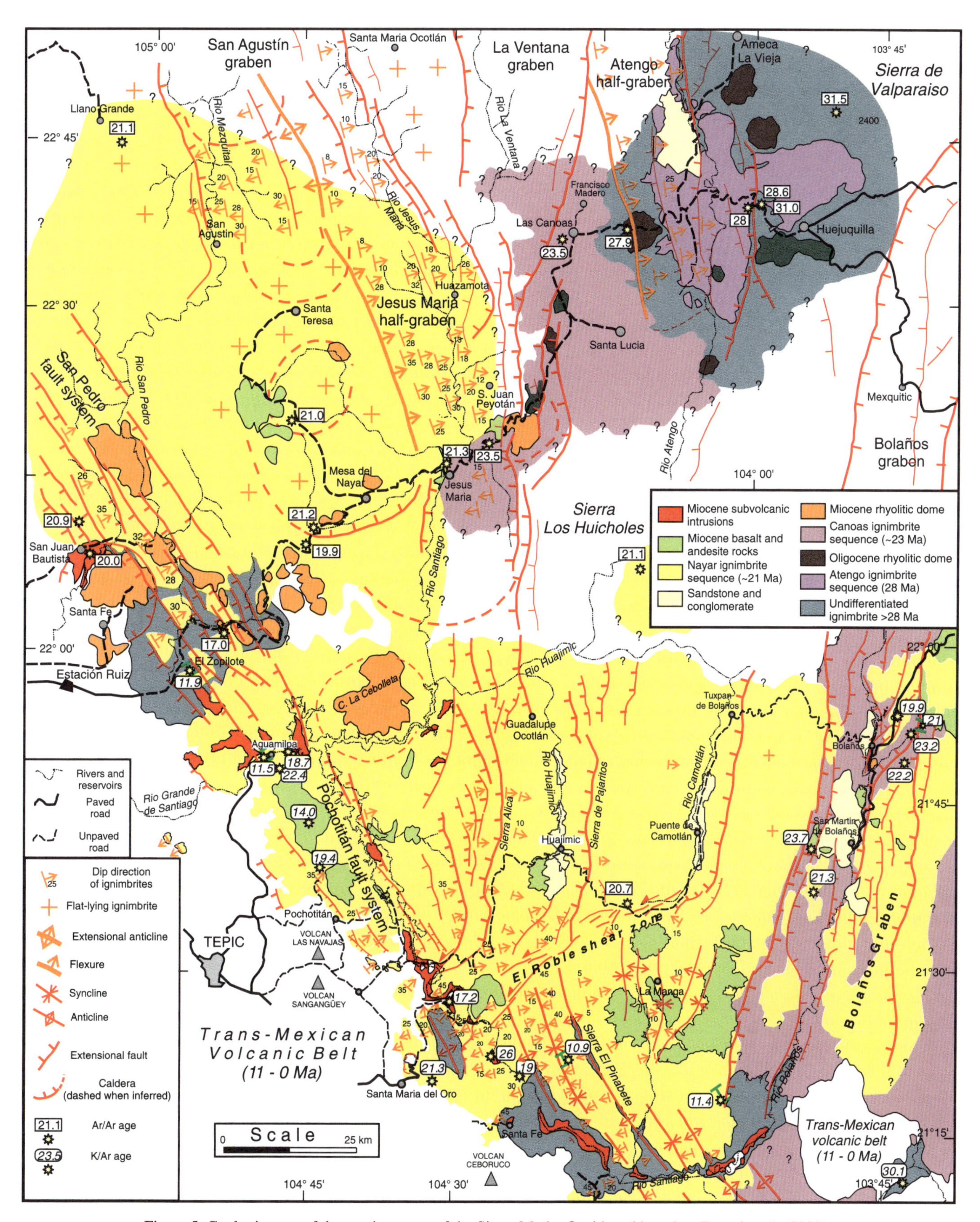

Figure 5. Geologic map of the southern part of the Sierra Madre Occidental based on Ferrari et al. (2002).

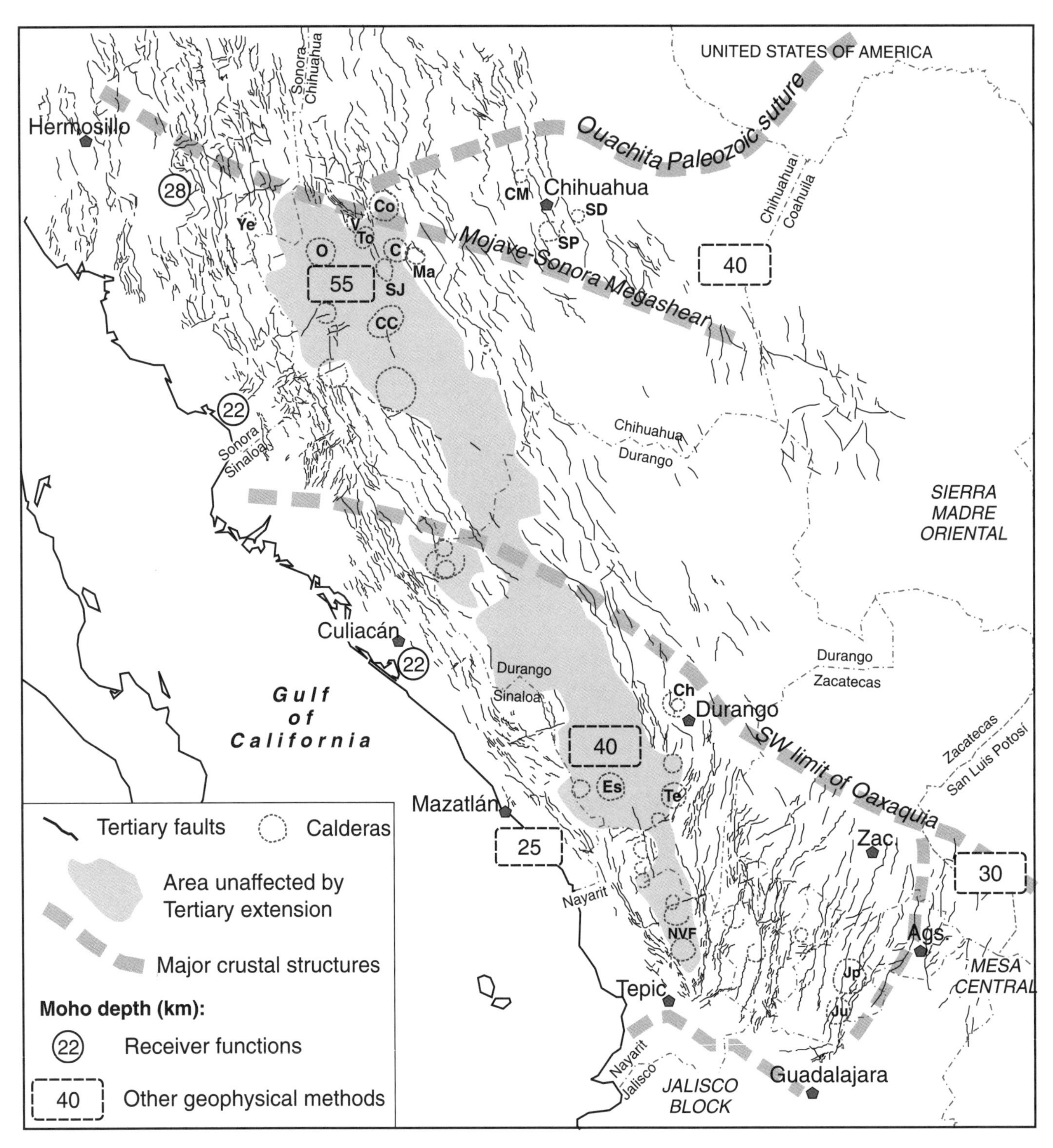

Figure 6. Tectonic map of the Sierra Madre Occidental. The map includes the main Tertiary faults reported in the literature (see text), integrated with an interpretation of a mosaic of orthorectified images of Landsat Enhanced Thematic Mapper (7, 4, and 2 bands) with a resolution of 14.25 m. The southwestern limit of Oaxaquia (after Lawlor et al., 1999) also corresponds to the limit of the continent at the end of the Paleozoic. The boundary between the Sierra Madre Occidental and the Mesa Central is according to Nieto-Samaniego et al. (1999). The main calderas are: SD—Santo Domingo (Megaw, 1986); SP—Sierra Pastoría (Megaw, 1990); CM—Caldera Majalca (Mauger, 1992; or San Marcos, *in* Ferriz, 1981); To—Tómochic and V—Las Varas (Wark et al., 1990); Co—Corralito and O—Ocampo (Swanson and McDowell, 1984); C—El Comanche; Ma—Manzanita; SJ—San Juanito and CC—Copper Canyon (Swanson et al., 2006); Ye—Yécora (Cochemé and Demant, 1991); Ch—Chupaderos (Swanson and McDowell, 1985); Te—Temoaya and Es—El Salto (Swanson and McDowell, 1984); NVF—Nayar caldera field (Ferrari et al., 2002); Ju—Juchipila and Jp—Jalpa (Webber et al., 1994). Other calderas inferred based on remote sensing. See text for reference on crustal thickness. Ags.—Aguascalientes; Zac.—Zacatecas.

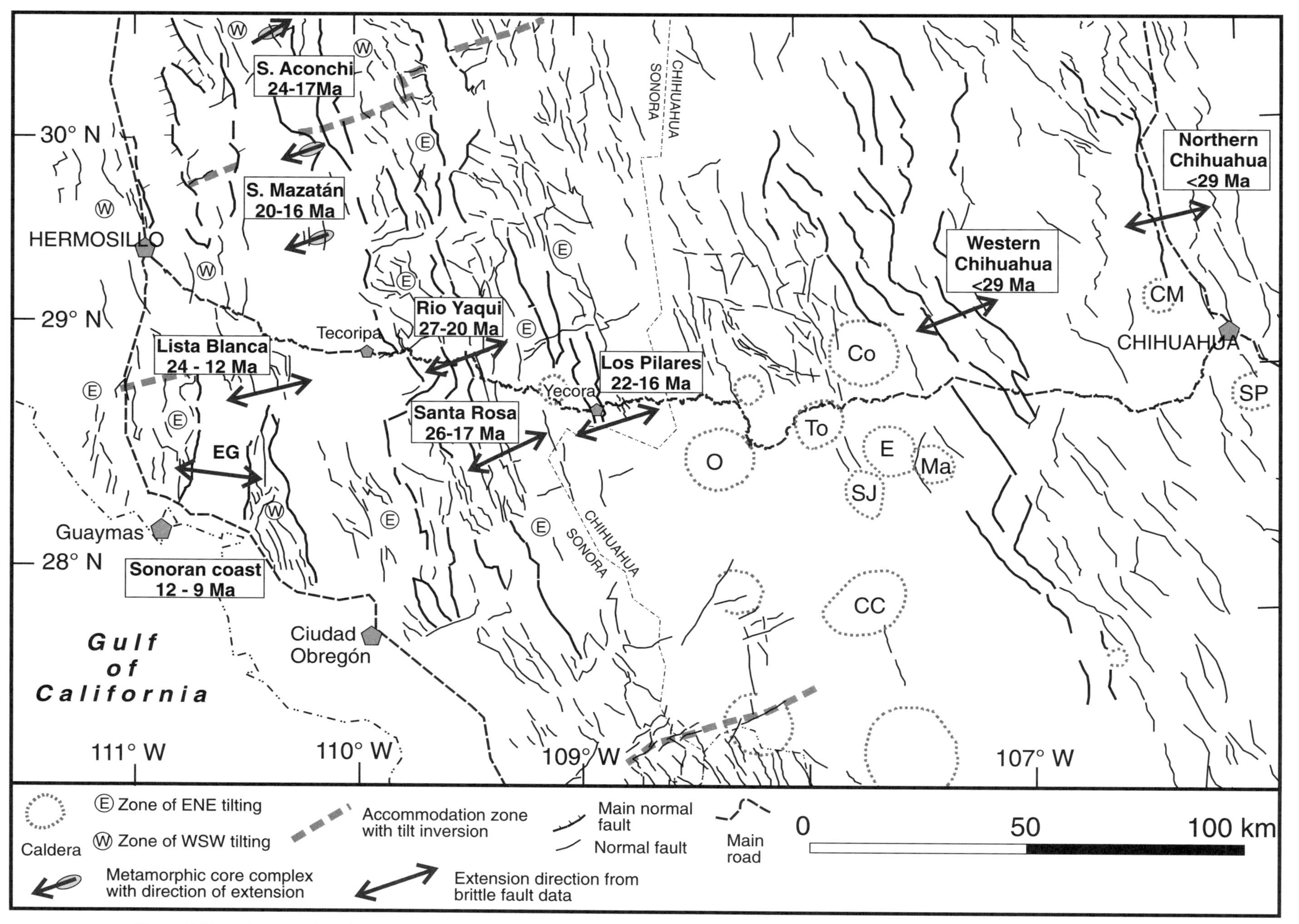

Figure 7. Tectonic map of northern part of the Sierra Madre Occidental showing orientation and age of extensional deformation (see text for details and references). Faults and calderas as in Figure 6. EG—Empalme graben.

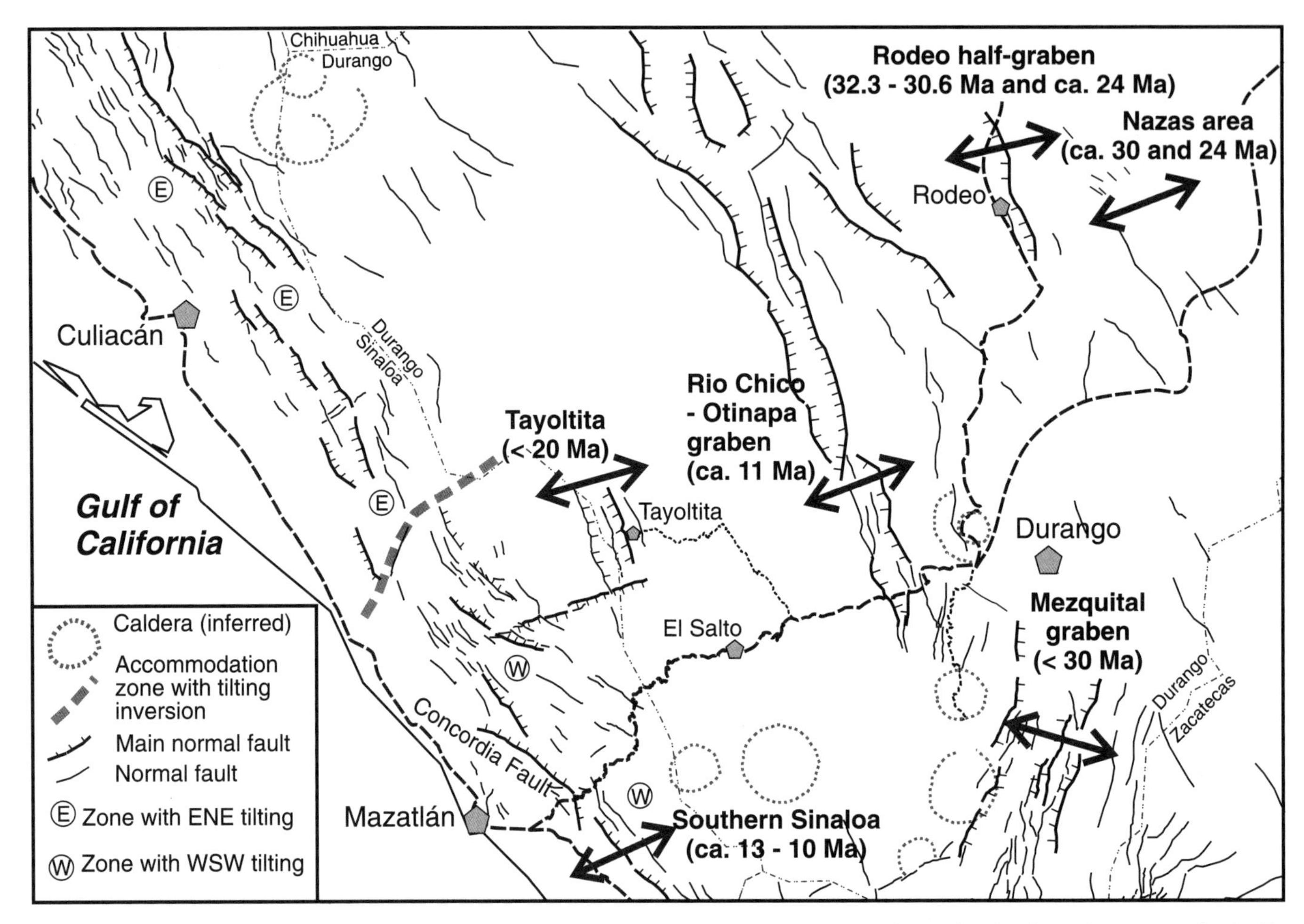

Figure 8. Tectonic map of central part of the Sierra Madre Occidental showing orientation and age of extensional deformation (see text for details and references). Faults and calderas as in Figure 6.

States) (Morán-Zenteno et al., 1999). The present work focuses on the geology within the Sierra Madre Occidental physiographic province, since the geology of the Mesa Central plateau is treated in a companion paper (Nieto-Samaniego et al., this volume).

The Sierra Madre Occidental comprises several different igneous rock assemblages emplaced while subduction of the Farallon plate occurred beneath the North American plate (Fig. 2), and includes: (1) plutonic and volcanic rocks of Late Cretaceous-Paleocene age; (2) Eocene andesitic and lesser dacitic-rhyolitic volcanic rocks; (3) silicic ignimbrites emplaced in two main pulses in the Early Oligocene and Early Miocene; (4) basaltic lavas erupted during the later stages of, and after, each ignimbritic pulse; and (5) repeated episodes of alkaline basaltic lavas and ignimbrites generally emplaced along the periphery of the Sierra Madre Occidental in the Late Miocene, Pliocene, and Quaternary. Assemblages 1–2 and 3 have been defined as the "Lower Volcanic Complex" and the "Upper Volcanic Supergroup," respectively (McDowell and Keizer, 1977). Mafic volcanic rocks of assemblage 4, in the northern part of the Sierra Madre Occidental, have been considered an extension of the "southern Cordillera basaltic andesite" belt, defined by Cameron et al. (1989). Basalts of assemblage 5 have been directly related to various episodes of extension and opening of the Gulf of California (Henry and Aranda-Gómez, 2000). The products of all these magmatic episodes, which are partly superimposed (Fig. 2), cover a heterogeneous and poorly exposed Precambrian, Paleozoic, and Mesozoic basement (James and Henry, 1993; McDowell et al., 1999; Dickinson and Lawton, 2001).

In order to summarize the regional stratigraphy of the Sierra Madre Occidental, one must take into account that our knowledge is significantly diminished by difficulties in access, the scarcity of outcrops of pre-Oligocene units, and the intense post-Eocene extensional faulting. In particular, the true extent and significance of Cretaceous-Eocene magmatism in the region is not readily appreciable (Figs. 2–5). For practical purposes, in describing the Cretaceous-Tertiary magmatism and tectonics, the Sierra Madre Occidental is divided into three sectors: northern (Sonora-Chihuahua), central (Sinaloa-Durango) and southern (Nayarit-Jalisco-Zacatecas) sectors; these sectors center on the three main highways that cross the Sierra Madre Occidental from west to east. The regional geology of these sectors is illustrated in

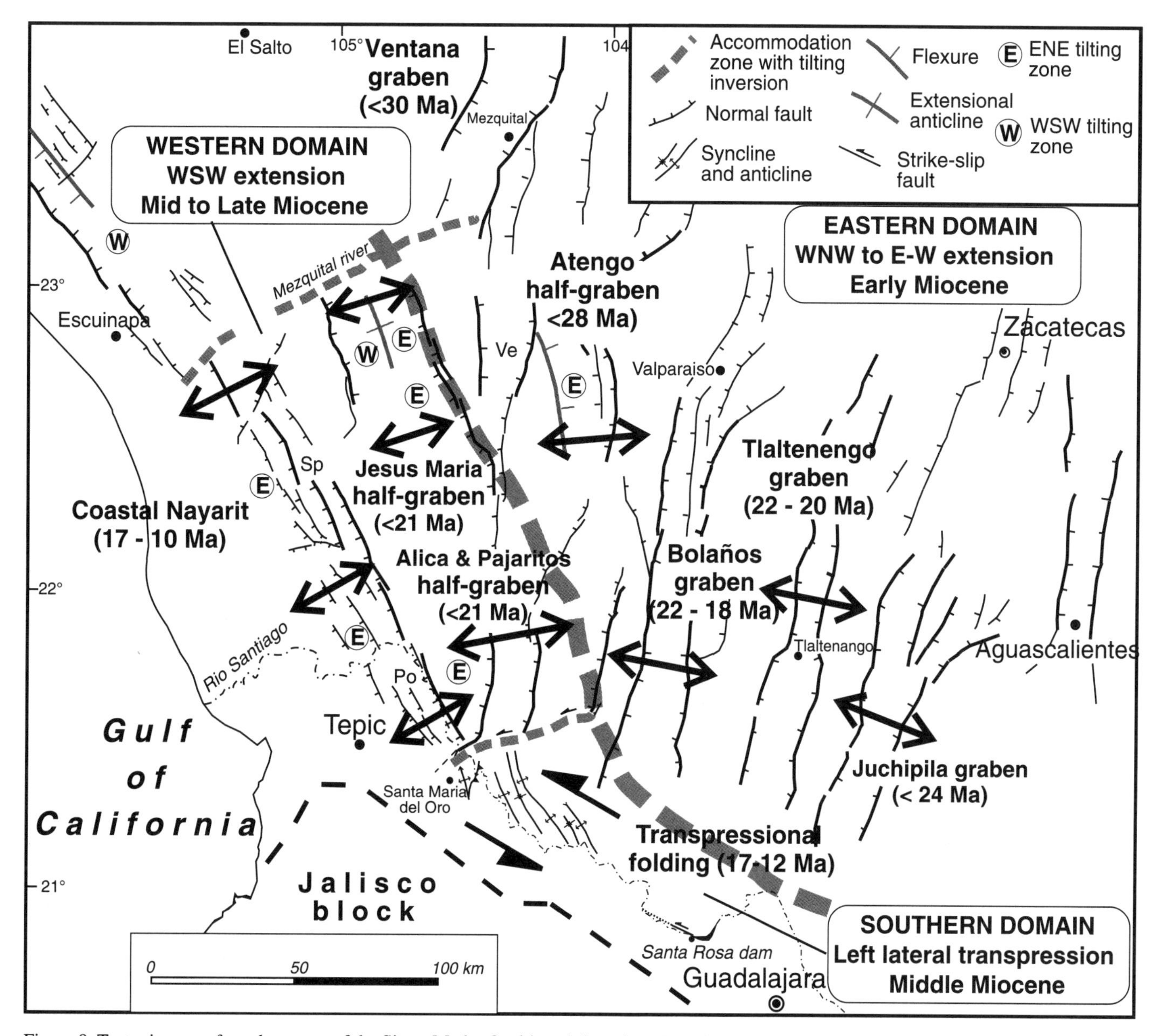

Figure 9. Tectonic map of southern part of the Sierra Madre Occidental (based on Ferrari et al., 2002) showing orientation and age of extensional deformation (see text for details and references). Faults and calderas as in Figure 6. Sp—San Pedro fault system; Ve—La Ventana graben.

the maps provided in Figures 3, 4, and 5. The first two maps are based on the 1:250,000 scale geological cartography published in the past 9 years by the former *Consejo de Recursos Minerales* (CRM), presently named *Servicio Geológico Mexicano* (SGM), and particularly the geological sheets for Hermosillo, Madera, Buenavista, Sierra Libre, Tecoripa, Chihuahua, Guaymas, Ciudad Obregón, San Juanito, Pericos, Santiago Papasquiaro, Culiacán, Durango, Mazatlán, and El Salto. These geologic maps were reinterpreted for this work and verified on the basis of the geochronologic data available in the literature. Although the geology of the SGM maps has been partly inferred, they represent the most detailed and updated map sheets covering the Sierra Madre Occidental province. For the southern sector of the Sierra Madre Occidental, the geologic map shown in Figure 5 is modified from those presented by Ferrari et al. (2002), as they provide more stratigraphic detail than SGM maps for the region.

Prevolcanic Basement

Paleo- and Mesoproterozoic

Abundant outcrops of Proterozoic rocks of the North American craton are recognized particularly in the basement of northwestern Sonora. Part of this basement was presumably displaced ~800 km southeast during the Mid to Late Jurassic, by a left lateral fault called the "Mojave-Sonora megashear" (Silver and Anderson, 1974; Anderson and Silver, 1979). Since its conception, the model of the Mojave-Sonora megashear has been controversial, however, it appears to be a fundamental feature in most tectonic reconstructions of México (e.g., Anderson et al., 2005; Sedlock et al., 1993). According to this model, the basement transported to the southeast, also defined as the Caborca terrane (Campa and Coney, 1983), is characterized by granitic plutons, gneiss, and schist, which yielded isotopic ages between 1.8 and 1.7 Ga. In contrast, the autochthonous basement to the north of the megashear is mostly characterized by metamorphosed clastic and volcanic rocks of the Pinal schist, with ages between 1.7 and 1.6 Ga (Anderson and Silver, 1979; Anderson and Schmidt, 1983). Recently proposed alternative models for the tectonic reconstruction of the Proterozoic basement of Sonora (Dickinson and Lawton, 2001; Iriondo et al., 2004a) still require an important shear zone, similar to that proposed by Anderson and Silver (1979).

Except in northeastern Sonora, Proterozoic crystalline rocks do not outcrop in the Sierra Madre Occidental province. To the east, granitic rocks of Grenvillian age (ca. 1.0 Ga) cut by amphibolitic dikes are recognized in east-central Chihuahua, in the Sierras of Los Filtros and El Carrizalillo (Ruiz et al., 1988b; McDowell and Mauger, 1994) (Fig. 3). Proterozoic rocks are known to occur at depth in northeastern and southeastern Chihuahua based on the occurrence of lower crustal xenoliths brought to the surface by recent alkali basalts exposed at El Potrerillo and La Olivina (Ruiz et al., 1988b; Cameron et al., 1992). Old North American basement is therefore considered to underlie the Sierra Madre Occidental and particularly the northern sector, and may have modified the final isotopic composition of the Cenozoic volcanic rocks. McDowell et al. (2001) and Albrecht and Goldstein (2000) used the Sr, Nd, and Pb isotopic compositions of Eocene-Oligocene ignimbrites to indirectly locate the boundary of the North American Proterozoic basement beneath the Sierra Madre Occidental in Chihuahua. In both studies, changes in whole-rock isotopic compositions defined a roughly WNW- or NW-trending boundary, located between the city of Chihuahua to the northeast and the Tómochic caldera to the southwest (Figs. 3 and 6). Isotopic studies of Laramide granitic rocks from northwestern Mexico suggest that this Proterozoic crustal boundary may be located further to the south, near the Sonora-Sinaloa-Chihuahua state borders (Valencia-Moreno et al., 2001, 2003) (Fig. 6).

Neoproterozoic and Paleozoic

Extensive sequences of marine sedimentary rocks cover the Proterozoic crystalline rocks of northern and northwestern Mexico. These sequences display a clear temporal continuity from the Neoproterozoic through most of the Paleozoic, and for this reason it is more convenient to treat them here as a single assemblage. A comprehensive descriptive inventory of known exposures of these rocks in Sonora was presented by Stewart and Poole (2002). In general, the sedimentary sequences represent two main geological environments: (1) a shallow-water marine platform, mostly occurring in western and central Sonora, and (2) a deep-marine basin extending further to the south and represented by the Paleozoic sedimentary sequences (e.g., Poole et al., 1991). The latter was considered part of the Cortéz terrane (Coney and Campa, 1987; Valencia-Moreno et al., 1999), however, other authors consider these sedimentary rocks to have been deposited in a basin fringing the North American continent, and therefore to be part of the same tectonic block that was transported in the Jurassic along the Mojave-Sonora megashear (e.g., Sedlock et al., 1993; Ortega-Gutiérrez et al., 1994). In eastern Sonora, the Neoproterozoic and Paleozoic rocks clearly continue beneath the Tertiary volcanic rocks of the Sierra Madre Occidental, and are observable west of Yécora and to the southeast of Sahuaripa (Stewart et al., 1999; Almazán-Vázquez, 1989), and along the western edge of the Sierra Madre Occidental (Fig. 3). In the northern part of Sinaloa, deformed deep-water marine rocks of Paleozoic age have also been recognized (Carrillo-Martínez, 1971; Mullan, 1978; Gastil et al., 1991), with exposures likely extending further north to near Álamos, in southern Sonora (Fig. 3).

The contact between these two Paleozoic environments is well known in central Sonora (e.g., Poole et al., 1991; Valencia-Moreno et al., 1999), but its extension to the east is hidden beneath the volcanic rocks of the Sierra Madre Occidental. East of the Sierra Madre Occidental, outcrops of Paleozoic rocks are relatively scarce, but they occur in northern Chihuahua where they are considered an extension of the Ouachita orogenic belt (Stewart, 1988) (Fig. 6). Further to the south in this region, Paleozoic low-grade metamorphic rocks and volcanic and volcaniclastic sequences of the Coahuila block, and associated with the Las Delicias arc (McKee et al., 1988; Sedlock et al., 1993; Dickinson and Lawton, 2001), are interpreted as the remnants of the accretion of Gondwana and Laurentia at the beginning of the Permian.

The southernmost exposures of pre-Mesozoic rocks in the Sierra Madre Occidental occur in northern Durango and northern Sinaloa. Limited exposures of muscovite schist outcrop beneath the Oligocene ignimbrites to the southwest of San Juan del Rio, Durango (Fig. 4). A minimum age for the associated metamorphic event has recently been assigned to the Permo-Triassic boundary, based on a $^{40}Ar/^{39}Ar$ dating of muscovite (Iriondo et al., 2003). Another metamorphic volcano-sedimentary sequence (El Fuerte Group) is widely exposed east of Culiacán, mainly in the valley of the San Lorenzo River (Fig. 4). The age of the El Fuerte Group is uncertain, however, Mullan (1978) considered this sequence to be Jurassic or possibly older.

Prevolcanic Mesozoic

Mesozoic rocks are abundant to the west of the Sierra Madre Occidental in Sonora, but they are less common toward the south of the state. In the east-central part of Sonora, the older Mesozoic rocks consist of a sequence of clastic continental and minor marine sedimentary rocks of Late Triassic–Early Jurassic age (Stewart and Roldán-Quintana, 1991; Valencia-Moreno et al., 1999). These rocks are locally known as the Barranca Group (Alencaster and de Cserna, 1961), which contains important coal horizons defining a middle member, as well as two conglomeratic members. The sediments of the Barranca Group were deposited in an E-W–oriented basin, whose northern limit abuts exposures of Paleozoic rocks (Valencia-Moreno et al., 1999). This basin extends to the western margin of the Sierra Madre Occidental, and presumably continues further east. It is considered to be genetically associated with a "pull-apart" type extensional rupture developed at the beginning of the Triassic (Stewart and Roldán-Quintana, 1991). Marine sedimentary rocks of similar age are also reported in northwestern Sonora as the Antimonio Group (González-León, 1997), but the relationship between this and the Barranca Group is poorly understood. The Antimonio Group is unconformably overlain by an interbedded sequence of clastic sedimentary and volcanic rocks, and both paleontologic and isotopic data indicate an age between Early and Late Jurassic. Although most commonly exposed in northwestern Sonora, exposures extend further east to north-central Sonora (Anderson and Silver, 1979; Rodríguez-Castañeda, 1996). Antimonio Group rocks and associated plutonic rocks have been interpreted as the products of a Jurassic continental arc. The presence of a single Late Triassic pluton exposed in the northwestern tip of Sonora has been used to suggest that magmatic arc activity in Sonora was relatively continuous from ca. 220–140 Ma (e.g., Anderson and Silver, 1979; Damon et al., 1981; Stewart, 1988). Igneous rocks of similar ages reported in southern Chihuahua and northern Durango have been interpreted as an extension of the Triassic-Jurassic arc, but as being displaced to the east by the Mojave-Sonora megashear (Grajales-Nishimura et al., 1992). Uncomformably overlying these Jurassic sequences are Late Jurassic to Early Cretaceous fluvio-deltaic and marine sedimentary rocks of the Bisbee Group that were deposited in a series of subsiding basins (González-León, 1994). Bisbee Group sedimentary rocks outcrop primarily in central and northern Sonora, but are also reported from Arivechi (Almazán-Vázquez, 1989) and Lampazos (González-León, 1988) in the eastern part of the state, and near Caborca, in northwestern Sonora (Jacques-Ayala, 1995).

At a more regional scale, similar sedimentary sequences have been reported east of the Sierra Madre Occidental, in the Chihuahua and Sabinas basins (Dickinson et al., 1989; Haenggi, 2002). Temporally equivalent rocks to the west correspond to the Alisitos Formation, exposed mainly in the northern part of the peninsula of Baja California, which represents the accreted remnants of a rifted oceanic arc terrane (e.g., Busby et al. [1998]). This Albian-Aptian formation consists of volcanic, volcaniclastic, and carbonate sequences deposited mainly in marine environments (Almazán-Vázquez, 1989; Dickinson and Lawton, 2001), and forms the basis for the Alisitos terrane (Campa and Coney, 1983) that was later redefined as the Yuma terrane (Sedlock et al., 1993). In the Late Cretaceous, syntectonic basins associated with the Laramide orogeny developed in northeastern Sonora, and accumulated fluvial and lacustrine sediments, followed by thick clastic sedimentary wedges (González-León and Lawton, 1995). Locally, these sedimentary rocks, which are collectively known as the Cabullona Group, contain horizons with abundant plant fragments, and invertebrate and vertebrate fossils, including dinosaur bones (Lucas et al., 1995).

In Sinaloa, orthogneisses, metasedimentary, and metavolcanic rocks are intruded by Cretaceous batholiths (Henry et al., 2003). The orthogneisses are intensely foliated and have been interpreted as Jurassic in age (Mullan, 1978). Keppie et al. (2006), however, recently interpret U-Pb ages zircons from the amphibolite facies Francisco gneiss in northern Sinaloa as Late Triassic. Henry and Mortensen (written communication, 2005) obtained a U-Pb zircon age near the Jurassic-Cretaceous boundary for other orthogneiss in central Sinaloa. The U/Pb age of two concordant fractions is 134.7 ± 0.4 Ma, which is indistinguishable from K-Ar hornblende dates obtained by the same authors from a layered gabbro exposed in the same area. The metasedimentary rocks consist of phyllite, quartzite, and quartz-muscovite-biotite schist of probable Jurassic age, as well as marble formed at the contact with the batholithic rocks (Henry et al., 2003). The marbles of Sinaloa are Albian in age, and are locally underlain by andesitic lavas and conglomerate (Bonneau, 1970). Scarce outcrops of amphibolite of pre-Albian age are also reported in southern Sinaloa (Henry et al., 2003).

In the southern part of the Sierra Madre Occidental, the existence of a pre-Cenozoic sedimentary basement is indicated by the presence of small outcrops of slate, greywacke, and limestone exposed in the canyon of the Santiago River prior to construction of the Aguamilpa reservoir (Fig. 5) (Gastil et al., 1978; Ferrari et al., 2000). These rocks are spatially associated with Oligocene to Early Miocene granitic intrusive bodies (Gastil et al., 1978; Nieto-Obregón et al., 1985; Ferrari et al., 2002). Older igneous rocks are not known in this region, but do outcrop further south of the Trans-Mexican Volcanic Belt in the Jalisco block (Fig. 2) (Ferrari et al., 2000).

Late Cretaceous–Paleocene Magmatism

Northern Sector

At the end of the Cretaceous and at the beginning of the Tertiary, magmatism in northern México was dominated by the activity of the so-called Laramide magmatic arc, since it was contemporaneous with Laramide deformation occurring in the western United States and Canada. Laramide-age magmatism produced significant volumes of plutonic and volcanic rocks, collectively grouped into the Lower Volcanic Complex by McDowell and Keizer (1977). Large composite batholiths vary in composition from diorite and quartz-diorite to alkaline granite (e.g., Roldán-Quintana, 1991; Valencia-Moreno et al., 2001), whereas coeval volcanic sequences are dominated by andesitic lavas, and locally known as the Tarahumara Formation (Wilson and Rocha, 1949). The Lower Volcanic Complex includes an upper member of rhyolitic and dacitic tuffs and lavas interbedded with sedimentary rocks that locally contain plant fossils (González-León et al., 2000; McDowell et al., 2001). According to Damon et al. (1983), the plutonic rocks of the Lower Volcanic Complex in northwestern México have ages between 90 and 40 Ma, and become progressively younger to the east. A more recent study of the volcanic rocks of the Tarahumara Formation, in the east-central part of Sonora, shows that eruptions occurred between 90 and 60 Ma

(McDowell et al., 2001). This suggests that Laramide magmatism in northern México may have been much more complex than indicated in earlier models, suggesting that a single magmatic arc migrated eastward (e.g., Coney and Reynolds, 1977; Damon et al., 1983). In general, it is accepted that igneous rocks of the Lower Volcanic Complex were emplaced as part of Cordilleran magmatic activity temporally associated with the Laramide orogeny. It should be noted, however, that the Lower Volcanic Complex of McDowell and Clabaugh (1979) also included older rocks of the Peninsular Ranges batholith of Baja California and its extension into Sinaloa (ca. 120–85 Ma).

In the westernmost portion of Sonora, the Lower Volcanic Complex is very well exposed even though considerable uplift associated with Cenozoic extension has resulted in substantial erosion of the associated volcanic sequences and exposure of the plutonic rocks. The rocks of the Tarahumara Formation are best preserved in the east-central and north-northeast portions of Sonora (e.g., González-León et al., 2000; McDowell et al., 2001). One of the most notable features of the Lower Volcanic Complex, besides its great extent of exposures, is that it has been host to the formation of a variety of ore deposits, largely exposed during Late Tertiary tectonic unroofing. Numerous porphyry copper deposits are distributed mostly in the eastern portion of the Lower Volcanic Complex belt (Damon et al., 1983; Staude and Barton, 2001), but especially in northeastern Sonora, where the world-class deposits of Cananea and La Caridad represent the best examples (Valencia-Moreno et al., 2001).

In Sonora, the Lower Volcanic Complex belt is considerably wider than its extension to the south into Sinaloa, which may reflect the greater effect of Neogene extension to the north (Damon et al., 1983). The Lower Volcanic Complex rocks disappear to the east under the ignimbritic volcanic province of the Sierra Madre Occidental, but there are indications that the complex extends into central Chihuahua. About 30 km northwest of the city of Chihuahua, Mauger (1981, 1983) described a >3000-m-thick volcanic sequence (Peñas Azules volcanics), which is interpreted as a complex of stratocone-related volcanic successions dated at 68.2 ± 1.6 Ma (K-Ar in plagioclase) and 67.5 ± 1.0 Ma (U-Pb in zircons) (McDowell and Mauger, 1994). Additionally, these authors reported small intrusive bodies and silicic tuffs with Paleocene ages in the region adjacent to the city of Chihuahua.

Central Sector

In the central sector, Cretaceous-Paleocene magmatism has been studied in more detail along the western margin of the Sierra Madre Occidental in Sinaloa, where crustal extension associated with the opening of the Gulf of California has exposed batholiths of the Lower Volcanic Complex. Cretaceous-Paleocene batholiths most likely underlie a large part of the Sierra Madre Occidental, given that Cretaceous dioritic intrusions are also reported in the Nazas area in western Durango (Aguirre-Díaz and McDowell, 1991). All known intrusive rocks are calc-alkaline in composition, and vary from diorite to granite, but granodioritic plutons are by far the dominant composition. Granitic rocks associated with the Sinaloa batholith have U-Pb and K-Ar ages between 101 and 46 Ma, and have been divided into two groups: pre- or syntectonic rocks and post-tectonic rocks (Henry and Fredrikson, 1987; Henry et al., 2003). Pre- or syntectonic intrusions are characterized by mineral foliations and lineations, suggesting emplacement prior to or during deformation that began ca. 85 Ma. Post-tectonic intrusions are more homogeneous and massive. The pre- and syntectonic rocks were emplaced along a belt close to the coast, whereas post-tectonic intrusive rocks are found further east, ~30 km away from the coast (Fig. 4).

Volcanic rocks of the Lower Volcanic Complex have received little study in the central part of the Sierra Madre Occidental, mainly due to the intense hydrothermal alteration. In general, they comprise a sequence of andesitic and rhyolitic lavas and silicic ignimbrites generally restricted to exposures in deep canyons. The main outcrops occur in the canyons of the Piaxtla and Presidio Rivers, as well as in the proximity to Pánuco and Copales, on the Mazatlán–El Salto road (Fig. 4) (Henry and Fredrikson, 1987).

Southern Sector

Isolated Cretaceous-Paleocene intrusive rocks outcrop along the eastern edge of the Sierra Madre Occidental in Zacatecas. In the area of La Tesorera-Zacatón, 20 km east of Zacatecas, the capital city of the state, Campanian K-Ar biotite ages (74 ± 6 Ma) were obtained from plutonic rocks of granodioritic composition (Mújica-Mondragón and Jacobo-Albarrán, 1983; Sole and Salinas, 2002). These plutonic bodies commonly intrude Lower Cretaceous marine sedimentary rocks (CRM, 1997). Further west in western Jalisco and Nayarit, there are no reports of pre-Cenozoic magmatic rocks. The scarcity of exposures of the Cretaceous-Paleocene igneous rocks in the southern sector of the Sierra Madre Occidental is largely due to the extensive cover of ignimbrites of Oligocene and particularly Early Miocene age, which attain their maximum areal extent in this region.

Eocene Magmatism

Previous reviews on the geology of Sierra Madre Occidental (e.g., McDowell and Clabaugh, 1979) included Eocene igneous rocks in the Lower Volcanic Complex along with the Cretaceous-Paleocene batholithic and volcanic rocks. However, later works (e.g., Aguirre-Díaz and McDowell 1991) have documented the space-time extent of this volcanism, which can be considered a distinct episode in the magmatic evolution of western México. For this reason, we discuss the Eocene igneous activity, which in several areas may be interpreted as a precursor to the Oligocene ignimbritic event, separately.

Northern Sector

Eocene volcanic rocks in the northern sector are exposed mainly in Chihuahua along the eastern edge of the Sierra Madre Occidental and in deeply incised canyons within the interior of the province (Fig. 3). In general, the first Eocene ignimbrites of the northern part of the Sierra Madre Occidental are porphyritic, crystal-rich, and commonly biotite-bearing (Magonthier, 1988). Eocene volcanic rocks near the city of Chihuahua are well exposed in a N-S belt that includes the sierras of El Gallego (Keller et al., 1982), del Nido, Sacramento, Pastorías, Las Palomas, Magistral (McDowell and Mauger, 1994 and references therein), Santa Eulalia (Megaw, 1990), and Los Arados (Iriondo et al., 2003) (Fig. 3). This episode began at 46 Ma after a period of scarce and intermittent magmatism, and continued almost without interruption until 27.5 Ma. Consequently, distinguishing between Eocene magmatism and the ignimbritic pulse of the "Upper Volcanic Supergroup" in this part of the Sierra Madre Occidental is extremely difficult. In central Chihuahua, the Eocene rocks comprise two sequences of silicic ignimbrites, which are separated by a thick sequence of massive lavas of intermediate to felsic composition (McDowell and Mauger, 1994 and references therein). A similar sequence, but less complete, is exposed further west in the area of Tómochic, in the interior of the Sierra Madre Occidental. In this region, Wark et al. (1990) reported a succession of andesitic lavas with ages between ca. 38–35 Ma, which are in turn covered by a 34-Ma-old ignimbritic sequence associated with the formation of Las Varas caldera. This sequence is partly covered by volcanic products from the two Oligocene calderas of Tómochic and Ocampo (Swanson and McDowell, 1985; Wark et al., 1990). The Las Varas sequence may extend further west into eastern Sonora, where Montigny et al. (1987) obtained a K-Ar age of 35.3 Ma on sanidine from a porphyritic ignimbrite exposed east of Yécora. In Sonora, however, Eocene volcanism is generally, distinctly older. In the area of Santa Rosa (Fig. 3), Gans (1997) obtained two $^{40}Ar/^{39}Ar$ ages on sanidine at 54.3 ± 0.2 Ma and 43.8 ± 0.2 Ma, from a several hundred-meters-thick sequence of ignimbrites and dacitic lavas. This sequence may be coeval with a granodioritic pluton dated at ca. 60.0 ± 0.5 Ma and exposed close to the town of Santa Rosa (Gans, 1997). These ages indicate the Santa Rosa sequences could be correlated with the upper, more felsic member of the Lower Volcanic Complex, which is exposed in an adjacent area to the west (McDowell et al., 2001). Additional geochronologic studies are needed to establish more precisely, the timing and spatial distribution of Eocene volcanism in the northern sector of the Sierra Madre Occidental, and its relationship with the Oligocene ignimbrite pulse.

Central Sector

In the central part of the Sierra Madre Occidental, the largest thickness of Eocene volcanic rocks occurs near the Sinaloa-Durango state boundary, particularly in the area of Tayoltita (Figs. 4, 8) (Henry and Fredrikson, 1987). In this area, the Piaxtla River has carved a deep canyon that exposes at least 1500 m of the Eocene sequence hosting important gold and silver mineralization (Horner, 1998; Horner and Enríquez, 1999; Enríquez and Rivera, 2001). The sequence, which is strongly tilted to the E-NE, consists of rhyolitic and andesitic lavas, and subvolcanic intrusions of dioritic composition that have yielded K-Ar ages from 39.9 to 36.6 Ma. The volcanic rocks cover, and the dioritic intrusion crosscuts, a granodioritic to dioritic batholith dated at 45.1 Ma (Enríquez and Rivera, 2001). Dating of adularia from mineralized veins yielded younger ages than the batholith (40.4 ± 0.4 Ma, Henry et al., 2003; 38.5–32.9 Ma, Enríquez and Rivera, 2001), and suggests a genetic relationship with the Eocene sequence. The Eocene rocks and mineralization are separated from an overlying Early Miocene ignimbritic sequence by a fluvio-lacustrine succession of conglomerates, sandstone, and shale (Horner and Enríquez, 1999). By analogy with the stratigraphy in the Piaxtla River valley, andesitic rocks reported in many areas elsewhere in Sinaloa beneath the Oligo-Miocene ignimbritic cover are interpreted here as Eocene in age (Fig. 4). However, there are no geochronological data supporting this correlation.

Another area where important Eocene magmatic activity has been documented occurs along the eastern edge of the Sierra Madre Occidental. In the region of Nazas (Fig. 4), two andesitic lavas have been dated at 48.8 and 40.3 Ma. They are separated by rhyolitic lavas and ignimbrites dated between 45.2 and 42.9 Ma, and with younger rhyolitic lavas dated at 34 Ma (Aguirre-Díaz and McDowell, 1991). Red-bed deposits also separate the Eocene volcanics from Oligocene rhyolitic ignimbrites. In the area of El Cuarenta, ~210 km NNW of the city of Durango (Fig. 4), an ignimbrite from the base of a rhyolitic volcanic succession has been dated by $^{40}Ar/^{39}Ar$ at 39.6 Ma (whole rock, Tuta et al., 1988). To the south of the city of Durango, in El Mezquital graben (Fig. 8), McDowell and Keizer (1977) also reported a thick sequence of andesitic lavas, from which they obtained a single whole-rock K-Ar age of ca. 52 Ma.

Southern Sector

Several exposures of Eocene volcanic rocks are known from the southern part of the Sierra Madre Occidental (Nieto-Samaniego et al., 1999). Silicic ignimbrites, rhyolitic domes, and andesitic lavas with K-Ar ages between 38 and 34 Ma locally outcrop in the area of Fresnillo and Sain Alto, Zacatecas (Ponce and Clark 1988; Lang et al., 1988; Tuta et al., 1988), whereas Nieto-Samaniego et al. (1996) reported a K-Ar age of 40.6 ± 1.0 Ma (sanidine) for a rhyolite exposed in the Cerro El Picacho, Aguascalientes. Further west, a sequence of pervasively altered andesitic lavas from which a K-Ar age of 48.1 ± 2.6. Ma (K-feldspar) was obtained, is exposed at the base of a Paleocene sequence in the area of Juchipila (Webber et al., 1994). This sequence is covered by reddish sandstone and conglomerate containing clasts of andesite volcanic rocks, which separate the Paleocene rocks from the Oligo-Miocene ignimbrites. Ferrari et al. (2000) reported a similar sequence in the area of Santa María del Oro (Fig. 5), but this section lacks absolute age constraints. Andesites dated by $^{40}Ar/^{39}Ar$ at ca. 51 Ma (plagioclase and feldspar) occur beneath the Trans-Mexican Volcanic Belt, being intersected in geothermal wells drilled in the San Pedro–Ceboruco graben (Ferrari et al., 2000, 2003), located immediately to the south of the Sierra Madre Occidental (Fig. 5).

Oligocene–Early Miocene Ignimbritic Pulses ("Ignimbrite Flare-Up")

As a volcanic province, the Sierra Madre Occidental is commonly associated with the huge ignimbritic succession, which reaches more than 1000 m in thickness and covers most of the western part of México (McDowell and Clabaugh, 1979). This sequence, known also as the Upper Volcanic Supergroup, was largely, unconformably emplaced on top of the Lower Volcanic Complex (McDowell and Keizer, 1977) and Eocene igneous rocks. The Upper Volcanic Supergroup comprises a thick sequence of rhyolitic ignimbrites, air-fall tuffs, silicic to intermediate lavas, and lesser mafic lavas that are particularly well exposed along the edges of the volcanic province (McDowell and Clabaugh, 1979; Cochemé and Demant, 1991). Averaging 250 km wide by 1200 km long and a present volume exceeding 300,000 km^3, the Sierra Madre Occidental volcanic province is the most notable geological feature of the Mexican subcontinent. In addition to its huge areal extent, another feature that must be emphasized is the relatively short time in which the ignimbrites were emplaced. Several geochronologic studies have shown that the first and more extensive ignimbritic pulse occurred at the beginning of the Oligocene with an impressive synchronicity across the entire province, while a second pulse in the Early Miocene was more restricted to the southwestern part of the Sierra Madre Occidental and central México beneath the Trans-Mexican Volcanic Belt. The characteristics of these explosive volcanic episodes are discussed further below.

Northern Sector

Several sections studied at different locations at the longitude of the city of Chihuahua contain a sequence of ignimbrites with K-Ar ages of ca. 33–30 Ma (McDowell and Mauger, 1994), overlain by peralkaline tuffs with a more restricted distribution and yielding ages from 30.5 to 29 Ma (Mauger, 1981). In San Buenaventura (Fig. 3), in northwestern Chihuahua, the volcanic sequence includes rhyolitic ignimbrites interbedded with dacite, rhyolite, and minor basalt lavas, which were emplaced onto Proterozoic basement (Albrecht and Goldstein, 2000). These authors reported an Rb/Sr age of 33.2 Ma for the lower part of the sequence. In the core of the province, where the ignimbritic sequence reaches its maximum thickness of ~1 km, age ranges for the exposed ignimbrite sections are no more than 3 Ma. Wark et al. (1990) reported ages from 31.8 to 31.4 Ma for the Rio Verde tuff, which is associated with the formation of the Tómochic cauldron (Figs. 4 and 6) and 29.0 Ma for the Cascada tuff, whose intracauldron facies form the spectacular waterfall of Basaseachic (Fig. 3). Further south, the thick ignimbritic sequence of Batopilas (Fig. 3) has been dated by K-Ar between 30.1 and 28.1 Ma (Lanphere et al., 1980). Ages obtained for the sequence of volcanic rocks exposed in the Barranca del Cobre also lie within this range, and are similar to the age of 29.3 Ma obtained in the area of Pito Real (Montigny et al., 1987) (Fig. 3). This indicates that over 1.5 km of ignimbrites were deposited between ca. 30 and ca. 29 Ma in the region of El Divisadero–Creel (Albrecht and Goldstein, 2000) for which several calderas have been recognized as sources for most of these ignimbrite sheets (Swanson et al., 2006).

The ignimbritic carpet of the Sierra Madre Occidental extends into eastern Sonora (Fig. 3), with the best-studied section located in the region of Yécora (Fig. 3), which was documented by Bockoven (1980), Cochemé and Demant (1991), and more recently by Gans (1997). In this region, the Sierra Madre Occidental volcanic rocks were unconformably deposited on an eroded section of the Lower Volcanic Complex, with a conglomeratic horizon containing clasts of an underlying granodioritic pluton occurring along the contact (Bockoven, 1980; Cochemé and Demant, 1991). Although in this part of Sonora the ignimbritic sequence remains to be systematically dated, an $^{40}Ar/^{39}Ar$ age of 33 Ma (sanidine) in Santa Rosa (Fig. 3) (Gans, 1997), plus two K-Ar ages of 33.5 ± 0.8 Ma (plagioclase) and 27.1 ± 0.9 Ma (K-feldspar) in the valley of the Yaqui River (Fig. 3) (McDowell et al., 1997), indicate a similar age range for ignimbrite eruptions to that in Chihuahua. However, the ignimbritic cover becomes considerably thinner in this part of the Sierra Madre Occidental than in the region of western Chihuahua. In the area of Tecoripa, McDowell et al. (2001) reported an average thickness of only 100 m for the rhyolitic ignimbrites erupted from volcanic centers located further east.

Central Sector

In the central sector of the Sierra Madre Occidental, the ignimbritic sequence outcrops mainly in the state of Durango, where it has been studied in some detail along the Durango-Mazatlán highway and in the area of Nazas (Fig. 4). In Nazas, Aguirre-Díaz and McDowell (1993) recognized two Oligocene ignimbritic sequences that reach a combined thickness of ~500 m, from which K-Ar ages of 32.2 ± 0.7 and 29.5 ± 0.6 Ma were obtained, respectively, on sanidine. $^{40}Ar/^{39}Ar$ dating in the area adjacent to Rodeo has yielded similar ages of 32.3 ± 0.7 and 30.6 ± 0.09 Ma (sanidine) (Luhr et al., 2001). In this same age range, a $^{40}Ar/^{39}Ar$ age from sanidine was obtained by Iriondo et al. (2004b) for an andesitic vitrophyre in the locality of Ignacio Ramírez, ~90 km to the south. Oligocene rocks also outcrop near the coast of Sinaloa (Fig. 4). Henry and Fredrikson (1987) reported K/Ar ages of 31.7 ± 0.4 Ma (biotite) for a quartz-dioritic dike in the area of Tayoltita and a 28.3 ± 0.7 Ma (biotite) for a faulted and tilted rhyolitic ignimbrite exposed north of Mazatlán.

An Oligocene ignimbrite sequence exposed around the city of Durango has an approximate thickness of 800 m, and has been related to the formation of the Chupaderos caldera (Swanson et al., 1978). K-Ar ages between 32.8 and 29.5 Ma (McDowell and Keizer, 1977; Swanson et al., 1978) were obtained from the Durango ignimbrite successions, however, ages obtained by $^{40}Ar/^{39}Ar$ on the same feldspar separates yielded a slightly more restricted age range between 32 and 30 Ma (Aranda-Gómez et al., 2003). To the southwest of the city of Durango, in the area between Mezquital and Milpillas (Fig. 4), three ignimbrites have been dated by K-Ar between ~28 (sanidine) and ca. 27 Ma (plagioclase) (Sole and Salinas, 2002).

Further southwest along the Durango-Mazatlán highway, the Durango ignimbrite sequence is covered beneath an extensive rhyolitic dome (Las Adjuntas) dated at 28 Ma, and by the Early Miocene ignimbritic sequence of the El Salto–Espinazo del Diablo (McDowell and Keizer, 1977) (Fig. 4). The El Salto–Espinazo del Diablo sequence has a total thickness of ~1000 m, consisting of four packages of ignimbrites with lesser rhyolitic and basaltic lavas. The K-Ar ages obtained from this section cluster at ca. 23.5 Ma (McDowell and Keizer, 1977). Iriondo et al. (2004b) obtained a nearly identical $^{40}Ar/^{39}Ar$ date on plagioclase from a vitrophyric ignimbrite from this sequence, sampled ~15 km west of El Salto. Early Miocene ignimbrites outcrop along most of the Durango-Mazatlán transect, however, it is not known how far they may extend to the north (Fig. 4). The northernmost localities where Early Miocene silicic rocks are known occur in the region of Tayoltita and Culiacán. In Tayoltita, ignimbrites and lavas that lie on the top of this sequence have been dated by K-Ar at 24.5 and 20.3 Ma (Enríquez and Rivera, 2001), whereas north of Culiacán, Iriondo et al. (2003) obtained a $^{40}Ar/^{39}Ar$ hornblende date of 23.2 ± 0.15 Ma for a subvolcanic granodioritic pluton. Sole and Salinas (2002) report a whole rock K-Ar age of 24.0 ± 1.0 Ma for a basaltic andesite lava that underlies one of the uppermost ignimbrites in the sequence exposed in the area of Milpillas (Fig. 4).

Southern Sector

The southern part of the Sierra Madre Occidental is covered by silicic ignimbrites that have traditionally been considered as Oligocene in age (e.g., McDowell and Clabaugh, 1979) by correlation with rocks exposed more to the east in the Mesa Central (Nieto-Samaniego et al., 1999). However, it has been recently shown that two distinct ignimbritic pulses occurred in this region of Early Oligocene and Early Miocene age (Ferrari et al., 2002). Oligocene ignimbrites dominate exposures in the eastern part through Aguascalientes, Zacatecas, and northern Jalisco (Nieto-Samaniego et al., 1999 and references therein). In the arca of Fresnillo, to the north of Zacatecas, rhyolite lavas and ignimbrites of the Sierra Valdecañas have yielded K-Ar ages between 29.1 and 27.5 Ma, whereas ages between 33.5 and 32.2 Ma were obtained from subvolcanic bodies associated with silver mineralization (Lang et al., 1988). To the southeast, good exposures of the Oligocene rocks occur in the Sierra de Morones, between Jalpa and Tlaltenango (Fig. 9), where Nieto-Obregón et al. (1981) reported a K-Ar date of 29.1 ± 0.6 Ma (sanidine) for one of the ignimbrites higher in the sequence. In this zone, the Oligocene sequence is composed of areally extensive, but relatively thin rhyolitic ignimbrites, which regularly do not exceed 10–20 m each. They are separated from underlying Eocene andesitic volcanics by beds of continental sandstone and red conglomerate. Further west, in the Huejuquilla–Estación Ruiz transect, ignimbrites and rhyolites with $^{40}Ar/^{39}Ar$ ages (sanidine) between 31.5 and 28 Ma constitute much of the Sierra de Valparaíso, and are also exposed to the west in the area of Huejuquilla and the Atengo half graben (Fig. 5) (Ferrari et al., 2002). A $^{40}Ar/^{39}Ar$ age of 27.9 ± 0.3 Ma on sanidine was obtained for a complex of rhyolitic exogenous domes in the western part of the half graben (Ferrari et al., 2002). These ages are almost identical to those observed for the sequence of Durango and the dome of Las Adjuntas, respectively, exposed ~80 km to the north (McDowell and Keizer, 1977).

West of Atengo and along all of the Bolaños-Tepic transect (Fig. 5), Early Miocene ignimbrites are dominant, although the Oligocene sequence could be underlying all the eastern part of the region (Fig. 5), since a package of ignimbrites with a K-Ar age of 30.1 Ma has been recognized in the southern part of the Bolaños graben (Ferrari et al., 2002). The Early Miocene ignimbritic sequence covers the Sierra Madre Occidental in Nayarit. Ferrari et al. (2002) recognized the Las Canoas and El Nayar sequences that represent two ignimbrite packages of different ages and provenance. The Las Canoas sequence is ~350 m thick and has been dated by K-Ar (Clark et al., 1981) and by $^{40}Ar/^{39}Ar$ (Ferrari et al., 2002) at 23.5 Ma, and temporally overlaps with the El Salto–Espinazo del Diablo sequences exposed ~80 km to the north (McDowell and Keizer, 1977). To the south, the age of the Las Canoas sequence overlaps that of the lower part of the succession exposed in the Bolaños graben (Fig. 5), where Scheubel et al. (1988) reported K-Ar ages of 23.7 and 23.2 Ma for an andesite and an ignimbrite, respectively. Similarly aged ignimbritic sequences outcrop in the area of Teúl to the southeast (ca. 23 Ma, Moore et al., 1994), in the area of the Santa Rosa dam (23.6 Ma, Nieto-Obregón et al., 1985), in Juchipila (ca. 24–23 Ma, Webber et al., 1994) (Fig. 9), and in the Sierra de Pénjamo (ca. 24 Ma, Castillo-Hernández and Romero-Ríos, 1991; Sole and Salinas, 2002).

The El Nayar sequence is a NNW-oriented belt, which has an average width of 75 km along the western edge of the Sierra Madre Occidental (Fig. 5). This sequence reaches its maximum thickness in the region of the Mesa del Nayar where Ferrari et al. (2002) reported the presence of a series of calderas and cauldrons that may represent the main eruptive sites (Fig. 5). In the interior of El Nayar caldera, 11 different ignimbritic units can be observed, with an average total thickness of ~1000 m (Ferrari et al., 2002). Seven $^{40}Ar/^{39}Ar$ ages were obtained from different stratigraphic levels of the El Nayar sequence, which partially overlap but define an age range between 21.2 and 19.9 Ma, with an average of 20.9 Ma for the sequence (Ferrari et al., 2002). Southwards, the sequence correlates with ignimbrites exposed in Santa María del Oro (21.3 Ma, Gastil et al., 1979), Aguamilpa (22.4 Ma, Damon et al., 1979), and in the upper part of the Bolaños graben (21.3–20.1 Ma, Scheubel et al., 1988) (Fig. 5). For the El Nayar sequence, Ferrari et al. (2002) estimated that a volume of ~4500 km^3 may have been erupted in ~1.4 m.y.

The sequence of El Nayar is truncated to the west by extensional faulting associated with the opening of the Gulf of California. However, in the southern part of the Baja California peninsula (areas of La Paz and Loreto), Hausback (1984) and Umhoefer et al. (2001) obtained K-Ar ages ranging between ca. 23 and ca. 17 Ma for different ignimbrites interbedded in the lower part of the Comondú Formation. They include the La Paz tuff, a sequence exposed near the city of La Paz, for which Hausback (1984) reported K-Ar ages from 21.8 ± 0.2 to 20.6 ± 0.2 Ma. Due to the large areal extent of volcanic rocks of El Nayar age, Ferrari et al. (2002) suggested that the La Paz tuff may be a distal deposit from a caldera-forming eruption in the region of La Mesa de El Nayar. The original distance between La Paz and the Mesa del Nayar is difficult to estimate, however, the edge of the El Nayar caldera is ~42 km from the coastal plain of Nayarit, whereas La Paz is located at a similar distance from the eastern coast of Baja California Sur. Therefore, the La Paz tuff may have been related to a caldera located less than 100 km further east.

The ignimbritic volcanism of the Sierra Madre Occidental does not continue further south of the Trans-Mexican Volcanic Belt in the Jalisco block. In contrast to what has been reported on several regional geological maps (e.g., Ortega-Gutiérrez et al., 1992; López-Ramos, 1995), ignimbrites of Eocene-Miocene age are not known in the Jalisco block. Different geochronologic studies have shown that the ignimbrites, which are widely exposed in the northern part of the block, have $^{40}Ar/^{39}Ar$ ages from 81 to 60 Ma (Wallace and Carmichael, 1989; Lange and Carmichael, 1991; Righter et al., 1995; Rosas-Elguera et al., 1997), and therefore most likely correlate with the Lower Volcanic Complex exposed in Sonora (McDowell et al., 2001). Oligocene and Early Miocene ignimbrites are reported south of the Trans-Mexican Volcanic Belt in Michoacán, to the south of the Chapala lake (31.8 Ma, Rosas-Elguera et al., 2003; 23.5 Ma, Ferrari et al., 2002) and to the south of Morelia (21 Ma; Pasquarè et al., 1991).

Postignimbritic Volcanism

After the ignimbritic pulse (the "ignimbrite flare up" of McDowell and Clabaugh, 1979), magmatism became more heterogeneous and dispersed in the Sierra Madre Occidental. Volcanism was generally bimodal and discontinuous, and tended to migrate west toward the region of the future Gulf of California. Among the mafic rocks, a group of basaltic-andesites has been distinguished, which was emplaced soon after the ignimbritic pulse, and a second group of more mafic and alkaline rocks that generally followed the ending of subduction. In the northern part of the Sierra Madre Occidental, the beginning of the second event was preceded by the emplacement of a series of alkaline ignimbrites with a distinctive character. In this section, we briefly synthesize the distribution and age of these rocks.

Postignimbritic Transitional Mafic Volcanism (Southern Cordillera Basaltic Andesite Province)

Across all of the Sierra Madre Occidental, basaltic-andesite lavas were emplaced discontinuously in the final stages of, and after, each ignimbritic episode. The main exposures of these lavas are distributed along a belt roughly oriented NNE between San Buenaventura and Chihuahua to the north, and Navojoa and Sinaloa to the south, and crossing Yécora (Fig. 3). These mafic volcanic rocks were grouped into the Southern Cordillera Basaltic Andesite Province, which was proposed by Cameron et al. (1989) for a regionally widespread assemblage that extends further north into Arizona and New Mexico, and is interpreted as representing an initial phase of extension in an intra-arc setting. The ages reported for these rocks in Chihuahua and Sonora vary between 33 and 17.6 Ma (Cameron et al., 1989 and references therein; McDowell et al., 1997; Paz-Moreno et al., 2003). However, most ages are Oligocene. In northeastern Sonora, some rhyolitic tuffs are interbedded with the basaltic-andesitic lavas from which a $^{40}Ar/^{39}Ar$ age of 25.4 Ma on sanidine was obtained (González-León et al., 2000). Volcanism became younger and progressively more silicic in composition to the west. In the Sierra Santa Úrsula north of Guaymas (Fig. 3), ignimbrites, andesites, and dacitic domes are reported with ages between ca. 23 and 15 Ma (Mora-Álvarez and McDowell, 2000). These rocks also outcrop to the northwest in the region of Bahía Kino (Fig. 3), where ca. 18 Ma andesitic rocks are reported (Gastil and Krummenacher, 1977).

In the central sector of the Sierra Madre Occidental, basaltic lavas dated between 30 and 29 Ma cover the ignimbritic sequence of Durango (Caleras basalts of Swanson et al., 1978). Due to their composition and age, Luhr et al. (2001) associated these rocks with the Southern Cordillera Basaltic Andesite Province. In addition, ca. 24 Ma basalts cover Oligocene ignimbrites in Nazas (Aguirre-Díaz and McDowell, 1993) and El Rodeo (Aranda-Gómez at al., 2003; Sole and Salinas, 2002). Cameron et al. (1989) included the Nazas basalts among the SCORBA-type. However, later studies by Luhr et al. (2001) and Aranda-Gómez et al. (2003) have shown that these alkaline rocks (hawaiites) have a geochemical signature more akin to intraplate basalts, and more typical of the Mexican Basin and Range Province.

For the southern part of the Sierra Madre Occidental, there are no geochemical and petrologic studies on postignimbritic mafic rocks, although they can be observed in different places. Basaltic lavas emplaced shortly after the Early Oligocene ignimbrites are exposed in the area of Huejuquilla (Fig. 5) and are reported in the 1:250,000 geological maps of the Consejo de Recursos Minerales in western Zacatecas and northern Jalisco. Basaltic lavas emplaced after the Early Miocene ignimbritic pulse have been mapped in the area of Milpillas (Fig. 4), from which whole rock K-Ar dates of ca. 21 Ma have been obtained (Sole and Salinas, 2002). Mafic volcanic rocks have also been observed in the area of the Mesa del Nayar and Jesús María (Fig. 5), and have been dated by $^{40}Ar/^{39}Ar$ at 21.3 Ma (whole rock; Ferrari et al., 2002). Similar basaltic lavas are also found in the Bolaños graben, with K-Ar ages between 21 and 19.9 Ma (Nieto-Obregón et al., 1981). Some of these basalts are intercalated with the ignimbrites in the upper part of the Bolaños section.

Alkaline and Peralkaline Volcanism

In the northwestern part of the Sierra Madre Occidental, post-subduction magmatism is marked by the eruption of a distinctive sequence of ignimbrites and peralkaline rhyolitic and rhyodacitic lavas, locally known as the Lista Blanca Formation. These rocks are widely exposed elsewhere in southwestern Sonora (Fig. 3) and have a relatively restricted age range between ca. 14 and 11 Ma (Gastil and Krummenacher, 1977; Bartolini et al., 1994; McDowell et al., 1997; Mora-Álvarez and McDowell, 2000; Oskin et al., 2003; Mora-Klepeis and McDowell, 2004; Vidal-Solano et al., 2005). Some of these ignimbrites have comenditic compositions and relatively high concentrations of iron and alkalis ($Na_2O + K_2O$ = 8–10 wt%). These alkaline geochemical signatures have been related to an asthenospheric mantle origin after an important period of crustal thinning during the initial opening of the Gulf of California, which promoted the up rise of convective mantle to the base of the continental crust (Vidal-Solano et al., 2005). Mora-Klepeis and McDowell (2004) interpreted the 12–11-Ma-old silicic rocks of the Sierra de Santa Úrsula as the first stage of postsubduction volcanism in Sonora, although it may more likely be a final magmatic phase related to intracontinental extension (Vidal-Solano et al., 2005). The 12.6 Ma San Felipe tuff is an important and widespread ignimbrite related to this event (Oskin et al., 2001; Oskin and Stock, 2003). It has been genetically associated with the Puertecitos volcanic province in Baja California, and has been correlated with different rock exposures along the Gulf of California up to coastal Sonora, and including Tiburón Island (Oskin et al., 2001).

In the rest of the Sierra Madre Occidental, alkaline volcanism essentially consists of alkali basalts that were emplaced as modest volumes of fissure-fed lavas, as well as large monogenetic volcanic fields. In general, these mafic rocks are found at the edges of the Sierra Madre Occidental, at the boundary with the Mesa Central, as well as in the eastern part of the Gulf of California. The alkali basalts are associated with three main extensional episodes that occurred during the Early Miocene (ca. 24–22 Ma), the Late Miocene (ca. 13–11 Ma), and the Plio-Quaternary (ca. 4–0 Ma) (Henry and Aranda-Gómez, 2000). At the eastern edge of the Sierra Madre Occidental, the main formations include: the Rodeo and Nazas basalts (Fig. 4) that were erupted from 24.1 to 23.3 Ma (Aranda-Gómez et al., 2003; Sole and Salinas, 2002) and at ca. 24 Ma (Aguirre-Díaz and McDowell, 1993), respectively; Metates Formation hawaiites in the southern part of the Rio Chico–Otinapa graben (Figs. 4 and 8), with ages of 12.7–11.6 Ma (McDowell and Keizer, 1977; Henry and Aranda-Gómez, 2000); the Camargo volcanic field, with ages from 4.7 Ma to the Holocene (Aranda-Gómez et al., 2003); and the Quaternary volcanic field of Durango (Smith et al., 1989). Extensive basaltic lavas exposed along the southern edge of the Sierra Madre Occidental in the area of La Manga (Fig. 5) have not been dated yet, but may correlate with this volcanic episode since mafic dikes exposed in nearby areas have been dated at ca. 11 Ma (Damon et al., 1979). Along the western edge of the Sierra Madre Occidental along the Gulf of California, postsubduction mafic rocks include: ca. 11–10 Ma tholeiitic basalts along the Sonora coast (Mora-Álvarez and McDowell, 2000; Mora-Klepeis and McDowell, 2004); the Pericos volcanic field to the north of Culiacán (this has not been studied, but is interpreted to be Latest Pliocene to Quaternary in age based on its young morphology); and the basalts of Punta Piaxtla and Mesa de Cacaxtla, exposed to the north of Mazatlán (3.2–2.1 Ma, Aranda-Gómez et al., 2003) (Fig. 5). Additionally, alkaline mafic dikes with a dominant NNW direction are common in the southwestern part of the Sierra Madre Occidental, from which Late Miocene ages (ca. 12–10 Ma) have been obtained in southern Sinaloa (Henry and Aranda-Gómez, 2000), as well as in Nayarit (Ferrari et al., 2002 and references therein). Available geochemical data are scarce but indicate that these rocks are also subalkaline in composition.

3. TECTONICS OF THE SIERRA MADRE OCCIDENTAL IGNEOUS COMPLEXES

Pre-Oligocene Deformation

Deformation that preceded Cenozoic extension has not been studied in detail. This is partly due to the scarcity of outcrops, the intense weathering of rocks, and the intense normal faulting that may obscure previous episodes of deformation in some areas. In Sonora, the Aptian-Albian marine succession is affected by folds and reverse faults but they did not produce any significant deformation in the overlying Lower Volcanic Complex rocks. In Sinaloa, Henry et al. (2003) recognized foliation and dynamic recrystallization textures in some tonalites and granodiorites of the coastal batholitic complex. The foliation is vertical on ENE-striking planes in the batholith and subparallel to the host rocks (orthogneiss, gabbro, and marble), suggesting that all these rocks were deformed at the same time (Henry 1986; Henry et al., 2003). K-Ar ages of different minerals and U-Pb ages of zircons from pre-, syn-, and post-tectonic plutons were interpreted by Henry et al. (2003) to indicate that deformation occurred between ca. 101 and ca. 89 Ma.

The Late Cretaceous to Paleocene volcanic succession of the Tarahumara Formation (ca. 90–60 Ma) is faulted and tilted, but folding and thrust faulting are not observed such that tilting may be related to Neogene normal faulting (McDowell et al., 2001). Further east in central Chihuahua, tilting of the ca. 68 Ma Peña Azules volcanic sequence was attributed to the Laramide orogeny by McDowell and Mauger (1994). However, the absence of reverse faulting or folding again raises the possibility that tilting was the result of Late Cenozoic normal faulting. East of Zacatecas, Campanian age intrusive rocks in the La Tesorera–Zacatón area are undeformed, as opposed to the Early Cretaceous marine succession that they intrude, which is involved in thrusts and open folds.

In summary, the few available data seem to indicate that between the Coniacian and Eocene, contractile deformation was negligible across most of the Sierra Madre Occidental, which was restricted to northwestern México, and in contrast to the western North American cordillera to the north. In Sonora and Sinaloa in the western part of the Sierra Madre Occidental, it is also common to find ENE-WSW–trending extensional fractures and faults that affected pre-Oligocene rocks (Horner and Enríquez, 1999; Staude and Barton, 2001). Most of the Cu-Mo porphyry deposits of the Sierra Madre Occidental are emplaced in highly fractured zones, developed concurrently, or at a late stage, during this phase of deformation (Barton et al., 1995; Horner and Enríquez, 1999). Geochronology studies of these deposits have consistently produced Paleocene and Eocene K-Ar ages (Damon et al., 1983; Staude and Barton, 2001), some of which have been recently confirmed by Re-Os dating (Barra et al., 2005). Horner and Enríquez (1999) interpreted the E-W– and ENE-WSW–trending structures as the result of a final phase of the Laramide shortening. However, the available data also permit these structures to be related to a different deformational episode that was transitional between the Laramide orogeny and the Oligo-Pliocene phase of extension.

Extensional Tectonics

Most of the Sierra Madre Occidental has been affected by different episodes of dominantly extensional deformation that began in the Oligocene and potentially at the end of Eocene. Extensional deformation has not affected the core of the Sierra Madre Occidental, which now represents a physiographic boundary between what has been defined as the "Mexican Basin and Range," to the east, and the "Gulf Extensional Province," to the west (Henry and Aranda-Gómez, 2000). In this work, we use the term "Basin and Range" essentially in a physiographic sense, with no genetic implications (i.e., in the sense of Dickinson, 2002). At the northern and southern ends of the Sierra Madre Occidental (i.e., northern Sonora and Chihuahua and Nayarit-Jalisco, respectively), these two provinces merge where extension has affected the entire width of the Sierra Madre Occidental (Fig. 6).

Northern Sector

Understanding the tectonic events that affected the northern part of the Sierra Madre Occidental is particularly complex since various extensional episodes overlap in space and time with igneous activity. It is evident that extension had begun in the Eocene, since a moderate angular unconformity exists between a 42–37 Ma succession and Oligocene ignimbrite south of Chihuahua (Megaw, 1990). In contrast, McDowell and Mauger (1994) considered that the transition between a contractile and an extensional deformation regime was at ca. 33 Ma, as recorded by the first occurrences of peralkaline ignimbrites and transitional basalts (SCORBA) in the region. The first extensional episode that can be documented regionally from structural evidence immediately follows the Oligocene silicic volcanism, whose maximum activity occurred between 34 and 29 Ma (McDowell and Clabaugh, 1979). Extension in Chihuahua is limited to Basin and Range structures that formed after 29 Ma based on different depositional relationships between pre- and post-29 Ma ignimbrites. No detailed structural studies are available for this region, but the presence of high-angle normal faults and the modest tilting of the volcanic successions suggest a limited amount of extension.

In Sonora, extension was much more intense and slightly younger than in Chihuahua. During a major episode of intracontinental deformation, rocks formed at middle crustal depths were locally exhumed along a wide belt subparallel to the Sierra Madre Occidental between Hermosillo and Tecoripa (Fig. 7) (Nourse et al., 1994; Vega-Granillo and Calmus, 2003; Wong and Gans, 2003, 2004). Along this belt are both high- and low-angle normal faults as well as metamorphic core complexes, which are characterized by detachment faults along which undeformed upper plate rocks are juxtaposed with lower plate milonite, gneiss, and peraluminous plutons (Davis and Coney, 1979; Nourse et al., 1994). Crustal extension during this episode is estimated to have locally exceeded 100% (Gans, 1997), and also formed tectonic troughs that were filled by clastic sediments with occasional accumulation of borate, lavas, and pyroclastic rocks (Fig. 3 and 7). Metamorphic core complexes are well known in the Magdalena (Nourse et al., 1994), Acónchi (Rodríguez-Castañeda, 1996; Calmus et al., 1996), Puerto del Sol (Nourse et al., 1994), and Mazatán areas (Vega-Granillo and Calmus, 2003; Gans et al., 2003; Wong and Gans, 2003) (Figs. 3 and 7). Peraluminous plutons exposed in these areas are thought to have been generated by partial melting of deformed crust (e.g., Nouse et al., 1994). Modeling $^{40}Ar/^{39}Ar$ ages in K-Feldspar, Gans et al. (2003) and Wong and Gans (2003) suggested an age between 20 and 16 Ma for the exhumation of the Mazatán core complex (Fig. 3), which is in agreement with an age of 18 ± 3 Ma obtained from apatite fission-track analysis for the same area (Vega-Granillo and Calmus, 2003). Published ages compiled by Nourse et al. (1994) for the different metamorphic core complexes in Sonora indicate an age range of ca. 25–15 Ma for this episode of crustal extension.

In addition to core complexes, extensional basins bounded by high-angle normal faults are common in central-eastern Sonora, and provide an additional constraint on the age of this extensional event. They are generally oriented NNW-SSE to ~N-S (Fig. 7), and contain thick successions of highly compacted conglomerate and sandstone that were initially defined as the Baucarit Formation (King, 1939). Basaltic to andesitic lavas ranging in age between 27 and 20 Ma are common toward the base of these clastic successions (McDowell et al., 1997; Paz-Moreno et al., 2003). The upper parts of these rift basin successions are less compacted and contain intercalated Middle Miocene tuff and rhyolite to rhyodacite lavas. Extension in these basins is much less than in the core complexes, indicating that extension may have been focused in areas where the basement had been previously weakened, and that the high values of stretching of ~100% obtained for some areas (Gans, 1997) is not representative for all of Sonora.

Along a coastal belt in Sonora, volcanic successions of Middle Miocene age are modestly tilted, with inclinations between 10° and 35° to the E or the W (McDowell et al., 1997; Mora-Álvarez and McDowell, 2000; MacMillan et al., 2003; Gans et al. 2003). The tilted blocks are covered in angular unconformity by flat-lying alkaline basalt lavas. In the Guaymas, Sierra Libre and Sierra del Bacatete areas, age dating constrains the boundary between the tilted rocks and flat-lying basalts to between ca. 12 and 10 Ma (Mora-Álvarez and McDowell, 2000; MacMillan et al., 2003) and between 10.7 and 9.3 Ma in the San Carlos–El Agujaje area (Gans et al., 2003) (Fig. 7). To the NW of Guaymas (Fig. 2), however, alkaline basalts dated at 8.3 Ma are cut by normal faults related to a younger extensional episode producing the "Empalme graben" described by Roldán-Quintana et al. (2004) (Fig. 7). The Empalme graben has been interpreted as marking the transition between Basin and Range–style block faulting and a strike-slip deformational regime associated with the beginning of the opening of the Gulf of California.

Although it is low in intensity, tectonic activity is still active in northeastern Sonora and northwestern Chihuahua. At least 64 historic earthquakes were recorded from 1887 to 1999 (Suter, 2001). The largest of these events is the Bavispe, Sonora, earthquake, of 3 May 1887, with $M_w = 7.4$. The earthquake rupture has been traced for over 100 km along three segments of a N-S fault active since the Miocene (Suter and Contreras, 2002).

Central Sector

In the central sector, extensional tectonics affected the Sierra Madre Occidental mainly along its borders, leaving a relatively unextended zone at its center (Fig. 6). Along its easternmost margin in Durango State, high-angle normal faults define basins much like those found in Chihuahua. The age of extensional deformation in this area dates back at least to the Oligocene and is characterized by a general ENE-WSW direction of elongation. In the Nazas area, ignimbrites dated by K-Ar at 29.9 ± 1.6 Ma are tilted up to 35° to the NE and are covered by flat-lying ignimbrites dated at 29.5 ± 0.6 Ma (Aguirre-Díaz and McDowell, 1993). In the Rodeo area of Durango State, Luhr et al. (2001) recognized an early episode of extension between 32.3 and 30.6 Ma that led to the formation of a NNW-trending half graben with an estimated ~3 km of vertical displacement. In addition, Aranda-Gómez et al. (2003) related the eruption of ca. 24 Ma alkaline lavas in both Nazas and Rodeo to a second extensional episode (Fig. 8). To the SSW of Durango city, the Mezquital graben is a 40-km-wide structure with a NNE orientation. The graben has not been studied in detail but the rocks cut by the high-angle–bounding faults are Oligocene in age, as the youngest is an ignimbrite dated at 27.0 ± 1.0 Ma (K-Ar age reported in Aranda-Gómez et al., 1997). In this area, these authors observed two generations of striae in the normal fault planes: the oldest one indicated extension oriented NW whereas the younger generation suggested a NE direction of extension.

To the west of Durango city, the Rio Chico–Otinapa graben is a 160-km-long and 20-km-wide extensional structure with a N-S to NNW-SSE orientation and a minimum vertical displacement estimated at 900 m (Aranda-Gómez et al., 2003) (Fig. 8). The high-angle normal faults bounding the graben cut the Oligocene ignimbrite sequence, which is covered by hawaiite lavas of the Metates Formation (Córdoba, 1963); the faults also acted as feeders for the hawaiite lavas. Amphibole separated from these lavas yielded ages of 12.7 ± 0.4 Ma by the K-Ar method (McDowell and Keizer, 1977) and 11.60 ± 0.07 Ma by the $^{40}Ar/^{39}Ar$ method (Henry and Aranda-Gómez, 2000). Based on these tectonic and stratigraphic relationships, it is concluded that the Rio Chico–Otinapa graben must have initiated before the eruption of the lavas, likely around ~12 Ma in response to WSW-ENE extension (Aranda-Gómez et al., 2003).

The western side of the Sierra Madre Occidental in Sinaloa has been profoundly affected by extensional faulting producing half grabens with a general NNW orientation. A NE-trending accommodation zone north of Tayoltita divides the region into two tilt domains, with the volcanic succession inclined to the ENE to the north, and to the WSW to the south (Fig. 8). The Concordia Fault is a major NW-trending structure to the east of Mazatlán (Fig. 8) that dips 40° to 70°NE, and with a vertical displacement of ~5 km (Aranda-Gómez et al., 2003). The Late Cretaceous to Paleocene Sinaloa batholith occurs in the footwall of the Concordia Fault, whereas the hanging wall consists of Oligocene to Early Miocene ignimbrites covered by poorly consolidated and weakly graded gravels that are also intruded by mafic dikes. Two mafic dikes have been dated at 10.7 ± 0.2 and 11.03 ± 0.16 Ma by $^{40}Ar/^{39}Ar$ on whole rock (Henry and Aranda-Gómez, 2000). Some dikes are also tilted indicating that extension continued after their emplacement. Extension in this region is estimated to range between 20% and 50% for a listric or plane geometry, respectively, to the Concordia Fault (Henry, 1989).

Close to the Sinaloa-Durango state boundary along the Durango-Mazatlán federal highway, the Early Miocene ignimbrite sequence is flat-lying and only minor faults are observed. However, to the north of the highway the same sequence is tilted by up to 30° due to normal faulting both in the Presidio River area (Aranda Gómez et. al, 2003) and in the valley of Piaxtla River close to Tayoltita (Horner and Enríquez, 1999; Enríquez and Rivera, 2001). This indicates that Middle to Late Miocene extension preceding the opening of the Gulf of California must have discontinuously affected the interior of the Sierra Madre Occidental.

Southern Sector

In the southern sector, all of the Sierra Madre Occidental has been affected by extensional tectonics (Figs. 6 and 9). At the northeastern boundary of this region in Fresnillo, Zacatecas State, an early phase of extension is evident with 39-Ma-old ignimbrites tilted up to 30° to the SW, which are covered by horizontal tuffs whose secondary alteration has been dated by K-Ar at 29.1 Ma (Lang et al., 1988). In the rest of the southern sector, however, there is no evidence of a pre-Oligocene phase of extension. In southern Zacatecas and northern Jalisco, several grabens are the continuation of extensional structures affecting the Central Mesa to the east (Nieto-Samaniego et al., 1999), whereas further west in Nayarit, half grabens are the dominant structures (Fig. 9). Ferrari et al. (2002) grouped the structure of the southern part of the Sierra Madre Occidental into three main domains (Fig. 9). An eastern domain is dominated by NNE- to N-S–oriented grabens between 40 and 120 km long that cut into Late Oligocene or Early Miocene ignimbrites. In the Tlaltenango graben, the youngest ignimbrite of the succession is 22 Ma (Moore et al., 1994) and is cut by normal faults with a minimum displacement of 400 m. Inside the graben, a shield volcano dated at 21 Ma (Moore et al., 1994) has fault scarps less than 50 m high, which suggests that some extension may have occurred between ca. 22 and ca. 20 Ma (Ferrari et al., 2002). In the Bolaños graben, which has a vertical displacement exceeding 1400 m, crosscutting relations among the different volcanic units indicate extension in the Early Miocene, which likely occurred in several phases (Ferrari et al., 2002; Lyons, 1988). For the other grabens, the present state of knowledge is insufficient to precisely constrain the inception of extension, but in all cases, normal faults related to WNW to E-W extension cut Early Miocene ignimbrites. In summary, extension leading to the development of grabens was generally synchronous across the eastern domain and began during the Early Miocene (Fig. 9).

The main structures of the western domain are the Alica, Pajaritos, and Jesús María half grabens, as well as the Pochotitán and San Pedro normal fault systems (Fig. 9). All these structures have a N-S to NNW-SSE orientation and systematically tilt blocks of Early Miocene ignimbrites to the ENE. An NE-trending accommodation zone across which there is a tilting reversal is inferred along the Mezquital River, such that to the north the tilt vergence is to the WSW (see also discussion for the northern sector). Master faults of the half grabens cut ignimbrites of the Nayar succession dated at ca. 21 Ma (Ferrari et al., 2002) but no minimum age can be inferred for fault activity. The San Pedro and Pochotitán fault systems were formed by an ENE- to NE-SW-directed extension and can be considered part of the Gulf Extensional Province. The Pochotitán fault system tilts rocks as young as 17 Ma, which are covered by flat-lying basaltic lavas dated at ca. 10 Ma (Ferrari and Rosas-Elguera, 2000). In this region, many mafic dikes intrude NNW normal faults or strike parallel to them. As in southern Sinaloa, the mafic dikes were intruded between 11.9 and 10.9 Ma, and were concurrent with the extensional tectonics (Ferrari et al., 2002).

The structural character of the southern domain contrasts with that of the rest of the southern sector as Oligocene and Early Miocene volcanic sequences of the Sierra Madre Occidental are characterized by open folds in an en echelon array, small thrusts and several left lateral faults all of which developed in the Middle Miocene (Ferrari, 1995). These structures are distributed in a belt with a WNW-ESE orientation at the southernmost boundary of the Sierra Madre Occidental with the Jalisco block basement. The folds are cut by vertical mafic dikes dated at ca. 11 Ma (Damon et al., 1979), providing a minimum age for the deformation. Ferrari (1995) interpreted these structures as a left lateral transpressional shear zone produced by the opposite motion between the Sierra Madre Occidental, during the waning of subduction of the Magdalena microplate, and the Jalisco block, beneath which subduction of the Cocos plate continued.

4. PETROLOGY AND GEOCHEMICAL CHARACTERISTICS OF THE SIERRA MADRE OCCIDENTAL MAGMATISM

Volcanic rocks of the Sierra Madre Occidental form a typical calc-alkaline association, characterized by intermediate to high potassium contents (Cochemé and Demant, 1991), together with a relatively low Fe enrichment (Cameron et al., 1980a). The main petrologic characteristics of these rocks are shown in Figure 10, which is based primarily on data from the northern sector of the Sierra Madre Occidental. The data indicate a broad range of compositions between ~49 and 78 wt% SiO_2, although igneous compositions are predominantly bimodal with silicic (~66–78 wt% SiO_2) and mafic (49–62 wt% SiO_2) groupings. Intermediate compositions (~62–66 wt% SiO_2) are rare. The Oligo-Miocene rhyodacite and rhyolite ignimbrites define the silicic compositional grouping whereas the mafic grouping includes andesite, basaltic andesite, and basalts of the Lower Volcanic Complex and postignimbritic transitional and alkaline mafic volcanism. The total alkalis-silica plot indicates a dominantly subalkaline association with a subordinate group of samples plotting in the alkaline field, which are mainly the postignimbritic mafic volcanic rocks. The postignimbritic mafic volcanic rocks also extend to subalkaline compositions and overlap the field of the southern Cordillera basaltic andesites (SCORBA), supporting the extension of this province to the northern part of the Sierra Madre Occidental as suggested by Cameron et al. (1989). Lower Volcanic Complex igneous compositions are broadly similar to the rest of the Sierra Madre Occidental rocks (Fig. 10), but are entirely subalkaline, have a more restricted range of silica contents with true basaltic compositions absent, and do not display a bimodal character (e.g., Valencia-Moreno et al., 2001, 2003). It is also worth comparing the Eocene to Oligocene and Neogene (ca. 25–10 Ma) rocks of Sierra Madre Occidental. Although only a few data are available for the Neogene volcanic rocks, they display a similar range of silica contents but rhyolitic compositions sensu stricto are absent. The limited trace element data available for the Sierra Madre Occidental rocks are indicative of a continental margin setting influenced by subduction. Linear trends in basalt-andesite-rhyolite suites in the Batopilas region for elements such as Rb, Sr, Nb, Y, Th, Zr, and REE were interpreted in early studies as the result of assimilation and fractional crystallization of mantle-derived mafic magmas (Cameron et al., 1980b; Bagby et al., 1981).

The mechanism responsible for the generation of the huge volume of silicic ignimbrites of the Sierra Madre Occidental has become highly debated since the first geochemical studies were published in the early 1980s (e.g., Cameron et al., 1980a and b; Cameron and Hanson., 1982; Verma, 1984; Ruiz et al., 1988a; Cameron et al., 1992; McDowell et al., 1999; Albrecht and Goldstein, 2000). Most approaches addressing this issue have relied on the whole-rock isotopic signatures of the volcanic rocks. Numerous studies have established that the initial $^{87}Sr/^{86}Sr$ ratios range between 0.7041 and 0.7070 and εNd between +2.3 and −3.2 for ignimbrites from several sites of the Sierra Madre Occidental. Lower εNd values (−5.2 to −5.8), and higher $^{87}Sr/^{86}Sr$ ratios (0.7089 and 0.7086) have been reported for areas of the northern Sierra Madre Occidental in the Tómochic and San Buenaventura areas (McDowell et al., 1999; Albrecht and Goldstein, 2000). Figure 11 shows the distribution of both εNd and $^{87}Sr/^{86}Sr$ isotopic data available for the Sierra Madre Occidental rocks. In general,

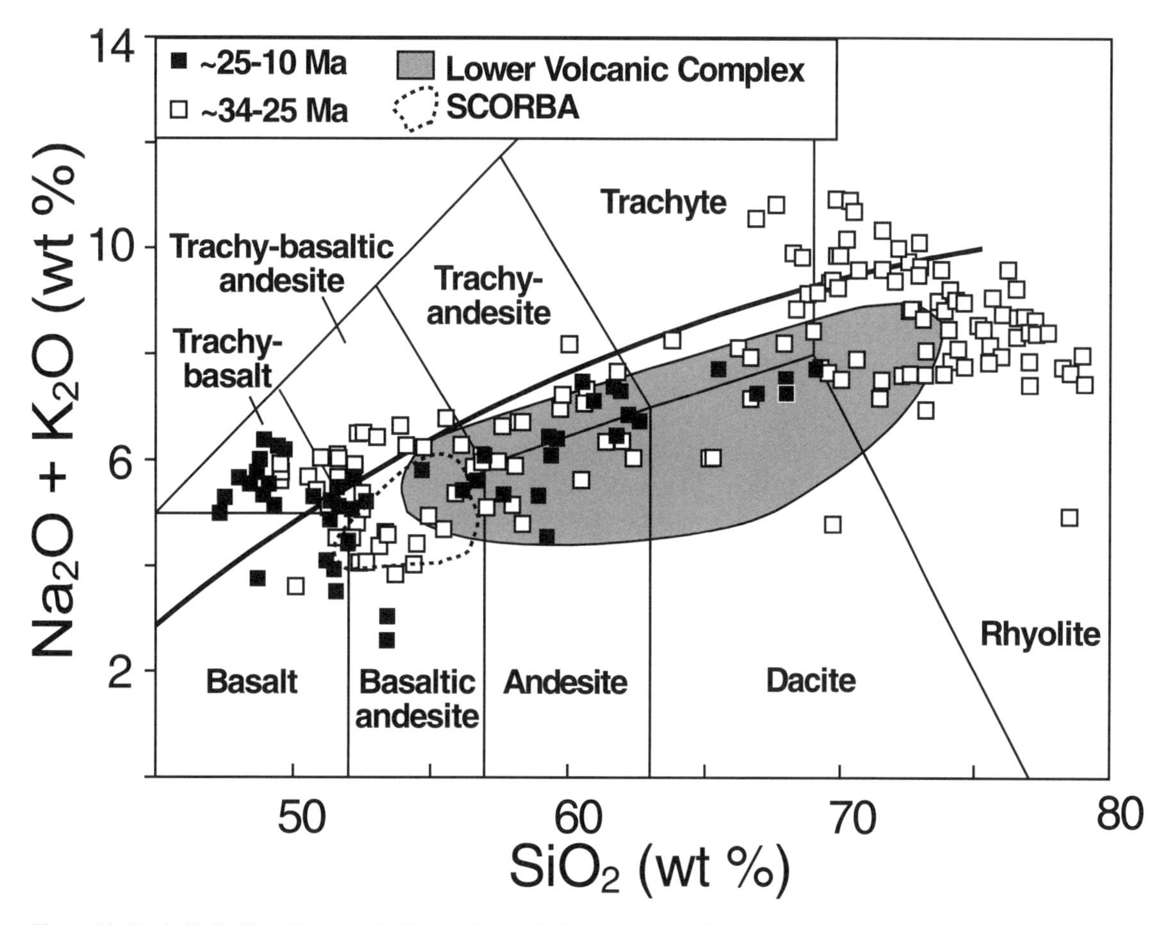

Figure 10. Total alkali-silica diagram (TAS) according to LeMaitre et al. (1989) for volcanic rocks of the northern Sierra Madre Occidental (open squares). The boundary between the alkaline and subalkaline fields (thicker line) is according to Irvine and Baragar (1971). The black squares represent volcanic rocks from the eastern Sierra Madre Occidental, within the Basin and Range Province. The field of the Southern cordillera basaltic andesites (SCORBA), based on Figure 5 of McDowell et al. (1997), is included here for comparison (dotted line). The field of magmatic rocks of the Lower Volcanic Complex (LVC) (shaded), based mainly on data from granitic intrusives, is also included. Sources: Sierra Madre Occidental—Cameron et al., 1980; Lanphere et al., 1980; Piguet, 1987; Wark, 1991; Gans, 1997; McDowell et al., 1997, 1999; Albrecht and Goldstein, 2000; González-León et al., 2000; Mora-Álvarez and McDowell, 2000. Post–Sierra Madre Occidental—Gastil and Krummenacher, 1977; Gastil et al., 1979; Bartolini et al., 1994; Gans, 1997; McDowell et al., 1997; González-León et al., 2000; Mora-Álvarez and McDowell, 2000; Henry et al., 2003. Lower Volcanic Complex—Bagby et al., 1981; Mora-Álvarez and McDowell, 2000; Roldán-Quintana, 1991; Valencia-Moreno et al., 2001, 2003; Henry et al., 2003.

many of the ignimbrites have isotopic compositions similar to the bulk Earth composition. The plot also shows a relatively continuous trend between samples with a more pronounced mantle signature (positive εNd and relatively primitive Sr values) to samples with negative εNd values and higher Sr ratios, suggesting significant crustal involvement in their petrogenesis.

To generate the rhyolites, Cameron et al. (1980a) proposed an AFC (assimilation and fractional crystallization) model for mantle-derived mafic magmas where the ignimbrites represented no more than 20% of the initial basaltic magma volume. This model implied that ~80% of the initial volume must have remained as residual material, adding some kilometers of new gabbroic crust beneath the Sierra Madre Occidental (Ruiz et al., 1988b; Cameron et al., 1992). Although isotopic studies are limited to a few areas, Cameron et al. (1986) suggested that the AFC model may be representative for the whole Sierra Madre Occidental and proposed that less than 25% crustal assimilation was involved in the generation of the ignimbrites. Verma (1984) put forward a similar genetic model for ignimbrites in Zacatecas and San Luis Potosí areas (Fig. 6), but also considered a second, shallower stage of fraction crystallization, which led to crustal assimilation being in the order of 80% before the final emplacement of the ignimbrites. In contrast, based on the modeling of many lower crustal xenoliths from areas to the east of the Sierra Madre Occidental, Ruiz et al. (1988b) proposed that most, if not all, of the volume of Sierra Madre Occidental ignimbrites may have been generated by partial melting of the lower continental crust since the isotopic compositions of the

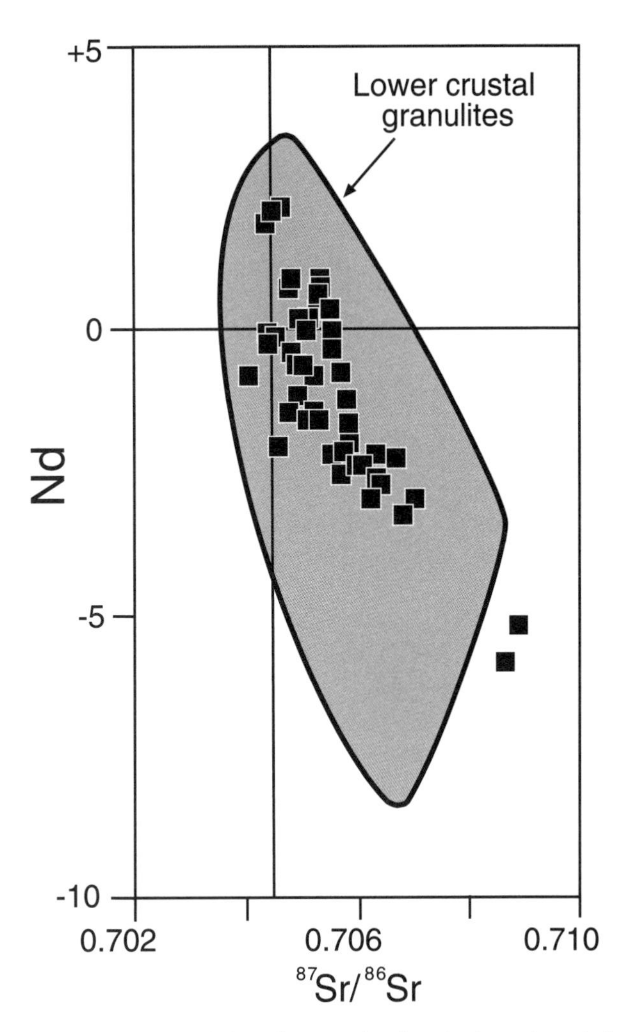

Figure 11. Sr-Nd correlation diagram showing the isotopic variation of volcanic rocks of the mid-Tertiary Sierra Madre Occidental. The cross hairs represent the bulk Earth composition. The field of the lower crustal granulites corresponds to intermediate and mafic orthognesiss xenoliths after Ruiz et al. (1988a). Sources: Wark, 1991; McDowell et al., 1999; Albrecht and Goldstein, 2000.

available crustal xenoliths (mafic to intermediate granulites of Paleozoic to Precambrian age) were identical to those of the Sierra Madre Occidental ignimbrites (Fig. 11). An important implication of this interpretation was that the magmatic episode that generated the Sierra Madre Occidental did not represent a period of significant crustal growth. Although mafic granulites/metaigneous xenoliths have the most appropriate chemical and isotopic compositions to be potential source materials for the Sierra Madre Occidental rhyolites, it remains unclear whether the xenoliths represent old lithosphere that was melted to generate the rhyolites (Ruiz et al., 1988b) or juvenile underplated material (i.e., cognate cumulate) essentially coeval with the rhyolites, and therefore representing several km of new crust that formed during the middle Tertiary beneath México (Cameron and Robinson, 1990).

5. THE CRUST AND THE MANTLE BENEATH THE SIERRA MADRE OCCIDENTAL: GEOPHYSICAL DATA

Despite its scientific and economic significance, few geophysical studies are available for the Sierra Madre Occidental. Most studies have a regional character and, although they define the major structures of the crust and upper mantle, and have not correlated the geophysical features with the detail of the surface geology.

The gross structure of the upper mantle beneath the Sierra Madre Occidental is mainly known from regional or global seismic tomography studies using different kinds of seismic waves and processing methods (Grand, 1994; Alsina et al., 1996; Van der Lee and Nolet, 1997; Bijwaard and Spakman, 2000; Ritzwoller et al., 2002; Ritsema et al., 2004). All of these studies have noted that mantle lithosphere beneath the Sierra Madre Occidental is thin or lacking. Although differing in details, all tomographic models show a low-velocity zone extending from ~80 km to ~250 km. This negative velocity anomaly extends from the Gulf of California to the Mesa Central, and from the U.S. Basin and Range Province to latitude 20° N in southern México. Estimations of the thermal structure based on tomographic models indicate that in this wide region, the mantle has a temperature ~500 °C higher than beneath the North America cratón to the east (Goes and van der Lee, 2002). These observations suggest that the lithospheric mantle has been mostly removed and replaced by asthenosphere.

Seismological data show a significant difference in crustal thickness between the Sierra Madre Occidental core and its margins. In a regional seismic study using surface and S-wave velocities, Gomberg et al. (1988) estimated an average thickness of 40 km for northern México. Given the ray trajectories used in this study, the authors also considered this crustal thickness as representative of the region to the east of the Sierra Madre Occidental, and the northern part of the Mesa Central.

More recently, Bonner and Herrin (1999) defined the crustal thickness of the northern part of the Sierra Madre Occidental by studying dispersion of surface waves generated by earthquakes occurring in the Gulf of California and received in Texas. Given this geographic array, the results may be considered representative of the central, less extended, part of the northern Sierra Madre Occidental. A crustal profile 55 km thick and with three layers: an ~5-km-thick layer with low velocity (Vs ~2.8 km/sec), an intermediate layer of ~20 km with Vs ~3.6 km/sec, and a lower layer of ~30 km characterized by high seismic velocity (Vs ~4.0 km/sec) provided the best fit to the seismic data (Bonner and Herrin, 1999).

In another recent study, Persaud (2003) used receiver functions at three points along the western side of the Sierra Madre Occidental to constrain the Moho depth in this part of the province where the crust has been thinned by extension that led to the formation of the Gulf of California. Moho depths reported in this study range from 28 km east of Hermosillo to 22 km both in southern Sonora (Navojoa) and northern Sinaloa (Culiacán) (Fig. 6).

Integrating gravimetric and seismic refraction data, Couch et al. (1991) estimated a minor contrast in crustal thickness in the central-southern part of the Sierra Madre Occidental where the crust at the center of the Sierra Madre Occidental is ~40 km but thins to 25 km at the coast south of Mazatlán (Fig. 6). Crustal thickness is also less to the east of the Sierra Madre Occidental beneath the Mesa Central; Fix (1975) estimated a value of ~30 km based on surface waves, whereas Campos-Enríquez et al. (1994) estimated a Moho depth of 33 km (Fig. 6).

Assuming a maximum Moho depth of 55 km for the unextended core of the northern Sierra Madre Occidental (Fig. 6) at the end of the ignimbritic pulse and an average of ~25 km for the Moho in Sonora and Sinaloa, Oligo-Miocene stretching of the crust on the western side of the Sierra Madre Occidental must have exceeded 100% if extension was uniform. This value is comparable with the extension calculated at surface for the region of core complexes in Sonora (e.g., Gans, 1997), but contrasts with geologic estimates in other areas of Sonora (McDowell et al., 1997) and Sinaloa (Henry, 1989), where less than 50% of extension and more likely in the order of ~20%–30% is indicated. The extreme extension estimated in certain areas of Sonora is a local feature where basement structures permitted the focusing of deformation, and/or there was decoupling between brittle and ductile crust, such that the latter may have flowed laterally during continental extension and opening of the Gulf of California (Persaud, 2003). This situation is not too different from in the western United States, where areas of highly extended crust are adjacent to areas of relatively unextended crust (Gans and Bohrson, 1998; Gans et al. 1989).

6. DISCUSSION

Space-Time Evolution of Convergent Margin Magmatism and Extensional Tectonics

The magmatic history of the Sierra Madre Occidental is intimately related to the evolution of the western margin of North America and the history of subduction of the Farallon plate. In a simplistic way, the magmatic evolution of the Sierra Madre Occidental fits into the pattern of inland migration and subsequent trenchward return of igneous activity already recognized for the southwestern part of the North America Cordillera between the Late Cretaceous and the Present (e.g., Coney and Reynolds, 1977; Damon et al., 1981; Damon et al., 1983) (Fig. 12A). According to this model, arc migration was primarily controlled by the variation in the dip of the Farallon plate being subducted beneath North America. At the beginning of the Late Cretaceous, the arc was located relatively close to the trench, with the Sierra Nevada and Peninsular Range batholiths (including Baja California) and the Lower Volcanic Complex of the Sierra Madre Occidental interpreted as the eroded remnants of this supra-subduction zone magmatic arc. The eastward migration of magmatism at the end of the Cretaceous has been related to the progressive decrease in slab dip associated with the Laramide orogeny. Once compression waned at the end of the Eocene, supra-subduction zone magmatism retreated westwards toward the trench as the dip of the subducted slab steepened.

Superficially, the space-time patterns of magmatism in México and the Sierra Madre Occidental show this type of pattern (Fig. 12). Eastward migration of magmatism is more evident in the United States and the northern Sierra Madre Occidental, where it reaches ~1000 km from the paleotrench (Damon et al. 1981), than in its central and southern parts. Henry et al. (2003) showed that magmatism in the central sector of the Sierra Madre Occidental only reached 400 km from the paleotrench and that the rate of eastward migration of magmatism took place at an order of magnitude lower (1–1.5 km/Ma) than the westward migration toward the trench. For the southern part of the Sierra Madre Occidental, the easternmost position of magmatism was in the Oligocene, when it reached a maximum of 600 km from the paleotrench (Nieto-Samaniego et al., 1999).

In the northern part of the Sierra Madre Occidental, eastward migration volcanism seems to postdate contractile deformation. The volcanic rocks of the Lower Volcanic Complex (ca. 90–60 Ma), both in Sonora (Tarahumara Fm.) and in Chihuahua (Peñas Azules volcanics), are only tilted and do not show clear evidence of having been affected by shortening. In the central part of the Sierra Madre Occidental, the ca. 101–89 Ma syntectonic plutons of Sinaloa (Henry et al., 2003) record deformation in the Lower Volcanic Complex that pre-dates Paleogene Laramide deformation of the Sierra Madre Oriental (Eguiluz de Antuñano et al., 2000). An additional point is that the Eocene rocks along the entire Sierra Madre Occidental do not show any evidence of contractile deformation.

The presence, during Late Cretaceous to Paleogene times, of magmatism located between the paleotrench and the deformation front in México precludes any model of flat subduction, as invoked to explain the Laramide orogeny in the United States (e.g., Coney and Reynolds, 1977, Bird, 1984, 1988; Saleeby, 2003), since the necessary coupling between subducting and upper plates would close the mantle wedge and shut-off arc magmatism. This is a similar situation for the Canadian sector of the Laramide orogen (English et al., 2003) and an alternative model is required to explain magmatism and deformation in México and Canada.

At a continental scale, volcanism during the Oligocene extended westwards toward the trench, although in detail, time-space patterns at the local scale may be more complex. Importantly, the Late Oligocene ignimbrite flare-up has essentially the same age across the entire Sierra Madre Occidental (ca. 32–28 Ma), and is distributed in a wide belt with a general NNW orientation (Fig. 2) without any apparent internal migration. In contrast, Early Miocene volcanism is clearly displaced toward the western half of the Sierra Madre Occidental and shows significant differences from north to south. In the northern Sierra Madre Occidental, Early Miocene volcanism is less abundant and was dominantly mafic in composition, whereas in the central and southern Sierra Madre Occidental, volcanism was more

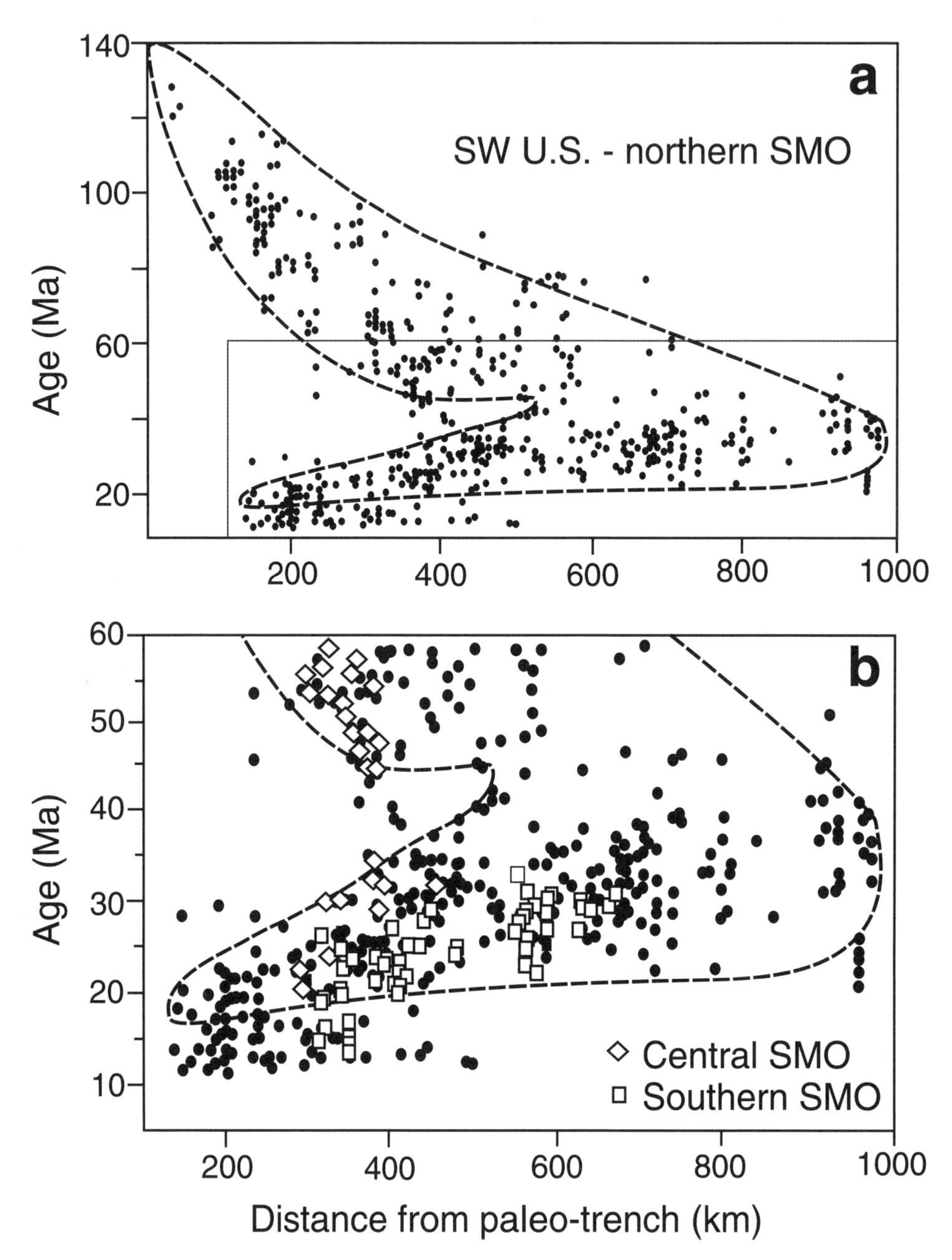

Figure 12. Distribution of available ages of Early Cretaceous to Miocene igneous rocks from México versus distance from the paleotrench. (A) Distribution of ages between ca. 140 and 10 Ma (modified from Damon et al., 1981) showing the inland migration of Cordilleran magmatism up to ~1000 km from the paleotrench between the Cretaceous and the end of Laramide orogeny (ca. 40 Ma). The inland migration was followed by an apparent rapid retreat of the magmatic activity toward the coast. The ages younger than 20 Ma and outside the enclosed field of ages mostly correspond to extension-related basalts. (B) Detailed time-space plot for ages between 60 and 10 Ma. Solid circles are data from Damon et al. (1981). The ages compiled by Nieto-Samaniego et al. (1999) for the southern part of the Sierra Madre Occidental (open squares), as well as those reported by Henry et al. (2003) for southern Sinaloa (open rhombs), are also shown.

bimodal with a second pulse of silicic ignimbrites and minor amounts of basalt. The eruption of andesite in the Early to Middle Miocene occurred further west in Baja California (the Comondú arc; Sawlan, 1991; Umhoefer et al., 2001) (Fig. 2), marking the first clear reestablishment of typical supra-subduction zone arc magmatism and ending the broad extension and migration of magmatism trenchward beginning ca. 40 Ma. In the southern part of the Sierra Madre Occidental, however, the pattern of migration appears more complex. Ferrari et al. (1999) showed that the orientation of the volcanic belt tended to rotate counterclockwise since the Oligocene from a NNW orientation to a WNW trend in the Early to Middle Miocene when volcanism extended to southern Baja California, but was absent south of latitude 20° N. In more detail, Ferrari et al. (2002, their Fig. 6) showed that during the Early Miocene, ignimbrite volcanism migrated to the ESE as far as longitude 101° W.

At the provincial scale, the onset of extension corresponds to the westward expansion of magmatic activity toward the trench. Data summarized section 3 indicate that extension, or more likely the extensional front, migrated episodically from east to west along the entire Sierra Madre Occidental (see also Stewart, 1998). Importantly, the onset of upper crustal extension seems to follow the first ignimbrite pulse. Extension probably began in the Eocene in the Mesa Central (Aranda-Gómez and McDowell, 1998; Nieto-Samaniego et al., 1999; Nieto-Samaniego et al., this volume), although more rapid extension began at ca. 30 Ma, after emplacement of the first rhyolites (Orozco-Esquivel et al., 2002). In the eastern part of the Sierra Madre Occidental, the earliest extension also occurs in mid-Oligocene times, after the emplacement of the first ignimbrite sheets ca. 30 Ma at Nazas and Rodeo, Durango (Aranda-Gómez et al., 2003), and between ca. 39 and ca. 29 Ma in Fresnillo, Zacatecas (Lang et al., 1988).

A second episode affected the central part of the Sierra Madre Occidental, with an extensional front moving westward between the end of the Oligocene and the Early Miocene. In western Chihuahua and Durango, high-angle normal faulting postdates ca. 28–27 Ma ignimbrites (see section 3 and Figures 4 and 5). Extension propagated northwestwards to eastern Sonora, where the main basins hosting the Báucarit Formation and *core complexes* developed mostly between ca. 26 and 16 Ma (Fig. 7). In the central part of the Sierra Madre Occidental, this episode is only expressed by the eruption of alkaline basalts in Rodeo and Nazas, Durango, at ca. 24 Ma (Aranda-Gómez et al., 2003). In the southern Sierra Madre Occidental, this extensional episode is expressed by the formation of various grabens with ~NNE-SSW to N-S orientation in southern Zacatecas and northwestern Jalisco (Fig. 9) between ca. 22 and 18 Ma, and after the beginning of the second ignimbritic pulse at 24–20 Ma.

At the end of the Middle Miocene (ca. 12 Ma), a third extensional episode affected the westernmost belt of the Sierra Madre Occidental. The age of this deformation, commonly referred to as the "Gulf Extensional Province" or "Proto-Gulf extension," is remarkably similar along the entire Sierra Madre Occidental occurring between ca. 12 and 9 Ma in Sonora (Gans et al., 2003), Sinaloa (Henry and Aranda-Gómez, 2000), and Nayarit (Ferrari and Rosas-Elguera, 2000; Ferrari et al., 2002). Although with less intensity, extension of this age occurred to the east of the Sierra Madre Occidental, with the formation of the Río Chico–Otinapa graben (Henry and Aranda-Gómez, 2000), other extensional structures in the Mesa Central, which are still seismically active in the Durango and Tepehuanes volcanic fields (Nieto-Samaniego et al., this volume), as well as with the eruption of alkaline lavas in the Los Encinos volcanic field in San Luis Potosí (Luhr et al. 1995).

The evolving geographic pattern of extensional tectonics summarized in the previous paragraphs clearly indicates a progression of extension through discrete episodes that discontinuously affected the entire region encompassing the Sierra Madre Occidental and Gulf of California (Fig. 1). Consequently, it is difficult to establish a limit between the Gulf Extensional Province and the Mexican Basin and Range, such that Henry and Aranda-Gómez (2000) proposed that the Gulf Extensional Province is part of the Basin and Range Province, and that the relatively unextended core of the Sierra Madre Occidental geographically separates the two provinces (Fig. 6). Therefore, continental extension was a widespread event that affected the entire southwestern part of the North American plate. These extensional processes, however, led to fundamentally different extended continental margins in the United States and México. In the western United States, Tertiary deformation eventually created an extensional province over 1000 km in width without rupturing the continental lithosphere. In México, after over 20 m.y., extension became localized in the Gulf of California region, resulting in the formation of a rift floored by oceanic crust since the Pliocene. In this framework, the region of Sonora—characterized by much higher extension than the rest of the Sierra Madre Occidental—marks the transition between thinned continental crust and oceanic crust flooring the Gulf of California.

At a continental scale, two causes may explain this differing behavior: (1) crustal rheologic contrasts, and (2) a thermal weakening of the zone in which the Gulf of California eventually developed. Crustal rheologic contrasts were proposed by Langenheim and Jachens (2003) who suggested that the Peninsular Range batholiths (together with its deeper mafic counterpart) represents a more rigid crustal block that controlled the localization of extensional deformation to the east, eventually leading to the separation of Baja California from mainland México. Thermal weakening to focus extension in the Gulf of California has been related to prior volcanic arc activity (i.e., the Comondú arc in southern Baja California) that occurred in this region before opening of the rift. Although Early to Middle Miocene rocks occur in Baja California (e.g., Sawlan, 1991; Stock and Lee, 1994), the older successions commonly comprise massive sedimentary deposits (fluvial conglomerate produced by the erosion of volcanics) and distal facies of ignimbrites (e.g., Hausback, 1984; Dorsey and Burns, 1994; Umhoefer et al., 2001), indicating minimal magmatism and therefore thermal input into the crust in this region. It was from the Middle to Late Miocene (ca. 15–12 Ma) that an andesitic arc was active and close to the coast of Baja California (see section 2 for Sonora and Umhoefer et al., 2001, for Baja California Sur). Thermal weakening of the crust may alternatively have been caused by the Oligo-Miocene volcanism in the region, and the Comondú arc may have only assisted in focusing extension in the region of the future Gulf of California.

Genesis of the Silicic Magmatism

Clearly, the most peculiar aspect of the Sierra Madre Occidental magmatism is the eruption of large volumes of silicic magma in a relatively short time. These ignimbrite pulses are not common, and large-volume silicic volcanism only characterizes volcanic arcs when undergoing extension, or rifting to form a backarc basin (e.g., Taupo Volcanic Zone), and the formation of volcanic rifted margins (Bryan et al., 1997, 2002). Discussions on the petrogenesis of the silicic magmas in the Sierra Madre Occidental have been recurrent in the literature and have generally been divided between two end-member models. The first model suggests that the rhyolites formed largely by partial melting of the crust (e.g., Huppert and Sparks, 1988) as a consequence of the arrival of large amounts of basaltic magmas from the mantle, which provide the heat necessary to melt the crust (Ruiz et al., 1988a, 1990; Albrecht and Goldstein, 2000; Ferrari et al., 2002). This model has been invoked for the generation of other silicic large igneous provinces (Pankhurst and Rapela, 1995; Ewart et al., 1998; Riley et al., 2001; Bryan et al., 2002). In the second model, the rhyolites are interpreted as the final product of the differentiation of basaltic magmas with small or negligible contributions from the crust (Cameron and Hanson, 1982; Cameron and Cameron, 1985; Cameron et al., 1980a, b; Cameron and Robinson, 1990; Wark, 1991; Smith et al., 1996). The basis for some of these models is limited as they have been formulated from data gathered from a few sites and representative of a limited time scale (e.g., 1–2 Ma), yet the results have been generalized to explain the origin of large volumes (>10^5 km^3) of rhyolite over a relatively large time span (10–20 m.y.). It is worth emphasizing that existing studies on the petrogenesis of the Sierra Madre Occidental rhyolites are restricted to Oligocene sections from four sites in western Chihuahua (Batopilas, Divisadero, San Buenaventura, and Tómochic), one site in Zacatecas at the southeastern margin of the Sierra Madre Occidental, and two sections located in the Mesa Central (La Olivina, Chihuahua, and San Luis Potosí).

In principle, discriminating between the two end-members (anatexis versus fractional crystallization) and assessing crustal contributions to rhyolitic magmatism may appear relatively straightforward, as mantle and continental crust are thought to have distinctive $^{87}Sr/^{86}Sr$ and εNd (and other isotopic) compositions. The use of whole rock geochemical and isotopic data and lower crustal xenoliths, which have provided a more direct sampling of the lower crust, have been the main approach used in the Sierra Madre Occidental (Cameron and Hanson, 1982; Cameron and Cameron, 1985; Cameron et al., 1980a, b; Ruiz et al., 1988a, 1990; Cameron and Robinson, 1990; Wark, 1991; Smith et al., 1996; Albrecht and Goldstein, 2000). The results, however, have frequently been ambiguous (see section 4), since the rhyolites and the scarce associated mafic lavas have Sr and Nd isotopic compositions intermediate between those inferred for a mantle metasomatized by subduction fluids, and the Paleozoic or Mesozoic lower crust though which they emplaced (e.g., εNd of ~+4 to −4, Fig. 11) (Wark, 1991; Johnson, 1991). Nevertheless, some crustal contribution to rhyolite magmatism is generally required from isotopic considerations (Cameron and Cameron, 1985; Wark, 1991).

The debate on a dominantly mantle versus crustal origin for the Sierra Madre Occidental rhyolitic ignimbrites has several implications that have been discussed in the literature for other silicic large igneous provinces. If rhyolite generation is due primarily to fractional crystallization of mantle-derived basaltic magmas, then significant material transfer and new crust formation is implied. Furthermore, if the rhyolites are the product of basalt differentiation only, then a fourfold volume of mafic cumulates in the crust is required (e.g., Cameron et al., 1980a; Cameron and Hanson, 1982; Ruiz et al., 1988a). In turn, if these mafic residues are formed in the proximity of the Moho, this may lead to convective instability at the base of the crust (e.g., Kay and Mahlburg-Kay, 1991; Kay et al., 1992; Meissner and Mooney, 1998; Jull and Kelemen, 2001). Alternatively, if cumulates formed at a different level in the crust, then a variation in the composition, thermal structure, and isostatic, seismic, and rheologic properties of the crust is expected (Gans, 1997; Gans et al., 1989; Klemperer, 1989; Glazner and Ussler, 1989; Johnson, 1991; Miller and Paterson, 2001). In contrast, if rhyolites formed primarily by partial melting of the crust, then a substantially lower volume of basaltic magma is required and residual cumulates produced (Ruiz et al., 1988a), resulting in a different crustal thickness and petrologic and rheologic crustal profiles.

Whether a basaltic magma contributes heat and/or mass to the generation of the rhyolites, an important question that can be addressed by geophysical data is the localization and the fate of the cumulate residual material (Smith et al., 1996; Jull and Kelemen, 2001; Ducea, 2002). The geophysical data synthesized in this work indicate that the crust in the unextended core of the Sierra Madre Occidental is thicker than along its margins (Fig. 6): in the northern Sierra Madre Occidental the unextended part is ~55 km versus ~40 km to the east and ~28–22 km along the coast of the Gulf of California. The thickness of the Sierra Madre Occidental core also appears thicker than that of other Precambrian or Paleozoic crustal sections (e.g., Mooney et al., 1998). Even for the central and southern part of the Sierra Madre Occidental where only post-Paleozoic basement is inferred to exist, the crust is ~10 km thicker than its counterpart in the Mesa Central to the east (see also Nieto-Samaniego et al., 1999). These anomalous thicknesses, coupled with the high seismic velocity contrast of the lower layer detected by Bonner and Herrin (1999) in the northern part of the Sierra Madre Occidental, suggest that the lower crust has been heavily intruded by mafic magmas.

The intrusion or underplating of mafic magma represents new additions to the continental crust and, by consequence, a lowering of the Moho, which is a process common in many active continental margins (Klemperer, 1989). The regional character of the seismic studies available for the Sierra Madre Occidental, together with a lack of detailed estimations of the volume of ignimbrites and their intrusive equivalent, currently prevent robust discrimination between the fractional crystallization and partial melting models and relative contributions of crust and mantle to Sierra Madre Occidental rhyolite generation. However, the presence of a significant volume of mafic intrusions in the lower crust is consistent with the cause of the ignimbrite pulses resulting from the arrival of large volumes of basaltic magma from the mantle.

A more general and important question is what controls the generation of large volumes of rhyolite or basalt to form a Large Igneous Province (LIP)? The fertility of the crust has been invoked as a determining factor in the generation of large volumes of rhyolites (Bryan et al., 2002), since many of the silicic LIPs are found on Phanerozoic crust, whereas most LIPs, as typified by the Mesozoic to Recent continental flood basalt provinces, are located on Archean crust and generally mafic. Numeric modeling of magma intrusion into the lower crust has demonstrated that the presence of hydrous minerals and basalts with high T and low water content promote crustal melting (Annen and Sparks, 2002). For the Sierra Madre Occidental, despite the presence of Precambrian crust in the northern part of the province, little involvement of Precambrian crust in rhyolite magma generation is indicated by whole-rock isotopic compositions of the ignimbrites, and from the observation that the Oligocene ignimbrite pulse produced similar volumes across the whole province. Nevertheless, the fertility of the lithosphere was likely increased by geologic events prior to the ignimbrite flare-up. Humphreys et al. (2003) proposed that sub-horizontal subduction of the Farallon plate during the Laramide orogeny may have hydrated and fertilized the Precambrian lithosphere of the North America plate. Once the Farallon plate was removed from the base of the North America plate in the Tertiary, the uprise of asthenospheric mantle led to crustal heating, partial melting, and the outburst of silicic volcanism. This general model, originally proposed to explain mid-Tertiary silicic volcanism in the southwestern United States, may be equally applicable to the Sierra Madre Occidental, which shares a similar tectonic history.

In conclusion, a range of possibilities exist to explain the generation of the Sierra Madre Occidental silicic magmas, with fractional crystallization and crustal anatexis as end-members. Geophysical information and petrologic studies are still too scarce to fully understand the processes of silicic magma generation and if fractional crystallization or partial melting played a dominant role. Nevertheless, the presence of (1) crustal isotopic signatures and isotopic differences between ignimbrites erupted through different types of crustal basement (McDowell et al., 1997; Albrecht and Goldstein, 2000; Valencia-Moreno et al., 2001); (2) Precambrian, Mesozoic, and Tertiary inheritance in magmatic zircons within the ignimbrites (McDowell et al., 1997; Bryan et al., 2006); and (3) the essentially bimodal character of the Oligocene and Early Miocene volcanic pulses suggest that assimilation and/or melting of the crust was significant in the Sierra Madre Occidental. We consider that the petrogenesis of the Sierra Madre Occidental ignimbrites was dominated by processes of large-scale magma mixing and/or AFC capable of producing large amounts of silicic magmas and lesser volumes of variably contaminated basalt and basaltic andesite. In detail, the locus of the crust involved in assimilation and melting likely varied with time with progressively shallower zones potentially affected as the intrusion of mafic magmas induced a densification of the lower crust and/or remelted earlier intruded magma.

Geodynamic Causes of Magmatism and Extension

Based on the considerations of the previous section, it seems clear that, regardless of the mechanism responsible for the generation of rhyolites, each ignimbrite pulse was related to the arrival of considerable amounts of mantle-derived mafic magma at the base of the crust. The remarkable synchronism and large volume of the first ignimbrite pulse extending across the entire Sierra Madre Occidental indicate that this phenomenon cannot be the result of a "normal" subduction regime but, rather, the consequence of a continental scale mechanism related to plate dynamics. The results of different seismic tomography studies indicate that the upper mantle beneath the Sierra Madre Occidental is characterized by a significant thermal anomaly (see section 5), which is interpreted to be asthenosphere. The tectono-magmatic evolution of the Sierra Madre Occidental indicates this thermal anomaly is the result of the removal of the Farallon plate from the base of the North American plate in mid-Tertiary times (e.g., Humphreys, 1995). In México, the removal of the subducted Farallon plate occurred at different stages, each one characterized by different mechanisms. Eocene volcanics, if interpreted as supra-subduction zone volcanism in the eastern part of the Sierra Madre Occidental, would indicate that by this time, fluid-induced melting of a mantle wedge was occurring after the foundering of the Farallon plate. This volcanic episode coincides with the first signs of a decrease in the velocity of convergence between the Farallon and North America plates between 43 and 39 Ma (Norton, 1995). The first ignimbrite pulse (ca. 32–28 Ma) of the Sierra Madre Occidental coincides with a second stage of decrease in convergence rates between ca. 33 and 25 Ma (Norton, 1995), and the first contact between the East Pacific Rise and North America occurring south of California at ca. 28 Ma (Atwater and Stock, 1998). The decrease in convergence velocity occurred when progressively younger and more buoyant oceanic crust arrived at the paleotrench and, eventually, the northern part of the Farallon plate lost some fragments which became the Monterey and Jasper plates between ca. 28.5 and ca. 27.5 Ma (Lonsdale, 2005). Decreases in subduction velocity typically produce an increase in slab dip. This, in turn, induces a strong convection as it "pulls" asthenospheric mantle into the opening mantle wedge. This process of bringing asthenospheric mantle to shallower depths and in contact with the base of the continental crust led to a sudden increase in partial melting of the mantle and the generation of an ignimbrite pulse within a short period of time.

The second ignimbrite pulse in the southern part of the Sierra Madre Occidental (ca. 24–20 Ma) mostly coincided with the formation of metamorphic core complexes in the northern Sierra Madre Occidental (and southern United States). During this period, the Farallon plate was still being subducted beneath parts of western México after the Pacific and North America plates had come into direct contact further north, in southern California. The interaction between the Pacific and North America plates, with a diverging relative motion, resulted in the formation of a slab window in front of the contact zone (Atwater and Stock, 1998; Dickinson, 1997, 2002). In this zone, the deeper part of the subducting Farallon slab must have detached from the shallower part of the plate. Ferrari et al. (2002) proposed that once initiated, the detachment may have propagated toward the south-southwest because of an increase in the slab pull on that part of the plate still attached, documented in other areas of the world (e.g., Wortel and Spakman, 2000). Consequently, there may have been asthenospheric flow into the mantle wedge. In the northern part of the Sierra Madre Occidental, the flow of the asthenospheric material, together with plate boundary forces (e.g., Sonder and Jones, 1999), may have produced the high-magnitude extension that formed the metamorphic core complexes. In the southern Sierra Madre Occidental, slab rupture may have induced a second episode of asthenospheric underplating, triggering crustal partial melting and the subsequent ignimbrite pulse.

The review of the magmatic and tectonic history of the Sierra Madre Occidental presented in this work indicates that this geologic province is closely connected to the evolution of the Cretaceous-Cenozoic subduction system of the Farallon beneath the North America plate. In particular, the Sierra Madre Occidental as an Oligo-Miocene silicic LIP was the result of events that occurred at the end of subduction of the Farallon plate. The Sierra Madre Occidental shares many characteristics with other silicic LIPs (Table 1), some of which preceded the formation of volcanic rifted margins (Bryan et al., 1997). In this sense, the ignimbrite pulses of the Sierra Madre Occidental are not the product of normal supra-subduction zone volcanism, but rather, the precursors of lithospheric rupture that eventually led to the formation of the Gulf of California.

ACKNOWLEDGMENTS

Some of the results reported in this work for the northern and southern part of the Sierra Madre Occidental are a product of grants CONACyT I29887 T and DGAPA/UNAM IN10660 (M. Valencia) and CONACyT P-0152 T (L. Ferrari). Centro de Geociencias, UNAM, and the Damon Wells Research Fellowship to S. Bryan at Yale University supported fieldwork related with this project in Chihuahua, Zacatecas, and Jalisco during 2004. We thank the editors of this special issue for encouraging us to carry out this review. We are also thankful for the detailed reviews of Chris Henry and Ángel Nieto-Samaniego and the informal reviews of Enrique González Torres and Gabriel Chávez Cabello, which helped improve the clarity of the final manuscript, and Peter Reiners for discussions on aspects of Sierra Madre Occidental magmatism.

REFERENCES CITED

Aguirre-Díaz, G.J., and Labarthe, G., 2003, Fissure ignimbrites: Fissure-source origin for voluminous ignimbrites of the Sierra Madre Occidental and its relationship with Basin and Range faulting: Geology, v. 31, p. 773–776, doi: 10.1130/G19665.1.

Aguirre-Díaz, G.J., and McDowell, F.W., 1991, The volcanic section at Nazas, Durango, Mexico, and the possibility of widespread Eocene volcanism within the Sierra Madre Occidental: Journal of Geophysical Research, v. 96, p. 13373–13388.

Aguirre-Díaz, G.J., and McDowell, F.W., 1993, Nature and timing of faulting and syn-extensional magmatism in the southern Basin and Range, central-eastern Durango, Mexico: Geological Society of America Bulletin, v. 105, p. 1435–1444, doi: 10.1130/0016-7606(1993)105<1435:NATOFA>2.3.CO;2.

Albrecht, A., and Goldstein, S.L., 2000, Effects of basement composition and age on silicic magmas across an accreted terrane-Precambrian crust boundary, Sierra Madre Occidental, Mexico: Journal of South American Earth Sciences, v. 13, p. 255–273, doi: 10.1016/S0895-9811(00)00014-6.

Alencaster, G., and de Cserna, Z., 1961, Paleontología del Triásico Superior de Sonora Parte 1, Estratigrafía del Triásico Superior de la parte central del Estado de Sonora: Universidad Nacional Autónoma de México, Instituto de Geología, Paleontología Mexicana, v. 11, 18 p.

Almazán-Vázquez, E., 1989, El Cámbrico-Ordovícico de Arivechi, en la región centro-oriental del Estado de Sonora: Universidad Nacional Autónoma de México, Instituto de Geología, Revista, v. 8, p. 58–66.

Alsina, D., Woodward, R.L., and Sneider, R.K., 1996, Shear wave velocity structure in North America from large-scale waveform inversions of surface waves: Journal of Geophysical Research, v. 101, p. 15969–15986, doi: 10.1029/96JB00809.

Anderson, T.H., and Schmidt, V.A., 1983, The evolution of Middle America and the Gulf of Mexico-Caribbean Sea region during Mesozoic time: Geological Society of America Bulletin, v. 94, p. 941–966, doi: 10.1130/0016-7606(1983)94<941:TEOMAA>2.0.CO;2.

Anderson, T.H., and Silver, L.T., 1979, The role of the Mojave-Sonora megashear in the tectonic evolution of northern Sonora, *in* Anderson, T.H., and Roldán-Quintana, J., eds., Geology of Northern Sonora: Guidebook-Field Trip 27, Annual meeting of the Geological Society of America, Universidad Nacional Autónoma de México, Instituto de Geología and the University of Pittsburgh, Pittsburgh, p. 59–68.

Anderson, T.H., Nourse, J.A., McKee, J.W., and Steiner, M.B., eds., 2005, The Mojave-Sonora Megashear Hypothesis: Development, Assessment, and Alternatives: Geological Society of America Special Paper 393, 693 p.

Annen, C., and Sparks, R.S.J., 2002, Effects of repetitive emplacement of basaltic intrusions on thermal evolution and melt generation in the crust: Earth and Planetary Science Letters, v. 203, p. 937–955, doi: 10.1016/S0012-821X(02)00929-9.

Aranda-Gómez, J.J., and McDowell, F.W., 1998, Paleogene extension in the southern basin and range province of Mexico: Syndepositional tilting of Eocene red beds and Oligocene volcanic rocks in the Guanajuato mining district: International Geology Review, v. 40, p. 116–134.

Aranda-Gómez, J.J., Henry, C.D., Luhr, J.F., and McDowell, F.W., 1997, Cenozoic volcanism and tectonics in NW Mexico—a transect across the Sierra Madre Occidental volcanic field and observations on extension related magmatism in the southern Basin and Range and Gulf of California tectonic provinces, *in* Aguirre-Díaz, G.J., Aranda-Gómez, J.J., Carrasco-Núñez, G., and Ferrari, L., eds., Magmatism and Tectonics in the Central and Northwestern Mexico—A Selection of the 1997 IAVCEI General Assembly Excursions, México, D.F., Universidad Nacional Autónoma de México, Instituto de Geología, p. 41.

Aranda-Gómez, J.J., Luhr, J.F., Housh, T.B., Connor, C.B., Becker, T., and Henry, C.D., 2003, Synextensional Pliocene-Pleistocene eruptive activity in the Camargo volcanic field, Chihuahua, México: Geological Society of America Bulletin, v. 115, p. 298–313, doi: 10.1130/0016-7606(2003)115<0298:SPPEAI>2.0.CO;2.

Atwater, T., and Stock, J.M., 1998, Pacific-North America plate tectonics of the Neogene southwestern United States: An update: International Geology Review, v. 40, p. 375–402.

Bagby, W.C., Cameron, K.L., and Cameron, M., 1981, Contrasting evolution of calc-alkalic volcanic and plutonic rocks of western Chihuahua, Mexico: Journal of Geophysical Research, v. 86, p. 10402–10410.

Bain, J.H.C., and Draper, J.J., eds., 1997, North Queensland Geology: Queensland Geology, vol. 9: Australian Geological Survey Organisation bulletin: Geological Survey of Queensland, Department of Mines and Energy, 600 p.

Barra, F., Ruiz, J., Valencia, V.A., Ochoa-Landín, L., Chesley, J.T., and Zürcher, L., 2005, Laramide porphyry Cu–Mo mineralization in northern Mexico: Age constraints from Re–Os geochronology in molybdenites: Economic Geology and the Bulletin of the Society of Economic Geologists, v. 100, p. 1605–1616.

Bartolini, C., Damon, P.E., Shafiqullah, M., and Morales-Montaño, M., 1994, Geochronologic contributions to the Tertiary sedimentary-volcanic sequences ("Baucarit Formation") in Sonora: Geofísica Internacional, v. 34, p. 67–77.

Barton, M.D., Staude, J.-M.G., Zürcher, L., and Megaw, P.K., 1995, Porphyry copper and other intrusion-related mineralization in Mexico, *in* Pierce, F.W., and Bolm, J.G., eds., Bootprints of the Cordillera: Arizona Geological Society Digest, v. 20, p. 487–524.

Bijwaard, H., and Spakman, W., 2000, Non-linear global P-wave tomography by iterated linearized inversion: Geophysical Journal International, v. 141, p. 71–82, doi: 10.1046/j.1365-246X.2000.00053.x.

Bird, P., 1984, Laramide crustal thickening event in the Rocky Mountain foreland and Great Plains: Tectonics, v. 3, p. 741–758.

Bird, P., 1998, Kinematic history of the Laramide orogeny in latitudes 35°–49° N, western United States: Tectonics, v. 17, p. 780–801, doi: 10.1029/98TC02698.

Bockoven, N.T., 1980, Reconnaissance geology of the Yecora-Ocampo area, Sonora and Chihuahua, Mexico [Ph.D. thesis]: Austin, University of Texas, 197 p.

Bonneau, M., 1970, Una nueva área cretácica fosilífera en el estado de Sinaloa: Boletín de la Sociedad Geológica Mexicana, v. 32, p. 159–167.

Bonner, J.L., and Herrin, E.T., 1999, Surface wave study of the Sierra Madre Occidental of northern Mexico: Bulletin of the Seismological Society of America, v. 89, p. 1323–1337.

Bryan, S.E., 2005a, Silicic Large Igneous Provinces: http://www.mantleplumes.org/SLIPs.html (November 2006).

Bryan, S.E., 2005b, The Early Cretaceous Whitsunday Silicic Large Igneous Province of eastern Australia: http://www.largeigneousprovinces.org/LOM.html, August 2005 Large Igneous Province (November 2006).

Bryan, S.E., Constantine, A.E., Stephens, C.J., Ewart, A., Schön, R.W., and Parianos, J., 1997, Early Cretaceous volcano-sedimentary successions along the eastern Australian continental margin: Implications for the break-up of eastern Gondwana: Earth and Planetary Science Letters, v. 153, p. 85–102, doi: 10.1016/S0012-821X(97)00124-6.

Bryan, S.E., Ewart, A., Stephens, C.J., Parianos, J., and Downes, P.J., 2000, The Whitsunday Volcanic Province, central Queensland, Australia: Lithological and Stratigraphic investigations of a silicic-dominated large igneous province: Journal of Volcanology and Geothermal Research, v. 99, p. 55–78, doi: 10.1016/S0377-0273(00)00157-8.

Bryan, S.E., Riley, T.R., Jerram, D.A., Leat, P.T., and Stephens, C.J., 2002, Silicic volcanism: an undervalued component of large igneous provinces and volcanic rifted margins, *in* Menzies, M.A., Klemperer, S.L., Ebinger, C.J., and Baker J., eds., Magmatic Rifted Margins: Geological Society of America Special Paper 362, p. 99–120.

Bryan, S.E., Holcombe, R.J., and Fielding, C.R., 2003, The Yarrol terrane of the northern New England Fold Belt: Forearc or backarc?: Reply: Australian Journal of Earth Sciences, v. 50, p. 271–293.

Bryan, S.E., Ferrari, L., Reisner, P.W., Allen, C.M., and Campbell, I.H., 2006. New insights into large volume rhyolite generation at the mid-Tertiary Sierra Madre Occidental Province, Mexico, revealed by U-Pb geochronology: GEOS, Boletín Unión Geofísica Mexicana, vol. 26, n. 1. p. 134.

Busby, C., Smith, D., Morris, W., and Fackler-Adams, B., 1998, Evolutionary model for convergent margins facing large ocean basins: Mesozoic Baja California, Mexico: Geology, v. 26, p. 227–230, doi: 10.1130/0091-7613(1998)026<0227:EMFCMF>2.3.CO;2.

Calmus, T., Pérez-Segura, E., and Roldán-Quintana, J., 1996, The Pb–Zn ore deposits of San Felipe (Sonora, Mexico): An example of detached mineralization in the basin and range Province: Geofísica Internacional, v. 35, p. 115–124.

Cameron, K.L., and Cameron, M., 1985, Rare earth element, ^{87}Sr/^{86}Sr, and ^{143}Nd/^{144}Nd compositions of Cenozoic orogenic dacites from Baja California, northwestern Mexico, and adjacent West Texas: Evidence for the predominance of a subcrustal component: Contributions to Mineralogy and Petrology, v. 91, p. 1–11, doi: 10.1007/BF00429422.

Cameron, K.L., and Hanson, G.N., 1982, Rare earth element evidence concerning the origin of voluminous mid-Tertiary rhyolitic ignimbrites and related volcanic rocks, SMO, Chihuahua, Mexico: Geochimica et Cosmochimica Acta, v. 46, p. 1489–1503, doi: 10.1016/0016-7037(82)90309-X.

Cameron, K.L., and Robinson, J.V., 1990, Nd-Sr isotopic compositions of lower crustal xenoliths—evidence for the origin of mid-Tertiary felsic volcanics in Mexico: Comment: Contributions to Mineralogy and Petrology, v. 104, p. 609–618, doi: 10.1007/BF00306668.

Cameron, M., Bagby, W.C., and Cameron, K.L., 1980a, Petrogenesis of voluminous mid-Tertiary ignimbrites of the Sierra Madre Occidental: Contributions to Mineralogy and Petrology, v. 74, p. 271–284, doi: 10.1007/BF00371697.

Cameron, K.L., Cameron, M., Bagby, W.C., Moll, E.J., and Drake, R.E., 1980b, Petrologic characteristics of mid-Tertiary volcanic suites, Chihuahua: Geology, v. 8, p. 87–91, doi: 10.1130/0091-7613(1980)8<87:PCOMVS>2.0.CO;2.

Cameron, K.L., Cameron, M., and Barreiro, B., 1986, Origin of voluminous mid-Tertiary ignimbrites of the Batopilas region, Chihuahua: Implications for the formation of continental crust beneath the Sierra Madre Occidental: Geofisica Internacional, v. 25, no. 1, p. 39–59.

Cameron, K.L., Nimz, G.J., Kuentz, D., Niemeyer, S., and Gunn, S., 1989, Southern Cordilleran basaltic andesite suite, southern Chihuahua, Mexico; a link between Tertiary continental arc and flood basalt magmatism in North America: Journal of Geophysical Research, v. 94, p. 7817–7840.

Cameron, K.L., Robinson, J.V., Niemeyer, S., Nimz, G.J., Kuentz, D.C., Harmon, R.S., Bohlen, S.R., and Collerson, K.D., 1992, Contrasting styles of Pre-Cenozoic and mid-Tertiary crustal evolution in northern Mexico: Evidence from deep crustal xenoliths from La Olivina: Journal of Geophysical Research, v. 97, p. 17353–17376.

Campa, M.F., and Coney, P.J., 1983, Tectono-stratigraphic terranes and mineral resource distribution in Mexico: Canadian Journal of Earth Sciences, v. 20, p. 1040–1051.

Campos-Enríquez, J.O., Kerdan, T., Morán-Zenteno, D.J., Urrutia-Fucugauchi, J., Sánchez-Castellanos, E., and Alday-Cruz, R., 1994, Estructura de la litósfera superior a lo largo del Trópico de Cáncer: GEOS, Boletín Union Geofisica Mexicana, v. 12, p, 75–76.

Camprubí, A., Ferrari, L., Cosca, M., Cardellach, E., and Canals, A., 2003, Age of epithermal deposits in Mexico: Regional significance and links with the evolution of Tertiary volcanism: Economic Geology and the Bulletin of the Society of Economic Geologists, v. 98, p. 1029–1037.

Carrillo-Martínez, M., 1971, Geología de la hoja San José de Gracia, Sinaloa (B.Sc. thesis): México, D.F., Universidad Nacional Autónoma de México, 154 p.

Castillo-Hernández, D., and Romero-Ríos, F., 1991, Estudio geológico-regional de Los Altos de Jalisco y El Bajío, Open-File Report 02-91, Comisión Federal de Electricidad, Gerencia de Proyectos Geotermoeléctricos, Departamento de Exploración, Morelia, Michoacán, México, 35 p.

Cather, S.M., Dunbar, N.W., McDowell, F.W., McIntosh, W.C., and Sholle, P.A., 2003, Early Oligocene global cooling, volcanic iron fertilization, and the ignimbrite flare-up of southwestern North America: Geological Society of America Abstracts with Programs, v. 35, p. 255.

Clark, K.F., Damon, P.E., Shafiquillah, M., Ponce, B.F., and Cárdenas, D., 1981, Sección geológica-estructural a través de la parte sur de la Sierra Madre Occidental, entre Fresnillo y la costa de Nayarit: Asociación Ingenieros de Minas, Metalurgistas y Geólogos de México, Memoria Técnica, v. XIV, p. 69–99.

Cochemé, J.J., and Demant, A., 1991, Geology of the Yécora area, northern Sierra Madre Occidental, Mexico, *in* Pérez-Segura, E., and Jacques-Ayala,

C., eds., Studies of Sonoran Geology: Geological Society of America Special Paper 254, p. 81–94.

Coney, P.J., and Campa, M.F., 1987, Lithotectonic Terrane Map of Mexico (west of the 91st meridian): Miscellaneous Field Studies, Map MF 1874-D.

Coney, P.J., and Reynolds, S.J., 1977, Cordilleran Benioff zones: Nature, v. 270, p. 403–406, doi: 10.1038/270403a0.

Consejo de Recursos Minerales (CRM), 1997, Carta geológico-minera Hoja Zacatecas F13-6, escala 1:250 000, Secretaria de Comercio y Fomento Industrial, 1 sheet.

Córdoba, D.A., 1963, Geología de la región entre Río Chico y Llano Grande, Municipio de Durango, Estado de Durango. Universidad Nacional Autónoma de México, Instituto de Geología, Boletín v. 71, Part 1, 21 p.

Couch, R.W, Ness, G.E., Sanchez-Zamora, O., Calderon-Riveroll, G., Doguin, P., Plawman, T., Coperude, S., Huehn, B., and Gumma, W., 1991, Gravity anomalies and crustal structure of the Gulf and Peninsular Province of the Californias, *in* Dauphin, J.P., and Simoneit, B.R.T., eds., The Gulf and the Peninsular Province of the Californias: American Association of Petroleum Geologists, Memoir, v. 47, p. 47–70.

Damon, P.E., Nieto-Obregón, J., and Delgado-Argote, L., 1979, Un plegamiento neogénico en Nayarit y Jalisco y evolución geomórfica del Río Grande de Santiago: Asociación Ingenieros de Minas, Metalurgistas y Geólogos de México, Memoria Técnica, v. XIII, p. 156–191.

Damon, P.E., Shafiqullah, M., and Clark, K.F., 1981, Age trends of igneous activity in relation to metallogenesis in the southern Cordillera: Arizona Geological Society Digest, v. 14, p. 137–154.

Damon, P.E., Shafiqullah, M., and Clark, K., 1983, Geochronology of the porphyry copper deposits and related mineralization in Mexico: Canadian Journal of Earth Sciences, v. 20, p. 1052–1071.

Davis, G.H., and Coney, P.J., 1979, Geological development of metamorphic core complexes: Geology, v. 7, p. 120–124, doi: 10.1130/0091-7613(1979)7<120:GDOTCM>2.0.CO;2.

De Silva, S.L., 1989a, Altiplano-Puna volcanic complex of the central Andes: Geology, v. 17, p. 1102–1106, doi: 10.1130/0091-7613(1989)017<1102:APVCOT>2.3.CO;2.

De Silva, S.L., 1989b, Geochronology and stratigraphy of the ignimbrites from the 21°30′S to 23°30′S portion of the central Andes of northern Chile: Journal of Volcanology and Geothermal Research, v. 37, p. 93–131, doi: 10.1016/0377-0273(89)90065-6.

Dickinson, W.R., 1997, Tectonic implications of Cenozoic volcanism in coastal California: Geological Society of America Bulletin, v. 109, p. 936–954, doi: 10.1130/0016-7606(1997)109<0936:OTIOCV>2.3.CO;2.

Dickinson, W.R., 2002, The Basin and Range Province as a composite extensional domain: International Geology Review, v. 44, p. 1–38.

Dickinson, W.R., and Lawton, T.F., 2001, Carboniferous to Cretaceous assembly and fragmentation of Mexico: Geological Society of America Bulletin, v. 113, p. 1142–1160, doi: 10.1130/0016-7606(2001)113<1142:CTCAAF>2.0.CO;2.

Dickinson, W.R., Fiorillo, A.R., Hall, D.L., Monreal-Saavedra, R., Potochnik, A.R., and Swift, P.N., 1989, Cretaceous strata of southern Arizona, *in* Jenney, J.P., and Reynolds, S.J., eds., Geologic Evolution of Arizona: Arizona Geological Society Digest, v. 17, p. 447–461.

Dorsey, R.J., and Burns, B.A., 1994, Regional stratigraphy, sedimentology and tectonic significance of Oligocene-Miocene sedimentary and volcanic rocks, northern Baja California, Mexico: Sedimentary Geology, v. 88, p. 231–251, doi: 10.1016/0037-0738(94)90064-7.

Ducea, M.N., 2002, Constraints on the bulk composition and root foundering rates of continental arcs—a California arc perspective: Journal of Geophysical Research, v. 107, p. 2304, doi: 10.1029/2001JB000643, doi: 10.1029/2001JB000643.

Eguiluz de Antuñano, S., Aranda-García, M., and Marrett, R., 2000, Tectónica de la Sierra Madre Oriental, México: Boletín de la Sociedad Geológica Mexicana, Tomo, v. LIII, no. 1, p. 1–26.

English, J., Johnston, S.T., and Wang, K., 2003, Thermal modelling of the Laramide orogeny: Testing the flat slab subduction hypothesis: Earth and Planetary Science Letters, v. 214, p. 619–632, doi: 10.1016/S0012-821X(03)00399-6.

Enríquez, E., and Rivera, R., 2001, Timing of magmatic and hydrothermal activity in the San Dimas District, Durango, Mexico, *in* New Mines and Mineral Discoveries in Mexico and Central America, Society of Economic Geologists Special Publication, n. 8, p. 33–38.

Ewart, A., Milner, S.C., Armstrong, R.A., and Duncan, A.R., 1998, Etendeka volcanism of the Goboboseb Mountains and Messum Igneous Complex, Namibia, Part II: Voluminous quartz latite volcanism of the Awahab magma system: Journal of Petrology, v. 39, p. 227–253, doi: 10.1093/petrology/39.2.227.

Ferrari, L., 1995, Miocene shearing along the northern boundary of the Jalisco block and the opening of the southern Gulf of California: Geology, v. 23, p. 751–754, doi: 10.1130/0091-7613(1995)023<0751:MSATNB>2.3.CO;2.

Ferrari, L., and Rosas-Elguera, J., 2000, Late Miocene to Quaternary Extension at the Northern Boundary of the Jalisco Block, western Mexico: The Tepic-Zacoalco rift revised, *in* Delgado-Granados, H., Aguirre-Díaz, G.J., and Stock, J.M., eds., Cenozoic Tectonics and Volcanism of Mexico: Geological Society of America Special Paper 334, p. 41–64.

Ferrari, L., López-Martínez, M., Aguirre-Díaz, G., and Carrasco-Núñez, G., 1999, Space-time patterns of Cenozoic arc volcanism in central Mexico: From the Sierra Madre Occidental to the Mexican Volcanic Belt: Geology, v. 27, p. 303–306, doi: 10.1130/0091-7613(1999)027<0303:STPOCA>2.3.CO;2.

Ferrari, L., Pasquarè, G., Venegas, S., and Romero, F., 2000, Geology of the Western Mexican Volcanic Belt and Adjacent Sierra Madre Occidental and Jalisco Block, *in* Delgado-Granados, H., Aguirre-Díaz, G.J., and Stock, J.M., eds., Cenozoic Tectonics and Volcanism of Mexico: Geological Society of America Special Paper 334, p. 65–84.

Ferrari, L., López-Martínez, M., and Rosas-Elguera, J., 2002, Ignimbrite flare-up and deformation in the southern Sierra Madre Occidental, western Mexico—implications for the late subduction history of the Farallon Plate: Tectonics, v. 21, doi: 10.1029/2001TC001302.

Ferrari, L., Petrone, C.M., Francalanci, L., Tagami, T., Eguchi, M., Conticelli, S., Manetti, P., and Venegas-Salgado, S., 2003, Geology of the San Pedro-Ceboruco graben, western Trans-Mexican Volcanic Belt: Revista Mexicana de Ciencias Geológicas, v. 20, p. 165–181.

Fix, J.E., 1975, The crust and upper mantle of central Mexico: Geophysical Journal of the Royal Astronomical Society, v. 43, p. 453–499.

Ferriz, H., 1981, Geología de la caldera de San Marcos, Chihuahua: Universidad Nacional Autónoma de México, Instituto de Geología, Revista, v. 5, p. 65–79.

Gans, P.B., 1997, Large-magnitude Oligo-Miocene extension in southern Sonora: Implications for the tectonic evolution of northwest Mexico: Tectonics, v. 16, p. 388–408, doi: 10.1029/97TC00496.

Gans, P.B., and Bohrson, W.A., 1998, Suppression of volcanism during rapid extension in the Basin and Range Province, United States: Science, v. 279, p. 66–68, doi: 10.1126/science.279.5347.66.

Gans, P.B., Mahood, G.A., and Schermer, E.R., 1989, Synextensional magmatism in the Basin and Range Province—a case study from the eastern Great Basin: Geological Society of America Special Paper 233, p. 1–53.

Gans, P.B., MacMillan, I., and Roldán-Quintana, J., 2003, Late Miocene (Proto Gulf) extension and magmatism on the Sonoran margin: EOS, Transactions AGU, v. 84(46), Fall Meeting Supplement, p. F1405.

Gastil, R.G., and Krummenacher, D., 1977, Reconnaissance geology of coastal Sonora between Puerto Lobos and Bahía de Kino: Geological Society of America Bulletin, v. 88, p. 189–198, doi: 10.1130/0016-7606(1977)88<189:RGOCSB>2.0.CO;2.

Gastil, R.G., Krummenacher, D., and Jensky, W.E., 1978, Reconnaissance geology of west-central Nayarit, Mexico: Geological Society of America, text to accompany Map and Chart Series, Map MC-24.

Gastil, R.G., Krummenacher, D., and Minch, J., 1979, The record of Cenozoic volcanism around the Gulf of California: Geological Society of America Bulletin, v. 90, p. 839–857, doi: 10.1130/0016-7606(1979)90<839:TROCVA>2.0.CO;2.

Gastil, R.G., Miller, R., Anderson, P., Crocker, J., Campbell, M., Buch, P., Lothringer, C., Leier Englehardt, P., DeLattre, M., Hoobs, J., and Roldán-Quintana, J., 1991, The relation between the Paleozoic strata on opposite sides of the Gulf of California, *in* Pérez-Segura, E., and Jacques-Ayala, C., eds., Studies of Sonoran Geology: Geological Society of America Special Paper 254, p. 7–18.

Glazner, A.F., and Ussler, W., III, 1989, Crustal extension, crustal density, and the evolution of Cenozoic magmatism and the Basin and Range of the western United States: Journal of Geophysical Research, v. 94, p. 7952–7960.

Goes, S., and van der Lee, S., 2002, Thermal structure of the North American uppermost mantle inferred from seismic tomography: Journal of Geophysical Research, v. 107, doi: 10.1029/2000JB000049, doi: 10.1029/2000JB000049.

Gomberg, J.S., Priestley, K.F., Masters, G., and Brune, J., 1988, The structure of the crust and upper mantle in northern Mexico: Geophysical Journal, v. 94, p. 1–20.

González-León, C.M., 1988, Estratigrafía y geología estructural de las rocas sedimentarias cretacicas del área de Lampazos, Sonora: Universidad Nacional Autónoma de México, Instituto de Geología, Revista, v. 7, p. 148–162.

González-León, C.M., 1994, Early Cretaceous tectono-sedimentary evolution on the southwestern margin of the Bisbee basin: Revista Mexicana de Ciencias Geológicas, v. 11, p. 139–146.

González-León, C.M., 1997, Sequence stratigraphy and paleogeographic setting of the Antimonio Formation (Late Permian-Early Jurassic), Sonora, Mexico: Revista Mexicana de Ciencias Geológicas, v. 14, p. 136–148.

González-León, C.M., and Lawton, T.F., 1995, Stratigraphy, depositional environments and origin of the Cabullona basin, northwestern Sonora, *in* Jacques-Ayala, C., González-León, C.M., and Roldán-Quintana, J., eds., Studies on the Mesozoic of Sonora and Adjacent Areas, Geological Society of America Special Paper 301, p. 121–143.

González-León, C.M., McIntosh, W.C., Lozano-Santacruz, R., Valencia-Moreno, M., Amaya-Martínez, R., and Rodríguez-Castaneda, J.L., 2000, Cretaceous and Tertiary sedimentary, magmatic, and tectonic evolution of north-central Sonora (Arizpe and Bacanuchi Quadrangles), northwest Mexico: Geological Society of America Bulletin, v. 112, p. 600–610, doi: 10.1130/0016-7606(2000)112<0600:CATSMA>2.3.CO;2.

Grajales-Nishimura, J.M., Terrell, D.J., and Damon, P.E., 1992, Evidencias de la prolongación del arco magmático cordillerano del Triásico Tardío-Jurásico en Chihuahua, Durango y Coahuila: Boletín de la Asociación Mexicana de Geólogos Petroleros, v. XLII, no. 2, p. 1–18.

Grand, S.P., 1994, Mantle shear structure beneath the Americas and surrounding oceans: Journal of Geophysical Research, v. 99, p. 11591–11621, doi: 10.1029/94JB00042.

Hausback, B.P., 1984, Cenozoic volcanic and tectonic evolution of Baja California Sur, Mexico, *in* Frizzell, A., ed., Geology of the Baja Peninsula: Society of Economic Paleontologists and Mineralogists Special Paper 39, p. 219–236.

Haenggi, W.T., 2002, Tectonic history of the Chihuahua trough, Mexico and adjacent USA, Part II: Mesozoic and Cenozoic: Boletín de la Sociedad Geológica Mexicana, tomo LV, p. 38–94.

Henry, C.D., 1986, East-northeast trending structures in western México: Evidence for oblique convergence in the late Mesozoic: Geology, v. 14, p. 314–317, doi: 10.1130/0091-7613(1986)14<314:ESIWME>2.0.CO;2.

Henry, C.D., 1989, Late Cenozoic Basin and Range structure in western Mexico adjacent to the Gulf of California: Geological Society of America Bulletin, v. 101, p. 1147–1156, doi: 10.1130/0016-7606(1989)101<1147: LCBARS>2.3.CO;2.

Henry, C.D., and Aranda-Gómez, J.J., 2000, Plate interactions control middle-late Miocene proto-Gulf and Basin and Range extension in the southern Basin and Range: Tectonophysics, v. 318, p. 1–26, doi: 10.1016/S0040-1951(99)00304-2.

Henry, C.D., and Fredrikson, G., 1987, Geology of part of southern Sinaloa, Mexico, adjacent to the Gulf of California: Geological Society of America Map and Chart Series, MCH063, 1 sheet, 14 p.

Henry, C.D., McDowell, F.W., and Silver, L.T., 2003, Geology and geochronology of the granitic batholithic complex, Sinaloa, México: Implications for Cordilleran magmatism and tectonics, *in* Johnson, S.E., Paterson, S.R., Fletcher, J.M., Girty, G.H., Kimbrough, D.L., and Martín-Barajas, A., eds., Tectonic Evolution of Northwestern México and the Southwestern USA: Geological Society of America Special Paper 374, p. 237–274.

Horner, H., 1998, Structural Geology and Exploration in the San Dimas District, Durango, Mexico—An Alternative Geologic Model: Faculty of Natural Sciences, University of Salzburg, 120 p.

Horner, J.T., and Enríquez, E., 1999, Epithermal precious metal mineralization in a strike-slip corridor: The San Dimas district, Durango, Mexico: Economic Geology and the Bulletin of the Society of Economic Geologists, v. 94, p. 1375–1380.

Houghton, B.F., Wilson, C.J.N., McWilliams, M., Lanphere, M.A., Weaver, S.D., Briggs, R.M., and Pringle, M.S., 1995, Chronology and dynamics of a large silicic magmatic system: Central Taupo Volcanic Zone, New Zealand: Geology, v. 23, p. 13–16.

Huppert, H.E., and Sparks, R.S.J., 1988, The generation of granitic magmas by intrusion of basalt into continental crust: Journal of Petrology, v. 29, p. 599–642.

Humphreys, E.D., 1995, Post-Laramide removal of the Farallon slab, western United States: Geology, v. 23, p. 987–990, doi: 10.1130/0091-7613(1995)023<0987:PLROTF>2.3.CO;2.

Humphreys, E.D., Hessler, E., Dueker, K., Erslev, E., Farmer, G.L., and Atwater, T., 2003, How Laramide-age hydration of North America by the Farallon slab controlled subsequent activity in the western U.S.: International Geology Review, v. 45, p. 575–595.

Iriondo, A., Kunk, M.J., Winick, J.A., and Consejo de Recursos Minerales, 2003, ^{40}Ar/^{39}Ar dating studies of minerals and rocks in various areas in Mexico: USGS/CRM Scientific Collaboration (Part I): U.S. Geological Survey Open-File Report 2003-020, 79 p.

Iriondo, A., Premo, W.R., Martínez-Torres, L.M., Budahn, J.R., Atkinson, W.W., Siems, D.F., and Guarás-González, B., 2004a, Isotopic, geochemical, and temporal characterization of Proterozoic basement rocks in the Quitovac region, northwestern Sonora, Mexico: Implications for the reconstruction of the southwestern margin of Laurentia: Geological Society of America Bulletin, v. 116, p. 154–170, doi: 10.1130/B25138.1.

Iriondo, A., Kunk, M.J., Winick, J.A., and Consejo de Recursos Minerales, 2004b, ^{40}Ar/^{39}Ar dating studies of minerals and rocks in various areas in Mexico: USGS/CRM Scientific Collaboration (Part II): U.S. Geological Survey Open-File Report, 2004-1444, 46 p.

Irvine, T.N., and Baragar, W.R.A., 1971, A guide of the chemical classification of common volcanic rocks: Canadian Journal of Earth Science, v. 8, p. 523–548.

Jacques-Ayala, C., 1995, Paleogeography and provenance of the Lower Cretaceous Bisbee Group in the Caborca-Santa Ana area, northwestern Sonora, *in* Jacques-Ayala, C.., González-León, C.M., and Roldán-Quintana, J., eds., Studies on the Mesozoic of Sonora and Adjacent Areas: Geological Society of America Special Paper 301, p. 79–98.

James, E.V., and Henry, C.D., 1993, Compositional changes in Trans-Pecos Texas magmatism coincident with Cenozoic stress realignment: Journal of Geophysical Research, v. 96, p. 13561–13575.

Johnson, C.M., 1991, Large-scale crust formation and lithosphere modification beneath Middle to Late Cenozoic calderas and volcanic fields, western North America: Journal of Geophysical Research, v. 96, p. 13,485–13,507.

Jull, M., and Kelemen, P.B., 2001, On the conditions for lower crustal convective instability: Journal of Geophysical Research, v. 106, p. 6423–6446, doi: 10.1029/2000JB900357.

Kay, R.W., and Mahlburg-Kay, S., 1991, Creation and destruction of lower continental crust: International Journal of Earth Sciences, v. 80, p. 259–278.

Kay, R.W., Mahlburg-Kay, S., and Arculus, R.J., 1992, Magma genesis and crustal processing, *in* Fountain, D.M., Arculus, R., and Kay, R.W., eds., Continental Lower Crust: Developments in Geotectonics, v. 23, p. 423–445.

Keller, P., Bockoven, N., and McDowell, F.W., 1982, Tertiary volcanic history of the Sierra del Gallego area, Chihuahua, Mexico: Geological Society of America Bulletin, v. 93, p. 303–314, doi: 10.1130/0016-7606(1982)93<303:TVHOTS>2.0.CO;2.

Keppie, J.D., Dostal, J., Millar, B.V., Ortega-Rivera, M.A., Roldan-Quintana, J., and Lee, J.W.K., 2006, Geochronology and Geochemistry of the Francisco Gneiss: Triassic Continental Rift Tholeiites on the Mexican Margin of Pangea Metamorphosed and Exhumed in a Tertiary Core Complex: International Geology Review, v. 48, p. 1–16.

King, R.E., 1939, Geologic reconnaissance in the northern Sierra Madre Occidental of Mexico: Geological Society of America Bulletin, v. 50, p. 1625–1722.

Klemperer, S.L., 1989, Deep seismic reflection profiling and the growth of the continental crust: Tectonophysics, v. 161, p. 233–244, doi: 10.1016/0040-1951(89)90156-X.

Lang, B., Steinitz, G., Sawkins, F.J., and Simmons, S.F., 1988, K-Ar Age studies in the Fresnillo Silver District, Zacatecas, Mexico: Economic Geology and the Bulletin of the Society of Economic Geologists, v. 83, p. 1642–1646.

Lange, R., and Carmichael, I.S.E., 1991, A potassic volcanic front in western Mexico: Lamprophyric and related lavas of San Sebastian: Geological Society of America Bulletin, v. 103, p. 928–940, doi: 10.1130/0016-7606(1991)103<0928:APVFIW>2.3.CO;2.

Langenheim, V.E., and Jachens, R.C., 2003, Crustal structure of the Peninsular Ranges batholith from magnetic data: Implications for Gulf of California rifting: Geophysical Research Letters, v. 30, p. 1597, doi: 10.1029/2003GL017159.

Lanphere, M.A., Cameron, K.L., and Cameron, M., 1980, Sr isotopic geochemistry of voluminous rhyolitic ignimbrites and related rocks, Batopilas area, western Mexico: Nature, v. 286, p. 594–596, doi: 10.1038/286594a0.

Lawlor, P.J., Ortega-Gutierrez, F., Cameron, K.L., Ochoa-Camarillo, H., López, R., and Sampson, D.E., 1999, U-Pb geochronology, geochemistry and provenance of the Grenvillian Huiznopala Gneiss of eastern Mexico: Precambrian Research, v. 94, p. 73–99.

Le Maitre, R.W., 1989, A Classification of Igneous Rocks and Glossary of Terms: Recommendations of the IUGS Subcommission on the Systematics of Igneous Rocks: London, Blackwell Scientific, 193 p.

Lonsdale, P., 2005, Creation of the Cocos and Nazca plates by fission of the Farallon plate: Tectonophysics, v. 404, p. 237–264, doi: 10.1016/j.tecto.2005.05.011.

López-Ramos, E., 1995, Carta geológica de los estados de Jalisco y Aguascalientes, escala 1:750,000, con resumen: Universidad Nacional Autónoma de México, Instituto de Geología, Cartas Geológicas Estatales, 1 sheet.

Lucas, S.G., Kues, B.S., and González-León, C.M., 1995, Paleontology of the Upper Cretaceous Cabullona Group, northeastern Sonora, *in* Jacques-Ayala, C.., González-León, C.M., and Roldán-Quintana, J., eds., Studies on the Mesozoic of Sonora and Adjacent Areas: Geological Society of America Special Paper 301, p. 143–165.

Luhr, J.F., Pier, J.G., Aranda-Gómez, J.J., and Podosek, F.A., 1995, Crustal contamination in early Basin-and-Range hawaiites of the Los Encinos Volcanic Field, central Mexico: Contribution to Mineralogy and Petrology, v. 118, p. 321–339, doi: 10.1007/s004100050018.

Luhr, J.F., Henry, C.D., Housh, T.B., Aranda-Gómez, J.J., and McIntosh, W.C., 2001, Early extension and associated mafic alkalic volcanism from the southern Basin and Range Province: Geology and petrology of the Rodeo and Nazas volcanic fields, Durango (Mexico): Geological Society of America Bulletin, v. 113, p. 760–773, doi: 10.1130/0016-7606(2001)113<0760:EEAAMA>2.0.CO;2.

Lyons, J.I., 1988, Geology and ore deposits of the Bolaños silver district, Jalisco, Mexico: Economic Geology and the Bulletin of the Society of Economic Geologists, v. 83, p. 1560–1582.

MacMillan, I., Gans, P.B., and Roldán-Quintana, J., 2003, Voluminous mid-Miocene silicic volcanism and rapid extension in the Sierra Libre, Sonora, Mexico: Geological Society of America Abstracts with Programs, v. 35, no. 4, p. 26.

Magonthier, M.C., 1988, Distinctive rhyolite suites in the mid-Tertiary ignimbritic complex of the Sierra Madre Occidental, western Mexico: Bulletin de la Société Géologique de France, Serie 8, v. 4, p. 57–68.

Mauger, R.L., 1981, Geology and petrology of the Calera-del Nido block, Chihuahua Mexico, *in* Goodell, P.C., ed., Uranium in Volcanic and Volcaniclastics Rocks: American Association of Petroleum Geologists, Studies in Geology, v. 13, p. 202–242.

Mauger, R.L., 1992, The mid-Eocene Majalca Canyon caldera, Chihuahua, Mexico, *in* Clark, K.F., et al., eds., Geology and Mineral Resources of the Northern Sierra Madre Occidental, Mexico: El Paso Geological Society 1992, Field Conference Guidebook, p. 127–132.

Mauger, R.L., 1983, Geologic map of the Majalca-Punta de Agua area, central Chihuahua, Mexico, *in* Clark, K.F., and Goodell, P.C., eds., Geology and Mineral Resources of North-central Chihuahua: Guidebook of the El Paso Geological Society Field Conference, p. 169–174.

Megaw, P.K.M., 1986, Geology and geological history of the Santa Eulalia Mining District, Chihuahua, Mexico, *in* Clark, K.F., et al., eds., Field Excursion Guidebook: Society of Economic Geology, p. 213–232.

Megaw, P.K.M., 1990, Geology and geochemistry of the Santa Eulalia mining district, Chihuahua, Mexico [Ph.D. thesis]: University of Arizona, Tucson, Arizona, 463 p.

McDowell, F.W., and Clabaugh, S.E., 1979, Ignimbrites of the Sierra Madre Occidental and their relation to the tectonic history of western Mexico, in Chapin, C.E., and Elston, W.E., eds., Ash-Flow Tuffs: Geological Society of America Special Paper 180, p. 113–124.

McDowell, F.W., and Keizer, R.P., 1977, Timing of mid-Tertiary volcanism in the Sierra Madre Occidental between Durango City and Mazatlán, Mexico: Geological Society of America Bulletin, v. 88, p. 1479–1487, doi: 10.1130/0016-7606(1977)88<1479:TOMVIT>2.0.CO;2.

McDowell, F.W., and Mauger, R.L., 1994, K-Ar and U-Pb zircon chronology of Late Cretaceous and Tertiary magmatism in central Chihuahua State, Mexico: Geological Society of America Bulletin, v. 106, p. 118–132, doi: 10.1130/0016-7606(1994)106<0118:KAAUPZ>2.3.CO;2.

McDowell, F.W., Roldán-Quintana, J.J., and Amaya-Martínez, R., 1997, Interrelationship of sedimentary and volcanic deposits associated with Tertiary extension in Sonora, Mexico, Geological Society of America Bulletin, v. 109, p. 1349–1360.

McDowell, F.W., Housh, T.B., and Wark, D.A., 1999, Nature of crust beneath west-central Chihuahua, Mexico, based upon Sr, Nd, and Pb isotopic compositions at the Tómochic volcanic center: Geological Society of America Bulletin, v. 111, p. 823–830, doi: 10.1130/0016-7606(1999)111<0823:NOTCBW>2.3.CO;2.

McDowell, F.W., Roldán-Quintana, J., and Connelly, J., N., 2001, Duration of Late Cretaceous–early Tertiary magmatism in east-central Sonora, Mexico: Geological Society of America Bulletin, v. 113, p. 521–531.

McKee, J.W., Jones, N.W., and Anderson, T.H., 1988, Las Delicias basin: a record of late Paleozoic arc volcanism in northeastern Mexico: Geology, v. 16, p. 37–40, doi: 10.1130/0091-7613(1988)016<0037:LDBARO>2.3.CO;2.

Meissner, R.O., and Mooney, W.D., 1998, Weakness of the lower continental crust; a condition for delamination, uplift, and escape: Tectonophysics, v. 296, p. 47–60, doi: 10.1016/S0040-1951(98)00136-X.

Miller, R.B., and Paterson, S.R., 2001, Influence of lithological heterogeneity, mechanical anisotropy, and magmatism on the rheology of an arc, North Cascades, Washington: Tectonophysics, v. 342, p. 351–370, doi: 10.1016/S0040-1951(01)00170-6.

Miranda-Gasca, M.A., and De Jong, K.A., 1992, The Magdalena mid-Tertiary extensional basin, *in* Clark, K.F., Roldán-Quintana, J., and Schmidt, R.H., eds., Geology and Mineral Resources of the Northern Sierra Madre Occidental, México: Guidebook of the El Paso Geological Society Field Conference, p. 377–384.

Miranda-Gasca, M.A., Gómez-Caballero, J.A., and Eastoe, C.J., 1998, Borate deposits of Northern Sonora, Mexico—Stratigraphy, tectonics, stable isotopes, and fluid inclusions: Economic Geology and the Bulletin of the Society of Economic Geologists, v. 93, p. 510–523.

Montigny, R., Demant, A., Delpretti, P., Piguet, P., and Cocheme, J.J., 1987, Chronologie K/Ar de sequences volcanique tertiaires du nord de la Sierra Madre Occidental, Mexique: Académie des Sciences Comptes Rendus, Paris: Ser. D, v. 304, p. 987–992.

Mooney, W.D., Laske, G., and Masters, G.T., 1998, Crust 5.1: A global crustal model at 5x5 degrees: Journal of Geophysical Research, v. 103, p. 727–747, doi: 10.1029/97JB02122.

Moore, G., Marone, C., Carmichael, I.S.E., and Renne, P., 1994, Basaltic volcanism and extension near the intersection of the Sierra Madre volcanic province and the Mexican Volcanic Belt: Geological Society of America Bulletin, v. 106, p. 383–394, doi: 10.1130/0016-7606(1994)106<0383:BVAENT>2.3.CO;2.

Mora-Álvarez, G., and McDowell, F.W., 2000, Miocene volcanism during late subduction and early rifting in the Sierra Santa Ursula of Western Sonora, Mexico, *in* Stock, J.M., Delgado-Granados, H., and Aguirre-Díaz, G., eds., Cenozoic Tectonics and Volcanism of Mexico: Geological Society of America Special Paper 334, p. 123–141.

Mora-Klepeis, G., and McDowell, F.W., 2004, Late Miocene calc-alkalic volcanism in northwestern Mexico: An expression of rift or subduction-related magmatism?: Journal of South American Earth Sciences, v. 17, p. 297–310, doi: 10.1016/j.jsames.2004.08.001.

Morales-Montaño, M., Bartolini, C., Damon, P.E., and Shafiqullah, M., 1990, K-Ar dating, stratigraphy, and extensional deformation of Sierra Lista Blanca, Central Sonora, Mexico: Geological Society of America Abstracts with Programs, v. 22, p. A364.

Moran-Zenteno, D.J., Tolson, G., Martiny, B., Martinez-Serrano, R.G., Schaaf, P., Silva Romo, G., Macias, C., Alba-Aldave, L., Hernandez Bernal, M.S., and Solis-Pichardo, G., 1999, Tertiary arc-magmatism of the Sierra Madre del Sur, Mexico, and its transition to the volcanic activity of the Trans-Mexican Volcanic Belt: Journal of South American Earth Sciences, v. 12, p. 513–535.

Mujica-Mondragón, M.R., and Jacobo-Albarrán, J., 1983, Estudio petrogenético de las rocas ígneas y metamórficas del Altiplano Mexicano: Instituto Mexicano del Petróleo, Subdirección Técnica de Exploración, Proyecto C-1156, 78 p. (Open-File Report.)

Mullan, H.S., 1978, Evolution of part of the Nevada orogen in northwest Mexico: Geological Society of America Bulletin, v. 111, p. 347–363.

Nieto-Obregón, J., Delgado-Argote, L., and Damon, P.E., 1981, Relaciones petrológicas y geocronológicas del magmatismo de la Sierra Madre Occidental el Eje Neovolcánico en Nayarit, Jalisco y Zacatecas: Asociación de Ingenieros de Minas, Metalurgistas y Geólogos de México, Memoria Técnica, v. XIV, p. 327–361.

Nieto-Obregón, J., Delgado-Argote, L., and Damon, P.E., 1985, Geochronologic, petrologic, and structural data related to large morphologic features between the Sierra Madre Occidental and the Mexican volcanic belt: Geofísica Internacional, v. 24, p. 623–663.

Nieto-Samaniego, Á.F., Macias-Romo, C., and Alaniz-Álvarez, S.A., 1996, Nuevas edades isotópicas de la cubierta volcánica cenozoica de la parte meridional de la Mesa Central, México: Revista Mexicana de Ciencias Geológicas, v. 13, p. 117–122.

Nieto-Samaniego, Á.F., Ferrari, L., Alaniz-Álvarez, S.A., Labarthe-Hernández, G., and Rosas-Elguera, J., 1999, Variation of Cenozoic extension and volcanism across the southern Sierra Madre Occidental volcanic province, Mexico: Geological Society of America Bulletin, v. 111, p. 347–363, doi: 10.1130/0016-7606(1999)111<0347:VOCEAV>2.3.CO;2.

Norton, I., 1995, Plate motions in the North Pacific: The 43 Ma nonevent: Tectonics, v. 14, p. 1080–1094, doi: 10.1029/95TC01256.

Nourse, J.A., Anderson, T.H., and Silver, L.T., 1994, Tertiary metamorphic core complexes in Sonora, northwestern Mexico: Tectonics, v. 13, p. 1161–1182, doi: 10.1029/93TC03324.

Orozco-Esquivel, M.T., Nieto-Samaniego, Á.F., and Alaniz-Álvarez, S.A., 2002, Origin of rhyolitic lavas in the Mesa Central, Mexico, by crustal melting related to extension: Journal of Volcanology and Geothermal Research, v. 118, p. 37–56, doi: 10.1016/S0377-0273(02)00249-4.

Ortega-Gutiérrez, F., Mitre-Salazar, L.M., Roldán-Quintana, J., and Aranda-Gómez, J.J., Morán- Zenteno, D., Alaniz-Álvarez S.A., and Nieto-Samaniego, Á.F., 1992, Carta geológica de la Republica Mexicana: Universidad Nacional Autónoma de México, scale 1:2,000,000, 1 sheet, with text of 74 p.

Ortega-Gutiérrez, F., Sedlock, R.L., and Speed, R.C., 1994, Phanerozoic tectonic evolution of Mexico, *in* Speed, R.C., ed., Phanerozoic Evolution of North American Continent-Ocean Transitions: Geological Society of America, Decade of North American Geology Summary Volume to accompany the DNAG Continent-Ocean Transects Series, p. 265–306.

Oskin, M., and Stock, J., 2003, Cenozoic volcanism and tectonics of the continental margins of the Upper Delfín basin, northeastern Baja California and western Sonora, *in* Johnson, S.E., Paterson, S.R., Fletcher, J.M., Girty, G.H., Kimbrough, D.L., and Martín-Barajas, A., eds., Tectonic Evolution of Northwestern México and the Southwestern USA: Geological Society of America Special Paper 374, p. 421–438.

Oskin, M., Stock, J.M., and Martín-Barajas, A., 2001, Rapid localization of Pacific–North America plate motion in the Gulf of California: Geology, v. 29, p. 459–462, doi: 10.1130/0091-7613(2001)029<0459:RLOPNA>2.0.CO;2.

Oskin, M., Iriondo, A., and Nourse, J., 2003, Geologic reconnaissance and geochronology of proto-Gulf of California extension, western Sonora: Geological Society of America Abstracts with Programs, 99th Cordilleran Meeting.

Pankhurst, R.J., and Rapela, C.R., 1995, Production of Jurassic rhyolite by anatexis of the lower crust of Patagonia: Earth and Planetary Science Letters, v. 134, p. 23–36, doi: 10.1016/0012-821X(95)00103-J.

Pankhurst, R.J., Leat, P.T., Sruoga, P., Rapela, C.W., Márquez, M., Storey, B.C., and Riley, T.R., 1998, The Chon Aike silicic igneous province of Patagonia and related rocks in Antarctica: A silicic large igneous province: Journal of Volcanology and Geothermal Research, v. 81, p. 113–136, doi: 10.1016/S0377-0273(97)00070-X.

Pankhurst, R.J., Riley, T.R., Fanning, C.M., and Kelley, S.R., 2000, Episodic silicic volcanism along the proto-Pacific margin of Patagonia and the Antarctic Peninsula: plume and subduction influences associated with the break-up of Gondwana: Journal of Petrology, v. 41, p. 605–625, doi: 10.1093/petrology/41.5.605.

Paterson, B.A., Stephens, W.E., Rogers, G., Williams, I.S., Hinton, R.W., and Herd, D.A., 1992, The nature of zircon inheritance in two granitic plutons: Transactions of the Royal Society of Edinburgh: Earth Sciences, v. 83, p. 459–471.

Pasquarè, G., Ferrari, L., Garduño, V.H., Tibaldi, A., and Vezzoli, L., 1991, Geological map of the central sector of Mexican Volcanic Belt, States of Guanajuato and Michoacan: Geological Society of America Map and Chart Series, MCH 072, 1 sheet, 20 p.

Paz-Moreno, F., Demant, A., Cochemé, J.J., Dostal, J., and Montigny, R., 2003, The Quaternary Moctezuma volcanic field: A tholeiitic to alkali basaltic episode in the central Sonoran Basin and Range Province, Mexico, *in* Johnson, S.E., Paterson, S.R., Fletcher, J.M., Girty, G.H., Kimbrough, D.L., and Martín-Barajas, A., eds., Tectonic Evolution of Northwestern México and the Southwestern USA: Geological Society of America Special Paper 374, p. 439–455.

Perry, F.V., DePaolo, D.J., and Baldridge, W.S., 1993, Neodymium isotopic evidence for decreasing crustal contributions to Cenozoic ignimbrites of the western United States: Implications for the thermal evolution of the Cordilleran crust: Geological Society of America Bulletin, v. 105, p. 872–882, doi: 10.1130/0016-7606(1993)105<0872:NIEFDC>2.3.CO;2.

Persaud, P., 2003, Images of early continental breakup in and around the Gulf of California and the role of basal shear in producing wide plate boundaries [Ph.D. thesis]: California Institute of Technology, 144 p.

Piguet, P., 1987, Contribution à l'étude de la Sierra Madre Occidental (Mexique): la séquence volcanoque tertiaire de la transversale Moctezuma-La Norteña [Ph.D. thesis]: Université de Droit, d'Économie et des Sciences D'Aix-Marseille, 300 p.

Ponce, B., and Clark, K., 1988, The Zacatecas mining district—a Tertiary caldera complex associated with precious and base metal mineralization: Economic Geology and the Bulletin of the Society of Economic Geologists, v. 83, p. 1668–1682.

Poole, F.G., Madrid, R.J., and Oliva-Becerril, J.F., 1991, Geological setting and origin of stratiform barite in central Sonora, Mexico, *in* Raines, G.L., Lisle, R.E., Schafer, R.W., and Wilkinson, W.H., eds., Geology and Ore Deposits of the Great Basin: Geological Society of Nevada, v. 1, p. 517–522.

Poole, F.G., Stewart, J.H., Berry, W.B.N., Harris, A.G., Repetski, J.E., Madrid, R.J., Ketner, K.B., Carter, C., and Morales-Ramírez, J.M., 1995, Ordovician ocean-basin rocks of Sonora, Mexico, *in* Cooper, J.D., Droser, M.L., and Finney, S.C., eds., Ordovician Odyssey: Short Papers for the Seventh International Symposium on the Ordovician System, Las Vegas: Fullerton, California: Pacific Section Society for Sedimentary Geology (SEPM), n. 77, p. 277–284.

Righter, K., Carmichael, I.S.E., and Becker, T., 1995, Pliocene-Quaternary faulting and volcanism at the intersection of the Gulf of California and the Mexican Volcanic Belt: Geological Society of America Bulletin, v. 107, p. 612–627, doi: 10.1130/0016-7606(1995)107<0612:PQVAFA>2.3.CO;2.

Riley, T.R., Leat, P.T., Pankhurst, R.J., and Harris, C., 2001, Origins of large volume rhyolitic volcanism in the Antarctic Peninsula and Patagonia by crustal melting: Journal of Petrology, v. 42, p. 1043–1065, doi: 10.1093/petrology/42.6.1043.

Ritsema, J., Van Heijst, H., and Woodhouse, J., 2004, Global transition zone tomography: Journal of Geophysical Research, v. 109, doi: 10.1029/2003JB002610.

Ritzwoller, M.H., Shapiro, N.M., Barmin, M.P., and Levshin, A., 2002, Global surface wave diffraction tomography: Journal of Geophysical Research, v. 107, p. 2335, doi: 10.1029/2002JB001777.

Rodríguez-Castañeda, J.L., 1996, Late Jurassic and mid-Tertiary brittle-ductile deformation in the Opodepe region, Sonora, México: Revista Mexicana de Ciencias Geológicas, v. 13, p. 1–9.

Roldán-Quintana, J., 1991, Geology and chemical composition of El Jaralito and Aconchi batholiths in east-central Sonora, *in* Pérez-Segura, E., and Jacques-Ayala, C., eds., Studies of Sonoran Geology: Geological Society of America Special Paper 254, p. 19–36.

Roldán-Quintana, J., Mora-Klepeis, G., Calmus, T., Valencia-Moreno, M., and Lozano-Santacruz, R., 2004, El graben de Empalme, Sonora, México: magmatismo y tectónica extensional asociados a la ruptura inicial del Golfo de California: Revista Mexicana de Ciencias Geológicas, v. 21, p. 320–334.

Rosas-Elguera, J., Ferrari, L., López-Martínez, M., and Urrutia-Fucugauchi, J., 1997, Stratigraphy and tectonics of the Guadalajara region and the triple junction area, western Mexico: International Geology Review, v. 39, p. 125–140.

Rosas-Elguera, J., Alva-Valdivia, L.M., Goguitchaichvili, A., and Urrutia-Fucugauchi, J., 2003, Counterclockwise rotation of the Michoacan block: Implications for the tectonics of western Mexico: International Geology Review, v. 45, p. 814–826.

Rossotti, A., Ferrari, L., López-Martínez, M., and Rosas-Elguera, J., 2002, Geology of the boundary between the Sierra Madre Occidental and the Trans-Mexican Volcanic Belt in the Guadalajara region, western Mexico: Revista Mexicana de Ciencias Geológicas, v. 19, p. 1–15.

Ruiz, J., Patchett, P.J., and Arculus, R.J., 1988a, Nd-Sr isotope composition of lower crustal xenoliths: Evidence for the origin of mid-Tertiary felsic volcanics in Mexico: Contributions to Mineralogy and Petrology, v. 99, p. 36–43, doi: 10.1007/BF00399363.

Ruiz, J., Patchett, P.J., and Ortega-Gutiérrez, F., 1988b, Proterozoic and Phanerozoic basement terranes of Mexico from Nd isotopic studies: Geological Society of America Bulletin, v. 100, p. 274–281, doi: 10.1130/0016-7606(1988)100<0274:PAPBTO>2.3.CO;2.

Ruiz, J., Patchett, P.J., and Arculus, R.J., 1990, Comments on Nd-Sr isotopic compositions of lower crustal xenoliths—Evidence for the origin of mid-Tertiary volcanics in Mexico: Reply: Contributions to Mineralogy and Petrology, v. 104, p. 615–618, doi: 10.1007/BF00306669.

Saleeby, J., 2003, Segmentation of the Laramide Slab—evidence from the southern Sierra Nevada region: Geological Society of America Bulletin, v. 115, p. 655–668, doi: 10.1130/0016-7606(2003)115<0655:SOTLSF>2.0.CO;2.

Sawlan, M.G., 1991, Magmatic evolution of the Gulf of California rift, *in* Dauphin, J.P., and Simoneit, B.A., eds., The Gulf and Peninsular Province of the Californias: American Association of Petroleum Geologists Memoir, 47, p. 301–369.

Scheubel, F.R., Clark, K.F., and Porter, E.W., 1988, Geology, tectonic environment, structural controls in the San Martin de Bolaños District, Jalisco: Economic Geology and the Bulletin of the Society of Economic Geologists, v. 83, p. 1703–1720.

Sedlock, R., Ortega-Gutiérrez, F., and Speed, R.C., 1993, Tectonostratigraphic terranes and tectonic evolution of Mexico: Geological Society of America Special Paper 278, 153 p.

Silver, L.T., and Anderson, T.H., 1974, Possible left-lateral early to middle Mesozoic disruption of the southwestern North America craton margin: Geological Society of America Abstracts with Programs, v. 6, p. 955–956.

Smith, J.A., Luhr, J.F., Pier, J.G., and Aranda-Gómez, J.J., 1989, Extension-related magmatism of the Durango volcanic field, Durango, Mexico: Geological Society of America Abstracts with Programs, v. 21, p. A201.

Smith, R.D., Cameron, K.L., McDowell, F.W., Niemeyer, S., and Sampson, D.E., 1996, Generation of voluminous silicic magmas and formation of mid-Cenozoic crust beneath north-central Mexico—evidence from ignimbrites, associated lavas, deep crustal granulites, and mantle pyroxenites: Contributions to Mineralogy and Petrology, v. 123, p. 375–389, doi: 10.1007/s004100050163.

Sole, J., and Salinas, J.C., 2002, Edades K-Ar de 54 rocas ígneas y metamórficas del occidente, centro y sur de México: GEOS, Boletín de la Unión Geofísica Mexicana, v. 22, n. 2, p. 260.

Sonder, L.J., and Jones, C.H., 1999, Western United States extension: How the West was widened: Annual Review of Earth and Planetary Sciences, v. 27, p. 417–462, doi: 10.1146/annurev.earth.27.1.417.

Staude, J.M., and Barton, M.D., 2001, Jurassic to Holocene tectonics, magmatism, and metallogeny of northwestern Mexico: Geological Society of America Bulletin, v. 113, p. 1357–1374, doi: 10.1130/0016-7606(2001)113<1357:JTHTMA>2.0.CO;2.

Stewart, J.H., 1988, Latest Proterozoic and Paleozoic southern margin of North America and the accretion of Mexico: Geology, v. 16, p. 186–189, doi: 10.1130/0091-7613(1988)016<0186:LPAPSM>2.3.CO;2.

Stewart, J.H., 1998, Regional characteristics, tilt domains, and extensional history of the later Cenozoic Basin and Range Province, western North America, *in* Faulds, J.E., and Stewart, J.H., eds., Accommodation Zones and Transfer Zones: The Regional Segmentation of the Basin and Range Province: Geological Society of America Special Paper 323, p. 47–74.

Stewart, J.H., and Poole, F.G., 2002, Inventory of Neoproterozoic and Paleozoic strata in Sonora, Mexico: U.S. Geological Survey Open-File Report 02-97, 50 p.

Stewart, J.H., and Roldán-Quintana, J., 1991, Upper Triassic Barranca Group; Nonmarine and shallow-marine rift-basin deposits of northwestern Mexico, *in* Pérez-Segura, E., and Jacques-Ayala, C., eds., Studies of Sonoran Geology: Geological Society of America Special Paper 254, p. 19–36.

Stewart, J.H., Anderson, R.E., Aranda-Gómez, J.J., Beard, L.S., Billingsley, G.H., Cather, S.M., Dilles, J.H., Dokka, R.K., Faulds, J.E., Ferrari, L., Grose, T.L.T., Henry, C.D., Janecke, S.U., Miller, D.M., Richard, S.M., Rowley, P.D., Roldán-Quintana, J., Scott, R.B., Sears, J.W., and Williams, V.S., 1998, Map showing Cenozoic tilt domains and associated structural features, western North America, *in* Faulds, J.E., and Stewart, J.H., eds., Accommodation Zones and Transfer Zones: The Regional Segmentation of the Basin and Range Province: Geological Society of America Special Paper 323, plate 1.

Stewart, J.H., Poole, F.G., Harris, A.G., Repetski, J.E., Wardlaw, B.R., Mamet, B.L., and Morales-Ramirez, J.M., 1999, Neoproterozoic (?) to Pennsylvanian inner-shelf, miogeoclinal strata in Sierra Agua Verde, Sonora, Mexico: Revista Mexicana de Ciencias Geológicas, v. 16, p. 35–62.

Stock, J.M., and Lee, J., 1994, Do microplates in subduction zones leave a geologic record?: Tectonics, v. 13, p. 1472–1487, doi: 10.1029/94TC01808.

Suter, M., 2001, The historical seismicity of northeastern Sonora and northwestern Chihuahua, Mexico (28–32° N, 106–111° W): Journal of South American Earth Sciences, v. 14, p. 521–532, doi: 10.1016/S0895-9811(01)00050-5.

Suter, M., and Contreras, J., 2002, Active Tectonics of Northeastern Sonora, Mexico (Southern Basin and Range Province) and the 3 May 1887 Mw 7.4 Earthquake: Seismological Society of America Bulletin, v. 92, p. 581–589, doi: 10.1785/0120000220.

Swanson, E.R., and McDowell, F.W., 1984, Calderas of the Sierra Madre Occidental volcanic field, western Mexico: Journal of Geophysical Research, v. 89, p. 8787–8799.

Swanson, E.R., and McDowell, F.W., 1985, Geology and geochronology of the Tómochic caldera, Chihuahua, Mexico: Geological Society of America Bulletin, v. 96, p. 1477–1482, doi: 10.1130/0016-7606(1985)96<1477:GAGOTT>2.0.CO;2.

Swanson, E.R., Keizer, R.P., Lyons, J.I., Jr., and Clabaugh, S.E., 1978, Tertiary volcanism and caldera development near Durango City, Sierra Madre Occidental, Mexico: Geological Society of America Bulletin, v. 89, p. 1000–1012, doi: 10.1130/0016-7606(1978)89<1000:TVACDN>2.0.CO;2.

Swanson, E.R., Kempter, K., McDowell, F.W., and McIntosh, W.C., 2006, Major ignimbrites and volcanic centers of the Copper Canyon area: A view into the core of Mexico's Sierra Madre Occidental: Geosphere, v. 2, p. 125–141, doi: 10.1130/GES00042.1.

Tuta, Z., Sutter, J.H., Kesler, S.E., and Ruiz, J., 1988, Geochronology of mercury, tin, and fluorite mineralization in northern Mexico: Economic Geology and the Bulletin of the Society of Economic Geologists, v. 83, p. 1931–1942.

Umhoefer, P.J., Dorsey, R., Willsey, S., Mayer, L., and Renne, P., 2001, Stratigraphy and geochronology of the Comondú Group near Loreto, Baja California Sur, Mexico: Sedimentary Geology, v. 144, p. 125–147, doi: 10.1016/S0037-0738(01)00138-5.

Valencia-Moreno, M., Ruiz, J., and Roldán-Quintana, J., 1999, Geochemistry of Laramide granitic rocks across the southern margin of the Paleozoic North American continent, Central Sonora, Mexico: International Geology Review, v. 41, p. 845–857.

Valencia-Moreno, M., Ruiz, J., Barton, M.D., Patchett, P.J., Zürcher, L., Hodkinson, D., and Roldán-Quintana, J., 2001, A chemical and isotopic study of the Laramide granitic belt of northwestern Mexico: Identification of the southern edge of the North American Precambrian basement: Geological Society of America Bulletin, v. 113, p. 1409–1422, doi: 10.1130/0016-7606(2001)113<1409:ACAISO>2.0.CO;2.

Valencia-Moreno, M., Ruiz, J., Ochoa-Landín, L., Martínez-Serrano, R., and Vargas-Navarro, P., 2003, Geology and Geochemistry of the Coastal Sonora Batholith, Northwestern Mexico: Canadian Journal of Earth Sciences, v. 40, p. 819–831, doi: 10.1139/e03-020.

van der Lee, S., and Nolet, G., 1997, Upper mantle S velocity structure of North America: Journal of Geophysical Research, v. 103, p. 22815–22838, doi: 10.1029/97JB01168.

Vega-Granillo, R., and Calmus, T., 2003, Mazatán metamorphic core complex (Sonora, Mexico): Structures along the detachment fault and its exhumation evolution: Journal of South American Earth Sciences, v. 16, 193–204.

Verma, S.P., 1984, Sr and Nd isotopic evidence for petrogenesis of mid-Tertiary felsic volcanism in the mineral district of Zacatecas, Zac. (Sierra Madre Occidental), Mexico: Isotope Geoscience, v. 2, p. 37–53.

Vidal-Solano, J.R., Paz-Moreno, F.A., Iriondo, A., Demant, A., and Cochemé, J.J., 2005, Middle Miocene peralkaline ignimbrites in the Hermosillo region (Sonora, México): Geodynamic implications: Compte Rendue Geosciences, v. 337, p. 1421–1430, doi: 10.1016/j.crte.2005.08.007.

Wallace, P., and Carmichael, I.S.E., 1989, Minette lavas and associated leucitites from the western front of the MVB: Petrology, chemistry and origin: Contribution to Mineralogy and Petrology, v. 103, p. 470–492, doi: 10.1007/BF01041754.

Ward, P.L., 1995, Subduction cycles under western North America during the Mesozoic and Cenozoic eras, *in* Miller, D.M., and Busby, C., eds., Jurassic Magmatism and Tectonics of the North American Cordillera: Geological Society of America Special Paper 299, p. 1–45.

Wark, D.A., 1991, Oligocene ash flow volcanism, northern Sierra Madre Occidental: Role of mafic and intermediate-composition magmas in rhyolite genesis: Journal of Geophysical Research, v. 96, p. 13389–13411.

Wark, D.A., Kempter, K.A., and McDowell, F.W., 1990, Evolution of waning, subduction-related magmatism, northern Sierra Madre Occidental, Mexico: Geological Society of America Bulletin, v. 102, p. 1555–1564, doi: 10.1130/0016-7606(1990)102<1555:EOWSRM>2.3.CO;2.

Webber, K.L., Fernández, L.A., and Simmons, W.B., 1994, Geochemistry and mineralogy of the Eocene-Oligocene volcanic sequence, southern Sierra Madre Occidental, Juchipila, Zacatecas, Mexico: Geofísica Internacional, v. 33, p. 77–89.

Wilson, C.J.N., Houghton, B.F., McWilliams, M.O., Lanphere, M.A., Weaver, S.D., and Briggs, R.M., 1995, Volcanic and structural evolution of Taupo Volcanic Zone, New Zealand: A review: Journal of Volcanology and Geothermal Research, v. 68, p. 1–28.

Wilson, F.I., and Rocha, S.V., 1949, Coal deposits of the Santa Clara district near Tónichi, Sonora, Mexico: U.S. Geological Survey Bulletin 962A, 80 p.

Wong, M., and Gans, P.B., 2003, Tectonic implications of early Miocene extensional unroofing of the Sierra Mazatán metamorphic core complex, Sonora, Mexico: Geology, v. 31, p. 953–956, doi: 10.1130/G19843.1.

Wong, M., and Gans, P.B., 2004, Evidence for Early crustal thickening, polyphase Oligo-Miocene Extension, and footwall rotation at the Sierra Mazatán metamorphic core complex, Sonora, Mexico: EOS, American Geophysical Union, Fall Meeting 2004, abstract T41E–1270.

Wortel, R., and Spakman, W., 2000, Subduction and slab detachment in the Mediterranean-Carpathian region: Science, v. 290, p. 1910–1917, doi: 10.1126/science.290.5498.1910.

Manuscript Accepted by the Society 29 August 2006

Printed in the USA

Geological Society of America
Special Paper 422
2007

Mesa Central of México: Stratigraphy, structure, and Cenozoic tectonic evolution

Ángel Francisco Nieto-Samaniego*
Susana Alicia Alaniz-Álvarez
Antoni Camprubí
Centro de Geociencias, Universidad Nacional Autónoma de México, Apartado postal 1-742, Querétaro, Qro. 76230, México

ABSTRACT

Mesa Central is an elevated plateau that can be divided into two regions. In the southern region, the topography is higher than 2000 masl, except for the Aguascalientes valley. This region is mostly covered by Cenozoic volcanic rocks. The northern region shows an advanced degree of erosion, and is below 2000 masl. The crust in Mesa Central is ~32 km thick, and it is bordered by the Sierra Madre Oriental, which has an average crustal thickness of ~37 km, and the Sierra Madre Occidental, which has an average crustal thickness of ~40 km. The presence of magmas below the crust is inferred, suggesting an underplating process. The oldest rocks are Triassic marine facies underlain by Jurassic continental rocks. Marine environment prevailed between the Oxfordian and the Cretaceous, forming three distinctive lithological sequences, from E to W: the Valles–San Luis Potosí Platform, the Mesozoic Basin of Central México, and marine volcanosedimentary Mesozoic rocks. All of the above rocks have plicative deformation and inverse faulting, which was produced during the Laramide orogeny. An angular unconformity separates these lithological sequences from the continental Cenozoic rocks. The bottom of the Cenozoic sequence consists of conglomerate with andesitic and rhyolitic volcanic rocks. These were followed by Oligocene topaz-bearing rhyolites, and the uppermost part of the Cenozoic sequence is Miocene-Quaternary alkaline basalt. The boundaries of Mesa Central are the Sector Transversal de Parras and major fault systems active during the Cenozoic to the E, W, and S. A major structure, the San Luis–Tepehuanes fault system, separates the northern and southern regions of Mesa Central. The majority of the mineral deposits found in Mesa Central or in its vicinities, especially epithermal deposits, is located on the traces of the major fault systems described above. The data available suggest that the structures associated with the major fault systems controlled the emplacement of both volcanic-hypabyssal rocks and mineral deposits.

Keywords: México, Mesa Central, Cenozoic stratigraphy, tectonics of México.

*afns@geminis.geociencias.unam.mx

Nieto-Samaniego, Á.F., Alaniz-Álvarez, S.A., and Camprubí, A., 2007, Mesa Central of México: Stratigraphy, structure, and Cenozoic tectonic evolution, *in* Alaniz-Álvarez, S.A., and Nieto-Samaniego, Á.F., eds., Geology of México: Celebrating the Centenary of the Geological Society of México: Geological Society of America Special Paper 422, p. 41–70, doi: 10.1130/2007.2422(02). For permission to copy, contact editing@geosociety.org.

INTRODUCTION

Mesa Central is located in central-northern México, a semi-arid region. Most geological studies in this region have focused on the exploration and exploitation of ore deposits, mainly precious metals. The works published in the nineteenth century, and in the beginnings of the twentieth, describe the stratigraphic record of mining districts such as Zacatecas, Fresnillo, Guanajuato, or Real de Catorce, to name a few (e.g., López-Monroy, 1888; Botsford, 1909).

Knowledge concerning Cenozoic geology and the processes that formed Mesa Central was scarce until the second half of the twentieth century. In the 1968 Geologic Map of the Mexican Republic (Hernández-Sanchez-Mejorada and López-Ramos, 1968), a "middle Cenozoic volcanic unit," which represents the undifferentiated Cenozoic volcanic cover, is located in the southern part of Mesa Central. Until that time, the rocks had not been divided into lithostratigraphic units. However, the 1968 map includes Cretaceous rocks and local outcrops of older rocks in the northern region of Mesa Central, in Zacatecas, Catorce, Ojo Caliente, and Charcas. The rest of the surface is covered by an extensive unit labeled "tuffs and residual material," which is given an age of Recent and Pleistocene and not formally separated into lithostratigraphic units.

Systematic topographic and geologic cartography, initiated in the 1970s by the National Institute of Statistics, Geography and Informatics (INEGI, 2004), constitutes the basis for the modern geological research in the area. There are two other regional series of geological maps that currently constitute the largest database of geologic information about Mesa Central. The first set of geologic maps is the geological cartography (scale 1:50,000) of the Institute of Geology of the Autonomous University of San Luis Potosí, published in the Technical Reports series. These maps cover the San Luis Potosí state and neighboring regions. The second set of geologic maps was published by the Mexican Geological Survey (formerly Council for Mineral Resources, Coremi). These maps cover the entire Mesa Central at a scale of 1:250,000 and many mining areas at a scale of 1:50,000. This information was gradually made available to the public during the 1970s, 1980s, and 1990s. This systematic cartographic program is still active. The geologic study of Mesa Central during the last decades of the twentieth century focused mainly on its southern region and Cenozoic rocks. Once the stratigraphic nomenclature was established, many volcanic rocks were dated, allowing structural studies to be done. The information provided in the past few decades has allowed researchers to propose models of tectonic evolution.

In this work, we present a critical analysis of the geologic knowledge of Mesa Central, mainly focusing on the evolution of Cenozoic rocks and structures. We also analyze the definition of this physiographic province, and review the known chronostratigraphic units and their interpretations. The structural analysis is made on the basis of location, geometry, and the age of the larger structures, as seen from a kinematic point of view. Additionally, we discuss the main Cenozoic tectonic models and, in conclusion, we propose a series of tectonic events that could have formed Mesa Central. Finally, we suggest that further research is necessary to close the major remaining scientific gaps concerning the geology of Mesa Central.

LOCATION AND EXTENSION OF MESA CENTRAL

Mesa Central was defined as a physiographic province by Raisz (1959), as "a basin surrounded by more elevated mountains. More high and flat than the Basin and Range province (located to the north). Instead of elongated ranges, this one has low elevated areas, mainly dissecting old volcanic rocks." If we observe the orographic configuration of the Mexican Republic, this province clearly stands out in the central part of the country (Fig. 1). Nevertheless, the modern knowledge of the orography and geology of that region, along with the possibility of observing México as a whole in satellite images and digital elevation models, necessitates a redefinition of the geologic boundaries of Mesa Central.

The boundaries of Mesa Central in Figure 1 were drawn according to the definitions of "physiographic province" proposed by Bates and Jackson (1987) and Lugo-Hubp (1989). We considered the specific morphologic and geologic characteristics of Mesa Central and identified its boundaries with neighboring regions. The configuration we drew matches the one proposed in the Digital Map of México by the National Institute of Statistics, Geography and Informatics (INEGI).

Mesa Central is a plateau located in the central part of México. More than half of its total area is found above 2000 masl, and the topographic highs within it are considered moderate, providing that typical height differences between the plateau and the topographic highs within are less than 600 m (Fig. 1). Mesa Central is bounded to the N and E by the Sierra Madre Oriental, toward the W by the Sierra Madre Occidental, and to the S by the El Bajío Basin. The physiographic provinces surrounding Mesa Central are distinctive in that they have a more abrupt relief. The Sierra Madre Oriental is a mountain chain formed during the Laramide orogeny (e.g., Eguiluz-de Antuñano et al., 2000) that lies N and E of Mesa Central. The Sierra Madre Oriental is topographically lower than 2000 masl in the eastern part and higher in its northern and northeastern parts, and is over 2800 masl in the higher mountains (Fig. 1). To the W of Mesa Central is the Sierra Madre Occidental, a mountain chain of volcanic origin (e.g., Aranda-Gómez et al., 2000). A N-S zone, which is ca. 130 km wide and has elevations over 2800 masl in many places, appears near the Mesa Central boundary (Fig. 1). Westward of that zone the elevations are less than 2000 masl and decrease gradually down to the coast. The southern section of Mesa Central is bounded by the El Bajío Basin, where elevations suddenly drop from more than 2000 masl to an average of 1800 masl.

Within Mesa Central, we recognize two regions: the southern region is higher in elevation than the northern; with the sole exception of the Aguascalientes valley, the southern region is higher than 2000 masl. It is covered by Cenozoic volcanic rocks

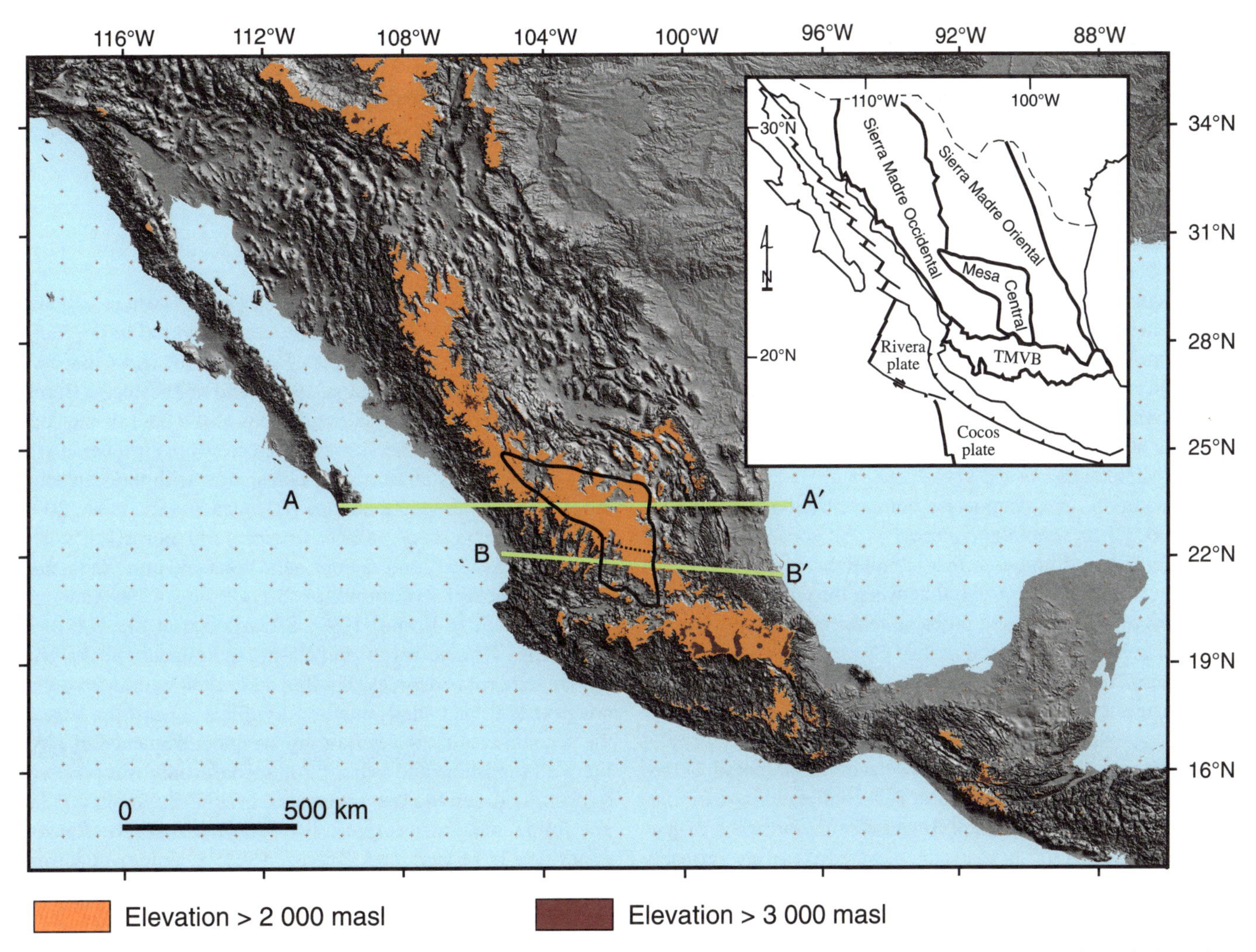

Figure 1. Digital elevation model of México and location of Mesa Central. TMVB—Trans-Mexican Volcanic Belt. The dotted line is the boundary between the northern and southern sectors of Mesa Central. The green lines indicate the location of sections in Figure 2.

(Nieto-Samaniego et al., 1999), shows immature stages of erosion compared to the northern region, and has normal faults with moderately eroded escarpments as its major structures. In an E-W profile through the southern region of Mesa Central and adjacent areas, it is evident that the average elevations of the Sierra Madre Occidental and the Sierra Madre Oriental are lower than the average elevation of Mesa Central. It is also remarkable that the relief within Mesa Central is flatter than it is within the Sierra Madre Occidental and Sierra Madre Oriental. These characteristics describe Mesa Central as an elevated plateau, more elevated than the surrounding regions (Fig. 1). In contrast, the northern region shows advanced stages of erosion with large alluvial and lacustrine continental basins, many of them endorheic. This region is lower in elevation, below 2000 masl, and flatter than the southern region (Fig. 1). Unlike the southern region, when an E-W profile is traced, Mesa Central appears as depressed and flattened in comparison to the surrounding mountains (Fig. 1), thus constituting a depressed plain.

The boundary between the northern and southern regions of Mesa Central is a NW-striking lineament over 1600 km long. Cenozoic normal faults have been documented along such lineaments in San Luis Potosí–San Luis de La Paz (Alaniz-Álvarez et al., 2001), Salinas de Hidalgo (La Ballena; Silva-Romo, 1996), and Nazas-Rodeo (Aranda-Gómez et al., 2003). In addition, there are local reports of NW normal faults along this lineament in the Zacatecas and Fresnillo mining districts (De Cserna, 1976; Albinson, 1988; Ponce and Clark, 1988).

THE CRUST AND LITHOSPHERE OF MESA CENTRAL

The crustal structure of Mesa Central and the Sierra Madre Occidental has been interpreted from seismic profiles reported by Meyer et al. (1958), Fix (1975), and Rivera and Ponce (1986). Kerdan (1992) made a model of the crustal structure based on a gravimetric profile through the Tropic of Cancer, from the Baja

California Peninsula to the Gulf of México, using the densities proposed in the aforementioned seismic studies.

The most probable crustal thickness of Mesa Central is ca. 32 km, in contrast with the crustal thicknesses determined for the Sierra Madre Oriental and the Sierra Madre Occidental, ca. 37 and 40 km, respectively (Fig. 2) (Campos-Enríquez et al., 1992; Kerdan, 1992; Nieto-Samaniego et al., 1999). Using a seismic model, Fix (1975) interpreted a zone with ca. 20% partial melting below the thinned crust of Mesa Central. The melting zone occurs between 34 and 260 km deep in the crust, and the relative degree of partial melting decreases gradually upward. Additional evidence for partial melting is found in the occurrence of mantle and crustal xenoliths in Quaternary basalts, which have been reported by Aranda-Gómez et al. (1993a). Indirect temperature data at the base of the crust were obtained by Hayob et al. (1989). These authors used granulitization (950–1125 °C), feldspar exsolution, and plagioclase homogenization temperatures (850–900 °C) in xenoliths collected in the southern region of Mesa Central, and proposed that granulitization occurred in the lower crust below Mesa Central between the Oligocene and the Quaternary. Thus, the most probable temperature at the base of the crust below Mesa Central is 850 °C.

Figure 2c displays the structure of the crust and the uppermost mantle below Mesa Central. Thus, the crust is relatively thin, topographically elevated, and bounded by two thicker and depressed crustal blocks. Below the crust of Mesa Central, bodies of partial melt occur in the mantle. This configuration resembles underplating processes, which, because of the presence of partially molten material beneath the elevated zone, are probably responsible for the crustal rise. As a consequence, the heated lower and middle crust shows risen isotherms.

STRATIGRAPHY

SW of San Juan del Río in Durango, near the edge of Mesa Central, Iriondo et al. (2003) reported small bodies of muscovite schist with metamorphism dated ca. 252 Ma ($^{40}Ar/^{39}Ar$ in muscovite; Fig. 3). There are no other mapped outcrops of Paleozoic rocks within Mesa Central. In contrast, Paleozoic rocks appear at the edges of Mesa Central and within the Sierra Madre Oriental (Ortega-Gutiérrez et al., 1992; Sánchez-Zavala et al., 1999).

Some authors attributed Paleozoic ages to rocks within Mesa Central, although convincing evidence was not provided. In Zacatecas, Burckhardt and Scalia (1906) considered a likely Paleozoic age for schist and phyllite that discordantly underlie a phyllite sequence with Carnian fossils; that Paleozoic age was questioned by McGehee (1976) because he could not identify the unconformity in his stratigraphic section. No additional information supporting the occurrence of Paleozoic rocks in Mesa Central has been published since then. In the Sierra de Catorce, Bacon (1978), Zárate-del Valle (1982), and Franco-Rubio (1999) considered a Paleozoic age for the base of the stratigraphic sequence. Such interpretation was mainly supported by the presence of Pennsylvanian spores (Bacon, 1978) and fragments of fossil plants (Franco-Rubio, 1999); but these ages have been questioned by Hoppe et al. (2002) and Barboza-Gudiño et al. (2004), because of sedimentary and stratigraphic evidence that indicates a more probable Upper Triassic age for the bottom of the sedimentary sequence at the Sierra de Catorce.

Mesozoic

Triassic

Late Triassic rocks within Mesa Central, whose age was established by their fossil content, were reported in the Sierra de Salinas (Peñón Blanco; Silva-Romo, 1996) and Charcas in San Luis Potosí (Cantú-Chapa, 1969), and in Zacatecas (Burckhardt and Scalia, 1906). The turbidites at the base of the sedimentary sequence in the Sierra de Catorce were attributed to the Triassic by means of lithologic correlation with the sequences at the other mentioned localities (Barboza-Gudiño et al., 2004; Fig. 3). In the Sierra de Salinas, Charcas, and Sierra de Catorce, Triassic rocks are found as thick sandstone and lutite sequences with characteristics of turbidites (Silva-Romo, 1996; Centeno-García and Silva-Romo, 1997; Silva-Romo et al., 2000). In Zacatecas, Triassic rocks are clastic sedimentary rocks with low metamorphic degrees. Phyllite and schist with interbedded conglomerate and sandstone are dominant toward the base of the sequence, and are overlain by an upper sequence of phyllite with sandstone and marble. An unconformity was proposed to occur between the two sequences, but this interpretation has not found unanimous agreement (McGehee, 1976; Ranson, et al., 1982; Monod and Calvet, 1991; Quintero-Legorreta, 1992). Mafic rocks, termed "green rocks" by several authors, intruded the above sedimentary rocks forming dikes, sills, and other intrusive bodies. Such rocks are younger than the metasedimentary sequences and have an attributed Cretaceous age (Centeno-García and Silva-Romo, 1997). However, some "green rock" outcrops were reported as pillow lavas interbedded with the Triassic sedimentary rocks (Burckhardt and Scalia, 1906; McGehee, 1976).

The Triassic turbidites located in the Sierra de Salinas, Sierra de Catorce, and Charcas in the State of San Luis Potosí were interpreted as evidence for the occurrence of a continental margin. The continent supplying clastic material would have been located toward the E-NE, with an ocean to the W-SW, which is represented by clastic rocks in Zacatecas (Fig. 4).

There are localities with volcanosedimentary rocks correlated by some authors with the Triassic unit at Zacatecas, but their age is a matter for debate. The existence of Triassic rocks in Fresnillo was proposed by De Cserna (1976), but there is no agreement about this age, as other authors consider them to be Cretaceous (Centeno-García and Silva-Romo, 1997). In La Manganita, 80 km NW of Zacatecas, similar volcanosedimentary rocks were reported along with "hybrid tuffs" (López-Infanzón, 1986), but no information about their age and composition is available. In the Sierra de Guanajuato, marine basin strata similar to the Zacatecas sequence were reported, but the most likely ages

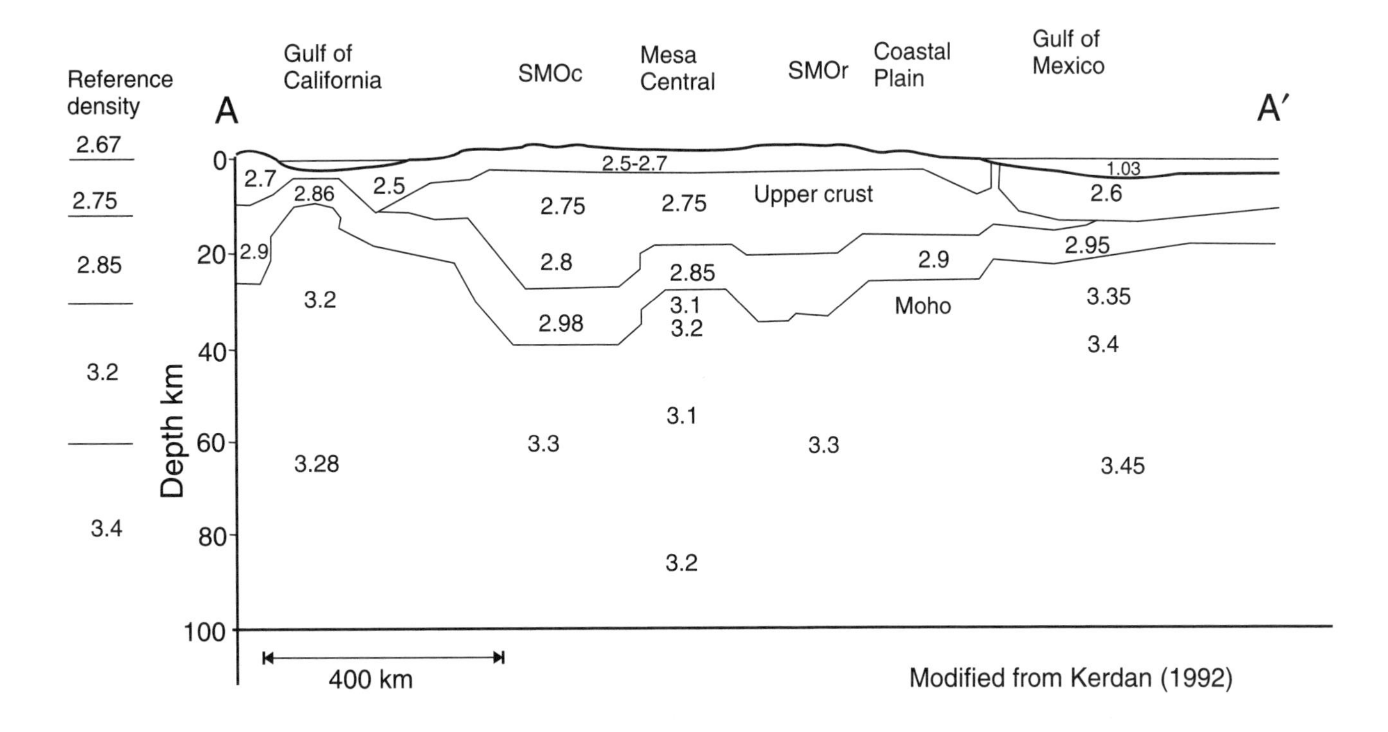

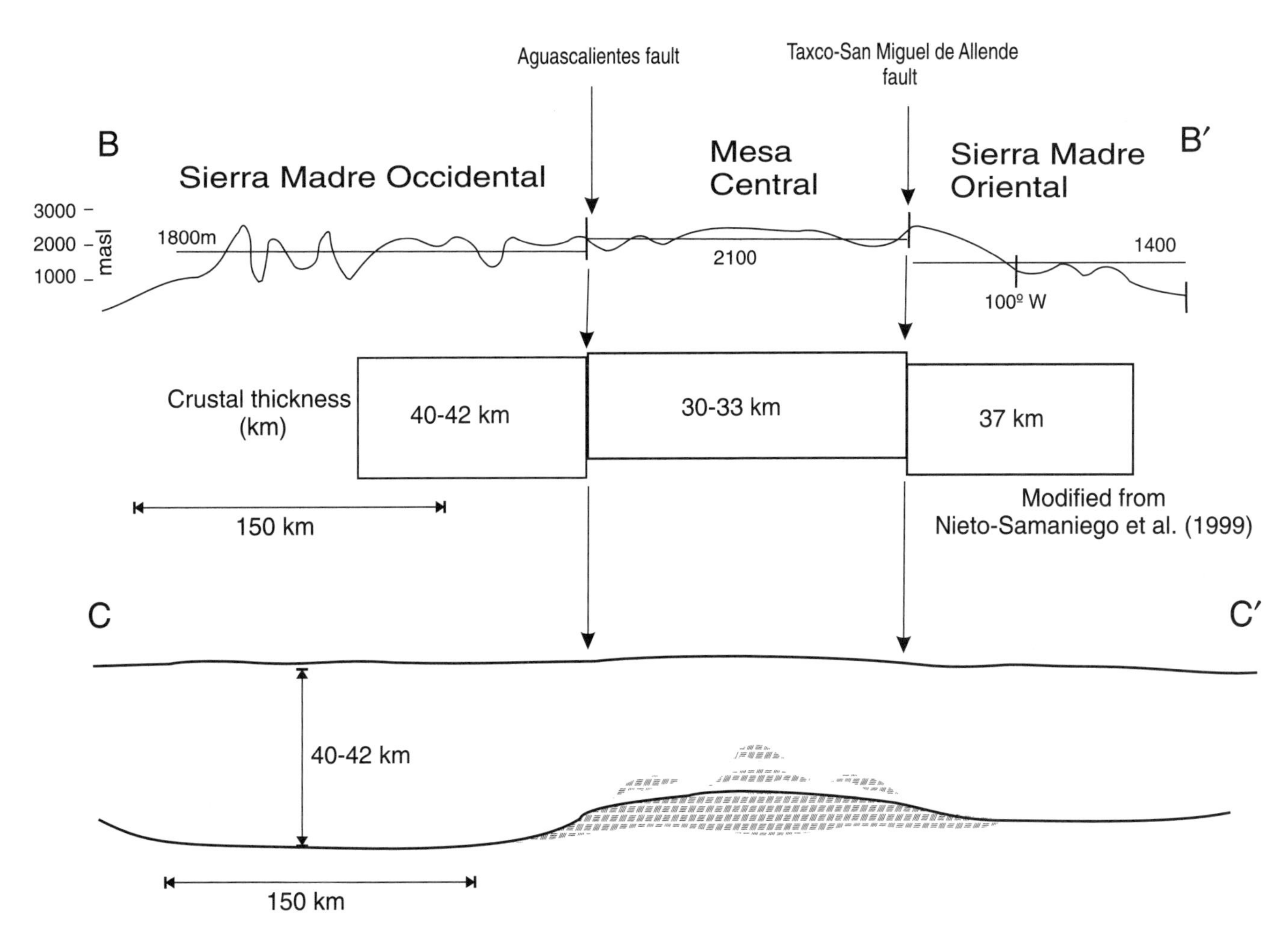

Figure 2. Crustal structure of Mesa Central, the location of sections is shown in Figure 1. Section A–A′ corresponds to the model obtained from gravimetric data and section B–B′ from hypsography and published crustal thickness (see text for complete discussion). Section C–C′ is an idealized model of the Mesa Central crustal structure proposed in this work. SMOc—Sierra Madre Occidental; SMOr—Sierra Madre Oriental.

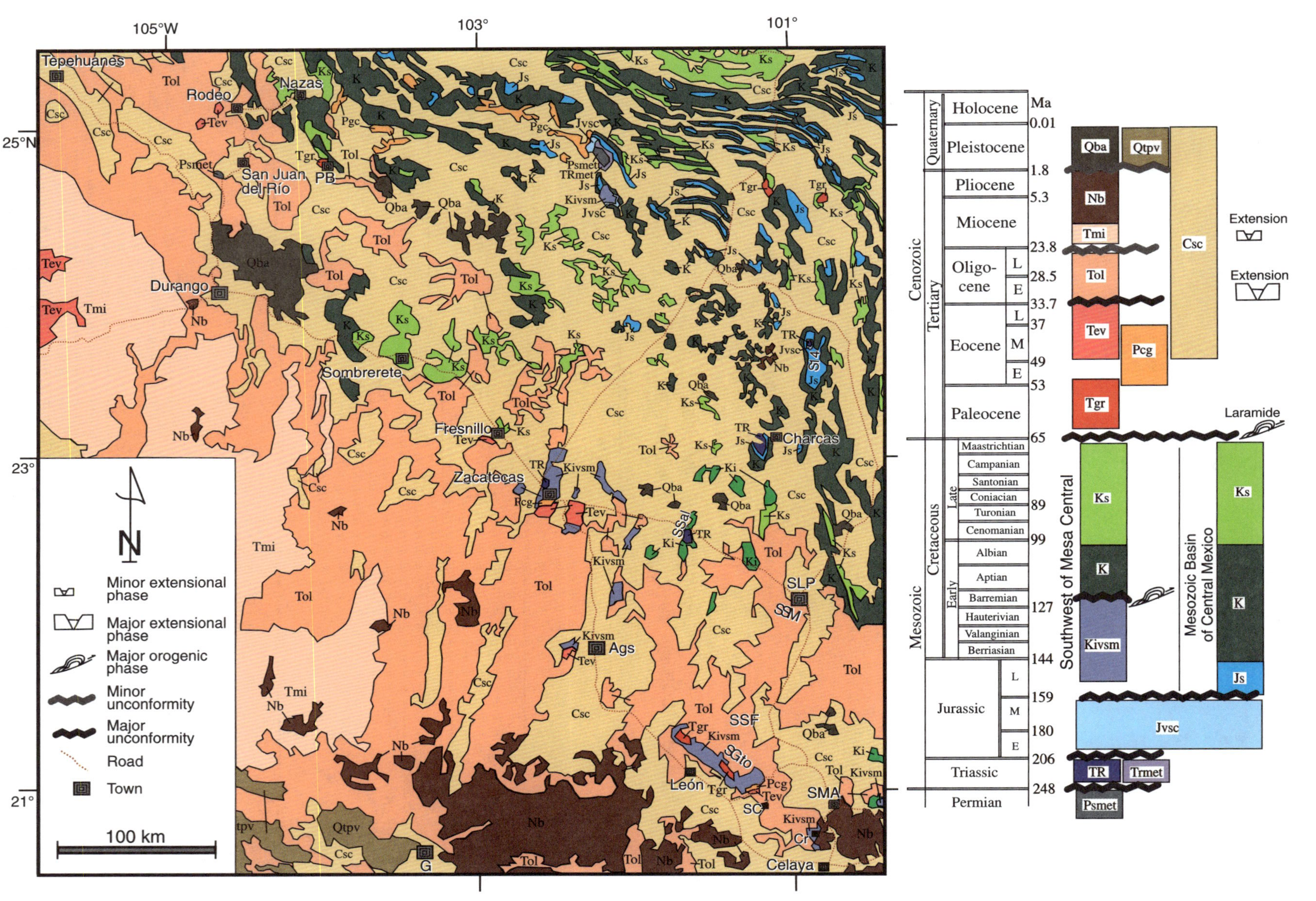

Figure 3. Geologic map indicating the localities mentioned in the text. Ags—Aguascalientes; Cr—Corrales; G—Guadalajara; PB—Peñón Blanco; S14—Sierra de Catorce; SGTO—Sierra de Guanajuato; SLP—San Luis Potosí; SMA—San Miguel de Allende; SSM—Sierra de San Miguelito; SSF—Sierra de San Felipe; SSal—Sierra de Salinas; SC—La Sauceda; Qba—Quaternary basalt; Qtpv—Quaternary-Pliocene volcanic rock; Nb—Neogene basalt; Tmi—Miocene ignimbrite; Csc—Cenozoic continental strata; Tol—Oligocene rhyolitic rocks; Tev—Eocene volcanic rocks; Pcg—Paleogene continental conglomerate; Tgr—Tertiary granite; Ks—Upper Cretaceous marine strata; K—Lower Cretaceous marine strata; Kivsm—Upper Jurassic–Lower Cretaceous volcanosedimentary marine strata; Js—Upper Jurassic marine strata; Jvsc—Middle Jurassic volcanosedimentary continental strata; TR—Triassic marine strata; Trmet—Triassic metamorphic rocks; Psmet—Paleozoic metamorphic rocks.

Late Triassic

T
25°N
continental crust
oceanic crust
turbidites
SLP
21°
101°W

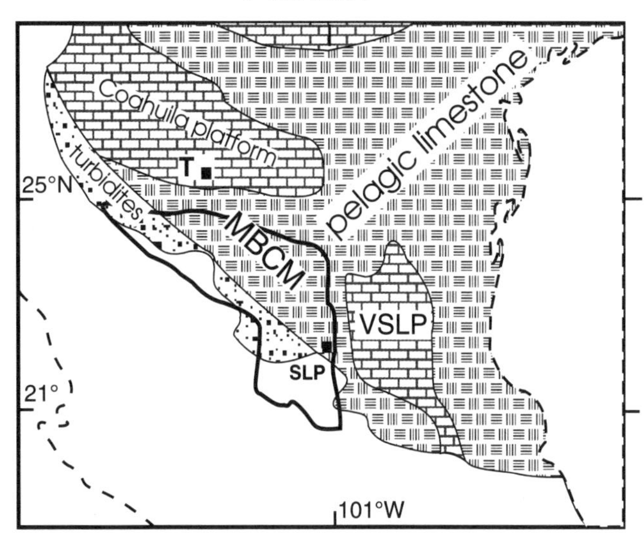

Figure 4. (A) Triassic (modified from Centeno-García and Silva-Romo, 1997) and (B) Albian (modified from Eguilez-de Antuñano et al., 2000) paleogeographic reconstruction of Mesa Central. VSLP—Valles–San Luis Potosí platform; MBCM—Mesozoic Basin of Central México; T—Torreón; SLP—San Luis Potosí.

for them are Late Jurassic to Early Cretaceous. Freydier et al. (1996) considered this to be the age of the basin strata because they are located near a volcanic arc sequence (magmatic sequence of Guanajuato) that is dated Early Cretaceous (K-Ar in whole-rock basalt, diorite, tonalite, and gabbro; Lapierre et al., 1992). An additional U-Pb age of 146.1 Ma in zircon was reported by Mortensen et al. (2003). There is direct stratigraphic evidence of the relationship between the volcanosedimentary sequence and the marine carbonate rocks near León in Guanajuato. The volcanosedimentary rocks in that locality underlie limestones with Aptian-Cenomanian fauna in angular unconformity (Chiodi et al., 1988; Quintero-Legorreta, 1992). The volcanic rocks show greenschist facies metamorphism, hydrothermal alteration, and two penetrative deformations. Quintero-Legorreta (1992) correlated those rocks with the Triassic rocks of Zacatecas because of their lithologic similarity; however, he did not report isotopic or paleontologic ages. A marine sedimentary sequence with basaltic rocks that resemble those of Guanajuato and Zacatecas crops out in Corrales, Guanajuato, but it is Early Cretaceous in age (Freydier et al., 1996).

Upper Triassic–Lower Jurassic Unconformity

There is an unconformity over Triassic rocks in Mesa Central that can be observed in Zacatecas (Zacatecas), Sierra de Salinas, Sierra de Catorce, and Charcas (San Luis Potosí). This unconformity has a regional extension, and is overlain by continental rocks that are widely distributed within Mesa Central, in the Sierra Madre Oriental (Carrillo-Bravo, 1971; Barboza-Gudiño et al., 1999), and in southern México (Morán-Zenteno et al., 1993).

Jurassic

The Jurassic comprises two lithologic units and one unconformity:

Lower-Middle Jurassic. The oldest Jurassic rocks are continental rocks (sandstone, conglomerate, and volcanic rocks), which record the emersion of Mesa Central. Their Lower-Middle Jurassic age is well known because these rocks discordantly overlie marine Triassic rocks and are discordantly underlain by Upper Jurassic marine rocks (Silva-Romo, 1996; Barboza-Gudiño et al., 2004). The occurrence of Lower-Middle Jurassic rocks testifies to the end of marine clastic sedimentation in the margin of the North American craton. Instead, continental sedimentation with subaerial volcanic activity occurred. The time lapse when Mesa Central was emerged is poorly constrained.

Middle Jurassic unconformity. An unconformity that covers the entire Mesa Central is found on the Lower-Middle Jurassic continental rocks (Pantoja-Alor, 1963; López-Infanzón, 1986; Silva-Romo, 1996). The stratigraphic interval contained in this unconformity is not well constrained because the ages of the overlying strata are not definite. The unconformity is located between the upper part of the Middle Jurassic and the base of the Upper Jurassic.

Although this unconformity is described in detail in several localities (Bacon, 1978; Silva-Romo, 1996), its existence was questioned in the Sierra de Catorce by Barboza-Gudiño et al. (2004). The pre-Oxfordian stratigraphic record of Mesa Central has been widely discussed in the literature, but the age of several lithostratigraphic units and the correlations between them are dubious. For complete revisions on this subject, see Barboza-Gudiño et al. (1999) and Bartolini et al. (1999).

Upper Jurassic. The second stratigraphic set of Jurassic rocks consists of calcareous marine rocks. The lower part of this unit is formed of platform limestone and dolomitic limestone (Zuloaga Formation), whereas the upper part is primarily formed of calcareous mudstone with chert layers (La Caja Formation). The age of this unit, Oxfordian to Tithonian, was defined by its fossil content (Imlay, 1938). These rocks record the sedimentation switch from the continental to the marine environment in the entire Mesa Central. The marine environment remained in the area during ~90 Ma, from the Oxfordian until the Upper Cretaceous (Caracol Formation).

Cretaceous

The marine environment initiated in the Upper Jurassic and produced three different marine strata in Mesa Central:

(1) The 4000-m-thick strata, corresponding to the Valles–San Luis Potosí Platform, were deposited in the eastern Mesa Central and toward the E (Carrillo-Bravo, 1971). Deposition initiated in the Upper Jurassic over continental red beds, and continued during the Cretaceous. These deposits consist of evaporite and reef limestone, fore reef and reef front facies, and the whole sequence was folded (Carrillo-Bravo, 1971; López-Doncel, 2003).

(2) A thick sequence of marine rocks that corresponds to open-sea and deep-water facies strata is located W of the Valles–San Luis Potosí platform, where it covers nearly the entire Mesa Central. These strata are known as the Mesozoic Basin of Central México (*Cuenca Mesozoica del Centro de México*) (Carrillo-Bravo, 1971). The stratigraphic record indicates the occurrence of a subsidence that resulted in the accumulation of up to 6000 m of sedimentary rocks. This accumulation initiated with the Upper Jurassic platform limestone that overlies red beds and continental volcanic rocks. During the Cretaceous, argillaceous limestone and calcareous shale were deposited and the sedimentation in the Mesozoic Basin culminated with detritic rocks, mainly sandstone containing fragments of volcanic rocks (Caracol Formation). The last facies in this basin precedes the end of the marine environment in Mesa Central during the Late Cretaceous. The age of the Caracol Formation is poorly known: it is considered Coniacian in Durango (Roldán-Quintana, 1968), based on lithological correlation and stratigraphic position criteria, and it is considered Maastrichtian in San Luis Potosí (Carrillo-Bravo, 1971), although no diagnostic criteria are specified. The rocks of the Mesozoic Basin of Central México have shortening deformation like those in the Valles–San Luis Potosí Platform.

(3) West of the Mesozoic Basin of Central México and located in the western and southern margins of Mesa Central, a marine volcanosedimentary sequence occurs. This sequence consists of a thick pile of pillow lava basalt, lava flows, and intrusive diabase bodies, with tuff, slate, chert, and radiolarite layers. The most prominent sedimentary unit consists of shale, sandstone, and limestone (Ranson et al. 1982; Martínez-Reyes, 1992; Centeno-García and Silva-Romo, 1997). These rocks are widespread in the western part of México, and to the N and S of Mesa Central (Freydier et al., 1996). The volcanosedimentary sequence has a strong shortening deformation and, contrary to the Mesozoic Basin of Central México and the Valles–San Luis Potosí Platform, shows metamorphism in the greenschists facies, a characteristic that is more easily observed in volcanic rocks. This sequence was interpreted to have formed in a marine volcanic arc whose age is not well constrained, between Late Jurassic and Early Cretaceous according to isotopic ages and paleontologic evidence (Corona-Chávez and López-Picos, 1988; Freydier et al., 1996; Mortensen et al., 2003).

The limit between the Valles–San Luis Potosí Platform and the Mesozoic Basin of Central México remarkably coincides with the eastern limit of Mesa Central and with the trace of the Taxco–San Miguel de Allende fault system (Alaniz-Álvarez et al., 2002). Also, the boundary between the Mesozoic Basin of Central México and the Mesozoic volcanosedimentary sequence approximately follows the trace of the San Luis–Tepehuanes fault system.

Neocomian unconformity. There is an unconformity on the Late Jurassic to Early Cretaceous volcanosedimentary sequence, whose extension and stratigraphic range is not well constrained. It was documented by Quintero-Legorreta (1992) close to León in Guanajuato as an angular unconformity between (1) clastic rocks with marine volcanic rocks of poorly defined Upper Jurassic-Neocomian age, and (2) calcareous rocks with Aptian-Cenomanian fossils. This unconformity has not been documented on rocks of the Mesozoic Basin of Central México, where marine rocks span from the Oxfordian to the Campanian-Maastrichtian with no major hiatus within the series. Thus, it can be inferred that the unconformity occurs only in the western part of Mesa Central, between the volcanosedimentary sequence and the calcareous rocks of the Mesozoic Basin of Central México, and that it includes the upper part of the Neocomian rocks. The presence of low-grade metamorphism and two phases of deformation in the volcanosedimentary sequence—in contrast with the rocks of the Mesozoic Basin of Central México, which had no metamorphism and only one shortening phase—led Quintero-Legorreta (1992) to infer a Nevadian orogeny phase for the volcanosedimentary sequence. The deformation would have occurred during the tectonic transport of the volcanosedimentary sequence toward the E-NE, and the corresponding time span would be represented by the described unconformity.

Cenozoic

The Mesozoic rocks are overlain by Cenozoic continental sedimentary and volcanic rocks, an environment that lasts until the present time in Mesa Central. Three regional unconformities were identified in the stratigraphic record:

Mesozoic-Tertiary Unconformity

The Mesozoic rocks, deformed and sometimes showing low-grade metamorphism, underlie an angular and erosional unconformity. This unconformity extends to all of Mesa Central and most of México. In Mesa Central, this unconformity represents a lapse in sedimentation from the Late Cretaceous to the Paleocene in its western part, and from the Maastrichtian to the Paleocene in the eastern part; however their temporal limits are not well defined. Munguía-Rojas et al. (2000) constrain the unconformity to the Santonian at 26° lat. N and 104° long. W, and it becomes older westward. This unconformity represents the sedimentary hiatus produced during the Laramide orogeny.

Continental Paleocene–Middle Eocene Strata

These rocks appear in the N and NW and as isolated outcrops in the center, S, and E of Mesa Central (Fig. 3). They are mainly sandstone and conglomerate, with mafic volcanic rocks in some localities, or less abundantly, felsic pyroclastic rocks. This unit is variable in thickness, changing from tens of meters to some hundreds in short distances. This variability indicates that the unit deposited in a surface with a high topographic relief, apparently associated with the development of grabens, as documented in Guanajuato (Edwards, 1955; Aranda-Gómez and McDowell, 1998), León (Martínez-Reyes, 1992), Zacatecas (Edwards, 1955), Fresnillo and Sombrerete (Albinson, 1988), some localities in the State of San Luis Potosí (Labarthe-Hernández et al., 1982), near Durango (Cordova, 1988), Rodeo (Aguirre-Díaz and McDowell, 1991), and Peñón Blanco (Roldán-Quintana, 1968). All of these deposits are younger than the Laramide shortening. The age of these rocks has been commonly assigned to the Paleocene-Eocene based on their stratigraphic position, and they have been dated in three localities: Guanajuato, San Luis Potosí, and Zacatecas. In Guanajuato, their Eocene age was determined using vertebrate fossils (Edwards, 1955) and an isotopic age from an andesite lava flow at the lower part of the sequence, at 49.3 ± 1 Ma (K-Ar, plagioclase; Aranda-Gómez and McDowell, 1998). Thus, these authors concluded that the deposit started to form during the early Eocene. The oldest rock that unconformably overlies the conglomerates was dated at 37 ± 3 Ma (Gross, 1975), thus constraining the minimum age of these rocks to the middle Eocene. In San Luis Potosí, palynologic analyses indicate a Paleocene-Eocene age for these rocks (Labarthe-Hernández et al., 1982), whereas in Zacatecas the upper part of the conglomerate is interbedded with pyroclastic rocks dated at 46.8 Ma (K-Ar, biotite; Ponce and Clark, 1988) of the Alamitos ignimbrite (Loza-Aguirre, 2005). Data indicate that the conglomerates deposited during the early Eocene and part of the middle Eocene, but a Paleocene age for the lower part of this sequence cannot be discarded.

Middle Eocene Unconformity

The Paleocene to middle Eocene rocks described above underlie local angular unconformities beneath a volcanic rock cover. In the southern, central, and western portions of Mesa Central, the stratigraphic position of the unconformities indicates that they span the upper part of the middle Eocene. In other localities where the Paleocene to lower Eocene rocks are absent, the middle Eocene unconformity joins the Mesozoic-Tertiary unconformity. In the Sierra de Guanajuato, the middle Eocene unconformity extends to the Oligocene and in the northeastern part of Mesa Central, it extends to the Recent, provided that there is no volcanic cover.

Middle Eocene Volcanic Rocks

The Paleocene-Eocene conglomerates underlie volcanic rocks, which are mainly mafic but also felsic. The localities where these rocks were mapped are Guanajuato, Aguascalientes, Zacatecas, Fresnillo, San Luis Potosí (Nieto-Samaniego et al., 1996), the Sombrerete-Colorada area (Albinson, 1988), Durango (Swanson et al., 1978), Nazas, and N (Aguirre-Díaz and McDowell, 1991) and S of Mesa Central (Morán-Zenteno et al., 2000). Thus, they form a discontinuous volcanic belt in western México whose ages span from the late Paleocene to the Eocene (Ferrari et al., 2005). Due to the middle Eocene unconformity, in some localities like Guanajuato, the volcanic rocks unconformably cover the Paleocene to Eocene conglomerates. However, in other localities like Zacatecas, these rocks are concordant to the underlying conglomerates. The isotopic ages of these volcanic rocks in Mesa Central span from 37 to 49 Ma, indicating a main middle Eocene age (Table 1).

Upper Eocene–Oligocene Unconformity

Between the middle Eocene and Oligocene rocks there is an angular unconformity documented in Zacatecas, Fresnillo, and Sombrerete (Albinson, 1988). In the Sierra de Guanajuato, this unconformity is inferred from the stratigraphic position of the Oligocene volcanic cover, which commonly directly overlies lower Eocene or Mesozoic rocks. This placement is also observed in the region between San Luis Potosí and Salinas de Hidalgo. This unconformity represents a hiatus in the volcanic activity and coincides with changes in the composition of the magmas. The Eocene volcanism has both andesitic and rhyolitic compositions, whereas the Oligocene volcanism is mostly rhyolitic.

Oligocene Volcanic Rocks

This group of rocks is more characteristic of the southern and western parts of Mesa Central. In the southern part of Mesa Central, the best geologic data set of the region was obtained and was used to establish a complete lithostratigraphic column (Labarthe-Hernández et al., 1982; Nieto-Samaniego et al., 1996; and references therein). These rocks are almost volcanic of rhyolitic composition, with compositional variations to latite and dacite. The lower part of this group of rocks consists mainly of effusive rocks, like lava flows and domes, with some pyroclastic rocks. The domes and lava flows form large volcanic complexes, over 400 m thick, which constitute the nuclei of the ranges within southern Mesa Central. Some examples are the Sierra de San Miguelito, Sierra de Guanajuato, Sierra de

TABLE 1. ISOTOPIC AGES OF MESA CENTRAL

Sample	Stratigraphic unit	State	Longitude (W)	Latitude (N)	Rock type	Method	Dated material	Age (Ma)	Error ± σ	Reference
H90-9	Pliocene and Quaternary	Durango	–105.33	24.91	Basalt	K/Ar	Whole rock	2.3	0.2	Aranda-Gómez et al., 1997
H90-13	Pliocene and Quaternary	Durango	–105.33	24.91	Basalt	K/Ar	Whole rock	2.5	0.2	Aranda-Gómez et al., 1997
H90-15a	Pliocene and Quaternary	Durango	–105.33	24.91	Basalt	K/Ar	Whole rock	2.3	0.2	Aranda-Gómez et al., 1997
-	Middle and upper Miocene	Guanajuato	–100.76	21	Basalt	K/Ar	Whole rock	10.7	0.7	Carranza-Castañeda et al., 1994
ZA 8	Middle and upper Miocene	Querétaro	–100.18	20.91	Andesite	K/Ar	Whole rock	10.9	0.5	Carrasco-Núñez et al., 1989
ES 135	Middle and upper Miocene	San Luis Potosí	–100.75	22.39	Basalt	K/Ar	Whole rock	11	1	Murillo-Muñetón and Torres-Vargas, 1987
H96-9	Middle and upper Miocene	Durango	–104.87	23.93	Basalt	Ar/Ar	Hornblende	11.59	0.05	Henry and Aranda-Gómez, 2000
H96-6	Middle and upper Miocene	Durango	–104.85	23.95	Basalt	Ar/Ar	Plagioclase	11.6	0.07	Henry and Aranda-Gómez, 2000
RK 12	Middle and upper Miocene	Durango	–104.87	23.92	Basalt	K/Ar	Amphibole	11.7	0.3	McDowell and Keizer, 1977
H96-8	Middle and upper Miocene	Durango	–104.86	23.94	Basalt	Ar/Ar	Hornblende	11.9	0.5	Henry and Aranda-Gómez, 2000
-	Middle and upper Miocene	Guanajuato	–101.37	21.02	Basalt	K/Ar	Whole rock	12.1	0.2	Aguirre-Díaz et al., 1997
RK 24	Middle and upper Miocene	Durango	–104.86	23.93	Basalt	K/Ar	Amphibole	12.4	0.4	McDowell and Keizer, 1977
-	Middle and upper Miocene	Guanajuato	–100.76	21	Basalt	K/Ar	Whole rock	12.5	0.9	Carranza-Castaneda et al., 1994
NA 41	Middle and upper Miocene	Durango	–104.3	25.42	Basalt	K/Ar	Whole rock	20.3	0.4	Aguirre-Díaz and McDowell, 1993
NA 41	Middle and upper Miocene	Durango	–104.3	25.42	Basalt	K/Ar	Whole rock	20.3	0.4	Aguirre-Díaz and McDowell, 1993
PZ 3	Middle and upper Miocene	Zacatecas	–101.53	22.4	Rhyolite	K/Ar	Sanidine	22	4	Murillo-Muñetón and Torres-Vargas, 1987
Na 55 a	Middle and upper Miocene	Durango	–104.2	25.19	Basalt	K/Ar	Whole rock	22.1	0.4	Aguirre-Díaz and McDowell, 1993
Na 55 a	Middle and upper Miocene	Durango	–104.2	25.19	Basalt	K/Ar	Whole rock	22.1	0.4	Aguirre-Díaz and McDowell, 1993
R 14	Middle and upper Miocene	Durango	–104.49	25.12	Basalt	K/Ar	Whole rock	22.4	0.4	Aguirre-Díaz and McDowell, 1993
R 14	Middle and upper Miocene	Durango	–104.49	25.12	Basalt	K/Ar	Whole rock	22.4	0.4	Aguirre-Díaz and McDowell, 1993
Mx 88-38	Oligocene volcanic rocks	Guanajuato	–101.22	20.93	Rhyolite	K/Ar	Sanidine	23	1	Pasquarè et al., 1991
PZ 2	Oligocene volcanic rocks	Zacatecas	–101.68	22.45	Rhyolite	K/Ar	Sanidine	23	2	Murillo-Muñetón and Torres-Vargas, 1987
FM291	Oligocene volcanic rocks	Zacatecas	–103.02	21.43	Ignimbrite	K/Ar	Whole rock	23.7	1.4	Webber et al., 1994
PZ 1	Oligocene volcanic rocks	Zacatecas	–101.44	22.22	Rhyolite	K/Ar	Sanidine	24	3	Murillo-Muñetón and Torres-Vargas, 1987
SL29	Oligocene volcanic rocks	Durango	–104.2	25.26	Basalt	K/Ar	Whole rock	24.3	0.5	Aguirre-Díaz and McDowell, 1993
SL29	Oligocene volcanic rocks	Durango	–104.2	25.26	Basalt	K/Ar	Whole rock	24.3	0.5	Aguirre-Díaz and McDowell, 1993
SOM-3	Oligocene volcanic rocks	Zacatecas	–103.47	23.47	Tuff	K/Ar	Sanidine	24.6	0.3	Tuta et al., 1988
HR460	Oligocene volcanic rocks	Zacatecas	–103.17	21.5	Ignimbrite	FT	Zircon	24.9	2.7	Webber et al., 1994
BQ483	Oligocene volcanic rocks	Zacatecas	–103.3	21.45	Ignimbrite	FT	Zircon	25.2	2.2	Webber et al., 1994
HR467	Oligocene volcanic rocks	Zacatecas	–103.18	21.5	Ignimbrite	FT	Zircon	25.3	2.4	Webber et al., 1994
SOM-6	Oligocene volcanic rocks	Zacatecas	–103.48	23.43	Tuff	K/Ar	Sanidine	25.6	0.3	Tuta et al., 1988
LP476	Oligocene volcanic rocks	Zacatecas	–103	21.51	Ignimbrite	FT	Zircon	25.9	2.5	Webber et al., 1994
SA 5	Oligocene volcanic rocks	Zacatecas	–103.13	23.55	Rhyolite	K/Ar	Sanidine	26	2	Murillo-Muñetón and Torres-Vargas, 1987
REA-5	Oligocene volcanic rocks	Guanajuato	–100	21.467	Rhyolite	K/Ar	Sanidine	26.7	0.3	Tuta et al., 1988

(continued)

TABLE 1. ISOTOPIC AGES OF MESA CENTRAL (*continued*)

Sample	Stratigraphic unit	State	Longitude (W)	Latitude (N)	Rock type	Method	Dated material	Age (Ma)	Error ± σ	Reference
-	Oligocene volcanic rocks	San Luis Potosí	−101.16	21.82	Ignimbrite	K/Ar	Whole rock	26.8	1.3	Labarthe-Hernández et al., 1982
IL 89 5	Oligocene volcanic rocks	Zacatecas	−103.48	21.42	Ignimbrite	K/Ar	Sanidine	27.11	0.05	Moore et al., 1994
REA-7	Oligocene volcanic rocks	Guanajuato	−100	21.467	Rhyolite	K/Ar	Sanidine	27.5	0.3	Tuta et al., 1988
SS-227	Oligocene volcanic rocks	Zacatecas	−102.94	23.12	Ignimbrite	K/Ar	Whole rock	27.5	1.2	Lang et al., 1988
SS-149	Oligocene volcanic rocks	Zacatecas	−102.93	23.06	Ignimbrite	K/Ar	Whole rock	27.7	0.6	Lang et al., 1988
TS 15	Oligocene volcanic rocks	Durango	−104.22	22.63	Rhyolite	Ar/Ar	Sanidine	27.9	0.3	Ferrari et al., 2002
IMP 619	Oligocene volcanic rocks	San Luis Potosí	−100.38	22.62	Granite	K/Ar	Biotite	28	2	Murillo-Muñetón and Torres-Vargas, 1987
TS 11	Oligocene volcanic rocks	Jalisco	−103.99	22.66	Ignimbrite	Ar/Ar	Sanidine	28	2	Ferrari et al., 2002
SN-2	Oligocene volcanic rocks	San Luis Potosí	−100.28	22.667	Granite	Ar/Ar	Biotite	28	0.4	Tuta et al., 1988
SA 6	Oligocene volcanic rocks	Zacatecas	−103.2	23.71	Ignimbrite	K/Ar	Sanidine	28	2	Murillo-Muñetón and Torres-Vargas, 1987
AMC-01	Oligocene volcanic rocks	Guanajuato	−101.53	21.33	Ignimbrite	K/Ar	Sanidine	28.2	0.7	Nieto-Samaniego et al., 1996
RK 4 P	Oligocene volcanic rocks	Durango	−104.67	23.83	Ignimbrite	K/Ar	Plagioclase	28.3	0.6	McDowell and Keizer, 1977
RK 3 F	Oligocene volcanic rocks	Durango	−104.67	23.92	Ignimbrite	K/Ar	Feldspar	28.3	0.6	McDowell and Keizer, 1977
REM-2	Oligocene volcanic rocks	Durango	−104.35	24.04	Rhyolite	K/Ar	Sanidine	28.3	0.3	Tuta et al., 1988
RK 1 F	Oligocene volcanic rocks	Durango	−104.88	23.93	Ignimbrite	K/Ar	Feldspar	28.5	1.5	McDowell and Keizer, 1977
TS 10	Oligocene volcanic rocks	Jalisco	−103.99	22.67	Ignimbrite	Ar/Ar	Sanidine	28.6	0.3	Ferrari et al., 2002
RK 11 F	Oligocene volcanic rocks	Durango	−104.87	23.92	Ignimbrite	K/Ar	Feldspar	28.6	0.6	McDowell and Keizer, 1977
JL 1 R	Oligocene volcanic rocks	Durango	−104.59	24.12	Basalt	K/Ar	Whole rock	28.6	0.5	McDowell and Keizer, 1977
RK 2 F	Oligocene volcanic rocks	Durango	−104.77	23.97	Ignimbrite	K/Ar	Feldspar	28.7	0.6	McDowell and Keizer, 1977
RK 15 F	Oligocene volcanic rocks	Durango	−104.95	23.9	Ignimbrite	K/Ar	Feldspar	28.8	1.1	McDowell and Keizer, 1977
Na.4	Oligocene volcanic rocks	Durango	−104.187	25.208	Tuff	K/Ar	Feldspar	28.8	1.9	Aguirre-Díaz and McDowell, 1993
3-71-1 F	Oligocene volcanic rocks	Durango	−104.97	23.89	Ignimbrite	K/Ar	Feldspar	28.9	0.6	McDowell and Keizer, 1977
-	Oligocene volcanic rocks	San Luis Potosí	−101.03	21.91	Ignimbrite	K/Ar	Whole rock	29	1.5	Labarthe-Hernández et al., 1982
2-71-2 F	Oligocene volcanic rocks	Durango	−104.97	23.89	Ignimbrite	K/Ar	Feldspar	29.1	0.6	McDowell and Keizer, 1977
SI-56	Oligocene volcanic rocks	Durango	−104.325	25.383	Tuff	K/Ar	Feldspar	29.1	1.8	Aguirre Díaz and McDowell, 1993
Ped Zac 3	Oligocene volcanic rocks	Zacatecas	−103.18	21.77	Ignimbrite	K/Ar	Sanidine	29.15	0.6	Nieto-Obregón et al. 1981
JAG 1	Oligocene volcanic rocks	San Luis Potosí	−101.27	22.04	Rhyolite	K/Ar	Biotite	29.2	0.8	Aguillon-Robles et al., 1994
JAG 1	Oligocene volcanic rocks	San Luis Potosí	−101.27	22.04	Rhyolite	K/Ar	Biotite	29.2	0.8	Aguillon-Robles et al., 1994
RK 14 F	Oligocene volcanic rocks	Durango	−104.95	23.9	Ignimbrite	K/Ar	Feldspar	29.2	0.7	McDowell and Keizer, 1977
-	Oligocene volcanic rocks	San Luis Potosí	−100.7	21.8	Ignimbrite	K/Ar	Whole rock	29.5	1.5	Labarthe-Hernández et al., 1982
JL BR SI R	Oligocene volcanic rocks	Durango	−104.57	24.08	Basalt	K/Ar	Whole rock	29.5	0.6	McDowell and Keizer, 1977
Na-33	Oligocene volcanic rocks	Durango	−104.386	25.409	Tuff	K/Ar	Feldspar	29.5	0.6	Aguirre-Díaz and McDowell, 1993
OCH-5	Oligocene volcanic rocks	Durango	−103.92	24.03	Tuff	K/Ar	Sanidine	29.6	0.4	Tuta et al., 1988
SOM-6	Oligocene volcanic rocks	Zacatecas	−103.48	23.43	Tuff	K/Ar	Whole rock	29.7	0.3	Tuta et al., 1988
JL BM LBR	Oligocene volcanic rocks	Durango	−104.22	24.53	Basalt	K/Ar	Whole rock	29.8	0.6	McDowell and Keizer, 1977
RK 6 F	Oligocene volcanic rocks	Durango	−104.67	23.95	Ignimbrite	K/Ar	Feldspar	29.9	0.7	McDowell and Keizer, 1977
-	Oligocene volcanic rocks	San Luis Potosí	−101.15	21.85	Rhyolite	K/Ar	Whole rock	30	1.5	Labarthe-Hernández et al., 1982
RK 5 F	Oligocene volcanic rocks	Durango	−104.67	23.93	Ignimbrite	K/Ar	Feldspar	30	0.7	McDowell and Keizer, 1977
AMC-06	Oligocene volcanic rocks	Guanajuato	−101.02	21.08	Rhyolite	K/Ar	Sanidine	30.1	0.8	Nieto-Samaniego et al., 1996
SOM-3	Oligocene volcanic rocks	Zacatecas	−103.47	23.47	Tuff	K/Ar	Whole rock	30.1	0.4	Tuta et al., 1988

(*continued*)

TABLE 1. ISOTOPIC AGES OF MESA CENTRAL (*continued*)

Sample	Stratigraphic unit	State	Longitude (W)	Latitude (N)	Rock type	Method	Dated material	Age (Ma)	Error ± σ	Reference
SOM-1	Oligocene volcanic rocks	Zacatecas	–103.47	23.47	Tuff	K/Ar	Sanidine+ Plagioclase	30.2	0.4	Tuta et al., 1988
77.4	Oligocene volcanic rocks	Durango	–104.567	24.35	Ignimbrite	K/Ar	Feldspar	30.4	0.9	Magonthier, 1988
REA-7	Oligocene volcanic rocks	Guanajuato	–100	21.467	Rhyolite	K/Ar	Whole rock	30.5	0.4	Tuta et al., 1988
-	Oligocene volcanic rocks	San Luis Potosí	–100.68	21.9	Rhyodacite	K/Ar	Whole rock	30.6	1.5	Labarthe-Hernández et al., 1982
2-71-1 F	Oligocene volcanic rocks	Durango	–104.67	24.05	Ignimbrite	K/Ar	Feldspar	30.6	0.9	McDowell and Keizer, 1977
AMC-08	Oligocene volcanic rocks	Guanajuato	–101.27	21.45	Rhyolite	K/Ar	Sanidine	30.7	0.8	Nieto-Samaniego et al., 1996
RK 23 B	Oligocene volcanic rocks	Durango	–104.83	23.99	Rhyolite	K/Ar	Biotite	30.7	0.5	McDowell and Keizer, 1977
JL-RD-A P	Oligocene volcanic rocks	Durango	–104.65	24.63	Rhyolite dike	K/Ar	Plagioclase	30.7	0.7	McDowell and Keizer, 1977
REA-5	Oligocene volcanic rocks	Guanajuato	–100	21.467	Rhyolite	K/Ar	Whole rock	30.8	0.3	Tuta et al., 1988
REA-8	Oligocene volcanic rocks	Guanajuato	–100	21.467	Rhyolite	K/Ar	Whole rock	30.9	0.4	Tuta et al., 1988
M-3	Oligocene volcanic rocks	San Luis Potosí	–100.38	22.66	Granite	K/Ar	Biotite	31	2.0	Murillo-Muñetón and Torres-Vargas, 1987
OCH-1	Oligocene volcanic rocks	Durango	–103.92	24.03	Rhyolite	K/Ar	Whole rock	31	0.4	Tuta et al., 1988
REM-2	Oligocene volcanic rocks	Durango	–104.35	24.04	Rhyolite	K/Ar	Ground mass	31	0.4	Tuta et al., 1988
RK 20 F	Oligocene volcanic rocks	Durango	–104.74	24.07	Ignimbrite	K/Ar	Feldspar	31	0.7	McDowell and Keizer, 1977
CUEVS-TF	Oligocene volcanic rocks	San Luis Potosí	–100.6	21.95	Rhyolite	K/Ar	Whole rock	31.1	0.3	Tuta et al., 1988
TS 5	Oligocene volcanic rocks	Jalisco	–103.99	22.66	Ignimbrite	Ar/Ar	Sanidine	31.1	0.4	Ferrari et al., 2002
OCH-5	Oligocene volcanic rocks	Durango	–103.92	24.03	Tuff	K-Ar	Ground mass	31.1	0.3	Tuta et al., 1988
RK 10 P	Oligocene volcanic rocks	Durango	–104.67	23.83	Ignimbrite	K/Ar	Plagioclase	31.2	1.5	McDowell and Keizer, 1977
RK 17 F	Oligocene volcanic rocks	Durango	–104.75	23.96	Ignimbrite	K/Ar	Feldspar	31.2	0.7	McDowell and Keizer, 1977
RK 22 F	Oligocene volcanic rocks	Durango	–104.5	23.87	Ignimbrite	K/Ar	Feldspar	31.4	0.7	McDowell and Keizer, 1977
K-LP-T F	Oligocene volcanic rocks	Durango	–104.63	24.12	Ignimbrite	K/Ar	Feldspar	31.4	0.7	McDowell and Keizer, 1977
77.9	Oligocene volcanic rocks	Durango	–104.421	24.593	Ignimbrite	K/Ar	Feldspar	31.5	1.0	Magonthier, 1988
TS 56	Oligocene volcanic rocks	Zacatecas	–103.72	22.89	Ignimbrite	Ar/Ar	Sanidine	31.6	0.3	Ferrari et al., 2002
-	Oligocene volcanic rocks	Guanajuato	–101.25	21.25	Rhyolite	K/Ar	Whole rock	32	1.0	Gross, 1975
JL-JE-A F	Oligocene volcanic rocks	Durango	–104.63	24.12	Ignimbrite	K/Ar	Feldspar	32.1	1.9	McDowell and Keizer, 1977
Na-81a	Oligocene volcanic rocks	Durango	–104.270	25.157	Tuff	K/Ar	Feldspar	32.2	0.7	Aguirre-Díaz and McDowell, 1993
SOM-1	Oligocene volcanic rocks	Zacatecas	–103.47	23.47	Tuff	Ar/Ar	Biotite	32.3	0.5	Tuta et al., 1988
Uaka 79-21	Middle Eocene volcanic rocks	Zacatecas	–102.55	22.75	Ignimbrite	K/Ar	Sanidine	36.8	0.8	Clark et al., 1981
-	Middle Eocene volcanic rocks	Guanajuato	–101.25	21.25	Ignimbrite	K/Ar	Whole rock	37	3	Gross, 1975
SLP 7	Middle Eocene volcanic rocks	San Luis Potosí	–100.68	22.43	Andesita	K/Ar	Hornblende	37.7	1.9	Nieto-Samaniego et al., 1999
S-2	Middle Eocene volcanic rocks	Zacatecas	–102.82	23.13	Ignimbrite	K/Ar	Whole rock	38.3	0.8	Lang et al., 1988
TOM-02	Middle Eocene volcanic rocks	Aguascalientes	–102.41	21.88	Rhyolite	K/Ar	Sanidine	40.6	1.0	Nieto-Samaniego et al., 1996
na	Middle Eocene volcanic rocks	Durango	–104.19	25.21	Tuff	K/Ar	Feldspar +Biotite	43	-	Aguirre-Díaz and McDowell, 1991
	Middle Eocene volcanic rocks	San Luis Potosí	–101.17	22.27	Andesita	K/Ar	Whole rock	44.1	2.2	Labarthe-Hernández et al., 1982

(continued)

TABLE 1. ISOTOPIC AGES OF MESA CENTRAL (*continued*)

Sample	Stratigraphic unit	State	Longitude (W)	Latitude (N)	Rock type	Method	Dated material	Age (Ma)	Error ± σ	Reference
CTO-01	Middle Eocene volcanic rocks	Aguascalientes	–102.41	21.9	Ignimbrite	K/Ar	Sanidine	47.2	1.2	Nieto-Samaniego et al., 1996
BFA	Middle Eocene volcanic rocks	Zacatecas	–103.15	21.65	Andesite	K/Ar	Feldspar	48.1	2.6	Webber et al., 1994
	Middle Eocene volcanic rocks	Guanajuato	–101.29	20.98	Andesite	K/Ar	Whole rock	49.3	1.0	Aranda-Gómez and McDowell, 1998
-	Vulcanosedimentary sequence	Guanajuato	–101.55	21.117	Ultramafic	K/Ar	Whole rock	112	-	Lapierre et al., 1992
-	Vulcanosedimentary sequence	Guanajuato	–101.55	21.117	Ultramafic	K/Ar	Actinolite	82	-	Lapierre et al., 1992
-	Vulcanosedimentary sequence	Guanajuato	–101.402	21.027	Diorite	K/Ar	Hornblende	122	-	Lapierre et al., 1992
-	Vulcanosedimentary sequence	Guanajuato	–101.402	21.027	Diorite	K/Ar	Whole rock	120	-	Lapierre et al., 1992
-	Vulcanosedimentary sequence	Guanajuato	–101.354	21.024	Dolerite (dike)	K/Ar	Whole rock	157–143	-	Lapierre et al., 1992
-	Vulcanosedimentary sequence	Guanajuato	–101.356	21.027	Basalt	K/Ar	Whole rock	108–66	-	Lapierre et al., 1992
SN-2	Intrusive bodies	San Luis Potosí	–100.28	22.667	Granite	K/Ar	Whole rock	32	0.4	Tuta et al., 1988
ES 323	Intrusive bodies	San Luis Potosí	–100.73	22.7	Granite	K/Ar	Biotite	36	3.0	Murillo-Muñetón and Torres-Vargas, 1987
2M-200	Intrusive bodies	San Luis Potosí	–100.71	23.66	Granodiorite	K/Ar	Biotite	36	3.0	Murillo-Muñetón and Torres-Vargas, 1987

Note: Some ages referred to within the text are not in this table because they do not have enough information; those ages should be consulted in the original source cited in the text.

Codornices, and Sierra de San Felipe (Fig. 3). The ages of the rhyolites in the southern Mesa Central are of early Oligocene, between 32 and 29 Ma (Nieto-Samaniego et al., 1996). In the region between Fresnillo and Sombrerete, there is no detailed stratigraphic information available, but the reported isotopic ages range from 30 to 27 Ma (Huspeni et al., 1984). The upper part of the Oligocene sequence consists of explosive volcanic rocks. It forms a cover with ignimbrites and other pyroclastic rocks of rhyolitic composition that overlie the rhyolitic domes and lava flows. The formal lithostratigraphic units for this type of cover were defined by Labarthe-Hernández et al. (1982). The ignimbrites are broadly distributed, but their thickness is moderate, only occasionally greater than 250 m. The ages reported in San Luis Potosí by Labarthe-Hernández et al. (1982) range from 29 to ~27 Ma. In the Sierra de Guanajuato region and in the Sierra de Codornices, Nieto-Samaniego et al. (1996) reported a ~25 Ma ignimbrite, the youngest rock in the pyroclastic group. Such ages have been further documented close to Celaya in Guanajuato (Ojeda-García, 2004) (Table 2).

In the southern part of Mesa Central, there is a remarkable absence of caldera-like structures associated with volcanic rocks, whereas in the western part of Mesa Central such structures have been documented by Swanson et al. (1978) and Ponce and Clark (1988). The absence of calderas in the southern part of Mesa Central and their scarcity in the eastern limit of the Sierra Madre Occidental led some authors to propose that the pyroclastic rhyolitic rocks in this region formed from fissure eruptions. This idea is supported by the observation of numerous pyroclastic dikes hosted by underlying rocks (Labarthe-Hernández et al., 1982; Aguirre-Díaz and Labarthe-Hernández, 2003).

Based on the distribution, stratigraphic position, and chemical and isotopic composition of the Oligocene volcanic rocks, Orozco-Esquivel et al. (2002) distinguished two groups of rocks with different origins. The lower group consists of rhyolitic lava flows, some domes, and pyroclastic material, which are associated with magmas that have evolved from the mantle. The upper group is a voluminous assemblage of rhyolite domes and lava flows, which forms the upper part of the sequence of the effusive volcanism and the totality of the pyroclastic cover. This second group was formed by the processes of disequilibrium causing partial melting in the crust, with a small contribution of mantle-derived magmas. The composition and main source of the magmas apparently changed at ~30 Ma.

Many private reports and some published works indicate that some volcanic rocks occur interbedded with sedimentary rocks, filling the continental basins of the southern part of Mesa Central (Martínez-Ruiz and Cuellar-González, 1978; Jiménez-Nava, 1993). A similar setting was reported by Cordova (1988) in the San Pablo Formation, ~20 km N of San Juan del Río in Durango, where the local volcanic rocks probably correspond to the Oligocene ignimbrites. Isotopic ages for the interbedded rocks have been reported only in the San Miguel de Allende–Guanajuato region (Nieto-Samaniego et al., 1996), but it is necessary to have a larger number of isotopic ages to establish the stratigraphic range of Oligocene volcanic rocks in Mesa Central. Data, however, indicate that the Oligocene continental sedimentary rocks deposited contemporaneously with volcanism in Mesa Central.

Middle and Upper Miocene Volcanic Rocks

The Oligocene volcanic rocks interbedded with the sedimentary rocks of the Cenozoic continental basins are unconformably overlain by Miocene mafic volcanic rocks. These rocks have been reported in many localities of Mesa Central, especially in its central, southern, and western parts.

Córdoba (1988) defined the Metates Formation at the NW border of Mesa Central as lava flows of olivine basalt and andesitic basalt that commonly contain xenoliths of peridotite, gneiss, and other metamorphic rocks. The Metates Formation was dated at ca. 12 Ma in its type locality, W of the city of Durango, outside Mesa Central (McDowell and Keizer, 1977; Henry and Aranda-Gómez, 2000).

In the Los Encinos region (Fig. 3), there is a volcanic alkaline basalt field with ages that range from 10.6 to 13.6 Ma (Luhr et al., 1995). In the State of San Luis Potosí, Labarthe-Hernández et al. (1982) defined a lithostratigraphic unit, which they called "Cabras Basalt." The age of this unit is unknown, but these authors attributed a late Oligocene age to it, because it overlies ignimbrites dated at ~27 Ma. However, a 13.2 ± 0.6 Ma K-Ar age in plagioclase was reported by Nieto-Samaniego et al. (1999) for andesitic rocks from the NE part of the State of San Luis Potosí, correlating with the Cabras Basalt.

TABLE 2. ISOTOPIC AGES OF THE SAN NICOLAS AND EL SALTO IGNIMBRITES

Sample	UTM	Lithostratigraphic unit	Mineral	%K	^{40}Ar* (mol/g)	%^{40}Ar*	Age (Ma)
A029MDA173	14Q285018 2295393	San Nicolás ignimbrite[†] [#]	Sanidine	5.93	2.560^{-10}	99	24.7 ± 1.1
A029MDA172	14Q284842 2295584	El Salto ignimbrite[#]	Sanidine	8.03	3.919^{-10}	95.9	28.0 ± 1.1

Note: Mineral separation was performed in the Centro de Geociencias and analysis in the Laboratorio Universitario de Geoquímica Isotópica, Instituto de Geología, Universidad Nacional Autónoma de México. UTM—Universal Transverse Mercator.

*Radiogenic Argon.

[†]Nieto-Samaniego (1992).

[#]Ojeda-García (2004).

South of the city of San Luis Potosí, in the upper parts of the Sierra de Guanajuato, the basalts and andesites are known as El Cubilete Basalt (Martínez-Reyes, 1992), which was dated at 13.5 Ma by Aguirre-Díaz et al. (1997). Near San Miguel de Allende, in the SE corner of Mesa Central, there are andesitic and basaltic lava flows and stratovolcanoes dated between 16 and ~10 Ma (Pérez-Venzor et al., 1996; Verma and Carrasco-Núñez, 2003). These volcanic rocks extend widely outside of Mesa Central along the northern zone of the Trans-Mexican Volcanic Belt.

Pliocene and Quaternary Volcanic Rocks

In the central part of Mesa Central, volcanic Quaternary to Pliocene rocks exist as small groups of lava and cinder cones, as well as maars, located in the State of San Luis Potosí (Labarthe-Hernández et al., 1982). In Durango, these rocks form a volcanic field with an overall extension of ~2000 km^2, which contains over 100 Quaternary cinder and lava cones and some maars (Swanson 1989; Aranda-Gómez et al., 2003). The Pliocene and Quaternary volcanic rocks of Mesa Central are characterized by the alkaline composition of the lava and commonly contain lerzolite xenoliths from the upper mantle and granulite xenoliths from the base of the crust. The geographic distribution, the volcanoes, and the chemistry of lavas and xenoliths were described by Swanson (1989) and Aranda-Gómez et al. (1993a, 1993b). For a complete study of the Miocene to Quaternary volcanic rocks of Mesa Central and adjacent areas, as well as a broad discussion about their origin, see Aranda-Gómez et al. (2005).

Oligocene-Quaternary Sedimentary Rocks

Within Mesa Central, mainly in its southern and western parts, the topography is dominated by relatively low mountain ranges surrounded by continental river basins, containing fluvial and lacustrine rocks. These rocks have not been studied much, and their stratigraphic range varies in different locations. In the Durango region, the lower part of the sequence has been named San Pablo Formation and the upper part, Guadiana Formation (Córdoba, 1988). The felsic and mafic volcanic rocks contained in this sequence correspond to Oligocene and Pliocene-Quaternary, respectively. In the States of San Luis Potosí, Zacatecas, and Guanajuato, the sedimentary rocks show interbedded Oligocene or Miocene felsic and mafic volcanic rocks. In addition, the fossil record indicates that the upper parts of the sedimentary sequence have ages that correspond up to the Quaternary (Montellano-Ballesteros, 1990; Carranza-Castañeda et al., 1994).

Intrusive Bodies

Small intrusive bodies of mainly Tertiary age crop out in Mesa Central. There are only a handful of specific studies about them and, in general, these come from nonconventional sources or are hard to find (theses, private reports for mining companies or government dependencies). The most important source of information about the location and age of these bodies are the geologic maps published by the Mexican Geological Survey, which are free and accessible electronically at http://www.coremisgm.gob.mx/. Three of the most important plutonic bodies in Mesa Central are described below:

Comanja Granite

This pluton was defined by Quintero-Legorreta (1992), and is found as discontinuous outcrops in most of the Sierra de Guanajuato (Fig. 5). According to Quintero-Legorreta, the granite body extends for ca. 160 km^2 and is formed by K-feldspar, quartz, plagioclase, and biotite.

The granite is hosted by the Mesozoic folded sequence. It is postorogenic because it was not folded, and it intrudes the Mesozoic rocks, producing a contact metamorphism aureola. There is a nonconformity between the granite and the Tertiary rocks. The granite underlies the continental conglomerates of the Continental Paleocene–middle Eocene rocks described previously, as well as the Oligocene rhyolitic ignimbrites (28.2 ± 0.7 Ma, K-Ar in sanidine; Nieto-Samaniego et al., 1996). Mujica-Mondragón and Albarrán-Jacobo (1983, in Quintero-Legorreta, 1992) obtained K-Ar ages in biotite of 55 ± 4 and 58 ± 5 Ma, evidence that the granite experienced cooling during the late Paleocene.

Peñón Blanco Intrusion, Zacatecas

The Peñón Blanco intrusive body is located in the Sierra de Salinas along with several plutonic and hypabyssal bodies, it is the largest body among them. The intrusive body is texturally porphyritic with phenocrysts in an equigranular groundmass. It is made up of quartz (ca. 35%), orthoclase (10%–15%), oligoclase (ca. 10%), and muscovite (3.5%–4%), with black tourmaline as a secondary mineral. The chemical and mineralogical compositions indicate that this intrusive body has a peraluminous granitic composition (Silva-Romo, 1996). The granite bodies intrude Triassic rocks, commonly in normal faults zones. K-Ar dating in muscovite for the Peñón Blanco yielded an age of 48 ± 4 Ma (Mujica-Mondragón and Albarrán-Jacobo, 1983, in Silva-Romo, 1996).

Palo Verde Intrusion (El Realito–El Refugio)

The Palo Verde intrusion consists of two main bodies and numerous small bodies (Labarthe-H. et al., 1989). Labarthe-Hernández et al. describe the intrusion as a subvolcanic gray to brown, holocrystalline to inequigranular rock with 20% of phenocrysts of orthoclase or sanidine, plagioclase and quartz sized between 2 and 3 mm. The matrix is an aggregate of feldspar and quartz with ferromagnesian mineral (biotite, hastingsite, and riebeckite) crystals up to 1 mm in diameter and forming 3–5% of the matrix. The larger bodies show lateral transitional changes to rhyodacite. The age of this intrusive body is unknown but an Oligocene age is inferred providing that these bodies crosscut ignimbrites aged 29.1 ± 0.3 Ma (K-Ar, whole rock; Labarthe-H. et al., 1989), and change transitionally to volcanic rocks that correlate with Oligocene volcanic units.

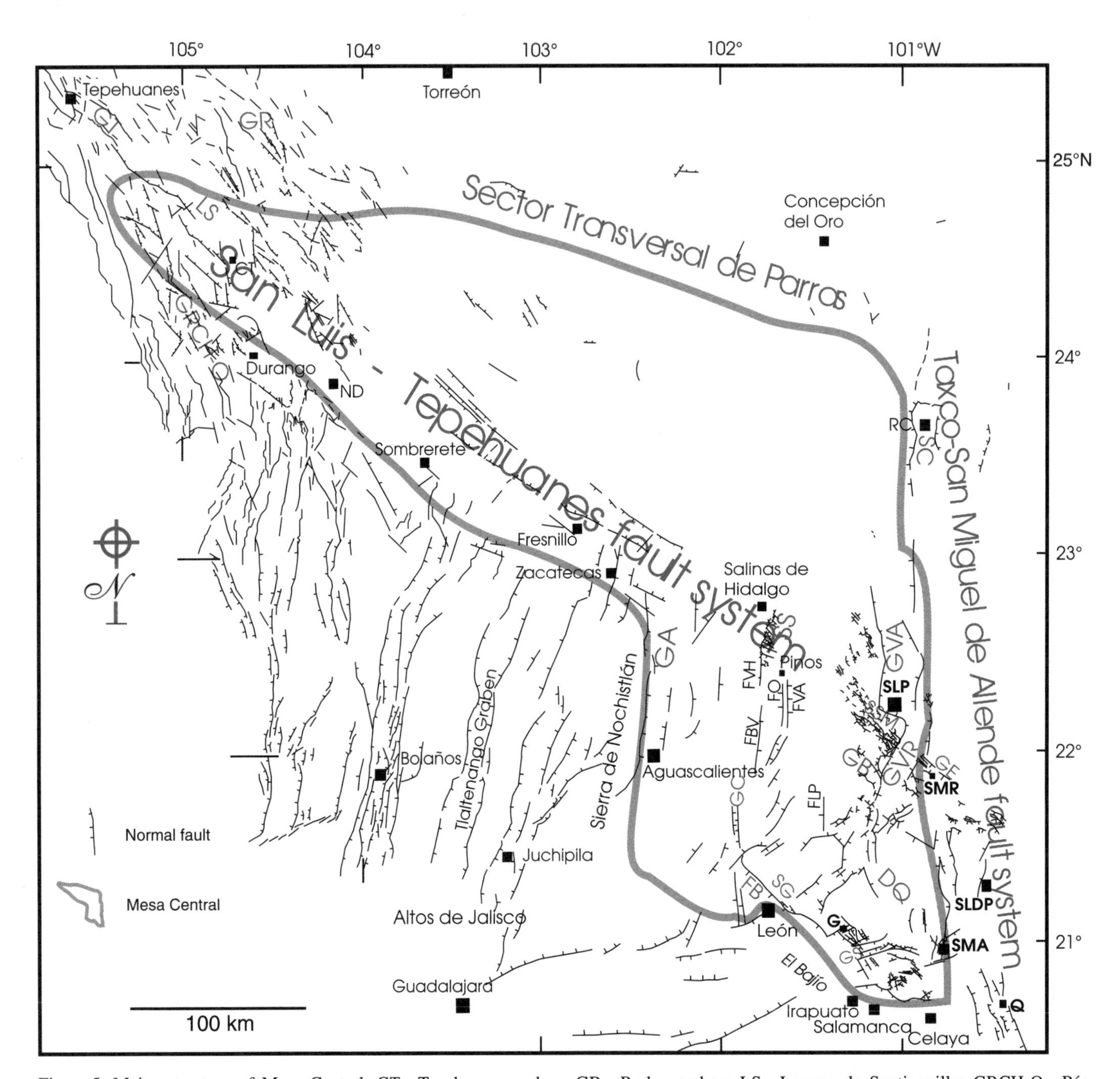

Figure 5. Major structure of Mesa Central. GT—Tepehuanes graben; GR—Rodeo graben; LS—Laguna de Santiaguillo; GRCH-O—Río Chico–Otinapa graben; GA—Aguascalientes graben; FVH—Villa Hidalgo fault; FBV—Buena Vista fault; FO—El Obraje fault; FVA—Villa de Arriaga fault; FLP—Los Pájaros fault; GC—El Cuarenta graben; FB—El Bajío fault; GS—La Sauceda graben; GB—Bledos graben; GVR—Villa de Reyes graben; GVA—Villa de Arista graben; GE—Enramadas graben; DQ—La Quemada depression; SSM—Sierra de San Miguelito; SG—Sierra de Guanajuato; SC—Sierra de Catorce; SS—Sierra de Salinas; RC—Real de Catorce; G—Guanajuato; SLP—San Luis Potosí; SMR—Santa María del Río; SLDP—San Luis de La Paz; ND—Nombre de Dios; SMA—San Miguel de Allende; Q—Querétaro.

CENOZOIC MAJOR STRUCTURES

Major Faults Bounding Mesa Central

El Bajío Fault

The southern boundary of Mesa Central is a normal fault that consists of two large segments. The easternmost segment extends from Celaya to Irapuato, in the State of Guanajuato; it strikes nearly E-W and forms the El Bajío Basin (Fig. 5). In the footwall of the El Bajío fault, there are two groups of faults, which strike NW-SE and ENE to NE and produce a zigzag pattern in the fault trace. The minimum throw that can be estimated is 250–350 m, which was obtained from the escarpment height, 150–250 m, plus the reported thickness of the El Bajío Basin filling near Celaya, ~100 m (Trujillo-Candelaria, 1985). Also, there are transverse faults covered by alluvial deposits up to 250 m thick in some parts of the hanging block (Trujillo-Candelaria, 1985). Thus, we estimate in Celaya a total throw of 350–500 m for the El Bajío fault.

The second segment of the El Bajío fault forms the SW escarpment of the Sierra de Guanajuato, which is located in the raised block of the fault. The segment is 80 km long, and extends from Irapuato to León in the State of Guanajuato. The fault exposed the Cretaceous marine volcanosedimentary sequence, which has been described in previous sections. This segment consists of a series of normal high-angle faults with downthrown blocks to the SW. The displacements of 850 m measured in Oligocene volcanic units (Quintero-Legorreta, 1992) and the thicknesses of the alluvial fillings in the neighborhoods of León which reach 500 m (Hernández-Laloth, 1991) resulted in a throw over 1,200 m long that accumulated after the early Oligocene. It is remarkable that the throw accumulated from the Miocene to Recent is ca. 500 m in Cerro El Cubilete, 20 km SE of León, and is obtained from the displacement of El Cubilete basalt, which is 13.5 Ma (Aguirre-Díaz et al., 1997). The Veta Madre fault is a NW-SE normal fault located near the city of Guanajuato, 10 km E of the El Bajío fault; the faults are parallel. The Veta Madre fault has a displacement of over 1,500 m, which occurred between the late Eocene and the early Oligocene. If we added the displacements of both faults, they would result in a total displacement of over 2 km and a throw of ~800 m for the Guanajuato area.

The early Paleocene-Eocene red conglomerates that crop out along the trace of the El Bajío fault indicate possible activity in the Eocene, although it has not been documented. It is not possible to dismiss the idea that the El Bajío fault is older than the above age. However, its individualization is not evident until the Oligocene, when it reached its peak activity, and it experienced successive reactivation phases until the early Miocene. The El Bajío fault crosscuts middle Miocene rocks in both the Guanajuato (El Cubilete basalts) and Salamanca regions (La Ordeña andesite; Ojeda-García, 2004). There is no clear evidence for Pliocene or younger activity in the El Bajío fault or associated faults. Volcanic rocks and volcanoes of that age in the Salamanca-Celaya region are not faulted.

The Taxco–San Miguel de Allende Fault System

The eastern boundary of Mesa Central is a N-S–trending normal fault system that extends from San Miguel de Allende and Querétaro to tens of kilometers to the N of the Sierra de Catorce, forming the boundary between Mesa Central and the Sierra Madre Occidental. These faults also correspond to the northern part of the Taxco–San Miguel de Allende fault system (Alaniz-Álvarez et al., 2002; Alaniz-Álvarez and Nieto-Samaniego, 2005). The fault system also coincides with the paleogeographic boundary between the Valles–San Luis Potosí Platform to the E, and the Mesozoic Basin of Central México to the W. Such paleogeographic elements were described in previous sections. The importance of the coincidence between the paleogeographic transition and the Taxco–San Miguel de Allende fault system lies in the bathymetric and probable crustal thickness differences between the paleogeographic domains in the Cretaceous. In the present time, there are similar differences in the crustal thickness on either side of this fault system. This suggests that most of the faults and the differences in crustal thickness are inherited characteristics of a Mesozoic or older structure, which constituted the transition zone from a marine platform to a marine basin, probably a system of normal faults similar to the Taxco–San Miguel de Allende fault system.

The fault system is segmented by several NW-SE lineaments in Mesa Central, some of which have been documented as fault systems. The larger lineaments are the San Luis–Tepehuanes fault system and the El Bajío fault (Fig. 5).

The Taxco–San Miguel de Allende fault system has been recognized in the following localities:

- In the Sierra de Catorce, the fault system strikes N-S, and is represented by W-dipping faults observable in the western flank. There is no detailed information on the magnitudes or age of the displacements, because there are no Tertiary rocks that allow measurement. However, considering the morphology and height of the escarpments, we infer that they are primarily normal displacements. The probable age of faulting is Paleogene, as it is only known that the escarpments crosscut NW-SE faults containing quartz-monzonitic dikes dated 53 ± 4 Ma (Mujica-Mondragón and Albarrán-Jacobo, 1983). The magnitude of the displacements is determined by the height of the escarpment, which goes as high as 1000 m. East of the Sierra de Catorce there is a normal fault with its hanging wall to the E; the wall is aligned with the western margin of the Valles–San Luis Potosí Platform (Barboza-Gudiño et al., 2004).
- The Villa de Arista graben (a northern continuation of the Villa de Reyes graben) consists of highly eroded escarpments of normal faults that form a graben filled with fluvial deposits (Moreira-Rivera et al., 1998). In the city of San Luis Potosí, the throw is ~500 m (Tristán-González, 1986), and the faults were active during the Oligocene (Nieto-Samaniego et al., 1997).

- In San Miguel de Allende and Querétaro, the Taxco–San Miguel de Allende fault system is a group of parallel normal faults that produced an E-W extension. In San Miguel de Allende, a throw of 450 m and two phases of activity between the Oligocene and the middle Miocene were documented. In Querétaro, the faults have a throw of ~100 m and activity ages corresponding to the late Miocene (Alaniz-Álvarez et al., 2001).

The San Luis–Tepehuanes Fault System

This fault system is observed in satellite images and digital elevation models as a NW-SE–striking major lineament that extends from San Luis de La Paz in the State of Guanajuato to Tepehuanes in the State of Durango. The lineament roughly coincides with the NE boundary between the Cenozoic volcanic rocks of the Sierra Madre Occidental and the outcrops of the Mesozoic volcanosedimentary sequences, and with the northern edge of the N-S grabens of the Sierra Madre Occidental (Figs. 3 and 5). The San Luis–Tepehuanes fault system can be divided into two segments:

The eastern segment extends from San Luis de La Paz to Salinas de Hidalgo and constitutes the boundary between the southern and northern regions of Mesa Central. Between San Luis de La Paz and Santa María del Río, a system of numerous normal faults trending NW-SE and dipping to the SW were mapped (Labarthe-Hernández and Tristán-González, 1980; Tristán-González, 1987; Alvarado-Méndez et al., 1997). In some places, the faults form grabens that cut early Oligocene rocks, indicating a maximum age of late Oligocene; the magnitudes of the displacements are unknown.

West of the Villa de Reyes graben, in the Sierra de San Miguelito (Figs. 3 and 5), the San Luis–Tepehuanes fault system was exhaustively studied by Labarthe-Hernández and Jiménez-López (1992, 1993, 1994), Nieto-Samaniego et al. (1997), and Xu et al. (2004). In that locality, the fault system consists of numerous faults trending N60°W to N20°W and dipping 45° to 75° to the SW. The faults form a domino-like system that tilts the Oligocene volcanic rock layers 20° to the NE on average. The deformation produced a 20% extension in the NE-SW direction, perpendicularly to the trend of the faults (Xu et al., 2004). The maximum age of the activity is unknown, but it occurred after the Oligocene and it reactivated several times. The main deformation occurred in the early Oligocene, and a second deformation of lesser magnitude occurred during the late Oligocene, which could have extended until the early Miocene (Nieto-Samaniego et al., 1997).

The next well-documented locality with normal faults of the San Luis–Tepehuanes fault system is the Sierra de Salinas, which is 10 km SSE of Salinas de Hidalgo (Figs. 3 and 5). The Sierra de Salinas is 30 km long from N to S and 5–10 km wide, the W side is a normal fault with the hanging wall to the W. The San Luis–Tepehuanes fault system crosscuts the Sierra de Salinas obliquely, its faults trend N50°W with slickensides that indicate a dominant normal movement, which forms a horst-and-graben system. Silva-Romo (1996) identified three phases of extensional deformation in the Sierra de Salinas. The oldest extensional phase activated N-S and WNW faults. The second phase reactivated the WNW faults that crosscut the range, corresponding to the activity of the San Luis–Tepehuanes fault system. Silva-Romo (1996) attributed both phases to the early to middle Eocene as they were coeval or previous to the emplacement of granite bodies dated 48 Ma, provided that there are many dikes hosted by faults of both phases. The last phase of deformation activated the N-S faults that bound the range westward (Silva-Romo, 1996). The age of such phase of deformation is not specified by Silva-Romo, but is younger than the previous phases, probably late Oligocene or early Miocene, because it affects the Villa Hidalgo ignimbrite, which correlates with Oligocene ignimbrites in Mesa Central (Labarthe-Hernández et al., 1982). The magnitudes of the displacement are unknown.

About 60 km W of the Sierra de Salinas, the San Luis–Tepehuanes fault system crosscuts the Aguascalientes graben, which ends up in the intersection zone and is not prolonged to the N. There is an apparent left jump of the W shoulder of the Aguascalientes graben, appearing again until Guadalupe, 15 km to the W (Figs. 3 and 5). In the Guadalupe zone, there are many N70°W faults located on the trace of the San Luis–Tepehuanes fault system with dominantly normal displacements. Although some smaller faults with lateral displacements were observed, no evidence for major faults with such displacement was found.

Two large fault systems were mapped in Zacatecas on the trace of the San Luis–Tepehuanes fault system (Ponce and Clark, 1988; Caballero-Martínez et al., 1999). The oldest consists of NW-SE faults that dip between 50° and 70° to the SW. The fault lengths vary from 4 to 16 km, and they host mineralized veins. These faults (or veins) include La Plomosa, Tajos de Pánuco, Veta Grande, Mala Noche, El Bote, Cantera, and San Rafael (Ponce and Clark, 1988). The displacement ages of these Cenozoic faults are not well known because they are located in Mesozoic rocks. The NW-SE faults were crosscut by N-S faults that limit the Sierra de Zacatecas forming a horst. No detailed data about the displacements of these faults are available. The size of the observed escarpments suggests, however, that the displacement must exceed an accumulated throw of 400 m. The faults crosscut middle Eocene volcanic rocks (namely the Los Alamitos ignimbrite, dated at 46.8 Ma, and the La Virgen formation, dated at 36.8 Ma; Ponce and Clark, 1988) and do not affect the Oligocene volcanic rocks (the Garabato ignimbrite, dated at 28.0 ± 0.8 Ma by K-Ar in sanidine, and the Sierra Fría rhyolite, dated at 27.0 ± 0.7 Ma by K-Ar in sanidine; Loza-Aguirre, 2005). Thus, the age of such faults is early Eocene-Oligocene. The N-S faults are displaced by new NW-SE faults, which we interpret as resulting from a reactivation of the San Luis–Tepehuanes fault system, which occurred in the late Oligocene, provided that they crosscut the Garabato ignimbrite and the Sierra Fría rhyolite.

Based on previous observations, two phases of activity for the San Luis–Tepehuanes fault system in Zacatecas can be envisaged: a first phase occurred between the late Eocene and the early

Oligocene coeval with the unconformity described in the stratigraphy section; and a second phase occurred at the end of the Oligocene, or in the Miocene.

Following the trace of the San Luis–Tepehuanes fault system westward, the Fresnillo and Sombrerete mining districts are found 50 and 150 km NW of Zacatecas, respectively. Cenozoic normal faults have been mapped in both localities. In Fresnillo, the longer faults are the Fresnillo and the Laguna Blanca, which strike N30° to 60°W dipping NE; there are minor normal faults parallel to these major ones (see Fig. 2 in De Cserna, 1976). The fault displacements are not well documented but we refer to the minimum 1000 m of displacement that De Cserna (1976) estimated for the Laguna Blanca fault. The faulting activity in Fresnillo occurred after the formation of the Fresnillo conglomerate, which lies in angular unconformity on Cretaceous rocks, does not present plicative deformation, and underlies volcanic rocks dated at 38.3 Ma (Albinson, 1988). Such volcanic rocks host some faults and associated hydrothermal alteration. Hydrothermal K-feldspars were dated at 29.1 Ma (Albinson, 1988). These observations indicate that faulting in the Fresnillo region occurred between the late Eocene and the early Oligocene, but previous faulting between the Paleocene and the middle Eocene might have been possible as well.

In the Sombrerete zone, there are two groups of NW-SE faults of differing age, as well as a small number of E-W and NE-SW faults. The oldest faults host veins formed between the late Eocene and the early Oligocene (Albinson 1988), provided that they crosscut early to middle Eocene rocks. However, the stratigraphic position within the volcanic sequence to which those ages correspond is unknown. This indicates that faulting occurred between the middle and the late Eocene. The youngest faults displace the veins and cut rhyolites that are probably Oligocene (30–25 Ma), as suggested by lithologic correlation with the Oligocene rhyolites in Fresnillo (Albinson, 1988).

Northwest of Sombrerete, following the trace of the San Luis–Tepehuanes fault system, there is a group of normal faults that forms continental basins with alluvial deposits that extend as far as the Tepehuanes vicinities (Fig. 5). There is no detailed information about such faults, but they are found in maps (*Cartas Geológico-Mineras,* published by the *Consejo de Recursos Minerales*: Fresnillo, Durango, and El Salto, scale 1:250,000, and Tepehuanes, scale 1:50,000). The Durango volcanic field of Quaternary age appears on the trace of the San Luis–Tepehuanes fault system between Villa Unión and Canatlán. Normal NW-SE faults within this volcanic field were reported by Aranda-Gómez et al. (2003). The fault system extends to the NW and forms the 40-km-long NW-SE graben that contains the Santiaguillo lagoon. During the past century, there were earthquakes and reports of local instrumental seismicity in the graben region (Yamamoto, 1993). The NW part of the fault system is the Tepehuanes graben, mapped at a scale of 1:50,000 (*Carta Geológico-Minera* of the *Consejo de Recursos Minerales*: Tepehuanes), but detailed information about the faults that form the graben is not available. In the Durango region, the San Luis–Tepehuanes fault system is bounded by large NNW normal faults and grabens. To the E, two phases of deformation were documented in the Rodeo graben, one during the early Oligocene (dated between 32.3 and 30.6 Ma), and the other during the early Miocene (dated at ~24 Ma). The Río Chico–Otinapa graben, which was active between 12 and 25 Ma, is located W of the Durango region (Aranda-Gómez et al., 2003).

The Aguascalientes Graben

This fault system bounds the southern sector to the W of Mesa Central, and separates it from the Sierra Madre Occidental (Fig. 5). It is a half graben with the main fault located to the W. The main fault trends N-S and is 150 km long, extending from the Altos de Jalisco region to its intersection with the San Luis–Tepehuanes fault system near Zacatecas. The displacement of this fault in Aguascalientes is ~900 m, as deduced from the topographic relief of ~400 m between the volcanic rocks in the raised (western) block and the level of the valley, plus ~500 m from the sediment thickness in the valley, obtained from a deep well (Jiménez-Nava, 1993). The western fault displaced volcanic rocks, mainly felsic ignimbrites whose stratigraphic range is not well known. The oldest rocks W of the city of Aguascalientes were dated middle Eocene, and the ignimbrites of the upper part of the Nochistlán range, located immediately to the W, were dated late Oligocene (Nieto-Samaniego et al., 1997). Additionally, in the deep well at Aguascalientes, there are basaltic rocks of unknown age interbedded with the strata. The nearest basalts that overlie the ignimbritic rocks have early Miocene ages (21.8 ± 1.0 Ma by K-Ar in the rock matrix; Moore et al., 1994), and are found in the Tlaltenango graben within the Sierra Madre Occidental. The main phase of faulting of the Aguascalientes graben occurred after the early Oligocene but the number of deformation events is unknown. The Cenozoic volcanic cover is present in Mesa Central as well as in the Sierra Madre Occidental, though the morphology and the structural style of these physiographic provinces are different. The Sierra Madre Occidental contains very long, narrow and parallel NNE to NNW horst-and-graben structures with little alluvial filling, whereas Mesa Central shows a complex pattern of normal faults with different trends, as well as horst-and-graben structures that commonly form rhombohedral arrangements with basins that contain thick sequences of alluvial and lacustrine continental rocks.

Sector Transversal de Parras of the Sierra Madre Oriental

The Mesa Central is bounded to the N by an ~E-W mountain range that consists of folded Mesozoic rocks of the Sierra Madre Oriental and is known as Sector Transversal de Parras. The relief in the Sector Transversal de Parras, above 2000 masl, is higher to the E than in the northern part. The folds have direction N70°W and they are large and narrow, with vergency to the NNE, though some folds have opposite vergency (Eguiluz-de Antuñano et al., 2000). Near Mesa Central, the folds are not affected by the N-S–trending Caballo, Almagre, or Juarez faults, which are Oligocene in age (Eguiluz-de Antuñano, 1984).

The San Marcos fault is located to the N of the northern Mesa Central edge; the San Marcos fault and associated faults are parallel to the Sector Transversal de Parras and were active between the late Miocene and the early Pliocene (Chávez-Cabello et al., 2005; Aranda-Gómez et al., 2005). The morphologic characteristics of the Sector Transversal de Parras suggest the existence of a major Cenozoic fault system, but there are no faults that form the northern boundary of Mesa Central or that clearly separate it from the Sierra Madre Oriental.

Fault Systems inside Mesa Central

In Mesa Central, there are many Cenozoic faults that commonly bound basins filled with fluvial and lacustrine deposits. Such faults can be found mainly in the southern sector of Mesa Central, where they cut Oligocene rocks; the faults strike N-S, E-W, NW-SE, and NE-SW. The diversity in the fault orientations imprints on this sector of Mesa Central a complex structural configuration (Fig. 5). The main structures in the region are the Villa de Reyes and the La Sauceda and the El Cuarenta grabens. The Villa de Reyes graben is oriented NNE-SSW, is ~100 km long, and shows throws that reach 500 m (Tristán-González, 1986). The La Sauceda graben is oriented ENE-WSW, is 25 km long, and shows throws ~400 m (Nieto-Samaniego, 1992). El Cuarenta graben is oriented N-S, is 40 km long, and shows throws up to 1000 m (Quintero-Legorreta, 1992). The major N-S normal faults are El Obraje, Villa Hidalgo, Los Pájaros, Buenavista, and Villa de Arriaga. The more important NW-SE structures are the Bledos and Enramadas grabens, and the La Quemada depression (Fig. 5).

GEOLOGIC EVOLUTION

The lithologic record suggests that the evolution of Mesa Central initiated in the Late Triassic with marine strata. The most important characteristic of these rocks is their variation: in the eastern part of Mesa Central (Sierra de Salinas, Sierra de Catorce, and Sierra de Charcas), they consist of turbiditic sequences, whereas in the western part of Mesa Central (Zacatecas), they consist of clastic sequences with a volcanic component and low-grade metamorphism (phyllites). This distribution of facies indicates the existence of a continental margin in Mesa Central, with the ocean toward the W. The shape of this margin cannot be reconstructed due to the scarcity of outcrops.

There are additional areas with marine basin strata of Mesozoic age, along with mafic volcanic rocks (Fig. 3). Their distribution shows that the marine basin sequences with volcanic rocks extend NW-SE along the western part of Mesa Central. Freydier et al. (1996) explain the presence of these rocks as part of what they inferred to be the Late Jurassic to Early Cretaceous age Arperos Basin, including the outcrops of Zacatecas and Guanajuato. In Guanajuato, Aptian to Albian marine rocks overlie unconformably the deformed sequence of the Arperos Basin, thus supporting the Late Jurassic to Early Cretaceous age assigned by Freydier et al. (1996). The occurrence of that unconformity led Quintero-Legorreta (1992) to infer a deformation phase at the end of the Neocomian. However, in the Sierra de Salinas, ca. 100 km N of the Sierra de Guanajuato and ca. 40 km E of Zacatecas, the stratigraphic record of the Mesozoic Basin of Central México contains a continuous record from the Oxfordian to the Turonian without the occurrence of either volcanic rocks or unconformities. For that reason, it is also possible to suppose a much older age than Early Cretaceous for the volcanosedimentary rocks of Sierra de Guanajuato under the unconformity. The age of the rocks is an unsolved problem in the stratigraphy and tectonic evolution of México.

It is remarkable that E and NE of Mesa Central there are Late Triassic red beds (Carrillo-Bravo, 1971; López-Infanzón, 1986). In Mesa Central, the unconformity on Triassic rocks and the continental origin and volcanosedimentary character of the overlying Middle Jurassic rocks indicate a significant change in the geologic environment: the conditions of Mesa Central change from marine to subaerial, and these continue until the Middle Jurassic with the development of a continental volcanic arc. The unconformity on the Middle Jurassic volcanosedimentary continental rocks, as well as the Upper Jurassic (Oxfordian-Tithonian) marine sequence's lack of any volcanic component, suggests that a marine transgression occurred as the volcanic activity ceased. In other words, the immersion of Mesa Central was produced at the same time as the westward migration of the volcanic zone. This theory is supported by the Late Jurassic to Early Cretaceous volcanosedimentary sequences that underlie the Aptian-Albian rocks in the southwestern border of Mesa Central, and also by the occurrence of Late Jurassic (Oxfordian) volcanic arc sequences in Sonora (Rangin, 1977) and the occurrence of Cretaceous volcanic rocks at the bottom of the Sierra Madre Occidental volcanic sequence (Ferrari et al., 2005). A simple explanation is that subducted plate increased the subduction angle during the Late Jurassic. That mechanism may have produced the westward migration of the volcanic arc and the extension and subsidence in the continental plate.

The marine conditions in Mesa Central continued during the Late Jurassic and Cretaceous forming the sequence of the Mesozoic Basin of Central México, which does not contain major unconformities that indicate the occurrence of orogenic phases. Only the upper units of the Mesozoic Basin of Central México show a change of facies, from limestone to clastic rocks with volcanic clasts in the younger strata. Such change has been interpreted as an indication of the beginning of the Laramide orogeny (Centeno-García and Silva-Romo, 1997). The ages of the clastic sequences with volcanic material in Mesa Central have been dated Cenomanian-Turonian after their fossil content (Tardy and Maury, 1973; Silva-Romo, 1996), and up to the Maastrichtian in the Parras region in the State of Coahuila (Tardy and Maury, 1973).

The unconformity on the Mesozoic rocks that is found in the entire Mesa Central region corresponds in time to the Laramide orogeny. The orogenic front migrated from W to E. The maximum age of the deformation is constrained by the youngest deformed rocks, which are Turonian in age in the western part of Mesa Central and Maastrichtian in the eastern part. Many of the deformed rocks in Mesa Central are covered by Neogene rocks, which means that the folded units may include younger rocks that have not been observed. The minimum age of the Laramide deformation is inferred from the ages of undeformed continental sedimentary or plutonic rocks. In the Sierra de Guanajuato, post-orogenic intrusions dated at 54 Ma (Quintero-Legorreta, 1992), as well as continental conglomerates with interbedded mafic lavas dated at 49 Ma (Aranda-Gómez and McDowell, 1998) were mapped. There are also volcanic rocks, with ages ranging from 51 to 37 Ma with no contractive deformation, that appear in the western part of Mesa Central as part of a continental volcanic arc (Fig. 3). Simultaneous with the volcanism, there was a NE-SW extension. It is important to consider that the Laramide orogenic front was still active in the Sierra Madre Oriental during the Eocene epoch (Eguiluz-de Antuñano et al., 2000). The NE-SW shortening in northern and eastern México was coeval or almost synchronic with the NE-SW extension in the western part of Mesa Central. The distribution of the deformation fields is not simple to explain. Considering that the orogenic phase was associated with a low subduction angle, the beginning of extension with associated arc volcanism (of unknown chemical signature) would have required that the subducted plate reach the necessary depth to cause partial melting in the mantle wedge. A simple explanation for the plate reaching that depth is that the subducted plate broke apart, and the segment near the trench subducted with a higher angle than the hanging segment, causing partial melting and the associated volcanism. A similar mechanism and geometry were proposed for this region by van der Lee and Nolet (1997) for the Miocene. Although this is still a matter of speculation, this hypothesis would explain the geologic record in the region.

It is reasonable to think that the concentration of the Eocene extension in the San Luis–Tepehuanes Fault System results from the existence of an inherited major weakness zone. The existence of a weakness zone is inferred from the distribution of oceanic basin rocks along the structure, indicating that such distribution corresponds to either (1) a thinned crust or (2) the continental crust edge for the Late Triassic, or for the Early Cretaceous according to Freydier et al. (1996). The juxtaposition zone of the Cretaceous marine volcanic arcs on the marine sequences of the Mesozoic Basin of Central México occurs along the San Luis–Tepehuanes fault system. Although an Eocene extension in a wide region of Mesa Central cannot be dismissed, such deformation has not been documented in the central and northeastern parts of Mesa Central and seems to have concentrated along the fault system.

The unconformity overlying Eocene rocks of Mesa Central marks a change in: (1) the composition and location of the volcanism, (2) the region affected by the extensional tectonics, and (3) the direction of extension that prevailed during the Oligocene. The Oligocene volcanic rocks cover the southern region of Mesa Central, forming large rhyolitic dome fields with a relatively thin cover of rhyolitic ignimbrites. The northern region of Mesa Central lacks Oligocene volcanic rocks except in the western part of Mesa Central along the boundary with the Sierra Madre Occidental. The Oligocene volcanism was coeval with extensional deformation; the principal extensions were~20% in the E-W direction and ~10% in the N-S direction. During this event, the following occurred: (1) crustal thinning and rise (Nieto-Samaniego et al., 1997), (2) a temperature increase with granulitization of the lower crust (Hayob et al., 1989), (3) the emission of large volumes of rhyolitic effusive rocks during 2 Ma with chemical signatures indicative of crustal melting (Orozco-Esquivel et al., 2002), and (4) the formation of many continental basins that were filled up by alluvial and lacustrine deposits.

Nieto-Samaniego et al. (1999) proposed the increase in the subduction rate as a feasible mechanism for producing the events described above. The increase in the subduction rate produces a difference between the rates of subduction and convergence, thus raising the critical value needed to produce extension in the overriding plate. The reasons why the northern region of Mesa Central was not affected by these events are unknown. Nieto-Samaniego et al. proposed that the increase in the subduction rate is related to different oceanic expansion rates to the N and S of the Shirley fracture of the Pacific plate. The rate of expansion was 20% higher to the S of the Shirley fracture than to the N during the Oligocene and the Miocene, and that fracture is aligned with the transition zone between the northern and southern regions of Mesa Central for the Oligocene (Nieto-Samaniego et al., 1999).

The volcanic arc and the extensional deformation migrated toward the W and the S of Mesa Central during the Miocene, and were located in the margins of Mesa Central during the late Miocene and Pliocene. The Miocene volcanic rocks in Mesa Central are located in the central (San Luis Potosí region) and western part of the province (Durango), and are represented by fissure basalts that form plateaus and belong to the Extensional Province of Northern México, and to the subprovince of the southern Basin-and-Range (see Aranda-Gómez et al., 2005). A second volcanic region, formed by volcanoes and fissure lava flows, occurs at the southern edge of Mesa Central and has been considered as a part of the Trans-Mexican Volcanic Belt. The volcanism was associated with normal faults, but the rates and magnitudes of deformation are much smaller than during the Oligocene event (Alaniz-Álvarez et al., 2001).

The youngest volcanic event, which culminates the magmatic history of Mesa Central and unconformably covers the geological units described above, consists of mafic alkaline Pliocene to Quaternary rocks, and belongs to the Extensional Province of Northern México and to the subprovince of the southern Basin-and-Range. The parental magmas originated in the mantle and contain xenoliths from the upper mantle and the lower crust. The petrologic and geochemical characteristics of such magmas, as well as the style of the resulting volcanism, indicate that they ascended quickly through the crust using deep faulted zones for their ascent (Aranda-Gómez et al., 2005). However, faults of significant length, magnitude, or deformation rates have not been mapped in association with these volcanic rocks.

MINERAL DEPOSITS IN MESA CENTRAL

One of the particular aspects of Mesa Central is that it contains a vast amount and a variety of ore deposits, with some of the richest in México among them. In this paper, we locate these deposits in space and time to determine the possible linkages between them and major geologic processes in Mesa Central.

Massive Sulfide Deposits (Volcanogenic and Sedex)

In the Mexican Mesa Central, the oldest ore deposits with some economic relevance are polymetallic massive sulfide bodies in Francisco I. Madero and El Salvador–San Nicolás in the Zacatecas region (Johnson et al., 1999; Miranda-Gasca, 2000; Olvera-Carranza et al., 2001). The Zn-Pb-Ag-(Cu) Francisco I. Madero deposit was interpreted to have formed through a sedimentary-exhalative model (Sedex), although some controversy about this explanation persists, whereas the Zn-Cu-(Ag-Au) El Salvador–San Nicolás deposit is a volcanogenic-hosted massive sulfide deposit (VMS). These mining districts also contain skarn and vein deposits (probably epithermal) formed during the Tertiary. In the Guanajuato district there are several small stratiform massive sulfide deposits, sized up to 1 Mt, which occur within the Sierra de Guanajuato Volcanosedimentary Complex (Miranda-Gasca, 2000), in a geologic context similar to that of the deposits found in Zacatecas. Both groups of massive sulfide deposits are hosted in volcanosedimentary sequences in submarine arcs and in back-arc basins that formed between the Upper Jurassic and the Lower Cretaceous (Corona-Chávez and López-Picos, 1988; Freydier et al., 1996).

Metalliferous Deposits in Skarns

The metalliferous skarn deposits are especially abundant in the border areas of Mesa Central, where the essential conditions for their formation occur. Thus, the Providencia–Concepción del Oro deposits in Zacatecas, dated 26.6 Ma, are found in the vicinity of the northwestern border of Mesa Central; the Mapimí deposits in Durango, dated 36.1 Ma, are found close to the northwestern border of Mesa Central; the Charcas (46.6 Ma) and Guadalcázar deposits in San Luis Potosí are found toward the eastern border, and the San Martín district in Zacatecas is located on the southeastern border of Mesa Central. The ages and main characteristics of these deposits were compiled by Megaw et al. (1988) from numerous sources. The San Martín district contains the largest skarn deposits in México (Aranda-Gómez, 1978; Rubin and Kyle, 1988); their formation results from the intrusion of a quartz-monzonitic stock on Cretaceous carbonate rocks of the Cuesta del Cura Formation, which yielded a K-Ar age in biotite of 46.2 ± 1 Ma (Damon et al., 1983). The Cerro San Pedro deposits in San Luis Potosí are hosted by the same formation, and also by the Soyatal and Tamaulipas Formations (Consejo de Recursos Minerales, 1996; Petersen et al., 2001). The skarn deposits in Mesa Central and neighboring areas, and the associated deposit types, have been historically mined for Ag, Au, Pb, Zn, Cu, Sn, Hg, As, Sb, Bi, and fluorite.

Epithermal Deposits

In Mesa Central, like many regions in México, the most frequently occurring deposit type is the epithermal type (Camprubí et al., 1998, 2003; Albinson et al., 2001; Camprubí and Albinson, 2006), since it is the most common source in México for Ag, as well as for Au, Bi, Se, Zn, Pb, Cu, Hg, As, Sb, and so forth. Many of the most famous mines in México were developed on this type of deposit. Mesa Central contains the deposits of Real de Asientos in Aguascalientes; Velardeña and Papanton in Durango; Guanajuato and Pozos in Guanajuato; Comanja de Corona in Jalisco; Santa María de La Paz and Real de Catorce in San Luis Potosí; Colorada, Fresnillo, Panuco, Pinos, Real de Ángeles, Saín Alto, Sombrerete, and Zacatecas in Zacatecas (González-Reyna, 1956; Petruk and Owens, 1974; Salas, 1975; Buchanan, 1981; Albinson, 1985, 1988; Gemmell et al., 1988; Gilmer et al., 1988; Lang et al., 1988; Pearson et al., 1988; Ponce and Clark, 1988; Ruvalcaba-Ruiz and Thompson, 1988; Simmons et al., 1988; Mango et al., 1991; Simmons, 1991; Consejo de Recursos Minerales, 1992, 1996; Rivera, 1993; Gunnesch et al., 1994; Randall et al., 1994; Albinson et al., 2001). Among the different deposit types found in Mesa Central, the epithermal deposits are probably the most studied, due to their economic importance. The epithermal deposits known to date in México are Tertiary in age, from Lutetian (middle Eocene) to Aquitainian (early Miocene) (Camprubí et al., 2003), and their space distribution was mostly determined by the evolution of volcanism in the Sierra Madre Occidental and the Sierra Madre del Sur (see Damon et al., 1981; Clark et al., 1982; Camprubí et al., 2003). The distribution of epithermal deposits can be divided into three main groups: (1) older than ~40 Ma, which comprises Real de Ángeles; (2) between ~40 and ~27 Ma, the preferential age range for the formation of these deposits in México (between 35 and 30 Ma, according to Albinson, 1988) and which comprises the rest of the deposits with known ages in Mesa Central; and (3) younger than ~23 Ma. The Real de Ángeles district is found SE of the city of Zacatecas, contains Pb-Zn-Ag-(Cd) polymetallic deposits with an estimated tonnage of 85 Mt, and thus constitutes the epithermal deposit in México with the largest tonnage. However, it is not the richest deposit because most of the ore occurs as low-grade disseminations in carbonate rocks, as well as veins and stockwork zones (Pearson et al., 1988; Megaw, 1999; Albinson et al., 2001). The total tonnage of this deposit is similar to that of the Pachuca–Real del Monte district in Hidalgo, which accounts for the largest historical production in the world, but average metal grades in this deposit are almost an order of magnitude higher than in Real de Ángeles. Other major world-class epithermal deposits in Mesa Central are Guanajuato and Fresnillo, with an estimated production of 40 and 7 Mt, respectively (the potential for Fresnillo may actually be higher), in high-grade veins and mantos (Albinson et al., 2001).

Iron Oxide and Apatite Deposits (Iron Oxide–Copper-Gold Type)

One of the best-known ore deposits in Mesa Central is the Cerro de Mercado iron deposit in Durango (Swanson et al., 1978; Lyons, 1988), which is famous for its gem-quality apatite crystals. This deposit formed ~31.5 Ma within silicic volcanic rocks of the Carpintero Group, which erupted from the Chupaderos caldera. The age of this deposit was obtained through fission tracks and (U-Th)/He in apatite (Young et al., 1969; McDowell and Keizer, 1977; Farley, 2000), and apatite from this deposit is commonly used as an international standard for both dating techniques. Lyons (1988) stated a volcanic origin for this deposit that, according to the present understanding on similar iron oxide and apatite deposits, would be a Phanerozoic equivalent to the iron oxide–copper-gold (or IOCG) type of deposit (Pollard, 2000).

Carbonate Replacement Deposits

This type of deposit was formed by the replacement of carbonate rocks that are not related to skarns, including both Mississippi Valley–type (MVT) deposits and low-temperature deposits whose type remains uncertain. Most of these deposits are found on the eastern border of Mesa Central and some are located in the Sierra Madre Oriental, but for the purposes of this study, it is convenient to consider them as a group. These deposits are the Las Cuevas and San José Tierras Negras (Wadley) in San Luis Potosí, and El Realito in Guanajuato (González-Reyna and White, 1947; González-Reyna, 1956; Ruiz et al., 1980; Consejo de Recursos Minerales, 1992, 1996; Levresse et al., 2003). Las Cuevas fluorite deposits are the largest and most productive in the world: the "G" body alone contains over 50 Mt with almost 99% fluorite, and the district contains over 150 Mt high-grade fluorite bodies. There is noticeable antimoniferous mineralization in Wadley, with Ag, Pb, Cu, Hg, and Zn, hosted by the Jurassic Zuloaga Formation and found within the Charcas–Real de Catorce trend. Some of the carbonate replacement deposits in Mesa Central cannot be assigned to a specific deposit type with certainty, and further study is necessary to better characterize their origin.

Other Deposit Types

Several tin mines and prospects, which are found near the northwestern border of Mesa Central, developed in Quaternary alluvial placers. The most important deposits are found in Sapiorís and América in Durango, although their prospective area runs from Sapiorís to Coneto de Comonfort. The mineralogy of these placers comprises cassiterite, topaz, durangite, and minor crisoberyl, emerald, gold, and silver (Fabregat-G, 1966). The source areas for the placer deposits are found in the neighboring ranges, which were formed by Tertiary rhyolites, where neumatolitic tin veins formed along breccia zones. Most of these tin veins are strictly found in the Sierra Madre Occidental. There are many small tin vein deposits in Ahualulco, Villa de Arriaga, and Villa de Reyes in San Luis Potosí, Ochoa in Durango, Sierra de Chapultepec in Zacatecas, Cosío in Aguascalientes, and Tlaquicheros in Guanajuato (Bracho-Valle, 1960; Fabregat-G, 1966; Salas, 1975; Consejo de Recursos Minerales, 1996). The most widely known topaz mineralization in rhyolitic domes in Mesa Central is found in Tepetate, San Luis Potosí (Aguillón-Robles et al., 1994), due to the gem quality of the topaz crystals in that locality. There are also gold and tin placers in Guadalcázar and El Realejo in San Luis Potosí (Consejo de Recursos Minerales, 1996). The wide variety of the mineralogenic environments in Mesa Central is further represented by the Montaña de Manganeso district in San Luis Potosí, which contains Mn-rich hydrothermal veins, stockworks, and jasperoids hosted by the Upper Cretaceous Caracol Formation (Consejo de Recursos Minerales, 1996). In addition, there are supergene alunite deposits in Comonfort and Santa Cruz de Galeana in Guanajuato (González-Reyna, 1956), although they are not metallogenically or economically relevant. There are also phosphate deposits, essentially formed by variscite, due to supergene alteration of rhyolitic tuffs at La Herradura, San Luis Potosí (Consejo de Recursos Minerales, 1996). Many supergene alteration zones or gossan deposits developed on metalliferous deposits, increasing the economic value or recoverability of the primary deposits, as in the Real de Catorce deposits with the formation of chlorargyrite and bromargyrite. Lastly, the Salinas de Hidalgo district in San Luis Potosí contains abundant prospects of many substances in different deposit types, although the area is better known for its recent evaporitic potash, as well as its sodium chloride, sulfate, and carbonate deposits (Consejo de Recursos Minerales, 1996).

General Considerations

Considering that the San Luis–Tepehuanes fault system has been active since the Oligocene, and that the majority of ore deposits in Mesa Central or neighboring areas concentrate along the fault system (Fig. 6), it seems likely that the formation of such deposits may have been influenced by the activity of this fault system. A similar grouping of ore deposits occurs along the Taxco–San Miguel de Allende fault zone and the El Bajío fault, though less abundant than in the previous case. In contrast, the northern border of Mesa Central is almost devoid of ore deposits, except for those that were not formed under the influence of or due to any magmatic activity. Additionally, since the deposit types grouped along the San Luis–Tepehuanes fault system (skarn, epithermal, IOCG, tin veins) have a genetic affinity with magmatic phenomena, it is reasonable to state that the fault system favored the channeling of magmas that, in turn, formed or contributed to the formation of ore deposits. The fault system reactivated several times in different segments, from middle Eocene to lower Miocene, and a well-known characteristic of such reactivation stages is that their occurrence coincides with the preferential age range for the formation of epithermal deposits in central México (Albinson, 1988; Camprubí et al., 2003), some skarn deposits (Megaw et al., 1988), and the Cerro de Mercado IOCG deposit

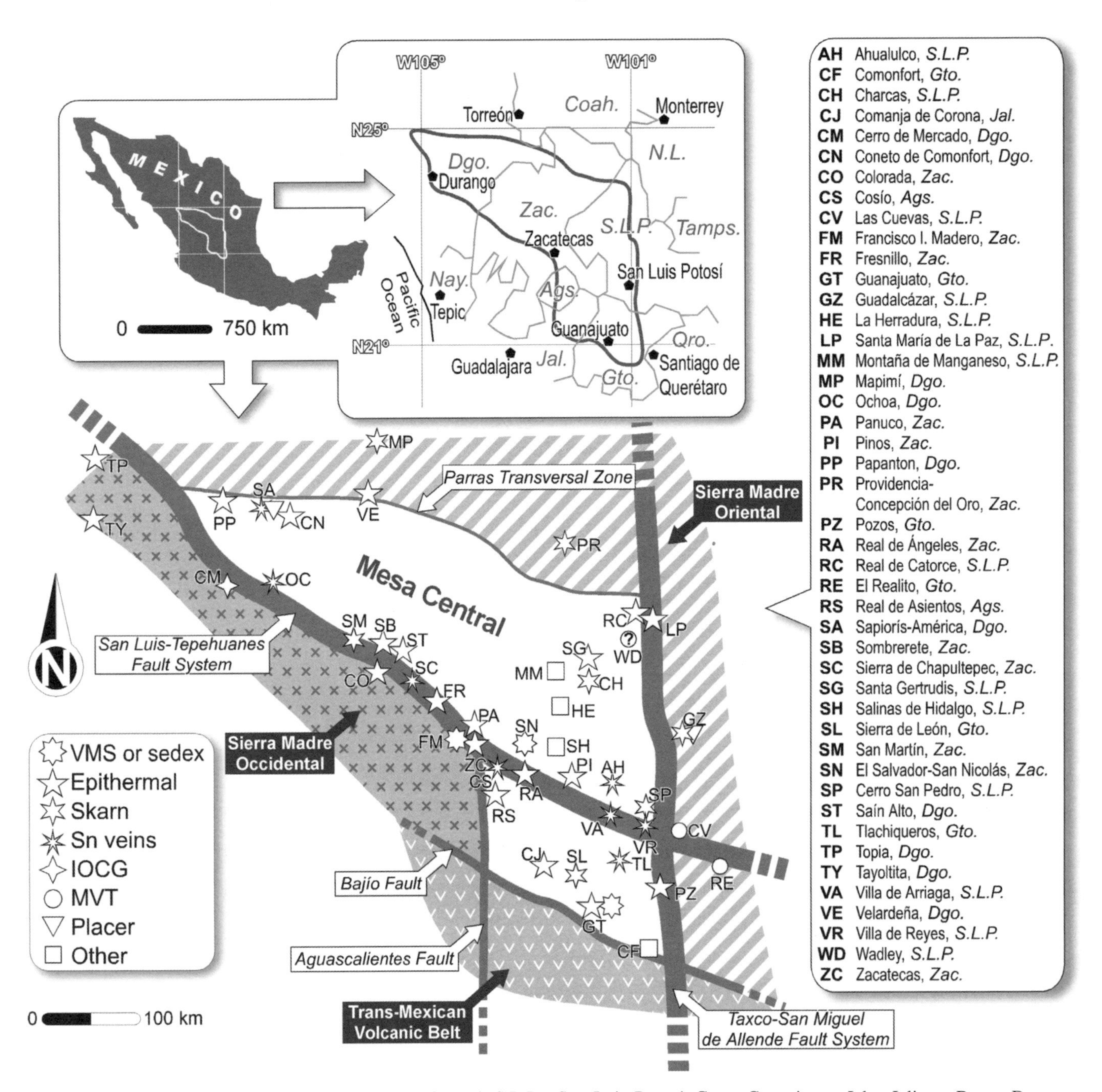

Figure 6. Distribution of mineral deposits in Mesa Central. S.L.P.—San Luis Potosí; Gto.—Guanajuato; Jal.—Jalisco; Dgo.—Durango; Zac.—Zacatecas; Ags.—Aguascalientes; VMS—volcanogenic hosted massive sulfide deposit; IOCG—iron oxide–copper-gold deposit; MVT—Mississippi Valley–type deposit.

(Lyons, 1988). The zones with structural weakness that include major fault zones like the San Luis–Tepehuanes and the Taxco–San Miguel de Allende or El Bajío fault may effectively favor the upwelling and emplacement of magmas and fluids, and thus the formation of associated mineral deposits. For example, as already described in previous sections, in the Zacatecas and Fresnillo mining districts (located on the San Luis–Tepehuanes fault system), normal faulting and magmatic activity have been documented during the late Eocene and the early Oligocene, when the epithermal deposits were forming (Albinson, 1988; Lang et al., 1988; Ponce and Clark, 1988). However, this does not imply that the emplacement of magmas and ore deposits actually occurred or must occur along the main faults. The formation of epithermal deposits, for instance, is controlled by regional-scale faults in zones with intense tensional fracturing (Mitchell and Balce, 1990; Nesbitt, 1990; Staude, 1993; Ponce and Glen, 2002). Such faults determine the location of deposits and channel ways for the emplacement of magmas that provide the necessary heat source to activate the hydrothermal activity (Hedenquist, 1986; Fournier, 1987). Although major faults directly control the emplacement of mineral deposits, these preferentially formed in subsidiary faults nonetheless (White and Hedenquist, 1990). In the case of Mesa Central, it is necessary to evaluate the role that subsidiary faults played in the formation of epithermal deposits in every case, as well as the association between major faults and these deposits, and between major faults and other deposit types. Based on the existing data, we can merely indicate that mineral deposits (especially epithermal) and major fault systems were active at the same time and in the same locations in Mesa Central, that the circulation of fluids through the crust is favored through large fault systems like the San Luis–Tepehuanes fault system, and thus that the abundance of mineral deposits along this fault system is not merely coincidental.

CONCLUSIONS

Mesa Central is a physiographic province characterized by a moderate topographic relief, located in the central-northern part of México, and surrounded by the Sierra Madre Occidental, Sierra Madre Oriental, and the Trans-Mexican Volcanic Belt. The Cenozoic extensional deformation was the main tectonic event that configured the province. The large fault systems present in Mesa Central had a fundamental role in the geologic evolution of the region.

Mesa Central is bound to the N by the Sector Transversal de Parras of the Sierra Madre Oriental, to the S by the El Bajío fault system, to the E by the Taxco–San Miguel de Allende fault system, and to the W by the Aguascalientes graben and the San Luis–Tepehuanes fault system. Within Mesa Central, two sectors can be distinguished: (1) the northern sector, in a more advanced state of erosion and with no significant evidence for Neogene tectonics, (2) and the southern sector, where major extensional tectonic events occurred during the Oligocene, and minor extensional events happened from the Miocene to the Quaternary. The boundary between these sectors, the San Luis–Tepehuanes fault system, roughly follows the western boundary of Triassic turbidites along the interpreted boundary of the Triassic continent.

The crust under Mesa Central is thinner than in the regions E and W of its limits. Recent zones of partial melting in the upper mantle under Mesa Central are inferred from seismic data, and the presence of alkaline basalts with mantle xenoliths is widely documented in the northern region of Mesa Central.

In the southern region of Mesa Central, detailed studies show that the crust was extended ~20% in an approximate E-W orientation mainly during the early Oligocene, but that minor deformation events occurred up to the Miocene. Intense silicic magmatism also occurred during the early Oligocene, mainly from partial melting of the crust.

In Mesa Central, magmatism was coeval with ore deposits, as noted in Mesa Central by several authors, as well as in other regions in México. However, the geographic distribution of the volcanic rocks in Mesa Central is wider than that of ore deposits, whereas in the case of volcanic rocks of the Sierra Madre Occidental, the associated deposits are mainly located at the northeastern limit. Most ore deposits are found on the trace of the San Luis–Tepehuanes fault system and their ages are congruent with the activity of this structure. Even though such space and time relations are evident, the role of large fault systems in the location of ore deposits is poorly studied and understood. It is reasonable to suppose that the weakness zones in the upper crust allow a better circulation of fluids, either magmatic or hydrothermal, thus favoring the formation of ore deposits with enhanced porosity if the country rocks are highly fractured.

Finally, we would like to point out/address the gaps in the geological knowledge concerning Mesa Central:

- More accurate and an increased number of age determinations of the volcanosedimenty marine sequences are necessary in order to separate the marine volcanic arcs of probable Late Jurassic–Early Cretaceous age from the Triassic sedimentary or volcanosedimentary sequences that correlate with the Zacatecas outcrops.
- The Cenozoic sedimentary rocks that fill continental basins have not been studied from a sedimentologic point of view. These studies would contribute fundamental information if they contain almost the totality of the aquifers in Mesa Central, and also contain part of the lithologic and paleontologic record about the time of emplacement of the volcanic sequences in the Sierra Madre Occidental.
- The Cenozoic volcanic stratigraphy is still incomplete and more radiometric age data from the associated rocks are needed. Systematic work is necessary to establish correlations between the outcrops in the different ranges separated by sedimentary basins. Until now, most correlations have been based on lithologic criteria, which is highly uncertain for volcanic rocks. The isotopic age determinations will help to reconstruct the evolution of structural systems and the tectonic history of the region, provided that most of the basins that separate mountain ranges are of tectonic origin.
- The migration of the volcanism during the Cenozoic toward the margins of Mesa Central has been documented in a general way. The details of that migration considering ages, volumes of rocks and styles, and composition of the volcanism, as well as the associated deformation remain unknown.
- The role that the major fault systems of Mesa Central played during the different tectonic events, which occurred during the Cretaceous and Cenozoic, is still obscure. Further systematic studies of the major structures and associated rocks are required.

ACKNOWLEDGMENTS

This work was financially supported by projects PAPIIT IN102602 and CONACYT 41044-F. The digital elevation model of Figure 1 was elaborated by Gabriel Origel. We thank Luca Ferrari for his comments and suggestions about the tectonics and stratigraphy of western México, as well as for the revision of the manuscript, and Rafael Barboza-Gudiño for his careful revision and suggestions, which helped to improve this work substantially.

REFERENCES CITED

Aguillón-Robles, A., Aranda-Gómez, J.J., and Solorio-Munguía, J.G., 1994, Geología y tectónica de un conjunto de domos riolíticos del Oligoceno medio en el sur del estado de San Luis Potosí, México: Revista Mexicana de Ciencias Geológicas, v. 11, p. 29–42.

Aguirre-Díaz, G.J., and Labarthe-Hernández, G., 2003, Fissure ignimbrites: Fissure-source origin for voluminous ignimbrites of the Sierra Madre Occidental and its relationship with Basin and Range faulting: Geology, v. 31, p. 773–776, doi: 10.1130/G19665.1.

Aguirre-Díaz, G.J., and McDowell, F., 1991, The volcanic section at Nazas, Durango, Mexico, the possibility of widespread Eocene volcanism in the Sierra Madre Occidental: Journal of Geophysical Research, v. 96, p. 13,373–13,388.

Aguirre-Díaz, G.J., and McDowell, F.W., 1993, Nature and timing of faulting and synextensional magmatism in the southern Basin and Range, central-eastern Durango, Mexico: Geological Society of America Bulletin, v. 105, p. 1435–1444, doi: 10.1130/0016-7606(1993)105<1435:NATOFA>2.3.CO;2.

Aguirre-Díaz, G.J., Nelson, S.A., Ferrari, L., and López-Martínez, M., 1997, Ignimbrites of the central Mexican Volcanic Belt, Amealco and Huichapan Calderas (Querétaro-Hidalgo), *in* Aguirre-Díaz, G.J., Aranda-Gómez, J.J., Carrasco-Núñez, G., and Ferrari, L., eds., Magmatism and Tectonics of Central and Northwestern Mexico—A Selection of the 1997 IAVCEI General Assembly Excursions: México, D.F., Universidad Nacional Autónoma de México, Instituto de Geología, Excursion 1, p. 1–39.

Alaniz-Álvarez, S.A., and Nieto-Samaniego, A.F., 2005, El sistema de fallas Taxco-San Miguel de Allende y la Faja Volcánica Transmexicana, dos fronteras tectónicas del centro de México activas durante el Cenozoico: Boletín de la Sociedad Geológica Mexicana, v. 57, p. 65–82.

Alaniz-Álvarez, S.A., Nieto-Samaniego, A.F., Reyes-Zaragoza, M.A., Orozco-Esquivel, M.T., Ojeda-García, A.C., and Vasallo-Morales, L.F., 2001, Estratigrafía y deformación de la región San Miguel de Allende-Querétaro: Revista Mexicana de Ciencias Geológicas, v. 18, p. 129–148.

Alaniz-Álvarez, S.A., Nieto-Samaniego, A.F., Orozco-Esquivel, M.T., Vasallo-Morales, L.F., and Xu Shunshan, 2002, El sistema de Fallas Taxco-San Miguel de Allende: implicaciones en la deformación post-Eocénica del centro de México: Boletín de la Sociedad Geológica Mexicana, v. 55, p. 12–29.

Albinson, T., 1985, Zoneamientos térmicos y su relación a la distribución mineral en algunos yacimientos epitermales en México, *in* XVI Convención Nacional, Memorias técnicas, Mazatlán, Sinaloa, Asociación de Ingenieros de Minas, Metalurgistas y Geólogos de México, 17 p.

Albinson, T., 1988, Geologic reconstruction of paleosurfaces in the Sombrerete, Colorada, and Fresnillo district, Zacatecas state, Mexico: Economic Geology and the Bulletin of the Society of Economic Geologists, v. 83, p. 1647–1667.

Albinson, T., Norman, D.I., Cole, D., and Chomiak, B., 2001, Controls on formation of low-sulfidation epithermal deposits in Mexico: Constraints from fluid inclusion and stable isotope data, *in* Albinson, T., and Nelson, C.E., eds., New mines and discoveries in Mexico and Central America. Littleton, Colorado, Society of Economic Geologists, Special Publication 8, p. 1–32.

Alvarado-Méndez, H., Sánchez-Garrido, E., Pérez-Vargas, M.A., and Caballero-Martínez, J.A., 1997, Carta geológico-minera Guanajuato F14-7: Consejo de Recursos Minerales, scale 1:250,000, second edition, 1 sheet.

Aranda-Gómez, J.J., 1978, Metamorphism, mineral zoning, and paragenesis in the San Martín mine, Zacatecas, Mexico [Ms. thesis]: Golden, Colorado, Colorado School of Mines, 90 p.

Aranda-Gómez, J.J., and McDowell, F., 1998, Paleogene extension in the southern Basin and Range province of Mexico: Syndepositional tilting of Eocene red beds and Oligocene volcanic rocks in the Guanajuato Mining District: International Geology Review, v. 40, p. 116–134.

Aranda-Gómez, J.J., Luhr, J.F., and Nieto-Samaniego, A.F., 1993a, Localidades recién descubiertas de xenolitos del manto y de la base de la corteza en el estado de San Luis Potosí, México, *in* Xenolitos del manto y de la base de la corteza en el estado de San Luis Potosí, México: Universidad Nacional Autónoma de México, Boletín del Instituto de Geología 106, part 2, p. 23–36.

Aranda-Gómez, J.J., Luhr, J.F., and Pier, J.G., 1993b, Geología de los volcanes cuaternarios portadores de xenolitos del manto y de la base de la corteza en el estado de San Luis Potosí, México, *in* Xenolitos del manto y de la base de la corteza en el estado de San Luis Potosí, México: Universidad Nacional Autónoma de México, Boletín del Instituto de Geología 106, part 1, p. 1–22.

Aranda-Gómez, J.J., Henry, C.D., Luhr, J.F., and McDowell, F.W., 1997, Cenozoic volcanism and tectonics in NW Mexico: A transect across the Sierra Madre Occidental volcanic field and observations on extension related magmatism in southern Basin and Range and the Gulf of California tectonic provinces, *in* Aguirre-Díaz, G.J., Aranda-Gómez, J.J., Carrasco-Núñez, G., and Ferrari, L., eds., Magmatism and tectonics of central and northwestern Mexico—A selection of the 1997 IAVCEI General Assembly excursions: México, D.F., Universidad Nacional Autónoma de México, Instituto de Geología, Excursion 1, p. 1–39.

Aranda-Gómez, J.J., Henry, C.D., and Luhr, J., 2000, Evolución tectonomagmática post-paleocénica de la Sierra Madre Occidental y de la porción meridional de la provincia tectónica de Cuencas y Sierras, México: Boletín de la Sociedad Geológica Mexicana, v. 53, p. 59–71.

Aranda-Gómez, J.J., Henry, C.D., Luhr, J., and McDowell, F.W., 2003, Cenozoic volcanic-tectonic development of northwestern Mexico—A transect across the Sierra Madre Occidental volcanic field and observations on extension-related magmatism in the southern Basin and Range and Gulf of California tectonic provinces, *in* Geologic transects across Cordilleran Mexico, Guidebook for field trips 99th Annual Meeting of the Cordilleran Section of the Geological Society of America, México, D.F., March 25–30, 2003: Universidad Nacional Autónoma de México, Instituto de Geología, Centro de Geociencias, Special publication 1, p. 71–121.

Aranda-Gómez, J.J., Housh, T.B., Luhr, J.F., Henry, C.D., Becker, T., and Chávez-Cabello, G., 2005, Reactivation of the San Marcos fault during mid- to late Tertiary extension, Chihuahua, México, *in* Nourse, J.A., Anderson, T.H., McKee, J.W., and Steiner, M.B., eds., The Mojave-Sonora Megashear Hypothesis: Development, Assessment, and Alternatives: Geological Society of America Special Paper 393, p. 509–522.

Bacon, R.W., 1978, Geology of the northern Sierra de Catorce, San Luis Potosí, México [Ms. thesis]: Arlington, Texas, University of Texas at Arlington, 124 p.

Barboza-Gudiño, R., Tristán-González, M., and Torres-Hernández, J.R., 1999, Tectonic setting of pre-Oxfordian units from central and northeastern Mexico: A review, *in* Bartolini, C., Wilson, J.L., and Lawton, T.F., eds., Mesozoic Sedimentary and Tectonic History of North-Central Mexico: Geological Society of America Special Paper 340, p. 197–210.

Barboza-Gudiño, J.R., Hoppe, M., Gómez-Anguiano, M., and Martínez-Macías, P.R., 2004, Aportaciones para la interpretación estratigráfica y estructural de la porción noroccidental de la Sierra de Catorce, San Luis Potosí, México: Revista Mexicana de Ciencias Geológicas, v. 21, p. 299–319.

Bartolini, C., Lang, H., and Stinnesbeck, W., 1999, Volcanic rock outcrops in Nuevo Leon, Tamaulipas and San Luis Potosí, Mexico: Remnants of the Permian-early Triassic magmatic arc?, *in* Bartolini, C., Wilson, J.L., and Lawton, T.F., eds., Mesozoic Sedimentary and Tectonic History of North-Central Mexico: Geological Society of America Special Paper 340, p. 347–355.

Bates, R.L., and Jackson, J.A., eds., 1987, Glossary of Geology, 3rd ed.: Alexandria, Va., American Geological Institute, 788 p.

Botsford, C.W., 1909, The Zacatecas district and its relation to Guanajuato and other camps: The Engineering and Mining Journal, v. 87, p. 1227–1228.

Bracho-Valle, F., 1960, Yacimientos de estaño en la Sierra de Chapultepec, Zac., La Ochoa, Dgo. y Cosío, Ags: Consejo de Recursos Naturales no Renovables, Boletín 48, 116 p., 31 sheets.

Buchanan, L.J., 1981, Precious metal deposits associated with volcanic environments in the Southwest, *in* Dickson, W.R., and Payne, W.D., eds., Relations of tectonics to ore deposits in the southern Cordillera: Arizona Geological Society Digest 14, 237–262.

Burckhardt, C., and Scalia, S., 1906, Géologie des environs des Zacatecas, *in* X International Geological Congress, guide excursion: Instituto Geológico de México 1(16), 26 p.

Caballero-Martínez, J.A., Isabel-Blanci, J., and Luévano-Pinedo, A., 1999, Carta geológico-minera Zacatecas F13–B58: Consejo de Recursos Minerales, escala 1:50,000, 1 sheet.

Campos-Enríquez, J.O., Kerdan, T., Morán-Zenteno, D.J., Urrutia-Fucugauchi, J., Sánchez-Castellanos, E., and Alday-Cruz, R., 1992, Estructura de la litósfera superior a lo largo del Trópico de Cáncer: Geos, v. 12, p. 75–76.

Camprubí, A., and Albinson, T., 2006, Los depósitos epitermales: revisión sobre el estado actual de su conocimiento, métodos de estudio y presencia en México: Boletín de la Sociedad Geológica Mexicana, v. 58, p. 27–81.

Camprubí, A., Prol-Ledesma, R.M., and Tritlla, J., 1998, Comments on "Metallogenic evolution of convergent margins: Selected ore deposit models" by S. E. Kesler: Ore Geology Reviews, v. 14, p. 71–76.

Camprubí, A., Ferrari, L., Cosca, M.A., Cardellach, E., and Canals, A., 2003, Ages of epithermal deposits in Mexico: Regional significance and links with the evolution of Tertiary volcanism: Economic Geology and the Bulletin of the Society of Economic Geologists, v. 98, p. 1029–1038.

Cantú-Chapa, C.M., 1969, Una nueva localidad Triásico Superior en México: Revista del Instituto del Petróleo, v. 1, p. 71–72.

Carranza-Castañeda, O., Petersen, M.S., and Miller, W.E., 1994, Preliminary investigation of the geology of northern San Miguel de Allende area, northeastern Guanajuato, Mexico: Brigham Young University, Geological Studies, v. 40, p. 1–9.

Carrasco-Núñez, G., Milán, M., and Verma, S.P., 1989, Geología de Volcán El Zamorano, México: Revista del Instituto de Geología, v. 8, p. 194–201.

Carrillo-Bravo, J., 1971, La plataforma de Valles-San Luis Potosí: Boletín de la Asociación Mexicana de Geólogos Petroleros, v. 23, p. 1–110.

Centeno-García, E., and Silva-Romo, G., 1997, Petrogenesis and tectonic evolution of central Mexico during Triassic-Jurassic time: Revista Mexicana de Ciencias Geológicas, v. 14, p. 244–260.

Chávez-Cabello, G., Aranda-Gómez, J.J., Molina-Garza, R.S., Cossío-Torres, T., Arvizu-Gutiérrez, I.R., and González-Naranjo, G., 2005, La Falla San Marcos: una estructura jurásica de basamento multirreactivada del noreste de México: Boletín de la Sociedad Geológica Mexicana, v. 57, p. 27–52.

Chiodi, M., Monod, O., Busnardo, R., Gaspar, D., Sanchez, A., and Yta, M., 1988, Une discordance ante Albienne datée par une faune d'ámmonites et de braquiopodes de type Téthysien au Mexique Central: Geobios, v. 21, p. 125–135.

Clark, K.F., Damon, P.E., Shafiquillah, M., Ponce, B.F., and Cardenas, D., 1981, Sección geológica-estructural a través de la parte sur de la Sierra Madre Occidental, entre Fresnillo y la costa de Nayarit: Asociación de Ingenieros Mineros, Metalúrgicos y Geólogos de México, Memoria Técnica, v. XIV, p. 69–99.

Clark, K.F., Foster, C.T., and Damon, P.E., 1982, Cenozoic mineral deposits and subduction-related magmatic arcs in Mexico: Geological Society of America Bulletin, v. 93, p. 533–544, doi: 10.1130/0016-7606(1982)93<533: CMDASM>2.0.CO;2.

Consejo de Recursos Minerales (Coremi), 1992, Monografía geológico-minera del estado de Guanajuato: Secretaría de Energía, Minas e Industria Paraestatal, Consejo de Recursos Minerales, Publicación M-6e, 136 p., 4 sheets.

Consejo de Recursos Minerales (Coremi), 1996, Monografía geológico-minera del estado de San Luis Potosí: México, Secretaria de Energía, Minas e Industria Paraestatal, Consejo de Recursos Minerales, Publicación M-7e, 217 p., 1 map, 2 sheets.

Córdoba, D.A., 1988, Estratigrafía de las rocas volcánicas de la región entre Sierra de Gamón y Laguna de Santiaguillo, estado de Durango: Revista del Instituto de Geología, v. 7, p. 136–147.

Corona-Chávez, P., and López-Picos, A., 1988, Análisis estratigráfico-estructural de la secuencia volcánico-sedimentaria metamorfizada de la Sierra de Guanajuato, *in* IX Convención Geológica Nacional abstracts: Sociedad Geológica Mexicana, p. 104.

Damon, P.E., Shafiqullah, M., and Clark, K.F., 1981, Evolución de los arcos magmáticos en México y su relación con la metalogénesis: Universidad Nacional Autónoma de México, Revista del Instituto de Geología, v. 6, p. 223–239.

Damon, P.E., Shafiqullah, M., and Clark, K.F., 1983, Geochronology of the porphyry copper deposits and related mineralization of Mexico: Canadian Journal of Earth Sciences, v. 20, p. 1052–1071.

De Cserna, Z., 1976, Geology of the Fresnillo area, Zacatecas, Mexico: Geological Society of America Bulletin, v. 87, p. 1191–1199, doi: 10.1130/0016-7606(1976)87<1191:GOTFAZ>2.0.CO;2.

Edwards, J.D., 1955, Studies of some early Tertiary red conglomerates of Central Mexico: U.S. Geological Survey, Professional Paper 264-H, 183 p.

Eguiluz-de Antuñano, S., 1984, Tectónica cenozoica del norte de México: Boletín de la Asociación Mexicana de Geólogos Petroleros, v. 36, p. 43–62.

Eguiluz-de Antuñano, S., Aranda-García, M., and Marrett, R., 2000, Tectónica de la Sierra Madre Oriental, México: Boletín de la Sociedad Geológica Mexicana, v. 53, p. 1–26.

Fabregat-G., F.J., 1966, Los minerales mexicanos 3. Durangita: Boletín Instituto de Geología 77, 113 p.

Farley, K.A., 2000, Helium diffusion from apatite: General behaviour as illustrated by Durango fluorapatite: Journal of Geophysical Research, v. 105, B2, p. 2903–2914, doi: 10.1029/1999JB900348.

Ferrari, L., López-Martínez, M., and Rosas-Elguera, J., 2002, Ignimbrite flare up and deformation in the southern Sierra Madre Occidental, western Mexico: Implications for the late subduction history of the Farallon plate: Tectonics, v. 21, p. 17-1–17-24.

Ferrari, L., Valencia-Moreno, M., and Bryan, S., 2005, Magmatismo y tectónica en la Sierra Madre Occidental y su relación con la evolución de la margen occidental de Norteamérica: Boletín de la Sociedad Geológica Mexicana, v. 57, p. 343–378.

Fix, J.E., 1975, The crust and upper mantle of central Mexico: Geophysical Journal of the Royal Astronomical Society, v. 43, p. 453–499.

Fournier, R.O., 1987, Conceptual models of brine evolution in magmatic-hydrothermal systems. *in* Decker, R.W., Wright, T.L., and Stauffer, P.H., eds., Volcanism in Hawaii, vol. 2: U.S. Geological Survey Professional Paper 1350, p. 1487–1506.

Franco-Rubio, M., 1999, Geology of the basement below the decollement surface, Sierra de Catorce, San Luis Potosí, México, *in* Bartolini, C., Wilson, J.L., and Lawton, T.F., eds., Mesozoic Sedimentary and Tectonic History of North-Central Mexico: Geological Society of America Special Paper 340, p. 211–227.

Freydier, C., Martínez, J.R., Lapierre, H., Tardy, M., and Coulon, C., 1996, The early Cretaceous Arperos Basin (western Mexico), Geochemical evidence for an aseismic ridge formed near a spreading center: Tectonophysics, v. 259, p. 343–367, doi: 10.1016/0040-1951(95)00143-3.

Gemmell, J.B., Simmons, S.F., and Zantop, H., 1988, The Santo Niño silver-lead-zinc vein, Fresnillo District, Zacatecas, Mexico; Part I: Structure, vein stratigraphy, and mineralogy: Economic Geology and the Bulletin of the Society of Economic Geologists, v. 83, p. 1597–1618.

Gilmer, A.L., Clark, K.F., Conde, J., Hernández, I., Figueroa, J.I., and Porter, E.W., 1988, Sierra de Santa María, Velardeña mining district, Durango, Mexico: Economic Geology and the Bulletin of the Society of Economic Geologists, v. 83, p. 1802–1829.

González-Reyna, J., 1956, Riqueza minera y yacimientos minerales de México, 3a edición: Banco de México, 497 p., 15 sheets.

González-Reyna, J., and White, D.E., 1947, Los yacimientos de antimonio de San José, Sierra de Catorce, estado de San Luis Potosí: Comité Directivo para la Investigación de los Recursos Minerales de México, Boletín 14, 36 p.

Gross, W.H., 1975, New ore discovery and source of silver-gold veins, Guanajuato, Mexico: Economic Geology and the Bulletin of the Society of Economic Geologists, v. 70, p. 1175–1189.

Gunnesch, K.A., Torres del Ángel, C., Cuba-Castro, C., and Sáez, J., 1994, The Cu-(Au) skarn and Ag-Pb-Zn vein deposits of La Paz, northeastern Mexico: Mineralogical, paragenetic, and fluid inclusion characteristics: Economic Geology and the Bulletin of the Society of Economic Geologists, v. 89, p. 1640–1650.

Hayob, J.L., Essene, E.J., Ruiz, J., Ortega-Gutiérrez, F., and Aranda-Gómez, J.J., 1989, Young high-temperature granulites from the base of the crust in central Mexico: Nature, v. 342, p. 265–268, doi: 10.1038/342265a0.

Hedenquist, J.W., 1986, Geothermal systems in the Taupo volcanic zone: Their characteristics and relation to volcanism and mineralization, *in* Smith, I.E.M., ed., Late Cenozoic volcanism in New Zealand: Royal Society of New Zealand Bulletin 23, p. 134–168.

Henry, C.D., and Aranda-Gómez, J.J., 2000, Plate interactions control middle-late Miocene, proto-Gulf and Basin and Range extension in the southern

Basin and Range: Tectonophysics, v. 318, p. 1–26, doi: 10.1016/S0040-1951(99)00304-2.

Hernández-Laloth, N., 1991, Modelo conceptual de funcionamiento hidrodinámico del sistema acuífero del valle de León, Guanajuato [BSc. thesis]: México, D. F., Universidad Nacional Autónoma de México, 129 p.

Hernández-Sánchez-Mejorada, S., and López Ramos, E., 1968, Carta Geológica de la República Mexicana escala 1:2,000,000: Comité de la Carta Geológica de México, Universidad Nacional Autónoma de México, Instituto de Geología, 1 sheet.

Hoppe, M., Barboza-Gudiño, J.R., and Schulz, H.M., 2002, Late Triassic submarine fan deposits in northwestern San Luis Potosí, México—Lithology, facies and diagenesis: Neues Jahrbuch füer Geologie und Paläontogie: Monatshefte, v. 12, p. 705–724.

Huspeni, J.R., Kesler, S.E., Ruiz, J., Tuta, Z., Sutter, J.F., and Jones, L.M., 1984, Petrology and geochemistry of rhyolites associated with tin mineralization in northern Mexico: Economic Geology and the Bulletin of the Society of Economic Geologists, v. 79, p. 87–105.

Imlay, R.W., 1938, Studies of the Mexican geosyncline: Geological Society of America Bulletin, v. 49, p. 1651–1694.

Instituto Nacional de Estadística, Geografía e Informática (INEGI), 2004, Mapa digital de México: México, Instituto Nacional de Estadística, Geografía e Informática, Sistemas Nacionales Estadístico y de Información Geográfica: http://galileo.inegi.gob.mx/website/mexico/viewer.htm?c=423.

Iriondo, A., Kunk, M.J., Winick, J.A., and Consejo de Recursos Minerales (Coremi), 2003, $^{40}Ar/^{39}Ar$ dating studies of minerals and rocks in various areas in Mexico: USGS/Coremi scientific collaboration (Part I): U.S. Geological Survey, Open-File Report (03–020), 79 p., online: http://pubs.usgs.gov/of/2003/ofr-03-020/.

Jiménez-Nava, F.J., 1993, Aportes a la estratigrafía de Aguascalientes mediante la exploración geohidrológica a profundidad, *in* Simposio sobre la geología del Centro de México, resúmenes y guía de excursión: Universidad de Guanajuato, Facultad de Minas, Metalurgia y Geología, Extensión Minera, abstracts book 1, p. 93.

Johnson, B., Montante, A., Kearvell, G., Janzen, J., and Scammell, R., 1999, Geology and exploration of the San Nicolás polymetallic (Zn-Cu-Au-Ag) volcanogenic massive sulphide deposit, *in* Ambor, J.L., ed., VMS and Carbonate-Hosted Polymetallic Deposits of Central Mexico: Vancouver, B.C.: British Columbia and Yukon Chamber of Mines, Cordillera Roundup, Special Volume, p. 45–54.

Kerdan, T.P., 1992, Estructura de la corteza y manto superior en el norte de México (a lo largo del Trópico de Cáncer desde Baja California hasta el Golfo de México) [MSc. thesis]: Universidad Nacional Autónoma de México, 347 p.

Labarthe-Hernández, G., and Jiménez-López, L.S., 1992, Características físicas y estructura de lavas e ignimbritas riolíticas en la Sierra de San Miguelito, S. L. P.: Universidad Autónoma de San Luis Potosí, Instituto Geología, Folleto Técnico 114, 31 p., 4 sheets.

Labarthe-Hernández, G., and Jiménez-López, L.S., 1993, Geología del domo Cerro Grande, Sierra de San Miguelito, S. L. P.: Universidad Autónoma de San Luis Potosí, Instituto Geología, Folleto Técnico 117, 22 p., 3 sheets.

Labarthe-Hernández, G., and Jiménez-López, L.S., 1994, Geología de la porción sureste de la Sierra de San Miguelito, S. L. P.: Universidad Autónoma de San Luis Potosí, Instituto Geología, Folleto Técnico 120, 34 p., 2 sheets.

Labarthe-Hernández, G., and Tristán-González, M., 1980, Cartografía geológica hoja Santa María del Río, San Luis Potosí: Universidad Autónoma de San Luis Potosí, Instituto de Geología y Metalurgia, Folleto Técnico 67, 32 p.

Labarthe-Hernández, G., Tristán-González, M., and Aranda-Gómez, J.J., 1982, Revisión estratigráfica del Cenozoico de la parte central del estado de San Luis Potosí: Universidad Autónoma de San Luis Potosí, Instituto de Geología, Folleto Técnico 85, 208 p., 1 sheet.

Labarthe-H., G., Tristán-G., M., Aguillón-R., A., Jiménez-L., L.S., and Romero, A., 1989, Cartografía geológica 1:50,000 de las hojas El Refugio y Mineral El Realito, estados de San Luis Potosí y Guanajuato: Universidad Autónoma de San Luis Potosí, Instituto de Geología, Folleto Técnico 112, 76 p., 4 sheets.

Lang, B., Steinitz, G., Sawkins, F.J., and Simmons, S.F., 1988, K-Ar age studies in the Fresnillo silver district, Zacatecas, Mexico: Economic Geology and the Bulletin of the Society of Economic Geologists, v. 83, p. 1642–1646.

Lapierre, H., Ortiz, L.E., Abouchami, W., Monod, O., Coulon, Ch., and Zimmermann, J.L., 1992, A crustal section of an intra-oceanic island arc: The late Jurassic-early Cretaceous Guanajuato magmatic sequence, central México: Earth and Planetary Sciences Letters, v. 108, p. 61–77, doi: 10.1016/0012-821X(92)90060-9.

Levresse, G., González-Partida, E., Tritlla, J., Camprubí, A., Cienfuegos-Alvarado, E., and Morales-Puente, P., 2003, Fluid characteristics of the world-class, carbonate-hosted Las Cuevas fluorite deposit (San Luis Potosí, Mexico): Journal of Geochemical Exploration, v. 78–79, p. 537–543, doi: 10.1016/S0375-6742(03)00145-6.

López-Doncel, R., 2003, La Formación Tamabra del Cretácico medio en la porción central del margen occidental de la Plataforma Valles-San Luis Potosí, centro-noreste de México: Revista Mexicana de Ciencias Geológicas, v. 20, p. 1–19.

López-Infanzón, M., 1986, Estudio petrogenético de las rocas ígneas en las formaciones Hizachal y Nazas: Boletín de la Sociedad Geológica Mexicana, v. 47, p. 1–42.

López-Monroy, P., 1888, Las minas de Guanajuato: Anales del Ministerio de Fomento (México) 10, 69 p.

Loza-Aguirre, I., 2005, Estudio estructural de la actividad cenozoica del sistema de fallas San Luis-Tepehuanes de la región Zacatecas-San José de Gracia [BSc. thesis]: Instituto Tecnológico de Ciudad Madero, 96 p.

Lugo-Hubp, J., 1989, Diccionario geomorfológico: México, D.F., Universidad Nacional Autónoma de México, Instituto de Geografía, 337 p.

Luhr, J.F., Pier, J.G., Aranda-Gómez, J.J., and Podosek, F.A., 1995, Crustal contamination in early Basin-and-Range hawaiites of the Los Encinos Volcanic Field, central Mexico: Contributions to Mineralogy and Petrology, v. 118, p. 321–339, doi: 10.1007/s004100050018.

Lyons, J.I., 1988, Volcanogenic iron oxide deposits, Cerro de Mercado and vicinity, Durango, Mexico: Economic Geology and the Bulletin of the Society of Economic Geologists, v. 83, p. 1886–1906.

Magonthier, M.C., 1988, Distinctive rhyolite suites in the mid-Tertiary ignimbritic complex of the Sierra Madre Occidental, western Mexico: Bulletin de la Société Géologique de France, Huitième Serie, v. 4, p. 57–68.

Mango, H., Zantop, H., and Oreskes, N., 1991, A fluid inclusion and isotope study of the Rayas Ag-Au-Cu-Pb-Zn mine, Guanajuato, Mexico: Economic Geology and the Bulletin of the Society of Economic Geologists, v. 86, p. 1554–1561.

Martínez-Reyes, J., 1992, Mapa geológico de la Sierra de Guanajuato con resumen de la geología de la Sierra de Guanajuato: Universidad Nacional Autónoma de México, Instituto de Geología, Cartas Geológicas y Mineras 8, 1 sheet.

Martínez-Ruiz, J., and Cuellar-González, G., 1978, Correlación de superficie y subsuelo de la cuenca geohidrológica de San Luis Potosí, S. L. P.: Universidad Autónoma de San Luis Potosí, Instituto Geología, y Metalurgia, Folleto Técnico 65, 25 p., 1 sheet.

McDowell, F.W., and Keizer, R.P., 1977, Timing of mid-Tertiary volcanism in the Sierra Madre Occidental between Durango City and Mazatlan, Mexico: Geological Society of America Bulletin, v. 88, p. 1479–1487, doi: 10.1130/0016-7606(1977)88<1479:TOMVIT>2.0.CO;2.

McGehee, R., 1976, Las rocas metamórficas del arroyo de La Pimienta, Zacatecas, Zac: Boletín de la Sociedad Geológica Mexicana, v. 37, p. 1–10.

Megaw, P.K.M., 1999, The high-temperature, Ag-Pb-Zn-(Cu) carbonate replacement deposits of Central Mexico, *in* Ambor, J.L., ed., VMS and Carbonate-Hosted Polymetallic Deposits of Central Mexico: Vancouver, B.C., British Columbia and Yukon Chamber of Mines, Cordillera Roundup, Special Volume, p. 25–44.

Megaw, P.K.M., Ruiz, J., and Titley, S.R., 1988, High-temperature, carbonate-hosted Ag-Pb-Zn-(Cu) deposits of Northern Mexico: Economic Geology and the Bulletin of the Society of Economic Geologists, v. 83, p. 1856–1885.

Meyer, R.P., Steinhart, J.S., and Woolard, G.P., 1958, Seismic determination of crustal structure in the central plateau of Mexico: Transactions—American Geophysical Union, v. 39, p. 525.

Miranda-Gasca, M.A., 2000, The metallic ore deposits of the Guerrero terrane, western Mexico: An overview: Journal of South American Earth Sciences, v. 13, p. 403–413, doi: 10.1016/S0895-9811(00)00032-8.

Mitchell, A.H.G., and Balce, G.R., 1990, Geological features of some epithermal gold systems, Philippines, *in* Hedenquist, J.W., White, N.C., and Siddeley, G., eds., Epithermal gold mineralization of the Circum-Pacific: Geology, geochemistry, origin and exploration, I: Journal of Geochemical Exploration, v. 35, p. 241–296.

Monod, O., and Calvet, P.H., 1991, Structural and Stratigraphic re-interpretation of the Triassic units near Zacatecas (Zac.), Central Mexico: Evidence of

Laramide nappe pile: Zentralblatt für Geologie und Paläontologie, Teil, v. 1, p. 1533–1544.

Montellano-Ballesteros, M., 1990, Una edad del Irvingtoniano al Rancholabreano para la fauna Cedazo del estado de Aguascalientes: Revista del Instituto de Geología, v. 9, p. 195–203.

Moore, G., Marone, C., Carmichael, I.S.E., and Renne, P., 1994, Basaltic volcanism and extension near the intersection of the Sierra Madre volcanic province and the Mexican Volcanic Belt: Geological Society of America Bulletin, v. 106, p. 383–394, doi: 10.1130/0016-7606(1994)106<0383:BVAENT>2.3.CO;2.

Morán-Zenteno, D.J., Caballero-Miranda, C.I., Silva-Romo, G., Ortega-Guerrero, B., and González-Torres, E., 1993, Jurassic-Cretaceous paleogeographic evolution of the northern Mixteca terrane, southern Mexico: Geofisica Internacional, v. 32, p. 453–473.

Morán-Zenteno, D.J., Martiny, B., Tolson, G., Solís-Pichardo, G., Alba-Aldave, L., Hernández-Bernal, M.S., Macías-Romo, C., Martínez-Serrano, R., Schaaf, P., and Silva-Romo, G., 2000, Geocronología y características geoquímicas de las rocas magmáticas terciarias de la Sierra Madre del Sur: Boletín de la Sociedad Geológica Mexicana, v. 53, p. 27–58.

Moreira-Rivera, F., Flores-Aguillón, G., Martínez-Rodríguez, L., Palacios-García, R., Chiapa-García, R., Olvera-Campos, A., Pérez-Benavidez, A., and González-Monsivais, P., 1998, Carta geológico-minera San Luis Potosí F14–4, México: Consejo de Recursos Minerales, escala 1:250,000, 1 sheet.

Mortensen, J.K., Hall, B.V., Bissig, T., Friedman, R.M., Danielson, T., Oliver, J., Rhys, D.A., and Ross, K.V., 2003, U-Pb zircon age and Pb isotopic constrains on the age and origin of volcanogenic massive sulfide deposits in the Guerrero terrane of central Mexico, *in* 99th Annual Meeting Cordilleran Section: Geological Society of America Abstracts with Programs, v. 35, no. 4, p. 61–62.

Mujica-Mondragón, M.R., and Albarrán-Jacobo, J., 1983, Estudio petrogenético de las rocas ígneas y metamórficas del Altiplano Mexicano: Instituto Mexicano del Petróleo, Proyecto C-1156, 78 p.

Munguía-Rojas, P., García-Padilla, J.L., Armenta-Román, R., and Camacho, J.M., 2000, Carta geológico-minera Santiago Papasquiaro G13–8, Durango, México: Consejo de Recursos Minerales, escala 1:250,000, 1 sheet.

Murillo-Muñetón, G., and Torres-Vargas, R., 1987, Mapa petrogenético y radiométrico de la República Mexicana: Instituto Mexicano del Petróleo, Subdirección de Tecnología de Exploración, informe del proyecto C-2010, 78 p.

Nesbitt, B.E., 1990. Fluid flow and chemical evolution in the genesis of hydrothermal ore deposits, *in* Nesbitt, B.E., ed., Short Course on Fluids in Tectonically Active Regimes of the Continental Crust: Toronto, Ontariom Mineralogical Association of Canada, Short Course Handbook 18, p. 261–297.

Nieto-Samaniego, Á.F., 1990(1992), Fallamiento y estratigrafía cenozoicos en la porción sudoriental de la Sierra de Guanajuato: Revista del Instituto de Geología, v. 9, p. 146–155.

Nieto-Samaniego, Á.F., Macías-Romo, C., and Alaniz-Álvarez, S.A., 1996, Nuevas edades isotópicas de la cubierta volcánica cenozoica de la parte meridional de la Mesa Central, México: Revista Mexicana de Ciencias Geológicas, v. 13, p. 117–122.

Nieto-Samaniego, Á.F., Alaniz-Álvarez, S.A., and Labarthe-Hernández, G., 1997, La deformación cenozoica poslaramídica en la parte meridional de la Mesa Central, México: Revista Mexicana de Ciencias Geológicas, v. 14, p. 13–25.

Nieto-Samaniego, Á.F., Ferrari, L., Alaniz-Álvarez, S.A., Labarthe-Hernández, G., and Rosas-Elguera, J., 1999, Variation of Cenozoic extension and volcanism across the southern Sierra Madre Occidental volcanic province, Mexico: Geological Society of America Bulletin, v. 111, p. 347–363, doi: 10.1130/0016-7606(1999)111<0347:VOCEAV>2.3.CO;2.

Ojeda-García, Á.C., 2004, Cartografía Geológica 1: 50,000 del límite El Bajío-Mesa Central, Porción Salamanca-Irapuato, estado De Guanajuato, Mexico [BSc. thesis]: Centro de Estudios Superiores del Estado de Sonora, 86 p., 1 sheet.

Olvera-Carranza, K., Centeno, E., and Camprubí, A., 2001, Deformation and distribution of massive sulphide deposits in Zacatecas, Mexico, *in* Pietrzynsi, A., et al., eds., Mineral Deposits at the Beginning of the 21st Century: Lisse, Swets & Zeitlinger Publishers, p. 313–316.

Orozco-Esquivel, M.T., Nieto-Samaniego, Á.F., and Alaniz-Álvarez, S.A., 2002, Origin of rhyolitic lavas in the Mesa Central, Mexico, by crustal melting related to extension: Journal of Volcanology and Geothermal Research, v. 118, p. 37–56, doi: 10.1016/S0377-0273(02)00249-4.

Ortega-Gutiérrez, F., Mitre-Salazar, L.M., Roldán-Quintana, J., Aranda-Gómez, J.J., Morán-Zenteno, D.J., Alaniz-Álvarez, S.A., and Nieto-Samaniego, Á.F., 1992, Carta geológica de la República Mexicana, quinta Edición: Universidad Nacional Autónoma de México-Secretaría de Energía, Minas e Industria Paraestatal, escala 1:2,000,000, 1 sheet.

Pantoja-Alor, J., 1963, Hoja San Pedro del Gallo, estado de Durango: México, D.F., Universidad Nacional Autónoma de México, Instituto de Geología, Carta Geológica de México series, escala 1:100,000, 1 sheet.

Pasquarè, G., Ferrari, L., Garduño, V.H., Tibaldi, A., and Vezzoli, L., 1991, Geologic map of the central sector of the Mexican Volcanic belt, States of Guanajuato and Michoacán: Geological Society of America Map and Chart series MCH072, 1 sheet, 20 p. text.

Pearson, M.F., Clark, K.F., and Porter, E.W., 1988, Mineralogy, fluid characteristics, and silver distribution at Real de Ángeles, Zacatecas, Mexico: Economic Geology and the Bulletin of the Society of Economic Geologists, v. 83, p. 1737–1759.

Pérez-Venzor, J.A., Aranda-Gómez, J.J., McDowell, F., and Solorio-Munguía, J.G., 1996, Geología del volcán Palo Huérfano, Guanajuato, México: Revista Mexicana de Ciencias Geológicas, v. 13, p. 174–183.

Petersen, M.A., Della Libera, M., Jannas, R.R., and Maynard, S.R., 2001, Geology of the Cerro San Pedro porphyry-related gold-silver deposit, San Luis Potosí, Mexico, *in* Albinson, T., and Nelson, C.E., eds., New Mines and Discoveries in Mexico and Central America: Littleton, Colorado, Society of Economic Geologists, Special Publication 8, p. 217–241.

Petruk, W., and Owens, D., 1974, Some mineralogical characteristics of the silver deposits in the Guanajuato mining district, Mexico: Economic Geology and the Bulletin of the Society of Economic Geologists, v. 69, p. 1078–1085.

Pollard, P.J., 2000, Evidence of a magmatic fluid and metal source for Fe-oxide Cu-Au mineralization, *in* Porter, T.M., ed., Hydrothermal Iron Oxide Copper-Gold & Related Deposits: A Global Perspective: Glenside, South Australia, Australia, Australian Mineral Foundation 1, p. 27–41.

Ponce, B.F., and Clark, K.F., 1988, The Zacatecas Mining District: A tertiary caldera complex associated with precious and base metal mineralization: Economic Geology and the Bulletin of the Society of Economic Geologists, v. 83, p. 1668–1682.

Ponce, D.A., and Glen, J.M.G., 2002, Relationship of epithermal gold deposits to large-scale fractures in Northern Nevada: Economic Geology and the Bulletin of the Society of Economic Geologists, v. 97, p. 3–9.

Quintero-Legorreta, O., 1992, Geología de la región de Comanja, estados de Guanajuato y Jalisco: Universidad Nacional Autónoma de México, Revista del Instituto de Geología, v. 10, p. 6–25.

Raisz, E., 1959, Landforms of Mexico: Cambridge, Massachussetts, U.S. Office of Naval Research, Geography Branch, scale ca. 1:3,000,000, 1 sheet.

Randall, J.A., Saldaña, E., and Clark, K.F., 1994, Exploration in a vulcano-plutonic center at Guanajuato, Mexico: Economic Geology and the Bulletin of the Society of Economic Geologists, v. 89, p. 1722–1754.

Rangin, C., 1977, Sobre la presencia de Jurásico Superior con amonitas en Sonora septentrional: Universidad Nacional Autónoma de México, Revista del Instituto de Geología, v. 1, p. 1–14.

Ranson, W.A., Fernández, L.A., Simmons, W.B., and Enciso-de la Vega, S., 1982, Petrology of the metamorphic rocks of Zacatecas, Mexico: Boletín de la Sociedad Geológica Mexicana, v. 41, p. 37–59.

Rivera, J., and Ponce, L., 1986, Estructura de la corteza al oriente de la Sierra Madre Occidental, México, basada en la velocidad del grupo de las ondas Rayleigh: Geofísica Internacional, v. 25, p. 383–402.

Rivera, R., 1993, Cocientes metálicos e inclusiones fluidas del distrito minero de Real de Asientos, Aguascalientes, *in* XX Convención Nacional, Memorias técnicas: Asociación de Ingenieros de Minas, Metalurgistas y Geólogos de México, p. 310–325.

Roldán-Quintana, J., 1968, Estudio geológico de reconocimiento de la región de Peñón Blanco, estado de Durango: Boletín de la Sociedad Geológica Mexicana, v. 31, p. 79–105.

Rubin, J.N., and Kyle, J.R., 1988, Mineralogy and geochemistry of the San Martín skarn deposit, Zacatecas, Mexico: Economic Geology and the Bulletin of the Society of Economic Geologists, v. 83, p. 1782–1801.

Ruiz, J., Kesler, S.E., Jones, L.M., and Sutter, J.F., 1980, Geology and geochemistry of the Las Cuevas fluorite deposit, San Luis Potosí, Mexico: Economic Geology and the Bulletin of the Society of Economic Geologists, v. 75, p. 1200–1209.

Ruvalcaba-Ruiz, D.C., and Thompson, T.B., 1988, Ore deposits at the Fresnillo mine, Zacatecas, Mexico: Economic Geology and the Bulletin of the Society of Economic Geologists, v. 83, p. 1583–1596.
Salas, G.P., 1975, Carta y provincias metalogenéticas de la República Mexicana: Consejo de Recursos Minerales, Publicación 21E, 242 p., 1 sheet.
Sánchez-Zavala, J.L., Centeno-García, E., and Ortega-Gutiérrez, F., 1999, Review of Paleozoic stratigraphy of Mexico and its role in the Gondwana-Laurentia connections, *in* Ramos, V.A., and Keppie, J.D., eds., Laurentia-Gondwana Connections before Pangea: Geological Society of America Special Paper 336, p. 211–226.
Silva-Romo, G., 1996, Estudio de la estratigrafía y estructuras tectónicas de la Sierra de Salinas, Edos. de S. L. P. y Zac. [MSc. thesis]: Universidad Nacional Autónoma de México, 139 p.
Silva-Romo, G., Arellano-Gil, J., Mendoza-Rosales, C., and Nieto-Obregón, J., 2000, A submarine fan in the Mesa Central Mexico: Journal of South American Earth Sciences, v. 13, p. 429–442, doi: 10.1016/S0895-9811(00)00034-1.
Simmons, S.F., 1991, Hydrologic implications of alteration and fluid inclusion studies in the Fresnillo District, Mexico. Evidence for a brine reservoir and a descending water table during the formation of hydrothermal Ag-Pb-Zn ore bodies: Economic Geology and the Bulletin of the Society of Economic Geologists, v. 86, p. 1579–1601.
Simmons, S.F., Gemmell, J.B., and Sawkins, F.J., 1988, The Santo Niño silver-lead-zinc vein, Fresnillo District, Zacatecas, Mexico; Part II. Physical and chemical nature of ore-forming solutions: Economic Geology and the Bulletin of the Society of Economic Geologists, v. 83, p. 1619–1641.
Staude, J.M.G., 1993, Gold, silver, and base metal epithermal mineral deposits around the Gulf of California, Mexico: Relationship between mineralization and mayor structures, *in* Scott, R.W., Jr., Detra, P.S., and Berger, B.S., eds., Advances related to United States and international mineral resources: Developing frameworks and exploration technologies: U.S. Geological Survey Bulletin 2039, p. 69–78.
Swanson, E.R., 1989, A new type of maar volcano from the State of Durango—The El Jagüey-La Breña complex reinterpreted: Revista del Instituto de Geología, v. 8, p. 243–248.
Swanson, E.R., Keizer, R.P., Lyons, J.I., and Clabaugh, S.E., 1978, Tertiary volcanism and caldera development near Durango City, Sierra Madre Occidental, Mexico: Geological Society of America Bulletin, v. 89, p. 1000–1012, doi: 10.1130/0016-7606(1978)89<1000:TVACDN>2.0.CO;2.
Tardy, M., and Maury, R., 1973, Sobre la presencia de elementos de origen volcánico en las areniscas de los flyschs de edad Cretácica superior de los estados de Coahuila y Zacatecas: Boletín de la Sociedad Geológica Mexicana, v. 34, p. 5–12.
Tristán-González, M., 1986, Estratigrafía y tectónica del graben de Villa de Reyes en los estados de San Luis Potosí y Guanajuato, México: Universidad Autónoma de San Luis Potosí, Instituto de Geología, Folleto Técnico 107, 91 p.
Tristán-González, M., 1987, Cartografía geológica Hoja Tierra Nueva, San Luis Potosí: Universidad Autónoma de San Luis Potosí, Instituto de Geología y Metalurgia, Folleto Técnico 109, 103 p.
Trujillo-Candelaria, J.A., 1985, Origen del fallamiento, *in* Flores-Núñez, J., ed., Fallamiento de terrenos en Celaya: Sociedad Mexicana de Mecánica de Suelos, p. 3–9.
Tuta, Z.H., Sutter, J.F., Kesler, S.E., and Ruiz, J., 1988, Geochronology of mercury, tin, and fluorite mineralization in northern Mexico: Economic Geology and the Bulletin of the Society of Economic Geologists, v. 83, p. 1931–1942.
van der Lee, S., and Nolet, G., 1997, Seismic image of the subducted trailing fragments of the Farallon plate: Nature, v. 386, p. 266–269, doi: 10.1038/386266a0.
Verma, S.P., and Carrasco-Núñez, G., 2003, Reappraisal of the geology and geochemistry of Volcán Zamorano, Central Mexico: Implications for the discrimination of the Sierra Madre Occidental and Mexican Volcanic Belt province: International Geology Review, v. 45, p. 724–752.
Webber, K.L., Fernández, L.A., and Simmons, B., 1994, Geochemistry and mineralogy of the Eocene-Oligocene volcanic sequence, southern Sierra Madre Occidental, Juchipila, Zacatecas, México: Geofísica Internacional, v. 33, p. 77–89.
White, N.C., and Hedenquist, J.W., 1990, Epithermal environments and styles of mineralization: Variations and their causes, and guidelines for exploration, *in* Hedenquist, J.W., White, N.C., and Siddeley, G., eds., Epithermal gold mineralization of the Circum-Pacific: Geology, geochemistry, origin and exploration, II: Journal of Geochemical Exploration, v. 36, p. 445–474.
Xu, S.S., Nieto-Samaniego, Á.F., and Alaniz-Álvarez, S.A., 2004, Vertical shear mechanism of faulting and estimation of strain in the Sierra de San Miguelito, Mesa Central, Mexico: Geologica Acta, v. 2, p. 189–201.
Yamamoto, J., 1993, Actividad microsísmica en el área de Canatlán, Durango y su relación con la geología regional: Geofísica Internacional, v. 32, p. 501–510.
Young, E.J., Myers, A.T., Munson, E.L., and Conklin, N.M., 1969, Mineralogy and geochemistry of fluorapatite from Cerro de Mercado, Durango, Mexico: U.S. Geological Survey Professional Paper, v. 650-D, p. D84–D93.
Zárate-del Valle, P.F., 1982, Geología y análisis metalogénico de la Sierra de Catorce: Boletín de la Sociedad Geológica Mexicana, v. 43, p. 1–21.

Manuscript Accepted by the Society 29 August 2006

Geological Society of America
Special Paper 422
2007

The Cenozoic tectonic and magmatic evolution of southwestern México: Advances and problems of interpretation

Dante Jaime Morán-Zenteno*
Instituto de Geología, Universidad Nacional Autónoma de México, Ciudad Universitaria, Delegación Coyoacán, 04510 México D. F.

Mariano Cerca
Centro de Geociencias, Universidad Nacional Autónoma de México, Campus Juriquilla, Querétaro, Qro. 76230, México

John Duncan Keppie
Instituto de Geología, Universidad Nacional Autónoma de México, Ciudad Universitaria, Delegación Coyoacán, 04510 México D. F.

ABSTRACT

Recent advances in the knowledge of the Cenozoic structure and stratigraphy of southern México reveal a geological evolution characterized by Upper Cretaceous orogenic deformation, followed by truncation of the continental margin and gradual extinction of arc magmatism in the Sierra Madre del Sur, prior to the onset of magmatism in the Trans-Mexican Volcanic Belt. Orogenic deformation began in the Late Cretaceous and was coeval with the Laramide orogeny with structures of similar orientation. Deformation consisted of E-W shortening that migrated to the east with time and with a general easterly vergence. Models that relate the Laramide deformation to a decrease in the angle of subduction of the Farallon plate, which was converging in western México, cannot be applied in southern México because Paleocene to upper Eocene arc magmatism occurs near the inferred paleo-trench. An alternative possible origin due to collision of an insular arc against the western margin of México suffers from an absence of features and petrogenetic associations indicating the closure of an oceanic basin.

In light of recent geochronological data, the general pattern of magmatic extinction from Upper Cretaceous–Paleocene in Colima and Jalisco to the middle Miocene in central and southeastern Oaxaca presents some variations inconsistent with a simple pattern of extinction toward the E-SE.

Maastrichtian to lower Paleocene plutonism recognized in the Jalisco block and Manzanillo areas is contemporaneous with a magmatic episode that has some documented adakitic affinities in the central part of the Sierra Madre del Sur. Magmatism from the Paleocene to middle Eocene seems to be concentrated in the Presa del Infiernillo area, although isolated centers existed in areas such as Taxco or the eastern Jalisco block. Finally, the main axis of magmatism between the middle Eocene and Oligocene developed along what is the present-day continental margin and extends

*dantez@servidor.unam.mx

Morán-Zenteno, D.J., Cerca, M., and Keppie, J.D., 2007, The Cenozoic tectonic and magmatic evolution of southwestern México: Advances and problems of interpretation, *in* Alaniz-Álvarez, S.A., and Nieto-Samaniego, Á.F., eds., Geology of México: Celebrating the Centenary of the Geological Society of México: Geological Society of America Special Paper 422, p. 71–91, doi: 10.1130/2007.2422(03). For permission to copy, contact editing@geosociety.org.

200 km inland as a broad band. In general, the geochemical characteristics of this magmatism indicate a low degree of continental crustal assimilation.

Two episodes of principally sinistral lateral faulting that activated NW-SE– and later N-S–oriented faults, with variations in time and space, have been documented during the Eocene and lower Oligocene. The N-S set of faults was active only in the north of the Sierra Madre del Sur, whereas the activity of the NW-SE set continued during the Oligocene along the Oaxaca continental margin. The recognition of these deformational episodes suggests that extensional directions related to lateral faulting changed from NNW-SSE to NE-SW, and locally produced normal displacements on preexisting discontinuities.

Fundamental problems still exist in the interpretation of the plate tectonic processes that produced the stress regimes acting on the different sets of faults, as well as in the determination of the factors influencing the migration of magmatism. Some of the arguments used to postulate the presence of the Chortis block off the southwestern Mexican continental margin during the early Cenozoic are uncertain. On the other hand, models that explain restricted displacements of the Chortis block with respect to the Maya block—without juxtaposition with the southwestern margin of México—suggest that continental truncation was essentially caused by subduction erosion and leave open the interpretation of the observed magmatic migration.

Keywords: southern México, arc magmatism, strike-slip tectonics, Cenozoic.

INTRODUCTION

Knowledge about the structure and Cenozoic evolution of southern México has experienced significant progress during the past few years. Although there are still areas lacking detailed geological description, general tectonic models have been generated based upon: (1) observations and data from key areas in the continental crust; and (2) inferences about the kinematics of the adjacent oceanic plates. Although there are important controversies about significant aspects of the geological evolution of this region, growth of the available data has allowed more profound discussion. One of the more important ideas initiated in the 1980s was the recognition of different pre-Mesozoic basement rocks, based on petrochemical and geochronological data, that define a mosaic of contrasting terranes (Ortega-Gutiérrez, 1981; Campa and Coney, 1983; Sedlock et al., 1993; Keppie, 2004) (Fig. 1), which allowed certain aspects of the Mesozoic and Cenozoic of southern México to be better understood. This was augmented by the recognition and general description of the Cenozoic stratigraphy of various areas of southern México (e.g., Fries, 1960, 1966; De Cserna and Fries, 1981; Ferrusquía-Villafranca, 1976, 1992; Martiny et al., 2000), associated tectonics (Ratschbacher et al., 1991; Tolson et al., 1993; Riller et al., 1992; Meschede et al., 1996; Nieto-Samaniego et al., 1995; Alaniz-Álvarez et al., 2002; Cerca et al., 2004), and geochemical and geochronological data that have permitted petrogenetic analysis and definition of space-time relationships of the magmatic activity (e.g., Herrmann et al., 1994; Schaaf et al., 1995; Morán-Zenteno et al., 1999, 2004; Martiny et al., 2000; Ferrari et al., 1999; Meza-Figueroa et al., 2003; González-Partida et al., 2003; Ducea et al., 2004a). Also important for the interpretation of the geology of southern México are studies about the structure, age, and kinematics of the adjacent oceanic plates (Fig. 1). This information forms the basis for Cenozoic tectonic models of this region (e.g., Malfait and Dinkelman, 1972; Anderson and Schmidt, 1983; Ross and Scotese, 1988; Pindell et al., 1988; Pindell and Barret, 1990; Keppie and Morán-Zenteno, 2005).

A wide variety of volcanic, plutonic, and continental sedimentary rocks in southern México is spatially related to extensional basins associated with transcurrent faults, some of which are reactivated faults. The process of uplift and erosion that initiated toward the end of the Cretaceous caused the deposition of continental sequences that are only locally preserved.

Structural and magmatic studies of rocks south and north of the Trans-Mexican Volcanic Belt show some similarities; however, there are important differences that appear to relate to the underlying basement and to Pacific plate tectonics. Although not studied in detail, differences in structural style during magmatism south and north of the Trans-Mexican Volcanic Belt imply tectonic differences during the Eocene and Oligocene (e.g., Nieto-Samaniego et al., 1999; Ferrari et al., 2002; Alaniz-Álvarez et al., 2002; Morán-Zenteno et al., 2004).

This paper reviews the present state of knowledge about the Cenozoic evolution of southern México, and establishes the main tectonic and magmatic episodes in the region. The review focuses mainly on the mountainous regions of the Sierra Madre del Sur between the Trans-Mexican Volcanic Belt and the Pacific Coast, and bounded on the east by the Isthmus of Tehuantepec and the Sierra de Juarez (Juarez terrane) (Fig. 1). It excludes the Trans-Mexican Volcanic Belt, which forms the topic of another paper (Gomez-Tuena et al., this book), and Chiapas and the Yucatan Peninsula, which are related to the opening of the Gulf of Mexico and require

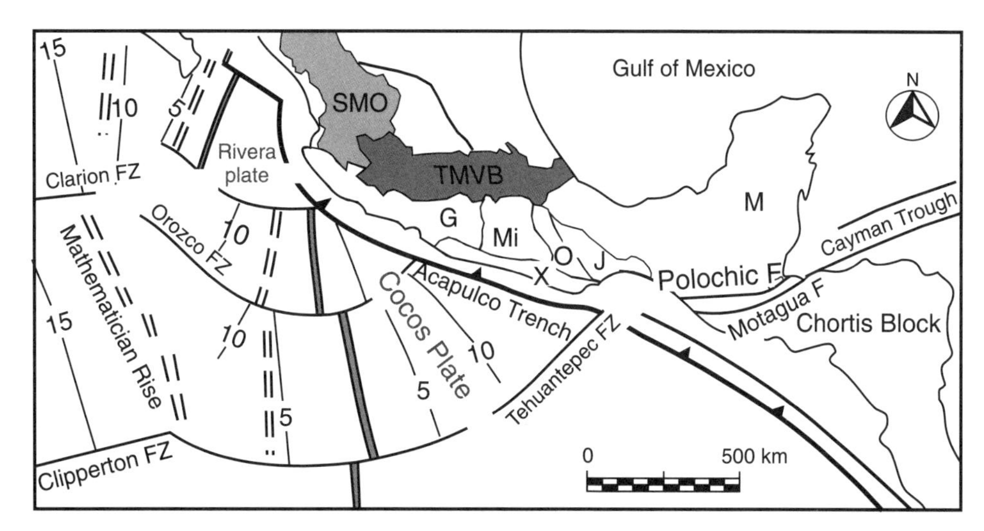

Figure 1. Main tectonic features of the present-day oceanic framework around southern México. The distribution of pre-Cenozoic tectonostratigraphic terranes in continental México (modified after Campa and Coney, 1983) and major volcanic arcs are also shown. SMO—Sierra Madre Occidental; TMVB—Trans-Mexican Volcanic Belt; G—Guerrero terrane; Mi—Mixteca terrane; O—Oaxaca terrane; J—Juarez terrane; X—Xolapa terrane; M—Maya terrane. FZ—fracture zone. Numbers 5, 10, and 15 indicate isochrones in Ma.

separate treatment. Although this paper is not an exhaustive review, it does review the main interpretations using relevant geochemical and age data found in various papers (i.e., Schaaf et al., 1995; Morán-Zenteno et al., 1999; Garduño-Monroy et al., 1999; Ferrari et al., 1999; and Western North American volcanic and intrusive rock database, http://navdat.geongrid.org/).

CRUSTAL STRUCTURE AND GENERAL TECTONIC FRAMEWORK

Based on petrologic, tectonic, and geochronological contrasts in the Precambrian and Paleozoic basement, as well as on the petrotectonic nature of Mesozoic terranes bordering the old continental nucleus of southern México, the region has been divided into five tectonostratigraphic terranes: Oaxaca, Mixteca, Guerrero, Xolapa, and Juarez (Campa and Coney, 1983) (Fig. 1). Sedlock et al. (1993) later introduced a similar division with some variations in the position of contacts between terranes. The tectonic nature of these contacts has been the matter of controversy (i.e., Sedlock et al., 1993; Centeno-García et al., 1993a; Lang et al., 1996; Freydier et al., 1996; Elías Herrera et al., 2000; Cabral-Cano et al., 2000a, 2000b; Keppie, 2004), however, there is a general consensus concerning the main lithological contrasts. Interpretations of the structure of the crust, based on geophysical data, are scarce. Modeling of the crustal structure, based on magnetotelluric profiles, shows some significant lateral changes in electric conductivity that do not necessarily coincide with proposed terrane boundaries (Jording et al., 2000).

The Guerrero and Juarez terranes, both containing a thick Cretaceous succession, mark the western and eastern flanks, respectively, of a pre-Mesozoic nucleus composed of the metamorphic basements of the Mixteca and Oaxaca terranes with a thinner Cretaceous cover. The N-S tectonic boundaries between these terranes are truncated to the south by the metamorphic-plutonic Xolapa terrane, which occupies the continental margin. One of the most striking tectonic features of southern México is the truncated and exhumed character of the continental margin (Bellon et al., 1982; Morán-Zenteno et al., 1996). The extensive exposures of plutons from Manzanillo to the Isthmus of Tehuantepec, as well as the presence of the Late Cretaceous Puerto Vallarta batholith in the Jalisco region (Fig. 2)—compared with the inland regions where Cretaceous and Cenozoic sedimentary, volcano-sedimentary, and volcanic sequences are extensively exposed—reflect the differential uplift of the continental margin. The distribution of arc plutonic rocks along the coast and offshore near the trench (Bellon et al., 1982) indicates not only the exhumed character of the margin but the relatively recent landward advance of the trench.

The truncation of the continental margin of southwestern México has generally been attributed to the detachment and displacement of the Chortis block (Malfait and Dinkelman, 1972; Ross and Scotese, 1988; Pindell et al., 1988; Schaaf et al., 1995) (Fig. 3A). Southward migration of the arc magmatism in the Sierra Madre del Sur and occurrence of NW-trending shear zones with left lateral components have been invoked in support of this interpretation (Herrmann et al., 1994; Schaaf et al., 1995; Morán-Zenteno et al., 1996). However, the average rotation pole of the Caribbean plate with respect to North America (Pindell et al., 1988) is incompatible with such displacement if the Chortis was attached to the Caribbean plate since Eocene-Oligocene time.

The low rate of sediment accumulation in the trench indicates active subduction erosion along the continental margin (Ducea

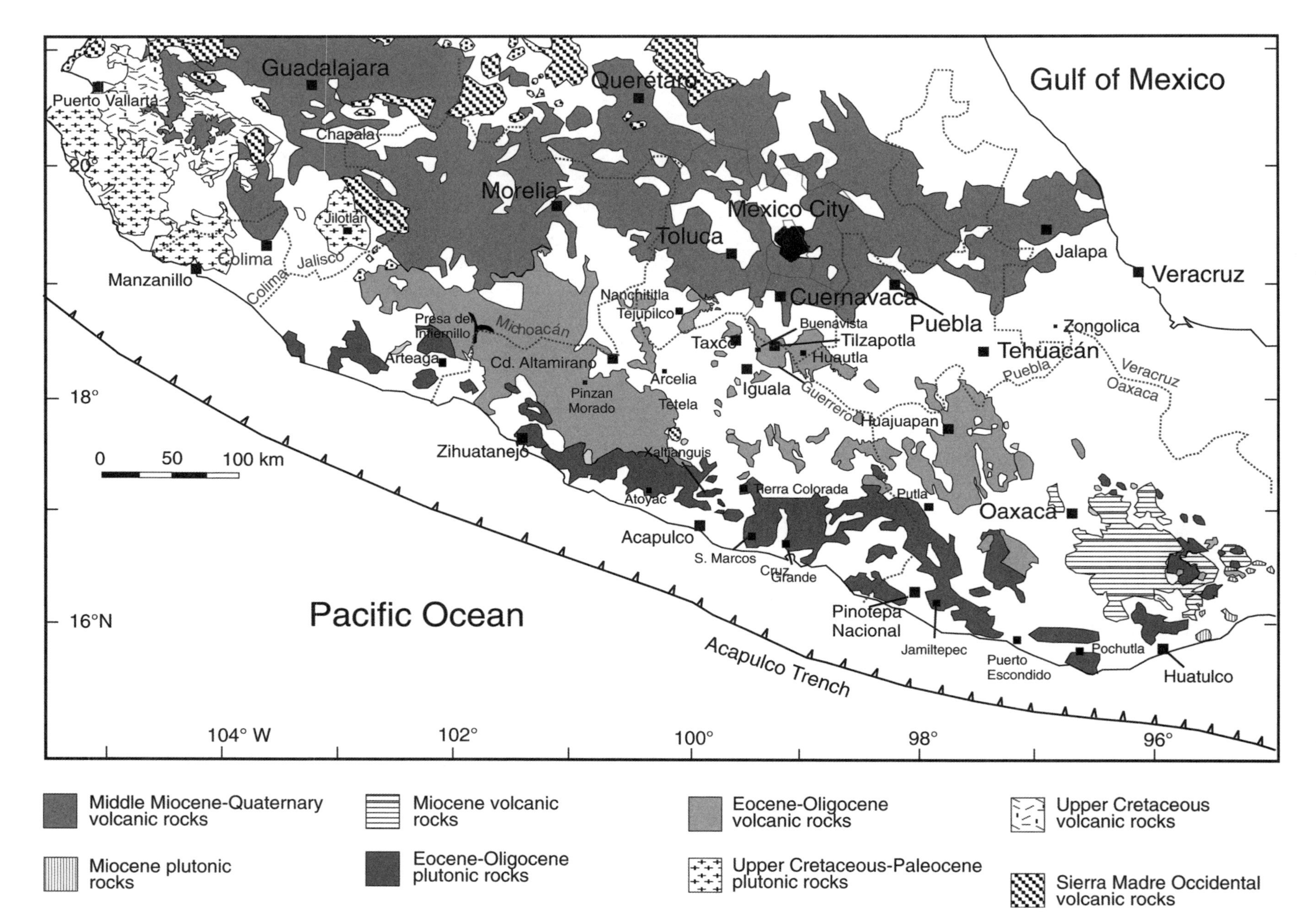

Figure 2. Schematic geologic map of southern México showing the distribution of the Cenozoic magmatic rocks (modified after Ortega-Gutiérrez et al., 1992, and Morán-Zenteno et al., 1999).

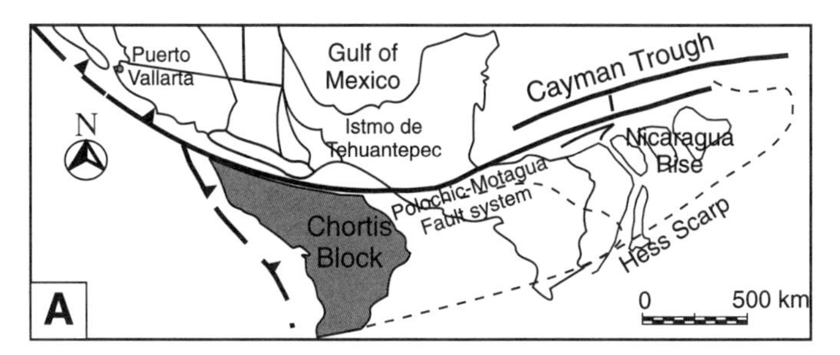

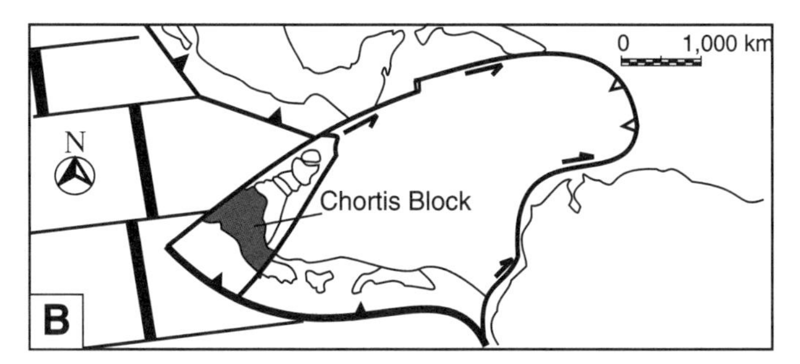

Figure 3. Tectonic restorations for the Eocene showing the alternative positions suggested for the Chortis block (in gray): (A) restoration based on the patterns of magmatic extinction documented by the isotopic age database proposed by Schaaf et al. (1995) using the restoration of Ross and Scotese (1988); and (B) alternative restoration proposed by Keppie and Morán-Zenteno (2005) based on an average rotation pole calculated for the Caribbean plate in the Eocene, which is located near the present-day position of Santiago de Chile (Pindell et al., 1988): The pole agrees with the general trace of the proposed limit between the Caribbean and North American plates. Note that the western end of the Polochic-Motagua fault system is characterized by concavity toward the north that clearly differs from the trace of the Cayman trough—and was probably caused by the recent (Neogene) transpressive interaction of the Chortis block and the Chiapas segment of the Yucatan block.

et al., 2004b). Modeling based on seismic profiles along the coast of eastern Guerrero and Oaxaca shows the continental crust thickness ranges from 15 to 20 km (Nava et al., 1988). Seismic models of the geometry of the subduction zone indicate a broad contact zone between the continental crust and the subducted slab of the Cocos plate, at least for 100 km inland from the coast (Valdés et al., 1986; GEOLIMEX Working Group, 1994; Pardo and Suárez, 1995). This interpretation suggests that significant changes in the geometry of subduction occurred after the extinction of the magmatic arc represented by the Cenozoic plutons of the continental margin.

The shallowing of the subduction zone and the landward advance of the trench imply the removal of a segment of lithospheric mantle and the lowest part of the continental crust. Gravity modeling carried out by Urrutia-Fucugauchi and Flores-Ruiz (1996), and seismic modeling by Valdés et al. (1986), estimates that the thickness of the continental crust between the coast and the Trans-Mexican Volcanic Belt varies from 20 to 45 km.

At present the Cocos and Rivera plates are subducting along the Acapulco trench (Fig. 1). The rotation pole of the Cocos plate with respect to North America is located in front of the Pacific Coast of Baja California. The convergence rate varies southeastward from 5.5 to 6.8 cm/yr (Pardo and Suárez, 1995; DeMets and Traylen, 2000). The Rivera plate subducts beneath the North American plate in the Jalisco segment about a rotation pole located at the mouth of the Gulf of California. The calculated rate of convergence of Manzanillo is 3.8 cm/yr (De Mets and Traylen, 2000). The trench is intersected by the Rivera, Orozco, and O'Gorman fracture zones (Fig. 1), the former representing the boundary of the Cocos and Rivera plates and producing inland extension parallel to the coast (Ferrari et al., 1994; Bandy et al., 1995). No tectonic features have been recognized along the inland projections of the Orozco and O'Gorman fracture zones. Off the Isthmus of Tehuantepec, the trench is intersected by the Tehuantepec fracture zone, which is close to the proposed diffuse triple junction between the Caribbean, North American, and Cocos plates. The dip of the subduction zone changes from 10 to 45° in this area (Pardo and Suárez, 1995). Beneath the Isthmus of Tehuantepec there is a well-defined linear zone of frequent seismicity perpendicular to the trench (Ponce et al., 1992). Cenozoic faults with left lateral displacement, which have been reported in southeastern Oaxaca and Chiapas, have been related to the Miocene to Recent evolution of the trench-trench-transform triple junction (Delgado-Argote and Carballido-Sánchez, 1990; Tolson-Jones, 1998; Guzmán-Speziale and Meneses-Rocha, 2000). In the same region, Barrier et al. (1998) identified a N-trending fault zone that they related to the influence of the subducted Tehuantepec fracture zone and the lateral change in the angle of subduction. These authors also relate the recent N-S extension in the Isthmus of Tehuantepec to the eastward displacement of the Caribbean plate.

Reconstructions based on the pattern and age of ocean floor magnetic anomalies show that the kinematics of the oceanic lithosphere segments off southwestern México underwent changes related to episodic jumps of different segments of the Pacific-Farallon ridge at 25, 12.5–11, and 6.5–3.5 Ma (Mammerickx and Klitgord, 1982). These episodes were related to changes in the rotation poles and produced the segmentation of the Farallon plate into the Cocos and Rivera plates. There is no unequivocal evidence that might support a connection between the fragmentation of the Farallon plate with a change in the geometry of subducted lithosphere beneath southwestern México, but the relationship with variations in the convergence rates suggests that they are somehow linked.

LATE CRETACEOUS OROGENIC DEFORMATION AND ESTABLISHMENT OF CONTINENTAL REGIME

During the Late Cretaceous, a major event of progressive shortening with eastward and northeastward tectonic transport produced folding and thrusting of Mesozoic marine and older sequences. Based on the time of deformation and structural style, this event has been generally related to the Laramide orogeny of the North American Cordillera (Campa et al., 1976; Campa and Ramírez, 1979; Salinas-Prieto et al., 2000; Dickinson and

Lawton, 2001). In the Sierra Madre del Sur, evidence of shortening is present in a broad area extending from Michoacán to Veracruz states, but it has been recently documented in more detail in the Guerrero state (Lang et al., 1996; Cabral-Cano et al., 2000a; Salinas-Prieto et al., 2000; Elías-Herrera et al., 2000).

The dominant structural style in southern México consists of N-S–trending, E-verging folds and low-angle thrusts associated with some strike-slip faults. There are also some structures with the opposite vergence in the Teloloapan region that have been interpreted as part of a process of progressive shortening (Salinas-Prieto et al., 2000). Opposite vergence structures in the Guerrero-Morelos platform have been considered to be the result of lateral variations in the competency between the Mesozoic marine sequences and the metamorphic Acatlán Complex (Cerca et al., 2004).

Differences in the structural style associated with lateral changes in the lithology and topography led to a division of the shortening structures in the Sierra Madre del Sur into three zones (Fig. 4). The western zone from the Arcelia area to the middle zone of the Guerrero Morelos platform, east of the Teloloapan thrust, includes N-S–trending low-angle structures. The main structures are expressed in the Cretaceous cover. In some specific zones, for example at Tejupilco, State of México (Elías-Herrera et al., 2000), at Pinzón-Morado, Guerrero (Montiel-Escobar et al., 2000), and at Arteaga, Michoacán (Centeno-García et al., 1993b), there are exposed pre-Cretaceous sedimentary and metamorphic rocks that constitute basement tectonic highs. The area west of Arcelia is made up of extensive terrestrial deposits that range in age from the latest Cretaceous to the Paleogene (Jansma and Lang, 1997). In the Tiquicheo region, Benammi et al. (2005) recently reported Late Cretaceous dinosaur remains in terrestrial deposits unconformably overlying folded sequences of the Guerrero terrane, indicating that orogenic deformation ended before the end of the Cretaceous in that region.

In contrast with the W, the central zone is related to extensive outcrops of a pre-Mesozoic metamorphic nucleus (Fig. 4). This zone extends from the eastern structures of the Guerrero Morelos

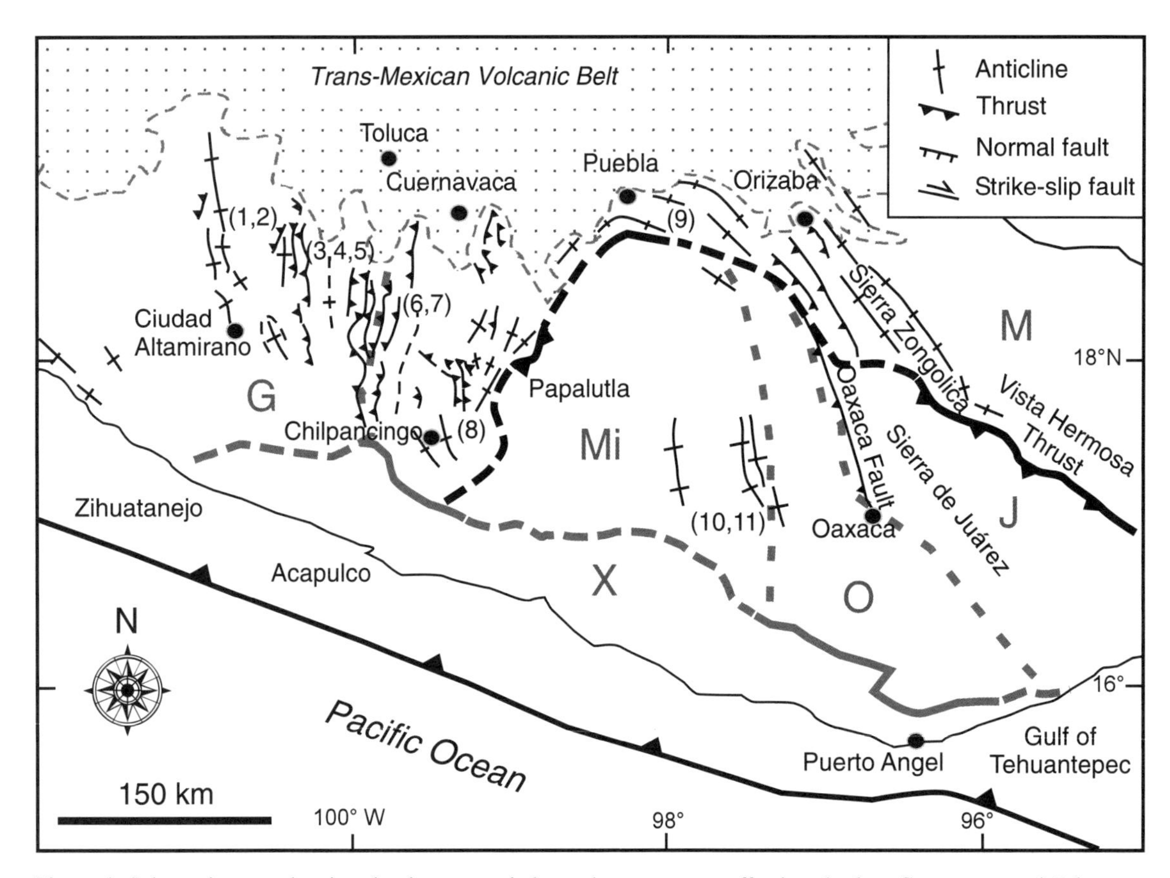

Figure 4. Schematic map showing the documented shortening structures affecting the late Cretaceous and Paleogene rocks in southern México (after Cerca et al., 2004). Gray discontinuous lines indicate the limits of the tectonostratigraphic terranes modified after Campa and Coney (1983) and Sedlock et al. (1993). Terrane nomenclature is indicated with capital letters: G—Guerrero; Mi—Mixteca; O—Oaxaca; J—Juarez; M—Maya; and X—Xolapa. Black dotted lines indicate the approximate limits of pre-Mesozoic rocks, whereas discontinuous lines indicate areas where Paleozoic rocks thrust over Mesozoic rocks. Structures were compiled from the following studies: (1) Montiel-Escobar et al. (2000); (2) Altamira-Areyán (2002); (3) Elías-Herrera et al. (2000); (4) Salinas-Prieto et al. (2000); (5) Cabral-Cano et al. (2000b); (6) Rivera-Carranza et al. (1998); (7) Campa-Uranga et al. (1997); (8) Cerca (2004); (9) Monroy and Sosa (1984); (10) Meneses-Rocha et al. (1994); and (11) González-Ramos et al. (2001), and references therein.

platform to metamorphic rocks exposed east of the Oaxaca fault. Anomalies in the pattern of N-S–oriented shortening structures observed in the central zone probably originated from vertical and lateral displacements during the orogenic deformation that, in this area, was Paleogene in age (Cerca et al., 2004). According to these authors, the mechanical contrast between the metamorphic block and adjacent sedimentary sequences concentrated the deformation at boundaries of this block. Furthermore, the geometry of the folds and thrusts delineates a north-convex arcuate promontory with structures displaying an outward radial vergence and a decrease in shortening toward the center of the block.

The eastern zone comprises shortening structures of the Sierra de Juárez, Sierra de Zongolica, and the Vistahermosa thrust (Fig. 4). The dominant structural style consists of a NW-trending fold-and-thrust belt with a northeastward vergence. Orogenic deformation in this zone lasted into the Eocene (Mossman and Viniegra, 1976; Meneses-Rocha et al., 1996; Eguiluz de Antuñano et al., 2000).

The Laramide deformation in the North American Cordillera has been bracketed within the 75–35 Ma interval (Dickinson et al., 1988; Bird, 1998) with a peak in deformation during the Eocene for the orogenic front of the Sierra Madre Oriental (Eguiluz de Antuñano et al., 2000, and references therein). In southern México, the beginning of shortening in the western zone has been associated with the terrigenous drowning of carbonate platforms at the Cenomanian-Turonian boundary and the appearance of siliciclastic sedimentation (ca. 93 Ma) (Lang and Frerichs, 1998). This is consistent with the age of the youngest folded volcanic rocks in the western zone, which range in age from 93 to 88 Ma in the Teloloapan and Arcelia areas (Delgado-Argote et al., 1992; Elías-Herrera et al., 2000; Cabral-Cano et al., 2000a).

Estimated ages for the end of the deformation show a migration from W to E. In the western zone, reported dates for post-orogenic magmatic rocks are as old as Maastrichtian in age (Ortega-Gutiérrez, 1980; González-Partida et al., 2003; Levresse et al., 2004; Cerca et al., 2004). Also in this zone, are some structures like the Tzitzio structure, whose kinematics and age seem to be anomalous with respect to the general style: Palinological data suggest that the continental beds involved in the Tzitzio fold are as young as Paleocene in age (Altamira-Areyán, 2002). To the E, in the Zongolica area and farther E, deformation continued until the late Eocene (Eguiluz de Antuñano et al., 2000). As Laramide deformation progressed to the E, sedimentary marine deposits gradually changed to continental deposits and subaerial volcanism accumulated in intermontane basins (Figs. 5 and 6).

MAASTRICHTIAN TO PALEOCENE MAGMATISM AND CONTINENTAL SEDIMENTATION

Recent geochronologic information indicates that, during the Maastrichtian and Paleocene, arc magmatism had a broader distribution in southwestern México than previously estimated. Two zones can be delineated for the magmatism of this age interval. One of them is represented by extensive batholiths and some volcanic units in a broad zone of Jalisco, Colima, and Michoacán regions (Schaaf et al., 1995; Morán-Zenteno et al., 1999, and references therein). To the E, there are reports of some intrusions and volcanic units between longitudes 99°W and 100°W in Guerrero and México states (Ortega-Gutiérrez, 1980; Morán-Zenteno et al., 1999; Meza-Figueroa et al., 2003; González-Partida et al., 2003; Cerca, 2004) (Fig. 7A).

The earliest descriptions of the basal beds of the Balsas Group considered them to be Late Cretaceous in age (Fries, 1960), an inference confirmed by Maastrichtian to late Eocene age dates in continental deposits intercalated with volcanic rocks of the Tetelcingo Formation (lowest part of the Balsas Group) in the western region of the Sierra Madre del Sur (Fries, 1960; Ortega-Gutiérrez, 1980) (Figs. 5 and 6). These deposits originated in continental basins bounded by Laramide folds and thrusts, as well as by strike-slip and normal faults active during deposition (Figs. 5 and 6). Volcanic and plutonic rocks ranging in age from 67 to 55 Ma are distributed from the Mezcala area to Acapulco (Ortega-Gutiérrez, 1980; Morán-Zenteno, 1992; Meza-Figueroa et al., 2003; González-Partida et al., 2003; Tritlla et al., 2003; Ducea et al., 2004a) (Fig. 7A). These rocks indicate that a magmatic arc episode developed in this region at the end of Laramide shortening.

In the Mezcala area (Fig. 7A), Late Cretaceous–Paleocene intrusive rocks are distributed along NW-trending lineaments that offset Laramide structures. ^{40}Ar-^{39}Ar (biotite) and U-Pb (zircon populations) dating carried out in these rocks yielded ages ranging from 68 to 60 Ma (Meza-Figueroa et al., 2003; Levresse et al., 2004). Some of the geochemical peculiarities of these rocks have been related to adakitic magmatism which resulted from the partial fusion of the subducted slab at the amphibolite-eclogite transition and its subsequent reaction with the mantle wedge after a period of subhorizontal subduction (González-Partida et al., 2003). Meza-Figueroa et al. (2003) also identified geochemical characteristics in these rocks, which indicate a contrasting petrogenesis with respect to later Cenozoic magmatism in the region. Similar ages have been reported for a volcano-sedimentary continental unit in the nearby Balsas River. Ortega-Gutiérrez (1980) named this unit the Tetelcingo Formation and divided the sequence into three members dominated by volcanic breccias and tuffs, lava flows and breccias, and volcanic breccias, respectively. There are also some intercalated conglomerates and lacustrine deposits. Lava varies in composition from andesite to rhyolite (De Cserna et al., 1980). Reported K-Ar ages for the base and the middle part of the sequence are 68.8 ± 2.4 and 66 ± 2.3, respectively (Ortega-Gutiérrez, 1980).

Similar ages have been reported in plutons of the Peña Colorada area, Colima (Tritlla et al., 2003, and references therein); in volcanic rocks of the northern Zihuatanejo area, Guerrero (Garduño-Monroy et al., 1999); in extensive plutons of the Jalisco block (Schaaf et al., 1995); and in some plutons of the Acapulco area (Morán-Zenteno, 1992; Ducea et al., 2004a) (Fig. 7A).

Due to the relatively restricted outcrops of Late Cretaceous–Paleocene rocks, the volume of magmatism is not well known,

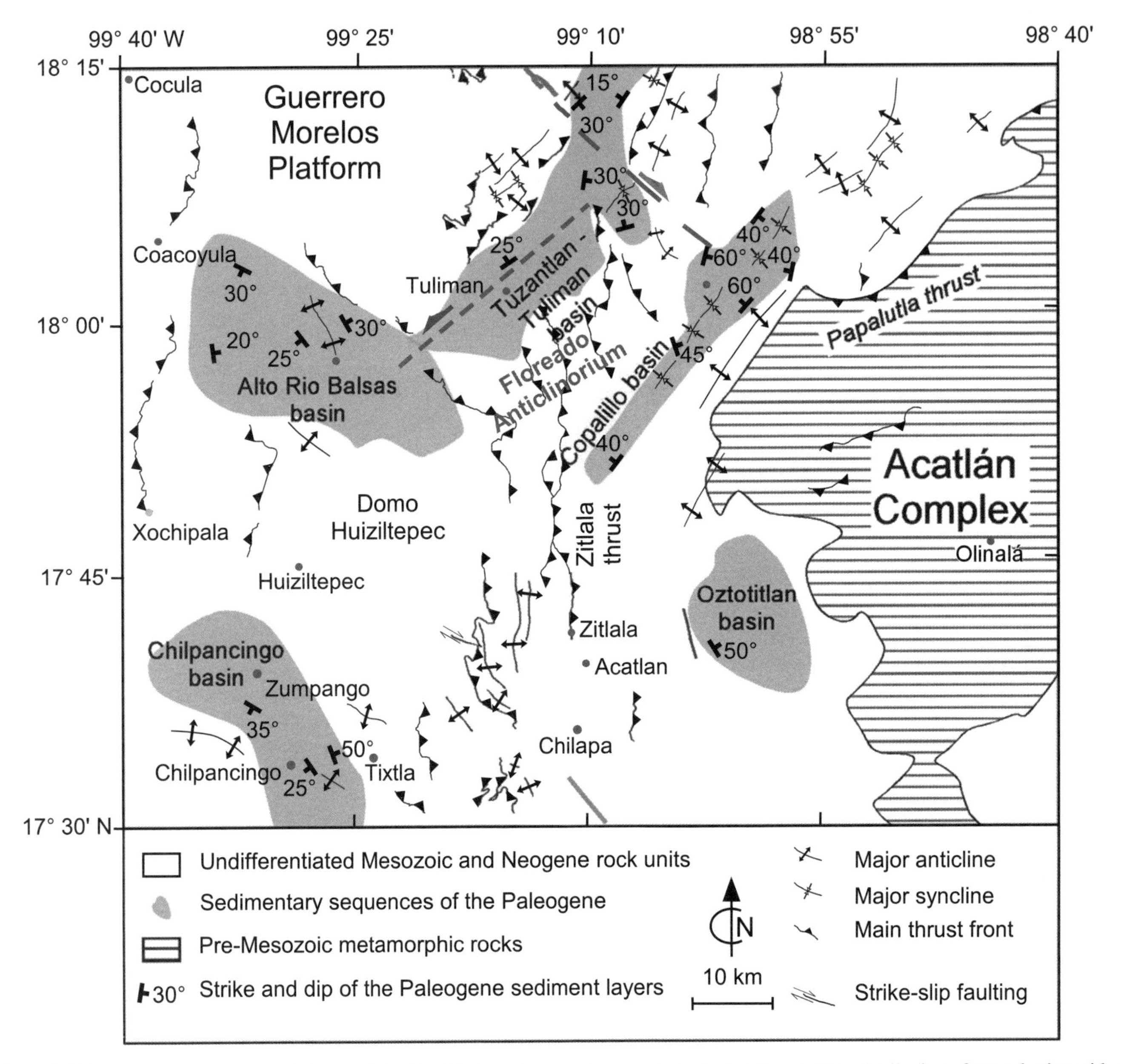

Figure 5. Schematic structural map of the zone W of the Mixteca terrane (Guerrero Morelos Platform). The distribution of some basins with a continental Paleogene sedimentary record is shown in gray. These basins are bounded by shortening structures and are affected by lateral displacement faults (modified after Cerca et al., 2004).

but the broad area where it is distributed suggests a regional magmatic episode. In the Mezcala area, the Paleocene intrusions define NW-trending lineaments (González-Partida et al., 2003), obliquely oriented with respect to N-S–trending Laramide shortening structures. Fracture zones with a similar orientation in the northern part of the Sierra Madre del Sur show left-lateral displacements of Eocene age (Morán-Zenteno et al., 1999; Alaniz-Álvarez et al., 2002). If the emplacement of volcanic rocks along strike-slip faults is a significant feature of the Maastrichtian and Paleocene time, it could have important implications for interpretation of terminal Laramide deformation. It is tentatively suggested that following the Laramide deformation, the increase of the subduction angle was associated with the segmentation of the subducting slab produced by the changes in the convergence rate. The subducted boundaries of the oceanic plate segments could have induced discontinuities in an orientation perpendicular to the trench in the overriding plate that, in part, could have controlled the emplacement of arc magmatic rocks (Cerca, 2004). However, an increase in the subduction angle does not satisfactorily explain the southeastward migration in the age of the plutons along the southwest coast since Late Cretaceous, nor the occurrence of two magmatic arc zones located at different distances from a pos-

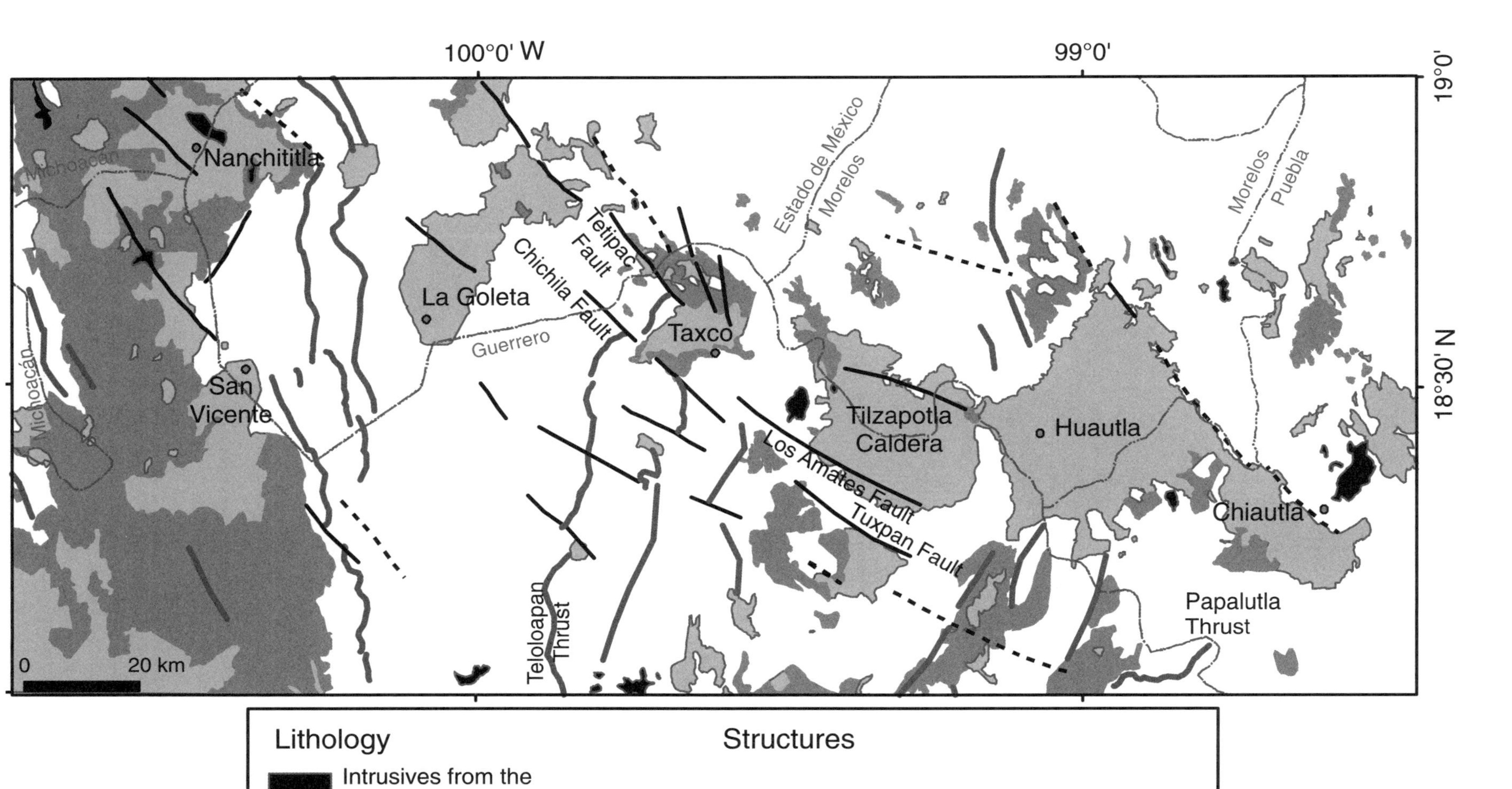

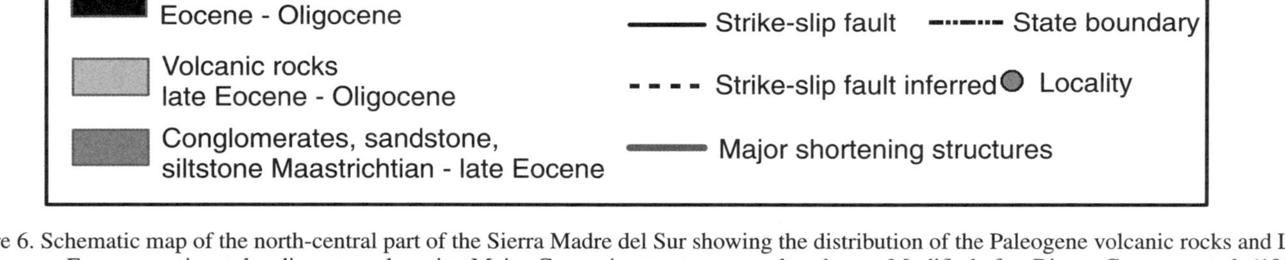

Figure 6. Schematic map of the north-central part of the Sierra Madre del Sur showing the distribution of the Paleogene volcanic rocks and Late Cretaceous–Eocene continental sedimentary deposits. Major Cenozoic structures are also shown. Modified after Rivera-Carranza et al. (1998), and includes data from Alaníz-Álvarez et al. (2002) and Morán-Zenteno et al. (2004).

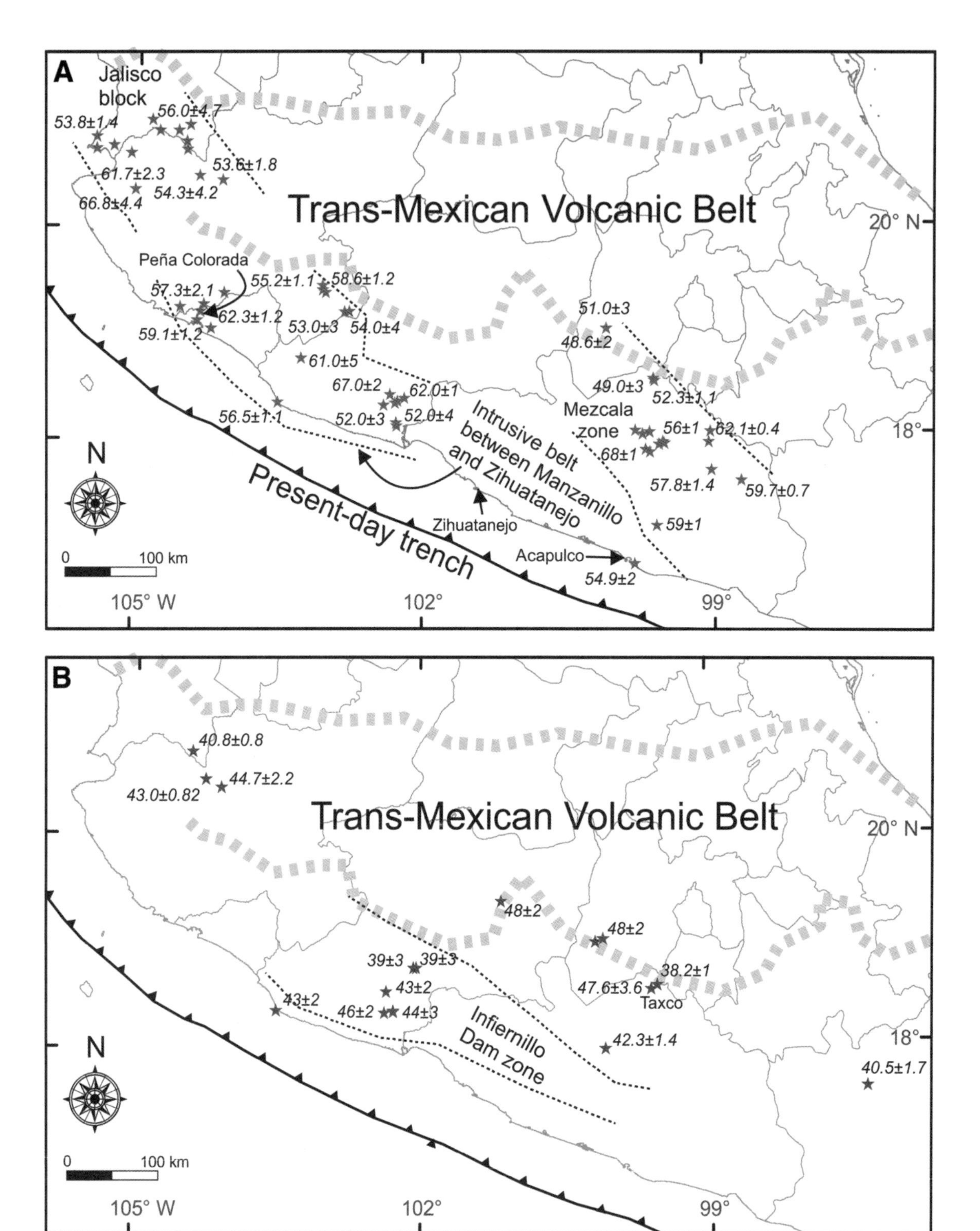

Figure 7 (*on this and following page*). Spatial and temporal distribution of the Tertiary magmatic rocks in southern México. The compiled age database is presented in intervals of (A) 68–48 Ma, (B) 48–38 Ma, and (C) 38–28 Ma. Available ages have been obtained by diverse methods, which have different interpretations and limitations (K-Ar, Ar-Ar, Rb-Sr, U-Pb). In spite of the limitations, analysis of the database allows delineation of general patterns of the distribution of the thermal-magmatic events in southern México during the Cenozoic. Some examples of the data are shown in the age labels. The compiled database was taken from previous compilations such as: Schaaf et al. (1995); Morán-Zenteno et al. (1999); Ferrari et al. (1999); Garduño-Monroy et al. (1999); Cerca (2004); NAVDAT (2005); and from the studies by Meza-Figueroa et al. (2003); Ducea et al. (2004a); Morán-Zenteno et al. (2004); and Levresse et al. (2004). Stars indicate localities with dated rocks, fine dashed lines delimit the probable areas of distribution of magmatic activity for every time interval.

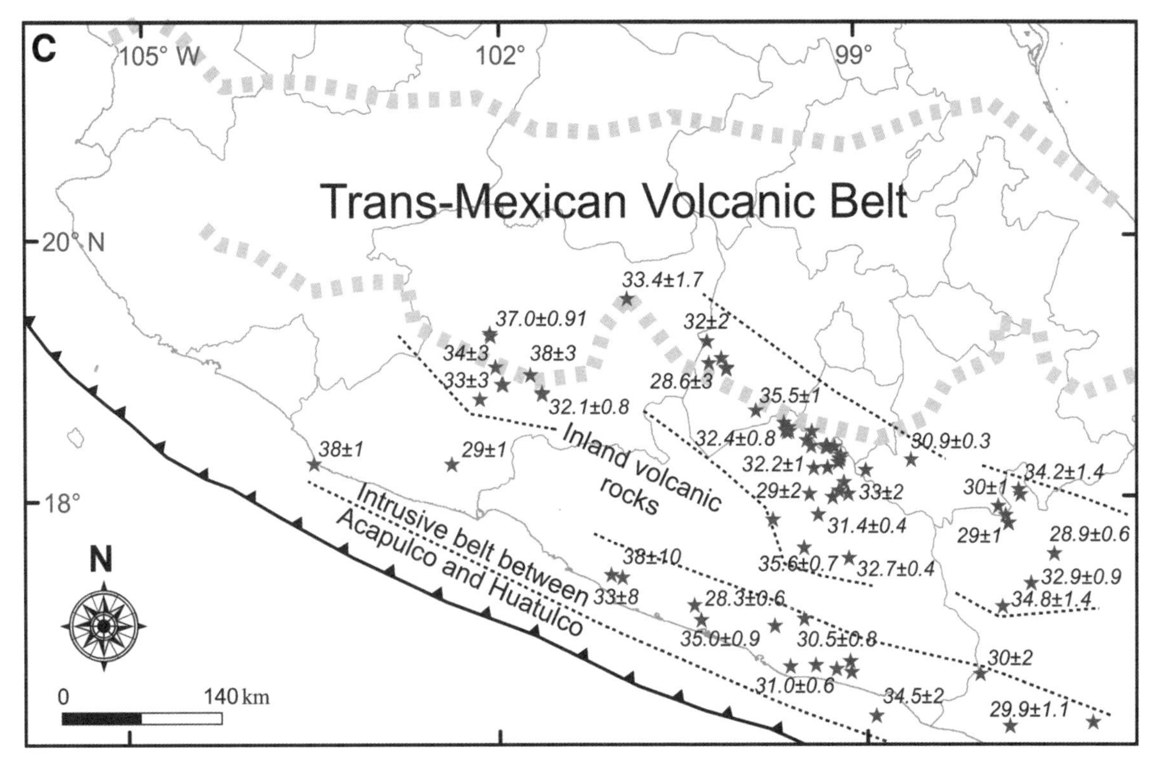

Figure 7 (*continued*).

sible N-S–oriented paleo-trench in the Paleocene–early Eocene (Fig. 7A). The occurrence of these two zones is more compatible with a NW-trending trench nearly parallel to the present trench, which would include the two zones within a discontinuous arc. This scenario raises some doubts about placing the Chortis block off southwestern México at this time. Furthermore, this alternative does not clearly explain the southeastward migration of the Cenozoic magmatism along the margin or the N-trending orientation of the Laramide structures.

CENOZOIC PLUTONISM AND EVOLUTION OF THE CONTINENTAL MARGIN

The Pacific coastal zone of southwestern México is characterized by extensive outcrops of Late Cretaceous and Paleogene plutons with abundant batholithic-sized bodies (Fig. 2). About 50% of the 100-km-wide coastal zone is occupied by outcrops of these plutons, which form a total outcrop area of up to 30,000 km^2. Information collected from drilling sites between the coast and the trench (Leg 66 of the Deep Sea Drilling Project) (Bellon et al., 1982), as well as from submarine samples collected off the Jalisco region (Mercier de Lépinay et al., 1997), indicate that similar plutons extend to the trench. There are also older deformed plutons that have been attributed to the prebatholitic Xolapa metamorphic complex (Ortega-Gutiérrez, 1981) whose age has been the subject of controversy (Robinson et al., 1989; Morán-Zenteno, 1992; Herrmann et al., 1994; Ducea et al., 2004a).

The composition of Cenozoic plutons along the coast is mostly silicic and subalkaline with granodioritc and tonalitic compositions dominating over granite compositions (Schaaf et al., 1995; Morán-Zenteno et al., 1999). The main accessory minerals are hornblende and biotite. In some segments, such as in the Manzanillo area, gabbroic and dioritic plutons are especially abundant (Schaaf, 1990) (Fig. 2). $^{87}Sr/^{86}Sr$, $^{143}Nd/^{144}Nd$, and common Pb isotopic variability of Cenozoic magmatic rocks in the Sierra Madre del Sur is relatively low compared with other arc magmatic rocks of the North American Cordillera. Initial $^{87}Sr/^{86}Sr$ typically range from 0.7035 to 0.7055 (Fig. 8) with most values falling between 0.7039 and 0.7046, whereas εNd vary typically from +1 to +3 (Morán-Zenteno et al., 1999, and references therein). Extreme high values of up to +5 have been reported for plutons located between Zihuatanejo and Manzanillo, as well as in the Tilzapotla volcanic center in the State of Morelos (Fig. 2) (Schaaf, 1990; Morán-Zenteno et al., 2004). Negative values up to –3 have been reported in the Puerto Escondido region (Hernández-Bernal and Morán-Zenteno, 1996). In contrast, the Cretaceous batholith contains granitiods with lower εNd Values (~–7) and higher $^{87}Sr/^{86}Sr$ values (Schaaf, 1990). Pb isotopic signatures of Cenozoic magmatic rocks in southwestern México typically vary within the following ranges: $^{206}Pb/^{204}Pb$ = 18.65–18.97, $^{207}Pb/^{204}Pb$ = 15.55–15.65, $^{208}Pb/^{204}Pb$ = 38.25–38.82 (Fig. 9) (Herrmann et al., 1994; Martiny et al., 2000; Martiny et al., 2002; Morán-Zenteno et al., 2004). Sr isotopic signa-

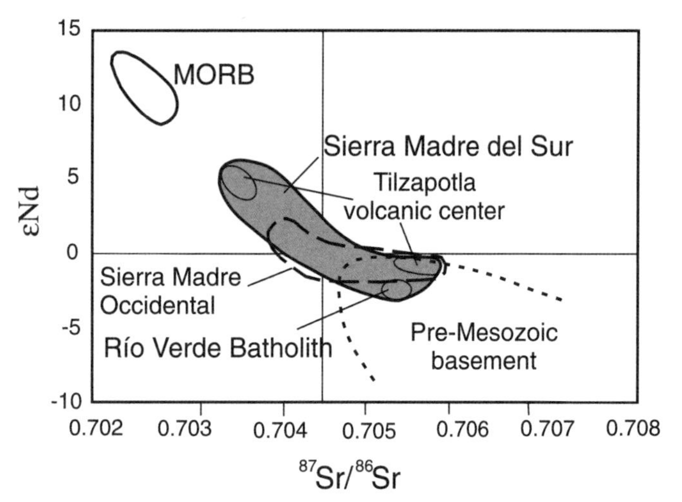

Figure 8. Distribution of the initial isotopic values of the magmatic rocks of southern México in a $^{87}Sr/^{86}Sr_{(i)}$ versus $\varepsilon Nd_{(i)}$ diagram. The typical distribution field of the volcanic rocks of the Sierra Madre Occidental and some areas in southern México with extreme values are shown for reference. Data compiled from Schaaf (1990), Schaaf et al. (1995, and references therein), Morán-Zenteno et al. (1999, 2004, and references therein), Martiny et al. (2000, and references therein). MORB—mid-oceanic-ridge basalt.

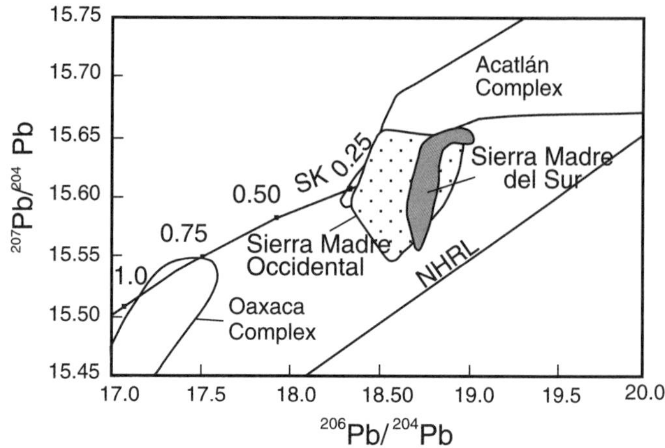

Figure 9. Distribution fields of $^{207}Pb/^{204}Pb$ and $^{206}Pb/^{204}Pb$ for the magmatic rocks of southern México. The typical distribution field of the Pb isotopic relations of the Oaxacan and Acatlán Complexes and the rocks of the Sierra Madre Occidental are shown for reference. Data compiled from Martiny et al. (2000), Smith et al. (1996), and Albrecht and Goldstein (2000). Reference lines include the two-stage lead evolution curve (Stacey and Kramers, 1975), marked by 250 Ma intervals (SK), and the Northern Hemisphere Reference Line (NHRL) (Hart, 1984).

tures do not show significant differences with respect to those reported in the Sierra Madre Occidental in northern México, however, Pb isotopic signatures of the Sierra Madre del Sur suggest more homogeneous crustal contamination components than those of the Sierra Madre Occidental. In general, the Cenozoic granitic rocks of the southwestern Pacific Coast of southern México are Type 1 granites or hornblende granites, in the sense of Chapell and White (1974) and Kemp and Hawkesworth (2004), respectively.

The restricted variability in the isotopic signatures of the Cenozoic magmatic rocks of southwestern México, and their tendency to relatively primitive values, suggest low degrees of assimilation of old continental crust, compared with other sectors of the North American Cordillera. The dominantly silicic composition and the great volume of the plutons occupying the continental margin suggest that the source of this magmatism could be related to a basalt underplating process. Given the petrogenetic impossibility of generating large volumes of granitic rocks in the mantle wedge, it has been proposed that this kind of voluminous silicic volcanism can be generated either in two stages (Pitcher, 1993), or through a continuous process (Kemp and Hawkesworth, 2004) that can involve: (1) underplating of basaltic magmas derived from the mantle wedge at the base of the continental crust, or the hybridization of the lower continental crust by voluminous intrusion of basaltic magmas derived from the mantle, and (2) partial fusion of the metasomatized mafic or hybridized crust by the thermal input of basalts derived from the mantle and mixing with melts that resulted from the partial fusion of the continental crust. There is evidence that the voluminous silicic magmatism that produced the extensive plutons along the continental margin between Acapulco and Huatulco was nearly coeval with some mafic intrusions. Pulses of basaltic magmas coming from the mantle seem to be a necessary condition to produce partial fusion by transferring heat to the lower mafic or hybridized crust. The occurrence of mafic and intermediate dike swarms crosscutting silicic plutons in the Tierra Colorada region seems to support this relationship. Magmatic and crystal-plastic foliation in granites parallel to the dikes margins suggests that the plutons still preserved some heat during emplacement.

Crystallization ages of plutons along the continental margin of southwestern México range from ~100 Ma (Albian), in the Puerto Vallarta batholith, to 29 Ma (Oligocene), in the Huatulco region, with the Cenozoic plutons (<65 Ma) distributed southeast from Manzanillo (Herrmann et al., 1994; Schaaf et al., 1995, and references therein; Ducea et al., 2004a) (Fig. 10). Age distributions indicate a southeastward migration in the time of magmatism. The rate of migration was lower in the Puerto Vallarta–Zihuatanejo sector (100–40 Ma) than in the Acapulco-Huatulco sector (34–29 Ma): an exception is the Acapulco pluton and other deformed plutons. Based on dates of individual zircons, Ducea et al. (2004a) concluded that the migration in the magmatism of the later sector is not evident and suggested a nearly simultaneous extinction. However, this conclusion is based on the geochronologic results of a restricted segment of the continental margin from Acapulco to Puerto Escondido where dates show a small but recognizable difference with a southeastward-decreasing age. K-Ar and Rb-Sr cooling ages also portray a southeastward-decreasing pattern (Schaaf et al., 1995) (Fig. 10).

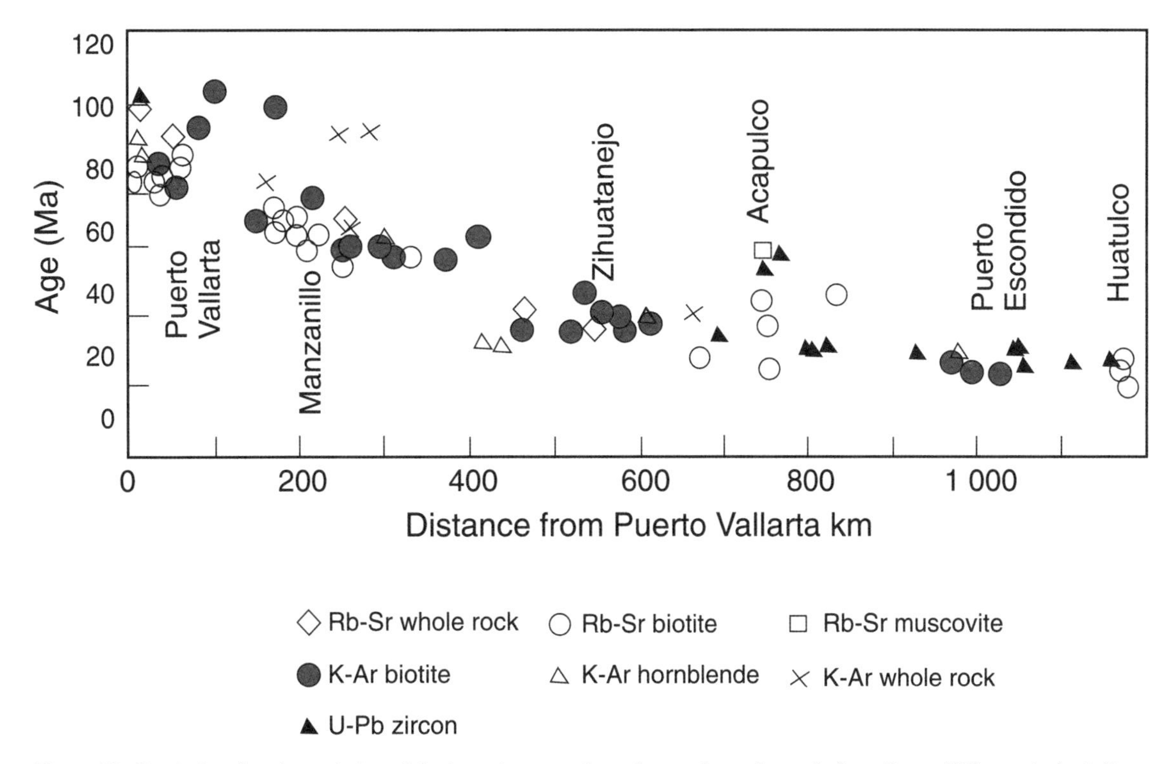

Figure 10. Graph showing the variation of the intrusive ages along the continental margin from Puerto Vallarata in the Jalisco block to Huatulco in the Oaxaca State. Modified after Schaaf et al. (1995) and includes data from Ducea et al. (2004a).

The Acapulco pluton, with U-Pb date of 54.9 ± 2 (Ducea et al., 2004a) and Rb-Sr and K-Ar cooling ages ranging from 43 to 50 Ma (López-Infanzón and Grajales-Nishimura, 1984; Guerrero-García, 1975; Schaaf et al., 1995), represents an anomaly in this migration pattern. Its age is more compatible with the Maastrichtian-Paleocene volcanism reported in the inland area at the same longitude (Ortega-Gutiérrez, 1980; Meza-Figueroa et al., 2003; Cerca, 2004; Levresse et al., 2004), which occurred prior to the late Eocene–Oligocene magmatic pulse.

Reported dates for the orthogneisses included in the metamorphic Xolapa Complex indicate emplacement and deformation ages ranging from the Late Jurassic to the Paleocene (Guerrero-García, 1975; Robinson et al., 1989; Morán-Zenteno, 1992; Herrmann et al., 1994; Ducea et al., 2004a). The fact that these rocks are intruded by undeformed late Eocene–early Oligocene plutons constrains the minimum age of the regional deformation and metamorphism of the Xolapa Complex.

In the Zihuatanejo-Huatulco sector, there are NW- and E-W–trending crystal-plastic and brittle shear zones overprinting the deformation and metamorphism of the Xolapa Complex (Ratschbacher et al., 1991; Riller et al., 1992; Tolson et al., 1993). Crystal-plastic deformation normally precedes the undeformed late Eocene–early Oligocene plutons. Kinematic features of the shear zones indicate normal and left-lateral components that have been related to the exhumation of the continental margin and coeval left-lateral deformation nearly parallel to the present Acapulco trench.

The NW-trending shear zone located south of Tierra Colorada is cut by a 34 Ma pluton, although there is evidence that deformation along the shear zone continued after the pluton emplacement (Riller et al., 1992). North of Huatulco, near the Gulf of Tehuantepec, Tolson-Jones (1998) reported that an E-W mylonitic zone forming the northern boundary of the Xolapa Complex crosscuts the 29 Ma Huatulco pluton. This supports the inference that crystal-plastic deformation along some of these shear zones continued after the voluminous, early Oligocene magmatic pulse.

The origin of the continental margin exhumation from Puerto Vallarta to the Gulf of Tehuantepec region is not clearly understood. Processes such as basalt underplating, combined with input from voluminous silicic magmas and subduction erosion, could have played an important role. Alternatively, Tolson-Jones (1998) interpreted that the uplift in the Huatulco region could have been produced by a southeastward-migrating lithospheric fold produced by differences in the age of the subducted oceanic crust and variations in trench orientations following the trench-trench-transform triple junction that accompanied the passage of the Chortis block. Based on Al_{tot} geobarometry in igneous hornblende of Paleogene plutons, Morán-Zenteno et al. (1996) calculated that uplift along the coast varied from 13 to 20 km. Ducea et al. (2004b) obtained fission tracks and (U-Th)/He dates indicating that the coastal ranges have been slowly decaying since the early Miocene, which implies that the main episode of uplift occurred in the Oligocene, after the main pulse of silicic plutonism.

VOLCANISM AND DEFORMATION IN THE INLAND REGION DURING THE EOCENE-MIOCENE

In the Paleogene, there was a migration in time of volcanism similar to that documented along the present-day coast region: early Eocene in the Michoacán region to middle Miocene in the southeastern Oaxaca region, although there are some exceptions to this general pattern (Figs. 7B and 7C).

Between Zihuatanejo and Ciudad Altamirano in eastern Michoacán and western Guerrero, there is a broadly distributed cover of volcanic rocks intercalated with terrestrial deposits that extends for more than 20,000 km^2 including the Infiernillo dam (presa del Infiernillo, Fig. 2). This cover represents a significant intermediate and silicic volcanic pulse, although detailed stratigraphic and tectonic studies are still lacking. K-Ar reported dates from dikes and lava flows vary from 61.2 ± 1.3 (Pantoja-Alor, 1986) to 33 Ma (Frank et al., 1992), but typically they range from 46 to 43 Ma (Fig. 7B) (Kratzeisen et al., 1991; Frank et al., 1992; Garduño-Monroy et al., 1999). There are also reports indicating the occurrence of some volcanic rocks with similar age (48–38 Ma) in northern Guerrero and southern Mexico State, which represent lower volume volcanism compared to that of the Zihuatanejo–Ciudad Altamirano region. North of this latter region, between longitude 101°W and 102°W, there are also some volcanic and intrusive rocks with younger ages (32–34 Ma) (Damon et al., 1981; Pasquaré et al., 1991) that represent an anomaly in the southeastward migration pattern of both the inland and coastal magmatism.

Late Eocene–early Oligocene volcanic activity developed in Morelos, Guerrero, and western Oaxaca regions between longitude 97°W and 100°W, and represents one of the major volcanic pulses in the inland region of southwestern México. It is concentrated in a NW-trending belt parallel and coeval with coastal plutons in similar longitude (Fig. 2).

In the Taxco-Huautla region, there is a series of dominantly silicic volcanic fields and minor intrusions with K-Ar and Ar-Ar ages ranging from 38 to 31 Ma (De Cserna and Fries, 1981; Alaniz-Álvarez et al., 2002; Morán-Zenteno et al., 2004), with the main volcanic pulse occurring at around 35 Ma. This volcanic activity is generally younger than the Paleogene volcanism in Michoacán and western Guerrero regions and is, in part, coeval with the plutonism in the Tierra Colorada region. Volcanic fields in the Taxco-Huautla region are made up of collapse calderas and rhyolitic domes. They are related in time and space to a series of NW-trending, left-lateral faults, especially recognized in areas like Taxco (Alaniz-Álvarez et al., 2002) and Tilzapotla (Morán-Zenteno et al., 2004). The kinematics of these faults implies a NNW-trending maximum extension. In Taxco, this extension produced NW-trending sedimentary basins with the accumulation of a thick terrestrial Eocene sequence (Chontalcuatlán Formation) that underlies rhyolites and ignimbrites. At the beginning of the Oligocene, a new tectonic regime activated N-S–trending left-lateral faults and produced a change in the kinematics in NW-trending faults from left to right lateral displacement, which implies a NE-SW extension.

W and NW of Taxco, in the La Goleta, Sultepec, Nanchititla, and San Vicente ranges (Fig. 6), there are also undated silicic volcanic fields (De Cserna, 1982) that show vestiges of exhumed collapse calderas. Their composition and exhumed character suggest that they were contemporaneous with the Taxco and Tilzapotla volcanic centers. SE of the Huautla range, around the Chiautla zone, there are reports of isolated silicic volcanic exposures that seem to be related to this volcanic pulse (Rivera-Carranza et al., 1998; Silva-Romo et al., 2001) (Fig. 6). The total volume of the exposed silicic volcanic rocks in this inland region is ~3000 km^3, which represents a minimum estimate of the original volume, taking into account the effects of erosion in a partially exhumed region. Between the Taxco volcanic region and the coastal plutons, the late Eocene–early Oligocene volcanic rocks are less abundant. In this region there are preliminary geochronologic reports also indicating late Eocene–early Oligocene silicic volcanic rocks (Cerca, 2004; Campa-Uranga et al., 2002).

In the western Oaxaca and southern Puebla region, there is an extensive Paleogene volcanic cover dominated by rocks of intermediate composition and minor units of silicic composition (Ferrusquía-Villafranca, 1976; Martiny et al., 2000). The volcanic succession transitionally overlies a succession of terrestrial and epiclastic beds that indicates an episode of continental sedimentation in intermontane and tectonic basins. The volcanic eruptive style is that of composite volcanoes and monogenetic volcanic fields that are now in an advanced state of erosion. Reported K-Ar dates for these rocks range mainly from 34 to 31 (Martiny et al., 2000), however, there are some reports indicating ages as young as 26 Ma (Ferrusquía-Villafranca, 1992). These ages confirm that the western Oaxaca volcanic rocks are in part coeval with the youngest volcanic units of the Taxco-Huautla region. Further to the E, in the Valley of Oaxaca zone, there is geochronologic evidence of volcanism as young as 19 Ma (Ferrusquía-Villafranca, 1992; Urrutia-Fucugauchi and Ferrusquía-Villafranca, 2001).

In western Oaxaca, early Oligocene volcanism was also coeval with episodes of strike-slip tectonics that were initially associated with NW- and E-W–oriented left-lateral faults followed by N-trending faults also with left-lateral displacement (Martiny et al., 2000; Silva-Romo et al., 2001). These episodes are not completely coeval with those of the Taxco-Huautla region, which suggests variations in time and space in the tectonic conditions for the northern Sierra Madre del Sur.

Nieto-Samaniego et al. (2006) interpreted the Oligocene as a time of NE-SW extension, probably related to the left-lateral displacement of N-S–trending faults in western Oaxaca. This phase of deformation involved the Valley of Oaxaca zone where it produced a NNW-trending half-graben. Accumulation of terrestrial sediments and volcanic rocks in this half-graben took place at least from the Oligocene until the early Miocene (Centeno-García, 1988; Nieto-Samaniego et al., 1995; Urrutia-Fucugauchi and Ferrusquía-Villafranca, 2001).

In southeastern Oaxaca, in the region located between Oaxaca City and Salina Cruz, there is an extensive Miocene volcanic sequence of intermediate lava flows and silicic pyroclastic rocks overlying terrestrial sandstone and conglomerate (Ferrusquía-Villafranca, 1999, 2001). These volcanic and sedimentary successions cover an area of ~6000 km^2 (Fig. 2). Reported K-Ar ages for the volcanic rocks range from 13 to 17 Ma (Ferrusquía-Villafranca, 1992).

TECTONIC SUMMARY RELATING TO THE PLATE TECTONICS KINEMATICS IN SOUTHWESTERN MÉXICO

The tectonic evolution of southwestern México since the Late Cretaceous is characterized first by an orogenic episode of shortening with general easterly vergence. Based on its age and deformation style, this episode has been generally correlated with the Laramide orogeny. Later, in the Eocene, the tectonic scenario changed to a tectonic regime characterized by strike-slip faults. Variations in the orientation and kinematics of different age groups of faults, as well as in the extension and shortening directions associated with the strike-slip tectonics, indicated changes in time and space of the stress regime.

The origin of the stress conditions that produced the Laramide shortening in this region is still subject to controversy. It is generally accepted that shortening initiated in the Late Cretaceous in the Guerrero terrane and propagated eastward into the Eocene in the Veracruz region (e.g., Salinas-Prieto et al., 2000; Eguiluz de Antuñano et al., 2000; Nieto-Samaniego et al., 2006). The most generally accepted hypothesis at a regional scale in the North American Cordillera is that mechanical coupling and traction were produced by the low-angle subduction beneath the North American plate. This low-angle geometry has been attributed to the increase in the rate of convergence between the Farallon and the North American plates (Engebretson et al., 1985; Dickinson et al., 1988; Bird, 1988; Bunge and Grand, 2000; Saleeby, 2003). The beginning of the shortening also coincides with a period of increase in the westward absolute motion of North America (Engebretson et al., 1985). There are important problems in extrapolating this model to explain the Laramide deformation in southern México. The occurrence of arc magmatism during and immediately after the deformation in zones located near the inferred trench in the Jalisco block suggests little change in the dip of the subduction zone, in contrast with the expected magmatic extinction suggested by recent thermal models (English et al., 2003). Another inconsistency that has recently emerged is the recognition of postorogenic magmatism W of the old metamorphic block (Oaxaca and Acatlán Complexes) coeval with the easternmost manifestations of shortening in the Sierra Madre del Sur (Eguiluz de Antuñano et al., 2000).

An alternative explanation suggested by some authors to explain the shortening in the Sierra Madre del Sur involves a more complex tectonic scenario with accretion of an island arc associated with westward-dipping subduction and the closure of a marginal basin (Campa and Coney, 1983; Monod et al., 1994; Salinas-Prieto et al., 2000; Freydier et al., 1996). Some objections to this model cite the lack of petrogenetic assemblages typical of suture zones in the Sierra Madre del Sur, as well as the clear continental contribution to sediments and geochemical signatures in magmatic rocks of the Guerrero terrane that may be related to the oldest terranes in southern México (Elías-Herrera and Ortega-Gutiérrez, 1998).

Additionally, the deformation style of the Laramide shortening in southwestern México involves thin-skinned deformation with patterns crossing some basement block or subterrane boundaries (e.g., Lang et al., 1996; Salinas-Prieto et al., 2000; Cabral-Cano et al., 2000a, 2000b; Elías-Herrera et al., 2000). This is at odds with an expected style of deformation strongly dependent on the geometry of the block boundaries. Typically, deformation concentrates along the boundaries of blocks, but some deformation can propagate into the blocks depending on intensity and duration of deformation, and rheologic properties.

The Late Maastrichtian–Paleocene magmatic episode postdates the Laramide deformation in the western and central Sierra Madre del Sur (González-Partida et al., 2003). Geochemical characteristics of these rocks indicate arc magmatism related to a convergent boundary whose orientation is not precisely known. Trends of fold-and-thrust structures suggest a N-S–oriented Pacific margin, however, the distribution of Maastrichtian-Paleocene arc magmatism is more compatible with a NW-SE orientation (Fig. 7A).

It is not precisely known when the Laramide deformation ended in different sectors of southwestern México, but there is evidence in late Eocene time that the strike-slip tectonics associated with NW-SE and E-W faults was already active. This has been recognized especially in the north-central part of the Sierra Madre del Sur and along the present continental margin between Zihuatanejo and Puerto Escondido (Figs. 6 and 10) (Ratschbacher et al., 1991; Morán-Zenteno, et al., 1999; Alaniz-Álvarez et al., 2002). At the beginning of the Oligocene, E-W–trending, left-lateral faults were still active in the Huatulco area (Tolson et al., 1993). In the north-central part of the Sierra Madre del Sur, however, there was a change in the kinematics of the deformation at the beginning of the Oligocene that reactivated N-S faults with left-lateral displacement. This has been documented from northern Guerrero to central Oaxaca regions (Alaniz-Álvarez et al., 2002; Silva Romo et al., 2001, Nieto-Samaniego et al., 2006). When considering the possible influence of the basement heterogeneities and preexisting tectonic discontinuities in the region, it is not possible to precisely infer the changes in the stress regime during this episode, but extension associated with the strike-slip tectonics changed from NNW to NE. According to Nieto-Samaniego et al. (2006), the last extension direction was responsible for reactivation of the Oaxaca fault and the formation of a half-graben structure (Fig. 11). Based on inversion of fault-slip data measured in different sectors of the Sierra Madre del Sur, Meschede et al. (1996) recognized different groups of faults that were associated with distinct paleo-stress tensors of different ages. These authors inferred that for the time between 70 and 40 Ma, σ_1 was nearly E-W with a subhorizontal σ_2 that produced a strike-slip deformational regime. The NW-trending shear zones

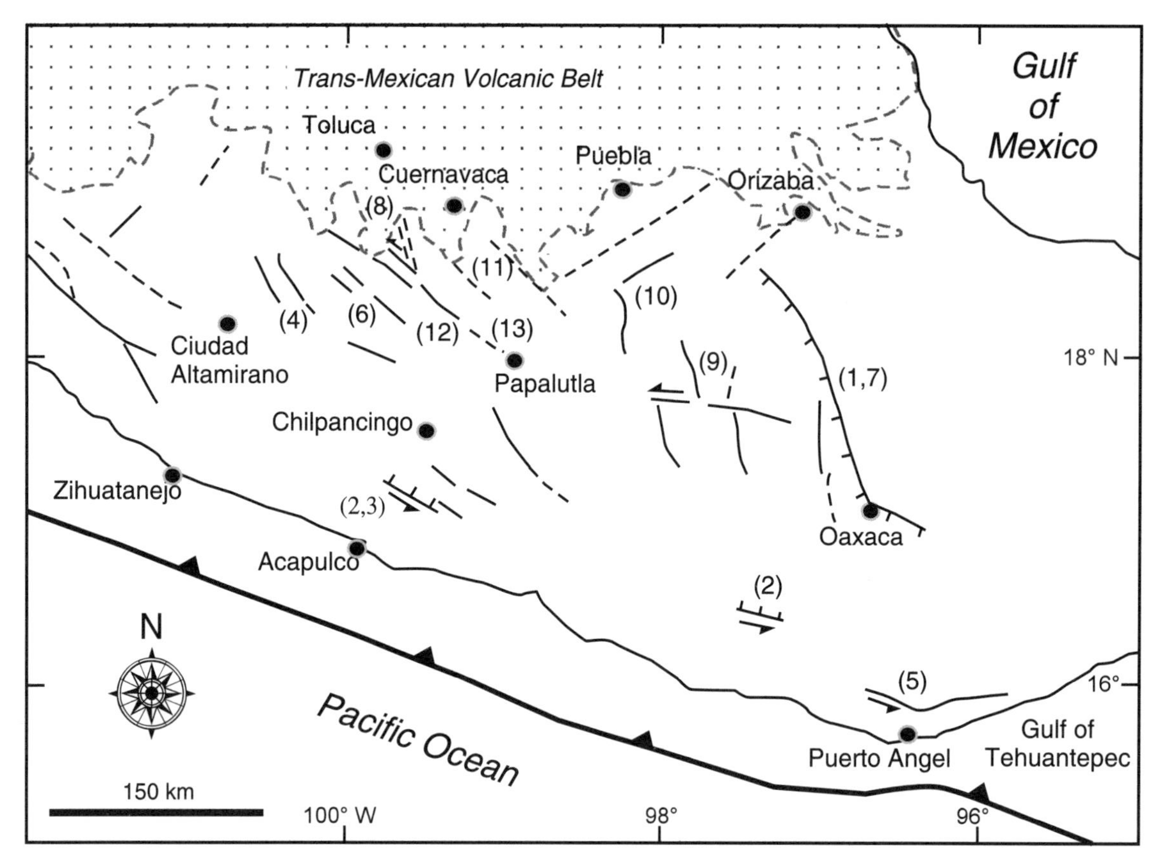

Figure 11. Spatial distribution of the main tectonic features recognized in southern México for the Cenozoic. The compilation includes features with a detailed description of their geometric and kinematic characteristics, as well as the most evident lineaments reported on the 1:250,000 scale maps of the Consejo de Recursos Minerales (Montiel-Escobar et al., 2000; Rivera-Carranza et al., 1998; Campa-Uranga et al., 1997; González-Ramos et al., 2001; Martínez-Amador et al., 2001), whose geometric and kinematic details are unknown. The indicators of the relative displacement of most of the lineaments are not shown because a complex kinematic history has been documented. Where shown, the relative displacement refers to the documented main motion for the late Eocene or early Oligocene. Numbers indicate the source of the information: (1) Centeno-García (1988); (2) Ratschbacher et al. (1991); (3) Riller et al. (1992); (4) Jansma and Lang (1997); (5) Tolson-Jones (1998); (6) Cabral-Cano et al. (2000b); (7) Alaniz-Álvarez et al. (1994); (8) Alaniz-Álvarez et al. (2002); (9) Martiny et al. (2002); (10) Silva-Romo et al. (2001); (11) Campa-Uranga et al. (1997); (12) Rivera-Carranza et al. (1998) and Morán-Zenteno et al. (2004); and (13) Cerca (2004). Refer to the key in Figure 4.

along the Pacific coastal region would be part of this episode. However, the chronology of the deformation and stress orientation inferred by Meschede at al. (1996) is not completely compatible with the chronology and kinematics of the large faults of the inland region, including those from Taxco to Huajuapan in the northern Sierra Madre del Sur for which significant activity has been documented for a time after 40 Ma (Alaniz-Álvarez et al., 2002; Martiny et al., 2002; Morán-Zenteno et al., 2004).

The truncation of the continental margin, the eastward migration of magmatism, and occurrence of NW- and E-W–trending shear zones with left-lateral kinematics have generally been related to the displacement of the Chortis block from a position between Puerto Vallarta and the Isthmus of Tehuantepec (Fig. 3A) (Schaaf et al., 1995; Meschede et al., 1996; Morán-Zenteno et al., 1999). The extinction of the Eocene-Oligocene magmatism of the Sierra Madre del Sur and its development in the Trans-Mexican Volcanic Belt since the Miocene have been related to a change in the geometry of the subducted slab. The prevailing interpretation of this change suggested it was the result of the passage of the trench-trench-transform triple junction that accompanied the displacement of the Chortis block, and that is also related to the exhumation of the continental margin (Schaaf et al., 1995; Morán-Zenteno et al., 1996; Tolson-Jones, 1998). The position of the Chortis block off southwestern México was previously interpreted by other authors as a geometric necessity in the paleogeographic reconstructions related to the evolution of the Caribbean plate (Malfait and Dinkelman, 1972; Anderson and Schmidt, 1983; Ross and Scotese, 1988; Pindell et al., 1988). Although this interpretation is compatible with the geochronological patterns in arc magmatism, it presents some problems when analyzed from the perspective of plate kinematics in the region and the anticipated effects along the continental margin of southwestern México. The main problems concern the lack of agreement between the displacement direction of the Chortis block and the rotation of the Caribbean plate since the Eocene, and the lack of correlation between old tectonic and petrotectonic features between Honduras and México. The northern transform boundary of the Caribbean plate is delineated by the Cayman trough and the Polochic-Motagua fault system. Based on the basement topography, depth, and patterns of magnetic anomalies in the Cayman trough, it has been estimated that the displacement of the Caribbean plate with respect to North America has been 1100 km since the middle Eocene (Rosencrantz and Sclater, 1986). The chronology of the displacement is also constrained by the change in the orientation of the northern boundary of the Caribbean plate after collision of Cuba against the Bahamas platform. The average pole rotation of the Caribbean plate since the Eocene, which is suggested by the slight concavity of the northern transform boundary of the Cayman trough, is located in the southern hemisphere (Pindell et al., 1988). On the other hand, the Polochic and Motagua faults display a slight concavity to the north.

In most reconstructions, the displacement of the Chortis block is to the SE, inconsistent with the easterly motion of the Caribbean plate. The predicted effects of the displacement of the Chortis block as part of the Caribbean plate would produce transpressive deformation along the southwestern margin of México. Transpressive features along this margin have only been documented for pre-Eocene tectonic episodes (Cerca et al., 2004) and for the Chiapas region since the Miocene (Meneses-Rocha, 1991; Guzmán-Speziale and Meneses-Rocha, 2000). However, the Eocene-Oligocene shear zones along the continental margin are more compatible with transtensional mechanisms (Ratschbacher, et al., 1991; Tolson et al., 1993; Riller et al., 1992; Herrmann et al., 1994; Tolson-Jones, 1998). Another problem with the prevailing model is the occurrence of an undeformed marine sequence in the Gulf of Tehuantepec ranging in age from the Late Cretaceous to the Holocene (Sánchez-Barreda, 1981); this is also incompatible with a transpressional deformation. A further significant problem is the limited 130–170 km total displacement documented for the Polochic–Motagua fault system (i.e., Burkart et al., 1987; Donnelly et al., 1990; Guinta et al., 2002, and references therein).

Keppie and Morán Zenteno (2005) evaluated the possibility for the Chortis block to be in a position to the W or SW of its present position, as suggested by the trace of the northern transform boundary of the Caribbean plate in the Cayman trough sector, which implies a clockwise rotation. Tectonic interactions since the Miocene between the Chortis and the Maya block would be responsible for the transpressional deformation (Guzmán-Speziale and Meneses-Rocha, 2000) and the concavity to the N of the western segments of the Polochic-Motagua fault system. The alternative trajectory of the Chortis block would allow collision of the Tehuantepec ridge with southwestern México after the Eocene, which would have influenced the slab geometry (Keppie and Morán-Zenteno, 2005). This model directly explains neither the truncation of the continental margin, nor the Eocene-Oligocene strike-slip faulting documented in the Guerrero and Oaxaca regions; therefore additional mechanisms were involved. Subduction erosion and oblique convergence could produce these effects. Keppie and Morán-Zenteno (2005) speculated that subduction erosion NW of the Isthmus of Tehuantepec would have been produced by the impingement of a sea mount province whose counterpart is now located between the Molokai and Clarion fracture zones. According to convergence vectors calculated by Schaaf et al. (1995), and based on rotation poles from Engebretson et al. (1985) and Pindell et al. (1988), there was an oblique convergence that would have caused the NW-trending left lateral faults. Another factor that might have contributed to the development of the left-lateral faulting would be the variations in the geometry of the subducted slab N and S of the present Trans-Mexican Volcanic Belt. North of the belt, the subducted slab initiated a rollback process in the early Oligocene (Nieto-Samaniego et al., 1999), whereas south of the belt, there was a process of shallowing in the geometry of the subduction zone. This could also have influenced the migration patterns of the magmatism.

Although accumulated observations and data do not completely permit the discard of traditional models on issues like the Late Cretaceous–early Paleogene Laramide shortening, or the Chortis block position off southwestern México in the Cenozoic, the discussion presented in this paper shows that alternative models require a more detailed analysis.

ACKNOWLEDGMENTS

M. Cerca is grateful for the support of the National Council for Science and Technology (CONACYT) for his Ph.D. research and to Dr. Luca Ferrari for stimulating discussion on the geology of southern México. This research was supported, in part, by the Program for Science and Technologic Innovation (PAPIIT) of the National University of Mexico (UNAM). Critical comments by two anonymous reviewers were very helpful and improved the final manuscript.

REFERENCES CITED

Alaniz-Álvarez, S.A., Nieto-Samaniego, Á.F., and Ortega-Gutiérrez, F., 1994, Structural evolution of the Sierra de Juárez mylonitic complex, state of Oaxaca, México: Revista Mexicana de Ciencias Geológicas, v. 11, p. 147–156.

Alaniz-Álvarez, S.A., Nieto-Samaniego, Á.F., Morán-Zenteno, D.J., and Alba-Aldave, L., 2002, Rhyolitic volcanism in extension zone associated with strike-slip tectonics in the Taxco region, southern Mexico: Journal of Volcanology and Geothermal Research, v. 118, p. 1–14, doi: 10.1016/S0377-0273(02)00247-0.

Albrecht, A., and Goldstein, S.L., 2000, Effects of basement composition and age on silicic magmas across an accreted terrane-Precambrian crust boundary, Sierra Madre Occidental, Mexico: Journal of South American Earth Sciences, v. 13, p. 255–273, doi: 10.1016/S0895-9811(00)00014-6.

Altamira-Areyán, A., 2002, Las litofacies y sus implicaciones de la cuenca sedimentaria Cutzamala-Tiquicheo, estados de Guerrero y Michoacán, México [M.Sc. thesis]: México, Universidad Nacional Autónoma de México, Instituto de Geología, Posgrado en Ciencias de la Tierra, 79 p.

Anderson, T.H., and Schmidt, V.A., 1983, The evolution of middle America and the Gulf of Mexico-Caribbean sea region during Mesozoic time: Geological Society of America Bulletin, v. 94, p. 941–966, doi: 10.1130/0016-7606(1983)94<941:TEOMAA>2.0.CO;2.

Bandy, W., Mortera-Gutiérrez, C., Urrutia-Fucugauchi, J., and Hilde, T.W.C., 1995, The subducted Rivera-Cocos plate boundary, where is it, what is it, and what is its relationship to the Colima rift?: Geophysical Research Letters, v. 22, p. 3075–3078, doi: 10.1029/95GL03055.

Barrier, E., Velasquillo, L., Chávez, M., and Goulon, R., 1998, Neotectonic evolution of the Isthmus of Tehuantepec (southeastern Mexico): Tectonophysics, v. 287, p. 77–96, doi: 10.1016/S0040-1951(98)80062-0.

Bellon, H., Maury, R.C., and Stephan, J.F., 1982, Dioritic basement, site 493. petrology, geochemistry, and geodynamics: Initial Reports of the Deep Sea Drilling Project, v. LXVI, p. 723–731.

Benammi, M., Centeno-García, E., Martínez-Hernández, E., Morales-Gámez, M., Tolson, G., and Urrutia-Fucugauchi, J., 2005, Presencia de dinosaurios en la Barranca Los Bonetes en el sur de México (Región de Tiquicheo, Estado de Michoacán) y sus implicaciones cronoestratigráficas: Revista Mexicana de Ciencias Geológicas, v. 22, no. 3, p. 429–435.

Bird, P., 1988, Formation of the Rocky Mountains, western United States: A continuum computer model: Science, v. 239, p. 1501–1507.

Bird, P., 1998, Kinematic history of the Laramide orogeny in latitudes 35°–49° N, western United States: Tectonics, v. 17, p. 780–801, doi: 10.1029/98TC02698.

Bunge, H.P., and Grand, S.T., 2000, Mesozoic plate-motion history below the northeast Pacific Ocean from seismic images of the subducted Farallon slab: Nature, v. 405, p. 337–340, doi: 10.1038/35012586.

Burkart, B., Deaton, B.C., Dengo, C., and Moreno, G., 1987, Tectonic wedges and offset Laramide structures along the Polochic fault of Guatemala and Chiapas, Mexico: Reaffirmation of large Neogene displacement: Tectonics, v. 6, p. 411–422.

Cabral-Cano, E., Lang, H.R., and Harrison, C.G.A., 2000a, Stratigraphic assessment of the Arcelia–Teloloapan area, southern Mexico: Implications for southern Mexico's post-Neocomian tectonic evolution: Journal of South American Earth Sciences, v. 13, p. 443–457, doi: 10.1016/S0895-9811(00)00035-3.

Cabral-Cano, E., Draper, G., Lang, H.R., and Harrison, C.G.A., 2000b, Constraining the late Mesozoic and early Tertiary tectonic evolution of southern Mexico: Structure and deformation history of the Tierra Caliente region, southern Mexico: The Journal of Geology, v. 108, p. 427–446, doi: 10.1086/314414.

Campa, M.F., and Coney, P.J., 1983, Tectono-stratigraphic terranes and mineral resource distributions in Mexico: Canadian Journal of Earth Sciences, v. 20, p. 1040–1051.

Campa, M.F., and Ramírez, J., 1979, La evolución geológica y la metalogénesis del noroccidente de Guerrero: Plan Piloto Proyecto de Recursos Naturales de Guerrero. Taxco, Universidad Autónoma de Guerrero, Serie Científica Técnica, v. 1, 100 p.

Campa, M.F., Oviedo, A., and Tardy, M., 1976, La cabalgadura laramídica del dominio volcánico-sedimentario (Arco de Alisitos–Teloloapán) sobre el miogeosinclinal mexicano en los límites de los estados de Guerrero y México, III Congreso Latinoamericano de Geología, Resúmenes. México, D. F., Universidad Nacional Autónoma de México, Instituto de Geología, 23 p.

Campa-Uranga, M.F., García-Díaz, J.L., Bustamante-García, J., Torreblanca-Castro, T. de J., Aguilera-Martínez, M.A., and Vergara Martínez, A., 1997, Carta geológico-minera de la hoja Chilpancingo E14–8: Pachuca, Hidalgo, Consejo de Recursos Minerales, scale 1:250,000, 1 sheet.

Campa-Uranga, M.F., Fitz-Díaz, E., and Martínez-Hernández, E., 2002, Revisión de la edad de la Formación Oapan y su significado en el graben de San Agustín Oapan y el sinclinorio de Zacango, Estado de Guerrero, XII Congreso Nacional de Geoquímica, Resúmenes. Actas INAGEQ v. 8, 165 p.

Centeno-García, E., 1988, Evolución estructural de la falla de Oaxaca durante el Cenozoico. México [M.Sc. thesis]: Universidad Nacional Autónoma de México, 156 p.

Centeno-García, E., García, J.L., Guerrero-Suástegui, M., Ramírez-Espinosa, J., Salinas-Prieto, J.C., and Talavera-Mendoza, O., 1993a, Geology of the southern part of the Guerrero terrane, Ciudad Altamirano-Teloloapan area, *in* Ortega-Gutiérrez, F., Centeno-García, E., Morán-Zenteno, D.J., and Gómez-Caballero, A., eds., First Circum-Pacific and Circum-Atlantic Terrane Conference, Terrane geology of southern Mexico. Guidebook of Field trip B. Guanajuato, Gto.: Universidad Nacional Autónoma de México, Instituto de Geología, p. 22–33.

Centeno-García, E., Ruiz, J., Patchett, P.J., and Ortega-Gutiérrez, F., 1993b, Guerrero terrane: Its role in the southern Cordillera from new geochemical data: Geology, v. 21, p. 419–423, doi: 10.1130/0091-7613(1993)021<0419:GTOMIR>2.3.CO;2.

Cerca, M., 2004, Deformación y magmatismo Cretácico tardío—Terciario temprano en la zona de la Plataforma Guerrero Morelos [Ph.D. thesis]: Juriquilla, Qro., Universidad Nacional Autónoma de México, 175 p.

Cerca, M., Ferrari, L., Bonini, M., Corti, G., and Manetti, P., 2004, The role of crustal heterogeneity in controlling vertical coupling during Laramide shortening and the development of the Caribbean—North American transform boundary in southern Mexico. Insights from analogue models, *in* Grocott, J., Taylor, G., and Tikoff, B., eds., Vertical coupling and decoupling in the lithosphere: Geological Society [London] Special Publication, v. 227, p. 117–140.

Chapell, B.W., and White, A.J.R., 1974, Two contrasting granite types: Pacific Geology, v. 8, p. 173–174.

Damon, P.E., Shafiqullah, M., and Clark, K.F., 1981, Geochronology of the porphyry copper deposits and related mineralization of Mexico: Canadian Journal of Earth Sciences, v. 20, p. 1052–1071.

De Cserna, Z., 1982, Hoja Tejupilco 14Q-g(9), con resumen de la hoja Tejupilco, estados de Guerrero, México y Michoacán: México, D. F., Universidad Nacional Autónoma de México, Instituto de Geología, Carta Geológica de México, Serie de 1:100,000, 1 sheet and text, 28 p.

De Cserna, Z., and Fries, C., Jr., 1981, Hoja Taxco 14Q-h(7), con resumen de la hoja Taxco, estados de Guerrero, México y Morelos: México, D. F., Universidad Nacional Autónoma de México, Instituto de Geología, Carta Geológica de México, Serie de 1:100,000, 1 sheet and text, 47 p.

De Cserna, Z., Ortega-Gutiérrez, F., and Palacios-Nieto, M., 1980, Reconocimiento geológico de la parte central de la cuenca del alto Río Balsas, estados de Guerrero y Puebla, *in* V Convención Geológica Nacional, Libro Guía de la excursión geológica a la parte central de la cuenca del alto Río Balsas, estados de Guerrero y Puebla. México, Sociedad Geológica Mexicana, p. 2–33.

Delgado-Argote, L., and Carballido-Sánchez, E., 1990, Análisis tectónico del sistema transpresivo neogénico entre Macuspana, Tabasco, y Puerto Ángel, Oaxaca: Revista del Instituto de Geología, Universidad Nacional Autónoma de México, v. 9, p. 21–32.

Delgado-Argote, L., López-Martínez, M., Cork, D., and Hall, M., 1992, Geologic framework and geochronology of ultramafic complexes of southern Mexico: Canadian Journal of Earth Sciences, v. 29, p. 1590–1604.

DeMets, C., and Traylen, S., 2000, Motion of the Rivera plate since 10 Ma relative to the Pacific and North American plates and the mantle: Tectonophysic, v. 318, p. 119–159, doi: 10.1016/S0040-1951(99)00309-1.

Dickinson, W.R., and Lawton, T.F., 2001, Carboniferous to Cretaceous assembly and fragmentation of Mexico: Geological Society of America Bulletin, v. 113, p. 1142–1160, doi: 10.1130/0016-7606(2001)113<1142:CTCAAF>2.0.CO;2.

Dickinson, W.R., Klute, M.A., Hayes, M.J., Janecke, S.U., Lundin, E.R., McKittrick, M.A., and Olivares, M.D., 1988, Paleogeographic and paleotectonic setting of Laramide sedimentary basins in the central Rocky Mountain region: Geological Society of America Bulletin, v. 100, p. 1023–1039, doi: 10.1130/0016-7606(1988)100<1023:PAPSOL>2.3.CO;2.

Donnelly, T.W., Horne, G.S., Finch, R.C., and López-Ramos, E., 1990, Northern central America; the Maya and Chortis blocks, *in* Dengo, G., and Case, J.E., eds., The Caribbean Region: Boulder, Colorado, Geological Society of America, Geology of North America, v. H, p. 37–76.

Ducea, M.N., Gehrels, G.E., Shoemaker, S., Ruiz, J., and Valencia, V.A., 2004a, Geologic evolution of the Xolapa Complex, southern Mexico: Evidence from U-Pb Zircon geochronology: Geological Society of America Bulletin, v. 116, p. 1016–1025, doi: 10.1130/B25467.1.

Ducea, M., Valencia, V.A., Shoemaker, S., Reiners, P.W., DeCelles, P.G., Campa, M.F., Morán-Zenteno, D., and Ruiz, J., 2004b, Rates of sediment recycling beneath the Acapulco trench: Constraints from (U-Th)/He thermochronology: Journal of Geophysical Research, v. 109(B9), B09404, doi 10.1029/2004JB003112, 11 p.

Eguiluz de Antuñano, S., Aranda-García, M., and Marrett, R., 2000, Tectónica de la Sierra Madre Oriental, México: Boletín de la Sociedad Geológica Mexicana, v. 53, no. 1, p. 1–26.

Elías-Herrera, M., and Ortega-Gutiérrez, F., 1998, The early Cretaceous Arperos oceanic basin (western Mexico): Geochemical evidence for an aseismic ridge formed near a spreading center—Comment: Tectonophysic, v. 292, p. 321–326, doi: 10.1016/S0040-1951(98)00051-1.

Elías-Herrera, M., Sánchez-Zavala, J.L., and Macías-Romo, C., 2000, Geologic and geochronologic data from the Guerrero terrane in the Tejupilco area, southern Mexico. New constraints on its tectonic interpretation: Journal of South American Earth Sciences, v. 13, p. 355–375, doi: 10.1016/S0895-9811(00)00029-8.

Engebretson, D.C., Cox, A., and Gordon, R.G., 1985, Relative plate motions between oceanic and continental plates in the Pacific basin: Boulder, Colorado: Geological Society of America Special Paper 206, 64 p.

English, J.M., Johnston, S.T., and Wang, K., 2003, Thermal modeling of the Laramide orogeny: Testing the flat-slab subduction hypothesis: Earth and Planetary Science Letters, v. 214, p. 619–632, doi: 10.1016/S0012-821X(03)00399-6.

Ferrari, L., Pasquarè, G., Venegas, S., Castillo, D., and Romero, F., 1994, Regional tectonics of western Mexico and its implications for the northern boundary of the Jalisco block: Geofísica Internacional, v. 33, p. 139–151.

Ferrari, L., López-Martínez, M., Aguirre-Díaz, G., and Carrasco-Núñez, G., 1999, Space-time patterns of Cenozoic arc volcanism in central Mexico: From the Sierra Madre Occidental to the Mexican Volcanic Belt: Geology, v. 27, p. 303–306, doi: 10.1130/0091-7613(1999)027<0303:STPOCA>2.3.CO;2.

Ferrari, L., López-Martínez, M., and Rosas-Elguera, J., 2002, Ignimbrite flare up and deformation in the southern Sierra Madre Occidental, western Mexico: Implications for the late subduction history of the Farallon plate: Tectonics, v. 21, no. 4, doi: 10.1029/2001TC001302, p. 17/1–17/24.

Ferrusquía-Villafranca, I., 1976, Estudios geológico-paleontológicos en la región Mixteca, Parte 1. Geología del área Tamazulapan-Teposcolula-Yanhuitlán, Mixteca Alta, estado de Oaxaca, México: Boletín del Instituto de Geología, Universidad Nacional Autónoma de México, v. 97, 185 p.

Ferrusquía-Villafranca, I., 1992, Contribución al conocimiento del Cenozoico en el sureste de México y de su relevancia en el entendimiento de la evolución regional: III Congreso Geológico de España y VIII Congreso Latinoamericano de Geología, Actas de las sesiones científicas. Salamanca, v. 4, p. 40–44.

Ferrusquía-Villafranca, I., 1999, Contribución al conocimiento de Oaxaca, México–El área Laollaga-Lachivizá: Boletín del Instituto de Geología, Universidad Nacional Autónoma de México, v. 110, 103 p.

Ferrusquía-Villafranca, I., 2001, Contribución al conocimiento geológico de Oaxaca. México-El área de Nejapa de Madero: Boletín del Instituto de Geología, Universidad Nacional Autónoma de México, v. 111, 100 p.

Frank, M., Kratzeisen, M., Negendank, J.F.W., and Boehnel, H., 1992, Geología y tectónica en el terreno Guerrero (México-Sur), en III Congreso Geológico de España y VIII Congreso Latinoamericano de Geología, Actas de las sesiones científicas. Salamanca, España, v. 4, p. 290–293.

Freydier, C., Martínez, J.R., Lapierre, H., Tardy, M., and Coulon, C., 1996, The early Cretaceous Arperos oceanic basin (western Mexico): Geochemical evidence for an aseismic ridge formed near a spreading center: Tectonophysics, v. 259, p. 343–367, doi: 10.1016/0040-1951(95)00143-3.

Fries, C., Jr., 1960, Geología del estado de Morelos y de partes adyacentes de México y Guerrero, región central meridional de México: Boletín del Instituto de Geología, Universidad Nacional Autónoma de México, v. 60, 236 p., 5 láms.

Fries, C., Jr., 1966, Hola Cuernavaca 14Q-h(8), con resumen de la geología de la hoja Cuernavaca, estados de Morelos, México, Guerrero y Puebla: México, Universidad Nacional Autónoma de México, Instituto de Geología, Carta Geológica de México, Serie de 1:100,000, 1 sheet with text.

Garduño-Monroy, V.H., Corona-Chávez, P., Israde-Alcántara, I., Mennella, L., Arreygue, E., Bigioggero, B, and Chiesa, S., 1999, Carta geológica del estado de Michoacán, escala 1:250,000. Morelia, Universidad Michoacana de San Nicolás de Hidalgo, 111 p., 1 sheet.

GEOLIMEX Working Group, 1994, Reflections from the subducting plate? First results of a Mexican geotraverse, *in* Miller, H., Rosenfeld, U., and Weber-Diefenbach, K., eds., 13. Symposium on Latin-America Geosciences: Zentralblatt für Geologie und Paläontologie, Teil 1. Allgemeine, Angewandte, Regionale und Historische Geologie, v. 1–2, p. 541–553.

González-Partida, E., Levresse, G., Carrillo-Chávez, A., Cheilletz, A., Gasquet, D., and Jones, D., 2003, Paleocene adakite Au-Fe bearing rocks, Mezcala, Mexico: Evidence from geochemical characteristics: Journal of Geochemical Exploration, v. 80, no. 1, p. 25–40, doi: 10.1016/S0375-6742(03)00180-8.

González-Ramos, A., Sánchez-Rojas, L.E., Mota-Mota, S., Arceo y Cabrilla, F.A., Onofre-Espinoza, L., Zárate-López, J., and Soto-Araiza, R., 2001, Carta geológico-minera y geoquímica de la hoja Oaxaca E14–9: Pachuca, Hidalgo, Consejo de Recursos Minerales, scale 1:250,000, 1 sheet.

Guerrero García, J.C., 1975, Contributions to paleomagnetism and Rb-Sr geochronology [Ph.D. thesis]: Dallas, University of Texas, 131 p.

Guinta, G., Beccaluva, L., Coltorti, M., Sienna, F., Mortellaro, D., and Cutrupia, D., 2002, The peri-Caribbean ophiolites: Structure, tectono-magmatic significance and geodynamic implications: Caribbean Journal of Earth Sciences, v. 36, p. 1–20.

Guzmán-Speziale, M., and Meneses-Rocha, J.J., 2000, The North America-Caribbean plate boundary west of the Motagua-Polochic fault system: A fault jog in southeastern Mexico: Journal of South American Earth Sciences, v. 13, p. 459–468, doi: 10.1016/S0895-9811(00)00036-5.

Hart, S.R., 1984, A large-scale isotope anomaly in the southern hemisphere mantle: Nature, v. 309, p. 753–757, doi: 10.1038/309753a0.

Hernández-Bernal, M.S., and Morán-Zenteno, D.J., 1996, Origin of the Rio Verde batholith, southern Mexico, as inferred from is geochemical characteristics: International Geology Review, v. 38, p. 361–373.

Herrmann, U.R., Nelson, B.K., and Ratschbacher, L., 1994, The origin of a terrane: U/Pb zircon geochronology and tectonic evolution of the Xolapa complex (southern Mexico): Tectonics, v. 13, p. 455–474, doi: 10.1029/93TC02465.

Jansma, P.E., and Lang, R., 1997, The Arcelia graben: New evidence for Oligocene Basin and Range extension in southern Mexico: Geology, v. 25, p. 455–458, doi: 10.1130/0091-7613(1997)025<0455:TAGNEF>2.3.CO;2.

Jording, A., Ferrari, L., Arzate, J., and Jödicke, H., 2000, Crustal variations and terrane boundaries in southern Mexico as imaged by magnetotelluric transfer functions: Tectonophysics, v. 327, p. 1–13, doi: 10.1016/S0040-1951(00)00166-9.

Kemp, A.I.S., and Hawkesworth, C.J., 2004, Granitic perspectives on generation and secular evolution of the continental crust, *in* Rudnick, R. L., ed., Treatise on Geochemistry, Volume 3, The Crust: Amsterdam, Elsevier, p. 349–410.

Keppie, J.D., 2004, Terranes of Mexico revisited: 1.3 billion year odyssey: International Geology Review, v. 46, p. 765–794.

Keppie, J.D., and Morán-Zenteno, D.J., 2005, Tectonic implications of alternative Cenozoic reconstructions for southern Mexico and the Chortis Block: International Geology Review, v. 47, p. 473–491.

Kratzeisen, M.J., Frank, M.M., Negendank, J.F.W., Boehnel, H., and Terrell, D., 1991, The continental margin of southern Mexico: Tectonic evolution during the Tertiary: Zentralblatt für Geologie and Paläontologie, Teil I: Allgemeine, Angewandte, Regionale and Historische Geologie, v. 6, p. 1545–1555.

Lang, H.R., and Frerichs, W.E., 1998, New planktic foraminiferal data documenting Conician age for Laramide orogeny onset and paleoceanography in southern Mexico: The Journal of Geology, v. 106, p. 635–640.

Lang, H.R., Barros, J.A., Cabral-Cano, E., Draper, G., Harrison, C.G.A., Jansma, P.E., and Johnson, C.A., 1996, Terrane deletion in northern Guerrero state: Geofísica Internacional, v. 35, p. 349–359.

Levresse, G., González-Partida, E., Carrillo-Chávez, A., Tritlla, J., Camprubí, A., Cheilletz, A., Gasquet, D., and Deloule, E., 2004, Petrology, U/Pb dating and (C-O) stable isotope constraints on the source and evolution of the adakite-related Mezcala Fe-Au skarn district, Guerrero, México: Mineralium Deposita, v. 39, p. 301–312, doi: 10.1007/s00126-003-0403-y.

López-Infanzón, M., and Grajales-Nishimura, M., 1984, Edades de K-Ar de rocas ígneas y metamórficas del estado de Guerrero, VII Convención Nacional, Resúmenes: México, Sociedad Geológica Mexicana, 215 p.

Malfait, B.T., and Dinkelman, M.G., 1972, Circum-Caribbean tectonic and igneous activity and evolution of the Caribbean plate: Geological Society of America Bulletin, v. 83, p. 251–272.

Mammerickx, J., and Klitgord, K.D., 1982, Northern east Pacific rise: Evolution from 25 m.y. B.P. to the present: Journal of Geophysical Research, v. 87, B8, p. 6751–6759.

Martínez-Amador, H., Zárate-Barradas, R., Loaeza-García, J.P., Saenz-Pita, R., and Cardosa-Vázquez, E.A., 2001. Carta geológica-minera de la hoja Orizaba. E14–4: Pachuca, Hidalgo, Consejo de Recursos Minerales, scale 1:250,000, 1 sheet.

Martiny, B., Martínez-Serrano, R., Morán-Zenteno, D.J., Macías-Romo, C., and Ayuso, R., 2000, Stratigraphy, geochemistry and tectonic significance of the Oligocene magmatic rocks in western Oaxaca, southern Mexico: Tectonophysics, v. 318, p. 71–98, doi: 10.1016/S0040-1951(99)00307-8.

Martiny, B., Silva-Romo, G., and Morán-Zenteno, D.J., 2002, Tertiary faulting and relationship with Eocene-Oligocene volcanism in western Oaxaca, southern Mexico, Annual Meeting, Cordilleran Section: Geological Society of America Abstracts with Programs, v. 34, no. 5, p. A-97.

Meneses-Rocha, J.J., 1991, Tectonic development of the Ixtapa graben, Chiapas, Mexico [Ph.D. thesis]: Austin, University of Texas, 120 p.

Meneses-Rocha, J.J., Monroy-Audelo, M.E., and Gómez-Chavarria, J.C., 1994, Bosquejo paleogeográfico y tectónico del sur de México durante el Mesozoico: Boletín de la Asociación Mexicana de Geólogos Petroleros, v. 44, p. 18–45.

Meneses-Rocha, J.J., Rodríguez-Figueroa, D., Tóriz-Gama, J., Banda-Hernández, J., Hernández de la Fuente, R., and Valdivieso-Ramos, V., 1996, Excursión geológica al cinturón plegado y cabalgado de Zongolica: México, Petróleos Mexicanos, Exploración y Producción, internal report.

Mercier de Lépinay, B., Michaud, F., Calmus, T., Burgois, J., Poupeau, G., and Saint Marc, P., 1997, Large Neogene subsidence event along the middle America Trench off Mexico (18°N–19°N): Evidence from submersible observations: Geology, v. 25, p. 387–390, doi: 10.1130/0091-7613(1997)025<0387:LNSEAT>2.3.CO;2.

Meschede, M., Frisch, W., Herrmann, U.R., and Ratschbacher, L., 1996, Stress transmission across an active plate boundary: An example from southern Mexico: Tectonophysics, v. 266, p. 81–100, doi: 10.1016/S0040-1951(96)00184-9.

Meza-Figueroa, D., Valencia-Moreno M.V., Valencia, V.A., Ochoa-Landín, L., Pérez-Segura, E., and Díaz-Salgado, C., 2003, Major and trace element geochemistry and 40Ar/39Ar geochronology of Laramide plutons associated with gold-bearing Fe skarn deposits in Guerrero state, southern Mexico: Journal of South American Earth Sciences, v. 16, p. 205–217.

Monod, O., Faure, M., and Thieblemont, D., 1994, Guerrero terrane of Mexico: Its role in the southern Cordillera from new geochemical data–Comment: Geology, v. 22(5), 477 p.

Monroy, M., and Sosa, A., 1984, Geología de la Sierra del Tentzo, Puebla, borde norte del terreno Mixteco: Boletín de la Sociedad Geológica Mexicana, v. 45, p. 43–72.

Montiel-Escobar, J.E., Segura de la Teja, M.A., Estrada-Rodarte, G., Cruz-López, D., and Rosales-Franco, E., 2000, Carta geológica-minera de la hoja Ciudad Altamirano E14–4: Pachuca, Hidalgo, Consejo de Recursos Minerales, scale 1:250,000. 1 sheet.

Morán-Zenteno, D.J., 1992, Investigaciones isotópicas de Rb-Sr y Sm-Nd en rocas cristalinas de la región Tierra Colorada–Acapulco–Cruz Grande, estado de Guerrero. México [Ph.D. thesis]: Universidad Nacional Autónoma de México, 186 p.

Morán-Zenteno, D.J., Corona-Chávez, P., and Tolson, G., 1996, Uplift and subduction erosion in southwestern Mexico since the Oligocene: Pluton geobarometry constraints: Earth and Planetary Science Letters, v. 141, p. 51–65, doi: 10.1016/0012-821X(96)00067-2.

Morán-Zenteno, D.J., Tolson, G., Martínez-Serrano, R.G., Martiny, B., Schaaf, P., Silva-Romo, G., Macías-Romo, C., Alba-Aldave, L., Hernández-Bernal, M.S., and Solís-Pichardo, G.N., 1999, Tertiary arc-magmatism of the Sierra Madre del Sur, Mexico, and its transition to the volcanic activity of the TMVB: Journal of South American Earth Sciences, v. 12, p. 513–535, doi: 10.1016/S0895-9811(99)00036-X.

Morán-Zenteno, D.J., Alba-Aldave, L.A., Solé, J., and Iriondo, A., 2004, A major resurgent caldera in southern Mexico: The source of the late Eocene Tilzapotla ignimbrite: Journal of Volcanology and Geothermal Research, v. 136, p. 97–119, doi: 10.1016/j.jvolgeores.2004.04.002.

Mossman, R.W., and Viniegra, F., 1976, Complex fault structures in Veracruz Province of Mexico: The Association of Petroleum Geologists Bulletin, v. 60, p. 379–388.

Nava, A., Núñez-Cornú, F., Córdoba, D., Mena, M., Ansorge, J., González, J., Rodríguez, M., Banda, E., Mueller, S., Udías, A., García-García, M., and Calderón, G., 1988, Structure of the middle America trench in Oaxaca, México: Tectonophysics, v. 154, p. 241–251, doi: 10.1016/0040-1951(88)90106-0.

NAVDAT, 2005, The Western North American Volcanic and Intrusive Rock Database. National Science Foundation; China Lake Naval Air, Weapons Station, Geothermal Program Office: http://navdat.geongrid.org (May 2005).

Nieto-Samaniego, Á.F., Alaniz-Álvarez, S.A., and Ortega-Gutiérrez, F., 1995, Estructura interna de la falla de Oaxaca (México) e influencia de las anisotropías litológicas durante su actividad cenozoica: Revista Mexicana de Ciencias Geológicas, v. 12, p. 1–8.

Nieto-Samaniego, Á.F., Ferrari, L., Alaniz-Álvarez, S.A., Labarthe-Hernández, G., and Rosas-Elguera, R., 1999, Variation of Cenozoic extension and volcanism across the southern Sierra Madre Occidental volcanic province, México: Geological Society of America Bulletin, v. 111, p. 347–363, doi: 10.1130/0016-7606(1999)111<0347:VOCEAV>2.3.CO;2.

Nieto-Samaniego, A.F., Alaniz-Álvarez, S.A., Silva-Romo, G., Eguiza-Castro, M.H., and Mendoza-Rosales, C.C., 2006, Latest Cretaceous to Miocene deformation events in the eastern Sierra Madre del Sur, Mexico, inferred from the geometry and age of major structures: Geological Society of America Bulletin, v. 118, p. 238–252, doi: 10.1130/B25730.1.

Ortega-Gutiérrez, F., 1980, Rocas volcánicas del Maestrichtiano en el área de San Juan Tetelcingo, estado de Guerrero, en V Convención Geológica Nacional, Libro guía de la excursión geológica a la parte central de la cuenca del alto Río Balsas, estados de Guerrero y Puebla: México: Sociedad Geológica Mexicana, p. 34–38.

Ortega-Gutiérrez, F., 1981, Metamorphic belts of southern Mexico and their tectonic significance: Geofísica Internacional, v. 20, p. 177–202.

Ortega-Gutiérrez, F., Mitre-Salazar, L.M., Roldán-Quintana, J., Aranda-Gómez, J.J., Morán-Zenteno, D.J., Alaniz-Álvarez, S.A., and Nieto-Samaniego, Á.F., 1992, Carta geológica de la República Mexicana, quinta edición escala 1:2,000,000: México, D. F., Universidad Nacional Autónoma de México, Instituto de Geología; Secretaría de Energía, Minas e Industria Paraestatal, Consejo de Recursos Minerales, 1 sheet.

Pantoja-Alor, J., 1986, Siete edades geocronométricas cenozoicas de la cuenca media del Río Balsas, en Primer Simposio Geología Regional de México, Resúmenes: México, D.F., Universidad Nacional Autónoma de México, Instituto de Geología, p. 60–61.

Pardo, M., and Suárez, G., 1995, Shape of the subducted Rivera and Cocos plates in southern Mexico: Seismic and tectonic implications: Journal of Geophysical Research, v. 100, B7, p. 12,357–12,374, doi: 10.1029/95JB00919.

Pasquarè, G., Ferrari, L., Garduño, V.H., Tibaldi, A., and Vezzoli, L., 1991, Geology of the central sector of Mexican Volcanic Belt, states of Guanajuato and Michoacán: Boulder, Colorado, Geological Society of America, Map and Chart Series MCH072, scale 1:300,000, 1 sheet, 22 p.

Pindell, J.L., and Barret, S.F., 1990, Geological evolution of the Caribbean region; a plate tectonic perspective, in Dengo, G., and Case, J.E., eds., The Caribbean Region: Boulder, Colorado: Geological Society of America, Geology of North America, v. H, p. 405–432.

Pindell, J.L., Cande, S.C.W., Pitman, W.C., Rowley, D.B., Dewey, J.F., Lebrecque, J., and Haxby, W., 1988, A plate-kinematic framework for models of Caribbean evolution: Tectonophysics, v. 155, p. 121–138, doi: 10.1016/0040-1951(88)90262-4.

Pitcher, W.S., 1993, The nature and origin of granite: London, Blackie Academic & Professional, 321 p.

Ponce, L., Gaulon, R., Suárez, G., and Lomas, E., 1992, Geometry and state of stress of the downgoing Cocos plate in the Isthmus of Tehuantepec, Mexico: Geophysical Research Letters, v. 19, p. 773–776.

Ratschbacher, L., Riller, U., Meschede, M., Herrmann, U., and Frisch, W., 1991, Second look at suspect terranes in southern Mexico: Geology, v. 19, p. 1233–1236, doi: 10.1130/0091-7613(1991)019<1233:SLASTI>2.3.CO;2.

Riller, U., Ratschbacher, L., and Frisch, W., 1992, Left-lateral transtension along the Tierra Colorada deformation zone, northern margin of the Xolapa magmatic arc of southern Mexico: Journal of South American Earth Sciences, v. 5, p. 237–249, doi: 10.1016/0895-9811(92)90023-R.

Rivera-Carranza, E., Tejada-Segura, M.A., Miranda-Huerta, A., Lemus-Bustos, O., Motolinía-García, O., León-Ayala, V., and Moctezuma-Salgado, M.D., 1998, Carta geológico-minera de la hoja Cuernavaca E14–5: Pachuca, Hidalgo, Consejo de Recursos Minerales, scale 1:250,000, 1 sheet.

Robinson, K.L., Gastil, R.G., Campa, M.F., and Ramírez-Espinosa, J., 1989, Geochronology of basement and metasedimentary rocks in southern Mexico and their relation to metasedimentary rocks in Peninsular California: Geological Society of America, Abstracts with Programs, v. 21(5), p. 135.

Rosencrantz, E., and Sclater, J.G., 1986, Depth and age in the Cayman trough: Earth and Planetary Science Letters, v. 79, p. 133–144, doi: 10.1016/0012-821X(86)90046-4.

Ross, M., and Scotese, C.R., 1988, A hierarchical tectonic model of the Gulf of Mexico and Caribbean region: Tectonophysics, v. 155, p. 139–168, doi: 10.1016/0040-1951(88)90263-6.

Saleeby, J., 2003, Segmentation of the Laramide slab—Evidence from the southern Sierra Nevada region: Geological Society of America Bulletin, v. 115, p. 655–668, doi: 10.1130/0016-7606(2003)115<0655:SOTLSF>2.0.CO;2.

Salinas-Prieto, J.C., Monod, O., and Faure, M., 2000, Ductile deformations of opposite vergence in the eastern part of the Guerrero terrane (SW Mexico): Journal of South American Earth Sciences, v. 13, p. 389–402, doi: 10.1016/S0895-9811(00)00031-6.

Sánchez-Barreda, L.A., 1981, Geologic evolution of the continental margin of the Gulf of Tehuantepec in southern Mexico [Ph.D. thesis]: Austin, University of Texas, 192 p.
Schaaf, P., 1990, Isotopengeochemishe untersuchugen an granitoiden gesteinen eines aktiven kontinentalrandes. Alter und herkufen der tiefengesteinskomplexe an der pazifikküste Mexikos zwischen Puerto Vallarta und Acapulco [Ph.D. thesis]: München, Facultät fur Geowissenschaten der Ludwig-Maximilians Universität, 202 p.
Schaaf, P., Morán-Zenteno, D., Hernández-Bernal, M.S., Solís-Pichardo, G., Tolson, G., and Köhler, H., 1995, Paleogene continental margin truncation in southwestern Mexico: Geochronological evidence: Tectonics, v. 14, p. 1339–1350, doi: 10.1029/95TC01928.
Sedlock, R.L., Ortega-Gutiérrez, F., and Speed, C., 1993, Tectonostratigraphic Terranes and Tectonic Evolution of Mexico: Geological Society of America Special Paper 278, 153 p.
Silva-Romo, G., Martiny, B., and Mendoza-Rosales, C., 2001, Formación de cuencas continentales en el sur de México y su cronología respecto al desplazamiento del bloque de Chortis: XI Congreso Latinoamericano de Geología, III Congreso Uruguayo de Geología, Memoria: Montevideo, Uruguay, 9 p.
Smith, R.D., Cameron, K.L., McDowell, F.W., Niemeyer, S., and Sampson, E., 1996, Generation of voluminous silicic magmas and formation of mid-Cenozoic crust beneath north-central Mexico: Evidence from ignimbrites, associated lavas, deep crustal granulites and mantle pyroxenites: Contributions to Mineralogy and Petrology, v. 123, p. 375–389, doi: 10.1007/s004100050163.
Stacey, J.S., and Kramers, J.D., 1975, Approximation of terrestrial south lead isotope evolution by a two-stage model: Earth and Planetary Science Letters, v. 26, p. 207–221, doi: 10.1016/0012-821X(75)90088-6.
Tolson, G., Solís-Pichardo, G., Morán-Zenteno, D.J., Victoria-Morales, A., and Hernández-Treviño, J.T., 1993, Naturaleza petrográfica y estructural de las rocas cristalinas de 1a zona de contacto entre los terrenos Xolapa y Oaxaca, región de Santa María Huatulco, Oaxaca, *in* Delgado-Argote, L., and Martín-Barajas, A., eds., Contribuciones a la tectónica del occidente de México: Ensenada, B. C., Unión Geofísica Mexicana, Monografía 1, p. 327–349.
Tolson-Jones, G., 1998, Deformación, exhumación y neotectónica de la margen continental de Oaxaca: datos estructurales, petrológicos y geotermobarométricos [Ph.D. thesis]: México, Universidad Nacional Autónoma de México, 98 p.
Tritlla, J., Camprubí, A., Centeno-García, E., Corona-Esquivel, R., Sánchez-Martínez, S., Gasca-Durán, A., Cienfuegos-Alvarado, E., and Morales-Puente, P., 2003, Estructura y edad del depósito de hierro de Peña Colorada (Colima), un posible equivalente fanerozoico e los depósitos de tipo IOCG: Revista Mexicana de Ciencias Geológicas, v. 20, p. 182–201.
Urrutia-Fucugauchi, J., and Ferrusquía-Villafranca, I., 2001, Paleomagnetic results for the Middle-Miocene continental Suchilquitongo Formation, Valley of Oaxaca, southeastern Mexico: Geofísica Internacional, v. 40, p. 191–206.
Urrutia-Fucugauchi, J., and Flores-Ruiz, J.H., 1996, Bouguer gravity anomalies and regional crustal structure in central Mexico: International Geology Review, v. 38, p. 176–194.
Valdés, C.M., Mooney, W., Singh, K., Meyer, S., Lomnitz, C., Luetgert, J., Hesley, C., Lewis, B., and Mena, M., 1986, Crustal structure of Oaxaca, Mexico, from seismic refraction measurements: Bulletin of the Seismological Society of America, v. 76, p. 547–563.

Manuscript Accepted by the Society 29 August 2006

Printed in the USA

Geological Society of America
Special Paper 422
2007

Late Cenozoic intraplate-type volcanism in central and northern México: A review

José Jorge Aranda-Gómez*
Departamento de Geología Económica, Instituto Potosino de Investigación Científica y Tecnológica, Apartado postal 3-74, San Luis Potosí, S. L. P., 78216, México

James F. Luhr†
Department of Mineral Sciences, Smithsonian Institution, P.O. Box 37012, NHB-119, Washington, DC 20013-7012, USA

Todd B. Housh
Department of Geological Sciences, University of Texas at Austin, Austin, Texas 78712, USA

Gabriel Valdez-Moreno
Posgrado en Ciencias de la Tierra, Centro de Geociencias, Universidad Nacional Autónoma de México, Campus Juriquilla, Apartado postal 1-742, Querétaro, Qro., 76001, México

Gabriel Chávez-Cabello§
Posgrado en Ciencias de la Tierra, Centro de Geociencias, Universidad Nacional Autónoma de México, Campus Juriquilla, Apartado postal 1-742, Querétaro, Qro., 76001, México

ABSTRACT

Intraplate-type volcanism (late Oligocene to Quaternary) occurs in México on both the North American and the Pacific plates. Oceanic localities are voluminous shield volcanoes on or near fossil-spreading ridges. Subaerial rocks of these volcanoes form either geochemically continuous and coherent rock series or bimodal suites. Low-pressure crystal fractionation of alkali basalt and assimilation of hydrothermally altered rocks from the volcanic pile determined the compositions of the mafic-intermediate rocks at Socorro. Trachytes at Socorro were formed by partial melting of alkali basalt, and rhyolites through crystal fractionation of parental trachytes. These felsic rocks also assimilated hydrothermally altered rocks.

Continental localities (late Oligocene to Quaternary) occur scattered north of the Trans-Mexican Volcanic Belt; their location is independent of older volcanic provinces and boundaries between today's geologic/tectonic provinces. Intraplate-type rocks

*Present address: Centro de Geociencias, UNAM, Campus Juriquilla, Querétaro, Qro. 76230, México; jjag@geociencias.unam.mx.
†deceased
§Present address: Facultad de Ciencias de la Tierra, Universidad Autónoma de Nuevo León, Linares, N.L. 67700, México.

Aranda-Gómez, J.J., Luhr, J.F., Housh, T.B., Valdez-Moreno, G., and Chávez-Cabello, G., 2007, Late Cenozoic intraplate-type volcanism in central and northern México: A review, *in* Alaniz-Álvarez, S.A., and Nieto-Samaniego, Á.F., eds., Geology of México: Celebrating the Centenary of the Geological Society of México: Geological Society of America Special Paper 422, p. 93–128, doi: 10.1130/2007.2422(04). For permission to copy, contact editing@geosociety.org.

have high TiO_2, Nb, and Ta contents, and host mantle and/or lower crust xenoliths ± megacrysts. Many fields are in the southern Basin and Range and the most extensive and voluminous were contemporaneous with normal faulting. The locations of some fields suggest that magma ascent was influenced by regional structures. However, normal faulting is minor or absent in many other Mexican fields.

The petrogenetic processes in late Oligocene to Miocene magmas differ from those in Plio-Quaternary magmas. Slow ascent of magmas formed during early stages of extension-favored assimilation-fractional crystallization and gravitational settling of xenoliths. Plio-Quaternary xenolith-bearing magmas traveled faster through cooler crust where brittle structures caused by extension were able to propagate deeper. Geochemical evidence of assimilation is more subtle in the younger volcanic rocks.

Keywords: alkalic volcanic rocks, extension, Basin and Range, peridotite xenoliths.

INTRODUCTION

Based on the type of crust from which they erupted, the intraplate-type volcanic fields in México can be divided into two groups: oceanic and continental. The oceanic fields are comprised of five islands or archipelagoes located on or near abandoned ocean ridges on what is now the Pacific plate. All but one of the continental localities are on the North American plate; the exception is San Quintín, whose underlying continental crust is also part of the Pacific plate. Four large magmatic provinces have been traditionally considered in the geological literature about México. These provinces are the Eastern Alkalic, Sierra Madre Occidental, Californian, and Trans-Mexican Volcanic Belt (see inset in Fig. 1). A series of late Tertiary and Quaternary volcanic fields located north of the belt have been documented in the past 20 yr. We propose that these localities should be considered as a fifth magmatic province overprinting the previous four provinces (e.g., Demant and Robin, 1975; Ortega-Gutiérrez et al., 1992; Morán-Zenteno, 1994). We have named this province the Northern Mexico Extensional Province. To systematize the data included in this paper we have divided the province into three regions: Baja California, Southern Basin and Range, and Eastern/Alkalic (Fig. 1).

Published information on both oceanic and continental intraplate-type volcanic fields is reviewed in this paper. The islands comprising the oceanic group are alkalic volcanoes younger than 15 Ma that formed on top of ocean-floor tholeiites. All but one of the continental volcanic fields are Miocene or younger (<24 Ma) and are located throughout the Northern Mexico Extensional Province. The one exception is the Bernal de Horcasitas volcanic neck (K-Ar: ca. 28 Ma). The rocks from continental localities are usually mafic, and plot in Miyashiro's (1974) diagram of SiO_2 versus FeO^T/MgO above the $FeO^T/MgO = (0.156 \times SiO_2) - 6.69$ line (i.e., outside of the calc-alkaline region). The volcanic products are commonly alkaline (i.e., with *nepheline* in the CIPW norm), although there are unaltered rocks in many of the volcanic fields with normative *hypersthene* or even *quartz*. Their high contents of TiO_2, Nb, and Ta, as well as their geologic occurrence away from plate margins, are consistent with their designation as products of intraplate magmatic activity. In general, this type of volcanism is monogenetic, with magmas usually erupted through central conduits forming cinder cones and associated lava flows, small continental lava shields, or isolated maar volcanoes. Small-volume lava flows produced by fissural activity occur in a few places. The intraplate activity in the Northern Mexico Extensional Province produced volcanic fields that cover small isolated areas and the volume of the material erupted is significantly smaller than in the four Mexican volcanic provinces proposed by Demant and Robin (1975).

Our initial interest in México's continental intraplate-type volcanism was triggered by the fact that these magmas commonly carried mantle or deep-crustal xenoliths to the surface (Aranda-Gómez and Ortega-Gutiérrez, 1987; Luhr and Aranda-Gómez, 1997), as well as complex megacryst assemblages that might be either accidental or cogenetic with the host magmas. The overall distribution of these volcanic fields and the occurrence of Cenozoic normal faults in central and northern México indicate that most of these fields are located in a region that was extended during the middle and late Cenozoic. This region has been considered by Henry and Aranda-Gómez (1992, 2000) as the southern part of the Basin and Range Province. However, mantle xenolith localities, such as San Quintín (northern Baja California), those in the Eastern/Alkalic region, and others in the Trans-Mexican Volcanic Belt, are clearly located outside of the Basin and Range Province (Fig. 1).

The localities discussed in this paper are first divided on the basis of whether they lie atop oceanic or continental crust. Then, for the continental localities, they are grouped on the basis of their age, as there is evidence of significant petrological contrasts between those that erupted at the beginning of the crustal extension (late Oligocene–Miocene) and those erupted later. The Plio-Quaternary localities (<5 Ma) are also classified in terms of their location in the Northern Mexico Extensional Province (Fig. 1). Systematic petrological and geochemical investigations have been published for only a few of these younger volcanic fields (e.g., Ventura–Espíritu Santo). For other fields, detailed investigations have been published for one or more volcanoes (e.g., Durango). And in still other regions we assume that a volcanic field is formed by intraplate-type magmatism based on its regional

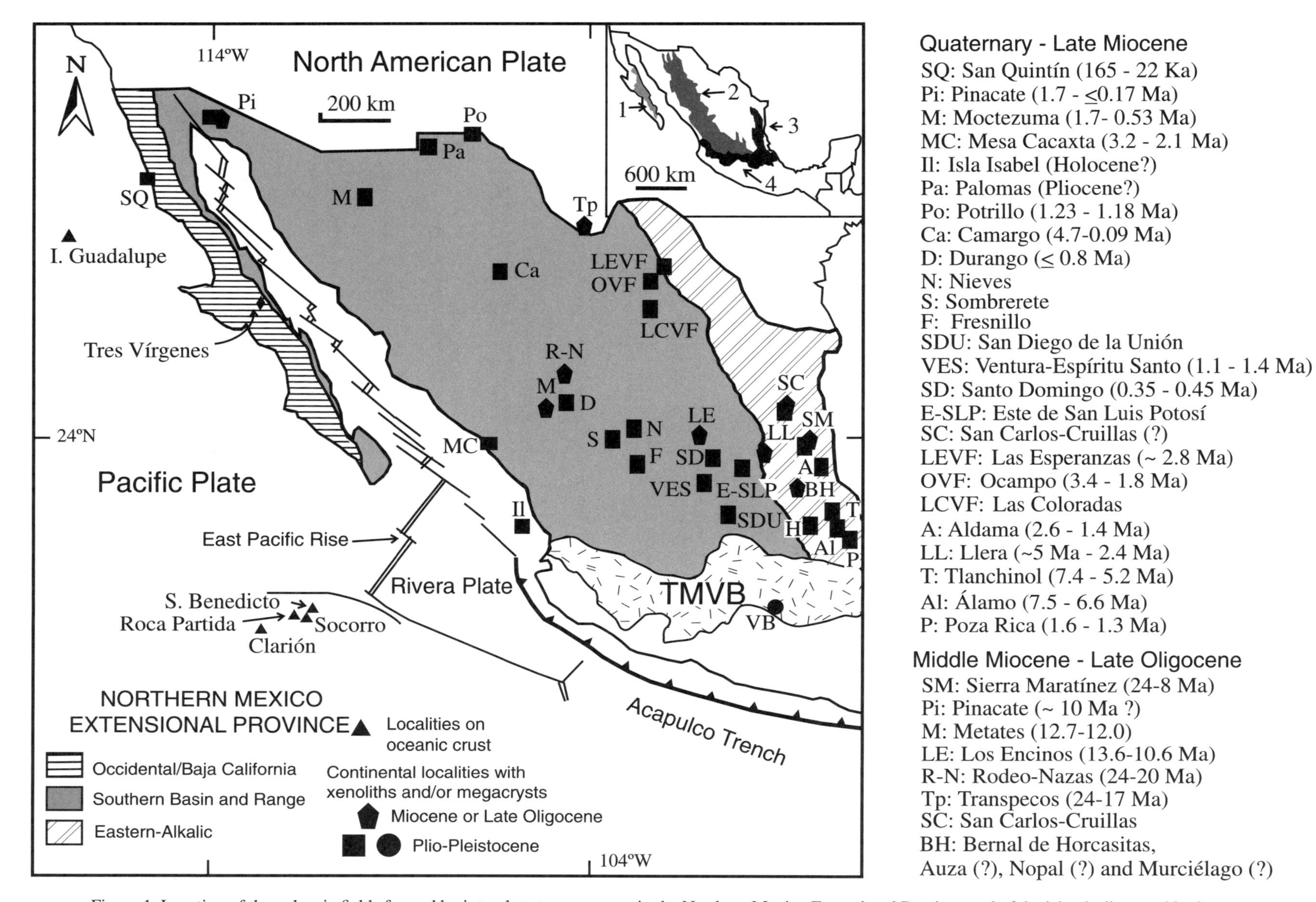

Figure 1. Location of the volcanic fields formed by intraplate-type magmas in the Northern Mexico Extensional Province and of the islands discussed in the text. All the continental fields mentioned in this paper are north of the Trans-Mexican Volcanic Belt and many contain xenoliths derived from the mantle and/or from deep portions of the crust, as well as megacryst assemblages. The localities with mantle xenoliths in the Trans-Mexican Volcanic Belt are Valle de Bravo (VB) and Palma Sola (outside of the map area). The host rock for the xenoliths in Valle de Bravo is an andesite (Blatter and Carmichael, 1998). This figure also shows the location of the Tres Vírgenes volcano, which is mentioned in the text. If this figure is compared with Figure 1 in Aranda-Gómez and Ortega-Gutiérrez (1987), the reader will note that a few xenolith localities are not included here. The reasons for their omission are discussed in the text. The inset shows the four volcanic provinces defined by Demant and Robin (1975); key: 1—Californian, 2—Sierra Madre Occidental, 3—Eastern Alkalic, and 4—Trans-Mexican Volcanic Belt.

setting, chemical analysis of a few samples (e.g., Camargo), and/or the occurrence of xenoliths (e.g., Fresnillo) and/or megacrysts (e.g., Metates) in the lavas and/or tephra deposits.

OCEANIC MAGMATISM

Guadalupe Island

The petrology of this locality was studied by Batiza (1977), and most of the information we summarize comes from his paper. Medina et al. (1989) and Siebert et al. (2002) provided additional information about recent volcanic activity on this island. Guadalupe is located ~300 km west of the coast of Baja California (Fig. 1), close to 29°N and 118°W, and it has a maximum elevation of 1100 masl. The island is elongated (N-S) and is formed by two eroded overlapping shield volcanoes; the volcano located in the northern part of the island is younger. Calderas mark the summits of both shields; the flanks of the central volcanoes, as well as the floors of the calderas, are partially covered by the products of late fissural activity. The island is on the axis of an abandoned fossil ridge flanked by magnetic anomaly 5B (ca. 15.5 Ma). The age of the oldest shield activity is unknown and the oldest rocks dated (K-Ar) are 7 ± 2 Ma. The youngest activity was considered by Medina et al. (1989) and Siebert et al. (2002) to have occurred in the Holocene. The shield-forming lavas, together with the products of the late fissural activity, constitute a complete and coherent alkali basalt–trachyte series (Fig. 2A). All the intermediate members, including hawaiite, mugearite, and benmoreite, are present (Fig. 2A). Although the volcanic rocks on Guadalupe are generally crystal-poor, with <10% phenocrysts, some are aphyric and others are strongly porphyritic. The mineralogy of the rocks varies according to rock composition. The observed mineral assemblage in basalts is: Ol + Pl + Aug + FTO (Table 1). The hawaiites, mugearites, and benmoreites all have similar mineral assemblages: Ol + Pl + Aug + Ap + FTO ± Krs (not present in the shield-forming lavas). Although their mineral compositions differ, plagioclase have higher albite contents and ferromagnesian minerals are richer in Fe in the more silicic rocks. The mineral assemblage in the trachytes is: Pl + Ol + Cpx + Bt + FTO ± Amph ± Kspar. The mafic alkalic rocks (Fig. 3A) contain 5%–10% normative *nepheline* and have Ti, K, Na, P, Rb, Sr, Zr, and LREE (light rare-earth element) abundances similar to those of typical oceanic island basalts. Batiza (1977) modeled whole-rock compositional variations in the series by fractional crystallization of mineral assemblages similar to those found in the rocks, starting from an alkali basalt parent at pressures between 2 and 10 kb. This model is consistent with the smooth chemical variations, the temporal succession of different rock types, and the relatively homogeneous $^{87}Sr/^{86}Sr$ values (0.70321–0.70330) observed in the series. The tectonic location of Isla Guadalupe on the axis of a fossil ridge has significant implications for the size of upper mantle domains with different chemical and isotopic compositions and the shape of the source areas that feed the magmatic systems beneath mid-ocean ridges and nearby off-axis ocean-island and seamount volcanoes.

Revillagigedo Archipelago

The Revillagigedo Archipelago comprises four volcanic islands—Socorro, San Benedicto, Clarión, and Roca Partida (Fig. 1)—located in the eastern Pacific, near the northern end of Mathematician Ridge. This topographic feature on the seafloor marks the location of a spreading center that was active 3.5 Ma (Mammerickx et al., 1988). The islands represent alkalic magmatic activity that occurred after the abandonment of the ridge. All but one of the Revillagigedo islands (Roca Partida) have volumes in the upper 1%–2% of Pacific seamounts and islands located on Pliocene age oceanic crust (Batiza, 1982). The youngest periods of volcanic activity in the archipelago are the 1952–1953 eruptions on San Benedicto Island (Richards, 1959) and a submarine eruption near Socorro in 1993 (McClelland et al., 1993; Siebe et al., 1995). Regional mantle heterogeneity is evident in the elemental (Figs. 4A and 5A) and isotopic data (Figs. 6A and 7A) for mildly alkalic transitional basalts from Socorro, mugearites from San Benedicto, and submarine basalts collected near Socorro (Bohrson and Reid, 1995).

Socorro Island

The initial geologic work at this locality was done by Bryan (1959, 1966, 1976). Recently, the island was studied by Bohrson and Reid (1995, 1997) and Bohrson et al. (1996). Our review is based on this information.

Socorro is a unique locality in the Pacific because its subaerial volcanic activity has been dominated (~80%) by silicic peralkaline magmas, although it is believed that the volcano is mostly basaltic. The total volume of the volcano is 2400 km^3, and less than 2% of the structure (40 km^3) is above sea level. A caldera (~4.5 × 3.8 km) is located at the summit of the volcano, but it is almost completely covered by younger volcanic products. The eruptive history of the volcano has been divided into three stages: pre-, syn-, and post-caldera. The pre- and syn-caldera rocks are lava flows and ignimbrites of trachytic and rhyolitic composition with ^{40}Ar-^{39}Ar ages between 540 and 370 ka. Post-caldera activity is dated from 180 ka and the youngest dated rock yielded an age of 15 ka. Post-caldera volcanic activity formed lava domes and flows of silicic composition. Cerro Evermann is a tephra cone with associated domes and its summit marks the highest point on the island (1050 masl). The Evermann Formation is post-caldera in age and is dominated by domes, peralkaline silicic lavas, and associated tephra, but it also contains alkali basalt lava flows. The post-caldera Lomas Coloradas Formation (180–15 ka) is exposed in the southeastern part of the island and is dominated by alkali basalt associated with less abundant peralkaline silicic rocks. No intermediate composition rocks (SiO_2 = 54–61%) have been found on the island (Fig. 2A). This compositional gap is also observed in other oxides and elements such as K_2O, TiO_2, P_2O_5, and Sr.

Post-Caldera Basalt Genesis. The mafic volcanic rocks from Socorro are mostly aphyric or contain only a few phenocrysts. The phenocryst assemblages observed are: Pl >> Ol ≈ Cpx or Pl >> Ol. The matrix has a pilotaxitic or intergranular texture and it

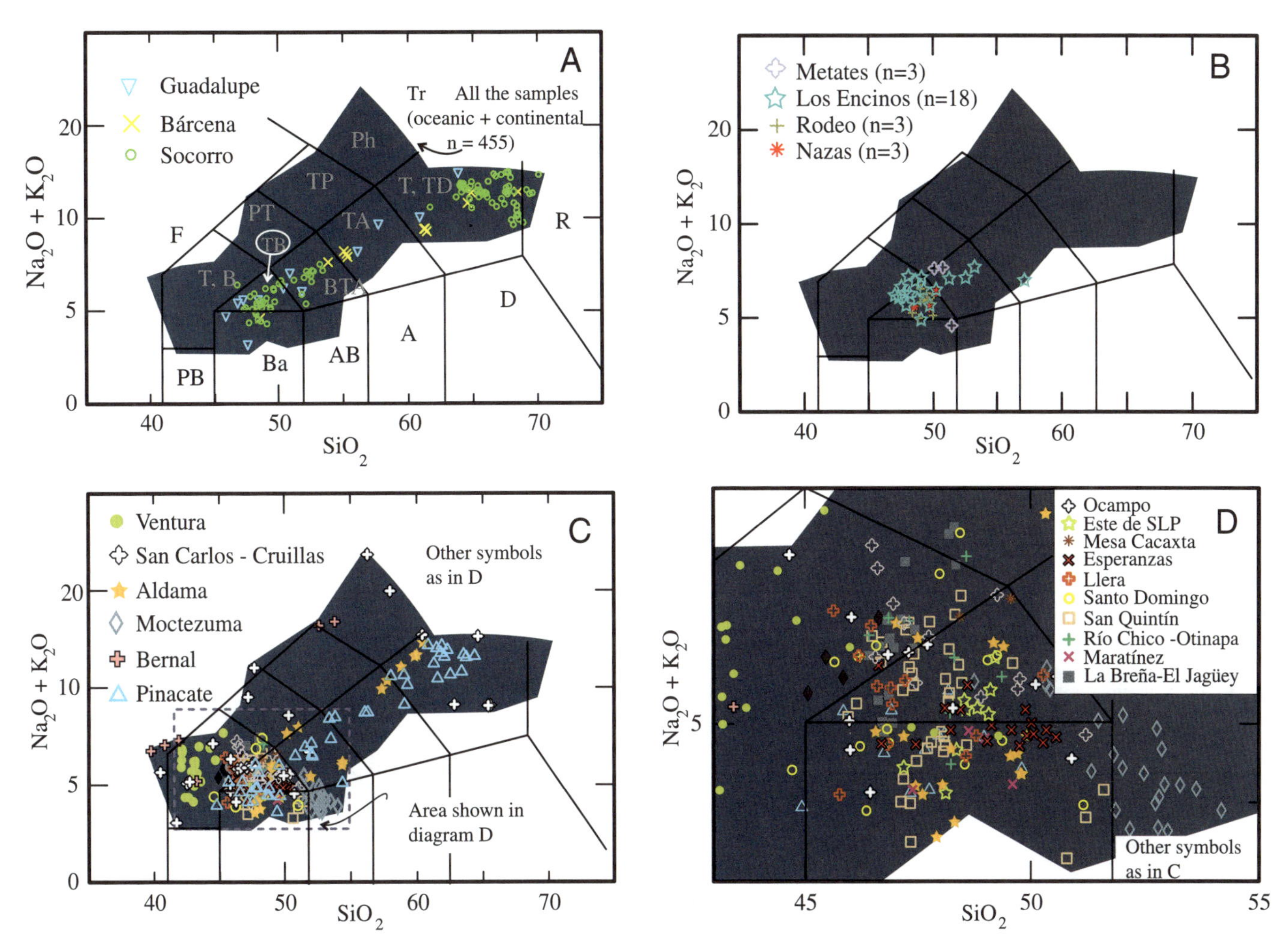

Figure 2. TAS classification (Le Maitre et al., 2002) of the intraplate-type volcanic rocks from México. SiO_2 and ($Na_2O + K_2O$) are in wt%. All the analyzed samples (n = 455), both from continental and oceanic localities, plot inside the blue area. (A) Samples collected at volcanoes formed on oceanic crust (n = 124). Socorro's samples (n = 90) are clearly bimodal. The compositional gap between these groups is covered by samples from Guadalupe (n = 13) and Bárcena (n = 13). (B) Samples from continental localities formed during the early stage of intraplate volcanism (Miocene). Most analyses (n = 35) plot in the compositional region where a large number of Plio-Quaternary rocks also plot. (C) Samples collected at Plio-Quaternary continental localities (n = 296). Note that with the exception of the volcanic fields listed in the legend, the rest of the analyses tend to concentrate near the hawaiite field. The most extreme chemical variations are found in the listed volcanic fields. (D) Despite the tendency of many of the samples to plot around the hawaiite field, samples from the same volcanic field tend to cluster in well-defined regions in the diagram (e.g., Ventura, Moctezuma, and Esperanzas). Abbreviations: F—foidite; PB—picro-basalt; T—tephrite (Ol < 10%); B—basanite (Ol > 10%); PT—phonotephrite; TP—tephriphonolite; Ph—phonolite; Ba—basalt; TB—trachybasalt (hawaiite or potassic trachybasalt); BTA—basaltic trachyandesite (mugearite or shoshonite); TA—trachyandesite (benmoreite or latite); Tr—trachyte (quartz < 20%) or TD—trachydacite (Qtz > 20%); Ba—basalt; A—andesite, D—dacite; R—rhyolite; SLP—San Luis Potosí; AB—basaltic andesite. Names inside of parentheses are applied based on whether Na_2O - 2 is greater or less than K_2O (Le Maitre et al., 2002).

is composed of Pl + Ol + Cpx + FTO ± Ap. The studied samples form a magma series dominated by alkali basalt, hawaiite, and mugearite (Figs. 2A and 3A) with a small amount of associated mildly alkaline transitional basalt. Most variations in major and high-field strength elements (HFSE) are consistent with low-pressure crystal fractionation of Pl + Cpx + Ol + FTO. If an alkali basalt parental magma is considered, it takes 50% fractional crystallization to produce a mugearite. The limited variation of Zr/Nb values (5.7–7.2) in Socorro's post-caldera alkali basalts suggests that the parental magmas are also the result of roughly the same amount of partial melting from a relatively homogenous mantle. The $^{87}Sr/^{86}Sr$ and $^{143}Nd/^{144}Nd$ data (Fig. 6A) show relatively narrow ranges (0.7031–0.7032, and 0.5128–0.5130, respectively). Pb isotope ratios (Fig. 7A) are $^{206}Pb/^{204}Pb$ = 18.74–19.16, $^{207}Pb/^{204}Pb$ = 15.56–15.65, and $^{208}Pb/^{204}Pb$ = 38.36–38.88, and are similar to those found in other alkalic seamounts near the East Pacific Rise. Many samples have negative Ce anomalies (Figs. 4A and 5A) and are enriched in P_2O_5, Ba, and intermediate REE (rare-earth elements). These features cannot be explained by the fractional crystallization model. Instead they have been explained by assim-

TABLE 1. SYMBOLS FOR MODAL MINERALS[†]

Ab	Albite	Di	Diopside	Ol	Olivine
Acm	Acmite	Fa	Fayalite	Opx	Orthopyroxene
Aeg*	Aegirine	FHd*	Ferrohedenbergite	Pgt	Pigeonite
Aen*	Aenigmatite	Fo	Forsterite	Pl	Plagioclase
Amph*	Amphibole	FTO*	Fe and Ti oxides	Po	Pyrrhotite
An	Anorthite	Grt	Garnet	Qtz	Quartz
Anor*	Anorthoclase	Hbl	Hornblende	Sa	Sanidine
Ap	Apatite	Hy*	Hypersthene	Sil	Sillimanite
Aug	Augite	Ilm	Ilmenite	Spl	Spinel
Bri*	Britholite	Krs	Kaersutite	TAug*	Titanaugite
Bt	Biotite	Kspar*	Potassium feldspar	TMt*	Titanomagnetite
Cha*	Chalcedony	Lct	Leucite		
Cpx*	Clinopyroxene	Ne	Nepheline		

[†]Symbols are after Kretz (1983), except those marked with an asterisk, as they are not included in his table for rock-forming minerals. The full name of normative components is given in the text in italics.

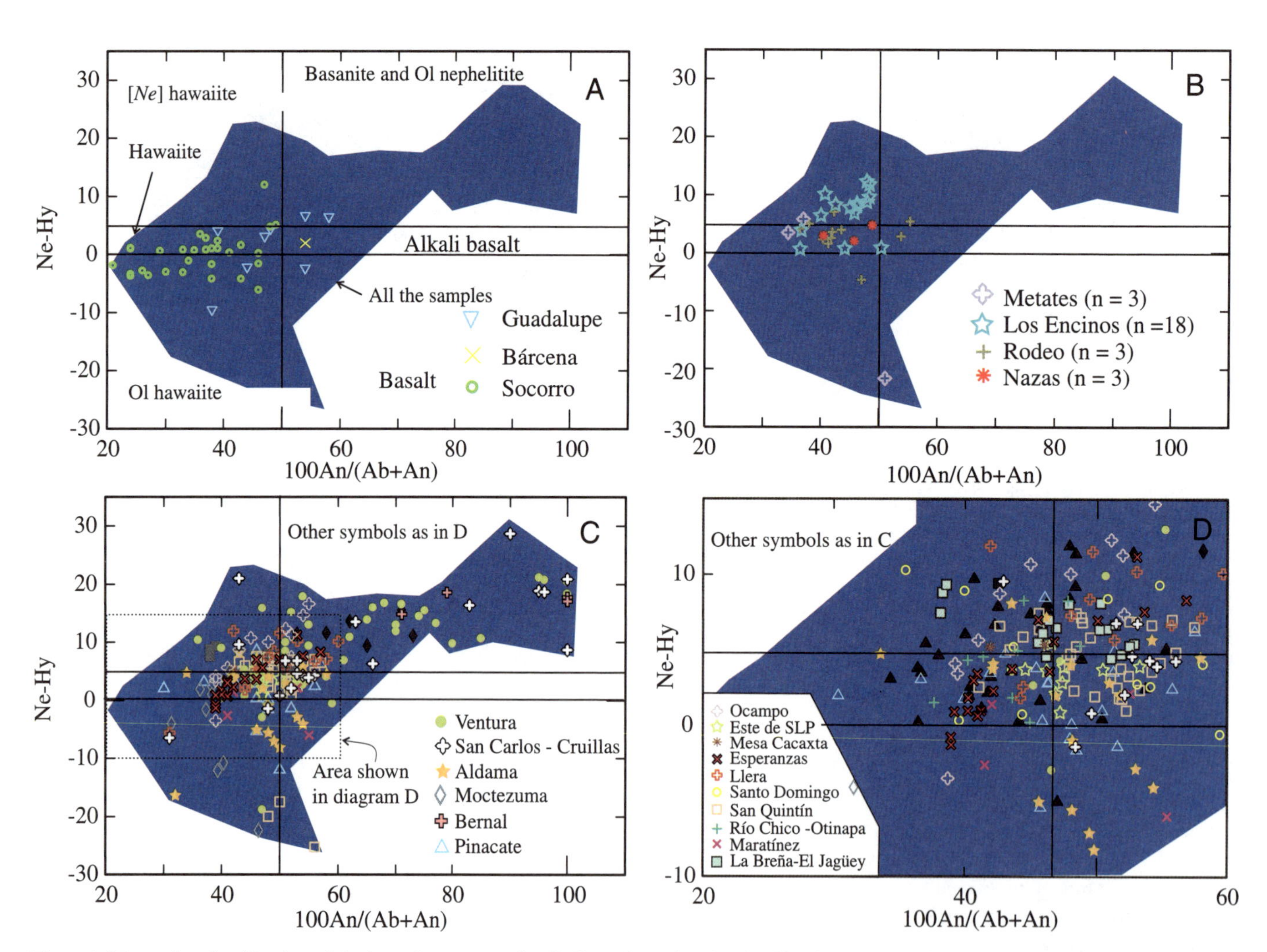

Figure 3. Normative classification of the intraplate-type mafic alkalic rocks, using the classification scheme of Luhr et al. (1995b). Normative composition of plagioclase is plotted against normative *nepheline* minus normative *hypersthene* in rocks with $SiO_2 < 52$ (wt%). All the analyzed samples (n = 313), both from continental and oceanic localities, plot inside of the blue area. (A) Samples collected at volcanoes formed on oceanic crust (n = 40). (B) Samples (n = 27) from continental localities formed during the early stage of intraplate volcanism (Miocene). (C) Samples collected at Plio-Quaternary continental localities (n = 246). The area inside the box is enlarged and composes (D). SLP—San Luis Potosí.

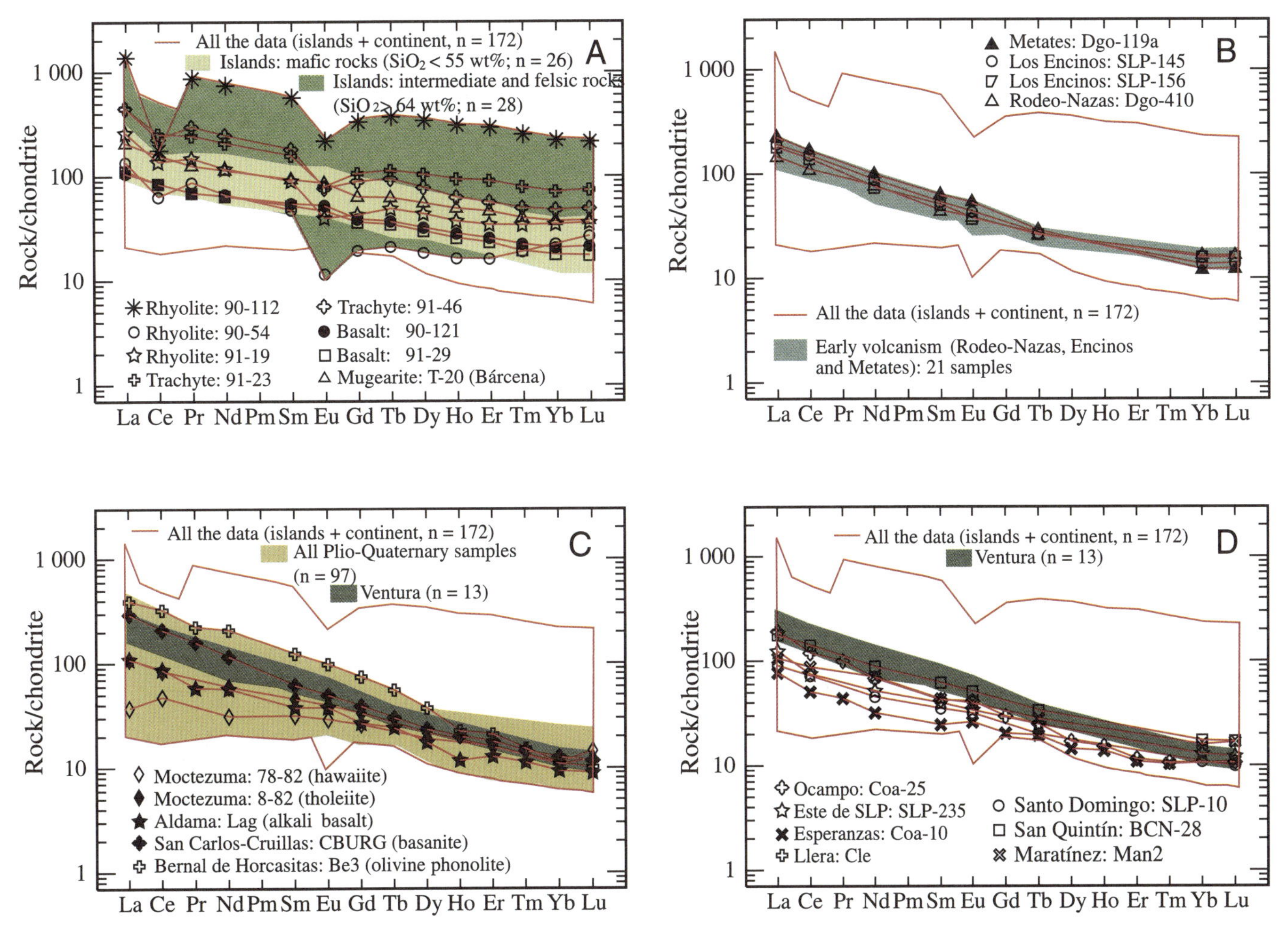

Figure 4. Chondrite-normalized (Sun and McDonough, 1989) REE plots. The solid red line in all the diagrams (A–D) borders the area where all the oceanic and continental samples plot. Each of the graphs shows a few representative examples from the localities discussed in the text. (A) Samples from the islands display different behavior as a function of SiO_2 content. Note that the curves for the mafic samples have similar behavior to the continental intraplate rocks. (B) Representative examples from continental localities formed during the early stage (Miocene) of intraplate volcanism. (C) All the samples from Plio-Quaternary continental volcanic fields plot inside the light green area. The Ventura samples, which we consider as the most extreme of the intraplate-type rocks with mantle xenoliths, plot in the dark green area. This graph includes one Oligocene sample (Bernal de Horcasitas) from the Eastern/Alkalic region. Unlike other rocks from the early stage of intraplate volcanism (Rodeo-Nazas, Los Encinos, and Metates), rocks from Bernal de Horcasitas do not contain megacrysts or xenoliths of feldspathic granulites from the deep crust. (D) Other examples from the Plio-Quaternary volcanic fields. SLP—San Luis Potosí.

ilation of oceanic crustal components, metalliferous sediments, and/or apatite formed in previous plutonic stages of the Socorro magma system. Anomalous values of radiogenic Sr (Fig. 6A) are attributed to interaction with seawater or hydrothermal fluids. The scale of the geochemical anomalies caused by assimilation and contamination is variable and is a function of the amount and type of material incorporated.

Origin of Peralkaline Trachytes and Rhyolites. The sample set (n = 51) studied by Bohrson and Reid (1997) included pre-, syn-, and post-caldera rocks collected from domes, lava flows, and ignimbrites. Samples collected from young lavas and domes are vitrophyres, while older lavas and ignimbrites are holocrystalline. Phenocrysts form 0–15% (volume) of the rocks, and the assemblage is Kspar ($Ab_{65}Or_{35}$–$Ab_{67}Or_{33}$) >> sodic pyroxene ± Fa ± FTO ± Aen. Mineralogy in the groundmass is similar to that in the phenocryst assemblage. All the studied rocks but one (a mugearite) are peralkaline trachytes or rhyolites with CIPW *quartz*, *acmite*, and *sodium silicate*, and enrichments (compared to metaluminous rocks) in Na_2O, K_2O, FeO_t, and in some HFSE. The rocks are depleted in Al_2O_3, Sr, and Ba. All the samples are enriched in LREE with La/Yb_N values between 2.9 and 8.9, and most have negative Eu anomalies. Some of the rocks show Ce anomalies, which are almost always negative (Fig. 5A). Nd and Pb isotope ratios range $^{143}Nd/^{144}Nd$ = 0.512869–0.512956, $^{206}Pb/^{204}Pb$ = 18.76–19.00, $^{207}Pb/^{204}Pb$ = 15.55–15.61, $^{208}Pb/^{204}Pb$ = 38.36–38.71, similar to the ratios found in Socorro's alkali basalts (Figs. 6A and 7A). The $^{87}Sr/^{86}Sr$ values, measured in feldspars, are 0.703086–0.704632. Although whole-rock $^{87}Sr/^{86}Sr$ values are considerably more vari-

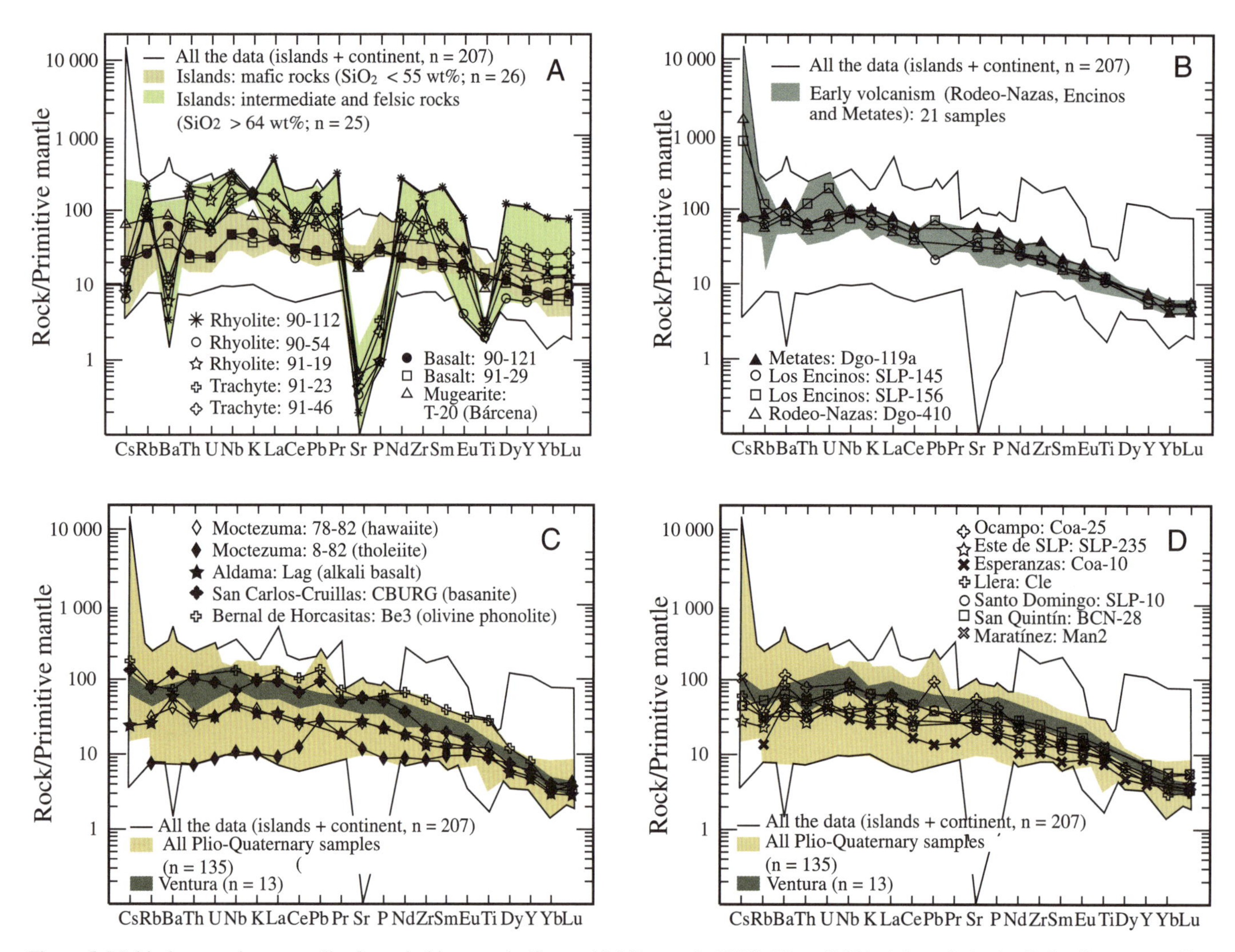

Figure 5. Multi-element plots normalized to primitive mantle (Sun and McDonough, 1989). The solid black boundaries in all the diagrams outline the area covered by all oceanic and continental analyses. Each of the panels shows a few representative examples from the localities discussed in the text. (A) Samples from the island volcanoes show different patterns as a function of SiO_2 content. (B) Representative examples from continental localities formed during the early stage (Miocene) of intraplate volcanism. (C) All the samples from Plio-Quaternary continental volcanic fields plot inside the light green area. The Ventura samples plot in the dark green area. This panel includes the Oligocene sample from Bernal de Horcasitas (Eastern/Alkalic region); see Figure 4 caption. (D) Other examples from the Plio-Quaternary volcanic fields. SLP—San Luis Potosí.

able and higher (0.703431–0.708621), the high values are found in rocks with very low Sr contents, which are susceptible to contamination; thus, the values are attributed to interaction with hydrothermal fluids dominated by seawater.

Based on the similarity between the Nd and Pb isotopes in the mafic and felsic rocks of Socorro (Fig. 7A), it was hypothesized that the trachytes could have been formed by fractional crystallization from a mildly alkaline mafic magma. This hypothesis was rejected, however, because attempts to model this process failed to mimic the major- and trace-element contents of the trachytes. Partial-melting models with previously erupted alkali basalts and associated cumulophyric rocks as sources suggest that 5–10% melting could produce the appropriate amounts of K_2O and some incompatible trace elements such as Zr and Nb in the trachytes. Unaltered rhyolites (i.e., without devitrification and consequent loss of Na_2O) can be modeled by fractional crystallization (low pressure, up to 80%) from a parental magma with trachytic composition. The mineral assemblage used in the model is mainly Kspar and Cpx + lesser Fa + Ilm + Ap. Negative anomalies of Ce and other REE elements (Fig. 4A) are interpreted as evidence for assimilation of variable amounts of hydrothermal sediments. The large variations in Sr isotope ratios—in samples that were not acid-leached—were attributed to the interaction between rocks with a low Sr content with a hydrothermal system dominated by seawater. Based on this investigation, Bohrson and Reid (1997) concluded that three factors lead to the formation of peralkaline magmas: (1) moderate extension rates, (2) shallow magma chambers, and (3) a parental magma with a mildly alkaline basaltic composition.

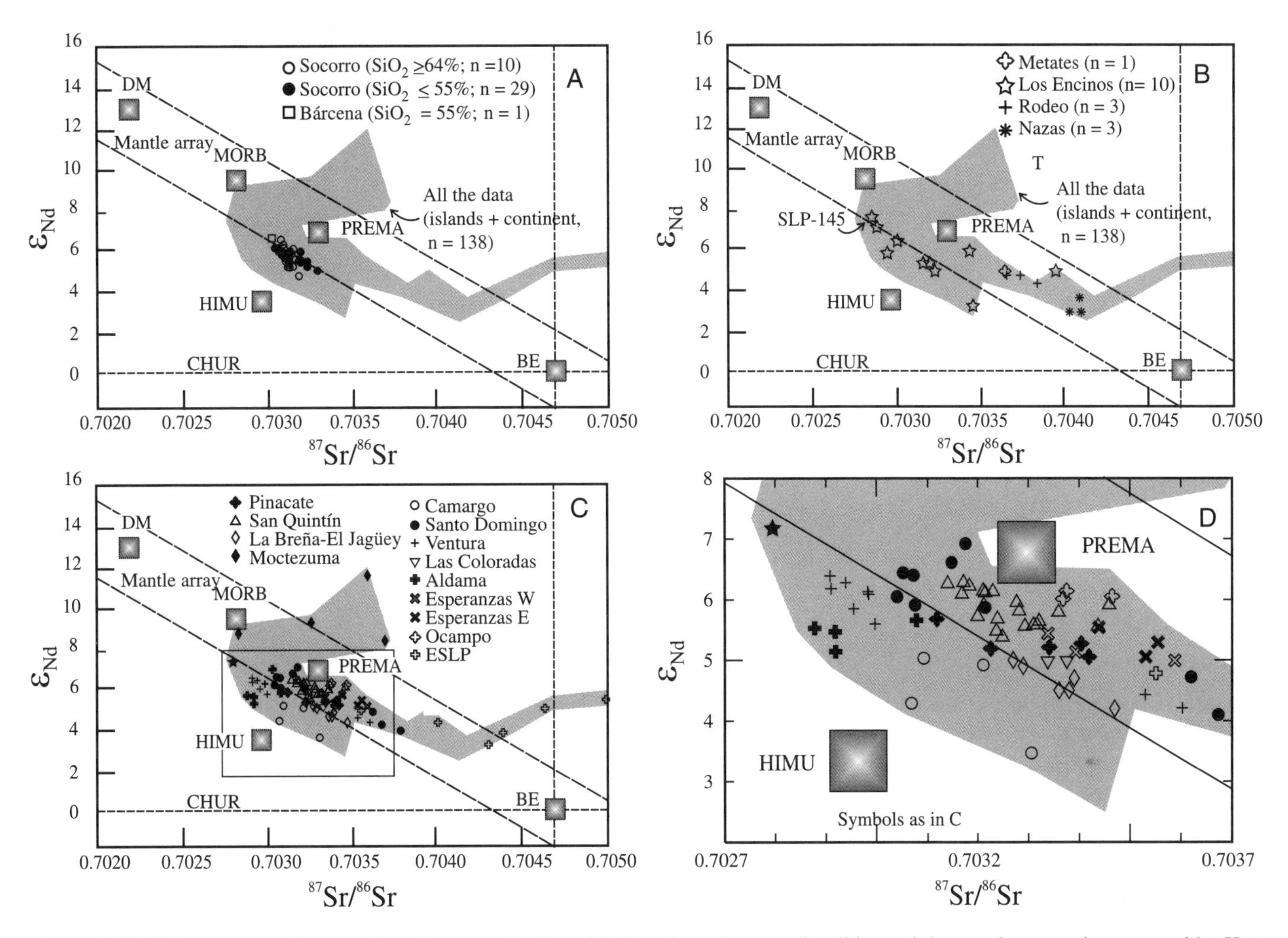

Figure 6. $^{87}Sr/^{86}Sr$ versus ε_{Nd} diagrams that show samples from Mexican intraplate-type localities and the mantle reservoirs proposed by Hart et al. (1986): HIMU (high U/Pb), MORB (mid-oceanic-ridge basalt), DM (depleted mantle), PREMA (prevalent mantle), and BE (bulk earth). CHUR—Chondrite Upper Reservoir. Most samples analyzed plot in the gray area shown in each panel. Only two samples are not shown because they plot off scale to the right: one from East San Luis Potosí and the other from El Pinacate have $^{87}Sr/^{86}Sr$ ~0.706 and ε_{Nd} ~+5. The values for Socorro correspond to feldspar analyses. (A) Oceanic localities. (B) Early continental magmas (initial values). (C) Plio-Quaternary continental samples (n = 81). The box in C corresponds to (D), where the bimodal character of the Ventura–Espíritu Santo and Santo Domingo volcanic fields may be seen. ESLP—East San Luis Potosí.

San Benedicto Island

Bárcena volcano is the most prominent topographic feature on San Benedicto, the southernmost island in the Revillagigedo archipelago. Bárcena volcano is located 350 km south of the tip of Baja California (Fig. 1) and was formed during eruptions that occurred in 1952–1953 (Richards, 1959, 1965, 1966). San Benedicto Island is elongated in a NE-SW direction. A series of trachyte domes occur in its northern part, while Bárcena volcano and a tephra cone (Montículo Cinerítico) are at its southern end. The Montículo Cinerítico, which is almost completely covered by volcanic products from Bárcena, is probably Holocene in age and is composed of mugearite and soda rhyolite. Bárcena is a tuff cone more than 300 m high, with a crater 700 m in diameter. Post-maar trachyte lava domes filled the crater bottom. A trachyte lava flow that erupted from the southeastern flank of the cone extended the shoreline 700 m seaward, creating a lava delta 1200 m long (Siebert et al., 2002). No detailed petrologic investigations have been published for San Benedicto. Bohrson and Reid (1995) reported a few chemical analyses of samples collected on the island (Figs. 2A and 6A).

CONTINENTAL MAGMATISM

Early Activity (Late Oligocene–Miocene): Southern Basin and Range Province

Pinacate Volcanic Field

Vidal-Solano et al. (2000) recently reported early volcanic activity (Miocene?) intimately associated with transitional alkalic basalts. This volcanic activity produced felsic rocks with (Na_2O +

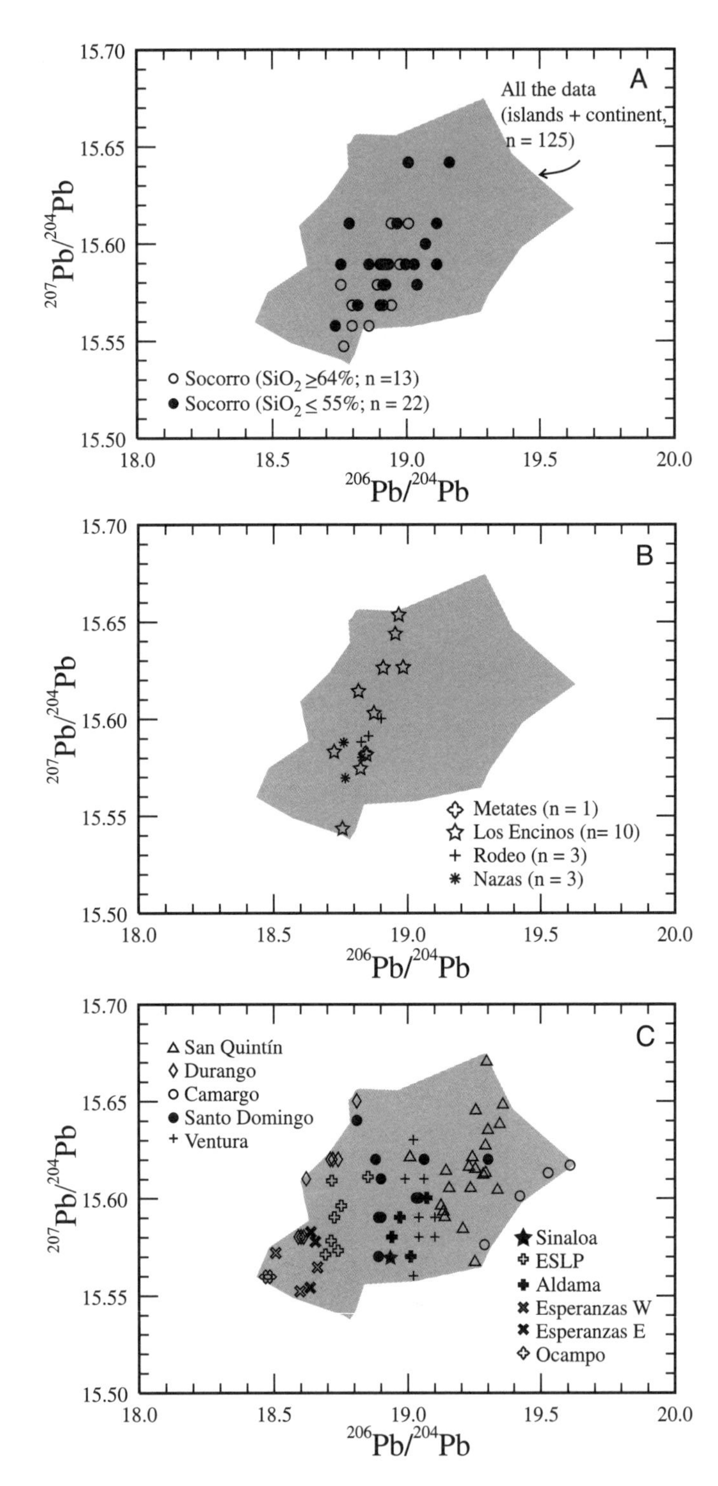

Figure 7. $^{206}Pb/^{204}Pb$ versus $^{207}Pb/^{204}Pb$ plots. All the analyzed samples, both from continental and oceanic localities, plot inside the gray area. (A) Islands. (B) Initial values for early continental intraplate-type volcanic rocks (Miocene). (C) Plio-Quaternary continental intraplate-type volcanic rocks (n = 73). ESLP—East San Luis Potosí.

$K_2O)/Al_2O_3$) = 1.2; the mineralogy of these pantellerites and comendites is sanidine, anorthoclase, fayalite, amphibole, ferrohedenbergite, aenigmatite, and titanomagnetite. Trace-element data show enrichments in Y, Zr, Zn, Ta, Nb, and Hf, plus REE patterns with a pronounced negative slope in the LREE, marked negative Eu anomalies, and flat HREE patterns. It is believed that these magmatic rocks are related to an early pulse of Tertiary crustal extension in the region (i.e., the Protogulf of California). Other localities in the Gulf of California Extensional Province where rocks with similar chemical composition have been found are near Hermosillo and in NW Sonora (Paz-Moreno et al., 2000).

Rodeo and Nazas, Durango

The first petrologic manifestation related to the beginning of crustal extension in northern México and the southwestern United States was the eruption of the southern Cordillera basaltic andesite (SCORBA) suite, which has trace-element contents and isotopic compositions similar to those found in the subduction-related basalt-andesite-rhyolite suites of the Sierra Madre Occidental, but with consistently low SiO_2 contents (Cameron et al., 1989). Even though these SCORBA rocks are not considered here as products of intraplate-type magmatism, their chemical characteristics were interpreted by Cameron et al. (1989) as indicating faster magma ascent compared with the older orogenic magmas that generally were more differentiated and viscous. SCORBA-type rocks are exposed immediately north of Durango City (Fig. 8), where they are known as the Caleras basalt (K-Ar, whole rock: 29–30 Ma). The oldest intraplate-type rocks documented in northern México are hawaiites at Rodeo and Nazas, Durango (Figs. 1 and 8). Field relations and radiometric ages were reported by Aguirre-Díaz and McDowell (1993) and Aranda-Gómez et al. (1997); the petrogenesis of these rocks was discussed by Luhr et al. (2001). The intraplate-type rocks of Rodeo and Nazas are similar in age and composition to some rocks exposed in Trans-Pecos, Texas (James and Henry, 1991). The volcanic vents at Rodeo are located near a breakaway fault zone that formed a significant half-graben (Fig. 8). Volcanic rocks are intercalated with graben-fill gravel deposits and gravel deposits and hawaiites are both cut by normal faults, therefore, intraplate-type volcanism was contemporaneous with a significant pulse of extension in the area. The hawaiites in Nazas comprise several undeformed lava fields that cover fault blocks. These blocks consist of tilted Oligocene felsic volcanic rocks, which are partially buried by flat-lying graben-fill gravel deposits. Some of the lava flows in Nazas can be followed all the way to their vents, which are deeply eroded cinder cones located either on or near the traces of normal faults.

Rodeo and Nazas hawaiites (SiO_2: 47.4–49.5 wt%) contain phenocrysts and microphenocrysts of plagioclase, olivine, clinopyroxene, and titanomagnetite. Nearly half of the samples studied by Luhr et al. (2001) also contain biotite microphenocrysts. Most samples contain xenocryst assemblages with Na-feldspar, Fe-rich olivine, Al-rich clinopyroxene, and a great variety of spinel crystals. Textures and chemical compositions at the borders of the megacrysts suggest that they were in disequilibrium

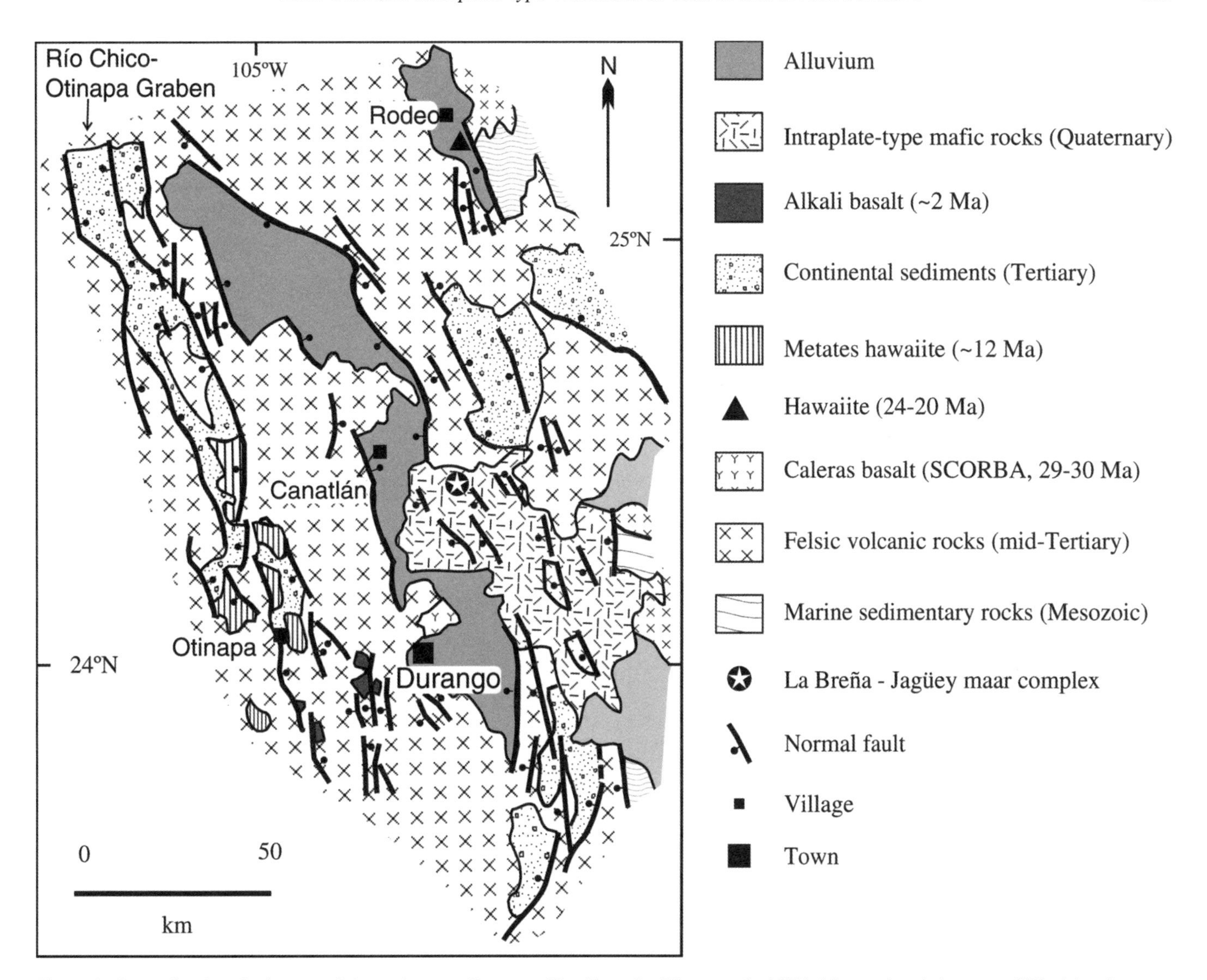

Figure 8. Generalized geologic map of the region near Durango City (Aranda-Gómez et al., 1997). Metates basalt is exposed W of the city, near the Río Chico–Otinapa graben.

with the host magma. For example, the cores of feldspar megacrysts have compositions in the range An_{26-51}, whereas the rims of the crystals are zoned with more calcic compositions (An_{57-65}). Likewise, the cores of clinopyroxene megacrysts have 7–9 wt% Al_2O_3, whereas the rims have only 1–5 wt% Al_2O_3. In both cases the compositions of megacryst borders are similar to those found in primary phenocrysts and microphenocrysts in the same rock. Some of the plagioclase cores are polycrystalline aggregates that give some idea of the mineral parageneses (Pl + Spl + Ap + Po; Pl + Spl + Cpx; Pl + Cpx + Ol) of the protoliths from which they were derived. Based on the chemical and mineralogical characteristics of these relicts, it is believed that the protoliths were derived from the lower crust beneath the volcanoes.

Rodeo and Nazas samples are moderately to poorly preserved and have elevated contents of TiO_2 (2.1–2.4 wt%), Nb (40–82 ppm), and Ta (2.1–4.2 ppm). According to Le Maitre et al. (2002)'s total alkali versus silica (TAS) classification scheme, the unaltered samples are hawaiites (Fig. 3). These samples also contain *nepheline* in the norm, and their MgO (5.5–7.1 wt%), Ni (48–83 ppm) and Cr (73–186 ppm) contents, and Mg# (52. 2–59.7) indicate that they are differentiated and do not represent primary magmas derived from mantle peridotites. Incompatible element abundances are similar to those found in rocks from many intraplate-type localities in the southern part the Basin and Range Province (Figs. 4B and 5B). Notable exceptions are four samples with a large Cs enrichment and Rb depletion (Fig. 9). It is interpreted that these geochemical characteristics were formed by interaction of these magmas with continental crust, preferentially incorporating Cs with respect to Rb. Therefore, their Rb/Cs ratio decreased. The isotopic compositions of the Rodeo and Nazas hawaiites are distinctive; although the variations are small, they form separate groups (Figs. 6B and 7B). Compared to the Nazas rocks, the Rodeo sam-

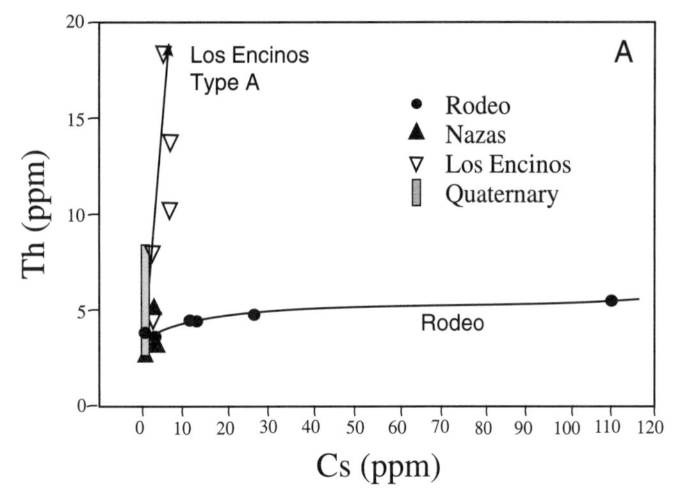

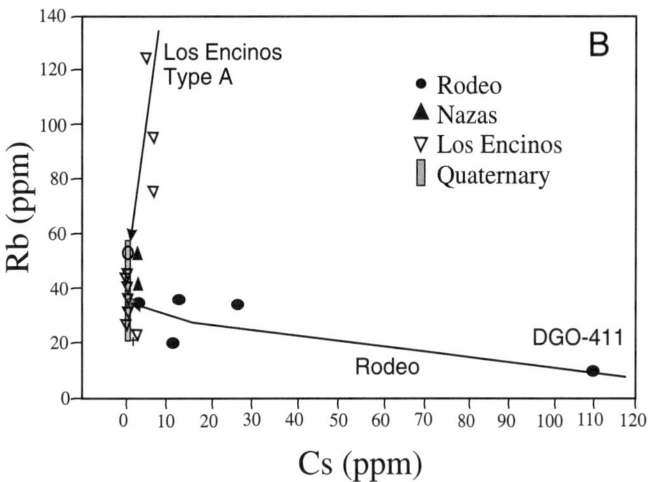

Figure 9. Whole-rock abundances of Cs versus Th and Rb (ppm) in early intraplate-type volcanic rocks (Miocene) from Los Encinos and Rodeo-Nazas fields. The compositions of 50 Plio-Quaternary rocks plot in the small areas defined by 0–2 ppm Cs, 0–8 ppm Th, and 20–58 ppm Rb. Contamination vectors for Rodeo and Los Encinos samples are shown as arrows (Luhr et al., 2001).

ples have lower $^{87}Sr/^{86}Sr$ (0.7037–0.7038 vs. 0.7040–0.7041), and higher ε_{Nd} (+4.2 to +4.8 vs. +2.8 to +3.5), $^{206}Pb/^{204}Pb$ (18.84–18.91 vs. 18.77–18.84), and $^{208}Pb/^{204}Pb$ (38.55–38.61 vs. 38.51–38.57). The values of $^{207}Pb/^{204}Pb$ in both areas overlap (15.57–15.60). The isotopic relations are interpreted as evidence of interaction of these magmas with continental crust.

Los Encinos Volcanic Field (San Luis Potosí)

Los Encinos volcanic field occupies an extensive region (>5000 km^2) of Mesa Central in northwest San Luis Potosí and northeast Zacatecas (Fig. 10). The petrogenesis of the intraplate-type volcanic rocks from Los Encinos and the significance of their inclusion assemblages are discussed in detail by Luhr et al. (1995b). Some tectonic implications of the vent distribution in the area are discussed by Henry and Aranda-Gómez (2000). The Los Encinos region is a high plateau with an elevation greater than 2000 masl, with isolated ranges that may be as high as 2600 masl. The eastern limit of the distribution of the Los Encinos vents is Sierra de Catorce (≤3800 masl), which is a N-S range bounded on the W by an important normal fault (Fig. 10A). Volcanic rock outcrops are scattered throughout a large region ~50 × 100 km, but they only cover a small portion of the area. There are also a few scattered remnants of Eocene–Oligocene calc-alkaline volcanic rocks in the region. The middle Miocene (K-Ar: 13.6–10.6 Ma; Luhr et al., 1995b) mafic intraplate-type volcanic rocks of Los Encinos occur as volcanic necks with conspicuous columnar jointing and as small mesas capped by lava flows. Some of the necks located immediately W of Sierra de Catorce are now being buried by active alluvial fans that have developed at the base of the range. This suggests the occurrence of postvolcanic faulting, as most of the volcanoes are on the downthrown side of the block. Likewise, in the same area, the vents define two rough lineaments (Fig. 10A) approximately parallel to the tectonic grain defined by Cenozoic normal faults in the southern part of Mesa Central (Fig. 10C). This is interpreted as evidence of a structural control on the volcanism and volcano-tectonic activity during the middle Miocene (Henry and Aranda-Gómez, 2000).

The stable mineral paragenesis in Los Encinos hawaiites (Fig. 3B) is: Pl (An_{53-64}) + Ol (Fo_{61-88}) + Cpx + TMt + Bt (tr). Without exception, the samples contain megacryst (1–3 cm) assemblages of feldspar (sanidine, anorthoclase, and/or plagioclase), kaersutite, clinopyroxene, and titanomagnetite. These may be accidental or co-genetic with the host magma. It is also common that the mafic volcanic rocks contain smaller (<1 cm) xenocrysts of apatite, olivine (Fo_{45-79}), and clinopyroxene. Ferromagnesian minerals are Fe-rich compared to primary phases crystallized from the magma. Quartz xenocrysts, invariably surrounded by reaction rims, are ubiquitous in all samples. Both megacrysts and xenocrysts are partially resorbed or are surrounded by remarkable reaction rims at their contacts with the matrix. Some of these volcanic rocks contain polycrystalline aggregates with mineral parageneses consistent with (1) feldspathic granulites (quartz- and sillimanite-bearing paragneisses) from the lower crust, or (2) gabbroic rocks crystallized at high pressure with whole-rock compositions similar to the host magmas, but more evolved. The chemical compositions of minerals in the gabbros are like those observed in the megacrysts. Therefore, it is believed that these are samples of the megacryst protoliths. Up to now, there have been no reports of mantle peridotites, or xenocrysts derived from them, in Los Encinos volcanic rocks.

The intraplate-type rocks from Los Encinos are hawaiites and *nepheline*-normative hawaiites (Fig. 3B), similar to those found in Trans-Pecos (Texas), Rodeo-Nazas, and Metates (Durango). Most samples display remarkable differences with respect to the Quaternary rocks from the nearby Santo Domingo and Ventura–Espíritu Santo volcanic fields, located farther south in Mesa Central (Fig. 10B). The Quaternary volcanic fields also contain hawaiites but these are associated with more primitive basanites and nephelinites. The compositional variations in Los Encinos samples are different from those in the Quaternary rocks (e.g., SiO_2: 57.0–46.9

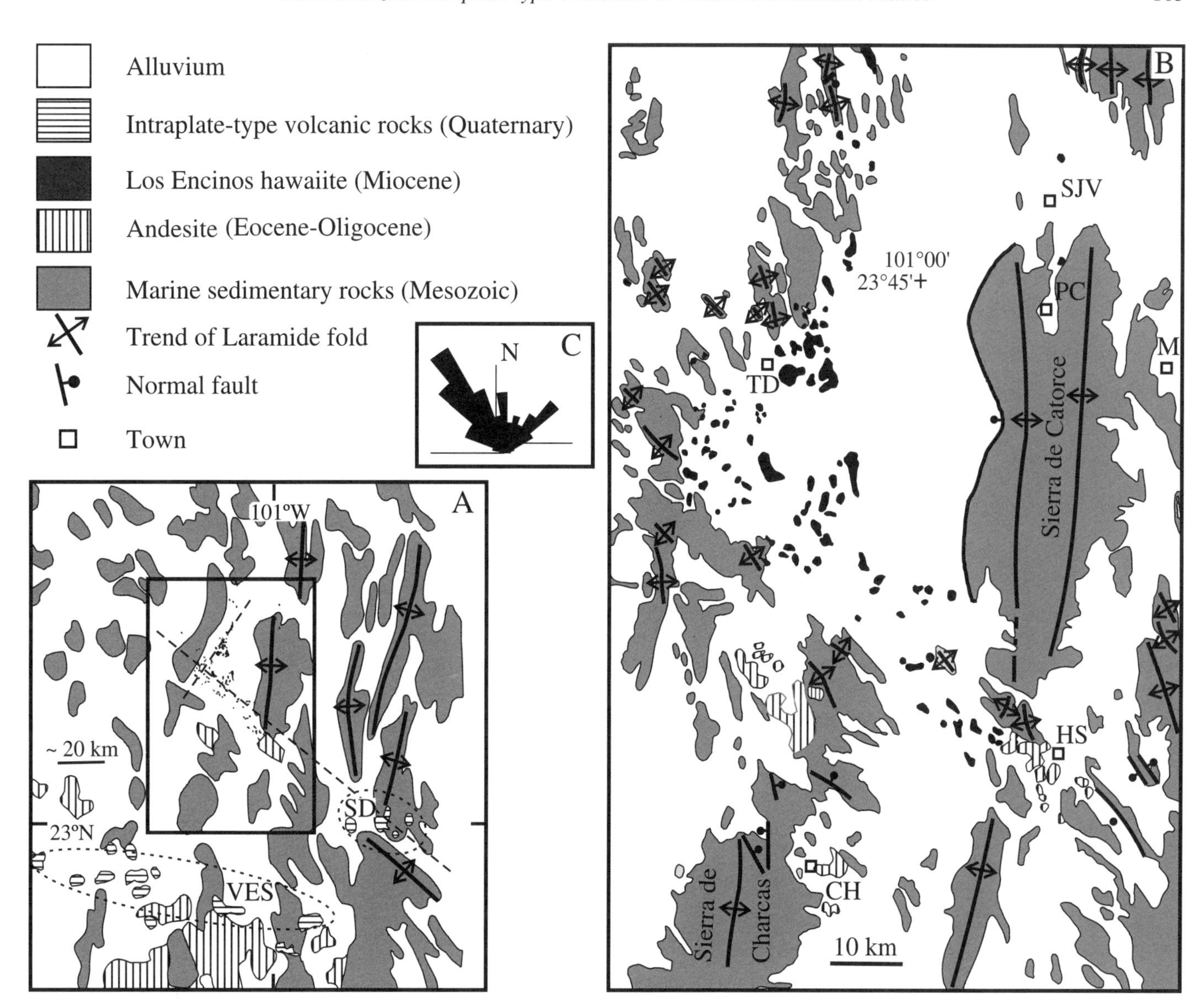

Figure 10. (A) Regional geologic setting of Los Encinos volcanic field. Its relationship to the intraplate-type Quaternary volcanic fields is shown in a diagrammatical way (SD—Santo Domingo and VES—Ventura–Espíritu Santo). Note that Santo Domingo is located along the same inferred basement structure as Los Encinos. The dashed lines show the lineaments discussed in the text. Key: SJV—San Juan de Vanegas; PC—Potrero de Catorce; M—Matehuala; TD—Tanque de Dolores; HS—Hacienda Solís; CH—Charcas. The box shows the location of the area that is enlarged and has become 10B. (B) Generalized geologic map of the Los Encinos volcanic field (Henry and Aranda-Gómez, 2000). Outcrops of Tertiary volcanic rocks are shown in relation to the location of the inferred basement structure. Note the change in the trend of Laramide folds along the lineament and its relationship to the location of mid-Tertiary and Miocene volcanic rocks in the region. (C) Rose diagram of mid- and late Tertiary normal faults mapped in the southern portion of Mesa Central, in the region between the cities of San Luis Potosí and Guanajuato. The main trends in the rose diagram correspond to the orientation of the lineaments in the Los Encinos volcanic field.

and MgO: 11.9–3.9 vs. SiO2: 51.2–41.8, MgO: 15.8–5.7 wt%, respectively). Trace-element contents in a few samples from Los Encinos are similar to those of Quaternary volcanic rocks (Figs. 5B and 5C), but other volcanic rocks from Los Encinos are enriched in Cs, Rb, Th, and U (Figs. 5B and 9). These enrichments are interpreted as evidence of contamination with lower crustal materials. This is because the high concentrations of these elements are independent from the concentrations of many other incompatible elements in the same rocks, which are interpreted to reflect control by their abundances in the parental rocks of the magmas and/or by the amount of partial melting or differentiation.

The samples studied by Luhr et al. (1995b) were separated into three groups based on their geochemistry. The U-type (uncontaminated) rocks are similar to Quaternary volcanic rocks, but are more evolved (compare sample SLP-145 [Fig. 5B] with the Ventura pattern in Fig. 5C). The isotopic compositions of the U-type samples (e.g., SLP-145, Figure 6B: $\varepsilon_{Nd} = +7.6$, $^{87}Sr/^{86}Sr = 0.70286$, and $^{206}Pb/^{204}Pb = 18.74$) are more extreme than those observed in the nearby Quaternary fields (Santo Domingo and Ventura–Espíritu Santo, Fig. 6C). The rocks in the other two groups are contaminated with crustal material, as indicated by their higher Yb values, higher $^{87}Sr/^{86}Sr$ (up to 0.7040) and $^{206}Pb/^{204}Pb$ (up to 18.98), and lower ε_{Nd} (up to + 3.1). This is a product of either bulk assimilation or assimilation and fractional crystallization (AFC) of feldspathic granulites, especially garnet paragneiss. A-type rocks have anomalous high values of Cs, Rb, Th, Sb, U, Pb, K, and Si (e.g., Figure 9 and SLP-156 in Fig. 5B). Those samples that do not clearly cluster with the U-types and A-types were classed as B-types.

Metates (Durango)

The Metates basalt (sensu lato) is exposed in and near the Río Chico–Otinapa graben (Córdoba, 1963), ~20 km west of Durango City. The age of this lithostratigraphic unit is 12.7 Ma (K-Ar, Hbl: McDowell and Keizer, 1977) and has been used as the main evidence for the age of a pulse of extension that affected the eastern part of the Sierra Madre Occidental, and is contemporaneous with the formation of the Protogulf of California (Henry and Aranda-Gómez, 2000). Stratigraphic and structural relations indicate that the graben formed a short time before the eruption of the Metates basalt. Close to the area where the Durango–Mazatlán highway crosses the graben, the oldest Metates basaltic lava flow rests atop a thin deposit composed of graben-fill gravels. In this particular outcrop, the lavas contain only a few megacrysts (feldspar and kaersutite). West of the graben, another basaltic lava flow, devoid of inclusions, has been displaced ~60 m by a normal fault. At the same locality, a basaltic dike without inclusions was apparently emplaced along the normal fault. This dike might be the feeder of the overlying lava flows. An outcrop of the Metates basalt, characterized by abundant megacrysts and some feldspathic granulite xenoliths, is located at km marker 44 of the highway. At that locality, the 12.0 Ma (K-Ar, Hbl) basalt overlies a 29.3 Ma (K-Ar, Sa) rhyolitic dome and associated pyroclastic deposits (McDowell and Keizer, 1977).

To date, no detailed petrologic study of the Metates basalt has been published. Aranda-Gómez et al. (1997) published chemical analyses of three rocks, emphasizing the similarities between the chemical compositions and megacryst assemblages of these hawaiites (Fig. 3B) with those for hawaiites from Los Encinos and Rodeo-Nazas (Luhr et al., 1995b, 2001).

Late Magmatism (Early Pliocene–Quaternary): Occidental Region/Baja California

San Quintín Volcanic Field

The San Quintín volcanic field is an exceptional place, as it is currently the only known locality with both mantle and crustal xenoliths in the peninsula of Baja California (Fig. 1). The San Quintín volcanic field consists of ten Quaternary volcanic complexes formed by small lava shields and cinder cones (^{40}Ar-^{39}Ar, groundmass: 126–90 ka, Luhr et al., 1995b, and ^{3}He-^{4}He: 165–22 ka, Williams, 1999). Compared to other localities in México, the San Quintín volcanic rocks and especially the mantle xenoliths have been extensively studied (e.g., Basu, 1977a, 1977b, 1978, 1979; Basu and Murthy, 1977; Bacon and Carmichael, 1973; Rogers et al., 1985; Cabanes and Mercier, 1988; Storey et al., 1989; Righter and Carmichael, 1993), making San Quintín the most intensely studied xenolith locality in México. San Quintín is located outside the extended areas related to the Basin and Range Province and the transtensional Gulf of California Province. The intraplate-type character of San Quintín's volcanic rocks has been attributed to (1) formation of a *no-slab window* that allowed access to hot asthenospheric mantle beneath this region (Rogers et al., 1985; Saunders et al., 1987), and (2) the short distance between San Quintín and the paleo-trench, which caused the mantle beneath this locality to be unaffected by subduction (Sawlan, 1991).

The most systematic study of the geology, geochemistry, and petrology of the volcanic rocks from San Quintín is that of Luhr et al. (1995a). Their results are summarized as follows: First, the stable mineral assemblage in the San Quintín volcanic rocks is Ol + Pl + Cpx + TMt + Ilm. Olivine phenocrysts usually contain spinel inclusions. These primary mineral phases are commonly accompanied by xenocrysts and megacrysts. The most common xenocrysts are those derived from mantle peridotites, and their abundance in the volcanic rocks is correlated with the abundance of xenoliths observed in the field. Nearly one half of the studied samples contain quartz xenocrysts. Some samples contain partially melted granulite xenoliths of gabbroic composition; those volcanoes where xenoliths are abundant also contain plagioclase and clinopyroxene megacrysts, up to 2 cm long. Second, almost all the studied samples (n = 63) have *nepheline* in the norm and were classified as hawaiite, *nepheline* hawaiite, alkali basalt, or basanite (Figs. 2D and 3D). These rock types all occur in approximately the same proportion in the set of studied samples. In addition to these types, there are three samples with normative *hypersthene* that have been classified as basalts or hawaiites (Fig. 3C). Third, the Mg# in the set of studied samples ranges from 51 to 67. The oldest volcanoes erupted primitive (Mg# > 64) lavas and pyroclasts with rare xenoliths of small size. As time went by, the volcanoes erupted more differentiated products with large, abundant xenoliths. The youngest volcanoes produced primitive lavas and tephra deposits devoid of xenoliths, but very rich in olivine phenocrysts. Fourth, there is a decrease in incompatible element content with time, which implies a larger degree of melting or the progressive depletion of these elements in the magma source. Fifth, most compositions of the evolved rocks can be modeled in a closed system from the most primitive rock composition in the series through fractionation of olivine, plagioclase, clinopyroxene, and spinel. Sixth, isotopic variability in the studied subset (n = 20) is small: $^{87}Sr/^{86}Sr$ = 0.70314–0.70346, ε_{Nd} = +5.4 – +6.3, and $^{206}Pb/^{204}Pb$ = 19.01–19.36 (Figs. 6D and 7C). The small variations in the Sr, Nd, and Pb isotope ratios observed in all rocks support the crystal-fractionation model. There are only two volcanic complexes where elemental and isotopic evidence of contamination with crustal material is evident. Seventh, an unusual feature of the primitive rocks from San Quintín, compared with other intraplate-type localities throughout the world, is that they have relatively high Al_2O_3 and Yb contents (Fig. 5D) accompanied by low La/Yb and CaO/Al_2O_3 contents (Fig. 11C). These trends of rising Al_2O_3 and falling CaO with increasing incompatible element abundance (Figs. 11A and 11B) are consistent with an origin for the San Quintín magmas of progressive partial melting at relatively low pressures in the mantle.

Late Magmatism (Early Pliocene–Quaternary): Southern Basin and Range Region

El Pinacate Volcanic Field

El Pinacate is an extensive volcanic field located a short distance NW of the Gulf of California's northern end. The volcanoes occur in an ~55 × 60 km area and their eruptive products cover nearly 1500 km^2. Early activity during the Quaternary (K-Ar: 1.7–1.1 Ma; Lynch et al., 1993) formed a 1200-m-high lava shield (Santa Clara volcano), which is comprised of alkaline series rocks that vary from basanite to trachyte in stratigraphic order (Figs. 2C and 2D). A large number of younger (K-Ar: <0.19 Ma; Lynch et al., 1993) monogenetic volcanoes (maars, tuff, and cinder cones) of basaltic composition lie atop Santa Clara volcano and in the surrounding desert. The Ives lava flow, which has an estimated volume between 0.25 and 0.50 km^3 (Lynch et al., 1993), occurs in the southern part of the volcanic field. This lava erupted from north–south fissures and covers 75 km^2, accounting for ~5% of the total area of El Pinacate. The pahoehoe surface morphology of the Ives flow is distinct from other lava flows. El Pinacate has a unique tectonic setting, as it is located only ~50 km from an active plate margin formed by spreading ridge segments and associated transform faults (Fig. 1).

Although a number of papers have been published on the physical volcanology (e.g., Arvidson and Mutch, 1974; Gutmann and Sheridan, 1978; Gutmann, 1979, 2002; Lutz and Gutmann, 1995) and some of the mineralogical aspects of the volcanic rocks of El Pinacate (e.g., Gutmann, 1974, 1986; Gutmann and Martin, 1976) and their xenoliths (Gutmann, 1986), no systematic study of the geochemistry and petrology of the El Pinacate volcanic rocks has been published. The data summarized in this paper are from the dissertation by Lynch (1981) and Lynch et al. (1993), which focused on 7 representative samples and reported their Sr and Nd isotopic compositions: early and late Quaternary samples, the Ives lava flow, and one spinel lherzolite xenolith from El Pinacate. Six mafic to intermediate samples (basanite, basalt, and basaltic trachyandesite) have $^{87}Sr/^{86}Sr$ in the range 0.70312–0.70342 and ε_{Nd} between +5.0 and +5.7 (Fig. 6C). A trachyte collected at the Santa Clara volcano has higher $^{87}Sr/^{86}Sr$ (~0.70611) and lower ε_{Nd} (+5.0), probably caused by contamination with crustal material (this sample is not shown in Fig. 6C). Despite the differences between the Ives lava flow and other rocks in the volcanic field (e.g., it is *hypersthene* normative), their Sr and Nd compositions are identical. This suggests that Ives and other magmas in the field were derived from the same source (or the same combination of sources) and that Ives's peculiar geochemical composition is related to a higher degree of partial melting. The isotopic composition of the volcanic

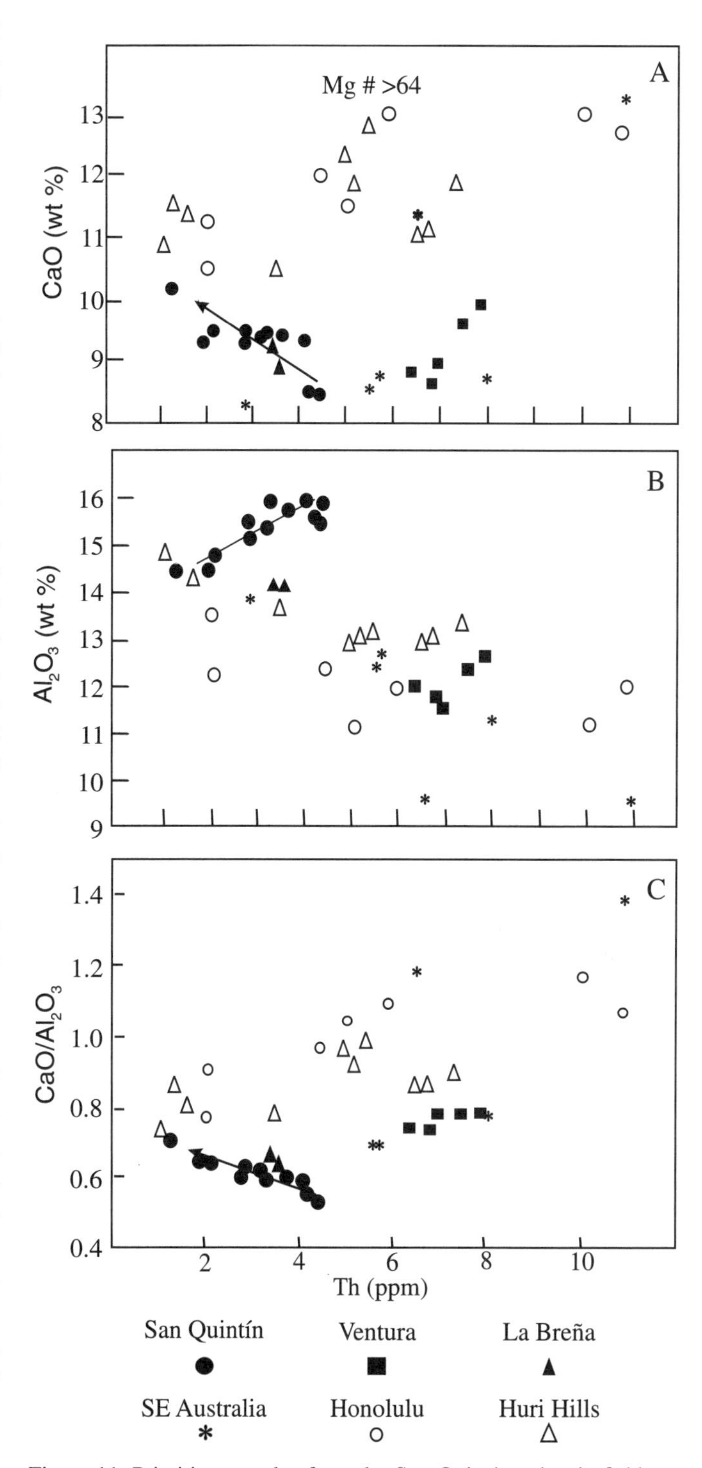

Figure 11. Primitive samples from the San Quintín volcanic field on plots of Th versus CaO, Al_2O_3, and CaO/Al_2O_3 (Luhr et al., 1995a). Other volcanic suites are shown for comparison (see references for these data sets in Luhr et al., 1995a).

rocks, as well as that of diopside separated from a spinel lherzolite xenolith, are similar to that of OIB (oceanic-island basalt) or an enriched MORB (mid-oceanic-ridge basalt). The data suggest that El Pinacate magmas were derived from a source dominated by asthenospheric mantle. Lynch et al. (1993) did not find compelling isotopic evidence in El Pinacate rocks for the influence of the nearby Gulf of California spreading ridges. However, they found a considerable similarity to other intraplate-type localities in the northern portion of the Basin and Range Province (e.g., Geronimo volcanic field and Kilbourne Hole; Menzies, 1989).

Moctezuma Volcanic Field

This volcanic field was studied by Paz-Moreno et al. (2003), and the information presented in this review was taken from that work. Moctezuma is located in Sonora, at the foothills of the northwestern end of the Sierra Madre Occidental (Fig. 1). The Moctezuma field, which is located in a NNW-trending semi-graben, is composed of volcanic rocks with normative *hypersthene* and/or *quartz* and other rocks with normative *nepheline* (Figs. 3C and 3D). A basalt (^{40}Ar-^{39}Ar, Pl: 22.3 Ma) interlayered with graben-fill fanglomerate indicates that extension in this region began during the early Miocene. The mafic volcanic rocks associated with the early phase of extension have a SCORBA-like geochemistry similar to that found by Cameron et al. (1989) in rocks from Chihuahua. After a long period of volcanic quiescence, mafic volcanism resumed in the Quaternary (^{40}Ar-^{39}Ar, whole rock: 1.7 Ma). Lava flows with *hypersthene* and/or *quartz* in the norm, from the initial pulses of Quaternary volcanism, were erupted from fissures located close to the master fault in the eastern part of the semi-graben and now cap extensive mesas. Younger volcanic activity produced mafic alkaline lavas (K-Ar: 0.53 Ma) that were erupted from central vents located near the center of the semi-graben. This late activity formed cinder cones and associated lava flows. The area covered by the Quaternary volcanic rocks is ~400 km^2 and the estimated volume is < 2 km^3. Unlike other intraplate-type localities, no upper mantle or lower crustal xenoliths—or crystals derived from their disaggregation—have been found in the Moctezuma rocks. Primary mineralogy in both the hypersthene- and nepheline-normative mafic rocks is: Ol + Pl + Cpx +Ilm. Silica-saturated rocks tend to be crystal-poor, whereas some alkaline rocks are phenocryst-rich (up to 30 vol.%). Olivine is the only mineral that forms phenocrysts in the silica-saturated rocks, whereas the hawaiites commonly contain glomeroporphyritic aggregates of Ol + Cpx + Pl. A significant mineralogical difference between the two rock suites is that clinopyroxene in the rocks with *hypersthene* and/or *quartz* in the norm is Ca-poor augite or pigeonite, whereas clinopyroxene in the *nepheline*-normative hawaiites is Ca-rich augite or diopside. The silica-saturated rocks are basalts and subalkaline basaltic andesites following the classification schemes of Le Maitre et al. (2002) and Irvine and Baragar (1971), whereas the alkaline rocks are hawaiites (Figs. 2C and 2D). Some of the samples from Moctezuma are transitional as they plot in the alkaline field, but they lack normative *nepheline*. The Mg# values in the whole Moctezuma data set range from 57 to 66, but are between 60 and 62 in most samples. These Mg#s, as well as the low contents of Ni (silica-saturated rocks: 154–220 ppm; hawaiites: 101–203 ppm) and Cr (silica-saturated rocks: 190–322 ppm; hawaiites: 171–272 ppm), led to the conclusion that these rocks do not represent primary magmas that have been derived from partial melting of typical upper mantle peridotite. Chondrite-normalized REE diagrams show contrasting patterns for the two series (Fig. 4C): Hawaiites with normative *nepheline* form a linear pattern with a pronounced negative slope (e.g., 78–82 in Fig. 4C) due to the pronounced enrichment in LREE, whereas the silica-saturated samples are less enriched in LREE and have a gentler slope to their patterns. HREE concentrations in rocks of the two suites are similar, and this feature is interpreted as evidence for the presence of garnet in the source rocks for both magma types. Primordial mantle–normalized multi-element plots (Fig. 5C) show OIB-like patterns for the *nepheline*-normative hawaiites; samples with *hypersthene* and/or *quartz* in the norm show patterns similar to enriched MORB or OIB tholeiites. The Quaternary volcanic products from Moctezuma have elevated values for ε_{Nd} (+8 to +11) and low $^{87}Sr/^{86}Sr$ = 0.7028–0.7036 (Fig. 6C). Based on these values and on their multi-element patterns, Paz-Moreno et al. (2003) suggested that the source region of the Moctezuma magmas is dominated by asthenospheric mantle.

Mesa Cacaxta

In the area located ~50 km north of Mazatlán (Sinaloa) there are several outcrops of mafic alkalic volcanic rocks. We refer to this volcanic field as Mesa Cacaxta (Fig. 1) since this locality contains the most extensive outcrop in the region (~500 km^2). Mesa Cacaxta is capped by a 3.2 Ma (^{40}Ar-^{39}Ar, whole rock and plagioclase: Aranda-Gómez et al., 1997) lava with plagioclase megacrysts. Punta Piaxtla, which is west of Mesa Cacaxta on the Gulf of California coast (see location in Figure 24 in Aranda-Gómez et al., 1997), is where *nepheline*-normative hawaiites (2.1 Ma, ^{40}Ar-^{39}Ar, whole rock and plagioclase: Aranda-Gómez et al., 1997) contain xenoliths of spinel lherzolite (Luhr and Aranda-Gómez, 1997), pyroxenite, and feldspathic granulite, as well as pyroxene and plagioclase megacrysts (Henry et al., 1987; Righter and Carmichael, 1993). Punta Prieta, Cerro Carey, Punta Los Labrados, and Punta Gruesa are additional localities in this area with similar volcanic rocks. Chemical analyses of two samples reported by Smith (1989) show that these rocks are *nepheline*-normative hawaiites (Fig. 3D). The geochemistry of these samples is compared with samples from the Durango volcanic field and with the Metates basalt in a variation diagram of Sr versus Mg# (Fig. 12).

It is remarkable that although these hawaiites are contemporaneous with sea-floor MORB generation along oceanic spreading ridges in the nearby Gulf of California, Mesa Cacaxta itself is unaffected by faulting or tilting. This is in sharp contrast with the Pliocene trans-tensional deformation documented by Umhoefer et al. (2002) on the western side of the Gulf, in a region where up to now there have been no reports of similar xenolith-bearing volcanic rocks.

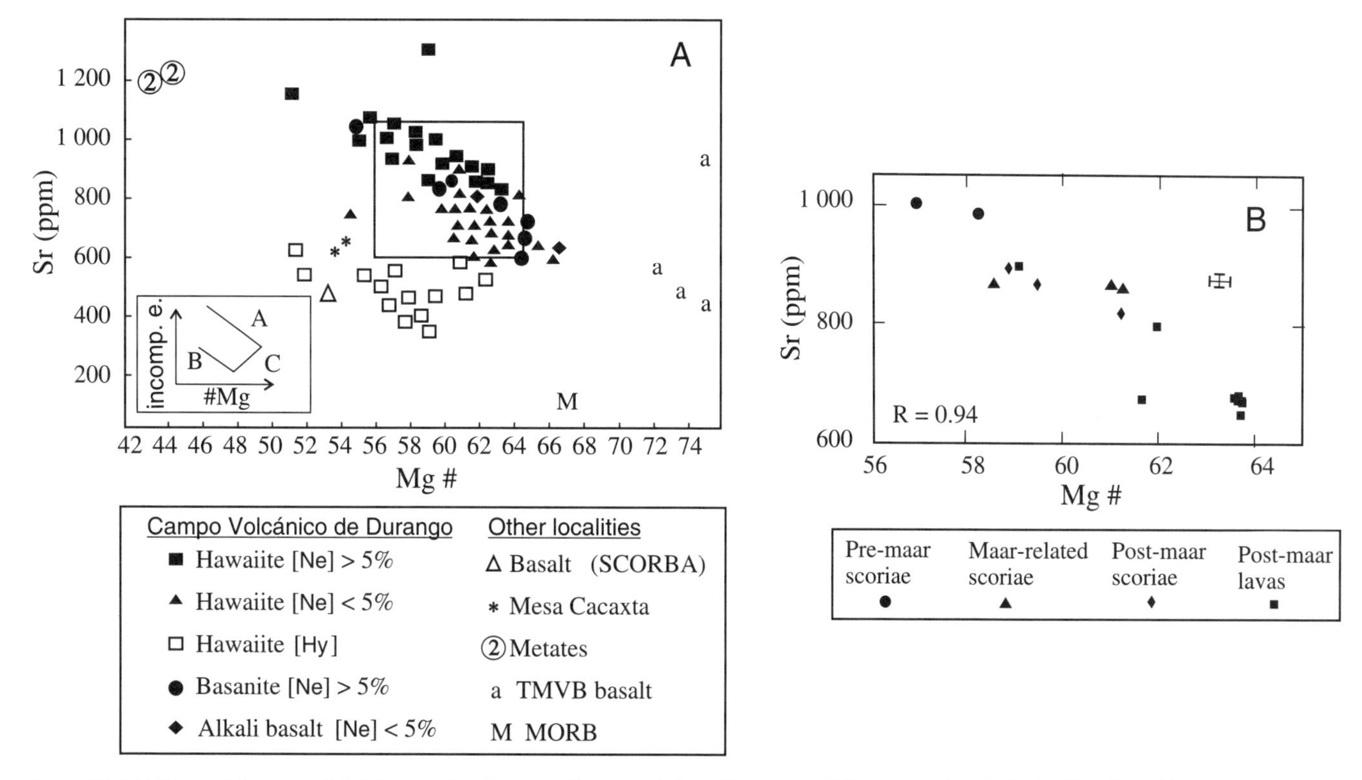

Figure 12. (A) Plots of Sr versus Mg# in samples from the Durango, Mesa Cacaxta, and Metates volcanic fields (Smith, 1989). Inset depicts the three linear trends (A, B, and C) identified by Smith in her Durango volcanic field data set. Other symbols used in the plot: a—arc basalts from the Trans-Mexican Volcanic Belt (samples MAS-21, JOR-44, 417A, and 426B taken from Luhr et al., 1989a); M—MORB (Wood et al., 1979). La Breña–El Jagüey samples (Durango) plot along linear trend A and its correlation coefficient is higher than for the full trend A in Figure 12A. Note that Mg# increased with time during the eruption (Pier et al., 1992). Inset in Figure 12A is shown in (B) for the La Breña–El Jagüey samples only. Incomp. e.—incompatible elements.

Isabel Island

Isabel is a small island (1.5 km long) located in the eastern part of the mouth of the Gulf of California, ~30 km from the coast of Nayarit (Fig. 1). The island is formed by a group of coalescent volcanoes that erupted on the continental shelf (Aranda-Gómez et al., 1999). Southeast of Isabel, on the mainland, and projecting toward it, is the NW-trending Tepic-Zacoalco rift zone (Allan, 1986; Allan et al., 1991), which is a major structural and volcanic zone in the western part of the Trans-Mexican Volcanic Belt. Most coastal exposures of Isabel are formed by pyroclastic-surge deposits related to the formation of tuff cones through magma-seawater interactions. Lava flows are also exposed in the sea cliffs and on the northwestern portion of the island, along with scoria deposits produced by strombolian activity. The chemical compositions of rock samples analyzed by Cabral-Cano (1988) indicate that they are alkali basalts (*sensu lato*), with 5–14 wt% *nepheline* in the norm. Thus, these rocks are broadly similar to many other intraplate-type rocks from the Mexican portion of the Basin and Range Province. Spinel-peridotite xenoliths, as well as some feldspar megacrysts and rare feldspathic-granulite xenoliths, occur on Isabel (Ortega-Gutiérrez and González-González, 1980). Based on the youthful morphology of some of the basaltic lava flows on Isabel, the island is assumed to be of Quaternary age.

Palomas Volcanic Field

The Palomas volcanic field, which covers an area of ~150 km^2, is located at the U.S. border, roughly 70 km west of Ciudad Juárez (Fig. 1). More than 30 cinder cones have been identified in this field. These cones commonly have a horseshoe-shaped morphology due to the presence of lava flows that have broken through and rafted away parts of them. N-S–trending linear dikes, as well as some curved dikes, also occur in the area. Some of the volcanoes are located at the intersections of dikes.

Palomas is on the western flank of the Río Grande rift and displays some significant contrasts to volcanic fields located inside the rift (Frantes and Hoffer, 1982): (1) Palomas seems to be older, as reflected in the exposures of dikes; (2) andesites and trachytes occur at Palomas, where these rocks are considered as differentiates from the *olivine*-normative basalts in the area; and (3) some pillow lavas are present within the field, thought to reflect eruption into a shallow lake. Field relations suggest that the andesites and trachytes are older than the basalts. Some of the olivine basalts brought mantle xenoliths to the surface. No isotopic ages are available, but based on a comparison with other volcanic fields in Luna County, New Mexico, Frantes and Hoffer (1982, p. 8) reached the conclusion that the volcanic field is "at least of Pliocene age."

Potrillo Volcanic Field

Potrillo maar is literally located on the border between México and the United States, ~40 km west of Ciudad Juárez (Fig. 1). This volcano is located at the southern end of Potrillo volcanic field, which covers ~4600 km^2 of Doña Ana County, New Mexico, in the southern part of the Río Grande rift. Alkali olivine-basalt was erupted during the middle and late Quaternary (ca. 1 Ma–8 ka; Williams, 2002) through numerous cinder cones, maars, and at least one lava shield volcano in a region crossed by at least three major faults: Fitzgerald, Robledo, and Aden. Active faulting (late Pleistocene–Holocene) has been documented for the first two structures (Hoffer, 1976). Pyroclastic deposits related to the Potrillo maar are cut by the Robledo fault (Williams, 2002). Two important xenolith localities are known in the Potrillo volcanic field: Kilbourne Hole and Potrillo maar, where mantle peridotites, feldspathic granulites, and kaersutite megacrysts occur (Aranda-Gómez and Ortega-Gutiérrez, 1987). Pb, Sr, and Nd isotopic relations in the amphibole crystals are similar to those found in OIB. This is interpreted to indicate that they were derived from an asthenospheric mantle source similar to that of enriched MORB or OIB (Ben-Othman et al., 1990). Rock samples collected in the northern part of the pyroclastic deposit of Potrillo maar, and lava associated with a cinder cone, yielded K-Ar ages of 1.23 ± 0.06 and 1.18 ± 0.03 Ma, respectively (Hawley, 1981).

Camargo Volcanic Field

The Camargo volcanic field is located in southeastern Chihuahua, close to the Coahuila state line (Figs. 1 and 13). To date, it is the largest (~3000 km^2) and most voluminous (~120 km^3) accumulation of intraplate-type volcanic products known in central and northern México. More than 300 volcanoes, mainly cinder cones and associated lavas, and a few maars have been identified in the field. The ^{40}Ar-^{39}Ar ages of rocks from Camargo (n = 23) vary between 4.7 and 0.09 Ma, and a systematic displacement of volcanic activity, from SW to NE, at an estimated rate of 15 mm/a, has been documented in the area. Unlike many other intraplate-type localities in México, a strong association between volcanism and normal faulting is evident at Camargo, as the lava field is cut by a complex graben (Fig. 13). Partial information on the ages of the faults suggests that there could also have been a northeastward migration of the deformation, coeval with the northeastward migration of volcanic activity (Aranda-Gómez et al., 2003). Camargo is located atop the buried trace of the San Marcos fault (Fig. 13), which is a structure that formed in Jurassic time (McKee et al., 1984, 1990; Cameron et al., 1991). This basement structure has been reactivated in every known pulse of deformation since its formation, including the mid- and late Cenozoic extensional events (Aranda-Gómez et al., 2005). To date, there has been no systematic study of the petrology of the mafic volcanic rocks in this field. The geochemistry of a few volcanoes located around La Olivina—an important mantle- and crustal-xenolith locality (Nimz et al., 1986, 1993, 1995; Rudnick and Cameron, 1991)—has been reported. On the basis of six samples collected around La Olivina, Nimz (1989) concluded: (1) The rocks are basanites with normative *nepheline* between 9.4 and 15.6 wt.% and Mg# between 56 and 66. (2) Variations in Mg# and other geochemical parameters are evidence of crystal fractionation of clinopyroxene and olivine. (3) Based on the chondrite-normalized REE patterns and on multi-element plots, the trace-element concentrations in Camargo rocks are believed to be similar to those found in other mantle-xenolith localities in the U.S. portion of the Basin and Range Province. However, the Camargo magmas require a greater garnet/pyroxene ratio in the mantle source compared with other localities. (4) The concentration and proportions of trace elements in the Olivina suite are similar to those found in alkalic rocks erupted in western Texas during the early and mid-Cenozoic. This observation is tentatively interpreted as evidence of a common regional source that has been chemically stable during the Cenozoic and was involved in the magma production associated with these volcanic events. (5) The isotopic ranges of La Olivina basanites (Figs. 6D and 7C) are ε_{Nd} ~+ 3.4 to + 5.2, $^{87}Sr/^{86}Sr$ ~0.7030–0.7033, $^{206}Pb/^{204}Pb$ ~19.3–19.6, and $^{207}Pb/^{204}Pb$ ~15.58–15.63. The Pb values are different from those found in other xenolith localities in the northern portion of the Basin and Range Province, both in the southwestern United States and in México (Fig. 7C). According to Nimz (1989), this difference may be related to contamination with crustal material, or to the presence of an important tectonic limit north of Camargo, or may signify that the mantle beneath North America is laterally heterogeneous. Nimz argued that significant crustal contamination is unlikely in these rocks, because the suite contains mantle xenoliths. The similarities between Camargo's crustal xenoliths and those found at Kilbourne Hole (e.g., Padovani and Carter, 1977), and other localities in the southwestern United States, led Nimz (1989) to conclude that an important tectonic limit does not exist north of La Olivina. Therefore, he interpreted the isotopic data as a product of heterogeneities in the mantle beneath North America. La Olivina basanites require a source with a Pb isotopic composition similar to the HIMU (high μ, high $^{238}U/^{204}Pb$ ratio) reservoir proposed by Zindler and Hart (1986).

Durango Volcanic Field

In the northwestern part of Mesa Central, east of the Sierra Madre Occidental (Fig. 1), there is a large lava field that covers ~2200 km^2 and has an estimated volume of 20 km^3. Approximately 100 volcanoes—mainly cinder and lava cones—have been identified in the field. The radiometric ages (K-Ar on groundmass) of two samples collected in the area are < 0.8 Ma (Smith, 1989). We obtained a zero age (^{40}Ar-^{39}Ar) from a sample collected at La Breña–El Jagüey maar complex, which represents the youngest (likely Holocene) volcanic activity in the area. However, our study of cone degradation throughout the field indicates that some cones in the area are considerably older, probably early Pleistocene or late Pliocene. The region has undergone several extensional pulses in the mid- and late Cenozoic. The youngest period affected the Durango volcanic field, as attested by NW-trending normal faults that cut some of the lavas. Cinder cone alignments within the field suggest that the magmas reached the

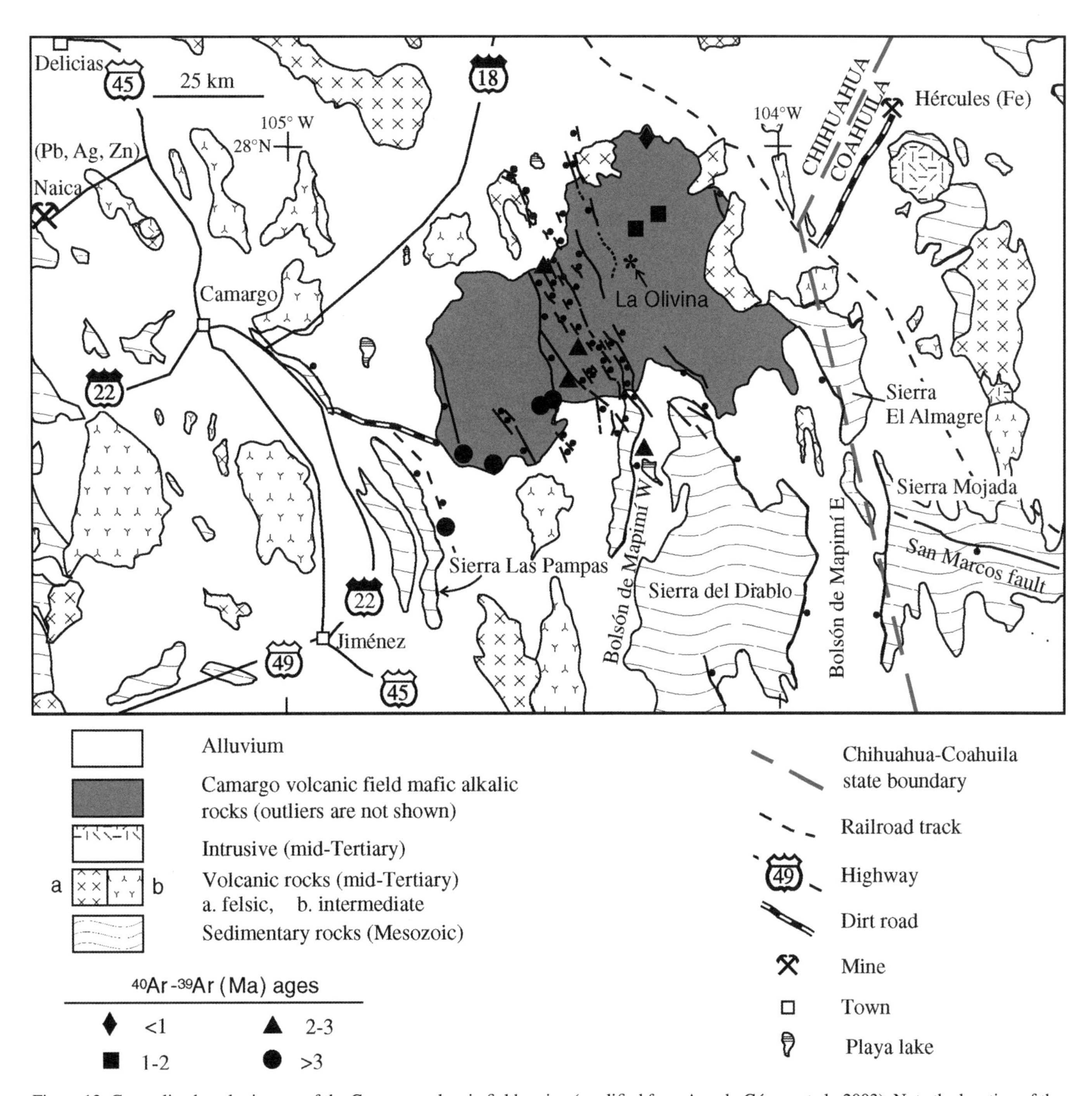

Figure 13. Generalized geologic map of the Camargo volcanic field region (modified from Aranda-Gómez et al., 2003). Note the location of the San Marcos fault in the SE corner of the map, the systematic northeastward younging in the age of the volcanism, and the occurrence of a central graben oriented NNW-SSE cutting across the volcanic field. The buried trace of the San Marcos fault lies below the Camargo volcanic field and is covered by alluvium and mid-Tertiary–Quaternary volcanic rocks (Aranda-Gómez et al., 2005).

surface through NW-trending tensional fractures, parallel to Quaternary normal faults and to older regional faults (Fig. 8).

Smith (1989) studied the petrology and geochemistry of 93 volcanic rock samples collected throughout the field and found that most of them (82%) contain normative *nepheline* (classified as hawaiites, alkali basalts, and basanites) whereas the rest (18%) are *hypersthene*-normative (hawaiites and basalts). The Smith (1989) data show that there is no geographic or stratigraphic pattern in the distribution of the two series. In one-third of the sampling localities, mantle and/or crustal xenoliths were found, including some *hypersthene*-normative host rocks. The Mg# of the rocks varies between 51 and 67. Variation diagrams of Mg#

versus incompatible elements show considerable dispersion (Fig. 12A), but three linear trends (A, B, and C) can be identified. Trends A and B are roughly parallel and have an inverse correlation with Mg#. Linear trend A includes rocks with *nepheline* in the norm: hawaiites, basanites, and alkali basalts. Rocks in trend B (hawaiites and basalts) have lower contents of incompatible elements and *hypersthene* in the norm. Trend C is weak and is formed only by *hypersthene*-normative hawaiites; it has a positive slope that cannot be explained by crystal fractionation or partial melting models. The chemical variations in *nepheline*-normative rocks can be modeled with fractionation of two different mineral assemblages: a high-pressure assemblage of Grt + Al-rich Aug + Krs + Ol or a low-pressure assemblage of Pl + TMt + Cpx + Ol. Smith (1989) preferred the high- pressure model based on the fact that these rocks contain mantle xenoliths, which indicates that the magmas rose rapidly from mantle depths. *Hypersthene*-normative rocks share many geochemical characteristics with subduction-related basalts and their chemical variations cannot be explained by crystal fractionation models. Thus, Smith (1989) postulated a metasomatized mantle component beneath the volcanic field, which is a remnant from mid-Tertiary subduction modification.

The results of a detailed study of 16 samples collected at the La Breña–El Jagüey maar complex show coherent chemical variations that suggest a cogenetic relationship (Aranda-Gómez et al., 1992; Pier et al., 1992). This is in sharp contrast to the results obtained in the regional study by Smith (1989).The linear trend shown in Figure 12B is similar to trend A (Fig. 12A) identified by Smith (1989) in the regional study. The maar complex is formed by two cinder cones and associated lava flows that underlie the hydrovolcanic sequence related to the maar volcanoes and by at least three spatter cones younger than the maars that grew inside La Breña crater (Aranda-Gómez et al., 1992). The rocks are hawaiites and basanites with content between 45.8 and 48.1 wt% SiO_2 and 4.9–9.7 wt% normative *nepheline* (Fig. 3D). The primary minerals are olivine, plagioclase, cliopyroxene, titanomagnetite, and ilmenite. Xenocrysts derived from mantle peridotites and feldspathic granulites are common in the samples. A remarkable feature in some of the postmaar samples is the presence of coarse-grained segregations of late-stage minerals. These minerals crystallized from a residual liquid rich in Fe and Ti that filled vesicles up to 1 cm in diameter. Mineral phases in the vesicle fillings are titanaugite and acicular crystals of ilmenite.

Mg# in La Breña–El Jagüey samples systematically increased during the evolution of the maar complex (Fig. 12B). Premaar samples have Mg# between 57.0 and 58.2; Mg# is 58.6–61.2 in scoriae samples interlayered with the phreatomagmatic surge beds, and 59.9–63.6 in postmaar lavas. Most analyzed elements show systematic variations with respect to Mg#. Incompatible elements define linear trends with negative correlations versus Mg#, which were interpreted by Pier et al. (1992) as evidence of crystal fractionation from a parental magma with a composition similar to the postmaar lavas, but with Mg# > 65. However, over small ranges of Mg# (e.g., 61–62 in Fig. 12B), most incompatible elements show variability greater than the analytical uncertainties. Therefore, the primary magmas appear to have had minor compositional differences related to slightly different degrees of partial melting. Chondrite-normalized REE trends are subparallel and show an increase of total content of REE with decreasing Mg#. Isotopic data are also consistent with a dominant role by crystal fractionation. The variations are small: $^{87}Sr/^{86}Sr$ = 0.70327–0.70347, ε_{Nd} = +4.2 to +5.0 (Fig. 6D), $^{206}Pb/^{204}Pb$ = 18.60–18.61 (Fig. 7C), $^{207}Pb/^{204}Pb$ = 15.58–15.65 (Fig. 7C), and $^{208}Pb/^{204}Pb$ = 38.19–38.58. Although petrographic evidence for contamination of these volcanic rocks with mantle and crustal material is compelling, these processes had little effect on the observed whole-rock elemental or isotopic compositions. The high concentrations of incompatible elements in the parental magmas render them relatively insensitive to such minor contamination.

San Luis Potosí

The central and western portions of the State of San Luis Potosí belong to a high plateau known as Mesa Central. Most mountain chains in the eastern part of the plateau are formed by sedimentary marine rocks of Cretaceous age that were folded during the Laramide Orogeny. The southern part of Mesa Central is covered by mid-Tertiary felsic volcanic rocks related to the Sierra Madre Occidental magmatic province. Today's physiography of Mesa Central was influenced by a complex set of middle and late Cenozoic normal faults that cut both the sedimentary marine rocks and the overlying volcanic sequence (e.g., Labarthe-Hernandez et al., 1982; Aranda-Gómez et al., 1989; Nieto-Samaniego et al. 1997). Scattered throughout Mesa Central are late Tertiary–Quaternary monogenetic volcanoes. These localities can be divided into two groups, based on the presence or absence of mantle xenoliths. Commonly the products of Plio-Quaternary volcanoes located west of 100°W longitude contain mantle and/or crustal xenoliths (Fig. 14). Those volcanoes located farther E generally lack such xenoliths and we refer to them as the East San Luis Potosí localities. Based on their geographic position, xenolith types, and the chemical composition of the volcanic rocks, the xenolith-bearing volcanoes are divided into the Ventura–Espíritu Santo and the Santo Domingo volcanic fields (Figs. 10A and 14).

Ventura–Espíritu Santo. This volcanic field is mostly formed by isolated cinder cones and associated lava flows that lie atop Mesozoic limestone, mid-Tertiary felsic volcanic rocks, or late Tertiary or Quaternary continental clastic deposits. The best-documented localities are three maars (Joya Honda, Joyuela, and Laguna de los Palau) located in the eastern part of the volcanic field (e.g., Labarthe-Hernández, 1978; Aranda-Gómez, 1982; Luhr et al., 1989b; Pier et al., 1989; Heinrich and Besch, 1992; Schaaf et al., 1994; Aranda-Gómez and Luhr, 1996). These maars are particularly important because of their abundance and variety of mantle and crustal xenoliths. Radiometric ages of the Joya Honda and Joyuela maars (K-Ar on groundmass) are 1.1 and 1.4 Ma, respectively (Aranda-Gómez and Luhr, 1996).

Compared with other intraplate-type localities (Fig. 3C), the Ventura–Espíritu Santo rocks have higher contents of normative *nepheline* (up to 28 wt%) and some samples contain

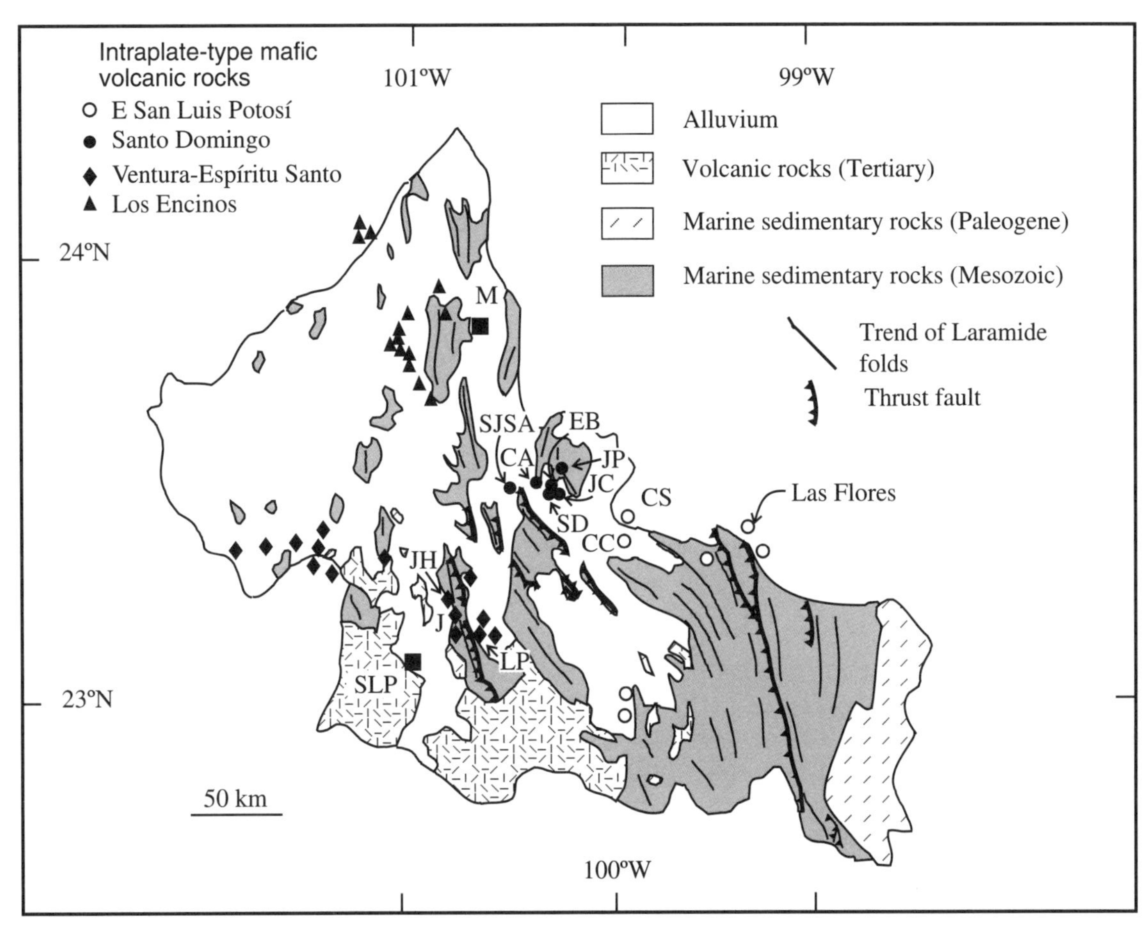

Figure 14. Generalized geological map of San Luis Potosí. The intraplate-type volcanic fields located east of longitude 100°W (shown as white circles) generally lack inclusions (xenoliths and/or megacrysts). Key to localities: M—Matehuala; SJSA—San Juan Sin Agua; CA—Cerro El Apaste; SD—Santo Domingo; JC—Joya de Los Contreras; EB—El Banco; JP—Joya Prieta; CS—Cerro Sarnoso; CC—Cerro Colorado; JH—Joya Honda; J—Joyuela; LP—Laguna de los Palau; SLP—San Luis Potosí.

CIPW-normative *leucite*. Excluding the rocks from the Bernal de Horcasitas, the volcanic products from this field are the most silica-undersaturated samples reported up to now in the Mexican portion of the Basin and Range Province.

Santo Domingo. These volcanoes are located NE of the Ventura–Espíritu Santo volcanic field and are separated from it by an ~50-km-wide zone without Quaternary volcanoes (Figs. 10A and 14). Vents in the Santo Domingo volcanic field are scarcer than in the Ventura–Espíritu Santo field. The best-studied localities in Santo Domingo are four maars: Santo Domingo, El Banco, Joya de los Contreras, and Joya Prieta (Labarthe-Hernández, 1978; Aranda-Gómez et al., 1993). The Santo Domingo field also contains several isolated lava flows, a well-preserved cinder cone (Cerro El Apaste), and three relatively large lava fields (San Juan sin Agua, Cerro Colorado, and Cerro Sarnoso, Fig. 14) that could be considered to be high-aspect-ratio lava shields. The Santo Domingo field is located in a region where there is a marked change in the trend of Laramide structures. South of Santo Domingo the fold axes are NW-trending; a few kilometers north of Santo Domingo the structures have N or NE trends (Fig. 10A). Therefore, it is believed that an important structure crosses the zone (a basement fault?) and that this structure influenced the style of deformation during the Laramide Orogeny in early Tertiary time. The same discontinuity projects south of Sierra de Catorce and crosses the central part of the Los Encinos volcanic field (Figs. 10A and 10B). Along this structural feature are volcanic vents of different ages (Eocene, Oligocene, Miocene, and Quaternary), suggesting that it repeatedly influenced the ascent of magma (Aranda-Gómez and Luhr, 1993). Most mantle xenoliths at the Santo Domingo volcanoes have mylonitic fabrics (Luhr and Aranda-Gómez, 1997), which suggest ductile deformation in the upper mantle, perhaps contemporaneous with the Quaternary volcanism. Radiometric ages (K-Ar on groundmass) obtained for lava samples collected at Cerro El Apaste and Joya de los Contreras are 0.35 and 0.45 Ma, respectively (Aranda-Gómez and Luhr, 1996). The most obvious difference between the Ventura–Espíritu Santo and Santo Domingo volcanoes is that at Santo Domingo xenoliths of hornblendite and hornblende-pyroxenites, as well as hornblende megacrysts, are very abundant, while this type of inclusion is not present in Ventura–Espíritu Santo. The Santo Domingo hornblende examined is high in Ti and classifies as kaersutite.

Petrology of the rocks from Ventura–Espíritu Santo and Santo Domingo. These rocks form a well-defined petrological series, which varies from very silica-undersaturated basanites and olivine nephelinites from Ventura–Espíritu Santo up to alkali basalts and hawaiites from Santo Domingo (Figs. 3C and 3D). Primary mineralogy in all the rocks is Ol + TAug + FTO ± Pl. Plagioclase is always present in the primary parageneses of rocks from Santo Domingo; it is present only in some rocks from Ventura–Espíritu Santo. As a general rule, the abundance of modal plagioclase decreases as the normative content of *nepheline* and *leucite* increases. In addition to the primary minerals, the rocks contain variable amounts of xenocrysts derived from the disaggregation of spinel peridotites (Ol + Opx + Cpx + Spl) and, to a lesser extent, of feldspathic granulites (Qtz + Pl). Some of the Ventura–Espíritu Santo localities contain rare pyroxene megacrysts derived from the mantle peridotites (Aranda-Gómez, 1982).

Silica content decreases in the sequence: hawaiite (51.2 wt%), alkali basalt, basanite, olivine-nephelinite (41.8 wt%). Normative contents of *nepheline* and *diopside* increase in the same order, as the normative contents of *albite* and *anorthite* are reduced. Many elements—Ti, K, Na, P, Rb, Sr, Zr, Nb, Ba, La, Ce, Nd, Sm, Eu, Hf, Ta, Th, and U—have incompatible behavior and progressively increase with CIPW *nepheline*. All Sr-Nd isotopic data for these two fields plot on or near the mantle array. Their distribution is bimodal (Figs. 6C and 6D), although both volcanic fields (Ventura–Espíritu Santo and Santo Domingo) include samples that fall in both clusters (Fig. 6D); this bimodal distribution in Sr and Nd isotopes contrasts with the smooth variations observed in other geochemical plots. If the clusters in the Sr and Nd data are individually analyzed, the Nd values for Santo Domingo are found to be more radiogenic than those for Ventura–Espíritu Santo. The Sr-Nd isotopic behavior is not mirrored by the Pb isotopes, which lack a bimodal distribution (Fig. 7C).

Many of the elemental variations documented in the suite can be explained by progressive partial melting of a garnet peridotite in the mantle. The Santo Domingo rocks in this model reflect higher degrees of partial melting compared with the Ventura–Espíritu Santo samples. An alternative interpretation of the data is that the chemical variations can be explained by the mixing of magmas from two different sources (Luhr et al., 1989b). The variability in the Sr and Nd isotopes cannot be explained using a simple partial-melting model involving magmas derived from a single source in the upper mantle. Its explanation by the assimilation of crustal material is also inadequate. Pier et al. (1989) concluded that the isotopic data require at least three different mantle reservoirs: (1) a depleted reservoir similar to the MORB source, (2) a Santa Helena–type source, and (3) a hydrated source associated with subduction. An alternative view is proposed here: The Sr and Nd data (see Figure 1A in Pier et al., 1989) can also be explained by invoking two mantle sources, one in the asthenosphere and the other in the lithosphere. The composition of group 1 could have an asthenospheric source, and group 2 could have been generated in the lithospheric mantle. Evidence of crustal contamination is also present in the rocks from Ventura–Espíritu Santo and Santo Domingo, and it is possible to envision group 2 as derived from magmas similar to those that originated in group 1 through this process. However, the marked differences between the groups indicate that the possibility of two mantle sources cannot be ruled out and that the isotopic compositions of some samples may reflect crustal contamination.

East San Luis Potosí. Throughout the eastern portion of the State of San Luis Potosí and in the neighboring State of Tamaulipas (Fig. 14), there exist isolated outcrops of hawaiite and picritic alkali basalt (Fig. 2D). The rocks in most of these localities lack xenoliths and/or megacrysts. An exception is a locality close to Cárdenas, S.L.P., where xenoliths of mantle peridotites occur. The most important locality in this broad region (~15,000 km^2) is Las Flores (Siebert et al., 2002), which lies in the area where the state boundaries of San Luis Potosí, Tamaulipas, and Nuevo León meet. The zone around the vent is covered by a lava field with an approximate area of ~200 km^2. Despite the humid, subtropical climate in the region, the lavas in this field have a youthful appearance, with ropy surfaces, lava tubes, collapse features, and pressure ridges; therefore, we assume that this field is very young. An exceptionally long lava flow was issued from Las Flores volcano. It flowed along the bottom of a synclinal valley and can be traced for nearly 80 km, all the way to the vicinity of Ciudad Valles, S.L.P. The results of a few chemical analyses of rocks from this field are shown in several graphs in Figure 23 in Aranda-Gómez et al. (1997). Based on these graphs, these authors concluded that the rocks from East San Luis Potosí are similar to other intraplate-type localities in México. The relation of these localities with the "volcanismo del altiplano," studied by Robin (1976) farther south in Hidalgo State, is unknown. We have analyzed the isotopic compositions of a few samples from East San Luis Potosí. Compared to the Santo Domingo and Ventura–Espíritu Santo samples, these rocks have higher $^{87}Sr/^{86}Sr$ (Figs. 6C and 6D) without a corresponding decrease in $^{143}Nd/^{144}Nd$. Therefore, we assume that these magmas may have been contaminated with the calcareous rocks of the Sierra Madre Oriental, which have high Sr contents, high $^{87}Sr/^{86}Sr$ values, and negligible Nd.

Coahuila

At least three localities with intraplate-type rocks occur in Coahuila State (Figs. 1 and 15): the Las Esperanzas volcanic field near Sabinas, Ocampo, in the central portion of the state, and Las Coloradas, which is located on the northeastern portion of the Coahuila Block, near the San Marcos fault. The geochemistry of the rocks in the first two localities was studied by Valdez-Moreno (2001). We recently analyzed the isotopic compositions of Sr and Nd in two samples from Las Coloradas.

Las Esperanzas volcanic field. A sample collected at Las Esperanzas yielded an ^{40}Ar-^{39}Ar age (groundmass) of ca. 2.8 Ma (Valdez-Moreno, 2001). The Esperanzas volcanic field can be

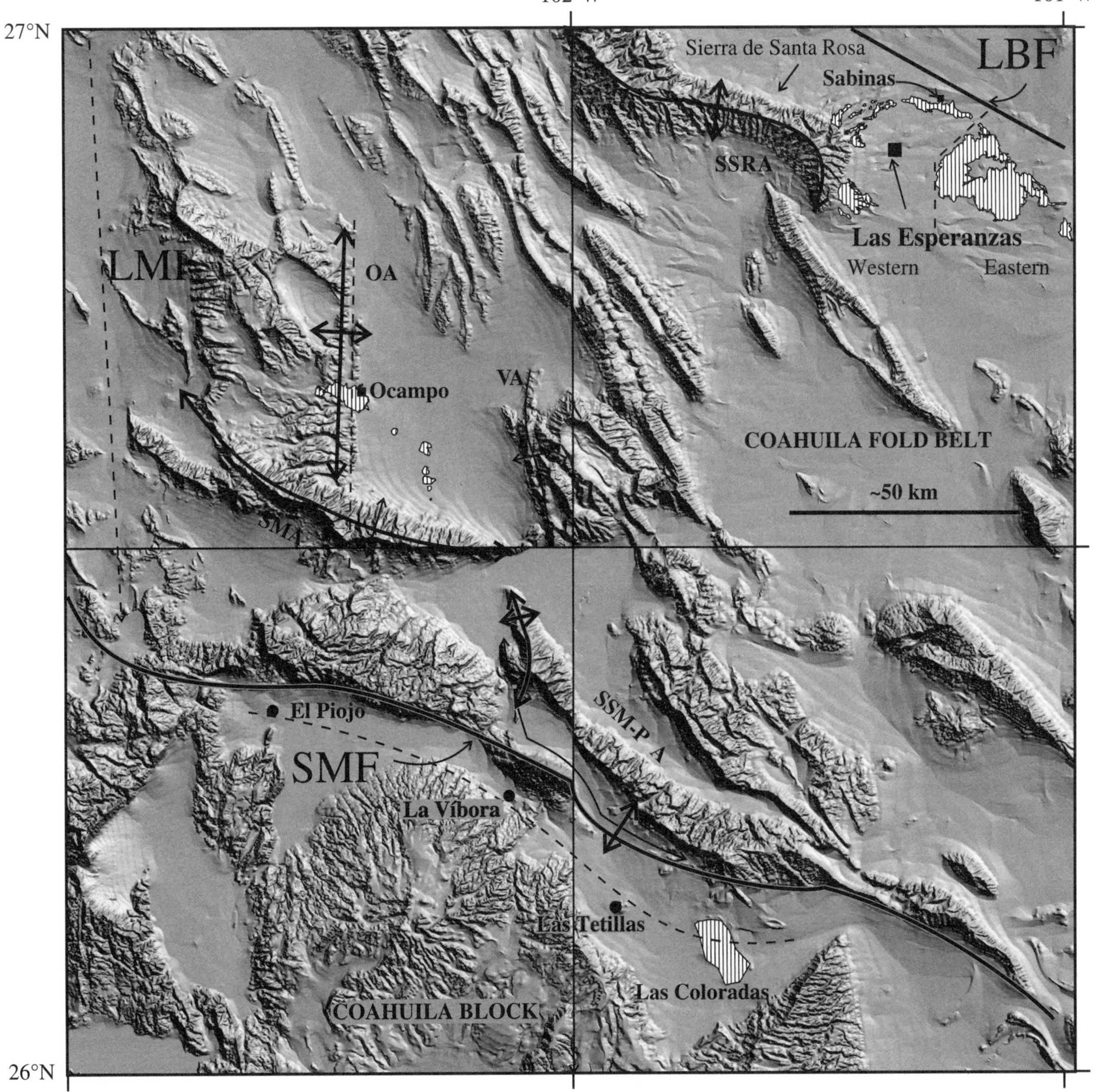

Figure 15. Digital elevation model of the central part of Coahuila state. Las Esperanzas, Ocampo, and Las Coloradas volcanic fields are shown, as well as the Las Tetillas intrusion and the isolated El Piojo and La Víbora volcanoes. The traces of the San Marcos (SMF) and La Babia (LBF) faults are shown along with other structural features mentioned in the text. The Coahuila fold belt is bounded by the San Marcos and La Babia faults and its geomorphology is controlled by the folds formed during the Laramide Orogeny (early Tertiary). Note the sharp contrast between the morphology at the folded belt and the neighboring Coahuila platform. Fold trends in the Coahuila fold belt often define orthogonal arrays or have sharp bends. These structural features are attributed to the influence of pre-Cretaceous basement faults during early Tertiary folding. La Mula Island (LMI), a Mesozoic paleogeographic feature, is probably limited by basement faults, as we infer from changes in the Laramide folds. OA—Ocampo Anticline; VA—La Vírgen Anticline; SMA—Sierra La Madera Anticline; SSRA—Sierra de Santa Rosa Anticline; SSM-P A—Sierras de San Marcos and Pinos Anticline.

divided into eastern and western zones. The western zone is formed by extensive lava flows, which were erupted from fissure vents located at the base of the Sierra de Santa Rosa anticline (Fig. 15). These lavas traveled up to 60 km away from the vent following fluvial paleochannels. The eastern zone of Las Esperanzas is formed by small continental lava shields with stacks of lavas that may be up to 35 m thick (COREMI, México [Consejo de Recursos Minerales, México], 1994). The primary mineral assemblage in the rocks from Las Esperanzas is: Ol + Cpx + Pl + FTO ± Ap. A few samples contain xenocrysts derived from mantle peridotites. Quartz xenocrysts surrounded by pyroxene coronas are also present in some rocks. According to the TAS classification (Le Maitre et al., 2002), Las Esperanzas rocks are basalts and hawaiites (Fig. 2D). Whole-rock Mg#s range from 60.5 to 66.5 (Fig. 16). One half of the studied samples are in the 63–73 range proposed by Green (1971) for primitive magmas. However, the Ni (118–263 ppm) and Cr (186–377 ppm) contents are relatively low for primary magmas; therefore it is believed that olivine and clinopyroxene were segregated from these magmas before they reached the surface. Variation diagrams of Mg# versus major and trace elements suggest that rocks from the western and eastern zones form two independent suites (Fig. 16). The geochemistry of the rocks suggests that some of the chemical variations reflect different degrees of partial melting and/or source-mantle heterogeneity.

Ocampo volcanic field. The Ocampo volcanic field is a zone with a total of ~600 km^2 of isolated areas covered by mafic lava flows and pyroclastic deposits (^{40}Ar-^{39}Ar, groundmass, 3.4–1.8 Ma; Valdez-Moreno, 2001). The most extensive outcrop occurs near the town of Ocampo (Fig. 15) and is formed by a stack of intracanyon lava flows, up to 80 m thick. These lavas were erupted from at least two scoria cones. A N-S–trending vent alignment formed by four cones and associated lava flows also occurs in the area (Fig. 15). The primary mineral assemblage in the samples collected at Ocampo is Pl + Ol + Cpx + FTO + Ap. A few samples contain quartz and/or potassium feldspar xenocrysts. According to the TAS classification, these rocks are basanites, hawaiites, and basalts (Fig. 2D). All but one sample contain normative *nepheline* (Fig. 3D). Whole-rock Mg# varies between 54 and 66 (Fig. 16), but most samples have Mg# > 60. Rocks with lower Mg# (54–57) were collected at the N-S vent alignment in the eastern portion of the volcanic field. Plots of major elements versus Mg# show rough linear trends, and Nb and Sr have negative correlations with Mg#. Incompatible elements such as Zr, Sr, and Nb have large variations within a narrow range of Mg# (Fig. 16). Other incompatible elements display similar behavior relative to Mg#, but the dispersion is smaller. Ni, Cr, and Co have positive correlations with Mg# with little scatter.

Petrology of the rocks from Las Esperanzas and Ocampo. Features common to the rocks from Las Esperanzas and Ocampo include enrichment in incompatible elements and the lack of negative anomalies in Nb and Ti (Fig. 5D). The LREE are enriched with respect to the HREE, which was interpreted as evidence of residual garnet in the parental magma. Initial $^{87}Sr/^{86}Sr$ varies

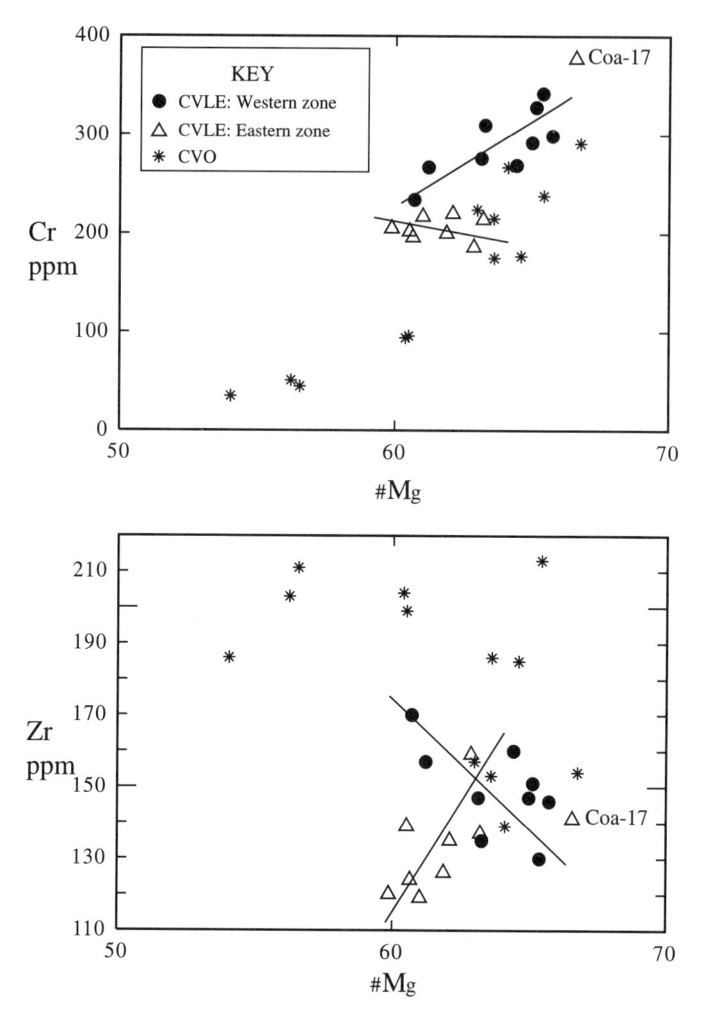

Figure 16. Variation diagrams for the rocks collected at the Las Esperanzas and Ocampo volcanic fields (Valdez-Moreno, 2001). Sample Coa-17 was collected in the eastern portion of Las Esperanzas and contains xenocrysts derived from mantle peridotites. In many variation diagrams, Coa-17 lies outside of the linear trends defined by other samples collected in the same area. Note that the slopes of the linear trends formed by samples collected at the eastern and western portions of the Las Esperanzas volcanic field are radically different. Therefore, the samples from the eastern and western parts of the field are considered as two separate suites. Likewise, note the poor correlation in the data sets.

between 0.70334 and 0.70359 for the samples collected at Las Esperanzas, and between 0.70337 and 0.70346 for the Ocampo samples. The ε_{Nd} values for Las Esperanzas are + 5.0 to + 5.5 and for Ocampo are +6.0 and +6.1 (Fig. 6D). The $^{206}Pb/^{204}Pb$ values are higher for Las Esperanzas (18.51–18.77) than for Ocampo (18.45–18.48). $^{207}Pb/^{204}Pb$ values are similar in both volcanic fields (Las Esperanzas: 15.55–15.58; Ocampo: 15.55–15.56) and $^{208}Pb/^{204}Pb$ is lower in Ocampo (38.08–38.18) than in Las Esperanzas (38.29–38.38). These results were interpreted by Valdez-Moreno (2001) as evidence of isotopic heterogeneity of the mantle and/or assimilation of crustal material.

Las Coloradas volcanic field. Las Coloradas, as well as the isolated vents of La Víbora and El Piojo, and the Las Tetillas intrusion—a subalkaline, subduction-related body (Cavazos-Tovar, 2004)—define a rough alignment, ~15 km south of and parallel to a basement structure known as the San Marcos fault (McKee et al., 1984, 1990) in central Coahuila (Fig. 15). A normal fault with the downthrown block to the S is related to La Víbora volcano. The intraplate nature of the Las Coloradas volcanic field is assumed from (1) the inferred age of the rocks (Plio-Quaternary), (2) the geographic location of the field, and (3) Sr and Nd isotopic data (Fig. 6D) for two samples we recently collected.

It is important to note that the normal faulting associated with the Río Grande rift ends abruptly in Coahuila (Gries, 1979; Seager and Morgan, 1979). All intraplate-type volcanism in the region appears to be related to the locations of old faults in the basement that bound large tectonic domains. For example, the Las Esperanzas volcanic field is located just at the base of the Sierra de Santa Rosa, which is a fault-bend fold that separates two regions with contrasting deformation styles: the Coahuila fold belt (Fig. 15) and the Burro-Peyotes Platform. The buried trace of La Babia fault (Charleston, 1974, 1981), an important basement structure covered by Cretaceous and Tertiary sedimentary sequences, crosses the region. The Ocampo volcanic field is in the eastern portion of a paleogeographic feature known as La Mula Island (Jones et al., 1984), and four volcanoes form a N-S alignment, which is parallel to the trend of the axes of Laramide folds in the region. Thus, the Ocampo volcanic field is also located on the trace of a basement structure. Our conclusion is that, at least in Coahuila, there is evidence of a close relationship between intraplate-type volcanism and basement faults formed during the Jurassic.

Late Magmatism (Early Pliocene–Quaternary): Eastern/Alkalic Region

These volcanic fields are located along the Gulf of Mexico coastal plain, and have been referred to as Mexico's Eastern Alkalic Province (Demant and Robin, 1975). This region has had a long and complex history of intermittent magmatic activity since the early Tertiary. The Eastern/Alkalic region is a NNW- to NS-trending belt (Fig. 1), roughly parallel to the Gulf of Mexico coastal plain, and in a few places appears to overlap parts of the Sierra Madre Oriental (Demant and Robin, 1975). Oligocene volcanic activity was roughly parallel to the Pacific paleo-trench (Ramírez-Fernández et al., 2000). The northern end of the Eastern/Alkalic region joins the Trans-Pecos Province in Texas. The southern end of the Eastern/Alkalic region is partially covered by or interacts in a complex way with the Trans-Mexican Volcanic Belt. Robin (1982) interpreted the volcanoes in the Tuxtlas region (Veracruz) as part of the Eastern Alkalic Province. The alkalic volcanism contemporaneous to the subduction of the Farallon plate in the northern part of this Eastern/Alkalic region has been interpreted as being associated with backarc extension (Demant and Robin, 1975). An alternative interpretation was proposed by Clark et al. (1982), who viewed it as the product of partial melting that occurred at greater depth while the arc was migrating to the east. A change in the chemical composition of the igneous rocks, from orogenic to intraplate-type, has been documented in the Trans-Pecos region of Texas. This change occurred in the period ca. 32–28 Ma and has been tied to the initial extension associated with the Basin and Range Province (James and Henry, 1991). Late Eocene–Oligocene magmatism at the Candela-Monclova intrusive complex and at the Sierra de San Carlos (Tamaulipas) has the geochemical signature of subduction. Younger intrusive rocks (ca. 30 Ma) and volcanic rocks exposed nearby have intraplate-type chemical compositions and were probably associated with extension (Viera-Décida, 1998; Ramírez-Fernández et al., 2000; Chávez-Cabello, 2005). Four Neogene (late Miocene–Quaternary) volcanic fields occur in the northern and central part of Veracruz: (1) Tlanchinol-Huautla, (2) Alamo–Sierra de Tantima, (3) Poza Rica, and (4) Chiconquiaco–Palma Sola (Orozco-Esquivel et al., 2003; Ferrari et al., 2005), which overlaps the eastern end of the Trans-Mexican Volcanic Belt. The volcanic fields in Veracruz, together with isolated volcanoes south of the Sierra de Tamaulipas (Bernal de Horcasitas and Cerros Auza, El Nopal, and El Murciélago; Fig. 17), were described by Robin (1976, p. 59) as the "volcanism of La Huasteca plains."

Sierra de San Carlos–Cruillas

Isolated outcrops of igneous rocks occur in an area of ~5000 km^2 in the central part of Tamaulipas (Fig. 1). Some of these localities have been studied by Nick (1988), Ramírez-Fernández et al. (2002), and Rodríguez-Saavedra (2003). The complex is composed of (1) mid-Tertiary (32–28 Ma; Bloomfield and Cepeda-Dávila, 1973) plutons (gabbro-syenite-monzonite) emplaced in Mesozoic limestone and shale, (2) sills (gabbroic porphyry) and dike swarms (basalt to phonolite), and (3) volcanic necks and associated lava flows of olivine basalt and trachybasalt. These researchers argued that the volcanic rocks are products of late activity associated with crustal extension. Some of these rocks contain mantle xenoliths (Treviño-Cázares, 2001). Petrological relations among the different parts of this igneous complex are not clear or direct. Mafic magmas are interpreted as mantle-derived, whereas intermediate and felsic rocks are interpreted as products of crystal fractionation in a stratified magma chamber (Ramírez-Fernández et al., 2002). Up to now, there have been no radiometric ages reported on the post-Oligocene intraplate-type rocks.

Sierra de Tamaulipas Magmatic Complex

Sierra de Tamaulipas is a N-S–trending anticlinorium developed in Cretaceous limestone, located between the Sierra Madre Oriental and the Gulf of Mexico coast (Fig. 17). Alkalic intrusions (gabbro to granite) are exposed in a ~3600 km^2 area (Ortega-Gutiérrez et al., 1992). The El Picacho complex is in the central part of the Sierra de Tamaulipas, and it is formed by olivine gabbro, kaersutite diorite, nepheline syenite, carbonatite dikes (Ramírez-Fernández et al., 2000), and apatite-, britholite-, and calc-rich hydrothermal veins (Elías-Herrera, 1984; Elías-

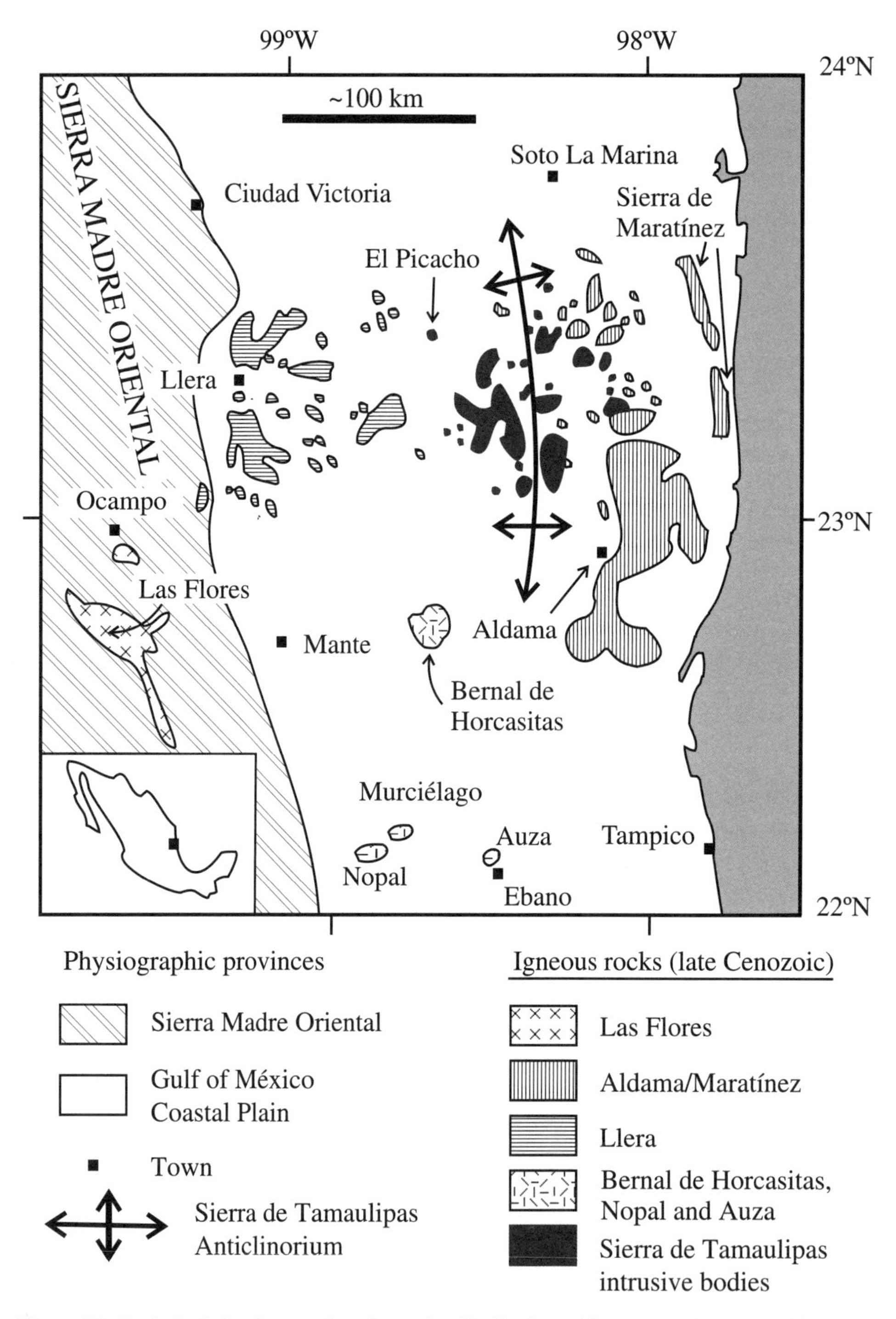

Figure 17. Geological sketch map that shows the distribution of igneous rocks around Sierra de Tamaulipas (modified from Treviño-Cázares et al., 2005). The Las Flores volcano is considered part of the East San Luis Potosí volcanic field in this review (see Fig. 14).

Herrera et al., 1990, 1991). The precise age of the complex is unknown, but other intrusive rocks from the Sierra de Tamaulipas yielded radiometric ages between 31.5 and 17.5 (Cantagrel and Robin, 1979; Seibertz, 1990; Camacho-Angulo, 1993). Ramírez-Fernández et al. (2000) interpreted the El Picacho complex as a typical carbonatite-ijolite–felsic alkalic rock association formed by mantle-derived magma in an intraplate-type environment. Based on the presence of interstitial crystals of calcite in the nepheline syenite and on the isotopic composition (C and O) of this phase, they believe that the carbonatite was formed from an immiscible carbonatitic liquid derived from a CO_2-rich silicate magma. The apatite, britholite, and calcite veins were interpreted by Ramírez-Fernández et al. (2000) as a second immiscible liquid formed from the carbonatitic magma. However, Elías-Herrera (1984) concluded that the veins have a hydrothermal origin.

Several outcrops of late Cenozoic volcanic rocks occur around the Sierra de Tamaulipas anticlinorium (Fig. 17). We grouped these outcrops into two volcanic fields: (1) Llera de Canales, west of the Sierra, and (2) Aldama–Sierra de Maratínez, east of the Sierra. Farther to the south of the Sierra are several other isolated volcanoes. At least two of these, located near Ébano, S.L.P., contain peridotite xenoliths (Cerros El Nopal and Auza; Treviño-Cázares, 2001).

Llera de Canales. A group of lava-capped mesas occur around the town of Llera de Canales, which is located just at the base of the Sierra Madre Oriental (Fig. 17). These mesas are nearly 200 m above the present-day height of the active Guayalejo riverbed. The capping lava flows are thin (<10 m). In the same area, several volcanic necks known in the region as "bernales" are found. Mafic rocks in the area are basanites, alkali basalts, and hawaiites (Fig. 3D). These mafic rocks have Mg# < 62, Ni $\simeq$ 210 ppm, and Cr < 345 ppm. Their REE contents were interpreted by Ramírez-Fernández (1997) as evidence for a garnet-amphibole-lherzolite mantle source. A hawaiite from Cerro La Clementina yielded a K-Ar age of 2.4 Ma (Camacho-Angulo, 1993). Aranda-Gómez et al. (2002) reported early Pliocene ^{40}Ar-^{39}Ar ages (ca. 5 Ma) for a number of rocks collected in the area. Several important mantle-xenolith localities exist in this volcanic field; some of them were first reported by Pettus (1979) and later studied in more detail by Treviño-Cázares (2001) and Treviño-Cázares et al. (2005).

Aldama and Sierra Maratínez. The Aldama volcanic field is in the coastal plain, immediately east of Sierra de Tamaulipas (Fig. 17). It has an estimated surface area of ~800 km^2 and contains ~20 volcanoes, mainly cinder cones and associated lava flows. The volcanic edifices are relatively well preserved in a humid and subtropical environment; therefore we believe that they are relatively young. Treviño-Cázares (2001) reported K-Ar ages (2.6 and 1.4 Ma) obtained by Camacho-Angulo (1993) on two samples collected in Aldama. Based on the degradation features observed in the cones, we believe that there might be younger volcanoes in the area. The chemical compositions of the Aldama samples range from trachyte to alkali basalt (Fig. 2C). To date, Aldama together with Palomas and Santa Clara volcano (El Pinacate) are the only reported Plio-Quaternary intraplate-type volcanic localities with felsic rocks in central and northern continental México; trachytes are common on the Mexican intraplate-type islands of Socorro, Bárcena, and Guadalupe (Luhr et al., 2006). Our unpublished isotopic analyses (Sr and Nd) for rocks from this volcanic field suggest that trachytes and alkali basalts are very similar (Fig. 6D) and probably cogenetic. The early eruptions in this field produced trachyte and phonolite and the late eruptions were basaltic (Siebert et al., 2002). Even though Vasconcelos et al. (2002) interpreted the Aldama rocks as a bimodal series, we note in the TAS diagram (Fig. 2C) that there is a continuous chemical variation in the analyses from basalt to trachyte. The mafic rocks include subalkaline basalts, alkali basalts, and hawaiites (Fig. 3D). The evolved samples are benmoreites and trachytes with primary mineral assemblages: Sa + Anor + Ne + Ol (Fa-rich) + Aug (Aeg-rich). The trachytes form two groups: one contains normative *nepheline* and *olivine* and the other contains normative *hypersthene* and *quartz*. The REE data obtained by Ramírez-Fernández (1997) display LREE enrichment and positive Eu anomalies in some of the samples with *hypersthene* and *quartz* in the norm, and negative Eu anomalies in *nepheline*- and *olivine*-normative rocks (Fig. 5C).

Sierra de Maratínez (Fig. 17), which is North of Aldama, is where several isolated outcrops of mafic volcanic rocks define a 60-km-long NNW-trending lineament (Ortega-Gutiérrez et al., 1992). The age of these rocks is unknown, but we note that they are deeply weathered and conclude that their age might be mid- to late Tertiary. Ramírez-Fernández (1996) analyzed four samples from this area, and found that they are alkali basalts. Neither Aldama nor Sierra de Maratínez rocks contain mantle xenoliths. We found feldspathic xenoliths in some of these localities, which were probably derived from the intrusive complex exposed at the Sierra de Tamaulipas.

Bernal de Horcasitas and Cerros Auza, El Nopal, and El Murciélago. Several outcrops of volcanic rocks are known south of Sierra de Tamaulipas (Fig. 17). The relation between these vents and the intrusive complex exposed in the Sierra is unknown. Bernal de Horcasitas is a large volcanic neck, ~750 m high, with a K-Ar age of ca. 28 Ma (Cantagrel and Robin, 1979). The neck is formed by olivine foidite with high contents of K_2O, TiO_2, and P_2O_5 (Ramírez-Fernández, 1996). On the coastal plain W of Tampico are three isolated hills: (1) Cerro El Murciélago is formed by sills of tinguaite emplaced in Mesozoic shale; (2) Cerro El Nopal is formed by a peralkaline phonolite; and (3) Cerro Auza is constituted by basanite. The occurrence of peridotite xenoliths in the latter two localities was reported by Treviño-Cázares (2001). The Nopal and Murciélago hills are small laccoliths with associated sills and their inferred age is Miocene (Robin, 1976). Cerro Auza is a subvolcanic dome (Robin, 1976).

Volcanic Fields in Northern and Central Veracruz

Orozco-Esquivel et al. (2003) reported K-Ar ages and a brief summary of the geochemical characteristics of igneous rocks scattered over a large region (~100 × 200 km) in Veracruz. The volcanic fields considered, from N to S, are (1) Tlanchinol-Huautla where 7.4–5.7 Ma basanites and hawaiites are exposed, (2) Alamo-Tantima with older alkali basalts (7.5–6.6 Ma), (3) Poza Rica with Quaternary alkali basalts and hawaiites (1.6–1.3 Ma), and (4) Chiconquiaco–Palma Sola with alkali basalts and hawaiites (7.0–2.0 Ma). The rocks in volcanic fields 1 and 2 are similar to OIB (Nb = 33–90 ppm, Ba/Nb = 6.7–8.1, Zr/Nb = 3.7–6.5, and La/Yb = 14.6–28.7) and can be modeled by low-degree partial melting from a lherzolite with residual garnet and hornblende. Subduction-related and OIB-like volcanic rocks are found in Poza Rica. According to Orozco-Esquivel et al. (2003), the Chiconquiaco–Palma Sola samples have a clear geochemical signature of subduction (Nb = 16–22, Ba/Nb = 10.7–23.3, Zr/Nb = 10.5–16.1, and La/Yb = 6.5), although we point out that the listed Nb values are well in excess of typical subduction-related rocks.

The geochemistry of the volcanic rocks from Palma Sola was interpreted by Gómez-Tuena et al. (2003) to be the product of interaction between diverse components related to subduction (oceanic crust, sediments, and fluids) in a mantle wedge composed by an enriched OIB-type source. The Gómez-Tuena et al. (2003) model for the origin of Palma Sola rocks infers that changes in the angle of the subducted slab controlled the type of material that was melted.

The isotopic ages for coastal-plain volcanic rocks in northern and central Veracruz reported by Orozco-Esquivel et al. (2003) are not consistent with the model proposed by Robin (1982) for the north-to-south migration of magmatic activity in his proposed Eastern Alkalic Province. In contrast, the radiometric data of Orozco-Esquivel el al. (2003) showed that there is no ordered succession of volcanic events in this region.

DISCUSSION

Oceanic Volcanism

The islands are located on or near fossil-spreading ridges and their products represent magmas formed after the production of MORBs along these ridges had ceased. The chemical compositions of the studied rocks vary from alkali basalts to rhyolites (Fig. 2A). Samples from Isla Guadalupe form a complete and coherent basalt–trachyte magma series, where all intermediate members—hawaiite, mugearite, and benmoreite—are represented. The chemical variations in the Guadalupe series can be successfully modeled by low- to intermediate-pressure (2–10 kb) crystal fractionation from a parental alkali basalt. The mineral assemblage used in the model is similar to that observed in the studied samples. The relatively homogenous $^{87}Sr/^{86}Sr$ values (0.70321–0.70330) are consistent with the crystal-fractionation hypothesis. The Socorro samples are clearly bimodal (Fig. 2A), with one group formed by basalts and hawaiites, and another group by trachytes and rhyolites. Compared with other oceanic islands, Socorro is an exceptional locality, as the exposed rocks are mostly (~80%) trachytes and rhyolites. The estimated volume of this alkalic volcano is twenty times larger than that of the Camargo volcanic field, which is up to now the most voluminous example known in the central and northern portion of continental México.

The geochemistry of Socorro's samples cannot be explained with a simple model. Elemental variations in the mafic rocks were modeled by low-pressure fractionation of Pl + Cpx + Ol + FTO from a parental alkali basalt. However, the Ce, P_2O_5, Ba, and intermediate REE concentrations (Figs. 4A and 5A) cannot be explained with this model. These anomalous values, as well as variations in radiogenic Sr (Fig. 6), were ascribed to contamination of the magma with components derived from the oceanic crust, metalliferous sediments, apatite formed in previous plutonic episodes, or hydrothermal fluids dominated by seawater. Socorro's felsic rocks cannot be modeled by crystal fractionation from parental alkali basalt. The trachytes can be explained by 5–10% partial melting of alkali basalt. Crystal fractionation of Cpx + Fe-rich Ol + Ilm + Ap may generate the rhyolites from a parental trachytic melt. As in the case of the mafic rocks, some Ce and REE anomalies in the felsic rocks (Figs. 4A and 5A) are attributed to assimilation of hydrothermal sediments and the variations in Sr isotopes are attributed to interaction with a hydrothermal system dominated by seawater.

Continental Localities

Detailed and systematic studies of a large number of the localities shown in Figure 1, such as Isla Isabel, Mesa Cacaxta, and East San Luis Potosí, are not available. Only a few analyses have been reported from these localities in graphs used to compare some chemical parameters in the intraplate-type rocks from México (e.g., Figure 23 in Aranda-Gómez et al., 1997). From these comparisons it is concluded that these localities represent intraplate-type volcanism. We have assumed that other volcanoes, such as those in Fresnillo, Sombrerete, and Nieves (Zacatecas), are intraplate-type localities, based on the occurrence of mantle xenoliths, geographic location, and regional geologic setting. A few mantle xenolith localities reported by Aranda-Gómez and Ortega-Gutiérrez (1987) are not included in Figure 1. We now know that the Jiménez, Lago Jaco, and La Olivina mantle xenolith localities are all part of the Camargo volcanic field. Likewise, the occurrence of mantle xenoliths at El Cardel, Puente Negro, and the Tuxtlas volcanic field could not be confirmed despite additional field work in these areas. Gómez-Tuena et al. (2003) and Blatter and Carmichael (1998) documented the presence of mantle xenoliths in volcanoes located in or near the Trans-Mexican Volcanic Belt. These two mantle xenolith localities, as well as volcanic fields with OIB-type rocks in the Trans-Mexican Volcanic Belt, were excluded from this review, as their proximity or close geographic and temporal association with continental subduction-related volcanism distinguishes them from other localities discussed in this paper.

Geographic Distribution and Association with Tectonic Provinces and/or Regional Tectonic Features

If it is assumed that all the continental volcanic fields shown in Figure 1 were formed by intraplate-type activity during the Cenozoic, it is evident that these phenomena occurred throughout the whole region north of the Trans-Mexican Volcanic Belt, independently of tectonic plates (Pacific or North American), older volcanic provinces, present-day boundaries between tectonic provinces, and tectonostratigraphic terranes proposed by several researchers (e.g., Campa and Coney, 1983; Sedlock et al., 1993).

Most magmas carrying mantle and deep-crustal xenoliths that erupted through continental crust are in the southern part of the Basin and Range Province (Fig. 1), but some volcanic fields are located in areas where normal faults are not obvious in the exposed geology. Some remarkable examples of this are (1) San Quintín, which lies on the "stable peninsula," W of the Main Gulf Escarpment (Gastil et al., 1975). This field is located in a place where the continental platform is very narrow and in an area where, based on the interpretation of geophysical surveys, there is a half-graben buried under a thick sequence of sediments (Espinosa-Cardeña et al., 1991; Almeida-Vega et al., 2000). (2) The Las Esperanzas and Ocampo volcanic fields (Coahuila) are in an area where Cenozoic normal faulting might be concealed or masked by the presence of thick evaporite deposits (e.g., Gries, 1979) interlayered in the Mesozoic sedimentary sequence accumulated in the Sabinas Basin. (3) The volcanic fields around Sierra de Tamaulipas are located on the Gulf of Mexico coastal plain. This region is presently under extension (Suter, 1991), and it has been argued that faulting associated with gravitational slip of sediments toward a depocenter located in the Gulf of Mexico must exist (Bryant et al., 1968). (4) The crustal extension implied in Robin's model (1982) for the origin of volcanism in the Eastern Alkalic Province has recently been challenged for the volcanic fields located in central and northern Veracruz. New data on the structural setting of these volcanic fields have shown an absence of normal faulting, parallel to the Gulf of Mexico coast, contemporaneous with the volcanism. Therefore, Ferrari et al. (2005) concluded that volcanism in the region cannot be associated with crustal rifting. They interpret the observed volcanic lineaments and elongation of volcanic necks as parallel to NE to ENE- and NW to NNW-trending basement faults that were formed during the opening of the Gulf of Mexico. It is believed that these structures controlled the magma ascent. According to Ferrari et al. (2005), partial melting was caused by detachment and sinking of the subducted Cocos plate.

Some of the most voluminous and extensive intraplate-type volcanic fields in northern México (e.g., Camargo and Durango: Smith, 1989, 1993; Aranda-Gómez et al., 2003) show evidence that volcanism and crustal extension were at least in part contemporaneous, because the volcanic sequences are cut by normal faults and segments of the fault traces are buried under younger volcanic products (Aranda-Gómez et al., 2003). The lengths of some of these faults exceed the dimensions of the intraplate-type volcanic fields and can be traced into adjacent areas. Both in Camargo and in Durango, it has been argued that the normal fault systems were formed in the Tertiary and later reactivated during the Quaternary (Aranda-Gómez et al., 1997, 2003). Volcanoes in Moctezuma (Sonora), Rodeo-Nazas, and Río Chico (Durango) are associated or aligned with regional normal faults, and the volcanic rocks are interlayered with graben-fill gravel deposits or conglomerates. This relationship is also evidence of contemporaneous normal faulting and intraplate-type volcanism (e.g., Aranda-Gómez et al., 1997, and Luhr et al., 2001). However, in most areas, evidence for a direct connection between intraplate-type volcanism and crustal extension is scant, as in the Ventura–Espíritu Santo and Los Encinos volcanic fields, where regional basement structures may have influenced magma ascent, but these structures are diffuse or have not been studied in detail.

The geographic location of some of the intraplate-type localities suggests that magma ascent was influenced by older regional faults, such as San Marcos (McKee and Jones, 1979; McKee et al., 1984, 1990), San Tiburcio (Mitre-Salazar, 1989), and the Santa María River fault systems (San Luis Potosí–Zacatecas). These structures limit regional tectonic domains in the basement and were reactivated during tectonic events after their formation (e.g., Aranda-Gómez et al., 2003, 2004).

Petrologic Considerations

Petrogenetic variations in a changing tectonic environment. Compared to the Plio-Quaternary rocks, the early intraplate-type volcanic rocks (late Oligocene–Miocene) display significant geochemical and petrologic differences. AFC processes were very important in the early stages of intraplate-type volcanism. This is shown by selective contamination of the magmas with elements derived from the continental crust (Figs. 5B and 9), coupled with clear evidence of high-pressure crystal fractionation. Observations that further support these conclusions include the following: (1) compared with younger intraplate-type volcanic rocks, the early volcanic rocks of intraplate type are more evolved (hawaiites) and usually contain complex sets of megacrysts and xenocrysts; (2) megacrysts may or may not be cogenetic with the host rock, yet they come from magmas similar in composition (but more evolved) than the host rocks; (3) the lava flows and tephra occasionally contain partially melted xenoliths of feldspathic granulites and/or gabbros crystallized at relatively high pressures; (4) mineral parageneses in the xenoliths are the same or similar to the megacryst/xenocryst assemblages; thus it is inferred that the latter crystals come from the disaggregation of these xenoliths; (5) the absence of peridotite xenoliths in the Oligocene-Miocene intraplate-type suites implies a lower ascent rate compared to the mantle xenolith-bearing Plio-Quaternary magmas; and (6) a few samples of primitive, uncontaminated rocks at Los Encinos have geochemical compositions similar to those of the Plio-Quaternary rocks, and plot at the extreme of "mixing vectors" in many variation diagrams (e.g., Fig. 9).

Luhr et al. (2001) proposed a model to explain the geochemical and petrological differences between the early and late magmas in the southern Basin and Range. The model is based on the inferred tectonic conditions and thermal state of the lithosphere. During the middle Oligocene (ca. 32 Ma), the eastern and southern portions of the Basin and Range Province underwent a transition from ENE compression to ENE extension (e.g., Henry and Aranda-Gómez, 1992). Generation and ascent of intraplate-type magmas, characterized by Ti, Nb, and Ta enrichments, may have started simultaneously or shortly after this change. Crustal extension began during the Sierra Madre Occidental ignimbrite flare-up (McDowell and Keizer, 1977; Aranda-Gómez et al., 2000), which caused a rise in the temperature of the crust and brought the brittle-ductile transition closer to the surface. These crustal conditions prevented the propagation of brittle structures deep into the crust, making intraplate-type magma ascent to the surface more difficult. As a consequence of this process, the first magmas generated were trapped and crystallized in the deep crust, forming gabbroic plutons and associated, more evolved cogenetic magma bodies. As the lithosphere was cooling down, the brittle–ductile transition in the crust descended, allowing the formation of volcanic conduits that facilitated the ascent of some hawaiites. As these 24–11 Ma magmas slowly traveled through the crust, they incorporated megacrysts and a few xenoliths from the intrusive bodies that had crystallized/differentiated from earlier intraplate-type magmas that never reached the surface. Likewise, the slow ascent rate of these magmas permitted cooling, crystal fractionation, and interaction with the crust, which caused selective contamination with incompatible elements, such as Cs, Th, U, and Rb (Figs. 5B, 5C, and 9) and the incorporation of xenocrysts of common crustal minerals (e.g., quartz). These processes combined to produce magmas more evolved than those erupted during the Plio-Quaternary. The younger intraplate-type rocks tend to be more primitive, and commonly brought large mantle and crustal xenoliths to the surface. This finding is consistent with faster ascent rates of the magmas through brittle structures that have developed during a more advanced stage of ENE extension. It is likely that some of the brittle structures penetrated more deeply in a cooler crust. This model is consistent with the data derived from late Oligocene hawaiites in Rodeo-Nazas (Durango) and with the middle Miocene hawaiites from Los Encinos (San Luis Potosí) and Metates (Durango). No other early intraplate-type volcanic fields are known, therefore it is not possible to evaluate if this model may be applied to other places in the southern Basin and Range Province.

The oldest alkalic rocks come from Bernal de Horcasitas (28 Ma). The chemical composition of samples from this volcanic neck is among the most extreme for all the volcanic fields in the Northern Mexico Extensional Province, comparable only with some samples from the Eastern/Alkalic region (Fig. 2C). The light- and middle-REE contents in rocks from Bernal de Horcasitas are considerably higher than those found in other continental localities, from both the early and late periods of volcanism (Figs. 4B and 4C). As far as we know, there are no megacrysts or mantle xenoliths at Bernal de Horcasitas, and in multi-element plots (Figs. 5B and 5C), the distinctive enrichments observed in some elements of the early volcanic rocks from the southern portion of the Basin and Range are not present. Therefore, we believe that the model outlined in this section may not be applicable to Bernal de Horcasitas.

Secular changes in the source regions of the intraplate magmas. Based on elemental and isotopic variations observed in intraplate-type rocks from the U.S. portion of the Basin and Range Province, it has been proposed that in the past 5–10 Ma, the source region for these magmas changed from the upper part of the lithospheric mantle to a deeper source in the asthenosphere (e.g., Perry et al., 1987; Fitton et al., 1991). Some of the Oligocene-Miocene intraplate volcanic rocks in México (i.e., Rodeo-Nazas, Los Encinos, and Metates) have trace-element contents that are commonly associated with asthenospheric magmas and isotopic composition that suggests a shallower source in the lithosphere. The evidence for significant contamination with crustal material documented by Luhr et al. (1995b, 2001) led those authors to argue that these rocks do not purely reflect the chemical compositions of their sources, because crustal-level processes associated with their transit to the surface obscured many of the geochemical signatures inherited from their sources. Luhr et al. (1995b, 2001) concluded that there is no clear evidence for a change in the magma source regions during the evolution of the southern portion of the Basin and Range Province. However, Paz-Moreno et al. (2003) argued in favor of a change of the source region for the mafic magmas from Moctezuma (Sonora), from a source in the lithospheric mantle in the early stages of extension (early Miocene) to an asthenospheric source during the Quaternary. This is similar to secular change observed in the United States. On the other hand, the bimodal distribution of the isotopic data in samples from Ventura–Espíritu Santo and Santo Domingo (Fig. 6D) may be compatible with two sources, one in the asthenosphere and the other in the lithospheric mantle, for both of these Quaternary volcanic fields in the southern part of the Basin and Range Province.

Extreme chemical compositions observed in the intraplate rocks. If Bernal de Horcasitas is excluded, the most silica-undersaturated volcanic rocks occur at the Ventura–Espíritu Santo region in San Luis Potosí. If it is assumed that both Santo Domingo and Ventura–Espíritu Santo have similar mantle-source rocks composed of garnet lherzolite, the chemical differences between these two sets of rocks may be explained by a higher degree of partial melting for the Santo Domingo magmas. A systematic study of the pressure and temperature of equilibration of peridotite xenoliths found in México has shown that the highest values correspond to the Ventura–Espíritu Santo samples (Luhr and Aranda-Gómez, 1997). Ventura–Espíritu Santo volcanic field is located in a region that has undergone several pulses of extension (Aranda-Gómez et al., 2000). However, its tectonic evolution is like that in other regions (e.g., Santo Domingo, S.L.P, and Durango) where intraplate-type volcanism of a similar age involved lesser degrees of silica undersaturation (e.g., Aranda-Gómez et al., 1997). Heinrich and Besch (1992), in a detailed investigation of the mantle xenoliths from Ventura–Espíritu Santo, argued for a thermal perturbation in this region caused by backarc extension related to the Quaternary Trans-Mexican Volcanic Belt. Up to now, no one has explored the possible effects of such a thermal pulse on the generation of the host magmas for the xenoliths.

Besides the Santa Clara volcano (El Pinacate, Sonora) trachytes and some samples from Palomas, the only other continental volcanic fields north of the Trans-Mexican Volcanic Belt that produced intraplate-type trachytes are San Carlos–Cruillas and Aldama, both in Tamaulipas (Fig. 2C). Our unpublished data on the Aldama rocks show that the isotopic ratios (Sr, Nd, and Pb) in the trachytes are similar to those in the associated mafic alkalic rocks (Figs. 6D and 7C). Presently it is unknown which factor favored the extreme differentiation in closed systems in this area. It is worth mentioning that the Eastern/Alkalic region has examples (San Carlos and Sierra de Tamaulipas) of large chemical variations within magma series of widely different ages (Oligocene-Quaternary) that have been attributed to crystal fractionation (Ramírez-Fernández, 1997). Likewise, it must be noted that there is no evidence in the exposed geology of significant crustal extension in these areas.

The magmatic activity in the Eastern/Alkalic region evolved over a long time span in a complex manner. The early phase of alkalic volcanism in the region was contemporaneous with orogenic volcanism that produced the Sierra Madre Occidental. This early alkalic activity has been interpreted either as a backarc equivalent of the orogenic volcanism (Demant and Robin, 1975) or as the product of an increase of the alkalinity of the magmas produced by partial melting near the Benioff zone at a greater depth (Clark et al., 1982). As the volcanic front of the continental arc was moving westward toward the paleo-trench, the magmatic activity changed to intraplate type and these rocks were erupted on top of subduction-related rocks formed during the previous event.

Pantellerites and comendites from Sonora are other extreme examples of differentiation in an alkalic magma series. However, at the present time, little is known about these localities, except that they are associated with transitional basalts and were probably formed during an early stage of extension in the region associated with formation of the Protogulf of California during the middle Miocene (Vidal-Solano et al., 2000).

Geographic and temporal association of mafic alkalic rocks with* hypersthene*-normative rocks in the continental intraplate-type localities. *Hypersthene*-normative rocks (Fig. 3C) occur in several volcanic fields (e.g., Santo Domingo, Durango, and San Quintín). These samples have a few geochemical characteristics that appear to be transitional to calc-alkaline rocks, even though there is no reasonable way to relate them to subduction. Field relations indicate that they were erupted during the same volcanic cycle as the mafic rocks with *nepheline* in the norm. As a general rule, this phenomenon has not been discussed in the literature nor have plausible hypotheses to explain their coexistence been proposed. The most obvious example of this association—because of the relative volumes involved—is Moctezuma. Paz-Moreno et al. (2003) found that the *hypersthene*-normative rocks are younger than those with *nepheline* in the norm. Likewise, they documented different eruptive styles for each magma type. Smith (1989) pointed out the following for the Durango volcanic field: (1) Some of the rocks with low to moderate contents of *hypersthene* in the norm contain mantle xenoliths. (2) Neither a clear pattern of distribution, nor a temporal variation of the occurrence of *hypersthene*-normative rocks was found. Linear trends B and C identified by Smith (1989) in many variation diagrams (e.g., Fig. 12A) correspond to *hypersthene*-normative rocks. Based on features observed in multi-element diagrams for some of the Durango volcanic field rocks and in a comparison with a sample from the Trans-Mexican Volcanic Belt, Smith (1989) concluded that *hypersthene*-normative rocks inherited some components from mantle domains that were influenced by earlier subduction.

A more striking example is that of a sample collected in an andesitic sill at the Camargo volcanic field. It yielded an ^{40}Ar-^{39}Ar age of ca. 14 Ma and its chemical composition corresponds to that of a mid-Tertiary subduction-related volcanic rock (Aranda-Gómez et al., 2003, p. 300). Emplacement of this sill occurred ca. 16 Ma after the end of subduction and the beginning of extension in the area and ca. 10 Ma after the first known occurrences on intraplate-type magmatism in northern México and the southern United States. A similar example is the Tres Vírgenes volcano (Fig. 1), where a calc-alkaline volcanic complex was formed during the Quaternary. Tres Vírgenes is located in a region under a transtensional tectonic regime associated with the Gulf of California Extensional Province. This calc-alkaline activity occurred at least 10 Ma after the end of subduction and has been interpreted as being produced by contamination with crustal material (Sawlan, 1986). Alternative hypotheses to explain the close association of *nepheline*- and *hypersthene*-normative magmas in intraplate environments, such as low-pressures or high degrees of partial melting, have not been explored.

Diversity of the petrogenetic processes in the intraplate-type rocks. As a general rule, whole-rock Mg# for the intraplate-type samples are too low for the magmas to have been in equilibrium with their entrained mantle peridotites (i.e., the peridotites are truly accidental fragments). Commonly, the Mg# and the contents of Ni and Cr are sufficiently low, making it likely that the magmas underwent some crystal fractionation of mafic phases prior to the incorporation of the xenoliths, or that the Mg# of the source rocks for the magmas was lower than that observed in the xenoliths (e.g., Luhr et al., 1989b, 1995a). Modeling performed by Smith (1989) with the Durango volcanic field data suggests that some of the chemical changes observed in the series may be consistent with crystal fractionation of either high- or low-pressure mineral assemblages. However, the presence of peridotite xenoliths in these rocks requires a rapid ascent rate and the absence of a shallow magma chamber underneath the volcanic field. Therefore, the modeled high-pressure mineral assemblage is considered more plausible. This points to a significant difference between the continental volcanic fields and the islands; in the latter, shallow magma chambers allow low-pressure crystal fractionation to generate significant volumes of trachytes.

There is no dominant petrogenetic process that explains the geochemistry of all the Plio-Quaternary continental volcanic fields shown in Figure 1. Chemical variations in some series appear to be influenced by different degrees of partial melting from source rocks with similar compositions and/or source heterogeneity in the mantle. Crystal fractionation appears to be the dominant process in other localities, such as La Breña (Pier et al., 1992). Changes in composition with time in San Quintín were attributed to an increase in the degree of partial melting or to the progressive depletion of incompatible elements in the source region.

Genetic Relationship between Alkalic Volcanism and Crustal Extension

A broad survey of the literature on the state of stress in the continental part of México, north of the Trans-Mexican Volcanic Belt, shows that the information is scarce (e.g., Suter, 1991; Zoback and Zoback, 1991). However, Suter (1991) concluded that the present-day state of stress in Chihuahua, Coahuila, Durango, Nuevo León, Tamaulipas, San Luis Potosí, and Querétaro is $S_V > S_N > S_E$ (where $_V$ = vertical, $_N$ = north, and $_E$ = east), and that the region is characterized by horizontal deviatoric tension. On the other hand, it is unknown how far back in time this interpretation can be extrapolated and if it can be applied over the entire continental region considered in this review.

The data summarized in this paper show that intraplate-type alkalic volcanism has occurred across the entire northern part of México, independent of the intensity of the deformation caused by crustal extension, which is larger in the Gulf of California Extensional Province and in the southern part of the Basin and Range Province than in the Gulf of Mexico coastal plain, where it might be absent. Therefore, we conclude that intraplate-type volcanism must be controlled by processes in the upper mantle rather than by the state of stress of the upper continental crust. Seismic data obtained by Grand (1994) and van der Lee and Nolet (1997) showed that most (or all?) of the continental region considered here is underlain by mantle characterized by anomalously slow S-wave velocities. Van der Lee and Nolet (1997) interpreted these results as evidence for upwelling of the asthenospheric mantle in the "slab window" created by subduction of the Farallon plate or by the presence of volatiles and partial melting in the same mantle region.

On the other hand, the state of stress or the presence of large discontinuities in the crust appears to have played an important role in determining whether the intraplate-type magmas reached the surface or not, and in the location of the volcanoes, or in the nature of some of the most obvious petrogenetic processes registered in the geochemistry of the rocks (e.g., AFC processes). However, the ultimate cause for the formation of these intraplate-type magmas is independent of the state of stress in the upper crust.

ACKNOWLEDGMENTS

This paper was written when J.J. Aranda-Gómez was on sabbatical leave at the Instituto Potosino de Investigación Científica y Tecnológica, A.C. Our work at the intraplate-type localities of México has been supported by several research grants from México's Consejo Nacional de Ciencia y Tecnología to Aranda-Gómez (3657PT, 37429, and 47071), National Science Foundation and Smithsonian Institution Scholarly Studies Program awards to J.F. Luhr, and the Posgrado en Ciencias de la Tierra (UNAM) to G. Chávez-Cabello and G. Valdez-Moreno. We thank all these institutions for their support. Soledad Medina-Malagón, librarian at the CGEO-UNAM, helped to compile the bibliographical materials and reviewed the reference list. Alejandro Morales and Cristina Morán prepared some of the figures. Rufino Lozano analyzed some of the rocks from Coahuila. Juan Tomás Vazquez and Crescencio Garduño prepared a large number of thin sections. José Luis Macías supported part of G. Valdez-Moreno's research in Coahuila.

REFERENCES CITED

Aguirre-Díaz, G.J., and McDowell, F.W., 1993, Nature and timing of faulting and synextensional magmatism in the southern Basin and Range, central-eastern Durango, Mexico: Geological Society of America Bulletin, v. 105, p. 1435–1444, doi: 10.1130/0016-7606(1993)105<1435:NATOFA>2.3.CO;2.

Allan, J.F., 1986, Geology of the northern Colima and Zacoalco grabens, southwest Mexico: Late Cenozoic rifting in the Mexican Volcanic Belt: Geological Society of America Bulletin, v. 97, p. 473–485, doi: 10.1130/0016-7606(1986)97<473:GOTNCA>2.0.CO;2.

Allan, J.F., Nelson, S.A., Luhr, J.F., Carmichael, I.S.E., Wopat, M., and Wallace, J.P., 1991, Pliocene-Recent Rifting in SW Mexico and Associated Volcanism: An Exotic Terrain in the Making: American Association of Petroleum Geologists Memoir 47, 425–445.

Almeida-Vega, M., Espinosa-Cardeña, J.M., and Ledesma-Vázquez, J., 2000, Structure of the San Quintín coastal valley, B.C. from aeromagnetic data analysis, *in* V Reunión Internacional sobre la Geología de la Península de Baja California, Memorias: Loreto, Baja California Sur, Universidad Autónoma de Baja California; Universidad Nacional Autónoma de México, Sociedad Geológica Peninsular, p. 64.

Aranda-Gómez, J.J., 1982, Ultramafic and high grade metamorphic xenoliths from central Mexico [Ph.D. thesis]: Eugene, University of Oregon, 228 p.

Aranda-Gómez, J.J., and Luhr, J.F., 1993, Geology of the Joya Honda and Santo Domingo maars, San Luis Potosí, Mexico—A visit to the mantle and crustal xenolith localities in the Sierra Madre Terrane, *in* Ortega-Gutiérrez, F., Centeno-García, E., Morán-Zenteno, D.J., and Gómez-Caballero, A., eds., First circum-Pacific and circum-Atlantic Terrane Conference. Pre-Mesozoic basement of NE Mexico, crust and mantle xenoliths of central Mexico, and northern Guerrero terrane. Guidebook of field trip A: Guanajuato, Gto., Universidad Nacional Autónoma de México, Instituto de Geología, p. 10–35.

Aranda-Gómez, J.J., and Luhr, J.F., 1996, Origin of the Joya Honda maar, San Luis Potosí, México: Journal of Volcanology and Geothermal Research, v. 74, p. 1–18, doi: 10.1016/S0377-0273(96)00044-3.

Aranda-Gómez, J.J., and Ortega-Gutiérrez, F., 1987, Mantle xenoliths in Mexico, *in* Nixon, P.H., ed., Mantle Xenoliths: New York, Wiley, p. 75–84.

Aranda-Gómez, J.J., Aranda-Gómez, J.M., and Nieto-Samaniego, A.F., 1989, Consideraciones acerca de la evolución tectónica durante el Cenozoico de la Sierra de Guanajuato y la parte meridional de la Meseta Central: Universidad Nacional Autónoma de México, Revista del Instituto de Geología, v. 8, no. 1, p. 33–46.

Aranda-Gómez, J.J., Luhr, J.F., and Pier, J.G., 1992, The La Breña-El Jagüey maar complex, Durango, México: I. Geological evolution: Bulletin of Volcanology, v. 54, p. 393–404, doi: 10.1007/BF00312321.

Aranda-Gómez, J.J., Luhr, J.F., and Pier, J.G., 1993, Geología de los volcanes cuaternarios portadores de xenolitos del manto y de la base de la corteza en el Estado de San Luis Potosí, México: Boletín del Instituto de Geología, Universidad Nacional Autónoma de México, no. 106, part 1, p. 1–22.

Aranda-Gómez, J.J., Henry, C.D., Luhr, J.F., and McDowell, F.W., 1997, Cenozoic volcanism and tectonics in NW Mexico: A transect across the Sierra Madre Occidental volcanic field and observations on extension related magmatism in the southern Basin and Range and Gulf of California tectonic provinces, *in* Aguirre-Díaz, G.J., Aranda-Gómez, J.J., Carrasco-Nuñez, G., and Ferrari, L., eds., Magmatism and Tectonics in the Central and Northwestern Mexico—A Selection of the 1997 IAVCEI General Assembly Excursions: México, D.F., Universidad Nacional Autónoma de México, Instituto de Geología, p. 41–84.

Aranda-Gómez, J.J., Luhr, J.F., and Housh, T.B., 1999, Reconnaissance geology of Isla Isabel, Nayarit, México: Geos, v. 19, no. 4, p. 320.

Aranda-Gómez, J.J., Henry, C.D., and Luhr, J.F., 2000, Evolución tectonomagmática post-paleocénica de la Sierra Madre Occidental y de la porción meridional de la provincia tectónica de Cuencas y Sierras, México: Boletín de la Sociedad Geológica Mexicana, v. 53, no. 1, p. 59–71.

Aranda-Gómez, J.J., Carranza-Castañeda, O., Luhr, J.F., and Housh, T.B., 2002, Origen de los sedimentos continentales en el valle de Jaumave: Tamaulipas: Geos, v. 22, no. 2, p. 327.

Aranda-Gómez, J.J., Luhr, J.F., Housh, T.B., Connor, C.B., Becker, T., and Henry, C.D., 2003, Synextensional, Plio-Pleistocene eruptive activity in the Camargo volcanic field, Chihuahua, México: Geological Society of America Bulletin, v. 115, p. 298–313, doi: 10.1130/0016-7606(2003)115<0298:SPPEAI>2.0.CO;2.

Aranda-Gómez, J.J., Luhr, J.F., Housh, T.B., Valdez-Moreno, G., and Chávez-Cabello, G., 2004, El vulcanismo intraplaca del Cenozoico tardío en el centro y norte de Méxco: una revisión, *in* Frías-Camacho, V.M., Silva-Corona, J.J., and Orozco-Esquivel, M.T., eds., IV Reunión Nacional de Ciencias de la Tierra, Libro de resúmenes: Geos, v. 24, no. 2, p. 147.

Aranda-Gómez, J.J., Housh, T.B., Luhr, J.F., Henry, C.D., Becker, T., and Chávez-Cabello, G., 2005, Reactivation of the San Marcos fault during mid- to late Tertiary extension, Chihuahua, México, *in* Nourse, J.A., Anderson, T.H., McKee, J.W., and Steiner, M.B., eds., The Mojave-Sonora Megashear Hypothesis: Development, Assessment, and Alternatives: Geological Society of America Special Paper 393, p. 509–522.

Arvidson, R.E., and Mutch, T.A., 1974, Sedimentary patterns in and around craters from the Pinacate Volcanic Field, Sonora, Mexico: Some comparisons with Mars: Geological Society of America Bulletin, v. 85, p. 99–104, doi: 10.1130/0016-7606(1974)85<99:SPIAAC>2.0.CO;2.

Bacon, C.R., and Carmichael, I.S.E., 1978, Stages in the P-T path of ascending basalt magma: An example from San Quintín, Baja California: Contributions to Mineralogy and Petrology, v. 41, p. 1–22, doi: 10.1007/BF00377648.

Basu, A.R., 1977a, Olivine-spinel equilibria in lherzolite xenoliths from San Quintín, Baja California: Earth and Planetary Science Letters, v. 33, p. 443–450, doi: 10.1016/0012-821X(77)90096-6.

Basu, A.R., 1977b, Textures, microstructures and deformation of ultramafic xenoliths from San Quintín, Baja California: Tectonophysics, v. 43, p. 213–246, doi: 10.1016/0040-1951(77)90118-4.

Basu, A.R., 1978, Trace elements and Sr isotopes in some mantle derived hydrous minerals and their significance: Geochimica et Cosmochimica Acta, v. 42, p. 659–668, doi: 10.1016/0016-7037(78)90084-4.

Basu, A.R., 1979, Geochemistry of ultramafic xenoliths from San Quintin, Baja California, *in* Boyd, F.R., and Meyer, H.O.A., eds., The Mantle Sample: Inclusions in Kimberlites and Other Volcanics: Proceedings of the Second International Kimberlite Conference: Washington, DC, American Geophysical Union, v. 2, p. 391–399.

Basu, A.R., and Murthy, V.R., 1977, Ancient lithospheric lherzolite xenoliths in alkalic basalt from Baja California: Earth and Planetary Science Letters, v. 35, p. 239–246, doi: 10.1016/0012-821X(77)90127-3.

Batiza, R., 1977, Petrology and chemistry of Guadalupe Island: An alkalic seamount on a fossil ridge crest: Geology, v. 5, p. 760–764, doi: 10.1130/0091-7613(1977)5<760:PACOGI>2.0.CO;2.

Batiza, R., 1982, Abundances, distributions, and sizes of volcanoes in the Pacific Ocean and the implications for the origin of non-hotspot volcanoes: Earth and Planetary Science Letters, v. 60, p. 195–206, doi: 10.1016/0012-821X(82)90003-6.

Ben Othman, D., Tilton, G.R., and Menzies, M.A., 1990, Pb, Nd and Sr isotopic investigations of kaersutite and clinopyroxene from ultramafic nodules, and their host basalts: The nature of the subcontinental mantle: Geochimica et Cosmochimica Acta, v. 54, p. 3449–3460, doi: 10.1016/0016-7037(90)90297-X.

Blatter, D.L., and Carmichael, I.S.E., 1998, Hornblende peridotite xenoliths from central Mexico reveal the highly oxidized nature of subarc upper mantle: Geology, v. 26, p. 1035–1038, doi: 10.1130/0091-7613(1998)026<1035:HPXFCM>2.3.CO;2.

Bloomfield, K., and Cepeda-Dávila, L., 1973, Oligocene alkaline igneous activity in NE Mexico: Geological Magazine, v. 110, p. 551–555.

Bohrson, W.A., and Reid, M.R., 1995, Petrogenesis of alkaline basalts from Socorro Island, Mexico: Trace element evidence for contamination of ocean island basalt in the shallow ocean crust: Journal of Geophysical Research, v. 100, p. 24,555–24,576, doi: 10.1029/95JB01483.

Bohrson, W.A., and Reid, M.R., 1997, Genesis of silicic peralkaline volcanic rocks in an ocean island setting by crustal melting and open-system processes; Socorro Island, México: Journal of Petrology, v. 38, p. 1137–1166, doi: 10.1093/petrology/38.9.1137.

Bohrson, W.A., Reid, M.R., Grunder, A.L., Heizler, M.T., Harrison, T.M., and Lee, J., 1996, Prolonged history of silicic peralkaline volcanism in the eastern Pacific Ocean: Journal of Geophysical Research, v. 101, p. 11,457–11,474, doi: 10.1029/96JB00329.

Bryan, W.B., 1959, High-silica lavas of Clarion and Socorro island, Mexico—Their genesis and regional significance [Ph.D. thesis]: Madison, University of Wisconsin, 164 p.

Bryan, W.B., 1966, History and mechanism of eruption of soda-rhyolite and alkali basalt Socorro Island, México: Bulletin of Volcanology, v. 29, p. 453–479.

Bryan, W.B., 1976, A basalt-pantellerite association from Isla Socorro, Islas Revillagigedo, Mexico, *in* Aoki, H., and Iizuka, S., eds., Volcanoes and Tectonosphere: Tokyo, Japan, Tokai University Press, p. 75–91.

Bryant, W.R., Antoine, J., Ewing, M., and Jones, B., 1968, Structure of Mexican continental shelf and slope, Gulf of Mexico: American Association of Petroleum Geologists Bulletin, v. 52, p. 1204–1228.

Cabanes, N., and Mercier, J.C.C., 1988, Insight into the upper mantle beneath an active extensional zone: The spinel-peridotite xenoliths from San Quintín (Baja California, Mexico): Contributions to Mineralogy and Petrology, v. 100, p. 374–382, doi: 10.1007/BF00379746.

Cabral-Cano, E., 1988, Paleomagnetismo y petrografía de la Isla Isabel, Nayarit [B.Sc. thesis]: México, Universidad Nacional Autónoma de México, Facultad de Ingeniería, 143 p.

Camacho-Angulo, F., 1993, Compilación geológica de la vertiente del Golfo de México: Veracruz, Ver., Comisión Federal de Electricidad, Subdirección Técnica, Gerencia de Estudios de Ingeniería Civil, Subgerencia de Estudios Geológicos, Departamento de Geología, Zona Golfo, 169 p.

Cameron, K.L., Nimz, G.J., and Kuentz, D., 1989, Southern cordillera basaltic andesite suite, southern Chihuahua, Mexico: A link between Tertiary continental arc and flood basalt magmatism in North America: Journal of Geophysical Research, v. 94, p. 7817–7840.

Cameron, K.L., Robinson, J.V., Niemeyer, S., Nimz, G.J., Kuentz, D.C., Harmon, R.S., Bohlen, S.R., and Collerson, K.D., 1992, Contrasting styles of pre-Cenozoic and mid-Tertiary crustal evolution in northern Mexico: Evidence from deep crustal xenoliths from La Olivina: Journal of Geophysical Research, v. 97, p. 17,353–17,376.

Campa, M.F., and Coney, P.J., 1983, Tectono-stratigraphic terranes and mineral resource distribution in Mexico: Canadian Journal of Earth Sciences, v. 20, p. 1040–1051.

Cantagrel, J.-M., and Robin, C., 1979, K-Ar dating on eastern Mexican volcanic rocks—Relations between the andesitic and the alkaline provinces: Journal of Volcanology and Geothermal Research, v. 5, p. 99–114, doi: 10.1016/0377-0273(79)90035-0.

Cavazos-Tovar, J.G., 2004, Petrografía y geoquímica del intrusivo Las Tetillas, Bloque Coahuila, México [B.Sc. thesis]: Linares, Nuevo León, Universidad Autónoma de Nuevo León, Facultad de Ciencias de la Tierra, 127 p.

Charleston, S., 1974, Stratigraphy, tectonics and hydrocarbon potential of the lower Cretaceous, Coahuila Series, Coahuila, Mexico [Ph.D. thesis]: Ann Arbor, University of Michigan, 268 p.

Charleston, S., 1981, A summary of the structural geology and tectonics of the state of Coahuila, Mexico, *in* Schmidt, C.I., and Katz, S.B., eds., Lower Cretaceous Stratigraphy and Structure, Northern México: Field Trip Guidebook: West Texas Geological Society Publication 81-74, p. 28–36.

Chávez-Cabello, G., 2005, Deformación y magmatismo cenozoicos en el sur de la Cuenca de Sabinas, Coahuila, México [Ph.D. thesis]: Juriquilla, Querétaro, Universidad Nacional Autónoma de México, Centro de Geociencias, Posgrado en Ciencias de la Tierra, 266 p., 1 map, 1 CD.

Clark, K.F., Foster, C.T., and Damon, P.E., 1982, Cenozoic mineral deposits and subduction-related magmatic arcs in Mexico: Geological Society of America Bulletin, v. 93, p. 533–544, doi: 10.1130/0016-7606(1982)93<533:CMDASM>2.0.CO;2.

Consejo de Recursos Minerales, México (COREMI, México), 1994, Inventario minero y exploración del carbón en el estado de Coahuila: México, Secretaría de Energía, Minas e Industria Paraestatal, Consejo de Recursos Minerales, 122 p.

Córdoba, D.A., 1963, Geología de la región entre Río Chico y Llano Grande, municipio de Durango, estado de Durango: Boletín del Instituto de Geología, Universidad Nacional Autónoma de México, no. 71, part. 1, p. 1–22, 1 map, 1 plate.

Demant, A., and Robin, C., 1975, Las fases del vulcanismo en México; una síntesis en relación con la evolución geodinámica desde el Cretácico: Revista del Instituto de Geología, Universidad Nacional Autónoma de México, v. 1, p. 70–83.

Elías-Herrera, M., 1984, Rocas alcalinas y mineralización de lantánidos en el área El Picacho, Sierra de Tamaulipas: Geomimet, v. 127, p. 61–75.

Elías-Herrera, M., Rubinovich-Kogan, R., Lozano-Santa Cruz, R., and Sánchez-Zavala, J. L., 1990, Petrología y mineralización de Tierras Raras del complejo ígneo El Picacho, Sierra de Tamaulipas: Boletín del Instituto de Geología, Universidad Nacional Autónoma de México, no. 108, p. 24–97, 2 plates.

Elías-Herrera, M., Rubinovich-Kogan, R., Lozano-Santa Cruz, R., and Sánchez-Zavala, J.L., 1991, Nepheline-rich foidolites and rare earth mineralization in the El Picacho Tertiary intrusive complex, Sierra de Tamaulipas, northeastern Mexico: Canadian Mineralogist, v. 29, p. 319–336.

Espinosa-Cardeña, J.M., Romo-Jones, J.M., and Almeida-Vega, M., 1991, Gravimetría y estructura del valle de San Quintín, B. C.: Geos, v. 2, no. 3, p. 10–15.

Ferrari, L., Takahiro, T., Mugihiko, E., Orozco-Esquivel, M.T., Petrone, C., Jacobo-Albarrán, J., and López-Martínez, M., 2005, Geology, geochronology and tectonic setting of late Cenozoic volcanism along the southwestern Gulf of Mexico: The Eastern Alkalic Province revisited: Journal of Volcanology and Geothermal Research, v. 146, p. 284–306, doi: 10.1016/j.jvolgeores.2005.02.004.

Fitton, J.G., James, D., and Leeman, W.P., 1991, Basic magmatism associated with late Cenozoic extension in the western United States: Compositional variations in space and time: Journal of Geophysical Research, v. 96, p. 13,693–13,711.

Frantes, T.J., and Hoffer, J.M., 1982, Palomas volcanic field, southern New Mexico and northern Chihuahua: New Mexico Geology, v. 4, no. 1, p. 6–8, 16.

Gastil, R.G., Phillips, R.P., and Allison, E.C., 1975, Reconnaissance Geology of the State of Baja California: Boulder, Colorado, Geological Society of America Memoir 140, 170 p., 3 maps.

Gómez-Tuena, A., LaGatta, A.B., Langmuir, C.H., Goldstein, S.L., Ortega-Guitérrez, F., and Carrasco-Nuñez, G., 2003, Temporal control of subduction magmatism in the eastern Trans-Mexican Volcanic Belt: Mantle sources, slab contributions and crustal contamination: Geochemistry, Geophysics, Geosystems, v. 4, p. 1–33.

Grand, S.P., 1994, Mantle shear structure beneath the Americas and surrounding oceans: Journal of Geophysical Research, v. 99, p. 11,591–11,621, doi: 10.1029/94JB00042.

Green, D.H., 1971, Composition of basaltic magmas as indicators of conditions of origin; applications to oceanic volcanism: Philosophical Transactions of the Royal Society of London. Series A: Mathematical and Physical Sciences, v. 268, p. 707–725.

Gries, J.C., 1979, Problems of delineation of the Rio Grande Rift into the Chihuahua tectonic belt of northeastern Mexico, *in* Riecker, R.E., ed., The Rio Grande Rift-Tectonics and Magmatism: Washington, DC, American Geophysical Union, p. 107–113.

Gutmann, J.T., 1974, Tubular voids within labradorite phenocrysts from Sonora, Mexico: The American Mineralogist, v. 59, p. 666–672.

Gutmann, J.T., 1979, Structure and eruptive cycle of cinder cones in the Pinacate volcanic field and the controls of Strombolian activity: The Journal of Geology, v. 87, p. 448–454.

Gutmann, J.T., 1986, Origin of four- and five-phase ultramafic xenoliths from Sonora, Mexico: The American Mineralogist, v. 71, p. 1076–1084.

Gutmann, J.T., 2002, Strombolian and effusive activity as precursors to phreatomagmatism: Eruptive sequence at maars of the Pinacate volcanic field, Sonora, Mexico: Journal of Volcanology and Geothermal Research, v. 113, p. 345–356, doi: 10.1016/S0377-0273(01)00265-7.

Gutmann, J.T., and Martin, R.F., 1976, Crystal chemistry, unit cell dimensions, and structural state of labradorite megacrysts from Sonora, Mexico: Schweizerische Mineralogische und Petrographische Mitteilungen, v. 56, p. 55–64.

Gutmann, J.T., and Sheridan, M.F., 1978, Geology of the Pinacate volcanic field, Sonora, Mexico, *in* Burt, D.M., and Peew, T.L., eds., Guidebook to the Geology of Central Arizona: Arizona Bureau of Geology and Mineral Technology Special Paper 2, p. 47–59.

Hart, S.R., Gerlach, D.C., and White, W.M., 1986, A possible new Sr-Nd-Pb mantle array and consequences for mantle mixing: Geochimica et Cosmochimica Acta, v. 50, p. 1551–1557, doi: 10.1016/0016-7037(86)90329-7.

Hawley, J.C., 1981, Pleistocene and Pliocene history of the international boundary area, southern New Mexico, *in* Hoffer, J-.M., and Hoffer, R.L., eds., Geology of the Border: Southern New Mexico-Northern Chihuahua: El Paso, Texas, El Paso Geological Society, p. 26–32.

Heinrich, W., and Besch, T., 1992, The upper mantle beneath a young back-arc extensional zone: Ultramafic xenoliths from San Luis Potosí, central México: Contributions to Mineralogy and Petrology, v. 111, p. 126–142, doi: 10.1007/BF00296583.

Henry, C.D., and Aranda-Gómez, J.J., 1992, The real southern Basin and Range: Mid-to late Cenozoic extension in Mexico: Geology, v. 20, p. 701–704, doi: 10.1130/0091-7613(1992)020<0701:TRSBAR>2.3.CO;2.

Henry, C.D., and Aranda-Gómez, J.J., 2000, Plate interactions control middle-late Miocene, proto-Gulf and Basin and Range extension in the southern Basin and Range: Tectonophysics, v. 318, no. 1–4, p. 1–26, doi: 10.1016/S0040-1951(99)00304-2.

Henry, C.D., Fredrikson, G., and Ames, J.T., 1987, Geology of part of southern Sinaloa, Mexico adjacent to the Gulf of California: Geological Society of America Map and Chart Series MCH063, scale 1:250,000, 1 sheet, 14 p. text.

Hoffer, J. M., 1976, Geology of Potrillo basalt field, south-central New Mexico: New Mexico Bureau of Geology & Mineral Resources, Circular, no. 149, 30 p., 1 plate.

Irvine, T.N., and Baragar, W.R.A., 1971, A guide to the chemical classification of the common volcanic rocks: Canadian Journal of Earth Sciences, v. 8, p. 523–548.

James, E.W., and Henry, C.D., 1991, Compositional changes in Trans-Pecos Texas magmatism coincident with Cenozoic stress realignment: Journal of Geophysical Research, v. 96, p. 13,561–13,575.

Jones, N.W., McKee, J.W., Márquez-D., B., Tovar, J., Long, L.E., and Laudon, T.S., 1984, The Mesozoic La Mula island, Coahuila, Mexico: Geological Society of America Bulletin, v. 95, p. 1226–1241, doi: 10.1130/0016-7606(1984)95<1226:TMLMIC>2.0.CO;2.

Kretz, R., 1983, Symbols for rock-forming minerals: The American Mineralogist, v. 68, p. 277–279.

Labarthe-Hernández, G., 1978, Algunos xalapascos en el estado de San Luis Potosí: Folleto Técnico del Instituto de Geología y Metalurgia, Universidad Autónoma de San Luis Potosí, no. 58, 17 p.

Labarthe-Hernandez., G., Tristán-G., M., and Aranda-Gómez, J.J., 1982, Revisión estratigráfica del Cenozoico de la parte central del Estado de San Luis Potosí: Folleto Técnico del Instituto de Geología y Metalurgia, Universidad Autónoma de San Luis Potosí, no. 85, 208 p., 1 map.

Le Maitre, R.W., ed., Streckeisen, A., Zanettin, B., Le Bas, M. J., Bonin, B., Bateman, P., Bellieni, G., Dudek, A., Efremova, S., Keller, J., Lameyre, J., Sabine, P.A., Schmid, R., Sorensen, H., and Woolley, A.R., 2002, Igneous Rocks: A Classification and Glossary of Terms: Recommendations of the International Union of Geological Sciences Subcommission on the Systematics of Igneous Rocks (2nd edition): Cambridge, Cambridge University Press, 236 p.

Luhr, J.F., and Aranda-Gómez, J.J., 1997, Mexican peridotite xenoliths and tectonic terranes: Correlations among vent location, texture, temperature, pressure, and oxygen fugacity: Journal of Petrology, v. 38, p. 1075–1112, doi: 10.1093/petrology/38.8.1075.

Luhr, J.F., Allan, J.F., Carmichael, I.S.E., Nelson, S.A., and Hasenaka, T., 1989a, Primitive calc-alkaline and alkaline rock types from the western Mexican Volcanic Belt: Journal of Geophysical Research, v. 94, p. 4515–4530.

Luhr, J.F., Aranda-Gómez, J.J., and Pier, J.G., 1989b, Spinel-lherzolite bearing, Quaternary volcanic centers in San Luis Potosí, México. I. Geology, Mineralogy, and Petrology: Journal of Geophysical Research, v. 94, p. 7916–7940.

Luhr, J.F., Aranda-Gómez, J.J., and Housh, T.B., 1995a, San Quintín Volcanic Field, Baja California Norte, México: Geology, Petrology and Geochemistry: Journal of Geophysical Research, v. 100, p. 10,353–10,380, doi: 10.1029/95JB00037.

Luhr, J.F., Pier, J.G., Aranda-Gómez, J.J., and Podosek, F., 1995b, Crustal contamination in early Basin-and-Range hawaiites of the Los Encinos volcanic field, central México: Contributions to Mineralogy and Petrology, v. 118, p. 321–339, doi: 10.1007/s004100050018.

Luhr, J.F., Henry, C.D., Housh, T.B., Aranda-Gómez, J.J., and McIntosh, W.C., 2001, Early extension and associated mafic alkalic volcanism from the southern Basin and Range Province: Geology and petrology of the Rodeo and Nazas volcanic fields, Durango (Mexico): Geological Society of America Bulletin, v. 113, p. 760–773, doi: 10.1130/0016-7606(2001)113<0760:EEAAMA>2.0.CO;2.

Luhr, J.F., Kimberly, P., Siebert, L., Aranda-Gómez, J.J., Housh, T.B., and Kysar, G., 2006, México's Quaternary volcanic rocks: Insights from the MEXPET petrological and geochemical database, *in* Siebe, C., Macías, J.L., and Aguirre-Díaz, G.J., eds., Neogene-Quaternary Continental Margin Volcanism: A Perspective from México: Geological Society of America Special Paper 402, p. 1–44.

Lutz, T.M., and Gutmann, J.T., 1995, An improved method of determining alignments of point-like features and its implications for the Pinacate volcanic field, Mexico: Journal of Geophysical Research, v. 100, p. 17,659–17,670, doi: 10.1029/95JB01058.

Lynch, D.J., 1981, Genesis and geochronology of alkaline volcanism in the Pinacate volcanic field northwestern Sonora, Mexico [Ph.D. thesis]: Tucson, University of Arizona, 265 p.

Lynch, D.J., Musselman, T.E., Gutmann, J.T., and Patchett, P.J., 1993, Isotopic evidence for the origin of Cenozoic volcanic rocks in the Pinacate volcanic field, northwestern Mexico: Lithos, v. 29, p. 295–302, doi: 10.1016/0024-4937(93)90023-6.

Mammerickx, J., Naar, D.F., and Tyce, R.L., 1988, The Matematician paleoplate: Journal of Geophysical Research, v. 93, p. 3025–3040.

McClelland, L.E., Venzke, E., and Goldberg, J., 1993, Socorro (Mexico) vesicular lava eruption from underwater vent 3 km W of the island: Smithsonian Institution: Bulletin of the Global Volcanism Network, v. 18, no. 1, p. 9–11.

McDowell, F.W., and Keizer, R.P., 1977, Timing of mid-Tertiary volcanism in the Sierra Madre Occidental between Durango City and Mazatlan, Mexico: Geological Society of America Bulletin, v. 88, p. 1479–1486, doi: 10.1130/0016-7606(1977)88<1479:TOMVIT>2.0.CO;2.

McKee, J.W., and Jones, N.W., 1979, A large Mesozoic fault in Coahuila, Mexico: Geological Society of America Abstracts with Programs, v. 11, p. 476.

McKee, J.W., Jones, N.W., and Long, L.E., 1984, History of recurrent activity along a major fault in northeastern Mexico: Geology, v. 12, p. 103–107, doi: 10.1130/0091-7613(1984)12<103:HORAAA>2.0.CO;2.

McKee, J.W., Jones, N.W., and Long, L.E., 1990, Stratigraphy and provenance of strata along the San Marcos fault, central Coahuila, Mexico: Geological Society of America Bulletin, v. 102, p. 593–614, doi: 10.1130/0016-7606(1990)102<0593:SAPOSA>2.3.CO;2.

Medina, F., Suárez, F., and Espíndola, J.M., 1989, Historic and Holocene volcanic centers in NW México: Bulletin of Volcanic Eruptions (Tokyo), v. 51, p. 91–93, doi: 10.1007/BF01197481.

Menzies, M.A., 1989, Cratonic, circumcratonic and oceanic mantle domains beneath the western United States: Journal of Geophysical Research, v. 94, p. 7899–7915.

Mitre-Salazar, L.M., 1989, La megafalla laramídica de San Tiburcio, estado de Zacatecas: Revista del Instituto de Geología, Universidad Nacional Autónoma de México, v. 8, p. 47–51.

Miyashiro, A., 1974, Volcanic rock series in island arcs and active continental margins: American Journal of Science, v. 274, p. 321–355.

Morán-Zenteno, D.J., 1994, Geology of the Mexican Republic: Tulsa, Oklahoma, American Association of Petroleum Geologists Studies in Geology, no. 39, 160 p.

Nick, K., 1988, Mineralogische, geochemische und petrographische untersuchungen in der Sierra de San Carlos (Mexiko) [Ph.D. thesis]: Karlsruhe, Geowissenschaften Universität Fakultät fur Biologie, 167 p.

Nieto-Samaniego, A.F., Alaniz-Alvarez, S.A., and Labarthe-Hernández, G., 1997, La deformación Cenozoica post-laramídica en la parte meridional de la Mesa Central, México: Revista Mexicana de Ciencias Geológicas, v. 14, p. 13–25.

Nimz, G.J., 1989, The geochemistry of the mantle xenolith suite from La Olivina, Chihuahua, Mexico [Ph.D. thesis]: Santa Cruz, University of California, 315 p.

Nimz, G.J., Cameron, K.L., Cameron, M., and Morris, S.L., 1986, The petrology of the lower crust and upper mantle beneath southeastern Chihuahua, Mexico: A progress report: Geofísica Internacional, v. 25, no. 1, p. 85–116.

Nimz, G.J., Cameron, K.L., and Niemeyer, S., 1993, The La Olivina pyroxenite suite and the isotopic compositions of mantle basalts parental to mid-Cenozoic arc volcanism of northern Mexico: Journal of Geophysical Research, v. 98, p. 6489–6509.

Nimz, G.J., Cameron, K.L., and Niemeyer, S., 1995, Formation of mantle lithosphere beneath northern Mexico: Chemical and Sr-Nd-Pb isotopic systematics of peridotite xenoliths from La Olivina: Journal of Geophysical Research, v. 100, p. 4181–4196, doi: 10.1029/94JB02776.

Orozco-Esquivel, M.T., Ferrari, L., Eguchi, M., Tagami, T., Petrone, C., and Albarran, J., 2003, The eastern alkaline province (Mexico) revised: Geology, geochronology and geochemistry of Neogene volcanism in Veracruz state, *in* 99ª Reunión Anual, Geological Society of America, Cordilleran Section, Puerto Vallarta, Jal., 1 a 3 de abril de 2003, Libro de resúmenes: México, Geological Society of America; Universidad Nacional Autónoma de México, Instituto de Geología, p. 58.

Ortega-Gutiérrez, F., and González-González, R., 1980, Nódulos de peridotita en la Isla Isabel, Nayarit: Revista del Instituto de Geología, Universidad Nacional Autónoma de México, v. 4, no. 1, p. 82–83.

Ortega-Gutiérrez, F., Mitre-Salazar, L.M., Roldán-Quintana, J., Aranda-Gómez, J.J., Morán-Zenteno, D., Alaníz-Álvarez, S.A., and Nieto-Samaniego, A.F., 1992, Carta geológica de la República Mexicana, escala 1:2,000,000, 5th ed.: México, Universidad Nacional Autónoma de México, Inst. Geología; Secretaría de Energía, Minas e Industria Paraestatal, Consejo de Recursos Minerales, 1 mapa.

Padovani, E.R., and Carter, J.L., 1977, Aspects of the deep crustal evolution beneath south central New Mexico, *in* Heacock, J.G., ed., The Earth's Crust: Its Nature and Physical Properties: Washington, D.C., American Geophysical Union, Monograph 20, 19–55.

Paz-Moreno, F., Demant, A., and Ornelas-Solís, R.E., 2000, Las ignimbritas hiperalcalinas neógenas de la región de Hermosillo, Sonora, México: mineralogía y geoquímica, *in* Calmus, T., and Pérez-Segura, E., eds., Cuarta Reunión sobre la Geología del Noroeste de México y Áreas Adyacentes, Libro de resúmenes: Hermosillo, Sonora, Universidad Nacional Autónoma de México, Instituto de Geología, Estación Regional del Noroeste; Universidad de Sonora, Departamento de Geología, p. 90–91.

Paz-Moreno, F.A., Demant, A., Cochemé, J.J., Dostal, J., and Montigny, R., 2003, The Quaternary Moctezuma volcanic field: A tholeiitic to alkali basaltic episode in the central Sonoran Basin and Range Province, México, *in* Johnson, S.E., Patterson, S.R., Fletcher, J.M., Girty, G.H., Kimbrough, D.L., and Martín-Barajas, A., eds., Tectonic Evolution of Northwestern México and the Southwestern USA: Geological Society of America Special Paper 374, p. 439–455.

Perry, F.V., Baldridge, W.S., and DePaolo, D.J., 1987, Role of asthenosphere and lithosphere in the genesis of late-Cenozoic basaltic rocks from the Rio Grande Rift and adjacent regions of the southwestern United States: Journal of Geophysical Research, v. 92, p. 9193–9213.

Pettus, D.S., 1979, Ultramafic xenoliths from Llera de Canales, Tamaulipas, Mexico [M.Sc. thesis]: Houston, University of Houston, 95 p.

Pier, J.G., Podosek, F., Luhr, J.F., Brannon, J., and Aranda-Gómez, J.J., 1989, Spinel-lherzolite-bearing, Quaternary volcanic centers in San Luis Potosí, México. II. Sr and Nd isotopic systematics: Journal of Geophysical Research, v. 94, p. 7941–7951.

Pier, J.G., Luhr, J.F., Podosek, F.A., and Aranda-Gómez, J.J., 1992, The La Breña-El Jaguey maar complex, Durango, Mexico: II. Petrology and geochemistry: Bulletin of Volcanology, v. 54, p. 405–428, doi: 10.1007/BF00312322.

Ramírez-Fernández, J.A., 1996, Zur petrogenesse des alkalikomplex der Sierra de Tamaulipas, NE-Mexiko [Ph.D. thesis]: Freiburg, Universität de Freiburg, 317 p.

Ramírez-Fernández, J.A., 1997, Volcanismo intraplaca típico de la planicie costera del Golfo de México; Sierra de Tamaulipas: Actas INAGEQ, v. 3, p. 323.

Ramírez-Fernández, J.A., Keller, J., and Hubberten, H.W., 2000, Relaciones genéticas entre las carbonatitas y las rocas nefelíniticas del complejo El Picacho, Sierra de Tamaulipas, NE de México: Revista Mexicana de Ciencias Geológicas, v. 7, no. 1, p. 45–55.

Ramírez-Fernández, J.A., Rodríguez-Saavedra, P., Jiménez-Boone, I., and Cossío-Torres, T., 2002, El complejo magmático Terciario de la sierra de San Carlos-Cruillas, estado de Tamaulipas: Geos, v. 22, no. 2, p. 250.

Richards, A.F., 1959, Geology of the Islas Revillagigedo, Mexico, 1. Birth and development of Volcan Barcena, Isla San Benedicto: Bulletin of Volcanology, v. 22, p. 73–124.

Richards, A.F., 1965, Geology of the Islas Revillagigedo, Mexico, 3. Effects of erosion on Isla San Benedicto 1952–61 following the birth of Volcan Barcena: Bulletin of Volcanology, v. 28, p. 381–403.

Richards, A.F., 1966, Geology of the Islas Revillagigedo, Mexico, 2. Geology and petrography of Isla San Benedicto: Proceedings of the California Academy of Sciences, v. 33, p. 361–414.

Righter, K., and Carmichael, I.S.E., 1993, Mega-xenocrysts in alkali basalts: Fragments of disrupted mantle assemblages: The American Mineralogist, v. 78, p. 1230–1245.

Robin, C., 1976, El vulcanismo de las planicies de la Huasteca (este de México): datos geoquímicos y petrográficos: Boletín del Instituto de Geología, Universidad Nacional Autónoma de México, no. 96, p. 55–92.

Robin, C., 1982, Relations volcanologie-magmatologie-geodynamique: Application au passage entre volcanismes alcalin et andesitique dans le sud Mexicain (Axe Trans-Mexicain et Province Alcaline Oriental)

[Ph.D. thesis]: Clermont-Ferrand, France, Université Clermont-Ferrand II.; U.E.R. de Recherche Scientifique et Technique, Annales Scientifiques de l'Université de Clermont-Ferrand II, Géologie, Minéralogie, 503 p.

Rodríguez-Saavedra, P., 2003, Petrografía y geoquímica de las rocas magmáticas de la Sierra de San Carlos-Cruillas [B.Sc. thesis]: Linares, Nuevo León, Universidad Autónoma de Nuevo León, Facultad de Ciencias de la Tierra, 134 p.

Rogers, G., Saunders, A.D., Terrell, D.J., Verma, S.P., and Marriner, G.F., 1985, Geochemistry of Holocene volcanic rocks associated with ridge subduction in Baja California, Mexico: Nature, v. 315, p. 389–392, doi: 10.1038/315389a0.

Rudnick, R.L., and Cameron, K.L., 1991, Age diversity of the deep crust in northern Mexico: Geology, v. 19, p. 1197–1200, doi: 10.1130/0091-7613(1991)019<1197:ADOTDC>2.3.CO;2.

Saunders, A.D., Rogers, G., Marriner, G.F., Terrell, D.J., and Verma, S.P., 1987, Geochemistry of Cenozoic volcanic rocks, Baja California, Mexico: Implications for the petrogenesis of post-subduction magmas: Journal of Volcanology and Geothermal Research, v. 32, p. 223–245, doi: 10.1016/0377-0273(87)90046-1.

Sawlan, M.G., 1986, Petrogenesis of late Cenozoic volcanic rocks from Baja California Sur, Mexico [Ph.D. thesis]: Santa Cruz, University of California, 174 p.

Sawlan, M.G., 1991, Magmatic evolution of the Gulf of California rift, *in* Dauphin, J. P., and Simoneit, B.A., eds., The Gulf and Peninsular Province of the Californias: Tulsa, Oklahoma, American Association of Petroleum Geologists, Memoir, no. 47, p. 301–369.

Schaaf, P., Heinrich, W., and Besch, T., 1994, Composition and Sm-Nd isotopic data of the lower crust beneath San Luis Potosí, central Mexico: Evidence from a granulite-facies xenolith suite: Chemical Geology, v. 118, p. 63–84, doi: 10.1016/0009-2541(94)90170-8.

Seager, W.R., and Morgan, P., 1979, Rio Grande Rift in southern New Mexico, west Texas, and northern Chihuahua, *in* Riecker, R.E., ed., The Rio Grande Rift, Tectonics and Magmatism: Washington, D.C., American Geophysical Union, p. 87–106.

Sedlock, R.L., Ortega-Gutiérrez, F., and Speed, R.C., 1993, Tectonostratigraphic terranes and tectonic evolution of Mexico: Geological Society of America Special Paper 278, 153 p.

Seibertz, E., 1990, El desarrollo cretácico del Archipiélago de Tamaulipas. II. Génesis y datación de un dique de basalto y su efecto en el ambiente deposicional medio-Cretácico de la Sierra de Tamaulipas (Cenomaniano-Turoniano): Actas de la Facultad de Ciencias de la Tierra, Universidad Autónoma de Nuevo León, v. 4, p. 99–123.

Siebe, C., Komorowski, J.C., Navarro, C., McHone, J.F., Delgado, H., and Cortes, A., 1995, Submarine eruption near Socorro Island, Mexico: Geochemistry and scanning electron microscopy studies of floating scoria and reticulite: Journal of Volcanology and Geothermal Research, v. 68, p. 239–271, doi: 10.1016/0377-0273(95)00029-1.

Siebert, L., Calvin, C., Kimberly, P., Luhr, J.F., and Kysar, G., 2002, Volcanoes of México: Washington, D.C., Smithsonian Institution, 1 CD.

Smith, J.A., 1989, Extension-related magmatism of the Durango volcanic field, Durango, Mexico [M.Sc. thesis]: Saint Louis, Missouri, Washington University, 290 p.

Smith, R.D., 1993, The Agua de Mayo mid-Cenozoic volcanic group and related xenoliths from La Olivina SE Chihuahua, Mexico [M.Sc. thesis]: Santa Cruz, University of California, 112 p.

Storey, M., Rogers, G., Saunders, A.D., and Terrell, D.J., 1989, San Quintin volcanic field, Baja California, Mexico: "Within plate" magmatism following ridge subduction: Terra Nova, v. 1, p. 195–202.

Sun, S.S., and McDonough, J.D., 1989, Chemical and isotopic systematics of oceanic basalts: Implications for mantle compositions and process, *in* Saunders, A.D., and Norry, M.J., eds., Magmatism in the Ocean Basins: Geological Society [London] Special Publication 42, p. 313–345.

Suter, M., 1991, State of stress and active deformation in Mexico and western Central America, *in* Slemmons, D.B., Engdahl, E.R., Zoback, M.D., and Blackwell, D.D., eds., Neotectonics of North America: Boulder, Colorado, Geological Society of America, Decade map, 1, 401–421.

Treviño-Cázares, A., 2001, Xenolitos del manto en la planicie costera del Golfo de México [B.Sc. thesis]: Linares, Nuevo León, Universidad Autónoma de Nuevo León, Facultad de Ciencias de la Tierra, 113 p.

Treviño-Cázares, A., Ramírez-Fernández, J.A., Velasco-Tapia, F., and Rodríguez-Saavedra, P., 2005, Mantle xenoliths and their host magmas in the Eastern Alkaline Province, Northeast Mexico: International Geology Review, v. 47, p. 1260–1286.

Umhoefer, P., Mayer, L., and Dorsey, B., 2002, Evolution of the margin of the Gulf of California near Loreto, Baja California peninsula, Mexico: Geological Society of America Bulletin, v. 114, p. 849–868, doi: 10.1130/0016-7606(2002)114<0849:EOTMOT>2.0.CO;2.

Valdez-Moreno, G., 2001, Geoquímica y petrología de los campos volcánicos Las Esperanzas y Ocampo, Coahuila, México [M.Sc. thesis]: México, Universidad Nacional Autónoma de México, Instituto de Geología, 104 p.

van der Lee, S., and Nolet, G., 1997, Upper mantle S velocity structure of North America: Journal of Geophysical Research, v. 102, p. 22,815–22,838, doi: 10.1029/97JB01168.

Vasconcelos, M., Ramírez-Fernández, J.A., and Viera-Décida, F., 2002, Petrología del vulcanismo traquítico del complejo volcánico de Villa Aldama, Tamps: Geos, v. 22, no. 2, p. 250–251.

Vidal-Solano, J.R., Paz-Moreno, F.A., and Demant, A., 2000, Estudio mineralógico y geoquímico de la fase volcánica hiperalcalina del evento miocénico pre-Pinacate, campo volcánico el Pinacate (NW Sonora, México), *in* Calmus, T., and Pérez-Segura, E., eds., Cuarta Reunión sobre la Geología del Noroeste de México y Áreas Adyacentes: Hermosillo, Sonora, Universidad Nacional Autónoma de México, Instituto de Geología, Estación Regional del Noroeste; Universidad de Sonora, Departamento de Geología, p. 142–143.

Viera-Décida, F., 1998, Delimitación, petrografía y geoquímica de los cuerpos intrusivos del Rancho El Salvador, Sierra de Tamaulipas [B.Sc. thesis]: Linares, Nuevo León, Universidad Autónoma de Nuevo León, Facultad de Ciencias de la Tierra, 96 p.

Williams, W.J.W., 1999, Evolution of Quaternary intraplate mafic lavas detailed using 3He surface exposure and ^{40}Ar/^{39}Ar dating, and elemental and He, Sr, Nd, and Pb isotopic signatures: Potrillo volcanic field, New Mexico, U.S.A., and San Quintin volcanic field, Baja California Norte, México [Ph.D. thesis]: El Paso, University of Texas at El Paso, 195 p.

Williams, W.J.W., 2002, Quaternary maar volcanism in the southern Rio Grande Rift, New Mexico, *in* Geological Society of America, 36th Annual Meeting, South-Central Section, Abstracts: http://gsa.confex.com/gsa/2002SC/finalprogram/abstract_32966.htm.

Wood, D.A., Tarney, J., Varet, J.J., Saunders, A.D., Bougault, H., Joron, J.L., Treuil, M., and Cann, J.R., 1979, Geochemistry of basalts drilled in the North Atlantic by IPOD Leg 49: Implications for mantle heterogeneity: Earth and Planetary Science Letters, v. 42, p. 77–97, doi: 10.1016/0012-821X(79)90192-4.

Zindler, A., and Hart, S., 1986, Chemical geodynamics: Annual Review of Earth and Planetary Sciences, v. 14, p. 493–571, doi: 10.1146/annurev.ea.14.050186.002425.

Zoback, M.D., and Zoback, M.L., 1991, Tectonic stress field of North America and relative plate motions, *in* Slemmons, D.B., Engdahl, E.R., Zoback, M.D., and Blackell, D.D., eds., Neotectonics of North America: Boulder, Colorado: Geological Society of America, Decade map, 1, 339–366.

Manuscript Accepted by the Society 29 August 2006

Printed in the USA

Geological Society of America
Special Paper 422
2007

Igneous petrogenesis of the Trans-Mexican Volcanic Belt

Arturo Gómez-Tuena
Ma. Teresa Orozco-Esquivel
Luca Ferrari
Centro de Geociencias, Universidad Nacional Autónoma de México, Juriquilla, Querétaro, 76230, México

ABSTRACT

The magmatic diversity of the Trans-Mexican Volcanic Belt is directly or indirectly controlled by two independent oceanic plates with differing geophysical and compositional parameters; by an extensional tectonic regime that operates with different intensities over the upper plate; by a continental basement with a diversity of ages, thicknesses, and compositions; and by a compositionally heterogeneous mantle wedge that has been modified to various extents by the slab-derived chemical agents. The convergent margin and the magmatic arc have not remained static throughout their geologic histories, but instead have shown significant changes in position, geometry, and composition. For these reasons, the Trans-Mexican Volcanic Belt is the result of one of the most complex convergent margins on the planet, the subject of more than a century of scientific investigations, and at the core of the most notorious debates on Mexican geology.

Keywords: Trans-Mexican Volcanic Belt, México, subduction, magmatic arc, igneous petrology, mantle, crust.

INTRODUCTION AND OBJECTIVES

The Trans-Mexican Volcanic Belt is arguably the best-studied geologic province in México. This attention is justified given the importance of fully understanding a geologically active region where most of the population and infrastructure of the country are concentrated. Nevertheless, and in spite of the copious publications about its tectonic framework, stratigraphy, petrology, geochemistry, and geophysics, it is still not fully understood. Each year the scientific contributions provide new data and hypotheses that gradually move us toward a general understanding of its origin, but only a few theories have awoken a generalized consensus among the scientific community. There is thus little doubt that the Trans-Mexican Volcanic Belt is one of the most interesting subjects of Mexican geology, and we are convinced that it will continue to be so for many years to come.

The main purpose of this contribution is to promote and facilitate the interdisciplinary study of the Mexican arc by providing an up-to-date revision of the data and ideas that have been proposed to explain its origin. Achieving such a goal is far from simple. Published research on Mexican volcanoes covers more than a century of investigations, and its sole compilation and analysis is likely to be a lengthy endeavor. Thus we emphasize that this paper will not be encyclopedic, and that we will dedicate more attention and space to those discoveries and problems that have been put forth in the past two decades, and which we consider more beneficial for most readers.

In the first part of this work, we will examine the current geologic and geophysical parameters of the magmatic arc, because these represent the framework upon which any petrogenetic interpretation must be constructed. We will then describe the compositional characteristics of the volcanic rocks by using a newly complied geochemical database with more than two thousand analyses (http://satori.geociencias.unam.mx/Centenario/Gomez-Tuena.xls). Finally, we will discuss the ideas and interpretations that have been proposed to explain the origin of the

Gómez-Tuena, A., Orozco-Esquivel, Ma.T., and Ferrari, L., 2007, Igneous petrogenesis of the Trans-Mexican Volcanic Belt, *in* Alaniz-Álvarez, S.A., and Nieto-Samaniego, Á.F., eds., Geology of México: Celebrating the Centenary of the Geological Society of México: Geological Society of America Special Paper 422, p. 129–181, doi: 10.1130/2007.2422(05). For permission to copy, contact editing@geosociety.org.

extraordinary magmatic diversity of the Mexican arc. We will critically confront the different hypotheses, emphasizing the premises and assumptions, without leaving behind the inconsistencies and unresolved observations. By doing this we expect to open new avenues for future research.

GEOLOGICAL AND GEODYNAMIC SETTING

It has been customary to define the Trans-Mexican Volcanic Belt as a continental magmatic arc constituted by nearly 8000 volcanic structures, and a few intrusive bodies, that extend from the Mexican Pacific coast in San Blás, Nayarit, and Banderas Bay, Jalisco, to the coast of the Gulf of Mexico in Palma Sola, Veracruz (Demant, 1978). The volcanic province is ~1000 km long and has an irregular width of ~80 to ~230 km. The arc follows an E-W orientation in its central and eastern sectors and a WNW-ESE trend in its western sector, and forms an angle of ~16° with the Middle America Trench (Fig. 1). This peculiar geometry led to the term *Trans-Mexican Volcanic Belt*, because it is transversally emplaced over most of the NNW-SSE–trending Mexican geologic provinces (Ortega-Gutiérrez et al., 1992).

The Trans-Mexican Volcanic Belt is also frequently divided into three different sectors with distinct geologic and tectonic features (Demant, 1978; Pasquaré et al., 1988) (Fig. 1): a western sector that is located between the Pacific Coast and the triple junction formed by the intersection of the Zacoalco, Chapala, and Colima rifts (Allan, 1986); a central sector that is placed between this triple junction and the Taxco–San Miguel de Allende fault system (Alaniz-Álvarez et al., 2002b); and an eastern sector located between these faults and the Gulf of Mexico. We consider this division useful and will use it in this paper.

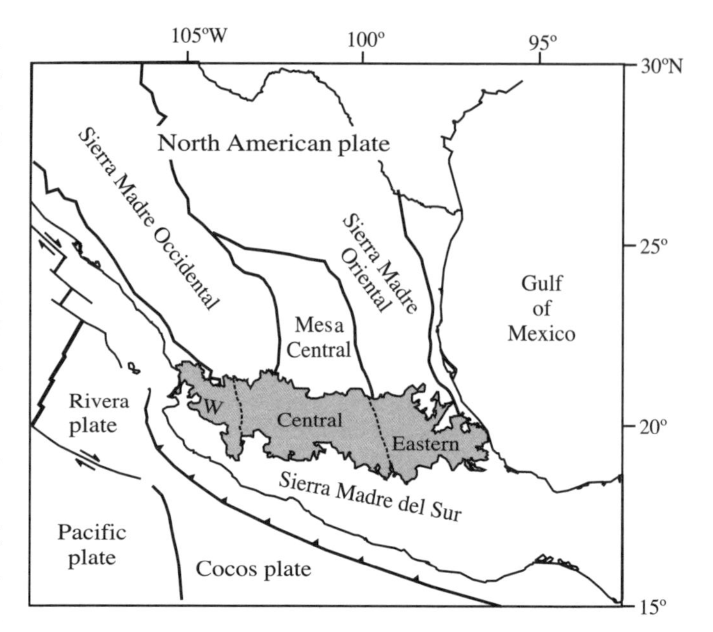

Figure 1. Location of the Trans-Mexican Volcanic Belt (gray) divided into the three sectors that are mentioned in the text. Also shown are the main geologic provinces of México and the current tectonic configuration. W—Western.

Geological Evolution of the Trans-Mexican Volcanic Belt

The space-time evolution of magmatism, and especially the transition between the Sierra Madre Occidental and the Trans-Mexican Volcanic Belt, was intensively debated for a long period of time (Mooser, 1972; Demant, 1978; Cantagrel and Robin, 1979; Demant, 1981; Robin and Cantagrel, 1982; Venegas et al., 1985; Nixon et al., 1987). However, the abundant isotopic ages obtained in the past two decades made it clear that the individualization of the Trans-Mexican Volcanic Belt as a distinctive geologic province dates back to the middle to late Miocene, as a result of a progressive counterclockwise rotation of the magmatic arc of the Sierra Madre Occidental (Ferrari et al., 1999).

Ferrari et al. (2005a) recently reported the first digital geologic map of the Trans-Mexican Volcanic Belt. This document, which includes a database of over 1300 ages and more than 2000 chemical analyses, constitutes the framework upon which a geologic history of the Trans-Mexican Volcanic Belt can be constructed (Plate 1, Fig. 2). The geologic evolution of the structure may be broken into four main episodes: (1) a middle to late Miocene arc of intermediate composition, (2) an episode of mafic volcanism of late Miocene age, (3) a latest Miocene silicic episode that becomes bimodal in the early Pliocene, and (4) the establishment of the modern arc with a large compositional variability since the late Pliocene.

Middle to Late Miocene Arc

The initial activity of the Trans-Mexican Volcanic Belt dates back to the middle Miocene when an arc with an ~E-W orientation began to form in central México between the location of the cities of Morelia and Querétaro and the Gulf of Mexico coast (Fig. 2A). The remnants of this ancestral Trans-Mexican Volcanic Belt are found close to the present volcanic front in the central sector of the arc: in the Sierra de Mil Cumbres and Sierra de Angangueo volcanic complexes in the State of Michoacán (Pasquaré et al., 1991; Capra et al., 1997); as well as in the Tenancingo and Malinalco areas in the State of México, and in the deepest part of the Mexico City basin (Ferrari et al., 2003a). Basaltic and andesitic lava flows of the Tenancingo and Malinalco areas in the State of México were considered to be late Miocene by García-Palomo et al. (2000), who correlated these rocks with a lava flow located ~20 km to the north that was dated at ca. 7.5 Ma (K-Ar age). However, Ferrari et al. (2003a) recently obtained five $^{40}Ar/^{39}Ar$ ages ranging between 19.5 and 16 Ma for the basal lavas of Malinalco and Tenancingo, which also correspond better to the age of ca. 21 Ma reported by García-Palomo et al. (2000) for a lava flow located just north of Malinalco. Given these new results, these sequences are considered to be part of the initial middle Miocene activity of the Trans-Mexican Volcanic Belt (Plate 1).

In a subsequent stage, volcanism migrated farther away from the trench, forming stratovolcanoes and lava cones with ages ranging between ca. 13 and 10 Ma. Among these are the Palo Huérfano, La Joya, and El Zamorano stratovolcanoes, located along the border of the States of Querétaro and Guanajuato (Carrasco-Núñez et al., 1989; Pérez-Venzor et al., 1996; Valdéz-Moreno et al., 1998; Verma and Carrasco-Núñez, 2003), the Cerro Grande stratovolcano in the state of Puebla (Carrasco-Núñez et al., 1997; Gómez-Tuena and Carrasco-Núñez, 2000), the Sierra de Guadalupe and other unnamed polygenetic volcanic centers to the northwest of Mexico City (Jacobo-Albarrán, 1986), as well as the Apan volcanic field, located between Mexico City and Pachuca, Hidalgo (García-Palomo et al., 2002) (Plate 1 and Fig. 2A). Toward the eastern end of the Trans-Mexican Volcanic Belt, in the Palma Sola area, the remnants of this volcanic episode are represented by plutonic and subvolcanic bodies of gabbroic to dioritic compositions with ages that vary between ca. 15 and 11 Ma (Gómez-Tuena et al., 2003; Ferrari et al., 2005b). Many of the middle and late Miocene volcanic rocks that were emplaced far away from the trench, from Querétaro to Palma Sola, display a geochemical composition akin to the "adakites" described by Kay (1978) and Defant and Drummond (1990). This led to the suggestion that these rocks may be the result of partial fusion of the subducted slab during a period of shallow dipping to subhorizontal subduction (Gómez-Tuena et al., 2003). This hypothesis is consistent with the gradual migration of the arc away from the trench since middle Miocene.

There is currently no evidence indicating the presence of a middle Miocene arc to the west of Morelia and Querétaro cities (Fig. 2A). The absence of middle Miocene volcanism does not appear to be related to covering by younger volcanic products, since late Miocene sequences in the Altos de Jalisco region, and in the Cotija area of Michoacán state are emplaced directly over Oligocene to early Miocene ignimbrites (Ferrari and Rosas-Elguera, 2000; Rosas-Elguera et al., 2003). Similarly, deep boreholes drilled in the Ceboruco area and in La Primavera caldera in Jalisco show that late Miocene lavas directly overlie Eocene andesites (Ferrari and Rosas-Elguera, 2000; Ferrari et al., 2003b). On the other hand, scattered volcanic centers of middle Miocene age are found in Nayarit in close proximity to the mouth of the Gulf of California (Fig. 2A) (Gastil et al., 1979; Ferrari et al., 2000a). This volcanism may be considered the southernmost end of the Comondú arc, better exposed along the eastern side of the Baja California Peninsula, with ages ranging between ca. 30 and 12 Ma (Umhoefer et al., 2001).

Late Miocene Mafic Episode

The relatively normal volcanic arc that developed during the middle Miocene experienced a sudden change in the late Miocene when mafic lavas were emplaced to the north of the previous arc, extending between Nayarit and Veracruz States, with a clear eastward migration pattern (Ferrari and Rosas-Elguera, 2000; Ferrari, 2004; Ferrari et al., 2005b) (Fig. 2B). This episode is mainly represented by fissural basaltic lava flows that frequently form widespread plateaus (*trapps*) with ages ranging between ca. 11 and 8.9 Ma in Nayarit (Righter et al., 1995; Ferrari et al., 2000a); ca. 11 and 8 Ma to the north of Guadalajara, in Los Altos de Jalisco, and in the Cotija area of Michoácan (Nieto-Obregón et al., 1981; Verma et al., 1985; Moore et al., 1994; Rosas-Elguera et al., 1997; Alva-Valdivia et al., 2000; Rossotti et al., 2002; Rosas-Elguera et al., 2003); and ca. 9–7 Ma in Querétaro State and in the Pathé area, Hidalgo (Pasquaré et al., 1991; Suter et al., 1995a; Aguirre-Díaz and López-Martínez, 2001). This mafic volcanism continues to the Tlanchinol-Huejutla area, and reaches the coast of the Gulf of Mexico in northern Veracruz (Tantima-Álamo) with ages that range between 7.5 and 6.5 Ma (Cantagrel and Robin, 1979; López-Infanzón, 1991; Ferrari et al., 2005b). Ferrari et al. (2000b) showed that even though these volcanic rocks have variable compositions, the late Miocene rocks emplaced between the Gulf of California and the western part of Hidalgo in the Pathé area have a clear subduction signature. In contrast, volcanism in the eastern part of Hidalgo State and northern Veracruz (Tlanchinol, Tantima, Álamo, and some of the Palma Sola plateau lava) tends to show an intraplate character (Orozco-Esquivel et al., 2007).

Latest Miocene Silicic Volcanism and Early Pliocene Bimodal Volcanism

By the end of the Miocene, volcanism decreased significantly and became more evolved. Latest Miocene to early Pliocene dacitic to rhyolitic dome complexes and voluminous ignimbrites were erupted from regional calderas, and emplaced in a belt located just south of the previous mafic episode (Fig. 2C). Significant volumes of rhyolitic flows and ignimbrites are exposed between Santa Maria del Oro and Plan de Barrancas (Jala Group; Ferrari et al., 2000a) in the western sector of the Trans-Mexican Volcanic Belt; whereas between the Santa Rosa dam and San Cristóbal to the north of Guadalajara exogenous domes and minor pyroclastic flows are the dominant products (Guadalajara group; Ferrari et al., 2000a; Rossotti et al., 2002). Although these silicic rocks span from 7.5 to ca. 3 Ma (Gilbert et al., 1985; Rossotti et al., 2002; Ferrari et al., 2003b, 2004), they essentially represent the only volcanic activity of the western Trans-Mexican Volcanic Belt until ca. 5 Ma.

After the silicic event, the first appearance of mafic volcanism occurred around Guadalajara and in the northern part of the Colima rift. Relatively low-volume alkaline lava flows with an intraplate character were emplaced since ca. 5.5 Ma in the Guadalajara area (Gilbert et al., 1985; Moore et al., 1994). Similar rocks continued to be emplaced during the early Pliocene along with some silicic dome complexes and ignimbrites (Moore et al., 1994; Ferrari et al., 2000a; Frey et al., 2004). It is worth noting that most of the intraplate lavas were emplaced in the rear part of the arc, where, together with the rhyolites, they constitute a typical bimodal magmatic association (Ferrari, 2004). In the northern part of the Colima rift, and in the Ayutla volcanic field, alkaline mafic volcanism with a weak subduction signature dominated the early Pliocene (Allan, 1986; Righter and Rosas-Elguera, 2001). The same kind of volcanism is found around Chapala Lake, although the oldest available ages indicate that this activity began at ca. 6 Ma (Delgado-Granados et al., 1995).

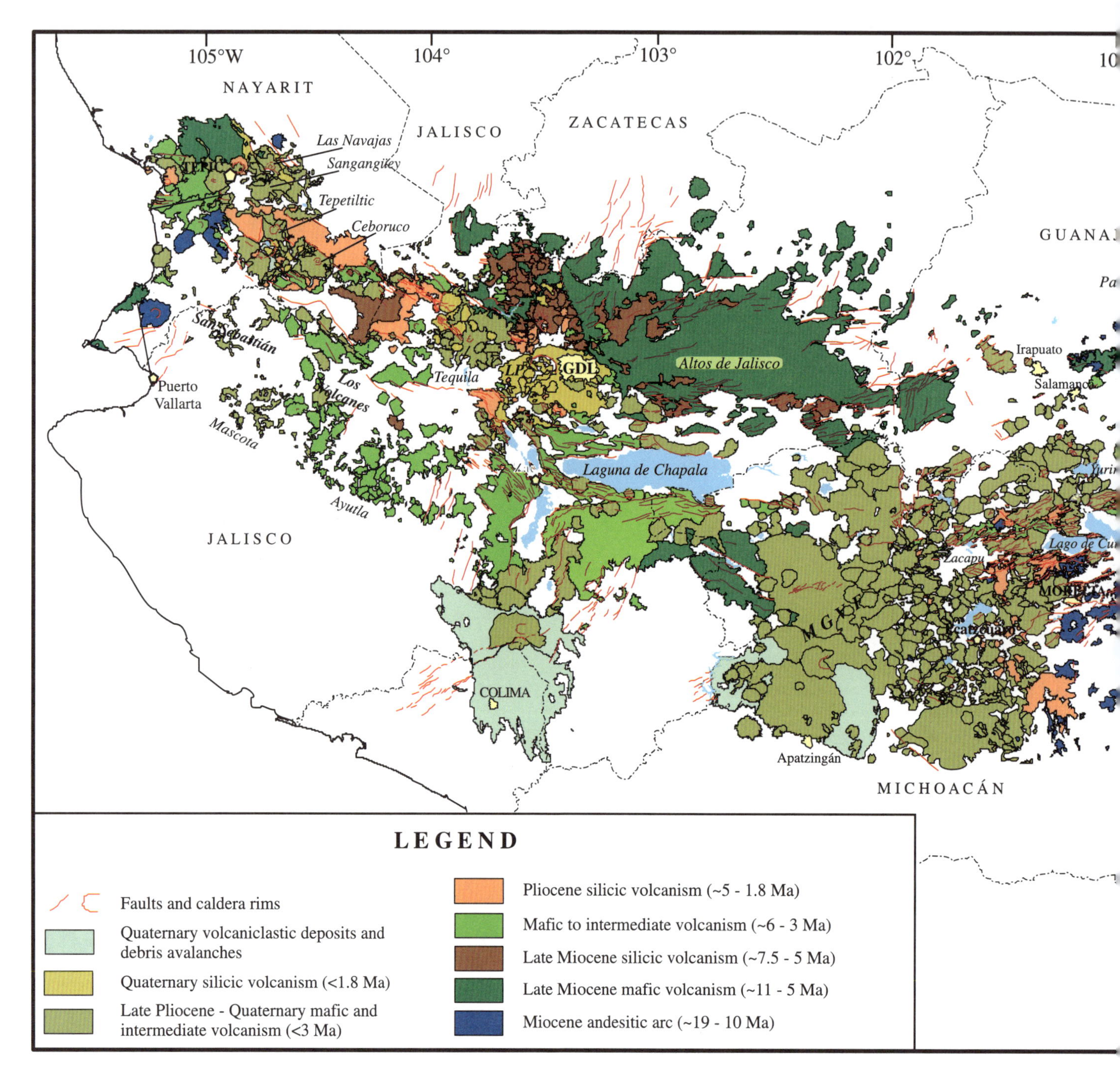

Plate 1. Simplified geologic map of the Trans-Mexican Volcanic Belt based on the compilation of Ferrari et al. (2005a). GDL—Guadalajara; Zac—Zacoalco; MGVF—Michoacán-Guanajuato volcanic field; NT—Nevado de Toluca; LP—La Primavera; Izta—Iztaccíhuatl; Popo—Popocatépetl; Pico—Pico de Orizaba; Cu—Las Cumbres; Cofre—Cofre de Perote. This plate is also available as item 2007086 in the GSA Data Repository, www.geosociety.org/pubs/ft2007.htm, or on request from editing@geosociety.org, Documents Secretary, GSA, P.O. Box 9140, Boulder, CO 80301, USA.

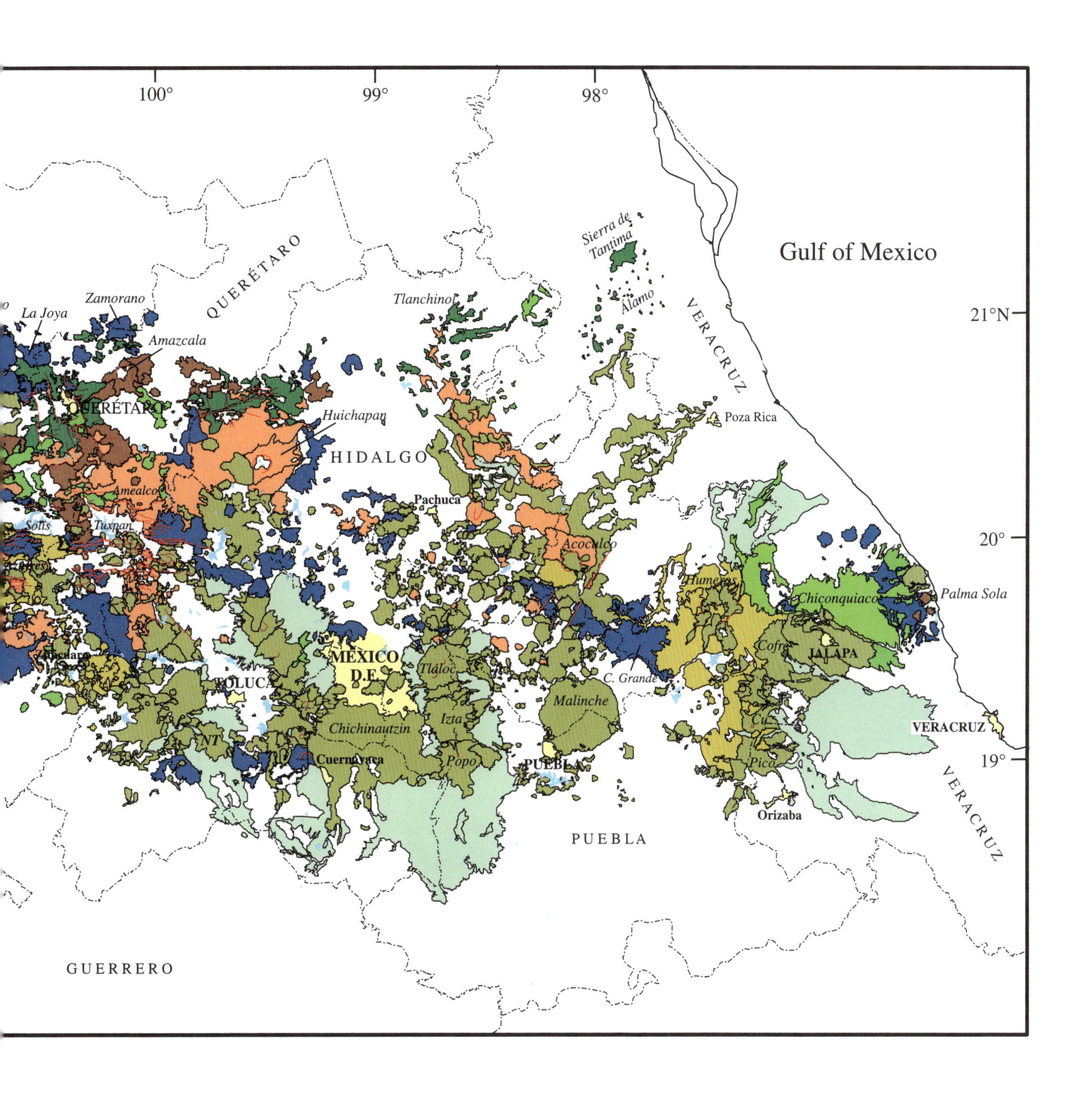

100°
99°
98°
Gulf of Mexico
21°N
20°
19°
QUERÉTARO
HIDALGO
VERACRUZ
PUEBLA
GUERRERO
Sierra de Tantima
Tlanchinol
Álamo
Zamorano
La Joya
Amazcala
QUERÉTARO
Huichapan
Poza Rica
Amealco
Pachuca
Solís
Tuxpan
Acoculco
Humeros
Chiconquiaco
Palma Sola
Cofre
JALAPA
MEXICO D.F.
Tláloc
C. Grande
TOLUCA
Malinche
Chichinautzin
Izta
VERACRUZ
NT
Cuernavaca
Popo
PUEBLA
Pico
Orizaba

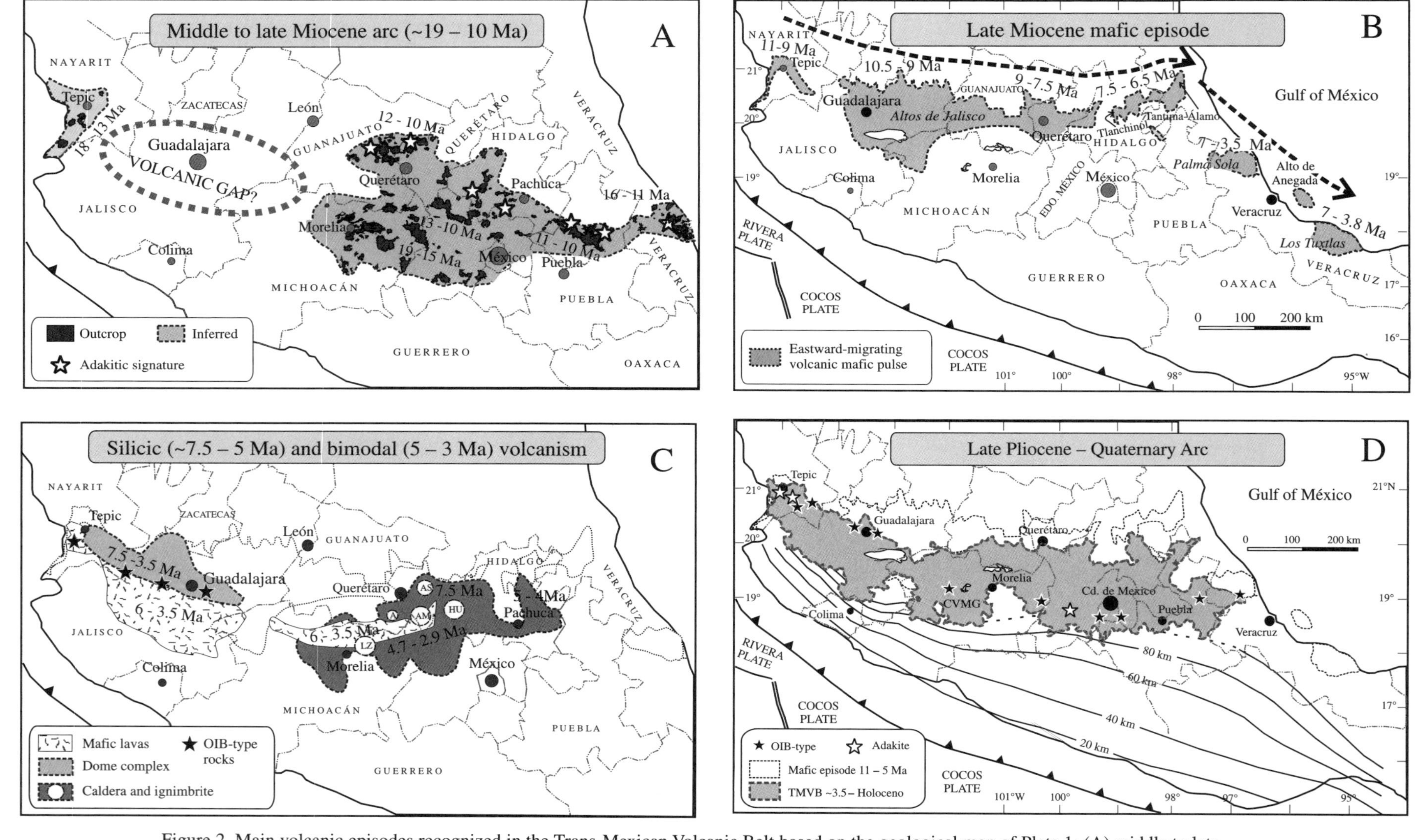

Figure 2. Main volcanic episodes recognized in the Trans-Mexican Volcanic Belt based on the geological map of Plate 1: (A) middle to late Miocene arc (ca. 19–10 Ma); (B) late Miocene mafic episode; (C) silicic (ca. 7.5–5 Ma) and bimodal (5–3 Ma) volcanism with important calderas identified (LZ—Los Azufres; AM—Amealco; AS—Amazcala; HU—Huichapan; A—Agustinos); and (D) late Pliocene–Quaternary arc. The Michoacán-Guanajuato volcanic field (CVMG) is also indicated. TMVB—Trans-Mexican Volcanic Belt.

The central sector of the Trans-Mexican Volcanic Belt is dominated by large caldera-forming ignimbrites and ash-flow tuffs with volumes of several tens of cubic kilometers distributed over thousands of square kilometers. The most prominent calderas are Amazcala (7.3–6.6 Ma; Aguirre-Díaz and López-Martínez, 2001), Amealco (4.7 Ma; Aguirre-Díaz and McDowell, 2000), Huichapan (4.7–3.4 Ma; Aguirre-Díaz et al., 1997), Los Azufres (ca. 6–3 Ma; Ferrari et al., 1991; Pradal and Robin, 1994), Zitácuaro (late Miocene–early Pliocene; Capra et al., 1997), and Apaseo and Los Agustinos (early Pliocene; Aguirre-Díaz et al., 1997). In the same region, mafic to intermediate lava flows of latest Miocene to early Pliocene (ca. 6–3.4 Ma) are exposed in a WSW-ENE–trending belt located north of the Zacapu basin, Cuitzeo lake, and Solís and Tuxpan dams (Ferrari et al., 1991; Pasquaré et al., 1991; Aguirre-Díaz, 1996) (Plate 1 and Fig. 2C). These lavas are often interbedded with the youngest caldera-forming ignimbrites that were mentioned above. Early Pliocene (4.9–4.5 Ma) pyroclastic flow deposits are also reported in the eastern sector of the Trans-Mexican Volcanic Belt between Pachuca and Tlanchinol, Hidalgo State, along the frontal part of the Sierra Madre Oriental orogenic belt. These rocks are interlayered and covered by basaltic lavas and predictably form a bimodal magmatic association (Cantagrel and Robin, 1979; Ochoa-Camarillo, 1997).

Late Pliocene to Quaternary Arc

By the early and late Pliocene, the silicic and bimodal volcanism was replaced by a basaltic to andesitic volcanic arc (Fig. 2D). In the western sector of the Trans-Mexican Volcanic Belt, a second pulse of intraplate volcanism started at ca. 3.6 Ma, although lavas with a more typical subduction signature were also emplaced in the northeastern part of the arc (Righter et al., 1995; Ferrari et al., 2000a). Since the late Pliocene, the volcanic front of the western Trans-Mexican Volcanic Belt is dominated by monogenetic volcanic fields (Mascota, Los Volcanes, San Sebastián, Atenguillo) with products that tend to form a lamprophyric association (Wallace and Carmichael, 1989; Lange and Carmichael, 1990, 1991; Righter and Carmichael, 1992; Carmichael et al., 1996). The large stratovolcanoes of the western sector of the Trans-Mexican Volcanic Belt were mostly built during the Quaternary. The Colima volcanic complex, located at the southern end of the homonymous rift, represents the largest volume of erupted volcanic material in the region by far (ca. 700 km³) (Robin et al., 1987). Other stratovolcanoes in the area have volumes of less than 100 km³ (Tequila, Ceboruco, Tepetiltic, Sangangüey, Las Navajas, and San Juan), and are aligned along a regional fault system with a WNW-ESE orientation that defines the northern boundary of the Jalisco block (Fig. 3): a Cretaceous batholith (ca. 100–75 Ma) (Schaaf et al., 1995) that apparently experienced uplift during the Paleogene (Rosas-Elguera et al., 1996).

Toward the central part of the Trans-Mexican Volcanic Belt, the Michoacán-Guanajuato volcanic field covers a wide region located between Chapala Lake and the western part of Querétaro State (Plate 1). Volcanism in this area began at ca. 2.8 Ma and continues to be active today as witnessed by the historic eruptions of Jorullo (1759–1774) and Parícutin (1943–1952) cinder cones. The volcanic field has over 1000 monogenetic cones and a few

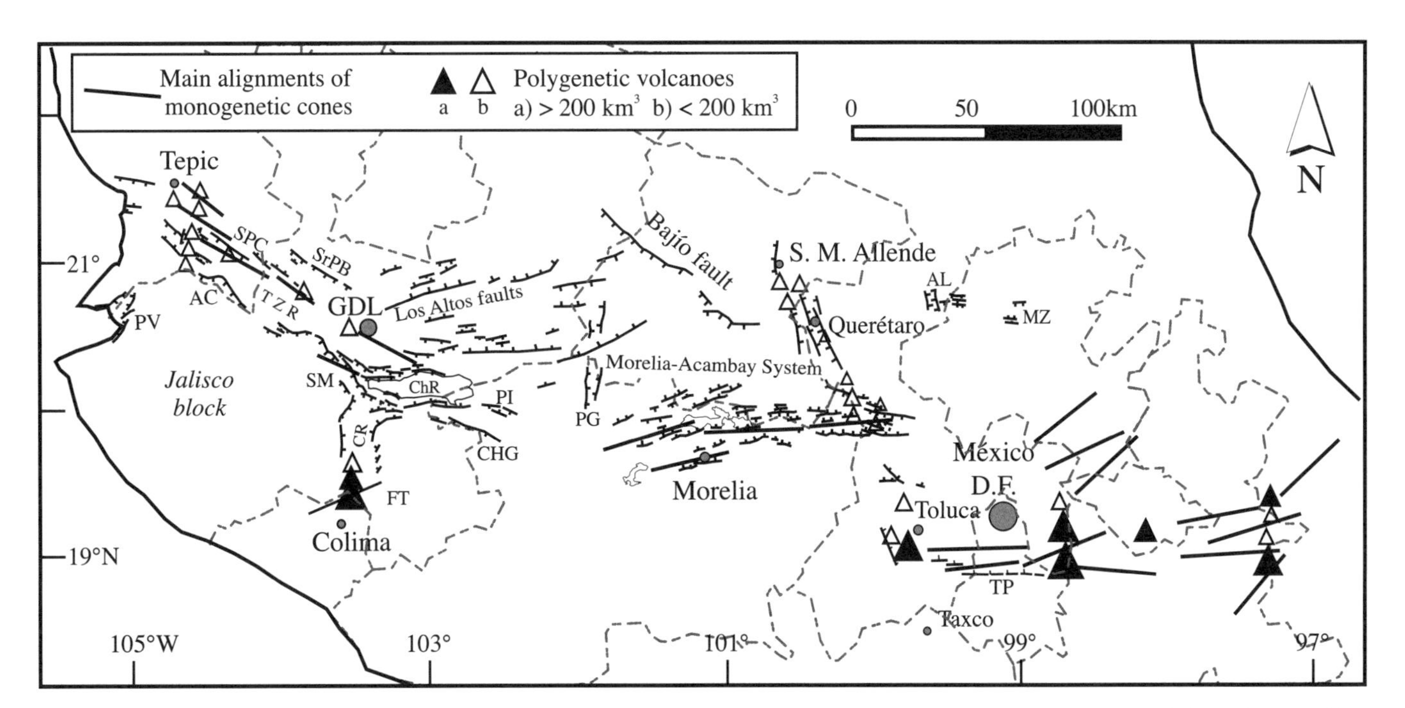

Figure 3. Relationships between the main Miocene-Quaternary fault systems and the Trans-Mexican Volcanic Belt. TZR—Tepic-Zacoalco rift; PV—Puerto Vallarta graben; SPC—San Pedro–Ceboruco graben; AC—Amatlán de Cañas half-graben; SrPB—Santa Rosa–Plan de Barrancas graben; SM—San Marcos fault; CR—Colima rift; FT—Tamazula fault system; ChR—Chapala rift; PI—Pajacuarán–Ixtlán de los Hervores fault; CHG—Cotija half-graben; PG—Penjamillo graben; AL—Aljibes half-graben; MZ—Mezquital graben; GDL—Guadalajara; TP—Tenango–La Pera fault system.

domes of intermediate to silicic composition (Hasenaka and Carmichael, 1985). It also includes more than 400 polygenetic centers, most of which are small shield volcanoes (Hasenaka, 1994). The only stratovolcano is Cerro Tancítaro, dated at ca. 0.5 Ma (Ban et al., 1992).

Volcanism is more scattered and slightly more evolved compositionally toward the east of the Michoacán-Guanajuato volcanic field. In the region of Maravatío-Zitácuaro–Valle de Bravo, middle Pleistocene to Holocene monogenetic cones are generally aligned along regional tectonic structures (Capra et al., 1997; Blatter et al., 2001), and some Quaternary dacitic dome complexes are emplaced in the outskirts of the Los Azufres and Zitácuaro calderas (Ferrari et al., 1991; Capra et al., 1997).

Excluding the alkaline basalts of the Palma Sola region, and the Pachuca-Tlanchinol bimodal sequence, a clear gap of volcanism is observed in the eastern sector of the Trans-Mexican Volcanic Belt between the end of the Miocene and the early Pliocene. Volcanism resumes at ca. 3.7 Ma to the west of Mexico City basin with the emplacement of polygenetic volcanic centers that formed the NNW-SSE–trending Sierra de las Cruces (Mora-Alvarez et al., 1991; Osete et al., 2000). Some medium-sized polygenetic centers with a similar age were also emplaced north of Mexico City in the Apan volcanic field (García-Palomo et al., 2002). Nonetheless, most of the volcanic rocks in the eastern sector of the Trans-Mexican Volcanic Belt have been emplaced during the Quaternary. Mafic volcanism is dominant in monogenetic volcanoes of the Sierra Chichinautzin volcanic field (Martin-Del Pozzo, 1982; Márquez et al., 1999c; Siebe et al., 2004b) and Apan (García-Palomo et al., 2002), as well as around Pico de Orizaba and Cofre de Perote stratovolcanoes (Plate 1) (Siebert and Carrasco-Núñez, 2002). More evolved volcanic products are found in the Acoculco (Verma, 2001a) and Los Humeros calderas (Ferriz and Mahood, 1984; Verma, 2000b), the Las Cumbres silicic center (Rodríguez et al., 2002; Rodríguez, 2005), and the massive isolated domes of Las Derrumbadas, Cerro Pinto, and Cerro Pizarro (Riggs and Carrasco-Núñez, 2004), all located in the State of Puebla. The Sierra Nevada, a N-S–trending volcanic chain composed of Cerro Tláloc, the Iztaccíhuatl volcanic complex, and Popocatépetl volcano formed on the eastern side of the Mexico City basin, with ages that become progressively younger toward the south (Nixon, 1989). Further to the east, La Malinche volcano and the ~N-S–trending Pico de Orizaba–Cofre de Perote volcanic chain are all younger than 1 Ma (Carrasco-Núñez and Ban, 1994).

Structural Geology and Neotectonics

The well-known obliquity of the Trans-Mexican Volcanic Belt to the Middle America Trench and to most of the Mexican geologic provinces (Fig. 1) inspired several researchers to suggest that an important tectonic structure might be concealed beneath the volcanic arc. The idea of a major crustal structure was originally proposed by von Humboldt (1808), and later reawakened by Mooser (1972) and Gastil and Jensky (1973), who suggested that the Trans-Mexican Volcanic Belt could be the expression of an ancient suture, or shear zone, that was reactivated during the Tertiary. This idea was then adopted by several authors who, in their reconstruction of the opening of the Gulf of Mexico and the tectonic evolution of the Caribbean, found themselves with the geometric need for an additional "megashear" located somewhere in central México (Silver and Anderson, 1974; Pindell, 1985; Ross and Scotese, 1988). Important advances in structural geology and tectonics performed during the past fifteen years have been able to define the geometry, kinematics, and age constraints of the main fault systems (Fig. 3). These studies have shown that volcanism within most of the Trans-Mexican Volcanic Belt has been closely associated with faulting, but a major crustal discontinuity has not been recognized.

Demant (1981) was the first to describe the three narrow tectonic basins arranged in a triple junction to the south of Guadalajara, on the western sector of the Trans-Mexican Volcanic Belt (Fig. 3). Demant (1981) defined these structures as the grabens of Tepic-Chapala (later defined as Tepic-Zacoalco), Colima, and Chapala. Allan (1986) later studied the volcanism around the triple junction in greater detail, and provided many isotopic ages that allowed him to establish that extensional faulting in the area had begun by the early Pliocene. The discovery of alkaline rocks associated with the extensional faulting led Luhr et al. (1985) to suggest an active rifting model for the Jalisco block, as a result of an eastward "jump" of a segment of the East Pacific rise. This model prevailed in the literature for more than a decade (Allan et al., 1991; Bourgois and Michaud, 1991; Michaud et al., 1991), and predicted a dextral lateral displacement along the Tepic-Zacoalco rift for Plio-Quaternary times, even though these authors did not carry out structural studies in the region. At the same time, some authors described the Tepic-Zacoalco rift as a series of grabens and pull-apart basins of Pliocene to Holocene age (Barrier et al., 1990; Allan et al., 1991; Garduño and Tibaldi, 1991), whereas others claimed that dextral lateral faulting was presently active in its eastern sector, around the Santa Rosa dam (Nieto-Obregón et al., 1985; Moore et al., 1994). However, by the early 1990s, several detailed structural studies began to show that the kinematics of the fault systems forming the Tepic-Zacoalco rift were essentially extensional during Plio-Quaternary times (Nieto-Obregón et al., 1992; Quintero-Legorreta et al., 1992; Ferrari et al., 1994; Rosas-Elguera et al., 1997), but that a strike-slip deformation also affected the region during the middle and late Miocene (Ferrari, 1995). In a detailed study of the main fault systems between Guadalajara and the mouth of the Gulf of California, Ferrari and Rosas-Elguera (2000) concluded that the Tepic-Zacoalco rift consists of a series of grabens and half-grabens that developed in different tectonic episodes since the late Miocene. The southernmost structures of the Tepic-Zacoalco rift (Puerto Vallarta graben, Amatlán de Cañas half-graben, San Marcos fault, and Zacoalco fault zone) show geologic (Ferrari et al., 1994; Rosas-Elguera et al., 1997) and seismologic evidence (Suárez et al., 1994; Pacheco et al., 1999) of neotectonic activity for the Quaternary. Based on the bathymetry of the marine floor off of Puerto Vallarta, Alvarez (2002) also proposed the existence of normal faulting in the Bahía de Banderas, which seems to be confirmed by the seismicity that has been registered over the past few years (Núñez-Cornú et al., 2002).

The Colima rift was defined as a Plio-Quaternary graben with a rough N-S orientation divided into three segments (north, central, and south) (Allan et al., 1991). However, other structural and geophysical studies disputed the existence of Plio-Quaternary extensional faulting in the southern segment of the Colima graben (south of Colima volcano), since the evidence suggested only a pre-Pliocene transpressive deformation (Serpa et al., 1992). Subsequently, some workers documented the existence of many NE-SW–trending transcurrent and normal faults (collectively called the "Tamazula fault") passing through the Colima volcanic complex and reaching the Pacific Coast at Manzanillo, Colima (Rosas-Elguera et al., 1996; Garduño-Monroy et al., 1998). These authors proposed that the Tamazula fault represents the southeastern boundary of the Jalisco block, located to the south of Colima volcano.

As a result of all of these studies, the model of Luhr et al. (1985) was revised by Rosas-Elguera et al. (1996) and Ferrari and Rosas-Elguera (2000), who proposed that the Tepic-Zacoalco and Colima rifts represent the continental boundaries of the Jalisco block. According to these authors, these boundaries were partially reactivated during Plio-Quaternary times with a mostly extensional motion as a consequence of the interaction between the Rivera and North American plates.

The eastern segment of the Guadalajara triple junction is formed by the Chapala rift, originally defined as an E-W–trending graben (Demant, 1978), which formed as a result of ~N-S extension that was active during the Plio-Quaternary (Garduño-Monroy et al., 1993). Subsequent studies demonstrated that this structure is in fact made of two half-grabens with opposite vergences: south on the western part and north on the eastern one (Urrutia-Fucugauchi and Rosas-Elguera, 1994; Rosas-Elguera and Urrutia-Fucugauchi, 1998). The master faults of these two half-grabens cut rocks as young as 3.4 Ma near the town of Chapala (Rosas-Elguera and Urrutia-Fucugauchi, 1998), and 3.3 Ma in the Pajacuarán–Ixtlán de los Hervores area (Rosas-Elguera et al., 1989), although the geomorphic expression of the fault escarpments suggests that the tectonic activity ended before the Quaternary. In contrast, a Quaternary faulting is suggested in the Citlala graben, a parallel structure located immediately to the south of Chapala Lake (Garduño-Monroy et al., 1993; Rosas-Elguera and Urrutia-Fucugauchi, 1998). Further to the south, the Cotija graben is an asymmetric extensional structure with a WNW-ESE orientation and SSW vergence cutting late Miocene rocks (Rosas-Elguera et al., 2003).

Toward the central sector of the Trans-Mexican Volcanic Belt, the widespread volcanism of the Michoacán-Guanajuato volcanic field prevents the recognition of possible pre-Pliocene faulting events. However, WNW-ESE– and WSW-ENE–trending normal faults cut Pliocene rocks at the western and eastern ends of the field, respectively (Plate 1). A statistical analysis of the orientation of cinder cones within the Michoacán-Guanajuato volcanic field shows that alignments of 3 to 6 cones are parallel to these fault systems (Connor, 1990). To the north of the volcanic field, a series of normal faults with a slight left-lateral component affects the late Miocene basalts of the Los Altos de Jalisco region (Plate 1) (Ferrari et al., 2000b). These faults have a general WSW-ENE orientation and are parallel to the alignment of some lava cones in the central part of the Los Altos plateau. Ferrari et al. (2000b) interpreted this to mean that these faults must have begun their activity during the final phase of the mafic volcanic episode at ca. 8 Ma.

Further to the east, and between the locations of León and Querétaro cities, the Plio-Quaternary volcanism of the Trans-Mexican Volcanic Belt is emplaced over a wide asymmetric basin bounded to the north by the Bajío normal fault system and to the south by the Morelia-Acambay system. The Bajío fault system has a length that exceeds 70 km, and a minimum vertical displacement of 2 km. These faults were active mainly during the Eocene and Oligocene times, although a displacement of at least 500 m took place after the middle Miocene (Nieto-Samaniego et al., 1999; Alaniz-Álvarez and Nieto-Samaniego, 2005). The Morelia-Acambay system was initially described by Martínez Reyes and Nieto-Samaniego (1990) and its central part by Pasquaré et al. (1988), Ferrari et al. (1990), and Pasquaré et al. (1991), as a ~30-km-wide belt of WSW-ENE–trending faults bounding the tectonic depressions of Zacapu, Cuitzeo, Morelia, and Acambay among others. Kinematic analyses on these faults indicate that the initial activity most likely occurred during the Pliocene, and was dominantly left lateral to oblique, but that it became progressively more extensional during the Quaternary (Ferrari et al., 1990; Suter et al., 1995b). Most of the faults in the western part of the system dip to the north and tilt to the south the Miocene to Pliocene volcanic successions. However, Garduño-Monroy et al. (2001) also reported evidence of Quaternary activity in the Morelia region. The eastern part of the system is an asymmetric graben system formed by the Epitacio Huerta and Acambay-Tixmadeje normal faults to the north, and the Venta de Bravo and Pastores faults to the south, showing a small left-lateral motion component (Suter et al., 1992, 1995b). This sector is the most active part of the system, as witnessed by the 1912 Acambay earthquake with a Ms = 6.9 (Urbina and Camacho, 1913; Suter et al., 1995b, 1996), for which a recurrent time of ca. 3600 yr has been estimated for the Holocene (Langridge et al., 2000). More detailed studies on the neotectonics of the whole Morelia-Acambay system have been recently provided by Suter et al. (2001) and Szynkaruk et al. (2004).

The Taxco–San Miguel de Allende fault system was first described by Demant (1981) as a group of transverse structures located between Taxco and San Miguel de Allende, and crossing the Trans-Mexican Volcanic Belt with a NNW orientation. Several segments of the fault system have been studied in detail during the past decade. In their synthesis, Alaniz-Álvarez et al. (2002b) described the system as a major continental structure over 500 km long, and up to 35 km wide, representing the boundary between crustal blocks with different thicknesses and topographies. Segments of the Taxco–San Miguel de Allende fault system have been reactivated with distinct kinematics since the Oligocene (Alaniz-Álvarez et al., 2002a, 2002b).

Deformation is older in the eastern sector of the Trans-Mexican Volcanic Belt. The Mexico City basin is a tectonic depression over 2 km in depth, which formed during the Oligocene or even the Eocene (Ferrari et al., 2003a; Alaniz-Álvarez and Nieto-Samaniego, 2005). The basin is bounded to the west by the Taxco–San Miguel de Allende fault system and to the south by the La Pera–Tenango fault system with an E-W orientation (García-Palomo et al., 2000; Ferrari et al., 2003a), which also seems to control monogenetic volcanism in the Sierra Chichinautzin volcanic field (Márquez et al., 1999c).

The most prominent tectonic structures of the eastern Trans-Mexican Volcanic Belt are the Aljibes half-graben, the Mezquital graben (Suter et al., 2001), and the normal faults of the Apan volcanic field (García-Palomo et al., 2002). The Aljibes half-graben is formed by four main normal faults with an E-W orientation located in the northern limit of the Trans-Mexican Volcanic Belt at ~140 km to the WNW of Mexico City (Suter et al., 1995a). The faults that dip to the south affect late Miocene basalts and are considered potentially active (Suter et al., 1995a, 1996). The Mezquital graben is an E-W–trending structure located at ~40 km to the east of the Aljibes half-graben. Gravimetric studies suggest that both structures are part of a single tectonic depression (Campos-Enríquez and Sánchez-Zamora, 2000). The Mezquital graben is bounded by the Cardonal fault, which cuts rocks older than 4.6 Ma (Suter et al., 2001). On the other hand, and despite the diffuse seismicity of the area, two earthquakes of magnitude (Mw) ~5 strongly indicate that this structure is potentially active (Suter et al., 1996; Quintanar et al., 2004). Finally, García-Palomo et al. (2002) described several NE-trending normal faults in the Apan volcanic field (Plate 1) that affect middle Miocene rocks and do not appear to be active.

In summary, the overall geologic evidence indicates that the Trans-Mexican Volcanic Belt is presently under an extensional tectonic regime, although the Quaternary faults of the central sector have a small and variable left-lateral component. This weak transtensional regime may be explained by considering that the convergence between the Cocos and North American plates is slightly oblique, and that a deformation partitioning at the plate boundary may produce a small, trench-parallel (sinistral) motion component that is accommodated by the upper plate (Ego and Ansan, 2002). Since the Trans-Mexican Volcanic Belt is the most important zone of crustal weakness north of the trench, it represents a suitable place to accommodate such a trench-parallel motion component (Ego and Ansan, 2002).

The relationship between magma emplacement mechanisms and tectonics along the Trans-Mexican Volcanic Belt has been the subject of strong debate. Several studies have suggested that the notable N-S alignment of the main stratovolcanoes of the Trans-Mexican Volcanic Belt should be related to some extensional fault systems that follow the same orientation (Cantagrel and Robin, 1979; Höskuldsson and Robin, 1993; Alaníz-Álvarez et al., 1998). Although this model might explain the alignment of stratovolcanoes along the Taxco–San Miguel de Allende fault system (Fig. 3), no study to date has reported extensional or strike-slip faults affecting the Popocatépetl-Iztaccíhuatl-Tláloc or the Cofre de Perote–Pico de Orizaba volcanic chains. It has also been suggested that the most primitive volcanic rocks of the eastern sector, generally associated with monogenetic volcanoes (e.g., Chichinautzin volcanic field), have been emplaced along faults and fissures that follow a preferred E-W orientation (Márquez et al., 1999c; García-Palomo et al., 2000). Indeed, fault systems with this orientation seems to favor a rapid ascent of primitive magmas in this sector, and also represent the dominant Quaternary deformation structures in the region (Suter et al., 2001). Following these observations, Alaníz-Álvarez et al. (1998) proposed a general model to explain the contrasting preferred orientation between monogenetic centers (parallel to the arc) and stratovolcanoes (transversal to the arc). These authors suggested that the oblique orientation of the arc with respect to the trench makes the faults parallel to the arc accommodate most of the extensional deformation, allowing fast magma ascent and the formation of monogenetic centers. In contrast, the arc transversal faults that accommodate only a small amount of extension have a low deformation rate, which favors magma trapping and differentiation, eventually leading to the formation of stratovolcanoes. Even though this model has been questioned by several authors (Contreras and Gómez-Tuena, 1999; Siebe et al., 1999; Suter, 1999), who pointed out several inconsistencies based on structural, petrologic, and field observations, no other model has been put forward to explain the unusual alignment of stratovolcanoes in the Trans-Mexican Volcanic Belt (Alaníz-Álvarez et al., 1999).

The Oceanic Plates

History and Geometry of the Subducting Oceanic Plates

The origin of the Trans-Mexican Volcanic Belt and its anomalous lack of parallelism with the Middle America Trench were extensively discussed for more than a century (von Humboldt, 1808; De Cserna, 1958; Mooser, 1972; Gastil and Jensky, 1973; Johnson and Harrison, 1990). Today, however, the geophysical evidence clearly indicates that the Cocos and Rivera oceanic plates are currently being subducted under the continent in the direction of the Trans-Mexican Volcanic Belt (Fig. 4) (Urrutia-Fucugauchi and Del Castillo, 1977; Urrutia-Fucugauchi and Böhnel, 1987; Pardo and Suárez, 1993, 1995). For these reasons, most researchers consider that magma genesis in the Trans-Mexican Volcanic Belt, as well as the oblique orientation to the trench, must be related in some way to the subduction process. Despite this evidence, some researchers find this relation questionable in light of the petrologic peculiarities of the arc, the unusual presence of an extensional tectonic regime accompanying primitive magmas, and the poorly defined Wadati-Benioff zone beneath most of the arc (Márquez et al., 1999a; Verma, 1999; Sheth et al., 2000; Verma, 2000a, 2002). It is clear that all of these factors need to be taken into account for a comprehensive evaluation of the origin of the Trans-Mexican Volcanic Belt; and an extensive discussion on these issues will be given later in this paper.

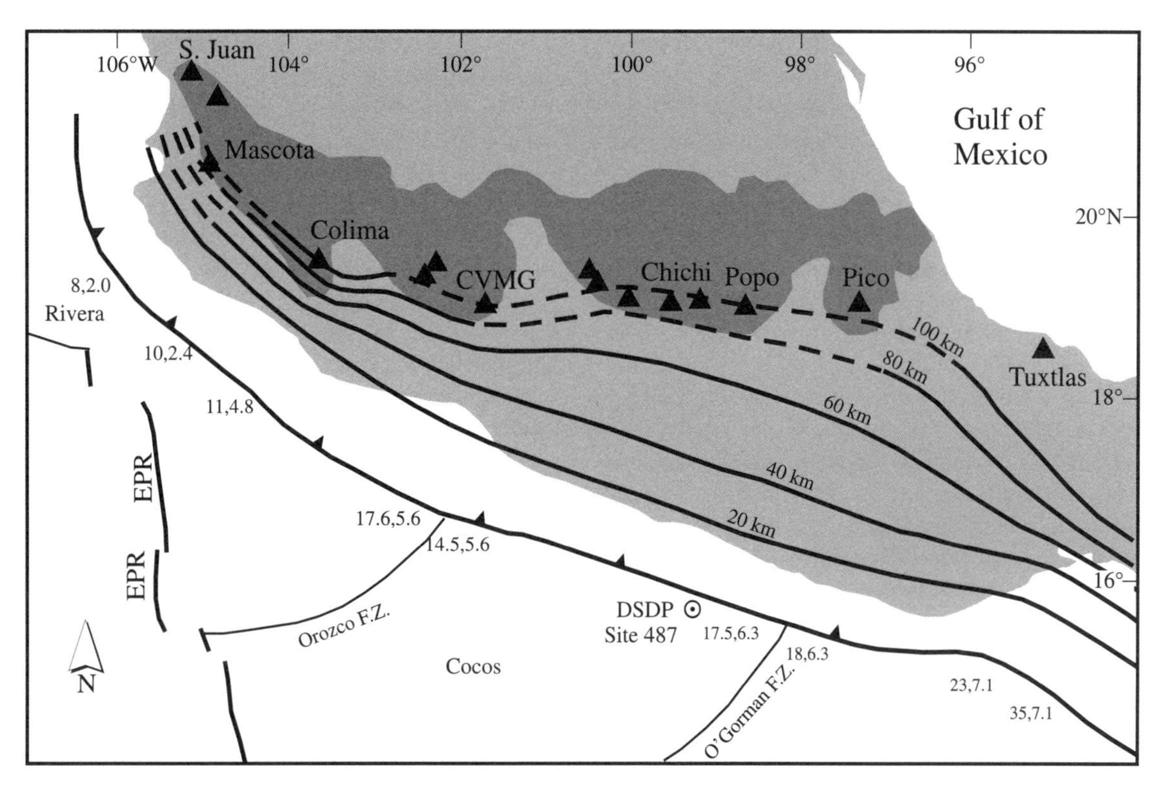

Figure 4. Generalized tectonic map of the Mexican subduction system (Pardo and Suárez, 1995) and the Trans-Mexican Volcanic Belt (gray field). Numbers separated by a comma indicate the age of the oceanic crust (in Ma) and the relative convergence velocity (in cm/a), respectively. Contour lines represent the depth of the top of the subducting plate (dashed when inferred). Site 487 of the Deep Sea Drilling Project Leg 66 is also indicated (DSDP Site 487). Main volcanic fields and edifices are included for reference: Los Tuxtlas volcanic field (Tuxtlas), Pico de Orizaba volcano (Pico), Popocatépetl volcano (Popo), Sierra Chichinautzin volcanic field (Chichi), Michoacán-Guanajuato volcanic field (CVMG), Colima volcanic complex (Colima), Mascota volcanic field (Mascota), San Juan volcano (S. Juan). EPR—East Pacific Rise.

The tectonic history of the oceanic plates in the Pacific realm was explored in a number of landmark publications analyzing the magnetic anomalies left on the present Pacific plate (Atwater, 1970; Menard, 1978; Mammerickx and Klitgord, 1982; Londsdale, 1991). These studies showed that the current tectonic configuration of the Mexican subduction zone is the result of successive fragmentation events that affected the ancient Farallon plate, as different segments of the East Pacific rise arrived in close proximity to the paleo-trench off of western North America. This process of fragmentation resulted in reorganization of the oceanic plate boundaries, and imposed important changes in the orientations and velocities of the convergence vector along the western margin of North America (Engebretson et al., 1985).

At the present time, the western sector of the Trans-Mexican Volcanic Belt is underlain by the Rivera plate, whereas the central and eastern sectors are underlain by the Cocos plate (Fig. 4). Although some minor discrepancies exist concerning the age and convergence velocities of these plates, the general consensus is that the Rivera plate is relatively younger and has a slower convergence rate than the Cocos plate (Nixon, 1982; Pardo and Suárez, 1993; DeMets et al., 1994; Kostoglodov and Bandy, 1995; Pardo and Suárez, 1995). The Rivera plate is between ca. 9 and ca. 13 Ma old at the trench off of Puerto Vallarta and Manzanillo, respectively, and converges with the North American plate at a rate that varies between 1.7 and 2.2 cm/yr (DeMets et al., 1994) or between 4 and 4.9 cm/yr (Kostoglodov and Bandy, 1995), depending on the model used. The age of the Cocos plate ranges between 12.7 and 16 Ma along the trench, and is older to the east. The convergence velocity also increases toward the east from ~4.7 to ~6.7 cm/yr (Pardo and Suárez, 1995).

Seismic studies of the subduction zone show that the Rivera plate dips at an angle of ~50° beneath the continent, and that earthquakes extend to a depth of ~120 km (Pardo and Suárez, 1993, 1995). In contrast, the subduction angle of the Cocos plate is more variable and the hypocenters do not exceed ~80 km depth (Fig. 4). In fact, the dip angle of the Cocos plate progressively decreases from its boundary with the Rivera plate to approximately 101°W long, after which it becomes almost horizontal in front of the western half of the eastern sector of the Trans-Mexican Volcanic Belt. East of the Pico de Orizaba volcano, the dip angle of the Cocos plate suddenly increases to an angle of 45–50° beneath the Tehuantepec isthmus, the Chiapas massif, and the Central American arc (Pardo and Suárez, 1995). This peculiar geometry is consistent with the obliquity of the arc, which is located farther away from the trench where the plate has the shallowest dip angle.

The evolution of the Mexican subduction system and the origin of its peculiar geometry are still poorly known. Global plate reconstructions indicate that the Cocos plate came into existence at ca. 23 Ma when the Farallon plate fragmented into two plates, which also created the Nazca plate to the south (Atwater and Stock, 1998; Londsdale, 2005). The Rivera plate branched off the Cocos plate at ca. 10 Ma, the oldest age for which a deformation can be observed to the north of the abandoned Mathematician Ridge (DeMets and Traylen, 2000).

It has also been proposed that the present Middle America Trench is the result of the eastern translation of the Chortís block, which was probably located at the longitude of Zihuatanejo in pre-Eocene times (Ratschbacher et al., 1991; Herrmann et al., 1994; Schaaf et al., 1995; Morán-Zenteno et al., 1996, 1999). According to these reconstructions, the western edge of the Chortís block was at the longitude of eastern Oaxaca at the end of the middle Miocene, implying that the trench in front of the Trans-Mexican Volcanic Belt had a geometry similar to the present one by this time (Morán-Zenteno et al., 1999). The presence of the middle Miocene arc in a location close to the Quaternary volcanic front of the central-eastern Trans-Mexican Volcanic Belt (Plate 1, Fig. 2A), and its gradual migration to the north, also suggests that the subhorizontal geometry of the Cocos plate dates back to this time (Gómez-Tuena and Carrasco-Núñez, 2000; Gómez-Tuena et al., 2003). In the western sector of the Trans-Mexican Volcanic Belt, the notable trenchward migration of the volcanic front starting at ca. 8.5 Ma led Ferrari et al. (2001) to propose that the Rivera plate began to rollback since the end of Miocene, a hypothesis that seems consistent with the sudden decrease in the convergence velocity relative to North America since ca. 9 Ma (DeMets and Traylen, 2000).

Thermal Structure of the Mexican Subduction Zone

The thermal structure of the Mexican subduction zone has been studied only recently, and even though the numerical models published to date are fairly sophisticated, it is still necessary to incorporate more complex features to properly describe the three-dimensional structure of the thermal state and to arrive at a satisfactory match with the geologic and petrologic record.

The limited measurements of heat flow in southern and central México are generally consistent with the typical structure of a subduction zone (Smith et al., 1979; Polak et al., 1985; Prol-Ledesma and Juárez, 1985; Ziagos et al., 1985). The lowest values (13–22 mWm^{-2}) are found in the forearc region, and increase significantly toward the magmatic arc (~100 mWm^{-2}). However, the quality and quantity of heat-flow measurement are still insufficient. Direct measurements are few and relatively shallow (<200 m); and the large variations in topography, geology, and hydrology add local complexities that are not easy to quantify and correct. For these reasons the main limitation for the present generation of thermal models of the Mexican subduction zone is the lack of sufficiently robust heat-flow measurements to independently constrain the numerical modeling results (Currie et al., 2002).

The main physical parameters considered in the thermal models of the Mexican subduction zone are similar to those used in other magmatic arcs around the world: age, geometry, and convergence rate of the subducting plate (Currie et al., 2002; Manea et al., 2004, 2005). However, other parameters such as the thermal conductivity of the oceanic and continental geologic units, the heat generated by radioactive decay, the shear heating at the plates' interface, and the viscosity of the mantle lithosphere are all important parameters that are difficult to estimate.

The thermal models of Currie et al. (2002), which were aimed at understanding the depth and extension of the seismogenic zone in the subduction zone, indicated that the oceanic plate is anomalously cold, despite its young age. This might be explained by considering a vigorous hydrothermal circulation that affects the oceanic plate (Prol-Ledesma et al., 1989), and the lack of a thick sedimentary cover insulating the basaltic crust. In these models, the seismogenic zone is restricted to temperatures of ~350 °C or less which, in the Mexican case, corresponds to <40 km depth. These models also predict a remarkably low temperature for the mantle wedge beneath the volcanic arc (<1000 °C), which is somehow inconsistent with the available heat-flow measurements and with the mere existence of volcanism.

A new generation of more refined thermal models became available recently. These new models take into account a wider zone of coupling between the two plates (~200 km) (Kostoglodov et al., 2003; Manea et al., 2004) and a temperature-dependent mantle viscosity (Manea et al., 2004). The models predict that the seismogenic zone is located between the isotherms of 150 °C and 450 °C. Additionally, plotting the results of their thermal model in a phase diagram for oceanic basalts and mantle peridotites, Manea et al. (2004) suggest that the eclogite facies assemblages would start to form at >450 °C and 1.3 GPa, conditions that are met precisely where the oceanic plate decouples from the continental lithosphere. In these models, the deepest seismic events occur at ~80 km, which coincides with the stability limit of hydrous minerals. On the other hand, the inclusion of a temperature-dependent mantle viscosity that makes use of experimentally determined nominal viscosity for olivine at upper mantle conditions (Hirth and Kohlstedt, 2003) allowed Manea et al. (2005) to predict partial melting of the subducted oceanic crust at depths as shallow as 50–60 km, as well as mantle wedge temperatures that are sufficiently high to promote its melting under hydrous conditions (>1200 °C). Yet these temperatures are still insufficient to generate melts from an anhydrous peridotite mantle wedge; thus the models are unable to explain the presence of intraplate lavas at the current volcanic front.

Composition of the Subducting Oceanic Plates

Even though the thickness and compositional variations of the subducted oceanic crust and its sedimentary cover are still poorly known, the studies carried out offshore of the coast of southern México, in the frame of the Deep Sea Drilling Project

(DSDP), as well as recent interpretations of gravimetric profiles across the trench (Manea et al., 2003), allow at least some tangible approximations to these parameters.

The study of Manea et al. (2003) allows us to interpret that the thickness of the sedimentary column over the Rivera plate should not exceed ~20 m, and that it gradually increases eastward along the trench. On the other hand, the DSDP carried out several seismic profiles and drillings along the Middle America Trench at a site known as Leg 66 (Moore et al., 1982). A drill hole at site 487, located ~11 km offshore of Guerrero State (lat 15°51.210′ N and long 99°10.518′ W), was the only one that penetrated the entire sedimentary column, and was used to recover samples of the underlying igneous oceanic crust (Figs. 4 and 5). The sedimentary column at this site is composed of ~100 m of Quaternary hemipelagic sediments which overlay ~70 m of late Miocene to Pliocene pelagic sediments. The oceanic crust at this site should be ca. 13 Ma according to the magnetic anomaly patterns.

The lithologic column of site 487 (Fig. 5) has been described and studied in great detail by a number of authors (Moore et al., 1982; Plank and Langmuir, 1998; Verma, 2000a; LaGatta, 2003). The hemipelagic sediments have a terrigenous character that indicates a provenance from the continental margin. This layer is mainly composed of gray mud with quartz, feldspar, and mica crystals—typical mineralogies observed in a belt of coastal plutons of southern México (Schaaf et al., 1995). Pelagic sediments consist essentially of yellowish, reddish, and black clays, with an origin associated with the hydrothermal activity of the mid-ocean ridges. The fragments of the oceanic crust recovered at this site consist of basaltic lavas with plagioclase and olivine crystals and, to a lesser extent, aphyric basalts (Verma, 2000a).

Although the compositions of sediments and igneous oceanic crust may vary significantly along the trench, rocks recovered at site 487 represent the only direct approximation of the materials that are presumably being subducted along this sector of the Middle America Trench. For this reason, a number of geochemical studies has been devoted to the samples collected at site 487 (Plank and Langmuir, 1998; Verma, 2000a; LaGatta, 2003). Figure 5 shows the trace-element and rare earth–element

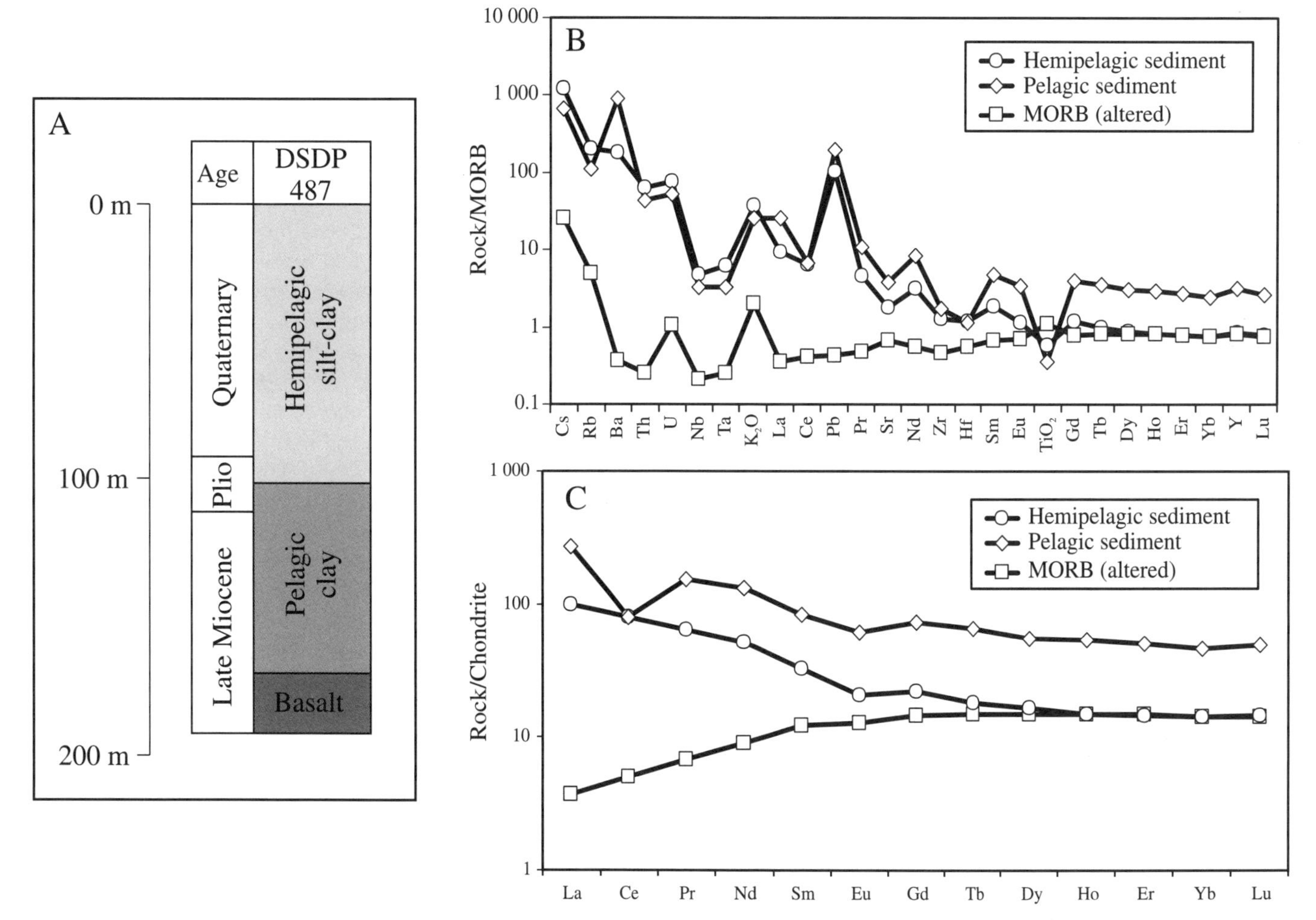

Figure 5. (A) Lithology and stratigraphy of the Cocos plate sampled at DSDP site 487. (B) Trace-element and (C) rare earth–element patterns of the different rock units (Legget, 1982; Plank and Langmuir, 1998; Verma, 2000a; LaGatta, 2003) were normalized to the N-MORB (Sun and McDonough, 1989) and CI-Chondrites (McDonough and Sun, 1995), respectively. MORB—mid-oceanic-ridge basalts.

(REE) patterns of the different sedimentary components and the altered basaltic crust (LaGatta, 2003). The pelagic and hemipelagic layers are geochemically distinct, the former being overall richer in trace elements. For instance, pelagic sediments have about five times more Ba, nearly double the light rare earth elements (LREE), whereas the heavy rare earth elements (HREE) are almost three times higher than in the hemipelagic sediments. The most conspicuous difference, however, is the *Ce anomaly,* which is very pronounced in the pelagic component. This anomaly is a common feature of marine sediments and results from the relative solubility of tetravalent Ce in seawater (Rollinson, 1993; Plank and Langmuir, 1998). Intriguingly, both sedimentary layers present a limited variation in the Sr and Nd isotopic composition (~0.7085 and ~0.5125, on average), but differ strongly in their Pb isotopes. Terrigenous sediments display upper-crustal-like radiogenic values on the $^{206}Pb/^{204}Pb$ ratio (~18.8, on average), whereas the pelagic layer has a MORB-like composition (~18.5, on average) (Verma, 2000a; LaGatta, 2003).

The Continental Plate

Crustal Structure

Information on the crustal structure along the Trans-Mexican Volcanic Belt has been provided essentially by gravimetric (Molina-Garza and Urrutia-Fucugauchi, 1993; De la Fuente et al., 1994; Urrutia-Fucugauchi and Flores-Ruiz, 1996; Flores-Ruiz, 1997; García-Perez and Urrutia-Fucugauchi, 1997; Campos-Enríquez and Sánchez-Zamora, 2000) and seismic studies (Urrutia-Fucugauchi, 1986; Valdés et al., 1986; Nava et al., 1988; Geolimex-Group, 1994; Campillo et al., 1996). Some additional information comes from aeromagnetic studies (Campos-Enríquez et al., 1990) and magneto-telluric profiles (Jording et al., 2000), and even though their coverage is still very broad, they are generally in agreement with seismic and gravimetric data.

The pattern of gravimetric anomalies along the arc defines a relatively simple structure consistent with a gradual increase in crustal thickness from the coasts to the continental interior (Fig. 6A). However, Figure 6A also distinguishes between domains by contrasting crustal thickness. The eastern sector of the arc has a relatively thicker continental crust, extending from the volcanic front up to the northern part of the State of Puebla. The thickest crust is found around the Toluca and Mexico City valleys (~47 km). A relatively sharp boundary in crustal thickness is observed in coincidence with the Taxco–San Miguel de Allende fault system (Alaniz-Álvarez et al., 2002b), which bounds the central sector of the Trans-Mexican Volcanic Belt and is characterized by a relatively thinner crust (< 40 km). The crust is thinner toward the Pacific and the Gulf of Mexico coastal areas (15–20 km). The crustal thickness structure estimated for the Trans-Mexican Volcanic Belt is consistent with the data obtained from the seismic profiles in southern México, where thickness increases inland from the Pacific Coast to reach ~47 km beneath the Oaxacan complex (Urrutia-Fucugauchi, 1986; Valdés et al., 1986).

The present topography of the Trans-Mexican Volcanic Belt shows a general correlation with gravimetric data and crustal thickness (Fig. 6B). The average elevation along the volcanic arc is ~2200 m.a.s.l., partly because it includes the country's highest peaks. In a general way, elevations tend to increase from the coasts to the continental interior, where the maximum elevations (>3,500 m) correspond to the large active stratovolcanoes of the eastern sector. A general correlation can also be observed between the concentration of the stratovolcanoes and the regions of higher crustal thickness. Elevations also tend to decrease gradually toward the north of the volcanic front.

Basement Geology

The geologic nature, age, and composition of the basement rocks beneath the Trans-Mexican Volcanic Belt are poorly known because they are covered by an extensive post-Mesozoic volcanic and sedimentary cover. For this reason, the geographic extension of the crystalline terranes emplaced in southern México and their correlations with the limited outcrops and the xenoliths that have been collected to the north of the arc have been the subject of debate (Keppie and Ortega-Gutiérrez, 1998; Ruiz et al., 1999).

The pioneer work of Campa and Coney (1983) proposed that most of the Mexican territory is constituted by an assemblage of crustal blocks with different geologic histories that were accreted to the North American plate during distinct tectonic episodes. Later Sedlock et al. (1993) and Ortega-Gutiérrez et al. (1994) proposed a revision of the tectonostratigraphic terranes schema of Campa and Coney (1983). The result was a rather complex subdivision that has been progressively simplified as new ages, structural studies, and petrologic data have become available for the crystalline terranes, xenoliths, and rocks recovered by deep bore holes. Taking into account the tectonic reconstructions and correlations proposed by Sedlock et al. (1993) and Ortega-Gutiérrez et al. (1994), and considering the recognition of the Oaxaquia Grenvillian microcontinent (Ruiz et al., 1988; Keppie and Ortega-Gutiérrez, 1995; Ortega-Gutiérrez et al., 1995), the Trans-Mexican Volcanic Belt might be emplaced on top of at least three distinct tectonostratigraphic terranes: Guerrero, Mixteco, and Oaxaquia (Fig. 7). Although these terranes all show a complex geologic evolution—and various aspects about their extension, boundaries, and composition are not completely understood—we will now summarize their main geologic features because assimilation of these crustal units may play a role in the petrogenesis of the Trans-Mexican Volcanic Belt.

The Guerrero terrane. Covering an area of ~700,000 km^2, the Guerrero terrane is the largest of the North American Cordillera. The terrane was originally defined by Campa and Coney (1983), and later modified by Sedlock et al. (1993), and has been the subject of various detailed geochemical and petrologic studies (Lapierre et al., 1992; Centeno-García et al., 1993; Freydier et al., 1996, 1997; Elías-Herrera and Orterga-Gutiérrez, 1997, 1998; Elías-Herrera et al., 1998). In a very general way, the Guerrero terrane is composed of volcanic and volcaniclastic rocks with both oceanic and continental affinity, and by a significant proportion of

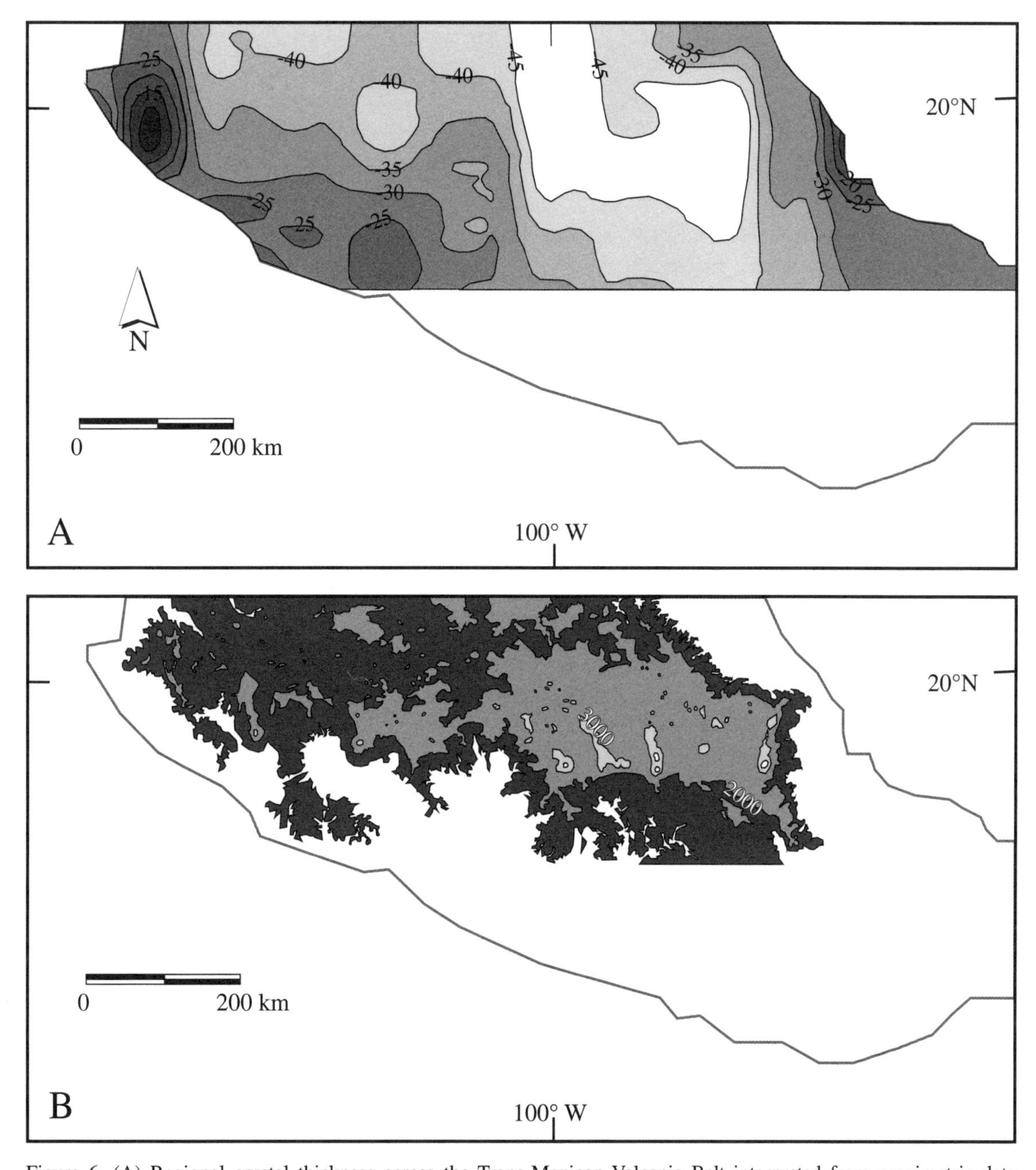

Figure 6. (A) Regional crustal thickness across the Trans-Mexican Volcanic Belt interpreted from gravimetric data (Molina-Garza and Urrutia-Fucugauchi, 1993; De la Fuente et al., 1994; Urrutia-Fucugauchi and Flores-Ruiz, 1996; Flores-Ruiz, 1997). (B) Hypsometric map of the Trans-Mexican Volcanic Belt. Contour lines are at 1000 m. Note the close match between crustal thickness and topography.

marine sedimentary rocks. Based on geochemical data, Centeno-García et al. (1993) suggested that the structural basement of the Guerrero terrane reflects a deep marine environment, although it is relatively close to the continent upon which an intra-oceanic island arc was constructed. The terrane was subsequently accreted to the North American plate during the Late Cretaceous.

Although most of the Guerrero terrane has been traditionally considered to consist of a series of oceanic arc sequences, other stratigraphic, geochronologic, and structural studies have questioned its alleged alloctonous nature. Some have proposed that the Guerrero terrane is a volcanic and sedimentary assemblage deposited upon a thinned North American crust that was later deformed by the Laramide orogeny (Cabral-Cano et al., 2000a, 200b). Similarly, recent geochemical data obtained on xenoliths found in Oligocene volcanic rocks suggest the presence of silicic continental rocks beneath the eastern part of the

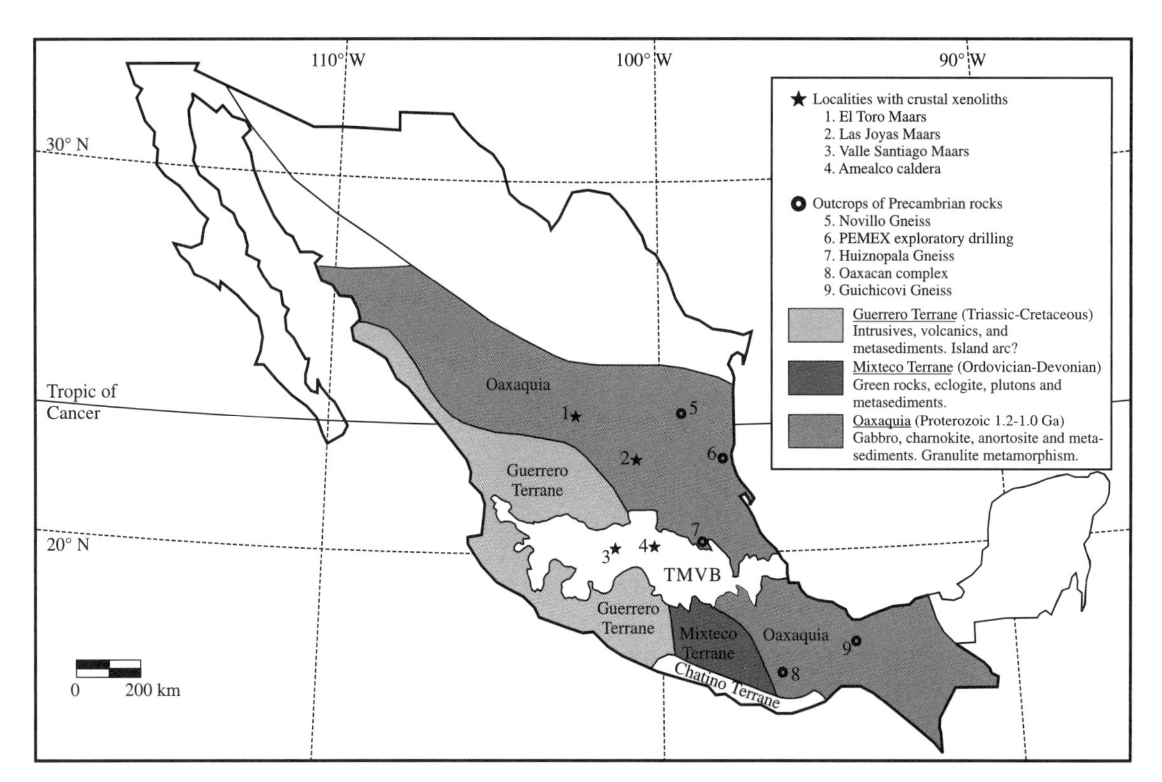

Figure 7. Simplified map of Mexican tectonostratigraphic terranes inferred to constitute the crustal basement of the Trans-Mexican Volcanic Belt (Campa and Coney, 1983; Ruiz et al., 1988; Sedlock et al., 1993; Ortega-Gutiérrez et al., 1994; Ortega-Gutiérrez et al., 1995).

Guerrero terrane (Elías-Herrera and Orterga-Gutiérrez, 1997, 1998; Elías-Herrera et al., 1998). In the same way, Nd model ages of the Jalisco block batholith (western Guerrero terrane) (Schaaf et al., 1995), as well as from a granulitic xenolith collected in the pyroclastic deposits of the Amealco caldera (central Guerrero terrane) (Aguirre-Díaz et al., 2002), suggest the presence of an older basement. This evidence indicates that the Guerrero terrane is probably emplaced above a continental crystalline basement that dates back to the Precambrian.

The Mixteco terrane. The basement of the Mixteco terrane is conformed by *Acatlán complex* (Yañez et al., 1991; Sedlock et al., 1993; Ortega-Gutiérrez et al., 1994). This complex is essentially made of metasedimentary and metaigneous rocks, with both oceanic and continental affinity, that were metamorphosed under amphibolite, eclogite, and greenschists facies and later intruded by granites and affected by migmatization.

The stratigraphy of the Acatlán complex includes the Acateco and Petlalcingo subgroups, the Tecomate Formation, the Totoltepec stock, and the San Miguel pluton. Subdivision of these units was established based on the different deformation styles and metamorphic facies. The Petlalcingo subgroup is made of pelitic schists, quartzite, metagraywacke, and gabbroic dikes that belong to the Chazumba Formation; and of amphibolite, quartzite, chert, calcareous schist, and greenrocks from the Cosoltepec Formation. The protolith of this subgroup has been interpreted as marine sedimentary rocks interlayered with mafic marine volcanics, which have been metamorphosed at high temperature and moderate pressure. The Acateco subgroup is composed of mylonitic green rocks, amphibolites, metagabbros, serpentinites, eclogites, quartzites, and metapelites of the Xayacatlán Formation; and granitic plutonic rocks, aplites, and pegmatites of Ordovician-Silurian age collectively known as the Esperanza granitoids (440–428 Ma). Eclogitic rocks reach a metamorphic peak of 15 Kb and 500–550 °C (Ortega-Gutiérrez, 1981). This sequence is discordantly covered by conglomerate, arkose, lutite, and limestone from the Devonian Tecomate Formation. The latter was subsequently deformed and metamorphosed before being covered by the Mississippian-Permian marine sediments from the Patlanoaya Formation, and the Pennsylvanian-Permian continental sediments of the Matzitzi Formation. The Acatlán complex has been intruded by the Totoltepec stock in late Pennsylvanian times (287 ± 2 Ma), and was later affected by a tectonothermal event during the Early to Middle Jurassic (205–170 Ma). The San Miguel intrusive and the Magdalena migmatites date back to this episode. The Paleozoic sequences are stratigraphically covered by: (1) volcanic and intrusive rocks of Triassic to Middle Jurassic age (Rosario Formation and San Miguel plutons), (2) marine sedimentary rocks (conglomerate, limestone, lutite, and sandstone) of Jurassic to Late Cretaceous age, and (3) Tertiary continental and volcanic sedimentary rocks (conglomerates, sandstone, lutite), silicic ignimbrites, volcaniclastic rocks, andesitic lavas, and lacustrine deposits (Morán-Zenteno et al., 1999).

The Mixteco terrane is bounded to the south by the Chatino terrane, to the west by the Guerrero terrane, and to the east by the Oaxaquia microcontinent along the Caltepec fault zone, a polydeformed tectonic boundary (Elías-Herrera et al., 2005). The volcanic cover of the Trans-Mexican Volcanic Belt masks the northern prolongation of this structure, so it is still unknown if it represents a zone of crustal weakness in central México.

Oaxaquia. The concept of the Oaxaquia microcontinent (Ortega-Gutiérrez et al., 1995) has drawn the attention of several workers in recent times. In a nutshell, Oaxaquia would represent a relatively large crustal block of Grenvillian age (ca. 1 Ga) with a geographic extension of ~1,000,000 km^2. The microcontinent is conformed by the Oaxaca, Juárez, Sierra Madre, and Maya terranes, as well as part of the Coahuila terrane of Campa and Coney (1983), or by their equivalence in the terminology of Sedlock et al. (1993): Zapoteco, Guachichil, Tepehuano, Maya, and part of the Cuahuiltecano terrane. However, the grouping of these terranes into a single Oaxaquia microcontinent is based on the similarity of the rocks forming the Middle Proterozoic (Grenvillian) basement, and does not necessarily include the rocks covering them.

The Oaxaquia microcontinent is defined by different outcrops located in the eastern part of México (Fig. 7): the Novillo Gneiss (Ciudad Victoria, Tamaulipas) (Ortega-Guitérrez, 1978); the Huiznopala Gneiss (Molango, Hidalgo) (Lawlor et al., 1999), the Oaxacan complex (Oaxaca) (Ortega-Gutiérrez, 1984); and the Guichicovi complex (La Mixtequita, Oaxaca) (Murrillo-Muñetón, 1994; Weber and Köhler, 1999). Additional locations come from the Nd model ages and Sm-Nd isochron ages found in lower crustal xenoliths carried by Cenozoic volcanic rocks at La Joya and La Olivina maars (San Luis Potosí) (Ruiz et al., 1988; Schaaf et al., 1994); and by Grenvillian age rocks found in a deep bore hole drilled by PEMEX in Tampico, Tamaulipas (Quezadas-Flores, 1961).

The Oaxacan complex is essentially composed of metapelites, quartzfeldspatic gneisses, calcsilicates, amphibolites, and marbles, commonly intruded by anorthosites, charnokites, and garnet-bearing mafic orthogneisses. The whole sequence is metamorphosed under granulite facies, and locally re-equilibrated to amphibolite facies conditions. Rocks from El Novillo, Huiznopala, and Guichicovi generally present a similar lithology to the Oaxacan complex, as well as similar metamorphic and deformation conditions. On the other hand, xenoliths found in Cenozoic volcanics from San Luis Potosí are dominantly mafic to intermediate in composition, and consist of granulite facies gabbros and tonalites.

IGNEOUS PETROGENESIS

In this section we will revise the chemical characteristics of volcanic rocks emplaced on the Trans-Mexican Volcanic Belt since the middle and late Miocene. We will make use of an extensive geochemical database compiled from the literature. For the sake of clarity we have divided the broad compositional variety of volcanic rocks into four main groups: Na-alkaline rocks, K-alkaline rocks, calc-alkaline rocks, and silicic magmatism. Although this division is largely arbitrary (because the chemical compositions of the different groups tend to overlap), we consider it to be useful because most petrological studies of the arc have traditionally focused on specific volcanic sequences, and also because it allows us to broadly explore the distinct petrogenetic processes that were involved in the sequences' formation.

Chemical Compositions of Volcanic Rocks

This work includes a new compilation of major- and trace-element contents, and isotopic compositions, for 2832 volcanic rocks collected on the Trans-Mexican Volcanic Belt and reported in the literature (http://satori.geociencias.unam.mx/Centenario/Gomez-Tuena.xls). The database includes the geographic coordinates for each sampling site, as well as the emplacement age determined by isotopic methods or estimated from stratigraphic field relations (Ferrari et al., 2005a).

Figures 8–13 show the relative abundances of major elements recalculated to 100% on a volatile-free basis, with the relative concentrations of FeO and Fe_2O_3 recalculated according to Middlemost (1989). It is possible that results obtained from this calculation schema do not coincide with those presented in the original publications, but a uniform normalization criterion is needed in order to make comparisons. We also emphasize that an exhaustive evaluation of the geochemical variations of the Trans-Mexican Volcanic Belt in time and space is beyond the scope of this work, and that we simply intend to illustrate the most relevant compositional characteristics of the arc. We also want to stress that even though samples included in this compilation are distributed along the whole magmatic arc, the reported data are not necessarily representative of the geochemical diversity of the entire Trans-Mexican Volcanic Belt. This is because some areas have received much attention in the literature, whereas data from other parts of the arc are still scarce.

Figure 8 shows the classic geochemical classification diagram based on total alkalis versus silica contents (Le Bas et al., 1986), as well as the line dividing alkaline from subalkaline rock fields (Irvine and Baragar, 1971). Subalkaline rocks are clearly dominant (83% of the samples) when compared to alkaline sequences. In the classification schema of Le Maitre (1989), most subalkaline rocks plot within the field of middle-K calc-alkaline (~86% of the samples) with the high-K calc-alkaline rocks less abundant (Fig. 9A). Alkaline rocks can be further subdivided into Na-rich and K-rich in the diagram proposed by Le Bas et al. (1986) (Fig. 9B). Na-alkaline rocks are relatively more abundant in this case (62% of the samples).

The geographic distribution of the different magmatic series is shown in Figure 10. These diagrams only include samples with MgO > 5% in order to filter out the effects of fractionation and/or contamination in the alkali contents. Calc-alkaline and alkaline rocks are broadly distributed along the Trans-Mexican Volcanic Belt, but alkaline rocks (sodic and potassic) tend to be more abundant in the western and eastern ends of the arc. Interestingly, K-alkaline rocks are preferentially distributed along the arc front

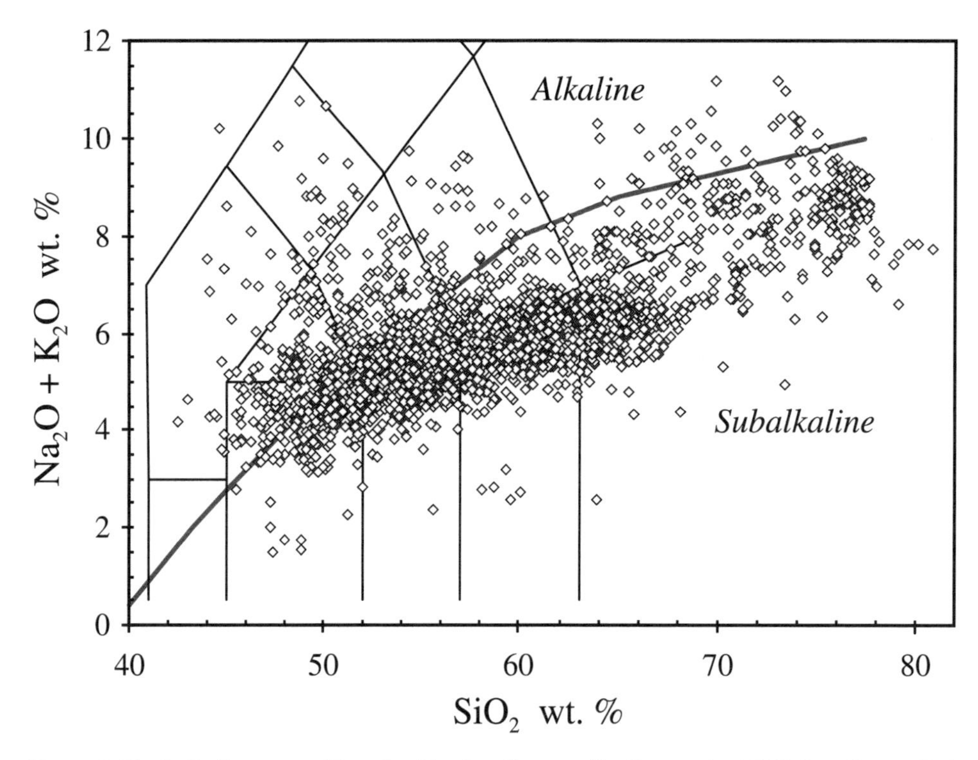

Figure 8. Total alkalis versus silica classification diagram (Le Bas et al., 1986) for all compiled analyses of the Trans-Mexican Volcanic Belt. Also shown is the line dividing subalkaline and alkaline fields (Irvine and Baragar, 1971).

of the western sector of the Trans-Mexican Volcanic Belt, and their relative abundance gradually decreases toward the east.

The major-element compositions of the three rock suites are also significantly different (Fig. 11). The alkaline rocks tend to have SiO_2 contents lower than 60%, whereas TiO_2 and P_2O_5 concentrations reach the highest values. The most obvious difference between K-alkaline and Na-alkaline rocks is the higher abundance of K_2O for the former and of Na_2O in the latter at similar SiO_2 contents. The less differentiated K-alkaline rocks also have the highest P_2O_5 contents. On the other hand, the highest SiO_2 contents are found in the subalkaline rocks, which also show less dispersion in major-element contents than the alkaline rocks.

The differences in trace-element contents between the three rock types are shown by the relative variations of Nb, Sr, and Ba (Fig. 12). Nb is markedly enriched in alkaline rocks, reaching slightly higher values in the Na-alkaline sequences. Most subalkaline rocks have low Nb contents (<20 ppm), although some samples have Nb values of up to 40 ppm. Very high Nb contents are also observed in rhyolitic rocks. K-alkaline rocks are characterized by very high Sr (up to 5109 ppm) and Ba (up to 4765 ppm), whereas Sr and Ba contents of Na-alkaline rocks are only slightly higher than those of subalkaline rocks.

Differences in the chemical compositions are especially marked between Na-alkaline and K-alkaline rocks, whereas data for subalkaline rocks tend to overlap with the other two series. This is exemplified in the Ba/Nb vs. TiO_2/K_2O diagram for samples with $MgO > 5\%$ (Fig. 13). Na-alkaline rocks have the lowest Ba/Nb and the highest TiO_2/K_2O ratios, with values that tend to overlap those of the field of ocean-island basalts (OIB). Yet not all Na-alkaline rocks plot within this field since some of them extend to higher Ba/Nb ratios. The opposite behavior is observed in the K-alkaline rocks, which are markedly enriched in the large ion lithophile elements (LILE) with respect to the high-field strength elements (HFSE) (i.e., low TiO_2/K_2O and high Ba/Nb ratios). Data for subalkaline rocks are more variable and overlap with the values observed on Na-alkaline and/or K-alkaline rocks, although they tend to concentrate at intermediate values between these rock types.

Sodium Alkaline Rocks and Intraplate Magmatism

Sodic-alkaline rocks with geochemical features that are similar to those of intraplate volcanic rocks (i.e., ocean islands and continental rifts) have been identified in several localities of the Trans-Mexican Volcanic Belt. Even though Na-alkaline rocks in the Trans-Mexican Volcanic Belt tend to have variable compositions, most of them do not exhibit geochemical signatures for significant contributions of the subducting plate or the continental crust. Yet rocks with such characteristics constitute a relatively small volume within the Trans-Mexican Volcanic Belt, and have been mostly emplaced in close space-time association with rocks showing appreciable subduction signatures.

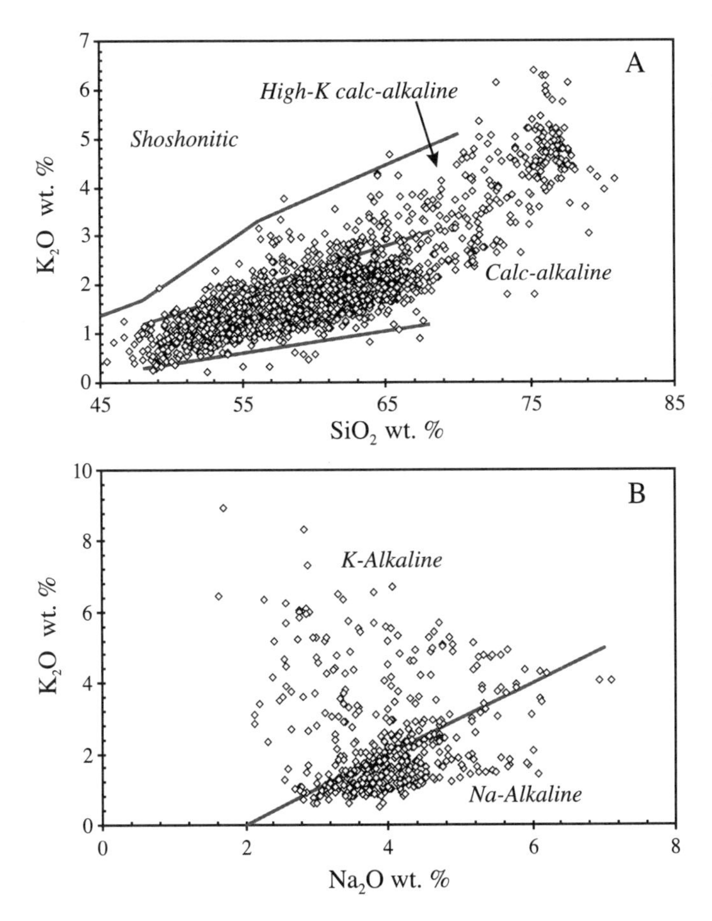

Figure 9. (A) Classification diagram for subalkaline rocks (Le Maitre, 1989). (B) Subdivision of the alkaline rocks into sodic and potassic series. Division line corresponds to Na_2O-2 = K_2O (Le Bas et al., 1986).

Intraplate-like rocks have been recognized in a great diversity of tectonic settings, and it is now clear that they are not necessarily restricted to regions away from active tectonic margins. Owing to their compositional similarity with ocean-island basalts (OIB), continental intraplate rocks have been considered to originate from partial melting of mantle domains that are relatively enriched in incompatible elements that are unlike those associated with the depleted mid-oceanic-ridge basalts (MORB). Nevertheless, the origin of this enrichment still remains unclear. Most workers agree that the enriched mantle domains that form the OIB are related to the recycling of ancient oceanic crusts (including their gabbroic and sedimentary portions), which have been incorporated into the convecting mantle by the global subduction process (Hofmann, 2003). However, other authors have suggested that the source of OIB-type magmas is located in a lithospheric mantle that has been metasomatized by asthenosphere-derived magmas that are enriched in volatiles and incompatible elements (Sun and McDonough, 1989; Halliday et al., 1995; Pilet et al., 2004). Delamination of a metasomatized continental lithosphere, and its incorporation into the asthenospheric mantle, has also been suggested for the origin of enriched mantle sources (McKenzie and O'Nions, 1995).

But despite these discussions concerning the origin of OIB mantle sources, the presence of intraplate-type rocks in the Trans-Mexican Volcanic Belt implies the existence of enriched mantle domains that have not been significantly modified by the subduction agents. It has long been argued that the contemporaneous eruption of intraplate-type magmas and magmas with subduction signatures does not agree with the classic models of magmatism at convergent margins, and many researchers have attempted to solve this apparent discrepancy. Yet it has been increasingly recognized that OIB-type rocks have erupted in several other volcanic arcs around the world (e.g., Hickey-Vargas et al., 1986; Carr et al., 2003; Strong and Wolff, 2003), and thus their presence is apparently much more common than previously thought.

Na-alkaline lavas constitute a relatively small volume in the Trans-Mexican Volcanic Belt, but their presence is significant in the western and eastern edges of the arc, and rocks of this kind are also found scattered in the monogenetic fields of Michoacán-Guanajuato (Hasenaka and Carmichael, 1987), Chichinautzin (Wallace and Carmichael, 1999; Verma, 2000a), Valle de Bravo–Zitácuaro (Blatter et al., 2001), and in the Los Humeros (Verma, 2000b) and Acoculco (Verma, 2001a) calderas.

The oldest Na-alkaline rocks of the Trans-Mexican Volcanic Belt were emplaced in the volcanic fields of Chiconquiaco–Palma Sola, Álamo, Sierra de Tantima, and Tlanchinol (Fig. 14), located toward the northern and at eastern limits of the arc. Almost all magmas emplaced between late Miocene (ca. 7.5 Ma) and early Pliocene (ca. 3 Ma) in these areas are Na-alkaline. They were mostly emplaced through fissures and formed extensive plateaus of lavas with a mafic character (Cantagrel and Robin, 1979; Negendank et al., 1985; López-Infanzón, 1991; Gómez-Tuena et al., 2003; Ferrari et al., 2005b). Na-alkaline rocks were also emplaced in this sector during the Pleistocene–Holocene, but their volumes are relatively small when compared to coeval calc-alkaline rocks (Negendank et al., 1985; López-Infanzón, 1991; Siebert and Carrasco-Núñez, 2002; Gómez-Tuena et al., 2003; Ferrari et al., 2005b). Rocks of similar age and composition have been also reported in the Los Tuxtlas volcanic field (Nelson and Gonzalez-Caver, 1992; Nelson et al., 1995). However, and in spite of the claims that the magmatism at Los Tuxtlas was also influenced by the subducted Cocos plate, the Los Tuxtlas volcanic field will not be included in this review because it lies outside the traditional boundaries of the Trans-Mexican Volcanic Belt (Plate 1, Fig. 1).

Early works by Demant and Robin (1975) and Robin (1976a) considered the alkaline rocks from the eastern sector of the Trans-Mexican Volcanic Belt as part of an independent magmatic province, related to extensional faulting parallel to the Gulf of Mexico coast. This volcanic province was named the "Eastern Alkaline Province" and extended from Sierra de San Carlos, Tamaulipas, to Los Tuxtlas, Veracruz. More recent studies have reported geochemical and isotopic evidence for significant contributions of subducted material to the alkaline magmas emplaced in Chiconquiaco–Palma Sola (Negendank et al., 1985; López-Infanzón, 1991; Gómez-Tuena et al., 2003). Also, Ferrari et al. (2005b) observed that alkaline volcanism initiated in the whole

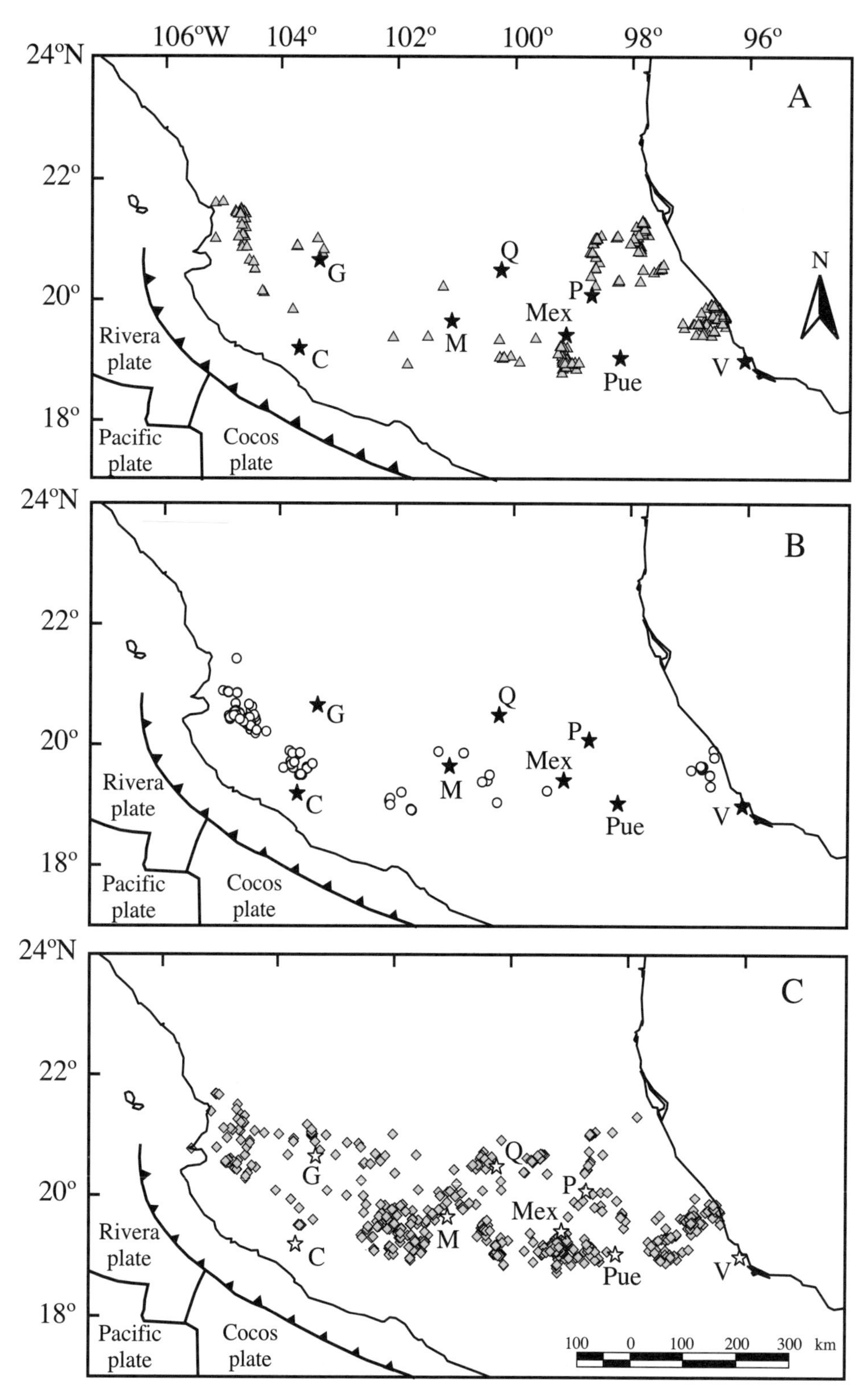

Figure 10. Geographic distribution of (A) Na-alkaline, (B) K-alkaline, and (C) subalkaline (calc-alkaline) rocks with more than 5 wt% MgO. Some cities are shown for reference, C—Colima; G—Guadalajara; M—Morelia; Q—Querétaro; Mex—Mexico City; P—Pachuca; Pue—Puebla; V—Veracruz.

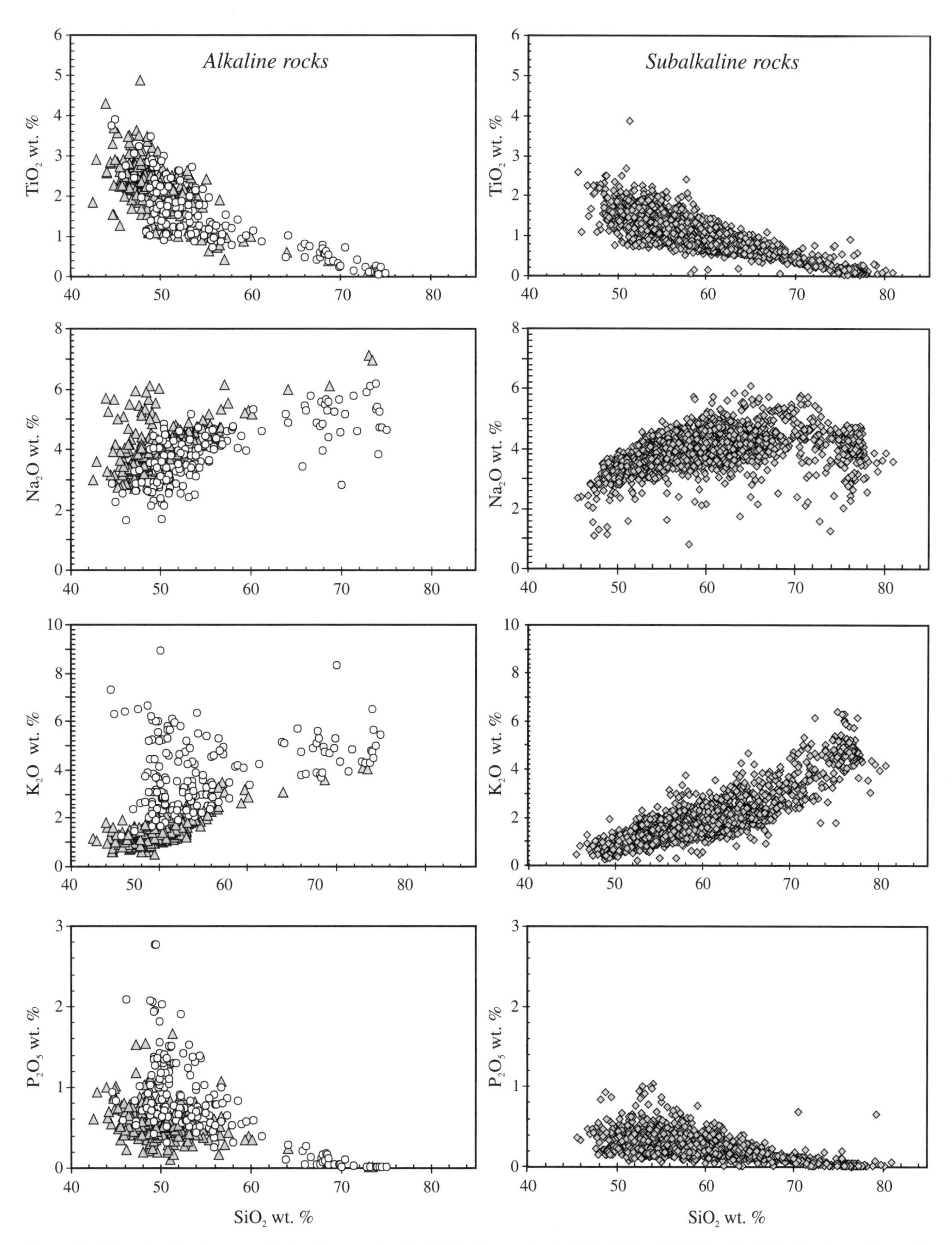

Figure 11. Selected major-element variation diagrams for Na-alkaline (gray triangles), K-alkaline (circles), and subalkaline (gray diamonds) rocks of the Trans-Mexican Volcanic Belt.

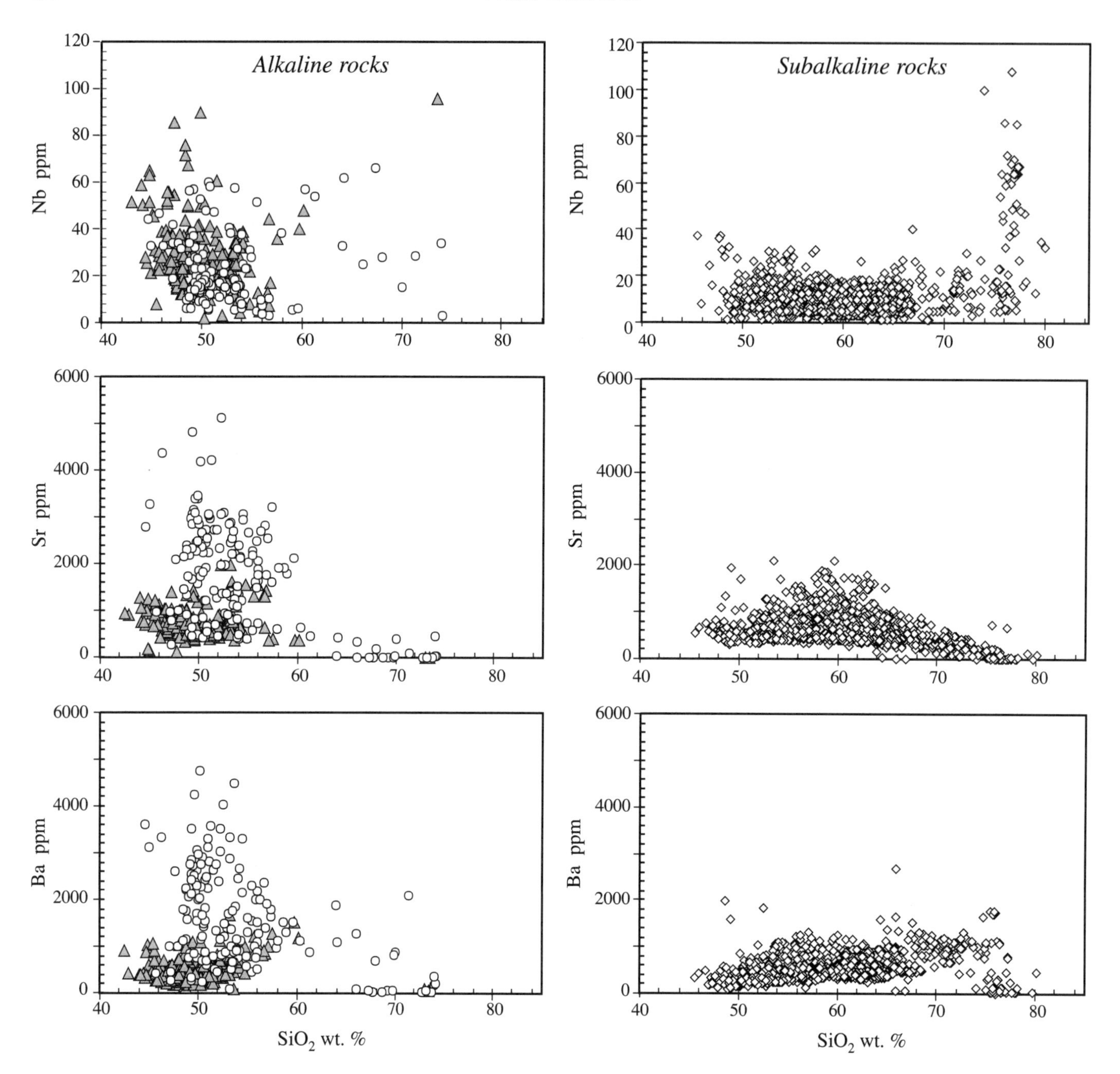

Figure 12. Selected trace-element variation diagrams against SiO_2 contents for Na-alkaline (gray triangles), K-alkaline (circles), and subalkaline (gray diamonds) rocks of the Trans-Mexican Volcanic Belt.

area almost simultaneously, which suggests a common regional mechanism for their genesis. Yet the authors discarded the rifting hypothesis because of the lack of late Tertiary normal faults parallel to the coast of the Gulf of Mexico, and the nonexistence of significant extensional faulting capable of promoting decompression melting in the mantle wedge. They also established that dikes, cone alignments, and cone elongations are dominantly oriented NE-SW and NNW-SSE, coinciding with preexisting crustal structures that could have served as pathways for the ascent of melts, but that were not responsible for inducing partial melting in the wedge. Therefore, the overall geochemical, geochronological, and geological evidence suggests that these volcanic fields should be part of the Trans-Mexican Volcanic Belt and were to some extent influenced by the Pacific subduction regime (Negendank et al., 1985; López-Infanzón, 1991; Gómez-Tuena et al., 2003; Ferrari et al., 2005b; Orozco-Esquivel et al., 2007).

In a geological and geochronological study of the northernmost volcanic fields of the eastern sector, Ferrari et al. (2005b) reported

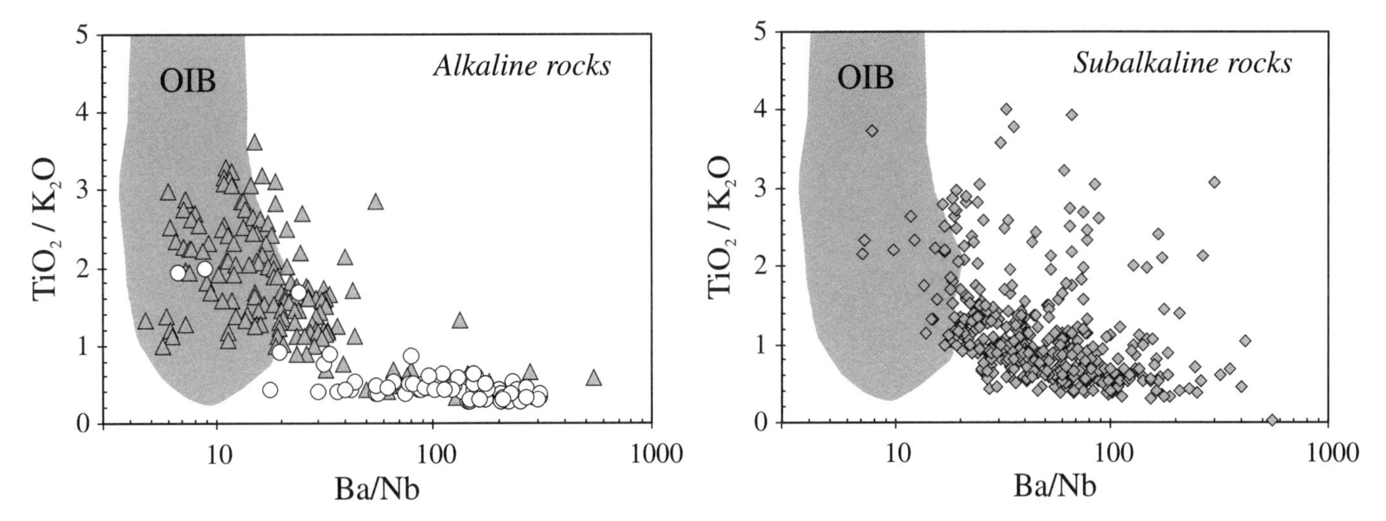

Figure 13. Ba/Nb versus TiO_2/K_2O variation diagram for rocks with more than 5 wt% MgO. Na-alkaline rocks (gray triangles), K-alkaline rocks (circles), subalkaline rocks (gray diamonds). The OIB field is the outline of 992 analyses from the GEOROC database (http://georoc.mpch-mainz.gwdg.de/georoc).

ages of 7.6–5.7 Ma for the Sierra de Tantima, the Álamo volcanic field, and the Tlanchinol lava flows. The lava-flow sequences in Sierra de Tantima and Tlanchinol vary in thickness from 700 m to 250 m, respectively. The elongation of Sierra de Tantima is considered to result from the emplacement along a NE-oriented fissure. The Álamo volcanic field is located to the south of Sierra de Tantima (Fig. 14), and is composed of at least 40 monogenetic cones for which only the volcanic necks or the feeding dikes are preserved. These structures are elongated and/or aligned to the NE-SW and to the NNW-SSE. Lavas in these three fields are classified as basanites, alkali basalts, hawaiites, and phonotephrites. Magmas from the monogenetic volcanoes of Álamo are more alkaline, with nepheline (*ne*) (2.6%–11.4%) in the CIPW norm, whereas those in Sierra de Tantima and Tlanchinol are *ne*-normative (1.4%–4.7%) and hypersthene (*hy*)-normative (2.9%–9.5%). The overall geochemical composition of these rocks is typical of intraplate magmas (Orozco-Esquivel, 1995; Ferrari et al., 2005b).

In the southern half of the eastern sector, the Chiconquiaco-Palma Sola area (Fig. 14) is characterized by a longer and more complex magmatic history during which calc-alkaline, transitional, Na-alkaline, and K-alkaline rocks were generated between the middle Miocene and the Quaternary. Successions of fissural lava flows of Na-alkaline composition form an 800-m-thick plateau along an area of ca. 1700 km^2. Lava emplacement was apparently controlled by a basement fault system oriented to the ENE, which facilitated the ascent of the primitive magmas (Ferrari et al., 2005b). This mechanism of emplacement, together with the presence of spinel-lherzolite xenoliths and clinopyroxene xenocrysts in some flows, indicates a rapid magma ascent from the mantle to the surface (Gómez-Tuena et al., 2003). Late Miocene to late Pliocene or Quaternary ages have been reported for the plateau (Cantagrel and Robin, 1979; López-Infanzón, 1991), but geochemical analyses of dated rocks indicate that the predominantly Na-alkaline event is restricted to an age interval between 6.9 and 3.2 Ma, the period during which most of the plateau was built. Small volumes of shoshonitic K-alkaline lavas were also emplaced in the late Pliocene (2.2–1.9 Ma) covering the plateau (Ferrari et al., 2005b).

Na-alkaline lavas in Chiconquiaco-Palma Sola have basanitic to mugearitic compositions, and are *ne* or *hy* normative (Gómez-Tuena et al., 2003; Ferrari et al., 2005b). Although some authors reported that all Na-alkaline lavas have subduction signatures (Negendank et al., 1985; López-Infanzón, 1991), later geochemical and isotopic studies performed by Gómez-Tuena et al. (2003) reported the interstratification of lavas enriched in Nb and Ta and with low LILE/HFSE ratios, similar to those observed in OIB, and lavas with moderate to high LILE/HFSE ratios with compositions that correlate with Pb, Nd, and Sr isotopic enrichments. These authors show that such variations are not related to crustal contamination processes, but reflect the inherited characteristics of an enriched mantle wedge that has been heterogeneously modified by the slab-derived chemical agents. According to Gómez-Tuena et al. (2003), the composition of rocks lacking subduction signatures (OIB type) are derived from melting of enriched mantle domains in the garnet stability field of peridotite. Rocks with subduction signatures display correlations in $^{206}Pb/^{204}Pb$ versus $^{207}Pb/^{204}Pb$ and $^{143}Nd/^{144}Nd$ versus Th/Nd diagrams, which are indicative of mixing processes between an enriched mantle and magmas derived from the partial fusion of subducted sediments. Nevertheless, contents of the most fluid-mobile elements (like Ba and Pb) in the rocks are lower than those expected from mixing with partial melts of the sediments sampled at DSDP site 487. These features indicate that the subducted sediments were significantly dehydrated prior to melting, an interpretation that is consistent with the experimental evidence (Johnson and Plank, 1999). In this model, the generation of magmas with such different geochemical features in close space-time association requires

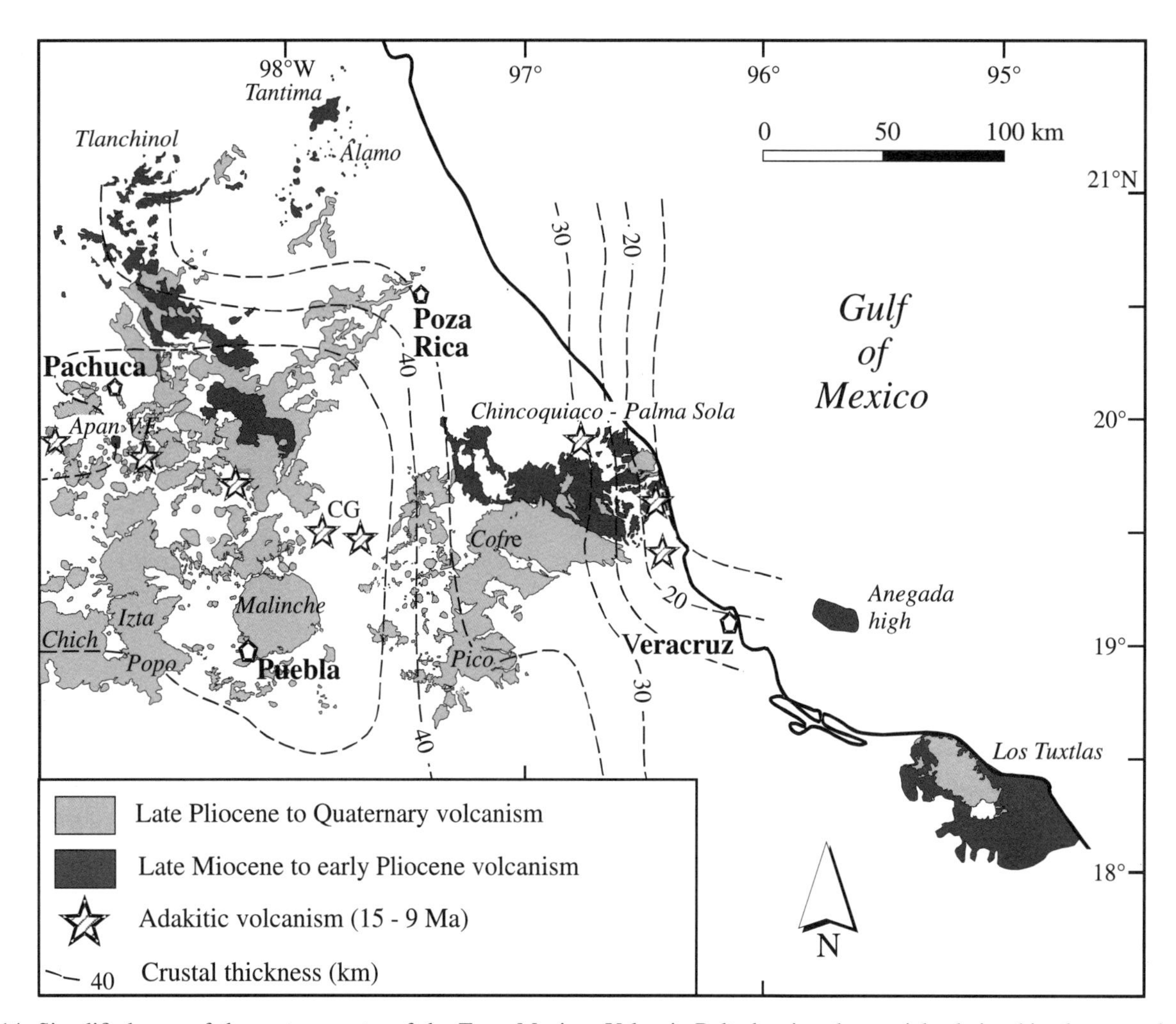

Figure 14. Simplified map of the eastern sector of the Trans-Mexican Volcanic Belt showing the spatial relationships between the mainly Na-alkaline late Miocene to early Pliocene volcanism, and the dominantly calc-alkaline late Pliocene to Quaternary volcanism. Some other volcanic fields are also shown: Apan—Apan volcanic field; CG—Cerro Grande volcanic field; Popo—Popocatépetl; Izta—Iztaccíhuatl; Chich—Sierra Chichinautzin; Pico—Pico de Orizaba; Cofre—Cofre de Perote. Figure modified from Ferrari et al. (2005b).

a complex tectonic scenario, in which (1) small portions of the mantle wedge, unmodified by subduction, generate OIB-type magmas by decompression melting, and (2) injection of subducted sediment melts into the wedge promotes melting of the mantle and generates magmas with typical arc features.

By the Quaternary, volcanism in the eastern sector is predominantly calc-alkaline, although scattered Na-alkaline volcanism is also present. Ferrari et al. (2005b) reported massive mafic lava flows that were emitted through fissures located at the front of the Sierra Madre Oriental to the west of Poza Rica, Veracruz (Fig. 14), and that traveled more than 90 km toward the coastal plains. These authors reported the first isotopic ages for those rocks (K-Ar in groundmass: 1.64 ± 0.06–1.31 ± 0.03 Ma). The lavas are *hy*- or quartz (*qz*)-normative alkaline basalts and hawaiites, with relatively depleted trace-element patterns and no clear subduction signatures (Orozco-Esquivel et al., 2007). Quaternary rocks in the Chiconquiaco–Palma Sola area are mainly calc-alkaline, although associated Na-alkaline rocks have also been reported (Gómez-Tuena et al., 2003). A remarkable example of this is the eruption in short succession of Na-alkaline (hawaiite, mugearite) and calc-alkaline basaltic lavas from the El Volcancillo paired vent at ca. 870 yr B.P. (Siebert and Carrasco-Núñez, 2002; Carrasco-Núñez et al., 2005). Both rock types show negligible contributions of subducted materials, and their compositional differences are apparently controlled by AFC processes as has been suggested for other Quaternary rocks emplaced in this area (Gómez-Tuena et al., 2003).

Na-alkaline volcanism in the western Trans-Mexican Volcanic Belt is represented by mildly alkaline rocks with geochemical features of intraplate magmas. Different from what is observed in the eastern sector, Na-alkaline volcanics in the western sector have a close relation with active extensional faulting (Demant, 1979; Luhr and Carmichael, 1985a; Ferrari et al., 2000b) (Fig. 15). The onset of intraplate volcanism also

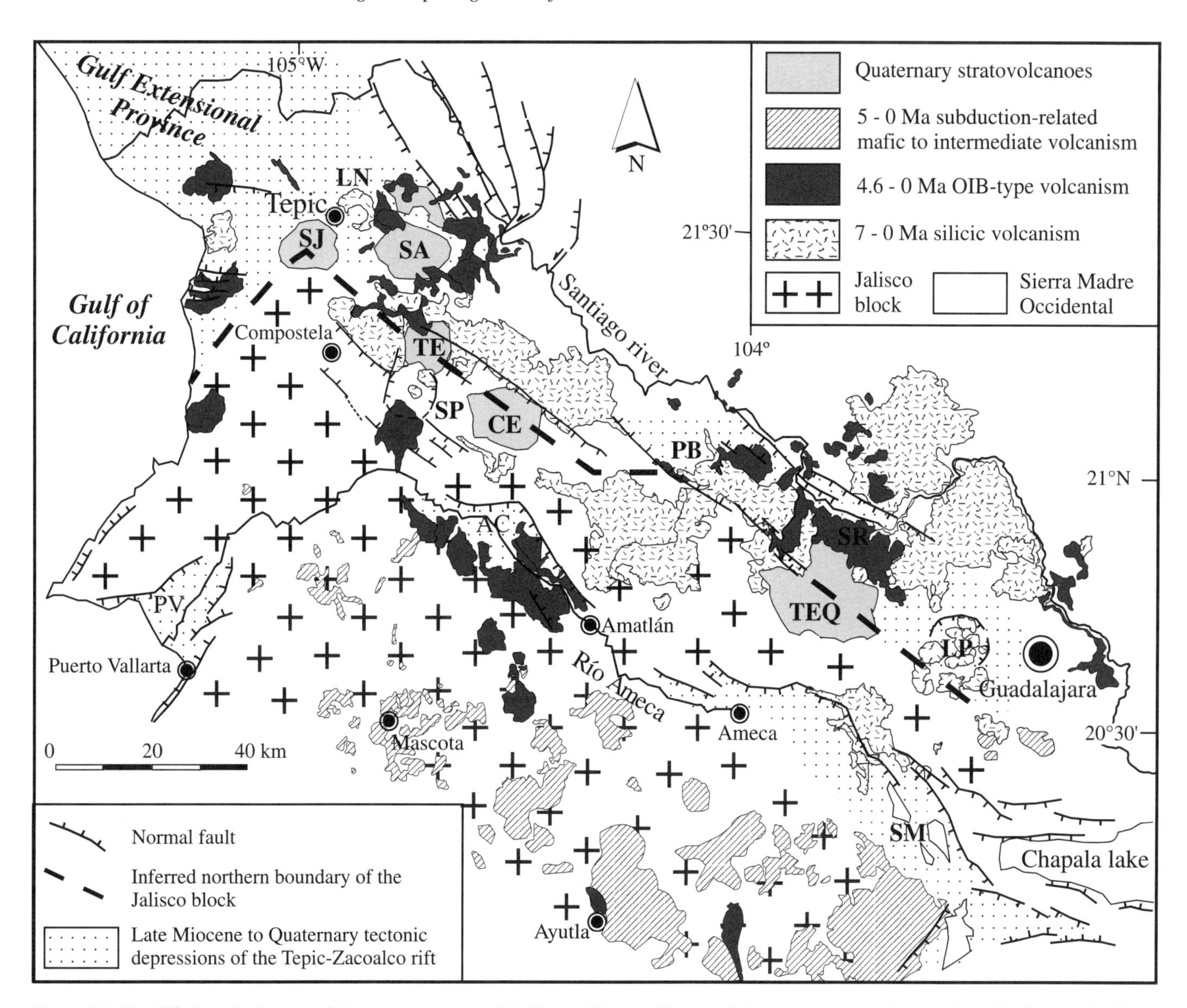

Figure 15. Simplified geologic map of the western sector of the Trans-Mexican Volcanic Belt showing the main extensional faults and the distribution of volcanism younger than 7 Ma. The main volcanic structures are: SJ—San Juan volcano; LN—Las Navajas volcano; SA—Sangangüey volcano; TE—Tepetiltic volcano; CE—Ceboruco volcano; TEQ—Tequila volcano; LP—La Primavera caldera. The main tectonic structures are PV—Puerto Vallarta graben; SP—San Pedro–Ceboruco graben; AC—Amatlán de Cañas; PB—Plan de Barrancas graben; SR—Santa Rosa graben; SM—San Marcos half-graben. Figure modified from Ferrari et al. (2003b).

occurred later in this area; it occurred during the early Pliocene. The oldest reported age of an intraplate alkaline rock in the western sector is a basalt from the Guadalajara region dated at 4.6 Ma (Moore et al., 1994), but intraplate volcanics continued to be intermittently emplaced until very recent times. The products of intraplate volcanic activity constitute only ~5% of the total volume of volcanic rocks emplaced in the western sector (Ferrari et al., 2001), and coexist spatially and temporally with calc-alkaline rocks and, in some localities, with K-alkaline rocks (Luhr et al., 1989b; Righter et al., 1995; Righter and Rosas-Elguera, 2001). Intraplate volcanic rocks also tend to be relatively more abundant away from the trench.

Na-alkaline rocks are found in the coastal area of Nayarit to the west of Tepic (Righter et al., 1995) in the Tepic-Zacoalco graben, including the Sangangüey volcano area (Nelson and Carmichael, 1984), in the San Pedro–Ceboruco graben (Ferrari et al., 2003b; Petrone et al., 2003), in the Plan de Barrancas–Santa Rosa graben, and the region to the northeast of Guadalajara (Moore et al., 1994). They have been reported within the Jalisco block in the Amatlán de Cañas graben (Atenguillo graben) (Righter and Carmichael, 1992; Ferrari and Rosas-Elguera, 2000), and also on the Ayutla volcanic field (Righter and Rosas-Elguera, 2001) (Fig. 15).

Intraplate magmatic activity has formed shield volcanoes, lava plateaus, and numerous cinder cones. The latter are often aligned within the grabens, paralleling the tectonic structures. This is exemplified by a ~30-km-long, NW-SE alignment of nearly 50 cones that passes through Sangangüey volcano (Nelson and Carmichael, 1984). In other areas, like Tepic and Ayutla, vents are more dispersed and do not appear to be associated with a clearly defined tectonic structure (Righter et al., 1995; Righter and Rosas-Elguera, 2001).

All Na-alkaline rocks from the western Trans-Mexican Volcanic Belt display a similar behavior in the major and trace elements. Lavas are classified as alkaline basalts, hawaiites, mugearites, benmoreites, and trachytes, generally having normative *ne* or *hy*, with typical features of OIB: high TiO_2, total Fe and Nb, and no enrichments of LILE and LREE with respect to the HFSE (Nelson and Carmichael, 1984; Righter and Carmichael, 1992; Moore et al., 1994; Righter et al., 1995; Righter and Rosas-Elguera, 2001; Petrone et al., 2003). The volcanic rocks are usually differentiated and do not represent primitive magmas. It is only in the Ayutla area, closer to the volcanic front, that some truly primitive magmas have been identified (Righter and Rosas-Elguera, 2001).

In the Sangangüey area, Nelson and Livieres (1986) interpreted the dispersion in geochemical data as resulting from ascending small and independent magma batches, which diversified through crystal fractionation and crustal assimilation during ascent. Nelson and Carmichael (1984) also noticed that cinder cones around Sangangüey erupting more differentiated lavas tend to be older than those generating alkali basalts, which suggests an incremental change in magma ascent velocity through time.

Although alkaline and subalkaline lavas were emplaced in close space-time association, works published on the western sector of the Trans-Mexican Volcanic Belt have concluded that no direct genetic relation between the series exists. The different magma types would have originated from different primary magmas, generated in a compositionally heterogeneous mantle wedge. This implies that the mantle beneath the western sector must include enriched asthenospheric domains, which have not been chemically modified by the subducted materials, and mantle domains that were strongly metasomatized by water-rich components released from the subducted slab. The existence of a heterogeneous mantle has been supported by diverse geochemical arguments, emphasizing the impossibility to generate calc-alkaline magmas from alkaline magmas through processes like crystal fractionation, magma mixing, or crustal assimilation (Nelson and Livieres, 1986; Verma and Nelson, 1989; Petrone et al., 2003). Also, pre-eruptive water contents in the magmas are significantly lower in alkaline rocks (1.5%–1.8%) than in calc-alkaline rocks (3%–5.8%), which would be consistent with an origin from a drier asthenospheric mantle for the former (Righter et al., 1995). Oxygen fugacities calculated relative to the nickel-nickel oxide buffer (ΔNNO) are also lower for intraplate alkaline rocks (−1 to +1) than for calc-alkaline rocks (+1 to +3) (Righter, 2000). These results indicate a close relationship between the oxidation state of the source and the type of rocks generated, with intraplate alkaline rocks originating in a more reduced asthenospheric mantle.

Most studies based on the abundance of trace elements locate the source of Na-alkaline magmas in an enriched mantle that is similar to the source of OIB. Yet the low La/Yb ratios and high Yb abundances in the rocks, as well as phase equilibria, indicate that this mantle source must be relatively shallow and within the stability field of spinel peridotite (Nelson and Livieres, 1986; Righter and Carmichael, 1992; Luhr, 1997; Righter and Rosas-Elguera, 2001).

Unfortunately, isotopic studies of Na-alkaline rocks from the western sector are still scarce. In a study of the Sangangüey area, Verma and Nelson (1989) reported relatively high $^{143}Nd/^{144}Nd$ values (0.512843–0.512964) and low and variable $^{87}Sr/^{86}Sr$ values (0.703003–0.703980) for Na-alkaline volcanics. Na-alkaline magmas in this area would have originated from an OIB-type source, and achieved the different isotopic compositions by variable degrees of crustal assimilation and fractional crystallization with a "hypothetical contaminant." In a more detailed geochemical study of the San Pedro–Ceboruco area, Petrone et al. (2003) reported $^{143}Nd/^{144}Nd$ (0.512946–0.512964) and $^{87}Sr/^{86}Sr$ (0.703195–0.703437) values that are within the range of those reported by Verma and Nelson (1989) for Sangangüey, but with less variations, as well as relatively high $^{206}Pb/^{204}Pb$ values (18.9011–19.0338) that are close to the values of an enriched mantle (EM) component. In this work, the authors present a model that considers the addition of a small amount (< 0.5%) of subducted component to the EM source to reproduce the composition of Na-alkaline magmas, documenting for the first time the need of a small amount of slab-derived contribution for the source of Na-alkaline magmas in the western sector of the Trans-Mexican Volcanic Belt.

Finally, although various authors have proposed the participation of crustal contamination processes in the genesis of the more evolved Na-alkaline rocks of the western sector (Verma and Nelson, 1989; Righter and Rosas-Elguera, 2001; Petrone et al., 2003), the identification and quantification of these contributions have been hindered by the lack of isotopic contrast between the mantle-derived magmas and the Jalisco block basement.

Potassium Alkaline Rocks

Even though K-alkaline rocks are considerably less abundant than the typical calc-alkaline sequences, these rocks have been the subject of numerous petrologic and geochemical investigations not only because of their unusual presence when compared to other magmatic arcs, but also because they have been emplaced in close space-time association with more typical calc-alkaline arc volcanics.

Potassic rocks are distinctly more abundant in the Jalisco block, but it is also possible to find them in small monogenetic cones, or isolated lava flows, in the Michoacán-Gunajuato volcanic field (Luhr and Carmichael, 1985b; Hasenaka and Carmichael, 1987), or in the Valle de Bravo–Zitácuaro region (Blatter et al., 2001). Some rocks with these characteristics are also found interstratified with Na-alkaline sequences in the late Pliocene volcanic successions of the Palma Sola volcanic field (Negendank et al., 1985; López-Infanzón, 1991; Orozco-Esquivel, 1995; Gómez-Tuena et al., 2003).

The geographic distribution of the Plio-Quaternary K-alkaline rocks clearly marks the existence of a potassic volcanic front in the western and central sectors of the arc (Lange and Carmichael, 1991), which gradually dilutes toward the east (if the Palma Sola region is excluded) (Fig. 10). Indeed, it is remarkable that these rocks practically disappear to the east of the Nevado de Toluca volcano, and their absence in the Chichinautzin volcanic field is somewhat surprising, given the relatively large abundance of primitive magmas in the region, and the conspicuous presence of an extensional tectonic regime (Márquez et al., 1999c; Wallace and Carmichael, 1999). Thus, even though the petrologic studies on these rocks have claimed a subduction origin for these sequences, a convincing explanation for their mysterious absence in the eastern sector of the arc is still lacking.

Potassic magmatism in the Jalisco block (and in other areas) has been coeval with the dominant calc-alkaline magmas, at least since the Pliocene. K-alkaline rocks have been studied in detail on several volcanic fields of the Jalisco block: San Sebastián (Lange and Carmichael, 1990, 1991), Los Volcanes (Wallace and Carmichael, 1989, 1992), the Atenguillo graben (Righter and Carmichael, 1992), the Ayutla and Tapalapa volcanic fields (Righter and Rosas-Elguera, 2001), and in the Colima rift (Luhr and Carmichael, 1981, 1982; Allan and Carmichael, 1984). Yet the best-studied sequences are those from Mascota, Jalisco (Carmichael et al., 1996).

Rocks from Mascota vary in composition from primitive absarokites and minettes to andesites and spessartites. Rocks with the highest K_2O contents also have the highest Mg#; therefore these variations cannot be the result of crystal fractionation or crustal contamination, and should instead reflect the inherited characteristic of a mantle wedge that has been extensively metasomatized by the subduction agents (Carmichael et al., 1996). The close space-time association of K-alkaline rocks with more typical calc-alkaline sequences in Mascota (and in other volcanic fields) strongly suggests a genetic relation between the different magmatic series. But how this relation occurs and which mechanisms govern the petrologic differences are questions that have been under discussion for years.

The work by Luhr (1997) probably contains the most comprehensive discussion about the origin of the K-alkaline rocks and their relation to the rest of the magmatic rocks emplaced in the western sector of the Trans-Mexican Volcanic Belt. Luhr (1997) discussed the chemical character of the most primitive magmas emplaced in the area, as defined by those rocks with MgO > 6%, Mg# > 62, in which $Fe^{2+} = 0.8*Fe^{total}$. Clearly, potassic rocks have high contents of alkaline metals ($K_2O/Na_2O>0.5$), but they also have extremely high contents of LILE and REE, and very high LILE/HFSE ratios, characteristics that point to a subduction origin. Given the high alkali contents, it is not surprising that the vast majority of the alkaline-potassic rocks contain nepheline, or even leucite, in the CIPW norm. In a typical total alkali versus silica variation diagram, these rocks are classified as trachybasalts and trachyandesites and some of them even reach the phonotephritic field. They can be further classified as shoshonites in the classic diagram of SiO_2 versus K_2O (Le Maitre, 1989), but Carmichael et al. (1996) and Luhr (1997) have emphasized that the potassic rocks of the Jalisco block should instead be classified as lamprophyres, in accordance with the mineralogical assemblies observed in thin sections. Indeed, the distinction between shoshonites and lamprophyres should be made in petrographic and not in geochemical terms: shoshonites have olivine, clinopyroxene, and plagioclase phenocrysts, whereas lamprophyres have amphibole and/or phlogopite phenocrysts but not feldspar. The presence of hydrous minerals, together with the conspicuous absence of plagioclase, is a clear sign of high pre-eruptive water content in the magmas, which has the effect of increasing the silica activity in the melt (Sekine et al., 1979). Lamprophyres also tend to have high and variable oxygen fugacities when compared to other magmatic series.

Sr, Nd, and Pb isotopic compositions of the lamprophyres are very similar to those observed in calc-alkaline sequences from the Jalisco block, indicating that the magma source region for both rock suites should be similar or even identical. Nevertheless Luhr (1997) indicated that the isotopic compositions of Pb suggest that the presubduction mantle wedge is distinctively radiogenic, and similar to what is observed in the Shimada Seamounts of the east Pacific.

The vast majority of researchers that have studied the K-alkaline rocks of the Jalisco block agree that their origin must be linked to the partial fusion of a modally metasomatized mantle source rich in phlogopite, amphibole, pyroxene, and apatite (Carmichael et al., 1996; Luhr, 1997). It has been suggested that these minerals are heterogeneously distributed inside a peridotitic mantle in the form of veins or discrete mantle domains. Low extents of partial melting of this mantle will unequally affect the veined regions and give rise to magmas with very high alkaline contents. In contrast, larger extents of partial melting will allow the incorporation of peridotitic lithologies into the melts, and consequently form the more typical calc-alkaline volcanic sequences.

Even though a modally metasomatized, veined mantle is the preferred model for explaining the origin of K-alkaline volcanics around the world (Foley, 1992), in the case of México there is still no consensus on how or when this modally metasomatic event took place. Some authors have suggested that these veins are an inherited characteristic of the long and complex process of subduction that affected the area as far back as the Jurassic and especially during the formation of the Sierra Madre Occidental during the Eocene-Oligocene (Lange and Carmichael, 1991; Hochstaedter et al., 1996; Righter and Rosas-Elguera, 2001). This metasomatized mantle could melt relatively easy when incorporated into the complex tectonic regime of the Jalisco block in which extension and subduction are coeval. In fact, this model could provide a partial explanation for why the potassic magmas gradually disappear in the eastern sector of the arc, since the influence of the Sierra Madre Occidental is restricted to the north and south limits of the Mexico City basin (Righter and Rosas-Elguera, 2001). Nonetheless, and even if this theory is tectonically appealing, it does not explain why K-alkaline rocks have essentially the same Sr isotopic composition as the calc-alkaline rocks. This peculiarity is not trivial because phlogopite, the mineral that is presumably playing a major role in the formation of these magmas, easily accepts Rb but does not incorporate large quantities of Sr into its mineral structure (Schmidt et al., 1999). Thus, if the metasomatic event that formed these veins occurred several million years ago, it seems inevitable that the phlogopite-rich veins will evolve to radiogenic $^{87}Sr/^{86}Sr$ ratios in a relatively short time, because the Rb/Sr ratios of the veins will be much higher than those of the peridotitic mantle.

Other authors have suggested that the formation of these veins must be related to the current subduction regime (Carmichael et al., 1996; Luhr, 1997). In these models, vein formation processes most likely occurred during amphibole breakdown in the subducted slab at ~80 km depth, the liberation of large amounts of fluids or melts, and their interaction with the mantle wedge. These veins and their surrounding mantle could then be gradually dragged down to deeper regions by viscous coupling and corner flow along with the subducting plate. At ~100 km depth, the phlogopite-rich veins could be re-melted forming lamprophyric magmas that ascend to the surface. In fact, Luhr (1997) suggested that lamprophyres must be regarded as the "essence" of subduction, even though they rarely reach the surface. In most cases, these magmas will lose water and freeze during ascent, and are most likely emplaced as dikes in the upper mantle and crust. Luhr (1997) further indicated that it is only because of the prevailing intricate tectonic regime in the Jalisco block, in which rifting and subduction act concurrently, that these magmas are able to erupt at the surface.

Potassic rocks from the Jalisco block have also been experimentally studied (Righter and Carmichael, 1996; Moore and Carmichael, 1998). These studies have demonstrated that the mineralogical diversity and crystallinity depend on several variables that could be reproduced experimentally: pressure, temperature, composition, oxygen fugacity, but especially water content. More recently, Hesse and Grove (2003) reproduced the equilibrium conditions of a primitive absarokite from the Jalisco block that must have equilibrated with mantle olivine at the time of segregation from its source. Their study indicates that the absarokite most likely segregated from a depleted lherzolite, or even a harzburgite, at pressures of 1.6–1.7 GPa (48–51 km depth) and temperatures of 1,400–1,300 °C, in which the variations in P and T depend on pre-eruptive water contents (1.7% to 5.1%). Notably, the experimental results do not predict the existence of nonperidotitic mineralogies in the mantle at the P-T of final equilibrium, indicating in any case that the phlogopite–amphibole-rich veins must be located at deeper mantle regions. Thus, Hesse and Grove (2003) support the idea that the hydrated minerals are formed by fluids derived from the subducted slab that are later dragged down to deeper regions along with the subducted slab. These minerals will become unstable at deeper levels in the subduction zone, liberate fluids that migrate to the mantle wedge, promote the wedge's partial melting, and form the K-rich magmas. Once formed, these magmas will constantly re-equilibrate with the asthenospheric mantle by dissolving peridotitic minerals. The last phase of re-equilibrium apparently occurs at ~50 km depth, and will not undergo additional modifications until reaching the surface.

Calc-alkaline Rocks

Rocks with calc-alkaline affinities are widely distributed along the arc and have been emplaced during all periods of activity (Fig. 10C). These rocks have also been emplaced by a variety of volcanic structures: cinder cones, shield volcanoes, domes, stratovolcanoes, maars, calderas, and also through fissures that are not related to a specific volcanic center. Even though the compositions of rocks derived from the different volcanic structures are broadly similar, the most primitive basaltic lavas are always associated with monogenetic cones and fissural lava flows, whereas stratovolcanoes and calderas are invariably more evolved.

For the sake of clarity and consistency with the petrologic and geochemical studies of the arc, in this contribution we have divided the calc-alkaline rocks into two large groups: monogenetic fields and stratovolcanoes. We will revise the characteristics of the two largest monogenetic fields of the arc, the Michoacán-Guanajuato and Chichinautzin, because they have both been extensively studied, and because together they encompass the entire petrologic diversity observed in the Trans-Mexican Volcanic Belt.

Surprisingly, the large Mexican stratovolcanoes have received comparatively less attention in the petrologic and geochemical investigations. Important exceptions are the Colima, Popocatépetl, and Iztaccíhuatl volcanoes, but there is still very little information on the Malinche, Pico de Orizaba, and Tancítaro volcanoes.

Monogenetic Fields

The Michoacán-Guanajuato volcanic field. This volcanic field has more than 1000 Quaternary eruptive centers distributed over an area of 40,000 km^2, and thus represents one of the areas with the highest concentrations of monogenetic cones on Earth. As opposed to what occurs in the Mexican stratovolcanoes, the life span of eruptive centers in the Michoacán-Guanajuato volcanic field is relatively short (<15 yr), and they rarely reactivate after concluding their eruptive cycle. This indicates that the magma supply rate is so low that it does not allow the formation of long-lived magma chambers (Hasenaka and Carmichael, 1985). Thus the processes of differentiation and mixing that have been thoroughly observed in the stratovolcanoes are somehow attenuated in the monogenetic systems.

The first petrologic studies on the Michoacán-Guanajuato volcanic field were focused on one of the most famous volcanoes in the world: Parícutin (1943–1952). Situated in a picturesque area of relatively easy access, the volcano attracted a large number of researchers from around the world who rapidly made Parícutin one of the best-studied volcanoes on the planet, and the subject of some classic petrologic studies (Wilcox, 1954;

Eggler, 1972). Given that its eruptive record was extraordinarily well documented since the beginning, Wilcox (1954) was able to recognize the systematic compositional evolution of the volcanic products. The first eruptive phase of Parícutin is made of basaltic andesites with olivine phenocrysts (55% SiO_2), whereas the culminating phase erupted hypersthene-bearing andesites (60% SiO_2). This differentiation trend was somewhat surprising to scientists at the time, because it was in clear contrast to what was observed in other volcanoes, such as Kilauea and Arenal, and it was very different from the classic conceptual model of a zoned magma chamber. Building upon the observations and hypothesis of Bowen (1928), Wilcox (1954) realized that the petrologic evolution of Parícutin was not controlled by the process of crystal fractionation alone, but also by the assimilation of an additional component derived from the local continental crust. These interpretations were later confirmed by more sophisticated studies that included the isotopic compositions of the volcanic rocks and the local basement (Reid, 1983; McBirney et al., 1987). These studies documented what is now considered a classic example of the petrologic process of assimilation-fractional crystallization (AFC), mathematically enunciated by DePaolo (1981).

The most comprehensive study to date on the Michoacán-Guanajuato volcanic field is the doctoral thesis of Toshiaki Hasenaka from the University of California at Berkeley (Hasenaka, 1986), and the formal publications that derived from this work (Hasenaka and Carmichael, 1985, 1987; Hasenaka, 1994). Hasenaka compiled an inventory of virtually all the volcanic centers in the field and made an exhaustive analysis of their morphology and relative ages, as well as the petrography and chemical composition of more than 200 rock samples. Given the high density of sampling, Hasenaka and Carmichael (1987) recognized three different petrologic associations in the Michoacán-Guanajuato volcanic field: calc-alkaline rocks with typical arc characteristics, K_2O-rich alkaline rocks with relatively high MgO contents, and TiO_2-rich alkaline rocks with relatively low MgO contents (Fig. 16). An important proportion of rocks could also be classified as transitional, because they clearly plot between the alkaline and calc-alkaline fields. Hasenaka and Carmichael (1987) observed that the compositional variations of the magmatic series cannot result from a single liquid line of descent from a common primitive magma, but would instead be governed by more complex and distinct petrogenetic processes. Nonetheless, they found some systematic variations in the temporal and geographic distribution of the volcanics. Hasenaka observed that virtually all volcanic centers with ages younger than 40,000 yr are emplaced in the southern part of the volcanic field (between 200 and 300 km from the trench), belong to the calc-alkaline series, and have slightly higher silica contents than older cones. In contrast, alkaline volcanoes appear to be morphologically older, although their exact ages and life spans have not yet been determined. Notably, alkaline cones with low MgO and high TiO_2 contents are preferentially located toward the northern sector of the field (between 350 and 400 km from the trench), whereas alkaline volcanoes with high MgO and K_2O are located closer to the trench (between 200 and 270 km from the trench). Therefore, volcanic centers in the Michoacán-Guanajuato volcanic field tend to decrease in MgO (as well as in other compatible elements like Ni and Cr) and increase in TiO_2 contents with increasing distance from the trench (Fig. 16). These variations, along with the inferences made about the stability fields of the different mineral species, indicate that magmas in the south tend to be emplaced more rapidly and efficiently, undergoing relatively less fractionation at higher pressures than magmas emplaced toward the north, which appear to stall at shallower crustal levels for longer periods of time. Interestingly, the systematic increase in K_2O contents (and in other incompatible elements), a feature that is commonly observed in many arcs with increasing distance from the trench (i.e., slab depth) (Dickinson and Hatherton, 1967), is not clearly observed in the Michoacán-Guanajuato volcanic field, and only becomes apparent if the high K-alkaline rocks, located toward the volcanic front, are excluded from the diagrams. In this sense, this volcanic field and the Mexican arc in general once again contradict the well-established paradigms of magma genesis in convergent margins.

Jorullo volcano, the other monogenetic cone with historical activity in the Michoacán-Guanajuato volcanic field, has also received some attention in geochemical and petrological studies. The work of Luhr and Carmichael (1985b) and, more recently, the work of Barclay and Carmichael (2004) documented the compositional variations of Jorullo and of a slightly older cone emplaced a few kilometers to the south known as Cerro La Pilita. Similar to what was observed in Parícutin, Jorullo lavas become more silicic with time, evolving from basalts (52% SiO_2) to andesites (55% SiO_2). Yet, and in contrast to Parícutin, this evolution cannot be explained by simple crystal fractionation, with or without assimilation, even though some of the lavas contain abundant granitic xenoliths from the local basement. Indeed, Luhr and Carmichael (1985b) observe that even if the major-element compositions of Jorullo volcano could be explained by crystallization of olivine, augite, plagioclase, and some spinel at lower crustal or upper mantle pressures, the HREE are anomalously depleted in more evolved rocks. The authors acknowledged that an additional petrogenetic process must have occurred, but they were not able to resolve its origin.

The chemical composition of Cerro La Pilita contrasts with that of Jorullo volcano. La Pilita erupted nepheline-normative trachybasalts with phenocrysts of hornblende, apatite, olivine, spinel, and augite, with very high K_2O, P_2O_5, Ba, and Sr contents (Luhr and Carmichael, 1985b). The authors show that magmas of Jorullo and La Pilita cannot be genetically related through simple differentiation, with or without crustal contamination, but should instead represent two different petrogenetic mechanisms that act almost concurrently in the volcanic front. Barclay and Carmichael (2004) further indicate that the mineralogical assembly of La Pilita can be experimentally reproduced if the magma had between 2.5 and 4.5% water and crystallized between 1040–970 °C at 50–150 MPa. The experiments demonstrate that a basaltic magma–crystallizing hornblende induces a marked increase in crystallinity (~15%–40%), which translates into higher viscosities, and therefore hinders its ability to reach the surface. Thus the scarcity

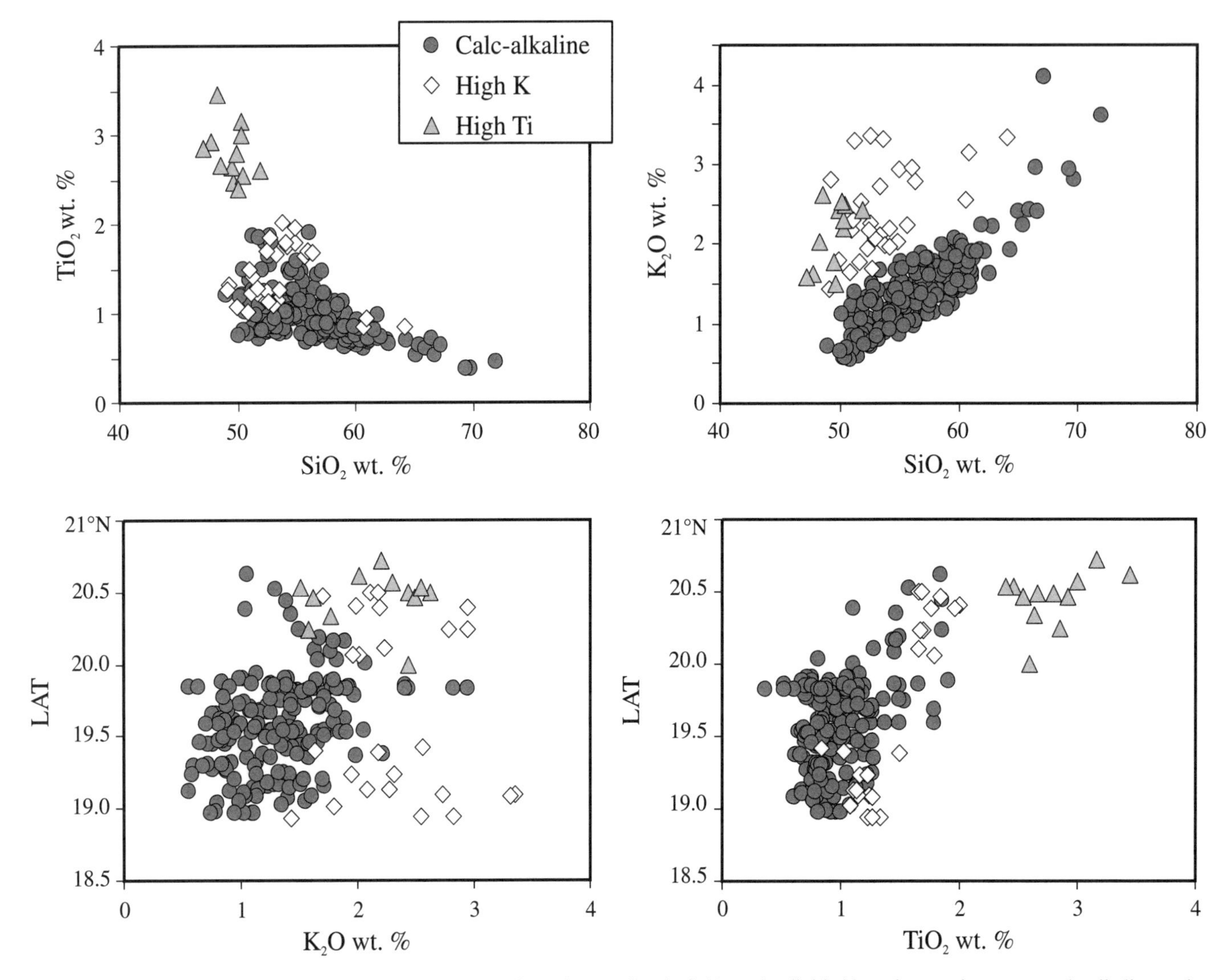

Figure 16. The compositional variations of the Michoacán-Guanajuato volcanic field can be divided into three main groups: calc-alkaline rocks, Ti-rich alkaline rocks, and K-rich alkaline rocks (Hasenaka, 1986; Hasenaka and Carmichael, 1987). Note that K_2O contents are uncorrelated with increasing distance from the trench (i.e., latitude), although Ti-rich rocks are always located faraway from the trench.

of hornblende-bearing basalts in volcanic arcs is probably due to the fact that these become trapped and stockpiled in the crust-mantle boundary (Barclay and Carmichael, 2004).

More recently, Verma and Hasenaka (2004) provided new Sr, Nd, and Pb isotopic compositions and additional major- and trace-element contents of a selection of samples that covers the compositional range of the Michoacán-Guanajuato volcanic field reported by Hasenaka (1986). The authors also discarded a simple differentiation process to explain the petrologic diversity of the volcanic field, and they further disregarded the possibility that it could be controlled by slab-derived chemical agents. Instead they adopted and modified the hypothesis put forth by Sheth et al. (2000), suggesting that the mantle wedge below México is compositionally heterogeneous and probably contains enriched metasomatic veins. The origin of these veins is interpreted to be related to a complex metasomatic process that occurs below an ancient crustal suture located beneath the proto–Trans-Mexican Volcanic Belt. As such, Verma and Hasenaka (2004) brought back to life the model of von Humboldt (1808) by suggesting that the Trans-Mexican Volcanic Belt is emplaced and governed by a first-order crustal discontinuity.

The Michoacán-Guanajuato volcanic field is also one of the few arc sequences in the world in which the Re-Os isotopic system has been employed to elucidate magma genesis (Lassiter and Luhr, 2001; Chesley et al., 2002). Re behaves as an incompatible element during magmatic processes whereas Os is highly compatible. Therefore, the continental crust has much higher Re/Os ratios than the mantle and will evolve to very radiogenic $^{187}Os/^{188}Os$ ratios through time (Shirey and Walker, 1998). Thus the Os isotopic compositions of volcanic rocks are ideal tracers to differentiate between crustal and mantle contributions to arc volcanics. Studies of the Re-Os system in the Michoacán-Guanajuato volcanic field confirm the participation of the continental crust in some evolved rocks of the volcanic field, but they also demonstrate that a good number of rocks with strong subduction signals (i.e., high LILE/HFSE ratios) have the low $^{187}Os/^{188}Os$ ratios that are typical of the mantle. These results have important implications in the petrogenesis of thick-crusted continental arcs like the Trans-Mexican Volcanic Belt because they demonstrate that some of the magmas have clearly escaped crustal contamination, and thus their chemical signals should be mainly mantle-derived and subduction-related phenomena.

The Chichinautzin volcanic field. This structure is an E-W elongated range that extends between the flanks of Popocatépetl and Toluca volcanoes. Occupying an area of nearly 2500 km^2, it is significantly smaller that the Michoacán-Guanajuato volcanic field, but it nonetheless has at least 220 cinder cones and shield volcanoes (Bloomfield, 1975; Martin-Del Pozzo, 1982). Paleomagnetic studies and the available isotopic ages have demonstrated that most of the volcanoes in the area are younger than ca. 0.78 Ma (Bloomfield, 1973; Mooser et al., 1974; Herrero and Pal, 1978; Siebe et al., 2004a). Even though there are no historical reports of volcanic activity, lava flows from Xitle volcano clearly buried and destroyed the pre-Hispanic settlement of Cuicuilco, which is now located within the southern edge of modern-day Mexico City (Siebe, 2000, and references therein). Detailed studies of the Chichinautzin volcanic field thus acquire additional social relevance because an eventual volcanic reactivation could affect one of the most densely populated areas in the world.

Petrographic and geochemical data from the Chichinautzin volcanic field have been reported by a number of authors, and there are also several publications that cover volcanologic and tectonic aspects (see Velasco-Tapia and Verma [2001] for a more extensive review on these issues). Nevertheless, and in spite of the large number of published studies about the area, there is still no consensus among researchers about its origin.

Wallace and Carmichael (1999) recognized that most of the volcanic rocks of this volcanic field are calc-alkaline andesites and dacites with typical arc geochemical features. However, it is rather unusual that true basalts have an alkaline affinity (Fig. 17). These alkaline basalts are not particularly abundant in this area, but they nonetheless represent some of the most primitive magmas emplaced in the arc, and their sole presence has produced great disputes among the scientific community. Wallace and Carmichael (1999) also observed that the least differentiated rocks in the Chichinautzin volcanic field (both alkaline and calc-alkaline) tend to be distributed in a N-S–trending band located in the center of the field, more or less along the 99.2°W meridian. Cinder cones emplaced toward the edges of the field are always more differentiated and tend to be chemically similar to the large stratovolcanoes.

Primitive rocks in the Chichinautzin volcanic field are generally olivine and plagioclase phyric, whereas pyroxenes and hornblende only occur in much more differentiated volcanics. A large number of rocks also contain ubiquitous quartz and sodic plagioclase xenocrysts in clear disequilibrium with the host magma. Wallace and Carmichael (1999) assumed that the most primitive calc-alkaline rocks are close to the primitive mantle-derived magmas that could have been formed by relatively low extents of partial melting (3.3% to 7.7%) of an incompatible element–depleted mantle source that has been enriched in water and fluid mobile elements derived from the subducted slab. In contrast, alkaline basalts should come from a compositionally heterogeneous mantle source, albeit one that is more enriched in incompatible elements, similar to the source of intraplate

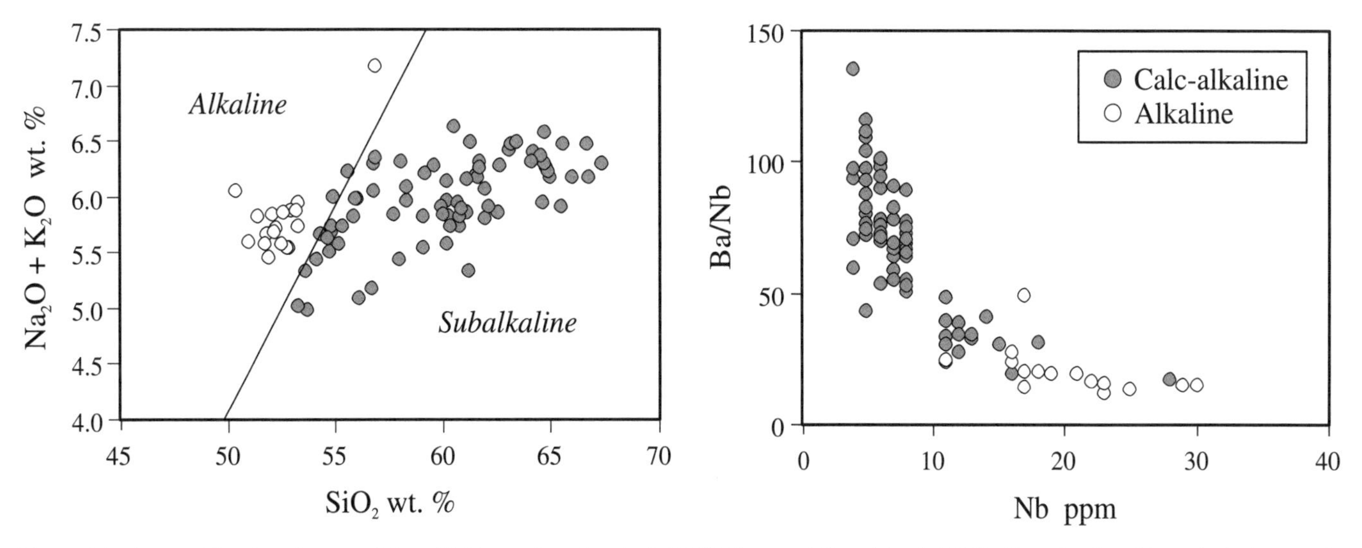

Figure 17. Compositional variations of the Chichinautzin volcanic field (Wallace and Carmichael, 1999). Note that basaltic rocks belong to the alkaline series, and have high Nb contents and very low subduction signals (i.e., low Ba/Nb ratios). Line dividing the alkaline and subalkaline fields is from McDonald and Katsura (1964).

basalts from the Mexican Basin and Range (Luhr et al., 1989a; Pier et al., 1992; Luhr, 1997). These authors argued that this enriched mantle domain could have been transferred from the backarc to the volcanic front by mantle convection induced by subduction, and that it underwent melting under nearly anhydrous conditions. Differences in the extent of mantle hydration thus generate magmas with various water contents and LILE/HFSE ratios, and confirm the complexity of magma generation processes in the Chichinautzin volcanic field (Cervantes and Wallace, 2003a, 2003b).

The works of Verma (1999, 2000a) challenged the conclusions reached by Wallace and Carmichael (1999). These studies recognized that the most primitive lavas in the Chichinautzin volcanic field have alkaline affinities, lack the chemical characteristics that are typical of arc volcanics, and are instead similar to ocean-island basalts (Fig. 17). Verma observed that more differentiated calc-alkaline rocks have lower concentrations of incompatible elements (HFSE and HREE) than more primitive rocks, a tendency that is difficult to explain by crystal fractionation or even crustal contamination. Primitive and differentiated rocks have similar Nd and Pb isotopic compositions, but more evolved rocks tend to have slightly more radiogenic Sr isotopes (Fig. 18). Verma indicated that this isotopic trend cannot be related to subduction because the data do not plot over the mixing trend made by the basaltic and sedimentary components of the subducted slab. Verma (1999, 2000a) suggested that more differentiated rocks are derived from partial melting of the lower continental crust, and that the more abundant intermediate volcanics represent simple mixtures between basaltic and evolved components.

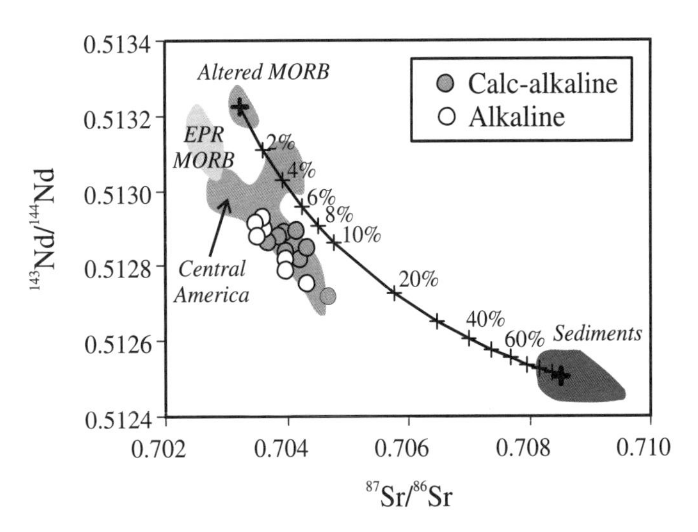

Figure 18. Sr and Nd isotopic compositions of rocks from the Chichinautzin volcanic field and the oceanic crust sampled at DSDP site 487 (Verma, 1999, 2000a). A mixing model between subducted sediments and altered oceanic crust is not able to reproduce the Chichinautzin data. This feature has been taken by Verma and co-workers as reason to disregard the participation of the subducted oceanic crust in the petrogenesis of the Chichinautzin volcanic field (and other parts of the arc) (Verma, 2002). Also shown is the isotopic variation field of the Central American arc (Feigenson and Carr, 1986; Feigenson et al., 2004). EPR-MORB—East Pacific Rise mid-oceanic-ridge basalts.

The geochemical arguments of Verma (1999, 2000a) were adapted by Márquez et al. (1999a), who associated the intraplate-like magmas from Chichinautzin with a mantle plume. In this model, the plume has been acting in conjunction with a propagating rift that has been gradually migrating toward eastern México since the late Miocene. Following a similar line of thought, the model by Sheth et al. (2000) envisages a continental rift that promotes mantle melting by decompression and the formation of primitive magmas without any influence from the subduction agents.

Márquez and De Ignacio (2002) later reported an extended database that includes whole-rock chemical data as well as mineral chemistry. Similar to Wallace and Carmichael (1999), they recognized the presence of two different mantle sources for the more mafic lavas: an essentially anhydrous but chemically enriched mantle source that gives rise to the OIB-like magmas, and a hydrated and metasomatized mantle source that forms calc-alkaline mafic rocks with subduction-related chemical signals. The intrusion of these mafic magmas into the continental crust could induce its partial fusion and generate evolved magmas with heterogeneous compositions. Magmas in the Chichinautzin volcanic field thus largely represent mixtures between various components from the mantle and crust. Nevertheless, and in contrast to Wallace and Carmichael (1999), Márquez and De Ignacio (2002) did not find compelling evidence indicating that the metasomatized mantle source could be related to the current subduction environment, and instead considered that metasomatism could have been induced by the effect of an "unrooted" mantle plume.

More recently, Siebe et al. (2004b) reported additional geochemical and detailed petrographic descriptions of three volcanic centers of the Chichinautzin volcanic field (Pelado, Guespalapa, and Chichinautzin), once again interpreting the isotopic and compositional heterogeneity of mafic rocks as a reflection of a compositionally diverse mantle source. Using a large variety of tectonic, thermodynamic, and petrologic arguments, the authors refuted the plume or rift hypothesis for their petrogenesis, instead supporting the model postulated by Wallace and Carmichael (1999). Similar to what is observed at Parícutin and Jorullo volcanoes from the Michoacán-Guanajuato volcanic field, rocks from these volcanic centers show the well-documented tendency to decrease HFSE and HREE contents with increasing SiO_2 contents. However, and in contrast to previous authors, Siebe et al. (2004a) explain these tendencies by polybaric crystal fractionation of accessory minerals, such as apatite, zircon, hornblende, and titaniferous phases, which act in combination with crustal assimilation.

In summary, all of these authors concur that mafic rocks in the Chichinautzin volcanic field reflect an extremely heterogeneous mantle source, but there is still no consensus about the origin of these heterogeneities: mantle plume, enriched mantle migration from the back arc, subduction-induced metasomatism, or a previously metasomatized lithospheric mantle. The origin of differentiated rocks in the volcanic field is also a matter of debate. All of these authors agree that simple fractionation of a typical low-pressure mineral assemblage is unable to explain their chemistry, yet the processes of crustal melting, assimilation, and crystallization of accessory mineral phases have not been thoroughly demonstrated with quantitative thermal or geochemical models. Thus, even though the Chichinautzin volcanic field is arguably the best-studied volcanic field in México, there is still no model that successfully reconciles the vast amount of petrologic, structural, and tectonic evidence that has been studied over the years.

Stratovolcanoes

México is a land of volcanoes and the most conspicuous manifestation of this is the large volcanic edifices that rise from coast to coast across the country (Plate 1). Reports on their activity go back to the codices and stelae of pre-Hispanic México, a wealth of valuable information that is still used today to establish the volcanoes' eruptive histories. It is thus not surprising that the indigenous volcano names allude to their persistent activity: Popocatépetl means "smoking mountain" and Colima is the name of "the place where the fire god dominates." Large volcanoes were considered to be deities in ancient Mexican cultures, and were worshiped in many rites that are still being carried out today albeit with Catholic nuances. Aztecs thought that the large volcanoes emerged in an age of fire after the oceans invaded the earth, in a time when the enormous lava flows and craters were formed. One of their legends is that since only birds were able to escape the wild fires, all men became birds with the exception of one who hid with his wife in a cave. This and other legends were compiled by Yarza de la Torre (1992) in her book *Volcanes de México.*

When the Spanish arrived in México they were awestruck by the huge volcanoes that majestically tower over the lowlands. They made numerous references to volcanoes in the chronicles of their conquest. In his *Historia General de las Cosas de la Nueva España* (Sahagún, 1992), Fray Bernardino de Sahagún (1499–1590) narrates:

> There is a mountain called Poyauhtécatl[1], near Auillizapán and Tecamachalco, that a few years ago began to burn on its top, and I saw it had the summit covered with snow, and I later saw when it began to blaze and the flames were there day and night and could be seen from twenty leagues, and now that the fire has wasted most of the mount's interior the fire can no longer be seen although it always burns.[2]

Petrologic and geochemical studies of the Mexican stratovolcanoes are not particularly abundant, yet it is encouraging to observe a certain level of consensus in the explanations about their origin. In fact, the vast majority of researchers have recognized that their petrologic varieties reflect some sort of mixing or mingling between magmatic components that reside in long-lived magma chambers. The mixing process allows for the effective homogenization of the magmatic end members, and results in the formation of compositionally uniform volcanic successions (Nelson, 1980; Luhr and Carmichael, 1982; Nixon, 1988a, 1988b; Robin et al., 1991; Straub and Martin-Del Pozzo, 2001; Schaaf et al., 2005). Compelling evidence for the mixing process is clearly recorded in the textures and mineral assemblies of the volcanic products. Indeed, it has been extensively documented that a large proportion of the volcanics erupted from stratovolcanoes' present antagonic mineralogies, like olivine and quartz, which could not have crystallized in equilibrium from a common primitive magma. Most phenocrysts also display textures of reaction-corrosion that demonstrate a complex transit in the magmatic system that is far from equilibrium. In addition, linear correlations in major-element variation diagrams are also typical of mixing, because they do not present the inflections that are commonly observed during crystal fractionation. Nonetheless, and even if mixing appears to be a very common process in the petrogenetic evolution of the stratovolcanoes, the specific character of the end members, their primary origin, and the physical mechanisms that allow for their effective homogenization are still difficult to understand.

The work by Nixon (1988a, 1988b) in Iztaccíhuatl volcano probably provides one of the best documentations of the magma-mixing process. Nixon explains that the magmatic evolution of Iztaccíhuatl is strongly controlled by a vigorous mixing process that acts in conjunction with crystal fractionation. In fact, Nixon demonstrated that the petrologic variations of Iztaccíhuatl display key aspects for understanding the theoretical models of a periodically refilled, tapped, and fractionated magma chamber (RTF) as postulated by O'Hara (1977) and O'Hara and Mathews (1981). Nixon called this process *dynamic* fractionation, because simple models of fractionation and mixing alone are not sufficient to explain the mineralogical and geochemical complexity of the volcanic successions. Nixon interpreted that the andesitic to dacitic lavas (58%–66% SiO_2) from Iztaccíhuatl always represent hybrid products, derived from a continuous and repetitive process of basaltic injections into a resident dacitic magma located in a long-lived magma chamber. Primary magmas feeding Iztaccíhuatl were considered to be basalts or basaltic andesites, with relatively high SiO_2 contents, but with MgO and Ni concentrations that are similar to primitive magmas. Yet these primitive magmas are far from homogeneous because they show important variations in Ba, Sr, and $^{87}Sr/^{86}Sr$ ratios. Nixon's calculations led him to suggest that the primitive magmas feeding Iztaccíhuatl could go from calc-alkaline hypersthene-normative basalts to alkaline basalts with nepheline in the norm, compositions that are also similar to basalts found in the Chichinautzin volcanic field and in the Mexico City basin. These features illustrate once more the highly heterogeneous character of the mantle wedge below the Trans-Mexican Volcanic Belt. On the other hand, the evolved magmatic component has a more homogenous dacitic composition, because it entails the last stage in the process of dynamic crystallization that should ultimately converge to equilibrium.

The best-studied volcano in México is undoubtedly Volcán Colima, the most productive central volcano in México and North America (Luhr and Carmichael, 1980, 1981, 1982, 1990; Robin et al., 1991; Moore and Carmichael, 1998; Luhr, 2002; Mora et al.,

[1]Ancient náhuatl name for Citlaltépetl or Pico de Orizaba.
[2]Free translation from ancient Spanish.

2002). Lavas from Volcán Colima somehow have a monotonous andesitic composition with phenocrysts of plagioclase ± orthopyroxene ± clinopyroxene ± titanomagnetite ± hornblende in a matrix made of glass and the same minerals, with the exception of hornblende. As with the rest of the stratovolcanoes, olivine is usually present with reaction coronas and other disequilibrium textures.

Luhr and Carmichael (1980) first recognized a certain eruptive cyclicity registered in the petrologic and geochemical variations of the volcanic products. Later studies indicated that during the second half of Colima's eruptive history, two similar volcanologic cycles can be identified: between 1818 and 1913, and between 1913 and the present (Luhr and Carmichael, 1990; Luhr, 2002). Each cycle is characterized by (1) an open crater that marks the final explosive event of each cycle; (2) a phase of dome building inside the conduit; (3) a phase of dome growth up to the crater rim, its eventual collapse, and the formation of block and ash flows; (4) brief periods of explosive activity in conjunction with phase 3; and (5) the culmination of the cycle with a major explosive event that produces widespread pyroclastic fall and flow deposits. These workers recognized that the volcano is currently approaching the end of a cycle, and predicted that a major explosive event will occur at some time during the next hundred years (Luhr and Carmichael, 1980). Eruptive cycles in Volcán Colima last about a century and appear to reflect the transit of discrete and compositionally zoned magmatic bodies that are enriched in SiO_2 toward the top. The last stage in the cycle is thus represented by the eruption of andesitic magmas that are relatively more primitive than those from previous stages.

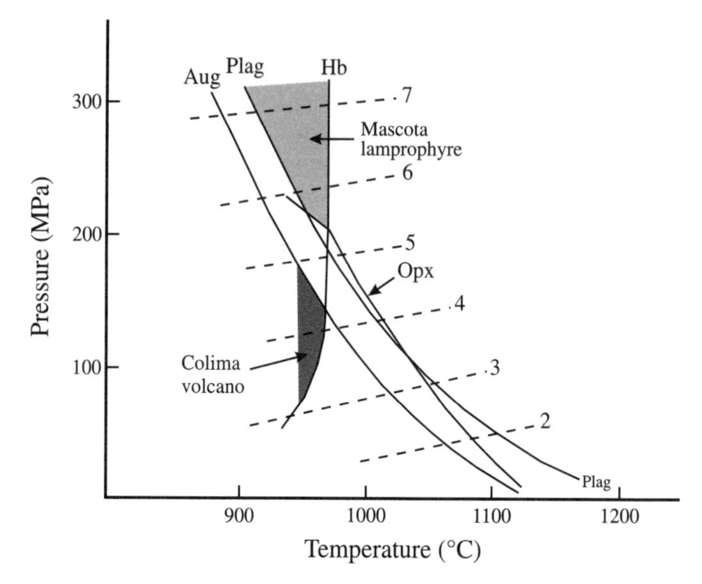

Figure 19. Water-saturated stability phase diagram for andesitic rocks from Colima volcano and spessartites from the Mascota volcanic field (Moore and Carmichael, 1998). Subhorizontal segmented lines are isopleths of wt% water content. Dark gray field represents the stability field of a typical mineral assembly from Colima volcano. Light gray field represents the *P-T* stability field of hornblende lamprophyres from the Mascota volcanic field in which plagioclase and pyroxene are absent. Aug—augite; Plag—plagioclase; Hb—hornblende; Opx—Orthopyroxene.

Similar to Iztaccíhuatl volcano, the petrologic evolution of Volcán Colima does not appear to be controlled by crystal fractionation or by a simple mixing mechanism. Instead, its mineralogical and geochemical features also point to a hybrid process in which crystallization is accompanied by mixing in an open magma chamber (Luhr and Carmichael, 1980). Yet experimental phase equilibria for andesitic magmas of Volcán Colima, and of lamprophyres from the Jalisco block, suggest a close genetic relation between them (Moore and Carmichael, 1998; Carmichael, 2002). Figure 19 shows the experimentally determined mineral stability fields of andesites in Vocán Colima and of lamprophyres from the Jalisco block. The authors documented that crystallinity and mineralogy in these rocks strongly depend on pre-eruptive water contents. Thus, the typical plagioclase-phyric andesite from Volcán Colima becomes so as a result of water loss and decompression-induced crystallization. The primary magma feeding the Colima system is represented by a hornblende-bearing lamprophyre with at least 6% water content, which is almost completely lost during decompression and ascent. Carmichael (2002) further observed that basalts and basaltic andesites in the Mexican arc are almost always poor in crystals, and that plagioclase, whose stability field is strongly controlled by dissolved water content, is not as abundant as in the andesites erupted from stratovolcanoes.

Popocatépetl volcano has also received some attention in the petrologic and geochemical research (Boudal and Robin, 1987; Straub and Martin-Del Pozzo, 2001; Schaaf et al., 2005). Similarly to other stratovolcanoes, Popocatépetl shows the previously mentioned features of magma mixing, and even though the periods of activity and rest are significantly longer than those for Volcán Colima, it has also been possible to infer some cyclicity in its eruptive behavior (Siebe et al., 1996).

The work by Schaaf et al. (2005) provides the most extensive geochemical data set available today for Popocatépetl and nearby cinder cones. Following a similar argument to that presented by Siebe et al. (2004a) for Chichinautzin, the authors suggested that the chemical variations observed in the cinder cones near Popocatépetl are mainly controlled by polybaric crystal fractionation in short-lived magma systems. In contrast, volcanic products from Popocatépetl are compositionally more homogenous and are the result of mingling primitive basalts with dacites in a long-lived magma chamber. Rocks from Popocatépetl usually show unequivocal evidence for magma-crust interactions in the form of xenoliths and xenocrysts. These interactions are also presumably observed in the isotopic compositions of the volcanic products. Rocks from Popocatépetl show higher $^{87}Sr/^{86}Sr$ ratios than those observed in nearby scoria cones at equivalent εNd. Schaaf et al. (2005) argued that these differences are the result of contamination with the Cretaceous limestone of the local basement.

On the other hand, the petrologic studies of Straub and Martin-Del Pozzo (2001) suggest that Popocatépetl andesites always represent hybrid magmas (Mg# 50–72) produced by mixing a mafic component (Mg# 68–86) with an evolved component (Mg# 35–40). The mafic component is considerably hydrated and is saturated with olivine and spinel at a temperature of 1170–1085 °C and at >12 kbar of pressure. This mafic magma crystallizes clinopyroxene during ascent through the crust, and also mixes with the evolved magmatic component that resides at 3–4 kbars at a temperature of ~950 °C. Using mass balance calculations and simple thermodynamic arguments, Straub and Martin-Del Pozzo (2001) disregarded the possibility that the rocks from Popocatépetl represent mixtures between Chichinautzin basalts and a siliceous magma derived from the partial fusion of the continental crust, or from a highly fractionated magma. They instead considered that the mafic component is actually a high MgO andesite (55%–62% SiO_2) with characteristics that are close to those of a primitive magma. As such, the evolved component could form by moderate extents of fractionation, although they do not dismiss the possibility that a small fraction of continental crust could have been assimilated.

Wallace and Carmichael (1994) did not find any evidence indicating that the primary magma feeding Tequila Volcano in western México is a basalt, and instead suggested that a primitive andesite interacts with a resident dacitic magma chamber. Observing the well-documented tendency of decreasing incompatible element contents with increasing silica in the volcanics, Wallace and Carmichael (1994) also dismissed fractional crystallization as the dominant petrogenetic process. The authors thus proposed the existence of distinct parental magmas that evolve independently, and allowed for some degree of crustal contamination during ascent.

The possibility that the primary magma that feeds the stratovolcanoes is not a basalt but a high-MgO andesite opens an interesting perspective for the petrogenesis of the Trans-Mexican Volcanic Belt and for arcs in general. Indeed, even though this hypothesis was put forth a few decades ago (Kushiro, 1972, 1974), more recent experimental and geochemical arguments restrict this possibility to extreme conditions—close to the limit of water saturation (Hirose, 1997)—that even today are difficult to reproduce in the most sophisticated laboratories (Parman and Grove, 2004). Yet the incorporation of hornblende-lherzolite nodules in a hydrous andesite from Zitácuaro could only have occurred while these magmas ascended through the mantle (Blatter and Carmichael, 1998a, 2001). This evidence along with that of the experimentally determined phase equilibria (Blatter and Carmichael, 2001), and the similarity of compositions between the andesitic lavas and the experimental water-saturated lherzolite melts (Hirose, 1997), strongly indicate a mantle origin for at least some of the Mexican andesites. It is noteworthy that andesites from Zitácuaro are compositionally very similar to those observed in the stratovolcanoes, and this could very well explain why there are simply no primitive basalts erupting from these volcanic systems.

More recently, a number of workers have put forth new evidence and arguments for an alternative explanation for the petrogenetic evolution of the Mexican arc (Luhr, 2000; Gómez-Tuena et al., 2003; Martínez-Serrano et al., 2004). These studies have shown that some of the intermediate rocks erupted from Miocene to Quaternary stratovolcanoes have chemical characteristics that are similar to those derived from the partial fusion of the subducted oceanic crust, including the sedimentary layers (Fig. 20). Even though these rocks are typical arc-related andesites, they also have very high Sr concentrations and low Y and HREE contents, characteristics that are difficult to explain with the differentiation of a parental basaltic magma (Defant and Drummond, 1990). These rocks also tend to have isotopic compositions that are close to MORB and thus cannot be formed by melting or assimilation of the continental crust. If the subducted oceanic crust below México is indeed melting, then it is possible to visualize the formation of high-MgO andesites through the interaction of these slab-derived melts with mantle peridotite during ascent (Kelemen et al., 2003). Therefore, if this hypothesis is confirmed, there would be no need to invoke water-saturated partial melting of the mantle to generate primitive andesites, because the high SiO_2 contents in the rocks could very well be an inherited characteristic of the melts derived from the partial fusion of the subducted oceanic crust (Rapp and Watson, 1995).

Silicic Magmatism

In a review paper on the silicic volcanism of the Trans-Mexican Volcanic Belt, Ferriz and Mahood (1986) identified five rhyolitic centers and some other isolated outcrops. Many more silicic centers have been recognized and studied since those pioneer studies, but most of the publications have focused on estab-

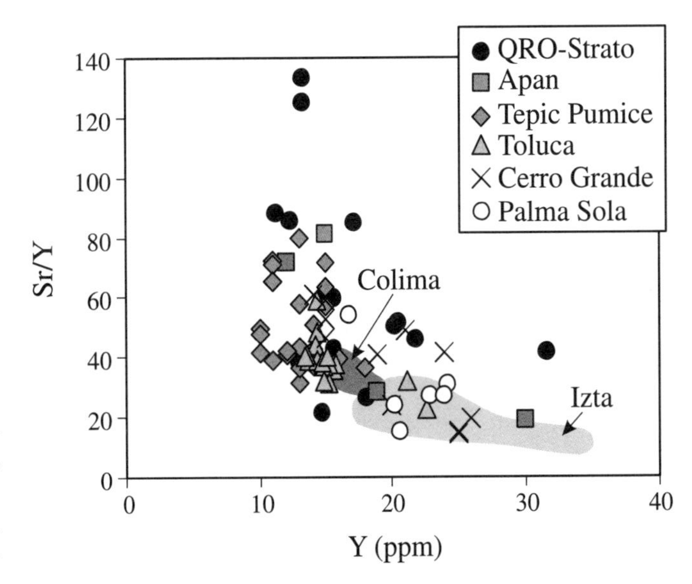

Figure 20. Sr/Y versus Y variation diagram for some volcanic fields of the Trans-Mexican Volcanic Belt. Adakite-like signals are observed in the Quaternary San Juan volcano (Luhr, 2000) and Nevado de Toluca volcano (Martínez-Serrano et al., 2004), as well as in the Miocene volcanic fields of Apan (García-Palomo et al., 2002), Cerro Grande (Gómez-Tuena and Carrasco-Núñez, 2000), Palma Sola, and Palo Huérfano–La Joya–Zamorano (Gómez-Tuena et al., 2003). Nonadakite-like rocks from the Colima (Luhr, 2002) and Iztaccíhuatl (Nixon, 1988b) volcanoes are also shown for comparison. QRO-Strato—Palo Huérfano–La Joya–Zamorano; Izta–Iztaccíhuatl.

lishing the stratigraphic relationships and the associated eruptive processes, whereas petrological and geochemical studies are still relatively scarce. On the basis of geographic distribution, stratigraphy, petrography, geochemistry, and emplacement mechanisms, we have distinguished three episodes of silicic volcanism associated with the evolution of the Trans-Mexican Volcanic Belt: (1) late Miocene, (2) early Pliocene, and (3) late Pliocene–Quaternary.

The first manifestations of rhyolitic volcanism appeared in the late Miocene (ca. 7 Ma) in the western and central sectors of the Trans-Mexican Volcanic Belt (Fig. 2C) (Aguirre-Díaz and López-Martínez, 2001; Ferrari et al., 2001; Rossotti et al., 2002). To the northwest and northeast of Guadalajara, in the western sector of the arc, some domes and pyroclastic flows with an estimated volume of 212 km^3 (Rossotti et al., 2002) have been dated at 7.15–5.2 Ma, and grouped into the *Guadalajara Group* (Gilbert et al., 1985). These rocks are mainly peraluminous rhyolites (molar $Al_2O_3/CaO + Na_2O + K_2O > 1$), with normative corundum, and high silica and K_2O contents (L. Ferrari, unpub.data) (Fig. 21). Toward the central sector, and to the northeast of Querétaro city, the Amazcala caldera was built between 7.3 and 6.6 Ma. The volcanic products are pumice-fall deposits, ignimbrites, and lava domes and flows of rhyolitic compositions and high K_2O contents (Aguirre-Díaz and López-Martínez, 2001). To the south of Querétaro, large rhyolitic dome complexes have been recognized, but only two ages have been reported (ca. 6 Ma; Aguirre-Díaz and López-Martínez, 2001; and 5.4 Ma; Ferrari et al., 1991). Some other rhyolitic domes have also been identified to the west of the Amazcala caldera with an age of 7.49 Ma (Jacobo-Albarrán, 1986). Volcanism related to this late Miocene rhyolitic episode was restricted to the northern portion of the Trans-Mexican Volcanic Belt, farthest from the trench, and occurred immediately after the late Miocene mafic episode (Fig. 2B–2C).

A considerable decrease in the magma extrusion rates occurred during the late Miocene rhyolitic episode of the western sector, as a result of a decrease in convergence rates between the Rivera and North American plates. The rhyolitic episode also coincides with the beginning of a regional trenchward migration of the volcanic front which has taken place since ca. 8.5 Ma. This has been interpreted to be the result of a gradual increase in the subduction angle of the Rivera plate (Ferrari et al., 2000a, 2001). This evidence indicates that plate dynamics apparently controlled the style and composition of volcanism in the early Trans-Mexican Volcanic Belt, but neither geochemical nor petrological data are available to provide further insight into their petrogenesis.

By the early Pliocene, rhyolitic rocks coexisted with mafic and intermediate rocks, and pyroclastic rocks became more prominent. Some ignimbrites contain dark glasses of basaltic andesitic to trachytic compositions that are mingled with colorless rhyolitic glasses (Gilbert et al., 1985; Aguirre-Díaz, 2001; Rossotti et al., 2002). In the Guadalajara area, such features were observed in deposits dated between 4.8 and 3.07 Ma (San Gaspar and Guadalajara ignimbrites; Gilbert et al., 1985), and in pyroclastic deposits in the *Chicharrón Group* (Rossotti et al., 2002). The persistence of rhyolitic volcanism in the Guadalajara area,

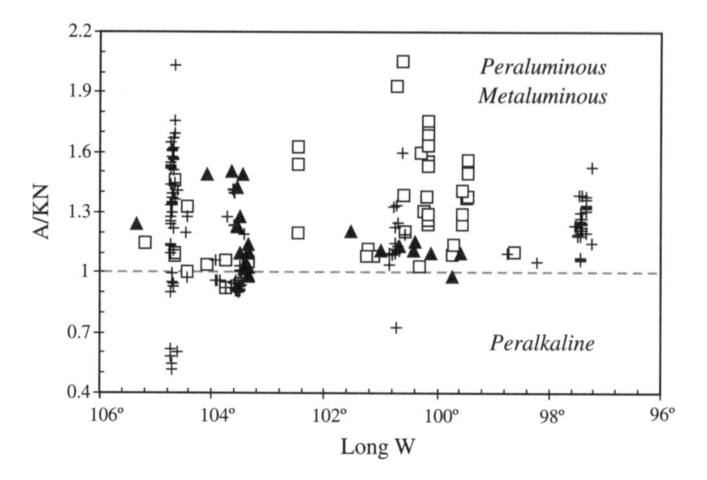

Figure 21. Geographic and temporal variations in the alumina saturation index for rhyolitic rocks (A/KN = molar $Al_2O_3/Na_2O + K_2O$). The stippled line divides the peraluminous-metaluminous and peralkaline fields. Late Miocene rocks (triangles), early Pliocene rocks (squares), late Pliocene–Quaternary rocks (crosses).

which initiated in the late Miocene and migrated to the south during Pleistocene times, has been also related to the presence of a shallow, long-lived magma chamber established at ca. 7 Ma. It has been proposed that the onset of intraplate alkaline volcanism in the area at 4.6 Ma (Moore et al., 1994) allowed the injection of mafic magmas into the silicic magma chamber, triggering explosive volcanism and causing magma mixing (Rosas-Elguera et al., 1997; Rossotti et al., 2002).

The early Pliocene rhyolitic volcanism of the central sector is associated with large calderas: Los Azufres, Amealco, and Huichapan (Plate 1). Volcanism is characterized by alternating events of mafic to intermediate and felsic composition, with a relative increase in the emission of mafic products when compared to the late Miocene episode. For instance, ~95% of the magma extruded from the Amealco caldera has an intermediate composition, and only the remaining 5% is rhyolitic (Aguirre-Díaz, 1996).

Intense faulting and a thick volcanic covering have complicated the definition of events responsible for the formation of the Los Azufres caldera. However, Ferrari et al. (1991) interpreted Los Azufres as a nested caldera formed by multiple collapses during late Miocene and Pliocene times. Pradal and Robin (1994) proposed, on the other hand, that the caldera was built at <3.4 Ma over a Miocene-Pliocene andesitic complex. The youngest volcanic products related to the caldera have been dated at 26.5 ka. According to the scarce geochemical data, volcanism occurred in alternating events of felsic and mafic composition. Felsic rocks are peraluminous dacites, rhyolites, and high-silica rhyolites that belong to the high-K calc-alkaline series (Cathalieau et al., 1987; Pradal and Robin, 1994). Geochemical and mineralogical evidence of magma mixing between rhyolitic and basaltic magmas has been reported as a mechanism to form dacites in this area (Cathalieau et al., 1987). The compositional variations of the extruded magmas have been related to the progressive sampling of a zoned magma chamber which was periodically replenished by more mafic magmas (Pradal and Robin, 1994). Differences in the $^{87}Sr/^{86}Sr$ ratios between a 0.6 Ma-old andesite (0.70357), a 0.026 Ma-old rhyolite (0.70553), and a recent basaltic cone (0.70367) were used to confirm this hypothesis (Pradal and Robin, 1994), although the authors did not discuss the processes by which the most differentiated rock acquired the most radiogenic isotopic composition. On the other hand, Ferrari et al. (1991) associated the generation of rhyolitic rocks with a compressive tectonic pulse that allowed magma stagnation and differentiation in the magma chamber. This pulse was followed by a transtensional tectonic phase that favored the ascent of mafic magmas, which replenished the magma chamber and mixed with the resident differentiated magmas. Large amounts of basaltic lavas were emplaced along extensional faults located outside the caldera rim during this phase.

The first products of the Amealco caldera (known as the Amealco tuff) were emplaced at 4.7 Ma. These rocks also present clear evidence of magma mingling (Aguirre-Díaz, 1996, 2001). In this case, mingling has been related to the repeated injection of less differentiated, hotter magma into a compositionally zoned magma chamber (Aguirre-Díaz, 2001). Five domes of low-silica rhyolitic composition were also emplaced within the caldera at 3.9 Ma, following the initial events that generated the Amealco tuff and other andesitic to dacitic domes and lava flows (Aguirre-Díaz, 1996). Some peraluminous and metaluminous high-silica rhyolitic domes were also emplaced in the periphery at 3.7 and 2.9 Ma (Aguirre-Díaz, 1996).

Early ignimbrite activity occurred in the Huichapan caldera at 4.8 and 4.4 Ma (Aguirre-Díaz et al., 1997). Silicic volcanism (dacite, trachyte, rhyolite) alternated with more mafic events (basaltic-andesite to andesite) in Huichapan; but as opposed to Amealco, the initial explosive event was exclusively rhyolitic (Verma, 2001b). Geochemical data indicate that the more differentiated rocks are high-K calc-alkaline, and predominantly peraluminous (corundum normative, molar A/CKN > 1) (Verma, 2001b).

Isotopic studies carried out by Verma et al. (1991) and Verma (2001b) in the Amealco and Huichapan calderas have shown that the volcanic products tend to have more enriched Sr, Nd, and Pb isotopic ratios as silica content increases, a feature consistent with crustal assimilation. The isotopic compositions of the Huichapan rocks were reproduced by a process of combined assimilation and fractional crystallization (AFC), considering a Miocene basalt as starting composition and a hypothetical crust with a similar isotopic composition to a very enriched rhyolite, that presumably originated by crustal melting during a previous volcanic event (Verma, 2001b). Similar to what is observed in the Los Azufres caldera, a cyclic recharge of more primitive magmas into a fractionated magma chamber also played an important role in the evolution of the Huichapan magmas. Interestingly, rhyolitic rocks in the Amealco caldera are only slightly more enriched in $^{87}Sr/^{86}Sr$ (0.70419–0.70424) and εNd (1.8–2.5) than the previously emplaced intermediate rocks ($^{87}Sr/^{86}Sr$ = 0.70396–0.70419; εNd = 2.1–2.6). Verma et al. (1991) explained the evolution of mantle-derived magmas to intermediate compositions through the assimilation of crustal material (up to 10%) at deeper crustal levels, or during early differentiation stages in a shallow magma chamber, but with very small to negligible crustal assimilation during further evolution to rhyolitic compositions.

In the eastern sector, and to the northeast of Pachuca, Hidalgo (Plate 1), extensive outcrops of pyoclastic deposits and rhyolitic lavas are closely associated with fissural emissions of moderately alkaline to transitional mafic lavas. These rocks have been dated at 4.3 and 4.4 Ma (Robin, 1976b, 1982). The origin of rhyolitic volcanism in this bimodal succession was briefly considered by Robin (1982) who reported mean values for the analyses of two rhyolite groups with a peraluminous composition. The composition of these rocks plots close to the ternary granite minimum in the Ab-Or-Q diagram, suggesting an anatectic origin (Robin, 1982). Based on the presence of Na-alkaline volcanism that evolves into a bimodal suite, this author proposed that volcanism in this area is related to an "Altiplano border sub-province," which would be associated with extensional processes parallel to the Gulf of Mexico coast, and unrelated to the subduction system of southern México. Nevertheless, this model has been reevaluated recently and the area is now seen as belonging to the Trans-Mexican Volcanic Belt (Ferrari et al., 2005b).

Late Pliocene to Quaternary rhyolitic volcanism is distinguished from the previous episodes by the emplacement of peralkaline magmas (molar $Al_2O_3/Na_2O + K_2O < 1$; normative acmite) (Fig. 21). Nonetheless, these have a relatively small volume and a restricted distribution when compared to the predominant metaluminous and peraluminous rhyolites. Peralkaline rocks are usually regarded as anorogenic, and their presence in the western and eastern sectors of the Trans-Mexican Volcanic Belt, where intraplate alkaline volcanism has been active (Tepic-Zacoalco graben, northeastern Hidalgo State), appears to indicate a genetic relationship between the two types of magmas.

In the western sector of the arc, peralkaline rhyolites have been described in the Las Navajas volcano (Nayarit), and in the Magdalena dome complex and Sierra La Primavera (Jalisco). The Las Navajas volcano (dated between <4 Ma and 200 ka; Nelson and Carmichael, 1984) is located at the western end of the Tepic-Zacoalco graben. This is the only place in the Trans-Mexican Volcanic Belt in which highly peralkaline rocks (pantellerites) have been reported in association with mildly peralkaline rocks (comendites). The early volcanic products at Las Navajas are mildly alkaline basalts to benmoreites, but magma compositions evolved through time to trachytes, comendites, and finally pantellerites. Peralkaline lavas and pyroclastic deposits display a mineral assembly that is typical of these kinds of rocks: quartz and K-feldspar, and subordinate amounts of fayalitic olivine, riebeckite, and aenigmatite in the comendites; or ferrohedenbergite, arfvedsonite, aenigmatite, and riebeckite in pantellerites (Nelson and Hegre, 1990). Pantellerites have low Al_2O_3 (8.2–9.4 wt%) and high Zr (839–2049 ppm) contents, suggesting an evolution from moderately alkaline basalts that underwent extreme fractional crystallization in a shallow magma chamber (Nelson and

Hegre, 1990). This implies that a considerable volume of alkaline mafic magma stagnated and differentiated in the continental crust, a phenomenon that contrasts with what is commonly observed in alkaline magmas from the western Trans-Mexican Volcanic Belt, which rapidly ascend to the surface through fractures and faults.

The Magdalena dome complex is constituted by 12 domes and rhyolitic flows (~35 km^3; SiO_2: 70%–76.5%) surrounding the andesitic Tequila Volcano. Rhyolitic rocks in this complex are moderately peralkaline, metaluminous, or peraluminous, and contain phenocrysts of sanidine, plagioclase, augite, and some hornblende (Demant, 1979; Harris, 1986). Rhyolitic magmas were emplaced at 1.12–0.24 Ma in close association with high-Ti basalts that predate the formation of Tequila Volcano (Harris, 1986; Lewis-Kenedi et al., 2005). These rocks are interpreted to be the result of discrete episodes of partial melting of the upper crust, probably driven by the emplacement of basaltic magmas that could, in part, ascend to the surface along northwest-southeast fractures and faults.

The Sierra La Primavera, located to the west of Guadalajara, has been better studied (Mahood, 1981a, 1981b; Mahood and Halliday, 1988). Rocks from La Primavera are domes, lava flows, and pyroclastic deposits with a rhyolitic composition (>75% SiO_2) that were emplaced between 145 and 30 ka. Most rocks typically contain sodic sanidine and quartz, and may contain small amounts (<3 vol%) of hedenbergite, fayalite, and ilmenite. Compositions vary from mildly peralkaline rhyolites (comendite) to subordinate metaluminous rhyolites. Early volcanic products are mildly peralkaline and are enriched in F, Rb, Na, Y, Zr, HREE, Hf, Ta, Pb, Th, and U. Later products are slightly metaluminous and enriched in Ca, LREE, and Ti. The magmatic evolution of La Primavera was related to sampling of progressively deeper levels of a zoned magma chamber (Mahood, 1981a, 1981b; Mahood and Halliday, 1988). Changes in temperature, volatile content, and peralkalinity with increasing depth apparently controlled the composition and abundance of fractionating phases and influenced the observed geochemical variations (Mahood, 1981a, 1981b; Mahood and Halliday, 1988). La Primavera rhyolites also have relatively primitive and constant $^{87}Sr/^{86}Sr$ (0.704–0.7048), εNd (4.5–5.8) and $\delta^{18}O$ (~6.6‰, in sanidine) isotopic compositions, which are consistent with an origin of fractional crystallization from a mantle-derived basalt with small to negligible contributions from the continental crust. Yet the absence of significant volumes of associated mafic and intermediate rocks in the area, and the observation that most siliceous magmas in México appear to be formed by crustal anatexis, led Mahood (1981a, 1981b) and Mahood and Halliday (1988) to suggest that the involvement of partial melts derived from intermediate to mafic plutonic rocks of the local Mesozoic basement (Guerrero terrane) that originally had a depleted mantle-like isotopic composition.

In the eastern sector, significant volumes of silicic lavas were emplaced during late Pliocene–Quaternary times (Plate 1). The most important volcanic centers are the Los Humeros and Acoculco calderas, and the domes of the Serdán-Oriental basin; but there are also some other volumetrically minor manifestations like the late Pliocene domes of Sierra Los Pitos (Zamorano-Orozco et al., 2002), and the Pleistocene Sierra Las Navajas in the State of Hidalgo (Geyne et al., 1963; Nelson and Lighthart, 1997).

Rhyolitic to dacitic domes and pyroclastic deposits were emplaced between 3 and 1.26 Ma in the Acoulco caldera, whereas the youngest activity was basaltic and occurred at 0.24 Ma (López-Hernández and Castillo-Hernández, 1997). A similar situation occurred at the Los Humeros caldera, where the early volcanic products were high-silica rhyolites to rhyodacites (ca. 0.47 Ma and ca. 0.22 Ma), followed by the emplacement of andesitic to rhyodacitic magmas (0.24 and 0.02 Ma), and culminated with the emplacement of small volumes of olivine basalt during the last 0.02 Ma. High-silica rhyolitic magmas volumetrically predominate over the rest of the sequences in Los Humeros. The more differentiated samples are metaluminous and belong to the high-K calc-alkaline series and some of them may contain biotite, hypersthene, or scarce hornblende phenocrysts (Ferriz and Mahood, 1984, 1987).

Following a similar line of thought to the one applied to Amealco and Huichapan calderas, Verma (2000b, 2001a) proposed that parent magmas for Acoculco and Los Humeros originated by partial melting of enriched domains (OIB-like) in the lithospheric mantle, and evolved to intermediate compositions through assimilation of crustal materials. Verma modeled the observed compositions by considering a hypothetical composition for the crustal component and assuming that the evolution from intermediate to rhyolitic compositions was controlled by fractional crystallization alone.

On the other hand, Ferriz and Mahood (1987) proposed a more complex evolution model for Los Humeros in which a zoned magma chamber is periodically replenished with mantle-derived basaltic magmas. This provides a thermal input to the reservoir and allowed magma mixing to occur. Mixing between basaltic and high-silica andesitic magmas would have produced basaltic andesites, which later evolved to rhyodacite and rhyolite compositions through fractional crystallization or AFC processes. In this model, the inverse zoning observed in single pyroclastic deposits resulted from tapping progressively deeper levels of a zoned magma chamber. The fact that high-silica rocks in Los Humeros have more enriched isotopic compositions ($^{87}Sr/^{86}Sr$ = 0.70414–0.70465; εNd = 1.4; $\delta^{18}O$ = 6.4–7.2‰) than younger basalts ($^{87}Sr/^{86}Sr$ = 0.70386; εNd = 4.1; $\delta^{18}O$ = 5.8‰) has been taken as evidence for mixing and/or crustal assimilation processes (Ferriz and Mahood, 1987). These authors also interpreted the temporal evolution to more mafic composition to an increase in the magma extraction rate that is accompanied by a decrease in the magma recharge rate into the reservoir, a process that would have prevented the reestablishment of a zoned magma chamber.

To the south of the Los Humeros caldera, a group of rhyolitic Quaternary domes was built following a rough N-S alignment inside the Serdán-Oriental basin: Las Derrumbadas, Cerro Águila, Cerro Pinto, and Cerro Pizarro (Ferriz and Mahood, 1986) (Plate 1). Cerro Pizarro was dated by Riggs and Carrasco-Núñez (2004) at 220,000, and Las Derrumbadas at 0.32 Ma by Yañez-García and Casique (1980), although Siebe and Verma (1988) estimated an age younger than 40,000 yr based on its morphology. To the east of this area, products related to the Las Cumbres volcanic complex include peraluminous high-silica rhyolitic domes dated at ca. 0.35 Ma, and an extensive rhyolitic pumice-fall deposit (Quetzalapa Pumice) produced by a plinian eruption that took place at 20 ka (Rodríguez et al., 2002; Rodríguez, 2005).

The Serdán-Oriental basin domes are mainly peraluminous rhyolites with phenocrysts of plagioclase, sanidine, quartz, biotite, magnetite, and ilmenite. Interestingly, Las Derrumbadas also contains garnet phenocrysts (Ferriz and Mahood, 1986). Even though all of the domes have similar silica contents, and were emplaced in a restricted area during a short period of time, their trace-element contents and isotopic compositions are unusually distinct. A number of workers (Negendank et al., 1985; Ferriz and Mahood, 1986, 1987; Besch et al., 1988, 1995; Riggs and Carrasco-Núñez, 2004) has shown that rocks from Las Derrumbadas are strongly depleted in MREE and HREE, and have a small, negative Eu anomaly and moderate Ba (770–889 ppm) and Sr (138–275 ppm) contents and the most enriched isotopic composition ($^{87}Sr/^{86}Sr = 0.70511$; $\varepsilon Nd = -2.8$) in the area. Cerro Pinto rhyolites are characterized by relatively flat REE patterns and a prominent, negative Eu anomaly, are strongly depleted in Ba (30–50 ppm) and Sr (21–22 ppm), and have isotopic ratios that are slightly less enriched than those from Las Derrumbadas ($^{87}Sr/^{86}Sr = 0.70506$; $\varepsilon Nd = -1.6$). Samples from Cerro Águila and Cerro Pizarro have LREE contents similar to those of Las Derrumbadas, but they are also strongly depleted in HREE and have a moderate negative Eu anomaly. Of all the domes studied in the area, Cerro Pizarro has the highest Ba abundances (1658–1776 ppm), relatively high Sr contents (176–205 ppm), and the least enriched isotopic ratios ($^{87}Sr/^{86}Sr = 0.70481$; $\varepsilon Nd = -1.4$). Taken together, the domes from the Serdán-Oriental basin have the most enriched isotopic compositions reported thus far for the Trans-Mexican Volcanic Belt (Fig. 22). The remarkable differences in mineralogy and REE abundances between localities prompted Ferriz and Mahood (1986, 1987) to interpret the origin of the rhyolitic domes as small-volume melts coming from a compositionally heterogeneous continental crust. Similarly, Besch et al. (1988) interpreted a residual source mineralogy with garnet, amphibole, and plagioclase for the Las Derrumbadas magmas; and with orthopyroxene, clinopyroxene, and plagioclase for Cerro Pinto. Alternatively, Siebe and Verma (1988) and Besch et al. (1995) considered rhyolite formation the result of crustal contamination from mantle-derived basaltic magmas, but they did not consider the well-known variations in trace elements among the different domes.

In summary, the main features of silicic volcanism in the Trans-Mexican Volcanic Belt can be described as follows: (1) An essentially effusive silicic episode of peraluminous rhyolites was established in the late Miocene toward the western and central sectors of the Trans-Mexican Volcanic Belt (Plate 1, Fig. 21). (2) By the early Pliocene, metaluminous and peraluminous rhyolites and ignimbrites coexisted with more mafic volcanics, either forming bimodal suites, or displaying evidence of magma mixing or mingling (Fig. 21). (3) During the late Pliocene(?) to Quaternary times, peralkaline, peraluminous, and metaluminous rhyolites were emplaced in the western sector in close space-time association with Na-alkaline volcanics, whereas peraluminous rhyolites were dominant in the central and eastern sectors (Fig. 21). Rhyolitic volcanism was also more restricted during this period and mainly located in the Tepic-Zacoalco graben (western sector), at the Los Azufres caldera (central sector), and at the Los Humeros caldera and Serdán-Oriental basin (eastern sector). (4) Rhyolites have been erupting in the Tepic-Zacoalco graben since the late Miocene, whereas rhyolitic volcanism in other areas has been migrating southward toward the trench (Plate 1). (5) The isotopic compositions (Fig. 22) indicate that crustal assimilation and/or anatexis have played an important role in the genesis of rhyolites in such places where the continental crust is thicker (i.e., the central and eastern sectors, Fig. 6). Nonetheless, crustal contributions to rhyolites from the western sector have not been clearly demonstrated, probably because the isotopic compositions of mantle-derived magmas and rocks from the Guerrero terrane are not very different.

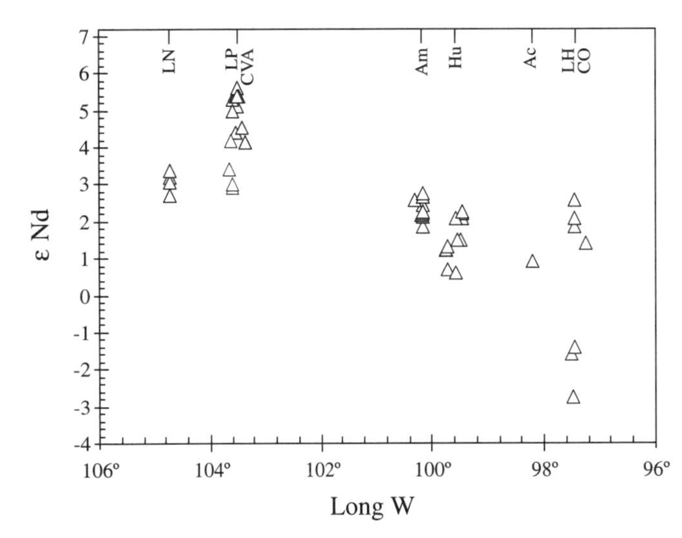

Figure 22. Geographic variations in the Nd isotopic compositions of rhyolitic rocks from the Trans-Mexican Volcanic Belt (Mahood and Halliday, 1988; Verma and Nelson, 1989; Verma et al., 1991; Besch et al., 1995; Verma, 2000b, 2001a, 2001b, 2001c; Maldonado-Sánchez and Schaaf, 2005). LN—Las Navajas; LP—La Primavera; CVA—Acatlán volcanic field; Am—Amealco caldera; Hu—Huichapan caldera; Ac—Acoculco caldera; LH—Los Humeros caldera; CO—Serdán-Oriental basin.

TECTONIC AND PETROGENETIC MODELS

The compositional diversity of a continental magmatic arc depends on various components and processes that are often difficult to identify and quantify. The current geologic and petrologic evidence indicates that, in the most complex scenario, arc volcanics could represent mixtures of a great number of geologic materials: (1) magmas derived from the partial fusion of the asthenospheric or lithospheric mantle that are intrinsically heterogeneous at all scales; (2) fluids released from the subducted basaltic crust, sediments, and underlying serpentinized mantle; (3) melts generated by partial fusion of the subducted materials; and (4) melts derived from a compositionally diverse continental crust. Even though interactions among these components could produce a plethora of compositions in the resulting magmas, their relative importance will strongly depend on the geologic and tectonic conditions that govern the thermal structure and deformation history of a subduction zone. Parameters, such as the age of the subducted slab, convergence rate, subduction dip angle, continental crustal thickness, and the state of stress that operates on the upper plate, are all fundamental in allowing some of the available components to be incorporated into the magmas to a lesser or greater extent. Indeed, while most workers agree that arc magmas represent some sort of recycling of materials derived from the mantle, the subducted crust, and the overriding plate, each arc has a unique geochemical imprint that must be attributed to subtle, or even prominent, variations in the parameters and components described above.

For these reasons the Trans-Mexican Volcanic Belt and the Mexican subduction zone have often been regarded as an anomaly among the magmatic arcs on Earth. In our view, it could hardly have turned out differently, since perhaps there is no other active continental margin on the planet with such a strong variability of geologic parameters operating at the same time. The current Mexican subduction zone is controlled by two independent oceanic plates with different ages, compositions, convergence velocities, and subduction dip angles that gradually vary along the Middle America Trench. These oceanic plates are also the result of a complex process of fragmentation and reorganization of the ancient Farallon plate. As a result, all of the parameters that govern the thermal structure of the subduction zone, the metamorphic histories of the subducted slabs, and thus the compositional characteristics of the slab-derived chemical agents that are presumably being delivered to the mantle wedge, should suffer modifications along the strike of the arc. Moreover, the arc is emplaced over a geologically diverse continental crust with different ages, thicknesses, and compositions. Each sector of the arc is also controlled by a complex state of stress and deformation history that involves extensional and strike-slip faulting that has been acting concurrently with magmatism. In addition, the convergent margin and the magmatic arc have not remained static throughout their histories, but instead have suffered notable modifications in terms of geometry and composition. With the help of an improved stratigraphic resolution coupled with high-quality geochemical data, it has also been recognized that different subduction components have been transferred to the magmatic products at different evolutionary stages.

Given all of these complexities, it is not surprising that the origin and petrologic evolution of the Trans-Mexican Volcanic Belt have been greatly debated for more than a century. These debates, which were particularly intense during the second half of the twentieth century, extend to today, and will most likely continue for many years to come. Indeed, in spite of the many theories proposed for the origin of the magmatism in the Trans-Mexican Volcanic Belt, researchers are far from reaching a consensus, which should perhaps be taken as evidence of the complexity that scientists have faced over the years.

The first scientific observations and interpretations of the Mexican volcanoes are generally attributed to the great German naturalist Alexander von Humboldt (1808). Humboldt described the alignment of stratovolcanoes along the parallel ~19° N as the surface expression of a major crustal discontinuity that was supposed to crosscut the continent from coast to coast. Even though more recent stratigraphic and tectonic studies have not recognized clear evidence for this great crustal discontinuity, Humboldt's hypothesis was very popular for a long period of time: it was adopted, modified, and extended by various authors during the 1950s, and it continued to find supporters until the 1970s. During those years, some researchers considered what they called the "Mexican Neovolcanic Axis" (the term used for the Trans-Mexican Volcanic Belt during the first half of the twentieth century) as the eastern extension of the Pacific Clarion fracture zone (Menard, 1955; Mooser and Maldonado-Koerdell, 1961). These researchers considered the Revillagigedo Islands as the westernmost expression of the volcanic arc. Later on, Mooser (1972) proposed the existence of a great pre-Paleozoic "geosuture" under the arc, which divided two cratonic blocks, that was later reactivated by the subduction regime along the Pacific margin. In this model, magmatism along the arc was related to differential heating along the great geosuture. Gastil and Jensky (1973) also considered a dextral strike-slip faulting system below the arc as a natural consequence of the transcurrent tectonics of the Gulf of California.

As plate tectonics took root as the paradigm of modern geology, most researchers considered magmatism in the Trans-Mexican Volcanic Belt to be the direct result of subduction of the Cocos and Rivera plates under the North American plate (Urrutia-Fucugauchi and Del Castillo, 1977; Urrutia-Fucugauchi and Böhnel, 1987; Pardo and Suárez, 1993, 1995). Many researchers today acknowledge that the nonparallel distribution of volcanoes with respect to the Middle America Trench is due to variations in the geometry of the subducted slabs, and especially to variations in the subduction angle (Menard, 1978; Nixon, 1982; Urrutia-Fucugauchi and Böhnel, 1987; Pardo and Suárez, 1993, 1995). Nonetheless, several authors have also recognized that a number of rocks emplaced along the arc do not show the petrological and chemical characteristics that would be expected in a typical convergent margin. Instead, these alkaline and transitional rocks have trace-element patterns and isotopic compositions that are similar to intraplate volcanics (oceanic-island basalts and continental rifts). Yet these rocks are not isolated, or involved in the formation of independent volcanic fields, but are always associated in time and space with more typical arc-like calc-alkaline and alkaline volcanics. Several lines of thought have been explored to explain the origin and genetic relationship of this peculiar petrologic diversity, but a unifying theory that reconciles all available observations has not been found.

On one hand, some workers have suggested a tight relation between subduction processes and local tectonics to explain the origin of both rock suites (Luhr and Carmichael, 1985a; Luhr, 1997; Wallace and Carmichael, 1999; Ferrari et al., 2001). In these models, magmas with high LILE/HFSE ratios are related to a compositionally heterogeneous mantle, yet mostly depleted and MORB-like, that has been intensively modified by fluids derived from the dehydration of the subducted Cocos and Rivera plates. In contrast, rocks with low LILE/HFSE ratios are related to the migration of an enriched OIB-like asthenospheric mantle that has not suffered intense modifications from the subduction zone, but that is able to melt due to the extensional tectonic regime that operates within the volcanic arc. It has been proposed that these enriched mantle domains could have been transferred from the backarc region by mantle convection and corner flow (Luhr, 1997; Wallace and Carmichael, 1999), because the compositions of the alkaline intraplate rocks are very similar to those emplaced in the Mexican Basin and Range (Luhr et al., 1989a). Consequently, these models suggest the existence of at least two compositionally different mantles below the Mexican arc that concurrently undergo partial fusion because of the interaction of two different melting mechanisms: adiabatic decompression (associated with extension) and fluid-fluxed melting (associated with subduction).

On the other hand, some authors have suggested that there is a heterogeneous mantle wedge, although enriched and similar to the source of OIB, that undergoes partial fusion by adiabatic decompression forming primitive magmas without appreciable subduction signatures (Márquez et al., 1999a; Verma, 1999; Sheth et al., 2000; Verma, 2000a, 2000b, 2002). In these models, more evolved rocks, which invariably present arc-like geochemical features, become so because of crustal contamination and/or direct anatexis, and therefore do not reflect the chemical imprints derived from the subduction environment. Some authors have even proposed that these OIB-like rocks could be associated with a mantle plume (Moore et al., 1994; Márquez et al., 1999a). Yet whereas Moore et al. (1994) restricted the activity of the plume to the Jalisco block, Márquez et al. (1999b) suggested that the whole length of the arc could be influenced by the plume and the concurrent activity of a continental rift that has been gradually propagating eastward since the Miocene. These models have been criticized by some workers who have shown that most of the OIB-like rocks of the western part of the arc display the negative Nb anomalies that are typical of arc volcanics, that there is no temporal progression of OIB volcanism toward the eastern part of México, and that there is no evidence for the prominent topographic uplift that could be expected to result from a mantle plume (Ferrari and Rosas-Elguera, 1999; Ferrari et al., 2001). Márquez et al. (1999a) defended their model, arguing that negative Nb anomalies do not necessarily reflect a subduction environment and that the absence of a prominent uplift is due to the lateral diffusion of the plume.

Even though the plume hypothesis has not attracted many experts (see for instance the extended discussion by Siebe et al. [2004a]), other authors still find it difficult to explain the petrologic diversity with the simple scenario of a magmatic arc. The detailed geochemical studies by Verma and co-workers on several volcanic centers of the Trans-Mexican Volcanic Belt, from the Miocene to the Quaternary, stand as the most prominent among these interpretations (Verma and Aguilar-y-Vargas, 1988; Verma et al., 1991; Verma, 1999, 2000a, 2000b, 2001a, 2000b, 2000c, 2002; Verma and Carrasco-Núñez, 2003; Verma and Hasenaka, 2004). In these publications, Verma and co-workers follow a similar line of thought, disregarding the influence of the subducted oceanic crust in the petrogenesis of the Trans-Mexican Volcanic Belt. The sudden disappearance of seismicity at ~80 km depth in the subducted Cocos plate and the consequent lack of a well-defined Wadatti-Beioff zone below the arc have been taken as evidence that the Trans-Mexican Volcanic Belt is not influenced by the subducted oceanic crust. In fact, Verma (2002) indicated that the extrapolations of the Cocos slab to ~100 km below the arc suggested by Pardo and Suárez (1995) are speculative if not completely erroneous. Verma (2002) argued that if the subducted slab remains at a shallower depth until reaching the arc front depths, it will be very difficult to achieve the thermal conditions for influencing magmatism in the Trans-Mexican Volcanic Belt. Yet volcanism in central México is clearly active and prolific, and it is associated with high heat flow. Verma and co-workers have also noted that the chemical compositions of some of the most primitive lavas in the arc do not present significant subduction signatures, and are in fact very different from what is observed in other mature arcs (Verma 2000a, 2002). Indeed, if the compositions observed in the Trans-Mexican Volcanic Belt are, for instance, compared to those from the Central American arc, it is very clear that both differ in their LILE/HFSE ratios, even though both segments of the Middle America Trench are subducted by the Cocos plate. But perhaps the most recurring argument used by Verma and co-workers is based on comparisons of the Sr, Nd, and Pb isotopic compositions displayed by the Mexican volcanics and a simple binary mixing line formed by the altered oceanic crust (AOC) and its overlying sediments, both sampled at DSDP site 487 (Fig. 18). Figure 18 shows that while rocks of the Central American arc volcanic front show a tendency toward this binary mixing model (Feigenson and Carr, 1986; Feigenson et al., 2004), the rocks in the Trans-Mexican Volcanic Belt tend to have lower $^{87}Sr/^{86}Sr$ ratios at similar $^{143}Nd/^{144}Nd$ values, plot away from the mixing trend of the subducted slab, and are in any case similar to the values observed in rocks from the Central American backarc region. This characteristic has been taken by Verma (1999, 2000a, 2002) as a strong argument for disregarding the participation of the subducted oceanic crust in the petrogenesis of the Trans-Mexican Volcanic Belt, and even though some authors have questioned the representativity of the AOC sampled at DSDP site 487 (Gómez-Tuena et al., 2003), it is also true that the divergence of the isotopic data from the mixing line is still difficult to explain by those who are looking for isotopic contributions of the subducted slab to the Mexican arc.

In any case, it is clear that Verma and co-workers do not consider the Trans-Mexican Volcanic Belt to be a continental volcanic arc because, in their view, it presents neither the tectonic nor the chemical compositions that should be present in a typical convergent margin. Verma and co-workers have instead suggested that magmatism in the Trans-Mexican Volcanic Belt is the product of an active continental rift that operates in central México with a structural expression denoted by the conspicuous presence of normal faults along the arc. The primitive magmatic products are thus related to partial melting of the mantle during adiabatic decompression (Verma and Aguilar-y-Vargas, 1988; Verma et al., 1991; Verma, 1999, 2000a, 2000b, 2001a, 2001b, 2001c, 2002; Verma and Carrasco-Núñez, 2003; Verma and Hasenaka, 2004). Primitive magmas will be compositionally similar to those observed at intraplate settings, whereas more evolved rocks, with the chemical characteristics of arc magmas, will become so through crustal contamination or crustal anatexis without any influence from the slab-derived chemical agents (Verma, 1999, 2002).

Even though the work of Verma and co-workers is an important contribution to our understanding of the Trans-Mexican Volcanic Belt, their rifting model has not been widely accepted (Ferrari and Rosas-Elguera, 1999; Siebe et al., 2004a). The vast majority of researchers conducting petrologic and geochemical studies on the Mexican arc are inclined to think that magmas are influenced to various extents by the subducted slabs. It has become clear nonetheless that integrating all the tectonic and petrologic peculiarities of the Trans-Mexican Volcanic Belt into a classic arc paradigm, as if it were, for instance, an island arc of the western Pacific, is irremediably condemned to failure because there is really no geologic condition that makes them similar, except for a subduction zone. For these reasons, it is noteworthy that the interpretations that have been inclined toward a subduction scenario are not exempt from limitations and controversy.

The petrologic studies done by the group at the University of California at Berkeley have provided some of the most compelling evidence to explain the origin of the Mexican magmatic sequences in terms of a volcanic arc. The more than two decades of studies performed by this group on the Mexican volcanoes, summarized in one of the most complete contributions about andesitic rocks in México (Carmichael, 2002), have shown that the mineral diversity and crystallinity of the volcanic rocks are directly linked to the amount of water dissolved in the magmas. Several lines of evidence indicate that at least some of this water must come from a source in the mantle. Indeed, Blatter and Carmichael (1998a) discovered the existence of water-rich andesitic magmas that contain rare hornblende-bearing mantle peridotite nodules. These findings clearly indicate that at least a portion of the mantle wedge below the arc has to be considerably hydrated. The presence of water and other volatile components (CO_2, S, Cl) in the mantle source of basalts has been confirmed by recent studies of melt inclusions in primitive olivines (Fo_{85-90}) (Cervantes and Wallace, 2003a, 2003b). These inclusions (tiny samples of basaltic magma that were trapped in the olivines during crystallization) have water contents that vary between 1.3% and 5.2%, incorporate large quantities of other fluid mobile elements, and have the high LILE/HFSE ratios that are typical of arc magmas. The presence of water and other volatiles in the most primitive melt inclusions is strong evidence for slab dehydration and recycling in the source of basaltic magmas (Cervantes and Wallace, 2003a, 2003b).

These studies seem to confirm the existence of a partially hydrated mantle wedge that is the source of magmas that are equally rich in water and other volatiles. This evidence, along with various experimental studies done of the Mexican rocks, led some researchers to propose that certain hydrous andesitic magmas could have been formed by direct partial fusion of the mantle with water contents that are close to saturation at 1 GPa (Blatter and Carmichael, 2001). Interpretations like this have enormous implications on our current knowledge of magma genesis in the Trans-Mexican Volcanic Belt, and of arcs in general, because they suggest that direct partial fusion of the mantle does not necessarily produce basaltic melts, as is generally assumed, but that SiO_2 rich liquids can also be formed provided that sufficient water is available. It might not be redundant to note that the only tectonic setting in which the mantle can be so hydrous is in a subduction zone. Therefore, if this hypothesis is correct, then the traditional strategy that we have been following to recognize the chemical imprints of the subduction environment (that is, analyzing the most primitive basalts) has allowed us to observe only a very small part of the problem, and to make interpretations that are, at best, incomplete.

The contributions of the continental crust as an essential part of magmatism in the Trans-Mexican Volcanic Belt have been documented since the classic studies of Wilcox (1954) in Parícutin. The presence of xenoliths and xenocrysts in clear chemical disequilibrium with the host magma (Blatter and Carmichael, 1998b), and the temporal evolution of some primitive magmas to more differentiated and isotopically enriched compositions (McBirney et al., 1987), constitute compelling evidence that mantle-derived magmas interact with crustal rocks at higher levels. Nonetheless, it is still unclear how much these crustal contributions influence the chemical compositions of the arc in general, and if these processes are able to obscure, or even erase, the chemical imprints from the subduction environment. Indeed, even though correlations between isotopic enrichment and differentiation indexes have been locally documented (McBirney et al., 1987; Chesley et al., 2002), these correlations are not always observed and there are just a few examples for which the origin and composition of the contaminant have been unequivocally identified. To overcome these limitations, the composition of the contaminant quite often has had to be inferred and the models adapted accordingly to coincide with a hypothetical assimilation process (Verma, 1999, 2000b, 2001a). On the other hand, other authors have simply disregarded the possibility that some of the evolved rocks in the arc could have their origin in a process of assimilation-fractional crystallization (AFC). For instance, Straub and Martin-Del Pozzo (2001) observed that the magmatic evolution of Popocatépetl volcano is related to magma mixing (as is the case for all large Mexican stratovolcanoes), but using simple thermodynamic and geochemical calculations, the authors reached the conclusion that a huge volume of ultramafic cumulates, equal to the volcano itself, would be needed in order to explain the origin of the most evolved rocks by the process of crustal contamination. In summary, even though interactions with the continental crust appear to be an inevitable phenomenon, it is still hard to define and quantify the crust's specific contributions to the arc magmas because we basically ignore the geological character of the crust below the arc, the structural depths at which contamination occurs, and the specific physicochemical mechanism that governs the process of contamination itself.

Some more recent studies have attempted not only to identify and quantify the specific geologic components derived from the subduction environment and the continental crust, but also to make inferences about the physical mechanisms of element recycling in the subduction zone. Comprehensive data sets on well-studied rock suites are essential for the accurate determination of these components and processes, because they allow people to make direct inferences about the tectonic process that governs the compositional variations of the arc in time and space. One such study was performed in the easternmost part of the Trans-Mexican Volcanic Belt, in the Palma Sola region (Gómez-Tuena et al., 2003). These authors recognized significant changes in the chemical characteristics of the subduction agents that have been injected into the mantle wedge from the middle Miocene to the Quaternary. They suggested that the earliest magmatic activity in the Palma Sola region was strongly influenced by melting of the subducted oceanic crust. Several middle to late Miocene volcanic successions in other parts of the Trans-Mexican Volcanic Belt also show the typical characteristics of adakites (Fig. 20), for which slab melting has been inferred (Kay, 1978; Defant and Drummond, 1990). Gómez-Tuena et al. (2003) also recognized that the subsequent magmatic event (late Miocene to Pleistocene) in the Palma Sola region derives from the partial fusion of a relatively deeper mantle region that was variably metasomatized by subducted sediment melts alone. In contrast, the most recent volcanics do not show important contributions from the subducted oceanic crust but have instead been influenced by contamination with the local continental crust. The temporal differences in the compositions of magmatic rocks have been explained by gradual modifications in the geometry of the subducted slab: The earliest magmas were formed during a phase of flat subduction, or a very shallow subduction angle, that has been gradually sinking until reaching its present position. A gradual change in the subduction angle influences the thermal and metamorphic path of the subducted oceanic crust, and allows different materials to be transferred from the slab into distinct portions of the mantle wedge. This process is also geologically expressed in a trenchward migration of the volcanic front.

The compositional diversity of the magmatic arc in time and space has provided the framework for more complex models that take into account not only the geochemical and petrologic evidence, but also the geologic and tectonic evolution of the convergent margin. A recent study by Ferrari (2004) proposed a novel hypothesis that, if confirmed, could reconcile some of the controversies that have been generated over the years. By making a detailed analysis of the regional geology and stratigraphy of the arc, Ferrari (2004) discovered an important mafic magmatic pulse that apparently migrated eastward from the Pacific Coast to the Gulf of Mexico between ca. 11.5 Ma and ca. 6 Ma. This magmatic event is geographically located toward the northern limit of the Trans-Mexican Volcanic Belt (Fig. 2B). Ferrari (2004) interpreted this mafic pulse as the volcanic expression of a slab-detachment event that started to develop at the mouth of the Gulf of California during late Miocene times, when the East Pacific Rise intersected the North American plate. The tear in the slab gradually propagated eastward to the Gulf of Mexico, paralleling the southern México trench system. The breaking of the subducted plate induced a sudden infiltration of deep, asthenospheric mantle, and produced a transitory increase of upper-mantle temperature promoting partial melting. The eruption of alkaline intraplate basalts 5–3 Ma after the inferred age of detachment in the western and central sectors of the arc is thus attributed to partial melting of a chemically enriched asthenospheric mantle that gradually filled the gap left by the missing slab. The mantle wedge below the current subduction zone may thus consist of enriched domains (the source of intraplate magmas) embedded in a matrix of depleted mantle (the source of subduction-related magmas), and melting of both domains may occur simultaneously.

SOME DIRECTIONS FOR FUTURE RESEARCH

Throughout this work we have tried to summarize most of the evidence that, in our view, is relevant to understanding the origin of the Trans-Mexican Volcanic Belt as a whole. We have combined geologic and stratigraphic information with geochemical and geophysical data, but we are also cognizant of the bias we have displayed by emphasizing the kind of studies that are most familiar to us: geochemistry and field geology. Far from being an exercise of autocriticism, we make this observation to emphasize that understanding such a complex region will require a true interdisciplinary approach. Research conducted on the Mexican arc has become highly specialized over the years; yet there are still very few scientists transcending their own fields and collaborating with researchers from other disciplines. We believe that the most fruitful perspective for the study of the Mexican arc will appear when geologists, geophysicists, petrologists, and geochemists start to cooperate and attack those questions that they have not been able to resolve alone. We will now highlight a few examples that could lead some of the future research.

The geochemical signals observed in arc magmas depend on the thermal structure of a subduction zone, which has a strong effect on the mineral stability of the subducted slab and its surrounding mantle. The subducted Cocos plate vanishes from seismic sight before reaching arc-front depths, and we are currently ignorant of whether the slab gradually sinks until the hypothetical ~100 km depth (Fig. 4), continues its subhorizontal trajectory, or is truncated below the arc. For this reason, a detailed seismic tomographic study of central México is of utter importance (see http://www.tectonics.caltech.edu/mase/index.html). Yet it is also important that the geologic and geochemical background is taken into consideration for data reduction and numerical modeling, if we aim to reconcile the experimental observations with the geophysical and field data.

An accurate discrimination between the contributions from the subducted oceanic plate and those from the continental crust in the chemistry of arc magmas requires a precise knowledge of the geological and chemical characteristics of the components involved. Nonetheless, our current knowledge of the composition and thickness of the slab that is presumably being subducted is restricted to one single deep core (DSDP site 487) and just a few gravimetric and seismic profiles. We know that the age of the slab and the thickness of the sedimentary cover vary widely along the trench, but we largely ignore the specific contributions of these components to the subduction zone. Similarly, the continental crust upon which the arc is emplaced is also extremely variable in terms of both thickness and composition, but accurate determinations of important parameters such as crustal thickness, age, composition, and geologic configuration of the basement rocks are for the most part missing. These are major limitations for all petrogenetic models that attempt to recognize the contributions of the continental and oceanic crusts to the compositions of arc magmas. And once again, the joint effort of those in all fields of earth sciences will be necessary to fill these gaps.

The recognition that the chemical diversity of primitive igneous rocks in the arc is related to a compositionally heterogeneous mantle wedge has been a recurring subject in all publications. Nonetheless, the origin of these heterogeneities and the methods that have been traditionally followed to recognize them are not exempt from limitations and controversy. For instance, it has been traditionally assumed that differences in the major-element contents of primitive rocks, as well as in the concentrations and ratios of fluid immobile elements (i.e., the HFSE), should reflect the inherited compositional characteristics of the mantle source. The isotopic compositions of the most primitive rocks have also been studied from this perspective. However, it has been increasingly recognized from experimental evidence and high-quality data sets of various arcs around the world that the major elements could vary as a function of a large number of parameters that are difficult to quantify, that the HFSE could be mobilized to some extent in the subduction flux if the mechanism of transport is a melt derived from the subducted slab and not a fluid, and that the traditional isotopic systems (i.e., Sr, Nd, and Pb) are particularly sensitive to contributions from the subducted slab and the continental crust. Therefore it is not at all clear that the customary geochemical tools that have been employed so far truly reflect mantle wedge heterogeneities since they could also be influenced by the subduction regime, the sublithospheric mantle, or the continental crust. Moreover, the possibility that at least some of the evolved andesites could have a direct mantle origin, or be related to slab melting, offers new perspectives that deserve more meticulous investigation. Resolving all these ambiguities is not a trivial endeavor since most of the tectonic theories that have been proposed to explain the origin of mantle heterogeneities below the Trans-Mexican Volcanic Belt (i.e., continental rift, mantle plume, mantle migration from back-arc, slab detachment, etc.) strongly depend upon correct characterization of the mantle.

The past few decades have been noteworthy for their increase in the determination of the chemical and mineralogical compositions of magmatic rocks in the Mexican arc. We currently are well acquainted with the behavior of the major elements, are aware of the most common mineral assemblages, and have a reasonable picture of some of the most important trace elements. We also have an extended database for the most common isotopic systems. By using this information, we have recognized that rocks in the Trans-Mexican Volcanic Belt depart from well-established arc paradigms, and we have also realized that if we aim to understand these discrepancies, we need to make use of the recent advances in petrology, stratigraphy, and field geology, but also take advantage of the possibilities that the new technologies provide. We should continue to construct large trace-element data sets with better precision, and include those elements that were traditionally difficult to analyze when using older methods and instruments (for instance, Li, Be, B, Tl, Ta, Hf, and Sb). We also need more and better data for the traditional isotopes, and especially for Pb since its true potential has been recognized only very recently. But, concurrently, we should start exploring other less conventional isotopic systems that have proved useful in other magmatic arcs and settings. Studies of the Trans-Mexican Volcanic Belt have been lagging behind in the analysis of U and Th decay series for determining residence times and ascent velocities of magmas. There are no data available thus far for the stable isotopes of Li, Cl, and B, and only a handful of volcanic rocks have been analyzed for O and Be isotopes. The same can be said of Hf isotopes and the noble gases. Detailed studies of these systems could provide invaluable information on mantle heterogeneities, and on the physical mechanisms of element recycling in the subduction zone and the continental crust.

But as with any other scientific discipline, our work must go far beyond the simple collection of new data and observations. The recent advances in experimental petrology, and the mathematical tools that describe the behavior of chemical elements in the magmatic processes, offer researchers the possibility of quantitatively modeling geochemical observations. Quantifying the specific contributions to magmatism, and modeling its evolution through time and space, should thus be used more frequently to validate our hypothesis.

ACKNOWLEDGMENTS

We thank Tania Norato and Carolina Muñoz who contributed by compiling and organizing the geochemical database, and Susanne Straub for giving us access to her own data compilation for Quaternary volcanoes. Careful and thoughtful reviews by Álvaro Márquez, Peter Schaaf, and Ángel Nieto-Samaniego helped to clarify several ideas and greatly improved the manuscript. L. Ferrari and Ma.T. Orozoco-Esquivel were supported by two bilateral CNR-CONACyT grants. A. Gómez-Tuena was financially supported by a young scientist CONACyT grant (39785).

REFERENCES CITED

Aguirre-Díaz, G., 1996, Volcanic stratigraphy of the Amealco caldera and vicinity, Central Mexican Volcanic Belt: Revista Mexicana de Ciencias Geológicas, v. 13, p. 10–51.

Aguirre-Díaz, G., 2001, Recurrent magma mingling in successive ignimbrites from Amealco caldera, central Mexico: Bulletin of Volcanology, v. 63, p. 238–251, doi: 10.1007/s004450100138.

Aguirre-Díaz, G., and López-Martínez, M., 2001, The Amazcala caldera, Querétaro, central Mexican Volcanic Belt, México. Geology and geochronology: Journal of Volcanology and Geothermal Research, v. 111, p. 203–218, doi: 10.1016/S0377-0273(01)00227-X.

Aguirre-Díaz, G., and McDowell, F., 2000, Volcanic evolution of the Amealco caldera, central Mexico, *in* Delgado-Granados, H., Stock, J., and Aguirre-Díaz, G., eds., Cenozoic Tectonics and Volcanism of Mexico: Geological Society of America Special Paper 334, p. 167–178.

Aguirre-Díaz, G., Nelson, S., Ferrari, L., and López, M., 1997, Ignimbrites of the Central Mexican Volcanic Belt, Amealco and Huichapan calderas (Querétaro-Hidalgo), *in* Aguirre-Díaz, G., Aranda-Gómez, J., Carrasco-Núñez, G., and Ferrari, L., eds., Magmatism and Tectonics of Central and Northwestern Mexico—A Selection of the 1997 IAVCEI General Assembly Excursions: México, DF, Instituto de Geología, Universidad Nacional Autónoma de México, p. 1–39.

Aguirre-Díaz, G., Dubois, M., Laureyns, J., and Schaaf, P., 2002, Nature and P-T conditions of the crust beneath the central Mexican Volcanic Belt based on a Precambrian crustal xenolith: International Geology Review, v. 44, p. 222–242.

Alaniz-Álvarez, S., and Nieto-Samaniego, A., 2005, El sistema de fallas Taxco-San Miguel de Allende y la Faja Volcánica Transmexicana, dos fronteras tectónicas del centro de México activas durante el Cenozoico: Boletín de la Sociedad Geológica Mexicana, v: Tomo, v. LVII, no. 1, p. 63–80.

Alaníz-Álvarez, S., Nieto-Samaniego, A., and Ferrari, L., 1998, Effects of strain rate in the distribution of monogenetic and polygenetic volcanism in the Transmexican volcanic belt: Geology, v. 26, no. 7, p. 591–594, doi: 10.1130/0091-7613(1998)026<0591:EOSRIT>2.3.CO;2.

Alaníz-Álvarez, S., Nieto-Samaniego, A., and Ferrari, L., 1999, Effects of strain rate in the distribution of monogenetic and polygenetic volcanism in the Transmexican volcanic belt: Reply: Geology, v. 26, no. 7, p. 591–594, doi: 10.1130/0091-7613(1998)026<0591:EOSRIT>2.3.CO;2.

Alaniz-Álvarez, S., Nieto-Samaniego, A., Morán-Zenteno, D., and Alba-Aldave, L., 2002a, Rhyolitic volcanism in extension zone associated with strike-slip tectonics in the Taxco region, Southern México: Journal of Volcanology and Geothermal Research, v. 118, p. 1–14, doi: 10.1016/S0377-0273(02)00247-0.

Alaniz-Álvarez, S., Nieto-Samaniego, A., Orozco-Esquivel, M., Vasallo-Morales, L., and Xu, S., 2002b, El Sistema de Fallas Taxco-San Miguel de Allende: Implicaciones en la deformación Post-Eocénica del Centro de México: Boletín de la Sociedad Geológica Mexicana, v. 55, p. 12–29.

Allan, J., 1986, Geology of the Colima and Zacoalco grabens, SW Mexico: Late Cenozoic rifting in the Mexican Volcanic Belt: Geological Society of America Bulletin, v. 97, p. 473–485, doi: 10.1130/0016-7606(1986)97<473:GOTNCA>2.0.CO;2.

Allan, J., and Carmichael, I., 1984, Lamprophyric lavas in the Colima Graben, SW Mexico: Contributions to Mineralogy and Petrology, v. 88, p. 203–216, doi: 10.1007/BF00380166.

Allan, J., Nelson, S., Luhr, J., Carmichael, I., Wopat, M., and Wallace, P., 1991, Pliocene–Recent rifting in SW Mexico and associated volcanism: An exotic terrain in the making, *in* Dauphin, J., and Simoneit, B., eds., Peninsular Province of the Californias, American Association of Petroleum Geologists Memoir, p. 425–445.

Alvarez, R., 2002, Banderas rift zone: A plausible NW limit of the Jalisco Block: Geophysical Research Letters, v. 29, no. 20, p. DOI:10.1029/2002GL016089.

Alva-Valdivia, L., Goguitchaichvili, A., Rosas-Elguera, J., Urrutia-Fucugauchi, J., Ferrari, L., and Zamorano, J., 2000, Paleomagnetic data from the trans-Mexican volcanic belt: Implications for tectonics and volcanic stratigraphy: Earth, Planets and Space, v. 52, p. 467–478.

Atwater, T., 1970, Implications of Plate Tectonics for the Cenozoic Tectonic Evolution of Western North America: Geological Society of America Bulletin, v. 81, p. 3513–3536.

Atwater, T., and Stock, J., 1998, Pacific–North America plate tectonics of the Neogene southwestern United States: International Geology Review, v. 8, p. 375–402.

Ban, M., Hasenaka, T., Delgado-Granados, H., and Takaoka, N., 1992, K-Ar ages of lavas from shield volcanoes in the Michoacán-Guanajuato volcanic field, Mexico: Geofísica Internacional, v. 31, p. 467–473.

Barclay, J., and Carmichael, I., 2004, A hornblende basalt from Western Mexico: Water-saturated phase relations constrain a pressure temperature window of eruptibility: Journal of Petrology, v. 45, no. 3, p. 485–506, doi: 10.1093/petrology/egg091.

Barrier, E., Borgois, J., and Michaud, F., 1990, Le système de rifts actifs du point triple de Jalisco: vers un proto-golfe de Jalisco: Comptes Rendus de l'Académie des Sciences, Paris, v. 310, p. 1513–1520.

Besch, T., Negendank, J., and Emmermann, R., 1988, Geochemical constraints on the origin of the calc-alkaline and alkaline magmas of the eastern Trans-Mexican Volcanic Belt: Geofísica Internacional, v. 27, p. 641–663.

Besch, T., Verma, S., Kramm, U., Negendank, J., Tobschall, H., and Emmermann, R., 1995, Assimilation of sialic crustal material by volcanics of the easternmost extension of the Trans-Mexican Volcanic Belt—Evidence from Sr and Nd isotopes: Geofísica Internacional, v. 34, p. 263–281.

Blatter, D., and Carmichael, I., 1998a, Hornblende peridotite xenoliths from central Mexico reveal the highly oxidized nature of subarc upper mantle: Geology, v. 26, p. 1035–1038, doi: 10.1130/0091-7613(1998)026<1035:HPXFCM>2.3.CO;2.

Blatter, D., and Carmichael, I., 1998b, Plagioclase-free andesites from Zitácuaro (Michoacán), México: Petrology and experimental constraints: Contributions to Mineralogy and Petrology, v. 132, p. 121–138, doi: 10.1007/s004100050411.

Blatter, D., and Carmichael, I., 2001, Hydrous phase equilibria of a Mexican high-silica andesite: A candidate for mantle origin?: Geochimica et Cosmochimica Acta, v. 65, p. 4043–4065, doi: 10.1016/S0016-7037(01)00708-6.

Blatter, D., Carmichael, I., Deino, A., and Renne, P., 2001, Neogene volcanism at the front of the central Mexican volcanic belt: Basaltic andesites to dacites, with contemporaneous shoshonites and high-TiO2 lava: Geological Society of America Bulletin, v. 113, no. 10, p. 1324–1342, doi: 10.1130/0016-7606(2001)113<1324:NVATFO>2.0.CO;2.

Bloomfield, K., 1973, The age and significance of the Tenango Basalt: Bulletin of Volcanology, v. 37, p. 586–595.

Bloomfield, K., 1975, A late-Quaternary monogenetic volcano field in central Mexico: Geologische Rundschau, v. 64, p. 476–497, doi: 10.1007/BF01820679.

Boudal, C., and Robin, C., 1987, Relation entre dynamismes éruptifs et réalimentations magmatiques d'origine profonde au Popocatepelt: Canadian Journal of Earth Sciences, v. 25, p. 955–971.

Bourgois, J., and Michaud, F., 1991, Active fragmentation of the North America plate at the Mexican triple junction area off Manzanillo: Geo-Marine Letters, v. 11, p. 59–65, doi: 10.1007/BF02431030.

Bowen, N., 1928, The Evolution of Igneous Rocks: Princeton, New Jersey, Princeton University Press, 334 p.

Cabral-Cano, E., Lang, H., and Harrison, C., 2000a, Stratigraphic assessment of the Arcelia-Teloloapan area, Southern Mexico: Implications for Southern Mexico's Post-Neocomian tectonic evolution: Journal of South American Earth Sciences, v. 13, p. 443–457, doi: 10.1016/S0895-9811(00)00035-3.

Cabral-Cano, E., Draper, G., Lang, H., and Harrison, C., 2000b, Constraining the late Mesozoic and early Tertiary tectonic evolution of southern Mexico: Structure and deformation history of the Tierra Caliente region, southern Mexico: The Journal of Geology, v. 108, p. 427–446, doi: 10.1086/314414.

Campa, M.F., and Coney, P.J., 1983, Tectono-stratigraphic terranes and mineral resource distributions in Mexico: Canadian Journal of Earth Sciences, v. 20, p. 1040–1051.

Campillo, M., Singh, S., Shapiro, N., Pacheco, J., and Herrmann, R., 1996, Crustal structure south of the Mexican volcanic belt, based on group velocity dispersion: Geofísica Internacional, v. 35, no. 4, p. 361–370.

Campos-Enríquez, J., and Sánchez-Zamora, O., 2000, Crustal structure across southern Mexico inferred from gravity data: Journal of South American Earth Sciences, v. 13, p. 479–489, doi: 10.1016/S0895-9811(00)00045-6.

Campos-Enríquez, J., Arroyo-Esquivel, M., and Urrutia-Fucugauchi, J., 1990, Basement, curie isotherm and shallow-crustal structure of the Trans-Mexican Volcanic Belt, from aeromagnetic data: Tectonophysics, v. 172, p. 77–90, doi: 10.1016/0040-1951(90)90060-L.

Cantagrel, J., and Robin, C., 1979, K-Ar dating on eastern Mexican volcanic rocks—Relations between the andesitic and the alkaline provinces: Journal of Volcanology and Geothermal Research, v. 5, p. 99–114, doi: 10.1016/0377-0273(79)90035-0.

Capra, L., Macías, J., and Garduño, V., 1997, The Zitácuaro Volcanic Complex, Michoacán, México: Magmatic and eruptive history of a resurgent caldera: Geofísica Internacional, v. 36, no. 3, p. 161–179.

Carmichael, I., 2002, The andesite aqueduct: Perspectives on the evolution of intermediate magmatism in west-central (105°-99°W) Mexico: Contributions to Mineralogy and Petrology, v. 143, p. 641–663.

Carmichael, I., Lange, R., and Luhr, J., 1996, Quaternary minettes and associated volcanic rocks of Mascota, western Mexico: A consequence of plate extension above a subduction modified mantle wedge: Contributions to Mineralogy and Petrology, v. 124, p. 302–333, doi: 10.1007/s004100050193.

Carr, M., Feigenson, M., Patino, L., and Walker, J., 2003, Volcanism and Geochemistry in Central America: Progress and Problems, *in* Eiler, J., ed., Inside the Subduction Factory: Washington, DC, American Geophysical Union, p. 153–179.

Carrasco-Núñez, G., and Ban, M., 1994, Geologic map and structural sections of the summit area of Citlaltépetl Volcano, México: Instituto de Geología, Universidad Nacional Autónoma de México.

Carrasco-Núñez, G., Milán, M., and Verma, S., 1989, Geología del volcán El Zamorano, Estado de Querétaro: Revista Instituto de Geología, v. 8, p. 194–201.

Carrasco-Núñez, G., Gómez-Tuena, A., and Lozano-Velázquez, L., 1997, Geologic map of Cerro Grande Volcano and surrounding area, central Mexico: Geological Society of America Maps and Charts series MCH081, scale 1:100 000, 1 sheet, 10 p. text.

Carrasco-Núñez, G., Righter, K., Chesley, J., Siebert, L., and Aranda-Gómez, J., 2005, Contemporaneous eruption of calc-alkaline and alkaline lavas in a continental arc (Eastern Mexican Volcanic Belt): Chemically heterogeneous but isotopically homogeneous source: Contributions to Mineralogy and Petrology, v. 150, no. 4, p. 423–440, doi: 10.1007/s00410-005-0015-x.

Cathalieau, M., Oliver, R., and Nieva, D., 1987, Geochemistry of volcanic series of the Los Azufres geothermal field (Mexico): Geofísica Internacional, v. 26, no. 2, p. 273–290.

Centeno-García, E., Ruíz, J., Coney, P.J., Patchett, P.J., and Ortega-Gutiérrez, F., 1993, Guerrero terrane of Mexico: Its role in the Southern Cordillera from new geochemical data: Geology, v. 21, p. 419–422, doi: 10.1130/0091-7613(1993)021<0419:GTOMIR>2.3.CO;2.

Cervantes, P., and Wallace, P., 2003a, Role of H2O in subduction-zone magmatism: New insights from melt inclusions in high-Mg basalts from central Mexico: Geology, v. 31, no. 3, p. 235–238, doi: 10.1130/0091-7613(2003)031<0235:ROHOIS>2.0.CO;2.

Cervantes, P., and Wallace, P., 2003b, Magma degassing and basaltic eruption styles: A case study of ~2000 year BP Xitle volcano in central Mexico: Journal of Volcanology and Geothermal Research, v. 120, p. 249–270, doi: 10.1016/S0377-0273(02)00401-8.

Chesley, J., Ruiz, J., Righter, K., Ferrari, L., and Gómez-Tuena, A., 2002, Source contamination versus assimilation: An example from the Trans-Mexican Volcanic Arc: Earth and Planetary Science Letters, v. 195, p. 211–221, doi: 10.1016/S0012-821X(01)00580-5.

Connor, C., 1990, Cinder cone clustering in the Trans-Mexican volcanic belt: Implications for structural and petrologic models: Journal of Geophysical Research, v. 95, p. 19,395–19,405.

Contreras, J., and Gómez-Tuena, A., 1999, Effect of strain rate in the distribution of monogenetic and polygenetic volcanism in the Transmexican volcanic belt: Comment: Geology, v. 27, p. 571–572, doi: 10.1130/0091-7613(1999)027<0571:EOSRIT>2.3.CO;2.

Currie, C., Hyndman, R., Wang, K., and Kostoglodov, V., 2002, Thermal models of the Mexico subduction zone: Implications for the megathrust seismogenic zone: Journal of Geophysical Research, v. 107, no. B12, p. doi: 10.1029/2001JB000886.

De Cserna, Z., 1958, Notes on the tectonics of Southern Mexico: American Association of Petroleum Geologists, v. 86, p. 523–532.

Defant, M., and Drummond, M., 1990, Derivation of some modern arc magmas by melting of young subducted lithosphere: Nature, v. 347, p. 662–665, doi: 10.1038/347662a0.

De la Fuente, M., Aitken, C., and Mena, M., 1994, Cartas gravimétricas de la República Mexicana, Carta de anomalía de Bouguer: Universidad Nacional Autónoma de México.

Delgado-Granados, H., Urrutia-Fucugauchi, J., Hasenaka, T., and Masso, B., 1995, Southwestward volcanic migration in the western Trans-Mexican Volcanic Belt during the last 2 Ma: Geofísica Internacional, v. 34, p. 341–352.

Demant, A., 1978, Características del Eje Neovolcánico Transmexicano y sus problemas de interpretación: Revista Instituto de Geología, v. 2, p. 172–187.

Demant, A., 1979, Vulcanologia y petrografia del sector occidental del Eje Neovolcanico: Revista Instituto de Geología, UNAM, v. 3, p. 39–57.

Demant, A., 1981, Interpretación geodinámica del volcanismo del Eje Neovolcánico Transmexicano: Revista Instituto de Geología, v. 5, p. 217–222.

Demant, A., and Robin, C., 1975, Las fases del vulcanismo en México, una síntesis en relación con la evolución geodinámica desde el Cretácico: Revista Instituto de Geología, v. 75, p. 66–79.

DeMets, C., and Traylen, S., 2000, Motion of the Rivera plate since 10 Ma relative to the Pacific and North American and the mantle: Tectonophysics, v. 318, p. 119–159, doi: 10.1016/S0040-1951(99)00309-1.

DeMets, C., Gordon, R., Argus, D., and Stein, S., 1994, Effects of recent revisions to the geomagnetic reversal time scale on estimates of current plate motions: Geophysical Research Letters, v. 21, p. 2191–2194.

DePaolo, D., 1981, Trace element and isotopic effects of combined wallrock assimilation and fractional crystallization: Earth and Planetary Science Letters, v. 53, p. 189–202, doi: 10.1016/0012-821X(81)90153-9.

Dickinson, W., and Hatherton, T., 1967, Andesitic volcanism and seismicity around the Pacific: Science, v. 157, p. 801–803.

Eggler, D., 1972, Water-saturated and undersaturated melting relations in a Paricutin andesite and an estimate of water content in the natural magma: Contributions to Mineralogy and Petrology, v. 34, p. 261–271, doi: 10.1007/BF00373757.

Ego, F., and Ansan, V., 2002, Why is the Central Trans-Mexican Volcanic Belt (102°–99°W) in transtensive deformation?: Tectonophysics, v. 359, p. 189–208, doi: 10.1016/S0040-1951(02)00511-5.

Elías-Herrera, M., and Ortega-Gutiérrez, F., 1997, Petrology of high-grade metapelitic xenoliths in an Oligocene rhyodacite plug—Precambrian crust beneath the southern Guerrero terrane, Mexico?: Revista Mexicana de Ciencias Geológicas, v. 14, no. 1, p. 101–109.

Elías-Herrera, M., and Ortega-Gutiérrez, F., 1998, The Early Cretaceous Arperos oceanic basin (western Mexico). Geochemical evidence for an aseismic ridge formed near a spreading center: Comment: Tectonophysics, v. 292, p. 321–326, doi: 10.1016/S0040-1951(98)00051-1.

Elías-Herrera, M., Ortega-Gutiérrez, F., and Lozano-Santa Cruz, R., 1998, Evidence for pre-Mesozoic sialic crust in the southern Guerrero terrane: Geochemistry of the Pepechuca high grade gneiss xenoliths: Actas INAGEQ, v. 4, p. 169–181.

Elías-Herrera, M., Ortega-Guitérrez, F., Sánchez-Zavala, J., Macías-Romo, C., Ortega-Rivera, M., and Iriondo, A., 2005, La falla de Caltepec: Raíces expuestas de una frontera tectónica de larga vida entre dos terrenos continentales del sur de México: Boletín de la Sociedad Geológica Mexicana, v. Volumen Conmemorativo del Centenario, no: Tomo LVII, Número, v. 1, p. 83–109.

Engebretson, A., Cox, A., and Gordon, R., 1985, Relative motions between oceanic and continental plates in the Pacific Basin: Geological Society of America Special Paper 206, 64 p.

Feigenson, M., and Carr, M., 1986, Positively correlated Nd and Sr isotope ratios of lavas from the Central American volcanic front: Geology, v. 14, p. 79–82, doi: 10.1130/0091-7613(1986)14<79:PCNASI>2.0.CO;2.

Feigenson, M., Carr, M., Maharaj, S., Juliano, S., and Bolge, L., 2004, Lead isotope composition of Central American volcanoes: Influence of the Galapagos plume: Geochemistry, Geophysics, Geosystems, v. 5, no. Q06001, doi: 10.1029/2003GC000621.

Ferrari, L., 1995, Miocene shearing along the northern boundary of the Jalisco block and the opening of the southern Gulf of California: Geology, v. 23, no. 8, p. 751–754, doi: 10.1130/0091-7613(1995)023<0751:MSATNB>2.3.CO;2.

Ferrari, L., 2004, Slab detachment control on mafic volcanic pulse and mantle heterogeneity in central Mexico: Geology, v. 32, no. 1, p. 77–80, doi: 10.1130/G19887.1.

Ferrari, L., and Rosas-Elguera, J., 1999, Alkalic (ocean-island basalt type) and calc-alkalic volcanism in the Mexican volcanic belt: A case for plume-related magmatism and propagating rifting at an active margin?: Comment: Geology, v. 27, p. 1055–1056, doi: 10.1130/0091-7613(1999)027<1055:AOIBTA>2.3.CO;2.

Ferrari, L., and Rosas-Elguera, J., 2000, Late Miocene to Quaternary extension at the northern boundary of the Jalisco block, western Mexico: The Tepic-Zacoalco rift revised, *in* Aguirre-Díaz, G., Delgado-Granados, H., and Stock, J., eds., Cenozoic Tectonics and Volcanism of Mexico, Geological Society of America Special Paper 334, p. 42–64.

Ferrari, L., Pasquaré, G., and Tibaldi, A., 1990, Plio-Quaternary tectonics of central Mexican Volcanic Belt and some constraints on its rifting mode: Geofísica Internacional, v. 29, p. 5–18.

Ferrari, L., Garduño, V., Pasquaré, G., and Tibaldi, A., 1991, Geology of Los Azufres caldera, Mexico, and its relations with regional tectonics: Journal of Volcanology and Geothermal Research, v. 47, p. 129–148, doi: 10.1016/0377-0273(91)90105-9.

Ferrari, L., Garduño, V., Innocenti, F., Manetti, P., Pasqueré, G., and Vaggelli, G., 1994, A widespread mafic volcanic unit at the base of the Mexican Volcanic Belt between Guadalajara and Querétaro: Geofísica Internacional, v. 33, p. 107–123.

Ferrari, L., Lopez-Martinez, M., Aguirre-Díaz, G., and Carrasco-Núñez, G., 1999, Space-time patterns of Cenozoic arc volcanism in central Mexico: From the Sierra Madre Occidental to the Mexican volcanic belt: Geology, v. 27, p. 303–306, doi: 10.1130/0091-7613(1999)027<0303:STPOCA>2.3.CO;2.

Ferrari, L., Pasquaré, G., Venegas, S., and Romero, F., 2000a, Geology of the western Mexican Volcanic Belt and adjacent Sierra Madre Occidental and Jalisco block, *in* Delgado-Granados, H., Aguirre-Díaz, G., and Stock, J., eds., Cenozoic Tectonics and Volcanism of Mexico: Geological Society of America Special Paper 334, p. 65–84.

Ferrari, L., Vaggelli, G., Petrone, C., Manetti, P., and Conticelli, S., 2000b, Late Miocene volcanism and intra-arc tectonics during the early development of the Trans-Mexican Volcanic Belt: Tectonophysics, v. 318, p. 161–185, doi: 10.1016/S0040-1951(99)00310-8.

Ferrari, L., Petrone, C., and Francalanci, L., 2001, Generation of oceanic-island basalt-type volcanism in the western Trans-Mexican volcanic belt by slab rollback, asthenosphere infiltration, and variable flux melting: Geology, v. 29, no. 6, p. 507–510, doi: 10.1130/0091-7613(2001)029<0507:GOOIBT>2.0.CO;2.

Ferrari, L., López-Martínez, M., González-Cervantes, N., Jacobo-Albarrán, J., and Hernández-Bernal, M., 2003a, Volcanic record and age of formation of the Mexico City basin: Reunión Anual de la Unión Geofísica Mexicana, v. 23, p. 120.

Ferrari, L., Petrone, C., Francalanci, L., Tagami, T., Eguchi, M., Conticelli, S., Manetti, P., and Venegas-Salgado, S., 2003b, Geology of the San Pedro-Ceboruco graben, western Trans-Mexican Volcanic Belt: Revista Mexicana de Ciencias Geológicas, v. 20, p. 165–181.

Ferrari, L., Rosas-Elguera, J., Orozco-Esquivel, M., Carrasco-Núñez, G., Norato-Cortez, T., and Gonzalez-Cervantes, N., 2005a, Digital geologic cartography of the Trans-Mexican Volcanic Belt (online), Digital Geosciences <http://satori.geociencias.unam.mx/digital_geosciences>, Universidad Nacional Autónoma de México.

Ferrari, L., Tagami, T., Eguchi, M., Orozco-Esquivel, M., Petrone, C., Jacobo-Albarrán, J., and López-Martínez, M., 2005b, Geology, geochronology and tectonic setting of late Cenozoic volcanism along the southwestern Gulf of Mexico: The Eastern Alkaline Province revisited: Journal of Volcanology and Geothermal Research, v. 146, p. 284–306, doi: 10.1016/j.jvolgeores.2005.02.004.

Ferriz, H., and Mahood, G., 1984, Eruption rates and compositional trends at Los Humeros Volcanic Center, Puebla, Mexico: Journal of Geophysical Research, v. 89, p. 8511–8524.

Ferriz, H., and Mahood, G., 1986, Volcanismo Riolítico en el Eje Neovolcánico Mexicano: Geofísica Internacional, v. 25, p. 117–156.

Ferriz, H., and Mahood, G., 1987, Strong compositional zonation in a silicic magmatic system: Los Humeros, Mexican Neovolcanic Belt: Journal of Petrology, v. 28, p. 171–209.

Flores-Ruiz, J., 1997, Estructura cortical de la Faja Volcánica Transmexicana [Ph.D. thesis]: Mexico, D.F., Universidad Nacional Autónoma de México, 150 p.

Foley, S., 1992, Vein-plus-wall-rock melting mechanisms in the lithosphere and the origin of potassic alkaline magmas: Lithos, v. 28, p. 435–453, doi: 10.1016/0024-4937(92)90018-T.

Frey, H., Lange, R., Hall, C., and Delgado-Granados, H., 2004, Magma eruption rates constrained by 40Ar/39Ar chronology and GIS for the Ceboruco–San Pedro volcanic field, western Mexico: Geological Society of America Bulletin, v. 116, p. 259–276, doi: 10.1130/B25321.1.

Freydier, C., Martinez, J., Lapierre, H., Tardy, M., and Coulon, C., 1996, The Early Cretaceous Arperos oceanic basin (western Mexico): Geochemical evidence for an aseismic ridge formed near a spreading center: Tectonophysics, v. 259, no. 4, p. 343–367, doi: 10.1016/0040-1951(95)00143-3.

Freydier, C., Lapierre, H., Briqueu, L., Tardy, M., Coulon, C., and Martinez-Reyes, J., 1997, Volcaniclastic sequences with continental affinities within the late Jurassic early Cretaceous Guerrero intra-oceanic arc terrane (western Mexico): The Journal of Geology, v. 105, no. 4, p. 483–502.

García-Palomo, A., Macías, J., and Garduño, V., 2000, Miocene to Recent structural evolution of the Nevado de Toluca volcano region, Central Mexico: Tectonophysics, v. 318, p. 281–302, doi: 10.1016/S0040-1951(99)00316-9.

García-Palomo, A., Macías, J., Tolson, G., Valdez, R., and Mora-Chaparro, J., 2002, Volcanic stratigraphy and geological evolution of the Apan region, east-central sector of the Transmexican Volcanic Belt: Geofísica Internacional, v. 41, p. 133–150.

García-Perez, F., and Urrutia-Fucugauchi, J., 1997, Crustal structure of the Arteaga Complex, Michoacán, southern Mexico, from gravity and magnetics: Geofísica Internacional, v. 36, p. 235–244.

Garduño, V., and Tibaldi, A., 1991, Kinematic evolution of the continental active triple junction of the western Mexican Volcanic Belt: Comptes Rendus de l'Académie des Sciences, Paris, v. II, p. 135–142.

Garduño-Monroy, V., Spinnler, J., and Ceragioli, E., 1993, Geological and structural study of the Chapala Rift, state of Jalisco, Mexico: Geofísica Internacional, v. 32, p. 487–499.

Garduño-Monroy, V., Saucedo-Girón, R., Jiménez, Z., Gavilanes-Ruiz, J., Cortés, A., and Uribe-Cifuentes, R., 1998, La falla Tamazula—límite suroriental del bloque Jalisco y sus relaciones con el complejo volcánico de Colima, México: Revista Mexicana de Ciencias Geológicas, v. 15, p. 132–144.

Garduño-Monroy, V., Arreygue-Rocha, E., Israde-Alcántara, I., and Rodríguez-Torres, G., 2001, Efectos de las fallas asociadas a sobreexplotación de acuíferos y la presencia de fallas potencialmente sísmicas en Morelia, Michoacán, México: Revista Mexicana de Ciencias Geológicas, v. 18, p. 37–54.

Gastil, G., and Jensky, W., 1973, Evidence of strike-slip displacement beneath the Trans-Mexican Volcanic Belt, *in* Kovach, R., and Nur, A., eds., Proceedings, Conference on Tectonic Problems of the San Andreas Fault System: Stanford, California, Stanford University Publications, Geological Sciences, p. 171–180.

Gastil, G., Krummenacher, D., and Jensky, A., 1979, Reconnaissance geology of west-central Nayarit, Mexico: Summary: Geological Society of America Bulletin, v. 90, p. 15–18, doi: 10.1130/0016-7606(1979)90<15:RGOWNM>2.0.CO;2.

Geolimex-Group, 1994, Reflections of the subducting plate? First results of a Mexican geotraverse, *in* Miller, H., Rosenfeld, U., and Weber-Diefenbach, K., eds., 13th Symposium on Latin-America Geosciences: Teil, Zentralblatt für Geologie und Paläontologie, p. 541–553.

Geyne, A., Fries, C., Segerstrom, K., Black, R., and Wilson, I., 1963, Geology and Mineral Deposits of the Pachuca-Real del Monte District, State of Hidalgo, Mexico: Consejo de Recursos Naturales no Renovables, Publicación 5E.

Gilbert, C., Mahood, G., and Carmichael, I., 1985, Volcanic stratigraphy of the Guadalajara area, Mexico: Geofísica Internacional, v. 24, p. 169–191.

Gómez-Tuena, A., and Carrasco-Núñez, G., 2000, Cerro Grande Volcano: The evolution of a Miocene stratocone in the early Transmexican Volcanic Belt: Tectonophysics, v. 318, p. 249–280, doi: 10.1016/S0040-1951(99)00314-5.

Gómez-Tuena, A., LaGatta, A., Langmuir, C., Goldstein, S., Ortega-Gutiérrez, F., and Carrasco-Núñez, G., 2003, Temporal control of subduction magmatism in the Eastern Trans-Mexican Volcanic Belt: Mantle sources, slab contributions and crustal contamination: Geochemistry, Geophysics, Geosystems, v. 4, no. 8, p. doi:10.1029/2003GC000524.

Halliday, A., Lee, D., Tommasini, S., Davis, G., Paslick, C., Fitton, J., and James, D., 1995, Incompatible trace elements in OIB and MORB and source enrichment in the sub-oceanic mantle: Earth and Planetary Science Letters, v. 133, p. 379–395, doi: 10.1016/0012-821X(95)00097-V.

Harris, J., 1986, Silicic volcanics of Volcan Tequila, Jalisco, Mexico [M.S. thesis]: Berkeley, University of California, 98 p.

Hasenaka, T., 1986, The cinder cones of Michoacán-Gunajuato, central Mexico [Ph.D. thesis]: Berkeley, University of California, 171 p.

Hasenaka, T., 1994, Size, distribution, and magma output rate for shield volcanoes of the Michoacan-Guanajuato volcanic field, Central Mexico:

Journal of Volcanology and Geothermal Research, v. 63, p. 13–31, doi: 10.1016/0377-0273(94)90016-7.
Hasenaka, T., and Carmichael, I., 1985, The cinder cones at Michoacan-Guanajuato, Central Mexico: Their age, volume, and distribution, and magma discharge rate: Journal of Volcanology and Geothermal Research, v. 25, p. 105–124, doi: 10.1016/0377-0273(85)90007-1.
Hasenaka, T., and Carmichael, I., 1987, The cinder cones of Michoacan-Guanajuato, Central Mexico: Petrology and chemistry: Journal of Petrology, v. 28, p. 241–269.
Herrero, E., and Pal, S., 1978, Paleomagnetic Study of Sierra de Chichinautzin: Geofísica Internacional, v. 17, p. 167–180.
Herrmann, U., Nelson, B., and Ratschbacher, L., 1994, The origin of a terrane: U/Pb zircon geochronology and tectonic evolution of the Xolapa complex (southern Mexico): Tectonics, v. 13, no. 2, p. 455–474, doi: 10.1029/93TC02465.
Hesse, M., and Grove, T., 2003, Absarokites from the western Mexican Volcanic Belt: Constraints on the mantle wedge conditions: Contributions to Mineralogy and Petrology, v. 146, p. 10–27, doi: 10.1007/s00410-003-0489-3.
Hickey-Vargas, R., Frey, F., Gerlach, D., and López-Escobar, L., 1986, Multiple sources for basaltic arc rocks from the southern volcanic zone of the Andes (34°-41°S): Trace element and isotopic evidence for contributions from subducted oceanic crust, mantle and continental crust: Journal of Geophysical Research, v. 91, p. 5963–5983.
Hirose, K., 1997, Melting experiments on lherzolite KLB-1 under hydrous conditions and generation of high-magnesian andesites: Geology, v. 25, p. 42–44, doi: 10.1130/0091-7613(1997)025<0042:MEOLKU>2.3.CO;2.
Hirth, G., and Kohlstedt, D., 2003, Rheology of the mantle wedge, *in* Eiler, J., ed., Inside the Subduction Factory: Washington, D.C., American Geophysical Union, p. 83–105.
Hochstaedter, A., Ryan, F., Luhr, J., and Hasenaka, T., 1996, On B/Be ratios in the Mexican Volcanic Belt: Geochimica et Cosmochimica Acta, v. 60, no. 4, p. 613–618, doi: 10.1016/0016-7037(95)00415-7.
Hofmann, A., 2003, Sampling mantle heterogeneity through oceanic basalts: Isotopes and trace elements, *in* Carlson, R., ed., Treatise on Geochemistry: Amsterdam, Elsevier, p. 61–101.
Höskuldsson, A., and Robin, C., 1993, Late Pleistocene to Holocene eruptive activity of Pico de Orizaba, Eastern Mexico: Bulletin of Volcanology, v. 55, p. 571–587, doi: 10.1007/BF00301810.
Irvine, T., and Baragar, W., 1971, A guide to the chemical classification of the common volcanic rocks: Canadian Journal of Earth Sciences, v. 8, p. 523–548.
Jacobo-Albarrán, J., 1986, Estudio petrogenético de las rocas ígneas de la porción central del Eje Neovolcánico, Reporte Interno de la Subdirección de Tecnología de Exploración: México, D.F., Instituto Mexicano del Petróleo, 47 p.
Johnson, C., and Harrison, C., 1990, Neotectonics in central Mexico: Physics of the Earth and Planetary Interiors, v. 64, p. 187–210, doi: 10.1016/0031-9201(90)90037-X.
Johnson, M., and Plank, T., 1999, Dehydration and melting experiments constrain the fate of subducted sediments: Geochemistry, Geophysics, Geosystems, v. 1, doi: 10.1029/1999GC000014.
Jording, A., Ferrari, L., Arzate, J., and Jodicke, H., 2000, Crustal variations and terrane boundaries in southern Mexico as imaged by magnetotelluric transfer functions: Tectonophysics, v. 327, p. 1–13, doi: 10.1016/S0040-1951(00)00166-9.
Kay, R., 1978, Aleutian magnesian andesites: Melts from subduction Pacific Oceanic crust: Journal of Volcanology and Geothermal Research, v. 4, p. 117–132, doi: 10.1016/0377-0273(78)90032-X.
Kelemen, P., Hanghøj, K., and Greene, A., 2003, One view of the geochemistry of subduction-related magmatic arcs, with emphasis on primitive andesite and lower crust, *in* Rudnick, R., ed., Treatise on Geochemistry: Amsterdam, Elsevier, p. 593–659.
Keppie, J., and Ortega-Gutiérrez, F., 1995, Provenance of Mexican Terranes: Isotopic Constrains: International Geology Review, v. 37, p. 813–824.
Keppie, J., and Ortega-Gutiérrez, J., 1998, Middle American Precambrian basement: A missing piece of the reconstructed 1 Ga orogen, *in* Ramos, V., and Keppie, J., eds., Laurentia-Gondwana Connections Before Pangea: Geological Society of America Special Paper 336, p. 199–210.
Kostoglodov, V., and Bandy, W., 1995, Seismotectonic constraints on the convergence rates between the Rivera and North American plates: Journal of Geophysical Research, v. 100, no. B9, p. 17,977–17,989, doi: 10.1029/95JB01484.
Kostoglodov, V., Singh, S., Santiago, J., Franco, S., Larson, K., Lowry, A., and Bilham, R., 2003, A large silent earthquake in the Guerrero seismic gap, Mexico: Geophysical Research Letters, v. 30, no. 15, doi: 10.1029/2003GL017219.
Kushiro, I., 1972, Effect of water on the compositions of magmas formed at high pressures: Journal of Petrology, v. 13, p. 311–334.
Kushiro, I., 1974, Melting of hydrous upper mantle and possible generation of andesitic magma: An approach from synthetic systems: Earth and Planetary Science Letters, v. 22, p. 294–299, doi: 10.1016/0012-821X(74)90138-1.
LaGatta, A., 2003, Arc magma genesis in the eastern Mexican volcanic belt [Ph.D. thesis]: Palisades, Columbia University, 365 p.
Lange, R., and Carmichael, I., 1990, Hydrous basaltic andesites associated with minette and related lavas in western Mexico: Journal of Petrology, v. 31, p. 1225–1259.
Lange, R., and Carmichael, I., 1991, A potassic volcanic front in western Mexico: The lamprophyric and related lavas of San Sebastian: Geological Society of America Bulletin, v. 103, p. 928–940, doi: 10.1130/0016-7606(1991)103<0928:APVFIW>2.3.CO;2.
Langridge, R., Weldon, R., Moya, J., and Suárez, G., 2000, Paleoseismology of the 1912 Acambay earthquake and the Acambay-Tixmadejé fault, Trans-Mexican Volcanic Belt: Journal of Geophysical Research, v. 105, p. 3019–3037, doi: 10.1029/1999JB900239.
Lapierre, H., Ortiz, L., Abouchami, W., Monod, O., Coulon, C., and Zimmerman, J., 1992, A crustal section of an intra-oceanic island arc: The Late Jurassic-Early Cretaceous Guanajuato magmatic sequence, central Mexico: Earth and Planetary Science Letters, v. 108, p. 61–77, doi: 10.1016/0012-821X(92)90060-9.
Lassiter, J., and Luhr, J., 2001, Osmium abundance and isotope variations in mafic Mexican volcanic rocks: Evidence for crustal contamination and constraints on the geochemical behavior of osmium during partial melting and fractional crystallization: Geochemistry, Geophysics, Geosystems, v. 2, no. 3, p. doi: 10.1029/2000GC000116.
Lawlor, P., Ortega-Gutiérrez, F., Cameron, K., Ochoa-Carrillo, H., Lopez, R., and Sampson, D., 1999, U/Pb Geochronology, Geochemistry and Provenance of the Grenvillian Huiznopala Gneiss of Eastern Mexico: Precambrian Research, v. 94, p. 73–99, doi: 10.1016/S0301-9268(98)00108-9.
Le Bas, M., Le Maitre, R., Streckeisen, A., and Zanettin, B., 1986, A chemical classification of volcanic rocks on the total alkali-silica diagram: Journal of Petrology, v. 27, no. 3, p. 745–750.
Legget, J., 1982, Geochemistry of Cocos Plate pelagic-hemipelagic sediment in Hole 487, DSDP Leg 66, *in* Watkins, J., and Moore, J., eds., Initial Reports of the Deep Ocean Drilling Project 66: Washington, D.C., U.S. Government Printing Office, p. 683–686.
Le Maitre, R., 1989, A Classification of Igneous Rocks and Glossary of Terms: Oxford, Blackwell, 193 p.
Lewis-Kenedi, C., Lange, R., Hall, C., and Delgado-Granados, H., 2005, The eruptive history of the Tequila volcanic field, western Mexico: Ages, volumes, and relative proportions of lava types: Bulletin of Volcanology, v. 67, p. 391–414, doi: 10.1007/s00445-004-0377-3.
Londsdale, P., 1991, Structural patterns of the Pacific floor offshore Peninsular California, *in* Dauphin, J., and Simoneit, B., eds., The Gulf and the Peninsular Province of the Californias: Tulsa, Oklahoma, American Association of Petroleum Geologists Memoir 47, p. 87–125.
Londsdale, P., 2005, Creation of the Cocos and Nazca plates by fission of the Farallon plate: Tectonophysics, v. 404, p. 237–264, doi: 10.1016/j.tecto.2005.05.011.
López-Hernández, A., and Castillo-Hernández, D., 1997, Exploratory drilling at Acoculco, Puebla, Mexico; a hydrothermal system with only nonthermal manifestations: Geothermal Research Council Transactions, v. 21, p. 429–433.
López-Infanzón, M., 1991, Petrologic study of the volcanic rocks in the Chiconquiaco-Palma Sola area, central Veracruz, Mexico [M.S. thesis]: New Orleans, Tulane University, 139 p.
Luhr, J., 1997, Extensional tectonics and the diverse primitive volcanic rocks in the western Mexican Volcanic Belt: Canadian Mineralogist, v. 35, no. 2, p. 473–500.
Luhr, J., 2000, The geology and petrology of Volcan San Juan Nayarit, Mexico and the compositionally zoned Tepic Pumice: Journal of Volcanology and Geothermal Research, v. 95, p. 109–156, doi: 10.1016/S0377-0273(99)00133-X.
Luhr, J., 2002, Petrology and geochemistry of the 1991 and 1998–1999 lava flows from Volcán de Colima, México: Implications for the end of the

current eruptive cycle: Journal of Volcanology and Geothermal Research, v. 117, p. 169–194.

Luhr, J., and Carmichael, I., 1980, The Colima Volcanic Complex, Mexico. I. Post-caldera andesites from Volcán Colima: Contributions to Mineralogy and Petrology, v. 71, p. 343–372, doi: 10.1007/BF00374707.

Luhr, J., and Carmichael, I., 1981, The Colima Volcanic Complex, Mexico. II. Late-Quaternary cinder cones: Contributions to Mineralogy and Petrology, v. 76, p. 127–147, doi: 10.1007/BF00371954.

Luhr, J., and Carmichael, I., 1982, The Colima Volcanic Complex, Mexico. III. Ash- and scoria-fall deposits from the upper slopes of Volcán Colima: Contributions to Mineralogy and Petrology, v. 80, p. 262–275, doi: 10.1007/BF00371356.

Luhr, J., and Carmichael, I., 1985a, Contemporaneous eruptions of calc-alkaline and alkaline magmas along the volcanic front of the Mexican Volcanic Belt: Geofísica Internacional, v. 24, p. 203–216.

Luhr, J., and Carmichael, I., 1985b, Jorullo Volcano, Michoacán, Mexico (1759–1774): The earliest stages of fractionation in calc-alkaline magmas: Contributions to Mineralogy and Petrology, v. 90, p. 142–161, doi: 10.1007/BF00378256.

Luhr, J., and Carmichael, I., 1990, Petrological monitoring of cyclical eruptive activity at Volcán Colima, México: Journal of Volcanology and Geothermal Research, v. 42, p. 235–260, doi: 10.1016/0377-0273(90)90002-W.

Luhr, J., Nelson, S., Allan, J., and Carmichael, I., 1985, Active rifting in southwestern Mexico: Manifestations of an incipient eastward spreading-ridge jump: Geology, v. 13, p. 54–57, doi: 10.1130/0091-7613(1985)13<54:ARISMM>2.0.CO;2.

Luhr, J., Aranda-Gómez, J.J., and Pier, J., 1989a, Spinel-lherzoilite-bearing quaternary volcanic centers in San Luis Potosi, Mexico. I. Geology, mineralogy, and petrology: Journal of Geophysical Research, v. 94, no. B6, p. 7916–7940.

Luhr, J., Allan, J., Carmichael, I., Nelson, S., and Hasenaka, T., 1989b, Primitive calc-alkaline and alkaline rock types from the western Mexican volcanic belt: Journal of Geophysical Research, v. 94, no. B4, p. 4515–4530.

Mahood, G., 1981a, A summary of the geology and petrology of the Sierra La Primavera, Jalisco, Mexico: Journal of Geophysical Research, v. 86, p. 10,137–10,152.

Mahood, G., 1981b, Chemical evolution of a Pleistocene rhyolitic center: Sierra La Primavera, Jalisco, Mexico: Contributions to Mineralogy and Petrology, v. 77, p. 129–149, doi: 10.1007/BF00636517.

Mahood, G., and Halliday, A., 1988, Generation of high-silica rhyolite: A Nd, Sr, and O isotopic study of Sierra La Primavera, Mexican Neovolcanic Belt: Contributions to Mineralogy and Petrology, v. 100, p. 183–191, doi: 10.1007/BF00373584.

Maldonado-Sánchez, G., and Schaaf, P., 2005, Geochemical and isotope data from the Acatlán Volcanic Field, western Trans-Mexican Volcanic Belt: Origin and evolution: Lithos, v. 82, p. 455–470, doi: 10.1016/j.lithos.2004.09.030.

Mammerickx, J., and Klitgord, K., 1982, North East Pacific Rise: Evolution from 25 m.y. B.P. to the present: Journal of Geophysical Research, v. 87, p. 6751–6759.

Manea, M., Manea, V., and Kostoglodov, V., 2003, Sediment fill in the Middle America Trench inferred from gravity anomalies: Geofísica Internacional, v. 42, no. 4, p. 603–612.

Manea, V., Manea, M., Kostoglodov, V., Currie, C., and Sewell, G., 2004, Thermal structure, coupling and metamorphism in the Mexican subduction zone beneath Guerrero: Geophysical Journal International, v. 158, p. 775–784, doi: 10.1111/j.1365-246X.2004.02325.x.

Manea, V., Manea, M., Kostoglodov, V., and Sewell, G., 2005, Thermomechanical model of the mantle wedge in Central Mexican subduction zone and a blob tracing approach for the magma transport: Physics of the Earth and Planetary Interiors, v. 149, no. 1–2, p. 165–186, doi: 10.1016/j.pepi.2004.08.024.

Márquez, A., and De Ignacio, C., 2002, Mineralogical and geochemical constraints for the origin and evolution of magmas in Sierra Chichinautzin, Central Mexican Volcanic Belt: Lithos, v. 62, p. 35–62.

Márquez, A., Oyarzún, R., Doblas, M., and Verma, S., 1999a, Alkalic (oceanic-island basalt type) and calc-alkalic volcanism in the Mexican volcanic belt: A case for plume-related magmatism and propagating rifting at an active margin?: Geology, v. 27, p. 51–54, doi: 10.1130/0091-7613(1999)027<0051:AOIBTA>2.3.CO;2.

Márquez, A., Oyarzún, R., Doblas, M., and Verma, S., 1999b, Alkalic (ocean-island basalt type) and calc-alkalic volcanism in the Mexican volcanic belt: A case for plume-related magmatism and propagating rifting at an active margin?: Reply: Geology, v. 27, p. 1056, doi: 10.1130/0091-7613(1999)027<1056:C>2.3.CO;2.

Márquez, A., Verma, S., Anguita, F., Oyarzun, R., and Brandle, J., 1999c, Tectonics and volcanism of Sierra Chichinautzin: Extension at the front of the central transmexican volcanic belt: Journal of Volcanology and Geothermal Research, v. 93, p. 125–150, doi: 10.1016/S0377-0273(99)00085-2.

Martin-Del Pozzo, A., 1982, Monogenetic volcanism in Sierra Chichinautzin, Mexico: Bulletin of Volcanology, v. 45, p. 9–24.

Martínez Reyes, J., and Nieto-Samaniego, A., 1990, Efectos geológicos de la tectónica reciente en la parte central de México: Revista Instituto de Geología, v. 9, p. 33–50.

Martínez-Serrano, R., Schaaf, P., Solís-Pichardo, G., Hernández-Bernal, M., Hernández-Treviño, T., Morales-Contreras, J., and Macías, J., 2004, Sr, Nd and Pb isotope and geochemical data from the Quaternary Nevado de Toluca volcano, a source of recent adakitic magmatism, and the Tenango Volcanic Field, Mexico: Journal of Volcanology and Geothermal Research, v. 138, p. 77–110, doi: 10.1016/j.jvolgeores.2004.06.007.

McBirney, A., Taylor, H., and Armstrong, R., 1987, Paricutín re-examined: A classic example of crustal assimilation in calc-alkaline magma: Contributions to Mineralogy and Petrology, v. 95, p. 4–20, doi: 10.1007/BF00518026.

McDonald, G., and Katsura, T., 1964, Chemical compositions of Hawaiian lavas: Journal of Petrology, v. 5, p. 82–133.

McDonough, W., and Sun, S., 1995, The composition of the earth: Chemical Geology, v. 120, no. 3–4, p. 223–253, doi: 10.1016/0009-2541(94)00140-4.

McKenzie, D., and O'Nions, R., 1995, The source regions of ocean island basalts: Journal of Petrology, v. 36, p. 133–159.

Menard, H., 1955, Deformation of the northeastern pacific basin and the west coast of North America: Geological Society of America Bulletin, v. 66, p. 1149–1168.

Menard, H., 1978, Fragmentation of the Farallon plate by pivoting subduction: The Journal of Geology, v. 86, p. 181–201.

Michaud, F., Quintero-Legorreta, O., Barrier, E., and Burgois, J., 1991, La frontière Nord du Bloc Jalisco (ouest Mexique): localisation et évolution de 13 Ma à l'actuel: Comptes Rendus de l'Académie des Sciences, Paris, II, v. 312, p. 1359–1365.

Middlemost, E., 1989, Iron oxidation ratios, norms and the classification of volcanic rocks: Chemical Geology, v. 77, p. 19–26, doi: 10.1016/0009-2541(89)90011-9.

Molina-Garza, R., and Urrutia-Fucugauchi, J., 1993, Deep crustal structure of central Mexico derived from interpretation of Bouger gravity anomaly data: Journal of Geodynamics, v. 17, p. 181–201, doi: 10.1016/0264-3707(93)90007-S.

Moore, G., and Carmichael, I., 1998, The hydrous phase equilibria (to 3 kbar) of an andesite and basaltic andesite from western Mexico: Constraints on water content and conditions of phenocryst growth: Contributions to Mineralogy and Petrology, v. 130, p. 304–319, doi: 10.1007/s004100050367.

Moore, G., Marone, C., Carmichael, I., and Renne, P., 1994, Basaltic volcanism and extension near the intersection of the Sierra Madre volcanic province and the Mexican Volcanic Belt: Geological Society of America Bulletin, v. 106, p. 383–394, doi: 10.1130/0016-7606(1994)106<0383:BVAENT>2.3.CO;2.

Moore, J., Watkins, J., Bachman, S., Beghtel, F., Butt, A., Didyk, B., Foss, G., Leggett, J., Lundberg, N., McMillan, N., Niitsuma, N., Shephard, L., Stephen, J., Shipley, T., and Strander, H., 1982, Facies belts of the Middle America Trench and forearc region, southern Mexico: Results from Leg 66 DSDP, *in* Leggert, J., ed., Trench-Forearc Geology: Sedimentation and tectonics on modern and ancient active plate margins: Oxford, Geological Society [London] Special Publication 10, p. 77–94.

Mooser, F., 1972, The Mexican volcanic belt structure and tectonics: Geofísica Internacional, v. 12, p. 55–70.

Mooser, F., and Maldonado-Koerdell, M., 1961, Tectónica penecontemporánea a lo largo de la costa mexicana del Océano Pacífico: Geofísica Internacional, v. 1, p. 3–20.

Mooser, F., Nairn, A., and Negendank, J., 1974, Palaeomagnetic investigations of the Tertiary and Quaternary igneous rocks: VIII. A palaeomagnetic and petrologic study of volcanics of the Valley of Mexico: Geologische Rundschau, v. 63, p. 451–483, doi: 10.1007/BF01820824.

Mora, J., Macías, J., Saucedo, R., Orlando, A., Manetti, P., and Vaselli, O., 2002, Petrology of the 1998–2000 products of Volcán de Colima, México: Journal of Volcanology and Geothermal Research, v. 117, p. 195–212, doi: 10.1016/S0377-0273(02)00244-5.

Mora-Alvarez, G., Cabellero-Miranda, C., Urrutia-Fucugauchi, J., and Uchiumi, S., 1991, Southward migration of volcanic activity in the Sierra de Las Cruces, basin of Mexico?—A preliminary K-Ar dating and palaeomagnetic study: Geofísica Internacional, v. 30, p. 61–70.

Morán-Zenteno, D., Corona-Chávez, P., and Tolson, G., 1996, Uplift and subduction erosion in southwestern Mexico since the Oligocene: Pluton geobarometry constraints: Earth and Planetary Science Letters, v. 141, p. 51–65, doi: 10.1016/0012-821X(96)00067-2.

Morán-Zenteno, D., Tolson, G., Martines-Serrano, R., Martiny, B., Schaaf, P., Silva-Romo, G., Macias-Romo, C., Alba-Aldave, L., Hernandez-Bernal, M., and Solis-Pichardo, G., 1999, Tertiary arc-magmatism of the Sierra Madre del Sur, Mexico, and its transition to the volcanic activity of the Trans-Mexican Volcanic Belt: Journal of South American Earth Sciences, v. 12, p. 513–535, doi: 10.1016/S0895-9811(99)00036-X.

Murrillo-Muñetón, G., 1994, Petrologic and geochronologic study of the Grenville-age granulaties and post-granulite plutons from the La Mixtequita area, State of Oaxaca, southern Mexico, and their tectonic significance [M.S. thesis]: Los Angeles, University of Southern California, 163 p.

Nava, A., Núñez-Cornu, F., Córdoba, D., Mena, M., Ansorge, J., González, J., Rodríguez, M., Banda, E., Müller, S., Udias, A., García-García, M., and Calderón, G., 1988, Structure of the Middle America trench in Oaxaca, México: Tectonophysics, v. 154, p. 241–251, doi: 10.1016/0040-1951(88)90106-0.

Negendank, J., Emmermann, R., Krawczyk, R., Mooser, F., Tobschall, H., and Wehrle, D., 1985, Geological and geochemical investigations on the eastern Trans-Mexican Volcanic Belt: Geofísica Internacional, v. 24, p. 477–575.

Nelson, S., 1980, Geology and petrology of Volcán Ceboruco, Nayarit, México: Geological Society of America Bulletin, v. 91, part 2, p. 2290–2431, doi: 10.1130/0016-7606(1980)91<639:GAPOVC>2.0.CO;2.

Nelson, S., and Carmichael, I., 1984, Pleistocene to Recent alkalic volcanism in the region of Sanganguey volcano, Nayarit, Mexico: Contributions to Mineralogy and Petrology, v. 85, p. 321–335, doi: 10.1007/BF01150290.

Nelson, S., and Gonzalez-Caver, E., 1992, Geology and K-Ar dating of the Tuxtla Volcanic Field, Veracruz, Mexico: Bulletin of Volcanology, v. 55, p. 85–96, doi: 10.1007/BF00301122.

Nelson, S., and Hegre, J., 1990, Volcán Las Navajas, a Pliocene-Pleistocene trachyte/peralkaline rhyolite volcano in the northwestern Mexican volcanic belt: Bulletin of Volcanology, v. 52, p. 186–204, doi: 10.1007/BF00334804.

Nelson, S., and Lighthart, A., 1997, Field excursion to the Sierra Las Navajas, Hidalgo, Mexico; a Pleistocene peralkaline rhyolite complex with a large debris avalanche deposit, II Convención sobre la Evolución Geológica de México y Recursos Asociados, Libro-Guía de las Excursiones Geológicas: Pachuca, México, Instituto de Investigaciones en Ciencias de la Tierra de la Universidad Autónoma de Hidalgo e Instituto de Geología, UNAM, p. 89–96.

Nelson, S., and Livieres, R., 1986, Contemporaneous calc-alkaline and alkaline volcanism at Sanganguey Volcano, Nayarit, Mexico: Geological Society of America Bulletin, v. 97, p. 798–808, doi: 10.1130/0016-7606(1986)97<798:CCAAVA>2.0.CO;2.

Nelson, S.A., Gonzalez-Caver, E., and Kyser, T.K., 1995, Constraints on the origin of alkaline and calc-alkaline magmas from the Tuxtla Volcanic Field, Veracruz, Mexico: Contributions to Mineralogy and Petrology, v. 122, no. 1–2, p. 191–211, doi: 10.1007/s004100050121.

Nieto-Obregón, J., Delgado-Argote, L., and Damon, P., 1981, Relaciones petrológicas y geocronológicas del magmatismo de la Sierra Madre Occidental y el Eje Neovolcánico en Nayarit, Jalisco y Zacatecas: Asociación de Ingenieros Mineros, Metalurgicos y Geólogos de México, Memoria Técnica, v. XIV, p. 327–361.

Nieto-Obregón, J., Delgado-Argote, L., and Damon, P., 1985, Geochronologic, petrologic and structural data related to large morphologic features between the Sierra Madre Occidental and the Mexican Volcanic Belt: Geofísica Internacional, v. 24, p. 623–663.

Nieto-Obregón, J., Urrutia-Fucugauchi, J., Cabral-Cano, E., and Guzman de la Campa, A., 1992, Listric faulting and continental rifting in western Mexico—A paleomagnetic and structural study: Tectonophysics, v. 208, p. 365–376, doi: 10.1016/0040-1951(92)90435-9.

Nieto-Samaniego, A., Ferrari, L., Alaniz-Alvarez, S., Labarthe-Hernández, G., and Rosas-Elguera, J., 1999, Variation of Cenozoic extension and volcanism across the southern Sierra Madre Occidental Volcanic Province, Mexico: Geological Society of America Bulletin, v. 111, p. 347–363, doi: 10.1130/0016-7606(1999)111<0347:VOCEAV>2.3.CO;2.

Nixon, G., 1982, The relationship between Quaternary volcanism in central Mexico and the seismicity and structure of subducted ocean lithosphere: Geological Society of America Bulletin, v. 93, p. 514–523, doi: 10.1130/0016-7606(1982)93<514:TRBQVI>2.0.CO;2.

Nixon, G., 1988a, Petrology of the younger andesites and dacites of Iztaccíhuatl Volcano, México: I. Disequilibrium phenocrysts assemblages as indicators of magma chamber processes: Journal of Petrology, v. 29, p. 213–264.

Nixon, G., 1988b, Petrology of the younger andesites and dacites of Iztaccíhuatl volcano, México: II. Chemical stratigraphy, magma mixing, and the composition of basaltic magma influx: Journal of Petrology, v. 29, p. 265–303.

Nixon, G., 1989, The Geology of Iztaccíhuatl Volcano and Adjacent Areas of the Sierra Nevada and Valley of Mexico: Geological Society of America Special Paper 219, 58 p.

Nixon, G., Demant, A., Armstrong, R., and Harakal, J., 1987, K-Ar and geologic data bearing on the age and evolution of the Trans-Mexican Volcanic Belt: Geofísica Internacional, v. 26, p. 109–158.

Núñez-Cornú, F., Rutz, L., Nava, F., Reyes-Dávila, G., and Suárez-Plascencia, C., 2002, Characteristics of seismicity in the coast and north of Jalisco Block, Mexico: Physics of the Earth and Planetary Interiors, v. 132, p. 141–155, doi: 10.1016/S0031-9201(02)00049-3.

Ochoa-Camarillo, H., 1997, Geología del anticlinorio de Huayacocotla, estado de Hidalgo, II convención sobre la Evolución Geológica de México y Recursos Asociados. Guía de excursión de campo: Mexico, D.F., Universidad Nacional Autónoma de México, Instituto de Geología.

O'Hara, M., 1977, Geochemical evolution during fractional crystallization of a periodically refilled magma chamber: Nature, v. 266, no. 503–507.

O'Hara, M., and Mathews, R., 1981, Geochemical evolution in an advancing, periodically replenished, periodically tapped, continuously fractionated magma chamber: Geological Society [London] Journal, v. 138, p. 237–277.

Orozco-Esquivel, M., 1995, Zur petrologie des Vulkangebietes von Palma-Sola, Mexiko. Ein Beispiel für den Übergang von anorogenem zu orogenem Vulkanismus [Ph.D. thesis]: Karlsruhe, Universität Karlsruhe, 167 p.

Orozco-Esquivel, M., Petrone, C., Ferrari, L., Tagami, T., and Manetti, P., 2007, Geochemical and isotopic variability controlled by slab detachment in a subduction zone with varying dip: The eastern Trans-Mexican Volcanic Belt: Lithos, v. 93, no. 1–2, p. 149–174.

Ortega-Guitérrez, F., 1978, El Gneiss Novillo y rocas metamórficas asociadas en los cañones del Novillo y la Peregrina, area de Ciudad Victoria, Tamaulipas: Revista Instituto de Geología, v. 2, p. 19–30.

Ortega-Gutiérrez, F., 1981, Metamorphic belts of southern Mexico and their tectonic significance: Geofísica Internacional, v. 20, p. 177–202.

Ortega-Gutiérrez, F., 1984, Evidence of Precambrian evaporites in the Oaxacan granulite Complex of Southern Mexico: Precambrian Research, v. 23, p. 377–393, doi: 10.1016/0301-9268(84)90051-2.

Ortega-Gutiérrez, F., Mitre-Salazar, L., Roldán-Quintana, J., Aranda-Gómez, J., Morán-Zenteno, D., Alaniz-Álvarez, S., and Nieto-Samaniego, A., 1992, Carta Geológica de la Republica Mexicana: Universidad Nacional Autónoma de México, Instituto de Geología; Secretaría de Minas e Industria Paraestatal, Consejo de Recursos Minerales, scale 1:2 000 000, 1 sheet, 74 p. text.

Ortega-Gutiérrez, F., Sedlock, R., and Speed, R., 1994, Phanerozoic tectonic evolution of Mexico, *in* Speed, R., ed., Phanerozoic Evolution of North American Continent-Ocean Transitions: Boulder, Colorado, Geological Society of America, p. 265–306.

Ortega-Gutiérrez, F., Ruiz, J., and Centeno-García, E., 1995, Oaxaquia, a Proterozoic microcontinent accreted to North America during the late Paleozoic: Geology, v. 23, no. 12, p. 1127–1130, doi: 10.1130/0091-7613(1995)023<1127:OAPMAT>2.3.CO;2.

Osete, M., Ruiz-Martínez, V., Cabellero, C., Galindo, C., Urrutia-Fucugauchi, J., and Tarling, D., 2000, Southward migration of continental volcanic activity in the Sierra de las Cruces, Mexico: Paleomagnetic and radiometric evidence: Tectonophysics, v. 318, p. 201–215, doi: 10.1016/S0040-1951(99)00312-1.

Pacheco, J., Mortera-Gutiérrez, C., Delgado-Granados, H., Singh, S., Valenzuela, R., Shapiro, N., Santoyo, M., Hurtado, A., Barrón, R., and Gutiérrez-Moguel, E., 1999, Tectonic significance of an earthquake sequence in the Zacoalco half-graben, Jalisco, Mexico: Journal of South American Earth Sciences, v. 12, p. 557–565, doi: 10.1016/S0895-9811(99)00039-5.

Pardo, M., and Suárez, G., 1993, Steep subduction geometry of the Rivera plate beneath the Jalisco Block in western Mexico: Geophysical Research Letters, v. 20, p. 2391–2394.

Pardo, M., and Suárez, G., 1995, Shape of the subducted Rivera and Cocos plate in southern Mexico: Seismic and tectonic implications: Journal of Geophysical Research, v. 100, p. 12,357–12,373, doi: 10.1029/95JB00919.

Parman, S., and Grove, T., 2004, Harzburgite melting with and without H2O: Experimental data and predictive modeling: Journal of Geophysical Research, v. 109, no. B02201, p. doi: 10.1029/2003JB002566.

Pasquaré, G., Garduño, V., Tibaldi, A., and Ferrari, M., 1988, Stress pattern evolution in the central sector of the Mexican Volcanic Belt: Tectonophysics, v. 146, p. 353–364, doi: 10.1016/0040-1951(88)90099-6.

Pasquaré, G., Ferrari, L., Garduño, V., Tibaldi, A., and Vezzoli, L., 1991, Geology of the central sector of the Mexican Volcanic Belt, States of Guanajuato and Michoacan: Geological Society of America Maps and Charts Series MCH072, scale 1:300 000, 1 sheet, 22 p. text.

Pérez-Venzor, J., Aranda-Gómez, J., McDowell, F., and Solorio Munguía, J., 1996, Geología del Volcán Palo Huérfano, Guanajuato, México: Revista Mexicana de Ciencias Geológicas, v. 13, no. 2, p. 174–183.

Petrone, C., Francalanci, L., Carlson, R., Ferrari, L., and Conticelli, S., 2003, Unusual coexistence of subduction-related and intraplate-type magmatism: Sr, Nd and Pb isotope and trace element data from the magmatism of the San Pedro–Ceboruco graben (Nayarit, Mexico): Chemical Geology, v. 193, p. 1–24, doi: 10.1016/S0009-2541(02)00229-2.

Pier, J., Luhr, J., Podosek, F., and Aranda-Gómez, J., 1992, The La Breña-El Jagüey Maar Complex, Durango, Mexico, II: Bulletin of Volcanology, v. 54, p. 405–428, doi: 10.1007/BF00312322.

Pilet, S., Hernandez, J., Bussy, F., and Sylvester, P., 2004, Short-term metasomatic control of Nb/Th ratios in the mantle sources of intraplate basalts: Geology, v. 32, p. 113–116, doi: 10.1130/G19953.1.

Pindell, J., 1985, Alleghenian reconstruction and subsequent evolution of the Gulf of Mexico, Bahamas, and Proto-Caribbean Sea: Tectonics, v. 4, no. 1, p. 133–156.

Plank, T., and Langmuir, C., 1998, The chemical composition of subducting sediment and its consequences for the crust and mantle: Chemical Geology, v. 145, p. 325–394, doi: 10.1016/S0009-2541(97)00150-2.

Polak, B., Kononov, V., Prasolov, E., Sharkov, I., Prol-Ledesma, R., González, A., Razo, A., and Molina-Berbeller, R., 1985, First estimations of terrestrial heat flow in the TMVB and adjacent areas based on isotopic compositions of natural helium: Geofísica Internacional, v. 24, p. 465–476.

Pradal, E., and Robin, C., 1994, Long-lived magmatic phases at Los Azufres volcanic center, Mexico: Journal of Volcanology and Geothermal Research, v. 63, p. 201–215, doi: 10.1016/0377-0273(94)90074-4.

Prol-Ledesma, R., and Juárez, G., 1985, Silica geotemperature mapping and thermal regime in the Mexican Volcanic Belt: Geofísica Internacional, v. 24, p. 609–622.

Prol-Ledesma, R., Sugrobov, V., Flores, E., Juarez, M., Smirov, Y., Gorshkov, A., Bondarenko, V., Rashidov, V., Nedopekin, L., and Gavrilov, V., 1989, Heat flow variations along the Middle America trench: Marine Geophysical Research, v. 11, p. 69–76, doi: 10.1007/BF00286248.

Quezadas-Flores, A., 1961, Las rocas del basamento de la cuenca Tampico-Misantla: Boletin de la Asociación Mexicana de Geólogos Petroleros, v. 13, p. 289–323.

Quintanar, L., Rodríguez-González, M., and Campos-Enríquez, O., 2004, A shallow crustal earthquake doublet from the Trans-Mexican Volcanic Belt (Central Mexico): Bulletin of the Seismological Society of America, v. 94, p. 845–855.

Quintero-Legorreta, O., Michaud, F., Bourgois, J., and Barrier, E., 1992, Evolución de la frontera septentrional del bloque Jalisco, México, desde hace 17 Ma: Revista Instituto de Geología, v. 10, p. 111–117.

Rapp, R., and Watson, E., 1995, Dehydration melting of metabasalt at 8–32 kb: Implications for continental growth and crust-mantle recycling: Journal of Petrology, v. 36, p. 891–931.

Ratschbacher, L., Riller, U., Meschede, M., Herrmann, U., and Frisch, W., 1991, Second look at suspect terranes in southern Mexico: Geology, v. 19, no. 12, p. 1233–1236, doi: 10.1130/0091-7613(1991)019<1233:SLASTI>2.3.CO;2.

Reid, M., 1983, Paricutín volcano revisited: Isotopic and trace element evidence for crustal assimilation: EOS, Transactions, American Geophysical Union, v. 64, no. 45, p. 907.

Riggs, N., and Carrasco-Núñez, G., 2004, Evolution of a complex isolated dome system, Cerro Pizarro, central México: Bulletin of Volcanology, v. 66, p. 322–335, doi: 10.1007/s00445-003-0313-y.

Righter, K., 2000, A comparison of basaltic volcanism in the Cascades and western Mexico: Compositional diversity in continental arcs: Tectonophysics, v. 318, p. 99–117, doi: 10.1016/S0040-1951(99)00308-X.

Righter, K., and Carmichael, I., 1992, Hawaiites and related lavas in the Atenguillo graben, western Mexican Volcanic Belt: Geological Society of America Bulletin, v. 104, p. 1592–1607, doi: 10.1130/0016-7606(1992)104<1592:HARLIT>2.3.CO;2.

Righter, K., and Carmichael, I., 1996, Phase equilibria of phlogopite lamprophyres from western Mexico: Biotite-liquid equilibria and P-T estimates for biotite-bearing igneous rocks: Contributions to Mineralogy and Petrology, v. 123, no. 1, p. 1–21, doi: 10.1007/s004100050140.

Righter, K., and Rosas-Elguera, J., 2001, Alkaline lavas in the volcanic front of the western Mexican volcanic belt: Geology and petrology of the Ayutla and Tapalpa volcanic fields: Journal of Petrology, v. 42, no. 12, p. 2333–2361, doi: 10.1093/petrology/42.12.2333.

Righter, K., Carmichael, I., Becker, T., and Renne, R., 1995, Pliocene to Quaternary volcanism and tectonics at the intersection of the Mexican Volcanic Belt and the Gulf of California: Geological Society of America Bulletin, v. 107, p. 612–626, doi: 10.1130/0016-7606(1995)107<0612:PQVAFA>2.3.CO;2.

Robin, C., 1976a, Présence simultanée de magmatismes de significations tectoniques opposées dans l'Est du Mexique: Bulletin de la Société Géologique de France, v. 18, p. 1637–1645.

Robin, C., 1976b, Las series volcánicas de la Sierra Madre Oriental (basaltos e ignombritas) descripción y caracteres químicos: Revista Instituto de Geología, v. 2, p. 12–42.

Robin, C., 1982, Relations volcanologie-magmatologie-géodynamique: application au passage entre volcanismes alcalin et andesitique dans le sud Mexicain (Axe Trans-mexicain et Province Alcaline Oriental) [PhD thesis]: Clermont-Ferrand, Annales Scientifiques de l'Université de Clermont-Ferrand II, 503 p.

Robin, C., and Cantagrel, J., 1982, Le Pico de orizaba (Mexique). Structure et evolution d' un grand volcan andesitique complexe: Bulletin of Volcanology, v. 45, p. 299–315.

Robin, C., Mossand, C., Camus, G., Cantagrel, J., Gourgaud, A., and Vincent, P., 1987, Eruptive history of the Colima volcanic complex: Journal of Volcanology and Geothermal Research, v. 31, p. 99–114, doi: 10.1016/0377-0273(87)90008-4.

Robin, C., Camus, G., and Gourgaud, A., 1991, Eruptive and magmatic cycles at Fuego de Colima volcano (México): Journal of Volcanology and Geothermal Research, v. 45, p. 209–225, doi: 10.1016/0377-0273(91)90060-D.

Rodríguez, S., 2005, Geology of Las Cumbres Volcanic Complex, Puebla and Veracruz states, Mexico: Revista Mexicana de Ciencias Geológicas, v. 22, no. 2, p. 181–199.

Rodríguez, S., Siebe, C., Komorowski, J., and Abrams, M., 2002, The Quetzalapa pumice: A voluminous late Pleistocene rhyolite deposit in the eastern Trans-Mexican Volcanic Belt: Journal of Volcanology and Geothermal Research, v. 113, p. 177–212, doi: 10.1016/S0377-0273(01)00258-X.

Rollinson, H., 1993, Using Geochemical Data: Essex, Longman Group, 352 p.

Rosas-Elguera, J., and Urrutia-Fucugauchi, J., 1998, Tectonic control on the volcano-sedimentary sequence of the Chapala graben, western Mexico: International Geology Review, v. 40, p. 350–362.

Rosas-Elguera, J., Urrutia-Fucugauchi, J., and Maciel, R., 1989, Geología del extremo oriental del Graben de Chapala; breve discusión sobre su edad: Zonas geotérmicas Ixtlan de Los Hervores-Los Negritos, México: Geotermia-Revista Mexicana de Geoenergía, v. 5, p. 3–18.

Rosas-Elguera, J., Ferrari, L., Garduño-Monroy, V., and Urrutia-Fucugauchi, J., 1996, Continental boundaries of the Jalisco Block in the Pliocene-Quaternary kinematics of western Mexico: Geology, v. 24, p. 921–924, doi: 10.1130/0091-7613(1996)024<0921:CBOTJB>2.3.CO;2.

Rosas-Elguera, J., Ferrari, L., Lopez-Martinez, M., and Urrutia-Fucugauchi, J., 1997, Stratigraphy and tectonics of the Guadalajara region and the triple junction area, western Mexico: International Geology Review, v. 39, p. 125–140.

Rosas-Elguera, J., Alva-Valdivia, L., Goguitchaichvili, A., Urrutia-Fucugauchi, J., Ortega-Rivera, M., and Archibald, D., 2003, Counterclockwise rotation of the Michoacan block: Implications for the tectonics of western Mexico: International Geology Review, v. 45, p. 814–826.

Ross, M., and Scotese, C., 1988, A hierarchical tectonic model of the Gulf of Mexico and Caribbean region: Tectonophysics, v. 155, p. 139–168, doi: 10.1016/0040-1951(88)90263-6.

Rossotti, A., Ferrari, L., López-Martínez, M., and Rosas-Elguera, J., 2002, Geology of the boundary between the Sierra Madre Occidental and the Trans-Mexican Volcanic Belt in the Guadalajara region, western Mexico: Revista Mexicana de Ciencias Geológicas, v. 19, p. 1–15.

Ruiz, J., Patchett, P., and Ortega-Gutiérrez, F., 1988, Proterozoic and Phanerozoic basement terranes of Mexico from Nd isotopic studies: Geological Society of America Bulletin, v. 100, p. 274–281, doi: 10.1130/0016-7606(1988)100<0274:PAPBTO>2.3.CO;2.

Ruiz, J., Tosdal, R., Restrepo, P., and Murillo-Muñetón, G., 1999, Pb Isotope evidence for Colombia-Southern Mexico connections in the Proterozoic, *in* Ramos, V., and Keppie, J., eds., Laurentia-Gondwana Connections before Pangea: Geological Society of America Special Paper 336, p. 183–197.

Sahagún, B., 1992, Historia General de Las Cosas de la Nueva España: Mexico, DF, Editorial Porrúa, 1093 p.

Schaaf, P., Heinrich, W., and Besch, T., 1994, Composition and Sm-Nd isotopic data of the lower crust beneath San Luis Potosí, central Mexico: Evidence from a granulite-facies xenolith suite: Chemical Geology, v. 118, p. 63–84, doi: 10.1016/0009-2541(94)90170-8.

Schaaf, P., Morán-Zenteno, D., Hernández-Bernal, M., Solís-Pichardo, G., Tolson, G., and Köhler, H., 1995, Paleogene continental margin truncation in southwestern Mexico: Geochronological evidence: Tectonics, v. 14, no. 6, p. 1339–1350, doi: 10.1029/95TC01928.

Schaaf, P., Stimac, J., Siebe, C., and Macías, J., 2005, Geochemical evidence for mantle origin and crustal processes in volcanic rocks from Popocatépetl and surrounding monogenetic volcanoes, Central Mexico: Journal of Petrology, v. 46, no. 6, p. 1243–1282, doi: 10.1093/petrology/egi015.

Schmidt, K., Bottazzi, P., Vannucci, R., and Mengel, K., 1999, Trace element partitioning between phlogopite, clinopyroxene and leucite lamproite melt: Earth and Planetary Science Letters, v. 168, p. 287–299, doi: 10.1016/S0012-821X(99)00056-4.

Sedlock, R., Ortega-Gutiérrez, F., and Speed, R., 1993, Tectonostratigraphic terranes and the tectonic evolution of Mexico: Geologial Society of America Special Paper 278, p. 153.

Sekine, T., Katsura, T., and Aramaki, S., 1979, Water-saturated phase relations of some andesites with application to the estimation of the initial temperature and water pressure at the time of eruption: Geochimica et Cosmochimica Acta, v. 43, no. 1367–1376.

Serpa, L., Smith, S., Katz, C., Skidmore, C., Sloan, R., and Pavlis, T., 1992, A geophysical investigation of the southern Jalisco block in the state of Colima, Mexico: Geofísica Internacional, v. 31, no. 247–252.

Sheth, H., Torres-Alvarado, I., and Verma, S., 2000, Beyond subduction and plumes: A unified tectonic-petrogenetic model for the Mexican volcanic belt: International Geology Review, v. 42, no. 12, p. 1116–1132.

Shirey, S., and Walker, R., 1998, Re-Os isotopes in cosmochemistry and high-temperature geochemistry: Annual Review of Earth and Planetary Sciences, v. 26, p. 423–500, doi: 10.1146/annurev.earth.26.1.423.

Siebe, C., 2000, Age and archaeological implications of Xitle volcano, southwestern Basin of Mexico City: Journal of Volcanology and Geothermal Research, v. 104, p. 45–64, doi: 10.1016/S0377-0273(00)00199-2.

Siebe, C., and Verma, S., 1988, Major element geochemistry and tectonic setting of Las Derrumbadas rhyolitic domes, Puebla, Mexico: Chemie der Erde, v. 48, p. 177–189.

Siebe, C., Abrams, M., Macías, J., and Obenholzner, J., 1996, Repeated volcanic disasters in Prehispanic time at Popocatépetl, central Mexico: Past key to the future?: Geology, v. 24, p. 399–402, doi: 10.1130/0091-7613(1996)024<0399:RVDIPT>2.3.CO;2.

Siebe, C., Quintero-Legorreta, O., García-Palomo, A., and Macías, J., 1999, Effect of strain rate in the distribution of monogenetic and polygenetic volcanism in the Transmexican volcanic belt: Comment: Geology, v. 27, p. 572–573.

Siebe, C., Rodríguez-Lara, V., Schaaf, P., and Abrams, M., 2004a, Geochemistry, Sr-Nd isotope composition, and tectonic setting of Holocene Pelado, Guespalapa and Chichinautzin scoria cones, south of Mexico City: Journal of Volcanology and Geothermal Research, v. 130, no. 3–4, p. 197–226.

Siebe, C., Rodríguez-Lara, V., Schaaf, P., and Abrams, M., 2004b, Radiocarbon ages of Holocene Pelado, Guespalapa, and Chichinautzin scoria cones, south of Mexico City: Implications for archaeology and future hazards: Bulletin of Volcanology, v. 66, p. 203–225, doi: 10.1007/s00445-003-0304-z.

Siebert, L., and Carrasco-Núñez, G., 2002, Late-Pleistocene to precolumbian behind-the-arc mafic volcanism in the eastern Mexican Volcanic Belt; implications for future hazards: Journal of Volcanology and Geothermal Research, v. 115, p. 179–205, doi: 10.1016/S0377-0273(01)00316-X.

Silver, L., and Anderson, T., 1974, Possible left-lateral early to middle Mesozoic disruption of the southwestern North America craton margin: Geological Society of America Abstracts with Programs, v. 6, p. 955–956.

Smith, D., Nuckels, C., Jones, R., and Cook, G., 1979, Distribution of heat flow and radioactive heat generation in northern Mexico: Journal of Geophysical Research, v. 84, no. 2371–2379.

Straub, S., and Martin-Del Pozzo, A., 2001, The significance of phenocryst diversity in tephra from recent eruptions at Popocatépetl Stratovolcano (central Mexico): Contributions to Mineralogy and Petrology, v. 140, p. 487–510, doi: 10.1007/PL00007675.

Strong, M., and Wolff, J., 2003, Compositional variations within scoria cones: Geology, v. 31, p. 143–146, doi: 10.1130/0091-7613(2003)031<0143:CVWSC>2.0.CO;2.

Suárez, G., García-Acosta, V., and Gaulon, R., 1994, Active crustal deformation in the Jalisco block, Mexico: Evidence for a great historical earthquake in the 16th century: Tectonophysics, v. 234, p. 117–127, doi: 10.1016/0040-1951(94)90207-0.

Sun, S., and McDonough, W., 1989, Chemical and isotopic systematics of oceanic basalts: Implications for mantle compositions and processes, *in* Saunders, A., and Norry, M., eds., Magmatism in the Ocean Basins: Geological Society [London] Special Publication 42, p. 313–345.

Suter, M., 1999, Effect of strain rate in the distribution of monogenetic and polygenetic volcanism in the Transmexican volcanic belt: Comment: Geology, v. 27, p. 571, doi: 10.1130/0091-7613(1999)027<0571:EOSRIT>2.3.CO;2.

Suter, M., Quintero, O., and Johnson, C., 1992, Active faults and state of stress in the central part of the Trans-Mexican Volcanic Belt. 1. The Venta de Bravo fault: Journal of Geophysical Research, v. 97, p. 11,983–11,994.

Suter, M., Carrillo-Martínez, M., López-Martínez, M., and Farrar, E., 1995a, The Aljibes half-graben—Active extension at the boundary between the trans-Mexican volcanic belt and the Basin and Range Province, Mexico: Geological Society of America Bulletin, v. 107, no. 6, p. 627–641, doi: 10.1130/0016-7606(1995)107<0627:TAHGAE>2.3.CO;2.

Suter, M., Quintero, O., López, M., Aguirre, G., and Ferrar, E., 1995b, The Acambay graben: Active intraarc extension in the Trans-Mexican Volcanic Belt: Tectonics, v. 14, p. 1245–1262, doi: 10.1029/95TC01930.

Suter, M., Carrillo-Martínez, M., and Quintero-Legorreta, O., 1996, Macroseismic study of shallow earthquakes in the central and eastern parts of the trans-Mexican volcanic belt, Mexico: Seismological Society of America Bulletin, v. 86, p. 1952–1963.

Suter, M., López-Martínez, M., Quintero-Legorreta, O., and Carrillo-Martínez, M., 2001, Quaternary intra-arc extension in the central Trans-Mexican volcanic belt: Geological Society of America Bulletin, v. 113, no. 6, p. 693–703, doi: 10.1130/0016-7606(2001)113<0693:QIAEIT>2.0.CO;2.

Szynkaruk, E., Garduño-Monroy, V., and Bocco, G., 2004, Active fault systems and tectono-topographic configuration of the central Trans-Mexican volcanic belt: Geomorphology, v. 61, p. 111–126, doi: 10.1016/j.geomorph.2003.10.006.

Umhoefer, P., Dorsey, R., Willsey, S., Mayer, S., and Renne, P., 2001, Stratigraphy and geochronology of the Comondú Group near Loreto, Baja California Sur, Mexico: Sedimentary Geology, v. 144, no. 1–2, p. 125–147, doi: 10.1016/S0037-0738(01)00138-5.

Urbina, F., and Camacho, H., 1913, La zona megasísmica de Acambay-Tixmadejé, Estado de México, conmovida el 19 de noviembre de 1912: Boletín del Instituto Geológico de México, v. 32, p. 125.

Urrutia-Fucugauchi, J., 1986, Crustal thickness, heat flow, arc magmatism, and tectonics of Mexico—Preliminary report: Geofísica Internacional, v. 25, no. 4, p. 559–573.

Urrutia-Fucugauchi, J., and Böhnel, H., 1987, Tectonic interpretation of the Trans-Mexican Volcanic Belt: Tectonophysics, v. 138, p. 319–323, doi: 10.1016/0040-1951(87)90047-3.

Urrutia-Fucugauchi, J., and Del Castillo, L., 1977, Un modelo del Eje Volcánico Mexicano: Boletín de la Sociedad Geológica Mexicana, v. 38, p. 18–28.

Urrutia-Fucugauchi, J., and Flores-Ruiz, J., 1996, Bouger gravity anomalies and regional crustal structure in central Mexico: International Geology Review, v. 38, p. 176–194.

Urrutia-Fucugauchi, J., and Rosas-Elguera, J., 1994, Paleomagnetic study of the eastern sector of Chapala Lake and implications for the tectonics of west-central Mexico: Tectonophysics, v. 239, p. 61–71, doi: 10.1016/0040-1951(94)90107-4.

Valdés, C., Mooney, W., Singh, S., Meyer, R., Lomnitz, C., Luetgert, J., Helsley, C., Lewis, B., and Mena, M., 1986, Crustal structure of Oaxaca, Mexico, from seismic refraction measurements: Bulletin of the Seismological Society of America, v. 76, p. 547–563.

Valdéz-Moreno, G., Aguirre-Díaz, G., and López-Martínez, M., 1998, El Volcán La Joya, Edos. de Querétaro y Guanajuato. Un estratovolcán antiguo del cinturón volcánico mexicano: Revista Mexicana de Ciencias Geológicas, v. 15, no. 2, p. 181–197.

Velasco-Tapia, F., and Verma, S., 2001, Estado actual de la investigación geoquímica en el campo monogenético de la Sierra de Chichinautzin: Análisis de información y perspectivas: Revista Mexicana de Ciencias Geológicas, v. 18, no. 1, p. 1–36.

Venegas, S., Herrera, F., and Maciel, R., 1985, Algunas características de la Faja Volcánica Mexicana y sus recursos geotérmicos: Geofísica Internacional, v. 24, p. 97–157.

Verma, S., 1999, Geochemistry of evolved magmas and their relationship to subduction-unrelated mafic volcanism at the volcanic front of the central Mexican Volcanic Belt: Journal of Volcanology and Geothermal Research, v. 93, p. 151–171, doi: 10.1016/S0377-0273(99)00086-4.

Verma, S., 2000a, Geochemistry of the subducting Cocos plate and the origin of subduction-unrelated mafic volcanism at the front of the central Mexican Volcanic Belt, *in* Delgado-Granados, H., Aguirre-Díaz, G., and Stock, J., eds., Cenozoic Tectonics and Volcanism of Mexico: Geological Society of America Special Paper 334, p. 1–28.

Verma, S., 2000b, Geochemical evidence for a lithospheric source for magmas from Los Humeros caldera, Puebla, Mexico: Chemical Geology, v. 164, p. 35–60, doi: 10.1016/S0009-2541(99)00138-2.

Verma, S., 2001a, Geochemical evidence for a lithospheric source for magmas from Acoculco caldera, eastern Mexican Volcanic Belt: International Geology Review, v. 43, p. 31–51.

Verma, S., 2001b, Geochemical and Sr-Nd-Pb isotopic evidence for a combined assimilation and fractional crystallisation process for volcanic rocks from the Huichapan caldera, Hidalgo, Mexico: Lithos, v. 56, p. 141–164, doi: 10.1016/S0024-4937(00)00062-1.

Verma, S., 2001c, Geochemical evidence for a rift-related origin of bimodal volcanism at Meseta San Juan, north-central Mexican Volcanic Belt: International Geology Review, v. 43, p. 475–493.

Verma, S., 2002, Absence of Cocos plate subduction-related basic volcanism in southern Mexico: A unique case on Earth?: Geology, v. 30, no. 12, p. 1095–1098, doi: 10.1130/0091-7613(2002)030<1095:AOCPSR>2.0.CO;2.

Verma, S., and Aguilar-y-Vargas, V., 1988, Bulk chemical composition of magmas in the Mexican Volcanic Belt (Mexico) and inapplicability of generalized arc-models: Chemie der Erde, v. 48, p. 203–221.

Verma, S., and Carrasco-Núñez, G., 2003, Reappraisal of the geology and geochemistry of Volcán Zamorano, Central Mexico: Implications for discriminating the Sierra Madre Occidental and Mexican Volcanic Belt Provinces: International Geology Review, v. 45, p. 724–752.

Verma, S., and Hasenaka, T., 2004, Sr, Nd, and Pb isotopic and trace element geochemical constraints for a veined-mantle source of magmas in the Michoacan-Guanajuato Volcanic Field, west-central Mexican Volcanic Belt: Geochemical Journal, v. 38, no. 1, p. 43–65.

Verma, S., López-Martínez, M., and Terrel, D., 1985, Geochemistry of Tertiary igneous rocks from Arandas-Atotonilco area, northeast Jalisco, Mexico: Geofísica Internacional, v. 24, p. 31–45.

Verma, S., and Nelson, S., 1989, Isotopic and trace element constraints on the origin and evolution of alkaline and calc-alkaline magmas in the northwestern Mexican volcanic belt: Journal of Geophysical Research, v. 94, p. 4531–4544.

Verma, S., Carrasco-Núñez, G., and Milán, M., 1991, Geology and geochemistry of Amealco caldera, Qro., Mexico: Journal of Volcanology and Geothermal Research, v. 47, p. 105–127, doi: 10.1016/0377-0273(91)90104-8.

von Humboldt, A., 1808, Essai politique sur le Royaume de la Nouvelle Espagne: Paris, F. Shoell, 904 p.

Wallace, P., and Carmichael, I., 1989, Minette lavas and associated leucitites from the western front of the Mexican Volcanic Belt: Petrology, chemistry and origin: Contributions to Mineralogy and Petrology, v. 103, p. 470–492, doi: 10.1007/BF01041754.

Wallace, P., and Carmichael, I., 1992, Alkaline and calc-alkaline lavas near Los Volcanes, Jalisco, Mexico: Geochemical diversity and its significance in volcanic arcs: Contributions to Mineralogy and Petrology, v. 111, no. 4, p. 423–439, doi: 10.1007/BF00320899.

Wallace, P., and Carmichael, I., 1994, Petrology of Volcán Tequila, Jalisco, México: Disequilibrium phenocryst assemblages and evolution of the subvolcanic magma system: Contributions to Mineralogy and Petrology, v. 117, p. 345–361, doi: 10.1007/BF00307270.

Wallace, P., and Carmichael, I., 1999, Quaternary volcanism near the Valley of Mexico: Implications for subduction zone magmatism and the effects of crustal thickness variations on primitive magma compositions: Contributions to Mineralogy and Petrology, v. 135, p. 291–314, doi: 10.1007/s004100050513.

Weber, B., and Köhler, H., 1999, Sm–Nd, Rb–Sr and U–Pb isotope geochronology of a Grenville terrane in Southern Mexico: Origin and geologic history of the Guichicovi complex: Precambrian Research, v. 96, p. 245–262, doi: 10.1016/S0301-9268(99)00012-1.

Wilcox, R., 1954, Petrology of Paricutin Volcano, Mexico: United States Geological Survey Bulletin, v. 965-C, p. 281–354.

Yañez, P., Ruiz, J., Patchett, J., Ortega-Gutiérrez, F., and Gehrels, G., 1991, Isotopic studies of the Acatlán Complex, southern México: Implications for Paleozoic North American tectonics: Geological Society of America Bulletin, v. 103, p. 817–828, doi: 10.1130/0016-7606(1991)103<0817: ISOTAC>2.3.CO;2.

Yañez-García, A., and Casique, J., 1980, Informe geológico del proyecto geotérmico los humeros-derrumbadas, Estados de Puebla y Veracruz: Comisión Federal de Electricidad, 97 p.

Yarza de la Torre, E., 1992, Volcanes de México: México, D.F., Instituto de Geografía, Universidad Nacional Autónoma de México, 173 p.

Zamorano-Orozco, J., Tenarro-García, L., Lugo-Hubp, J., and Sánchez-Rubio, G., 2002, Evolución geológica y gemorfología del complejo dómico Los Pitos, norte de la Cuenca de México: Revista Mexicana de Ciencias Geológicas, v. 19, p. 66–79.

Ziagos, J., Blackwell, D., and Mooser, F., 1985, Heat flow in southern Mexico and the thermal effects of subduction: Journal of Geophysical Research, v. 90, p. 5410–5420.

Manuscript Accepted by the Society 29 August 2006

Printed in the USA

Geological Society of America
Special Paper 422
2007

Geology and eruptive history of some active volcanoes of México

José Luis Macías*
Departamento de Vulcanología, Instituto de Geofísica, Universidad Nacional Autónoma de México, Coyoacán 04510, México D.F.

ABSTRACT

Most of the largest volcanoes in México are located at the frontal part of the Trans-Mexican Volcanic Belt and in other isolated areas. This chapter considers some of these volcanoes: Colima, Nevado de Toluca, Popocatépetl, Pico de Orizaba (Citlaltépetl), and Tacaná. El Chichón volcano is also considered within this group because of its catastrophic eruption in 1982. The volcanic edifice of these volcanoes, or part of it, was constructed during the late Pleistocene or even during the Holocene: Colima 2500 yr ago, Pico de Orizaba (16,000 yr), Popocatépetl (23,000 yr), Tacaná (~26,000 yr), and Nevado de Toluca (>45,000). The modern cones of Colima, Popocatépetl, Pico de Orizaba, and Tacaná are built inside or beside the remains of older caldera structures left by the collapse of ancestral cones. Colima, Popocatépetl, and Pico de Orizaba represent the youngest volcanoes of nearly N-S volcanic chains. Despite the repetitive history of cone collapse of these volcanoes, only Pico de Orizaba has been subjected to hydrothermal alteration and slope stability studies crucial to understand future potential events of this nature.

The magmas that feed these volcanoes have a general chemical composition that varies from andesitic (Colima and Tacaná), andesitic-dacitic (Nevado de Toluca, Popocatépetl, and Pico de Orizaba) to trachyandesitic (Chichón). These magmas are the result of several magmatic processes that include partial melting of the mantle, crustal assimilation, magma mixing, and fractional crystallization. So far, we know very little about the deep processes that occurred between the upper mantle source and the lower crust. However, new data have been acquired on shallower processes between the upper crust and the surface. There is clear evidence that most of these magmas stagnated at shallow magma reservoirs prior to eruption; these depths vary from 3 to 4 km at Colima volcano, 4.5–6 km at Nevado de Toluca, and ~6–8 km at Chichón volcano.

Over the past 15 years, there has been a surge of studies dealing with the volcanic stratigraphy and eruptive history of these volcanoes. Up to the present, no efforts have achieved integration of the geological, geophysical, chemical, and petrological information to produce conceptual models of these volcanoes. Therefore, we still have to assume the size and location of the magma chambers, magma ascent paths, and time intervals prior to an eruption. Today, only Colima and

*macias@geofisica.unam.mx

Macías, J.L., 2007, Geology and eruptive history of some active volcanoes of México, *in* Alaniz-Álvarez, S.A., and Nieto-Samaniego, Á.F., eds., Geology of México: Celebrating the Centenary of the Geological Society of México: Geological Society of America Special Paper 422, p. 183–232, doi: 10.1130/2007.2422(06). For permission to copy, contact editing@geosociety.org.

Popocatépetl have permanent monitoring networks, while Pico de Orizaba, Tacaná, and Chichón have a few seismic stations. Of these, Popocatépetl, Colima, and Pico de Orizaba have volcanic hazard maps that provide the basic information needed by the civil defense authorities to establish information programs for the population as well as evacuation plans in case of a future eruption.

Keywords: geology, eruptive chronology, active volcanoes, México.

INTRODUCTION

The first reports regarding volcanic activity in México appeared in the náhuatl codes; the volcanoes that deserved such attention were Pico de Orizaba (Citlaltépetl) and Popocatépetl. Their eruptions were shown in these codes as hills with a smoking top. The best record belongs to Popocatépetl, which means in náhuatl language "the smoking mountain." The Aztecs reported the occurrence of eruptions in 1363, 1509, 1512, and 1519–1528 (De la Cruz-Reyna et al., 1995). After the Spanish conquest, several volcanoes received more attention mostly focused on the activity of Popocatépetl volcano. However, it is only until the beginning of the nineteenth century that some scientific observations were carried out, such as those by Humboldt in 1804, Del Río in 1842, Del Castillo in 1870, and Sánchez in 1856 (in De la Cruz-Reyna et al., 1995). During this time several travelers, and among them some painters, produced detailed images of the volcanoes: Echeverría in 1793, Rugendas between 1831 and 1834, Baptiste in 1834, Egerton in 1834, Pieschel in 1856, Sattler in 1862, and White between 1862 and 1863. Their paintings provide information regarding the activity of several of the largest volcanoes in México, but they do not represent scientific studies aimed to understand their behavior. It was not until 1890 that two European travelers visited México and pointed out the distribution of the largest volcanoes (Felix and Lenk, 1890). In their report, they proposed that volcanoes like Popocatépetl, Cerro del Ajusco, Nevado de Toluca, Patzcuaro, Patambán, and La Bufa de Mascota (Bufa de Real Alto) were genetically linked to the presence of a single fissure. They also mentioned the existence of aligned volcanoes in secondary fissures, such as Pico de Orizaba–Cofre de Perote, el Telapón, Tlamacas, Iztaccíhuatl, Popocatépetl, and Ceboruco–Cerro de la Bufa. Without a specific purpose except to understand the nature of the Mexican volcanoes, the observations of Felix and Lenk (1890) produced great skepticism amongst the small Mexican geological community who began to investigate the origin, age and distribution of the volcanoes (Ordoñez, 1894). At that time, only general geological studies had been performed on some volcanoes, and some details of the relative age of certain volcanic chains were known. México had a reduced number of geologists who were mostly appointed to the National Geological Survey and to the Geological Commission of México. Ordoñez (1894), then a member of the Geological Commission and one of the nation's most prestigious geologists, reviewed the work by Felix and Lenk (1890) and denied the existence of a large fracture that produced all these volcanoes. He discussed the differences between the truncated crater of Nevado de Toluca and the Quaternary eruptions of Popocatépetl, the Ajusco volcano that did not have a crater, and the contemporaneity and common genesis of the volcanoes. Analyzing the existence of N-S faults, Ordoñez considered that the pyroxene-andesites of Iztaccíhuatl volcano and others in México were related to volcanoes that were destroyed in the past—in other words, old volcanoes, as opposed to the hornblende-andesites of Popocatépetl, which had a young crater and notorious activity. In this way, Ordoñez rejected the existence of secondary N-S fissures. Despite of this erroneous conclusion, Ordoñez demonstrated that he had a broad knowledge of the Mexican volcanoes, as reflected in his studies of Iztaccíhuatl (Ordoñez, 1894), Popocatépetl (Aguilera and Ordoñez, 1895), Colima and Ceboruco (Ordoñez, 1898), Cofre de Perote (Ordoñez, 1905), and Nevado de Toluca (Ordoñez, 1902). In these studies, he defined the morphologic features of the volcanic edifices, the petrographic nature of the rocks, and stratigraphic descriptions of the deposits. His work established the basic frame for new geological studies of the Mexican volcanoes, which started at the end of the nineteenth century and the beginning of the twentieth century. The first meteorological and volcanological observatory in México was founded in 1893 in Zapotlán, Jalisco, by priests Arreola and Diaz, who performed systematic observations of Colima volcano between 1893 and 1905. In 1895, Arreola founded the Colima Volcanological Observatory (Díaz, 1906), a surprising fact considering that the first volcanological observatory in the world was established at Vesubio volcano (Italy) by Palmieri only 54 years earlier, after a series of eruptions.

The first decade of the twentieth century had a large impact in the geological community worldwide, due to the occurrence of several volcanic eruptions in Latin America. On 8 May 1902, Mount Pelée on the island of Martinique erupted, producing a pyroclastic surge that completely destroyed the city of St. Pierre, killing more than 25,000 people (Perret, 1937). A day before, on 7 May 1902, Soufrière volcano devastated the island of St. Vincent, with a total death toll of 2000 people (Anderson and Flett, 1903). Finally, in October of the same year, Santa Maria volcano, Guatemala, awoke violently, producing a Plinian eruption that covered the NW portion of Guatemala with pumice and sent ash into central México. This eruption generated 12 km^3 of magma, which represents one of the largest eruptions in the twentieth century (Williams and Self, 1983). The catastrophic eruption of Mount Pelée attracted the attention of geologists worldwide, who studied it and subsequent eruptions in detail and established the Volcanological Observatory of Morne des Cadets in 1903.

These eruptions opened a new course in the study of active volcanoes in México, such as those of Böse (1903) about Tacaná volcano in the State of Chiapas. Böse concluded that Tacaná volcano consisted of three peaks that represented the remains of old edifices. In addition, he performed detailed observations of the summit, where he described small craters produced by recent eruptions. Waitz (1909) concluded that Nevado de Toluca volcano represented the ruins of an andesitic stratovolcano, with an extinct central dome affected by glacier activity, due to the presence of moraine deposits.

During the second decade of the twentieth century, two important volcanic eruptions took place in México that influenced the development of volcanological studies. These events were the 1913 eruption of Colima volcano and the 1913–1927 eruption of Popocatépetl. Despite the fact that these eruptions occurred during the troubled times of the Mexican Revolution, both attracted the attention of national and international geologists. This was followed by two decades of relative quietness, but this ended on 20 February 1943, when there was news of the birth of a volcano in a field corn of the State of Michoacán, close to San Juan Parícutin. This volcano, called Parícutin, reached an elevation of 424 m and buried the villages of Parícutin and San Juan Parangaricutiro (Flores-Covarrubias, 1945). This eruption was closely followed by volcanologists (Wilcox, 1954; González-Reyna, 1956; Segerstrom, 1956), and Parícutin became the most studied volcano in México (Luhr and Simkin, 1993). Today, Parícutin forms part of education programs at the elementary level and is a classic example of the birth of a volcano worldwide. In 1956, a spectacular eruption occurred on San Benedict Island in the Revillagigego archipelago, giving rise to the Bárcena volcano. Because of its remote location, this eruption was studied by only a small group of volcanologists (Richards, 1959, 1965). In 1962, the crater of Colima volcano, formed in 1913, was filled to the brim, beginning the emission of lava and pyroclastic flows on its flanks, with several eruptions occurring in 1962, 1976, 1981, 1987, 1991, 1994, 1998–2000, and 2002–2005.

The event that accelerated volcanological studies in México was the 28 March 1982 catastrophic eruption of El Chichón volcano in the State of Chiapas (Sigurdsson et al., 1984). The reactivation of this practically unknown volcano took place after 550 years of quiescence (Tilling et al., 1984; Macías et al., 2003), taking scientists by surprise. This eruption had a death toll of 2000 people, destroyed nine villages, and caused important global effects in the planet with the emission of 7 Mt of SO_2 into the atmosphere and the reduction of the planet temperature of ~0.5 °C during several months (Espíndola et al., 2002). The El Chichón eruption still represents the worst volcanic disaster in México in historic times. As a consequence of this eruption and the 1985 earthquake that devastated México City, the National Civil Protection System was created. During the late eighties, the volcanic activity continued with small events, such as the 1986 phreatic eruption of Tacaná volcano that ended with the emission of a fumarole. This event warned the state authorities to begin monitoring Tacaná volcano. During the 1990s, the small submarine eruption of Everman volcano in the Revillagigedo archipelago occurred (Siebe et al., 1995a). Another important event was the reactivation of Popocatépetl volcano occurred on 21 December 1994; it not only represented a breakthrough in volcanological studies, but for hazard mitigation studies as well. Immediately after the crisis began, a scientific committee was formed to evaluate the volcanic crisis at Popocatépetl. This committee deemed it pertinent to construct a volcanic hazards map for civil protection authorities. It delineated emergency plans that included evacuation routes, meeting points, shelters, etc. The Popocatépetl hazards map was published in February 1995 (Macías et al., 1995). This map was followed by the creation of the Colima volcano (Martín Del Pozzo et al., 1995; Navarro et al., 2003) and Pico de Orizaba (Sheridan et al., 2002) hazard maps.

LOCATION OF THE ACTIVE VOLCANOES IN MÉXICO

The highest concentration of volcanoes in México is located in the Trans-Mexican Volcanic Belt, where a large variety of volcanic landforms occurred as monogenetic volcanic fields, majestic stratovolcanoes with elevations higher than 4000 m above sea level, shield and composite volcanoes, calderas, fissural lava flows and domes (Fig. 1). Inside the Trans-Mexican Volcanic Belt there are several N-S to NE-SW volcanic chains made of stratovolcanoes and composite volcanoes, where the volcanic activity has migrated during the past 2 m.y., toward the frontal part of the volcanic arc. In other words, the active volcanoes are located in the southward end of this volcanic chain. These volcanic chains are the Cántaro–Nevado de Colima–Colima, Tláloc–Telapón–Iztaccíhuatl–Popocatépetl, and Cofre de Perote–Las Cumbres–Pico de Orizaba–Sierra Negra. In this work, some of the large active volcanoes in México are considered, such as Colima, Nevado de Toluca, Popocatépetl, Pico de Orizaba, and Tacaná. Due to its 1982 catastrophic eruption, El Chichón is also considered within this group. In the following section, there is a description of each volcano, a compilation of the available information, and then an analysis and reflection of the advances and limitations of such studies, as well as another studies yet to be accomplished.

COLIMA VOLCANO

Colima volcano or Fuego de Colima (19°30′45″, 103°37′) has an elevation of 3860 m above sea level and is the eighth highest peak in México (Fig. 2). The volcano is located 100 km south of the city of Guadalajara and 30 km north of the city of Colima. Colima volcano belongs to an N-S volcanic chain formed by the Cántaro, Nevado de Colima, and Colima volcanoes.

Previous Studies

The first geological studies of Colima volcano were carried out by Waitz (1906, 1915, 1935). Waitz made general observations of the volcano and described the generation of pyroclastic

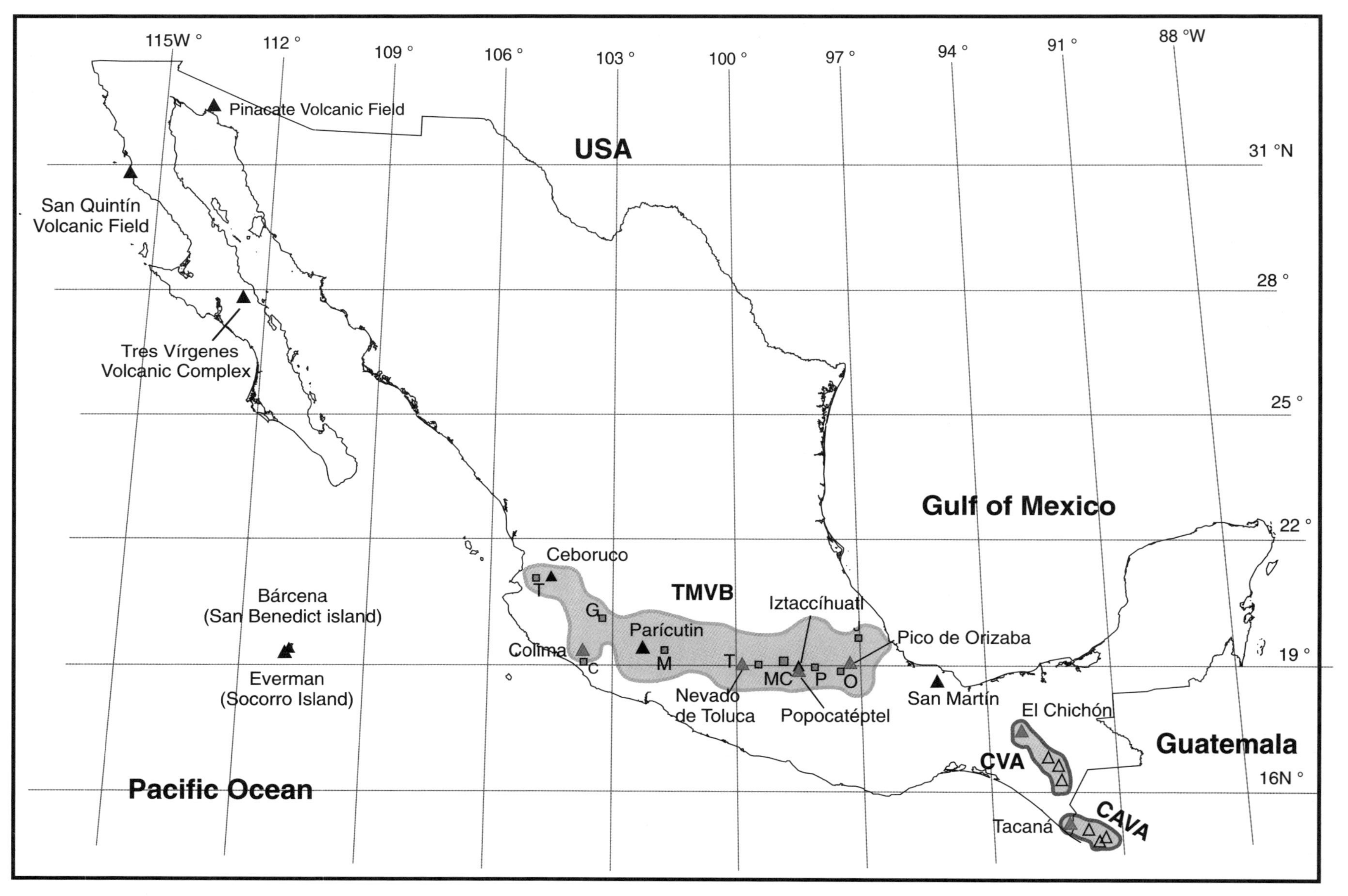

Figure 1. Location of main volcanoes in México (black triangles) that are grouped along the 19–20°N parallel of latitude forming the Trans-Mexican Volcanic Belt (TMVB). There are other regions of isolated volcanism, such as the Revillagigedo archipelago, the Tres Vírgenes Volcanic Complex, Los Tuxtlas Volcanic Complex, and the Chiapanecan Volcanic Arc. The volcanoes treated in this work appear as gray triangles. CVA—Chiapanecan Volcanic Arc; CAVA—Central America Volcanic Arc; C—Colima; G—Guadalajara; J—Jalapa; M—Morelia; MC—México City; O—Orizaba; P—Puebla; T—Tepic.

Figure 2. Panoramic view of the southern flank of Colima volcano taken in December 1998. Visible are some blocks collapsed from the summit that roll down the steep slopes of the volcano.

flows during the 1913 eruption. From that time until the end of the 1950s, Colima volcano's activity remained restricted to the crater's interior. In 1962, the crater was completely filled with lava, causing the emission of lava flows on the volcano flanks, attracting the attention of volcanologists who studied the morphology and some general features of the volcanic complex (Mooser, 1961), its Merapi type pyroclastic flows (Thorpe et al., 1977), and the evolution of the volcano (Demant, 1979). The 1980s were the starting point of the modern studies of Colima volcano. After the 1981 eruption, Medina-Martínez (1983) presented the first analysis of eruptive recurrence of the volcano during the past 400 yr. In addition, the first chemical and petrological studies of the volcano and adventitious cones were completed (Luhr and Carmichael, 1980, 1981, 1982, 1990a) as well as a study of the collapse of the volcanic edifice (Robin et al., 1987). This latter study created an enormous interest in the volcano and led to a surge of new studies and discussion over the subsequent years (Luhr and Prestegaard, 1988; Stoopes and Sheridan, 1992; Komorowski et al., 1997; Capra and Macías, 2002; Cortés, 2002). The 1990s were distinguished by the generation of several small eruptions at Colima that produced lava and pyroclastic flows. These eruptions were studied by different specialists, from seismic studies (Núñez-Cornu et al., 1994; Jiménez et al., 1995; Domínguez et al., 2001; Zobin et al., 2002), petrology (Robin et al., 1990; Robin and Potrel, 1993; Connor et al., 1993; Luhr, 2002; Macías et al., 1993; Mora et al., 2002; Valdez-Moreno et al., 2006), gas chemistry (Taran et al., 2002), and stratigraphy (Martín Del Pozzo et al., 1987; Rodríguez-Elizarrarás et al., 1991; Rodríguez-Elizarrarás, 1995; Navarro-Ochoa et al., 2002; Saucedo et al., 2002; Saucedo et al., 2004a). During the past 25 years, several geological maps of Colima volcano have been completed (Demant, 1979; Luhr and Carmichael, 1990b; Rodríguez-Elizarrarás, 1991; Cortés et al., 2005), as well as several types of hazards maps (Sheridan and Macías, 1995; Martín Del Pozzo et al., 1995; Navarro et al., 2003; Saucedo et al., 2004b).

Evolution of the Volcanic Chain

The NW part of the Trans-Mexican Volcanic Belt is subjected to the subduction of the Rivera plate underneath the North America plate and to the triple point junction formed by the Tepic-Zacoalco rift to the NW, the Chapala rift to the E, and the Colima graben to the S (Luhr et al., 1985; Garduño and Tibaldi, 1990) (Fig. 3). The first two rifts bound to the N and to the E of the so-called Jalisco block and are considered to be old cortical structures reactivated by forces acting at the plate boundaries (Rosas-Elguera et al., 1996). According to Cortés et al. (2005), the Colima Volcanic Complex is built upon the lower Cretaceous andesites and volcaniclastic deposits of the Tecalitlán Formation, sandstones and claystones of the Encino Formation, massive limestones of the Tepames Formation, red beds of the upper Cretaceous Coquimatlan Formation, Cretaceous intrusive rocks, and a Tertiary volcanic sequence made of basaltic and andesitic lava flows, dacitic volcanic breccias, and ignimbrites (Fig. 4).

The Quaternary volcanic activity in the graben began ca. 1.6 Ma with formation of the Cántaro stratovolcano (Allan, 1986). Allan and Carmichael (1984) reported K-Ar ages for this volcano that vary from 1.66 ± 0.24, 1.52 ± 0.20, and 1.33 ± 0.20 Ma. Cántaro volcano consists of andesitic lava flows followed by dacitic domes (Luhr and Carmichael, 1990b). The activity of Cántaro ended ca. 1.0 Ma.

Nevado de Colima Volcano

Afterward, the volcanic activity migrated ~15 km to the south to form the ancient Nevado de Colima volcano, which had a complex eruptive history composed of either two eruptive phases (Mooser, 1961), four phases (Robin et al., 1987), or six eruptive periods (Cortés et al., 2005). According to the latter authors, these eruptive periods were influenced by the activity of the Tamazula fault (Garduño et al., 1998); three of them are associated with the failure of the volcano and the generation of debris avalanche deposits. These eruptive periods are described below:

1. During the first eruptive period, ca. 53 Ma, a volcanic edifice 25 km in diameter was formed (Robin et al., 1987). This stratovolcano was composed of andesitic lavas, pyroclastic flows, and fallout deposits reaching a volume higher than 300 km^3.

2. The formation of a second edifice smaller than the first made of andesitic lava and pyroclastic flow and fall deposits of unknown age.

3. The construction of a third edifice composed of andesitic lava and pyroclastic flow and fall deposits. This edifice is associated with a 4-km-wide semicircular caldera open to the SE, produced by the collapse of the flank that caused a debris avalanche deposit exposed to the NE of the El Platanar village.

4. The activity continued with the building of a new volcanic edifice with the emission of 17 km long andesitic lava flows,

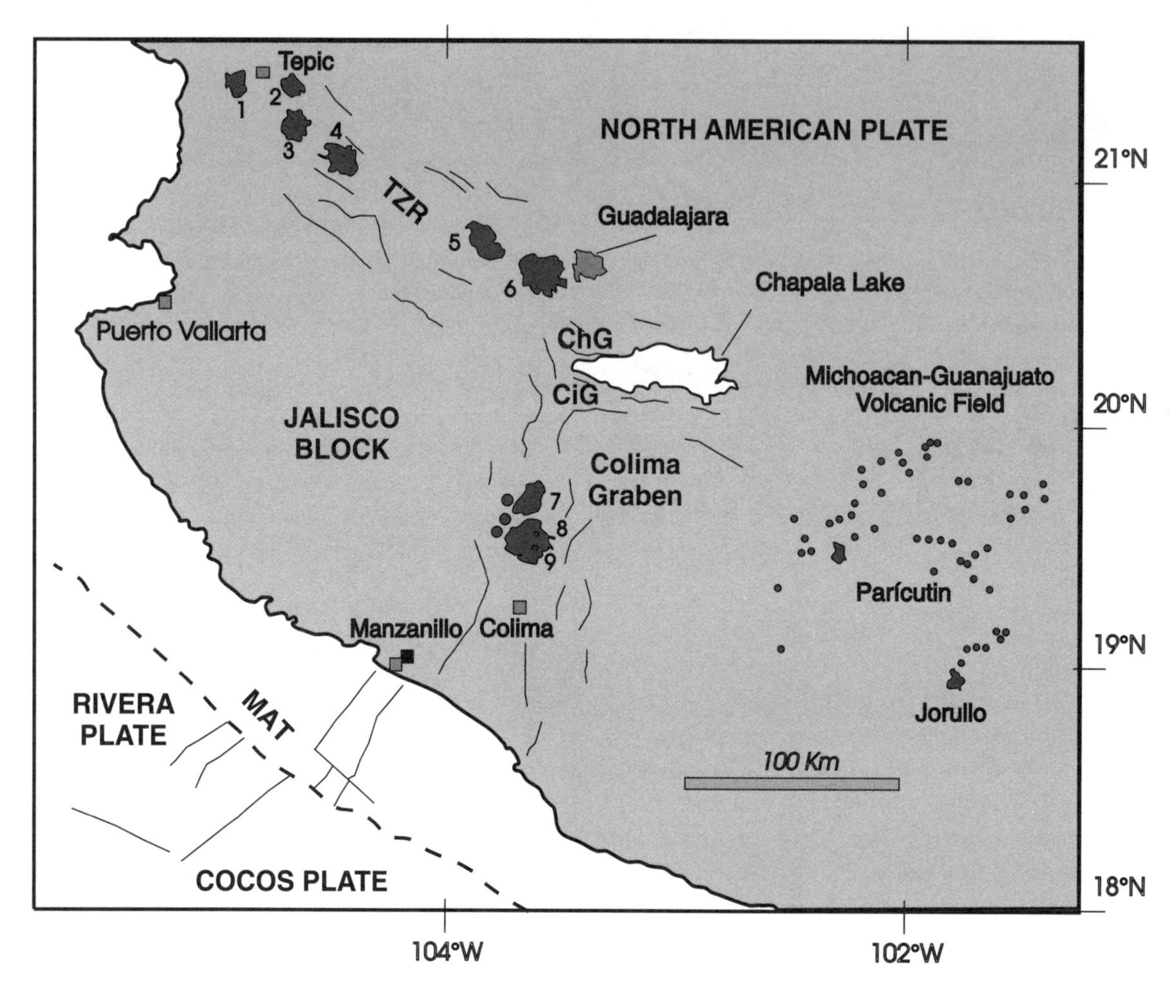

Figure 3. Tectonic setting of the western portion of the Trans-Mexican Volcanic Belt, after DeMets and Stein (1991), Lange and Carmichael (1991), and Luhr et al. (1989). The stratovolcanoes are 1—San Juan; 2—Sangangúey; 3—Tepetiltic; 4—Ceboruco; 5—Tequila; 6—La Primavera caldera. Colima Volcanic Complex: 7—Cántaro; 8—Nevado de Colima; 9—Colima volcano. ChG—Chapala Graben; CiG—Citala Graben; MAT—Middle America Trench (discontinuous line); TZR—Tepic-Zacoalco Rift. Thin lines represent fractures.

pyroclastic flows, and fallouts. Some of these lava flows have an age of 0.35 Ma (Robin and Boudal, 1987). The activity of the volcano ended with the lateral collapse of the edifice toward the SE, originating a second debris avalanche deposit exposed up to 25 km to the SE of the crater at the Beltrán and Tuxpan-Naranjo gullies. The precise age of the collapse is still unknown.

5. Renewed activity began at the interior of the caldera with the emplacement of thick lava flows and pyroclastic deposits exposed at the Atenquique gully; these deposits built a fifth volcanic edifice whose activity ended with an explosive event that produced block-and-ash flows traveling 17 km to the ESE from the crater to the Atenquique area and leaving a semicircular caldera opened to the ENE (Cortés et al., 2005). These deposits discordantly overlie the fluviatile deposits of the Atenquique Formation (Mooser, 1961) that Robin and Boudal (1987) dated at 0.38 and 0.26 Ma. However, this caldera has been associated by other authors with the gravitational collapse of Nevado de Colima, which occurred 18,500 yr B.P. and generated a third debris avalanche deposit (Robin et al., 1987; Stoopes and Sheridan, 1992; Capra, 2000; Capra and Macías, 2002). This debris avalanche deposit caused an enormous interest in the scientific community since Stoopes and Sheridan (1992) concluded that this avalanche had traveled along the Tuxpan-Naranjo and Salado rivers all the way to the Pacific Coast across a distance of 120 km from the summit, at that time the largest described in the world. However, Capra (2000), Pulgarín et al. (2001), and Capra and Macías (2002) stated that this event, in fact, began with the flank failure of Nevado de Colima that produced a debris avalanche that traveled 25 km up to the Naranjo River. The deposit dammed the river, forming a temporary lake that, after a few days, broke to produce a gigantic debris flow, which moved along the Tuxpan-Naranjo rivers up to the Pacific Ocean.

6. This event was followed by a period of volcanic quiescence of Nevado de Colima, which allowed the development of

soils and occasional explosive eruptions, forming extended yellow ash flows with pumice (Robin and Boudal, 1987). One of these deposits was dated at 17,960 yr B.P. at the NW edge of the Quesería village, whose age is very close to the younger debris avalanche of the volcano. It has been assumed that these may be associated events or closely spaced in time. In addition, there was a series of Plinian eruptions that occurred between 8000 and 2000 yr B.P. that emplaced fallouts and pyroclastic surges at the caldera interior (Navarro-Ochoa and Luhr, 2000; Luhr and Navarro-Ochoa, 2002). The eruptive activity of Nevado de Colima came to an end with andesitic lava flows contained inside the caldera walls and the emplacement of the El Picacho dome, which is the present summit of the volcano.

Paleofuego Volcano

At the same time as the last stages of activity of Nevado de Colima, Paleofuego began its construction 5 km south of the ancient cone of Colima volcano (Robin and Boudal, 1987) (Fig. 4). This volcanic edifice is now represented by a 5-km-wide caldera open to the south. The projection of the caldera walls suggests that the volcano had an approximate elevation of 4100 m above sea level (masl) (Luhr and Prestegaard, 1988). The 300-m-thick northern walls are made of andesitic lava flows alternated with block-and-ash flow deposits (Cortés et al., 2005). These lava flows moved 17.5 km to the SW and 31 km to the SE. A sequence composed of a pyroclastic flow, lahar, and lacustrine beds was dated at 38,400 yr B.P. by Komorowski et al. (1993), who associated it with Paleofuego volcano. Therefore, this age represents the minimum age of formation of Paleofuego. Waitz (1906) was the first author to describe the Paleofuego caldera as a maar; Demant (1979) associated its formation with a series of cyclic eruptions of scoria and ash; Robin et al. (1987) and Luhr and Prestegaard (1988) interpreted it as the remains of a structure from a Mount St. Helens–type collapse. Luhr and Prestegaard (1988) estimated an area of 1550 km^2 for the debris avalanche deposit, located to the south of the volcano, and dated charcoal material at its base at 4280 ± 110 yr B.P. However, Robin et al. (1987) assigned an older age of 9370 ± 400 yr B.P. to the deposit and considered it to be composed of several deposits. Later, this deposit was mapped by Komorowski et al. (1997), who concluded that the deposit represented the youngest debris avalanche of the volcano, which traveled 30 km to the south, covered a surface of ~1200 km^2 and occurred 2500 yr B.P. More recently, Cortés et al. (2005) reported that Paleofuego volcano collapsed at least five times during its eruptive history, with deposits covering an area of 5000 km^2 (Fig. 4). The following summary of events and deposits uses the abbreviations presented by Cortés et al. (2005):

The first debris avalanche (CVP3) crops out 40 km south of the volcano, in the vicinity of Coquimatlán. This deposit develops a series of small hummocks aligned in the flow direction. The deposit covers a surface of ~445.5 km^2 and has an average thickness of 20 m and a volume of 8.9 km^3. It is covered by a sequence of fluviatile and lahar deposits, divided by a paleosol dated at 16650 ± 135 yr B.P.

The second debris avalanche deposit (CVP4) outcrops near the town of Mazatán and is located to the west of the volcano, atop the Cerro Grande limestones. The deposit is covered by a lacustrine sequence dated at 6390 and 7380 yr B.P. (Komorowski et al., 1997).

The third deposit (CVP5) covers ~586 km^2 and has a volume of ~30 km^3. Outcrops from 0 to 15 km from the crater consist of block facies (hummocks) with close depressions forming lakes (El Jabalí, Carrizalillos, etc.). Here the blocks consist of red and gray lava with jigsaw-fit structure and scarce matrix. From 15 to 30 km from the volcano, the deposit developed a matrix facies with a morphology of smooth hills. This facies consists of angular blocks (up to 2 m in diameter) set in a poorly consolidated fine ash matrix. Organic material found inside the deposit was dated at 6990 ± 130 yr B.P. (Cortés and Navarro-Ochoa, 1992).

The fourth debris avalanche deposit (CVP6) extends up to 20 km to the SW, from the present Paleofuego caldera wall, and covers ~40 km^2. This deposit consists of megablocks assimilated from different volcanic and fluvio-lacustrine deposits (Cortés, 2002) and was dated at 3600 yr B.P. (Komorowski et al., 1997). The fifth debris avalanche is represented by the youngest deposit dated at 2500 yr B.P. by Komorowski et al. (1997).

Colima Volcano

After the last collapse of Paleofuego volcano 2500 yr ago, the volcanic activity migrated to the south inside the caldera floor of Paleofuego. This activity formed the modern Colima volcano, which has an approximate volume of 10 km^3 and has grown at a rate of 0.002 km^3/yr (Luhr and Carmichael, 1990a, 1990b). This stratovolcano is composed of an interlayering of andesitic lava flows and pyroclastic deposits (fall, surge and flow). One of the main features of Colima volcano is the generation of pyroclastic flow deposits that have reached 15 km from the summit, like those produced during the 1913 eruption. During the past 400 yr, Colima volcano has experienced ~43 eruptions, which places it among the most active volcanoes in North America (De la Cruz-Reyna, 1993; Saucedo et al., 2004b).

The most detailed studies summarizing the eruptive activity of Colima volcano are those by Medina-Martínez (1983), De la Cruz-Reyna (1993), Saucedo and Macías (1999), and Bretón et al. (2002). The oldest eruptions occurred in the sixteenth and seventeenth centuries (Tello, 1651), including the 13 December 1606 (Arreola, 1915) and the 15 April 1611 (Bárcena, 1887) eruptions, as well as those that took place in 1690 (De la Cruz-Reyna, 1993) and in 1771 (Bárcena, 1887), and many other minor eruptions. A detailed historical record began with the 15 February 1818 eruption (Sartorius, 1869), which destroyed a lava dome (Dollfus and Monserrat, 1867) and dispersed ash and scoria to the cities of Guadalajara, Zacatecas, Guanajuato, San Luis Potosí, and México (Bárcena, 1887; Arreola, 1915). After the 1818 eruption, Colima volcano had a 500-m-wide, funnel-shaped crater with 50–230-m-deep vertical walls. On 12 July 1869, the formation of the parasitic vent "El Volcancito" began; it ended in 1872 (Sartorius, 1869; Bárcena, 1887). Orozco et al. (1869) reported

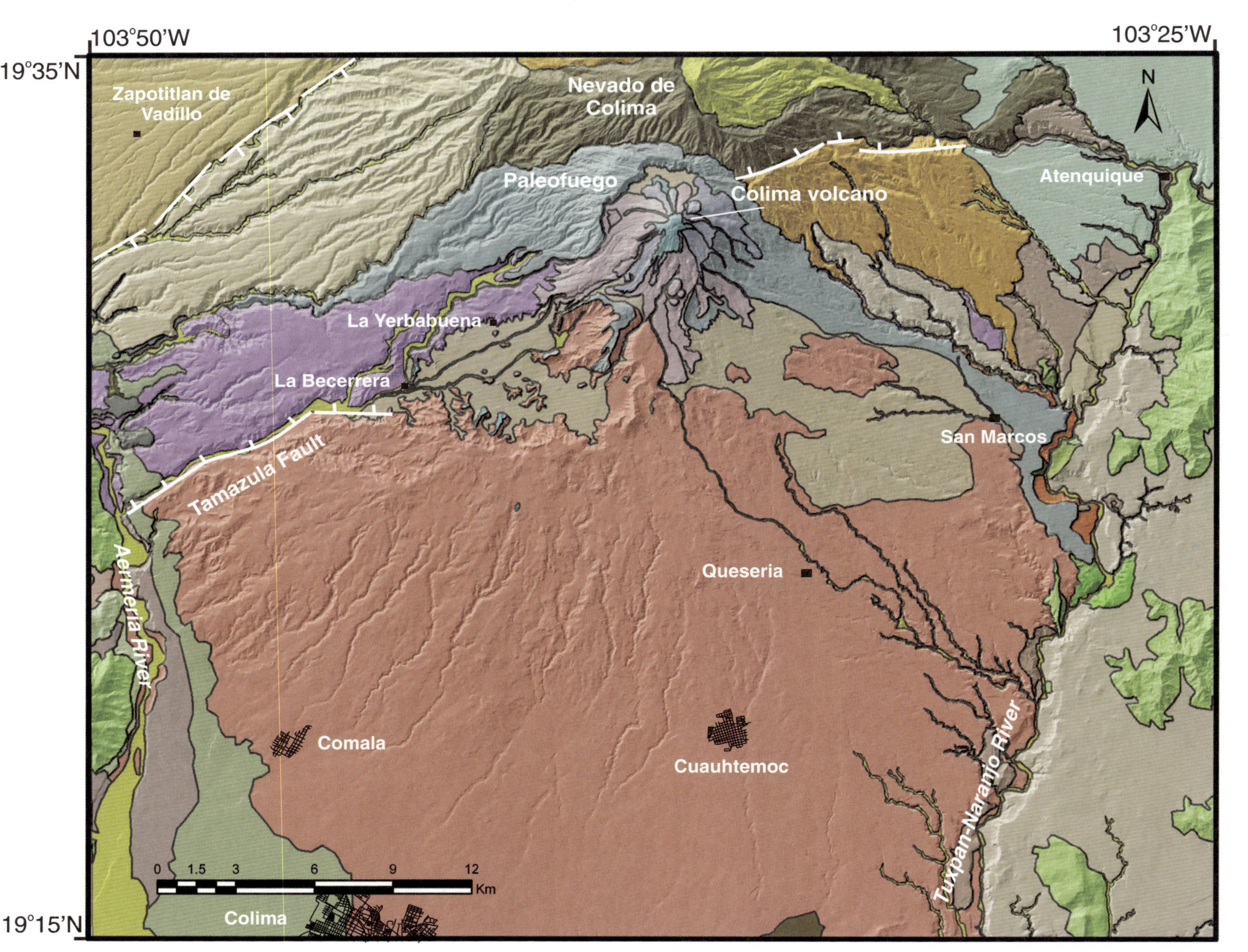

Figure 4 (*legend on following page*). Simplified geologic map of the southern portion of the Colima Volcanic Complex (taken from Cortés et al., 2005) superimposed on a digital elevation model. For the sake of simplicity, only the mapped units of Colima volcano and a portion of Nevado de Colima are labeled.

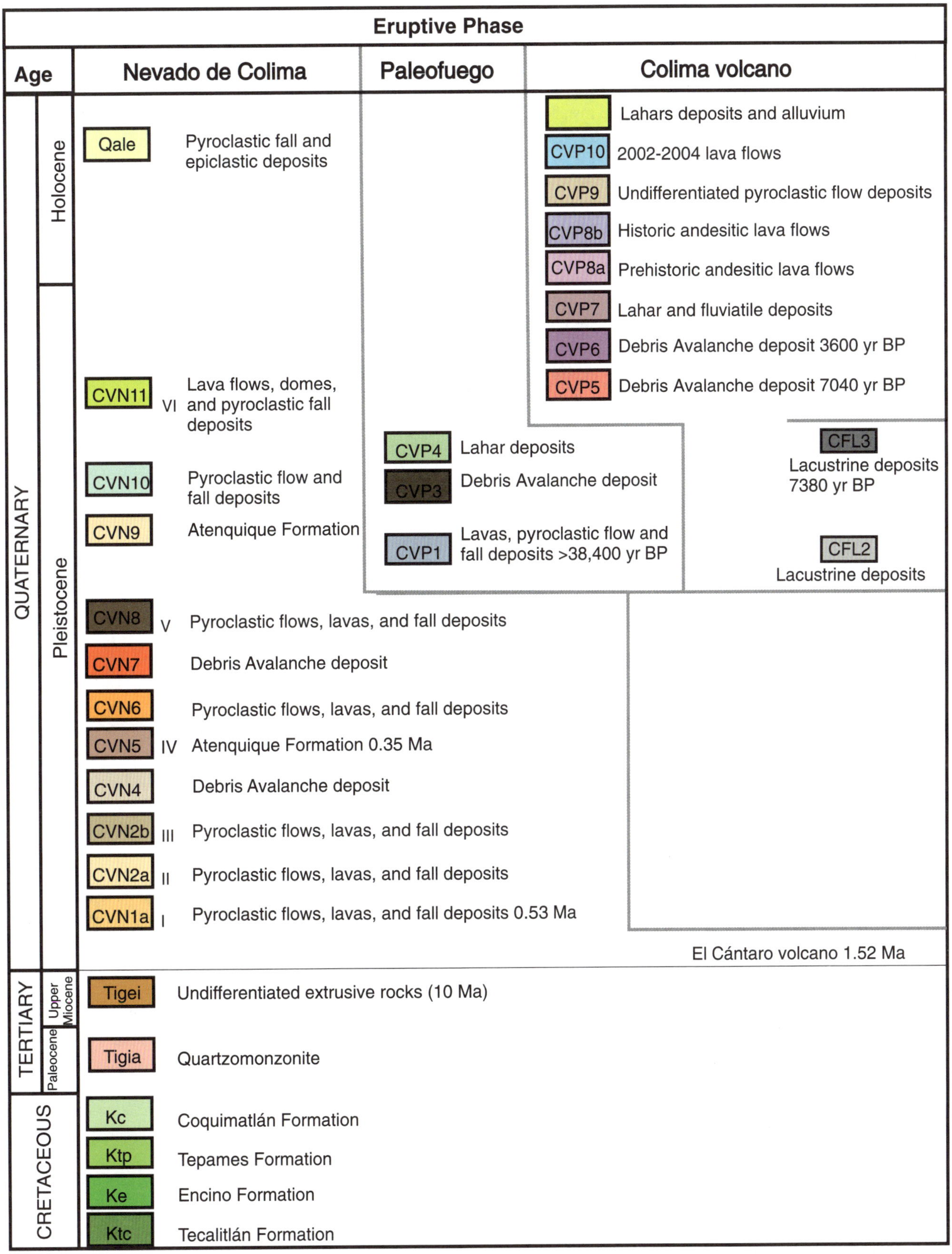

Eruptive Phase

Age	Nevado de Colima	Paleofuego	Colima volcano

QUATERNARY — Holocene

Nevado de Colima:
- Qale — Pyroclastic fall and epiclastic deposits

Colima volcano:
- Lahars deposits and alluvium
- CVP10 — 2002-2004 lava flows
- CVP9 — Undifferentiated pyroclastic flow deposits
- CVP8b — Historic andesitic lava flows
- CVP8a — Prehistoric andesitic lava flows
- CVP7 — Lahar and fluviatile deposits
- CVP6 — Debris Avalanche deposit 3600 yr BP
- CVP5 — Debris Avalanche deposit 7040 yr BP

QUATERNARY — Pleistocene

Nevado de Colima:
- CVN11 — VI — Lava flows, domes, and pyroclastic fall deposits
- CVN10 — Pyroclastic flow and fall deposits
- CVN9 — Atenquique Formation
- CVN8 — V — Pyroclastic flows, lavas, and fall deposits
- CVN7 — Debris Avalanche deposit
- CVN6 — Pyroclastic flows, lavas, and fall deposits
- CVN5 — IV — Atenquique Formation 0.35 Ma
- CVN4 — Debris Avalanche deposit
- CVN2b — III — Pyroclastic flows, lavas, and fall deposits
- CVN2a — II — Pyroclastic flows, lavas, and fall deposits
- CVN1a — I — Pyroclastic flows, lavas, and fall deposits 0.53 Ma

Paleofuego:
- CVP4 — Lahar deposits
- CVP3 — Debris Avalanche deposit
- CVP1 — Lavas, pyroclastic flow and fall deposits >38,400 yr BP

- CFL3 — Lacustrine deposits 7380 yr BP
- CFL2 — Lacustrine deposits

El Cántaro volcano 1.52 Ma

TERTIARY

- Upper Miocene: Tigei — Undifferentiated extrusive rocks (10 Ma)
- Paleocene: Tigia — Quartzomonzonite

CRETACEOUS

- Kc — Coquimatlán Formation
- Ktp — Tepames Formation
- Ke — Encino Formation
- Ktc — Tecalitlán Formation

Figure 4 (*legend*).

that "El Volcancito" flared up at an elevation of 3200 m, reaching 300 m high for a maximum elevation of 3500 m. Between 1893 and 1903, the priest Arreola in Colima, Colima, and his colleague Castellanos in Zapotlán, Jalisco (Cuidad Guzmán today), started systematic observations of the volcano with the installation of volcanological observatories. The results of these observations were published in the Central Meterological Observatory bulletin in México City (Arreola, 1915). These observations were very important because the volcano was heading toward a major eruption, which took place in 1913. Prior to the eruption, the volcano had a summit crater filled by a dome.

The eruption began on 17 January 1913 (Ortíz-Santos, 1944) with a series of explosions that generated dense ash columns rich in vapor. On 20 January, the eruption continued with the formation of a 21-km-high Plinian column that was dispersed to the northeast by stratospheric winds (Saucedo, 1997). The column deposited 15 cm of pumice at Zapotiltán, Jalisco, and 4 cm at La Barca, Jalisco, according to the *El Impacial* journal published on 21 January. The ash of this column reached the city of Saltillo, located 700 km northeast of the volcano. The column collapsed, producing 15-km-long pyroclastic flows that descended to the southern flank of the volcano (Waitz, 1915). As a result of the eruption, the volcano summit had lost 100 m and a new 400-m-wide jogged crater was formed. The depth of this crater oscillated between 50 and 100 m (Waitz, 1935). In 1958, this crater was almost completely occupied by a blocky lava dome (Mooser, 1961) that, between 1961 and 1962, started to spill over the northern flank of the crater, forming a lava flow that reached the location known as El Playón. The 1975–1976 eruptions generated lava flows and, for the first time since the 1913 eruption, some pyroclastic flows (Thorpe et al., 1977). Since this event, Colima has emitted lava domes, lava flows, and different types of pyroclastic flows. Pyroclastic flows have been produced by the gravitational collapse of the external parts of the summit dome such as those that occurred in 1982 (Luhr and Carmichael, 1990b) and 1991 (Rodríguez-Elizarrarás et al., 1991; Saucedo et al., 2004a), by the collapse of the frontal parts of advancing lava flows as in 1998 (Saucedo et al., 2002), by the collapse of low-altitude ash columns in 1991 (Sauceo et al., 2002), and by strong explosions produced at the crater in 2005 (Macías et al., 2006).

NEVADO DE TOLUCA VOLCANO

Nevado de Toluca volcano (99°45′W, 19°09′N) is located 23 km to the SW of the city of Toluca (Fig. 5). Its highest peak has an elevation of 4680 masl (the fourth highest mountain in México). Nevado de Toluca is also known as "Xinantécatl" (naked man in náhuatl language), although García-Martínez (2000) concluded that the proper náhuatl name is "Chicnauhtécatl" meaning "nine hills." The volcanic edifice is built upon a volcano-sedimentary metamorphic basement of Late Jurassic–Early Cretaceous age (Bonet, 1971; Bloomfield et al., 1977; Campa et al., 1974; Demant, 1981), Eocene rhyolitic ignimbrites, and Miocene andesitic lava flows (García-Palomo et al., 2002) (Fig. 6).

Nevado de Toluca is a composite volcano of late Pleistocene-Holocene age, made up of calc-alkaline andesites and dacites (Bloomfield and Valastro, 1974; Cantagrel et al., 1981; García-Palomo et al., 2002). The Nevado de Toluca crater has the remains of two ancient scars, located on the SE and NE flanks of the volcano, that are related to the partial collapse of the edifice. The northern flank of Nevado de Toluca has a relative elevation of 2015 m with respect to the Lerma river basin, and its southern flank has a relative elevation of 2900 m with respect to the Ixtapan de la Sal village (Fig. 5). The crater of Nevado de Toluca is truncated (Fig. 7A) and has a 2 × 1.5-km-wide elliptical crater with its main axis aligned to an E-W direction and opened to the east. The craters interior holds a dacitic central dome called the navel "El Ombligo" that separates the Sun and Moon lakes whose surfaces stand at an elevation of 4200 m (Fig. 7B). The water of these two lakes contains diatoms (Caballero, 1996), and has alkaline composition (Armienta et al., 2000). Prehispanic obsidian blades and pottery shards are found dispersed on the surface of the Ombligo dome and at the bottom of the lakes, since these sites were used first by the Matlazincas, followed by the Aztec people to perform religious ceremonies (Quezada-Ramírez, 1972). The surface of Nevado de Toluca was carved by glacier activity during late Pleistocene-Holocene time, as evidenced by moraine deposits and rock glaciers (Heine, 1976, 1988; Vázquez-Selem and Heine, 2004).

Previous Studies

The first studies of Nevado de Toluca described morphological and petrographic aspects of the volcano (Ordoñez, 1902; Otis, 1902; Hovey, 1907; Flores, 1906; Waitz, 1909). After that, the volcano remained unstudied until the 1970s, when the first geological and volcanological studies were accomplished (Bloomfield and Valastro, 1974; Bloomfield and Valastro, 1977; Bloomfield et al., 1977). These authors described two Plinian eruptions that produced the Lower Toluca Pumice, dated at ca. 24,000 yr B.P., and the Upper Toluca Pumice, dated at 11,600 yr B.P. In addition, during these studies, the authors established part of the modern eruptive history of Nevado de Toluca, as well as the calc-alkaline nature of its products (Whitford and Bloomfield, 1977). In 1982, Cantagrel et al. (1981) divided the evolution of the volcano in two main stages. The oldest stage was composed of andesitic lava flows that form the ancient edifice dated with the K-Ar method at 1.60 ± 0.12 and 1.23 ± 0.15 Ma. The youngest stage was composed of a complex sequence of volcaniclastic deposits that in the southern flank of the volcano had a minimum thickness of 100 m and an age of 100,000 yr. According to these authors, the activity between these two stages was mainly volcaniclastic.

During the 1990s, a new surge of studies at Nevado de Toluca began to understand the structural setting of the volcano (García-Palomo et al., 2000), the geology and eruptive history (Macías et al., 1997a; García-Palomo et al., 2002), the deposits produced during Plinian eruptions (Arce et al., 2003; Arce et al., 2005b; Capra et al., 2006), the collapse events (Capra

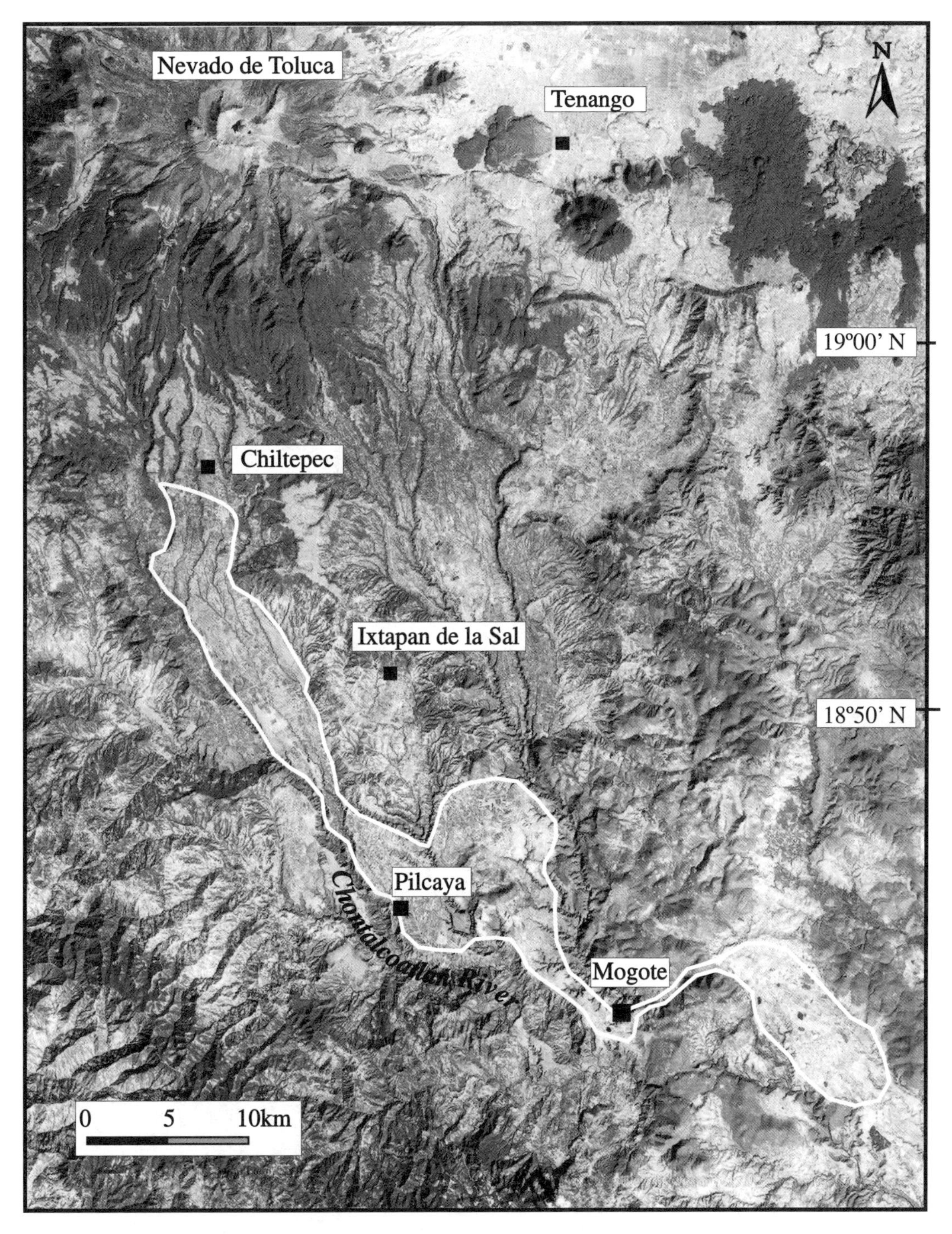

Figure 5. LANDSAT image showing Nevado de Toluca and the maximum extent of Pilcaya debris flow and the Mogote lahar deposits (Capra and Macías, 2002).

and Macías 2000), geomorphological aspects (Norini et al., 2004), the paleosols preserved between the volcanic deposits (Sedov et al., 2001, 2003; Solleiro-Rebolledo et al., 2004), and paleoenvironmental deposits in the Upper Lerma Basin (Metcalfe et al., 1991; Newton and Metcalfe, 1999; Caballero et al., 2001, 2002; Lozano-García et al., 2005).

Evolution of Nevado de Toluca

Nevado de Toluca was built upon the intersection of three fault systems with NW-SE, NE-SW, and E-W orientations (García-Palomo et al., 2000). This structural geometry favored the formation of coalescent pyroclastic fans, with smooth mor-

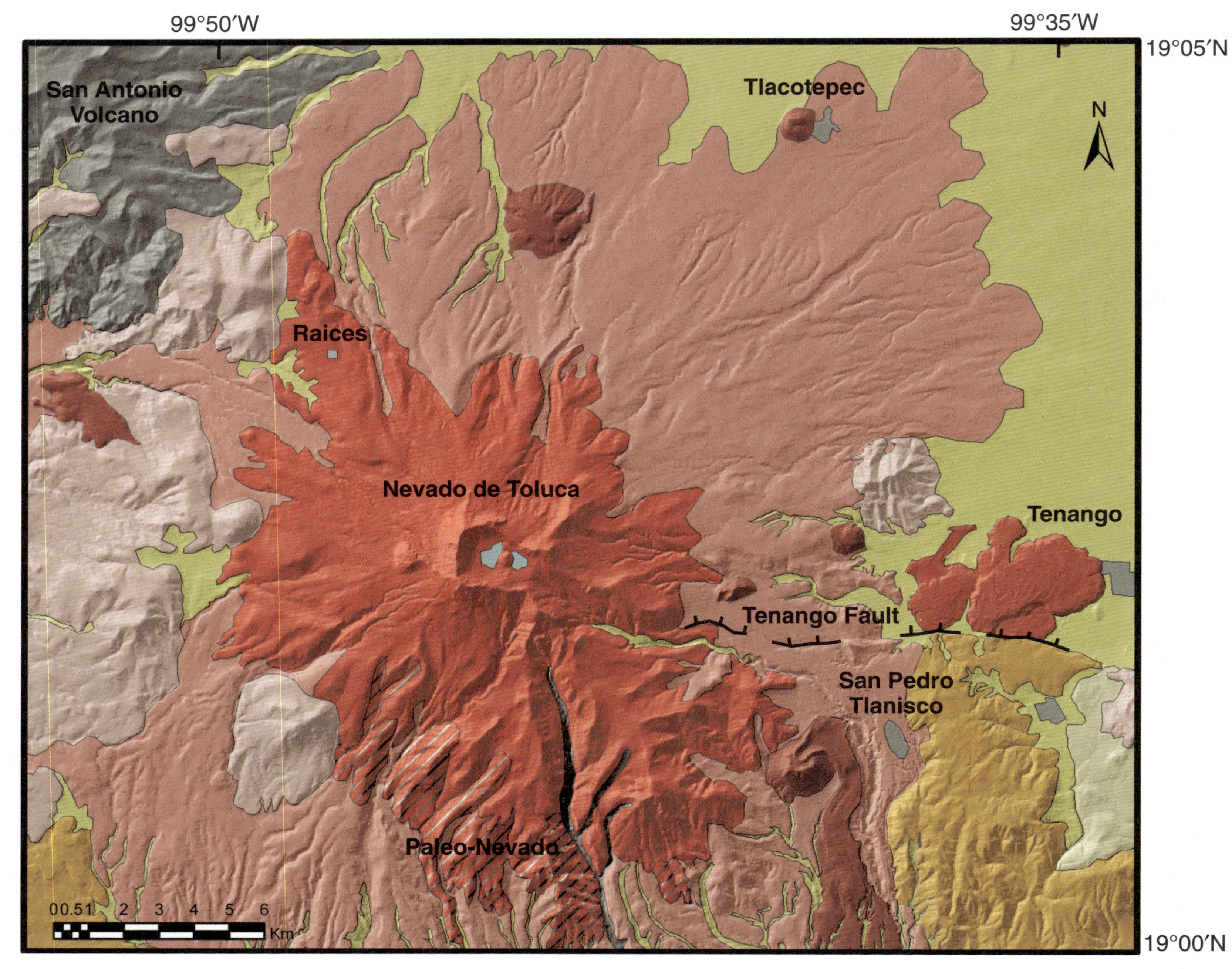

Figure 6 (*legend on following page*). Simplified geologic map of Nevado de Toluca volcano from García-Palomo et al. (2002) showing the basal sequence, San Antonio volcano, the units of Nevado de Toluca volcano, and the western portion of the Chichinautzin Volcanic Field.

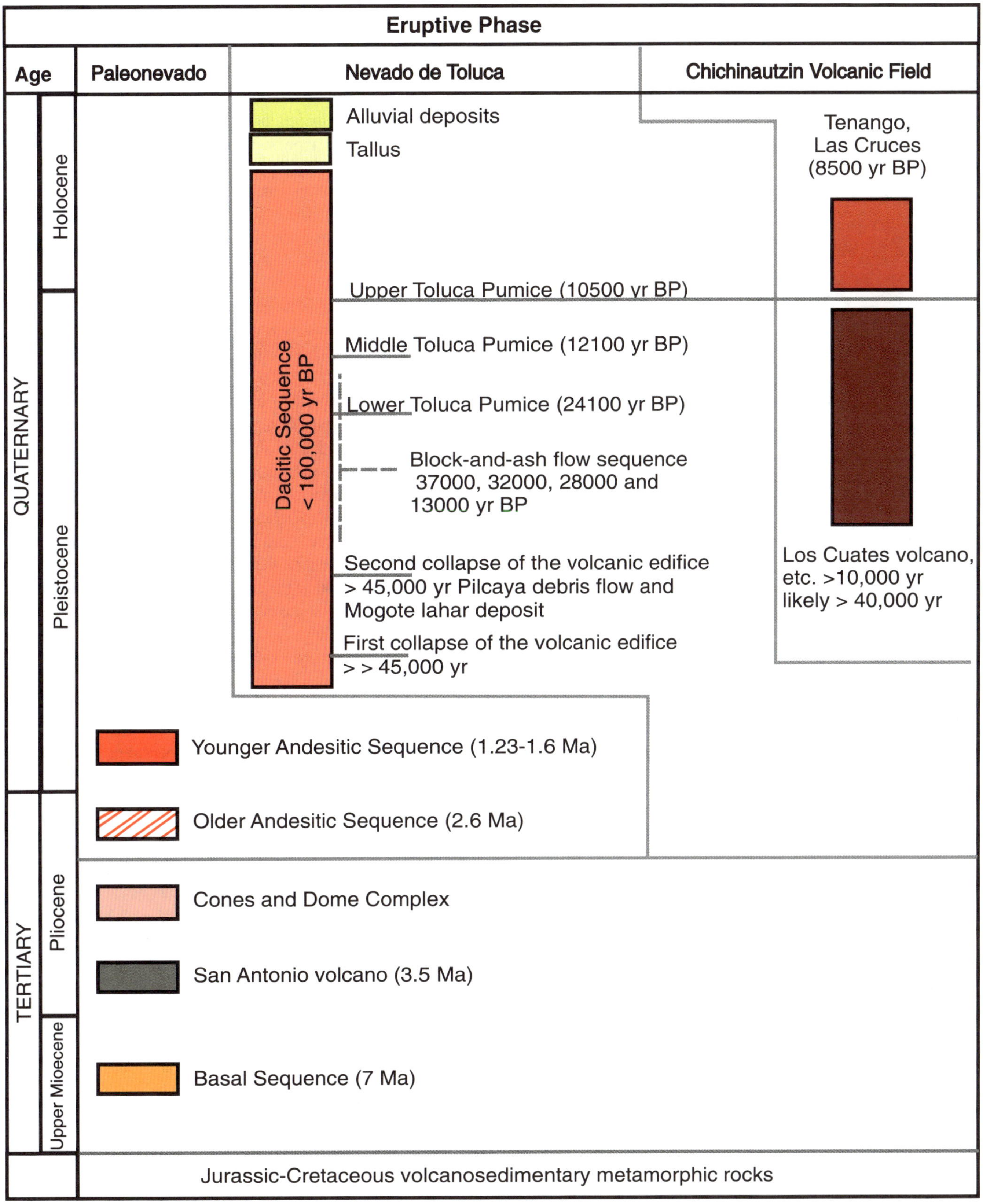

Eruptive Phase

Age | **Paleonevado** | **Nevado de Toluca** | **Chichinautzin Volcanic Field**

QUATERNARY

Holocene

Pleistocene

Alluvial deposits

Tallus

Dacitic Sequence
< 100,000 yr BP

Upper Toluca Pumice (10500 yr BP)

Middle Toluca Pumice (12100 yr BP)

Lower Toluca Pumice (24100 yr BP)

Block-and-ash flow sequence
37000, 32000, 28000 and
13000 yr BP

Second collapse of the volcanic edifice
> 45,000 yr Pilcaya debris flow and
Mogote lahar deposit

First collapse of the volcanic edifice
> > 45,000 yr

Tenango,
Las Cruces
(8500 yr BP)

Los Cuates volcano,
etc. >10,000 yr
likely > 40,000 yr

Younger Andesitic Sequence (1.23-1.6 Ma)

TERTIARY

Pliocene

Upper Mioecene

Older Andesitic Sequence (2.6 Ma)

Cones and Dome Complex

San Antonio volcano (3.5 Ma)

Basal Sequence (7 Ma)

Jurassic-Cretaceous volcanosedimentary metamorphic rocks

Figure 6 (*legend*).

Figure 7. (A) View of the western flank of Nevado de Toluca volcano showing the truncated crater with its maximum elevation represented by El Pico del Fraile. Photograph taken along the Toluca–Valle de Bravo highway by Jorge Neyra on 19 May 2001. (B) View to the southeast of the interior of Nevado de Toluca crater covered by snow in January 1992. Notice the dacitic El Ombligo dome and the frozen Lake of the Sun with its surface located at an elevation of 4200 m. In the background appears the volcano's crater open to the east.

phology on the northern flanks and the infilling of graben type structures in the southern flank (García-Palomo, 1998; García-Palomo et al., 2002). The volcanic activity in the region began 1.6–1.3 Ma, with the formation of the ancient edifice called Paleonevado, which was located to the S-SE of the present edifice (Cantagrel et al., 1981), although recent studies suggest that the activity began 2.6 Ma (Norini et al., 2004). The Paleonevado edifice was constructed through the emplacement of andesitic lavas 1.2 Ma (García-Palomo et al., 2002; Norini et al., 2004). Between 1.2 Ma and 0.1 Ma, Paleonevado was subjected to intense erosive activity (Cantagrel et al., 1981) that generated debris avalanches, debris flows, and fluviatile deposits (Macías et al., 1997a; Capra and Macías, 2000) (Fig. 8). The formation of the present edifice of Nevado de Toluca started ca. 0.1 Ma, with the emission of dacitic products that have given place to explosive eruptions (Macías et al., 1997a; García-Palomo et al., 2002). During the late Pleistocene, the southern flank of Nevado de Toluca collapse twice, originating debris avalanche deposits that were transformed into debris flows with distance (Capra and Macías, 2000; Scott et al., 2001). The scars produced by these collapses have disappeared due to subsequent volcanic and glacier activity.

The oldest collapse produced a light brown, partially lithified debris avalanche deposit (described as DAD1 by Macías et al., 1997a). It has blocks with jigsaw-fit structures set in a coarse sand matrix with an approximate thickness of 10 m. This deposit has been found up to a distance of 35 km from the volcano summit (Fig. 5). The collapse was followed by the emplacement of lahars and fluviatile and lacustrine sediments, deposited in small, closed depressions over several thousands of years, although no specific dates are available. A second collapse of Nevado de Toluca also occurred during the late Pleistocene along with intense hydrothermal alteration of the volcanic edifice (Capra and Macías, 2000). This event generated a debris avalanche deposit, which was immediately transformed into a debris flow called "Pilcaya Debris Flow" that traveled 55 km from the source (Capra, 2000; Capra and Macías, 2002). The deposit is light-brown massive heterolithologic, with blocks immersed in a coarse sand matrix. This deposit rests upon a light brown paleosol and overlies a lahar deposit called "El Mogote," which was originated by the Pilcaya debris flow (Capra and Macías, 2002). According to their stratigraphic position, an age of >45,000 yr has been assigned to these deposits.

The activity of Nevado de Toluca during the late Pleistocene continued with two large explosive eruptions that produced two pink and white pumiceous pyroclastic flows, respectively, in all directions around the volcano. One of these deposits contains charcoal material dated at 42,000 yr. These deposits are generally exposed at the base of several gullies around the volcano. They are covered by deposits produced by the partial or total destruction of dacitic central domes and the generation of Plinian columns, as described in the following section.

Destruction of Dacitic Domes

Block-and-ash flows, produced by the partial or total destruction of domes, are recognized as "old lahar assemblages" by Bloomfield and Valastro (1974, 1977). These authors estimated that the age of this deposit is 28,000 yr B.P., as they dated a paleosol on top. However, Macías et al. (1997a) identified two block-and-ash flow deposits, the oldest one containing disseminated charcoal that yielded an age of 37,000 ± 1125 yr and therefore did not correlate with the Bloomfield and Valastro (1977) deposit. Macías et al. (1997a) correlated this deposit with the "gray lahar" deposit of Heine (1988), dated at 35,600 +2600/–800 yr, and a deposit overlying a paleosol, dated at 38,000 yr by Cantagrel et al. (1981). The younger block-and-ash flow deposit was dated at 28,140 +865/–780 and 28,925 +625/–580 yr B.P., with charcoal found inside the deposits that clearly correlated with the age of 27,580 ± 650 yr B.P. obtained by Bloomfield and Valastro (1974, 1977). Subsequent studies have described at least five gray massive block-and-ash flow deposits dated at ca. 37, 32, 28, 26, and

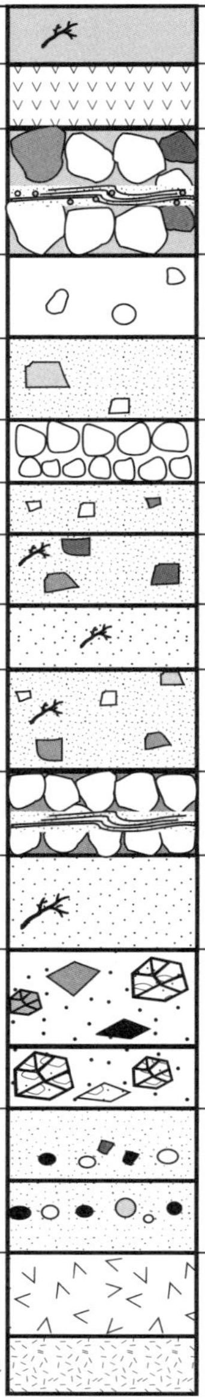

Thickness (meters)	Unit	Age (yr. B.P.)	Deposit	Description
1.5		~3.3 ka		Gray cross-stratified surge and brown ash flow deposits with disseminated charcoal.
100		8.5 ka		Andesite lava flows (Tenango) of the Chichinautzin Formation.
20	**UTP**	10.5 ka		Upper Toluca Pumice. Fall deposit composed of three members, interbedded with thick pyroclastic flow and surge beds.
20	**MTP**	~12.1 ka		Sequence of pumice fall, surge and two white pumice flow deposits rich in subrounded dacitic pumice and biotite crystals.
5	**BAF**	~14 ka		Gray massive block-and-ash flow deposits, ash flows and surge layers with accretionary lapilli. The deposits contain juvenile dacitic clasts and pumice.
3	**LTP**	21.7 ka		Lower Toluca Pumice. Inversely graded fallout bed rich in yellow pumice and a few schist clasts from the local basement capped by surge beds.
10	**BAF**	~26.5 ka		Gray massive block-and-ash flow deposit. There are scarce outcrops of this deposit.
10	**BAF**	~28 ka		Gray massive block-and-ash flow deposit, composed of at least three units and interbedded surges. It contains juvenile dacitic clasts, pumice and red altered dacite clasts.
5	**BAF**	~32 ka		Pale brown ash flow deposit, composed of several flow units, interbedded with surge deposits.
10	**BAF**	~37 ka		Gray block-and-ash flow sequence composed of three main massive units and minor flow and surge layers. Consists of juvenile gray dacite clasts, red altered dacites, and rare pumice
3.5	Ochre Pumice Fall	~36-39 ka		Ochre pumice fall deposit, composed of three layers interbedded with surge deposits and caped by a massive pyroclastic flow rich in pinkish pumice fragments and charcoal.
~4		~42 ka		Pink pumice flow deposit, composed of several flow units. Clasts include subrounded dacitic pumice and few andesitic fragments, set in a sandy matrix.
40	**PDF**			Heterolithologic cohesive debris flow deposit (Pilcaya deposit), composed of dacitic clasts and exotic components (basalt, limestone, rhyolite, sandstone) embedded in an indurated sandy matrix
15	**DAD1**			Monolithologic debris avalanche deposit, composed of dacitic clasts set in a sandy matrix.
200	Older Sequence			Interbedded sequence of debris flows, hyperconcentrated flows, fluviatile beds, and minor lacustrine deposits "Older Lahars" from Nevado.
150		1.2-1.6 Ma		Primitive andesitic-dacitic lava flows of Nevado de Toluca.
100		2.6 ± 0.2 Ma		Light-gray porphyritic lava flows

Figure 8. Stratigraphic column of Nevado de Toluca after Macías et al. (1997a) and García-Palomo et al. (2002). The PDF (Pilcaya debris flow) and DAD1 (Debris Avalanche Deposit 1) are two debris avalanche-debris flow deposits produced by ancient collapses of the volcano. The young dacitic sequence of Nevado de Toluca rests on top of these deposits. BAF—block-and-ash flow; LTP—Lower Toluca Pumice; MTP—Middle Toluca Pumice; UTP—Upper Toluca Pumice.

14 ka (García-Palomo et al., 2002). The youngest (14 ka) deposit also appears in the stratigraphic record of the Lake Chiconahuapan (Newton and Metcalfe, 1999; Caballero et al., 2001, 2002). These five deposits have been widely distributed around Nevado de Toluca, and some of them (37, 28, and 14 ka B.P.) reach minimum distances of 25 km from the summit, still with considerable thicknesses (Fig. 9). The deposits contain juvenile dacitic lithics with silica concentrations varying from 65 to 67 wt%.

Plinian Eruptions

The first volcanological studies of Nevado de Toluca were dedicated to the deposits of two Plinian eruptions that were exposed on the flanks of the volcano and as far as the Lerma Basin. These deposits were dubbed by Bloomfield and Valastro (1974, 1977) as the Lower Toluca Pumice of ca. 24,500 yr B.P. and the Upper Toluca Pumice of ca. 11,600 yr B.P. These eruptions were dated with charcoal material below the deposits or the underlying paleosols, but never from material carbonized during the eruption. Recent studies suggest that these eruptions had younger ages, ca. 21,700 yr B.P. for the Lower Toluca Pumice (Capra et al., 2006) and ca. 10,500 yr B.P. for the Upper Toluca Pumice (Arce, 2003; Arce et al., 2003). However, the stratigraphy of Nevado de Toluca had some other deposits associated with Plinian- or subplinian-type activity. Macías et al. (1997a) discovered another deposit that yielded an age of ca. 12,100 yr B.P., dubbed the "White Pumice Flow," that was later described as the Middle Toluca Pumice by Arce et al. (2005b). In addition, García-Palomo et al. (2002) recognized another fallout dubbed as the Ochre Pumice dated between ca. 36,000 and 37,000 yr B.P. We next summarize the characteristics of the three youngest fall deposits of Nevado de Toluca:

The Lower Toluca Pumice (ca. 21,700 yr) was originated from a Plinian column that reached an elevation of 24 km above the crater and was later dispersed to the NE by stratospheric winds. The eruption was then followed by subplinian pulses and hydromagmatic eruptions that produced a total volume of 2.3 km^3 (0.8 km^3 dense rock equivalent [DRE]). This eruption has a peculiar place within the eruptive history of Nevado de Toluca, because it incorporated schist fragments from the basement, and the magma has a chemical composition that varies from andesitic 55wt% SiO_2 to dacite 65wt% SiO_2. The pumice of the Lower Toluca Pumice represents the more basic magma emitted by Nevado de Toluca during the past 50,000 yr (Capra et al., 2006).

Figure 9. El Refugio quarry located 15 km northeast of Nevado de Toluca crater showing an exposure of the 37,000 yr B.P. block-and-ash flow deposits.

The Middle Toluca Pumice (ca. 12,100 yr) began with a 20-km-high Plinian column, dispersed to the NE by stratospheric winds. The column was interrupted by hydromagmatic explosions and then an 18–19-km-high subplinian column established for some time prior to wane. Afterward, another subplinian column formed, and was interrupted by hydromagmatic explosions that produced pyroclastic surge, causing the column to collapse, triggering two pyroclastic flows rich in white pumice that were designated the "White Pumice Flow" by Macías et al. (1997a). This eruption generated 1.8 km^3 (DRE) of dacitic magma, with a homogeneous composition of 63.54–65.06 wt% in SiO_2. The mineral assemblage of this deposit is formed by phenocrysts of plagioclase > orthopyroxene > hornblende ± ilmenite and titanomagnetite and xenocrysts of biotite, immersed in a rhyolitic groundmass (70–71 wt% in SiO_2). The xenocrysts of biotite were in reaction with the groundmass; some of these were dated with the $^{40}Ar/^{39}Ar$ method and yielded ages of 0.8 Ma, indicating that they were assimilated from the magma chamber or other rocks (Arce et al., 2005a).

The Upper Toluca Pumice (ca. 10,500 yr) was caused by a complex event that formed four Plinian eruptions, PC0, PC1, PC2, and PC3, reaching heights of 25, 39, 42, and 28 km above the crater, respectively, that were dispersed toward the NE by dominant winds (Fig. 10). The last three columns were interrupted by hydromagmatic explosions at the crater that dispersed pyroclastic surges and produced the collapse of the columns, forming pumiceous pyroclastic flows. Fallouts PC1 and PC2 covered a minimum area of 2000 km^2, an area that is presently occupied by the cities of Toluca and México, and generated a volume of 14 km^3 (~6 km^3 DRE). In the basin of México, the Upper Toluca Pumice was initially described as the pumice triple layer (Mooser, 1967). The composition of the magma erupted by the Upper Toluca Pumice has a homogeneous composition of 63–66 wt% in silica (Arce et al., 2003).

Depth of the Magma Chamber

The homogeneous chemical composition and mineral assemblage of the magmas produced during the eruptions of 14,000 yr B.P. (block-and-ash flow), 12,100 yr B.P. (Middle Toluca Pumice), and 10,500 yr B.P. (Upper Toluca Pumice), suggest that the magma ejected by these eruptions might come from a single large reservoir filled with a dacitic magma (Arce et al., 2005b). With this assumption, these authors analyzed the mineral chemistry of the products of these three eruptions, which

Figure 10. Outcrop located 15 km northeast of Nevado de Toluca crater that reveals the 14,000 yr B.P. block-and-ash flow deposits. BAF14—block-and-ash Flow 14; UTP—Upper Toluca Pumice.

resulted to be very similar and consist of plagioclase (An_{30-59}) > orthopyroxene (En_{56-59}) > hornblende (edenita-hornblende mainly) >> Fe-Ti oxides + rare apatite (in orthopyroxene) + biotite, embedded in a rhyolitic groundmass (72–76 wt% in silica). For the ilmenite-titanomagnetite pair found in the samples, these authors estimated a pre-eruptive temperature of 850 °C and oxygen fugacity of −11. Hydrothermal experiments run on Upper Toluca Pumice samples found stability conditions for these magmas at pressures of 160–210 MPa and depths of 4.5–6 km, beneath the Nevado de Toluca crater (Fig. 11).

The Holocene stratigraphic record of Nevado de Toluca consists of the extrusion of the Ombligo dacitic dome inside the crater. The dome surface has signs of glacier polish dated at 9100 ± 500 yr B.P., determined with the ^{36}Cl method (Arce et al., 2003). Therefore, it is clear that this dome was emplaced at the end of the Upper Toluca Pumice event or sometime later. The last eruption at Nevado de Toluca took place somewhere in the central crater ca. 3300 ^{14}C yr ago (Macías et al., 1997a). This eruption emitted a pyroclastic flow and surge on the NE flank of the volcano. Finally, a series of yellow lahar deposits are widely distributed around the volcano flanks. Nevado de Toluca should be considered as an active dormant volcano.

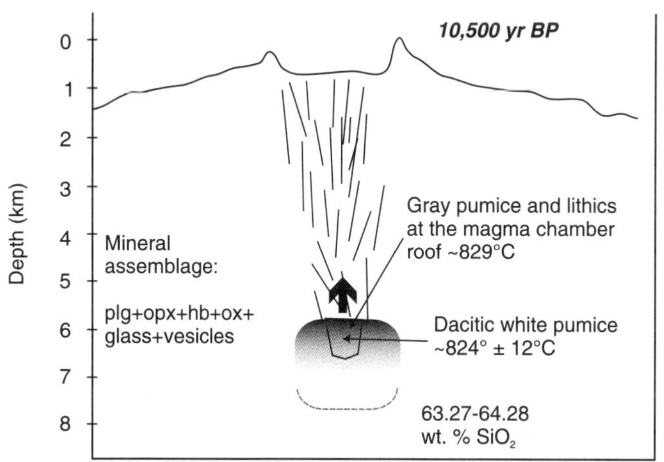

Figure 11. Cartoon showing the position of the magma chamber below Nevado de Toluca crater ca. 10,500 yr B.P. prior to the Plinian eruption that deposited the Upper Toluca Pumice (UTP). The UTP dacitic magma stagnated 4.5–6 km below the crater at a temperature of 860 °C.

POPOCATÉPETL VOLCANO

Popocatépetl volcano is located 65 km SE of México City and 45 km to the west of the city of Puebla (Figs. 1 and 12). More than one million people live in a radius of 40 km from the summit. The volcano forms the southern end of the Sierra Nevada range composed of the Tláloc, Telapón, Teyotl, Iztaccíhuatl, and Popocatépetl volcanoes. Popocatépetl means in náhuatl language "the smoking mountain," referring to the fact that during prehispanic times, the Aztecs observed several eruptions such as those that occurred in 1363, 1509, 1512, and 1519–1528; the last one was also described by the priest Bernal Díaz y Gomarra (De la Cruz-Reyna et al., 1995). During the Colonial period, several descriptions were written about Popocatépetl eruptions, such as those that occurred in 1530, 1539, 1540, 1548, 1562–1570, 1571, 1592, 1642, 1663, 1664, 1665, 1697, and 1720. Throughout the nineteenth century, several scientists visited the volcano, among them von Humboldt in 1804 (Humboldt, 1862), the geologists Del Río in 1842, Del Castillo in 1870, and Sánchez in 1856, who carried out general descriptions of the volcano's morphology. Aguilera and Ordóñez (1895) identified Popocatépetl as a truncated cone made of an alternating sequence of pyroclastic and lava deposits, with hypersthene-hornblende andesite composition. These authors pointed out the existence of seven fumaroles inside Popocatépetl's crater with temperatures <100 °C and a blue-greenish lake formed by thaw water with variable temperatures of between 28 and 52 °C. Weitzberg (1922) made a detailed study of the glacier. In 1906, the bottom of the crater had a funnel-shaped with a small lake and vertical walls. Popocatépetl volcano reawakened in February 1919. In March-April, several inhabitants of Amecameca described that in the bottom of the crater there was an accumulation of steaming rocks, like a turned over stewing pan (Atl, 1939). On 11 October 1920, Waitz

Figure 12. Panoramic view of the southern flank of Popocatépetl volcano with snow covering up to the tree line at ~4000 m. The photo shows the surface of a debris avalanche deposit with hummocky topography.

visited the crater and described a lava dome at the bottom (Waitz, 1921). On 15 November 1921, Camacho and Friedlaender photographed the crater's interior, showing a lava dome surrounded by abundant gas emissions; they also captured the occurrence of brief explosions from the center of the dome (Friedlaender, 1921; Camacho, 1925). In January 1922, Camacho observed that a small crater was occupying the center of the dome. Atl (1939) described the eruption and evolution of this crater; according to his observations, the eruption ended in 1927. Popocatépetl remained calm for 67 yr; it reawakened on 21 December 1994.

Previous Studies

The few geological studies of Popocatépetl volcano prior to its reactivation in 1994 were those of Heine and Heide-Weise (1973), Miehlich (1984), Robin (1984), Carrasco-Núñez (1985), and Boudal and Robin (1989), as well as some petrologic studies (Boudal, 1985; Boudal and Robin 1987; Kolisnik, 1990). These studies defined Popocatépetl as a stratovolcano and presented part of the stratigraphic record. The geologic evolution of Popocatépetl can be summarized by the following stages: (1) the activity started with the formation of Nexpayantla volcano (Mooser et al., 1958), the ancestral volcano (Robin, 1984), through the emission of andesitic to dacitic lava flows. An eruption that occurred ~200,000 yr ago promoted the volcano collapse and the formation of a caldera. Inside this caldera began the construction of El Fraile volcano through the emplacement of andesitic and dacitic lava flows. This last volcano collapsed <50,000 yr B.P. (Boudal and Robin, 1989), due to a Bezymiany-type eruption that destroyed the southern flank of the cone. These authors estimated a volume of 28 km^3 for this deposit. The eruption generated a debris avalanche that moved to the S-SW from the crater and was followed by the formation of a Plinian eruption that deposited a white pumice-fall layer toward the south of the volcano as well as pyroclastic flows. After this event, the formation of the modern cone known as Popocatépetl began. A large number of studies regarding geological (Siebe et al., 1995b, 1995c, 1996a, 1996b, 1997; Martín-Del Pozzo et al., 1997; Panfil et al., 1999; Espinasa-Pereña and Martín-del Pozzo, 2006) and geomorphological aspects (Palacios, 1996; Palacios et al., 2001) carried out during the last decade were driven by the reactivation of Popocatépetl. The stratigraphic studies can be summarized in the following sections.

Destruction of the Ancient Cone of Popocatépetl

Some 23,000 yr ago, a lateral eruption larger than the 18 May 1980 Mount St. Helens eruption produced the lateral collapse (to the south) of the ancient Popocatépetl cone (Fig. 13). The explosion generated a debris avalanche deposit that reached up to 70 km from the summit. The decompression of the magmatic system, due to the flank collapse, originated a lateral blast that emplaced a pyroclastic surge deposit and allowed the establishment of a Plinian column (Fig. 14). The column deposited a thick pumice-fall layer that is widely distributed on the southern flanks of the volcano. The column then collapsed and formed an ash flow that charred everything in its path. The deposit reached up to 70 km from the summit, covers an area of 900 km^2, and if we assign an average thickness of 15 m, a volume of 9 km^3 is obtained. This deposit overlies a paleosol that has charred logs dated at 23,445 ± 210 yr; disseminated charcoal found in the ash flow deposit yielded an age of 22,875 +915/−820 yr. Therefore the age of this eruption is ca. 23,000 yr B.P.

There are at least four debris avalanche deposits around Popocatépetl volcano. The oldest comes from the failure of the SE flank of Iztaccíhuatl volcano, and the other three come from the flank collapse of paleo-Popocatépetl (Siebe et al., 1995b; García-Tenorio, 2002), the youngest being the 23,000 yr B.P. deposit.

Construction of the Present Cone

Popocatépetl's present cone started growing 23,000 yr B.P.; it has a maximum elevation of 5472 masl and a relative elevation regarding the surrounding ground of 3000 m. The volcano has been built through the emission of alternating sequences of pyroclastic deposits and lava flows of andesitic-dacitic composition. The rocks consist of phenocrysts of plagioclase, hypersthene, augite, olivine, and rare hornblende in a glassy microcrystalline matrix. During the past 20,000 yr the explosive activity of Popocatépetl has been characterized by four major events (14,000, 5000, 2150, and 1100 yr B.P.) and four minor events (11,000, 9000, 7000 and 1800 yr B.P.) (Siebe et al., 1997; Siebe and Macías, 2006). The geologic history of Popocatépetl during the past 20,000 yr can be summarized in the following sections.

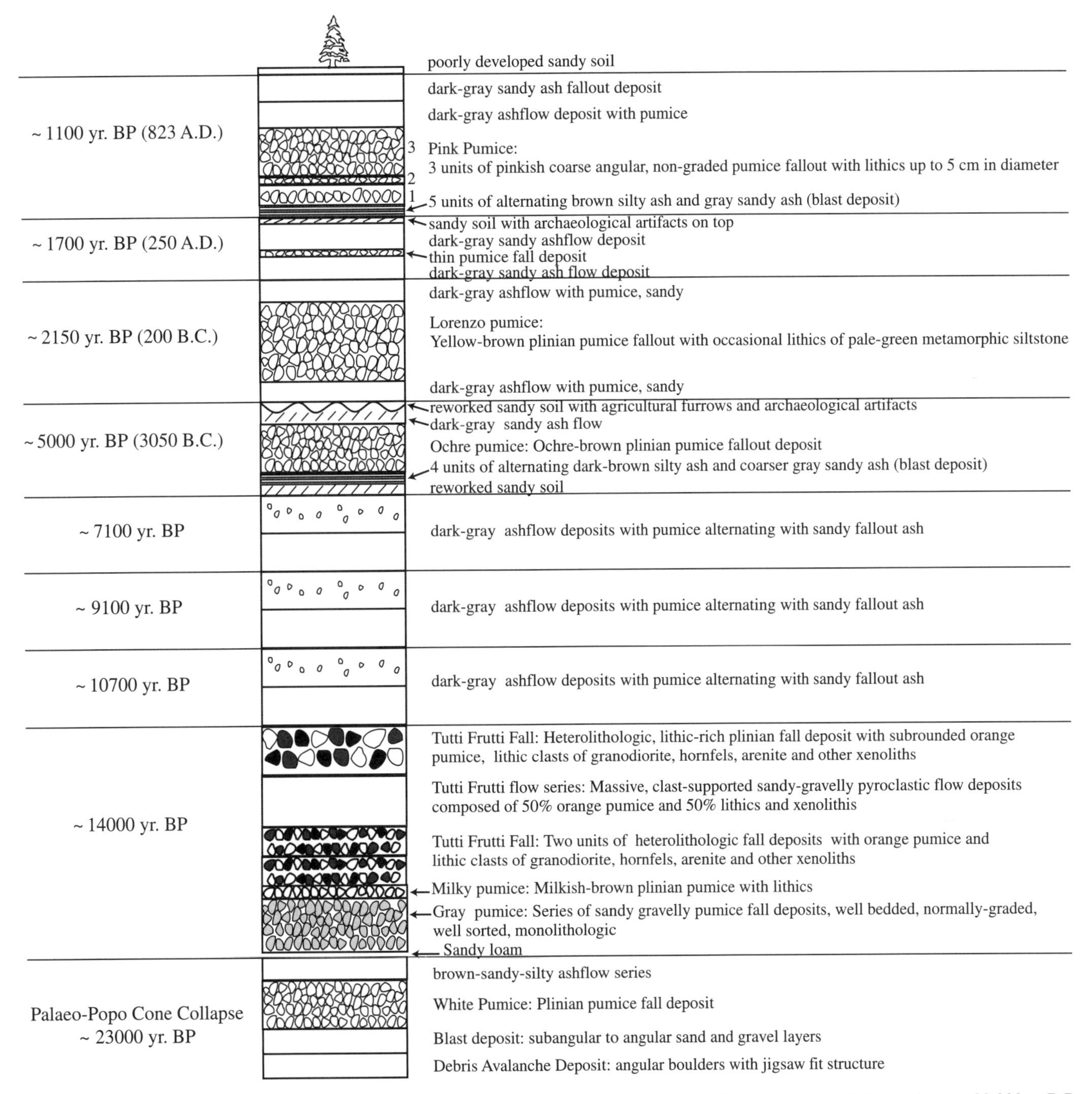

Figure 13. Simplified stratigraphic column of Popocatépetl volcano summarizing the pyroclastic deposits erupted during the past 23,000 yr B.P. (Siebe et al., 1995c; Siebe and Macías, 2006).

Phreatoplinian Eruption ca. 14,000 yr B.P. (Pómez con Andesita or Tutti Frutti)

A large magnitude eruption began with the emission of a gray ash fallout around the volcano followed by a series of proximal pyroclastic flows and surges that culminated with the formation of a Plinian column. This was dispersed by the stratospheric wind to the N-NW, toward the present area occupied by México City (Siebe et al., 1995b, 1997). This fallout layer is heterolithologic; it contains ochre dacitic pumice, gray granodiorite, metamorphose limestones, skarns, and other fragments from the basement.

This fallout layer was described in the Basin of México as the "Pómez con andesita" or pumice with andesite by Mooser (1967), with a thickness of 5 cm. This deposit is widely exposed around the volcano, but it is barren of charcoal material except at

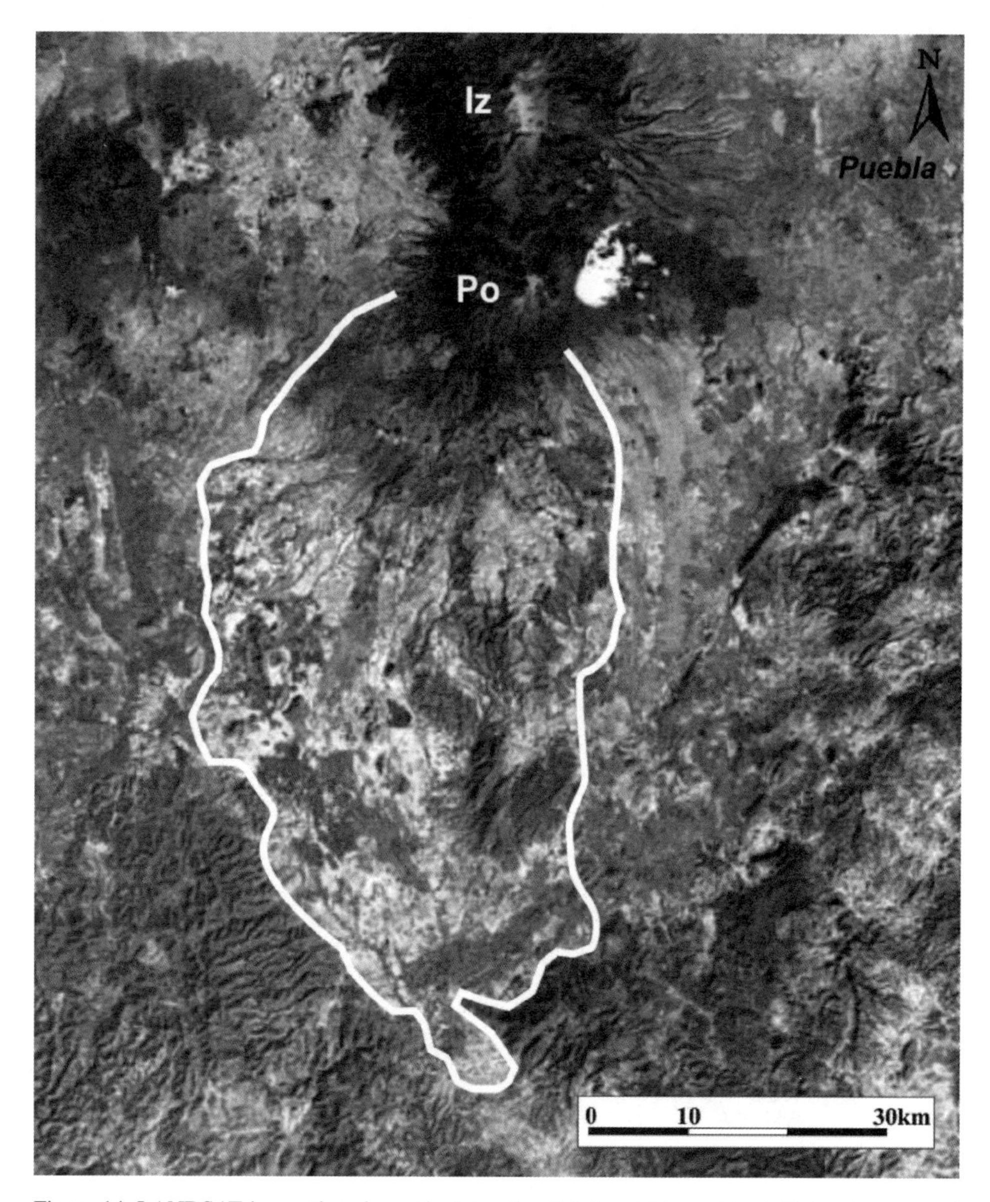

Figure 14. LANDSAT image that shows the Iztaccíhuatl (Iz) and Popocatépetl (Po) volcanoes and the distribution of the Popocatépetl debris avalanches according to Siebe et al. (1995b) (from Capra et al., 2002).

one site, where it yielded an age of ca. 14,000 yr B.P. (Siebe et al., 1997). This implies that at the time of the eruption Popocatépetl cone was covered by a glacier and therefore there was little vegetation on its flanks.

Recent Plinian Eruptions

At least three main Plinian fall deposits have been identified in the stratigraphic record of Popocatépetl volcano during the past 5000 yr. These deposits are dated at 5000, 2150, and 1100 yr B.P. (Fig. 15). The events that occurred at 5000 and 1100 yr B.P. had a similar evolution; they began with hydromagmatic explosions that dispersed wet pyroclastic surges up to 20 km from the summit. These explosions opened the magmatic conduit, decompressed the magmatic system, and formed >25-km-high Plinian columns (Siebe et al., 1996a, 1996b). These plumes were transported by the stratospheric winds to the N (5000 yr ago), to the E (2150 yr ago), and to the N-NE-E (1150 yr ago). Once these columns were fully established, their collapse was likely due to the consumption of the magma in the chamber, which generated pyroclastic flows that destroyed everything in their paths.

These eruptions blocked the hydrographic network of Popocatépetl and Iztaccíhuatl volcano located to the north. Springs and meteoric waters saturated the unconsolidated material produced by the pyroclastic flows to produce lahars that flooded the Basin of Puebla.

Figure 15. Photograph of the pumice fall deposits emplaced during the 5000 (A), 2150 (B), and 1100 (C) yr B.P. eruptions at Popocatépetl volcano. C. Siebe points to the 2150 yr B.P. deposit.

Popocatépetl has also produced effusive activity through the emission of lava flows from the central event or from lateral fissures (Schaaf et al., 2005; Espinasa-Pereña and Martín-del Pozzo, 2006), such as those flows located in the vicinity of San Nicolás de Los Ranchos, which were produced during the eruption 2150 yr ago.

Effects of the Plinian Eruptions

The Popocatépetl Plinian eruptions had a large impact on prehispanic settlements through the direct emplacement of hot pyroclastic flows and thick fallouts or by the secondary emplacement of lahars that flooded extensive areas (Siebe et al., 1996a, 1996b). The last three Plinian eruptions of Popocatépetl coincide with three important events in Mesoamerican history: The 5000 yr B.P. (3195–2830 B.C.) eruption coincides with the 3114 B.C. beginning of the Mesoamerican Calendar; the 2150 yr B.P. (800–215 B.C.) eruption coincides with the transition from the Preclassic to the Classic period; the last eruption, which occurred 1100 yr B.P. (675–1095 A.C., likely 823 A.C.), coincides with the Classic-Postclassic periods transition (Fig. 16).

The 2150 yr B.P. Plinian eruption produced a yellow, pumice-rich fallout that was deposited on the NE flank of the volcano. The eruption emplaced a 1-m-thick fallout, burying prehispanic settlements in the region (Seele, 1973). Detailed archaeological excavations suggest that the inhabitants of this village, now called Tetimpa, were not able to get their belongings prior to their escape from the volcano's eruption (Plunket and Uruñuela, 1999, 2000, 2005).

The 1100 yr B.P. Plinian eruption also directly impacted prehispanic populations; however, the largest damage was produced due to remobilization of primary materials by lahars that moved along depth gullies. Thick sequences of lahars have been found surrounding Classic period ceremonial centers, such as Cholula, Cacaxtla, and Xochitécatl. The lahar deposits consist of the same constituents as the primary pyroclastic flow, surge-and-fall deposits exposed close to the volcano, and pottery shards, obsidian blades, and charcoal. The charcoal found in the lahar deposits yielded ages that correlate with the Plinian eruption at Popocatépetl volcano (675–1095 A.C.), with an approximated age of 823 A.C. Suárez-Cruz and Martínez-Arreaga (1993) concluded that the ceremonial center of Cholula was temporarily abandoned ca. 800 A.C.; that in a way coincides with the age range obtained by Siebe et al. (1996a, 1996b), therefore indicating that this abandonment of Cholula was caused by lahar flooding.

The 1994–2007 Eruption

After a quiescent period of 67 yr, Popocatépetl volcano resumed activity on 21 December 1994 with an increase in seismic activity and the emissions of 2–3-km-high columns above the crater, composed of ash, water vapor, and gases. The emissions were generated from small vents located at the base of the SE crater wall, inside the crater. These events continued sporadically until March 1995, when a sudden decrease in activity took place. From this time until early March 1996 the volcano remained calm with minor fluctuation in the seismic activity. On 29 March 1995, a lava flow at the base of the crater appeared for the first time. Small explosions launched lapilli-size fragments on the volcano flanks and in some villages. One of these explosions ended the

Figure 16. View of an archaeological excavation located on the northeastern slopes of Popocatépetl volcano. This small house forms part of the Tetimpa village covered by 1 m of lapilli-size pumice fall during the 2150 yr B.P. eruption of Popocatépetl.

lives of five mountaineers who had climbed the crater on 30 April 1996. This explosion sent juvenile material from the dome (1–2 cm in diameter) to the villages of Xalitzintla, San Nicolás de los Ranchos, and others located at 12 km on the NE flank of the volcano. On 10 June 1996, the dome had a thickness of 50 m and had completely covered the crater formed in 1922. Between April 1996 and June 1997, the volcano had extruded at least three lava domes. On 30 June 1997, a strong explosion preceded by two hours of volcanotectonic earthquakes took place. The eruption formed an 8-km-high column that was transported by the winds toward México City, producing ash fall that caused the international airport to be closed. The next day, a 12-km-long lahar was produced that reached the town of Santiago Xalitzintla, partially flooding one house. From July 1997 to November 2000, four domes were emplaced at the crater; these domes were followed by stronger explosions that launched ballistic projectiles up to 5 km from the summit. From 12–16 December 2000, Popocatépetl activity increased notably, causing the evacuation of more than 40,000 inhabitants. At that time, Popocatépetl's crater was almost completely filled with lava, and the 18 December 2000 eruption had extensive coverage in the media. These vulcanian-type explosions projected dome rocks >5 km from the summit, triggering forest fires. After these events, an updated hazards map of the volcano was presented (Sheridan et al., 2001). From April 1996 to the present, the volcano has extruded more than 20 domes that were subsequently destroyed by vulcanian-type explosions. One of the last large explosions took place on 21 January 2001, producing a small column that suddenly collapsed to produce a pyroclastic flow with a maximum extent of 6.5 km, reaching to the edge of the tree line. The pyroclastic flow diverted into several tongues, one of which scoured the glacier forming, and lahars reached the town of Xalitzintla, located 15 km from the summit (Capra et al., 2004). After 11 yr of activity, Popocatépetl's crater is almost filled to the brim, thus overcoming the magnitude of the 1919–1927 eruption (Macías and Siebe, 2005) (Fig. 17).

PICO DE ORIZABA VOLCANO (CITLALTÉPETL)

Pico de Orizaba volcano represents the highest mountain in México (19°01′N, 97°16′W, 5675 m) (Fig. 18A). This volcano is also known as Citlaltépetl, which means in náhuatl language "Mountain of the Star," and is located in the eastern portion of the Trans-Mexican Volcanic Belt (Fig. 1). Its summit sets the boundary between the states of Veracruz and Puebla. There are several reports of the historic activity of Pico de Orizaba; the most important event occurred in 1687 (Mooser, et al., 1958), and reports of other small events have been summarized by De la Cruz-Reyna and Carrasco-Núñez (2002). Today, there are a few signs of minor activity at the volcano, such as weak emissions of SO_2 and sulfur sublimates in the inner crater walls (Waitz, 1910–1911). For this reason, it is considered an active but dormant volcano. The present edifice has a central oval crater with 500 × 400 m diameter and 300 m vertical walls (Fig. 18B). The crater's northern part is covered by a glacier (Heine, 1988). The cone has a symmetric shape with steep slopes that reach up to 40° (Fig. 18A). The volcano rises 2900 m with respect to the Serdán–Oriental Basin to the west and 4300 m with regard to the Coastal Gulf Plain to the east (Carrasco-Núñez, 2000).

Figure 17. Detailed view to the east of Popocatépetl crater's interior. The photograph was taken on 24 October 2004 by Jorge Neyra. At the center appears a 340-m-wide internal crater. In the background to the right is Pico de Orizaba volcano.

Figure 18. (A) Photograph of the western flank of Pico de Orizaba volcano seen from Tlalchichuca on 4 February 2004. Photograph by Jorge Neyra. (B) View from the summit of the E-NE internal wall of Pico de Orizaba crater. Photograph by Jorge Neyra taken on 8 November 1998.

Previous Studies

The first general observations of the volcano were made by Waitz (1910–1911) during a geologic field trip; after that, the volcano remained unstudied until the middle of the twentieth century, when it was considered an active volcano by Mooser et al. (1958). The first regional geological studies of the volcano were carried out by Yáñez-García and García-Durán (1982) and Negendank et al. (1985). The first stratigraphy of Pico with a general evolution model was presented by Robin and Cantagrel (1982). This work was followed by petrologic studies (Kudo et al., 1985; Singer and Kudo, 1986; Calvin et al., 1989), stratigraphic and eruptive chronology studies (Cantagrel et al., 1984; Hoskuldsson, 1992; Hoskuldsson et al., 1990; Hoskuldsson and Robin, 1993; Carrasco-Núñez, 1993, 1997; Carrasco-Núñez et al., 1993; Siebe et al., 1993; Carrasco-Núñez and Rose, 1995; Gómez-Tuena and Carrasco-Núñez, 1999; Rossotti and Carrasco-Núñez, 2004), and the completion of a geological map (Carrasco-Núñez and Ban, 1994; Carrasco-Núñez, 2000). All of this information allowed scientists to produce a volcano hazards map (Sheridan et al., 2002, 2004), applying different programs, such as FLOW3D (Kover, 1995) and LAHARZ (Iverson et al., 1998). Finally, Díaz-Castellón (2003) and Zimbelman et al. (2004) evaluated the stability of the volcanic edifice.

Eruptive History

Pico de Orizaba is a Quaternary volcano built upon Cretaceous limestones and claystones (Yáñez-García and García-Duran, 1982). Its eruptive history shows several episodes of construction and destruction of ancient cones. In fact, Robin and Cantagrel (1982) and Hoskuldsson (1992) proposed that the volcano was built during three main stages. However, subsequent stratigraphic studies indicated that Pico de Orizaba was built during four eruptive phases (Carrasco-Núñez and Ban, 1994; Carrasco-Núñez, 2000) (Fig. 19). These phases were dubbed from older to younger as the Torrecillas cone, the Espolón de Oro cone, silicic peripheral domes, and the Citlaltépetl cone, as described in the following:

1. The Torrecillas cone began its formation 0.65 ± 0.71 Ma (Hoskuldsson, 1992) with the emission of the Pilancon olivine basaltic andesites, followed by the Jamapa andesitic and dacitic lavas and the complex Torrecillas two pyroxene andesitic lavas, breccias, amphibole-bearing dacites, dated at 0.29 ± 0.5 Ma (Carrasco-Núñez, 2000), with two pyroxene andesites atop. The Torrecillas stratovolcano reached a total volume of 270 km^3 (Carrasco-Núñez, 2000).

The constructive stage of this volcano ended with the collapse of the edifice, which generated a debris avalanche (Hoskuldsson et al., 1990) called Jamapa (Carrasco-Núñez and Gómez-Tuena, 1997) ca. 0.25 Ma. The debris avalanche traveled 75 km to the east of the crater, along the Jamapa River (Carrasco-Núñez and Gómez-Tuena, 1997) (Fig. 20). This phase produced a caldera, whose remnants are exposed to the south of the present cone (Robin and Cantagrel, 1982; Carrasco-Núñez, 1993).

2. The Espolón de Oro cone was constructed to the north of the Torrecillas crater walls; the remnants of this structure are represented by amphibole-dacite dated at 0.21 ± 0.04 Ma (Carrasco-Núñez, 2000). The Espolón de Oro cone began its formation with the emission of plagioclase + amphibole andesitic lavas dubbed Paso de Buey; these were followed by the Espolon de Oro amphibole dacitic lavas and block-and-ash flows. The activity of the volcano continued first on the western flank with the lateral emission of El Carnero olivine basaltic andesites and on the northern flank with the emplacement of the Alpinahua sequence composed of pyroclasts and andesitic lavas between 0.15 and 0.09 Ma. This sequence ended with the emplacement of aphanitic andesitic lavas interbedded with breccias and a welded ignimbrite with fiamme structures. At this point, the Espolón de Oro cone had an approximated volume of 50 km^3 (Carrasco-Núñez, 1997). A flank collapse finished with the construction of the Espolón de Oro cone ca. 16,500 yr B.P. (Carrasco-Núñez et al., 2005). This collapse originated the Tetelzingo debris avalanche deposit; that downslope transformed into a cohesive debris flow (10%–16% clay) that traveled 85 km from the source, covered and area of 143 km^2, and had a volume of 1.8 km^3 (Carrasco-Núñez et al., 1993). These authors concluded that the collapse was due to hydrothermal alteration of the volcanic edifice and the presence of a summit glacier.

3. Silicic peripheryal domes: The domes Tecomate to the NE and Colorado to the SW (Fig. 20) were extruded during the construction of the Espolón de Oro cone. The Tecomate dome consists of rhyolitic obsidian lavas, while the Colorado dome is formed by dacitic lavas and associated pyroclastic flows. These domes were followed by the emplacement of the Sillatepec and Chichihuale dacitic domes to the NW of the cone. This stage ended with the emplacement of the Chichimeco dome complex, composed of domes and amphibole andesitic lavas covered by a scoria pyroclastic flow dated at 8630 ± 90 yr B.P. (Carrasco-Núñez, 1993).

4. The construction of the Citlaltépetl cone started ca. 16,500 yr B.P. within the Espolón de Oro caldera wall remnants. The cone began its construction with effusion of the hornblende dacite Malacara lavas that flowed 13 km to the SE flank. These lavas were followed by the emission toward the NE flank of the crater of the Vaquería, andesitic lavas, and finally by the emission of the thick Orizaba dacites toward the SW and NE flanks. The present cone has a volume of 25 km^3 (Carrasco-Núñez, 1997). The historic lavas that erupted in 1537, 1545, 1566, and 1613 (Crausaz, 1994) are well exposed around the summit (Carrasco-Núñez, 1997).

The formation of the Citlaltépetl cone is not solely related to the quiet effusion of andesitic-dacitic lavas on its flanks, but it is also related to explosive activity (Siebe et al., 1993; Hoskuldsson and Robin, 1993; Carrasco-Núñez and Rose, 1995; Rossotti and Carrasco-Núñez, 2004). The stratigraphic record shows at least three major explosive events: (1) an eruption produced pyroclastic pumice flows on the eastern flanks of the volcano ca. 13,000 yr B.P.; (2) the Citlaltépetl eruptive sequence that produced diverse pumice fallouts and pyroclastic flows between 8500 and 9000 yr B.P. (Carrasco-Núñez and Rose, 1995); and (3) a series of dome destruction events that emplaced block-and-ash flow deposits on the western and southeastern flanks of the volcano occurred ca. 4100 yr B.P. (Siebe et al., 1993; Carrasco-Núñez, 1999). Other pyroclastic flow deposits have been dated between 8170 and 1730 yr B.P., and six pyroclastic fall deposits have been dated between 10,600 and 690 yr B.P. (De-la Cruz-Reyna and Carrasco-Núñez, 2002). Among all the deposits described above, only the 8500–9000 and 4100 yr B.P. eruptions left traceable deposits around the crater. Because of their importance in the stratigraphic record of Pico de Orizaba, these are described in the following sections.

Citlaltépetl Ignimbrite (8500–9000 yr B.P.)

This sequence represents the most explosive event during the Holocene record of Pico de Orizaba. The sequence was originally described as two members composed of pyroclastic flows separated by pumice fallout (Carrasco-Núñez and Rose, 1995). These authors concluded that the lower member was formed by four pyroclastic flow units with charcoal dated at 8795 ± 57 yr B.P. (average of six dates), a lahar, and a poorly developed paleosol. The upper member was composed of a fall deposit at the base and a pyroclastic flow deposit with charcoal dated at 8573 ± 79 yr B.P. (average of ten dates). Carrasco-Núñez and Rose (1995) proposed that these eruptions occurred between 8500 and 9000 yr B.P. at the Citlaltépetl cone. The pyroclastic flows (Citlaltépetl Ignimbrite) produced during this event were dispersed around the crater up to a distance of 30 km. The deposits consist of andesitic scoria, minor pumice, and lithics set in a fine ash matrix and have a volume of 0.26 km^3 (Carrasco-Núñez and Rose, 1995). These authors concluded that the Citlaltépetl Ignimbrite was triggered by a magma mixing event between andesitic (58–59 wt% SiO_2) and (62–63 wt% SiO_2) dacitic magmas and emplaced through a boiling-over mechanism. Afterward, Gómez-Tuena and Carrasco-Núñez (1999) studied the pyroclastic flow deposit found in the lower member of the Citlaltépetl Ignimbrite. According to its distribution, granulometry, and textural features, these authors concluded that the ignimbrite settled through an accretion process in proximal facies and in masse in distal facies. Recent studies of the Holocene deposits of Pico de Orizaba have recognized at least 10 fallout layers, interbedded with four pyroclastic flows and three paleosols (Rossotti and Carrasco-Núñez, 2004; Rossotti, 2005). Carrasco-Núñez and Rose (1995) dated three samples of organic material in this sequence that gave ages falling in the 8500–9000 yr B.P. period, the same period proposed for the Citlaltépetl Ignimbrite. Based on new stratigraphic columns, correlation with previous work, and radiometric dating, Rossotti (2005) concluded that the Citlaltépetl sequence occurred in four eruptive phases, dated at ca. 9000–8900, ca. 8900–8800, ca. 8800–8700, and ca. 8700–8500 yr B.P. This means that in a 500 yr period, Citlaltépetl volcano produced four eruptions that emplaced fallouts to the NE and pyroclastic flows up to a distance of 30 km from the crater.

Dome Destruction Event (4100 yr B.P.)

This event was characterized by the destruction of a central dome that originated block-and-ash flow deposits that formed a pyroclastic fan in the outskirts of the Avalos village west to the crater (Siebe et al., 1993). The flows traveled ~16 km and had an H/L (elevation loss/lateral extent ratio) of 0.186. Siebe et al. (1993) estimated a volume of 0.048 km^3 for these deposits and obtained radiocarbon dates of 4040 ± 80 and 4060 ± 120 yr B.P. In 1999, Carrasco-Núñez described a sequence of block-and-ash flow deposits forming terraces to the southeast of the crater. These flows traveled ~28 km from the crater and have an H/L of 0.153. Carrasco-Núñez (1999) determined a volume of 0.162 km^3 and obtained a radiocarbon date of 4130 ± 70 yr B.P. Because both deposits have similar ages and a homogeneous chemical composition of the juvenile lithics (dacite 62.7–63.95 wt% SiO_2), Carrasco-Núñez (1999) proposed that these deposits, with a total volume of 0.27 km^3, were emplaced by the same eruption ca. 4100 yr B.P.

EL CHICHÓN VOLCANO

El Chichón volcano (17°21′N, W93°41′W; 1100 masl) is located in the NW portion of the State of Chiapas, at ~20 km to the SW of the Pichucalco village. El Chichón is the active volcano in the Chiapanecan Volcanic Arc (Damon and Montesinos, 1978). This arc straddles the State of Chiapas from the southeast at the Boquerón volcano to the northwest at El Chichón volcano. The Chiapanecan Volcanic Arc is composed of low-relief, small volume volcanoes of Pliocene to Recent age (Capaul, 1987) and is located between the Trans-Mexican Volcanic Belt and the Central America Volcanic Arc (Fig. 1). From a tectonic point of view, the volcano sits on the northern part of the strike-slip Motagua-Polochic fault Province (Meneses-Rocha, 2001). El Chichón is built upon Late Jurassic–Early Cretaceous evaporites and limestones, Early to Middle Cretaceous dolomites, and Tertiary claystones and limestones (Canul and Rocha, 1981; Canul et al., 1983; Duffield et al., 1984; García-Palomo et al., 2004) (Fig. 21). These rocks are folded in a NW-SE direction, forming the Catedral anticline and La Unión and Caimba synclines (Macías et al., 1997b; García-Palomo et al., 2004). In addition, these folded rocks are dissected by E-W left lateral faults, such as the San Juan Fault that cuts across the volcano and N45°E-trending Chapultenango normal faults, dipping to the NW (García-Palomo et al., 2004).

El Chichón consists of the 1.5 × 2 km wide Somma crater, which has a maximum elevation of 1150 masl (Fig. 22). This crater is a dome ring structure breached in its N, E, and SW flanks that has subvertical inner walls and gentler outer slopes. The relative elevation of the Somma crater with respect to the surrounding terrain is ~700 m to the east and 900 m to the west. The Somma crater is composed of trachyandesites with an age of 0.2 Ma (Damon and Montesinos, 1978; Duffield et al., 1984).

The Somma crater was disrupted by younger structures, such as the Guayabal tuff cone in the SE, and intruded by two trachyandesitic domes of unknown age exposed in the SW and NW flanks (Macías, 1994). Prior to the 1982 eruption, this Somma crater was occupied by a trachyandesitic dome with two peaks, the higher having an elevation of 1235 masl. The 1982 eruption destroyed this dome, forming a 1-km-wide crater with a maximum elevation of 1100 masl (Fig. 23). This crater has vertical walls up to 140 m deep, and its floor is at an elevation of 860 m, hosting a lake, fumaroles, mud, and boiling water ponds (Taran et al., 1998; Tassi et al., 2003; Rouwet et al., 2004). The water in the lake normally has a temperature of 32 °C (Armienta et al., 2000), and the fumaroles have temperatures of 100 °C (Taran et al., 1998; Tassi et al., 2003), with organic components (Capaccioni et al., 2004) connected to an active hydrothermal system (Rouwet et al., 2004) (Fig. 24).

Previous Studies

The first mention of El Chichón volcano was in 1930, when local inhabitants heard and felt noises on a small hill known as Cerro la Union or Chichonal (Müllerried, 1933). Müllerried visited the hill and described that, in fact, it was an active volcano composed of a crater (Somma crater) and a central dome. In addition, he found a set of fumaroles with temperatures ~90 °C and a small lake between the SE part of the dome and the crater (Somma crater), thus showing that it was an active volcano. El Chichón appeared later in the catalogue of active volcanoes of the world (Mooser et al., 1958). For several decades, the volcano remained unstudied until, in the 1970s, the National Power Company (Comisión Federal de Electricidad, CFE) began a geothermal prospection study of the area (González-Salazar, 1973; Molina-Berbeyer, 1974). During a mineral prospecting study of the State of Chiapas, Damon and Montesinos (1978) visited the volcano; they dated the eastern wall of the Somma crater at 0.209 ± 0.019 Ma (K-Ar method) and considered it to be an active

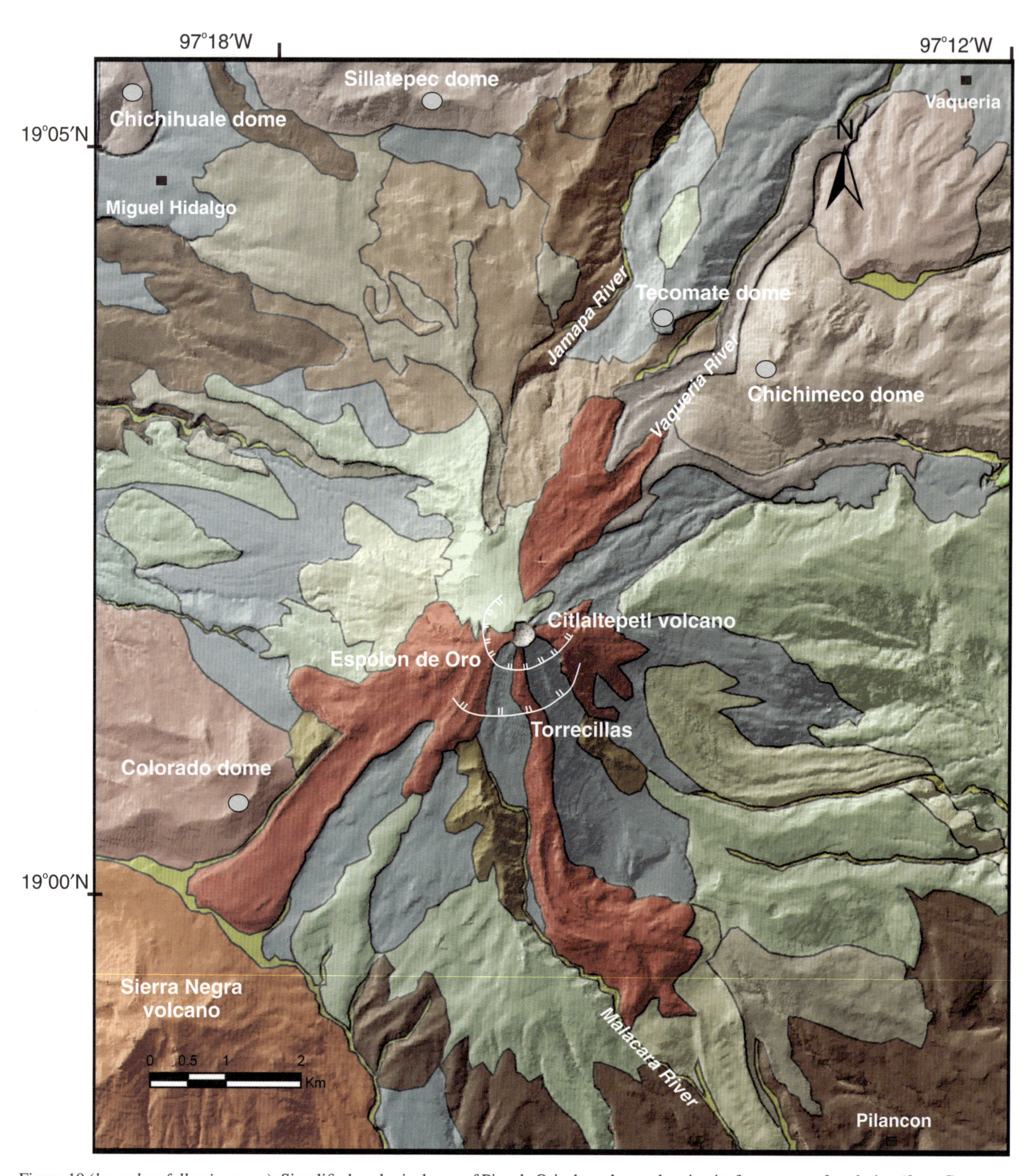

Figure 19 (*legend on following page*). Simplified geological map of Pico de Orizaba volcano showing its four stages of evolution (from Carrasco-Núñez, 2000).

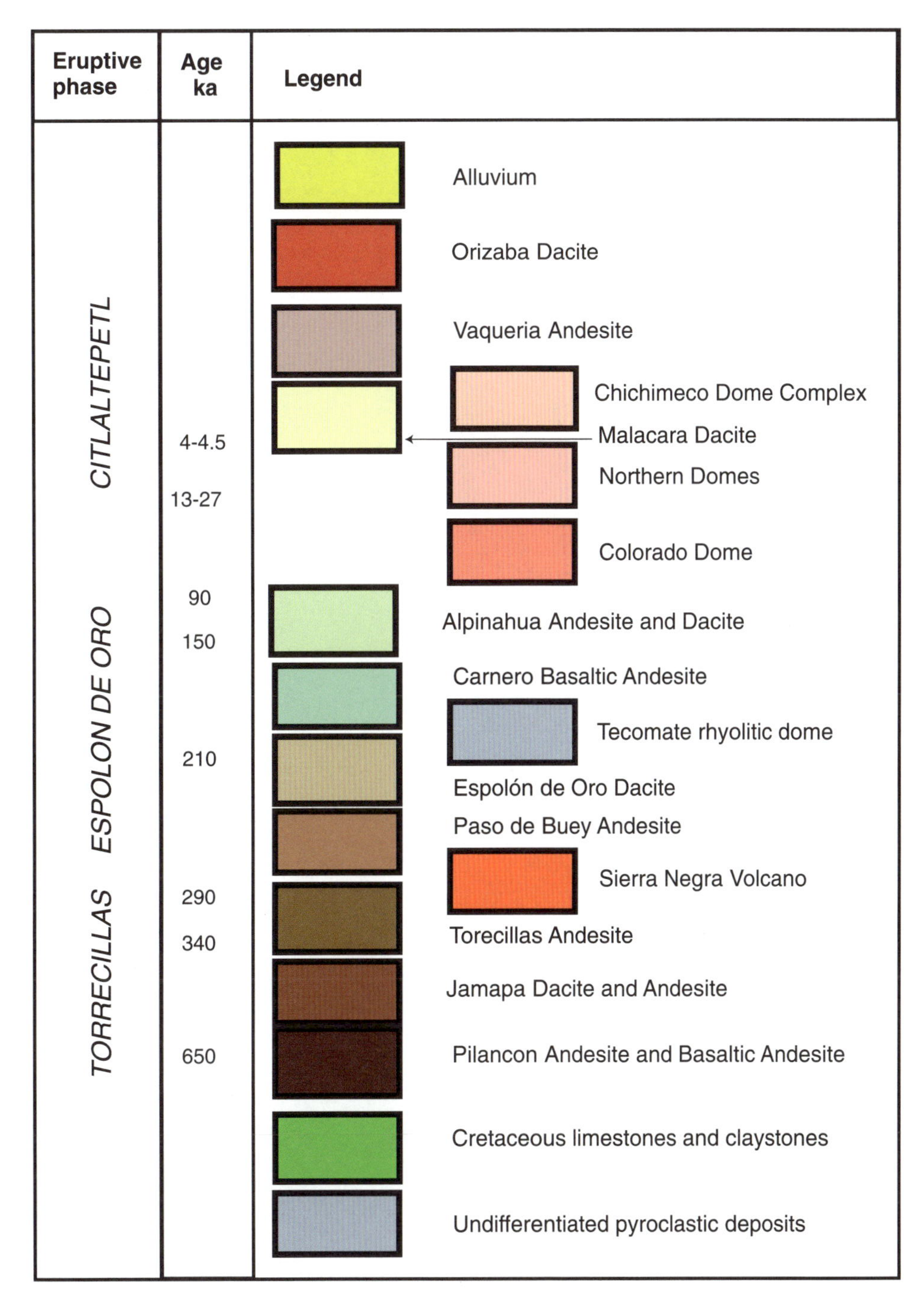

Figure 19 (*legend*).

volcano. At the beginning of the 1980s, the CFE started a geological reconnaissance (Canul and Rocha, 1981) and a fluid study (Templos, 1981) of the area. In their geological study, Canul and Rocha (1981) recognized older deposits rich in charcoal related to previous eruptions. While pursuing fieldwork in 1981, they felt and heard strong explosions at the volcano and realized that El Chichón might reactivate in the near future, as they stated in their internal report to the CFE.

The 1982 Eruption

Prior to the 1982 eruption, El Chichón consisted of the Somma crater and a central dome known as Chichón or Chichonal (Fig. 22). The eruption of El Chichón came as a surprise to the local population and the distant scientific community in México City, despite the increasing premonitory signs given by the volcano since late 1981.

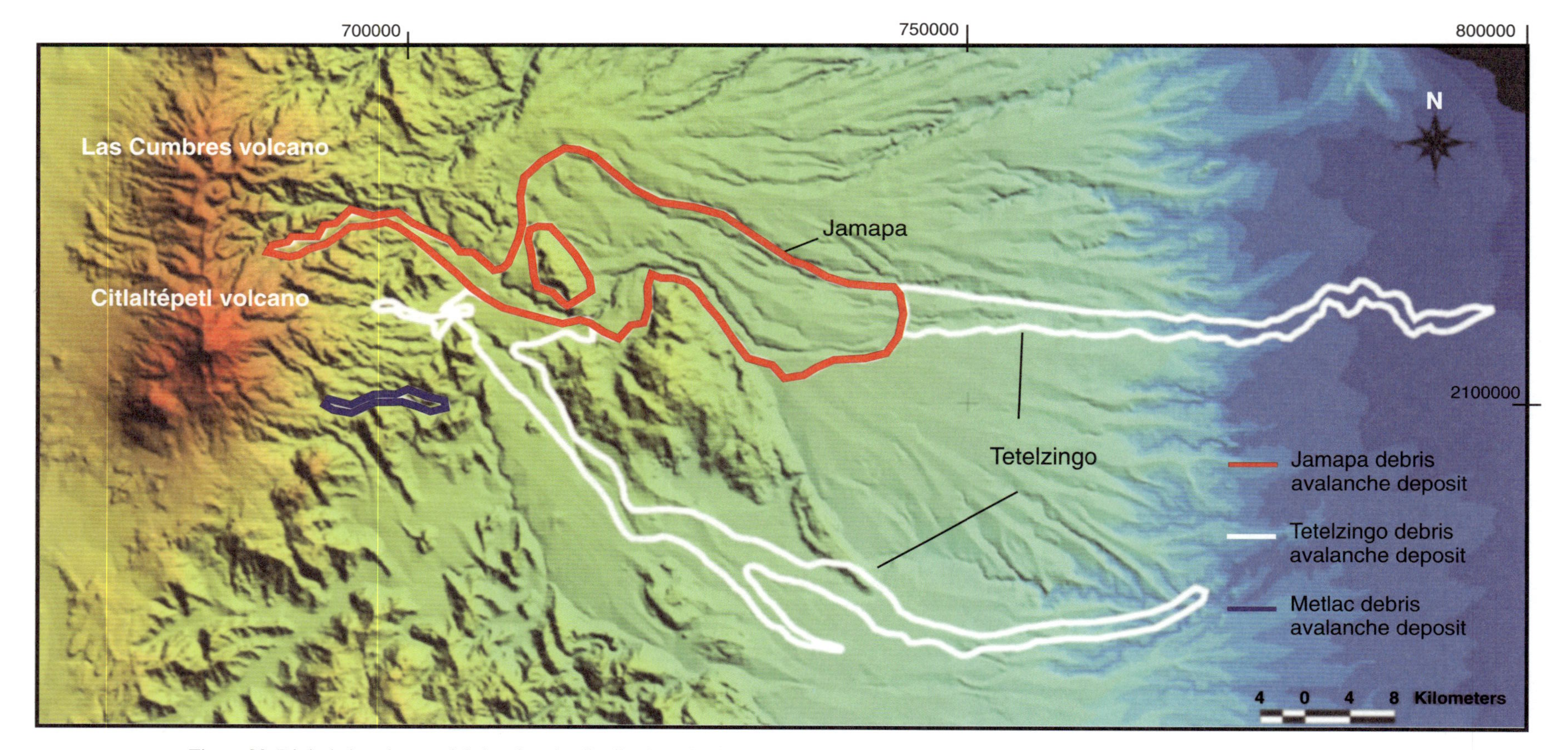

Figure 20. Digital elevation model showing the distribution of debris avalanches and debris flows produced at Pico de Orizaba volcano (Carrasco-Núñez et al., 2006). Topographic database with UTM coordinates obtained with Geosensing Engineering and Mapping.

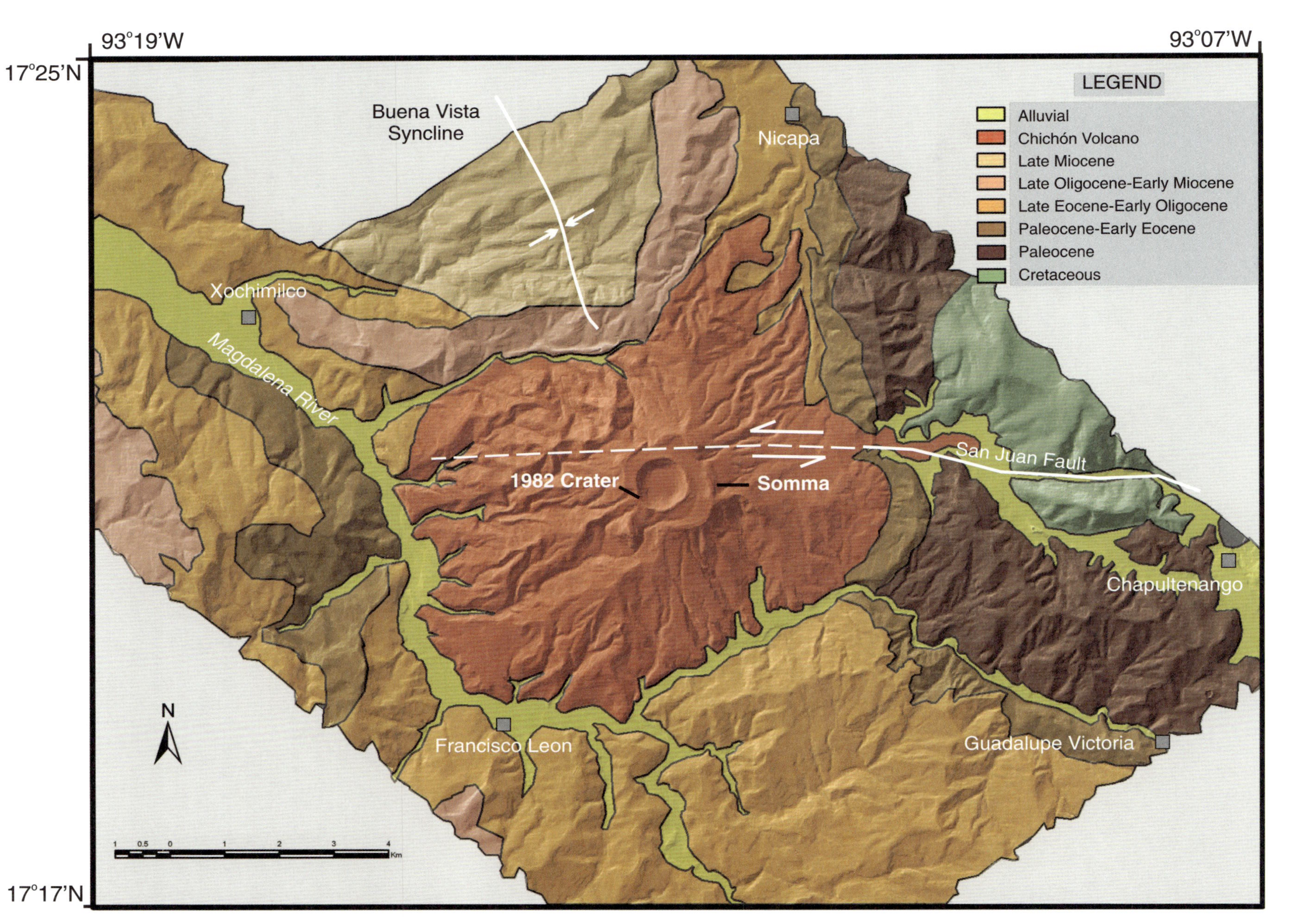

Figure 21. Simplified geological map of El Chichón volcano showing the Buena Vista syncline and the left-lateral San Juan Fault (taken from García-Palomo et al., 2004).

Figure 22. View to the west of El Chichón volcano with the Somma crater (arrow) and the central dome prior to the 1982 eruption. Photograph courtesy of Rene Canul.

Figure 23. Aerial view of El Chichón volcano after the 1982 eruption showing the crater that was formed in 1982 and an internal lake with fumaroles and sulfur precipitation. Photograph courtesy of Reynaldo Mota.

Figure 24. Panoramic view from the SE of El Chicón crater lake taken in May 2004. Fumarolic activity is observed on the righthand side of the picture.

The eruption was preceded by fumarolic activity and earthquakes; the latter were recorded at a seismic network installed at the Chicoasén hydroelectric plant, under construction at that time by the CFE (Espíndola et al., 2002). Jiménez et al. (1999) concluded that the seismic activity started in late 1980, increased through 1981, and peaked in March 1982. On Sunday night, 28 March 1982, the seismic activity developed into harmonic tremor of changing amplitude, followed by an hour of complete calm. Then, the beginning of the eruption was recorded as large amplitude tremor. The magmatic eruption opened a 150–180-m-wide crater in the central dome and formed a 27-km-high Plinian column (Medina-Martínez, 1982; Sigurdsson et al., 1984; Carey and Sigurdsson, 1986). This column was dispersed to the NE by the stratospheric winds, depositing fall layer A. The ash fall provoked confusion amongst the inhabitants, and they fled toward the nearest towns and cities of Pichucalco and Tuxtla Gutiérrez in Chiapas State and Villahermosa in Tabasco State. The next day, the area was closed by the army, which immediately applied its emergency plan called DN-III-E (Secretaría de la Defensa Nacional, 1983), during which most of the villagers were evacuated. From 29 March to 2 April, the volcano's activity decreased despite the occurrence of minor explosions and constant seismic activity. On Saturday, 3 April 1982, the military authorities allowed villagers to return to their homes, mainly to the Francisco Leon village, located in the southwestern part of the volcano, where a geologist and an army convoy spent the night. Unfortunately, the more violent explosion occurred that night (Yokoyama et al., 1992). The trachyandesitic magma came into contact with groundwater and the hydrothermal system, producing a series of hydromagmatic eruptions that dispersed pyroclastic surges (S1) up to 8 km from the crater, destroying nine towns and killing ~2000 people (Sigurdsson et al., 1984; Carey and Sigurdsson, 1986; Sigurdsson et al., 1987; Macías et al., 1997c; Macías et al., 1998; Scolamacchia and Macías, 2005) (Fig. 25). These explosions completely destroyed the central dome, first allowing the formation of a lithic-rich pyroclastic flow (F1) that partially flattened the rugged terrain, then the establishment of a 32-km-high phreatoplinian column depositing layer fall B, rich in lithics, the collapse of the column with the generation of pyroclastic flows and surges rich in pumice (F2 and S2), and finally a series of flow and surge events restricted to 2 km from the crater (UI). The volcano stayed relatively calm for four hours, until dawn on 4 April, when a 29-km-high Plinian column developed. The plume was dispersed to the NE by the stratospheric winds, which deposited layer C. After this, hydromagmatic explosions took place and emplaced pyroclastic surges (S3) up to 4 km from the crater. The eruptive plumes contained water vapor, as proved by the aggregation of ash (Varekamp et al., 1984) and the formation of different types of ash aggregates (Scolamacchia et al., 2005). The eruption left a 1-km-wide crater with four small inner craters with lakes. The walls of the crater exposed a stratigraphic record that clearly suggested this crater had been active in the past. After 4 April, El Chichón activity decreased drastically to small explosions, until September 1982.

The April 1982 eruption emitted 1.5 km^3 of magma (DRE) and abruptly modified the hydrologic network around the volcano. The pyroclastic deposits blocked all gullies around the volcano, especially the Susnubac-Magdalena and Platanar riverbeds (Riva-Palacio Chiang, 1983; Secretaría de la Defensa Nacional, 1982). Pyroclastic flows (F1 and F2) produced on the night of 3 April formed a 25–75-m-thick dam at the Magdalena River (Macías et al., 2004a). The water coming from the Susnubac River and rainfall during the months of April and May accumulated behind the dam, forming a lake (Fig. 26). The pyroclastic material in the lake substrate had a temperature of ca. 300 °C, and the lake water almost reached the boiling point. At the end of April, the lake was 4 km long, 300–400 m wide, and had a volume of hot water of 26 × 106 m^3, and in early May it had achieved a volume of 40×10^6 m^3 (Medina-Martínez, 1982). In the last days of May, the army evacuated 1288 inhabitants living downstream from the dam (Báez-Jorge et al., 1985) because the water from the lake started overtopping the dam. On 26 May 1982 at 1:30 a.m., the dam broke, producing two subsequent debris flows that rapidly were transformed into a single hyperconcentrated flow (Macías et al., 2004a). At 10 km from the dam, the flow had temperature of 82 °C, and reached the town of Ostuacán, burning coffee and cacao plantations. Downstream, the flow became a laden sediment flow, and at the junction between the Magdalena and Grijalva rivers, it diluted, promoting the rise of the water column. After 7 km, the flow front reached the wall of the Peñitas hydroelectric plant of CFE, at that time under construction. At the Peñitas dam, the surface of the river rose 7 m, with a surge of hot water (50 °C) that killed one worker, injured three others, and destroyed machinery.

The 1982 eruption devastated more than 100 km^2 of tropical forest and in distant places, such as Ostuacán and Pichucalco, caused churches and house roofs to collapse. The eruption injected seven million tons of SO_2 into the atmosphere, forming aerosols (Krueger, 1983; Matson, 1984). The aerosols formed a cloud that circumnavigated the planet and dropped the global temperature by ~0.5 °C. The juvenile products of the magma contained anhydrite, a mineral that had never been reported as primary mineral in magmas (Luhr et al., 1984; Rye et al., 1984). All these ingredients attracted the attention of a large number of specialists, who studied the crater lake (Casadevall et al., 1984), the pyroclastic deposits (Tilling et al., 1984; Sigurdsson et al., 1984, 1987; Rose et al., 1984), the ash and plume dispersion (Varekamp et al., 1984; Carey and Sigurdsson, 1986), the sulfur content of the magma (Devine et al., 1984; Carroll and Rutherford, 1987), and the petrology of the products (Luhr et al., 1984).

Holocene Eruptions

The 1982 eruption stripped all vegetation cover around the volcano and exposed the stratigraphic record of El Chichón. The first stratigraphic studies of this volcano revealed pyroclastic deposits with charcoal produced by eruptions that occurred at 550, 1250, and 1650 yr B.P. (Rose et al., 1984; Tilling et al., 1984). Surprisingly, Tilling et al. (1984) found abundant Mayan-type pottery shards embedded in the 1250 yr B.P. deposits, suggesting that the volcano had been inhabited by Mayan groups since that time. Later, Macías (1994) found the deposits of other two eruptions occurring at the volcano, dated at 900 and 1400 yr B.P. In the first systematic stratigraphic study of El Chichón, Espíndola et al. (2000) discovered at least 11 deposits related to the same number of eruptions during the past 8000 yr at 550, 900, 1250, 1400, 1700, 1800, 2000, 2400, 3100, 3700, and 7500 yr B.P. (Fig. 27). The 550, 1250, and 1450 yr B.P. eruptions were larger in magnitude than the 1982 eruption, having a volcanic explosivity index (VEI) of at least 4 (Newhall and Self, 1982). In fact, the 550 yr B.P. Plinian eruption ejected 1.4 km^3 of magma through the emplacement of a single fall layer that is thicker if compared to the total fallouts A, B, and C, produced by the 1982 eruption at the same distance (Macías et al., 2003). Strikingly, the 2400 yr B.P. pyroclastic flow also contained pottery shards, suggesting that the volcano was inhabited during the past 2500 yr. Unfortunately, these shards are homemade artifacts, and therefore it was not possible to assign them an Olmec or Maya origin, two of the oldest cultures in the region. The El Chichón eruptions not only had a local impact but also a regional one through the emplacement of falling ash at the Maya Low Lands in Guatemala during the Classic period (Ford and Rose, 1995). The inhabitants of this region used volcanic ash as a temper to prepare pottery. The chemical composition of the crystals and glass found in the pottery shards is similar to the content of juvenile materials located at El Chichón and Tacaná volcanoes in México and other volcanoes in Guatemala such as Cerro Quemado (Ford and Rose, 1995).

The repose period between the Holocene eruptions of El Chichón varies from 100 to 600 yr, compared with the 1982 eruption, which occurred after 550 yr of quiescence (Tilling

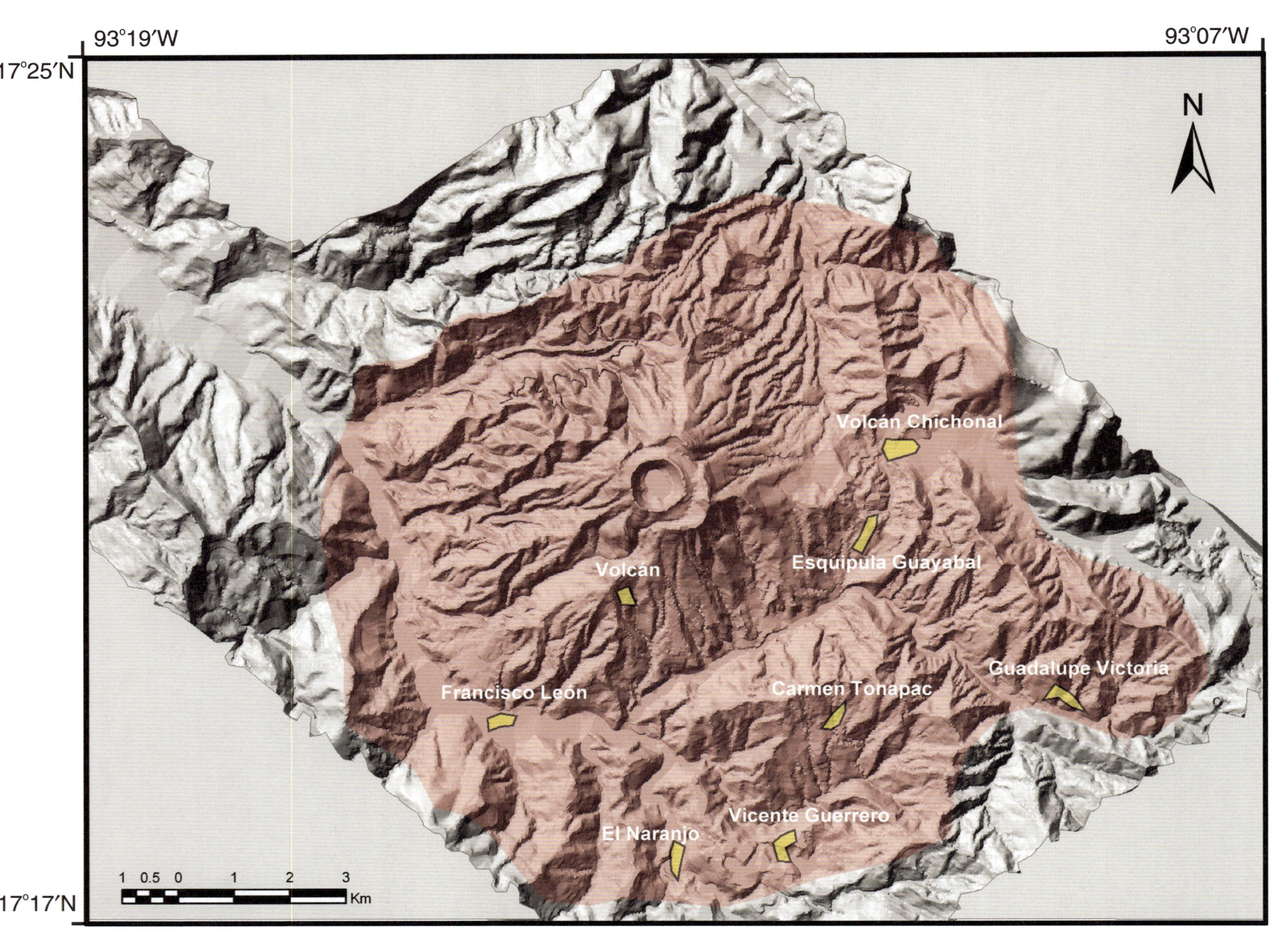

Figure 25. Digital elevation model of El Chichón volcano displaying the distribution of the pyroclastic surges around the volcano (after Scolamacchia and Macías, 2005). The towns in yellow were destroyed during the 1982 eruption.

et al., 1984; Espíndola et al., 2000). Therefore, today we cannot discount that the volcano might erupt under the present circumstances with an open crater occupied by a lake. In fact, the 900, 2000 and 2400 yr B.P. eruptions occurred under open vent conditions.

Chemical Composition of the Magmas

The magmas erupted by El Chichón during the past 8000 yr have a trachyandesitic composition (Duffield et al., 1984; Rose et al., 1984; McGee et al., 1987; Espíndola et al., 2000; Macías et al., 2003), with a similar mineral association of plagioclase > amphibole > augite, with magnetite, sphene, pyrrothite, biotite, and apatite as accessory minerals (Luhr et al., 1984; Duffield et al., 1984; Espíndola et al., 2000). The chemical composition and the mineral association are very similar to those described for the Chiapanecan Volcanic Arc rocks (Capaul, 1987). The 1982 trachyandesites were rich in sulfur and crystals (~53 vol.%) including ~2 vol.% of anhydrite (Luhr et al., 1984). Prior to the eruption, this magma had temperatures between 750 and 880 °C (Luhr et al., 1984) and stagnated at depths ~6 km (2 kilobars, Luhr, 1990). However, the analysis of the seismic record indicates that the magma chamber at the time of the eruption was at a depth between 7 and 13 km (seismic gap) below the volcano (Jiménez et al., 1999), suggesting a deeper reservoir in the sedimentary Jurassic-Cretaceous substrate where assimilation of sulfur from the rocks could have taken place (Rye et al., 1984). Prior to the 550 yr B.P. eruption, the magma had a temperature of 820–830 °C, it was water saturated (5–6 wt% H_2O), and it resided at a depth of ~6–7.5 km (2–2.5 kilobars) below the volcano (Macías et al., 2003).

The homogeneous chemical composition of the Holocene Chichón products, suggests that the magmatic system has been relatively stable. However, the presence of mafic enclaves (trachybasalts and trachybasaltic andesites) hosted by the trachyandesites of the Somma crater and the isotopic variations found in the plagioclase phenocrysts of El Chichón show that the volcano magmatic system has had the input of repeated recharges of deeper mafic magmas (Espíndola et al., 2000; Tepley et al., 2000; Davidson et al., 2001; Macías et al., 2003) (Fig. 28).

TACANÁ VOLCANO

Tacaná volcano (15°08′N, 92°09′W) lies in the State of Chiapas in southern México and in the San Marcos Department in Guatemala. Shared by the two countries, the summit delineates the international boundary. The volcano is located at the most northern part of the Central American Volcanic Arc (Figs. 1 and 29). For several small towns and coffee plantations, this volcano is a serious hazard to their population and economic activity. The city of Tapachula, Chiapas, México (~300,000 inhabitants) is located 30 km SW of the volcano's summit.

Previous Studies

Von Humboldt (1862) first described the volcano as Soconusco, the most northwestern volcano of Central America, while Dollfus and de Monserrat (1867) referred to it as Istak volcano. Later, Sapper (1896, 1899) clarified that Soconusco was synonym of Tacaná, the name used by Böse during his studies (1902, 1903, 1905). Finally, Waitz (1915) con-

Figure 26. View of the temporal lake formed in April to May 1982 in the southwestern flank of El Chicón volcano. The water in contact with the pyroclastic deposits was nearly to the boiling point. Photograph courtesy of Servando de la Cruz.

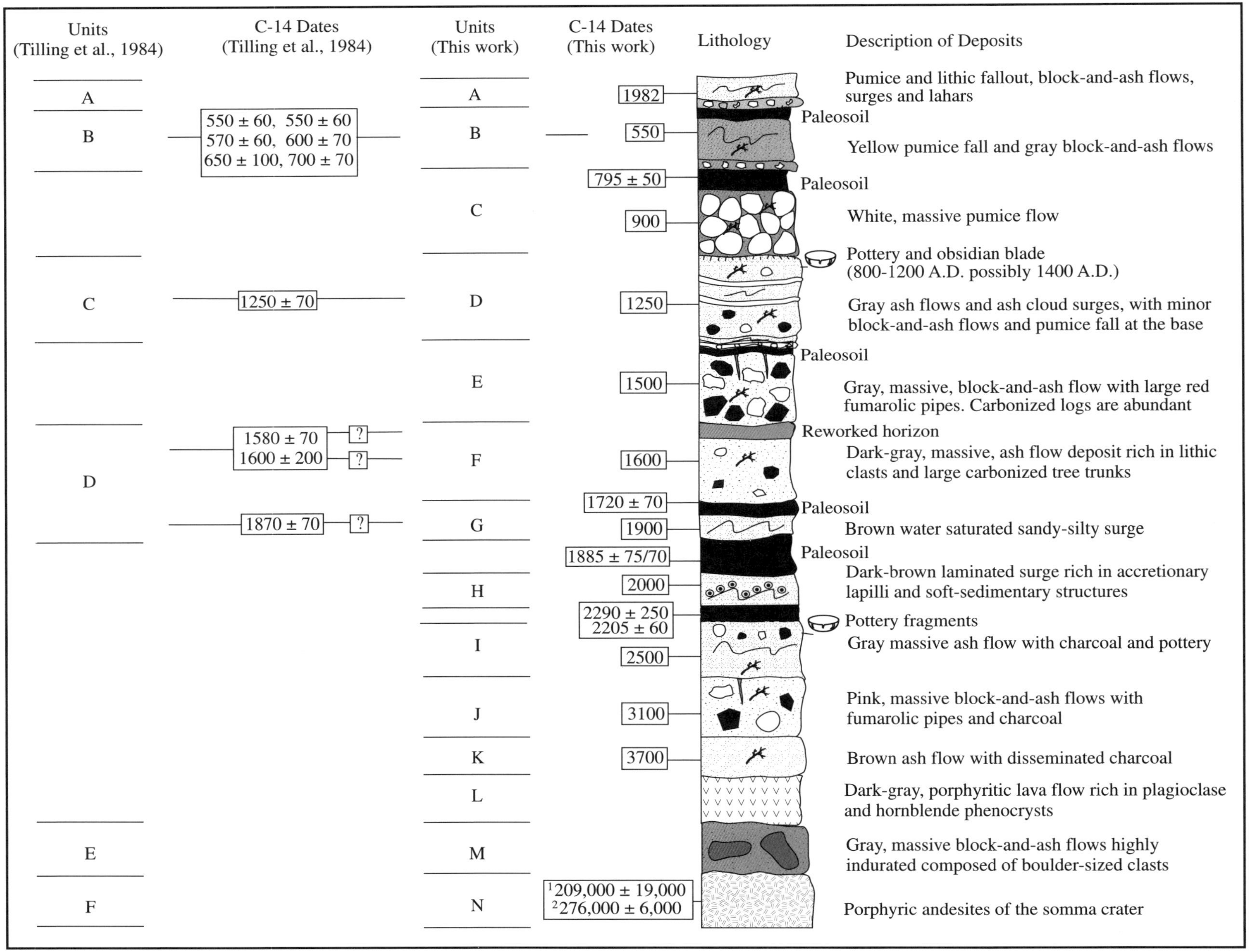

Units (Tilling et al., 1984)	C-14 Dates (Tilling et al., 1984)	Units (This work)	C-14 Dates (This work)	Description of Deposits
A		A	1982	Pumice and lithic fallout, block-and-ash flows, surges and lahars
				Paleosoil
B	550 ± 60, 550 ± 60 570 ± 60, 600 ± 70 650 ± 100, 700 ± 70	B	550	Yellow pumice fall and gray block-and-ash flows
			795 ± 50	Paleosoil
		C	900	White, massive pumice flow
				Pottery and obsidian blade (800-1200 A.D. possibly 1400 A.D.)
C	1250 ± 70	D	1250	Gray ash flows and ash cloud surges, with minor block-and-ash flows and pumice fall at the base
				Paleosoil
		E	1500	Gray, massive, block-and-ash flow with large red fumarolic pipes. Carbonized logs are abundant
				Reworked horizon
D	1580 ± 70 [?] 1600 ± 200 [?]	F	1600	Dark-gray, massive, ash flow deposit rich in lithic clasts and large carbonized tree trunks
			1720 ± 70	Paleosoil
	1870 ± 70 [?]	G	1900	Brown water saturated sandy-silty surge
			1885 ± 75/70	Paleosoil
		H	2000	Dark-brown laminated surge rich in accretionary lapilli and soft-sedimentary structures
			2290 ± 250 2205 ± 60	Pottery fragments
		I	2500	Gray massive ash flow with charcoal and pottery
		J	3100	Pink, massive block-and-ash flows with fumarolic pipes and charcoal
		K	3700	Brown ash flow with disseminated charcoal
		L		Dark-gray, porphyritic lava flow rich in plagioclase and hornblende phenocrysts
E		M		Gray, massive block-and-ash flows highly indurated composed of boulder-sized clasts
F		N	[1]209,000 ± 19,000 [2]276,000 ± 6,000	Porphyric andesites of the somma crater

Figure 27. Simplified stratigraphic column of El Chichón volcano with the Holocene deposits described by Espíndola et al. (2000). References are [1]Damon and Montesinos (1978) and [2]Duffield et al. (1984).

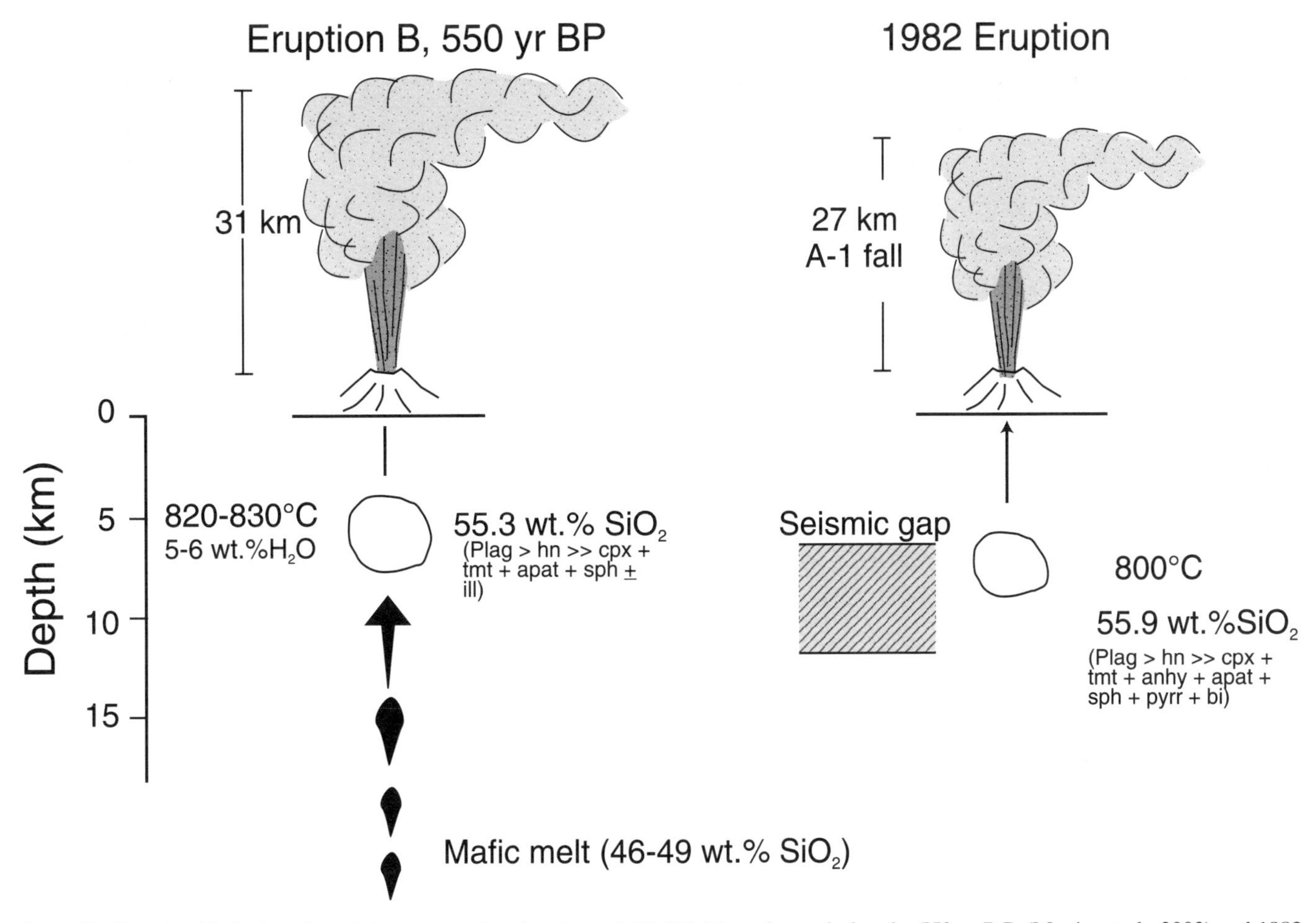

Figure 28. Sketch with the location of the magma chamber beneath El Chichón volcano during the 550 yr B.P. (Macías et al., 2003) and 1982 eruptions (Jiménez et al., 1999; Luhr, 1990).

cluded that Soconusco and Istak were synonyms of Tacaná (De Cserna et al., 1988). The first general descriptions of Tacaná were made by Sapper (1896, 1899). Bergeat (1894) made the first petrographic description of the Tacaná rocks, which he described as augite andesites. Detailed studies of the volcano began after a series of earthquakes on 22 September 1902 and the 24 October 1902 eruption of Santa Maria volcano in Guatemala. These events captured the attention of several geologists who visited the region at that time. The first detailed account of Tacaná was made by Böse (1902, 1903), who reported that the volcano base was located at an elevation of 2200 masl on top of a granitic basement. Böse also concluded that Tacaná consisted of three terraces located at elevations of 3448, 3655, and 3872 m (the uppermost crater hosted by a lava dome). The rocks collected by Böse were then petrographically analyzed by Ordóñez, who classified them as hypersthene-hornblende andesites. Böse (1902, 1903) also described an explosion crater with an elliptical shape 50 m in diameter and 5 m deep, located SW of the main summit, from which sulfur waters and fumes emanated at that time. Sapper (1897) pointed out that after the 12 January 1855 earthquake, there was formation of fissures on the flanks of Tacaná, with brief emission of fumes (Mooser et al., 1958). This activity might have been related to accounts by local inhabitants, who described an eruption at the volcano summit that ejected ash from fan-shaped holes and may account for another eruption in 1878. Böse (1902, 1903) and Waibel (1933) considered Tacaná as a dormant but not extinct, volcano.

The 1949 Eruption

On 22 September 1949 an earthquake took place at Tacaná volcano, after which the residents observed white columns and ash fall on the outskirts of Unión Juárez. This event alerted the local authorities, who asked for help from the Geology Institute of the Universidad Nacional Autónoma de México (UNAM). In January 1950 the geologist Müllerried visited the area and prepared a detailed account of the eruption (Müllerried, 1951).

Mülleried observed that the Tacaná summit crater was 70 m below the volcano summit (third step of Böse, 1902) and that it was open to the N-NW. He also described the other two steps described by Böse (1902): one located 160 m below the sum-

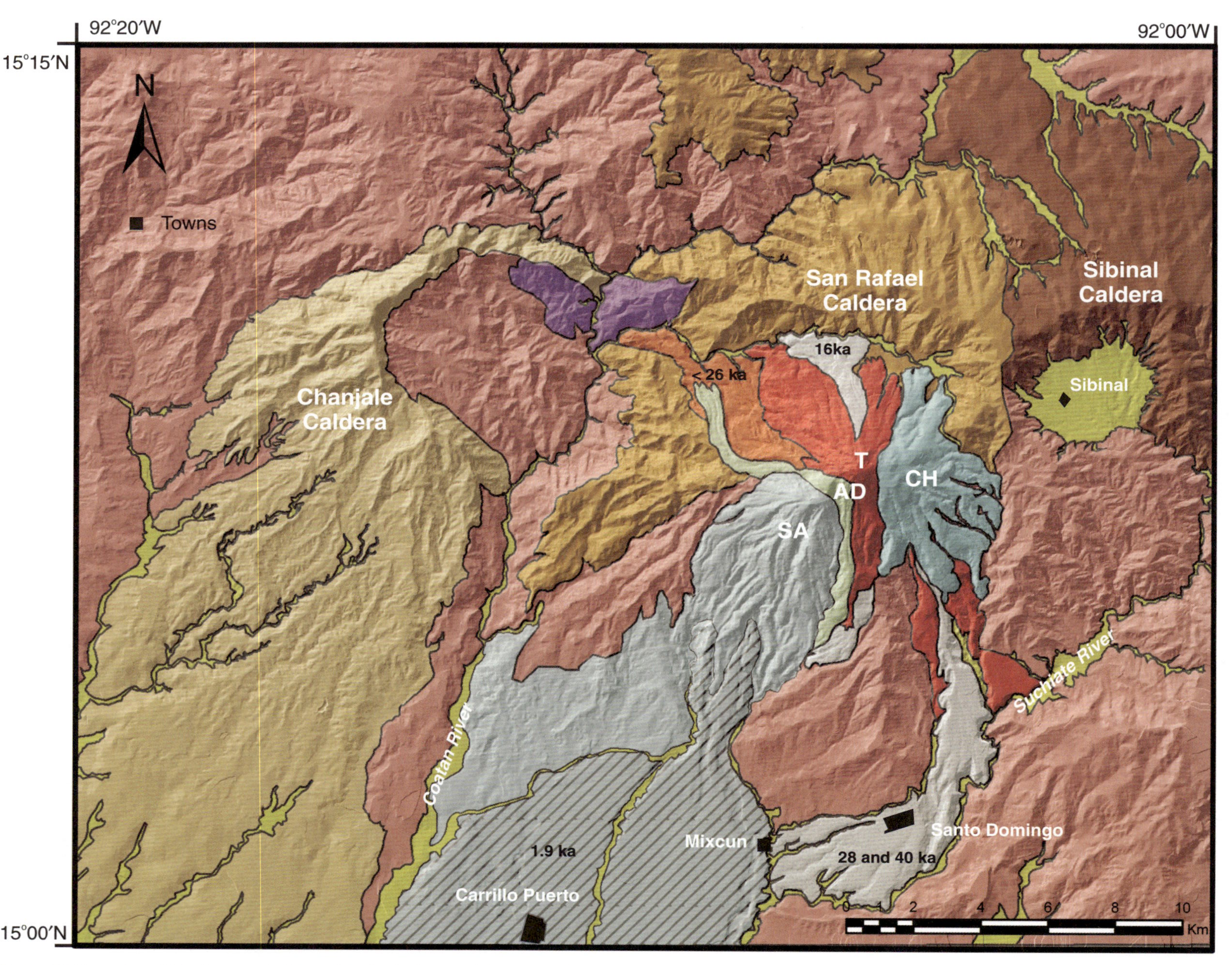

Figure 29 (*legend on following page*). General geological map of the Tacaná Volcanic Complex that is hosted inside the remains of the 1 Ma San Rafael caldera. The map shows the extent of some deposits younger than 50,000 yr B.P. from Tacaná volcano and the 1950 yr B.P. Mixcun deposits (after García-Palomo et al., 2006). CH—Chichuj; T—Tacaná; AD—Ardillas Dome; SA—San Antonio.

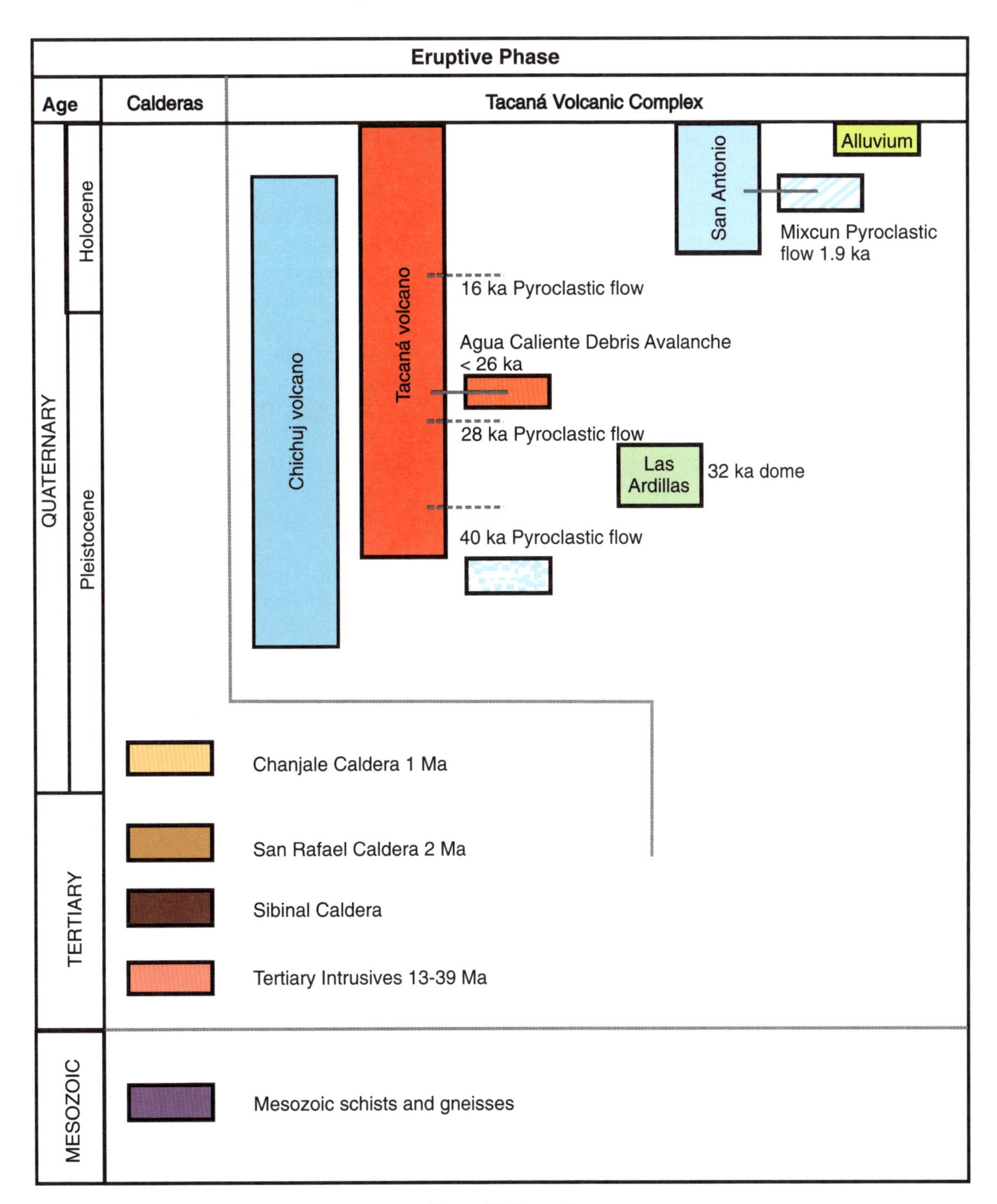

Figure 29 (*legend*).

mit with a dry lake, and the other 230 m below the summit at the lake site. According to Müllerried (1951), the 1949 activity occurred at 16 vents situated in three areas southwest of the summit. The area of these vents coincides with the suspected area of the nineteenth century eruptions mentioned by Böse (1902, 1903). The 1949 eruptive vents were 2–4 m wide and up to 4 m deep (fumarole 1; Müllerried, 1951). White to transparent fumes escaped from all the vents and were visible from the City of Tapachula. Some of these fumaroles emitted sulphurous gases and sulphur, and chlorine precipitations at their vents. The 1949 eruption pushed Tacaná into the catalogue of active volcanoes of the world (Mooser et al., 1958).

The 1986 Eruption

On December 1985, after 35 yr of repose, Tacaná volcano awoke with earthquakes and noises that continued thorough January 1986, when a seismic network was deployed at the volcano.

The most important seismic event occurred on 3 February 1986, producing damage to adobe homes at the Ixchiguan village in the San Marcos Department, Guatemala, located 25 km ENE from the crater. The seismic activity increased until 7 May, when an earthquake swarm provoked panic among the population. On 8 May, when the frequency of "felt" earthquakes by the population was two per minute, a phreatic explosion occurred outside the summit crater, almost along the international border, at an elevation of 3600 m (De la Cruz-Reyna et al., 1989). The explosion produced an 8-m-wide vent from which rose a 1-km-high column rich in water vapor and gases. Afterward, the seismic activity declined notably, and two days later this activity reached levels such as those recorded in April 1986. The fumarole was enriched in water vapor, without magmatic components (Martíni et al., 1987).

Prior to this eruption, CFE had begun a series of studies to evaluate the geothermic potential of the volcano using geological reconnaissance (Medina, 1985; De-la Cruz and Hernández, 1985) and the chemistry of the water springs around the volcano (Medina-Martínez, 1986). The latter pointed out that the spring water of Tacaná contains acid sulfate and reported fumaroles located 3200 and 3600 m to the S-SW from the summit, with temperatures of 82 and 94 °C. Today, the spring waters of Tacaná are rich in CO_2, with a composition that can be interpreted as a mixture of a deep source rich in SO_4-HCO_3-Cl and dilute meteoric water (Rouwet et al., 2004). The total emission of volatiles at Tacaná is ~50 t/d of SO_2, a typical value for passive degassing volcanoes.

De la Cruz and Hernández (1985) made the first geological map of the volcano at a scale of 1:120,000 and constructed the first stratigraphic column of the area with the volcano sitting atop a granitic basement and Tertiary andesites. In addition, these authors and Saucedo and Esquivias (1988) mapped three pyroclastic fans, dubbed Qt1, Qt2, and Qt3, which they associated with the formation of three calderas. Later, De Cserna et al. (1988) presented a photogeological map of the volcano at a 1:50,000 scale, defined 14 stratigraphic units, and summarized previous studies. In their study, these authors concluded that Tacaná is a polygenetic stratovolcano composed of three NE-SW aligned volcanoes, formed during the eruptive periods named Talquian, Tacaná, and El Águila. It is likely that these volcanoes correspond to the three calderas proposed by De la Cruz and Hernández (1985). Similarly, Mercado and Rose (1992) produced a photogeological map of Tacaná, hazard maps for the different types of hazards, and defined the chemistry of the rocks as calc-alkaline andesites. Recent studies have pointed out that Tacaná is actually a volcanic complex composed of four aligned structures in a NE-SW direction, as suggested by De Cserna et al. (1988), whose activity migrated from NW to the SE. These structures are, from oldest to youngest: Chichuj volcano (Talquian; De Cserna et al., 1988), Tacaná (Tacaná; De Cserna et al., 1988), las Ardillas dome, and San Antonio (El Águila: De Cserna et al., 1988; Macías et al., 2000; García-Palomo et al., 2006) (Fig. 30).

Regional Geology

Tacaná is located close to the triple point junction between the North America, Caribbean, and Cocos plates (Burkart and Self, 1985; Guzmán-Speziale et al., 1989). The boundary between the North America and Caribbean plates is given by the left lateral Motagua-Polochic Fault System, which separates the Maya block to the north and the Chortis block to the south (Ortega-Gutiérrez et al., 2004), where Tacaná volcano is located (Fig. 29). Previous studies suggest that Tacaná was built upon a Paleozoic basement (Mooser et al., 1958; De Cserna et al., 1988; De la Cruz and Hernández, 1985). However, due to its tectonic location within the Chortis block, its basement rocks, represented by schists and gneisses, should be younger, of Mesozoic age (García-Palomo et al., 2006). The Tacaná area was then affected by two intrusion episodes of granites, granodiorites, and tonalities, occurring at 29–35 Ma and 13–20 Ma (Mugica, 1987; García-Palomo et al., 2006). The volcanic activity began 2 Ma with the formation of

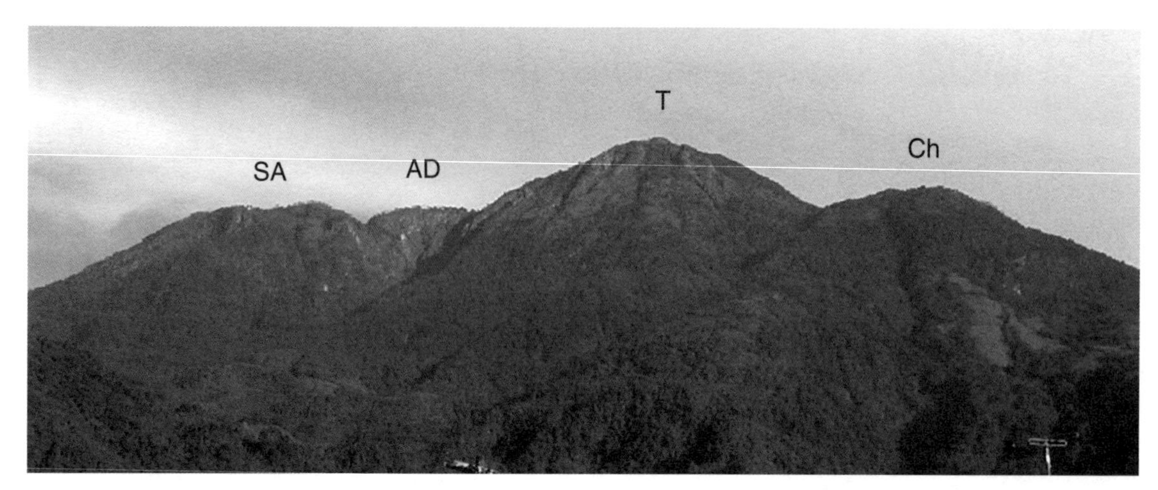

Figure 30. Panoramic view to the north of the Tacaná Volcanic Complex from the Unión Juárez village. The edifices are Chichuj (Ch), Tacaná (T), Las Ardillas Dome (AD), and San Antonio (SA).

the San Rafael caldera and continued 1 Ma with the formation of the Chanjale caldera and the Sibinal caldera (García-Palomo et al., 2006). The Tacaná Volcanic Complex began its construction 100,000 yr B.P. inside the San Rafael caldera.

Eruptive History

The first radiometric dates in charcoal of Tacaná volcano were obtained by Espíndola et al. (1989). These authors dated a block-and-ash flow deposit exposed at La Trinidad village at 42,000 yr B.P. This deposit corresponded to the oldest Qt3 pyroclastic fan of De la Cruz and Hernández (1985). Based on this age, Espíndola et al. (1989) considered that these deposits could correspond to some of the oldest deposits of Tacaná. Later on, Espíndola et al. (1993) obtained another radiometric age for the same deposit, yielding an age of 38,000 yr B.P., which confirmed the age of the deposit. In addition, these authors described another block-and-ash flow deposit near the Monte Perla locality that gave a date of 30,000 yr B.P.; this deposit was exposed within the pyroclastic fan Qt3. This new age suggested that the pyroclastic fan Qt3 consisted of several deposits associated with eruptions produced at Tacaná volcano, the tallest structure of the Tacaná Volcanic Complex. Later, Macías et al. (2000) identified deposits related to two eruptions dated at ca. 10,000 and 1950 yr B.P. The 1950 eruption, however, was originated at San Antonio volcano, the youngest volcano of the Tacaná Volcanic Complex. This eruption produced a later explosion that destroyed the summit dome, generating some pyroclastic surges but also a block-and-ash flow deposit that traveled up to 14 km from the summit along the Cahuacán and Mixcun ravines. The deposit is well exposed on the outskirts of the Mixcun village from which it takes its name (Macías et al., 2000) and corresponds to pyroclastic fan Qt2 (De la Cruz and Hernández, 1985). At that time, the main cultural center of the region was Izapa, a ceremonial center located ~17 km from the Tacaná summit. The archaeological excavations at Izapa have suggested that the center had been temporarily abandoned during the first century A.D. (Lowe et al., 1982). In this part of the excavation, Lowe et al. found objects with features similar to pottery produced in Guatemala and El Salvador, a fact that led these authors to propose that villagers of Izapa abandoned the city to conquer such lands. However, the 1950 yr B.P. eruption of San Antonio Volcano was dated within the period of the abandonment at Izapa (Macías et al., 2000) defined by Lowe et al. (1982), suggesting that the cause of the abandonment was the eruption of Tacaná itself. The pyroclastic flow produced by the 1950 yr B.P. eruption did not directly impacted Izapa; however, secondary hyperconcentrated and debris flows caused extensive damage in the area, destroying crops and isolating Izapa from Central México (Macías et al., 2000). This phenomenon would have caused the villagers of Izapa to migrate to Central America. This phenomenon is not uncommon in Chiapas, since tropical storms and hurricanes have isolated Tapachula from México and Central America during the last two decades.

The eruptions that occurred at 40,000, 30,000, and 1950 yr B.P. caused the partial or total destruction of andesitic central domes, and this generated block-and-ash flows able to move and infill ravines as far as 15 km around the volcano. Recent studies have identified at least other six eruptions, occurring at 32,000, 28,000, <26,000, 16,000, 7500, and 6500 yr B.P. (Macías et al., 2004b; Mora et al., 2004; García-Palomo et al., 2006), in addition to the 40,000, 30,000, 10,000, and 1950 yr B.P. eruptions. Some of these eruptions have not been considered in previous hazard zonation studies of the volcano (Mercado and Rose, 1992; Macías et al., 2000). The eruptions that occurred 40,000, 30,000, 16,000 and 1950 yr B.P. were produced by the partial destruction of andesitic domes from Tacaná and San Antonio volcanoes.

The 32,000 yr B.P. eruption formed a Plinian column that was dispersed to the NE into the Guatemala lands and that deposited a pumice fall layer. Prior to this report, there was a brief note of the presence of 2-m-thick fall deposits close to the Sibinal village, Guatemala, by Mercado and Rose (1992).

The 26,000 yr B.P. eruption was caused by the collapse of the NW part of Tacaná volcano. At that time, the crater was occupied by an andesitic central dome. The collapse produced a debris avalanche followed by a series of block-and-ash flows that traveled at least 8 km to the Coatán River. The debris avalanche has an H/L of 0.35, covers a minimum area of 8 km^2, and has a volume of 1 km^3 (Macías et al., 2004b; García-Palomo et al., 2006). The eruption ended with the generation of debris flows along the San Rafael and Chocab rivers.

The eruptions dated at ca. 7500 and 6500 yr B.P. produced ash flows and pyroclastic surges, respectively, forming a thin blanket that covers the Tacaná cone.

Tacaná volcano has mainly erupted two pyroxene andesites and minor dacites. The andesites are composed of plagioclase, augite, hypersthene + FeO oxides, and rare hornblende. The chemical composition of these rocks is very homogeneous (Mercado and Rose, 1992; Macías et al., 2000; Mora, 2001; Mora et al., 2004). Therefore, Chichuj volcano has been edified by andesitic lava flows and domes (59–63 wt% SiO_2), Tacaná volcano by rare basaltic andesite enclaves (56–61 wt% SiO_2), andesitic lava flows and dacitic domes (61–64 wt% SiO_2), and pyroclastic flows with juvenile andesitic lithics (60–63 wt% SiO_2). San Antonio volcano has formed andesitic lava flows and dacitic domes (58–64 wt% SiO_2).

DISCUSSION AND CONCLUSIONS

Most of the large volcanoes in México are located along the Trans-Mexican Volcanic Belt, a continental arc that transects Central México between the 19° and 20° parallel latitude north (Fig. 1). Volcanism also appears concentrated in small areas, such as the Tres Vírgenes Volcanic Complex and the San Quintín Volcanic Field in Baja California, the Pinacate Volcanic Field in Sonora, the Revillagigedo archipelago in the Pacific, Los Tuxtlas Volcanic Field in Veracruz, the Chiapanecan Volcanic Arc, and the northwestern edge of the Central America Volcanic Arc in

Chiapas. Several authors have concluded that the Trans-Mexican Volcanic Belt is formed by the subduction of the Rivera and Cocos beneath the North America plate at the Middle American Trench (Ponce et al., 1992; Singh and Pardo, 1993; Pardo and Suárez, 1993, 1995). The Trans-Mexican Volcanic Belt has an oblique position of ~15° with respect to the Middle American Trench; this is a strange feature for continental volcanic arcs. This feature caused some authors to propose that the Trans-Mexican Volcanic Belt is related to a megashear transecting the central part of México (Cebull and Shurbet, 1987) or to active extension like a rift zone (Sheth et al., 2000).

The Central America Volcanic Arc extends from western Panama to southeastern México. It runs parallel to the Middle America Trench, and at the México-Guatemala border volcanism becomes discontinuous and migrates inland along with the seismic isodepth curves (Fig. 31). The discontinuous volcanism appears as a scatter of landforms in the state of Chiapas forming the Chiapanecan Volcanic Arc, which has mainly erupted calc-alkaline products but also K-rich alkaline rocks at el Chichón, located 350 km from the trench, and Na-rich alkaline products at Los Tuxtlas Volcanic Field, located 400 km from the trench. Nixon (1982) proposed that the alkaline volcanism of El Chichón and Los Tuxtlas was due to extensional tectonics related to the triple point junction of the North America, Caribbean, and Cocos plates. However, other authors have concluded that the volcanism of the region is related to the subduction of the Cocos plate beneath the North America plate (Stoiber and Carr, 1973; Thorpe, 1977; Burbach et al., 1984; Bevis and Isacks, 1984; Luhr et al., 1984; García-Palomo et al., 2004).

Finally, the volcanism reappears in Central México at parallel 19°N at the Trans-Mexican Volcanic Belt, extending from the coast of Veracruz to the coast of Nayarit predominantly as calc-alkaline volcanism but with small regions of alkaline volcanism, such as the Colima graben and the Chichinautzin Volcanic Field (Fig. 31). The thickness of the continental crust increases from west to east on the Trans-Mexican Volcanic Belt: at Colima, the crust is 20–22 km thick; at Nevado de Toluca, it is 40 km thick; the crust at Popocatépetl is 47 km thick; and at Pico de Orizaba, it is >50 km thick (Molina-Garza and Urrutia-Fucugauchi, 1993; Urrutia-Fucugauchi and Flores-Ruiz, 1996). These variations are reflected in the $^{87}Sr/^{86}Sr$ versus εNd isotopic relationships of the volcanoes (Nelson et al., 1995; Macías et al., 2003; Martínez-Serrano et al., 2004; Schaaf et al., 2004) (Fig. 32). Colima is the closest volcano to the trench with the most primitive isotopic relationships compared to Nevado de Toluca, Popocatépetl, and Pico de Orizaba, which are located farthest from the trench with a thicker continental crust below them. Because of to this, the volcanoes show clear evidence of crustal contamination, as observed through Osmium isotopes (Chesley et al., 2000; Lassiter and Luhr, 2001) and the presence of crustal xenoliths (Valdez-

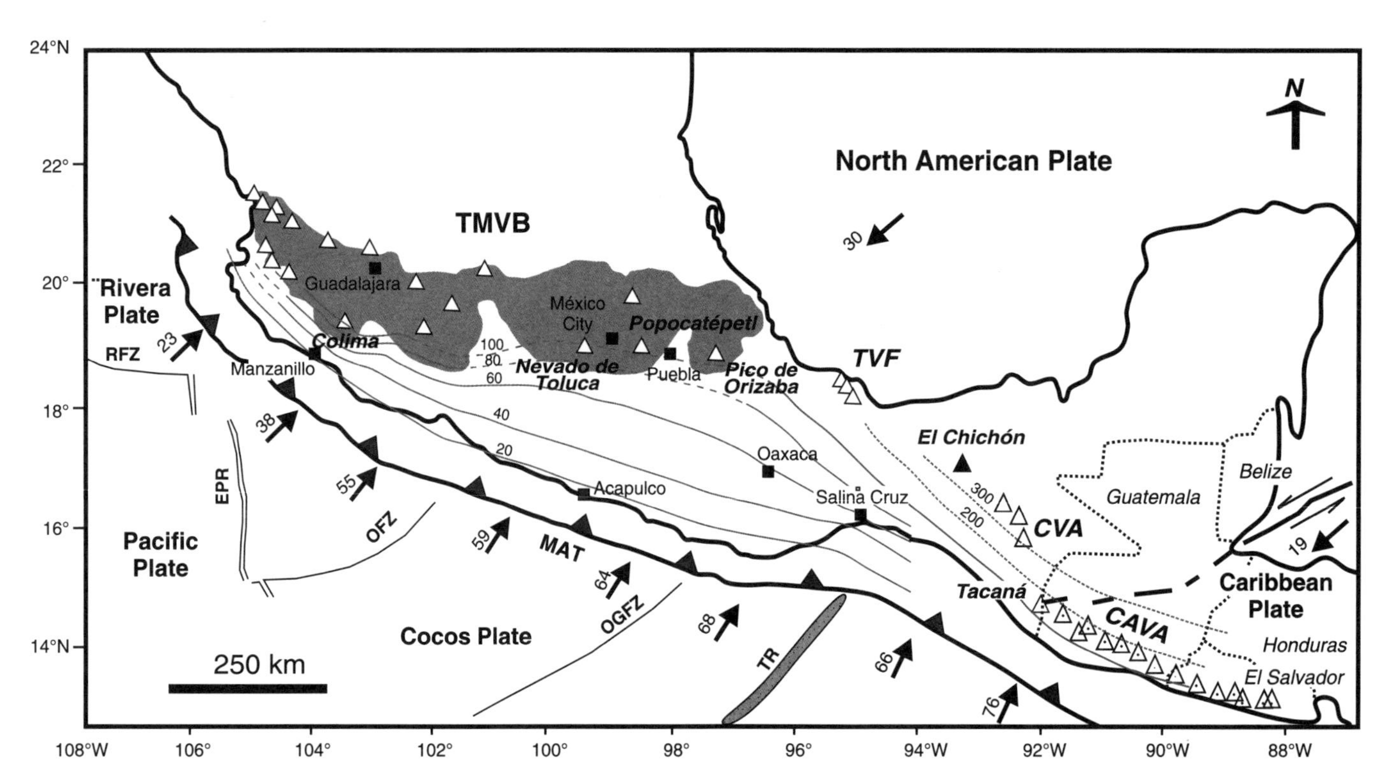

Figure 31. Tectonic setting of the southern portion of México with the location of the Trans-Mexican Volcanic Belt (TMVB), Los Tuxtlas Volcanic Field (TVF), Chiapanecan Volcanic Arc (CVA), and the Central America Volcanic Arc (CAVA) with respect to the plates. The gray lines are the isodepth curves of earthquakes. EPR—East Pacific Rise; MAT—Middle America Trench; OFZ—Orozco Fracture Zone; OGFZ—O'Gorman Fracture Zone; RFZ—Rivera Fracture Zone; TR—Tehuantepec Ridge (after García-Palomo et al., 2004).

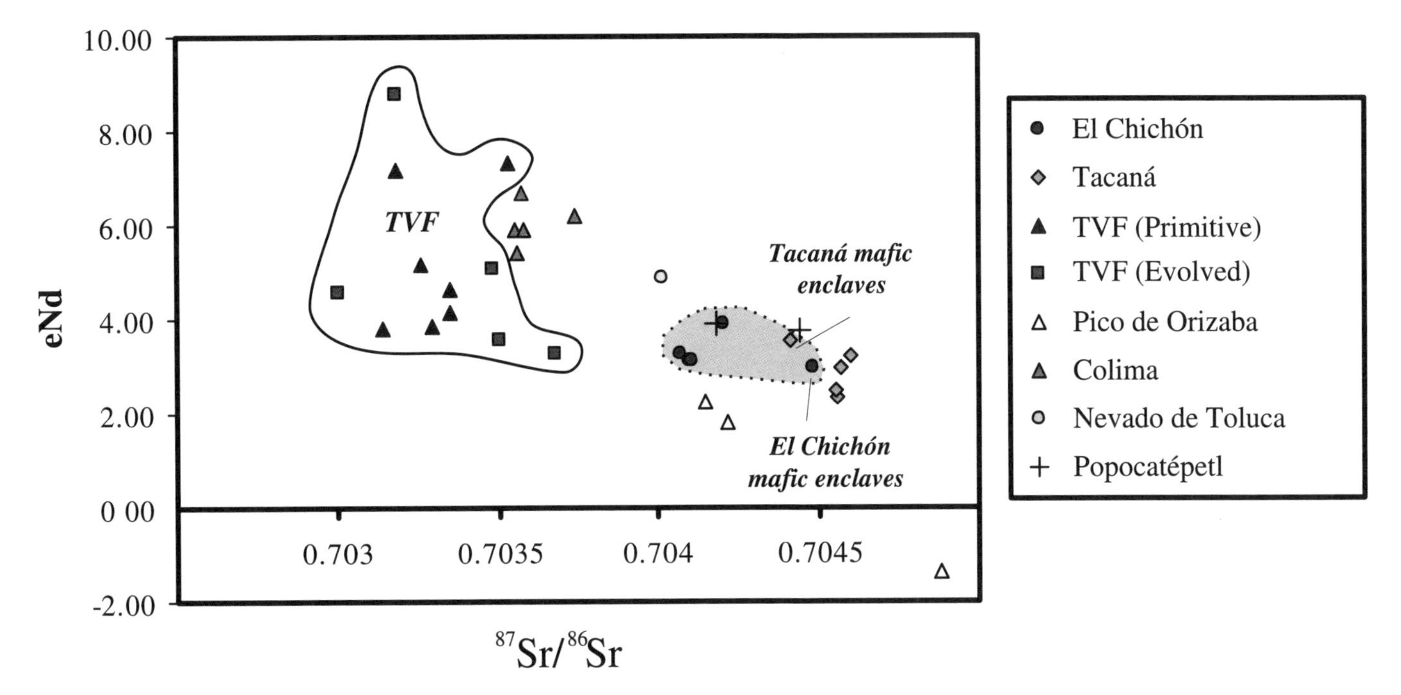

Figure 32. Isotopic ratios of $^{87}Sr/^{86}Sr$ versus epsilon neodymium of rocks from Colima, Nevado de Toluca, Popocatépetl, Pico de Orizaba, Tuxtlas Volcanic Field (TVF), Chichón, and Tacaná volcanoes.

Moreno et al., 2006). The degree of crustal assimilation is related to the age and composition of the continental crust through which these magmas interact as well as the storage time at the base of the crust and their subsequent ascent to the surface. Schaaf et al. (2004) concluded that the primitive isotopic relationships of Colima volcano are due to the presence of a young and primitive continental crust while the more evolved isotopic relationships of Popocatépetl and Pico de Orizaba are due to the presence of the Morelos Formation and Grenvillian rocks, respectively.

In southeastern México, there are volcanic products with different characteristics (Fig. 32). Tacaná volcano is located 100 km from the trench, atop a 40-km-thick continental crust (Rebollar et al., 1999). This volcano has isotopic ratios with a higher degree of crustal assimilation ($^{87}Sr/^{86}Sr$: 0.70441–0.70459; εNd: 2.26–3.57) (Mora et al., 2004), with respect to the isotopic ratios of Chichón volcano (Macías et al., 2003), located 250 km from the trench, and several volcanoes of Los Tuxtlas Volcanic Field, located 350 km from the trench (Nelson et al., 1995). The volcanic rocks erupted in southern México have a calc-alkaline signature with medium K contents, negative anomalies of Nb, Ti, and P, and enrichments in light rare earth elements, typical of subduction environments.

The compositional features (major, trace, rare earth elements, some isotope ratios) of Mexican volcanoes have been established during the past 30 yr. Despite these advances, we still known very little about the processes of magma genesis, the partial fusion of the mantle that occurred at depths of 75 km or deeper, the assimilation and contamination processes at depths of <50 km, magma storage, and other processes (fractional crystallization, magma mixing, etc.) occurring at shallower depths. There have been some recent advances in experimental petrology studies of our volcanoes attempting to understand the pre-eruptive conditions of magmas (i.e., pressure, temperature, and water saturation) of some eruptions. For instance, the magma erupted during the 1998 eruption of Colima volcano had a pre-eruptive temperature of 840–900 °C at depths of 3–7 km below the crater (Mora et al., 2002). During the April 1996 through February 1998 eruptions at Popocatépetl (Straub and Martín-Del Pozzo, 2001), the volcano erupted a hybrid magma originated by the mixing of an olivine-spinel–saturated andesitic magma (55% SiO_2, 1170–1085 °C) probably from the Moho and a clynopyroxene-orthopyroxene-plagioclase dacitic magma (62% SiO_2, ~950 °C) stored between ~4 and 13 km below the crater. Between 14,000 and 10,500 yr B.P., three eruptions occurred at Nevado de Toluca volcano that erupted magmas with the same chemical and mineral association that Arce et al. (2005a) interpreted as the same magma stagnated at a temperature of 814–840 °C and at depths of 6 km beneath the crater. Wallace and Carmichael (1999) concluded that some olivine basalts at the Chichinautzin Volcanic Field had temperatures of 1200–1290 °C prior to eruption. In the eastern Trans-Mexican Volcanic Belt, recent lavas erupted at two vents of Volcancito had pre-eruptive temperatures of 1198 and 1236 °C (calc-alkaline basalt Rio Nolinco) and 1166–1175 °C (hawaiite Toxtlacuaya) (Carrasco-Núñez et al., 2005). All these studies represent a considerable advance in our knowledge of shallow processes; however, our knowledge of deeper processes is still lacking and may improve with the study of mantle xenoliths (Blatter and Carmichael, 1998) and fluid inclusions (gas and melt inclusions) trapped in the magmas. There are already some studies in this field, such as those accomplished for Parícutin volcano (Luhr, 2001) and the Chichinautzin Volcanic Field (Cervantes and Wallace, 2003).

Other studies that could allow better understanding of the magmatic processes use Boron isotopes to estimate the influence of sediments in the subduction process (Hochstaedter et al., 1996), hydrogen isotopes to understand the isotopic characteristics of the magmatic water (Taran et al., 2002), and Sr-Nd-Pb isotopic ratios to evaluate petrologic process and the source of the magmas (Martínez-Serrano et al., 2004; Schaaf et al., 2005; Valdéz-Moreno et al., 2006).

As stated in previous sections, there has been a considerable advance in the study of the stratigraphic record of the Mexican volcanoes, which is necessary to better comprehend their eruptive history and to design hazard maps. In most Mexican volcanoes, the stratigraphic record dates back to 50,000 yr B.P. as covered by the radiocarbon dating method. With this period of time, we can document the eruptive record of Paleofuego and Colima volcano but not the stratigraphic record of Nevado de Colima. In other words, most of our volcanoes began their formation beyond the ^{14}C limit ca. 1–2 Ma. In order to completely understand the eruptive behavior of these volcanoes, other radiometric methods are needed, such as K-Ar, Ar-Ar in minerals and rocks, thermoluminiscense in minerals, ^{36}Cl in polished surfaces of moraine deposits, etc.

Geologic studies (petrology, geochemistry, volcanology, etc.) and the historic record are very important to understanding the past behavior of our active volcanoes and to forecast future activity. For instance, the eruptive record of Popocatépetl shows that during the past 23,000 yr, the volcano has had four major Plinian eruptions and that, therefore, a large future event might be of this type. The eruptive records of Paleofuego and Colima volcanoes indicate that these cones have produced debris avalanches and many small dome collapse events. Minor activity at Colima will generated small-volume pyroclastic flows like those produced during the past 20 yr. It is clear that based upon the eruptive record of a volcano, we must construct hazard maps that show the areas affected by eruptions with different magnitudes in the past. To date, México has hazard maps of Colima, Popocatépetl, and Pico de Orizaba volcanoes and preliminary hazard maps of Tres Vírgenes, Nevado de Toluca, Chichón, and Tacaná. These maps represent a basic source of information for the Civil Protection authorities, who should use them to create preventive emergency plans, education programs to the population, etc. (Macías and Capra, 2005).

The integration of the geologic information (eruptive history, historic record, hazard maps, chemistry of the erupted products, etc.) and the geophysical information (seismology, gravimetry, magnetometry, etc.) is key to understanding the present behavior of our volcanoes. The 1994–present eruption of Popocatépetl has allowed the gathering of this information in order to establish a monitoring network to forecast future events. Despite of all this development, we are not yet able to analyze and prepare a conceptual model of Popocatépetl with the depth, shape, and volume of the magma chamber, ascent times of the magma to the surface, and its relationship to degassing activity at the surface.

During the next decades, volcanologic studies in México should focus on completing the eruptive record of the volcanoes, to determine the magmatic processes that take place at depth, the storage conditions in the upper crust, the chemical evolution of the products, the times of magma ascent to the surface, and to update and create hazards maps of all the active volcanoes. Finally, all of this information should be available to the population living around the volcanoes in order to avoid a future volcanic disaster such as the 1982 eruption of El Chichón. An eruption cannot be stopped, but its consequences can be reduced considerably.

ACKNOWLEDGMENTS

This work has benefited from funding of diverse projects, among them Consejo Nacional de Ciencia y Tecnología (CONACYT, 47226) and Dirección General de Asuntos del Personal Académico de la Universidad Nacional Autónoma de México (DGAPA-UNAM, IN114206-3). Most of the observations and fieldwork carried out in these volcanoes were made with friends and colleagues: J.M. Espíndola at El Chichón and Tacaná volcanoes; C. Siebe at Popocatépetl; R. Saucedo, M.I. Bursik, J.C. Gavilanes, and A. Cortés at Colima volcano; C. Siebe and G. Carrasco at Pico de Orizaba volcano; and A. García, L. Capra, J.L. Arce, R. Saucedo, J.C. Mora, T. Scolamacchia, K. Scott, S. Hughes, and D. Sarocchi at all these volcanoes. All satellite imagery was courtesy of Michael Abrams from the Jet Propulsion Laboratory–National Aeronautics and Space Administration, USA. G. Carrasco and M. Ort provided careful review of this manuscript. I am thankful to C. Limon for producing the geologic maps of the volcanoes, C. Ordaz for looking at dark, old manuscripts in the UNAM library and putting together the reference list, Sara Solís for correcting the final version of the reference list, and A.M. Rocha for her technical support. I am grateful to several people who provided logistic support at these volcanoes, such as E. Segre and A. Fernández at Popocatépetl volcano and S. Hernández at Chichón and Tacaná volcanoes. Finally, I would like to give my most sincere appreciation to Reynaldo Mota, Servando de la Cruz, Robert I. Tilling, Jorge Neyra, R. Díaz, and G. Carrasco for the material they have provided to me during the course of the last decade.

REFERENCES CITED

Aguilera, J.G., and Ordoñez, E., 1895, Expedición Científica al Popocatépetl, México, Oficina de la Secretaría de Fomento: Comisión Geológica Mexicana, 48 p.

Allan, J., 1986, Geology of the northern Colima and Zacoalco grabens, southwest México: Late Cenozoic rifting in the Mexican Volcanic Belt: Geological Society of America Bulletin, v. 97, p. 473–485, doi: 10.1130/0016-7606(1986)97<473:GOTNCA>2.0.CO;2.

Allan, J., and Carmichael, I.S.E., 1984, Lamprophyric lavas in the Colima graben, SW México: Contributions to Mineralogy and Petrology, v. 88, p. 203–216, doi: 10.1007/BF00380166.

Anderson, T., and Flett, J.S., 1903, Report on the eruption of the Soufrière Vincent in 1902: Royal Society of London Philosophical Transactions, v. 200, p. 353–553.

Arce, J.L., 2003, Condiciones pre-eruptivas y evolución de la erupción pliniana Pómez Toluca Superior, volcán Nevado de Toluca [Ph.D. thesis]: México, D.F., Universidad Nacional Autónoma de México, 177 p.

Arce, J.L., Macías, J.L., and Vázquez-Selem, L., 2003, The 10.5 ka Plinian eruption of Nevado de Toluca volcano, Mexico: Stratigraphy and hazard implications: Geological Society of America Bulletin, v. 115, p. 230–248, doi: 10.1130/0016-7606(2003)115<0230:TKPEON>2.0.CO;2.

Arce, J.L., Macías, J.L., Gardner, J.E., and Layer, P.W., 2005a, Petrologic, ^{40}Ar-^{39}Ar and experimental constraints on the Nevado de Toluca magma chamber, Mexico, around 10.5 ka: Journal of Petrology, v. 47, no. 3, p. 457–479.

Arce, J.L., Cervantes, K.E., Macías, J.L., and Mora, J.C., 2005b, The 12.ka Middle Toluca Pumice: A dacitic Plinian–sub-Plinian eruption of Nevado de Toluca in central México: Journal of Volcanology and Geothermal Research, v. 147, p. 125–143, doi: 10.1016/j.jvolgeores.2005.03.010.

Armienta, M.A., De la Cruz-Reyna, S., and Macías, J.L., 2000, Chemical characteristics of the crater lakes of Popocatépetl, El Chichón, and Nevado de Toluca volcanoes, Mexico: Journal of Volcanology and Geothermal Research, v. 97, p. 105–125, doi: 10.1016/S0377-0273(99)00157-2.

Arreola, J.M., 1915, Catálogo de las erupciones antiguas del Volcán de Colima: Memorias de la Sociedad Científica Antonio Alzate, t. 32, p. 443–481.

Atl, Dr., 1939, Volcanes de México, La Actividad del Popocatépetl: México D.F., Polis, 1, 67 p.

Báez-Jorge, F., Rivera-Balderas, A., and Arrieta-Fernández, P., 1985, Cuando ardió el cielo y se quemó la tierra; condiciones socioeconómicas y sanitarias de los pueblos Zoques afectados por la erupción del volcán Chichonal: México: Instituto Nacional Indigenista, Colección INI, Serie de Investigaciones Sociales 14, 310 p.

Bárcena, M., 1887, Informe sobre el estado actual del volcán de Colima: Anales del Ministerio de fomento de México, p. 328–365.

Bergeat, A., 1894, Zur Kenntnis der jungen Eruptivgesteine der Republik Guatemala: Zeitschrift der Deutschen Zeitschrift der Deutschen Geologischen Gesellschaft, v. 46, p. 131–157.

Bevis, M., and Isacks, B.L., 1984, Hypocentral Trend Surface Analysis: Probing the geometry of Benioff Zones: Journal of Geophysical Research, v. 89, B7, p. 6153–6170.

Blatter, D.L., and Carmichael, I.S.E., 1998, Hornblende peridotite xenoliths from central Mexico reveal the highly oxidized nature of subarc upper mantle: Geology, v. 26, p. 1035–1038, doi: 10.1130/0091-7613(1998)026<1035:HPXFCM>2.3.CO;2.

Bloomfield, K., and Valastro, S., 1974, Late Pleistocene eruptive history of Nevado de Toluca volcano, central Mexico: Bulletin of Volcanology, v. 85, p. 901–906.

Bloomfield, K., and Valastro, S., 1977, Late Quaternary tephrochronology of Nevado de Toluca volcano, central Mexico: Overseas Geology and Mineral Resources, 46, 15 p.

Bloomfield, K., Sánchez Rubio, G., and Wilson, L., 1977, Plinian eruptions of Nevado de Toluca: Geologische Rundschau, v. 66, p. 120–146, doi: 10.1007/BF01989568.

Bonet, F., 1971, Espeleología de la región de Cacahuamilpa, Estado de Guerrero, México: Boletín del Instituto de Geología, Universidad Nacional Autónoma de México, 90, 88 p.

Böse, E., 1902, Breve noticia sobre el estado actual del Volcán Tacaná, Chiapas: Memorias y Revista de la Sociedad Científica "Antonio Alzate," v. 18, p. 266–270.

Böse, E., 1903, Los temblores de Zanatepec, Oaxaca, a fines de septiembre de 1902 y el estado actual del volcán Tacaná: Perergones del Instituto Geológico de México, v. 1, no. 1, 25 p.

Böse, E., 1905, Reseña acerca de la geología de Chiapas y Tabasco: Instituto Geológico de México, Boletín 20, 113 p.

Boudal, C., 1985, Petrologie d'un grand volcan andesitique mexicain: Le Popocatépetl, le rôle des mélanges de magmas dans les dynamismes de l'activité récente [Ph.D. thèse]: Clermont-Ferrand, France, U.E.R. de Sciences Exactes et Naturelles de l'Université de Clermont-Ferrand, France, 140 p.

Boudal, C., and Robin, C., 1987, Relations entre dynamismes éruptifs et réalimentations magmatiques d'origine profonde au Popocatépetl: Canadian Journal of Earth Sciences, v. 25, p. 955–971.

Boudal, C., and Robin, C., 1989, Volcan Popocatépetl: Recent eruptive history, and potential hazards and risks in future eruptions, *in* Latter, J.H., ed., Volcanic Hazards, International Association of Volcanology and Chemistry of the Earth's Interior Proceedings in Volcanology: Berlin, Springer-Verlag, 1, p. 110–128.

Bretón, G., Ramírez, J., and Navarro, C., 2002, Summary of the historical eruptive activity of Volcan de Colima México 1519–2000: Journal of Volcanology and Geothermal Research, v. 117, p. 21–46, doi: 10.1016/S0377-0273(02)00233-0.

Burbach, V.G., Frohlich, C., Pennington, D.W., and Matumoto, T., 1984, Seismicity and Tectonics of the subducted Cocos plate: Journal of Geophysical Research, v. 89, B9, p. 7719–7735.

Burkart, B., and Self, S., 1985, Extension and rotation of crustal blocks in northern Central America and effect on the volcanic arc: Geology, v. 13, p. 22–26, doi: 10.1130/0091-7613(1985)13<22:EAROCB>2.0.CO;2.

Caballero, M., 1996, The diatom flora of two acid lakes in central Mexico: Diatom Research, v. 11, p. 227–240.

Caballero, M., Macías, J.L., Lozano-García, S., and Urrutia Fucugauchi, J., 2001, Late Pleistocene-Holocene volcanic stratigraphy and palaeoenvironments of the Upper Lerma Basin, Mexico: International Association of Sedimentologists Special Publication 30, p. 247–261.

Caballero, M., Valadez, F., Ortega-Guerrero, B., Metcalfe, S., Macías, J.L., and Sugiura, Y., 2002, Sta Cruz Atizapan: A record of lake level changes and human occupation in the Upper Lerma Basin, Central Mexico: Palaeogeography, Palaeoclimatology, Palaeoecology, v. 186, p. 217–235.

Calvin, E.M., Kudo, A.M., Brookins, D.G., and Ward, D.B., 1989, Strontium isotope and trace element geochemistry of Pico de Orizaba-Trans-Mexican Volcanic Belt: Comparison phases II and III, *in* Continental magmatism [abs.]: New México Bureau of Mines & Mineral Resources Bulletin, v. 131, p. 38.

Camacho, H., 1925, Resultados de la visita al cráter del Popocatépetl el 9 de marzo de 1922: Anales del Instituto Geológico, v. 2, no. 1–3, p. 38–67.

Campa, M.F., Campos, M., Flores, R., and Oviedo, R., 1974, La secuencia mesozoica volcano-sedimentaria metamorfizada de Ixtapan de la Sal, México-Teloloapan, Guerrero: Boletín de la Sociedad Geológica Mexicana, v. 35, p. 7–28.

Cantagrel, J.M., Robin, C., and Vincent, P., 1981, Les grandes etapes d'evolution d'un volcan andesitique composite: Exemple du Nevado de Toluca: Bulletin of Volcanology, v. 44, p. 177–188.

Cantagrel, J.M., Gourgaud, A., and Robin, C., 1984, Repetitive mixing events and Holocene pyroclastic activity at Pico de Orizaba and Popocatepetl (Mexico): Bulletin of Volcanology, v. 47-4, no. 1, p. 735–748.

Canul, R.F., and Rocha, V.L., 1981, Informe geológico de la zona geotérmica de "El Chichonal," Chiapas: Comisión Federal de Electricidad, Morelia, Michoacán, México, Informe 32-81, 38 p.

Canul, R.F., Razo, A.M., and Rocha, V.L., 1983, Geología e historia volcanológica del volcán Chichonal, Estado de Chiapas, *in* El Volcán Chichonal: Universidad Nacional Autónoma de México, Instituto de Geología, p. 3–22.

Capaccioni, B., Tarán, Y., Tassi, F., Vaselli, O., Mangani, F., and Macías, J.L., 2004, Source conditions and degradation processes of light hydrocarbons in volcanic gases: An example from El Chichón volcano, Chiapas State of México: Chemical Geology, v. 206, p. 81–96, doi: 10.1016/j.chemgeo.2004.01.011.

Capaul, W.A., 1987, Volcanoes of the Chiapas Volcanic Belt, Mexico [Master's thesis]: Houghton, USA, Michigan Technological University, 93 p.

Capra, L., 2000, Colapsos de edificios volcánicos: Transformación de avalanchas de escombros en flujos de escombros cohesivos: Los casos del Monte Santa Elena-EEUU, Nevado de Toluca y Nevado de Colima [Ph.D. thesis]: México, D.F., Universidad Nacional Autónoma de México, 177 p.

Capra, L., and Macías, J.L., 2000, Pleistocene cohesive debris flows at Nevado de Toluca volcano, central Mexico: Journal of Volcanology and Geothermal Research, v. 102, no. 1–2, p. 149–167, doi: 10.1016/S0377-0273(00)00186-4.

Capra, L., and Macías, J.L., 2002, The cohesive Naranjo debris-flow deposit (10 km^3): A dam breakout flow derived from the Pleistocene debris avalanche deposit of Nevado de Colima volcano (Mexico): Journal of Volcanology and Geothermal Research, v. 117, p. 213–235, doi: 10.1016/S0377-0273(02)00245-7.

Capra, L., Macías, J.L., Scott, K.M., Abrams, M., and Garduño-Monroy, V.H., 2002, Debris avalanches and debris flows transformed from collapses in the Trans-Mexican volcanic belt, Mexico—Behavior, and implications for hazard assessment: Journal of Volcanology and Geothermal Research, v. 113, p. 81–110, doi: 10.1016/S0377-0273(01)00252-9.

Capra, L., Poblete, M.A., and Alvarado, R., 2004, The 1997 and 2001 lahars of Popocatépetl volcano (central Mexico): Textural and sedimentological constraints on their origin and hazards: Journal of Volcanology and Geothermal Research, v. 131, p. 351–369, doi: 10.1016/S0377-0273(03)00413-X.

Capra, L., Carreras, L., Arce, J.L., and Macías, J.L., 2006, The Lower Toluca Pumice: A ~21,700 yr B.P, Plinian eruption of Nevado de Toluca volcano, Mexico, *in* Siebe, C., Macías, J.L., and Aguirre-Díaz, G.J., eds., Neogene-Quaternary continental margin volcanism: A perspective from México: Geological Society of America Special Paper 402, Penrose Conference Series, p. 155–173.

Carey, S.N., and Sigurdsson, H., 1986, The 1982 eruptions of El Chichón volcano, Mexico: 2. Observations and numerical modeling of tephra fall distribution: Bulletin of Volcanology, v. 48, p. 127–141, doi: 10.1007/BF01046547.

Carrasco-Núñez, G., 1985, Estudio geológico del Volcán Popocatépetl [B.S. thesis]: México D.F., Facultad de Ingeniería, Universidad Nacional Autónoma de México, 138 p.

Carrasco-Núñez, G., 1993, Structure-eruptive history and some major hazardous events at Citlaltépetl volcano (Pico de Orizaba), México [Ph.D. thesis]: Houghton, USA, Michigan Technological University, 182 p.

Carrasco-Núñez, G., 1997, Lava flow growth inferred from morphometric parameters—A case study of Citlaltépetl volcano, Mexico: Geological Magazine, v. 134, p. 151–162, doi: 10.1017/S0016756897006614.

Carrasco-Núñez, G., 1999, Holocene block-and-ash flows from summit cone activity of Citlaltépetl volcano, eastern Mexico: Journal of Volcanology and Geothermal Research, v. 88, p. 47–66, doi: 10.1016/S0377-0273(98)00110-3.

Carrasco-Núñez, G., 2000, Structure and proximal stratigraphy of Citlaltépetl volcano (Pico de Orizaba), Mexico, *in* Delgado-Granados, H., Aguirre Diaz, G., and Stock, J.M., eds., Cenozoic Volcanism and Tectonics of Mexico: Geological Society of America Special Paper 334, p. 247–262.

Carrasco-Núñez, G., and Ban, M., 1994, Geologic map and structure sections of the summit area of Citlaltépetl volcano, Mexico with summary of the geology of the Citlaltépetl volcano summit area: México D.F., Universidad Nacional Autónoma de México, Instituto de Geología, Cartas Geológicas y Mineras, 9, scale 1:40,000, 1 sheet.

Carrasco-Núñez, G., and Gómez-Tuena, A., 1997, Volcanogenic sedimentation around Citlaltépetl (Pico de Orizaba) volcano and surroundings, Velacruz, México: IAVCEI General Assembly excursions: México, D.F., Universidad Nacional Autónoma de México, Instituto de Geología, Excursión 16, p. 131–157.

Carrasco-Núñez, G., and Rose, W.I., 1995, Eruption of a major Holocene pyroclastic flow at Citlaltépetl volcano (Pico de Orizaba), Mexico, 8.5–9.0 ka: Journal of Volcanology and Geothermal Research, v. 69, p. 197–215, doi: 10.1016/0377-0273(95)00023-2.

Carrasco-Núñez, G., Vallance, J.W., and Rose, W.I., 1993, A voluminous avalanche-induced lahar from Citlaltépetl volcano, Mexico: Implications for hazard assessment: Journal of Volcanology and Geothermal Research, v. 59, p. 35–46, doi: 10.1016/0377-0273(93)90076-4.

Carrasco-Núñez, G., Righter, K., Chesley, J., Siebert, L., and Aranda-Gomez, J.J., 2005, Contemporaneous eruption of calc-alkaline and alkaline lavas in a continental arc (Eastern Mexican Volcanic Belt): Chemically heterogeneous but isotopically homogeneous source: Contributions to Mineralogy and Petrology, v. 150, no. 4, p. 423–440, doi: 10.1007/s00410-005-0015-x.

Carrasco-Núñez, G., Díaz, R., Siebert, L., Hubbard, B., Sheridan, M., and Rodríguez, S., 2006, Multiple edifice-collapse events in the Eastern Mexican Volcanic Belt: The role of sloping substrate and implications for hazard assessment: Journal of Volcanology and Geothermal Research, vol. 158, p. 151–176.

Carroll, M.R., and Rutherford, M.J., 1987, The stability of igneous anhydrite: Experimental results and implications for sulfur behavior in the 1982 El Chichon trachyandesite and other evolved magmas: Journal of Petrology, v. 28, p. 781–801.

Casadevall, T.J., De la Cruz-Reyna, S., Rose, W.I., Bagley, S., Finnegan, D.L., and Zoller, W.H., 1984, Crater Lake and Post-Eruption Hydrothermal Activity, El Chichon Volcano, Mexico: Journal of Volcanology and Geothermal Research, v. 23, p. 169–191, doi: 10.1016/0377-0273(84)90061-1.

Cebull, S.E., and Shurbet, D.H., 1987, Mexican Volcanic Belt: An intraplate transform?: Geofísica International, v. 26, p. 1–14.

Cervantes, P., and Wallace, P., 2003, Magma degassing and basaltic eruption styles: A case study of ~2000 year B.P. Xitle volcano in central Mexico: Journal of Volcanology and Geothermal Research, v. 120, p. 249–270, doi: 10.1016/S0377-0273(02)00401-8.

Chesley, J.T., Ruiz, J., and Righter, K., 2000, Source versus crustal contamination in arc magmatism: Evidence for lower crustal assimilation in the Trans-Mexican Volcanic Belt: Eos (Transactions, American Geophysical Union), Fall Meeting, San Francisco, USA, v. 81, no. 48, p. F1269.

Connor, C., Luhr, J., and Martin del Pozzo, A., 1993, Structure and petrology of the March 1991 dome, volcan Colima, México: Possible transition toward explosive eruption: Geophysical Research Letters, v. 98, no. 11, p. 19713–19722.

Cortés, A., and Navarro-Ochoa, C., 1992, Las avalanchas volcánicas del complejo volcánico de Colima: Una nueva y desconcertante datación C-14: Unión Geofísica Mexicana, 12(5), 66 p.

Cortés, A., Garduño, V.H., Navarro-Ochoa, C., Komorowski, J.C., Saucedo, R., Macías, J.L., and Gavilanes, J.C., 2005, Carta geológica del Complejo Volcánico de Colima, con Geología del Complejo Volcánico de Colima: México, D.F., Universidad Nacional Autónoma de México, Instituto de Geología, Cartas Geológicas y Mineras 10, scale 1:100,000, 1 sheet, 37 p. text.

Cortés, C.A., 2002, Depósitos de avalancha de escombros originados hace 3,600 años por el colapso del Volcán de Colima [M.S. thesis]: México, D.F., Posgrado en Ciencias de la Tierra, Instituto de Geofísica, Universidad Nacional Autónoma de México, 135 p.

Crausaz, W., 1994, Pico de Orizaba o Citlaltépetl: Geology, archaeology, history, natural history and mountaineering routes: Ohio, USA, Geopress International, 594 p.

Damon, P., and Montesinos, E., 1978, Late Cenozoic volcanism and metallogenesis over an active Benioff Zone in Chiapas, México: Arizona Geological Society Digest, v. 11, p. 155–168.

Davidson, J., Tepley, F., Palacz, Z., and Meffan-Main, S., 2001, Magma recharge, contamination and residence times revealed by in situ laser ablation isotopic analysis of feldspar in volcanic rocks: Earth and Planetary Science Letters, v. 184, p. 427–442, doi: 10.1016/S0012-821X(00)00333-2.

De Cserna, Z., Aranda-Gómez, J.J., and Mitre Salazar, L.M., 1988, Mapa fotogeológico preliminar y secciones estructurales del volcán Tacaná: Universidad Nacional Autónoma de México, Instituto de Geología, Cartas Geológicas y Mineras 7, scale 1:50,000, 1 Sheet.

De la Cruz, V., and Hernández, R., 1985, Estudio geológico a semidetalle de la zona geotérmica del Volcán Tacaná, Chiapas: México, Comisión Federal de Electricidad, 41/85, 28 p.

De la Cruz-Reyna, S., 1993, The historical eruptive activity of Colima Volcano, México: Journal of Volcanology and Geothermal Research, v. 55, p. 51–68, doi: 10.1016/0377-0273(93)90089-A.

De la Cruz-Reyna, S., and Carrasco-Núñez, G., 2002, Probabilistic hazard analysis of Citlaltépetl (Pico de Orizaba) Volcano, eastern Mexican Volcanic Belt: Journal of Volcanology and Geothermal Research, v. 113, p. 307–318, doi: 10.1016/S0377-0273(01)00263-3.

De la Cruz-Reyna, S., Armienta, M.A., Zamora, V., and Juárez, F., 1989, Chemical changes in spring waters at Tacaná Volcano, Chiapas, México: Journal of Volcanology and Geothermal Research, v. 38, p. 345–353, doi: 10.1016/0377-0273(89)90047-4.

De la Cruz-Reyna, S., Quezada, J.L., Peña, C., Zepeda, O., and Sánchez, T., 1995, Historia de la actividad del Popocatépetl (1354–1995), Volcán Popocatépetl: Estudios realizados durante la crisis de 1994–1995: Centro Nacional de Prevención de Desastres, Universidad Nacional Autónoma de México, 3–22.

Demant, A., 1979, Vulcanología y petrografía del sector occidental del Eje Neovolcanico: Revista Instituto de Geología, Universidad Nacional Autónoma de México, v. 3, p. 39–57.

Demant, A., 1981, L'axe Néo-volcanique Transmexicain Etude Volcanologique et Pétrografique Significacion Géodynamique [Ph.D. thesis]: France, Université de Marseille, 459 p.

DeMets C., and Stein, R.G., 1991, Present-day kinematics of the Rivera plate and implications for tectonics in southwestern Mexico: Journal of Geophysical Research, v. 95, no. 21, p. 21,931–21,948.

Devine, J.D., Sigurdsson, H., Davis, A.N., and Self, S., 1984, Estimates of sulfur and chlorine yield to the atmosphere from volcanic reports and potential climatic effects: Journal of Geophysical Research, v. 89, no. B7, p. 6309–6325.

Díaz, S., 1906, Efemérides del volcán de Colima según las observaciones practicadas en los observatorios de Zapotlán y Colima de 1893 a 1905: Imprenta y Fototipia de la Secretaria de Fomento, 199 p.

Díaz-Castellón, R., 2003, Análisis de la estabilidad de edificios volcánicos del flanco oriental de la Sierra de Citlaltépetl-Cofre de Perote [Master's thesis]: México D.F., Universidad Nacional Autónoma de México, 135 p.

Dollfus, A., and Monserrat, E. de, 1867, Arcive de la commission scientifique du Mexique Ministére de L´instruction Publique: El Renacimiento, v. 1, p. 451–457.

Domínguez, T., Zobin, V.M., and Reyes-Davila, G.A., 2001, The fracturing in volcanic edifice before an eruption: The June–July 1998 high frequency earthquake swarm at Volcán de Colima, México: Journal of Volcanology and Geothermal Research, v. 105, p. 65–75, doi: 10.1016/S0377-0273(00)00243-2.

Duffield, W.A., Tilling, R.I., and Canul, R., 1984, Geology of El Chichón volcano, Chiapas, México: Journal of Volcanology and Geothermal Research, v. 20, p. 117–132, doi: 10.1016/0377-0273(84)90069-6.

Espinasa-Pereña, R., and Martín-del Pozzo, A.L., 2006. Morphostratigraphic evolution of Popocatépetl volcano, México, *in* Siebe, C., Macías, J.L., Aguirre-Díaz, G., eds., Neogene-Quaternary continental margin volcanism: A perspective from México, Geological Society of America Special Paper 402, Penrose Conference Series, p. 115–137.

Espíndola, J.M., Medina, F.M., and De los Ríos, M., 1989, A C-14 age determination in the Tacaná volcano (Chiapas, México): Geofísica Internacional, v. 28, p. 123–128.

Espíndola, J.M., Macías, J.L., and Sheridan, M.F., 1993, El Volcán Tacaná: Un ejemplo de los problemas en la evaluación del Riesgo Volcánico, *in* Proceedings, Simposio Internacional sobre Riesgos Naturales e Inducidos en los Grandes Centros Urbanos de America Latina: México D.F., Centro Nacional de Prevención de Desastres, p. 62–71.

Espíndola, J.M., Macías, J.L., Tilling, R.I., and Sheridan, M.F., 2000, Eruptive history of el Chichón volcano (Chiapas, México) and its impact on human activity: Bulletin of Volcanology, v. 62, p. 90–104, doi: 10.1007/s004459900064.

Espíndola, J.M., Macías, J.L., Godínez, L., and Jiménez, Z., 2002, La erupción de 1982 del Volcán Chichónal, Chiapas, México, *in* Lugo, H.J., and Inbar, M., eds., Desastres Naturales en América Latina: México, D.F., Fondo de Cultura Económica, p. 37–65.

Felix, J., and Lenk, H., 1890, Beitraege zur Geologie und Palaeontologie: Leipzig: A. Felix; Stuttgart: E. Schweizerbartsche Verlagshandlung, 142 p.

Flores, T., 1906, Le Xinantecatl ou Volcan Nevado de Toluca: México, D.F., Rep. 10th Session of International Geological Congress, Excursion Guide, México, 9, p. 1–22.

Flores-Covarrubias, L.C., 1945, Cálculos para la determinación de la altura del cono del volcán del Parícutin, *in* El Paricutín: México, D.F., Universidad Nacional Autónoma de México, 19–20.

Ford, A., and Rose, W.I., 1995, Volcanic ash in ancient Maya ceramics of the limestone lowlands, implications for prehistoric volcanic activity in the Guatemala highlands: Journal of Volcanology and Geothermal Research, v. 66, p. 149–162, doi: 10.1016/0377-0273(94)00068-R.

Foshag, W.F., and Gonzalez-Reyna, J., 1956, Birth and development of Parícutin volcano: U.S. Geological Survey Bulletin, v. 965-D, p. 355–489.

Friedlaender, I., 1921, La erupción del Popocatépetl: Memorias y Revista de la Sociedad: Antonio Alzate, v. 40, p. 219–227.

García-Martínez, B., 2000, Los nombres del Nevado de Toluca: Arqueología Mexicana, v. 8, no. 43, p. 24–26.

García-Palomo, A., 1998, Evolución estructural en las inmediaciones de Volcán Nevado de Toluca, Estado de México [Master's thesis]: México, D.F., Universidad Nacional Autónoma de México, 150 p.

García-Palomo, A., Macías, J.L., and Garduño, V.H., 2000, Miocene to Holocene structural evolution of the Nevado de Toluca volcano region, central Mexico: Tectonophysics, v. 318, p. 281–302, doi: 10.1016/S0040-1951(99)00316-9.

García-Palomo, A., Macías, J.L., Arce, J.L., Capra, L., Espíndola, J.M., and Garduño, V.H., 2002, Geology of Nevado de Toluca volcano and surrounding areas, central Mexico: Geological Society of America Map and Chart Series 89, 1 sheet, 26 p. text.

García-Palomo, A., Macías, J.L., and Espíndola, J.M., 2004, Strike-slip faults and K-Alkaline volcanism at El Chichón volcano, southeastern México: Journal of Volcanology and Geothermal Research, v. 136, p. 247–268, doi: 10.1016/j.jvolgeores.2004.04.001.

García-Palomo, A., Macías, J.L., Arce, J.L., Mora, J.C., Hughes, S., Saucedo, R., Espíndola, J.M., Escobar, R., and Layer, P., 2006, Geological evolution of the Tacaná Volcanic Complex, México-Guatemala, *in* Rose, W.I., Bluth, G.J.S., Carr, M.J., Ewert, J., Patino, L.C., and Vallance, J., eds., Volcanic hazards in Central America: Geological Society of America Special Paper 412, p. 39–57, doi: 10.1130/2006.2412(03).

García-Tenorio, F., 2002, Estratigrafía y petrografía del complejo volcánico Iztaccíhuatl [B.S. thesis]: México, D.F., Instituto Politécnico Nacional, 149 p.

Garduño, M.V., and Tibaldi, A., 1990, Kinematic evolution of the continental active triple junction of Western Mexican Volcanic Belt: Compte Rendus Academie Sciences, Paris, t. 312, 1–6.

Garduño, V.H., Saucedo, R., Jiménez, S., Gavilanes, J.C., Cortés, A., and Uribe, R.M., 1998, La Falla Tamazula, limite suroriental del bloque Jalisco, y sus relaciones con el complejo volcánico de Colima, México: Revista Mexicana de Ciencias Geológicas, v. 15, p. 132–144.

Gómez-Tuena, A., and Carrasco-Núñez, G., 1999, Fragmentation, transport and deposition of a low-grade ignimbrite: the Citlaltépetl ignimbrite, eastern Mexico: Bulletin of Volcanology, v. 60, no. 6, p. 448–464, doi: 10.1007/s004450050245.

González-Salazar, A., 1973, Informe preliminar de la zona Geotérmica del Volcán Chichonal, Chiapas: México, Comisión Federal de Electricidad, Reporte Interno.

Guzmán-Speziale, M., Pennington, W.D., and Matumoto, T., 1989, The triple junction of the North America, Cocos, and Caribbean Plates: Seismicity and tectonics: Tectonics, v. 8, p. 981–999.

Heine, K., 1976, Blockgletscher und blockzungen generationen am Nevado de Toluca: Mexiko: Die Erde, v. 107, p. 330–352.

Heine, K., 1988, Late Quaternary Glacial Chronology of the Mexican Volcanoes: Die Geowissenschaften, v. 6, p. 197–205.

Heine, K., and Heide-Weise, H., 1973, Jungquartäre Förderfolgen des Malinche-Vulkans und des Popocatépetl (Sierra Nevada, México) und ihre Bedeutung fur die Glazialgeologie, Palaeoklimatologie und Archaeologie: Münstersche Forschungen in der Paläontologie und Geologie, v. 31/32, p. 303–322.

Hochstaedter, A.G., Ryan, J.G., Luhr, J.F., and Hasenaka, T., 1996, On B/Be ratios in the Mexican Volcanic Belt: Geochimica et Cosmochimica Acta, v. 60, no. 4, p. 613–628, doi: 10.1016/0016-7037(95)00415-7.

Hoskuldsson, A., 1992, Le complexe volcanique Pico de Orizaba-Sierra Negra-Cerro Las Cumbres (Sud-Est Mexicain): Structure, dymamismes eruptifs et evaluations del areas [Ph.D. thesis]: Clermont-Ferrand, France, Université Blaise Pascal, 210 p.

Hoskuldsson, A., and Robin, C., 1993, Late Pleistocene to Holocene eruptive activity of Pico de Orizaba, eastern Mexico: Bulletin of Volcanology, v. 55, p. 571–587, doi: 10.1007/BF00301810.

Hoskuldsson, A., Robin, C., and Cantagrel, J.M., 1990, Repetitive debris avalanche events at Volcano Pico de Orizaba, Mexico, and their implications for future hazard zones: *in* International Association of Volcanology and Chemistry of the Earth's Interior International Volcanological Congress, Mainz, Germany, Abstract volume, 47 p.

Hovey, E.O., 1907, Volcanoes of Colima, Toluca, and Popocatépetl: Annals of the New York Academy of Sciences, 25, 646 p.

Humboldt, A., 1862, Cosmos: A sketch of the physical description of the universe: London, H.G. Bohn, 4, 575 p.

Iverson, R.M., Schilling, S.P., and Vallance, J.W., 1998, Objective delineation of lahar inundation hazard zones: Geological Society of America Bulletin, v. 110, p. 972–984, doi: 10.1130/0016-7606(1998)110<0972:ODOLIH>2.3.CO;2.

Jiménez, Z., Reyes, G., and Espíndola, J.M., 1995, The July 1994 episode of seismic activity at Colima Volcano, Mexico: Journal of Volcanology and Geothermal Research, Short Communication, v. 64, p. 321–326, doi: 10.1016/0377-0273(94)00118-Z.

Jiménez, Z., Espíndola, V.H., and Espíndola, J.M., 1999, Evolution of the Seismic Activity from the 1982 Eruption of El Chichón Volcano, Chiapas, México: Bulletin of Volcanology, v. 61, p. 411–422, doi: 10.1007/s004450050282.

Kolisnik, A.M., 1990, Phenocryst zoning and heterogeneity in andesites and dacites of Volcán Popocatépetl, México [M.S. thesis]: Kingston, Ontario, Canada, Queen's University, 247 p.

Komorowski, J.C., Navarro, C., Cortés, A., Siebe, C., and Rodríguez-Elizarrarás, S., 1993. Multiple collapse of Volcán Colima, México, since 10000 yr BP—Implications for eruptive style, magma yield, edifice stability and volcanic risk: International Association of Volcanology and Chemistry of the Earth's Interior, General Assembly, Camberra, Australia, abstract, 60 p.

Komorowski, J.C., Navarro, C., Cortés, A., Saucedo, R., and Gavilanes, J.C., 1997, The Colima Complex: Quaternary multiple debris avalanche deposits, historical pyroclastic sequences (pre-1913, 1991 and 1994), International Association of Volcanology and Chemistry of the Earth's Interior, Puerto Vallarta, México, 1997, Plenary Assembly, Excursion guidebook: Guadalajara, Jalisco, Gobierno del Estado de Jalisco, Secretaría General, Unidad Editorial, p. 1–38.

Kover, T., 1995, Application of a digital terrain model for the modelling of volcanic flows: A tool for volcanic hazard determination [Master's thesis]: Buffalo, State University of New York at Buffalo, 62 p.

Krueger, A.J., 1983, Sighting of El Chichon sulfur dioxide clouds with the Nimbus 7 Total Ozone Mapping Spectrometer: Science, v. 220, p. 1377–1378.

Kudo, A.M., Jackson, M.E., and Husler, J.M., 1985, Phase chemistry of recent andesite, dacite, and rhyodacite of volcan Pico de Orizaba, Mexican Volcanic Belt: Evidence for xenolithic contamination: Geofisica Internacional, v. 24, p. 679–689.

Lange, R., and Carmichael, I.S., 1991, A potassic volcanic front in western México: Lamprophyric and related lavas of San Sebastian: Geological Society of America Bulletin, v. 103, p. 928–940, doi: 10.1130/0016-7606(1991)103<0928:APVFIW>2.3.CO;2.

Lassiter, J., and Luhr, J., 2001, Osmium abundance and isotope variations in mafic Mexican volcanic rocks: Evidence for crustal contamination and constraints on the geochemical behavior of osmium during partial melting and fractional crystallization: Geochemistry Geophysics Geosystems, v. 2, paper number 2000GC000116.

Lowe, G.W., Lee, T.A., and Martínez-Espinosa, E., 1982, Izapa: An Introduction to the Ruins and Monuments: Provo, Utah, Papers of the New World Archaeological Foundation, 349 p.

Lozano-García, S., Sosa-Najera, S., Sugiura, Y., and Caballero, M., 2005, 23,000 yr of vegetation history of the Upper Lerma a tropical high altitude basin in Central Mexico: Quaternary Research, v. 64, p. 70–82, doi: 10.1016/j.yqres.2005.02.010.

Luhr, J.F., 1990, Experimental phase relations of water-and sulfur-saturated arc magmas and the 1982 eruptions of El Chichon volcano: Journal of Petrology, v. 31, p. 1071–1114.

Luhr, J.F., 2001, Glass inclusions and melt volatile contents at Parícutin Volcano, México: Contributions to Mineralogy and Petrology, v. 142, p. 261–283.

Luhr, J., 2002, Petrology and geochemistry of the 1991, 1998–1999 lava flows from the Volcan de Colima, México: Implications for the end of the current eruptive cycle: Journal of Volcanology and Geothermal Research, v. 117, p. 169–194, doi: 10.1016/S0377-0273(02)00243-3.

Luhr, J.F., and Carmichael, I.S.E., 1980, The Colima Volcanic Complex, México. Part I. Post-Caldera andesites from Volcán Colima: Contributions to Mineralogy and Petrology, v. 71, p. 343–372, doi: 10.1007/BF00374707.

Luhr, J.F., and Carmichael, I.S.E., 1981, The Colima Volcanic Complex, México. Part II, Late Quaternary cinder cones: Contributions to Mineralogy and Petrology, v. 76, no. 2, p. 127–147, doi: 10.1007/BF00371954.

Luhr, J.F., and Carmichael, I.S.E., 1982, The Colima Volcanic Complex, México. Part III. Ash and scoria fall deposits from the upper slopes of Volcán Colima: Contributions to Mineralogy and Petrology, v. 80, p. 262–275, doi: 10.1007/BF00371356.

Luhr, J., and Carmichael, I.S.E., 1990a, Geology of Volcán de Colima: Boletín del Instituto de Geología, no.107, p. 101, + plates.

Luhr, J.F., and Carmichael, I.S.E., 1990b, Petrological monitoring of cyclic eruptive activity at Volcán Colima, México: Journal of Volcanology and Geothermal Research, v. 42, p. 235–260, doi: 10.1016/0377-0273(90)90002-W.

Luhr, J.F., and Navarro-Ochoa, C., 2002, Excursión al Volcán Nevado de Colima, *in* VIII Reunión Internacional, Volcán de Colima, Enero, Abstracts, p. 22–25.

Luhr, J.F., and Prestegaard, K.L., 1988, Caldera formation at Volcán de Colima, México, by a large Holocene volcanic debris avalanche: Journal of Volcanology and Geothermal Research, v. 35, p. 335–348, doi: 10.1016/0377-0273(88)90027-3.

Luhr, J.F., and Simkin, T., 1993, Parícutin—The volcano born in a Mexican cornfield: Smithsonian Institution, 427 p.

Luhr, J.F., Carmichael, I.S.E., and Varekamp, J.C., 1984, The 1982 eruptions of El Chichón Volcano, Chiapas, Mexico: Mineralogy and petrology of the anhydrite-bearing pumices: Journal of Volcanology and Geothermal Research, v. 23, p. 69–108, doi: 10.1016/0377-0273(84)90057-X.

Luhr, J.F., Nelson, S.A., Allan, J.F., and Carmichael, I.S.E., 1985, Active rifting in southwestern Mexico: Manifestations of an incipient eastward spreading-ridge jump: Geology, v. 13, p. 54–57, doi: 10.1130/0091-7613(1985)13<54:ARISMM>2.0.CO;2.

Luhr, J.F., Allan, J.F., Carmichael, I.S.E., Nelson, S.A., and Asean, T., 1989, Primitive calc-alkaline and alkaline rock types from the Western Mexican Volcanic Belt: Journal of Geophysical Research, v. 94, no. 4, p. 4515–4530.

Macías, J.L., 1994, Violent short-lived eruptions from small-size volcanoes: El Chichón, Mexico (1982) and Shtyubel', Russia (1907) [Ph.D. thesis]: Buffalo, State University of New York at Buffalo, 193 p.

Macías, J.L., and Capra, L., 2005, Los volcanes y sus peligros: Situación actual en México y Latinoamérica: México D.F., Fondo de Cultura Económica, 132 p.

Macías, J.L., and Siebe, C., 2005, Popocatépetl's crater filled to the brim: Significance for hazard evaluation: Journal of Volcanology and Geothermal Research, v. 141, p. 327–330, doi: 10.1016/j.jvolgeores.2004.10.005.

Macías, J., Capaccioni, B., Conticelli, S., Giannini, M., Martini, M., and Rodríguez, S., 1993, Volatile elements in alkaline and calc-alkaline rocks from the Colima graben, Mexico: Constraints on their genesis and evolution: Geofísica Internacional, v. 32, no. 4, p. 575–589.

Macías, J.L., Carrasco, G., Delgado, H., Martin Del Pozzo, A.L., Siebe, C., Hoblitt, R., Sheridan, M.F., and Tilling, R.I., 1995, Mapa de peligros volcánicos del Popocatépetl: Universidad Nacional Autónoma de México, Instituto de Geofísica, scale 1:25,000, 1 sheet.

Macías, J.L., García-Palomo, A., Arce, J.L., Siebe, C., Espíndola, J.M., Komorowski, J.C., and Scott, K.M., 1997a, Late Pleistocene–Holocene cataclysmic eruptions at Nevado de Toluca and Jocotitlán volcanoes, central Mexico, *in* Kowallis, B.J., ed., Proterozoic to recent stratigraphy, tectonics, and volcanology, Utah, Nevada, southern Idaho and central Mexico: Brigham Young University, Geology Studies, p. 493–528.

Macías, J.L., Espíndola, J.M., Taran, Y., and García, P.A., 1997b, Explosive volcanic activity during the last 3,500 years at El Chichón volcano, Mexico, *in* International Association of Volcanology and Chemistry of the Earth's Interior, Puerto Vallarta, México Plenary Assembly, Fieldtrip guidebook: Guadalajara, Jalisco, Gobierno del Estado de Jalisco, Secretaria General, Unidad Editorial, p. 1–53.

Macías, J.L., Sheridan, M.F., and Espíndola, J.M., 1997c, Reappraisal of the 1982 eruptions of El Chichón volcano, Chiapas, Mexico: New data from proximal deposits: Bulletin of Volcanology, v. 58, p. 459–471, doi: 10.1007/s004450050155.

Macías, J.L., Espíndola, J.M., Bursik, M., and Sheridan, M.F., 1998, Development of lithic-breccias in the 1982 pyroclastic flow deposits of El Chichón Volcano, Mexico: Journal of Volcanology and Geothermal Research, v. 83, p. 173–196, doi: 10.1016/S0377-0273(98)00027-4.

Macías, J.L., Espíndola, J.M., García-Palomo, A., Scott, K.M., Hughes, S., and Mora, J.C., 2000, Late Holocene Peléan style eruption at Tacaná Volcano, Mexico-Guatemala: Past, present, and future hazards: Geological Society of America Bulletin, v. 112, no. 8, p. 1234–1249, doi: 10.1130/0016-7606(2000)112<1234:LHPASE>2.3.CO;2.

Macías, J.L., Arce, J.L., Mora, J.C., Espíndola, J.M., Saucedo, R., and Manetti, P., 2003, The ~550 bp Plinian eruption of el Chichon volcano, Chiapas, Mexico: Explosive volcanism linked to reheating of a magma chamber: Journal of Geophysical Research, v. 108, B12, p. 2569, doi: 10.1029/2003JB002551.

Macías, J.L., Capra, L., Scott, K.M., Espíndola, J.M., García-Palomo, A., and Costa, J.E., 2004a, The May 26, 1982, breakout flow derived from failure of a volcanic dam at El Chichón Volcano, Chiapas, Mexico: Geological Society of America Bulletin, v. 116, p. 233–246, doi: 10.1130/B25318.1.

Macías, J.L., Arce, J.L., Mora, J.C., and García-Palomo, A., 2004b, The Agua Caliente Debris Avalanche deposit a northwestern sector collapse of Tacaná volcano, México-Guatemala: International Association of Volcanology and Chemistry of the Earth's Interior General Assembly, Pucon, Chile, Symposium 11a-07.

Macías, J.L., Saucedo, R., Gavilanes, J.C., Varley, N., Velasco-García, S., Bursik, M., Vargas-Gutiérrez, V., and Cortes, A., 2006, Flujos piroclásticos asociados a la actividad explosive del Volcán de Colima y perspectivas futures: GEOS (Boletín Informativo de la Unión Geofísica Mexicana, A.C.), v. 25, no. 3, p. 340–351.

Martín Del Pozzo, A.L., Romero, V.H., and Ruiz-Kitcher, R.E., 1987, Los flujos piroclásticos del Volcán de Colima, México: Geofísica Internacional, v. 26, no. 2, p. 291–307.

Martín Del Pozzo, A.L., Sheridan, M.F., Barrera, D., Hubp, J.L., and Vázquez, L., 1995, Mapa de peligros, Volcán de Colima, México: Universidad Nacional Autónoma de México, Instituto de Geofísica, scale 1:25,000, 1 sheet.

Martín-Del Pozzo, A.L., Espinasa-Pereña, R., Lugo, J., Barba, I., López, J., Plunket, P., Uruñuela, G., and Manzanilla, L., 1997, Volcanic impact in central México: Puerto Vallarta, México, International Association of Volcanology and Chemistry of the Earth's Interior General Assembly Field Trip Guidebook, 31 p.

Martínez-Serrano, R.G., Schaaf, P., Solís-Pichardo, G., Hernández-Bernal, M.S., Hernández-Treviño, T., Morales-Contreras, J.J., and Macías, J.L., 2004, Sr, Nd, and Pb isotope and geochemical data from the Quaternary Nevado de Toluca volcano, a source of recent adakitic magmatism, and the Tenango volcanic field, México: Journal of Volcanology and Geothermal Research, v. 138, p. 77–110, doi: 10.1016/j.jvolgeores.2004.06.007.

Martíni, M., Capaccioni, B., and Giannini, L., 1987, Ripresa dell'attivita sismica e fumarolica al Vulcano di Tacana (Chiapas, Messico) dopo un quarantennio di quiescenza: Bollettino del Grupo Nazionale per la Vulcanologia, p. 467–470.

Matson, M., 1984, The 1982 El Chichon volcanic eruptions—A satellite perspective: Journal of Volcanology and Geothermal Research, v. 23, p. 1–10, doi: 10.1016/0377-0273(84)90054-4.

McGee, J.J., Tilling, R.I., and Duffield, W.A., 1987, Petrologic characteristics of the 1982 and pre-1982 eruptive products of El Chichón volcano, Chiapas, México: Geofisica Internacional, v. 26, p. 85–108.

Medina, H.A., 1985, Geoquímica de aguas y gases del Volcán Tacaná, Chiapas, Geotermia: Revista Mexicana de Geoenergía, v. 2, p. 95–110.

Medina-Martínez, F., 1982, El Volcán Chichón: GEOS, Boletín de la Unión Geofísica Mexicana, v. 2, no. 4, p. 19.

Medina-Martínez, F., 1983, Analysis of the eruptive history of the Volcán de Colima, México, 1560–1980: Geofísica Internacional, v. 22, p. 157–178.

Medina-Martínez, F., 1986, Análisis de las columnas eruptivas del volcán Chichón, marzo-abril, 1982, Velocidad de salida, presión de la cámara magmática y energía cinética asociada: Geofísica Internacional, v. 25, p. 233–249.

Meneses-Rocha, J.J., 2001, Tectonic evolution of the Ixtapa Graben, an example of a strike-slip basin of southeastern México: Implications for regional petroleum systems, *in* Bartolini, C., Buffler, R.T., and Cantú-Chapa, A., eds., The western Gulf of México Basin: Tectonics, Sedimentary Basins, and Petroleum Systems: American Association of Professional Geologists Memoir 75, p. 183–216.

Mercado, R., and Rose, W.I., 1992, Reconocimiento geológico y evaluación preliminar de peligrosidad del Volcán Tacaná, Guatemala/México: Geofísica Internacional, v. 31, p. 205–237.

Metcalfe, S.E., Street-Perrott, F.A., Perrott, F.A., and Harkness, D.D., 1991, Palaeolimnology of the upper Lerma Basin, central Mexico: A record of climatic change and anthropogenic disturbance since 11,600 yr B.P: Journal of Paleolimnology, v. 5, no. 3, p. 197–218, doi: 10.1007/BF00200345.

Miehlich, G., 1984, Chronosequenzen und anthropogene Veränderungen andesitischer Vulkanascheböden eines randtropischen Gebirges (Sierra Nevada, México) [Ph.D. thesis]: Hamburg, Germany, Universität Hamburg, 417 p.

Molina-Berbeyer, R., 1974, Informe preliminar de geoquímica de los fluidos geotérmicos del Volcán Chichonal, Chiapas: Comisión Federal de Electricidad, Internal Report.

Molina-Garza, R., and Urrutia-Fucugauchi, J.F., 1993, Deep crustal structure of central Mexico derived from interpretation of Bouguer gravity anomaly data: Journal of Geodynamics, v. 17, no. 4, p. 181–201, doi: 10.1016/0264-3707(93)90007-S.

Mooser, F., 1961, Los volcanes de Colima: Universidad Nacional Autónoma de México, Boletín del Instituto Geología, v. 61, p. 49–71.

Mooser, F., 1967, Tefracronología de la Cuenca de México para los últimos treinta mil años: Boletín Instituto Nacional de Antropología e Historia, no. 30, p. 12–15.

Mooser, F., Meyer-Abich, H., and McBirney, A.R., 1958, Catalogue of the active volcanoes of the world, including Solfatara fields. Part VI, Central America: Napoli International Volcanological Association, p. 26–30.

Mora, J.C., 2001, Studio vulcanologico e geochimico del Vulcano Tacana, Chiapas, Messico [Ph.D. Thesis]: Firenze, Italy, Università degli Studi di Firenze, 147 p.

Mora, J.C., Macías, J.L., Saucedo, R., Orlando, A., Manetti, P., and Vaselli, O., 2002, Petrology of the 1998–2000 products of Volcán de Colima, México: Journal of Volcanology and Geothermal Research, v. 117, p. 195–212, doi: 10.1016/S0377-0273(02)00244-5.

Mora, J.C., Macías, J.L., García-Palomo, A., Espíndola, J.M., Manetti, P., and Vaselli, O., 2004, Petrology and geochemistry of the Tacaná Volcanic Complex, México-Guatemala: Evidence for the last 40,000 yr of activity: Geofísica Internacional, v. 43, p. 331–359.

Mugica, M.R., 1987, Estudio petrogenético de las rocas ígneas y metamórficas en el Macizo de Chiapas: Instituto Mexicano del Petróleo, México, C-2009, 47 p.

Müllerried, F.K.G., 1933, El Chichón, único volcán en actividad descubierto en el estado de Chiapas: Memorias de la Sociedad Científica "Antonio Alzate," v. 54, p. 411–416.

Müllerried, F.K.G., 1951, La reciente actividad del Volcán de Tacaná, Estado de Chiapas, a fines de 1949 y principios de 1950: Tuxtla Gutiérrez, Departamento de Prensa y Turismo, Sección Autográfica; Universidad Nacional Autónoma de México, Instituto de Geología, 25 p.

Navarro, C., Cortés, A., and Téllez, A., 2003, Mapa de peligros del Volcán de Fuego de Colima: Universidad de Colima, México, scale 1:100,000, 1 sheet.

Navarro-Ochoa, C., and Luhr, J., 2000, Late-Holocene Tephrochronology at the Colima Volcanic Complex, México, *in* Séptima Reunión Internacional Volcán de Colima, Abstracts 6–10.

Navarro-Ochoa, C., Gavilanes, J.C., and Cortés, A., 2002, Movement and emplacement of lava flows at Volcán de Colima, México: November 1998–February 1999: Journal of Volcanology and Geothermal Research, v. 117, p. 155–167, doi: 10.1016/S0377-0273(02)00242-1.

Negendank, J.F.W., Emmerman, R., Krawczyc, R., Mooser, F., Tobschall, H., and Werle, D., 1985, Geological and geochemical investigations on the Eastern Trans-Mexican Volcanic Belt: Geofisica Internacional, v. 24, p. 477–575.

Nelson, S.A., Gonzalez-Caver, E., and Kyser, T.K., 1995, Constrains on the origin of alkaline and calc-alkaline magmas from the Tuxtla Volcanic Field, Veracruz, México: Contributions to Mineralogy and Petrology, v. 122, p. 191–211, doi: 10.1007/s004100050121.

Newhall, C.G., and Self, S., 1982, The Volcanic Explosivity Index (VEI): An estimate of explosive magnitude for historical volcanism: Journal of Geophysical Research, v. 87, C2, p. 1231–1237.

Newton, J.A., and Metcalfe, S.E., 1999, Tephrochronology of the Toluca Basin, central México: Quaternary Science Reviews, v. 18, p. 1039–1059, doi: 10.1016/S0277-3791(98)00043-2.

Nixon, G.T., 1982, The relationship between Quaternary volcanism in central México and the seismicity and structure of the subducted ocean lithosphere: Geological Society of America Bulletin, v. 93, p. 514–523, doi: 10.1130/0016-7606(1982)93<514:TRBQVI>2.0.CO;2.

Norini, G., Groppelli, G., Capra, L., and De Beni, E., 2004, Morphological analysis of Nevado de Toluca volcano (Mexico): New insights into the structure and evolution of an andesitic to dacitic stratovolcano: Geomorphology, v. 62, p. 47–61, doi: 10.1016/j.geomorph.2004.02.010.

Núñez-Cornu, F., Nava, A., De la Cruz, S., Jiménez, Z., Valencia, C., and García, R., 1994, Seismic activity related to the 1991 eruption of Colima Volcano, Mexico: Bulletin of Volcanology, v. 56, p. 228–237, doi: 10.1007/s004450050032.

Ordoñez, E., 1894, Nota acerca de los ventisqueros del Iztaccíhuatl: Memorias de la Sociedad Científica "Antonio Alzate," v. 8, p. 31–42.

Ordoñez, E., 1898, Les Volcans Colima et Ceboruco: Memorias de la Sociedad Científica "Antonio Alzate," v. 11, p. 325–333.

Ordoñez, E., 1902, Le Xinantecatl ou Volcan Nevado de Toluca: Memorias de la Sociedad Científica "Antonio Alzate," v. 18, p. 83–112.

Ordoñez, E., 1905, El nauhcampatepetl o Cofre de Perote: Boletín de la Sociedad Geológica Mexicana, 1, 151 p.

Ortega-Gutiérrez, F., Solari, L.A., Solé, J., Martens, U., Gómez-Tuena, A., Morán-Ical, S., Reyes-Salas, M., and Ortega-Obregón, C., 2004, Polyphase, High-Temperature Eclogite-Facies Metamorphism: in the Chuacús Complex, Central Guatemala: Petrology, Geochronology, and Tectonic Implications: International Geology Review, v. 46, p. 445–470.

Ortíz-Santos, G., 1944, La zona volcánica "Colima" del estado de Jalisco: monografía: Guadalajara, Universidad de Guadalajara, Instituto de Geografía, 49 p.

Otis, H.E., 1902, Volcanoes of Colima, Toluca and Popocatépetl: Science, v. 25, p. 646 p.

Palacios, D., 1996, Recent geomorphologic evolution of a glacio-volcanic active stratovolcano: Popocatépetl (Mexico): Geomorphology, v. 16, p. 319–335, doi: 10.1016/0169-555X(96)00003-7.

Palacios, D., Zamorano, J.J., and Gómez, A., 2001, The impact of present lahars on the geomorphologic evolution of proglacial gorges: Popocatépetl, Mexico: Geomorphology, v. 37, p. 15–42, doi: 10.1016/S0169-555X(00)00061-1.

Panfil, M.S., Gardner, T.W., and Hirth, K.G., 1999, Late Holocene stratigraphy of the Tetimpa archaeological sites, northeast flank of Popocatépetl volcano, central México: Geological Society of America Bulletin, v. 111, p. 204–218, doi: 10.1130/0016-7606(1999)111<0204:LHSOTT>2.3.CO;2.

Pardo, M., and Suárez, G., 1993, Steep subduction geometry of the Rivera Plate beneath the Jalisco Block in Western Mexico: Geophysical Research Letters, v. 320, p. 2391–2394.

Pardo, M., and Suárez, G., 1995, Shape of the subducted Rivera and Cocos plates in southern Mexico: Seismic and tectonic implications: Journal of Geophysical Research, v. 100, p. 12,357–12,373, doi: 10.1029/95JB00919.

Perret, F.A., 1937, The eruption of Mt. Pelée 1929–1932: Carnegie Institution Washington Publication, v. 458, p. 1–26.

Pieschel, C., 1856, Die vulkane der republik mexiko: Berlin, Reimer, 1 p., 18 plates.

Plunket, P., and Uruñuela, G., 1999, Preclassic household patterns preserved under volcanic ash at Tetimpa, Puebla, México: Latin American Antiquity, v. 9, no. 4, p. 287–309.

Plunket, P., and Uruñuela, G., 2000, The archaeology of a Plinian eruption of the Popocatépetl volcano, *in* McGuire, W.J., Griffiths, D.R., Hancock, P.L., and Stewart, I.S., eds., The archaeology of geological catastrophes: Geological Society [London] Special Publication 171, p. 195–204.

Plunket, P., and Uruñuela, G., 2005, Recent research in Puebla prehistory: Journal of Archaeological Research, v. 13, p. 89–127, doi: 10.1007/s10804-005-2485-5.

Ponce, L., Gaulon, R., Suárez, G., and Lomas, E., 1992, Geometry and state of stress of the downgoing Cocos plate in the Isthmus of Tehuantepec, Mexico: Geophysical Research Letters, v. 19, p. 773–776.

Pulgarín, B., Capra, L., Macías, J.L., and Cepeda, H., 2001, Depósitos de flujos de escombros gigantes (10 km^3) asociados a colapsos de edificios volcánicos, los casos de los volcanes Nevado del Huila (Colombia) y Nevado de Colima (México): Revista de Geofísica, Instituto Panamericano de Geografía e Historia, v. 55, p. 51–75.

Quezada-Ramírez, N., 1972, Los Matlatzincas: Época prehispánica y época colonial hasta 1650, Instituto Nacional de Antropología e Historia, 22, 142 p.

Rebollar, C.J., Espíndola, V.H., Uribe, A., Mendoza, A., and Pérez-Vertti, A., 1999, Distribution of stress and geometry of the Wadati-Benioff zone under Chiapas, Mexico: Geofísica Internacional, v. 38, p. 95–106.

Richards, A.F., 1959, Geology of the Islas Revillagigedo, Mexico. 1. Birth and development of Volcan Barcena, Isla San Benedicto: Bulletin of Volcanology, v. 22, p. 73–124.

Richards, A.F., 1965, Geology of the Islas Revillagigedo, 3. Effects of erosion on Isla San Benedicto 1952–1961 following the birth of Volcan Barcena: Bulletin of Volcanology, v. 28, p. 381–419.

Riva Palacio-Chiang, R., 1983, Informe y comentarios acerca del Volcán Chichonal, Chiapas, *in*: VI Convención Geológica Nacional, El Volcán Chichonal: México, D.F., Universidad Nacional Autónoma de México, Instituto Geología, 49–56.

Robin, C., 1984, Le volcan Popocatépetl (Mexique): Structure, evolution pétrologique et risques: Bulletin of Volcanology, v. 47, p. 1–23, doi: 10.1007/BF01960537.

Robin, C., and Boudal, C., 1987, A gigantic Bezymianny-type event at the beginning of modern Volcan Popocatépetl: Journal of Volcanology and Geothermal Research, v. 31, p. 115–130, doi: 10.1016/0377-0273(87)90009-6.

Robin, C., and Cantagrel, J.M., 1982, Le Pico de Orizaba (Mexique): Structure et evolution d'un grand volcan andésitique complexe: Bulletin of Volcanology, v. 45, p. 299–315.

Robin, C., Komorowski, J.C., Boudal, C., and Mossand, P., 1990, Evidence of magma mixing in juvenile fragments from pyroclastic surge deposits associated with debris avalanche deposits at Colima volcanoes, México: Bulletin of Volcanology, v. 52, p. 391–403, doi: 10.1007/BF00302051.

Robin, J., and Potrel, A., 1993, Multi-stage magma mixing in the pre-caldera series of Fuego de Colima volcano: Geofísica Internacional, v. 32, p. 605–615.

Robin, J., Mossand, P., Camus, G., Cantagrel, J.M., Gourgaud, A., and Vincent, P.M., 1987, Eruptive history of the Colima volcanic complex (México): Journal of Volcanology and Geothermal Research, v. 31, p. 99–113, doi: 10.1016/0377-0273(87)90008-4.

Rodríguez-Elizarrarás, S., 1991, Geología del Volcán de Colima, estados de Jalisco y Colima [Master's thesis]: México, D.F., Universidad Nacional Autónoma de México, Instituto de Geología, 110 p.

Rodríguez-Elizarrarás, S.R., 1995, Estratigrafía y estructura del Volcán de Colima, México: Revista Mexicana de Ciencias Geológicas, v. 12, p. 22–46.

Rodríguez-Elizarrarás, S., Siebe, C., Komorowski, J., Espíndola, J., and Saucedo, R., 1991, Field observations of pristine block and ash flow deposits emplaced April 16–17, 1991 at Volcán de Colima, México: Journal of Volcanology and Geothermal Research, v. 48, p. 399–412, doi: 10.1016/0377-0273(91)90054-4.

Rosas-Elguera, J., Ferrari, L., Garduño-Monroy, V.G., and Urrutia-Fucugauchi, J., 1996, Continental boundaries of the Jalisco Block and their influence in the Pliocene-Quaternary kinematics of western México: Geology, v. 24, p. 921–924, doi: 10.1130/0091-7613(1996)024<0921:CBOTJB>2.3.CO;2.

Rose, W.I., Bornhorst, T.J., Halsor, S.P., Capaul, W.A., Plumley, P.S., De la Cruz, S.R., Mena, M., and Mota, R., 1984, Volcán El Chichón, México: Pre-1982 S-rich eruptive activity: Journal of Volcanology and Geothermal Research, v. 23, p. 147–167, doi: 10.1016/0377-0273(84)90060-X.

Rossotti, A., 2005, Reconstrucción de la historia eruptiva de la Pómez Citlaltépetl (Volcán Pico de Orizaba) [Ph.D. thesis]: México, Juriquilla-Querétaro, Universidad Nacional Autónoma de México, 142 p.

Rossotti, A., and Carrasco-Núñez, G., 2004, Stratigraphy of the 8.5–9.0 ka B.P. Citlaltépetl pumice fallout sequence: Revista Mexicana de Ciencias Geológicas, v. 21, no. 3, p. 353–370.

Rouwet, D., Taran, Y., and Varley, N., 2004, Dynamics and mass balance of El Chichón crater lake, México: Geofísica Internacional, v. 43, p. 427–434.

Rye, R.O., Luhr, J.F., and Wasserman, M.D., 1984, Sulfur and oxygen isotopes systematics of the 1982 eruptions of El Chichon volcano, Chiapas, México: Journal of Volcanology and Geothermal Research, v. 23, p. 109–123, doi: 10.1016/0377-0273(84)90058-1.

Sapper, K., 1896, La geografía física y la geografía de la Península de Yucatán: Instituto Geológico de México, 3, 58 p.

Sapper, K., 1897, Ueber die räumliche Anordnung der mittelamerikanischen Vulkane: Berlin: Zeitschrift der Deutschen Geologischen Gesellschaft, v. 1897, p. 672–682.

Sapper, K., 1899, Ueper Gebirsbau und Boden des noerdlichen Mittelamerika: Petermanns Geographische Mitteilungen, 127, 119 p.

Sartorius, C., 1869, Eruption of the Volcano of Colima in June 1869: Smithsonian Report, 423 p.

Saucedo, G.R., 1997, Reconstrucción de la erupción de 1913 del Volcán de Colima [Master's thesis]: México D.F., Universidad Nacional Autónoma de México, 185 p.

Saucedo, G.R., and Esquivias, H., 1988, Evaluación del riesgo volcánico en el área del Volcán Tacaná, Chiapas [B.S. thesis]: México, D.F., Instituto Politécnico Nacional, Escuela Superior de Ingenieria y Arquitectura, 142 p.

Saucedo, R., and Macías, J.L., 1999, La Historia del Volcán de Colima: Tierra Adentro, v. 98, p. 8–14.

Saucedo, R., Macías, J., Bursik, M., Mora, J., Gavilanes, J., and Cortes, A., 2002, Emplacement of pyroclastic flows during the 1998–1999 eruption of Volcán de Colima, Mexico: Journal of Volcanology and Geothermal Research, v. 117, p. 129–153, doi: 10.1016/S0377-0273(02)00241-X.

Saucedo, R., Macías, J., and Bursik, M., 2004a, Pyroclastic flow deposits of the 1991 eruption of Volcán de Colima, México: Bulletin of Volcanology, v. 66, p. 291–306, doi: 10.1007/s00445-003-0311-0.

Saucedo, G.R., Macías, J.L., Sheridan, M.F., Bursik, I., and Komorowski, J.C., 2004b, Modeling of pyroclastic flows of Colima Volcano, México: Implications for hazard assessment: Journal of Volcanology and Geothermal Research, v. 139, p. 103–115, doi: 10.1016/j.jvolgeores.2004.06.019.

Schaaf, P., Martínez-Serrano, R., Siebe, C., Macías, J.L., Carrasco, G., Castro, R., and Valdez, G., 2004, Heterogeneous magma compositions of Transmexican Volcanic Belt stratovolcanoes—Geochemical and isotopic evidence for different basement compositions, *in* Aguirre, J., Macías, J.L., and Siebe, C., eds., Neogene-Quaternary continental margin volcanism, Proceedings of the Geological Society of America, Penrose Conference Series, Special Publication, Instituto de Geología, UNAM [abs.], p. 68.

Schaaf, P., Stimac, J., Siebe, C., and Macías, J.L., 2005, Geochemical evidence for mantle origin and crustal processes in volcanic rocks from Popocatépetl and surrounding monogenetic volcanoes, central México: Journal of Petrology, v. 46, p. 1243–1282, doi: 10.1093/petrology/egi015.

Scientific Even Alert Network, 1982, Volcanic events: El Chichón volcano: Smithsonian Institution: Bulletin of Volcanology, v. 7, no. 5, p. 2–6.

Scolamacchia, T., and Macías, J.L., 2005, Diluted pyroclastic density currents of the 1982 eruptions of El Chichon volcano, Chiapas, México: Revista Mexicana de Ciencias Geológicas, v. 22, no. 2, p. 159–180.

Scolamacchia, T., Macías, J.L., Sheridan, M.F., and Hughes, S., 2005, Cylindrical ash aggregates: A new type of particle in wet pyroclastic surge deposits: Bulletin of Volcanology, v. 68, no. 2, p. 171–200, doi: 10.1007/s00445-005-0430-x.

Scott, K.M., Macías, J.L., Naranjo, J. A., Rodriguez, S., and McGeehin, J.P., 2001, Catastrophic Debris Flows Transformed from Landslides in Volcanic Terrains, Mobility, Hazard Assessment, and Mitigation Strategies: U.S. Geological Survey, Professional Paper, 1630, 67 p, 19 figs, 9 tabs.

Secretaría de la Defensa Nacional, 1983, El Plan DN-III-E y su aplicación en el área del Volcán Chichonal, *in* VI Contención Geológica Nacional de la Sociedad Geológica Mexicana: México, D.F., Universidad Nacional Autónoma de México, Instituto de Geología, p. 90–100.

Sedov, S., Solleiro-Rebolledo, E., Gama-Castro, J.E., Vallejo-Gómez, E., and González-Velázquez, A., 2001, Buried palaeosols of the Nevado de Toluca; an alternative record of late Quaternary environmental change in central México: Journal of Quaternary Science, v. 16, p. 375–389, doi: 10.1002/jqs.615.

Sedov, S., Solleiro-Rebolledo, E., and Morales-Puente, P., J., Arias-Herrera, A., Vallejo-Gomez, E., and Jasso-Castaneda, C., 2003, Mineral and organic components of the buried paleosols of the Nevado de Toluca, Central México as indicators of paleoenvironments and soil evolution: Quaternary International, v. 106–107, p. 169–184.

Seele, E., 1973, Restos de milpas y poblaciones prehispánicas cerca de San Buenaventura Nealtic: Estado de Puebla, Comunicaciones, v. 7, p. 77–86.

Segerstrom, K., 1956, Erosion studies at Paricutin, State of Michoacán, México: Bulletin of the Geological Survey, Bulletin 965-A, 164 p., 7 plates.

Sheridan, M.F., and Macías, J.L., 1995, Estimation of risk probability for gravity-driven pyroclastic flows at Volcán Colima, México: Journal of Volcanology and Geothermal Research, v. 66, p. 251–256, doi: 10.1016/0377-0273(94)00058-O.

Sheridan, M.F., Hubbard, B., Bursik, M.I., Siebe, C., Abrams, M., Macías, J.L., and Delgado, H., 2001, Gauging short-term potential volcanic hazards at Popocatépetl, México: Eos (Transactions, American Geophysical Union), v. 82, no. 16, p. 187–189.

Sheridan, M.F., Carrasco-Nuñez, G., Hubbard, B.E., Siebe, C., and Rodríguez-Elizarrarás, S., 2002, Mapa de Peligros del Volcán Citlaltépetl (Pico de Orizaba): Universidad Nacional Autónoma de México, Instituto de Geología, scale 1:25,000, 1 Sheet.

Sheridan, M.F., Hubbard, B., Carrasco-Núñez, G., and Siebe, C., 2004, Pyroclastic flow hazard at Volcán Citlaltépetl: Natural Hazards, v. 33, p. 209–221, doi: 10.1023/B:NHAZ.0000037028.89829.d1.

Sheth, H.C., Torres-Alvarado, I.S., and Verma, S., 2000, Beyond subduction and plume: A unified tectonic-petrogenetic model for the Mexican Volcanic Belt: International Geology Review, v. 42, p. 1116–1132.

Siebe, C., and Macías, J.L., 2006, Volcanic hazards in the Mexico City metropolitan area from eruptions at Popocatépetl, Nevado de Toluca, and Jocotitlán stratovolcanoes and monogenetic scoria cones in the Sierra Chichinautzin Volcanic Field, in Seibe, C., Macías, J.L., and Aguirre-Díaz, G.J., eds., Neogene-Quaternary continental margin volcanism: A perspective from México: Geological Society of America Special Paper 402, p. 253–329.

Siebe, H.C., Abrams, M., and Sheridan, M., 1993, Major Holocene block-and-ash fan at the western slope of ice-capped Pico de Orizaba volcano, México: Implications for future hazards: Journal of Volcanology and Geothermal Research, v. 59, p. 1–33, doi: 10.1016/0377-0273(93)90075-3.

Siebe, C.G., Komorowski, J.C., Navarro, C., McHone, J., Delgado, H., and Cortés, A., 1995a, Submarine eruption near Socorro Island, Mexico: Geochemistry and scanning electron microscopy studies of floating scoria and reticulite: Journal of Volcanology and Geothermal Research, v. 68, p. 239–272, doi: 10.1016/0377-0273(95)00029-1.

Siebe, H.C., Abrams, M., and Macías, J.L., 1995b, Derrumbes gigantes, depósitos de avalancha de escombros y edad del actual cono del volcán Popocatépetl: Volcán Popocatépetl, estudios realizados durante la crisis de 1994–1995: Centro Nacional de Prevención de Desastres, México, p. 195–220.

Siebe, C., Macías, J.L., Abrams, M., Rodríguez, S., Castro, R., and Delgado, H., 1995c, Quaternary explosive volcanism and pyroclastic deposits in east central Mexico: Implications for future hazards: Geological Society of America Field Trip Guidebook, v. 1, p. 1–48.

Siebe, C., Abrams, M., Macías, J.L., and Obenholzner, J., 1996a, Repeated volcanic disasters in Prehispanic time at Popocatépetl, central Mexico: Past key to the future?: Geology, v. 24, no. 5, p. 399–402, doi: 10.1130/0091-7613(1996)024<0399:RVDIPT>2.3.CO;2.

Siebe, C., Macías, J.L., Abrams, M., and Obenholzner, J., 1996b, La destrucción de Cacaxtla y Cholula: Un suceso en la historia erupiva del Popocatépetl: Revista Ciencias, Facultad de Ciencias, Universidad Nacional Autónoma de México, v. 43, p. 36–45.

Siebe, C., Macías, J.L., Abrams, M., Rodríguez, S., and Castro, R., 1997, Catastrophic prehistoric eruptions at Popocatépetl and Quaternary explosive volcanism in the Serdán-Oriental basin, east-central Mexico: Fieldtrip Guidebook, Pre-meeting Excursion, v. 4, 1997 International Association of Volcanology and Chemistry of the Earth's Interior General Assembly, Puerto Vallarta, México, 88 p.

Sigurdsson, H., Carey, S.N., and Espíndola, J.M., 1984, The 1982 eruptions of El Chichón volcano, Mexico: Stratigraphy of pyroclastic deposits: Journal of Volcanology and Geothermal Research, v. 23, p. 11–37, doi: 10.1016/0377-0273(84)90055-6.

Sigurdsson, H.C., Carey, S.N., and Fisher, R.V., 1987, The 1982 eruptions of El Chichón volcano, Mexico (3): Physical properties of pyroclastic surges: Bulletin of Volcanology, v. 49, p. 467–488, doi: 10.1007/BF01245474.

Singer, B.S., and Kudo, A.M., 1986, Origin of andesites and dacites from Pico de Orizaba, Mexican Volcanic Belt: Sr isotope and phase chemistry: Geological Society of America Abstracts with Programs, v. 18, no. 2, p. 186.

Singh, S.K., and Pardo, M., 1993, Geometry of the Benioff Zone and state of stress in the overriding plate in central Mexico: Geophysical Research Letters, v. 20, p. 1483–1486.

Solleiro-Rebolledo, E., Macías, J.L., Gama-Castro, J.E., Sedov, S., and Sulerzhitsky, L.D., 2004, Quaternary pedostratigraphy of the Nevado de Toluca volcano: Revista Mexicana de Ciencias Geológicas, v. 21, p. 101–109.

Stoiber, E.R., and Carr, M.J., 1973, Quaternary volcanic and tectonic segmentation of Central America: Bulletin of Volcanology, v. 37, p. 1–22.

Stoopes, G.R., and Sheridan, M.F., 1992, Giant debris avalanches from the Colima Volcanic Complex, Mexico: Implications for long-runout landslides (<100 km) and hazard assessment: Geology, v. 20, p. 299–302, doi: 10.1130/0091-7613(1992)020<0299:GDAFTC>2.3.CO;2.

Straub, S.M., and Martín-Del Pozzo, A.L., 2001, The significance of phenocrysts diversity in tephra from recent eruptions at Popocatépetl volcano (central Mexico): Contributions to Mineralogy and Petrology, v. 140, p. 487–510, doi: 10.1007/PL00007675.

Suárez-Cruz, S., and Martínez-Arreaga, S., 1993, Monografía de Cholula: Puebla, Offset Mabek, 43 p.

Taran, Y., Fisher, T.P., Pokrovsky, B., Sano, Y., Armienta, M.A., and Macías, J.L., 1998, Geochemistry of the volcano-hydrothermal system of El Chichón volcano, Chiapas, Mexico: Bulletin of Volcanology, v. 59, p. 436–449, doi: 10.1007/s004450050202.

Taran, Y., Gavilanes, J.C., and Cortés, A., 2002, Chemical and isotopic composition of fumarolic gases and the SO_2 flux from Volcán de Colima, México, between the 1994 and 1998 eruptions: Journal of Volcanology and Geothermal Research, v. 117, no. 1–2, p. 105–119, doi: 10.1016/S0377-0273(02)00239-1.

Tassi, F., Vaselli, O., Capaccioni, B., Macías, J.L., Nencetti, A., Montegrossi, G., Magro, G., and Buccianti, A., 2003, Chemical composition of fumarolic gases and spring discharges from El Chichón volcano, Mexico: Causes and implications of the changes detected over the period 1998–2000: Journal of Volcanology and Geothermal Research, v. 123, no. 1–2, p. 105–121, doi: 10.1016/S0377-0273(03)00031-3.

Tello, F.A., 1651, Libro segundo de la Crónica Miscelánea en que se trata de la Conquista Espiritual y temporal de la Santa Provincia de Jalisco en el Nuevo Reino de la Galicia y Nueva Vizcaína y Descubrimiento del Nuevo México, Guadalajara, Imprenta de la Republica Literaria.

Templos, L.A., 1981, Observaciones geoquímicas de la zona geotérmica del Chichonal, Chiapas: Comisión Federal de Electricidad, Informe.

Tepley, F.J., Davidson, J.P., Tilling, R.I., and Arth, J.G., 2000, Magma mixing, recharge and eruption histories recorded in plagioclase phenocrysts from El Chichón volcano, Mexico: Journal of Petrology, v. 41, p. 1397–1411, doi: 10.1093/petrology/41.9.1397.

Thorpe, R.S., 1977, Tectonic significance of alkaline volcanism in eastern Mexico: Tectonophysics, v. 40, p. T19–T26, doi: 10.1016/0040-1951(77)90064-6.

Thorpe, R.S., Gibson, I.L., and Vizcaíno, J.S., 1977, Andesitic pyroclastic flows from Colima Volcáno: Nature, v. 265, p. 724–725, doi: 10.1038/265724a0.

Tilling, R.I., Rubin, M., Sigurdsson, H., Carey, S., and Duffield, W.A., 1984, Prehistoric eruptive activity of El Chichón volcano, México: Science, v. 224, p. 747–749.

Urrutia-Fucugauchi, J., and Flores-Ruiz, J.H., 1996, Bouguer gravity anomalies and regional crustal structure in central México: International Geology Review, v. 38, p. 176–194.

Valdéz-Moreno, G., Schaaf, P., Macías, J.L., and Kusakabe, M., 2006, New Sr-Nd-Pb-O isotope data for Colima volcano and evidence for the nature of the local basement, *in* Siebe, C., Macías, J.L., and Aguirre-Díaz, G.J., eds., Neogene-Quaternary continental margin volcanism: A perspective from México: Geological Society of America Special Paper 402, p. 45–63.

Varekamp, J.C., Luhr, J.F., and Prestegaard, K.L., 1984, The 1982 eruptions of El Chichón volcano (Chiapas, Mexico): Character of the eruptions, ash-fall deposits, and gas phase: Journal of Volcanology and Geothermal Research, v. 23, p. 39–68, doi: 10.1016/0377-0273(84)90056-8.

Vázquez-Selem, L., and Heine, K., 2004, Late Quaternary glaciation of México, *in* Ehlers, J., and Gibbard, P.L., eds., Quaternary glaciations—Extent and chronology, part III: Amsterdam, Elsevier, p. 233–242.

Waibel, L., 1933, Die Sierra Madre de Chiapas: Mitteilungen der Geographuchen Gesellschaft in Hamburg, v. 43, p. 13–162.

Waitz, P., 1906, Le Volcan de Colima *in* X Congreso Geológico Internacional, Libreto-Guía: México, v. 13, p. 27.

Waitz, P., 1909, Excursión Geológica al Nevado de Toluca: Boletín de la Sociedad Geológica Mexicana, v. 6, p. 113–117.

Waitz, P., 1910–1911, Observaciones geológicas acerca del Pico de Orizaba: Boletín de la Sociedad Geológica Mexicana, v. 7, p. 67–76.

Waitz, P., 1915, El estado actual de los volcanes de México y la última erupción del Volcán de Colima (1913): Revista Volcanológica, p. 259–268.

Waitz, P., 1921, Popocatépetl again in activity: American Journal of Science, v. 1, p. 81–87.

Waitz, P., 1935, Datos históricos y bibliográficos acerca del Volcán de Colima: Memorias de la Sociedad Científica Antonio Alzate, v. 53, p. 349–383.

Wallace, P.J., and Carmichael, I.S.E., 1999, Quaternary volcanism near the Valley of México: Implications for subduction zone magmatism and the effects of crustal thickness variations on primitive magma compositions: Contribution to Mineralogy and Petrology, v. 135, p. 291–314, doi: 10.1007/s004100050513.

Weitzberg, F., 1922, El Ventisquero del Popocatépetl: Memorias de la Sociedad Científica Antonio Alzate, v. 41, p. 65–90.

Whitford, D.J., and Bloomfield, K., 1977, Geochemistry of Late Cenozoic Volcanic Rocks from the Nevado de Toluca area, México: Carnegie Institution of Washington, Yearbook, v. 75, p. 207–213.

Wilcox, R.E., 1954, The petrology of Parícutin volcano, México: U.S. Geological Survey Bulletin, v. 965C, p. 281–353.

Williams, S.N., and Self, S., 1983, The October 1902 Plinian eruption of Santa Maria volcano, Guatemala: Journal of Volcanology and Geothermal Research, v. 16, p. 33–56, doi: 10.1016/0377-0273(83)90083-5.

Yáñez-García, C., and García Durán, S., 1982, Exploración de la región geotérmica Los Humeros-Las Derrumbadas, estados de Puebla y Veracruz: Comisión Federal de Electricidad, Reporte Técnico, 96 p.

Yokoyama, I., De la Cruz-Reyna, S., and Espíndola, J.M., 1992, Energy partition in the 1982 eruption of El Chichón volcano, Chiapas, Mexico: Journal of Volcanology and Geothermal Research, v. 51, p. 1–21, doi: 10.1016/0377-0273(92)90057-K.

Zimbelman, D., Watters, R., Firth, I., Breit, G., and Carrasco-Núñez, G., 2004, Stratovolcano stability assessment methods and results from Citlaltépetl, Mexico: Bulletin of Volcanology, v. 66, p. 66–79, doi: 10.1007/s00445-003-0296-8.

Zobin, V.M., Luhr, J.F., Taran, Y., Bretón, M., Cortés, A., De La Cruz-Reyna, S., Domínguez, T., Galindo, I., Gavilanes, J.C., and Muñiz, J.J., 2002, Overview of the 1997–2000 activity of Volcán de Colima, México: Journal of Volcanology and Geothermal Research, v. 117, no. 1–2, p. 1–19, doi: 10.1016/S0377-0273(02)00232-9.

Manuscript Accepted by the Society 29 August 2006

Printed in the USA

The Geological Society of America
Special Paper 422
2007

The Mojave-Sonora megashear: The hypothesis, the controversy, and the current state of knowledge

Roberto S. Molina-Garza*
Centro de Geociencias, Universidad Nacional Autónoma de México, Campus Juriquilla, Querétaro, 76230, México

Alexander Iriondo
Centro de Geociencias, Universidad Nacional Autónoma de México, Campus Juriquilla, Querétaro, 76230, México, and *Department of Geological Sciences, University of Colorado, Boulder, Colorado 80309, USA*

ABSTRACT

The Mojave-Sonora megashear model, which implies left-lateral strike-slip motion of northern México in Jurassic time, remains one of the most influential ideas concerning the geology of México. A comprehensive review of the literature related to this topic does not yet allow resolution of the controversy over the validity of this hypothesis. A clear conclusion is that the original hypothesis was based on a relatively simplistic model of the geology of Sonora, as the basement of the Caborca terrane is not simply a fragment of the Mojave Precambrian basement province of eastern California. Attempts to use quantitative techniques in testing the model have yielded results contrary to the hypothesis, such as clockwise rotations indicated by paleomagnetic data, and the diversity and complexity of the basement of Caborca indicated by geochemical and geochronological data. Other quantitative methods such as zircon provenance studies in quartzites of the sedimentary cover yield inconclusive results. The main conclusion of the studies of detrital zircons is that Grenvillean zircons are relatively abundant, but that their presence cannot be attributed solely to sources in the Grenville province in a fixist model. Stratigraphic correlations of upper Paleozoic and Mesozoic rocks in Caborca with similar sequences in California and Nevada do not provide convincing arguments of large displacement, but should be evaluated in more detail. Elements that have the potential to test the hypothesis with greater certainty include detailed studies of basement rocks, a refined stratigraphy of the Jurassic volcanic and volcaniclastic arc rocks south of the inferred fault trace, and an increased understanding of depositional trends in the miogeoclinal sequence. Structural studies are sparse in this region. It is particularly important to gain a better understanding of the effects in time and space of Late Cretaceous–Tertiary contractional deformation. A tectonic evolution model that does not conflict with the existing data is the proposal that displacement of a para-autochthonous Caborca terrane may have occurred in the late Paleozoic. Nonetheless, available data and geologic relations in the Caborca region do not require Late Jurassic slip of several hundred kilometers.

Keywords: Mojave-Sonora megashear, Caborca terrane, northwestern México, Proterozoic.

*rmolina@geociencias.unam.mx

Molina-Garza, R.S., and Iriondo, A., 2007, The Mojave-Sonora megashear: The hypothesis, the controversy, and the current state of knowledge, *in* Alaniz-Álvarez, S.A., and Nieto-Samaniego, Á.F., eds., Geology of México: Celebrating the Centenary of the Geological Society of México: Geological Society of America Special Paper 422, p. 233–259, doi: 10.1130/2007.2422(07). For permission to copy, contact editing@geosociety.org.

RESUMEN

El modelo de la megacizalla Mojave-Sonora, el cual implica desplazamiento lateral izquierdo en el norte de México durante el Jurásico, permanece como una de las ideas más influyentes en la geología del país. Una revisión general de la literatura relacionada con el tema no permite aún resolver la controversia sobre la validez de la hipótesis, pero una conclusión clara es que la hipótesis original estaba basada en un modelo relativamente simplista de la geología de Sonora, ya que el basamento del terreno Caborca no es un simple fragmento de la corteza Mojave del este de California. Intentos de utilizar métodos cuantitativos han dado resultados contrarios a la hipótesis, como el de las rotaciones horarias indicadas por el paleomagnetismo y la diversidad de basamentos en Caborca que sugieren la geocronología y geoquímica; otros métodos producen resultados indeterminados, como la proveniencia de circones en las cuarcitas de la cobertura del terreno Caborca. La conclusión más relevante de esos estudios es la abundancia de circones de edad Grenvilleana, pero su presencia no puede simplemente atribuirse a fuentes en la Provincia Grenville en un modelo fijista. Las correlaciones estratigráficas entre secuencias Paleozoico tardío y Mesozoico en Caborca y secuencias similares en California y Nevada no producen argumentos convincentes a favor de grandes desplazamientos, pero deben considerarse con datos más detallados. Elementos que podrían evaluar la hipótesis con mayor contundencia son estudios más detallados del basamento, una estratigrafía fina del arco volcánico Jurásico y de las rocas volcanoclásticas al sur de la traza inferida de la falla y un mejor conocimiento de la secuencia miogeosinclinal. Son pocos los estudios estructurales en la región y en particular un problema importante es resolver en tiempo y espacio los efectos de la deformación compresional Cretácico-Terciario. Un modelo que no entra en conflicto con la evidencia existente es la propuesta de que el desplazamiento del terreno parautóctono Caborca haya ocurrido en el Paleozoico tardío. Sin embargo, los datos existentes y las relaciones geológicas en la región de Caborca, no requieren de un desplazamiento de cientos de kilómetros en el Jurásico Tardío.

Palabras clave: megacizalla Mojave-Sonora, terreno Caborca, noroeste de México, Proterozoico.

INTRODUCTION

The Megashear Hypothesis

The Mojave-Sonora megashear hypothesis is one of the most influential models concerning the geology of México. The hypothesis proposed in the early 1970s (Silver and Anderson, 1974) suggests that there is a NW-SE–oriented system of left-lateral strike-slip faults extending from the Mojave Desert, in California, across northern México. As known today, this fault system connects with a hypothetical spreading center in the present-day Gulf of México (Anderson and Schmidt, 1983). The fault system was active in Middle to Late Jurassic time. The original hypothesis was based on the observation of the apparent juxtaposition of two Precambrian provinces of different age and geologic history in northwest Sonora, in the Caborca region ("This zone ... disrupts two northeasterly trending orogenic and magmatic belts of Precambrian age" [Anderson et al., 1979. p. 59]). These two provinces were, however, defined on the basis of U-Pb zircon ages determined in a few localities, two of them located to the north of the Mojave-Sonora megashear trace (Anderson and Silver, 1981, 2005; Fig. 1).

The original argument of the juxtaposition of Precambrian basement provinces is more complex than originally perceived. Basement rocks in northwest Sonora, south of the hypothetical trace of the megashear, belong to both the Mojave and Yavapai provinces (Iriondo and Premo, 2003), and they are indeed different from basement rocks exposed in northeast Sonora. Rocks in northeast Sonora belong to the Mazatzal province (Conway and Silver, 1989). But to this day, the age, tectonic limits, and affinity of Sonoran basement rocks are not known in sufficient detail. This is not only because the analytical data of the original publication (Silver and Anderson, 1974) were few and have become available only recently (Anderson and Silver, 2005), but also because only a few other studies have been conducted since that time. The most recent publications on this topic (Iriondo and Premo, 2003; Iriondo et al., 2004; Nourse et al., 2005) indicate that the basement of a miogeoclinal Precambrian-Paleozoic sequence in the Caborca region is composed of various blocks of differing affinity. This basement includes elements of the Mojave and Yavapai

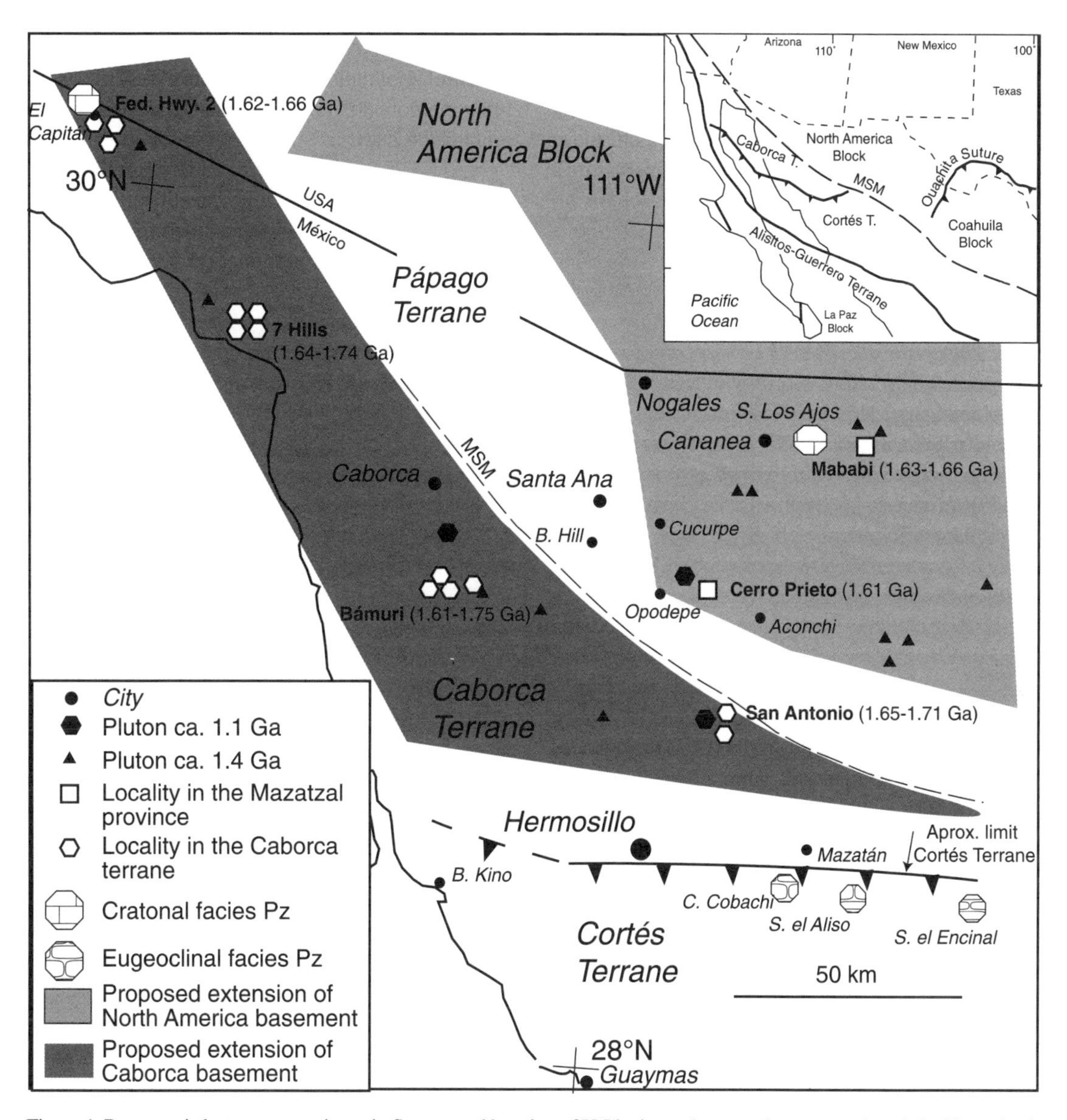

Figure 1. Proterozoic basement provinces in Sonora and location of U-Pb zircon dates used to present the original hypothesis for the Mojave-Sonora megashear. Modified after Anderson and Silver (1979). Also shown are the Paleozoic cratonal deposits in northeastern Sonora and the eugeosinclinal rocks from south-central Sonora. The inset shows tectono-stratigraphic terranes and other tectonic features in NW México. MSM—Mojave-Sonora megashear; S.—Sierra; C.—Cerro.

provinces, as well as rocks with characteristics of the transition between these provinces (Iriondo et al., 2004). This conclusion is solid as it is based on crystallization ages, model ages, and the age of metamorphism of these basement rocks.

The timing of activity and details of the displacement along the Mojave-Sonora megashear were defined in the Caborca region in northwest Sonora. Various authors recognized the miogeoclinal Neoproterozoic and lower Paleozoic sequence (Gamuza beds and overlying units), and they proposed that the sequence can be correlated, formation by formation, with the miogeoclinal sequence of eastern California and southern Nevada in the Inyo and San Bernardino Mountains (Eells, 1972; Stewart et al., 1984; Stewart, 2003). The correlation of the Caborca sequence with the section exposed in the Mojave Desert has been used to estimate ~800 km of displacement along the Mojave-Sonora megashear. The approximate age of the fault is based on the apparent juxtaposition of Middle Jurassic rocks north of the hypothetical trace of the megashear against rocks of the miogeoclinal sequence south of the trace (Anderson and Silver, 1979).

Geotectonic Significance

Since originally proposed, the Mojave-Sonora megashear hypothesis has been controversial. Yet, the hypothesis was readily adopted by researchers interested in late Paleozoic paleogeographic reconstructions of western equatorial Pangea, because it offers an elegant solution to the problem posed by the overlap between South America and most of México when closing the Gulf of Mexico (Pindell and Dewey, 1982; Pilger, 1978; and others). The overlap is inevitable and requires repositioning most of México, but as Figure 2 shows, reconstructions proposed more recently do not require large-scale displacement of northern México. The overlap is only evident between northern South America and southeast México. The hypothesis as viewed today has been presented in an elaborate model for the evolution of the circum-Gulf region (Anderson and Schmidt, 1983). In this model, cogenetic faults along the Trans-Mexican Volcanic Belt and along the Mesoamerican trench accommodated motion of cortical blocks in central and southern México in a similar direction (southeastward) as the Mojave-Sonora megashear (Fig. 3).

Other lineaments in northern México with a general northwest-southeast orientation may or may not be linked to the Mojave-Sonora megashear hypothesis (Fig. 4). Authors like De Cserna (1971, 1976) have suggested the existence of such structures; one example is the Torreón-Saltillo lineament, which follows the trend of faults and folds along the front of the Sierra Madre Oriental. Of these lineaments, the San Marcos fault (McKee et al., 1984, 1990), and perhaps the La Babia fault (Charleston, 1981; Padilla y Sánchez, 1986), both in Coahuila and north of the trace of the megashear, are the best documented. La Babia fault, also known as the Boquillas lineament, marks the limit between the Sabinas basin and the Burro-Peyotes platform. A component of left-lateral slip along this fault is inferred by the apparent displacement between the Burro-Peyotes platform and the Tamaulipas arch (Charleston, 1981; Padilla y Sánchez, 1986). The fault strikes S55°E. The San Marcos fault, which is also known as the Las Delicias arc, is located along the southern margin of the Sabinas basin, and separates the Sabinas basin from Coahuila Island. The fault was active in Late Jurassic and Early Cretaceous time as a normal fault (Chávez-Cabello et al., 2005). Yet another structure linked to the Mojave-Sonora megashear, at least conceptually, is the Texas lineament. This structure is manifested as a series of east-west fractures and Tertiary volcanism, exemplified by the Balcones fault, across southwest Texas.

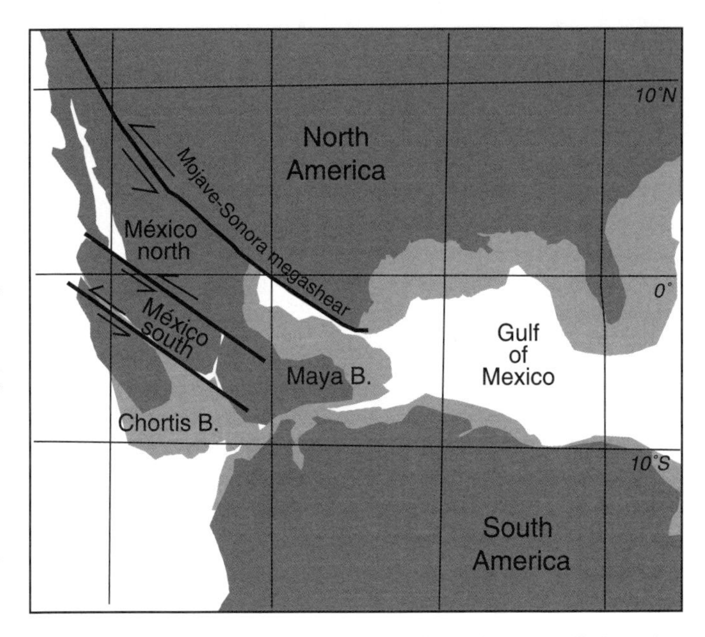

Figure 3. Paleogeographic reconstruction of the Circum-Gulf of Mexico after Anderson and Schmidt (1983). In more recent reconstructions (Lottes and Rowley, 1990), today's Gulf region is occupied by the Yucatan block and the continental masses are much closer. The continental platforms without attenuation effects are shown in light gray.

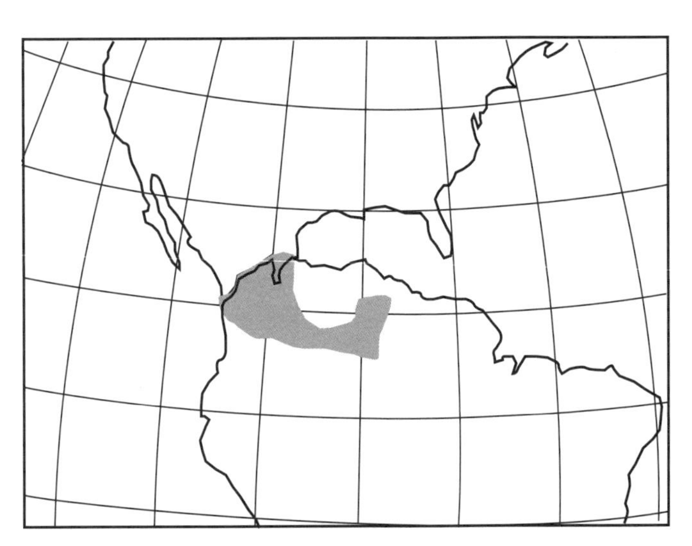

Figure 2. Reconstruction of continental masses in North America and South America on the Pangea configuration of Lottes and Rowley (1990) showing the overlap (gray) of South America over parts of southern México.

The eastward continuation of the megashear is inferred from scarce exposures of pre-Oxfordian rocks in central México (i.e., Jones et al., 1995). The trace is projected eastward below the volcanic cover of the Sierra Madre Occidental, along the southern margin of Coahuila Island, to the south of exposures of the upper Paleozoic Las Delicias arc, and continues eastward over the Laramide-age Parras basin. In this region, the Torreón-Saltillo or the Torreón-Monterrey lineaments (De Cserna, 1976) may represent the continuation of the Mojave-Sonora megashear. Nonetheless, these structures are associated with Laramide thrusting of the frontal section of Sierra Parras and Sierra Jimulco, as well as the Monterrey curvature, over the Parras–La Popa basin. The Cenozoic cover of the coastal plain masks the possible continuation of the Mojave-Sonora megashear east of the Sierra Madre Oriental. The San Tiburcio lineament in northern Zacatecas, with a general northwest-southeast orientation, has been interpreted as a regional fault with left-lateral slip (Mitre-Salazar, 1989) perhaps linked to the megashear.

Similarly, the hypothetical trace of the Mojave-Sonora megashear is projected to the northwest, to the desert region of southeast California and southern Nevada. Various authors have noticed that the proposed trace of the megashear in the Mojave

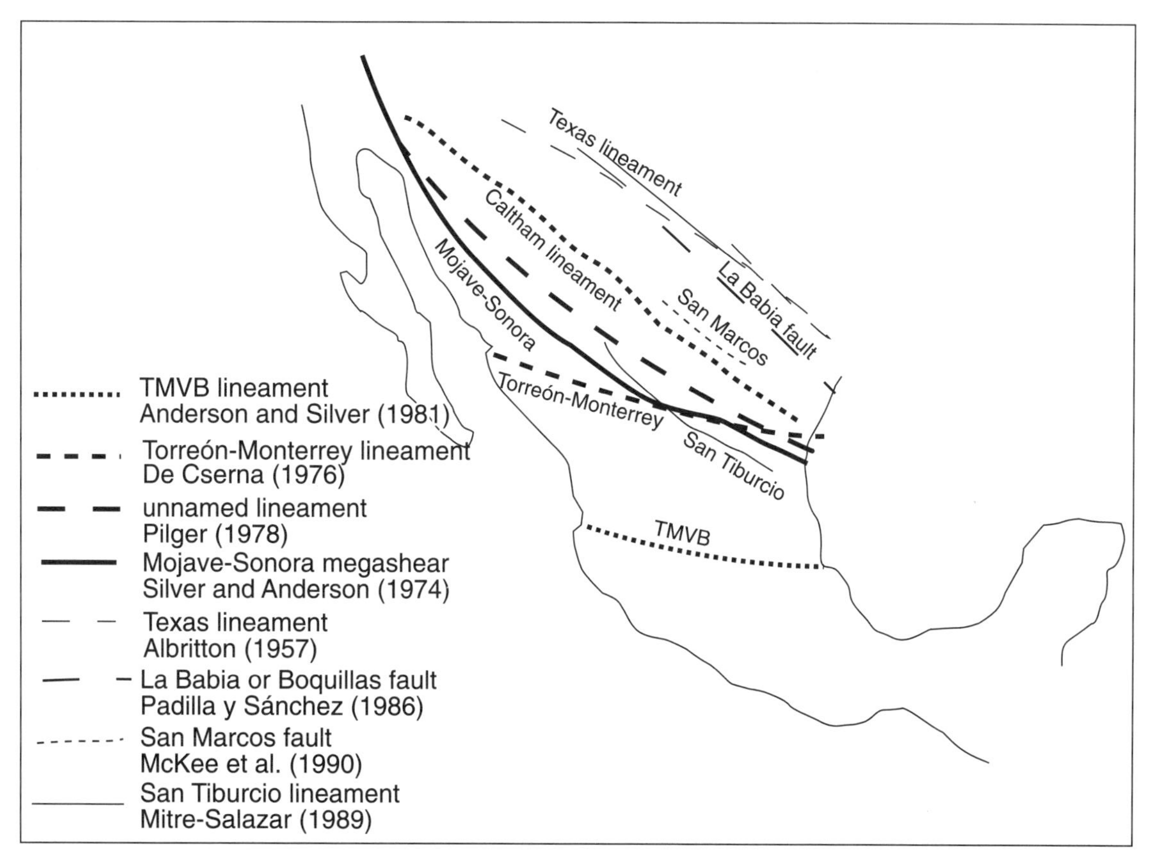

Figure 4. Regional structures and lineaments in northern México that may have a bearing on the Mojave-Sonora megashear hypothesis. TMVB—Trans-Mexican Volcanic Belt.

Desert cuts through apparently continuous late Paleozoic–Triassic stratigraphic trends and patterns (such as isopachs and facies changes; Cameron, 1981; Walker and Wardlaw, 1989). Stone and Stevens (1988) and Walker (1988) proposed instead that the Cordilleran margin of California was truncated in late Paleozoic time in the context provided by the Sonoman Orogeny. In the Mojave Desert, the evidence of continental margin truncation includes sedimentation patterns in Mississippian and Pennsylvanian rocks, and the apparent abrupt termination of stratigraphic trends in lower Paleozoic strata in the region of southwestern Nevada (Stevens et al., 1992). Even if the Cordilleran continental margin was truncated by lateral slip, the activity of the Mojave-Sonora megashear in Middle–Late Jurassic time is an element of the initial hypothesis that has proved difficult to demonstrate.

The purpose of this article is to present a compilation of existing information on the topic of the Mojave-Sonora megashear since the idea was first introduced. We attempt to weigh the evidence that favors and contradicts the hypothesis. A definitive conclusion on the validity of the megashear hypothesis, from the point of view of the authors, is not yet within reach. We also present alternatives to the megashear model that explain general stratigraphic, structural, geophysical, and geochemical observations.

GEOLOGY OF SONORA

Crystalline Precambrian basement in northwest México was first reported by Cooper and Arellano (1946), who identified a Middle Cambrian fauna in the sedimentary sequence of Caborca overlying a metamorphic basement. Damon et al. (1962) reported the first radiometric dates (K-Ar) in Paleoproterozoic basement rocks, also identifying Grenvillean age magmatism (ca. 1.0 Ga). Several years later, Anderson and Silver (1977) described the basement rocks as a Paleoproterozoic metamorphic series assigned to the Bámuri Complex. These authors depicted the metamorphic complex as a volcano-sedimentary sequence in amphibolite to greenschist facies, metamorphed ca. 1.65 Ga, and intruded by 1.4 and 1.1 Ga granitoides.

The first geological map and a general stratigraphic framework for northwest Sonora were published in studies conducted by Merriam (1972) and Merriam and Eells (1979); whereas a detailed stratigraphy and correlation of this stratigraphy with the known Neoproterozoic and Paleozoic miogeoclinal sequence of Nevada were established by Stewart et al. (1984) and Longoria and Pérez (1979). In general, the pre-Jurassic geology of Sonora allows distinction between three provinces: northern Sonora, the region of Caborca to Hermosillo, and the region of south-central Sonora. These provinces have been incorporated into the modern literature in the context of tectonostratigraphic terranes (Fig. 1, inset), corresponding to the North America, Caborca (or Seri), and Cortés terranes, respectively (Campa and Coney, 1983; Sedlock et al., 1993).

Caborca Terrane

The most characteristic lithologies of the Caborca-Hermosillo region belong to the Neoproterozoic and Paleozoic. Here the pre-Mesozoic geology consists of carbonate and siliciclastic rocks, considered part of the cordilleran miogeocline, exposed along the western margin of the craton. The sequence rests disconformably on Proterozoic basement. It has an estimated thickness of 3.3 km, and was deposited in shallow-marine environments on the passive margin of Laurentia that developed after Rodinia's breakup in the Neoproterozoic (Stewart et al., 2002).

Sequences of the mid-Paleozoic (Siluro-Devonian) are only locally preserved in the Caborca terrane. Also, a hiatus that includes most of the Silurian is evident (Poole et al., 2000), and an apparent disconformity in the Hermosillo region separates the Permian from the rest of the Paleozoic sequence (Stewart et al., 1997). The Paleozoic and Neoproterozoic sequences are folded—folds have an east to northeast vergence and locally overthrust Mesozoic sequences (De Jong et al., 1988). This has led some authors to suggest that the trace of the Mojave-Sonora megashear is regionally covered by a Cretaceous age fold-and-thrust belt.

Upper Paleozoic rocks in the Caborca terrane include platform carbonate rocks of Devonian, Mississippian, and Pennsylvanian age (Brunner, 1976; Stewart et al., 1997). At Sierra Santa Teresa, close to Hermosillo (Fig. 5), the upper Paleozoic is over 2000 m thick. A deep-water siliciclastic sequence ~600 m in thickness, assigned to the Lower Permian, may represent foredeep to foreland deposits of an orogen developing to the south (in the present paleogeography), or development of a marginal basin concurrent with transtensional faulting (Stewart et al., 1997).

Overlying the Paleozoic sequence rests, disconformably, a Mesozoic succession of transitional to deep-marine environments (Santa Rosa Formation and stratigraphic equivalents such as the Antimonio Formation; González-León, 1980). The disconformity between Paleozoic and Mesozoic strata is not well exposed except at localities east of Cerro Pozos de Serna (Calmus et al., 1997; Lucas et al., 1999), to the south of Cerro El Rajón (Stewart et al., 1984), in Sierra Santa Rosa (Hardy, 1981), and perhaps in the area of Rancho Placeritos (Poole et al., 2000, Fig. 5).

As originally defined, the Antimonio Formation has a thickness of ~3–4 km (González-León, 1980). The unit was subdivided by this author into a Triassic lower member and a Jurassic upper member. From the lower member, Lucas and Estep (1999) have identified an ammonoid fauna of the Spathian (Lower Triassic) to the Norian. In their work, Lucas and Estep (1999) revise the stratigraphy of the Antimonio Formation, and restrict the use of Antimonio Formation to the lower member of González-León (1980), with a stratigraphic range from the Spathian to the upper Carnian. This interval is dominated by siliciclastics (red mudstone and fine-grained sandstone) with intercalations of bioclastic limestone. This unit also contains conglomerate with clasts of quartzite, and sandstone petrology suggests a crystalline continental source (Stanley and González-León, 1995). Rocks of the redefined Antimonio Formation were deposited in deep-marine environments that become shallower in the upper part of the sequence; conglomerates are interpreted as shelf-slope deposits. Lucas and Estep (1999) assigned shallow-marine facies of the uppermost part of the lower member sensu González-León (1980) to the Río Asunción Formation, which is marked by intervals of limestone and sandstone with faunas from the Norian to the Lower Jurassic. Marine to transitional environments are typical of the Río Asunción Formation. High-energy fluvial and alluvial fan environments may apply to a distinct conglomeratic interval in the Río Asunción Formation containing clasts of quartzite, chert, metamorphic quartz, and granitic porphyry. A Jurassic sequence that was originally assigned to the upper member of the Antimonio Formation is reassigned by Lucas and Estep (1999) to the Santa Rosa Formation as defined by Hardy (1981) in Sierra Santa Rosa. This sequence contains a diverse, abundant Liassic fauna, possibly reaching the Pliesbachian (Stanley and González-León, 1995). The Santa Rosa Formation, sensu Lucas and Estep (1999), was derived from the erosion of a volcano-plutonic terrane, and consists of basinal facies dominated by siliciclastics.

The Jurassic sequence is well exposed at Sierra del Álamo. Equivalent strata are exposed at Pozos de Serna (Calmus et al., 1997; Lucas et al., 1999) and Sierra Santa Rosa (Hardy, 1981). These facies are characterized by turbidites that indicate deep-marine environments. Possibly correlative lower Mesozoic strata crop out at Sierra López, Sierra Santa Teresa, and Sierra la Flojera (in the Hermosillo region; Lucas and González-León, 1994; Stewart et al., 1997), and at Cerro El Rajón southeast of Caborca (Fig. 5; Longoria and Pérez, 1979). At this later locality the volcanic component is relatively important.

The tectonic environment of the Santa Rosa Formation at Sierra del Álamo and Sierra Santa Rosa is probably a fore-arc basin developed between a continental volcanic arc and a trench complex (González-León, 1997). The apparent lack of a fore-arc basin and trench-related rocks south of the trace of the hypothetical Mojave-Sonora megashear was interpreted by Silver and Anderson (1974) as evidence supporting truncation of the Jurassic continental arc of northern Sonora by the megashear. This interpretation must be reconsidered.

Cortés Terrane

In central and eastern Sonora, Paleozoic strata consist of eugeoclinal deep-water facies, which include siliciclastic and carbonate rocks, chert, and less abundant volcanic rocks (Stewart et al., 1990). The eugeoclinal sequence is well exposed in the Barita de Sonora, Sierra el Aliso, Cerro Cobachi, and Sierra el Encinal areas (Fig. 1). Eugeoclinal facies have also been recognized at Tiburón Island and in Baja California. The basement of the sequence is not

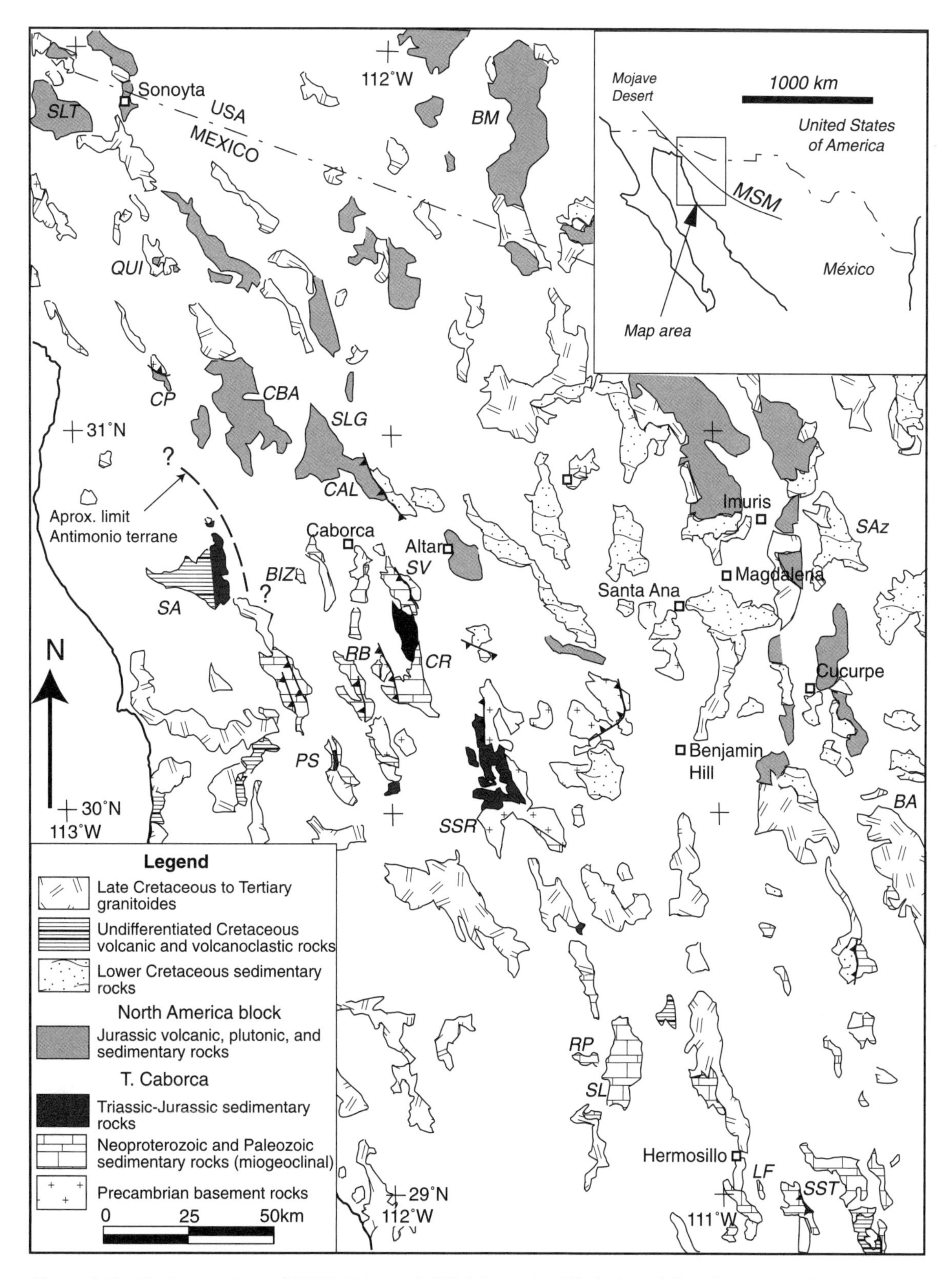

Figure 5. Pre-Tertiary geology of NW México and SW Arizona (modified after Molina Garza and Geissman, 1998). Abbreviations for localities: SLT—Sierra Los Tanques; QUI—Quitovac; SLG—Sierra La Gloria; CAL—Cerro Álamo; CP—Cerro Prieto; SA—Sierra del Álamo; CR—Cerro El Rajón; SSR—Sierra Santa Rosa; BIZ—Bizani; RB—Rancho Bámuri; SL—Sierra de López; SST—Sierra Santa Teresa; BA—Aconchi batholith; PS—Sierra Pozos Serna; LF—Sierra La Flojera; RP—Rancho Placeritos; SAz—Sierra Azul; SV—Sierra la Víbora; CBA—Cerro la Basura; BM—Baboquivari Mountains; MSM—Mojave-Sonora megashear.

known; the oldest strata are Ordovician in age. Deep-water facies crop out in a belt ~140 km long and ~50 km wide, roughly oriented east-west (Stewart et al., 1990). This belt consists of at least two facies associations, which are both similar and different from eugeoclinal facies of the North America western cordillera. In the Barita de Sonora region, the sequence has a minimum thickness of 700 m. It consists of (1) Lower to Middle Ordovician graptolitic shale and deep-water limestone; (2) banded chert, shale, and dolomite from the Upper Ordovician; (3) Devonian chert, shale, sandstone and barite; (4) Lower Mississippian limestone; (5) Upper Mississippian conglomerate, shale, and chert; and (6) Pennsylvanian silt, sandstone, chert, and barite (Poole et al., 2000, 2005).

The abrupt change from shallow-water platform facies to deep-water facies in the Lower Permian, overlying the miogeoclinal sequence at Sierra Santa Teresa near Hermosillo, has been interpreted as evidence of the tectonic emplacement of what has been called the Sonoran allochthon (Stewart et al., 1997; Poole et al., 2005). This event marks the accretion of eugeoclinal rocks of the Cortés terrane over the margin of the Caborca terrane. The work of Valencia-Moreno et al. (1999) on the isotopic signature of Laramide granitoides suggests that the Cortés terrane is thrusted over the Caborca terrane.

While juxtaposition of the Cortés and Caborca terranes is well documented in central Sonora, exposures of deep-water Permian strata (Monos Formation) in northwest Sonora at Sierra del Álamo are anomalous in the sense that similar facies have not been identified in other ranges of this area. The westward continuation of the Sonoran allochthon is obscured by younger tectonics and plutonism in both western Sonora and Baja California.

The eugeoclinal sequence is covered in a regional angular unconformity by Upper Triassic strata of the Barranca Group. The Barranca Group thus postdates juxtaposition and thrusting of rocks from the Cortés terrane over miogeoclinal rocks. The Barranca Group comprises a continental red beds unit assigned to the Arrayanes Formation, a coal-bearing marine-transition marine cyclothemic unit assigned to the Upper Carnian Santa Clara Formation, and a thick unit of quartz-cobble conglomerate of the Coyotes Formation (Stewart and Roldán-Quintana, 1991). The Barranca Group was deposited in fluvial (Arrayanes Formation), fluvio-deltaic and shallow marine environments (Santa Clara Formation), and alluvial fans (Coyotes Formation). An erosional surface separates the Santa Clara and Coyotes Formations. The Coyotes Formation may provide a stratigraphic link between post-Triassic strata in Sonora. Exposures of quartz-cobble conglomerate crop out from Sonoita in northwest Sonora to the oriental limit of exposures of the Barranca Group in east-central Sonora.

North America Block

In northeast Sonora, Paleoproterozoic basement crops out only locally. It consists primarily of schists and associated metavolcanic rocks, intruded by Mesoproterozoic granitoids (ca. 1.4 and 1.0 Ga). The sedimentary cover in northeastern Sonora is thinner than in the Caborca region; it consists mostly of shallow cratonal facies, and outcrops are also restricted (Krieger, 1961; Anderson and Silver, 1977). The oldest strata include the Bolsa Quartzite and carbonate rocks of the Abrigo Formation. The overlying upper Paleozoic succession is analogous to the well-known Paleozoic rocks of northern Chihuahua and southern Arizona. Overlying Paleozoic and Precambrian rocks is a Middle to Upper Jurassic volcanic sequence, coeval Jurassic granitoids intrude this sequence. These rocks are considered an extension of the Jurassic arc of southern Arizona. The region of southwestern Arizona and northwestern Sonora is characterized by the lack of exposures of basement rocks, and this region has been assigned to the Papago terrane (Haxel et al., 1984; Tosdal et al. 1989), more recently defined as the Papago Domain by Anderson et al. (2005). The volcanic sequence includes pyroclastic and volcaniclastic rocks, tuffs, and lava flows with relatively evolved compositions. Locally, volcanic clast conglomerates, sandstone, and carbonate rocks are present. Volcanic rocks range from andesitic to rhyolitic compositions; the volcanic arc they embody was built on continental crust. These rocks have been mapped in Sonora by Corona (1979), Nourse (1995), Rodríguez-Castañeda (1984, 1994) and Iriondo (2001), while Haxel et al. (1984) and Tosdal et al. (1989) mapped these rocks in Arizona. The arc sequence crops out extensively in the ranges north of Caborca and Santa Ana, although they are affected by intense deformation and metamorphism in the Late Cretaceous and again in the mid-Tertiary. According to Anderson and Nourse (1998, 2005), parts of the Jurassic arc rocks were affected by NW-SE–trending faults in the Late Jurassic. These faults, parallel to the trace of the Mojave-Sonora megashear, formed elongated basins filled by syntectonic conglomerate and marine sedimentary rocks of the Lower Cretaceous (Bisbee Group and correlative strata in Sonora).

Post-Jurassic Stratigraphy and Tectonics

In northern Sonora, overlying rocks of both the North America and Caborca blocks, there is a nonmarine, transitional, and shallow-marine sequence of Upper Jurassic and Lower Cretaceous strata assigned to the Bisbee Group and partly equivalent stratigraphic units (such as Cerro del Oro, El Represo, and Arroyo Sasabe Formations; González-León and Lucas, 1995; Jacques-Ayala et al., 1990; and others). The Bisbee Group is widely distributed in Sonora, Arizona, and New Mexico, and it is interpreted as a sequence deposited in extensional backarc environments. Common lithologies include volcanic rock conglomerate, red sandstone and mudstone, carbonate rocks, and locally volcanic rocks. Notable is the lack of both crystalline basement– and Neoproterozoic strata–derived clasts (Jacques-Ayala, 1995) in spite of studies that indicate north-directed paleocurrents or southerly sediment sources. Overlying Lower Cretaceous strata intermediate composition volcanic flows are common (Jacques-Ayala et al., 1990); similar rocks crop out extensively in central and eastern Sonora, where they are assigned to stratigraphic units such as the Palma and Tarahumara Formations (Amaya-Martínez et al., 1993; García-Barragán and Jacques-Ayala, 1993).

Bisbee Group strata and older units of the Caborca and North America blocks are intruded by plutonic rocks. Studies have recognized stocks and subvolcanic bodies that range from small to intermediate sizes, as well as Late Cretaceous to early Tertiary batholiths. These rocks were included in the Laramide Sonora batholith of Damon et al. (1983), and were emplaced between ca. 90 and 40 Ma. The vast majority of these intrusions are granitoides of the calc-alkaline series, and include peraluminous two-mica granites. The batholiths, typified by the Aconchi batholith (Fig. 5), crop out primarily along N-S–oriented ranges, which are a product of Tertiary extensional tectonics.

De Jong et al. (1988) recognize a Mid-Cretaceous compressional event, which does not affect the granodiorite intrusion dated ca. 80 Ma at Sierra la Víbora east of Caborca (Fig. 5). Coeval deformation in western North America is assigned to the Sevier Orogeny. Late Cretaceous to early Tertiary deformation has been identified in southern Arizona, in northwestern Sonora, and in the Altar region (Hayama et al., 1984; Damon et al., 1962; Iriondo, 2001; Iriondo et al., 2005). Late Cretaceous–early Tertiary Laramide deformation produced greenschist facies metamorphism and ductile deformation affecting the Jurassic volcanic arc sequence, the Bisbee Group strata, and Late Cretaceous granites. Metamorphic rocks resulting from this event have been assigned to the Altar Schist, and they could represent an eastward extension of the Orocopia Schist in California (Jacques-Ayala and De Jong, 1996). In fact, the area between Altar and Sonoita (Fig. 5) corresponds to a low-grade metamorphic belt, for which a unique association with Jurassic activity along the Mojave-Sonora megashear was once theorized. Nonetheless, an Upper Cretaceous protolith for the Altar Schist invalidates this interpretation (García-Barragán et al., 1998). The Altar-Sonoita metamorphic belt is of economic importance as it is host to "orogenic" gold mineralization (Clark, 1998; Jacques-Ayala and Clark, 1998; Iriondo and Atkinson, 2000; Iriondo, 2001). In the eastern portion of Sonora State in the region of Sahuaripa, Pubellier et al. (1995) recognize a Mid-Cretaceous thrust system. This deformation is partly contemporaneous with deposition of the Tarahumara Formation and correlative strata. In northeast Sonora in the Cabullona basin, compressive deformation affected Upper Cretaceous strata. Deformation is similar in style to the thick-skin Laramide deformation of the Rocky Mountains (González-León and Lawton, 1995).

The main events affecting the region during Tertiary time are (1) emplacement of silicic volcanic rocks of the Sierra Madre Occidental; (2) development of metamorphic core complexes (Nourse, 1990, 1995; Nourse et al., 1994); and (3) Basin and Range–style extension—linked too to the opening of the Gulf of California extensional province (Henry and Aranda-Gómez, 2000). Although the Tertiary geology is outside the scope of this article, it is important to mention, for example, that the tectonic fabric imposed on Precambrian through Mesozoic rocks by Late Cretaceous contraction and Tertiary extension may complicate the interpretation of sequences exposed along the hypothetical trace of the Mojave-Sonora megashear, to the north, northwest and east of Caborca, as well as in the Opodepe region.

DISCUSSION OF THE GEOLOGICAL EVIDENCE RELATING TO THE MOJAVE-SONORA MEGASHEAR HYPOTHESIS

Juxtaposition of Precambrian Basement Provinces in Northwestern Sonora

As we mentioned earlier, metavolcanic and metasedimentary rocks from the Bámuri igneous-metamorphic complex (greenschist to amphibolite facies) are intruded by calc-alkaline plutonic rocks with crystallization ages between 1.71 and 1.75 Ga (Anderson and Silver, 1981, 2005). All of these rocks were subsequently metamorphosed at ca. 1.66 Ga and later intruded by Mesoproterozoic ca. 1.4 and ca. 1.1 Ga granitoids that present neither deformation nor metamorphism (Anderson and Silver, 1981, 2005). This basement is quite different from the present one north of the Mojave-Sonora megashear in Sonora, where crystallization ages for the oldest rocks range between 1.6 and 1.7 Ga and were later affected by a magmatic pulse at ca. 1.4 Ga (Anderson and Silver, 1981, 2005) (Fig. 1). A simplistic model for the distribution of Precambrian basement provinces with respect to the Mojave-Sonora megashear for northern México is presented in Figure 6.

The metamorphic basement rocks present in Sonora have not been studied in detail, but similar, better-characterized rocks from southern Arizona and New Mexico (Yavapai and Mazatzal provinces) have been interpreted to have originally represented volcanic island arc sequences that were accreted to the North America craton during Paleoproterozoic time. In the case of the Mojave province in southern California, the arc has a conti-

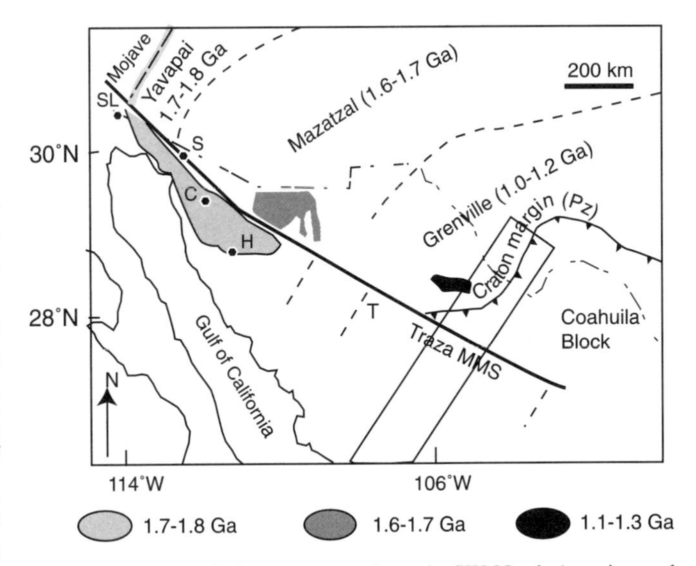

Figure 6. Proterozoic basement provinces in SW North America and localities of interest in northern México. A rectangular box depicts the area of geological transect by Cameron et al. (1989). Abbreviations: H—Hermosillo; C—Caborca; S—Sonoita; SL—San Luis Río Colorado; T—Tomochic Caldera; MMS—Mojave-Sonora megashear. Modified after McDowell et al. (1999).

nental nature with magmas incorporating the crustal signature from preexisting Archean crust. There is no doubt this is one of the key aspects for testing any potential correlation between the Paleoproterozoic Caborca terrane rocks with the ones present in the Mojave province. The first study to test that possible correlation was attempted south of the megashear in the Quitovac region (Fig. 5; Iriondo, 2001; Iriondo et al., 2004). These authors report U-Pb zircon crystallization ages for Caborca igneous rocks between 1.78 and 1.69 Ga. The inheritance from older crust is not evident in dated zircons for these granitoids (Iriondo 2001; Iriondo et al., 2004), but whole-rock Sm-Nd isotopic determinations on the same lithologies provide evidence for the interaction of these magmas with slightly older crust as inferred by the Nd model ages (2.07 and 1.88 Ga) and the εNd values (+0.6 to +2.6) determined for these rocks. However, Pb-Pb isotopic determinations for these rocks from Quitovac plot away from the field of the Mojave province rocks (Iriondo 2001; Iriondo et al., 2004). These authors conclude that the Caborca basement in the Quitovac region might be correlated with the Yavapai province, as it has been defined in northwestern Arizona, or perhaps to a transitional crust between the Mojave and Yavapai provinces, but not directly correlative to the Mojave crust.

In sharp contrast to the Caborca terrane, the basement from northeastern Sonora, also known as the North America terrane or block, is correlated with the Mazatzal province represented in southern Arizona as the Pinal Schist (Silver, 1965). Outcrops of this Precambrian basement are scattered along northern Sonora; some of the localities include areas around Cananea, Sierra San Jose, Sierra Los Ajos, Las Mesteñas, Nacozari, and Anibacochi (Anderson and Silver, 2005). The predominant lithology is a muscovite schist that locally is intruded by ca. 1.4 Ga granitoids (i.e., Herrera-López et al., 2005). The basement in northern Sonora is even less understood than the Bámuri igneous-metamorphic complex south of the Mojave-Sonora megashear, but it is assumed to have rocks with crystallization ages between 1.6 and 1.7 Ga, similar to rocks from southern Arizona (Anderson and Silver, 1977, 1981, 2005; Karlstrom and Williams, 1995).

In Quitovac, the Caborca terrane (Yavapai? province) is juxtaposed against the North America terrane (Mazatzal province) along Laramide ductile thrust faults (Iriondo, 2001; Iriondo et al., 2004, 2005). The age range for rocks from this terrane, based on U-Pb zircon geochronology, is 1.71–1.66 Ga. The age overlap between the Caborca and North America terranes for Quitovac rocks presented by Iriondo et al. (2004) proves that the distinction of one province from the other cannot be achieved based solely on crystallization ages, as originally suggested by Silver and Anderson (1974) when stating the Mojave-Sonora megashear hypothesis. Iriondo et al. (2004) suggest that geochemical and isotopic determinations (specifically Sm/Nd systematics) will be necessary to distinguish them. In the case of the North America terrane in Quitovac, the Nd model ages (1.8–1.74 Ga) and εNd values between +3.4 and +3.9 indicate that the basement rocks are more juvenile, that is, less enriched than their counterparts on the Caborca terrane. These characteristics were used by Iriondo et al. (2004) to propose that the North America terrane in Quitovac could be correlated to the Mazatzal province as defined in southern Arizona.

The presence of at least two distinct, juxtaposed Precambrian basement provinces in northwestern Sonora is indisputable. However, the Caborca terrane rocks in the Quitovac area do not appear to correlate well with the Mojave province; instead these Precambrian rocks have geochemical and isotopic signatures that resemble the rocks from the Yavapai province of northern Arizona or rocks with a transitional signature between those in the Mojave and Yavapai provinces. In addition, it is important to emphasize that the juxtaposition between these basement blocks in northwestern Sonora appears to be a thrust-related to Laramide compressional tectonism, instead of Mid- to Late Jurassic strike-slip displacement. Future work is needed to demonstrate if the Caborca terrane is a composite terrane made up of several of the aforementioned provinces present in southwestern North America (Fig. 7; i.e., Iriondo and Premo, 2003).

Displacement of the Cordilleran Miogeoclinal Sequence

The cordilleran miogeoclinal sequence is widely exposed in the Caborca region (Stewart et al., 1984). It is assigned to 14 formations (Fig. 8) consisting primarily of carbonate rocks, ultramature quartzite, and fine siliciclastic rocks. The Neoproterozoic sedimentary sequence lacks age diagnostic fossils that would permit precise age determinations; it rests disconformably on Precambrian crystalline basement near Aibó ranch southwest of Caborca (Aibó granite; 1.1 Ga; Damon et al., 1962; Anderson and Silver, 1977, 2005; Iriondo et al., 2003a; Farmer et al., 2005). At Aibó ranch the Neoproterozoic sequence overlies well-exposed diabase dikes with K-Ar radiometric dates ca. 900 Ma (Damon et al., 1962), thus indicating the maximum age of the sequence.

Lower Paleozoic rocks include the Proveedora Quartzite and the Puerto Blanco Formation with Lower to Middle Cambrian faunas (Cooper et al., 1952). The existence of the Proveedora Quartzite in northwestern and central Sonora permits the correlation of the Caborca sequence with the miogeoclinal section of California and Nevada, where it is equivalent to the Zabriskie Quartzite (Fig. 8). Similarly, an informal unit within the Johnnie Formation, the Johnnie oolite, of the Neoproterozoic has been identified in the Clemente Formation of the Caborca terrane. Ketner (1986) also recognized an equivalent to the Middle Ordovician Eureka Quartzite of California and Nevada in the Peña Blanca quartzite (informal) in the region of Cerro Cobachi in the Cortés terrane of central Sonora (Fig. 1). The notable similarity in grain size and accessory mineral composition between the Eureka and Peña Blanca quartzites supports the notion that they are indeed the same unit. It is important to note, however, that the age of the Peña Blanca quartzite has not been precisely determined. The abrupt disruption of isopachs of the Eureka Quartzite in eastern California supports the hypothesis that, after deposition, this unit was removed tectonically from its original position. Ketner (1986) suggests that the Eureka Quartzite–displaced

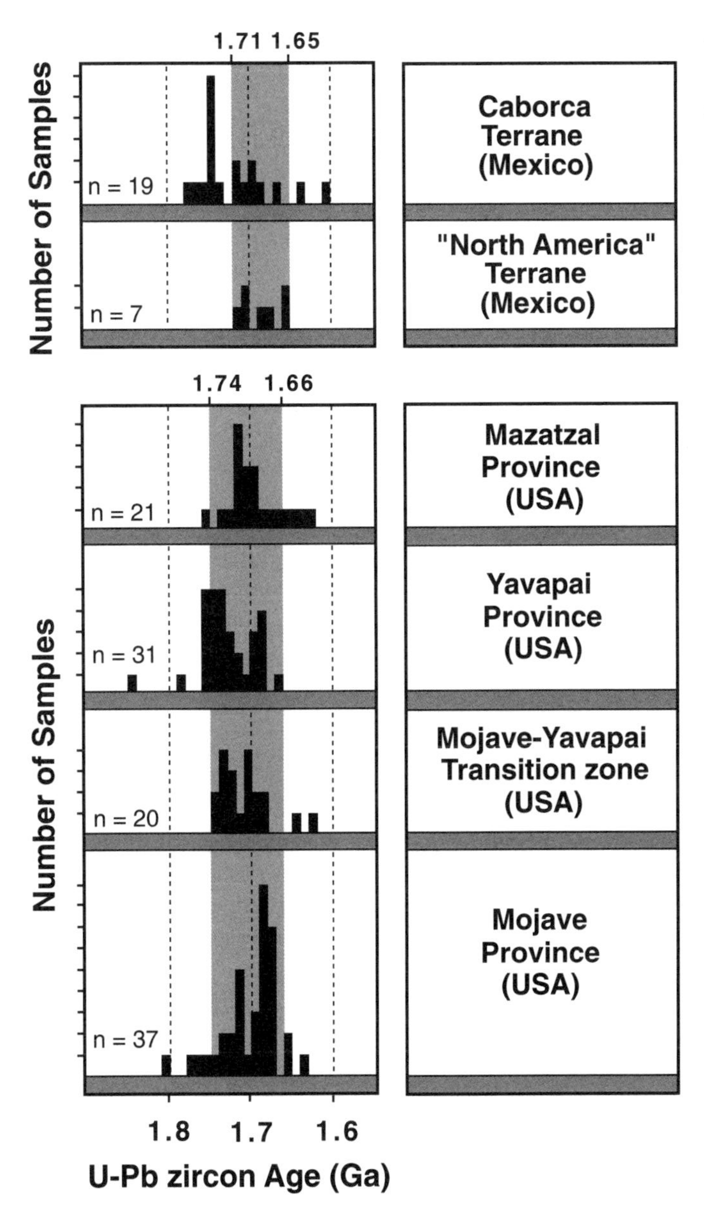

Figure 7. Histograms for U-Pb zircon ages of igneous rocks from the Proterozoic basement provinces in SW North America. Modified after Iriondo et al. (2004).

equivalent is located in the Cortés terrane. According to this model, the correlation of these units indicates a southeastward displacement of the Cortés terrane (and by inference the Caborca terrane) of ~1200 km.

In summary, not only are the miogeoclinal successions of Caborca and eastern California correlative at a gross scale, but also comparable individual stratigraphic units exist in both successions. Stewart (2003) points out other elements that link the Caborca sequence to the rest of the cordillera, such as the distribution of Neoproterozoic diamictites, the distribution of Lower Cambrian archaeocyathids, and stratified barite deposits of the Ordovician and Devonian in both the Roberts Mountains and in the Sonoran allochthon (Fig. 9). An essential test to demonstrate that the miogeoclinal sequence has been displaced, and does not represent a sequence that borders and surrounds a salient in the geography of the craton, is determining the trends in stratigraphic patterns in the Caborca terrane (facies changes, thickness variations, paleocurrents, etc.). In the western Cordillera, stratigraphic patterns trend in a general north-south (actual) orientation, but stratigraphic data for the Caborca sequence is insufficient to determine those trends with reasonable precision. Stewart (2003) suggests that existing data for the Caborca terrane indicate a north-south trend of stratigraphic patterns, supporting the notion that the miogeoclinal sequence was displaced. Nevertheless, restoration of the ~40° post–Middle Jurassic rotation of the Caborca block determined by paleomagnetic data (Molina-Garza and Geissman, 1999) changes the trends to a NW-SE orientation, supporting the hypothesis that the sequence is autochthonous, wrapping a salient of the Laurentia margin (i.e., Poole et al., 2005).

The observation of (1) thin shallow-water cratonal facies in the extreme northeast of Sonora, (2) thick miogeoclinal platformal facies in central and northwestern Sonora, and (3) deep-water eugeoclinal facies of variable thickness in south-central Sonora is common to the western cordillera (Stewart et al., 1984). This suggests a paleogeography that does not require large-scale modification of the southwestern margin of North America. However, transitional facies between shallow-cratonal strata and platform strata have not yet been recognized in Sonora. In those regions of Sonora that have been mapped with sufficient detail, eugeoclinal facies are found structurally below the miogeoclinal strata (Ketner and Noll, 1987), or the sequences are juxtaposed by normal faults. The juxtaposition of the migeoclional and eugeoclinal sequences in Sonora occurred before depositon of the Upper Triassic Barranca Group, possibly in the Early Permian (Stewart et al., 1997).

The stratigraphy of the miogeoclinal sequence of Caborca allows direct comparisons with California and Nevada, but Stewart et al. (1990) points out that correlations are not perfect. The Upper Cambrian Dunderberg Shale, with ample distribution in eastern California and southern Nevada, has no correlative equivalent in the Caborca sequence. The youngest Cambrian units in the Caborca terrane are the Abreojos and el Tren Formations of the Middle Cambrian.

Silurian platform facies in west-central Sonora (i.e., Rancho Placeritos area, west of Hermosillo) are significant to the discussion of the allochthoneity of the Caborca terrane because they could represent the westward continuation of a belt of Silurian platform facies that wraps around the southern North America craton from south Texas through Chihuahua to Sonora (Poole et al., 2000, 2005). Reconstruction of the Caborca terrane to the northwest of its present position, according to the Mojave-Sonora megashear model, would result in a wide gap of Silurian platform facies west of exposures of these rocks in Chihuahua.

Recent work in northern Sonora and southern Arizona suggests that the Cambrian succession of Caborca is autochthonous. According to Strickland and Middleton (2000), the

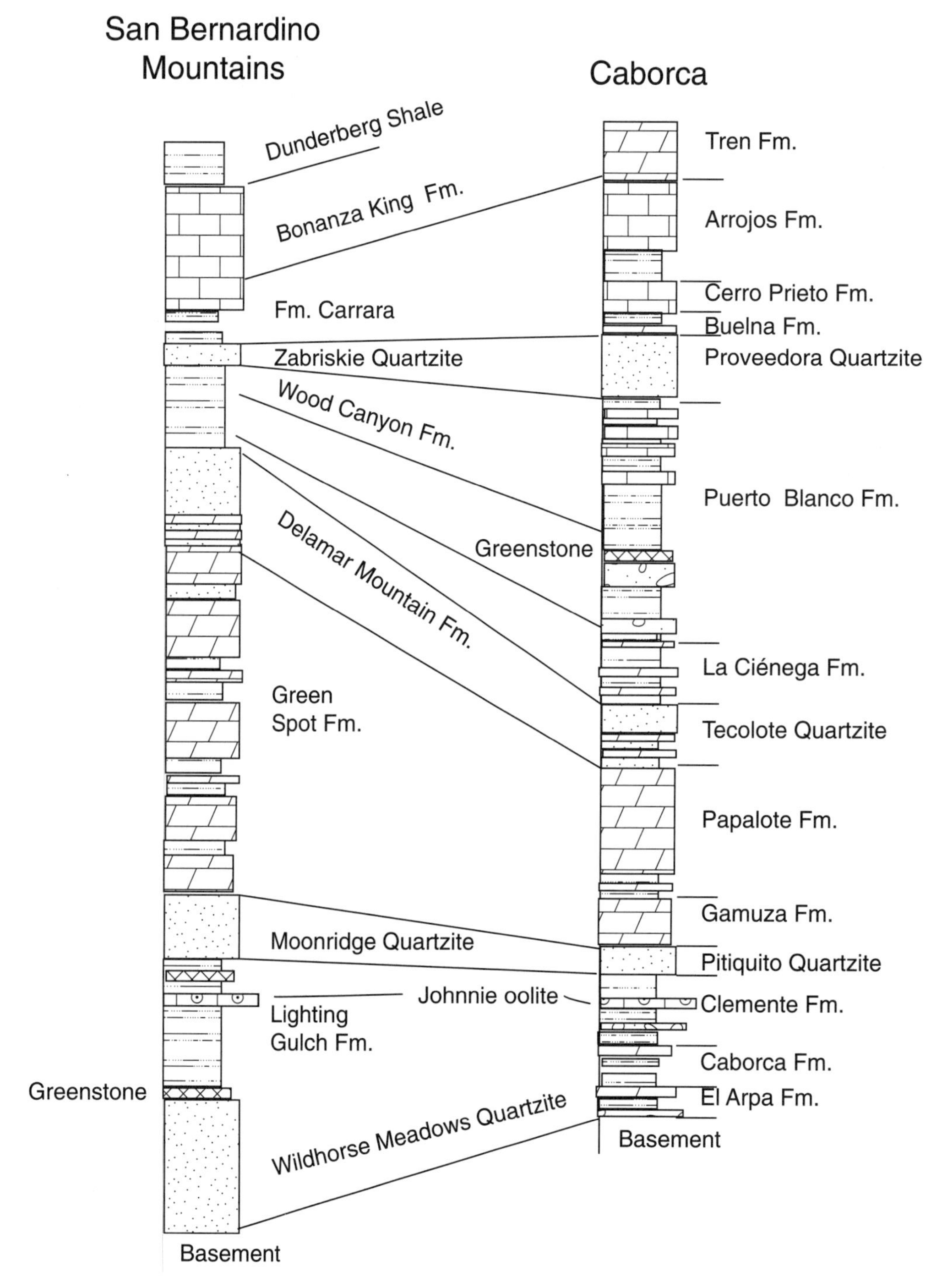

Figure 8. Comparison of stratigraphic columns for Neoproterozoic-Paleozoic rocks from Caborca in Sonora and the Death Valley region in California. Modified after Stewart et al. (1984).

Cambrian sequence represents deposition along a recess in an otherwise continuous continental margin wrapping the southwestern portion of the craton. The general distribution of the Bolsa Quartzite and the Abrigo Formation, in northern Sonora, suggests that this recess opened to the southwest during Mid-Cambrian times. Cambrian strata in the Caborca region in this paleogeography represent continuation of that trend to somewhat deeper waters (Strickland and Middleton, 2000). It is important to note that detrital zircon populations in the Bolsa and Proveedora Quartzites are indistinguishable (Kurtz et al., 1998). In particular, these units share the dominance of zircons ca. 1.1 Ga (n = 14 out of 21 zircons analyzed in the Bolsa

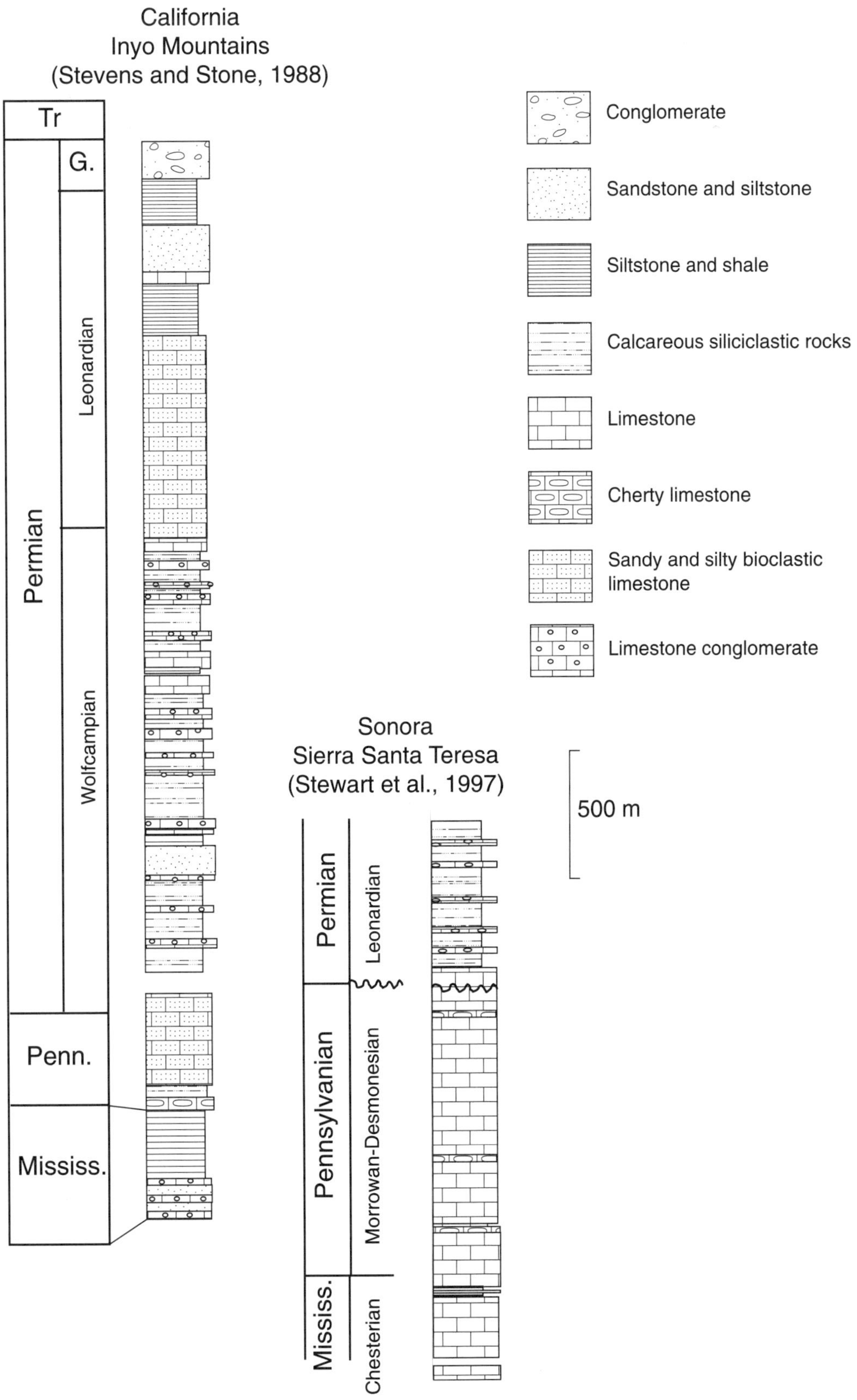

Figure 9. Comparison of stratigraphic columns for upper Paleozoic rocks from the southern Inyo Mountains in California (Stone and Stevens, 1988) and Sierra Santa Teresa in Sonora (Stewart et al., 1997).

Quartzite, and three grains ca. 1.2 Ga); this observation makes the Bolsa and Proveedora Quartzites distinct from similar strata in California, and suggests that the Caborca terrane is autochthonous with respect to the North America terrane.

A sedimentary sequence that is difficult to incorporate into the models that displaced the Caborca miogeoclinal sequence to the northwest of its present position crops out at Ejido Aquiles Cerdán near San Luis Río Colorado in the most northwestern section of Sonora (Fig. 6). This sequence essentially represents cratonal facies of the Colorado plateau (Fig. 6), including the Supai Group, Coconino Sandstone, and Kaibab Limestone (Leveille and Frost, 1984). The sequence is metamorphosed to greenschist facies, but the age of metamorphism is not well understood. If this sequence is autochthonous, as it suggests its remarkable similarity to the sequence of the Colorado plateau, it implies that the trace of the Mojave-Sonora megashear should lie further to the south of this locality.

Zircon Provenance Studies from Quartzites

One important method for testing the correlation between miogeocline rocks from Caborca with the ones from the rest of the cordillera is the zircon provenance studies from Paleozoic quartzites (Gehrels and Stewart, 1998). Such studies are not definitive but can be used to obtain some preliminary conclusions that are noteworthy. Zircons from the Cambrian Proveedora Quartzite located in the Caborca block were primarily derived from crystalline rocks ca. 1.1 Ga, 1.40–1.45 Ga, and 1.6–1.8 Ga (Gehrels and Stewart, 1998). The most abundant population is ca. 1.1 Ga (20 out of 35 zircons studied) with a few grains at ca. 1.24 Ga. This is very important because there is a lack of abundant igneous rocks in the Mojave Desert in eastern California and southwestern Nevada that could have served as the source of these zircons. Gehrels and Stewart (1998) note that the presence of 1.1 Ga plutonic rocks in the Caborca region (i.e., Aibó Granite) could indicate a local source for the detrital zircons present in the Proveedora Quartzite. However, we cannot discard the possibility that some of these zircons could have been derived from the nearby Grenville province in south Texas and Chihuahua.

Detrital zircons from Mid-Ordovician miogeocline rocks present in Sierra López just west of Hermosillo (Fig. 5) appear to have distal sources with ages between 1.77 and 2.07 Ga (9 grains), as well as ages between 2.47 and 2.90 Ga (5 grains). An upper Devonian quartzite from the miogeocline section at Agua Verde, north of Mazatán in central Sonora (Fig. 1), has detrital zircons with ages at ca. 1.43 Ga (10 grains), between 1.62 and 1.78 Ga (14 grains), and two more grains at 2.07 and 2.47 Ga. These provenance results are very intriguing but are not definitive as far as assessing the paleogeographic position of the Caborca block when the miogeocline rocks were formed. Perhaps future zircon provenance studies of more miogeocline rocks will help clarify if these Caborca rocks were formed in the Mojave Desert or instead were formed close to where we find them today in Sonora.

Eugeoclinal Sequence—The Sonoran Allochthon

A notable inconsistency in the Mojave-Sonora megashear model noted first by Poole and Madrid (1988), and later developed by Stewart et al. (1990), involves the stratigraphy and timing of deformation of the eugeoclinal sequence of central Sonora (Cortés terrane, Fig. 10). This sequence is allochthonous, and shares several characteristics with other eugeoclinal sequences along the North America cordillera. This includes general facies associations and zircon provenance data, with a high abundance of zircons with ages > 2.4 Ga. Based on current knowledge of eugeoclinal sequences in Sonora and the western United States, finding arguments that clearly support or invalidate the Mojave-Sonora megashear hypothesis is difficult at best. The age of thrusting of the Roberts Mountains allochthon in Nevada is not well constrained and the Antler Orogeny itself may have occurred diachronously along the cordilleran margin. Nonetheless, in contrast to eugeoclinal sequences in Nevada, which were juxtaposed in the mid-Paleozoic, the eugeoclinal and miogeoclinal sequences of central Sonora were juxtaposed in Permian time (Stewart et al., 1990, 1997). Poole and Madrid (1988) point out that there are some similarities between the Sonoran eugeoclinal allochthon and the Ouachita orogen of western Texas. The most notable of these is the general southwestern tendency of stratigraphic trends in Texas. These authors also draw attention to the apparent westward continuation of the Ouachita front into Sonora.

In a recent article, Poole et al. (2005) further developed the hypothesis that the Ouachita-Marathon belt can be extended westward from exposures in the Big Bend region of western Texas toward central Sonora. These authors interpret sedimentary facies and structures along the southern margin of Laurentia as the result of diachronous oblique collision, so that the foredeep facies and foreland basin depocenters become progressively younger toward the west. According to the collisional model of Poole et al. (2005), the deformation along the margin initiated in the Mississippian, and culminated in the late Pennsylvanian in the Ouachita segment of the orogen, in the Early Permian in the Marathon segment, and lastly, in the Late Permian in the Sonoran segment. The age difference is interpreted as the result of the progressive counterclockwise rotation of Gondwana during collision. In turn, this model requires that Sonora remain in a paleogeography similar to the paleogeography present during the late Paleozoic.

Comparison of the Upper Paleozoic Marine Sequences of East-Central California and the Caborca Terrane

During the late Paleozoic, carbonate platforms developed along the southwest margin of North America. In northeast Sonora, these facies are represented by the Escabrosa, Horquilla, El Tigre, Earp, and Epitah Formations from the Carboniferous to the Permian. These formations are well known in northern Chihuahua, southern Arizona, and southwestern New Mexico. In the Caborca terrane, the sequence is represented by a thick succes-

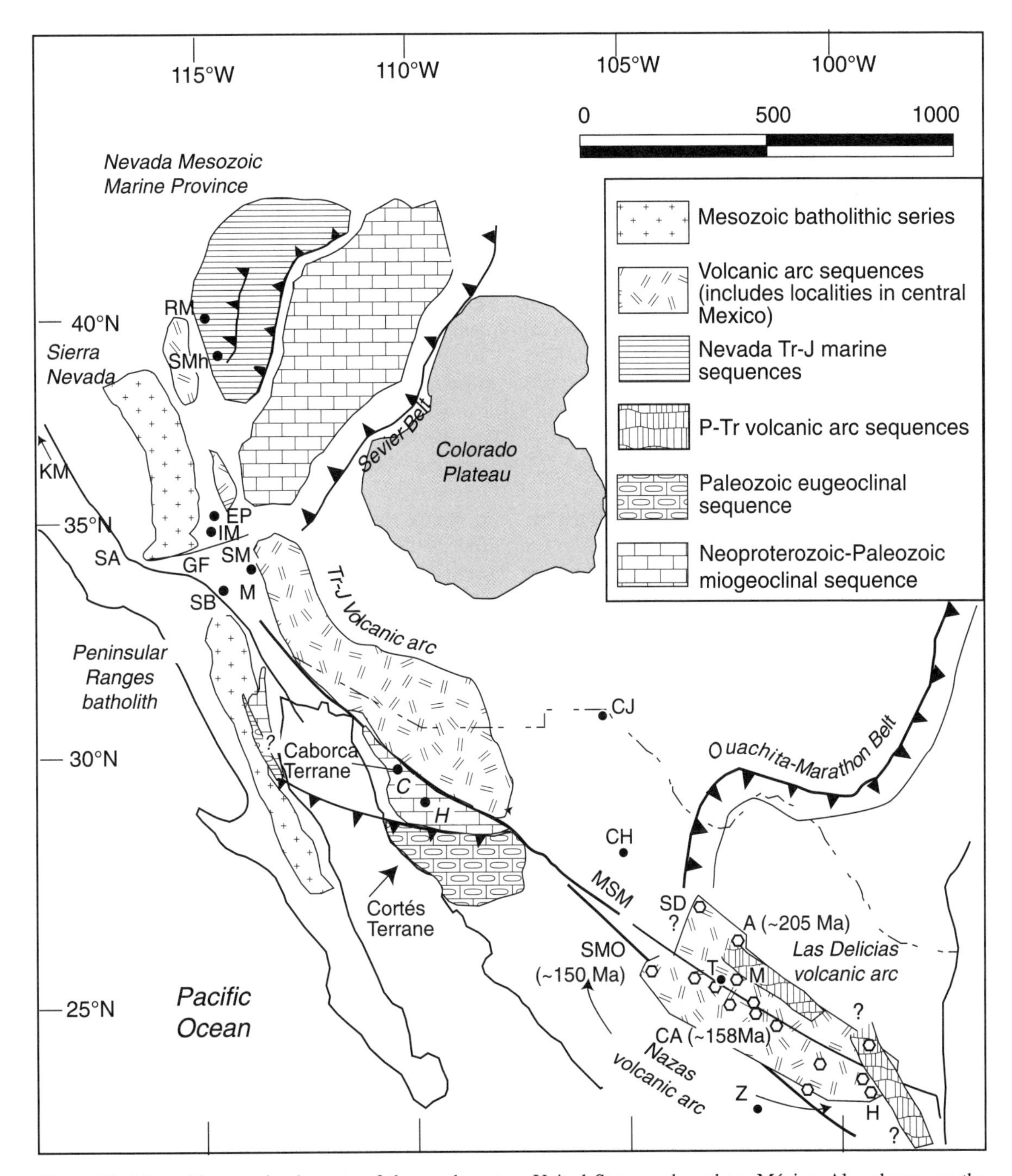

Figure 10. Map with tectonic elements of the southwestern United States and northern México. Also shown are the localities for the Triassic-Jurassic magmatic arc of north-central México (open hexagons). Abbreviations: *Localities in the Mojave Desert*: SA—San Andreas fault; SB—San Bernardino Mountains; EP—El Paso Mountains; SM—Soda Mountains; GF—Garlock fault; IM—Inyo Mountains. *Localities in México*: C—Caborca; H—Hermosillo; CH—Chihuahua; SMO—Santa María del Oro; CJ—Ciudad Juárez; SD—Sierra del Diablo; A—Acatita, Sierra Las Delicias; T—Torreón; CA—Caopas-Pico de Tera; Z—Zacatecas; H—Valle de Huizachal; M—Pozo Mayrán I; MSM—Mojave-Sonora megashear. *Localities in Nevada*: RM—Roberts Mountains; SMh—Shoshone Mountains. TR-J—Triassic-Jurassic; KM—Klamath Mountains; P-TR—Paleozoic-Triassic.

sion of seven informal units exposed at Sierra Santa Teresa near Hermosillo (Stewart et al., 1997). The Sierra Santa Teresa upper Paleozoic section is ~1500 m thick; it includes shallow-marine carbonate platform facies (mudstone, wackestone, and packstone) of the Upper Mississippian to the mid-Pennsylvanian, a hiatus from the late Pennsylvanian to Wolfcampian, 100 m of fusulinid packstone and briozoan fragments, and 600 m of calcareous siltstone, fine sandstone, and calcarenita with Leonardian to Guadalupian fauna. This sequence bears no resemblance to the upper Paleozoic sequence of the Inyo Mountains (Stevens and Stone, 1988). The Inyo Mountains sequence is characterized by the presence of Mississippian shales, Pennsylvanian calcareous sandstones and bioclastic calcareous siltstone, and an expanded Wolfcampian section of calcareous limolite (nearly 2000 m thick), and by calcarenita, bioclastic limestone, and limestone conglomerate of the Leonardian, covered by siliciclastic rocks (Fig. 9).

Correlation of the Marine Triassic Section of Northwest Sonora with the Mesozoic Marine Province of Nevada

Stanley and González-León (1995) have suggested that the Antimonio and Río Asunción Formations of Triassic to lowermost Jurassic age (González-León, 1980; Lucas and Estep, 1999), and exposed at Sierra del Álamo Muerto west of Caborca (Figs. 6 and 10), may be correlated with the Triassic-Jurassic sequence of central Nevada (Luning and Gabbs Formations; Oldow, 1984; Oldow et al., 1993).

At Sierra del Álamo, the Antimonio Formation overlies the Monos Formation, which consists of limestones bearing giant fusulinids characteristic of allochthonous terranes of the cordillera, such as the Klamath terrane of northern California. González-León and Stanley (1993) proposed including rocks of the Antimonio and Monos Formations in the suspect Antimonio terrane. The Antimonio terrane has been interpreted as allochthonous with respect to the miogeoclinal sequence of Caborca (González-León and Stanley, 1993; Molina-Garza and Geissman, 1996). The allochthoneity of the terrane cannot be clearly demonstrated as there is no clear evidence of a suture, but Lucas and Estep (1999) show the Río Asunción and Santa Rosa Formations in thrust contact over the Antimonio Formation. Based on the correlation of the marine Triassic sequences, González-León and Stanley (1993) infer that the Antimonio terrane is a fragment of the marine Mesozoic basin of Nevada, which has been tectonically transported ~1000 km southward until reaching its present position.

The basal coarse siliciclastic member of the Río Asunción Formation (quartzite and metamorphic quartz boulder conglomerate) is inserted between two marine horizons with ammonoid faunas; Hettangian below and Sinemurian above. This conglomeratic member has been recognized at Pozos de Serna and Sierra Santa Rosa (Fig. 5) and it is possibly equivalent to the Coyotes Formation of the Barranca Group of central Sonora. The Barranca Group overlies miogeoclinal and eugeoclinal strata in that region. Outcrops of the Liassic conglomerate are exposed in a band ~300 km possibly from Quitovac, in northwestern Sonora, to the Barranca Group basin in central Sonora. This correlation suggests that the inferred allochthoneity of the Antimonio terrane may be only valid for pre–Middle Jurassic time. This is because the Liassic boulder conglomerate overlies sequences characteristic of the Caborca terrane at various localities (i.e., Calmus et al., 1997).

The correlation of the Lunning Formation with Triassic rocks of Sierra del Álamo in Sonora is based, however, on a diverse subtropical reef fauna of the Norian, which lacks diagnostic provincialism. This fauna is present in a biostromal interval at the top of the Antimonio Formation, and contains coral, bivalve, sponge, and chephalopod taxa with affinities that allow this fauna to be equally associated with Norian faunas of the Mesozoic Marine province of west-central Nevada, northern Peru, or the western Thethys region; although the Antimonio fauna share more taxa with rocks from Nevada. Rocks from Sierra del Álamo also share taxa with the east Klamath and Stikinia terranes. Stanley and González-León (1995) suggest that thickness and lithofacies at Sierra del Álamo compare favorably with the Nevada sequence. For instance, in Sonora as in Nevada limestones dominate during Norian time. The succession in Nevada contains, however, expanded Lower and Middle Norian sections with much thicker carbonate intervals than at Sierra del Álamo. Also, at Sierra del Álamo, Early Norian rocks are dominated by siliciclastics.

Recent reviews of the ammonoid sequence of Sierra del Álamo (Gómez-Luna and Martínez-Cortés, 1997; Lucas and Estep, 1999) do not support the correlation of Stanley and González-León (1995). Oldow et al. (1993) show that the oldest rocks in the Lunning interval are from the Middle Triassic (Ladinian), while Lucas and Estep (1999) recognize ammonoids from the Smithian and Spathian stages (Lower Triassic) in basal strata of the Antimonio Formation. The ammonoid fauna in the Sierra del Álamo section also includes components of the *Dilleri* biozone (Carnian), the biozones Suessi, Columbianus, Cordilleranus (Upper Norian), and the Crickmayi zone of the latest Triassic (Rhaetian). According to Stanley and González-León (1995), this fauna resembles that reported for the middle part of the Gabbs Formation in west-central Nevada. We note that the Mesozoic marine sequence from Nevada contains a unit that may be correlated with the coarse siliciclastic interval (of quartz-boulder conglomerate) of the Río Asunción Formation. However, lithologically, the Dunlap Formation and the Sierra del Álamo sections do not resemble each other; there is a greater abundance of carbonate facies in the Liassic shallow-water facies from Nevada, whereas these facies are nearly absent in northwestern Sonora.

In their correlation of sequences from Nevada and the Antimonio terrane, Stanley and González-León (1995) point out that the ammonoid fauna characteristic of the Crickmayi zone resembles one reported for the Gabbs Formation; when these authors refer to ammonoid fauna that conforms to the Dilleri zone, they find affinity with the fauna present in the Hosselkus Limestone of California. Ammonoides of the Dilleri zone of the Late Carnian occur at Sierra del Álamo in thin micrite facies, bioclastic limestone, and fine-grained sandstone, interpreted as deep-water facies (González-León et al., 1996). Overlying this interval, the section includes sandstone, siltstone, and scarce limestone beds, capped by an interval of impure carbonates (biostromes) interstratified with shale. Near the base of this section there is a fauna of the Middle and Upper Norian (Columbianus or Cordilleranus zone of the lower Upper Norian). In turn, overlying this package there are general representatives of the Amoenum zone. These strata are covered by bioclastic limestone, impure limestone, and siltstone with an Upper Norian fauna (Crickmayi zone), and above this interval ammonoids of the Upper Hettangian were reported.

Rocks of the Lunning association were originally defined in the central region of the Pilot Mountains in Nevada as a sequence of dark dolomite, limestone, argillite, and shale, interstratified with coarse clastic rocks. In the Shoshone Mountains, three ammonoid associations have been recognized: the "Carnites" fauna of the Carnian is concordantly overlain by a fauna characteristic of the Tropites zones (Tropites welleri subzone) of the Late Carnian. The third association corresponds to the Guembelites fauna of the Late Carnian–Early Norian (Muller et al., 1936). The Guembelites zones were referred by Silberling and Tozer (1968) to the Kerri zone. In turn, the Gabbs Formation is defined as successions of shale and siltstone interstratified with black limestone overlain by bioclastic sandy and shaley limestone. The fauna in the lowest strata is from the Late Norian, assigned by Silberling and Tozer (1968) to the Suessi zone. Tozer (1980) indicates that the fauna from this zone occurs throughout the Late Norian, and he assigned instead basal strata of the Gabbs Formation to the Amoenum zone. Faunas forming the middle and upper members of the Gabbs Formation were correlated with the Crickmayi zone (Tozer, 1980).

In the standard zonation, the Dilleri zone underlies the Welleri zone. Thus, there is no correspondence between the Nevada ammonoid succession and that from Sierra del Álamo. Furthermore, the sequence in Nevada at the Shoshone Mountains is much more complete than in Sonora, including, in ascending order, the Welleri, Macrolabatus, Kerri, and Magnus zones (Gómez-Luna and Martínez-Cortés, 1997). Gómez-Luna and Martínez-Cortés noted that biostromes in Nevada at the Pilot Mountains that share coral and sponge taxa with Sonora occur between the Kerri and Magnus zones, while in Sonora they are restricted to a younger interval between the Columbianus and Cordilleranus zones. The later zones have no equivalent in Nevada. Thus on the basis of the Cepahlopode zonation, Gómez-Luna and Martínez-Cortés (1997) find no support for the correlation of Stanley and González-León (1995).

Ductile Deformation with NW-SE Orientation Representing the Juxtaposition of a Triassic Intrusive against Jurassic Volcanics Close to Sonoita, Northwest Sonora

Campbell and Anderson (1998, 2003) report the presence of a ductile deformation zone in Sierra Los Tanques, close to the town of Sonoita in NW Sonora. They interpret this zone as the original trace of the Mojave-Sonora megashear (Fig. 5). Ductile fabrics are represented by banded ultramylonite and coarse-grained mylonite with NW-SE orientations and a variable thickness from 1 to 3 km. Triassic age granitoids (225 Ma; U-Pb zircon, reported in Stewart et al., 1986) are juxtaposed against Jurassic volcanic arc rocks. Subhorizontal lineations present along the ductile zone suggest lateral displacements; however, our own assessment of kinematic indicators in the classic outcrops of the Mojave-Sonora megashear in Sierra Los Tanques was inconclusive in determining the real sense of displacement along the ductile fault zone.

Caudillo-Sosa et al. (1996) propose that Proterozoic basement rocks and Jurassic volcanic arc rocks in the Quitovac area just SW of Sonoita were deformed during a transpressional event directly associated with the Jurassic Mojave-Sonora megashear. This age constraint is based on the argument that Late Cretaceous granitoids in the area were not affected by this ductile deformation event. However, more detailed work by Iriondo (2001) and Iriondo et al. (2005) suggests that these intrusive rocks in Quitovac, dated with U-Pb zircon geochronology at ca. 79 Ma, are affected by the thrust-related, ductile deformation event. Supporting this idea, Jurassic volcanic and volcanoclastic rocks from the Comobabi and Baboquivari Mountains in Arizona, ~100 km north of Quitovac, present a similar style of deformation and metamorphism that has been proposed to be of Late Cretaceous to early Tertiary age (Haxel et al., 1984). Similar fabrics are also found at Cerro Prieto, ~50 km south of Quitovac, where Calmus and Sosson (1995) found Proterozoic rocks thrusting Jurassic arc volcanic rocks.

Campbell and Anderson (1998, 2003) report mylonites with subvertical foliations from Sierra Los Tanques. In addition, some of the lineations are locally subvertical with kinematic indicators suggesting both NE and SW thrusting. These compressional structures, forming a positive flower structure, are interpreted by these authors to represent transpression during Jurassic strike-slip movement associated with the Mojave-Sonora megashear.

There is general agreement about the geometry of the structures, but the main problem is defining the age of the deformational event forming them. While Campbell and Anderson (1998, 2003) propose a Jurassic age for the ductile fabrics in Sierra Los Tanques, Iriondo (2001), Iriondo et al. (2005), and Haxel et al. (1984) propose a Late Cretaceous age for the fabrics in NW Sonora and SW Arizona. In particular, a structural and geochronological study by Iriondo (2001) and Iriondo et al. (2005) in the Quitovac area suggests that the ductile deformation present in the region, including the mylonite zone from Sierra Los Tanques, is the result of Laramide compressional tectonism (Late Cretaceous–early Tertiary). The different thrust units proposed in Quitovac were later tilted as a result of movement along Basin-and-Range normal faults (Iriondo, 2001; Iriondo et al., 2005). It is possible that some of the ductile fabrics in the region are the result of reactivation of older fabrics, but studies by Iriondo and colleagues basically proved that the ductile fabrics are not all exclusively associated with Jurassic tectonism in connection with the Mojave-Sonora megashear as previously proposed.

Ductile Deformation in the Opodepe Region: Additional Structural Evidence

In the Opodepe region northeast of Hermosillo (Fig. 5), ductile and fragile deformation has been reported to be associated with the hypothetical Jurassic Mojave-Sonora megashear (Rodríguez-Castañeda, 1996). Proterozoic rocks with NE-SW foliation, and variable dips (~60°), present kinematic indicators suggesting dextral slip along a ductile structure. This zone of deformation in Opodepe has been interpreted as an antithetic structure to the main Mojave-Sonora megashear fault. The deformation is considered pre–ca. 36 Ma because an alkaline granite of that age intrudes the ductile fabrics; however, the pluton itself presents ductile fabrics with the same NE-SW orientation. The proposed Jurassic age for the deformation in Opodepe is based solely on "regional tectonic considerations" (Rodríguez-Castañeda, 1996).

It is important to consider the possibility that this mylonite zone in Opodepe could be related to late Tertiary metamorphic core complex deformation. However, Calmus et al. (1997) question the presence of a metamorphic core complex in Opodepe, and instead propose that the ductile fabric may be of Proterozoic age. Obviously, the presence of ductile fabrics in Sonora does not imply that the resulting mylonites have any particular age.

Even more speculative is the inference made by Araiza-Martínez (1998) that a gneissic fabric present in Proterozoic rocks in Estacion Llano, ~150 km north of Hermosillo, is attributed to the Jurassic Mojave-Sonora megashear based solely on its proximity to the proposed trace of the Mojave-Sonora megashear. Precambrian gneisses assigned to the San Francisco Group overthrust Jurassic volcanoclastic rocks with fairly low metamorphic grade. Locally, in the San Francisco gold mine, deformation with the same characteristics is associated with metamorphism in amphibolite facies, perhaps indicating that the deformation could be Proterozoic. There is no evidence that metamorphic grade associated with Late Cretaceous–early Tertiary compressional deformation, or for that matter Jurassic deformation, exceeded greenschist facies in the Altar region (Hayama et al., 1984). The age interpretations presented by Araiza-Martínez (1998) are a good example of the influence of the Jurassic Mojave-Sonora megashear hypothesis on many geologists working in México.

Interpretation of the Barranca Group as the "Displaced" Deltaic Facies of the Chinle Group from Southwest North America

Marzolf and Anderson (1996, 2000) have suggested that the Triassic Barranca Group, in central Sonora, was displaced from its original position in southern Nevada via the Mojave-Sonora megashear. The Barranca Group was interpreted earlier as a continental rift basin fill (i.e., Stewart and Roldán-Quintana, 1991). The rift would have developed over a late Paleozoic orogen that deformed the Sonora miogeocline and eugeocline sequences. Deposition of the Barranca Group in an elongated basin bordered by normal faults, or a series of basins with these characteristics, is difficult to demonstrate. Furthermore, most of the sandstones in the Barranca sequence are relatively mature with abundant quartz and the section lacks volcanic rocks typical of continental rifts; only the Santa Clara Formation of the Barranca Group contains immature arkoses. Paleocurrent indicators suggest south-flowing streams, and these are not consistent with the expected east-west flow that the orientation proposed for the hypothetical Barranca rift basin (Stewart and Roldán-Quintana, 1991).

Detrital zircons in sandstones of the Santa Clara Formation (Gehrels and Stewart, 1998) form groups with ages ca. 1.42 Ga (17 grains), 250–280 Ma (14 grains), and 225–235 Ma (3 grains). The abundance of Permo-Triassic zircons indicates a relatively close magmatic source of that age, and the probable source is Permo-Triassic rocks of the continental arc of Chihuahua, Coahuila, and northeast México (Torres-Vargas et al., 1999; McKee et al., 1990). Triassic zircons can also be derived form the Coahuila block to the east. Sources of magmatic zircons with those ages in the Mojave Desert are scarce but do exist (Miller et al., 1992; Barth et al., 1997). González-León (2005, personal communication) notes that igneous rocks in the age range from 240 to 280 Ma are common in southeast California and west-central Nevada. Triassic plutons have been also recognized near Sonoita (Stewart et al., 1986; Campbell and Anderson, 1998). A test for the proposed correlation of Triassic rocks in Sonora with the Chinle Group would be the presence of zircons with ages between 500 and 525 Ma, which are characteristic of the Osobb Formation (of the Auld Lang Syne Group in Nevada) in what has also been interpreted as the westward continuation of the Chinle Group drainages (Riggs et al., 1996).

Apparent Truncation of the Jurassic Continental Arc Sequence of Northern Sonora and Southern Arizona in the Caborca Region

Anderson and Schmidt (1983) and Anderson et al. (1979) have suggested that the Sonoran Jurassic continental arc was truncated by the Mojave-Sonora megashear. This segment of the Jurassic arc is typically included in the Papago terrane of Haxel et al. (1984) or Papago Domain of Anderson et al. (2005). The interpretation of truncation of the arc is based on the premise that outcrops of Jurassic volcanic rocks do not exist south of the inferred trace of the megashear. This argument is not entirely valid and needs revision. For instance, the arc sequence north of Caborca at Cerro la Basura and Cerro Álamo (Fig. 5) contains marine rocks (Corona, 1979) with Sinemurian ammonites (García-Barragán and Jacques-Ayala, 1993); this indicates the presence of a Liassic marine basin south of the arc.

In the mountains of Sierra Santa Rosa, Cerro El Rajón, Sierra del Álamo, and Pozos de Serna, there are outcrops of marine sequences with similar Liassic fauna (Hardy, 1981; Lucas et al., 1999; Stanley and González-León, 1995; Longoria and Pérez, 1979), thus correlative with rocks in the southern part of the arc sequence. Besides sharing an ammonoid fauna, rocks in those ranges show a continuity that suggests the structural integrity of the basin in which they were deposited. These sequences have been interpreted as fore-arc basin deposits (González-León, 1997). The sequence, as mentioned earlier, is characterized by turbidites and tuffaceous horizons; lateral facies changes are also important. Locally, there are conglomerates with abundant clasts of volcanic rocks and granite. Sandstones of the Santa Rosa Formation have a sedimentary petrology that indicates the rock components were derived from a volcanic arc (Stanley and González-León, 1995). For this reason, the present distribution of the arc in northern Sonora and a fore-arc basin in the Caborca region do not require displacement between these two elements. We must add that several outcrops of volcanic rocks of Jurassic age exist south of the trace of the megashear. Jurassic porphyry crops out overlying the Paleoproterozoic Bámuri Metamorphic Complex near Cerro Bámuri (P. Castiñeiras, 2004, personal communication), Jurassic andesites have also been reported for Cerro Chino (Longoria and Pérez, 1979), and Jurassic rhyolites are exposed near Cabo Tepoca (Anderson et al., 2005). These authors entertain the idea that these rocks were thrust southward, although thrusts with this vergence have not been reported.

Other authors have proposed that rocks of the volcanic arc extended, at least partially, south of the trace of the hypothetical megashear (Jacques-Ayala, 1995). This interpretation is based on observations of conglomerates of the Bisbee Group along the trace of the Mojave-Sonora megashear with abundant clasts of volcanic rocks. Paleocurrent indicators in the conglomerates suggest sources to the south. This, combined with the low frequency of clasts derived from the miogeoclinal sequence suggests, as Jacques-Ayala (1995) inferred, that rocks of the volcanic arc covered the mioclinal sequence during deposition of the Bisbee conglomerates.

Yet another important contribution to the debate over truncation of the Jurassic arc is the observation of a Middle Jurassic volcanic sequence in tectonic contact with Precambrian basement at a locality south of the trace of the Mojave-Sonora megashear (Calmus and Sosson, 1995). This locality, east of Puerto Peñasco in northwest Sonora (Fig. 5), clearly demonstrates that the Mojave-Sonora megashear, if it exists, must be located south of these exposures.

Jones et al. (1995) interpret volcanic rocks in north-central México in northern Zacatecas and Durango as the "displaced" counterpart of the volcanic arc of Sonora. Volcanic and volcanoclastic rocks in north-central México include the Caopas schist and the Nazas Formation (Grajales-Nishimura et al., 1992). Although the age and correlation of the Nazas Formation present a complex stratigraphic problem (Barboza-Gudiño et al., 1999), the presence of a Triassic-Jurassic volcanic arc in north-central México in southeast Chihuahua, southern Coahuila, and southern Tamaulipas (Fig. 10) is reasonably well documented. Rhyolites at the Mayran-1 well in Coahuila produce a two point Rb-Sr isochron with an age of 220 ± 20 Ma (Grajales-Nishimura et al. [1992]). Also, a K-Ar date on intrusive rocks at Santa María del Oro in Durango is also Jurassic (ca. 150 Ma; Grajales-Nishimura et al., 1992). U-Pb and Ar-Ar data for intrusive rocks at Sierra Las Delicias in Coahuila give dates ca. 215 Ma (Molina-Garza, 2005). Finally, a discordant U-Pb date on zircons from an intrusive related to the Caopas schists in northern Zacatecas is of 158 ± 4 Ma (Jones et al., 1995).

The hypothesis that rocks in north-central México were displaced by the megashear is inconsistent with the age of the basement in Durango, Zacatecas, and San Luis Potosí (Rudnick and Cameron, 1991). This hypothesis is also inconsistent with the association of volcanic rocks of the Nazas Formation, in Durango, with a thick continental sequence of red beds, and not with Liassic marine strata as rocks of the Sonora segment of the volcanic arc at Cerro la Basura (Fig. 5). Furthermore, the volcanic sequence in north-central México overlies Upper Triassic marine strata of the Zacatecas and La Ballena Formations, submarine fans with a restricted stratigraphic range. Similar strata are unknown in the arc of northern Sonora.

Rocks of the Nazas Formation and equivalent strata in the Mexican Altiplane are overlain by an Oxfordian transgressive sequence (Zuloaga and La Gloria Formations and other Jurassic strata). In the region of Caopas, Anderson et al. (1991) reported a deformation event that they assign to the Late Jurassic. A northwest-striking nappe is interpreted to be the result of transpression related to the Mojave-Sonora megashear. The nappe involves rocks of the Caopas schist and the Nazas Formation. Anderson et al. (1991) suggest that some deformation affects the Zuloaga Formation and interpret the deformation event as contemporaneous with deposition of this unit. We note that the Caopas schist is difficult to correlate with the Jurassic arc of northern Sonora, because the age of the protolith (a quartz porphyry) is unknown. Also, phillites in the region record a Cretaceous age of metamorphism (ca. 79 Ma; Iriondo et al., 2003b). Mylonites in the same region also suggest Laramide ages, but their Ar-Ar spectra are disturbed. The Jurassic age (158 ± 4 Ma) of the granite mentioned earlier, presumably related to the Caopas schist, is also problematic; the age is too close to the depositional age of the Oxfordian transgressive sequence.

Finally, in seeking to fit the model of displacement of the volcanic arc, Jones et al. (1995) draw the trace of the Mojave-Sonora megashear to the south of the Coahuila paleo-island (Las Delicias arc). This excludes exposures of calc-alkaline igneous rocks of the Acatita series at Sierra Las Margarita (Molina-Garza, 2005) and rhyolites of pre-Cretaceous age at Sierra del Diablo (McKee, et al., 1990; Fig. 10).

Transtension and Pull-Apart Basins along the Trace of the Mojave-Sonora Megashear in Sonora and Arizona

Early Cretaceous basins in Sonora and southern Arizona record an interval of nonmarine to marine deposition that followed activity along the hypothetical Mojave-Sonora megashear and associated faults. According to Nourse (1995, 2001), deformation linked to left-lateral shear along a broad region resulted in areas of localized subsidence, defined by outcrops of the Bisbee Group. According to this author, the margins of these areas of localized subsidence are NW- and E-trending faults that can be delineated by deposition of the Glance Conglomerate, the basal unit of the Bisbee Group. Thus, the basins have been interpreted as pull-aparts developed at releasing bends along the megashear. More recently, Anderson and Nourse (2005) elaborate on this hypothesis. Nourse et al. (1994) and Nourse (1995) have postulated the metamorphic core complexes in the same region developed along those areas that were stretched and fractured by transtensional faulting in the Late Jurassic and Early Cretaceous. These authors also suggest that the same discontinuities created by transtensional faulting accommodated compressional deformation in the Late Cretaceous.

McKee and Anderson (1999) have interpreted a Lower Cretaceous sequence near Sierra Azul north of Cucurpe (Fig. 5) as gravity-slide deposits emplaced in the Early Cretaceous in a basin with high-relief margins; they interpreted the basin as a pull-apart basin linked to Jurassic strike-slip. Although the basin in question is some 50 km north of the trace of the Mojave-Sonora megashear, these authors suggested that a transtensional regime existed in northern Sonora and southern Arizona. Other basins that have been incorporated into this model include the McCoy basin, in SW Arizona and deposits in the Altar–Santa Ana region of Sonora (Nourse, 2001). Similarly, Rodríguez-Castañeda (1994) and other authors have suggested that deposition of the Bisbee Group was controlled by the Mojave-Sonora megashear in northern Sonora, even if activity on the fault had presumably ceased by this time.

Interpreting the Bisbee Group as the fill of pull-apart basins is not straightforward. For instance, gravity-slide deposits in the Agua Prieta region studied by McKee and Anderson (1999) had been previously interpreted as a Paleozoic sequence thrusted on rocks of the Bisbee Group (Rangin, 1982). An important criticism of the model suggesting that NW-SE–trending faults affecting the region of the continental Jurassic arc of northern Sonora and southern Arizona are linked to the Mojave-Sonora megashear, as well as the models suggesting that these faults are concurrent with pull-apart basins and deposition of the Bisbee Group, is that it has been demonstrated that faults with this orientation (i.e., Sawmill Canyon fault in Arizona) are active in the Late Cretaceous and early Tertiary. The faults affect rocks as young as Late Cretaceous, producing rotations about vertical axes (Hagstrum and Sawyer, 1989; Sosson, 1990). However, one could argue that the faults were established in the Late Jurassic, but were reactivated during the Late Cretaceous. Jurassic slip on these faults has been suggested by Drewes (1996).

The Basement in the Northern Mexican Altiplane

Cameron et al. (1989) undertook an isotopic study of mid-Tertiary andesites along a 700-km NE-SW transect through Chihuahua. These andesitic lavas, with relatively low K/P values (<7) and variable Ba/Nb (50 versus 18), show similar isotopic values throughout the transect, suggesting that their isotopic signatures are not controlled by assimilation of significant amounts of crustal material. Along the transect, basaltic rocks have εNd and $^{87}Sr/^{86}Sr$ values close to bulk earth and isotopic relationships for $^{206}Pb/^{204}Pb$ and $^{207}Pb/^{204}Pb$ that plot on the 1.7 Ga pseudo-isochron. This NE-SW transect crosses the trace of the inferred Mojave-Sonora megashear. At this latitude, the megashear is considered a lithospheric discontinuity that separates Proterozoic basement to the NE from Phanerozoic basement to the SW. The geochemical/isotopic changes across the inferred fault are very smooth and appear to be gradational rather than abrupt. This isotopic uniformity for Sr and Nd determinations across the Mojave-Sonora megashear could be interpreted in at least three ways: (1) The subcontinental lithosphere for both the Proterozoic and the Phanerozoic basement blocks is indistinguishable for their Sr and Nd isotopic values. (2) The Mojave-Sonora megashear is not a lithospheric boundary separating two different basements at least not in the study area of the transect. (3) The isotopic signatures were acquired at the asthenosphere and not from the subcontinental lithosphere. McDowell et al. (1999) presented a similar conclusion based on the study of rocks from the Tomóchic volcanic field in central Chihuahua (Fig. 6). These authors suggested that a simplistic hypothesis in which Laurentian basement was displaced along the megashear was not possible.

Paleomagnetic Data

Paleomagnetic data for the Caborca terrane were cited as evidence in support of the Mojave-Sonora megashear. The data cited include preliminary data for the Triassic-Jurassic Antimonio Formation (3 sites) published by Cohen et al. (1986). More recently, Molina-Garza and Geissman (1996, 1999) reported paleomagnetic data for the Antimonio Formation, for the rocks of the Neoproterozoic-Paleozoic mioclinal sequence, and for rocks of the Jurassic volcanic arc assigned to the Fresnal Canyon sequence. Molina-Garza and Geissman (1999) also present paleomagnetic data for Lower Cretaceous strata and for intrusive rocks of the Laramide age Sonoran batholith.

The paleomagnetic data for the Antimonio Formation and for an unnamed yet correlative sequence at Barra Los Tanques (Fig. 5), ~50 km west of type locality of the Antimonio Formation at Sierra del Álamo, clearly show that the study of Cohen et al. (1986) cannot be used to support the Mojave-Sonora megashear hypothesis. First, the characteristic magnetization of the Antimonio Formation is of secondary origin; the age of the magnetization is pre–Early Cretaceous, but is not Triassic. Secondly, there is structural complexity at Sierra del Álamo, which Cohen et al. (1986) failed to recognize, such as the westward tilt of Cretaceous rocks overlying the previously folded Triassic-Jurassic sequence.

The mean paleomagnetic direction reported by Molina-Garza and Geissman (1996) for the Antimonio Formation was interpreted in terms of the Jurassic accretion of the Antimonio terrane to the margin of North America or the para-autochthonous Caborca terranes. The mean direction (D = 186.7.0°, I = -28.4°; n = 14 sites; k = 28.2; α95 = 7.5°) is discordant; it does not resemble the expected direction from Early Jurassic to Early Cretaceous cratonic reference poles. This magnetization is interpreted as the record of accretion of the Antimonio terrane to Caborca in the Late Jurassic. Alternatively, this magnetization may reflect deformation in an essentially autochthonous fore-arc basin. The declination of the characteristic magnetization indicates post–Middle Jurassic clockwise rotation of the Antimonio terrane with respect to North America. Similar clockwise rotations have been observed in other regions of Sonora and are explained below.

The mean paleomagnetic direction for the mioclinal sequence is also a secondary magnetization restricted by field and other evidence to date from the Middle to Early Jurassic (Molina-Garza and Geissman, 1999). The observed direction (D = 15.0°, I = 10.0°; n = 28 sites; k = 23.0; α95 = 5.8°) indicates that the Caborca terrane experienced a clockwise rotation with respect to North America. According to paleomagnetic data for Lower Cretaceous strata, the rotation occurred in Middle to Late Jurassic. These paleomagnetic data also argue against latitudinal displacement of the Caborca terrane from north to south (Fig. 11).

The Fresnal Canyon volcanic sequence from localities north of Caborca has a characteristic magnetization for which the mean (D = 15.0°, I = 4.0°; n = 10 sites; k = 12.4; α95 = 14.3°) is essentially indistinguishable from the direction observed in mioclinal rocks of the Caborca terrane. The magnetization of the volcanic rocks indicates a similar clockwise rotation and suggests that no relative displacement exists between the Caborca terrane and the Jurassic continental arc. The paleomagnetic data for the volcanic rocks at Cerro la Basura and Cerro Álamo predate deformation of the sequence, which suggest but do not prove that this is a primary magnetization. These rocks also pass a conglomerate test (Molina-Garza and Geissman, 1998), further supporting the interpretation of the magnetization as primary.

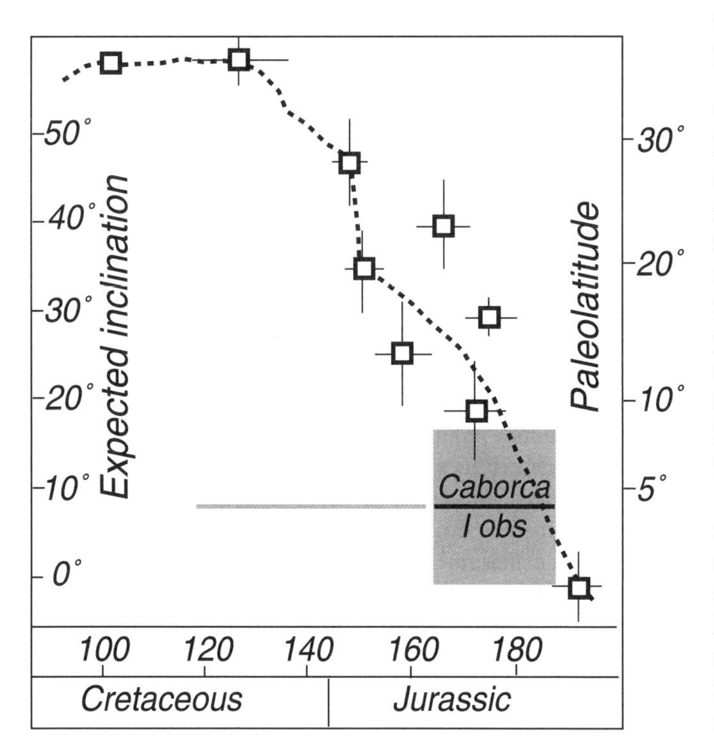

Figure 11. Expected (dotted line) and observed (I obs) inclinations for the Jurassic and Cretaceous times in the Caborca region. Modified after Molina-Garza and Geissman (1999).

The mean declinations of both the secondary magnetization in mioclinal rocks and of the primary magnetization in rocks of the Jurassic arc indicate a clockwise rotation of Sonora with respect to the craton. It is difficult to give a tectonic interpretation to secondary magnetization, unless the timing of remanence acquisition can be established with some certainty. In the region of Caborca, both Lower Cretaceous sedimentary rocks and Late Cretaceous intrusive rocks contain concordant magnetizations. That is, these rocks indicate no rotation or latitudinal displacement of the Caborca area with respect to North America. The youngest rocks of the Caborca terrane containing a discordant magnetization are latest Early Jurassic; thus the age of the regional remagnetization event is constrained between the Middle and Late Jurassic.

If the regional remagnetization event that affects rocks of the Caborca terrane had occurred in the Late Jurassic, because the observed inclination is more shallow than expected, it would indicate a latitudinal displacement from south to north of ~10° ± 7° (opposite of the Mojave-Sonora megashear proposed displacement). The observed declination would indicate a clockwise rotation of up to 50° for both the Jurassic volcanic arc and the Caborca terrane. On the other hand, if the remagnetization event affecting rocks of the Caborca terrane is Middle Jurassic, the interpretation is more uncertain. The low latitude track of the apparent polar wander path proposed by May and Butler (1986) predicts latitudes for a site in NW Sonora near the equator (1°S to 14°N), following the Jurassic northward drift of North America. The apparent polar wander path at high latitudes proposed by Van Fossen and Kent (1990) predicts latitudes of 1°S to 22°N. The observed paleolatitude of ~5°N (Fig. 11) may indicate displacement from north to south, but only if the remagnetization event is early Middle Jurassic. However, this displacement must apply to both the Caborca terrane and the Jurassic volcanic arc NE of the hypothetical Mojave-Sonora megashear. The evidence of a primary magnetization in rocks of the Jurassic volcanic arc north of Caborca (Molina-Garza and Geissman, 1998) and the reported isotopic ages for rocks within the arc (ranging from 153 to 170 Ma; Stewart et al., 1986; Anderson et al., 2005) suggest, however, that the age of the magnetization is late Middle Jurassic.

Together, the observation of pre-Cretaceous clockwise rotation indicated by NNE-directed declinations, the lack of evidence for north-south latitudinal displacement indicated by relatively shallow inclinations, and the observation of similar directions for localities north and south of the inferred trace for the megashear do not support reconstructions of the Caborca terrane north of its present position with respect to North America. The paleomagnetic data for relatively distant localities are similar, suggesting that the Caborca terrane behaves as a rigid block. In our view, paleomagnetic data provide perhaps the strongest evidence against the Mojave-Sonora megashear hypothesis.

Geology of the Mojave Desert Region

Upper Paleozoic and Lower Mesozoic rocks from the Mojave Desert region are very important to consider when assessing the hypothetical Mojave-Sonora megashear. The proposed trace is located west of the Soda Mountains in the central part of the Mojave Desert in California. Cameron (1981) was the first to note the apparent continuity of the Neoproterozoic and Paleozoic miogeosyncline and cratonal facies in the region (Stewart and Poole, 1975) that would hamper the presence of the hypothetical fault in the Mojave Desert. The apparent relationship between miogeosynclinal facies east and west of the Mojave Desert could be still consistent with the Mojave-Sonora megashear model only if the rocks from the San Bernardino Mountains were tectonically translated along the megashear at a later time; however, there is no evidence for such a geological relationship.

Following the Mojave-Sonora megashear model, Upper Paleozoic and Triassic sequences from the Soda Mountains and surrounding areas (Fig. 10) should have been present in Sonora if they were displaced along the megashear. However, rocks of these ages in the Caborca terrane are quite distinct from the ones in the Mojave Desert (Walker and Wardlaw, 1989). Late Permian rocks are scarce in Sonora, and Triassic rocks from the Barranca Group correspond to shallow continental to transitional facies with coal layers. In contrast, the lower Triassic in the Mojave Desert is represented by deep-water facies rocks (Silver Lake Formation) and other deep-marine units in the southern Sierra Nevada. In the Soda Mountains region, the Triassic rocks are represented by subaereal volcanic rocks, volcanoclastic sandstones, and eolian quartzites.

ALTERNATIVES AND CONCLUSIONS

In summary, evidence that clearly contradicts the Mojave-Sonora megashear model includes the paleomagnetic results (Molina-Garza and Geissman, 1999), the stratigraphic relationships between the Jurassic volcanic arc and the pre-arc basin present just south of the volcanic arc (poorly documented in the literature), and the impossibility of tracing the hypothetical fault through the Mojave Desert. However, we cannot ignore the apparent stratigraphic correlation between the miogeocline sequence in Caborca and the paleogeography of eastern California. Although, these correlations have been recently questioned in Sonora because of the apparent continuity between the Bolsa Quartzite (NE Sonora-Mazatzal province) and the Cambrian Proveedora Quartzite from the Caborca terrane, the correlation of Caborca miogeocline rocks with rocks of the Mojave Desert is a solid observation.

The apparent impossibility to trace the Mojave-Sonora megashear through the Mojave Desert in the lower Paleozoic suggests that the possible displacement of the miogeocline sequence occurred in the late Paleozoic as proposed by Stone and Stevens (1988) and Walker and Wardlaw (1989). The evidence that correlates the Triassic stratigraphic sequence from Nevada and the sequences from the Barranca Group and/or the Antimonio Formation in Sonora appears to be more equivocal and probably requires additional work to be tested. Another line of evidence to support the Mojave-Sonora megashear that requires further study is the correlation between Paleoproterozoic basement from Sonora and the basement from eastern California and southern Arizona. In addition, the hypothesis that a transpressional tectonic environment controlled the deposition of the Bisbee Group in southern Arizona and northern Sonora should be studied in greater detail. Even if these hypotheses were proven correct, this would not demonstrate the required, large latitudinal displacements of the Caborca terrane in the Mojave-Sonora megashear model. Stratigraphic, paleomagnetic, structural, and geochronological data that question the existence of the Mojave-Sonora megashear, however, do not preclude displacements of the miogeocline rocks during the late Paleozoic or even older displacements of the Proterozoic basement prior to deposition of the miogeocline sequence.

Some of the important observations that prop up the hypothesis for the Pennsylvanian to lower Permian truncation of the SW margin of Laurentia include the strike change for the continental margin, and the abrupt interruption of sedimentation patterns for lower Paleozoic rocks (Stevens et al., 1992, 2005). In Nevada, both the upper Paleozoic continental margin and the calcareous platform have a southwestern strike whereas this direction changes to southeast in southeastern California in the Mojave Desert region. In the lower Permian, eastern California was exposed to an extensional event with important subsidence (associated turbidite deposits) that was contemporaneous to local thrusting. This combination of contrasting deformations could be explained by transtension associated with a system of strike-slip faulting similar to the younger Jurassic Mojave-Sonora megashear. Stone and Stevens (1988) proposed that displacement of the Neoproterozoic and lower Paleozoic miogeocline rocks from the Caborca terrane occurred sometime in the late Paleozoic instead of Jurassic time. A similar reasoning for the tectonic evolution of SW North America is presented by Dickinson and Lawton (2001), where they propose Triassic displacement of the Caborca miogeocline rocks along what they call the California-Coahuila transform fault. These authors also proposed that the California-Coahuila fault system connected the convergence event occurring along the Sonora orogen in California and Nevada with the subduction taking place in east-central México at approximately the same time.

In conclusion, the unbiased evaluation presented here for the different sets of data, including stratigraphic relationships, structure, geochemistry, age constraints, and paleomagnetism, does not favor the original model of the Caborca terrane displaced along the Jurassic Mojave-Sonora megashear; however, these results cannot totally discredit the hypothesis. The impossibility of demonstrating latitudinal displacements during Middle to Late Jurassic time is perhaps, as shown in the paleomagnetic studies, the main weakness of the Mojave-Sonora megashear model. It is true that paleogeographic reconstructions of the Circum-Gulf of

Mexico require a different position for the Mexican subcontinent with respect to North America, but the Jurassic Mojave-Sonora megashear model or a similar set of faults are not the only alternatives for such reconstructions.

ACKNOWLEDGMENTS

This study has been supported by PAPIIT-UNAM project IN-1221002 and partly financed through a CONACyT project T-3286. We are thankful to T. Calmus, C. González-León, and T.H. Anderson for their reviews, which greatly improved an older version of this manuscript.

REFERENCES CITED

Albritton, C.C., and Smith, J.F., 1957, The Texas lineament, *in* XX Congreso Geológico Internacional, Relaciones entre la tectónica y la sedimentación: México, D. F., International Geological Congress, 2, p. 501–518.

Amaya-Martínez, R., González-León, C., and Moises, M., 1993, Las rocas volcánicas de la Formación Tarahumara y su relación con las secuencias volcanosedimentarias en la porción centro oriental del Estado de Sonora, *in* González-León, C., and Vega-Granillo, E.L., eds., III simposio de la Geología de Sonora y areas adyacentes: Instituto de Geología, Universidad Nacional Autónoma de México, 3, p. 1–2.

Anderson, T.H., and Nourse, J.A., 1998, Releasing bends of the sinistral Late Jurassic Mojave-Sonora fault system; principal structures within the southwestern borderland of North America: Geological Society of America, Rocky Mountain Section, 50th Annual Meeting: Abstracts with Programs, v. 30, no. 6, p. 2.

Anderson, T.H., and Nourse, J.A., 2005, Pull-apart basins at releasing bends of the sinistral Late Jurassic Mojave-Sonora fault system, *in* Anderson, T.H., Nourse, J.A., McKee, J.W., and Steiner, M.B., eds., The Mojave-Sonora Megashear Hypothesis: Development, Assessment, and alternatives: Geological Society of America Special Paper 393, p. 97–122, doi: 10.1130/2005-2393(03).

Anderson, T.H., and Schmidt, V.A., 1983, The evolution of Middle America and the Gulf of Mexico–Caribbean Sea region during Mesozoic time: Geological Society of America Bulletin, v. 94, p. 941–966, doi: 10.1130/0016-7606(1983)94<941:TEOMAA>2.0.CO;2.

Anderson, T.H., and Silver, L.T., 1977, Geochronometric and stratigraphic outlines of the Precambrian rocks of northwestern Mexico: Geological Society of America Abstracts with Programs, v. 9, no. 7, p. 880.

Anderson, T.H., and Silver, L.T., 1979, The role of the Mojave-Sonora megashear in the tectonic evolution of northern Sonora, *in* Anderson, T.H., and Roldán-Quintana J., eds., Geology of Northern Sonora: Geological Society of America Field Trip 27 Guidebook: University of Pittsburgh, Pennsylvania, p. 59–69.

Anderson, T.H., and Silver, L.T., 1981, An overview of Precambrian rocks in Sonora: Revista Instituto de Geología, Universidad Nacional Autónoma de México, v. 5, no. 2, p. 131–139.

Anderson, T.H., and Silver, L.T., 2005, The Mojave-Sonora megashear—Field and analytical studies leading to the conception and evolution of the hypothesis, *in* Anderson, T.H., Nourse, J.A., McKee, J.W., and Steiner, M.B., eds., The Mojave-Sonora Megashear Hypothesis: Development, Assessment, and Alternatives: Geological Society of America Special Paper 393, p. 1–50, doi: 10.1130/2005-2393(01).

Anderson, T.H., Eells, J., and Silver, L.T., 1979, Geology of Precambrian and Paleozoic rocks, Caborca-Bamori region, *in* Anderson, T.H., and J. Roldán-Quintana, J., eds., Geology of Northern Sonora: Geological Society of America Field Trip 27 Guidebook, University of Pittsburgh, Pennsylvania, p. 1–22.

Anderson, T.H., McKee, J.W., and Jones, N.W., 1991, A northwest trending, Jurassic fold nappe, northernmost Zacatecas, Mexico: Tectonics, v. 10, p. 383–401.

Anderson, T.H., Rodríguez-Castañeda, J.L., and Silver, L.T., 2005, Jurassic rocks in Sonora, Mexico: Relations to Mojave-Sonora Megashear and its inferred northwestward extension, *in* Anderson, T.H., Nourse, J.A., McKee, J.W., and Steiner, M.B., eds., The Mojave-Sonora Megashear hypothesis: Development, Assessment, and Alternatives, Geological Society of America Special Paper 393, p. 51–96.

Araiza-Martínez, H., 1998, The San Francisco gold mine municipality of Santa Ana, Sonora, Mexico, *in* Clark, K.F., ed., Gold Deposits of Northern Sonora, México: Guidebook Series—Society of Economic Geologists 30, p. 49–58.

Barboza-Gudiño, J.R., Tristán-González, M., and Torres Hernández, J.R., 1999, Tectonic setting of pre-Oxfordian units from central and northeastern Mexico: A review, *in* Bartolini, C., Wilson, J.L., and Lawton, T.F., eds., Mesozoic Sedimentary and Tectonic History of North-Central Mexico: Geological Society of America Special Paper 340, p. 197–210.

Barth, A.P., Tosdal, R.M., Wooden, J.L., and Howard, K.A., 1997, Triassic plutonism in Southern California: Southward younging of arc initiation along a truncated continental margin: Tectonics, v. 16, p. 290–304, doi: 10.1029/96TC03596.

Brunner, P., 1976, Litología y biostratigrafía del Misisípico en el área El Bisani, Caborca, Sonora: Revista Instituto Mexicano del Petróleo, v. 8, p. 7–41.

Calmus, T., and Sosson, M., 1995, Southwestern extension of the Papago terrane into the Altar Desert region, northwestern Sonora, and its implications, *in* Jacques-Ayala, C., González-León, CM., and Roldán-Quintana, J., eds., Studies on the Mesozoic of Sonora and Adjacent Areas: Geological Society of America Special Paper 301, p. 99–110.

Calmus, T., Pérez-Segura, E., and Stinnesbeck, W., 1997, La structuration de la marge pacifique nord-américaine et du "terrane Caborca": Apports de la découverte d'une faune du Jurassique inférieur et moyen dans la série de Pozos de Serna (Sonora, Mexique): Comptes Rendus de l'Académie des Sciences, t. 323, série IIa, p. 795–800.

Cameron, C.S., 1981, Geology of the Sugarloaf and Delamar mountain areas, San Bernardino Mountains, California [unpublished Ph.D. thesis]: Cambridge, Massachusetts Institute of Technology, 399 p.

Cameron, K.L., Nimz, G.J., Kuentz, D., Niemeyer, S., and Gunn, S., 1989, Southern Cordilleran basaltic andesite suite, southern Chihuahua, Mexico: A link between Tertiary continental arc and flood basalt magmatism in North America: Journal of Geophysical Research, v. 94, B6, p. 7817–7840.

Campa, M.F., and Coney, P.J., 1983, Tectonostratigraphic terranes and mineral resource distribution in Mexico: Canadian Journal of Earth Sciences, v. 20, p. 1040–1051.

Campbell, P.A., and Anderson, T.H., 1998, Structure and kinematics along a Jurassic Plate Boundary Transform, the Mojave-Sonora Megashear, *in* Clark, K.F., ed., Gold Deposits of Northern Sonora, Mexico: Society of Economic Geologists, Guidebook Series, 30: Fort Collins, Colorado, p. 177–186.

Campbell, P.A., and Anderson, T.H., 2003, Structure and kinematics along a segment of the Mojave-Sonora megashear: A strike-slip fault that truncates the Jurassic continental magmatic arc of southwestern North America: Tectonics, v. 22, no. 6, 16:1–21, doi: 10.1029/2002TC001367.

Caudillo-Sosa, G., Oviedo-Lucero, L.F., and Rodríguez-Castañeda, J.L., 1996, Falla Quitovac—Resultado de un evento de transpresión del "Mojave-Sonora megashear," noroeste de Sonora, México: Revista Mexicana de Ciencias Geológicas, v. 13, no. 2, p. 140–151.

Charleston, S., 1981, Lower Cretaceous stratigraphy and structure, northern Mexico: A field trip guidebook, West Texas Geological Society, Midland, Texas, 81–74.

Chávez-Cabello, G., Aranda-Gómez, J.J., Molina-Garza, R.S., Cossío-Torres, T., Arvizu-Gutiérrez, I.R., and González-Naranjo, G., 2005, La Falla San Marcos: Una estructura jurásica de basamento multireactivada del noreste de México: Volumen Centenario SGM—Grandes Fronteras Tectónicas de México: Boletín de la Sociedad Geológica Mexicana, Tomo, v. LVII, no. 1, p. 27–52.

Clark, K.F., 1998, Gold deposits of northwestern Sonora, Mexico: Society of Economic Geologists, Guidebook Series, 30: Fort Collins, Colorado, 252 p.

Cohen, K.K., Anderson, T.H., and Schmidt, V.A., 1986, A paleomagnetic test of the proposed Mojave-Sonora megashear in northwest Sonora, Mexico: Tectonophysics, v. 131, p. 23–51, doi: 10.1016/0040-1951(86)90266-0.

Conway, C.M., and Silver, L.T., 1989, Early Proterozoic rocks (1710–1615) in central and southeastern Arizona, *in* Jeney, J.P., and Reynolds, S.J., eds., Geologic Evolution of Arizona: Arizona Geological Digest, 17, p. 211–238.

Cooper, G.A., and Arellano, A.R.V., 1946, Stratigraphy near Caborca, northwest Sonora, Mexico: American Association of Petroleum Geologists Bulletin, v. 30, no. 4, p. 606–610.

Cooper, G.S., Okulich, V.J., Stoyanow, A., Villabos-Figueroa, A., Arellano, A.R.V., Johnson, J.H., and Lochman-Balk, C., 1952, Cambrian stratigraphy and paleontology near Caborca, northwestern Sonora, Mexico, Smithsonian Miscellaneous Collections, 119(1), 183 p.

Corona, F.V., 1979, Preliminary reconnaissance geology of Sierra La Gloria and Cerro Basura, northwestern Sonora, Mexico, *in* Anderson, T.H., and Roldán-Quintana, J., eds., Geology of Northern Sonora: Geological Society of America Field Trip 27 Guidebook, University of Pittsburgh, Pennsylvania, p. 59–69.

Damon, P.E., Livingston, D.E., Mauger, R.L., Gilletti, B.J., and Alor, P., 1962, Edad del Precámbrico Anterior y de otras rocas del zócalo de la región de Caborca-Altar de la parte noroccidental del Estado de Sonora: Boletín Instituto de Geología, Universidad Nacional Autónoma de México, v. 64, p. 11–64.

Damon, P.E., Shafiqullah, M., Roldán-Quintana, J., and Jacques-Ayala, C., 1983, El batolito Larámide (90–40 Ma) de Sonora: Asociación Ingenieros de Minas, Metalurgistas y Geólogos de México, Memoria Técnica, v. XV, p. 63–96.

De Cserna, Z., 1971, Development and structure of the Sierra Madre Oriental of Mexico: Geological Society of America Abstracts with Programs, v. 3, no. 6, p. 377–378.

De Cserna, Z., 1976, Mexico: Geotectonics and Mineral Deposits: New Mexico Geological Society Special Publication 6, p. 18–25.

De Jong, K.A., Escárcega, J.A., and Damon, P.E., 1988, Eastward thrusting, southwestward folding, and westward back-sliding in the Sierra de Víbora, Sonora, Mexico: Geology, v. 16, p. 904–907, doi: 10.1130/0091-7613(1988)016<0904:ETSFAW>2.3.CO;2.

Dickinson, W.R., and Lawton, T.F., 2001, Carboniferous to Cretaceous assembly and fragmentation of Mexico: Geological Society of America Bulletin, v. 113, p. 1142–1160, doi: 10.1130/0016-7606(2001)113<1142:CTCAAF>2.0.CO;2.

Drewes, H., 1996, A ramp-end strike-slip fault near Paradise, Chiricahua Mountains, southeastern Arizona: Geological Society of America Abstracts with Programs, v. 28, no. 5, p. 63.

Eells, J.L., 1972, Geology of the Sierra de la Berruga, northwestern Sonora, Mexico [master's thesis]: San Diego, San Diego State University, 77 p.

Farmer, G.L., Bowring, S.A., Matzel, J., Espinosa-Maldonado, G., Fedo, C., and Wooden, J., 2005, Paleoproterozoic Mojave province in northwestern Mexico? Isoptopic and U-Pb circón geochronologic studies of Precambrian and Cambrian crystalline and sedimentary rocks, Caborca, Sonora, *in* Anderson, T.H., Nourse, J.A., McKee, J.W., and Steiner, M.B., eds., The Mojave-Sonora Megashear Hypothesis: Development, Assessment, and Alternatives: Geological Society of America Special Paper 393, p. 183–198, doi: 10.1130/2005-2393(05).

García-Barragán, J.C., and Jacques-Ayala, C., 1993, Stratigraphy and depositional setting of a red conglomerate in the Cerro la Gloria, Caborca, Sonora, *in* González-León, C., and Vega-Granillo, E.L., eds., III Simposio de la Geología de Sonora y áreas adyacentes: Instituto de Geología, Universidad Nacional Autónoma de México, v. 3, p. 33–34.

García-Barragan, J.C., Jacques-Ayala, C.J., and DeJong, K.A., 1998, Stratigraphy of the Cerros El Amol, Altar, Sonora, México, *in* Clark, K.F., ed., Gold Deposits of Northern Sonora, México: Guidebook Series—Society of Economic Geologists 30, p. 101–116.

Gehrels, G.E., and Stewart, J.H., 1998, Detrital zircon U-Pb geochronology of Cambrian to Triassic miogeosinclinal and eugeosinclinal strata of Sonora, Mexico: Journal of Geophysical Research, B, Solid Earth and Planets, v. 103, no. 2, p. 2471–2487, doi: 10.1029/97JB03251.

Gómez-Luna, M.E., and Martínez-Cortés, A.M., 1997, Relaciones y diferencias de la sucesión de amonoideos Triásicos del noroeste de Sonora y de Nevada centro-oeste, *in* González-León, C., and Stanley, G., eds., Publicaciones Ocasionales Estación Regional del Noroeste: Instituto de Geología, Universidad Nacional Autónoma de México, v. 1, p. 19–22.

González-León, C.M., 1980, La Formación Antimonio (Triásico Superior-Jurásico Inferior) en la Sierra del Alamo, Estado de Sonora: Revista Instituto de Geología, Universidad Nacional Autónoma de México, v. 4, p. 13–18.

González-León, C.M., 1997, Sequence stratigraphy and paleogeographic setting of the Antimonio Formación (Late Permian-Early Jurassic), Sonora, México: Revista Mexicana de Ciencias Geológicas, v. 14, p. 136–148.

González-León, C.M., and Lawton, T.F., 1995, Stratigraphy, depositional environments and origin of the Cabullona basin, northwestern Sonora, *in* Jacques-Ayala, C., González-León, C.M., and Roldán-Quintana, J., eds., Studies on the Mesozoic of Sonora and Adjacent Areas: Geological Society of America Special Paper 301, p. 121–143.

González-León, C.M., and Lucas, S.G., 1995, Stratigraphy and paleontology of the Early Cretaceous Cerro de Oro Formación, central Sonora, *in* Jacques-Ayala C., González-León, C.M., and Roldán-Quintana, J., eds., Studies on the Mesozoic of Sonora and Adjacent Areas: Geological Society of America Special Paper 301, p. 41–48.

González-León, C., and Stanley, G.D., Jr., 1993, The Antimonio Terrane of western Sonora and its paleogeographic significance, *in* González-Leon, C., and Vega-Granillo, E.L., eds., III Simposio de la Geología de Sonora y áreas adyacentes: Instituto de Geología, Universidad Nacional Autónoma de México, 3, p. 41–43.

González-León, C.M., Taylor, D.G., and Stanley, G.D., Jr., 1996, The Antimonio Formation in Sonora, Mexico, and the Triassic-Jurassic boundary: Canadian Journal of Earth Sciences, v. 33, p. 418–428.

Grajales-Nishimura, J.M., Terrell, D.J., and Damon, P.E., 1992, Evidencias de la prolongación del arco magmático Cordillerano del Triásico Tardío-Jurásico en Chihuahua, Durango y Coahuila: Boletín Asociación Mexicana de Geólogos Petroleros, v. XLII, p. 1–18.

Hagstrum, J.T., and Sawyer, D.A., 1989, Late Cretaceous paleomagnetism and clockwise rotation of the Silver Bell Mountains, south central Arizona: Journal of Geophysical Research, B, Solid Earth and Planets, v. 94, no. 12, p. 17,847–17,860.

Hardy, L.R., 1981, Geology of the central Sierra de Santa Rosa, northwest Sonora, Mexico, *in* Ortlieb, L., and Roldán-Quintana, J., eds., Geology of Northwestern Mexico and Southern Arizona: Geological Society of America, Cordilleran Section Annual Meeting, Field Guides and Papers, p. 73–98.

Haxel, G.B., Tosdal, R.M., May, D.J., and Wright, J.E., 1984, Latest Cretaceous and early Tertiary orogenesis in south-central Arizona: Thrust faulting, regional metamorphism, and granitic plutonism: Geological Society of America Bulletin, v. 95, p. 631–653, doi: 10.1130/0016-7606(1984)95<631:LCAETO>2.0.CO;2.

Hayama, Y., Shibata, K., and Takeda, H., 1984, K-Ar ages of the low-grade metamorphic rocks in the Altar massif, northwest Sonora, Mexico: Journal of the Geological Society of Japan, v. 90, p. 586–596.

Henry, C.D., and Aranda-Gómez, J.J., 2000, Plate interactions control middle-late Miocene, proto-Gulf and Basin and Range extension in the southern Basin and Range: The influence of plate interaction on post-Laramide magmatism and tectonics in Mexico: Tectonophysics, v. 318, no. 1–4, p. 1–26, doi: 10.1016/S0040-1951(99)00304-2.

Herrera-López, P., Iriondo, A., and Rodríguez-Castañeda, J.L., 2005, Preliminary time constraints and geochemistry of the Proterozoic basement from Sierra Los Ajos and Cerros Las Mesteñas, NE Sonora, México: GEOS, Unión Geofísica Mexicana, A.C: Resúmenes y Programa, v. 25, no. 1, p. 87–88.

Iriondo, A., 2001, Proterozoic basements and their laramide juxtaposition in NW Sonora, Mexico: Tectonic constraints on the SW margin of Laurentia [Ph.D. thesis]: Boulder, University of Colorado at Boulder, 222 p.

Iriondo, A., and Atkinson, W.W., Jr., 2000, Orogenic gold mineralization along the proposed trace of the Mojave-Sonora megashear: Evidence for the Laramide Orogeny in NW Sonora, Mexico: Geological Society of America Abstracts with Programs, v. 32, no. 7, p. A393.

Iriondo, A., and Premo, W.R., 2003, The Caborca Block; an inhomogeneous piece of Paleoproterozoic crust in Sonora: Geological Society of America, Cordilleran Section, 99th Annual Meeting, Abstracts with Programs, v. 35, no. 4, p. 67.

Iriondo, A., Miggins, D., and Premo, W.R., 2003a, The Aibo-type (~1.1 Ga) granitic magmatism in NW Sonora, Mexico: Failed continental rifting of Rodinia?: Geological Society of America Cordilleran Section Abstracts with Programs, v. 35, n. 4, p. 84.

Iriondo, A., Kunk, M.J., Winick, J.A., and CRM, 2003b, ^{40}Ar/^{39}Ar dating studies of minerals and rocks in various areas in Mexico: USGS/CRM Scientific Collaboration Part I, U.S. Department of the Interior, U.S. Geological Survey, Open-File Report 03-020, 79 p.

Iriondo, A., Premo, W.R., Martínez-Torres, L.M., Budahn, J.R., Atkinson, W.W., Jr., Siems, D.F., and Guaras-González, B., 2004, Isotopic, geochemical, and temporal characterization of Proterozoic basement rocks in the Quitovac region, northwestern Sonora, Mexico: Implications for the reconstruction of the southwestern margin of Laurentia: Geological Society of America Bulletin, v. 116, no. 1/2, p. 154–170, doi: 10.1130/B25138.1.

Iriondo, A., Martínez-Torres, L.M., Kunk, M.J., Atkinson, W.W., Jr., Premo, W.R., and McIntosh, W.C., 2005, Northward Laramide thrusting in the Quitovac region, northwestern Sonora, Mexico: Implications for the juxtaposition of Paleoproterozoic basement blocks and the Mojave-Sonora megashear hypothesis, in Anderson, T.H., Nourse, J.A., McKee, J.W., and Steiner, M.B., eds., The Mojave-Sonora Megashear Hypothesis: Development, Assessment, and Alternatives: Geological Society of America Special Paper 393, p. 631–669, doi: 10.1130/2005-2393(24).

Jacques-Ayala, C., 1995, Geology and provenance of Lower Cretaceous Bisbee Group in the Caborca-Santa Ana area, northwest Sonora, *in* Jacques-Ayala, C., González-León, C.M., and Roldán-Quintana, J., eds., Studies on the Mesozoic of Sonora and Adjacent Areas: Geological Society of America Special Paper 301, p. 79–98.

Jacques-Ayala, C., and Clark, K.F., 1998, Lithology, structure and gold deposits of northwestern Sonora, Mexico, *in* Clark, K.F., ed., Gold Deposits of Northern Sonora, Mexico: Society of Economic Geologists—Guidebook Series 30, p. 203–234.

Jacques-Ayala, C., and De Jong, K.A., 1996, Extension of the Orocopia Schist belt into northwestern Sonora, Mexico: A working hypothesis: Geological Society of America Abstracts with Programs, v. 28, no. 5, p. 78.

Jacques-Ayala, C., García-Barragán, J.C., and De Jong, K.A., 1990, Caborca-Altar Geology: Cretaceous Sedimentation and Compression, Tertiary Uplift and Extension: Arizona Geological Survey Special Paper 7, v. 165–182.

Jones, N.W., McKee, J.W., Anderson, T.H., and Silver, L.T., 1995, Jurassic volcanic rocks in northeastern Mexico: A possible remnant of a Cordilleran magmatic arc, *in* Jacques-Ayala, C., González-León, C.M., and Roldán-Quintana, J., eds., Studies on the Mesozoic of Sonora and Adjacent Areas: Geological Society of America Special Paper 301, p. 179–190.

Karlstrom, K.E., and Williams, M.L., 1995, The case for simultaneous deformación, metamorphism and plutonism: An example from Proterozoic rocks in central Arizona: Journal of Structural Geology, v. 17, p. 59–81, doi: 10.1016/0191-8141(93)E0025-G.

Ketner, K.B., 1986, Eureka quartzite in Mexico?—Tectonic implications: Geology, v. 14, p. 1027–1030, doi: 10.1130/0091-7613(1986)14<1027:EQIMI>2.0.CO;2.

Ketner, K.B., and Noll, J.H., 1987, Preliminary geologic map of the Cerro Cobachi area, Sonora, Mexico: U.S. Geological Survey Misc, Field Studies Map MF-1980, scale 1: 20 000, 1 sheet.

Krieger, M.H., 1961, Troy Quartzite (younger Precambrian) and Bolsa and Abrigo Formations (Cambrian), northern Galiuro Mountains, southeastern Arizona: Article 207, U.S. Geological Survey Professional Paper, C160–C164.

Kurtz, J.R., Gehrels, G.E., and Stewart, J.H., 1998, Detrital zircons geochronology of the Bolsa Quartzite (Cambrian) in Sonora, Mexico: Geolological Society of America Abstracts with Programs, v. 30, no. 5, p. 49.

Leveille, G.P., and Frost, E.G., 1984, Deformed upper Paleozoic-lower Mesozoic cratonic strata, El Capitan, Sonora, Mexico: Geological Society of America, 97th Annual Meeting: Abstracts with Programs, v. 16, no. 6, p. 575.

Longoria, J.F., and Pérez, V.A., 1979, Bosquejo geológico de los Cerros Chino y Rajón, cuadrángulo Pitiquito-La Primavera (NW de Sonora): Universidad de Sonora, Boletín Departamento de Geología, v. 1, no. 2, p. 119–144.

Lottes, A.L., and Rowley, D.B., 1990, Early and Late Permian reconstructions of Pangaea, Palaeozoic palaeogeography and biogeography: Geological Society [London] Memoir, v. 12, p. 383–395.

Lucas, S.G., and Estep, J.W., 1999, Permian, Triassic, and Jurassic stratigraphy, biostratigraphy and sequence stratigraphy in the Sierra del Alamo Muerto, Sonora, Mexico, *in* Bartolini, C., Wilson, J.L., and Lawton, T.F., eds., Mesozoic Sedimentary and Tectonic History of North-Central Mexico: Geological Society of America Special Paper 340, p. 271–286.

Lucas, S.G., and González-León, C.M., 1994, Marine Upper Triassic strata at Sierra la Flojera, Sonora, México: Neues Jahrbuch fuer Geologie und Palaeontologie Monatshefte, v. H-1, p. 34–40.

Lucas, S.G., Estep, J.W., and Molina-Garza, R.S., 1999, Early Jurassic ammonites at Cerro Pozos de Serna, Sonora, Mexico, Neues Jahrbuch fuer Geologie und Palaeontologie Monatshefte, v. 6, p. 357–371.

Marzolf, J.E., and Anderson, T.H., 1996, Early Mesozoic paleogeography and the Mojave Sonora Megashear: Geological Society of America, 28th Annual Meeting, Abstracts with Programs, v. 28, no. 7, p. 114.

Marzolf, J.E., and Anderson, T.H., 2000, Restoration of the Caborca Block: Alternative models: Geological Society of America, Annual Meeting: Abstracts with Programs, v. 32, no. 7, p. 46.

May, S.R., and Butler, R.F., 1986, North American Jurassic apparent polar wander: Implications for plate motion, paleogeography and cordilleran tectonics: Journal of Geophysical Research, v. 91, p. 11,519–11,544.

McDowell, F.W., Housh, T.B., and Wark, D.A., 1999, Nature of the crust beneath west-central Chihuahua, Mexico, based upon Sr, Nd, and Pb isotopic compositions at the Tomóchic volcanic center: Geological Society of America Bulletin, v. 111, no. 6, p. 823–830, doi: 10.1130/0016-7606(1999)111<0823:NOTCBW>2.3.CO;2.

McKee, J.W., and Anderson, T.H., 1999, Mass-gravity deposits and structures in the Lower Cretaceous of Sonora, Mexico: Geological Society of America Bulletin, v. 110, no. 12, p. 1516–1529.

McKee, J.W., Jones, N.W., and Long, L.E., 1984, History of recurrent activity along a major fault in northeastern Mexico: Geology, v. 12, no. 2, p. 103–107, doi: 10.1130/0091-7613(1984)12<103:HORAAA>2.0.CO;2.

McKee, J.W., Jones, N.W., and Long, L.E., 1990, Stratigraphy and provenance of strata along the San Marcos fault, central Coahuila, Mexico: Geological Society of America Bulletin, v. 102, p. 593–614, doi: 10.1130/0016-7606(1990)102<0593:SAPOSA>2.3.CO;2.

Merriam, R., 1972, Reconnaissance geologic map of the Sonoita quadrangle, northwest Sonora, Mexico: Geological Society of America Bulletin, v. 95, p. 631–653.

Merriam, R., and Eells, J.L., 1979, Reconnaissance geologic map of the Caborca quadrangle, Sonora, Mexico: Boletín Departamento de Geología, Universidad de Sonora, v. 1, no. 2, p. 87–94.

Miller, C.F., Wooden, J.L., and Gerber, M.E., 1992, Plutonism at a tectonically evolving continent margin: Mesozoic granitoids of the eastern Mojave Desert, California, USA: 29th International Geological Congress; Abstracts, 29, p. 519.

Mitre-Salazar, L.M., 1989, La megafalla laramídica de San Tiburcio, Estado de Zacatecas: Revista Instituto de Geología, Universidad Nacional Autónoma de México, v. 8, p. 47–51.

Molina-Garza, R.S., 2005, Paleomagnetic data for the Late Triassic Acatita intrusives, Coahuila, Mexico: Tectonic implications: Geofísica Internacional, v. 44, p. 197–210.

Molina-Garza, R.S., and Geissman, J.W., 1996, Timing of deformacion and accretion of the Antimonio terrane, Sonora, from paleomagnetic data: Geology, v. 24, p. 1131–1134.

Molina Garza, R.S., and Geissman, J.W., 1998, Paleomagnetic evidence against Jurassic left-lateral (southeastward) displacement of the Caborca Terrane, *in* Clark, K.F., ed., Gold Deposits of Northern Sonora, Mexico: Guidebook Series—Society of Economic Geologists 30, p. 187–201.

Molina Garza, R.S., and Geissman, J.W., 1999, Paleomagnetic data from the Caborca terrane, Mexico: Implications for Cordillera tectonics and the Mojave-Sonora megashear: Tectonics, v. 18, p. 293–325, doi: 10.1029/1998TC900030.

Muller, S., Ferguson, W., and Gardiner, H., 1936, Triassic and Jurassic Formations of west-central Nevada: Geological Society of America Bulletin, v. 47, no. 2, p. 241–251.

Nourse, J.A., 1990, Tectonostratigraphic development and strain history of the Magdalena metamorphic core complex, northern Sonora, Mexico: Arizona Geological Survey Special Paper, v. 7, p. 155–164.

Nourse, J.A., 1995, Jurassic-Cretaceous paleogeography of the Magdalena region, northern Sonora, and its influence on the positioning of Tertiary metamorphic core complexes, *in* Jacques-Ayala, C., González-León, C.M., and Roldán-Quintana, J., eds., Studies on the Mesozoic of Sonora and Adjacent Areas: Geological Society of America Special Paper 301, p. 59–78.

Nourse, J.A., 2001, Tectonic Insights from an Upper Jurassic-Lower Cretaceous Stretched-Clast Conglomerate, Caborca-Altar Region, Sonora, Mexico: Journal of South American Earth Sciences, v. 14, no. 5, p. 453–474, doi: 10.1016/S0895-9811(01)00051-7.

Nourse, J.A., Anderson, T.H., and Silver, L.T., 1994, Tertiary metamorphic core complexes in Sonora, northwestern Mexico: Tectonics, v. 13, p. 1161–1182, doi: 10.1029/93TC03324.

Nourse, J.A., Premo, W.R., Iriondo, A., and Stahl, E.R., 2005, Contrasting Proterozoic basement complexes near the truncated margin of Laurentia, northwestern Sonora–Arizona international border region, *in* Anderson, T.H., Nourse, J.A., McKee, J.W., and Steiner, M.B., eds., The Mojave-Sonora Megashear Hypothesis: Development, Assessment, and Alternatives: Geological Society of America Special Paper 393, p. 123–182, doi: 10.1130/2005-2393(04).

Oldow, J.S., 1984, Evolution of a late Mesozoic back-arc fold and thrust belt, northwestern Great: Tectonophysics, v. 102, no. 1–4, p. 245–274, doi: 10.1016/0040-1951(84)90016-7.

Oldow, J.S., Satterfield, J.I., and Silberling, N.J., 1993, Jurassic to Cretaceous transpressional deformacion in the Mesozoic marine province of the northwestern Great Basin, *in* Lahren, M.M., Trexler, J.H., Jr., and Spinosa, C., eds., Crustal Evolution of the Great Basin and the Sierra Nevada: University of Nevada, Reno, p. 129–166.

Padilla y Sánchez, R.J., 1986, Post-Paleozoic tectonics of northeast Mexico and its role in the evolution of the Gulf of Mexico: Geofisica Internacional, v. 25, p. 157–206.

Pilger, R.H., 1978, A closed gulf of Mexico, pre-Atlantic Ocean plate reconstructions and the early rift history of the Gulf and North Atlantic: Gulf Coast Association of Geological Societies Transactions, v. 28, p. 383–393.

Pindell, J., and Dewey, J.F., 1982, Permo-Triassic reconstruction of western Pangea and the evolution of the Gulf of Mexico/Caribbean region: Tectonics, v. 1, no. 2, p. 179–211.

Poole, F.G., and Madrid, R.J., 1988, Allochthonous Paleozoic eugeosinclinal rocks of Barita, Sonora mine area, central Sonora, Mexico, *in* Rodríguez, T., ed., Libreto Guía Excursión Geológica, II Simposio sobre Geología y Minerales del Estado de Sonora, Excursión de Campo, Instituto de Geología, Universidad Nacional Autónoma de México y Universidad de Sonora, Hermosillo, Sonora, p. 32–41.

Poole, F.G., Amaya-Martínez, R., and Page, W.R., 2000, Silurian and Devonian carbonate shelf rocks and Lower Jurassic sequence near Rancho Placeritos, west-central Sonora: Field Guide for Field Trip 2, IV Symposium on the Geology of Northwest Mexico and Adjacent Areas: Center for the Arts, Universidad de Sonora, Hermosillo, México, 24 p.

Poole, F.G., Perry, W.J., Madrid, R.J., and Amaya-Martínez, R., 2005, Tectonic synthesis of the Ouachita-Marathon-Sonora orogenic margin of southern Laurentia: Stratigraphic and structural implications for timing of deformational events and plate tectonic model, *in* Anderson, T.H., Nourse, J.A., McKee, J.W., and Steiner, M.B., eds., The Mojave-Sonora Megashear Hypothesis: Development, Assessment, and Alternatives: Geological Society of America Special Paper 393, p. 543–596, doi: 10.1130/2005-2393(21).

Pubellier, M., Rangin, C., Rascon, B., Chorowicz, J., and Bellon, H., 1995, Cenomanian thrust tectonics in the Sahuaria region, Sonora: Implications about northwestern Mexico megashears, *in* Jacques-Ayala, C., González-León, C.M., and Roldán-Quintana, J., eds., Studies on the Mesozoic of Sonora and Adjacent Areas: Geological Society of America Special Paper 301, p. 111–120.

Rangin, C., 1982, Contribution à l'étude géologique du système Cordillerain du Nord-Ouest du Mexique [thèse de doctorat d'état]: Paris, Université Pierre et Marie Curie, 588 p.

Riggs, N.R., Lehman, T.M., Geherels, G.E., and Dickinson, W.R., 1996, Detrital zircon link between headwaters and terminous of the Upper Triassic Chinle-Dockum paleoriver system: Science, v. 273, p. 97–100.

Rodríguez-Castañeda, J.L., 1984, Geology of Tuape region, north-central Sonora, Mexico [M.S. thesis]: Pittsburgh, University of Pittsburgh, 157 p.

Rodríguez-Castañeda, J.L., 1994, Geología del área El Teguchi, Estado de Sonora, México: Revista Mexicana de Ciencias Geológicas, v. 11, no. 1, p. 11–28.

Rodríguez-Castañeda, J.L., 1996, Late Jurassic and mid-Tertiary brittle-ductile deformación in the Odepe region, Sonora, México: Revista Mexicana de Ciencias Geológicas, v. 13, no. 1, p. 1–9.

Rudnick, R.L., and Cameron, K.L., 1991, Age diversity of the deep crust in northern Mexico: Geology, v. 19, no. 12, p. 1197–1200.

Sedlock, R.L., Ortega-Gutiérrez, F., and Speed, R.C., 1993, Tectonostratigraphic terranes and tectonic evolution of Mexico: Geological Society of America Special Paper 278, 153 p.

Silberling, N.J., and Tozer, E.T., 1968, Biostratigraphic Classification of the Marine Triassic in North America: Geological Society of America Special Paper 110, 63 p.

Silver, L.T., 1965, Mazatzal Orogeny and Tectonic Episodicity: Geological Society of America Special Paper 82, p. 185–186.

Silver, L.T., and Anderson, T.H., 1974, Possible left-lateral early to middle Mesozoic disruption of the southwestern North America craton margin: Geological Society of America Abstracts with Programs, v. 6, no.7, p. 955–956.

Sosson, M., 1990, Le passage des Cordillères nord-américaines aux Sierras Madres mexicaines, sud de l'Arizona (Etats-Unis) et nord du Sonora (Mexique) [thèse d'habilitation]: Université de Nice Sophia-Antipolis, Nice, France, 428 p.

Stanley, G.D., and González-León, C., 1995, Paleogeographic and tectonic implications of Triassic fossils and strata from the Antimonio Formación, northwestern Sonora, *in* Jacques-Ayala, C., González-León, C., and Roldán-Quintana, J., eds., Studies of the Mesozoic of Sonora and Adjacent Regions: Geological Society of America Special Paper 301, p. 1–16.

Stevens, C.H., and Stone, P., 1988, Early Permian thrust faults in east-central California: Geological Society of America Bulletin, v. 100, p. 552–562, doi: 10.1130/0016-7606(1988)100<0552:EPTFIE>2.3.CO;2.

Stevens, C.H., Stone, P., and Kistler, R.W., 1992, A speculative reconstruction of the middle Paleozoic continental margin of southwestern North America: Tectonics, v. 11, p. 405–419.

Stevens, C.H., Stone, P., and Miller, J.S., 2005, A new reconstruction of the Paleozoic continental margin of southwestern North America: Implications for the nature and timing of continental truncation and the possible role of the Mojave-Sonora megashear, *in* Anderson, T.H., Nourse, J.A., McKee, J.W., and Steiner, M.B., eds., The Mojave-Sonora Megashear Hypothesis: Development, Assessment, and Alternatives: Geological Society of America Special Paper 393, p. 597–618, doi: 10.1130/2005-2393(22).

Stewart, J.H., 2003, Mojave-Sonora megashear: Evidence from Neoproterozoic to Lower Jurassic strata: Geological Society of America, Cordilleran Section, 99th Annual Meeting, Abstracts with Programs, v. 35, no. 4, p. 13.

Stewart, J.H., and Poole, F.G., 1975, Extension of the Cordilleran miogeocline belt to the San Andreas fault, southern California: Geological Society of America Bulletin, v. 86, p. 205–212, doi: 10.1130/0016-7606(1975)86<205:EOTCMB>2.0.CO;2.

Stewart, J.H., and Roldán-Quintana, J., 1991, Upper Triassic Barranca Group: Nonmarine and shallow-marine rift-basin deposits of northwestern Mexico, *in* Pérez-Segura, E., and Jacques-Ayala, C., eds., Studies of Sonoran Geology: Geological Society of America Special Paper 254, p. 19–35.

Stewart, J.H., McMenamin, A.S., and Morales-Ramírez, J.M., 1984, Upper Proterozoic and Cambrian rocks in the Caborca region, Sonora, México—Physical stratigraphy, biostratigraphy, paleocurrent studies, and regional relations: U.S. Geological Survey Professional Paper 1309, 32 p.

Stewart, J.H., Anderson, T.H., Haxel, G.B., Silver, L.T., and Wright, J.E., 1986, Late Triassic paleogeography of the southern Cordillera: The problem of a source for voluminous volcanic detritus in the Chinle Formation of the Colorado Plateau region: Geology, v. 14, p. 567–570, doi: 10.1130/0091-7613(1986)14<567:LTPOTS>2.0.CO;2.

Stewart, J.H., Poole, F.G., Ketner, K.B., Madrid, R.J., Roldán-Quintana, J., and Amaya-Martínez, R., 1990, Tectonics and Stratigraphy of the Paleozoic and Triassic southern margin of North America, Sonora, Mexico, *in* Gehrels, G.E., and Spencer, J.E., eds., Geologic Excursions through the Sonoran Desert Region, Arizona and Sonora: Arizona Geological Society Special Paper 7, p. 183–202.

Stewart, J.H., Amaya-Martínez, R., and Stamm, R.H., 1997, Stratigraphy and regional significance of Mississippian to Jurassic rocks in Sierra Santa Teresa, Sonora, México: Revista Mexicana de Ciencias Geológicas, v. 14, p. 115–135.

Stewart, J.H., Poole, F.G., Harris, A.G., Repetski, J.E., Wardlaw, B.R., Mamet, B.L., and Morales-Ramirez, J.M., 1999, Neoproterozoic(?) to Pennsylvanian inner-shelf, miogeosinclinal strata in Sierra Agua Verde, Sonora, México: Revista Mexicana de Ciencias Geológicas, v. 16, no. 1, p. 35–62.

Stewart, J.H., Amaya-Martínez, R., and Palmer, A.R., 2002, Neoproterozoic and Cambrian strata of Sonora, Mexico: Rodinian supercontinent to Laurentian Cordilleran margin, *in* Barth, A., ed., Contributions to Crustal Evolution of the Southwestern United States: Geological Society of America Special Paper 365, p. 5–48.

Stone, P., and Stevens, C.H., 1988, Pennsylvanian and Early Permian paleogeography of east-central California: Implications for the shape of the continental margin and the timing of continental truncation: Geology, v. 16, p. 330–333, doi: 10.1130/0091-7613(1988)016<0330:PAEPPO>2.3.CO;2.

Strickland, J.M., and Middleton, L.Y., 2000, Sedimentology, stratigraphy and paleogeographical implications of the Cambrian Bolsa Quartztite and Abrigo Formación, southeastern Arizona: Geological Society of America Abstracts with Programs, v. 32, no. 7, p. A308–A309.

Torres-Vargas, R., Ruiz, J., Patchett, P.J., and Grajales-Nishimura, M., 1999, Permo-Triassic continental arc in eastern Mexico: Tectonic implications for reconstructions of southern North America, *in* Bartolini, C., Wilson, J.L., and Lawton, T.F. eds., Mesozoic Sedimentary and Tectonic History of North-Central Mexico: Geological Society of America Special Paper 340, p. 191–196.

Tosdal, R.M., Haxel, G.B., and Wright, J.E., 1989, Jurassic geology of the Sonoran Desert region, southern Arizona, southeastern California, and northernmost Sonora: Construction of a continental-margin magmatic arc, *in* Jenney, J.P., and Reynolds, S.J., eds., Geologic Evolution of Arizona: Arizona Geological Society Digest 17, p. 397–434.

Tozer, E.T., 1980, New genera of Triassic Ammonoidea: Current Research, Part A, Paper—Geological Survey of Canada, 80–1A, p. 107–113.

Valencia-Moreno, M., Ruiz, J., and Roldán-Quintana, J., 1999, Geochemistry of Laramide granitic rocks across the southern margin of the Paleozoic North American Continent, central Sonora, Mexico: International Geology Review, v. 41, p. 845–857.

Van Fossen, M.C., and Kent, D.V., 1990, High latitude paleomagnetic poles from Middle Jurassic plutons and Moat Volcanics in New England and the controversy regarding Jurassic apparent polar wander for North America: Journal of Geophysical Research, v. 95, p. 17,503–17,516.

Walker, J.D., 1988, Permian and Triassic rocks of the Mojave Desert and their implications for timing and mechanisms of continental truncation: Tectonics, v. 7, p. 685–709.

Walker, J.D., and Wardlaw, B.R., 1989, Implications of Paleozoic and Mesozoic rocks in the Soda Mountains, northeastern Mojave Desert, California, for late Paleozoic and Mesozoic Cordilleran orogenesis: Geological Society of America Bulletin, v. 101, p. 1574–1583, doi: 10.1130/0016-7606(1989)101<1574:IOPAMR>2.3.CO;2.

Manuscript Accepted by the Society 29 August 2006

Printed in the USA

Geological Society of America
Special Paper 422
2007

The San Marcos fault: A Jurassic multireactivated basement structure in northeastern México

Gabriel Chávez-Cabello*
Posgrado en Ciencias de la Tierra, Centro de Geociencias, Universidad Nacional Autónoma de México, Campus Juriquilla, Apartado postal 1-742, Querétaro, Qro., 76001, México, and
Facultad de Ciencias de la Tierra, Universidad Autónoma de Nuevo León, Apartado postal 104, Kilómetro 8, Carretera Linares-Cerro Prieto, Linares, N.L., 67700, México

José Jorge Aranda-Gómez
Departamento de Geología Económica, Instituto Potosino de Investigación Científica y Tecnológica, Apartado postal 3-74, San Luis Potosí, S.L.P., 78216, México, and
Centro de Geociencias, Universidad Nacional Autónoma de México, Campus Juriquilla, Apartado postal 1-742, Querétaro, Qro., 76001, México

Roberto S. Molina-Garza
Centro de Geociencias, Universidad Nacional Autónoma de México, Campus Juriquilla, Apartado postal 1-742, Querétaro, Qro., 76001, México

Tomás Cossío-Torres
Facultad de Ciencias de la Tierra, Universidad Autónoma de Nuevo León, Apartado postal 104, Kilómetro 8, Carretera Linares-Cerro Prieto, Linares, N.L., 67700, México

Irving R. Arvizu-Gutiérrez
Posgrado en Ciencias de la Tierra, Centro de Geociencias, Universidad Nacional Autónoma de México, Campus Juriquilla, Apartado postal 1-742, Querétaro, Qro., 76001, México

Gildardo A. González-Naranjo
Posgrado en Ciencias de la Tierra, Facultad de Ciencias de la Tierra, Universidad Autónoma de Nuevo León, Apartado postal 104, Kilómetro 8, Carretera Linares-Cerro Prieto, Linares, N.L., 67700, México

ABSTRACT

The San Marcos fault is a regional structure in northeast México with a minimum length of 300 km, which separates the Coahuila block from the Coahuila fold belt; the fault dips north-northeast and its trend is west-northwest. The San Marcos fault is a basement structure that has been reactivated multiple times, and along its trace there is stratigraphic and structural evidence of intermittent activity since at least the Late Jurassic to the Pliocene-Quaternary. The structural evidence analyzed in this work suggests that the San Marcos fault accommodated mainly north-northeast crustal

*gabchave@hotmail.com

Chávez-Cabello, G., Aranda-Gómez, J.J., Molina-Garza, R.S., Cossío-Torres, T., Arvizu-Gutiérrez, I.R., and González-Naranjo, G.A., 2007, The San Marcos fault: A Jurassic multireactivated basement structure in northeastern México, *in* Alaniz-Álvarez, S.A., and Nieto-Samaniego, Á.F., eds., Geology of México: Celebrating the Centenary of the Geological Society of México: Geological Society of America Special Paper 422, p. 261–286, doi: 10.1130/2007.2422(08). For permission to copy, contact editing@geosociety.org.

extension in pre-Tithonian and Neocomian pulses of activity. This extension may have contributed to development and growth of the Sabinas basin to the north. We found no evidence to support previous proposals of large lateral offset across the fault in Late Jurassic time, but we document a small component of right-lateral slip.

At least four reactivation events have been recognized along the San Marcos fault. The first, in Neocomian time, was normal and triggered deposition of the San Marcos Formation. The second reactivation of the San Marcos fault involved reverse slip during Paleogene time, and it must include minor movements along secondary faults associated with the San Marcos fault. Interpretation of the reactivation event of the San Marcos fault as a reverse fault is based on (1) the occurrence of drape folds and minor tectonic transport to the south-southwest along the main trace of the fault; (2) the occurrence of a nearly perpendicular fold axis of different generation in the southwest sector of the Sabinas basin; (3) uplift of progressively older rocks toward the northeast within the San Marcos Valley; and (4) the existence of near perpendicular directions of tectonic transport determined for different structures within the San Marcos Valley (e.g., faults in the western sector of the valley record tectonic transport to the west and faults in the southwest sector of the valley record tectonic transport to the south-southwest). Secondary faults associated with the San Marcos fault vary in orientation from nearly east-west to nearly north-south, and are best represented by the El Caballo and El Almagre faults exposed in western Coahuila and southeastern Chihuahua. Reactivation of the San Marcos fault as a reverse fault occurred late, relative to an earlier stage of detachment (locally duplicating the stratigraphic sequence) in localities over the Coahuila platform and in the Sabinas basin itself. The relative importance and scale of these detachment folds need to be explored in further detail.

The third reactivation event was normal with a left-lateral component (late Miocene–early Pliocene), and the fourth and last event is dominantly normal (Pliocene-Quaternary). These last two reactivation events along the San Marcos fault were recognized along the segment of the fault buried by volcanic products of the Camargo volcanic field in southeast Chihuahua State. These late events might also be present along the San Marcos fault in Coahuila; the lack of Cenozoic sequences atop the fault trace makes their recognition difficult.

Keywords: San Marcos fault, Sabinas basin, Coahuila fold belt, Coahuila.

RESUMEN

La falla San Marcos es un lineamiento estructural regional con más de 300 km de largo, rumbo WNW y que se inclina hacia el NNE, separando el bloque de Coahuila del Cinturón Plegado de Coahuila en el noreste de México. La falla San Marcos es una estructura de basamento multirreactivada que, en superficie, muestra evidencias estratigráficas y estructurales que documentan su actividad intermitente por lo menos desde el Jurásico Tardío hasta el Plioceno-Cuaternario. Las evidencias estructurales más antiguas reconocidas en este trabajo documentan actividad de la falla San Marcos durante tiempos pre-Titoniano y Neocomiano, sugiriendo que la falla San Marcos acomodó principalmente extensión de la corteza en dirección NNE. Esta extensión contribuyó al crecimiento de la cuenca de Sabinas; con lo anterior, se pone en duda la existencia de grandes desplazamientos laterales a través de la falla San Marcos por lo menos para estos tiempos.

Se han reconocido al menos cuatro eventos de reactivación de la falla San Marcos. El primero fue con componente normal en el Neocomiano y causó el depósito de la Formación San Marcos. El segundo evento de reactivación fue inverso en el Paleógeno y debió incluir a fallas menores asociadas a la falla San Marcos. Se interpreta que el segundo evento de reactivación está representado por (1) la ocurrencia

de plegamiento tipo *drape* y transporte tectónico menor hacia el sur-suroeste sobre la traza principal de la falla San Marcos, (2) la ocurrencia de relaciones perpendiculares entre los ejes de pliegues en la parte suroeste de la cuenca de Sabinas, (3) el levantamiento de rocas más antiguas progresivamente hacia el noreste dentro del Valle San Marcos y, (4) la existencia de direcciones perpendiculares de transporte tectónico determinadas para diferentes estructuras en el Valle San Marcos (e.g., fallas en el sector oeste del Valle San Marcos registran transporte hacia el oeste y fallas en el sector suroeste registran transporte hacia el sur-suroeste). Las fallas menores asociadas a la falla San Marcos presentan orientaciones desde E-W hasta cercanamente N-S como las fallas El Caballo y El Almagre expuestas al oeste de Coahuila y sureste de Chihuahua. Este evento de reactivación inverso de la falla San Marcos es tardío con respecto a una fase anterior de despegues (duplicación de la secuencia por fallas) en localidades de la plataforma de Coahuila y la cuenca de Sabinas. La importancia y escala de los despegues debe ser explorado con mayor detalle en futuros trabajos.

La tercera reactivación es normal con componente lateral izquierda (Mioceno tardío-Plioceno temprano) y, la cuarta y última, predominantemente normal (Plioceno–Cuaternario). Estas reactivaciones fueron reconocidas sobre la traza de la falla San Marcos sepultada por productos del Campo Volcánico de Camargo, al sureste de Chihuahua. Los dos últimos eventos parecen estar presentes sobre los segmentos de la falla San Marcos en Coahuila; sin embargo, aquí no afectan a rocas jóvenes por lo que no es posible establecer sus edades.

Palabras clave: la falla San Marcos, la cuenca de Sabinas, Cinturón Plegado de Coahuila, Coahuila.

INTRODUCTION

The San Marcos fault was first identified by Charleston (1981), and it is the only basement structure in northeast México for which stratigraphic and structural evidence compiled in the field document its existence in a convincing way. The trace of the San Marcos fault has a general west-northwest trend and a minimal length of 300 km; it dips north-northeast, dividing the Coahuila State almost in half. The fault is a regional structural boundary between the Coahuila block and the Coahuila fold belt (Fig. 1). It has been suggested that the San Marcos fault can be extended for another 300 km toward the west-northwest from the Sierra Mojada area (Fig. 1). However, this western segment of the fault is buried by a thick sequence of Paleogene and Neogene volcanic rocks and/or alluvium (Aranda-Gómez et al., 2005a). The San Marcos fault has also been referred to as the "Sierra Mojada–China lineament" by Padilla y Sánchez (1982, 1986).

In contrast with other basement faults in NE México such as the Mojave-Sonora megashear (Anderson and Schmidt, 1983) and the La Babia fault (Charleston, 1981), also known as the Boquillas Del Carmen–Sabinas lineament (Padilla y Sánchez, 1982), or the Sabinas fault (Alfonso, 1978), the San Marcos fault trace is well defined. The trace can be followed from Potrero La Gavia, through San Marcos Valley, to Potrero Colorado and to the Sierra Mojada (Fig. 1) area. The fault is distinguished by (1) contrasting structural styles between the Coahuila block and the Coahuila folded belt, (2) the existence of a pre-Tithonian and Neocomian major sedimentary clastic wedge related to its activity and deposited in the hanging wall of the fault (Fig. 2; McKee et al., 1990), and (3) the existence of fault contacts between Permian, Jurassic, and Cretaceous rocks in the San Marcos Valley and Potrero Colorado areas, where the fault dips north-northeast with a high angle. Furthermore, in those areas where the San Marcos fault is now buried, such as the Camargo volcanic field, its hidden trace may be inferred from structural features in the surface such as local uplift associated with folding and/or lateral displacement. It has also been suggested that the buried trace of the fault channeled Plio-Quaternary basaltic magmas to the surface, and allowed the formation of the unusually large Camargo volcanic field (Aranda-Gómez et al., 2003, 2005b).

Despite the significance of the San Marcos fault and other regional structures such as the La Babia fault and the Mojave-Sonora megashear (Fig. 1) for understanding the tectonic evolution of northeast México, previous research focusing on the structural data is restricted to regional mapping mostly based on the interpretation of aerial photographs and satellite imagery. These techniques were used to define the structural styles in the Coahuila block and the Coahuila folded belt (Charleston, 1981; Padilla y Sánchez, 1982). The caveat is that the majority of the structural features discernible in aerial and satellite imagery are Paleogene structures formed during the Laramide orogeny and are not related to the Late Jurassic deformation.

In this work we describe the structural features that document the kinematics of the San Marcos fault since its pre-Tithonian formation until its last event of reactivation in Neogene-Quaternary time, as recorded in central Coahuila. We make a

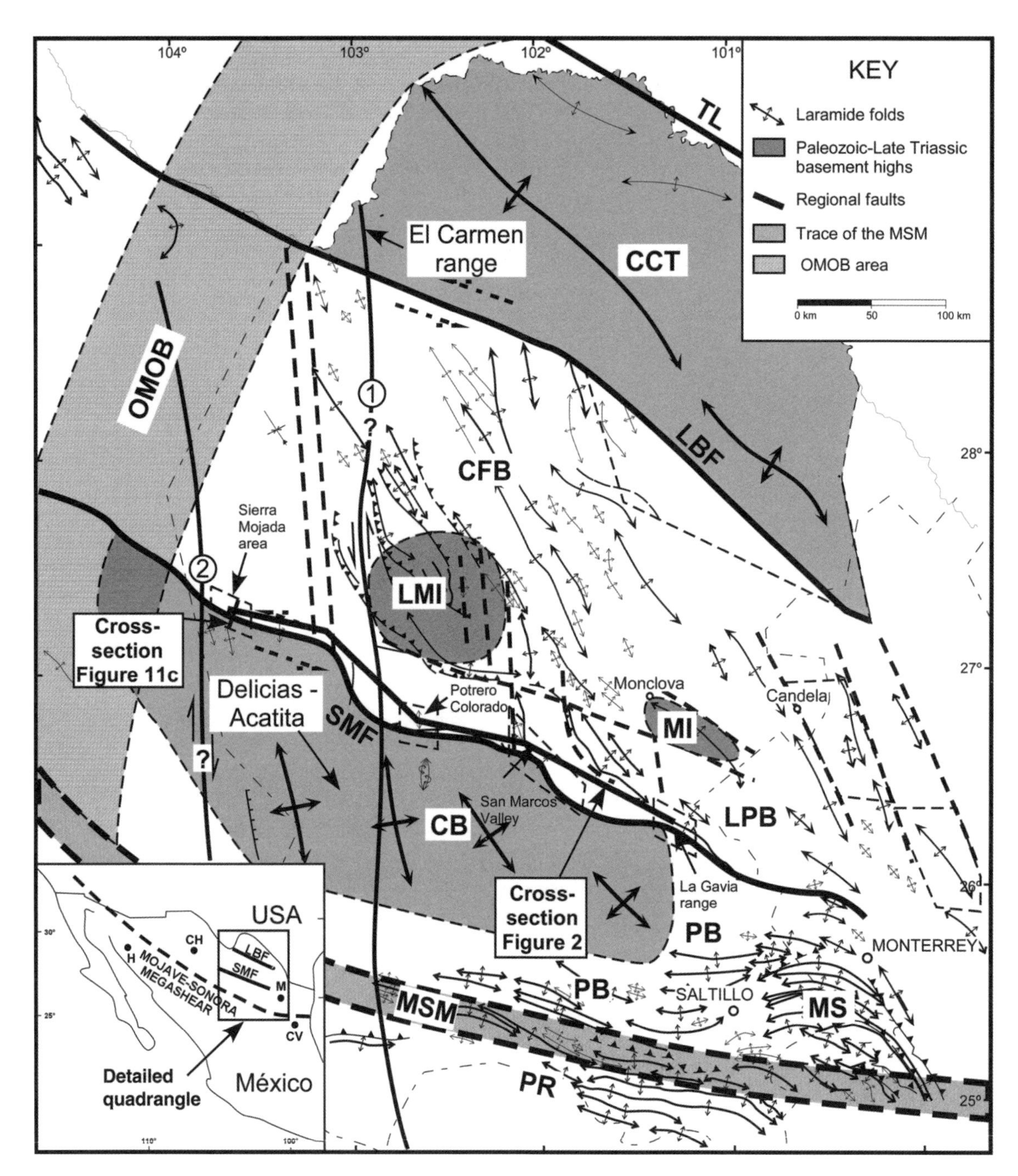

Figure 1. Generalized geologic map for Coahuila State and adjacent regions. It shows locations for major basement structures including the Texas lineament (TL), La Babia fault (LBF), San Marcos fault (SMF), and the inferred trace for the Mojave-Sonora fault (MSF). The La Babia and San Marcos faults separate regions with different deformation styles, as the Coahuila-Texas craton (CCT) from the Coahuila fold belt (CFB), which is also separated from the Coahuila block (CB). Second-order basement structures also delimit important paleogeographic features including La Mula Island (LMI) and Monclova Island (MI). Others morph structural features such as the Monterrey salient (MS), Parras range (PR), and the La Popa (LPB) and Parras foreland basins (PB). Almagre (1) and El Caballo (2) faults are also depicted, as are several localities cited in the text. Dashed zone OMOB represents the buried prolongation of the Ouachita marathon orogenic belt. The thick gray, dotted line between the Coahuila block and the Coahuila-Texas craton depicts the extension of the Sabinas basin according to Eguiluz (2001). H—Hermosillo; CH—Chihuahua; M—Monterrey; CV—Ciudad Victoria; MSM—Mojave-Sonora megashear.

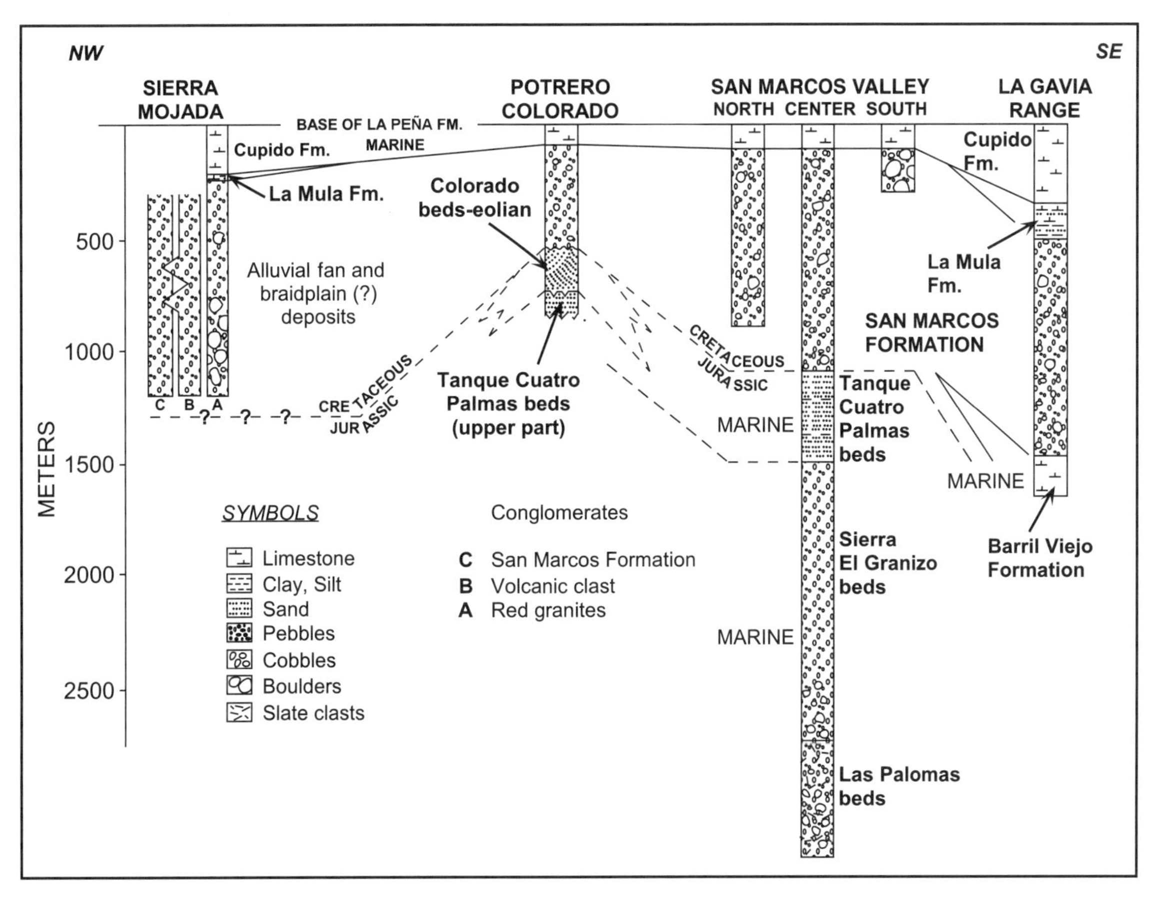

Figure 2. Correlation between clastic packages deposited on the hanging-wall block north of the San Marcos fault since Jurassic to Neocomian time (from McKee et al., 1990).

comparative study between the fault sets from localities in San Marcos Valley, Potrero Colorado, and Sierra Mojada. These are the same localities where McKee et al. (1990), using stratigraphic evidence, documented two reactivation events of the San Marcos fault (Neocomian and Paleogene). On the other hand, data collected bearing on the Neogene-Quaternary reactivation of the fault in Coahuila are compared with data for coeval structures reported by Aranda-Gómez et al. (2005a) for Sierras La Herradura and Aguachile and for the Camargo volcanic field, all in Chihuahua. The information obtained was used to regionally analyze the different pulses of deformation that reactivated the San Marcos fault. We recognized four periods of activity in the fault; the earliest pulse occurred in pre-Tithonian times and the youngest was Plio-Quaternary. Each of these events is described and discussed in the following sections.

STRATIGRAPHIC SEQUENCES RELATED TO THE SAN MARCOS FAULT ACTIVITY

The best evidence of the early activity of the San Marcos fault is stratigraphic. Faulting was inferred from the existence of a Jurassic (?), pre-Tithonian, sedimentary marine clastic wedge, deposited on the hanging-wall block north of the San Marcos fault along the southern margin of the Sabinas basin. The Jurassic clastic wedge is ~2000 m thick, and comprises three informal lithostratigraphic units recognized by McKee et al. (1990): Las Palomas, Sierra El Granizo, and Tanque Cuatro Palmas beds (Fig. 2). The Tanque Cuatro Palmas beds, which correlate with the Casita Formation (Kimmeridian–Lower Berriasian)—a widespread unit characteristic of the stratigraphy of northeastern México—are believed to represent sedimentation during a time of tectonic quiescence after a major pulse of activity along the San Marcos fault. This earlier pulse controlled the accumulation of the Las Palomas and Sierra El Granizo beds. These two units comprise 1600 m of syntectonic sedimentation, and the upper 400 m correspond to the Tanque Cuatro Palmas beds. Resting atop the Tanque Cuatro Palmas beds there are ~1000 m of continental conglomerate, fluvial sandstone, and mudstone from the San Marcos Formation. The San Marcos Formation represents the first reactivation event of the San Marcos fault in Neocomian time (Fig. 2). There are convincing observations indicating that the clasts in the Jurassic strata were derived from the Coahuila block to the south of the San Marcos fault. The clasts include fragments of proximal, low-grade metavolcanic rocks from the Las Delicias arc (Pennsylvanian-Permian) accumulated in Las Delicias basin, and the plutonic rocks that intrude the arc sequence (McKee et al., 1984, 1988, 1999).

Since diagnostic fossils are absent from the Las Palomas and Sierra El Granizo beds, we cannot discard the possibility that these beds are correlative with the clastic sequence of the Huizachal Group in northeastern México. The Huizachal Group was deposited during the rifting stage of the opening of the Gulf of Mexico. If this correlation were correct, then the formation of the San Marcos fault may have occurred as early as Late Triassic or Early Jurassic time. However, the age of the clastic wedge can only be securely established as pre-Tithonian in age. At the Potrero Colorado area, the clastic wedge is represented by four units: (1) a hematitic sequence of intercalated red sandstone and mudstone that may represent the upper part of Tanque Cuatro Palmas beds (the name we use throughout this article); (2) an unnamed 10-m-thick conglomerate; (3) a 191-m-thick package of eolian sandstones, which McKee et al. (1990) defined as Colorado beds; and (4) the San Marcos Formation (Fig. 2). Finally, in the Sierra Mojada area there are three Lower Cretaceous conglomeratic units that have been correlated with the San Marcos Formation (McKee et al., 1990).

FORMATION OF THE SAN MARCOS FAULT

The shallow-marine clastic sediments of the Tanque Cuatro Palmas beds give a minimum age of pre-Tithonian for the initiation of the San Marcos fault activity (McKee et al., 1990). The syntectonic Las Palomas and Sierra El Granizo beds cannot be used to date the early activity of the fault as they lack fossils.

This structural analysis focused on the study of faults and both regional and local fold axes related to the San Marcos fault. The faults were split into several sets corresponding to specific stratigraphic levels (i.e., Jurassic, Neocomian, and younger Cretaceous rocks). The independent study of each data set allowed analysis of the kinematics of the fault during its formation and during subsequent reactivations (in the Neocomian and Paleogene). To infer the orientation of the principal maximum and minimal compression directions for each faulting event, we applied the INVD software of Angelier (1990) to each fault data set.

The analysis of the kinematics of the initial activity of the San Marcos fault is complicated, as the Permo-Triassic basement rocks that were initially sheared by the fault are not clearly exposed and the fault zone is wide. According to McKee et al. (1990), near the San Marcos fault trace in the San Marcos Valley there are small outcrops of Triassic basement granitoids. However, Molina-Garza et al. (2003) and Arvizu-Gutiérrez (2003), based on paleomagnetic studies, concluded that these outcrops are not in situ basement. This interpretation is explained below.

On the other hand, due to the composition and structural characteristics of the Upper Jurassic clastic rocks deposited on the hanging-wall block of the San Marcos fault, only isolated structural evidence can be used to infer the kinematics of the early activity in the fault. Even so, we found independent areas where fold and fault data sets were collected. These data sets are discussed in the following paragraph.

Late Jurassic Folding (?) in the San Marcos Valley

McKee et al. (1990) initially recognized an anticline and syncline pair (Fig. 3) in Jurassic rocks exposed in the San Marcos Valley. These structures have NNW-SSE to north-south trends with fold axes plunging 20–30° northward. These folds, as well as high-angle shear zones in the marine Cretaceous sedimentary rocks, document intense deformation near the trace of the San Marcos fault; the intensity of deformation decreases away from the fault zone. However, the importance of these folds is that they are in a bend zone of the fault, and these are the only macrostructures exposed in Jurassic rocks along the San Marcos fault recognized up to now. Based on these relations, this area was interpreted as a restraining bend zone, and the folds as having formed during left-lateral shearing along the fault (McKee et al., 1990; Fig. 3). Intrigued by this interpretation, Arvizu-Gutiérrez (2003) and Molina-Garza et al. (2003) conducted a local paleomagnetic study to evaluate rotations and to date magnetizations in both the folded rocks and the granitoids of the area. The granitoids were interpreted by Jones et al. (1982) and McKee et al. (1990) as in situ basement with microstructural evidence of left-lateral movement for the San Marcos fault during its early activity.

The paleomagnetic results obtained by Arvizu-Gutiérrez and Molina-Garza et al. (2003) suggest that the outcrops of Triassic granitoids (242 ± 2 Ma, Rb-Sr model age in muscovite: McKee et al.., 1990) are not part of a coherent body of rocks partially buried in San Marcos Valley because the declination and inclination data are not related to the Triassic stable pole of North America, and site mean directions are not consistent. They suggest that these outcrops represent large blocks which fell down from a fault escarpment during the early activity of the San Marcos fault.

On the other hand, for the folds developed in the Las Palomas beds in the San Marcos Valley, Arvizu-Gutiérrez and Molina-Garza et al. (2003) established that these structures record an east-directed shallow-dipping magnetization, which if compared with the North American polar wander path, defines a clockwise rotation of 80–90° (Arvizu-Gutiérrez, 2003). These authors conclude that the clockwise rotation must be local, because paleomagnetic data for the slightly older Acatita series and Nazas Formation (Fig. 4) exposed in the central and southern parts of the Coahuila block, respectively, do not record significant rotation. Both the Acatita series and the Nazas Formation yield poles near what is expected for the stable North American craton.

It must be emphasized that the clockwise rotation of the Las Palomas beds is not consistent with the deformation associated with a restraining bend in a left-lateral strike-slip system at any time during the different periods of deformation in the area. Therefore, paleomagnetic data conflict with the McKee et al. (1990) hypothesis of significant left-lateral movement during the early stages of the San Marcos fault. An alternative explanation

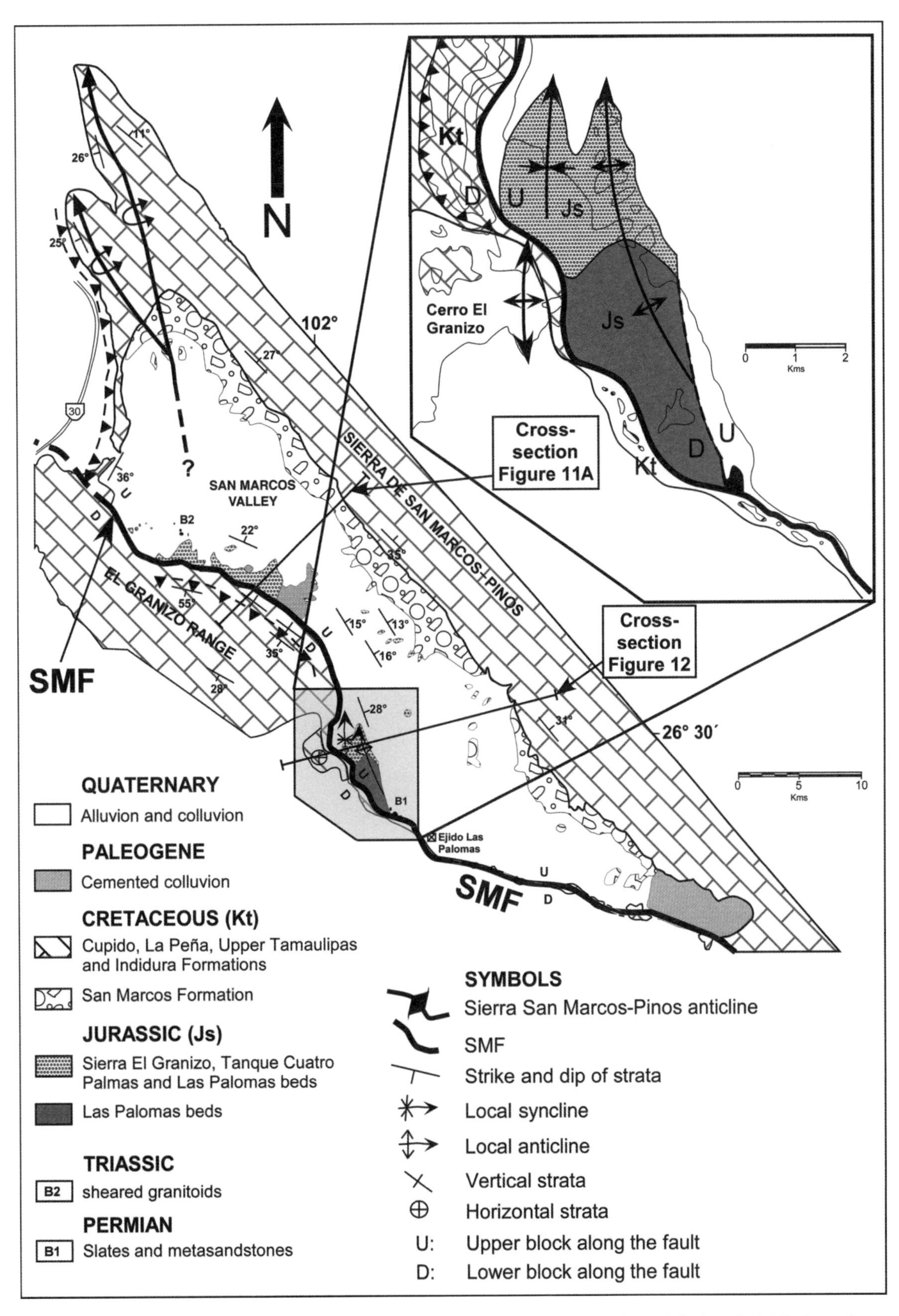

Figure 3. Simplified geologic and structural chart from the San Marcos Valley (see Figure 1 for location). The inset shows a detailed view from the bend zone of the San Marcos fault (SMF) that was studied by Molina-Garza et al. (2003) and Arvizu-Gutiérrez (2003). Faults in this area have high angles, as discussed in the text.

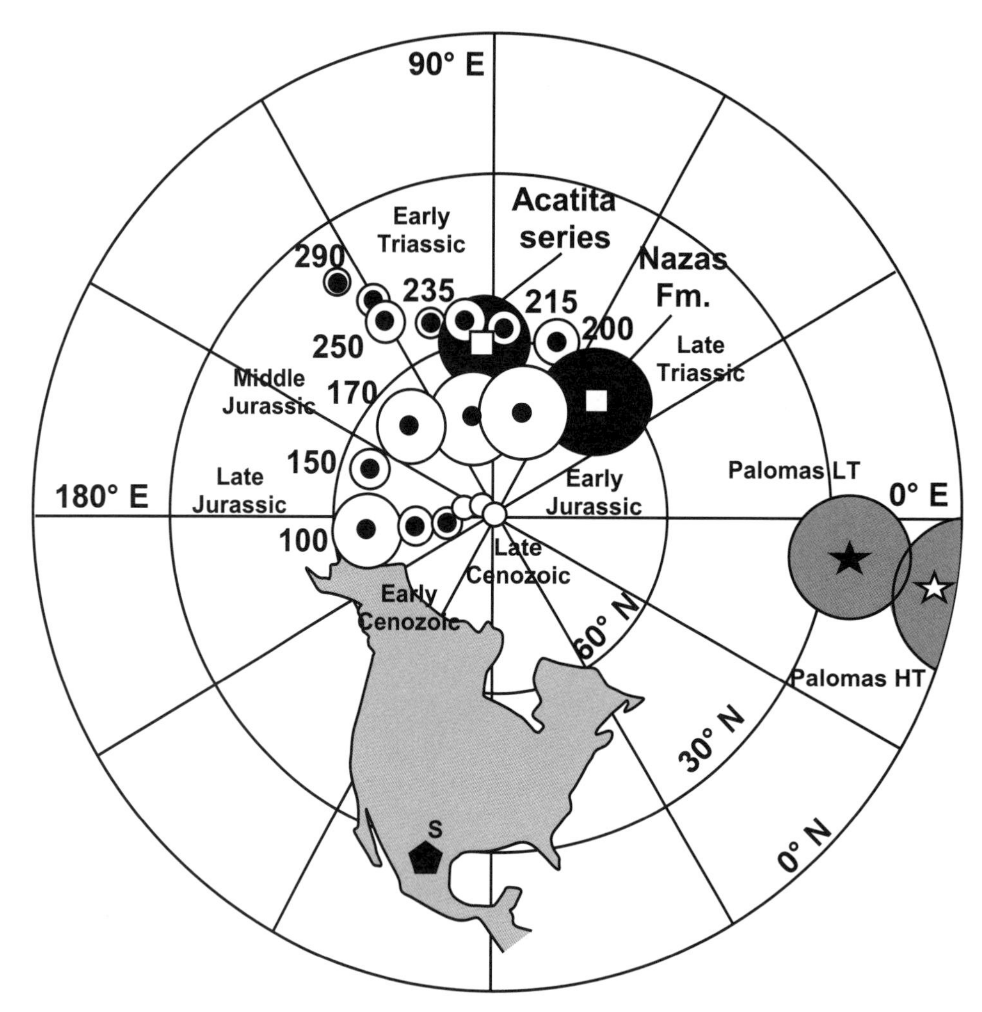

Figure 4. White circles represent the American polar wander path with its confidence intervals (α_{95}). Ages for reference poles are indicated by numbers. The white squares are paleomagnetic poles for a Mesozoic series in northern México (Molina-Garza, 2005) with its corresponding confidence intervals (α_{95}). The black and white stars are poles for the in situ Jurassic series Las Palomas beds, which have structural correction (black star: low temperature, and white star: high temperature); major circles represent the confidence interval for each one (α_{95}). The pentagon shows the sample location for Las Palomas beds in Coahuila, México.

for the origin of folds in Jurassic rocks present in San Marcos Valley, and the folds' clockwise rotation, is that after the folds developed, the rocks rotated as the result of complex deformation during the Paleogene Laramide orogeny.

Chávez-Cabello (2005) proposed that the Laramide orogeny was developed in two phases: an early phase produced fault-bend folding along the San Marcos fault, and a younger phase caused reverse reactivation of basement faults. An important observation is that the anticline-syncline pair in the Jurassic rocks occurs in an area bounded by two faults dipping NE, parallel to the bend zone in this segment of the San Marcos fault. It is inferred that they were reactivated with a reverse component during the late phase of the Laramide orogeny. Further support for this model could be derived from additional structural and paleomagnetic studies away from the bend zone, and would thus establish whether the rotation occurred only in this particular bend area or if it affected a larger area; additional data could also show whether or not deformation is consistent with other Laramide structures in the region.

Late Jurassic Faulting in Potrero Colorado

The structural evidence in Potrero Colorado, which is used here to interpret a Late Jurassic deformation event, is a set of northwest-southeast–trending normal faults exposed in the core of the La Fragua anticline. These structures are oblique to the San Marcos fault (here ~east-west). The normal faults affect a sequence composed of limolite and sandstone (correlated with the Tanque Cuatro Palmas beds), and also crosscut eolianites of the Colorado beds (Fig. 5A). Another set of faults found in this area has minor (<1m) dextral displacements. Apparently, this set

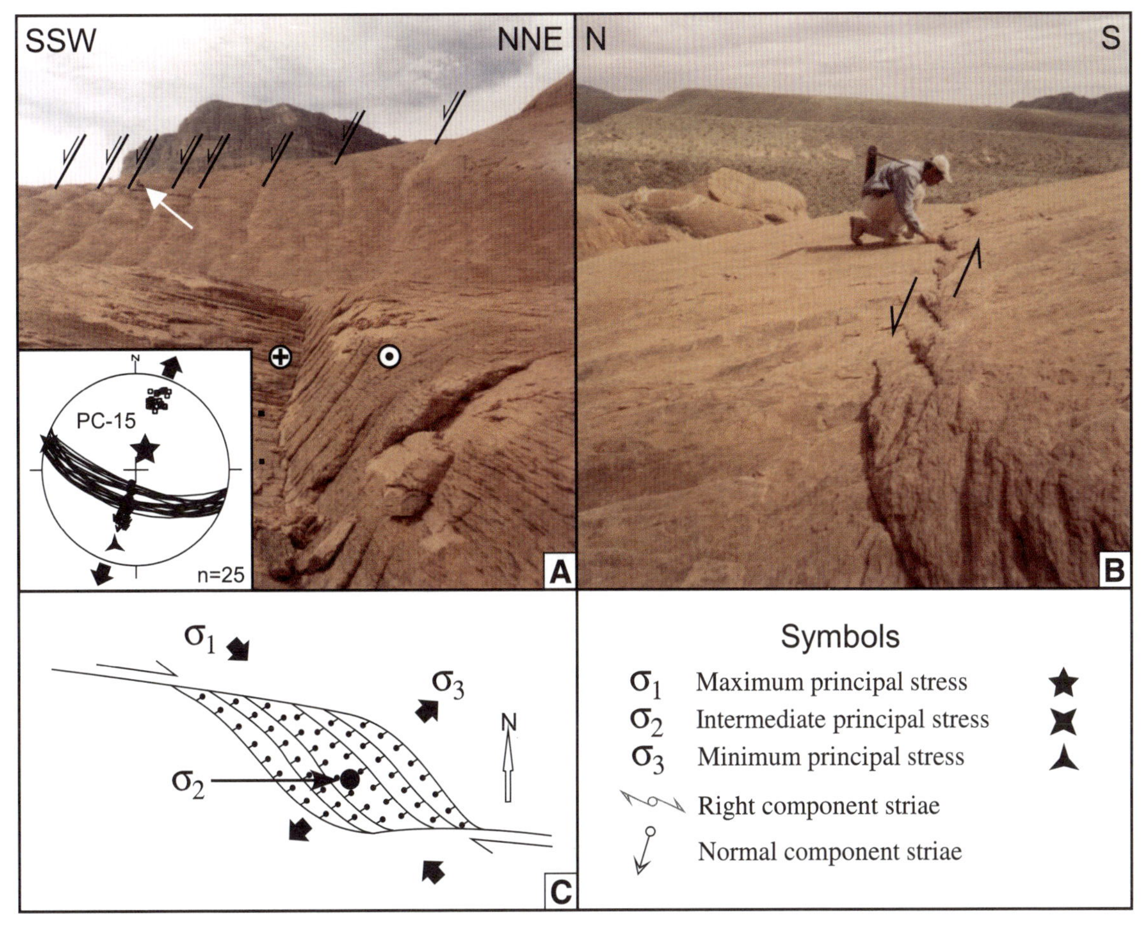

Figure 5. (A) The photograph shows a domino fault set. The dip in these faults is typical for high-angle normal faults; however it has kinematic indicators typical for younger dextral strike-slip faults. We interpret these relationships as initial normal faulting reactivated as dextral strike-slip faulting; in the stereographic projection we can see the arrangement of the stress axes estimated for the normal faulting. White arrow indicates the position of one person for scale in the horizon, the key PC-15 is the location where fault data were collected (Table 1, estimated stress tensor) and n: number of data collected. (B) En echelon fractures which suggest minimal left-lateral movement in the rock. For some strike-slip faults, striate directions were inferred using Riedel fractures in the fault surface; therefore the data must be used with limited confidence and (C) model of releasing bend which could be congruent for normal and dextral Jurassic faulting in Potrero Colorado.

reactivated the older normal faults present in the Colorado beds (Fig. 5A), and generated en echelon fractures with a left-lateral component (Fig. 5B). Readily interpretable structures that document the early phase of activity of the San Marcos fault (Late Jurassic) were only found in the Potrero Colorado area (Fig. 6).

The Colorado beds display large-scale cross bedding. They lie atop, through an abrupt contact (Charleston, 1973), a package of fluvial conglomerate and red sandstone ~10 m thick (Fig. 2). The eolianites underlay, through an erosive contact, conglomerate and sandstone of the San Marcos Formation (McKee et al., 1990). At Potrero Colorado, the San Marcos Formation is ~500 thick, significantly less than in the San Marcos Valley (Fig. 2). The fluvial conglomerates under the eolianite rest atop, through an erosive contact, a nonfossiliferous sequence composed of limolite and quartz sandstone ~100 m thick (Fig. 2). Initially Charleston (1973) considered the limolite and sandstone beds at Potrero Colorado as Cretaceous. However, McKee et al. (1990) later assigned them to the Upper Jurassic, and correlated them with the Tanque Cuatro Palmas beds described in the San Marcos Valley (Fig. 2). These beds display mud cracks, rain imprints, fossil plants, and ichnofossils that could be dinosaur footprints.

Because of the descriptions of McKee et al. (1990), it is clear that the inferred Late Jurassic age of the eolianites hinges on the correlation of the underlying sandstone-mudstone sequence with the Tanque Cuatro Palmas beds; a Tithonian age has been determined biostratigraphically for these beds in the San Marcos Valley, and the unit is covered by the Lower Cretaceous San Marcos Formation. Thus, the structural importance of the Tanque Cuatro Palmas and Colorado beds at Potrero Colorado is that they are faulted, and the faults' kinematic indicators such as en echelon and Riedel fractures constrain the Late Jurassic strain tensor for central Coahuila. Fault surfaces do not

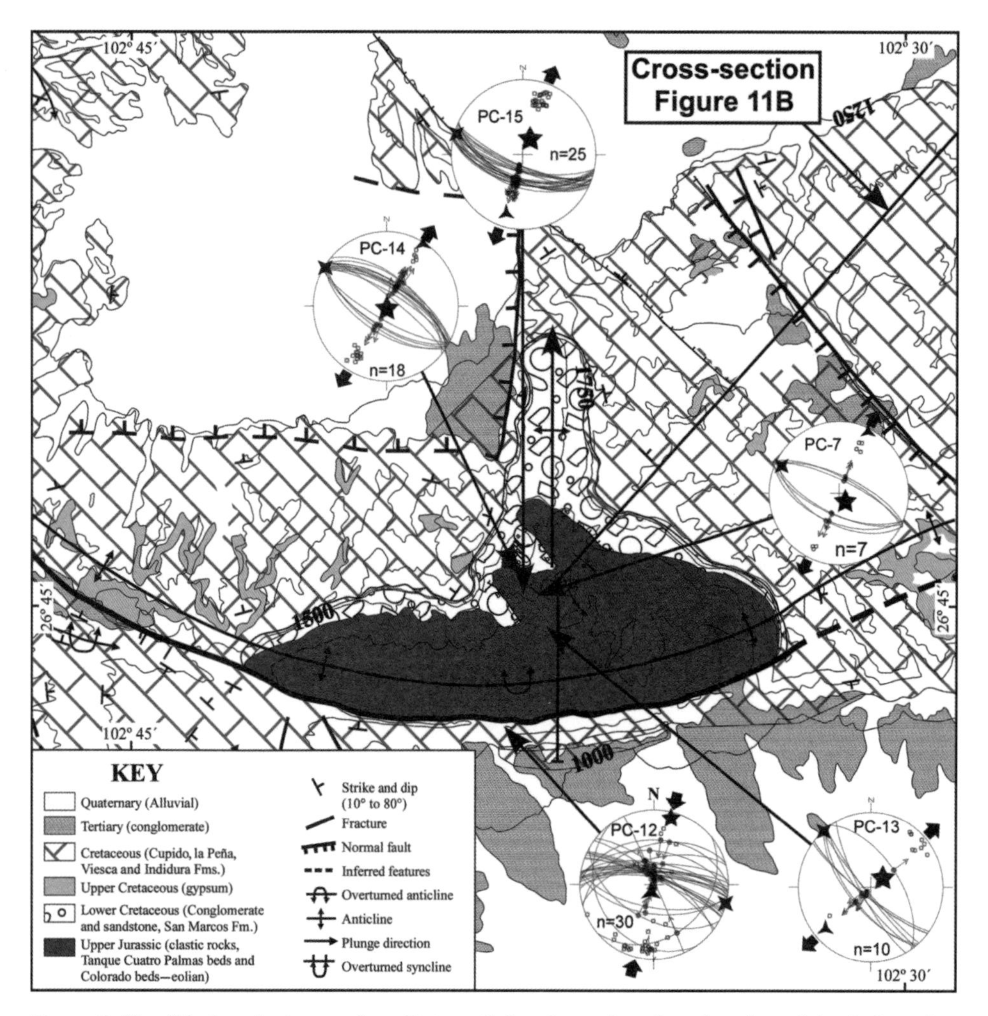

Figure 6. Simplified geologic map from Potrero Colorado used to show location of the fault station documented in this area. Location for part of this area (Fig. 11B cross section) is depicted in Figure 1. Please refer to the key in the Figure 5 caption.

contain slickensides, as the sandstone is poorly consolidated; in these cases the direction and sense of shear were approximately inferred from en echelon and Riedel fractures crosscutting the fault surface. We stress the fact that these faults do not affect younger rocks. Dextral-slip faults reactivated earlier normal faults in Jurassic rocks, and were not found to affect the San Marcos Formation. They thus predate this unit.

The Upper Jurassic limolite and sandstone sequence (a Tanque Cuatro Palmas beds equivalent) is affected by the same system of normal faults (Fig. 7A) as the eolianites. We interpret the Tanque Cuatro Palmas and Colorado beds as being deposited contemporaneously with active normal and minor strike-slip faulting in Potrero Colorado. This interpretation is supported by increases in bed thickness in the hanging-wall block; also, in some places the sediments present evidence of syndepositional liquefaction (Fig. 7A and 7B). The outcrop photographed in Figure 7A displays a paleosurface crosscut by normal faults. This feature was later covered by quartz sandstone, which now shows local differential compaction. The older faults in the limolite caused late propagation of fractures through the sandstone after its deposition (Fig. 7A). Some synsedimentary fault zones (i.e., dextral-slip faults) show structures produced by liquefaction in sediments. These structures suggest that faulting occurred while the sediments were still saturated with water and/or were poorly lithified (Fig. 7A). Faults with liquefaction features have dextral slip, and it is believed that they could have interacted alternatively with the normal faults (Fig. 7B).

The eolianite shows well-defined features that detail the three-dimensional geometry of Late Jurassic normal faults. We recognized a domino-style set of northwest-southeast–trending faults with the downthrown blocks to the southwest (Fig. 5A); we inferred that the fault zone must had been cemented and later reactivated, because the faults show well-developed en echelon and Riedel fractures that constrain minor dextral slip (Fig. 5A).

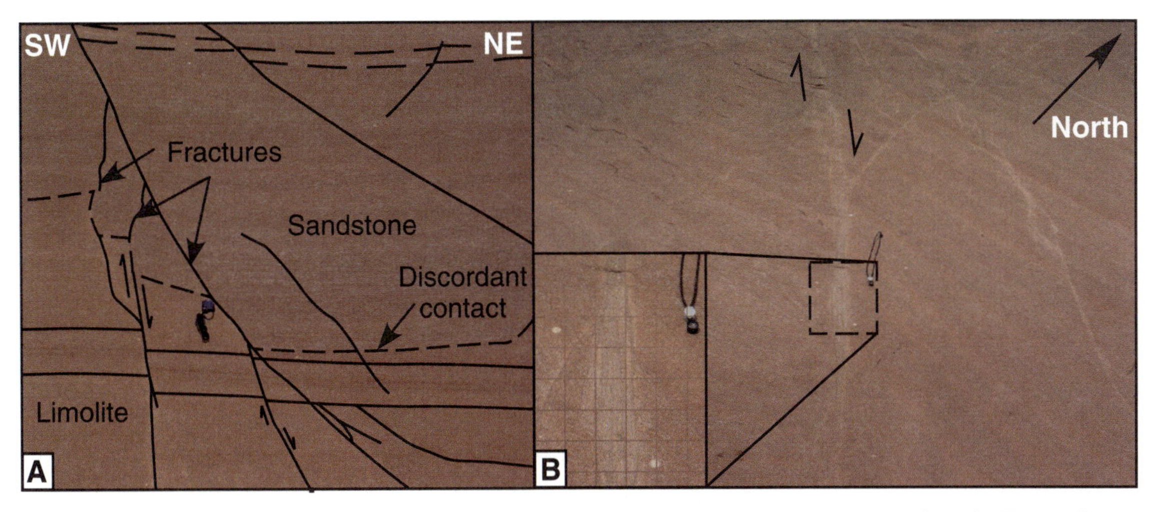

Figure 7. (A) Minor normal fault that crosscuts the contact between limolite and quartz sandstone from the Tanque Cuatro Palmas beds in Potrero Colorado; the sandstone was deposited after normal faulting and later minor fracture propagation in the sandstone occurred, which initiated from the older normal faults and minor sediment compaction; pocketknife for scale. (B) Dextral strike-slip fault in the limolite and quartz sandstone from the Tanque Cuatro Palmas beds showing features that suggest liquefaction (synsedimentary faulting prior to rock formation).

Additionally, in the eolianite we observed several minor fractures with en echelon arrangement (Fig. 5B), suggesting local left-lateral displacement. However, the dextral faults are far more abundant in the fault zone.

The dextral fault set, formed at the expense of older normal faults (Fig. 5A), cannot be used to estimate a stress tensor as the orientation of the fault planes was established during a previous faulting event. However, we infer that a local northwest-southeast shortening could have reactivated the normal faults with dextral shear (Fig. 5A and 5C). The reactivated fault planes with a dextral component are abundant in the eolianite and the underlying Tanque Cuatro Palmas beds. Additionally, in the central portion of Potrero Colorado near the domino-style faults, we found an exposure where the eolianite displays a synsedimentary set of horsts and grabens which was buried by undeformed younger eolian sediments. This exposure is evidence of Late Jurassic north-northeast–south-southwest extension in Potrero Colorado (Fig. 8A and 8B).

The outcrop depicted in Figure 8 shows that faulting and sedimentation were coeval, and that the faults do not cut the upper contact with the San Marcos Formation. A similar relation was documented in the older limolite and quartz sandstone. The faults in Figure 8 are consistent with a north-northeast–south-southwest direction of minimal compression. However, many faults were cut by younger faults or reactivated with minor dextral components. This was observed both in the Tanque Cuatro Palmas and in the eolianite of the Colorado beds. Therefore, it is concluded that Late Jurassic faulting was complex, producing normal faults that were cemented. These faults were later reactivated with centimetric to metric dextral slip, developing Riedel fractures in the early stages of this second pulse of deformation. The alternating activity of these fault systems can be explained as the result of pre-Tithonian or Late Jurassic deformation in a minor restraining bend zone or in a relay in a normal fault zone (González-Naranjo, 2006) in Potrero Colorado. The orientation of the inferred maximum compression axis was ~northwest-southeast and minimum compression was northeast-southwest (Figs. 5C and 6).

The relations among normal and dextral strike-slip faults found in Jurassic rocks of Potrero Colorado are better exposed than the sedimentary structures and folds in the San Marcos Valley, which were used to infer a restraining bend in a left-lateral system along the San Marcos fault during the Late Jurassic (McKee et al., 1990). Our interpretation of the structures in Potrero Colorado as a product of a minor releasing bend zone with dextral movement along the fault or as a normal relay fault zone argues against the previous interpretation.

We note that the bend of the San Marcos fault in Potrero Colorado has geometry similar to the bend in the San Marcos Valley. However, the normal- and dextral-slip faults documented in Potrero Colorado suggest opposing kinematics when compared with the folds recognized in the San Marcos Valley. Therefore, it cannot be argued that the same stress conditions in the crust generated both sets of structures at any time. In this study we propose that the normal and right-lateral faulting that acted during deposition of the Tanque Cuatro Palmas and Colorado beds in Potrero Colorado defines the timing and kinematics of deformation more clearly than the structures in the San Marcos Valley. Therefore, we conclude that the San Marcos fault had a pulse of activity with a minor right-lateral component in pre-Neocomian time, and that the folds with clockwise rotation of 80° to 90° (Arvizu-Gutiérrez, 2003; Molina-Garza et al., 2003) in the San Marcos Valley were formed during complex deformation caused by the Laramide orogeny in the Paleogene.

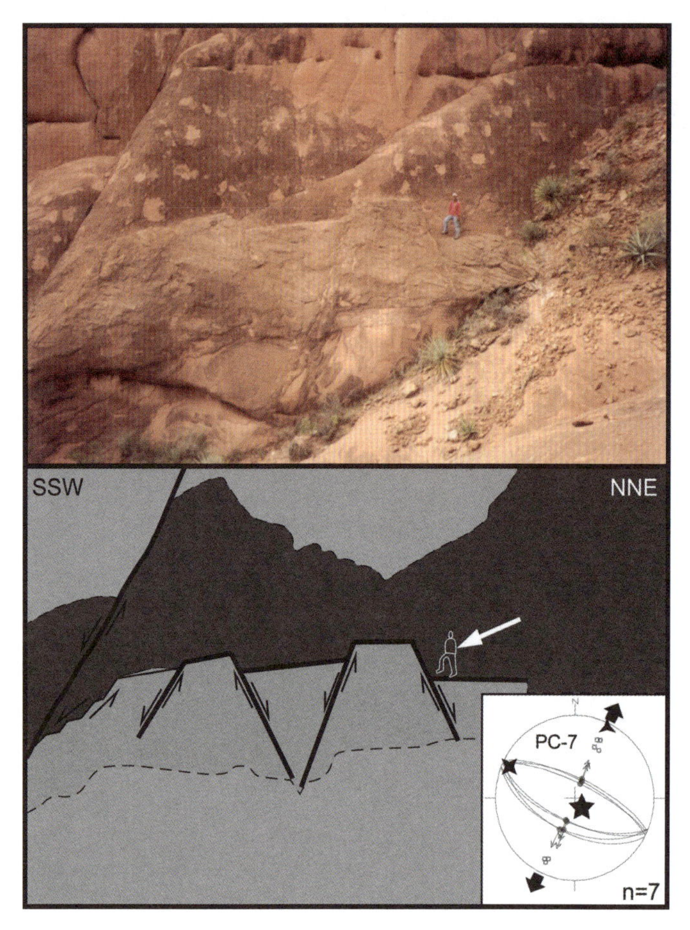

Figure 8. Grabens and horsts in the Upper Jurassic Colorado beds from Potrero Colorado. The upper picture shows the outcrop and the lower the structural interpretation; the inset shows the stereographic projection of the normal faults, which define a north-northeast direction of minimal compression. Faults in the center of the picture do not crosscut the younger beds, however the major fault in the left side crosscuts all the sequences. This major fault is a Neocomian fault that crosscuts the contact between the eolianite beds of the Colorado beds and the San Marcos Formation, whereas the minor faults, in the center of the picture, are interpreted as Late Jurassic synsedimentary faulting. The white arrow indicates a person for scale. The differing shades of gray show internal contacts between different directions of cross bedding in the Colorado beds. PC-7 indicates the site where fault samples were collected and n the number of data. Please refer to the key in the Figure 5 caption.

REACTIVATIONS OF THE SAN MARCOS FAULT

First Stage: Early Cretaceous Reactivation

The first major reactivation event of the San Marcos fault occurred during the Neocomian, and was identified by McKee et al. (1990) based on the deposition of a continental clastic package represented by the San Marcos Formation. The San Marcos Formation is composed of fluvial sediments, and in Potrero Colorado has eolian to fluvial deposits at its base. Neocomian lateral facies changes are also evident because of alluvial plain deposits in Sierra Mojada (Fig. 2). The occurrence of marked facies changes in the San Marcos Formation led McKee et al. (1990) to propose that the San Marcos fault experienced normal reactivation at a regional scale during the Neocomian. They suggested isostatic adjustment as the cause for this reactivation. This reactivation event was coeval with a global eustatic sea level fall recognized in northeastern México (Goldhammer, 1999; Eguiluz de Antuñano, 2001), producing second-order supersequences in the stratigraphy of both the Coahuila block and the Sabinas basin. In this study we document normal faults that acted during the Neocomian, which supports the McKee et al. (1990) hypothesis of major uplift drastically modifying sedimentation in central Coahuila.

As in the case of the Late Jurassic faulting, the Neocomian normal faults related to the accumulation of the San Marcos Formation were documented only in Potrero Colorado, as the contact between Upper Jurassic strata and the conglomerates of the San Marcos Formation is not exposed in either Sierra Mojada or the San Marcos Valley (Fig. 9).

The contact between the Colorado beds and the San Marcos Formation was carefully inspected, and some normal faults with vertical displacements of the order of more than 10 m (Fig. 9A) to a few meters (Fig. 9B) were found. On the other hand, a set of domino-style faults with the downthrown block toward the southwest was also found (Fig. 9C). These faults cut across the contact between the Colorado beds and San Marcos Formation without affecting the upper contact between the San Marcos Formation and the marine limestone (reef facies) of the Cupido Formation (Hauterivian-Barremian; Figs. 6 and 9), as these faults were buried during deposition of the upper package of the San Marcos Formation. Based on these relations, these rocks constrain the time of this pulse of normal faulting of the San Marcos fault to the Neocomian.

The northwest-southeast–trending normal faults crosscutting the contact between the Colorado beds and San Marcos Formation define a principal direction of minimal compression northeast-southwest (Fig. 9; localities PC-13 and PC-14: Figure 6 and Table 1). Our observations are consistent with the proposal of McKee et al. (1990) of a period extension in the Early Cretaceous, which reactivated the San Marcos fault as a normal fault, as suggested by their stratigraphic study. Our study provides structural arguments that strengthen the hypothesis of an important extension event that contributed to the growth of the Sabinas basin during Neocomian time.

Second Stage: Paleogene Laramide Reactivation

The structures that attest to Paleogene deformation are widespread along the trace of the fault and in the Coahuila fold belt (Fig. 1). This deformation event was coeval with the Laramide or Hidalgoan orogeny, and reactivated the San Marcos fault, as well as other secondary or subsidiary structures, with a reverse direction. During this period of deformation the Sabinas basin was tectonically inverted.

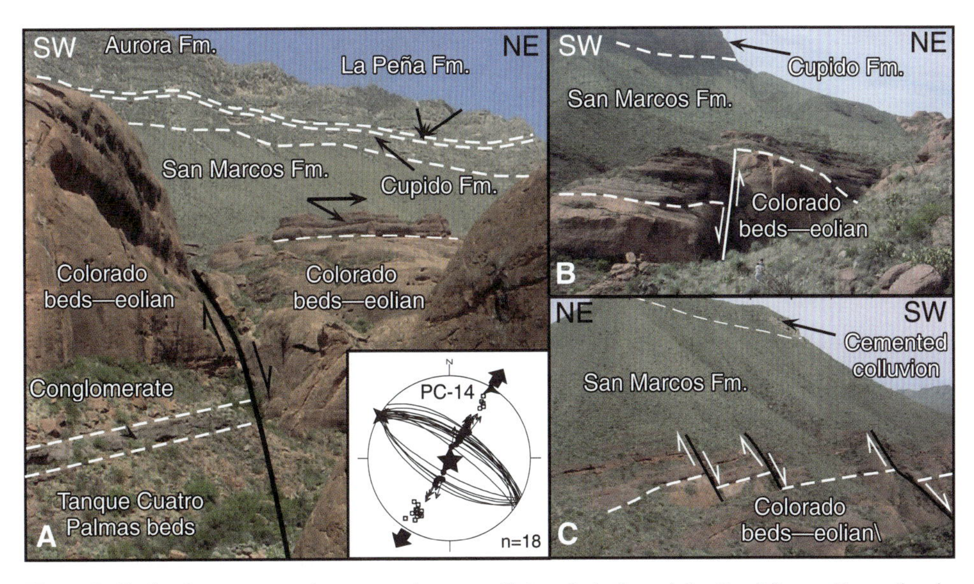

Figure 9. Faults that crosscut the contact between Colorado beds and the San Marcos Formation in northwestern Potrero Colorado. (A) Normal fault with vertical displacement >10 m and hanging-wall block toward the northeast, (B) normal fault with hanging-wall block toward the southwest (it and the former define a graben structure), and (C) domino arrangement of normal faulting with vertical displacement of a few meters and hanging-wall blocks toward the southwest; none of the faults crosscut the contact between the San Marcos and Cupido Formations. White dashed lines delimit lithologic contact. The stereographic projection in 9A shows a north-northeast minimum direction of compression for this set of faults. Please refer to the key in the Figure 5 caption.

Laramide Folding

Folding of Mesozoic marine and continental sedimentary rocks along the San Marcos fault trace was complex, as shown by the common occurrence of nearly perpendicular fold axes in southern Coahuila (Fig. 10). Drape-type folds are ubiquitous in the region (Figs. 10, 11A, and 11B). One exception exists immediately north-northwest of the San Marcos Valley at the Sierra San Marcos–Pinos. This range is formed by two north-northwest–plunging anticlines that change from steeply inclined to moderately inclined folds toward the west-southwest. These anticlines are separated by a tight syncline and their axes merge in the southeast part of the structure to conform a single major structure in the San Marcos Valley, where the San Marcos Formation is exposed at its core (Fig. 3).

We envision that the south-southeast portion of the Sierra de San Marcos–Pinos anticline was truncated during reverse reactivation of the San Marcos fault during a late phase of the Laramide orogeny (Fig. 3; Chávez-Cabello, 2005). Details about the origin and styles resulting from the early phase of Laramide deformation in the region are beyond the scope of this article and will not be discussed in detail here. We only address the principal crosscutting relationships among major structures in the area, pointing out the type of activity along the San Marcos fault and the structures generated by the fault at this time.

The El Granizo range (Fig. 3) is formed by a major stack of Upper Cretaceous calcareous rocks deformed by west-northwest–trending fault-bend folds with tectonic transport to the north-northeast. These structures were formed during the early phase of Laramide deformation. Like the Sierra San Marcos–Pinos anticline, these fault-bend folds were truncated by the San Marcos fault during its Early Tertiary reverse reactivation (Chávez-Cabello, 2005). This pulse of activity of the San Marcos fault tectonically juxtaposed Upper Jurassic clastic rocks with Cretaceous calcareous rocks at the base of the El Granizo range (Figs. 3, 11A, and 12). The reverse reactivation of the fault could have caused a tilt of nearly 30° toward the northeast of all Jurassic and Cretaceous sequence exposed in the San Marcos Valley and Sierra San Marcos–Pinos (Figs. 3, 11A, and 12). Additionally, intense simple shear was developed between Jurassic and Cretaceous rocks in northeastern El Granizo range, tilting the complete sequence toward the southwest (Fig. 11A).

On the other hand, the La Fragua range is formed by a WNW-ESE–trending asymmetric anticline (a drape fold; Fig. 11B), and it is practically uneroded at its core. The southeastern, upright, frontal fold limb dips 30° to the east and abruptly changes to vertical or inverted in the northwest. If inspected in detail, the southeastern limb is complex; as observed in the El Granizo range, there is at least one fault-bend fold that locally duplicates the sequence (i.e., El Mimbre canyon; Fig. 10B). Faults and folds suggest north-northeast tectonic transport, and were interpreted as produced during an early phase of Laramide deformation (Chávez-Cabello, 2005). The distal part of the northeastern limb of the Sierra La Fragua anticline dips gently north-northeast (~15°). The core of the anticline is exposed at Potrero Colorado, where sandstone and limolite of the Upper Jurassic are present as the oldest sedimentary rocks in the structure (McKee et al., 1990).

TABLE 1. VALUES FOR THE PRINCIPAL STRESS AXES OF THE STRESS TENSORS ESTIMATED BY INVERSE ANALYSIS OF FAULT SLIP DATA USING THE *INVD* PROGRAM OF ANGELIER (1990)

Locality	Formation	Fault type	Site	T	σ_1	σ_2	σ_3	ϕ	N	RUP	Q
PC	TCPB	Normal	PC-13	1	73/051	01/319	17/228	0.267	10	21	A
PC	CB	Normal	PC-15	1	71/019	04/289	18/199	0.024	25	10	A
PC	CB	Normal	PC-7	1	78/143	10/295	05/026	0.840	7	30	A
PC	CB-SM	Normal	PC-14	2	87/173	02/301	02/031	0.766	18	27	A
PC	TS	Reverse	PC-12	3	16/012	02/102	74/199	0.673	30	28	A
SMV	TS	Reverse	LP-1	3	12/014	03/105	78/208	0.621	29	33	A
SMV	Indidura	Reverse	CG	3	03/036	06/306	83/150	0.559	22	24	A
SMV	TS	Reverse	SG	3	16/191	13/284	69/051	0.623	24	38	A
SMV	Cupido	Lateral/ Reverse	SMM	3	05/296	25/029	64/196	0.687	10	46	B
SMV	Cupido	Reverse	FCW	3	05/287	13/018	76/175	0.730	8	9	A
Sierra Mojada	La Mula	Reverse oblique	SM-12	3	00/234	23/324	67/144	0.628	41	31	A
Sierra Mojada	La Mula	Reverse oblique	SM-4	3	04/069	15/338	74/175	0.831	36	34	A
Delicias	Las Uvas and basement	Lateral	D-1	4	07/077	52/177	37/342	0.243	11	23	A
Tetillas	Outer unit	Lateral	LT-1	4	08/265	70/152	18/358	0.651	9	32	A
Tetillas	Outer unit	Lateral	LT-2	4	27/063	04/160	45/251	0.670	11	40	B
Delicias	Las Uvas and basement	Normal	D-2	5	70/328	11/090	16/184	0.296	23	38	B
Tetillas	Outer unit	Normal	LT-3	5	67/355	15/227	17/132	0.285	17	32	B
Tetillas	Outer unit	Normal	LT-4	5	72/020	10/141	15/234	0.342	29	25	B

Note: Fault slip data were collected in structures above or near the San Marcos Fault. T: Chronology of faulting (1: Late Jurassic, neoformed left-lateral, 2: Early Cretaceous, neoformed normal, 3: Paleogene, inherited reverse developed during Laramide event, 4: Middle to Late Tertiary, inherited left-lateral, 5: Plio-Quaternary, inherited normal); σ_1, σ_2, and σ_3: principal axes of the stress tensor (plunge/trend); ϕ: $(\sigma_1-\sigma_3)/(\sigma_1-\sigma_3)$; N: Number of data; RUP: estimator of coherence of Angelier's program (1990) for data with RUP<75%; Q: Quality (A: good, B: intermediate). TCPB: Tanque Cuatro Palmas Beds, CB: Colorado beds, SM: San Marcos, PC: Potrero Colorado, SMV: San Marcos Valley, and TS: Tamaulipas Superior.

The anticline forming Sierra Mojada (Figs. 1 and 13) is also complex. The Sierra Planchada (north of the Sierra Mojada locality) is the northeast limb and dips gently (<30°) in the same direction. The southwestern limb of the anticline (i.e., the Sierra Mojada proper) shows intense faulting and the Lower Cretaceous sequence (i.e., the conglomerate of the San Marcos Formation, shale and sandstone of the La Mula Formation, and limestone of the Cupido Formation) is thrusted over Upper Cretaceous rocks (i.e., Tamaulipas Superior Formation; McKee et al., 1990; Fig. 11C). More detailed structural work is needed to establish the geometry of this fold.

The anticlines exposed at Sierra La Fragua, San Marcos–Pinos, and probably at Sierra Mojada are the result of complex faulting and folding, which developed in two phases of deformation during the Laramide orogeny. As a first approximation, both the La Fragua and San Marcos–Pinos ranges appear to be monoclines/drape-type anticlines, where the dip change between the limbs occurs abruptly, from a gentle dip in a limb that forms a large portion of the ranges to a nearly vertical dip in the other limb (Fig. 11A and 11B). The drape-type folds are commonly associated with inversion of basement faults (Harding et al., 1985; Buchanan and McClay, 1991). In the study area, faults in the basement and in the Jurassic rocks suggest major normal and minor right-lateral displacements for the Jurassic and major normal displacement for the Early Cretaceous. After sedimentation in the basin occurred, a phase of décollement and formation of fault-bend folds developed (Chávez-Cabello, 2005); this phase predated tectonic inversion in the region. Tectonic inversion was characterized by reactivation of former normal faults as reverse orientation of older basement faults controlled the generation of drape-type folds with variable trends.

Laramide Faulting in the San Marcos Valley

Among the areas studied along the trace of the San Marcos fault, the southeastern portion of the San Marcos Valley is the area with the largest variety of rock types with different ages.

Figure 10. (A) Arrangement of nearly perpendicular fold axes in the southwestern area of the Coahuila fold belt (CFB). Dashed line subparallel to the San Marcos fault south of it indicates the trend of a magmatic lineament defined by the volcanic neck of La Víbora, Las Tetillas pluton, and Las Coloradas volcanic field. Isotopic ages of the magmatic bodies are not known, but we infer that they could be Pliocene because the relief is young. An exception is the Las Tetillas pluton which may be late Eocene in age. SLFA—La Fragua anticline; SMPA—San Marcos–Pinos anticline; SLMA—La Madera anticline; OA—Ocampo anticline; LVA—La Virgen anticline; and CB—Coahuila block. Please refer to the key in the Figure 5 caption. (B) Fault bend fold in Cañón El Mimbre, which was cross-cut by late reactivation of the San Marcos fault.

This is an areas where juxtaposition of different lithostratigraphic units is attained by high-angle faults dipping northeast. This set of faults suggests that the San Marcos fault is not a simple fault plane but a wide zone of deformation. The high-angle faults in the San Marcos Valley are tectonic boundaries between blocks that juxtapose progressively older rocks toward the center of the valley (Figs. 3 and 12). Almost all of the secondary faults found in this area have a steep dip, which corresponds with the requirement of high-angle basement faults for developing drape-type folds through tectonic inversion. In the

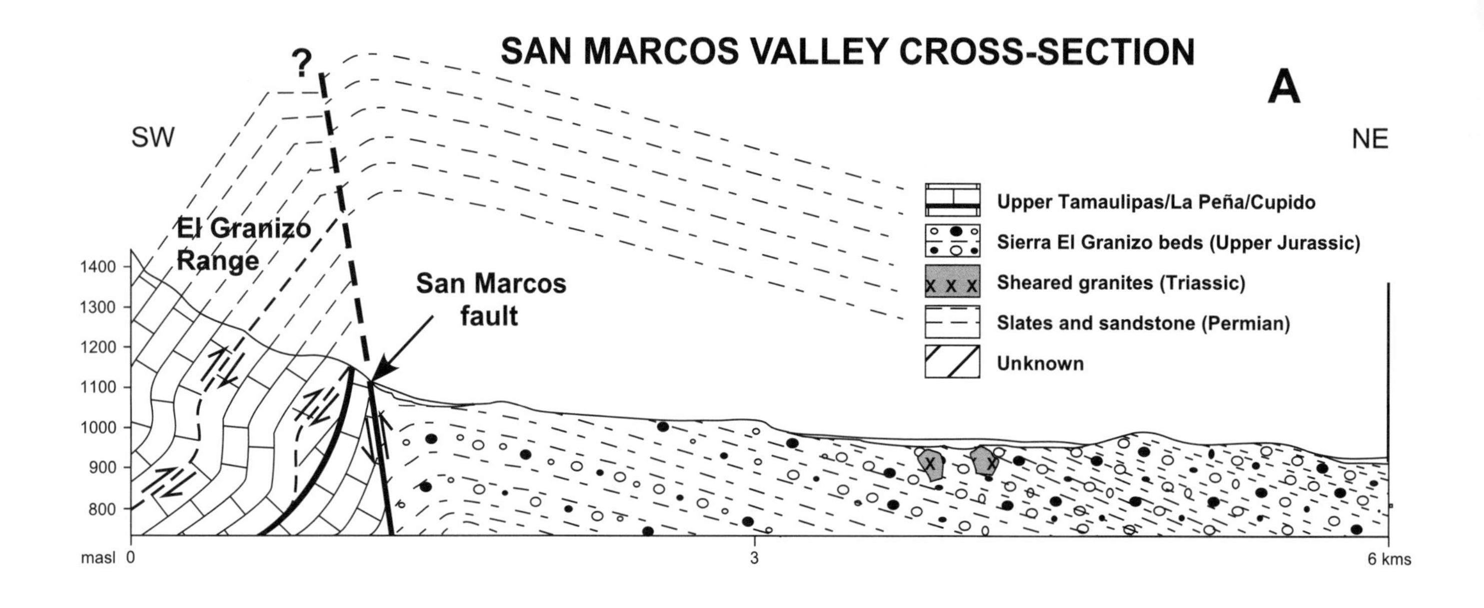
SAN MARCOS VALLEY CROSS-SECTION
A
SW
NE
?
El Granizo Range
San Marcos fault
Upper Tamaulipas/La Peña/Cupido
Sierra El Granizo beds (Upper Jurassic)
Sheared granites (Triassic)
Slates and sandstone (Permian)
Unknown
1400
1300
1200
1100
1000
900
800
masl 0
3
6 kms

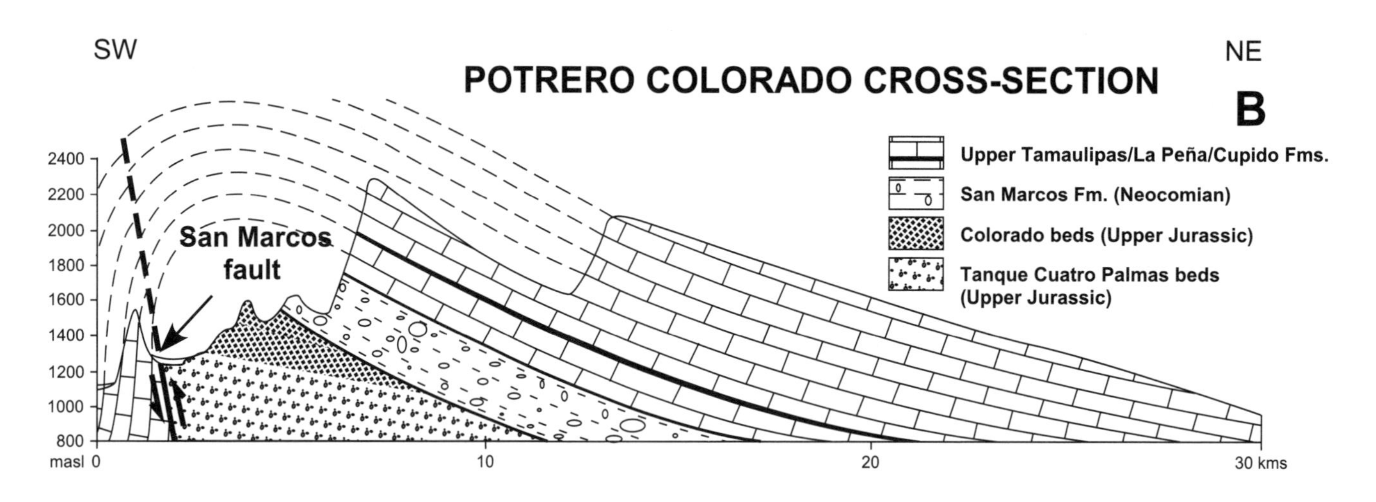
SW
NE
POTRERO COLORADO CROSS-SECTION
B
San Marcos fault
Upper Tamaulipas/La Peña/Cupido Fms.
San Marcos Fm. (Neocomian)
Colorado beds (Upper Jurassic)
Tanque Cuatro Palmas beds (Upper Jurassic)
2400
2200
2000
1800
1600
1400
1200
1000
800
masl 0
10
20
30 kms

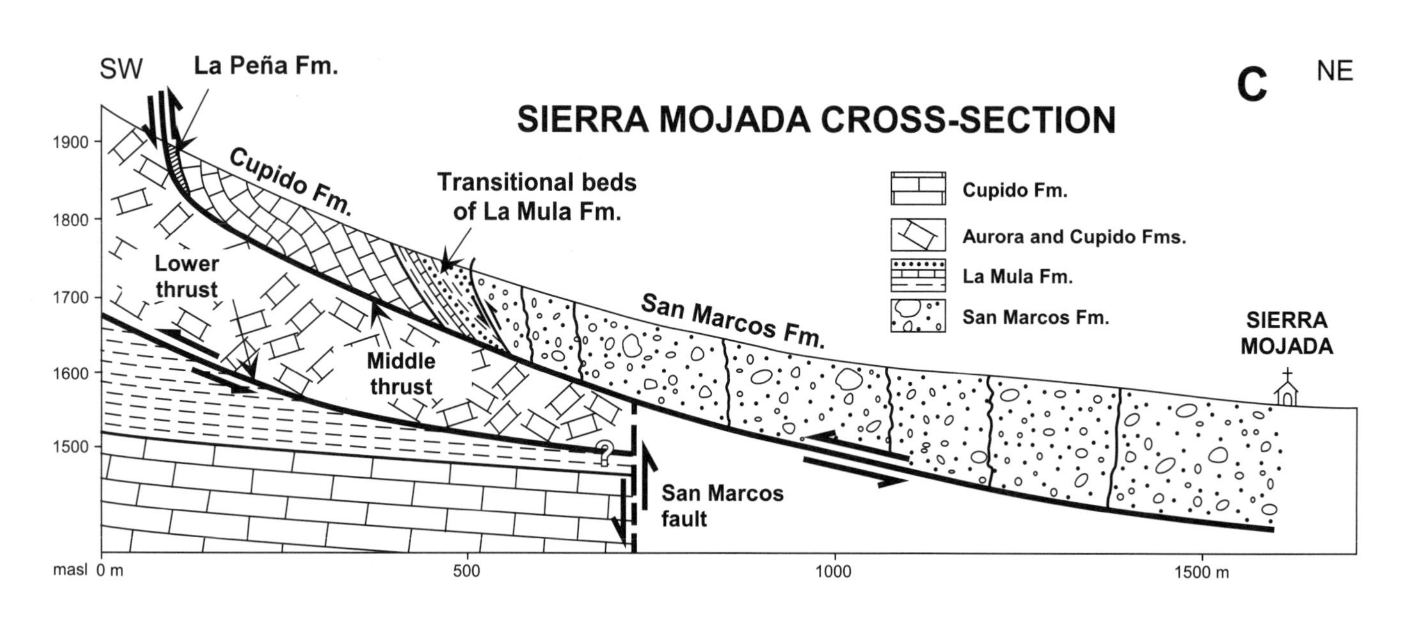
SW
La Peña Fm.
C
NE
SIERRA MOJADA CROSS-SECTION
Cupido Fm.
Transitional beds of La Mula Fm.
Lower thrust
Middle thrust
San Marcos Fm.
San Marcos fault
SIERRA MOJADA
Cupido Fm.
Aurora and Cupido Fms.
La Mula Fm.
San Marcos Fm.
1900
1800
1700
1600
1500
masl 0 m
500
1000
1500 m

Figure 11. Simplified geological cross sections that show the geometry of asymmetric anticlines developed over the San Marcos fault trace. We interpret that this geometry was controlled by tectonic inversion of the fault during the Laramide orogeny in the region. (A) Middle part of the San Marcos Valley, line of section depicted in Figure 3, (B) cross section of the asymmetric anticline of the La Fragua range through Potrero Colorado, line of section depicted in Figure 6, and (C) cross section of the Sierra Mojada area (from McKee et al., 1990), line of section depicted in Figure 1. Key: masl—meters above sea level.

southwestern area of the San Marcos Valley, two faults support the idea that the valley was uplifted above the Coahuila block during the Laramide orogeny in the Paleogene.

The fault located at the innermost part of the San Marcos Valley (Figs. 3 and 12) juxtaposes Upper Paleozoic slate with limestone of the Albian Tamaulipas Superior Formation, and also with Upper Jurassic clastic rocks of the Las Palomas beds. The next fault south juxtaposes Upper Jurassic clastic rocks of the Las Palomas beds with limestone of the Albian Tamaulipas Superior Formation, and finally, the last fault in the southwest juxtaposes limestones of the Tamaulipas Superior Formation with fine-grained marine clastic sediments of the Upper Cretaceous Indidura Formation close to El Granizo hill (see inset in Fig. 3 and Fig. 12).

In addition to the faults discussed in the previous paragraph, this work documented a set of minor faults along the San Marcos fault that were used to infer the kinematics of the second major reactivation of the fault during the Laramide orogeny. In the following paragraphs we discuss the evidence used to estimate the maximum compression direction from several fault localities studied in the San Marcos Valley.

North-northeast to northeast maximum compression. Along the northern foothills of the El Granizo range we studied faults in three localities and found that the data are consistent with north-northeast to northeast maximum compression (SG, LP-1, and CG; Fig. 13 and Table 1). Only locality CG yielded northeast compression, and this result was obtained from faults in the Indidura Formation ~200 m south from the bend zone of the San Marcos fault in the southeastern portion of the El Granizo range (Fig. 13 and Table 1).

The faults studied at locality LP-1, south of Ejido Las Palomas, form a conjugated system of reverse faults developed in limestone of the Tamaulipas Superior Formation (Fig. 13). These minor faults trend parallel to the San Marcos fault. The faults are WNW-ESE–trending with a nearly dip-slip movement of blocks. The inferred stress direction for locality LP-1 is consistent with the stress tensor orientation for an ideal set of conjugated reverse faults (σ_1 nearly horizontal, σ_2 parallel to the trend of faults, and σ_3 nearly vertical; Fig. 13, Table 1). It is similar to Anderson's theoretical model (1951) for faults formed in undeformed rocks, or fault movement along preexisting surfaces in the rocks which exerted minimal influence during deformation.

A second locality where faults were documented is between the El Granizo range and El Granizo hill (station CG, Fig. 13 and Table 1), south of the fault bend zone of the San Marcos fault. The structures affect shale and fine-grained calcareous sandstone of the Indidura Formation. In this locality we measured data from a set of minor faults ~200 m south of the younger vertical reverse fault that juxtaposes the limestone of the Tamaulipas Superior Formation with the shale and sandstone of the Indidura Formation. As in the LP-1 site, the trend of faults is west-northwest–east-southeast with nearly dip-slip movement of blocks in conjugate reverse faults, but here the faults dipping southwest are less abundant than those inclined to the northeast (Fig. 13, CG site).

The estimated principal stress directions have a clockwise rotation of 22° with respect to the stress tensor orientation obtained at site LP-1 with σ_1 nearly horizontal and σ_3 more verti-

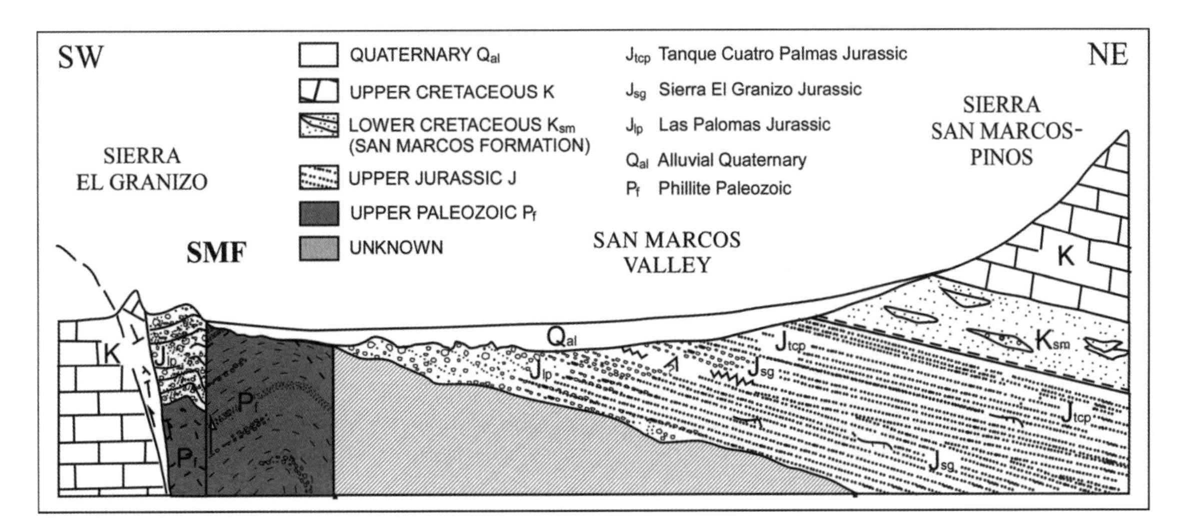

Figure 12. Geologic cross section that illustrates the vertical faulting in southwestern San Marcos Valley. Upper Paleozoic rocks were juxtaposed against Upper Jurassic rocks and the latter against Upper Cretaceous rocks (modified from McKee et al., 1990), line of section depicted in Figure 3. SMF—San Marcos fault.

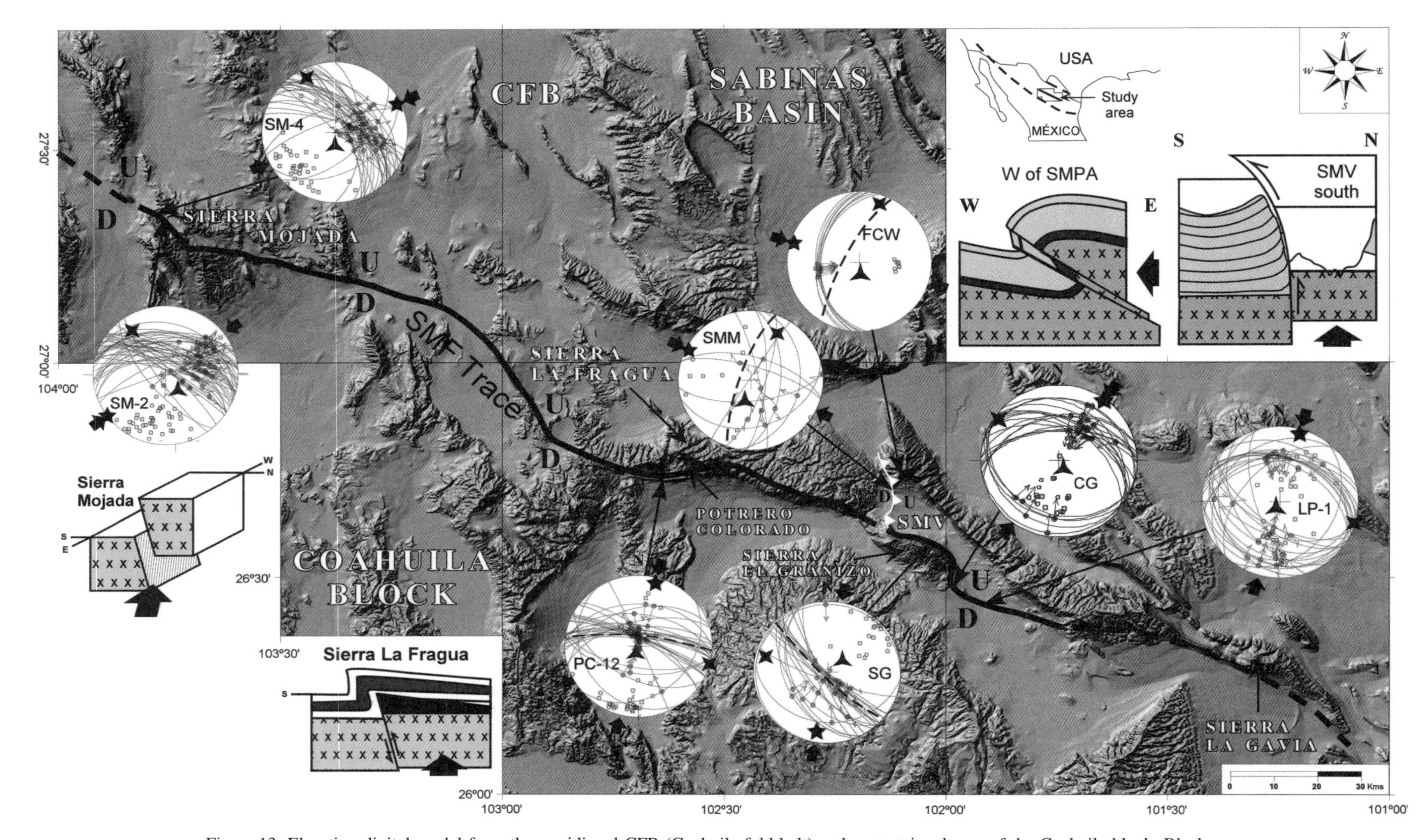

Figure 13. Elevation digital model from the meridional CFB (Coahuila fold belt) and septentrional area of the Coahuila block. Black arrows show the location of sites and the plots are the results of fault data analysis along the San Marcos fault (SMF). The data indicate north-northeast to northeast maximum compression during Paleogene time. The sketches in the upper right and lower left areas dilucidate the kinematics of the hanging-wall block of the San Marcos fault during the Laramide event. U—uplifted block on the fault plane during laramidic reverse reactivation; D—descended block; SMV—San Marcos Valley; SMPA—San Marcos Pinos Anticline. Please refer to the key in the Figure 5 caption.

cal (Fig. 13 and Table 1). We interpret this effect as the result of the interaction with the fault bend zone in the San Marcos fault.

Finally, a third fault station was measured northwest of the CG locality at the San Marcos fault, in the northern foothills of the El Granizo range (Fig. 13, site SG, and Table 1), in a place where McKee et al. (1990) identified an outcrop of the fault. The fault is marked by juxtaposition of conglomeratic rocks of Sierra El Granizo beds with limestone of the Albian Tamaulipas Superior Formation. In this zone, the La Peña Formation is intensely sheared and separates steeply dipping beds of the Cupido Formation exposed in the valley from beds of the Tamaulipas Superior Formation that dip ~45°SW. The faults measured at the SG site are nearly vertical and parallel to the vertical shear zone of the San Marcos fault and to faults parallel to the bedding in the contact between the Cupido and La Peña Formations.

We interpret the faults in LP-1 and CG, and the low-angle faults measured in SG, as having formed during the early shortening stage of the Laramide deformation in the zone, which in the Coahuila block was characterized by detachment controlled by stratigraphic units and by fault-bend folding at the boundary between the block and the Sabinas basin. The reverse high-angle faults observed in site SG were probably formed during the second stage of the Laramide deformation which inverted the San Marcos fault. In both cases the estimated maximum principal stress is similar (Fig. 13 and Table 1).

West-northwest maximum compression. In order to understand the causes of the trend change from west-northwest to ~north-south of the fold that forms the Sierra San Marcos–Pinos (Figs. 3 and 10), and the abrupt dip variation between the strata exposed in the northeastern part of Sierra La Fragua (horizontal) and the western part of Sierra San Marcos–Pinos (vertical or overturned), we searched for faults in the western limb of the anticline of the Sierra San Marcos–Pinos. Two fault stations were studied; the stress tensor obtained has σ_1 nearly horizontal and oriented west-northwest, nearly perpendicular to the maximum principal stress direction estimated in the southwestern part of the San Marcos Valley (i.e., El Granizo range, where σ_1 is north-northeast; Fig. 13 and Table 1).

Almost every fault observed in station SMM is oblique to bedding, and the structures are exposed in the access to a small exploration pit where the contact between the Cupido and San Marcos Formations is exposed (Fig. 13 and Table 1). The estimated maximum principal stress has a west-northwest direction with only 5° plunge. In addition, the area shows tight folds. The trend and dip of bedding is 190°/55° (Fig. 3). In this area where the range narrows, a major north-south–trending vertical breccia zone was recognized, and we interpret it as the structure that caused the change of trend of the Sierra San Marcos–Pinos from west-northwest to nearly north-south (Figs. 3 and 10).

The second fault station (FCW, Fig. 13 and Table 1) shows mainly minor thrust faults, which are oblique to bedding. These structures have north-south trends and dip 30° to 40°W. The orientation of the estimated stress tensor inferred from these faults is similar to that obtained in station SMM (Fig. 13). Other faults observed from the distance in the western limb of the Sierra San Marcos–Pinos anticline in the Tamaulipas Superior Formation are not discussed in this work as they are in an inaccessible area. Those steep faults have an approximate north-south trend and dip to the east, suggesting westward tectonic transport.

The paleostress directions inferred from fault stations in the San Marcos Valley are perpendicular (north-northeast in the south and west-northwest in the west) and show small plunges for σ_1 and σ_2 (Fig. 13 and Table 1). The origin of this perpendicular relationship between stress axes is discussed next.

Laramide Faulting in Potrero Colorado

North-northeast maximum compression. North-northeast to northeast is the most common direction for maximum compression obtained along the trace of the San Marcos fault in central Coahuila. We collected data for a fault station at the entrance to the Potrero Colorado ranch, in the southern part of that area, obtaining data from faults parallel or oblique to bedding in the Tamaulipas Superior Formation (station PC-12, Fig. 13 and Table 1). The faults dips are nearly vertical and their trend is ~east-west, similar to that in faults documented in station SG in the San Marcos Valley (Fig. 13). A vertical fault contact between a limolite and sandstone sequence of the Upper Jurassic Tanque Cuatro Palmas beds with Cretaceous limestone and gypsum was observed in Potrero Colorado (Fig. 11B). The fault data were also collected near the southern access point to Potrero Colorado in an Albian sequence of limestone and gypsum, where gypsum laterally flowed during deformation. It is believed that this sequence corresponds to a lateral facies change near the border of the Coahuila calcareous platform. Other regions of the platform show occurrences of the Acatita Formation (lagoonal facies), which is mainly composed of gypsum and is coeval with the Viesca (reef facies) and Tamaulipas Superior (basin facies) Formations.

The inferred maximum principal stress plunges 16° and trends N12°E (Fig. 13 and Table 1), σ_2 is horizontal and parallel to the faults, and σ_3 is nearly vertical. A few meters northeast of fault station PC-12, a major vertical reverse fault juxtaposes Upper Jurassic clastic rocks with Albian limestone and gypsum. This fault zone, steeply inclined to the north, is interpreted here as the main trace of the San Marcos fault, the one responsible for the regional drape-folding in Sierra La Fragua during the late phase of Laramide deformation.

Laramide Faulting in Sierra Mojada

Northeast maximum compression. Sierra Mojada was the third and last locality where Laramide fault data were measured along the trace of the San Marcos fault. Two fault localities were documented along the Palomas Negras canyon (SM-2 and SM-4), northwest of the town of Sierra Mojada. The estimated maximum principal stress trends northeast, and was constrained using some west-northwest–trending faults with dips of ~50° (SM-2 station) and other NW-trending faults inclined more than 60° (SM-4 station). Most structures measured at Sierra Mojada are interpreted as minor oblique reverse faults, as their pitch is ~60°E (Fig. 13).

The fault data collected in Palomas Negras canyon correspond to oblique reverse faults that cut sandstone and other clastic marine rocks of La Mula Formation. The estimated tensors show nearly horizontal maximum compression axes, oblique to the west-northwest–trending San Marcos fault; σ_2 is horizontal and parallel to the faults, and σ_3 is nearly vertical (Fig. 13 and Table 1). However, the reverse faulting shows dips >50°. We use this information to argue that these faults could have been generated during reverse reactivation of the San Marcos fault during the late phase of the Laramide deformation in the area.

Third Stage: Late Miocene–Early Pliocene Faulting

Faulting that postdates Laramide structures in Coahuila is scarce, and it is not as obvious as in neighboring Chihuahua State. Aranda-Gómez et al. (2005a) identified two post-Laramide reactivation events probably associated with the buried trace of the San Marcos fault in southeastern Chihuahua. The older pulse is late Miocene–early Pliocene in age (14–5 Ma), and is inferred from the occurrence of broad north-northwest–trending folds plunging to the south-southeast. The folded rocks are volcanic with isotopic ages between ~32 and 14 Ma. The folded structures and the Camargo volcanic field lie atop the buried trace of the San Marcos fault in southeastern Chihuahua. The folds are exposed northwest (Aguachile range) and southeast (La Herradura range), respectively, of the Camargo volcanic field.

Aranda-Gómez et al. (2005a) suggested that the San Marcos fault may be prolonged at least 300 km northwest into Chihuahua to an area close to Aldama, where other syncline in volcanic rocks was identified ~5 km west from Sierra El Morrión at Sierra Cuesta del Infierno.

Aranda-Gómez et al. (2005a) identified faults and folds formed during a pulse of northwest Basin-and-Range extension, which occurred in the period between 14 and 5 Ma. This pulse of extension has been documented in other regions of northwest México and the southwestern United States. According to their interpretation, this phase of extension caused oblique movement along the San Marcos fault (normal, with a minor left-lateral component). Due to the orientation of the basement fault with respect to the minor left-lateral component, shortening developed in some areas. Isolated structures congruent with lateral movements, such as Aguachile and La Herradura synclines, resulted from this deformation. In an attempt to document this deformation period in Coahuila, we explored: (1) areas where Laramide structures could have been affected by this reactivation event of the San Marcos fault, (2) fault traces in the Coahuila block that could have been reactivated between late Miocene and early Pliocene time, and (3) Tertiary intrusions and younger volcanic rocks (Pliocene?) emplaced after the end of the Laramide deformation. We expected these units to have recorded the reactivation event identified along the postulated concealed continuation of the San Marcos fault in southeastern Chihuahua.

The Laramide structures in central Coahuila are not affected by faults formed or reactivated during a Neogene extensional event. The only studied locality where breccia zones crosscut Laramide structures is El Mimbre canyon in the eastern part of the La Fragua anticline (Figs. 10 and 14). In this area a major vertical breccia zone ~200 m thick (Fig. 14A and 14B), divides almost symmetrically the hinge zone of La Fragua anticline and follows its trend (ESE-WNW). Major lineaments that crosscut Laramide structures parallel to the trend of the breccia zone are not observed in the region. In addition, this wide breccia zone cannot be traced to the east into the Sierra San Marcos–Pinos.

Structures that may suggest some displacement of the San Marcos fault and some Laramide structures are the El Caballo (left lateral) and Almagre faults (right lateral, Figure 1; Eguiluz de Antuñano, 1984), which with the Juárez fault define a set of three north-northwest–trending regional lineaments west of Potrero Colorado (Fig. 1). We infer that these faults cut earlier Laramide structures and may be basement faults reactivated during the late stage of the Laramide orogeny. These faults cut across major folds and thrust faults in the western part of the Sabinas basin, and can be traced all the way south to Sierra de Parras as indicated by Eguiluz de Antuñano (1984). Based on the Chávez-Cabello (2005) observations near Cuatro Ciénegas (Coahuila), we speculate that, like the San Marcos fault and subsidiary faults, these lineaments could have been produced by reactivation of basement faults during the late stage of the Laramide orogeny. The precise age of the activity of these Tertiary faults is uncertain, but we observe that some of them, like other faults in central Coahuila, channeled Tertiary magmas. However, it remains to be established whether these faults crosscut these rocks or if they only influenced the location of volcanic vents (Fig. 1).

The major breccia zone at the El Mimbre canyon displays a clear vertical displacement near its border and cuts across older Laramide structures (Fig. 14C). However, we interpret that these vertical displacements were generated during the same Laramide event in the region, but in its late stage.

In reference to the reactivation of other basement faults in the Coahuila block, important results were obtained from the study of a major basement fault near Nuevo Delicias in the El Agua Grande canyon. This fault plane trends northwest and dips southwest, showing two sets of striae. The older striations show left-lateral kinematic indicators (Riedel fractures). The younger striations correspond to mineral fibers, which are consistent with normal displacement along the fault. This fault affects the Upper Cretaceous marine sequence exposed in the area, which in turn suggests that at least the normal reactivation may be post-Laramide (Fig. 15A and 15B).

Another locality where minor lateral faults occur is the late Eocene subvolcanic Las Tetillas intrusion (35.13 ± 0.10 Ma; Chávez-Cabello, 2005). This intrusive body was emplaced in the Coahuila block ~15 km south of the San Marcos fault, near Ejido Estanque Palomas. The intrusion shows some outcrops with conjugate lateral faults, and many cooling fractures that accommodated minor lateral displacements (Fig. 15C and 15D, respectively). Based on the data obtained at Nuevo Delicias, Las

Figure 14. (A) Zone of 200-m-thick vertical breccia in limestone of Viesca Formation, the square at the lower right side represents the area of 14B, (B) details in the deformation zone, and (C) shear parallel to bedding (1) that was crosscut and displaced by a minor vertical brecciate zone, and (2) in the south-southwest part of the major brecciate zone. We interpret that these structures represent minor structures related to the first (1) and the second (2) phase of deformation in the Sabinas basin.

Tetillas, and from the isotopic age of Las Tetillas, we suggest that these minor left-lateral and normal faults present in the northern portion of the Coahuila block may be coeval with the late Miocene–early Pliocene reactivation event proposed by Aranda-Gómez et al. (2005a) in southeastern Chihuahua.

Fourth Stage: Pliocene-Quaternary Normal Faulting

Aranda-Gómez et al. (2005a) proposed a Pliocene-Quaternary extensional event in southeastern Chihuahua. Faults of this event cut across the Plio-Pleistocene lava fields of the Camargo volcanic field (~5–0.09 Ma). Apparently late Miocene faults exposed near the volcanic field were also reactivated and new faults that crosscut the La Herradura and Aguachile synclines were generated. This northeast to east-northeast extensional event is younger than 5 Ma, and is coeval with intense intraplate mafic magmatism above of a major bend of the San Marcos fault such that it may have operated as a pull-apart basin.

In Coahuila, mafic intraplate-type magmatism occurred in several localities inside the Sabinas basin, such as the Las Esperanzas and Ocampo volcanic fields (Valdez-Moreno, 2001). Similarly, on the Coahuila block the Las Coloradas volcanic field (Fig. 1) was formed. However, there is no clear evidence that this magmatism was coeval with normal faulting, like in the Camargo volcanic field (Aranda-Gómez et al., 2005a). Conversely, the magmas in the volcanic fields of Coahuila may have ascended through major discontinuities in the basement like the La Babia and San Marcos faults and the tectonic (?) borders of La Mula Island, which is in the central part of the Sabinas basin (Aranda-Gómez et al., 2005b).

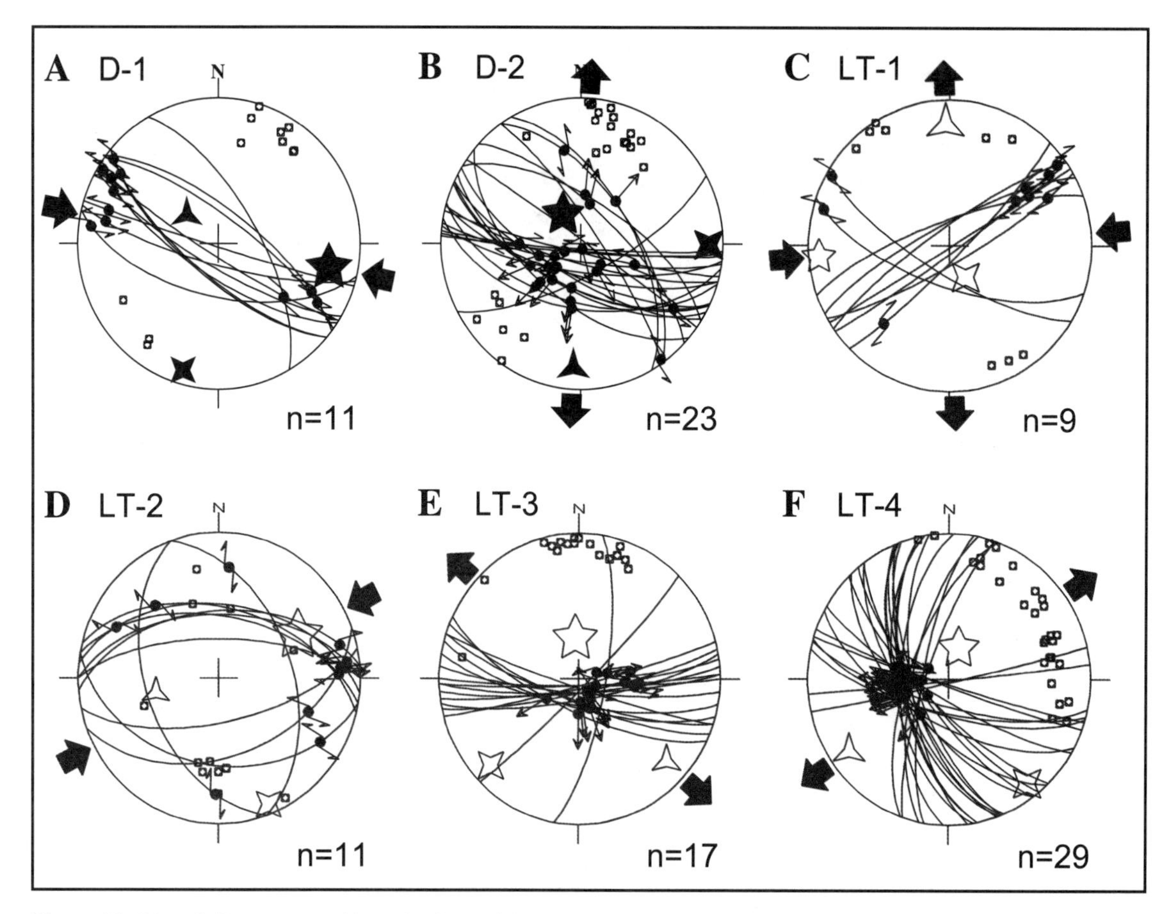

Figure 15. (A) and (B) stereographic projections of faults with stress axes for faults collected in the Nuevo Delicias area; see Figure 10A for site location. (A) Left-lateral faults in granites from the Coahuila block; (B) normal faults in the Upper Cretaceous sequence; faults in (A) and (B) are parallel to regional west-northwest basement lineaments, (C), (D), (E), and (F) come from the Las Tetillas pluton, see Figure 10A for location of sites: LT-1, LT-2, LT-3, and LT-4; (C) conjugated faults with minor lateral displacement collected inside the late Eocene Las Tetillas pluton; (D) cooling fractures which accommodate minor lateral displacement; (E) normal faults with oblique striations toward the east, which are in agreement with the northwest direction of minimum compression; (F) normal faults with pitch of striae close to 90° and highly variable strike of faults which define a northeast direction of minimum compression, we interpret that planes of faults reactivate preexisting planes of fractures. This is an example of where Angelier's INVD program failed to estimate stress tensor in faulted rocks, because of faulted blocks of rocks interacting to accommodate extension (Nieto-Samaniego and Alaniz-Álvarez, 1997). Diagrams (C), (D), and (F) were taken from García-Quintero (2004). Please refer to the key in the Figure 5 caption.

We note that a few northwest-trending normal faults with minor vertical displacements occur in the Coahuila block and in the Sabinas basin (Fig. 10). Additionally, we identified a lineament defined by magmatic features. The lineament suggests a structural element subparallel to but lying south of the San Marcos fault. The lineament is defined by the La Víbora volcanic neck (between Los Alamitos and El Granizo ranges), Las Tetillas intrusion (east of Los Alamitos range), and Las Coloradas volcanic field (located between Los the Alamitos, Paila, and San Marcos–Pinos ranges). Only the La Víbora volcanic neck presents evidence of normal faulting with the downthrown block south, coeval with the magmatic activity (Fig. 10).

On the other hand, in the El Agua Grande canyon near Nuevo Delicias, we documented fault planes with kinematic indicators that suggest normal displacement. This normal movement reactivated a preexisting basement fault, and accommodates north-northeast extension (Fig. 15B). Another locality where normal faulting with minor displacement was identified is the Las Tetillas intrusion. Here, the normal movements between blocks followed preexisting cooling fractures and, sometimes, reactivated older lateral faults with minor displacements (Fig. 15E and 15F). Finally, in the northern margins of the La Fragua and Los Alamitos anticlines there are northwest-southeast faults with minor normal displacements that suggest minor northeast-southwest extension (Fig. 10). In this work, we suggest that these lateral and normal post–late Eocene faulting events may correspond with the lateral and normal faulting found by Aranda-Gómez et al. (2005a) in and around the Camargo volcanic field for late Miocene–early Pliocene and Pliocene-Quaternary, respectively.

DISCUSSION

The Origin of the San Marcos Fault

McKee et al. (1984) inferred that the San Marcos fault is part of the Mojave-Sonora megashear system (or a branch of it) postulated by Anderson and Schmidt (1983), even though large-magnitude lateral slip along the fault can be neither proved nor excluded. This inference is based on the size of the fault, its orientation, its location, and the chronology of activity. A few years later, these same authors suggested that the San Marcos fault may have been part of a system of transform faults, which connected oceanic ridges from the Pacific and the Atlantic Oceans accommodating left-lateral slip in the Late Jurassic (McKee et al., 1990). In either model, the San Marcos fault is inferred to have accommodated large left-lateral displacement. Nevertheless, such displacements remain largely unproven. In fact, there is ongoing debate about the existence and origin of the Mojave-Sonora megashear system (e.g., Molina-Garza and Geissman, 1996, 1999; Iriondo, 2001; Molina-Garza and Iriondo, this volume). Admittedly, the Mojave-Sonora megashear system implies larger displacement than what can be accounted for along the San Marcos fault.

On the other hand, three tectonic models have been proposed for the origin of the Sabinas basin. This feature has been interpreted as: (1) a pull-apart basin (Longoria, 1984; Santamaría et al., 1991), (2) an intracratonic basin (an aulacogen or aborted rift arm [Charleston, 1973; Alfonso, 1978]); and (3) an intracratonic basin formed as a proto-rift (Salvador, 1991; Eguiluz de Antuñano, 2001). In each of these models, the San Marcos fault would act as a master fault along the southern margin of the basin. The first model would imply that the fault was established as a strike-slip fault, the other models require the fault to act as a normal fault.

The structural data collected in the course of this research cannot rule out the possibility that the fault may have originated as a strike-slip fault. Large vertical-axis rotations were interpreted from a small set of paleomagnetic data for Upper Jurassic strata in the San Marcos Valley (Molina-Garza et al., 2003; Arvizu-Gutiérrez, 2003) and they could be related to strike-slip tectonics, however the timing of the rotations is not necessarily Late Jurassic. These rotations may have occurred during the Paleogene Laramide orogeny, which in the Sabinas basin presents two phases of deformation (Chávez-Cabello, 2005). The models of Arvizu-Gutiérrez (2003) must be explored in greater detail with more complete structural and paleomagnetic studies in the San Marcos Valley and other localities where the fault trace is exposed, such as in Potrero Colorado. What is clear at Potrero Colorado is that the dominant faulting mechanism, that is synsedimentary deformation with significant displacement (tens of meters), is normal faulting. In this locality, northwest-trending normal fault planes were reactivated (in the Late Jurassic) to accommodate minor right-lateral slip (<1m) in Late Jurassic time.

Yet another characteristic of faults affecting Jurassic rocks at Potrero Colorado is intense syntectonic sedimentation, particularly in red beds correlated by McKee et al. (1990) with the Tanque Cuatro Palmas beds of the San Marcos Valley. A conglomerate identified by McKee et al. (1990) separating the red beds assigned to the Tanque Cuatro Palmas beds and the overlying Colorado beds may reflect minor rejuvenation of the relief, although it may also reflect decreasing sea level.

In general, the faults we documented that were affecting the red beds assigned to the Tanque Cuatro Palmas and the Colorado beds indicate that, from the Late Jurassic to the earliest Cretaceous (Neocomian), extension was significantly more important than lateral slip in central Coahuila. From this observation, we deduce that the Sabinas basin is more likely to have originated as an extensional basin delimited by the San Marcos fault on the southwest (via pulses of NNE-SSW–directed extension), than as a pull-apart basin delimited by a major strike-slip fault. This inference is further supported by the fact that the faults that border the Sabinas basin, San Marcos and La Babia (Fig. 1), have parallel orientations and do not join at either extreme of the basin. On a regional scale, this argues against the pull-apart model for the origin of basin.

Reactivations of the San Marcos Fault

McKee et al. (1990) suggested that the first event of reactivation of the San Marcos fault occurred in the Early Cretaceous, and that it resulted from isostatic adjustment. This interpretation is based on the existence of a clastic wedge up to 1000 m thick represented by the San Marcos Formation. Deposition of the San Marcos Formation marks a change from shallow-marine deposition (Tanque Cuatro Palmas) to sedimentation in continental environments. In this work we present the structural evidence that documents this first event of reactivation. As we showed earlier, in the Potrero Colorado area, normal faults observed at the proposed contact between the Colorado beds and the San Marcos Formation are characterized by larger vertical displacements than those observed in Upper Jurassic strata.

It appears that faulting in the Early Cretaceous exercised ample control in rejuvenation of the relief in the area. Faults documented here show that major north-northeast–directed extension occurred during that time. It is important to emphasize that there are no significant differences in the inclination and orientation of the Late Jurassic faults and those affecting the Neocomian sequence. As mentioned above, marine beds bearing Tithonian fossils represented by the Tanque Cuatro Palmas beds are covered by the San Marcos Formation, which was deposited in continental environments. Moreover, in this work we recognized that the oldest right lateral strike-slip faults affected only the Jurassic rocks, and not Neocomian strata where all of the faults are normal. Late Jurassic and Neocomian extension occurred in intermittent pulses that contributed to the development of the Sabinas basin. However, we should not underestimate right-lateral reactivation of the oldest normal faults present in the Colorado beds (Fig. 5A), because these could imply significant lateral displacements that cannot yet be estimated.

Regarding the event of the San Marcos fault reactivation in the Paleogene, it is clear that there was an inversion in the direction of the original down-dip component of movement. Here we document northwest-southeast to west-northwest–east-southeast inverse faults along the trace of the San Marcos fault, especially on the frontal zone of drape folds formed by inversion of the fault. The directions of maximum compression calculated using faults are very consistent along the trace of the San Marcos fault, varying from northeast at Sierra Mojada to north-northeast in the San Marcos Valley. In addition, the juxtaposition of older rocks toward the north-northeast via high-angle inverse faults in the valley (McKee et al., 1990) and at Potrero Colorado (Chávez-Cabello, 2005) strongly supports this hypothesis.

The nearly orthogonal relationship between fold axes of major folds in the southwestern part of the Sabinas basin, together with a change in strike from northwest-southeast to north-south for the San Marcos–Pinos range (Fig. 10), and the perpendicular directions of tectonic transport observed along the southwestern wall and the western walls of the ranges that form the San Marcos Valley all suggest that secondary faults, oblique to the trace of the San Marcos fault, were reactivated as well during the Laramide event. Basement faults that may have controlled those changes in the orientation of structures and tectonic transport may have been generated during the event of isostatic adjustment proposed for the Early Cretaceous by McKee et al. (1990), or even during the Late Jurassic. This is because there is a clear change in sedimentary facies among San Marcos Valley, Potrero Colorado, and Sierra Mojada for this time (McKee et al., 1990).

The inversion of secondary basement faults with respect to the San Marcos fault trace explains the orientations of fold axes nearly perpendicular to each other, although the direction of regional shortening did not change between what has been interpreted to be two phases of Laramide deformation in this area (Chávez-Cabello, 2005).

The reactivation of the San Marcos fault in the late Miocene–early Pliocene proposed by Aranda-Gómez et al. (2005a) in southeast Chihuahua, which accommodated minor left-lateral slip during a short period of northwest-southeast extension, manifests itself in a subtle way along the trace of the fault in Coahuila. During the course of this research we identified some fault planes within the zone of major vertical breccia at El Mimbre canyon; kinematic indicators show left-lateral slip. These faults did not, however, accommodate significant displacement. On the other hand, strike-slip faults observed in the late Eocene Las Tetillas pluton (García-Quintero, 2004), as well as reactivation of a basement fault near Nuevo Delicias, suggest that minor reactivations occurred south of the trace of the San Marcos fault.

Normal faulting recognized by Aranda-Gómez et al. (2005a) over the buried trace of the San Marcos fault in southeast Chihuahua does not manifest clearly along the trace of the fault in central Coahuila. We note, however, that ~15 km south of the trace of the fault there is a lineament of eruptive centers, which is roughly parallel to the fault. In one of the volcanic centers, magmatism and normal faulting coexisted (at Cerro La Víbora). This is similar to the activity described by Aranda-Gómez et al. (2003) for the Camargo volcanic field. Furthermore, there are small normal displacements in the Las Tetillas intrusion (Fig. 10) and in the basement fault identified near Nuevo Delicias. This suggests that post-Laramide tectonic and magmatic activity in central Coahuila appears to have migrated from the trace of the San Marcos fault toward the south, in contrast to what has been interpreted for Chihuahua. Another explanation for the relatively minor manifestations of Neogene normal faults in Coahuila, even though this area was under an extensional regime (Suter, 1991), is the presence of evaporites that could have inhibited or masked faulting. This assumption implies that the evaporitic horizons (Olvido Formation) absorbed a large proportion of the deformation, behaving as ductile layers.

There is no doubt that in the central part of Coahuila basement structures such as the San Marcos fault and the secondary faults related to it controlled the inclination and orientation of the structures generated since the Early Cretaceous. In the region quite a few fold axes and traces of faults have west-northwest to northwest orientations, similar to the general strike of the San Marcos fault and an unnamed basement fault south of Nuevo Delicias. This greatly complicates the study of reactivation, as at basement discontinuities younger tectonic events superimpose their characteristics on preexisting structures. As a result, the analysis of the tectonic evolution of the region requires detailed studies not only of the structures and crosscutting relationships, but also of sedimentation patterns, facies distribution, and spatial changes in the thickness of the different units exposed. On the other hand, the original geometry of the San Marcos fault and other basement features buried by hundreds of meters of sediment is difficult to discern with precision. For those studies based only on the analysis of features exposed at the surface, there will always be doubts as to whether the original geometry was preserved or gradually modified in later deformation events.

The data presented in this work on the reactivation of the San Marcos fault for the southern margin of the Sabinas basin are somewhat limited, and they do not allow choosing between the different hypotheses about the origin of the basin. Again, the data are solely for the southern margin. This information, however, underlines the relevance of faults with normal displacement and restricted to the Late Jurassic–Early Cretaceous—the timing for the formation of the basin.

It is evident that the Laramide orogeny is the event with better expression in the region's structures, because of the number of folds and faults with that origin in the localities we visited. There is also stratigraphic and sedimentological evidence of the early activity of the fault, preserved in clastic strata of the Upper Jurassic (Las Palomas, Sierra El Granizo, and Tanque Cuatro Palmas beds) and the lowermost Cretaceous (San Marcos Formation). Additional work is needed to better understand directions of shortening and extension that postdate the Laramide event. More detailed work will allow researchers to define the mechanisms involved and evaluate the importance of post-Laramide reactivations of the fault in central Coahuila, or else explain the reason for their nonexistence.

Finally, one important aspect that must be explored in greater detail is how Laramide folding evolved in the southern part of the basin. There is evidence suggesting that there were two phases of deformation. This is of pivotal importance from an economic point of view, because of the hydrocarbon, mineral, and deep-groundwater resources that can potentially be harvested and used to improve economic development in the region.

CONCLUSIONS

The San Marcos fault is a major basement structure, which is the best known in northeast México. This structure shows the most straightforward stratigraphic and structural evidence of intermittent activity from the Late Jurassic to the Pliocene. The structural and stratigraphic features recognized indicate that between the Late Jurassic and the Neocomian the fault accommodated primarily north-northeast–directed crustal extension, which we surmise contributed to the development of the Sabinas basin. Based on this information, we question earlier inferences of large strike-slip displacement (tens of km) on this fault.

On the other hand, reactivation of the San Marcos fault as a high-angle reverse fault in the Paleogene is responsible for the generation of drape folds in the area between San Marcos Valley and Sierra Mojada (Fig. 1). This reactivation appears to postdate an initial phase of deformation dominated by bedding parallel detachments with north-northeast tectonic transport in both the Coahuila Platform and the Sabinas basin. Further work is needed to demonstrate this more convincingly. The nearly perpendicular relationships between fold axes in the southwestern part of the Sabinas basin, the uplift of progressively older rocks toward the north-northeast within the San Marcos Valley, and the perpendicular directions of tectonic shortening determined in the San Marcos Valley area suggest that, in addition to the San Marcos fault, secondary basement faults must have been reactivated during the Paleogene Laramide orogeny.

Finally, the left-lateral late Miocene–early Pliocene event of reactivation along the hypothetical (buried) trace of the San Marcos fault, as well as a younger reactivation event as a normal fault evident in Chihuahua, have a very subtle expression in central Coahuila. Some deformation with this style is manifest in the Coahuila block, south of the San Marcos fault, and in other areas such as the Alamitos range and the northern segment of Sierra La Fragua.

Additional structural, geophysical, geochronologic, and paleomagnetic work is needed to understand vertical-axis rotations inferred from preliminary studies, and to determine whether these rotations occurred in the Jurassic or are the product of Paleogene Laramide deformation.

ACKNOWLEDGMENTS

This work is part of the Ph.D. dissertation of G. Chávez-Cabello, who thanks Universidad Autónoma de Nuevo León (PROMEP) for a grant for his graduate studies. Additional financial support to Chávez was provided by a PAEP grant from the Graduate Program of Earth Sciences at UNAM. Field and laboratory expenses for Chávez and I.R. Arvizu's theses were covered by CONACYT (grants 37429-T and 47071 to J. Aranda, R. Molina, and G. Chávez). We thank Samuel Eguiluz for a helpful review of an earlier version of the manuscript and are grateful to Ángel Nieto and Randall Marrett for their helpful comments, which improved the final version of the manuscript.

REFERENCES CITED

Alfonso, Z.J., 1978, Geología regional del sistema sedimentario Cupido: Boletín de la Asociación Mexicana de Geólogos Petroleros, v. 30, n. 1–2, p. 1–55.

Anderson, E.M., 1951, The Dynamics of Faulting, 2nd edition: Edinburgh, Oliver & Boyd, 206 p.

Anderson, T.H., and Schmidt, V.A., 1983, The evolution of Middle America and the Gulf of Mexico-Caribbean Sea region during Mesozoic time: Geological Society of America Bulletin, v. 94, p. 941–966, doi: 10.1130/0016-7606(1983)94<941:TEOMAA>2.0.CO;2.

Angelier, J., 1990, Inversion of field data in fault tectonics to obtain the regional stress, III: A new rapid direct inversion method by analytical means: Geophysical Journal International, v. 103, p. 363–376.

Aranda-Gómez, J.J., Luhr, J.F., Housh, T.B., Connor, C.B., Becker, T., and Henry, C.D., 2003, Synextensional Plio-Pleistocene eruptive activity in the Camargo volcanic field, Chihuahua, México: Geological Society of America Bulletin, v. 115, p. 298–313, doi: 10.1130/0016-7606(2003)115<0298:SPPEAI>2.0.CO;2.

Aranda-Gómez, J.J., Housh, T.B., Luhr, J.F., Henry, C.D., Becker, T., and Chávez-Cabello, G., 2005a, Reactivation of the San Marcos Fault during mid-to-late Tertiary extension, Chihuahua, Mexico, *in* Anderson, T.H., Nourse, J.A., McKee, J.W., and Steiner, M.B., eds., The Mojave-Sonora Megashear Hypothesis: Development, Assessment and Alternatives: Geological Society of America Special Paper 393, p. 509–521.

Aranda-Gómez, J.J., Luhr, J.F., Housh, T.B., Valdez-Moreno, G., and Chávez-Cabello, G., 2005b, El vulcanismo de intraplaca del Cenozoico tardío en el centro y norte de México: una revisión: Boletín conmemorativo del Centenario de la Sociedad Geológica Mexicana: Tomo, v. LVII, no. 3, p. 187–225.

Arvizu-Gutiérrez, I.R., 2003, Estudio paleomagnético de los granitoides permotriásicos y de las Capas Las Palomas (Jurásico Tardío), Valle San Marcos, Coahuila, México [B.S. thesis]: Facultad de Ciencias de la Tierra, Universidad Autónoma de Nuevo León, Linares, Nuevo León, 116 p.

Buchanan, P.G., and McClay, K.R., 1991, Sandbox experiments of inverted listric and planar faults systems, *in* Cobbold, P.R., ed., Experimental and Numerical Modelling of Continental Deformation: Tectonophysics, v. 188, p. 97–115.

Charleston, S., 1973, Stratigraphy, tectonics and hydrocarbon potential of the lower Cretaceous, Coahuila series, Coahuila, México [Ph.D. thesis]: Ann Arbor, University of Michigan, 268 p.

Charleston, S., 1981, A summary of the structural geology and tectonics of the State of Coahuila, Mexico, *in* Schmidt, C.I., and Katz, S.B., eds., Lower Cretaceous Stratigraphy and Structure, Northern Mexico: West Texas Geological Society Field Trip Guidebook, Publication 81–74, p. 28–36.

Chávez-Cabello, G., 2005, Deformación y Magmatismo Cenozoicos en el Sur de la Cuenca de Sabinas, Coahuila, México [Ph.D. thesis]: Centro de Geociencias, Universidad Nacional Autónoma de México, Juriquilla, Querétaro, 226 p.

Eguiluz de Antuñano, S., 1984, Tectónica cenozoica del norte de México: Boletín de la Asociación Mexicana de Geólogos Petroleros, v. 34, p. 41–62.

Eguiluz de Antuñano, S., 2001, Geologic evolution and gas resources of the Sabinas basin in northeastern México, *in* Bartolini, C., Buffler, R.T., and Cantú-Chapa, A., eds., The Western Gulf of México Basin: Tectonics, Sedimentary Basins, and Petroleum Systems: American Association of Petroleum Geologists Memoir 75, p. 241–270.

García-Quintero, J.J., 2004, Cartografía y análisis estructural del intrusivo Las Tetillas, Coahuila, México [B.S. thesis]: Facultad de Ciencias de la Tierra, Universidad Autónoma de Nuevo León, Linares, Nuevo León, 101 p.

Goldhammer, R.K., 1999, Mesozoic sequence stratigraphy and paleogeographic evolution of northeast of Mexico, *in* Bartolini, C., Wilson, J.L., and Lawton, T.F., eds., Mesozoic Sedimentary and Tectonic History of North-Central Mexico: Geological Society of America Special Paper 340, p. 1–58.

González-Naranjo, G.A., 2006, Análisis estructural y estudio paleomagnético en el área Potrero El Colorado, porción occidental de la sierra La Fragua, Coahuila, México [M.S. thesis]: Facultad de Ciencias de la Tierra, Universidad Autónoma de Nuevo León, Linares, Nuevo León, 221 p.

Harding, T.P., Vierbuchen, R.C., and Christie-Blick, N., 1985, Structural styles, plate tectonic settings, and hydrocarbon traps of divergent (transtensional) wrench faults, *in* Biddle, K., and Christie-Blick, N., eds., Strike-Slip Deformation, Basin Formation, and Sedimentation: Society of Economic Paleontologists and Mineralogists Special Publication 31, p. 51–75.

Iriondo, A., 2001, Proterozoic basements and their Laramide juxtaposition in NW Sonora, Mexico [Ph.D. thesis]: University of Colorado, Boulder, 222 p.

Jones, N.W., Dula, F., Long, L.E., and McKee, J.W., 1982, An exposure of a fundamental fault in Permian basement granitoids, Valle San Marcos, Coahuila, México: Geological Society of America Abstracts with Programs, v. 14, no. 7, p. 523–524.

Longoria, J.F., 1984, Stratigraphic studies in the Jurassic of northeastern Mexico: Evidence of the origin of the Sabinas basin, *in* Ventres, W.P.S., Bebout, D.G., Perkins, B.F., and Moore, C.H., eds., The Jurassic of the Gulf Rim: Austin Society of Economic Paleontologists and Mineralogists Foundation, Gulf Coast Section, Proceedings of the Third Annual Research Conference, p. 171–193.

McKee, J.W., Jones, N.W., and Long, L.E., 1984, History of recurrent activity along a major fault in northeastern Mexico: Geology, v. 12, p. 103–107, doi: 10.1130/0091-7613(1984)12<103:HORAAA>2.0.CO;2.

McKee, J.W., Jones, N.W., and Anderson, T.H., 1988, Las Delicias basin: A record of Late Paleozoic arc volcanism in northeastern Mexico: Geology, v. 16, p. 37–40, doi: 10.1130/0091-7613(1988)016<0037:LDBARO> 2.3.CO;2.

McKee, J.W., Jones, N.W., and Long, L.E., 1990, Stratigraphy and provenance of strata along the San Marcos fault, central Coahuila, Mexico: Geological Society of America Bulletin, v. 102, p. 593–614, doi: 10.1130/0016-7606(1990)102<0593:SAPOSA>2.3.CO;2.

McKee, J.W., Jones, N.W., and Anderson, T.H., 1999, Late Paleozoic and early Mesozoic history of the Las Delicias terrane, Coahuila, México, *in* Bartolini, C., Wilson, J.L., and Lawton, T.F., eds., Mesozoic Sedimentary and Tectonic History of North-Central México: Geological Society of America Special Paper 340, p. 161–189.

Molina-Garza, R.S., 2005, Paleomagnetism and geochronology of the Late Triassic Acatita intrusives, Coahuila, Mexico: Geofísica Internacional, v. 44, p. 197–210.

Molina-Garza, R.S., and Geissman, J.W., 1996, Timing of deformation and accretion of the Antimonio terrane, Sonora, from paleomagnetic data: Geology, v. 24, p. 1131–1134, doi: 10.1130/0091-7613(1996)024<1131: TODAAO>2.3.CO;2.

Molina-Garza, R.S., and Geissman, J.W., 1999, Paleomagnetic data from the Caborca terrane, Mexico: Implications for Cordillera tectonics and the Mojave-Sonora megashear hypothesis: Tectonics, v. 18, p. 293–325, doi: 10.1029/1998TC900030.

Molina-Garza, R.S., Arvizu-Gutiérrez, I.R., and Chávez-Cabello, G., 2003, Paleomagnetismo de la Formación Palomas (Jurásico) y granitoides permo-triásicos, sur de Coahuila: Implicaciones Tectónicas: GEOS, v. 23, no. 2, p. 112.

Nieto-Samaniego, A.F., and Alaniz-Álvarez, S.A., 1997, Origin and tectonic interpretation of multiple fault patterns: Tectonophysics, v. 270, p. 197–206.

Padilla y Sánchez, R.J., 1982, Geologic evolution of the Sierra Madre Oriental between Linares, Concepción del Oro, Saltillo and Monterrey, México [Ph.D. thesis]: Austin, University of Texas at Austin, 217 p.

Padilla y Sánchez, R.J., 1986, Post Paleozoic tectonics of northeast México and its role in the evolution of the Gulf of México: Geofísica Internacional, v. 25, p. 157–206.

Salvador, A., 1991, Origin and development of the Gulf of Mexico basin, *in* Salvador, A., ed., The Gulf of Mexico Basin: Boulder, Colorado, Geological Society of America, Geology of North America, v. J, p. 389–444.

Santamaría, O.D., Ortuño, A.F., Adatte, T., Ortíz, U.A., Riba, R.A., and Franco, N.S., 1991, Evolución geodinámica de la Cuenca de Sabinas y sus implicaciones petroleras, Estado de Coahuila: Instituto Mexicano del Petróleo, unpublished internal report.

Suter, M., 1991, State of stress and active deformation in México and western Central America, *in* Slemmons, D.B., Engdahl, E.R., Zoback, M.D., and Blackwell, D.D., eds., Neotectonics of North America, Decade Map I: Boulder, Colorado, Geological Society of America, p. 401–421.

Valdez-Moreno, G., 2001, Geoquímica y petrología de las rocas ígneas de los Campos Volcánicos de Las Esperanzas y Ocampo, Coahuila, México [M.S. thesis]: Instituto de Geología, Universidad Nacional Autónoma de México, México D.F., 128 p.

Manuscript Accepted by the Society 29 August 2006

Printed in the USA

The Geological Society of America
Special Paper 422
2007

Right-lateral active faulting between southern Baja California and the Pacific plate: The Tosco-Abreojos fault

François Michaud*
Géosciences Azur, La Darse, BP48, 06235 Villefranche/Mer, France

Thierry Calmus
Instituto de Geología, Universidad Nacional Autónoma de México (UNAM), 83000, Hermosillo, México

Jean-Yves Royer
Centre National de la Recherche Scientifique (CNRS) Domaines Océaniques,
Institut Universitaire Européen de la Mer (IUEM), F-29280, Plouzané, France

Marc Sosson
Géosciences Azur, La Darse, BP48, 06235 Villefranche/Mer, France

Bill Bandy
Carlos Mortera-Gutiérrez
Instituto de Geofísica, Universidad Nacional Autónoma de México (UNAM), México D.F., 04510, México

Jérôme Dyment
Institut Physique du Globe de Paris, 4 place Jussieu, 75252, Paris, France

Florence Bigot-Cormier
Anne Chabert
Jacques Bourgois
Géosciences Azur, La Darse, BP48, 06235 Villefranche/Mer, France

ABSTRACT

At 12.5 Ma, after subduction below the North American plate stops, right-lateral transform motion occurs along the margin between the Pacific and North American plates. The Tosco-Abreojos fault zone, located along the western margin of southern Baja California, has been interpreted as the main transform boundary between both plates until early Pliocene, when the plate boundary was transferred to the Gulf of California, leading to the capture of Baja California Peninsula by the Pacific plate. However, the morphology and the seismic activity of the Tosco-Abreojos fault zone suggest this right-lateral strike-slip motion is still active. The Tosco-Abreojos fault

*micho@geoazur.obs-vlfr.fr

Michaud, F., Calmus, T., Royer, J.-Y., Sosson, M., Bandy, B., Mortera-Gutiérrez, C., Dyment, J., Bigot-Cormier, F., Chabert, A., and Bourgois, J., 2007, Right-lateral active faulting between southern Baja California and the Pacific plate: The Tosco-Abreojos fault, *in* Alaniz-Álvarez, S.A., and Nieto-Samaniego, Á.F., eds., Geology of México: Celebrating the Centenary of the Geological Society of México: Geological Society of America Special Paper 422, p. 287–300, doi: 10.1130/2007.2422(09). For permission to copy, contact editing@geosociety.org.

zone is characterized by bathymetric scarps and asymmetric basins filled by recent sediments which are deformed. These observations are compatible with the hypothesis that the motion of the Pacific plate with respect to the North American plate is partitioned, as indicated by kinematic data (GPS versus global models) between the still active Tosco-Abreojos fault zone and the Gulf of California where most of the motion is accommodated. The Baja California Peninsula can thus be considered as an independent block limited to the west by the Tosco-Abreojos and San Benito fault zones and to the east by the Gulf of California transform boundary.

Keywords: Tosco-Abreojos fault, seafloor morphology, plate boundary, Baja California, Pacific plate.

INTRODUCTION

The geological evolution of northwestern México is related to interactions between the North American plate and the adjacent oceanic plates (Fig. 1). Presently the Gulf of California, which includes a complex transform fault system and incipient spreading centers (Aragon-Arreola et al., 2005), links the East Pacific rise to the San Andreas fault. Global kinematic models predict Pacific–North American relative motion ranging from 56 mm/yr (Minster and Jordan, 1978) to 50 mm/yr (NUVEL 1; DeMets et al., 1990) to 48.8 ± 1.8 mm/yr (DeMets, 1995). In the northern part of the Gulf, the correlation between upper Miocene volcanic rocks from both sides of the Gulf (29°N lat) led Oskin and Stock (2003) to suggest a 276 ± 13 km displacement during the last 6 m.y., which corresponds to a velocity of 46.0 ± 0.2 mm/yr. In the southern part of the Gulf, magnetic anomalies indicate plate motion velocity in the range of 58 and 48 mm/yr during the past 4 m.y. (Humphreys and Weldon, 1991; Ness et al., 1991; Lonsdale, 1995; DeMets and Traylen, 2000). This points out a discrepancy between the Pacific–North American plate instantaneous motion predicted by global kinematic models and the plate velocity based on geological evidence from the Gulf, suggesting that part of the plate motion takes place outside of the Gulf of California.

Moreover, Pacific–North American plate motion has been steady since 3.6 Ma, while southern Gulf of California divergence of the Baja California from North America has been significantly slower than the Pacific–North American motion (DeMets, 1995). This implies that Baja California may have moved relative to the Pacific plate up until Present (DeMets and Dixon, 1999; Dixon et al., 2000). These sets of data can be reconciled if we assume that the Baja California block is moving relative the Pacific plate, i.e., that Baja California has not been rigidly coupled to the Pacific plate since 3.6 Ma.

Along the western Baja California margin, subduction is assumed to stop after 12 Ma when a transform boundary initiated between the Pacific and North American plates. This plate boundary, known as the Tosco-Abreojos fault zone between 23°N and 28°N lat, is parallel to the Pacific margin of Baja California and is believed to have been active until 5–3.6 Ma when the transform motion migrated to the Gulf of California (Spencer and Normark, 1979, 1989; Stock and Lee, 1994; Stock and Hodges, 1989; Lonsdale, 1989). Consequently since 5–3.6 Ma Baja California is assumed to have transferred to the Pacific plate (Fig. 1).

Faults along the western margin of Baja California presumably accommodated some or all of this motion (Fletcher and Munguía, 2000). Among them, the Tosco-Abreojos fault zone would have been active between 12 Ma and 5 Ma (Spencer and Normark, 1989). However, coarse bathymetric data and single-channel seismic reflection profiles show that the fault zone controls angular bathymetric scarps and in some places cuts younger seafloor strata (Spencer and Normark, 1979; Normark et al., 1987; Dauphin and Ness, 1991; Ness et al., 1991). Moreover some seismicity occurs on the continental borderland (Fletcher and Munguía, 2000).

During the FAMEX cruise (R/V *l'Atalante*, 2002) new swath bathymetric data and seismic profiles were collected across the Tosco-Abreojos fault zone. The seismic profiles were shot using 300 inch^3 GI guns tuned in harmonic mode and a streamer with six 50-m-long channels. The 10 s pop rate at a speed of 10 knots allowed a 3-fold coverage with a record length of 4 s. These new data suggest that modern Pacific–North American plate motion is partitioned between the Gulf of California and active faults along the Pacific margin of Baja California.

SWATH BATHYMETRY

Swath bathymetric data between 27° and 24.5°N lat (Fig. 2A and 2B) show three 35–40°W–trending elongated basins located along the middle slope (Michaud et al., 2004). These basins define the Tosco-Abreojos trough (Spencer and Normark, 1979). The southern and central basins are 45 km long and reach their maximum depth at 1700 m; they are asymmetric. Westward these basins are bounded by NW40°- to NW35°-trending, fresh and steep scarps. On the eastern flank of the central basin, linear canyons are abruptly cut, suggesting that the scarp corresponds to an active fault. The northern basin (15 km wide, 70 km long) reaches its maximum depth at 2600 m (Fig. 2B and 2C). It exhibits a strong asymmetry with a linear 750-m-high steep scarp to the west and a less steep 1500-m-high scarp to the east. This eastern scarp exhibits a seaward arcuate topographic expression. The western scarp, which is less important than the eastern one, is

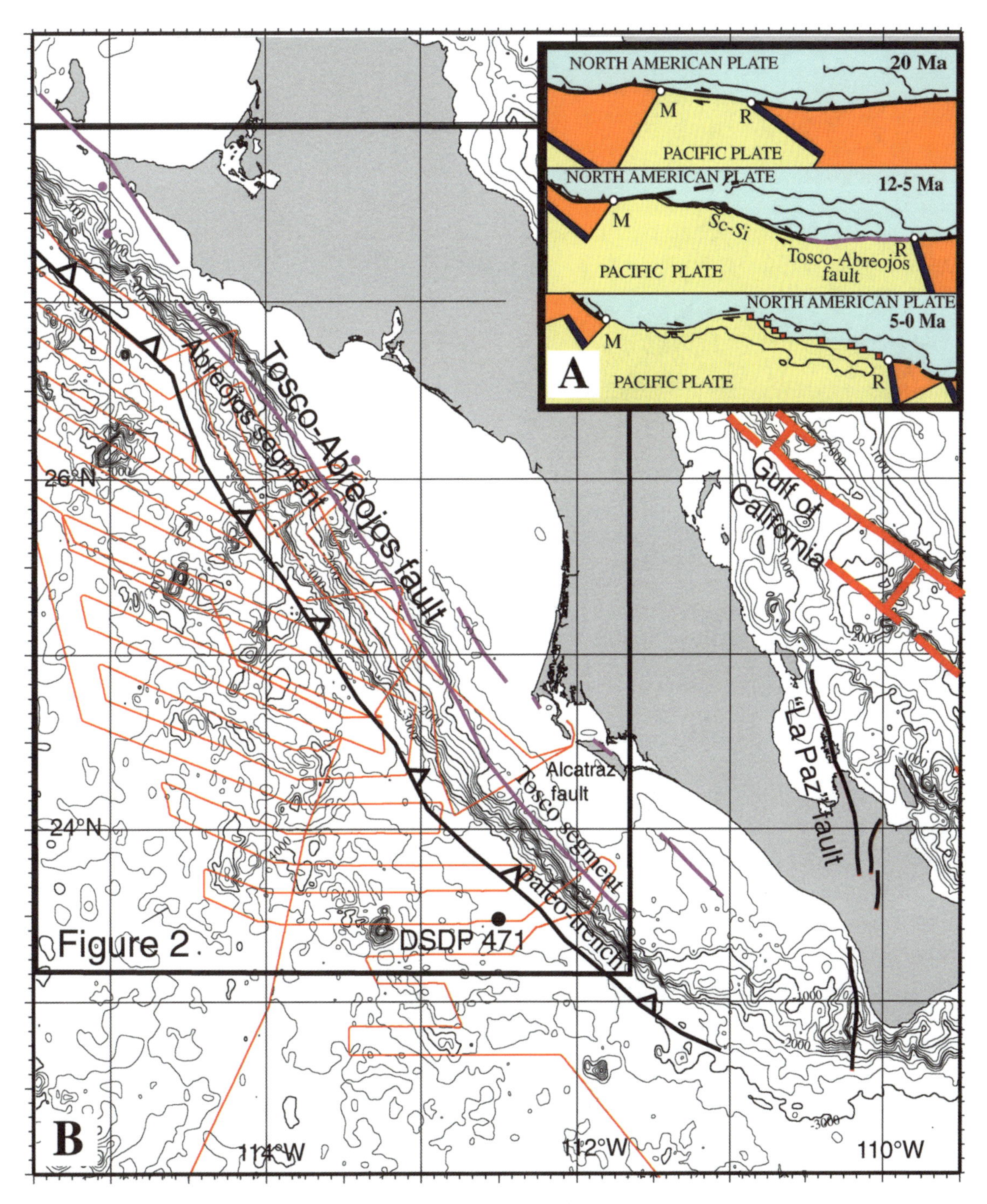

Figure 1. (A) Schematic evolution of the Pacific–North American plate margin during the last 20 Ma (after Dickinson and Snyder, 1979). R—Rivera triple junction; M—Mendocino triple junction; Sc-Si—San Clemente–San Isidro fault. (B) Bathymetric map of the Baja California margin and part of the Gulf of California (predicted bathymetry from Smith and Sandwell, 1997). Along the Baja California margin, the Tosco-Abreojos fault system (TAFS) as main active faults of the Magdalena seismic zone (MSZ) (Dixon et al., 2000) and of the Baja California tip are drawn. Light purple lines represent active faults along the Southern Baja California margin. Thin red lines: FAMEX cruise. Location of the Tosco-Abreojos fault system from Spencer and Normark (1989); location of "La Paz" fault from Fletcher and Munguía (2000). Black dots represent earthquakes (magnitude > 4) (from USGS, 1973–2005). Thick red lines denote principal faults related to the Gulf of California opening.

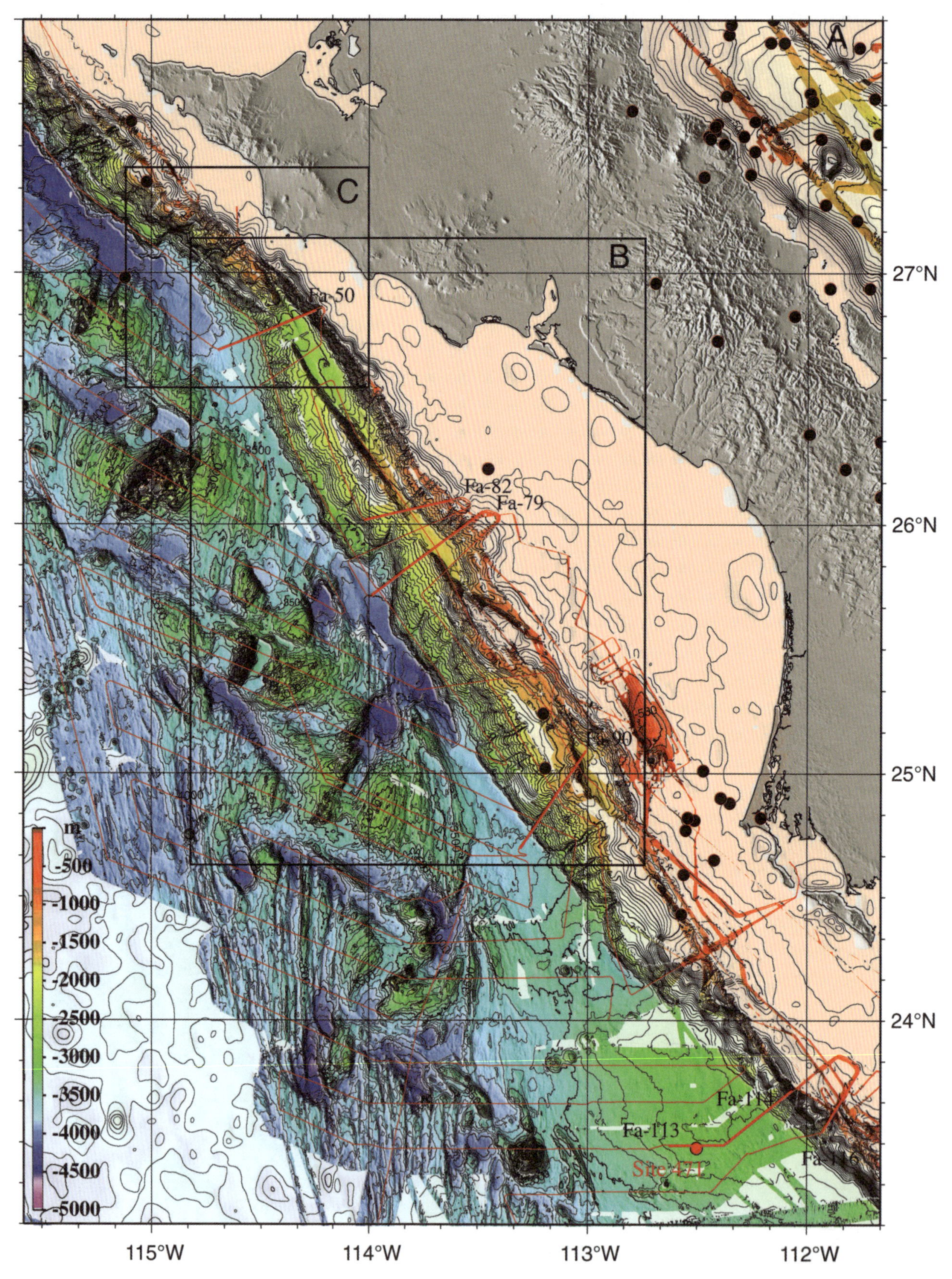

Figure 2 (*on this and following two pages*). (A) Swath bathymetry of the Baja California margin between 23° and 28°N (contour interval = 100 m). The data are from the FAMEX cruise with data principally from R/V *Marion Dufresne* and R/V *Melville*. This map shows three basins along the Tosco Abreojos trough and the associated scarps related to the Tosco-Abreojos fault system. Soft red lines represent FAMEX profiles; thick red lines underline the seismic profiles across the margin shown in this paper.

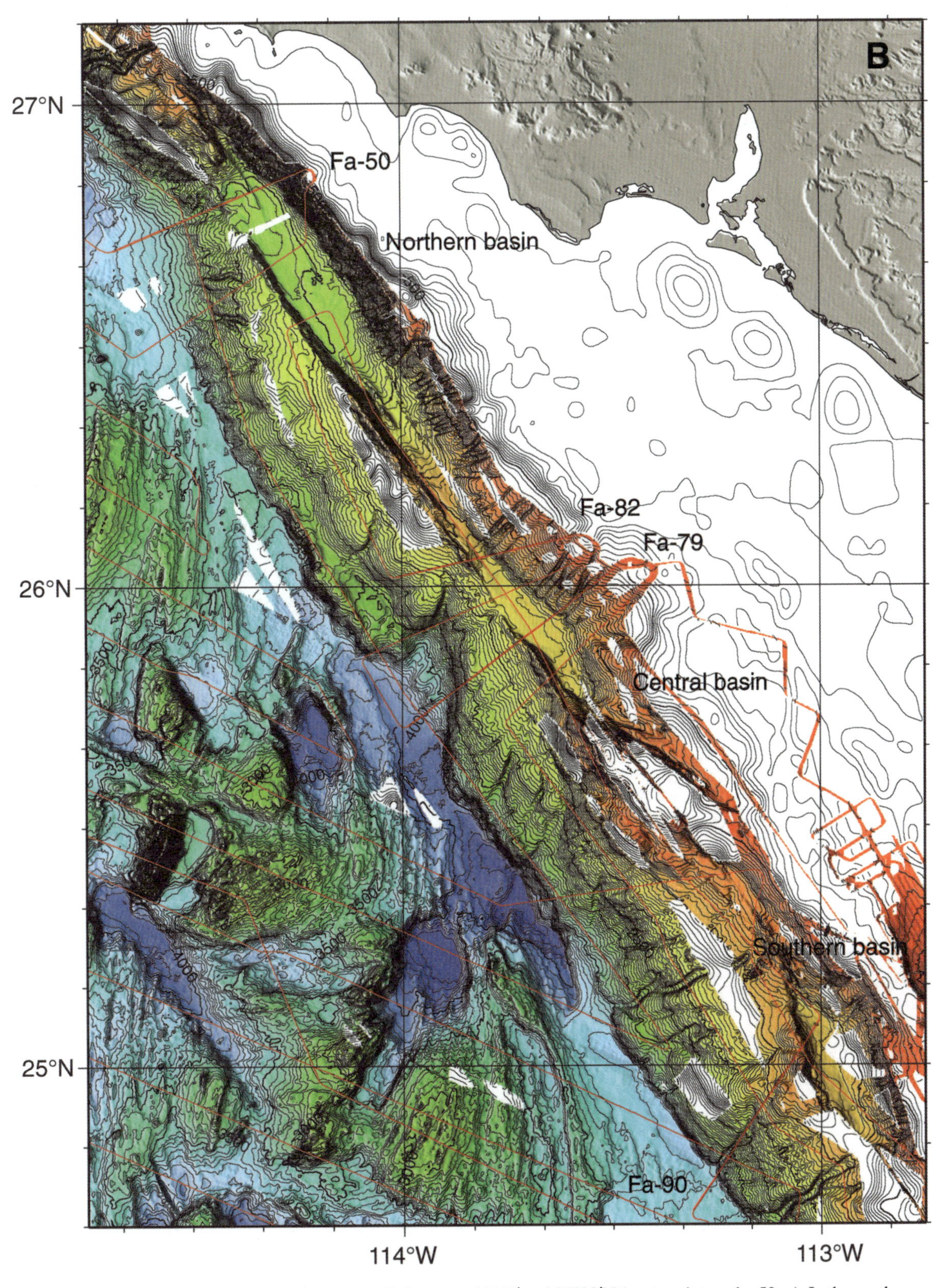

Figure 2 (*continued*). (B) Bathymetric map details between 24°40′ and 27°20′N (contour interval = 50 m). In the southern part of the figure, the northern basin exhibits a strong morphological asymmetry with a linear 750-m-high steep scarp to the west and a less steep 1500-m-high scarp to the east. The seafloor bottom is affected by a small scarp trending NW50°.

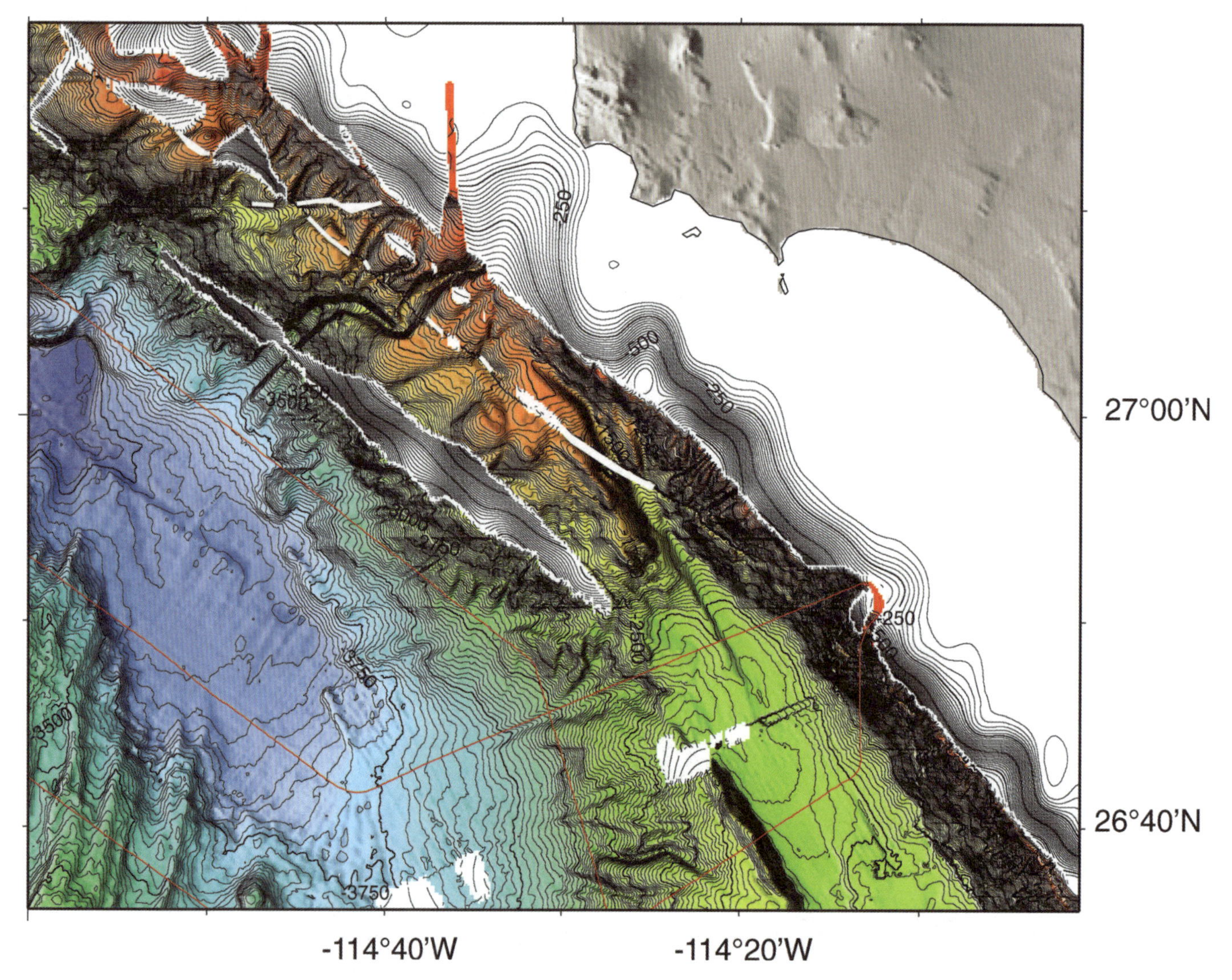

Figure 2 (*continued*). (C) North of 27°N, note the abrupt change in the canyon trend which suggests right lateral transform motion along possible recent active faults (contour interval = 25 m).

a high-angle fault dipping toward the continent. These tectonic features are evidence for active faulting. In some places, the flat floor of the basin shows NW30°-trending, 50-m-high rectilinear scarps which correspond to a Riedel-type fault in a transcurrent fault system (Fig. 2B and 2C).

SEISMIC DATA

The seismic profiles between 23° and 27°N across the Tosco Abreojos fault zone confirm the observations made from the swath bathymetry data. The seismic profiles (Figs. 3–7) show that the northern part of the Tosco Abreojos fault system is active (Figs. 4–7) as is the southern part (Fig. 3). The scarps observed from the bathymetric data correspond to faults at depth which deform the most recent sediments (Fig. 7). The eastern and western faults which limit the northern basin are dipping 30–35° landward and 40° seaward, respectively. This suggests that these faults intersect at depth (Fig. 5). Moreover, based on our bathymetric data set, these faults connect at the southward and northward limits of the basin and consequently define a flower structure. Such flower structures are also present along the central and southern basins (Fig. 6). Seismic profile Fa-90 (Fig. 4) which crosses the Tosco Abreojos fault zone in the southern basin (at 25°N) exhibits a sedimentary infill deformed by flower structures characterized by the coexistence of normal and thrust faults, which define a synclinal structure. We thus find evidence that the Tosco Abreojos fault is active along its entire length, and not only along its southern part, as postulated by Spencer and Normark (1989). Seismic profiles across the margin toe, such as profiles Fa-113 and Fa-114 (Fig. 3) exhibit several acoustic units (labeled A, B, C, and D). To the toe, the two oldest acoustic units correspond to the sediments of the oceanic plate (C) and of the paleo-trench (B). We tenta-

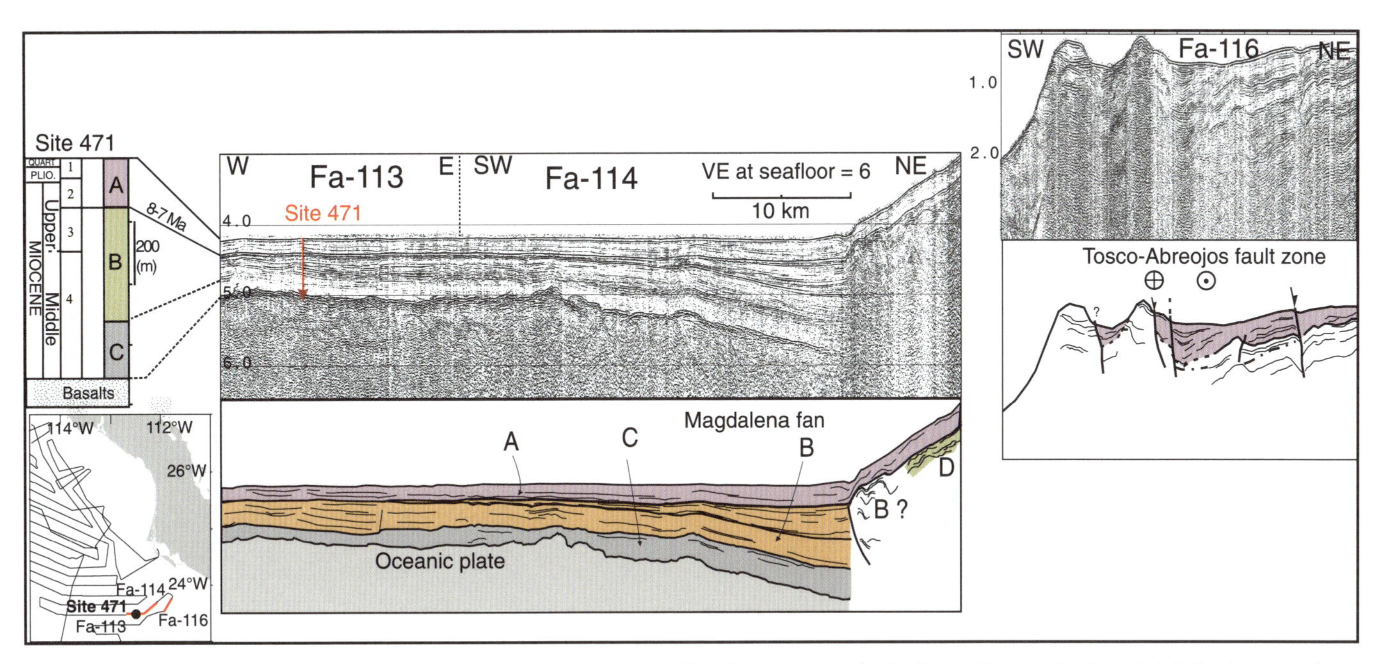

Figure 3. Seismic profiles Fa-113, Fa-114, and Fa-116 and corresponding line drawing; A, B, C, and D are seismic unit labels. Ages are from correlation with Site 471 DSDP 63 (location on Figs. 1 and 2) crossing by seismic line Fa-113. The DSDP Site 471 stratigraphic column is based on information in Yeats and Haq (1981): numbers 1–4 are the lithologic units from Marsaglia (2004). Seismic unit C, which corresponds to the oceanic sediment of the oceanic plate, could be followed beneath the upper plate. At the foot of the margin, seismic unit B shows a compressional deformation sealed by 8–7 Ma seismic unit A. Seismic unit A is intensely faulted by the Tosco-Abreojos system. VE—vertical exaggeration.

tively correlate unit B temporally with seismic unit on the margin (Figs. 5 and 6). Unit B shows compressional structures mainly located close to the margin toe (Fig. 3) and it is overthrusted by unit D (Figs. 3 and 7), corresponding to the frontal part of the upper plate. Unit A unconformably overlies these compressional structures and extends from the Pacific oceanic plate to the lower slope of the continental margin. Northward, similar compressional features are observed (Figs. 4–7; profiles Fa-90, Fa-79, and Fa-82). This observation suggests that the compressional features related to the thrusting deformation developed during the deposit of unit B. Profiles Fa-113 and Fa-114 cross Deep Sea Drilling Project (DSDP) Site 471 (Fig. 3) and can be used to project the DSDP results (Yeats and Haq, 1981; Yeats et al., 1981) to the margin toe. According to the microfossil assemblage at DSDP Site 471, the base of unit A is ~8–7 m.y. It unconformably overlies the compressional front and the seaward prolongation of the trench fill sediment (units B and C).

TECTONIC INTERPRETATION

The correlation between acoustic units observed along the margin with dated sediments sampled at DSDP Site 471 enables us to establish a chronology of the tectonic evolution of the western margin of Baja California Sur. The presence of acoustic unit C within the upper part of the slab indicates that the subduction was still active during middle Miocene, probably until 12.5 Ma, when a major kinematic reorganization occurred with the end of the subduction (Klitgord and Mammerickx, 1982; Mammerickx and Klitgord, 1982; Lonsdale, 1991). Unit B, whose age extends from 13.4 to 8–7 Ma, overlies unit C within the oceanic domain and unit D (the lateral equivalent to unit C) within the marginal domain. Nevertheless, seismic profiles Fa-50 and Fa-79 (Michaud et al., 2004) show that unit B is affected by reverse faulting in the subduction zone, suggesting a compressional component of the deformation until 8–7 Ma at the base of the margin. Unit A corresponds to sediments younger than 8–7 Ma which were deposited on the oceanic plate, the trench and the continental margin without apparent discontinuity. Unit A unconformably lies on unit B because it does not record the compressional deformation observed within unit B along the trench. Unit A is characterized by variations in thickness with an increase within basins. The deformation of unit A is exclusively located along the Tosco-Abreojos fault zone as shown on seismic profiles Fa-50, Fa-9, Fa-82, and Fa-90 located respectively across the northern, central, and southern basins. Along these profiles unit A is faulted and folded close to the normal faults fringing basins and within the basins themselves. East of the Tosco-Abreojos fault zone unit B is faulted. Seismic profile Fa-90 indicates that the southern basin is limited to the east by a low-angle normal fault, and to the west by a normal fault reactivated in the thrust fault. Folded sediments outline a general synclinal structure with vertical and reverse faults in the central part of the structure.

Seismic reflection and bathymetry data suggest that between 12.5 and 8–7 Ma the deformation is located at the foot of the

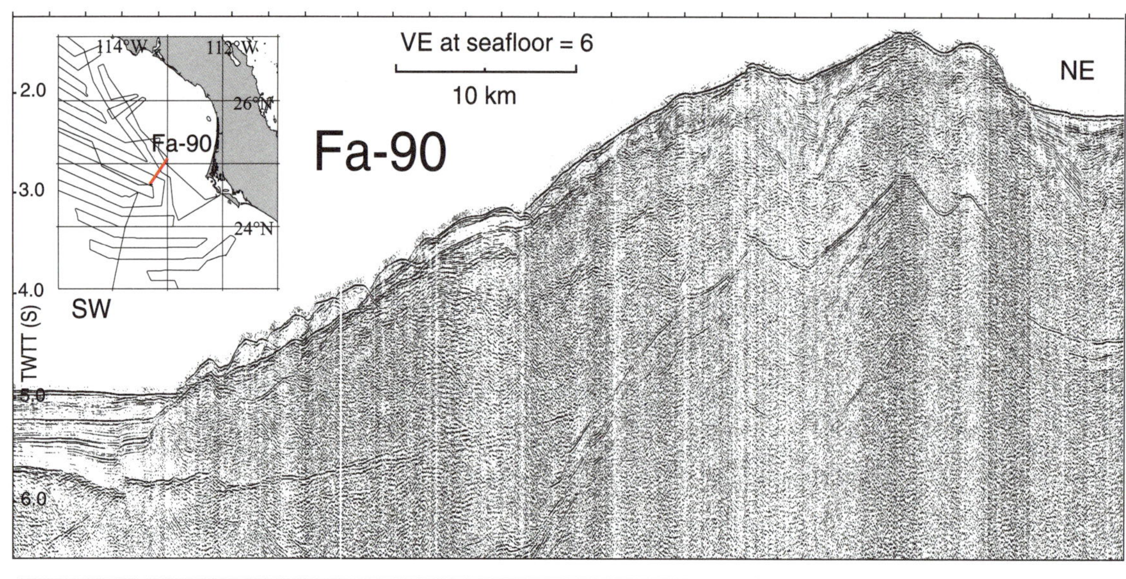

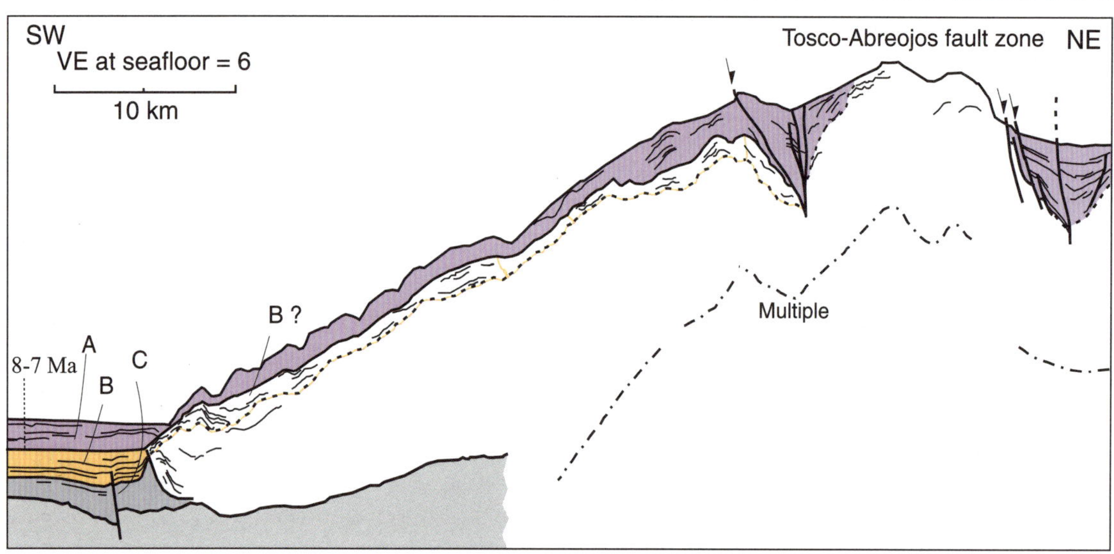

Figure 4. Seismic line Fa-90 (location on Fig. 2) and corresponding line drawing. As along the Fa-79 (Fig. 5) and Fa-50 (Fig. 7) profiles, this profile shows compressional deformation at the foot of the margin sealed by 8–7 Ma seismic unit A. The Tosco-Abreojos system is composed of two fault zones separated by a bathymetric high. VE—vertical exaggeration; TWTT—two-way traveltime.

margin. A compressive component probably associated with the beginning of shear motion along the margin originated reverse faulting up to the paleo-trench, as suggested by some seismic profiles showing thrusting of unit D onto the lowest part of unit B. After 8–7 Ma, deformation of the margin underwent a change from transpressive to transtensive regime which created some pull-apart basins along the Tosco-Abreojos fault zone. Nevertheless the data cannot tell exactly when the formation of the basins began along the Tosco-Abreojos fault zone. It is possible that this tectonic change along the western margin of Baja California Sur is related to the transfer of the main boundary between the Pacific and North American plates to the early Gulf of California during late Miocene and Pliocene. The deformation related to the Tosco-Abreojos fault zone was still active during the Pliocene, suggesting a progressive change of plate boundary.

The seismic activity recorded within the Magdalena seismic zone (Dixon et al., 2000) corroborates the theory that the Tosco-Abreojos fault zone is presently active. Two earthquakes of magnitude greater than 5 on the trace of the Tosco fault at the latitude of Vizcaino peninsula, and another of magnitude 4.4 located a few kilometers east of the Abreojos fault, indicate a right-lateral handed focal mechanism with a normal component along the N150 ± 5° direction consistent with the mean direction of the Tosco-Abreojos fault zone (Fig. 8). Thus, the Tosco-Abreojos fault zone is the best candidate to link the San Clemente–San Isidro–San Benito fault system (Krause, 1965), parallel to

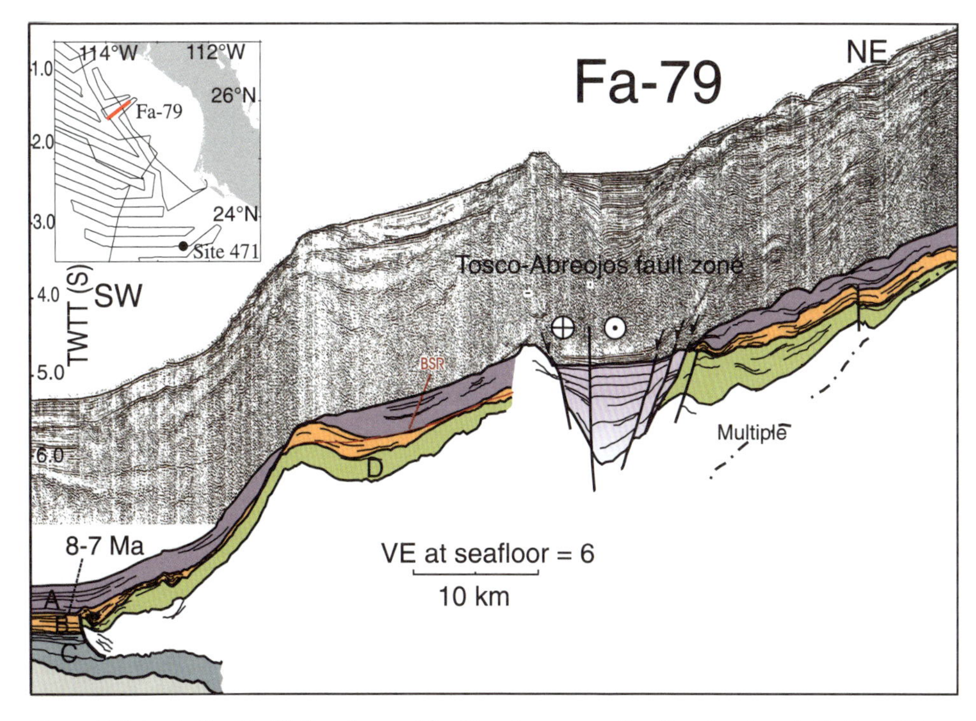

Figure 5. Seismic line Fa-79 (location on Fig. 2) and corresponding line drawing; A, B, C, and D are seismic unit labels and are the same as in Figure 3. Seismic unit C, which corresponds to the oceanic sediment of the oceanic plate, could be followed beneath the upper plate. VE—vertical exaggeration; TWTT—two-way traveltime.

the northern Baja California margin, with the Magdalena seismic zone to the south, where some active faults are described as Alcatraz fault located north of Santa Margarita Island (Yeats and Haq, 1981), Santa Margarita fault (Normark et al., 1987), Todos Santos and Carrizal faults (Ramos-Velázquez, 1998), located east of the La Paz-Los Cabos area—the fault in the Tinajas canyon, and, finally, a fault parallel to the eastern coast characterized by a 5.3 M earthquake in 1969 (Fletcher and Munguía, 2000).

The occurrence of the Tosco-Abreojos fault zone very close to the trench suggests that when the subduction ceased, the lowest part of the inner wall corresponded to a weak zone possibly because of (1) the right-lateral reactivation of a previous left-lateral fault related to the oblique subduction of the Magdalena plate, and (2) a thermal anomaly associated with the slab break below the thinned margin of the upper plate (Michaud et al., 2006).

DISCUSSION

Most of the relative motion between the Pacific and North American plates occurs in the Gulf of California. Magnetic anomalies created at the spreading centers in the southern part of the Gulf record a 10–15% increase in the spreading rates between Baja California and the North American plate (DeMets, 1995) from 44.8 mm/yr at 3.58 Ma (anomaly 2An.3) to 49.8 mm/yr at 0.78 Ma (anomaly 1n). Nevertheless the rate of motion of the Pacific plate with respect to the North American plate seems constant since 3.16 Ma, suggesting that part of the motion is accommodated along structures west of the Gulf of California, as the Tosco-Abreojos fault zone, or faults of the Magdalena seismic zone (Dixon et al., 2000; Fletcher and Munguía, 2000). The fact that the motion has been the same since 3.16 Ma and that the motion rate of Baja California with respect to the North American plate increased until 0.78 Ma also suggests that the motion rate along the Tosco-Abreojos fault zone decreased during the same period and that Baja California moved with respect to the Pacific plate before 0.78 Ma.

GPS motion vectors for some localities of Baja California peninsula with respect to North America (Fig. 9) are smaller than the motion of the Pacific plate with respect to the North American plate (DeMets and Dixon, 1999; Dixon et al., 2000). For example, the CONC locality (26.62°N, 111.81°W; Fig. 9) off the Bahía Concepción on the eastern coast of Baja California Sur moves at a rate of 43.6 ± 2.2 mm/yr with respect to North America, slower than the predicted velocity of 50.7 mm/yr of the Pacific with respect to North America (Dixon et al., 2000; Argus and Gordon, 2001). Dixon et al. (2000) interpret this difference as a consequence of elastic deformation accumulation along probable active faults near the CONC locality or as lateral motions along submarine faults located along the western margin of Baja California. The motion vector of the CONC locality has a direction ranging from 308.4° ± 2.2 (Dixon et al., 2000) to 303.9° ± 3.2 (Antonelis et al., 1999). Both results are slightly oblique with respect to the Tosco-Abreojos fault zone direction (15° in the first case). If we consider the hypothesis that there is no active

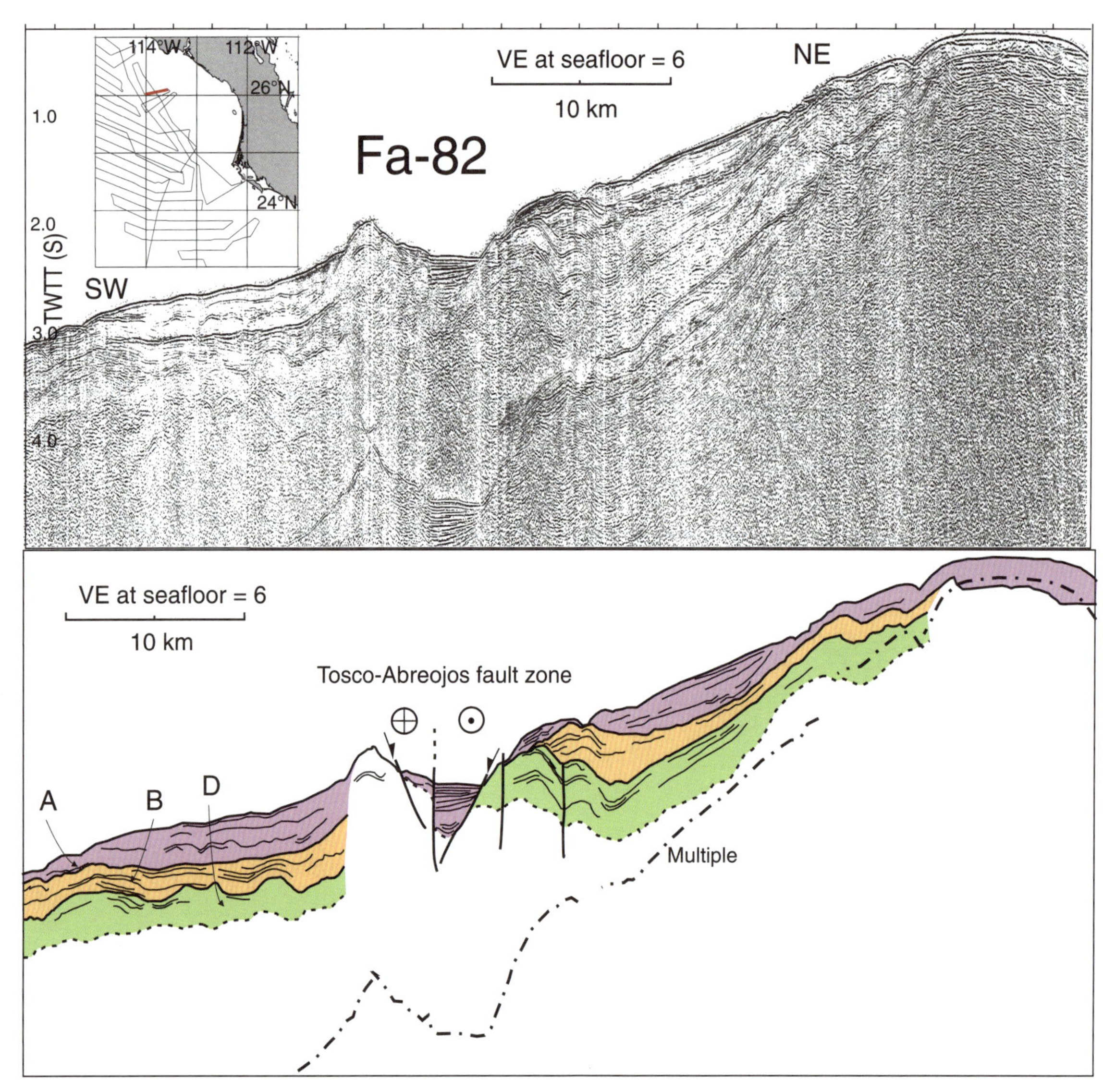

Figure 6. Seismic line Fa-82 (location on Fig. 2) and corresponding line drawing (see text for interpretation); A, B, and D are seismic unit labels and are the same as in Figure 3. VE—vertical exaggeration; TWTT—two-way traveltime.

fault between the CONC locality and the Tosco-Abreojos fault zone, the present motion along the fault corresponds to the difference between predicted and observed motion at the CONC locality, i.e., 7.1 mm/yr (Fig. 9). Due to the obliquity between the motion direction and the strike of the Tosco-Abreojos fault zone, the motion can be decomposed in a right-handed lateral component of 6.9 mm/yr with an extensional component of 1.8 mm/yr. This result coincides with our bathymetric and structural observations along the fault zone. Our estimates correspond with upper bounds and would be smaller if there were some additional active faults within the peninsula, for example along the Main Escarpment of the Gulf, or in areas where volcanism has been widespread during late Pliocene and Quaternary. The steadiness of the relative motion between the Pacific and North American plates since 3.16 Ma and the recent increase of oceanic spreading within the Gulf of California indicate that the lateral motion rate and extension components have been greater before 1 Ma, 12 mm/yr, and 3 mm/yr, respectively. For before 3.16 Ma, Atwater and Stock (1998) consider a faster motion between both plates of 56 mm/yr from 10.9 to 3.2 Ma, the value also established at the CONC locality. If we consider that the motion within the Gulf of

Figure 7. Seismic profile Fa-50 (location on Fig. 2) and corresponding line drawing (see text for interpretation; A, B, C, and D are seismic unit labels and are the same as in Figure 3. VE—vertical exaggeration; TWTT—two-way traveltime.

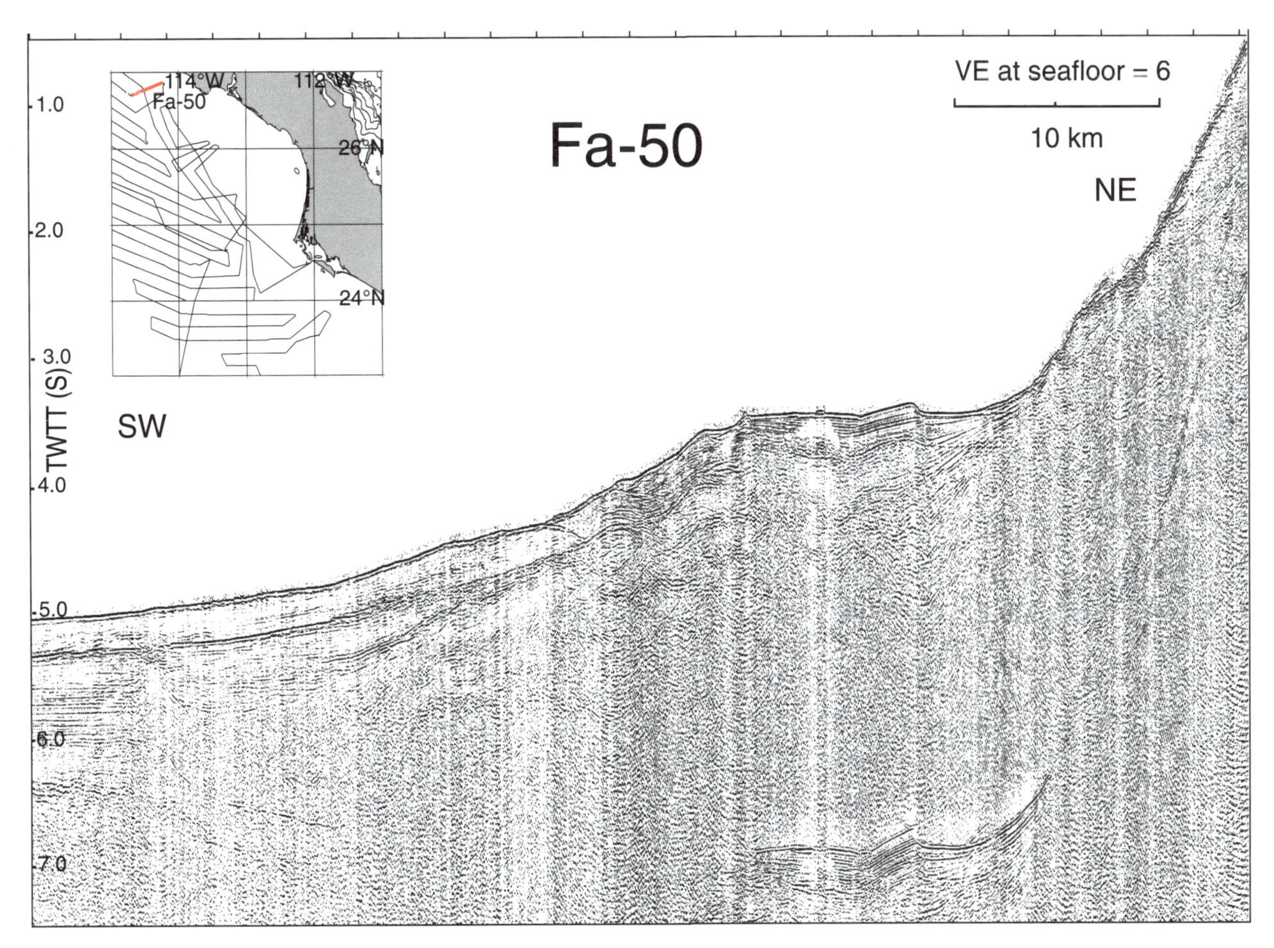

Fa-50
VE at seafloor = 6
10 km
NE
SW
TWTT (S)
1.0
2.0
3.0
4.0
5.0
6.0
7.0
114°W
112°W
Fa-50
26°N
24°N

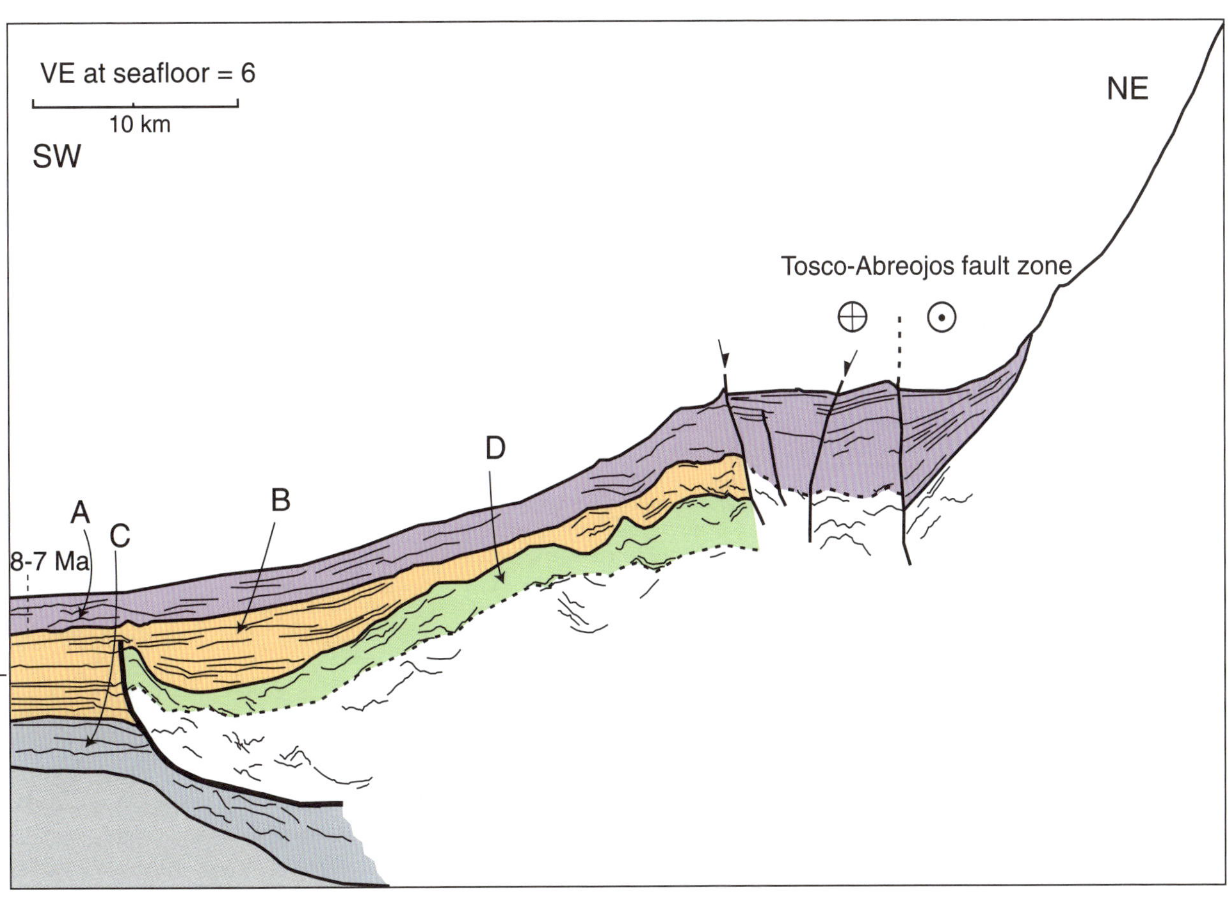

VE at seafloor = 6
10 km
SW
NE
Tosco-Abreojos fault zone
A
B
C
D
8-7 Ma

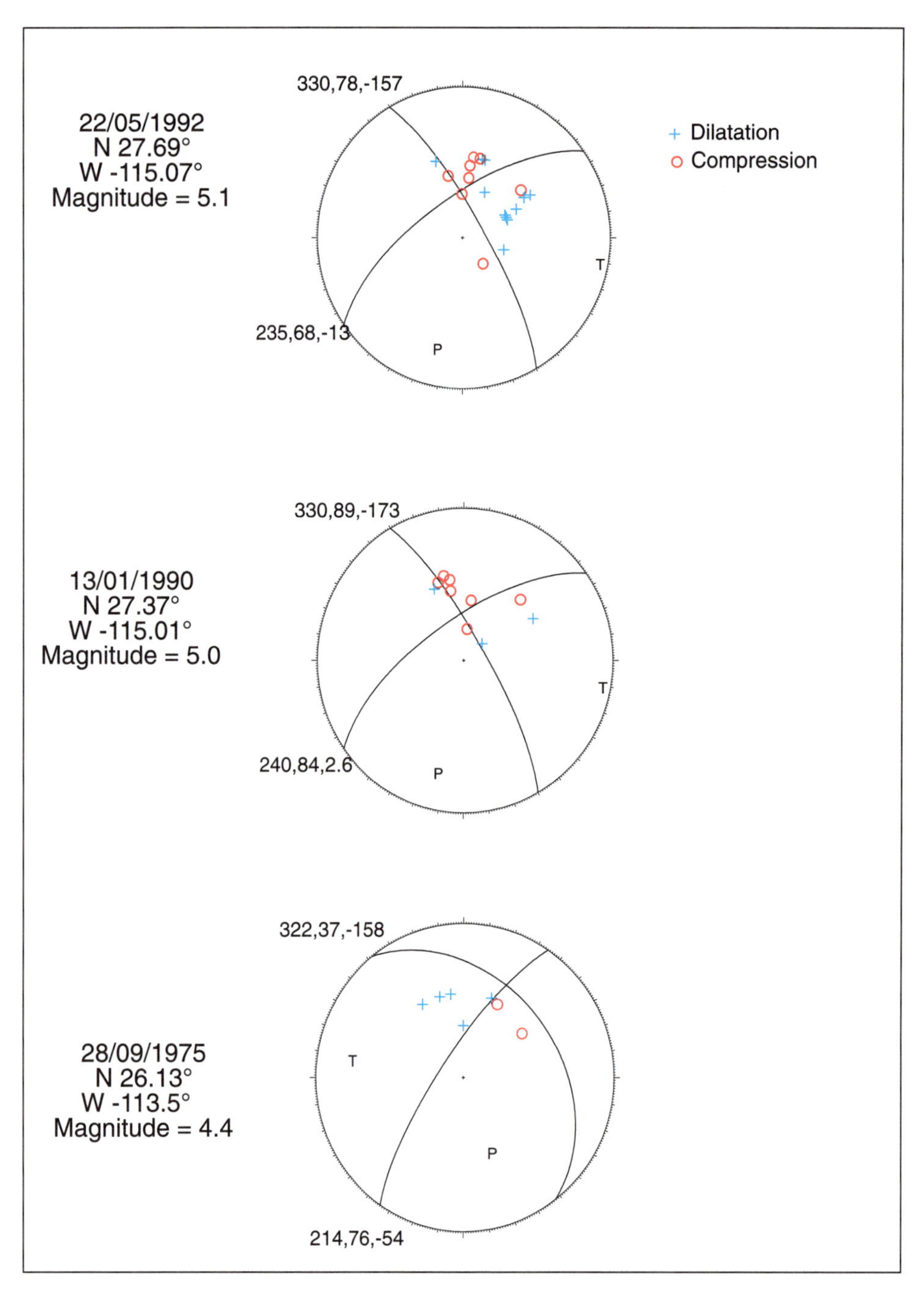

Figure 8. Focal mechanisms of earthquakes located along the Baja California margin near the Tosco-Abreojos fault system. These focal mechanisms show lateral transform motion with normal components. This is compatible with kinematics models and morpho-structural analysis (Michaud et al., 2004).

California was slower than or the same as the 45 mm/yr initial rate of spreading for the 2A magnetic anomaly (DeMets, 1995), the relative motion between Baja California and the Pacific plate must have been at least the same and likely faster than the present motion rate. The kinematic evolution of the region suggests that the Tosco-Abreojos fault zone remained active during the last 10 m.y. with a progressive decrease of lateral motion velocity.

CONCLUSIONS

Swath bathymetry and seismic reflection data provide new insights about the tectonic evolution of the Tosco-Abreojos fault zone. This fault system is presently active with a right-lateral motion and a transtensional component in agreement with the kinematic prediction of 2.9–6.9 mm/yr. Kinematic models sug-

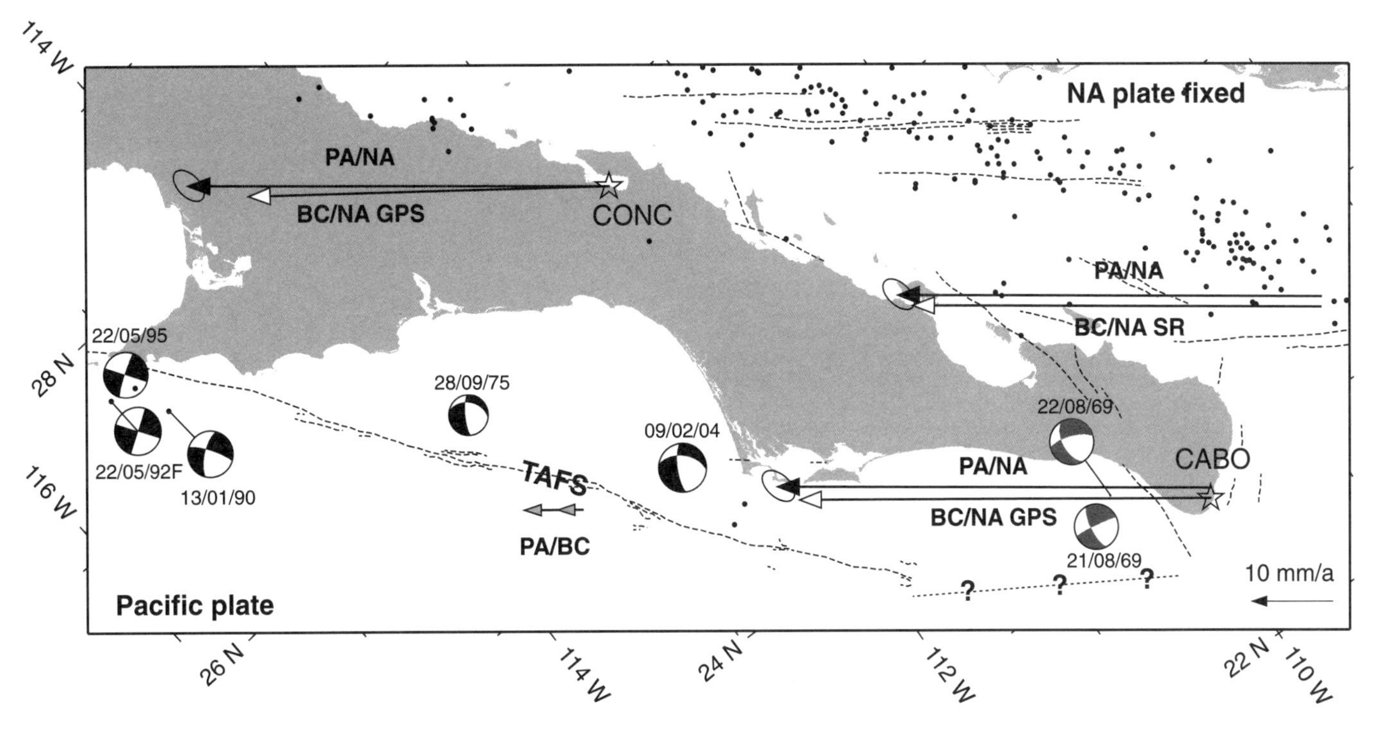

Figure 9. Kinematic predictions and observations in southern Baja California relative to the North American plate (NA) reference frame. Filled arrows with 95% uncertainties are the Pacific plate (PA) motion vectors derived from the PA/NA geodetic model of Argus and Gordon (2001). Open arrows represent the observed motion of Baja California (BC) relative to NA measured by GPS (CONC and CABO sites; Dixon et al., 2000) and from the central seafloor spreading magnetic anomalies at the Gulf rise (0.78 Ma; DeMets, 1995). The gray arrows south of the Tosco-Abreojos fault system (TAFS) represent the minimum and maximum Pacific/Baja California velocities (3–7 mm/yr) derived by subtracting the BC/NA from PA/NA motion vectors at the CONC site (see text). The orientation and direction of motion are compatible with the right-lateral strike-slip motion with a small extensional component observed along the Tosco-Abreojos fault system and with Harvard catalog focal mechanisms available in this area. Mercator oblique projection is relative to the PA/NA rotation pole. Black dots are earthquake epicenters with magnitude (mb) larger than 4.5 from the Harvard and NEIC catalogs. Focal mechanisms are in black from the Harvard moment tensor catalog and are in dark gray from Molnar (1973). Faults (dashed lines) are after the chart of Spencer and Normark (1989) except for the Tosco-Abreojos fault system trace derived from this study. SR—spreading rate.

gest that Pacific plate motion with respect to the North American plate is not exclusively located within the Gulf of California but that it is partitioned (DeMets, 1995; DeMets and Dixon, 1999; Dixon et al., 2000; Fletcher and Munguía, 2000). A significant amount of movement is partitioned across a wide zone (Mann et al., 1983) the westernmost and easternmost limits of which are the Tosco-Abreojos fault zone and probably the margin of Sonora, respectively. The Tosco-Abreojos fault zone presently records a right-lateral displacement of the Baja California peninsula relative to the Pacific plate. Baja California thus forms an independent block limited to the west by the Tosco-Abreojos fault zone and, to the east, by the Gulf of California transform system, composed of transform faults, pull-apart basins, and narrow spreading centers. To the north, the Tosco-Abreojos fault zone is connected with the submarine San Isidro–San Clemente fault system, along the margin of northern Baja California, México, and California, USA (Legg et al., 1991; Humphreys and Weldon, 1991). The continuation of the Tosco-Abreojos fault zone to the south is still unclear, although there are several linear fault escarpments parallel to the Tosco-Abreojos fault zone, such as the Alcatraz fault (Yeats and Haq, 1981), the San Margarita fault (Normark et al., 1987), and the normal fault system of the Los Cabos block as the Todos Santos and Carrizal faults (Fletcher and Munguía, 2000).

REFERENCES CITED

Antonelis, K., Johnson, D.J., Miller, M.M., and Palmer, R., 1999, GPS determination of current Pacific-North American plate motion: Geology, v. 27, p. 299–302, doi: 10.1130/0091-7613(1999)027<0299:GDOCPN>2.3.CO;2.

Aragon-Arreola, M., Morandi, M., Martin-Barajas, A., Delgado-Argote, L., and Gonzalez-Fernandez, A., 2005, Structure of the rifts basins in central Gulf of California: Kinematic implications for oblique rifting: Tectonophysics, v. 409, p. 19–38, doi: 10.1016/j.tecto.2005.08.002.

Argus, D.F., and Gordon, R.G., 2001, Present tectonic motion across coastal ranges and San Andreas fault system in central California: Geological Society of America Bulletin, v. 113, p. 1580–1592, doi: 10.1130/0016-7606(2001)113<1580:PTMATC>2.0.CO;2.

Atwater, T., and Stock, J., 1998, Pacific-North America Plate Tectonics of the Neogene Southwestern United States: An Update: International Geology Review, v. 40, p. 375–402.

Dauphin, J.P., and Ness, G.E., 1991, *in* Dauphin, J.P., and Simoneit, B.R.T., eds., The Gulf and Peninsular Province of the Californias, American Association of Petroleum Geologists, Memoir 47, p. 21–24.

DeMets, C., 1995, A reappraisal of seafloor spreading lineations in the Gulf of California: Implications for the transfer of Baja California to the Pacific plate and estimates of Pacific-North America motion: Geophysical Research Letters, v. 22, p. 3545–3548, doi: 10.1029/95GL03323.

DeMets, C., and Dixon, T.H., 1999, New kinematic models for Pacific-North America motion from 3 Ma to present, I: Evidence for steady motion and biases in the NUVEL-1A model: Geophysical Research Letters, v. 26, p. 1921–1924, doi: 10.1029/1999GL900405.

DeMets, C., and Traylen, S., 2000, Motion of the Rivera plate since 10 Ma relative to the Pacific and North American plates and the mantle: Tectonophysics, v. 318, p. 119–160, doi: 10.1016/S0040-1951(99)00309-1.

DeMets, C., Gordon, R.G., Argus, D.F., and Stein, S., 1990, Current plate motions: Geophysical Journal International, v. 101, p. 425–478.

Dixon, T., Farina, F., DeMets, C., Suarez-Vidal, F., Fletcher, J., Marquez-Azua, B., Miller, M., Sanchez, O., and Umhoefer, P., 2000, New kinematic models for Pacific-North America motion from 3 Ma to Present II: Evidence for a "Baja California shear zone": Geophysical Research Letters, v. 27, p. 3961–3964, doi: 10.1029/2000GL008529.

Fletcher, J.M., and Munguía, L., 2000, Active continental rifting in southern Baja California, Mexico: Implications for plate motion partitioning and the transition to seafloor spreading in the Gulf of California: Tectonics, v. 19, p. 1107–1123, doi: 10.1029/1999TC001131.

Humphreys, E.D., and Weldon, R.J., II, 1991, Kinematic constraints on the rifting of Baja California, *in* Dauphin, J.P., and Simoneit, B.R.T., eds., The Gulf and Peninsular Province of the Californias: American Association of Petroleum Geologists, Memoir 47, p. 217–229.

Klitgord, K.D., and Mammerickx, J., 1982, Northern East Pacific Rise: Magnetic anomaly and bathymetric framework: Journal of Geophysical Research, v. 87, p. 6725–6750.

Krause, D.C., 1965, Tectonics, bathymetry, and geomagnetism of the southern continental borderland west of Baja California, Mexico: Geological Society of America Bulletin, v. 76, p. 617–650.

Legg, M.R., Wong, O., and Suarez, V., 1991, Geologic structure and Tectonics of the inner continental borderland of northern Baja California, *in* Dauphin, J.P., and Simoneit, B., eds., The Gulf and Peninsular Province of the Californias: American Association of Petroleum Geologists Memoir 47, p. 145–196.

Lonsdale, P., 1989, Geology and tectonic history of the Gulf of California, *in* Winterer, E.L., Hussong, D.M., and Decker, R.W., eds., The Eastern Pacific Ocean and Hawaii, The Geology of North America, Geological Society of America, v. N., p. 499–521.

Lonsdale, P., 1991, Structural patterns of the Pacific floor offshore of peninsular California, *in* Dauphin, J.P., and Simoneit, B.R.T., eds., The Gulf and Peninsular Province of the Californias, American Association of Petroleum Geologists Memoir 47, p. 87–125.

Lonsdale, P., 1995, Segmentation and disruption of the East Pacific Rise in the mouth of the Gulf of California: Marine Geophysical Researches, v. 17, p. 323–359, doi: 10.1007/BF01227039.

Mammerickx, J., and Klitgord, K.D., 1982, Northern East Pacific Rise: Evolution from 25 m.y. B.P. to the Present: Journal of Geophysical Research, v. 87, p. 6751–6759.

Mann, P., Hempton, M.R., Bradley, D.C., and Burke, K., 1983, Development of pull-apart basins: Journal of Geology, v. 91, p. 529–554.

Marsaglia, K.M., 2004, Sandstone detrital modes support Magdalena Fan displacement from the mouth of the Gulf of California: Geology, v. 32, p. 45–48, doi: 10.1130/G20099.1.

Michaud, F., Sosson, M., Royer, J.-Y., Chabert, A., Bourgois, J., Calmus, T., Mortera, C., Bigot-Cormier, F., Bandy, W., Dyment, J., Pontoise, B., and Sichler, B., 2004, Motion partitioning between the Pacific plate, Baja California and the North America plate: The Tosco-Abreojos fault revisited: Geophysical Research Letters, v. 31, L08604, doi: 1029/2004GL019665.

Michaud, F., Royer, J.-Y., Bourgois, J., Dyment, J., Calmus, T., Bandy, W.L., Sosson, M., Mortera, C., Sichler, B., Rebolledo-Vieyra, M., and Pontoise, B., 2006, Oceanic-ridge subduction vs. slab break off: Plate tectonic evolution along the Baja California Sur continental margin since 15 Ma: Geology, v. 34, p. 13–16, doi: 10.1130/g22050.1.

Minster, J.B., and Jordan, T.H., 1978, Present-day plate motions: Journal of Geophysical Research, v. 83, p. 5331–5354.

Molnar, P., 1973, Fault plane solutions of earthquakes and direction of motion in the Gulf of California and on the Rivera Fracture Zone: Geological Society of America Bulletin, v. 84, p. 1651–1658, doi: 10.1130/0016-7606(1973)84<1651:FPSOEA>2.0.CO;2.

Ness, G.E., Lyle, M.E., and Couch, R.W., 1991, Marine magnetic anomalies and oceanic crustal isochrones of the Gulf and Peninsular Province of the Californias, *in* Dauphin, J.P., and Simoneit, B.R.T., eds., The Gulf and Peninsular Province of the Californias, American Association of Petroleum Geologists, Memoir 47, p. 47–69.

Normark, W., Spencer, J., and Ingle, J., 1987, Geology and Neogene history of the Pacific margin of Baja California Sur, *in* Scholl, D., Grantz, A., and Vedder, J., eds., Geology and Resource Potential of the Continental Margin of Western North America, Earth Science Series 6, Circum-Pacific Counsel for Energy and Mineral Resources, Houston, Texas, p. 449–472.

Oskin, M., and Stock, J., 2003, Pacific-North America plate motion and opening of the Upper Delfín basin, northern Gulf of California, Mexico: Geological Society of America Bulletin, v. 115, p. 1173–1190, doi: 10.1130/B25154.1.

Ramos-Velázquez, E., 1998, Características de la deformación en las rocas cristalinas cretácicas del este sureste de la ciudad de La Paz, BCS, México: Tesis de Maestría, Centro de Investigación Científica y de Educación Superior de Ensenada, Ensenada, México, p. 1–122.

Smith, W.H.F., and Sandwell, D.T., 1997, Global sea floor topography from satellite altimetry and ship depth soundings: Science, v. 277, p. 1957–1962.

Spencer, J.E., and Normark, W., 1979, Tosco-Abreojos fault zone: A Neogene transform plate boundary within the Pacific margin of southern Baja California, Mexico: Geology, v. 7, p. 554–557, doi: 10.1130/0091-7613(1979)7<554:TFZANT>2.0.CO;2.

Spencer, J.E., and Normark, W., 1989, Neogene plate-tectonic evolution of the Baja California Sur continental margin and the southern Gulf of California, Mexico, *in* Winterer, E.L., Hussong, D.M., and Decker, R., eds., The Eastern Pacific Ocean and Hawaii: Boulder, Colorado, Geological Society of America, Geology of North America, v. N, p. 489–497.

Stock, J.M., and Hodges, K.V., 1989, Pre-Pliocene extension around the Gulf of California and the transfer of Baja California to the Pacific plate, Tectonics, v. 28, p. 99–115.

Stock, J.M., and Lee, J., 1994, Do microplates in subduction zones leave a geological record?: Tectonics, v. 13, p. 1472–1487, doi: 10.1029/94TC01808.

Yeats, R.S., and Deep Sea Drilling Project Leg 63 Scientific Staff, 1978, Site 471: Offshore Magdalena Bay, *in* Yeats, R.S., Haq, B.U., et al., Initial Reports of the Deep Sea Drilling Project, Volume 63: Washington, D.C., U.S. Government Printing Office, p. 269–323.

Yeats, R.S., and Haq, B.U., 1981, Deep-Sea drilling off the Californias: Implications of Leg 63, Initial Reports of the Deep Sea Drilling Project, Volume 63, Washington, D.C., U.S. Government Printing Office, p. 949–961.

Manuscript Accepted by the Society 29 August 2006

Geological Society of America
Special Paper 422
2007

The Taxco–San Miguel de Allende fault system and the Trans-Mexican Volcanic Belt: Two tectonic boundaries in central México active during the Cenozoic

Susana A. Alaniz-Álvarez*
Ángel Francisco Nieto-Samaniego
Centro de Geociencias, Universidad Nacional Autónoma de México, Campus Juriquilla, Apartado Postal 1-742, 76001 Querétaro, Qro., México

ABSTRACT

The Trans-Mexican Volcanic Belt has been recognized as a major volcanic arc, which crosses México from the Pacific Coast to the Gulf of México, that has displayed normal faulting and volcanism since the Miocene. In this work we present the deformation events that have been recorded N and S of the belt in order to establish when the crustal discontinuity originated and also to determine the deformation field precursor of the volcanic arc emplacement. In Mesa Central, the post-Laramide deformation occurred in three extensional events during the Eocene, Oligocene, and Miocene-Recent. The three events produced extension in two horizontal directions and shortening in a vertical direction. The direction of the principal extension in the Eocene is not well known. A 20% extension in an ~ENE-WSW direction is recorded for the Oligocene event. The most recent event, active since the middle Miocene, has developed in the Trans-Mexican Volcanic Belt and along its northern boundary. In the Sierra Madre Oriental, Cenozoic deformation has been minimal. In the Taxco region, there were two post-Laramide deformation events, mainly a result of NW-SE and N-S lateral faults. The first one occurred in the late Eocene with a NNW-SSE horizontal extension direction. The second event was early Oligocene with a maximum extension to the NE-SW. It is concluded that since the Eocene, the deformation style has been different in Mesa Central and in the Sierra Madre del Sur, which implies the presence of a detachment zone between these provinces.

Keywords: Trans-Mexican Volcanic Belt, major fault, Cenozoic, reactivation.

INTRODUCTION

Since Humboldt's time, it has been recognized that in central México a volcanic belt, including many active volcanoes, crosses México from Tepic on the Pacific Coast to Veracruz on the Gulf of México. This belt is known as the Trans-Mexican Volcanic Belt. There are many hypotheses about its origin; the most common is that it is located in a weakness zone through which the magma has ascended to form the actual volcanic arc. This hypothesis results from its oblique position with respect to the Mesoamerican Trench (Mooser, 1972) (Fig. 1).

*alaniz@geociencias.unam.mx

Alaniz-Álvarez, S.A., and Nieto-Samaniego, Á.F., 2007, The Taxco–San Miguel de Allende fault system and the Trans-Mexican Volcanic Belt: Two tectonic boundaries in central México active during the Cenozoic, *in* Alaniz-Álvarez, S.A., and Nieto-Samaniego, Á.F., eds., Geology of México: Celebrating the Centenary of the Geological Society of México: Geological Society of America Special Paper 422, p. 301–316, doi: 10.1130/2007.2422(10). For permission to copy, contact editing@geosociety.org.

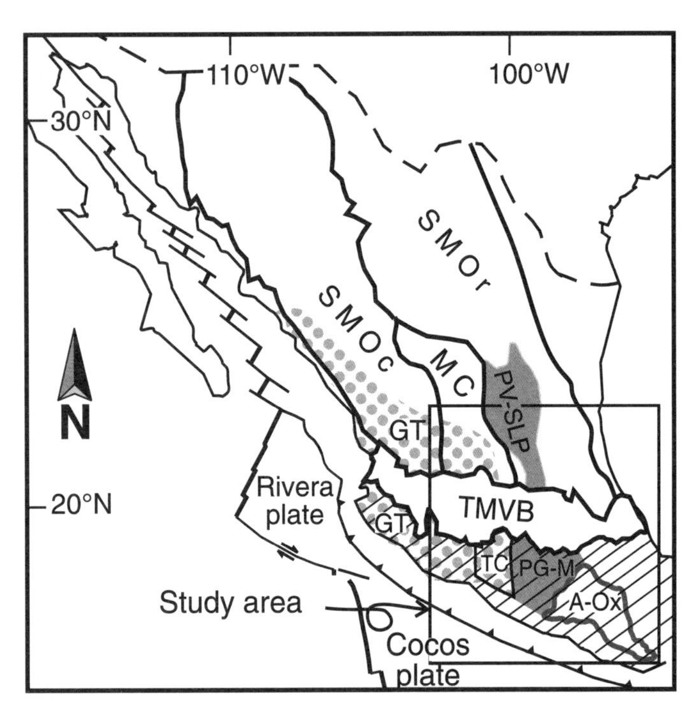

Figure 1. Location of the study area. SMOc—Sierra Madre Occidental volcanic province; SMOr—Sierra Madre Oriental; TMVB—Trans-Mexican Volcanic Belt; MC—Mesa Central; PG-M—Guerrero-Morelos Platform; TC—Tierra Caliente metamorphic complex; GT—Guerrero terrane, Mesozoic volcanic arcs; A-Ox—crustal block containing the Acatlán and Oaxaca complexes; PV-SLP—Valles–San Luis Potosí Platform. Dashed lines—Sierra Madre del Sur; dots—Guerrero terrane.

There were many papers written about the opening and evolution of the Gulf of México during the end of the 1970s and the beginning of the 1980s. In order to accommodate the lithospheric blocks during Jurassic time (between 180 and 140 Ma), the existence of several megashears along and across México was proposed. One of these proposed megashears was located along the actual position of the Trans-Mexican Volcanic Belt (Anderson and Silver, 1974; Pindell, 1985; Ross and Scotese, 1988). The idea of the emplacement of the Trans-Mexican Volcanic Belt along a crustal or lithospheric megashear was previously proposed by Mooser (1972), who suggested that the belt was located along a Paleozoic suture reactivated during the Tertiary. The Trans-Mexican Volcanic Belt has been the subject of many papers, many of which are related to volcanism and magma origin, however, the hypothesis of an ancient suture has yet to be proven.

It has been documented that major faults bound most of the major basins in central México and reach depths of hundreds or thousands of meters, as in the case of the México City basin (Venegas-Salgado et al., 1985; Pérez-Cruz, 1988; Silva-Romo et al., 2002, Siebe et al., 2004), the San Pedro Ceboruco graben (Ferrari et al., 2003a), and the El Bajío zone (Alaniz-Álvarez et al., 2001). If the Trans-Mexican Volcanic Belt is located along one or many tectonic depressions, it is essential to know what deformation event generated them. We focus on the main deformation events that occurred in the adjacent blocks in the central part of the structure, supported by studies along the Taxco–San Miguel de Allende fault, which crosses the belt from N to S and has been considered a weakness zone that limits different crustal zones (Nieto-Samaniego et al., 1999; Alaniz-Álvarez et al., 2002a).

The structural trend of the Trans-Mexican Volcanic Belt is ~E-W (Fig. 1), thus we analyzed the transversal structural features in order to recognize the continuity of the features N and S of the belt and the lateral displacements along it. In this paper, we describe the deformation events that have occurred in central México since the Albian, after which the pre-Mesozoic basements in México have been in their present-day position and have not experienced major displacements (Morán-Zenteno et al., 1988). With these considerations we define the tectonic character of the Trans-Mexican Volcanic Belt during the Cenozoic.

GEOLOGICAL SETTING

In central México there are many physiographic provinces: the Sierra Madre Occidental, Mesa Central, the Sierra Madre Oriental, the Trans-Mexican Volcanic Belt, and the Sierra Madre del Sur (Raisz, 1959) (Fig. 1). In this work we focus only on the sectors of these provinces whose geological constitution permits us to easily differentiate among them.

Structural Features Located N of the Trans-Mexican Volcanic Belt

The Sierra Madre Occidental is a volcanic province that is primarily composed of siliceous ignimbrites emplaced during the Oligocene-Miocene in western México. It is 1200 km long and extends from the United States–México border to Jalisco (Aranda-Gómez et al., 2000). In Mesa Central, the volcanism was emplaced by rhyolitic domes during the Oligocene; they cover a surface larger than 10,000 km^2 (Orozco-Esquivel et al., 2002). The Mesozoic basement underlies the silicic volcanism in Mesa Central; it is composed mainly of basin marine strata, forming the "Mexican geosyncline" (Imlay, 1938), which is also known as the "Central Mexico Mesozoic Basin" (Carrillo-Bravo, 1971). In the border between Mesa Central and the Sierra Madre Occidental, the outcrops of the basement are constituted by a Late Jurassic–Early Cretaceous volcanosedimentary sequence (Martínez-Reyes, 1992; Quintero-Legorreta, 1992; Centeno-García and Silva Romo, 1997; Freydier et al., 2000).

The Sierra Madre Oriental is a cordillera located in eastern México from Parral, Chihuahua, to Tuxtepec, Oaxaca. It is made up of Mesozoic marine strata that were contracted by the Laramide orogeny. The Mesozoic marine deposition initiated in the Late Triassic but from the Jurassic until the Late Cretaceous marine sediments were constantly deposited with contrasted facies distribution (Eguiluz et al., 2000). The Valles–San Luis Potosí Platform is one of the paleogeographic elements of this cordillera (Fig. 1); it is characterized by a thin layer of Upper Jurassic strata that underlies a 4000-m-thick layer of evaporitic rocks and Lower and Upper Cretaceous reef and back-reef limestone (Carrillo-Bravo, 1971). The Laramide orogeny produced, during the end of the Cretaceous to the Eocene, folds and thrusts of E-NE vergence, migrating the tectonic front from W to E (Eguiluz et al., 2000).

North of the Trans-Mexican Volcanic Belt, from E to W, outcrop Upper Jurassic–Cretaceous rocks belonging to the following paleogeographic environments: a platform located within the Sierra Madre Oriental, a marine basin located within Mesa Central, and further to the west, in the boundary between Mesa Central and the Sierra Madre Occidental, marine volcanic arcs.

Structural Features Located S of the Trans-Mexican Volcanic Belt

The Guerrero-Morelos Platform is located in the central part of the Sierra Madre del Sur. It contains a shallow-marine sedimentary sequence that spans from Neocomian to Turonian (Huitzuco Anhydrite, Morelos Formation, Cuautla Formation) (Hernández-Romano et al., 1997; Aguilera-Franco, 2003); it has continental red beds with limestone strata interdigited at its base (Zicapa Formation; De Cserna et al., 1980).

The Guerrero-Morelos Platform rocks were contracted with vergence to the E, and the age of deformation migrated to the E. Also, there have been reports of structures with vergence in the opposite direction to the Teloloapan thrusts at the western border of this platform (Salinas-Prieto et al., 2000), and in the Papalutla thrust along the eastern border (Cerca et al., 2004). Although the Guerrero-Morelos Platform has not been considered a part of the Sierra Madre Oriental, the same Early Cretaceous paleogeography continues N and S of the Trans-Mexican Volcanic Belt; it is evidenced by the presence of a platform just S of the Valles–San Luis Potosí Platform (Fig. 1).

The Tierra Caliente metamorphic complex is located W of the Guerrero-Morelos Platform between Arcelia and Teloloapan (Fig. 2). It is made up of a Cretaceous age volcanosedimentary marine sequence with low-grade metamorphism (Ortega-Gutiérrez, 1981). This sequence contains andesitic pillow lavas, ignimbrite, tuffs, sandstone, and limestone that belong to the Taxco Viejo Schist and Almoloya Phyllite (Cabral-Cano et al., 2000). The Tierra Caliente metamorphic complex belongs to the Guerrero terrane (Campa and Coney, 1983). In this work we focus only on the predominantly volcanic character of this sequence, which contrasts with the marine sedimentary predominance without volcanics content of the Guerrero-Morelos Platform. Thus, for Early Cretaceous time S of the Trans-Mexican Volcanic Belt, there is a platform to the E and marine volcanic arcs W of the Taxco–San Miguel de Allende fault system. This contrasts with the area N of the belt where a marine basin is separated by these two marine environments.

The magmatic province of the Sierra Madre del Sur is made up of Tertiary volcanic and plutonic rocks; these rocks were emplaced during the Paleogene to Miocene (Morán-Zenteno et al., 2000). The plutonic rocks, mainly granitic, are located parallel and adjacent to the Pacific Ocean Coast margin, while the volcanic rocks were emplaced in the internal part of the Sierra Madre del Sur. There are different basements underlying this province: the Guerrero terrane in the W and the Acatlán and Oaxaca complexes to the E of the Guerrero-Morelos Platform.

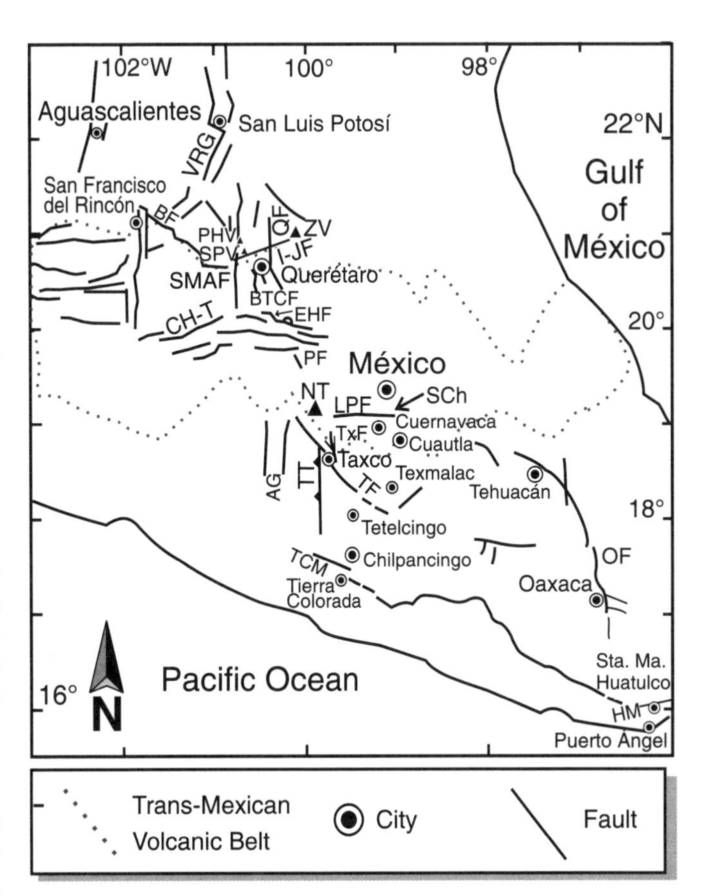

Figure 2. Location of major faults, volcanoes, and cities mentioned in the text. VRG—Villa de Reyes graben; QF—Querétaro fault; BTCF—San Bartolome, Tlacote, La Cuesta, and Lagunillas-Huimilpan faults; BF—El Bajío fault; I-JF—Ixtla–La Joya fault system; SMAF—San Miguel de Allende fault; CH-T—Chapala-Tula fault system; TF—Tetipac, Chichila, and Amates faults; LPF—La Pera fault system; TxF—Taxco, San Gregorio, Coapango, and Acamixtla faults; EHF—Epitacio Huerta fault; PF—Perales fault; AG—Arcelia graben; TT—Teloloapan thrust; OF—Oaxaca fault; TCM—Tierra Colorada mylonite; HM—Huatulco mylonite; SCh—Sierra de Chichinautzin; ZV—Zamorano volcano; PHV—Palo Huérfano volcano; SPV—San Pedro volcano; NT—Nevado de Toluca.

AGE OF THE LARAMIDE OROGENY

It is known that the Laramide orogeny migrated from W to E in the study area (De Cserna et al., 1980; Eguiluz et al., 2000). In order to estimate the deformation age in a specific locality, we considered the youngest rock affected by the orogeny and the oldest rock that was not affected. These were compared to the contractile deformation age within the Valles–San Luis Potosí and the Guerrero-Morelos Platforms.

In the Guanajuato range (101°–102°W), the Laramide deformation age was obtained between the Early Cenomanian, the maximum age of the La Perlita Limestone, which is slightly folded, and 54–58 Ma, the age of the Comanja Granite (Quintero-Legorreta, 1992). These data are consistent with the 49 Ma age of the Guanajuato Conglomerate, which records the post-Laramide extension in the Guanajuato range (Aranda-Gómez and McDowell, 1998). In the southern sector of the Valles–San Luis Potosí Platform, the age of the Laramide deformation is constrained by the Campanian-Maastrichtian Cárdenas Formation (Carrillo-Bravo, 1971), which contains the younger rocks affected by contraction, with 43 Ma, which is the age of the younger reported post-tectonic granite emplaced over the folded sediments (Vassallo et al., 2004).

In the Tetelcingo area (99°35′W) within the Guerrero-Morelos Platform, the age of the Laramide deformation was constrained by the Turonian-Coniacian, the age of the fossil content of the Mexcala Formation in this area, and 66 ± 2.3 to 68.8 ± 2.4 Ma (K-Ar biotite and whole rock), the age of the Tetelcingo Formation (Ortega-Gutiérrez, 1980). On the other hand, in Texmalac in the eastern part of the platform, the maximum age of the contractile deformation was Maastrichtian, based on the fossil content (Alencaster, 1980). The obtained ranks of age for the Laramide orogeny allow us to establish that the migration of the tectonic front toward the E happened in a similar way in platforms both N and S of the Trans-Mexican Volcanic Belt.

TAXCO–SAN MIGUEL DE ALLENDE FAULT SYSTEM

Demant (1978) named a structural feature that crosses the Trans-Mexican Volcanic Belt with a NNW-SSE trend the "Taxco–San Miguel de Allende lineament." He proposed that this structure had a dextral-strike displacement during the Miocene, based on the apparent displacement of this volcanic arc along this structure. In the past decade many structural data have been obtained along the Taxco–San Miguel de Allende lineament. Although its limits are beyond the original definition; we decided to keep *Taxco* and *San Miguel de Allende* in the term we use because they are well known in the Mexican geologic community. Thus, we have applied the term *Taxco–San Miguel de Allende fault system* to the set of faults trending N-S and NNW-SSE that crosses central México. These faults separate crustal blocks with different geologic histories, crustal thickness, and topographies (Alaniz-Álvarez et al., 2002a). The fault system's main geophysical characteristic is a pronounced magnetic anomaly (Urrutia-Fucugauchi and Flores-Ruiz, 1996).

Recent studies along the Taxco–San Miguel de Allende fault system show that its length is greater than 450 km, including from north of the Sierra de Catorce, S.L.P., to the State of Guerrero; this fault system has wider amplitude in the region of Querétaro (Alaniz-Álvarez et al., 2002a) (Fig. 2). The Taxco–San Miguel de Allende fault system includes faults with lengths from 15 to 50 km. The kinematics, displacement, and age of the major faults of this system have been carefully documented by diverse authors: Villa de Reyes graben by Tristán-González (1986) and San Miguel de Allende fault by Nieto-Samaniego and Alaniz-Álvarez (1994) in Mesa Central; the San Bartolome, Tlacote, and Querétaro faults by Alaniz-Álvarez et al. (2001) in the transition between Mesa Central and the Trans-Mexican Volcanic Belt; the Cuesta and Lagunillas-Huimilpan faults by Dávalos-Álvarez et al. (2005), Epitacio Huerta and Perales faults by Suter et al. (1995), the N-S–trending faults in the El Oro–Tlalpuhahua region by Flores (1920), and the horst and graben system near the Nevado de Toluca by García-Palomo et al. (2000) within the Trans-Mexican Volcanic Belt; the Taxco, San Gregorio, Coapango, and Acamixtla faults by Alaniz-Álvarez et al. (2002b) and the Teloloapan thrust by Cabral-Cano et al. (2000) in the Sierra Madre del Sur. The trending of this structure is N-S outside the Trans-Mexican Volcanic Belt and NNW-SSE within the volcanic arc.

Within the belt, the presence of the Taxco–San Miguel de Allende fault system is evidenced by normal faults located in the regions of Querétaro, Acambay, and Toluca, and the alignment of 12 polygenetic volcanoes along the trace of this structure (Alaniz-Álvarez et al., 1998). The fault system continues N of the city of San Luis Potosí forming the limit of a great tectonic depression that constitutes the N prolongation of the Villa de Reyes graben until the northern section of the Sierra de Catorce (Moreira-Rivera et al., 1996). Toward the S, the western limit of the Guerrero-Morelos Platform seems to be part of the Taxco–San Miguel de Allende fault system. It is constituted by the Teloloapan fault and many other parallel faults that are located along the limit between the Mesozoic volcanosedimentary marine sequence of the Tierra Caliente complex, and the marine Mesozoic sequence without volcanic material from the Guerrero-Morelos Platform (Cabral-Cano et al., 2000; Salinas-Prieto et al., 2000) (Figs. 1 and 2).

The kinematics and age of the faults of the Taxco–San Miguel de Allende fault system have obeyed the kinematics of the structural province to which each fault belongs; thus in Mesa Central and the Trans-Mexican Volcanic Belt, the main component of movement has been normal and has occurred in two different events in Oligocene and Miocene-Recent times (Alaniz-Álvarez et al., 2002a), whereas in the region of Taxco the main component was strike-slip, with senses of movement as much dextral as left, since they were subject to two different phases of deformation which occurred in late Eocene and early Oligocene (Alaniz-Álvarez et al., 2002b).

The width of the Taxco–San Miguel de Allende fault system varies depending on the number of faults that accommodate the deformation. In the region of Querétaro, the zone of faulting has a width maximum of nearly 30 km, which includes the area from the San Miguel de Allende to the Querétaro faults, whereas in other localities it is narrower.

The Taxco–San Miguel de Allende Fault System as a Boundary of Paleogeographic Elements

Alaniz-Álvarez et al. (2002a) mentioned that the Taxco–San Miguel de Allende fault system is located along the boundary of two crustal blocks with different crustal thicknesses, topography, and geological features. On a regional scale, it is located between Mesa Central and the Sierra Madre Oriental, between the central and eastern sectors of the Trans-Mexican Volcanic Belt with different volcanic and structural styles, and between the Tierra Caliente metamorphic complex and the Guerrero-Morelos Platform. There is a major Mesozoic paleogeographic element that it is located under the Taxco–San Miguel de Allende fault system, the western margin of the Valles–San Luis Potosí, and the Guerrero-Morelos Platforms. The eastern boundary of the Mesozoic marine volcanosedimentary outcrops that coincide with the fault system is located between San Miguel de Allende and Teloloapan. In San Miguel de Allende there is contact between the Soyatal Formation with mid-Turonian age (Hernández-Jáuregui et al., 2000) and the Mesozoic volcanosedimentary sequence; however in the Sierra de los Cuarzos, E of San Miguel de Allende and 20 km N of Querétaro City, the volcanosedimentary sequence outcrops again between the San Miguel de Allende and Querétaro faults (Alaniz-Álvarez et al., 2001). This contact deviates from San Miguel de Allende toward the NW (Centeno-García and Silva-Romo, 1997). In the El Oro–Tlapuhahua region in Michoacán, Mesozoic volcanic rocks outcrop along the western side of the Taxco–San Miguel de Allende fault system (Flores, 1920; Pasquaré et al., 1991); they form the only basement outcrop within the Trans-Mexican Volcanic Belt. To the S of the structure, Cabral-Cano et al. (2000) studied the history of deformation of the Tierra Caliente metamorphic complex between Arcelia and Teloloapan in order to determine the nature of its eastern limit with the Guerrero-Morelos Platform. They concluded that this contact is constituted by parallel N-S thrusts that formed during a contractive deformation, which occurred at the end of the Cretaceous and therefore correlates with the Laramide orogeny. The Teloloapan thrust does not present major deformation features such as high-pressure metamorphism or intense brittle deformation. Cabral-Cano et al. (2000) conclude that the Teloloapan thrust is not a tectonic boundary between terranes, as was proposed previously, but the manifestation of an old relief exposed to a contractional deformation. The western margin of the Cretaceous continental platforms Valles–San Luis Potosí and Guerrero-Morelos coincides with the Taxco–San Miguel de Allende fault system trace from Sierra de Catorce to Teloloapan. Therefore, the N-S trace of the Taxco–San Miguel de Allende fault system is located throughout the transition between a zone of continental platform and one of greater bathymetry; the slope change and the differences in existing crustal thicknesses during the Cretaceous contribute to the origin that we propose for the discontinuity with a nearly N-S direction that controlled the location of the faults belonging to the fault system during Cenozoic time.

The previous hypothesis explains some characteristics that appear along this lineament: the persistence of NNW-SSE–trending faults even though there are fault systems with other directions; the gravimetric anomaly that indicates different crustal thicknesses from both sides of the Taxco–San Miguel de Allende fault system; the widening of the fault zone between San Miguel de Allende and Querétaro where the Valles–San Luis Potosí Platform is narrowed; the location of the eastern limit of the Guerrero terrane in the S and the continuation of the fault system N of the city of San Luis Potosí on the limit between the Valles–San Luis Potosí Platform and the Mesozoic basin of central México.

CENOZOIC DEFORMATION EVENTS

There are few studies on southern México that focus on the Cenozoic deformations and practically all of them are concentrated to the E of the Taxco–San Miguel de Allende fault system with the exception of the studies in the Colima and Arcelia grabens, which are located to the W of this structure. In Mesa Central and the southern part of the Sierra Madre Occidental, the Cenozoic deformation is well documented (e.g., Nieto-Samaniego et al., 1999), whereas in the zone located to the E of the fault system in the Sierra Madre Oriental little information is available but apparently Cenozoic deformation is minimal or absent.

Paleocene Deformation Events

One of the least-studied aspects of the region included in this study is the Paleocene deformation, which includes the final stages of the Laramide orogeny and the postorogenic deformation related to the deposit of red beds associated with lateral or normal faults.

Although it is not possible to infer differences between the Paleocene deformation N and S of the Trans-Mexican Volcanic Belt, we consider it necessary to mention some data published on possible events of deformation during that time.

In the central part of the Sierra Madre del Sur, a contractile deformation event has been proposed to occur subsequent to the Laramide contractile climax, which produced wide folds and thrust with vergence toward the W. Salinas-Prieto et al. (2000) and Cabral-Cano et al. (2000) propose that these structures affected the volcanosedimentary sequence of Arcelia as well as the Guerrero-Morelos Platform, and that they were more poorly developed than the structures with vergence toward the E. In both papers it is proposed that this phase developed from the gravitational collapse of the accumulated sedimentary strata during the shortening of the Laramide orogeny. Cerca-Martínez (2004) also reported a series of wide folds and thrusts with opposite vergences and posteriors to the structures generated by the Laramide orogeny in the eastern margin of the Guerrero-Morelos Platform. He documented that many of the early Tertiary basins are associated with the contractile structures with vergences to the west, and he proposed that the origin of these structures is related to a post-Laramide transpressive event.

The contractive structures with vergence toward the W only affected some portions of southern México; there are few studies about their age and kinematics and there is no consensus on their origin. The evidence that we have indicates that these structures were formed immediately after the peak of the Laramide deformation (Cabral-Cano et al., 2000; Salinas-Prieto et al., 2000; Cerca-Martínez, 2004), which opens up the possibility that it is a delayed event of that orogeny, which occurred in the Paleocene.

The deposit of the post-Laramide red beds occurred as much in Mesa Central as in the Sierra Madre del Sur; these strata were recognized by Edwards (1955) as postorogenic continental conglomerates. Studies in the Sierra de Guanajuato and Taxco show that these deposits are of Paleocene-Eocene age and that they are not related to the Paleocene contractive event. The Guanajuato Conglomerate (Conglomerado Guanajuato), located in Mesa Central, was dated at Eocene by Aranda-Gómez and McDowell (1998) and they considered that this conglomerate was formed during an extensional event. In southern México, the Taxco Red Conglomerate (Conglomerado Rojo de Taxco) underlies the Acamixtla Ignimbrite of 38 Ma (K-Ar in sanidine; Alaniz-Álvarez et al., 2002b). Although it is possible that these deposits originated during a Paleocene-Eocene event of extensional deformation, similar to the one documented in the Guanajuato region, we do not have enough information to document it.

Eocene Deformation Events

The first post-Laramide extensional event occurred during the Eocene in Mesa Central, whereas S of the Trans-Mexican Volcanic Belt, during the late Eocene, a transcurrent deformation event took place (Fig. 3). In Mesa Central, Aranda-Gómez and McDowell (1998) documented an extensional event between 49 and 37 Ma. The horizontal main extensional directions have not been established; the NW- and NE-trending faults that liberate this deformation event were reactivated later during the Oligocene in a much greater event. Aranda-Gómez and McDowell (1998) propose that the Guanajuato Conglomerate registered two phases of Eocene extension: the first one activated NW faults with extension toward the NE-SW and the second one activated NE faults with an extension toward the NW-SE.

In the eastern part of the Sierra Madre del Sur, the Eocene deformation was liberated through left-lateral faults trending NW, parallel to the coast (Fig. 3). This event was registered from near the Pacific Ocean margin in mylonitic rocks, and to the southern edge of the Trans-Mexican Volcanic Belt in the region of Taxco in brittle faults. The Tierra Colorada mylonites with left-lateral kinematics have been related to the displacement of the Chortis block toward the southeast (Schaaf et al., 1995). In the Taxco region, the N45°W-trending vertical faults (Tetipac and Chichila faults) had left-lateral movement between 36 and 33 Ma (Alaniz-Álvarez et al., 2002b). Morán-Zenteno et al. (2004) report that on the Amates fault, whose direction is N45°W and which is part of the system of faults that includes the Tetipac fault, there was intruded a dike that contains striae and kinematic indicators of left-lateral motion. The dike contains material that fed the Tilzapotla caldera, which is dated at 34 Ma. In southern México there are structures documented with similar ages for this event, in localities with the same W long located from the coast to the southern margin of the Trans-Mexican Volcanic Belt (Fig. 3): in Tierra Colorada, there are mylonites with an age >34 Ma (Riller et al., 1992); brittle faults formed between 38 and 34 Ma in Taxco (Alaniz-Álvarez et al., 2002b) and at 34 Ma in Tilzapotla. The maximum extension direction documented in Taxco is NNW-SSE (Alaniz-Álvarez et al., 2002b, Fig. 6A).

Oligocene Deformation Events

During the Oligocene (from 32 to 24 Ma) extensional deformation occurred in Mesa Central, it was liberated through a complex system of normal faults with main directions of NW-SE, NE-SW, and N-S. This event of deformation was the main one that occurred in that region and had an extension of 20% to N79°E and 11% to N11°W (Nieto-Samaniego et al., 1999). In the region of San Miguel de Allende, it was documented that these three systems of faults were active during the Oligocene event; the main faults of these systems are the El Bajío fault, the Villa of Reyes graben, and the San Miguel de Allende fault (Alaniz-Álvarez et al., 2001).

In the S, in the region of Taxco, an event of transcurrent deformation was registered during the Oligocene, reactivating the preexisting faults (Alaniz-Álvarez et al., 2002b). The N-S faults had left-lateral displacement whereas the faults of Tetipac and Chichila with N45°W direction had right-lateral displacement. The direction of the maximum horizontal extension in the Oligocene was toward the NE-SW, whereas the maximum horizontal shortening is toward the NW-SE (Fig. 3); it is remarkable that the main directions of deformation changed in this region in a short time between the late Eocene and the early Oligocene.

Miocene-Recent Events

From Miocene to Recent times, the deformation in the central part of México has been concentrated throughout the Trans-Mexican Volcanic Belt and it has been liberated through normal faults. Martínez-Reyes and Nieto-Samaniego (1990) mapped the faults along the west side of the Taxco–San Miguel de Allende fault system in the Trans-Mexican Volcanic Belt, those that Suter et al. (1992, 1995) designated as the Chapala-Tula fault system

→

Figure 3. Scheme illustrating the primary deformation events. (A) Main faults that had activity at different times. The areas in beige and red have elevations greater than 2000–3000 masl, respectively; those levels emphasize the limits of the Trans-Mexican Volcanic Belt and Mesa Central. Note the different elevations on both sides of the Taxco–San Miguel de Allende fault; N of the Trans-Mexican Volcanic Belt the greater elevations are located toward the W, and S of the structure toward the E. (B) The shaded areas are regions that had deformation at different times. Arrows indicate the directions of the infinitesimal strain principal axes inferred from the kinematics of the faults that released this deformation. (C) This shows the horizontal finite strain ellipses inferred for the Trans-Mexican Volcanic Belt from the deformations that occurred N and S of it. A small deformation is assumed. Question marks indicate sites for which we do not have information. Red circle shows the zone without deformation. Lines are passive markers drawn to illustrate the extension and rotation expected. The star marks the location of the México City basin.

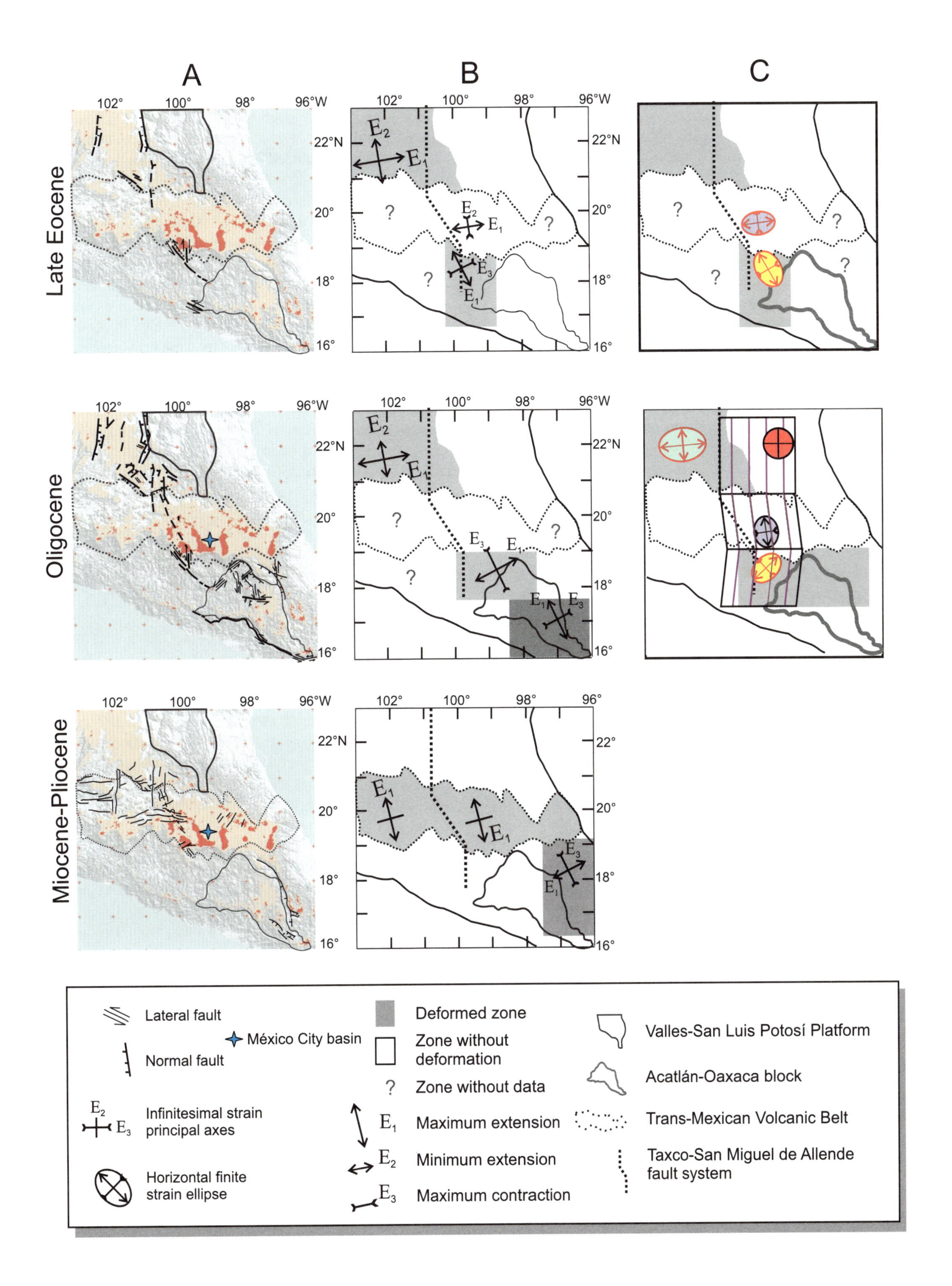

A
B
C
Late Eocene
Oligocene
Miocene-Pliocene
102°
100°
98°
96°W
22°N
20°
18°
16°
E2
E1
E3
?
Lateral fault
México City basin
Normal fault
Infinitesimal strain principal axes
Horizontal finite strain ellipse
Deformed zone
Zone without deformation
Zone without data
Maximum extension
Minimum extension
Maximum contraction
Valles-San Luis Potosí Platform
Acatlán-Oaxaca block
Trans-Mexican Volcanic Belt
Taxco-San Miguel de Allende fault system

(Fig. 2). The faults of this system are normal with a subordinated left lateral component. Suter et al. (2001) calculated an extension smaller than 3% toward the NNW-SSE during the Quaternary for the Chapala-Tula system.

To the east of the Taxco–San Miguel de Allende fault system, in the region of Apan north of México City, a system of tectonic horsts and grabens was formed with a NNE-SSW trend, affecting volcanic rocks from Miocene to Holocene age. García-Palomo et al. (2002) inferred a WNW-ESE extension direction for the faults in this period.

The limit between the Trans-Mexican Volcanic Belt and Mesa Central is discernible by the outcrop of the Mesozoic basement throughout the El Bajío and Ixtla–La Joya faults. In these faults activity has been documented during the Oligocene and Miocene events (Alaniz-Álvarez et al., 2002a), the former predominated in Mesa Central, whereas the second in the Trans-Mexican Volcanic Belt. In the northern margin of the central part of the belt, two systems of faults intersect—those that belong to the Chapala-Tula fault system with an ENE-WSW direction, and those that belong to the Taxco–San Miguel de Allende fault system with a NNW-SSE direction. In the region between Celaya and Querétaro, Alaniz-Álvarez et al. (2002a) documented that these three systems of faults, with directions N-S, NW-SE and NE-SW, activated contemporarily during the same deformation events. A detailed study made by Dávalos-Álvarez et al. (2005) in the region of Huimilpan, 30 km S of Querétaro, documented that the three systems of faults activated during the Pliocene time in the same event of deformation but in an alternated way.

In the Sierra Madre del Sur, the Miocene deformation was registered in Oaxaca, where an ignimbrite of the Suchilquitongo Formation of 19 Ma is cut by normal faults trending N15°W, suggesting an event of extension toward the ENE-WSW (Urrutia-Fucugauchi and Ferrusquía-Villafranca, 2001).

The detail of the geological cartography of the Mesa Central and Querétaro region have allowed us to recognize the presence of Miocene volcanic rock cut by normal faults with small displacements (Orozco-Esquivel et al., 2002). These indicate that the Miocene deformation existed as much in the southern part of Mesa Central as in Oaxaca, although most of it was concentrated throughout the Trans-Mexican Volcanic Belt.

MAJOR FAULTS THAT BORDER THE CENTRAL SECTION OF THE TRANS-MEXICAN VOLCANIC BELT

The activity of major faults plays an important role in deformation since faults not only have the capacity to release a great amount of displacement, but also have a high potential of reactivation and have control of the type of deformation, reactivating or producing neo-formation faults associated with them.

In Mesa Central it has been observed that the three systems of main faults present parallel the major faults (Nieto-Samaniego et al., 1999); this is evident as much for the Miocene deformation as for the Oligocene deformation event. The analogical models developed by Dubois et al. (2002) indicate that in a second phase of overlapping extension and oblique to a set of structures developed in a previous phase, the structures formed in the first stage will be reactivated and new parallel faults to these will be formed.

A normal fault with a length of more than 40 km implies that its depth can reach the limit between the brittle and ductile regimes. In the area considered in this study, the faults more than 40 km long in Mesa Central are the El Bajío fault, the Ixtla–La Joya fault system, the Villa de Reyes graben (in its northern part only), the San Miguel de Allende fault, and the Querétaro fault (Fig. 2). These faults constitute the limit between structural domains and different stratigraphic sequences.

In this work we describe the faults that constitute the boundaries of the Trans-Mexican Volcanic Belt; this includes the La Pera fault system, which is the meridional boundary of the México City and Toluca basins. A fault is considered part of the structural limit of the belt if it has the following characteristics: (1) it exposes Mesozoic rocks, (2) its trace can be followed for more than 40 km, (3) it has been activated in the Oligocene and Miocene events, (4) its displacement has been calculated to be more than 300 m, and (5) its direction is parallel or oblique to the greater axis of the Trans-Mexican Volcanic Belt.

El Bajío Fault

The El Bajío fault represents one of the northern segments of the limit of the Trans-Mexican Volcanic Belt since the Mesozoic basement is located in the foot wall; the Oligocene silicic volcanism characteristic of Mesa Central is located toward the N, whereas in the hanging-wall block there is Miocene-Recent mafic volcanism characteristic of the Trans-Mexican Volcanic Belt. This fault constitutes the northern limit of El Bajío, a region characterized by a low topographic zone that goes from San Francisco del Rincón to Querétaro City and in which are located the cities of Celaya, Salamanca, Irapuato, León, and Lagos de Moreno. This fault has acted as a normal fault; its orientation is N45°W dipping SW, but in its oriental part the direction changes to E-W dipping S (Fig. 2).

The hypsographic characteristic of this fault is prominent (Fig. 3), the continuous escarpment measures 70 km long. The total displacement of this fault is poorly known and it has been estimated only in two places. One of them is the region of León, Guanajuato, where its displacement of more than 850 m is obtained from the displacement of the Oligocene rocks located in the upper part of the Cuatralba range (Quintero-Legorreta, 1992); and furthermore if the 500 m thickness of the sediments within the El Bajío is considered (Hernández-Laloth, 1991), the throw would exceed 1200 m. The second place where its displacement is known is at Silao, Guanajuato, where the El Bajío fault displaced 500 m to the Cubilete Basalt dated in 13.5 Ma by Aguirre-Díaz et al. (1997).

This fault has had at least three episodes of activity. The first of them is of Eocene age as inferred by Aranda-Gómez and McDowell (1998) from the tilting of the basal layers of the Guanajuato Conglomerate in the raised block of this fault. The second event occurred in the Oligocene and was the most important. The fault that contains the Veta Madre in Guanajuato parallels the El Bajío fault and is located less than 10 km from this fault; it displaced more than 1500 m of volcanic rocks ca. 32 Ma in age. The vein that fills up the fault has an age constrained between 29–27 Ma (Gross, 1975). If the Miocene displacement mentioned above at the Silao region is added to the Veta Madre displacement, we would have a total displacement greater than 2 km in the region of Guanajuato.

The Ixtla–La Joya Fault System

The Ixtla–La Joya fault system constitutes the SE margin of a NE-trending structure with a width of 20 km, located among the San Pedro–Palo Huérfano volcanoes and the La Joya volcano. Nevertheless the morphologic attribute is very tenuous; its importance is inferred because along its trace are emplaced the oldest (middle Miocene) and the northernmost stratovolcanoes in the central Trans-Mexican Volcanic Belt, and there are also outcropping Mesozoic rocks along this structure. The structures that constitute the Ixtla–La Joya fault system are an alignment of rhyodacitic domes, and N48°E to N68°E fractures and faults dipping SE. It is inferred that these faults had a Oligocene throw of 250 m since the Mesozoic rocks have been found in wells more than 350 m deep in the Querétaro graben, and the faults of this graben have less than 100 m of Miocene displacement (Alaniz-Álvarez et al., 2001). The post-Miocene displacements are small and they are only observed in a few segments of fault; most of the Ixtla–La Joya fault system is covered by basalt lava flow from 7.5 and 6.2 Ma (Valdez-Moreno et al., 1998).

La Pera Fault System

The southern limit of the central portion of the Trans-Mexican Volcanic Belt is hidden by volcanic deposits, including volcaniclastic deposits from the Nevado de Toluca and Popocatépetl volcanoes and from the volcanic field of the sierra de Chichinautzin. It has been proposed that the sierra de Chichinautzin is located throughout an E-W–trending fault because of the numerous alignments of monogenetic cones with that direction (e.g., Márquez et al., 1999) and the unevenness between the outcrops of calcareous Mesozoic rocks located in Cuernavaca to an altitude of 1840 masl and the reported ones in the base of wells of the México City basin; the unevenness is between 1200 and 3775 m, the hanging wall is the one to the N (Pérez-Cruz, 1988). Márquez et al. (1999) documented that the alignment of more than 15 cones has orientations E-W and NE-SW. The La Pera fault system represents this E-W fault system. The absence of faults with E-W direction within the sierra de Chichinautzin is explained because most of the extension was released by the dikes that fed the volcanic field. Crisp (1984) proposed that the basic magma under the surface represents 3–10 times the material that arrives at the surface. Additionally, the magma deformation inhibited the normal faults formation (Parsons and Thompson, 1991). The major fault that displaced the basement is buried; the gravimetric anomalies reported in Urrutia-Fucugauchi and Flores-Ruiz (1996) and Ferrari et al. (2002) support the existence of this structure. The extension toward the N of this system corresponds to normal faults with a northern inclination found in Malinalco (González-Cervantes, 2004) and Tenango (Márquez et al., 1999; García-Palomo et al., 2000).

FAULTING-VOLCANISM RELATIONSHIPS

The relationships between volcanism and faulting have been studied in Mesa Central (Orozco-Esquivel et al., 2002), in the central part of the Trans-Mexican Volcanic Belt (Alaniz-Álvarez et al., 1998; García-Palomo et al., 2000), and in the Taxco region (Alaniz-Álvarez et al., 2002b; Morán-Zenteno et al., 2004). In Mesa Central, a volcanic field of rhyolite domes that cover an area of more than 10,000 km^2 formed during the Oligocene deformation event; this event produced elongation of more than 20% extension in an ~E-W direction (Xu et al., 2005).

In the Trans-Mexican Volcanic Belt, the monogenetic volcanoes, cinder cone alignments, and dykes are statistically parallel to the normal faults and orthogonal to the maximum extensional strain (Alaniz-Álvarez et al., 1998); in the region of Taxco in the Sierra Madre del Sur, where the system of faults is left-lateral, the emplacement of rhyolitic domes was documented in a pull-apart zone (Alaniz-Álvarez et al., 2002b), and, in a similar way, the caldera Tilzapotla was located in the left overlap of two left-lateral faults (Morán-Zenteno et al., 2004), indicating in all of the cases that the magmatism emplacement occurred in the maximum extensional strain zones.

In Mesa Central and in the Trans-Mexican Volcanic Belt, volcanism and faulting were contemporaneous; the mafic and felsic magmas had been emplaced in zones of maximum strain and the volcanism was emplaced in normal and lateral fault regimes. These features are consistent with the analogical and theoretical models about the upward transport of magmas (Secor and Pollard, 1975; Takada, 1994). Takada (1989, 1994), Watanabe et al. (1999), and others proposed that, in order for the volcanism to occur, it needed a magma supply and a buoyancy force, and that there should be a stress state favorable with $\sigma 3$ horizontal. The relations between volcanism and faulting could be quite varied and interdependent. In the regions studied, practically all of the Cenozoic events of deformation had taken place synchronously with the volcanism; but we have also seen that there have been periods of low contribution of magma such as the volcanic hiatus between the pulses of the Sierra Madre Occidental and the Trans-Mexican Volcanic Belt; in these cases we cannot establish the maximum age of faulting. For example, we know that the Oligocene volcanism occurred synchronous with faulting in Taxco, however it is not known if the deformation continued, since there are no younger rocks covering the faults.

AGE OF VOLCANISM COMMENCEMENT IN THE CENTRAL PART OF THE TRANS-MEXICAN VOLCANIC BELT

Hiatus between the volcanism of the Sierra Madre Occidental and the central part of the Trans-Mexican Volcanic Belt has been established in three sites. Near the San Miguel de Allende fault, in a westward direction, this hiatus comprises the period from 24 to 16 Ma. The first age corresponds to the San Nicolás Ignimbrite, which is the youngest rock recognized in the meridional part of Mesa Central (Nieto-Samaniego et al., 1996; Ojeda-García, 2004), whereas the second age corresponds to the oldest pulses of the Palo Huérfano volcano, which is one of the first stratovolcanoes in the Trans-Mexican Volcanic Belt (Pérez-Venzor et al., 1996). Similar ages (from 22 to 14.6 Ma) for this hiatus were obtained by Cerca-Martínez et al. (2000) in the Sierra de Guanajuato and between 22 and 10 Ma by Verma and Carrasco-Núñez (2003) for the Zamorano region (100°10′-20°55′).

The Teneria Formation is the youngest volcanic rock in the Taxco region, which corresponds to the central-northern part of the magmatic province of southern México; it has a K-Ar age of 31.6 Ma (Alaniz-Álvarez et al., 2002b) and is made up of domes and pyroclastic flows both of rhyolitic composition. In the region of Toluca and Malinalco, within the Trans-Mexican Volcanic Belt, the older volcanic rock is the San Antonio Andesite with an age of 21.6 Ma (García-Palomo et al., 2000). The basaltic volcanism has been reported in the meridional boundary of the Trans-Mexican Volcanic Belt with ages from 20.5 to 16.7 Ma (Ferrari et al., 2003b). Thus, in the Taxco-Tepoztlán region, the volcanic hiatus between the predominantly rhyolitic volcanism of the Sierra Madre del Sur and the basaltic volcanism of the Trans-Mexican Volcanic Belt would have been between ca. 31 and 21 Ma.

North and south of the Trans-Mexican Volcanic Belt, as well as east and west of the Taxco–San Miguel de Allende fault system, a volcanic hiatus that lasted between 6 and 12 Ma is registered in several localities. It is necessary to note that there could have been volcanic activity during the time we consider as the volcanic hiatus (Ferrari et al., 1999; García-Palomo et al., 2002; Cerca-Martínez, 2004), but its relative volumetric importance is much smaller. The previous statement seems to indicate that the geologic event that originated the Trans-Mexican Volcanic Belt was separated, in time, from the Sierra Madre Occidental and the Sierra Madre del Sur.

The older Cenozoic rocks reported from wells perforated in México City correspond to Oligocene basalts and tuffs. Using seismic reflection profiling and data from these wells, Pérez-Cruz (1988) documented that the base of the México City basin is irregular and that the Mesozoic basement goes to depths of 1600 m in the Mixhuca-1 well and more than 3200 m in the Rome-1 well (Fig. 4). Also, five ages of volcanic rocks have been reported, which range from 31.4 to 23 Ma in samples obtained in depths between 1000 and 2000 m (Marsal and Graue, 1969; Oviedo, 1970; Pérez-Cruz, 1988).

DISCUSSION

In this paper we seek to establish the conditions and the time under which the Trans-Mexican Volcanic Belt has become a tectonic barrier, by which we mean the limit that divides crustal blocks with different geologic or structural history. It is possible to establish different scales for a tectonic barrier; since a barrier represents a region that accommodates the deformation between different structural domains, a tectonic barrier can be as ample as a geologic province (Trans-Mexican Volcanic Belt), or can be formed by a systems of parallel faults (Taxco–San Miguel de Allende fault system), or be as discreet as a greater fault (e.g., El Bajío fault). Tectonic barriers have had an important influence on the stratigraphic record of the regions in which they are located.

The Trans-Mexican Volcanic Belt has been established as a volcanic arc since middle Miocene, even though its history as a tectonic barrier began long before then. Based on studies of central México, in the Cretaceous there were few differences between the northern and southern sides of the Trans-Mexican Volcanic Belt with regard to the structure and lithologies. To the W, there was the same paleogeographic environment with volcanic arcs, and the Laramide orogeny acted at the same time and with the same style and age of deformation.

The observations and data presented above do not let us establish significant differences for the Paleocene. The presence of contractive structures with vergence toward the W was documented solely in the northern-central part of the Sierra Madre del Sur. This deformation took place by the end of the Cretaceous or the beginning of the Cenozoic, and is present only in a few sites. In Tierra Caliente, Salinas-Prieto et al. (2000) documented that this deformation was of low intensity, whereas in the eastern margin of the Guerrero-Morelos Platform it was more intense (Cerca-Martínez, 2004). This variation in the vergence of the contractive structures, which has not been documented N of the Trans-Mexican Volcanic Belt, is not sufficiently important when considering that the structure acted like a tectonic border during that event of deformation.

During middle and late Eocene, remarkable differences were recorded in the deformations that occurred N and S of the Trans-Mexican Volcanic Belt. In the Sierra de Guanajuato, Aranda-Gómez and McDowell (1998) documented a post-Laramide extensional event liberated by normal faults which took place during most of the Eocene. In Taxco, S of the Trans-Mexican Volcanic Belt, the origin of the post-Laramide red beds cannot be established. During late Eocene (between 36 and 33 Ma), a phase of strike-slip faulting that generated horizontal extension toward the NNW was documented. This phase is characterized by the left-lateral displacement of coast-parallel faults and is documented in Taxco (Alaniz-Álvarez et al., 2002b), Tierra Colorada, Guerrero (Riller et al., 1992), Tilzapotla (Morán-Zenteno et al., 2004), and Huatulco, Oaxaca (Tolson, 2005); the deformation is younger toward the E. This is the first event documented in which the structural domains S and N of the Trans-Mexican Volcanic Belt are different.

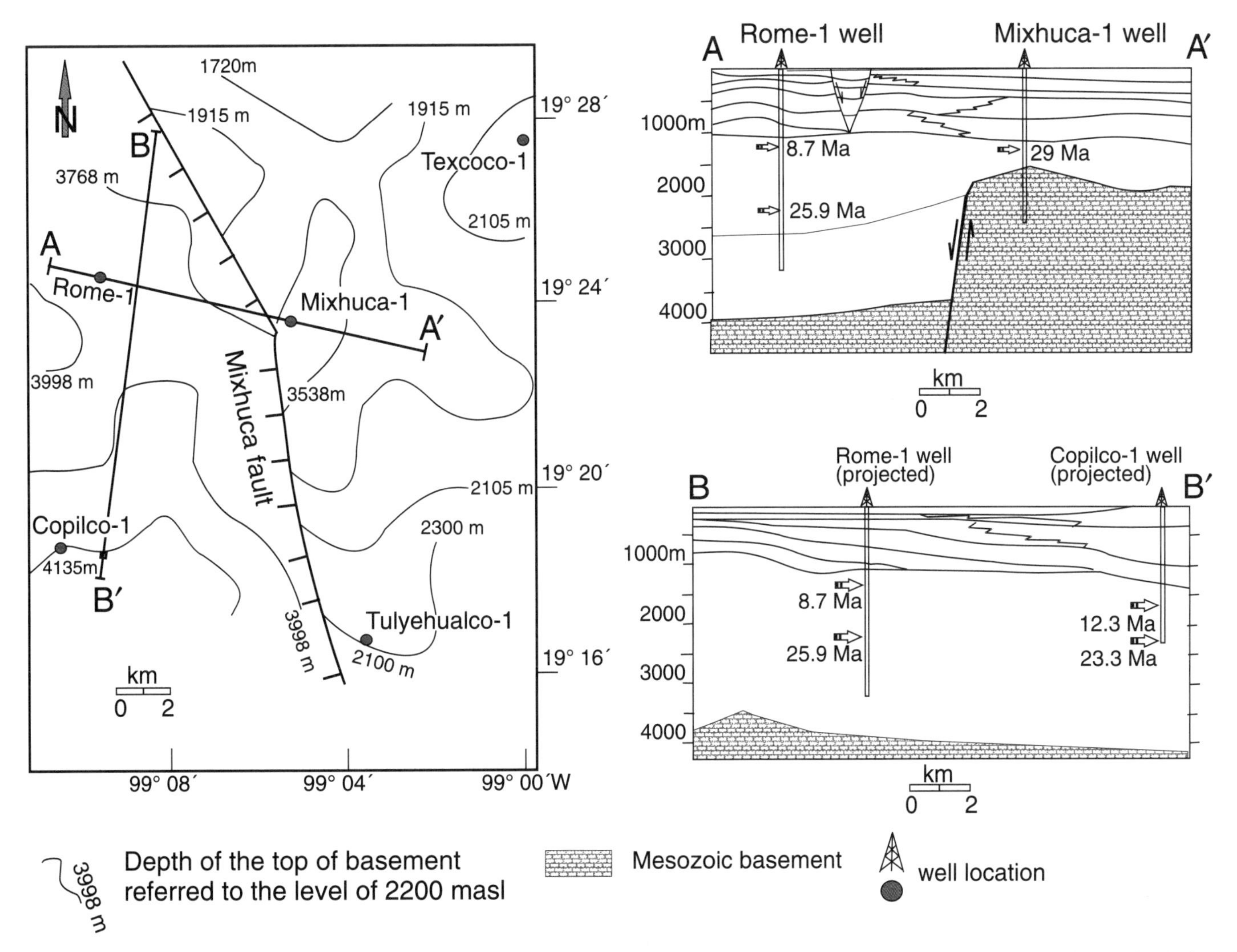

Figure 4. Geological sections and a map with lines of equal depth at the top of the Mesozoic basement in the México City basin, obtained with data from wells and seismic lines of reflection. Modified from Pérez-Cruz (1988).

During the Oligocene the difference between the deformation style occurring in Mesa Central and in the Sierra Madre del Sur continued; whereas there was extensional deformation N of the Trans-Mexican Volcanic Belt, transcurrent deformation occurred in the S. In Mesa Central, Nieto-Samaniego et al. (1999) documented three-dimensional deformation: The maximum extension direction was ~20% toward the ENE-WSW, there was 11% horizontal extension in the NNW-SSE direction, and shortening occurred vertically. Nieto-Samaniego et al. (1999) propose that the deformation, which migrated to the W and SE, was controlled by changes in the speed of accretion in the Pacific plate and its angle of subduction below the North American plate.

On the other hand, in the Taxco region it was documented that the Oligocene transcurrent deformation generated maximum horizontal extension toward the NE-SE and shortening toward the NW-SE. This event migrated toward the E and was registered mainly by normal faulting in the graben formed by the Oaxaca fault whose age of activity varies from before the Oligocene in Tehuacán to early Miocene in the region of Etla, near the city of Oaxaca (Nieto-Samaniego et al., 2006).

Since middle Miocene, the extensional deformation throughout the central portion of the Trans-Mexican Volcanic Belt has been syntectonic with a great magmatic pulse. The deformation has been liberated mainly by normal faults trending ENE-WSW in the Chapala-Tula system and NE-SE faults in the region of Apan, Hidalgo, although normal faults trending NNW-SSE belonging to the Taxco–San Miguel de Allende fault system have also been activated.

From analysis of these events it is inferred that, from the late Eocene, a crustal discontinuity was generated in the site occupied by the central section of the Trans-Mexican Volcanic Belt (Fig. 3). This idea arises when considering that during the transcurrent deformation that occurred in the Sierra Madre del Sur during the late Eocene and Oligocene-Miocene, in the Valles–San

Luis Potosí Platform there was no important deformation; it is also noteworthy that the migration of the deformations was in opposite directions, N of the Trans-Mexican Volcanic Belt it migrated toward the W and S of the belt it migrated toward the E.

It is necessary to consider that between the regions located N and S of the belt a zone had to exist to accommodate the incompatibility in deformation. The evidence available, which was previously mentioned, indicates that in the México City basin, the zone covered by volcanic rocks of Miocene-Holocene age should hide the faults formed during the Eocene-Oligocene. It seems reasonable to infer the presence of an adjustment zone from the same time in the western part of the Trans-Mexican Volcanic Belt, where Rosas-Elguera et al. (1996) and Ferrari and Rosas-Elguera (2000) considered that there is a crustal boundary between the Jalisco block and the Sierra Madre Occidental, which was individualized in the Eocene.

Deformation in the Initial Stages of the Trans-Mexican Volcanic Belt

The type of the deformation that occurred in the initial stages of the formation of the Trans-Mexican Volcanic Belt is difficult to establish, because most of the structures produced at that time are covered by a volcanic layer up to 4 km deep. Fortunately, there is information about the region below the surface from deep wells drilled by PEMEX (Petróleos Mexicanos, a Mexican petroleum company) in the México City basin (Marsal and Graue, 1969; Oviedo, 1970; Pérez-Cruz, 1988). The fact that the México City basin reaches more than 4000 m deep, and that it is affected by faults, indicates that its formation was related to one or several deformation events.

The unevenness between the level of the outcrop of marine Cretaceous calcareous rocks in Cuernavaca and its presence in the subsoil of the México City basin is the departure point for reconstructing the evolution of this basin. Within the México City basin, under the surface, the configuration of the contact between Mesozoic rocks and the continental basin-fill strata indicates the existence of a normal fault trending NNW-SSE, dipping W, known as the Mixhuca fault. This fault displays a throw of between 1600 and 2000 m, which is measured by the thickness of the basin-fill continental strata, and is buried by rocks of Miocene age. The structure was interpreted based on the seismic reflection profiles and lithologies nucleus recorded from the Rome-1, Mixhuca-1, Copilco-1, and Tulyehualco-1 wells (Pérez-Cruz, 1988), as well as the Texcoco well (Oviedo, 1970) (Fig. 4).

The lithological record from the wells shows that the oldest rocks in the fill of the México City basin are previous to late Oligocene. In the Texcoco well, the stratigraphic column reported by Oviedo (1970) shows a volcaniclastic sequence that goes from 480 to 1965 m deep; toward 1500 m he reports rocks with ages of 30 and 31.4 Ma, which were obtained by Marsal and Graue (1969). Underneath that sequence, ~100 m of continental anhydrite appear, which overlay the Texcoco conglomerate, a continental conglomerate whose age is not known, but which contains limestone clasts of the Upper Cretaceous. The thickness of the conglomerate was obtained considering the depth of the basement reported by Pérez-Cruz (1988); in the zone of the Texcoco it is a little greater than 2105 m, which would indicate a thickness of the conglomerate near 50 m. In the Rome-1 well the oldest age is 25.9 Ma and it is located at 2200 m deep; underneath there are ~1700 m of fill-basin rocks, measured from this site to the top of Mesozoic rocks. In the Copilco-1 well, the oldest age obtained is 23.3 Ma and it is located to 2250 m of depth; underneath is ~1900 m of basin-filling strata. In the Mixhuca well, the oldest volcanic rocks are 29 Ma and it is located to 1200 m of depth; underneath is ~220 m of basin-fill strata above Mesozoic rocks. The differences in thickness of the basin-fill strata of the México City basin between the wells in the foot-wall block of the Mixhuca fault (Texcoco, Mixhuca-1, and Tulyehualco-1) and those wells located in the hanging-wall block (Rome-1 and Copilco) show that the hanging-wall block has a thicker filling of around 2 km. Additionally the well data indicate that the activity of this structure was previous to early Miocene since strata of that age buried the fault.

The age of the deepest continental strata is not known; nevertheless as an approach (corrections by compaction are omitted) we can consider that in the Rome-1 well the upper 2200 m of basin-fill strata were deposited at a rate near 80 m per m.y. and in the Copilco-1 well a rate of 98 m per m.y. If we considered similar rates of sedimentation for the deepest filling (below 2200 m and as deep as 4100 m), we could suppose that its deposit includes a lapse between 15 and 20 Ma. We suppose that that deposit had to delay at least 10 Ma, which would indicate that the age at which basin filling began is late Eocene. This estimation is consistent with the ages obtained in the Texcoco and Mixhuca-1 wells, where early Oligocene rocks were reported above 200 m of basin-fill continental strata. Additionally, it should be considered that the stratigraphic record of the Texcoco well shows anhydrite and conglomerate in the lowest 150 m, whose rates of sedimentation were probably minor to those of the volcaniclastic filling.

Another important observation is that in the hanging wall as in the foot wall of the Mixhuca fault, the configuration of the top of Mesozoic rocks in the México City basin shows a slope with southerly inclination (Fig. 4) (Pérez-Cruz, 1988, Fig. 15). The top of Mesozoic rocks in the Rome-1 and Copilco-1 wells in the hanging wall of the Mixhuca fault shows an unevenness of 210 m dipping toward the S, whereas the top of Mesozoic rocks in the Mixhuca-1 and Tulyehualco-1 wells in the foot wall of the Mixhuca fault shows an unevenness of 525 m dipping toward the S (Fig. 4).

The distribution of Mesozoic rocks and the unevenness between their outcrops to the S of the Trans-Mexican Volcanic Belt in Cuernavaca, and beneath the surface of the México City basin, indicate the presence of faults trending E-W located toward the southern limit of this basin. These faults have been proposed by some authors, designating them as part of the La Pera fault system and locating them under the sierra de Chichinautzin (Ferrari et al., 2002; Siebe et al., 2004; Delgado-Granados et al.,

1995). The seismic-structural configuration reported by Pérez-Cruz (1988, Fig. 16) shows a limit between the "lower volcanic" and "upper volcanic" units, whose age was located at the end of middle Miocene according to the ages reported in that work. A set of fault scarps trending E-W can be observed in this limit and on the basin; the Miocene sequence has a general slope toward the S. These data are compatible with the existence of a major fault located S of the basin, whose activity was posterior to the displacement of 1600 m of the Mixhuca fault, since it tilts similarly to the hanging wall and the foot wall of that structure.

From the above data, we inferred the following sequence of events in the México City basin region: During the early Paleocene (perhaps including the upper part of the Maastrichtian in accordance with the ages of the Mexcala Formation in Texmalac) the Laramide orogeny folded Mesozoic marine sediments. Later the deposit of continental molasses, represented by the Texcoco conglomerate with limestone clasts but without volcanic content, which were observed in the deep wells drilled in the México City basin, took place; this followed a stage of minimum tectonic and volcanic activities during which the anhydrites were deposited. The abrupt beginning of volcanism and the tectonic activity in the Mixhuca fault occurred before early Oligocene time, and led to an accumulation of more than 2000 m of volcaniclastic material in the hanging-wall block of the Mixhuca fault. The volcanic accumulation continued until the Recent and the tectonic activity was focused on the La Pera fault system since the Oligocene. In concordance with previous observations, two main events of deformation took place: one during the late Eocene during which deformation was concentrated mainly along the Mixhuca fault with horizontal extension oriented E-W or WSW-ENE, and the other during Oligocene-Miocene for which deformation was liberated mainly in the La Pera fault system, with extension N-S or NNW.

If we compare the description of events of deformation in the México City basin with those documented N and S of the Trans-Mexican Volcanic Belt (Fig. 3), we can emphasize that for the Eocene-Miocene, the deformation was insignificant in the Sierra Madre Oriental (in any case the little deformation present would be with normal faulting and ~E-W maximum extension direction); whereas in the Sierra Madre del Sur there were two phases of deformation, both with transcurrent faults; one of Eocene–early Oligocene age with a maximum horizontal extension direction ~NW and shortening to ~NE-SW, and the second one of Oligocene-Miocene age with extension direction ~NE-SW and shortening to ~NW-SE.

Considering all of the regions studied, it is clear that the deformation should be heterogeneous (Fig. 3) and that the resulting maximum extension in the shear zone located between both structural domains in the La Pera fault region will depend on the magnitudes of the imposed vorticity and the extension in the region located S of the main shear zone. For the Oligocene–early Miocene, the direction of maximum extension predicted within the Trans-Mexican Volcanic Belt is located in the NW-SE direction (Fig. 3C); to a smaller amount of rotation, the direction of maximum extension will come near the N-S direction, which would produce vertical displacement in the La Pera fault system (Fig. 3C).

In the case of the deformations that occurred during late Eocene, when the normal displacement in the Mixhuca fault took place, it is inferred that throughout the La Pera fault zone the sliding had a right-lateral component due to the E-W extension documented within the basin, and the magnitudes of the displacement will depend mainly on the magnitude of that extension.

From the information that we have at the moment, and although this is still documented in only a few places, the Trans-Mexican Volcanic Belt is viewed as a great accommodation zone between incompatible deformations in the crustal blocks located N and S of this volcanic arc.

CONCLUSIONS

Study of the deformations which occurred in Mesa Central, Sierra Madre Oriental, Sierra Madre Occidental, the Trans-Mexican Volcanic Belt, and Sierra Madre del Sur, in central México, allow us to conclude that:

- The Taxco–San Miguel de Allende fault system is located at the site of a transition between a zone of continental platforms and one of greater bathymetry. This paleogeographic discontinuity, which is of Cretaceous age, controlled the location of Cenozoic faulting, trending N-S and NNW-SSE, which crosses the Trans-Mexican Volcanic Belt.
- Since the Eocene, the extensional deformation of Mesa Central and the transcurrent deformation of the Sierra Madre del Sur indicate that a zone of detachment between these provinces existed in the site occupied at the moment by the Trans-Mexican Volcanic Belt.
- In the Oligocene, the activity of the lateral faults in southern México implied shortening in a NW-SE direction and extension to the NE-SW; in Mesa Central three-dimensional deformation occurred with extension along two directions: ~E-W and ~N-S. In the Sierra Madre Oriental, north of the Trans-Mexican Volcanic Belt, significant Cenozoic deformation did not take place.
- From Miocene to Recent, deformation has been concentrated in the Trans-Mexican Volcanic Belt.
- The correlation of the data collected in deep wells and the studies of seismic reflection, reported by Pérez-Cruz (1988) for the México City basin, as well as the analysis of deformation events in this work, indicate that from the late Eocene, the deformation which occurred in the meridional part of the Trans-Mexican Volcanic Belt was an effect of the deformations in the blocks that are located N and the S of this volcanic arc.
- In the México City basin during the late Eocene, most of the activity was located on the Mixhuca fault, liberating E-W or WSW-ENE horizontal extension. During the Oligocene-Miocene the deformation was mainly liberated along the La Pera fault system with extension N-S or NNW-SSE.

ACKNOWLEDGMENTS

The authors want to thank Dante J. Morán Zenteno, Alvaro Márquez, and Luca Ferrari, whose revisions improved the manuscript. The financial support for this investigation was provided by the Consejo Nacional de Ciencia y Tecnología with project CONACYT 41044-F and by the Universidad Nacional Autónoma de México with project PAPIIT-IN102602.

REFERENCES CITED

Aguilera-Franco, N., 2003, Cenomanian–Coniacian zonation (foraminifers and calcareous algae) in the Guerrero–Morelos basin, southern México: Revista Mexicana de Ciencias Geológicas, v. 20, p. 202–222.

Aguirre-Díaz, G.J., Nelson, S.A., Ferrari, L., and López-Martínez, M., 1997, Ignimbrites of the central Mexican Volcanic Belt, Amealco and Huichapan Calderas (Querétaro-Hidalgo), *in* International Association of Volcanology and Chemistry of the Earth Interior General Assembly 1997, Field Trip #1 Guidebook: Puerto Vallarta, México, 54 p.

Alaniz-Álvarez, S.A., Nieto-Samaniego, Á.F., and Ferrari, L., 1998, Effect of the strain rate in the distribution of monogenetic and polygenetic volcanism in the Transmexican Volcanic Belt: Geology, v. 26, p. 591–594, doi: 10.1130/0091-7613(1998)026<0591:EOSRIT>2.3.CO;2.

Alaniz-Álvarez, S.A., Nieto-Samaniego, A.F., Reyes-Zaragoza, M.A., Orozco-Esquivel, M.T., Ojeda-García, A.C., and Vassallo-Morales, L.F., 2001, Estratigrafía y deformación de la región San Miguel de Allende-Querétaro: Revista Mexicana de Ciencias Geológicas, v. 18, p. 129–148.

Alaniz-Álvarez, S.A., Nieto-Samaniego, A.F., Orozco-Esquivel, M.T., Vassallo-Morales, L.F., and Xu, S.S., 2002a, El Sistema de Fallas Taxco-San Miguel de Allende: Implicaciones en la deformación post-Eocénica del centro de México: Boletín de la Sociedad Geológica Mexicana, v. 55, p. 12–29.

Alaniz-Álvarez, S.A., Nieto-Samaniego, A., Morán-Zenteno, D.J., and Alba-Aldave, L., 2002b, Rhyolitic volcanism in extension zone associated with strike-slip tectonics in the Taxco region, Southern México: Journal of Volcanology and Geothermal Research, v. 118, p. 1–14, doi: 10.1016/S0377-0273(02)00247-0.

Alencaster, G., 1980, Moluscos del Maastrichtiano de Texmalac, Guerrero: Sociedad Geológica Mexicana, Libro guía de la excursión geológica a la cuenca del alto Río Balsas, Guidebook, p. 39–42.

Anderson, T.H., and Silver, L.T., 1974, Late Cretaceous plutonism in Sonora, Mexico, and its relationship to circum-Pacific magmatism: Geological Society of America Rocky Mountain Section Annual Meeting Program [abs.], p. 484.

Aranda-Gómez, J.J., and McDowell, F.W., 1998, Paleogene extension in the southern Basin and Range Province of Mexico: Syn-depositional tilting of Eocene red beds and Oligocene volcanic rocks in the Guanajuato mining district: International Geology Review, v. 40, p. 116–134.

Aranda-Gómez, J.J., Henry, C.D., and Luhr, J.F., 2000, Evolución tectonomagmática post-paleocénica de la Sierra Madre Occidental y de la porción meridional de la provincia tectónica de Cuencas y Sierras, México: Boletín de la Sociedad Geológica Mexicana, v. 53, p. 59–71.

Cabral-Cano, E., Draper, G., Lang, H.R., and Harrison, C.G.A., 2000, Constraining the late Mesozoic and Early Tertiary tectonic evolution of southern Mexico: Structure and deformation history of the Tierra Caliente region, southern Mexico: Journal of Geology, v. 108, p. 427–446, doi: 10.1086/314414.

Campa, M.F., and Coney, P.J., 1983, Tectono-stratigraphic terranes and mineral resource distributions of México: Canadian Journal of Earth Sciences, v. 20, p. 1040–1051.

Carrillo Bravo, J., 1971, La Plataforma de Valles-San Luis Potosí: Boletín de la Asociación Mexicana de Geólogos Petroleros, v. 23, p. 1–110.

Centeno-García, E., and Silva-Romo, G., 1997, Petrogenesis and tectonic evolution of central Mexico during Triassic-Jurassic time: Revista Mexicana de Ciencias Geológicas, v. 14, p. 244–260.

Cerca, M., Ferrari, L., Bonini, M., Corti, G., and Manetti, P., 2004, The role of crustal heterogeneity in controlling vertical coupling during Laramide shortening and the development of the Caribbean–North America transform boundary in southern Mexico: Insights from analogue models, *in* Grocott, J., Taylor G., and Tikoff, B., eds., Vertical Coupling and Decoupling in the Lithosphere: Geological Society [London] Special Publication 227, p. 117–140.

Cerca-Martínez, M., 2004, Deformación y magmatismo Cretácico Tardío-Terciario temprano en la zona de la Plataforma Guerrero-Morelos [Ph.D. thesis]: México, D.F., Universidad Nacional Autónoma de México, 175 p.

Cerca-Martínez, L.M., Aguirre-Díaz, G., and López-Martínez, M., 2000, The geologic evolution of the southern sierra de Guanajuato, México: A documented example of the transition from the Sierra Madre Occidental to the Mexican Volcanic Belt: International Geology Review, v. 42, p. 131–151.

Crisp, J.A., 1984, Rates of magma emplacement and volcanic output: Journal of Volcanology and Geothermal Research, v. 20, p. 177–211, doi: 10.1016/0377-0273(84)90039-8.

Dávalos-Álvarez, O.G., Nieto-Samaniego, A.F., Alaniz-Álvarez, S.A., and Gómez-González, J.M., 2005, Las fases de deformación cenozoica en la region de Huimilpan, Queretaro, y su relacion con la sismicidad local: Revista Mexicana de Ciencias Geológicas, v. 22–2, p. 129–147.

De Cserna, Z., Ortega-Gutiérrez, F., and Palacios-Nieto, M., 1980, Reconocimiento geológico de la parte central de la cuenca del alto Río Balsas, Estados de Guerrero y Puebla, *in* Libro guía de la excursión geológica a la cuenca del alto Río Balsas: México, D.F., Sociedad Geológica Mexicana, Guidebook, p. 1–33.

Delgado-Granados, H., Nieto-Obregón, J., Silva-Romo, G., Mendoza-Rosales, C.C., Arellano-Gil, J., Lermo-Samaniego, J.F., and Rodríguez-González, M., 1995, La Pera detachment fault system: Active faulting south of México City (II): Geological evidence: Geos, v. 15, p. 64.

Demant, A., 1978, Características del Eje Neovolcánico Transmexicano y sus problemas de interpretación: Universidad Nacional Autónoma de México, Instituto de Geología, Revista, v. 2, p. 172–187.

Dubois, A., Odonneb, F., Massonnatc, G., Lebourgd, T., and Fabred, R., 2002, Analogue modelling of fault reactivation: Tectonic inversion and oblique remobilisation of grabens: Journal of Structural Geology, v. 24, p. 1741–1752, doi: 10.1016/S0191-8141(01)00129-8.

Edwards, J.D., 1955, Studies of some Early Tertiary red conglomerates of Central Mexico: U.S. Geological Survey Professional Paper 264-H, 179 p.

Eguiluz de Antuñano, S., Aranda-García, M., and Marrett, R., 2000, Tectónica de la Sierra Madre Oriental: Boletín de la Sociedad Geológica Mexicana, v. 53, p. 1–26.

Ferrari, L., and Rosas-Elguera, J., 2000, Late Miocene to Quaternary extension at the northern boundary of the Jalisco block, western Mexico: The Tepic-Zacoalco rift revised, *in* Delgado-Granados, H., Aguirre-Días, G.J., and Stock, J.M., eds., Cenozoic Tectonics and Volcanism of Mexico: Geological Society of America Special Paper 334, p. 42–64.

Ferrari, L., Lopez-Martinez, M., Aguirre-Diaz, G., and Carrasco-Nuñez, G., 1999, Space-time patterns of Cenozoic arc volcanism in central Mexico: From the Sierra Madre Occidental to the Mexican Volcanic Belt: Geology, v. 27, p. 303–306, doi: 10.1130/0091-7613(1999)027<0303:STPOCA>2.3.CO;2.

Ferrari, L., Mena, M., López-Martínez, M., Jacobo-Albarrán, J., Silva-Romo, G., Mendoza-Rosales, C.C., and González-Cervantes, N., 2002, Estratigrafía y Tectónica de la cuenca de la Ciudad de México y áreas colindantes: Geos, v. 22, p. 150.

Ferrari, L., Petrone, C.M., Francalanci, L., Tagami, T., Eguchi, M., Conticelli, S., Manetti, P., and Venegas-Salgado, S., 2003a, Geology of the San Pedro-Ceboruco graben, western Trans-Mexican Volcanic Belt: Revista Mexicana de Ciencias Geológicas, v. 20, p. 165–181.

Ferrari, L., López-Martínez, M., González-Cervantes, N., Jacobo-Albarrán, J., and Hernández-Bernal, M.S., 2003b, Volcanic record and age of formation of the Mexico City basin: Geos, v. 23, p. 120.

Flores, T., 1920, Estudio Geológico Minero de los distritos de El Oro y Tlalpuhahua, Boletín del Instituto de Geología, v. 37, 40 p.

Freydier, C., Lapierre, H., Ruíz, J., Tardy, M., Martínez, R.J., and Coulon, C., 2000, The Early Cretaceous Arperos basin: An oceanic domain dividing the Guerrero arc from nuclear Mexico evidenced by geochemistry of the lavas and sediments: Journal of South American Earth Sciences, v. 13, p. 325–336, doi: 10.1016/S0895-9811(00)00027-4.

García-Palomo, A., Macías, J.L., and Garduño, V.H., 2000, Miocene to Recent structural evolution of the Nevado de Toluca volcano region, central México: Tectonophysics, v. 318, p. 281–302, doi: 10.1016/S0040-1951(99)00316-9.

García-Palomo, A., Macías, J.L., Tolson, J., Valdéz, G., and Mora, J.C., 2002, Volcanic stratigraphy and geological evolution of the Apan region, east-central sector of the Trans-Mexican Volcanic Belt: Geofísica Internacional, v. 41, p. 133–150.

González-Cervantes, N., 2004, Tectónica extensional y volcanismo neogénico en la región de Malinalco y la cuenca de la Cd. de México [Bacherol thesis]: Tampico, Tamaulipas, Instituto Tecnológico de Ciudad Madero, 108 p.

Gross, W.H., 1975, New ore discovery and source of silver-gold veins, Guanajuato, Mexico: Economic Geology and the Bulletin of the Society of Economic Geologists, v. 70, p. 1175–1189.

Hernández-Jáuregui, R., Valencia-Islas, J.J., and González-Casildo, V., 2000, Facies turbidíticas relacionadas a movimientos orogénicos en el centro oriente del Estado de Querétaro, México: Geos, v. 20, p. 156.

Hernández-Laloth, N., 1991, Modelo conceptual de funcionamiento hidrodinámico del sistema acuífero del valle de León, Guanajuato [Bacherol thesis]: México, D.F., Universidad Nacional Autónoma de México, Facultad de Ingeniería, 129 p.

Hernández-Romano, U., Aguilera-Franco, N., Martínez-Medrano, M., and Barceló-Duarte, J., 1997, Guerrero-Morelos Platform drowning at the Cenomanian–Turonian boundary, Huitziltepec area, Guerrero State, southern Mexico: Cretaceous Research, v. 18, p. 661–686, doi: 10.1006/cres.1997.0078.

Imlay, R.W., 1938, Studies of the Mexican Geosyncline: Bulletin of the Geological Society of America, v. 49, p. 1661–1694.

Márquez, A., Verma, S.P., Anguita, F., Oyarzun, R., and Brandle, J.L., 1999, Tectonics and volcanism of Sierra Chichinautzin: Extension at the front of the Central Trans-Mexican Volcanic belt: Journal of Volcanology and Geothermal Research, v. 93, p. 125–150, doi: 10.1016/S0377-0273(99)00085-2.

Marsal, R.J., and Graue, R., 1969, El subsuelo del lago de Texcoco, *in* Carrillo Nabor, El hundimiento de la Ciudad de México y Proyecto Texcoco: México, D.F., Secretaría de Hacienda y Crédito Público, p. 167–202.

Martínez-Reyes, J., 1992, Mapa geológico de la sierra de Guanajuato, con Resumen de la geología de la sierra de Guanajuato: México, D.F., Instituto de Geología, Universidad Nacional Autónoma de México, scale 1:100 000, 1 sheet.

Martínez-Reyes, J., and Nieto-Samaniego, A.F., 1990, Efectos geológicos de la tectónica reciente en la parte central de México: Universidad Nacional Autónoma de México, Instituto de Geología, Revista, v. 9, p. 33–50.

Mooser, F., 1972, The Mexican Volcanic Belt: Structure and tectonics: Geofísica Interacional, v. 12, p. 55–70.

Morán-Zenteno, D.J., Urrutia-Fucugauchi, J., Böhnel, H., and González-Torres, E., 1988, Paleomagnetismo de rocas jurásicas del norte de Oaxaca y sus implicaciones tectónicas: Geofísica Internacional, v. 27, p. 485–518.

Morán-Zenteno, D.J., Martiny, B., Tolson, G., Solís-Pichardo, G., Alba-Aldave, L., Hernández-Bernal, M.S., Macías-Romo, C., Martínez-Serrano, R.G., Schaaf, P., and Silva Romo, G., 2000, Geocronología y características geoquímicas de las rocas magmáticas terciarias de la Sierra Madre del Sur: Boletín de la Sociedad Geológica Mexicana, v. 53, p. 27–58.

Morán-Zenteno, D.J., Alba-Aldave, L.A., Solé, J., and Iriondo, A., 2004, A major resurgent caldera in southern Mexico: The source of the late Eocene Tilzapotla ignimbrite: Journal of Volcanology and Geothermal Research, v. 136, p. 97–119, doi: 10.1016/j.jvolgeores.2004.04.002.

Moreira-Rivera, F., Martínez-Rodríguez, L., Palacios-García, R., Maldonado-Lee, J.M., Olvera-Campos, A., Mata-Pérez, F., Pérez-Benavides, A., and González-Monsivais, P., 1996, Carta geológico-minera Matehuala F14–1: Pachuca, Hidalgo, México, Consejo de Recursos Minerales, scale 1:250 000, 1 sheet.

Nieto-Samaniego, A.F., and Alaniz-Álvarez, S.A., 1994, La Falla de San Miguel de Allende: Características y evidencias de su actividad cenozoica: Tercera Reunión Nacional de Geomorfología, Sociedad Mexicana de Geomorfología, Guadalajara, Jal., México [abstract], p. 139–142.

Nieto-Samaniego, A.F., Macías-Romo, C., and Alaniz-Álvarez, S.A., 1996, Nuevas edades isotópicas de la cubierta volcánica cenozoica de la parte meridional de la Mesa Central, México: Revista Mexicana de Ciencias Geológicas, v. 13, p. 117–122.

Nieto-Samaniego, A.F., Ferrari, L., Alaniz-Álvarez, S.A., Labarthe-Hernández, G., and Rosas-Elguera, J., 1999, Variation of Cenozoic extension and volcanism across the southern Sierra Madre Occidental Volcanic Province, México: Geological Society of America Bulletin, v. 111, p. 347–363, doi: 10.1130/0016-7606(1999)111<0347:VOCEAV>2.3.CO;2.

Nieto-Samaniego, A.F., Alaniz-Álvarez, S.A., Silva-Romo, G., Eguiza-Castro, M.H., and Mendoza-Rosales, C.C., 2006, Latest Cretaceous to Miocene deformation events in the eastern Sierra Madre del Sur, Mexico, inferred from the geometry and age of major structures: Geological Society of America Bulletin, v. 118, p. 238–252, doi: 10.1130/B25730.1.

Ojeda-García, A.C., 2004, Cartografía geológica 1:50,000 del límite El Bajío-Mesa Central, porción Salamanca-Irapuato, Estado de Guanajuato, México [B.A. thesis]: Hermosillo, Sonora, Centro de Estudios Superiores del Estado de Sonora, 86 p.

Orozco-Esquivel, M.T., Nieto-Samaniego, A.F., and Alaniz-Álvarez, S.A., 2002, Origin of rhyolitic lavas in the Mesa Central, Mexico, by crustal melting related to extension: Journal of Volcanology and Geothermal Research, v. 118, p. 37–56, doi: 10.1016/S0377-0273(02)00249-4.

Ortega-Gutiérrez, F., 1980, Rocas volcánicas del Maestrichtiano en el área de San Juan Tetelcingo, Estado de Guerrero, in Libro-Guía, Excursión Geol. III: Reunión Nacional Geotectonia-Geotermia, México, p. 34–38.

Ortega-Gutiérrez, F., 1981, Metamorphic belts of southern Mexico and their tectonic significance: Geofísica Internacional, v. 20, p. 177–202.

Oviedo de León, A., 1970, El Conglomerado Texcoco y el posible origen de la Cuenca de México: Revista del Instituto Mexicano del Petróleo, v. 2, p. 5–20.

Parsons, T., and Thompson, G.A., 1991, The role of magma overpressure in suppressing earthquakes and topography: Worldwide examples: Science, v. 253, p. 1399–1402.

Pasquaré, G., Ferrari, L., Garduño, V.H., Tibaldi, A., and Vezzoli, L., 1991, Geologic map of the central sector of the Mexican Volcanic belt, States of Guanajuato and Michoacán: Geological Society of America Map and Chart series MCH072, 1 sheet, 20 p. text.

Pérez-Cruz, G.A., 1988, Estudio sismológico de reflexión del Subsuelo de la Ciudad de México [master's thesis]: México, D.F., Universidad Nacional Autónoma de México, 83 p.

Pérez-Venzor, J.A., Aranda-Gómez, J.J., McDowell, F., and Solorio-Munguía, J.G., 1996, Geología del volcán Palo Huérfano, Guanajuato, México: Revista Mexicana de Ciencias Geológicas, v. 13, p. 174–183.

Pindell, J.L., 1985, Alleghenian reconstruction and subsequent evolution of the Gulf of Mexico, Bahamas, and proto-Caribbean: Tectonics, v. 4, p. 1–39.

Quintero-Legorreta, O., 1992, Geología de la región de Comanja, estados de Guanajuato y Jalisco: Universidad Nacional Autónoma de México, Instituto de Geología, Revista, v. 10, p. 6–25.

Raisz, E., 1959, Landforms of Mexico: Cambridge, Massachusetts, Geography branch of the Office of Naval Research, scale 1:3 000 000, 1 map with text.

Riller, U., Ratschbacher, L., and Frisch, W., 1992, The Tierra Colorada deformation zone: Left-lateral transtension along the northern margin of the Xolapa complex, southern Mexico: Journal of South American Earth Sciences, v. 5, p. 237–249, doi: 10.1016/0895-9811(92)90023-R.

Rosas-Elguera, J., Ferrari, L., Garduno-Monroy, V.H., and Urrutia-Fucugauchi, J., 1996, Continental boundaries of the Jalisco Block and their influence in the Pliocene-Quaternary kinematics of western Mexico: Geology, v. 24, p. 921–924, doi: 10.1130/0091-7613(1996)024<0921:CBOTJB>2.3.CO;2.

Ross, M.I., and Scotese, C.R., 1988, A hierarchical tectonic model of the Gulf of Mexico and Caribbean region: Tectonophysics, v. 155, p. 139–168, doi: 10.1016/0040-1951(88)90263-6.

Salinas-Prieto, J.C., Monod, O., and Faure, M., 2000, Ductile deformations of opposite vergence in the eastern part of the Guerrero Terrane (SW Mexico): Journal of South American Earth Sciences, v. 13, p. 389–402, doi: 10.1016/S0895-9811(00)00031-6.

Schaaf, P., Morán-Zenteno, D.J., Henández-Bernal, M.S., Solís-Pichardo, G., Tolson-Jones, G., and Köhler, H., 1995, Paleogene continental margin truncation in southwestern Mexico: Geochronological evidence: Tectonics, v. 14, p. 1339–1350, doi: 10.1029/95TC01928.

Secor, D.T., and Pollard, D.D., 1975, On the stability of open hydraulic fractures in the Earth's crust: Geophysical Research Letters, v. 2, p. 510–513.

Siebe, C., Rodriguez-Lara, V., Schaaf, P., and Abrams, M., 2004, Geochemistry, Sr-Nd isotope composition, and tectonic setting of Holocene Pelado, Guespalapa and Chichinautzin scoria cones, south of Mexico City: Journal of Volcanology and Geothermal Research, v. 130, p. 197–226, doi: 10.1016/S0377-0273(03)00289-0.

Silva-Romo, G., Mendoza-Rosales, C.C., Nieto-Samaniego, A.F., and Alaniz-Alavarez, S.A., 2002, La paleocuenca de Aztlán, antecesora de la Cuenca de México: Geos, v. 22, p. 149–150.

Suter, M., Quintero, O., and Johnson, C.A., 1992, Active faults and state of stress in the central part of the Trans-Mexican volcanic belt, Mexico, 1, The Venta de Bravo fault: Journal of Geophysical Research, v. 97, p. 11983–11993.

Suter, M., Quintero, O., López, M., Aguirre, G., and Farrar, E., 1995, The Acambay graben: Active intraarc extension in the trans-Mexican volcanic belt, Mexico: Tectonics, v. 14, p. 1245–1262, doi: 10.1029/95TC01930.

Suter, M., López-Martínez, M., Quintero Legorreta, O., and Carrillo-Martínez, M., 2001, Quaternary intra-arc extension in the central Trans-Mexican Volcanic Belt: Geological Society of America Bulletin, v. 113, p. 693–703, doi: 10.1130/0016-7606(2001)113<0693:QIAEIT>2.0.CO;2.

Takada, A., 1989, Magma transport and reservoir formation by a system of propagating cracks: Bulletin of Volcanology, v. 52, p. 118–126, doi: 10.1007/BF00301551.

Takada, A., 1994, The influence of regional stress and magmatic input on styles of monogenetic and polygenetic volcanism: Journal of Geophysical Research, v. 99, p. 13563–13573, doi: 10.1029/94JB00494.

Tolson, G., 2005, La falla de Caltepec: Raíces expuestas de una frontera tectónica de larga vida entre dos terrenos continentales del sur de México: Boletín de la Sociedad Geológica Mexicana, v. 57, p. 83–109.

Tristán-González, M., 1986, Estratigrafía y tectónica del Graben de Villa de Reyes en los Estados de San Luis Potosí y Guanajuato, México: San Luis Potosí, Universidad Autónoma de San Luis Potosí, Instituto de Geología, Folleto técnico 107, 91 p.

Urrutia-Fucugauchi, J., and Ferrusquía-Villafranca, I., 2001, Paleomagnetic results for the Middle-Miocene continental Suchilquitongo Formation, Valley of Oaxaca, southeastern Mexico: Geofísica Internacional, v. 40, p. 191–205.

Urrutia-Fucugauchi, J., and Flores-Ruiz, J.H., 1996, Bouguer gravity anomalies and regional crustal structure in central Mexico: International Geology Review, v. 38, p. 176–194.

Valdez-Moreno, G., Aguirre-Díaz, G.J., and López-Martínez, M., 1998, El volcán La Joya, estados de Querétaro y Guanajuato—Un estratovolcán miocénico del Cinturón Volcánico Mexicano: Revista Mexicana de Ciencias Geológicas, v. 15, p. 181–197.

Vassallo, L.F., Sousa, J.E., and Olalde, G., 2004, Time magmatism and mineralization at the central part of Mexico: 49th Annual Meeting of the Geological Association of Canada, St. Catharines, Canada, Geological Association of Canada [abs.], p. 385.

Venegas-Salgado, S., Herrera-Franco, J., and Maciel-Flores, R., 1985, Algunas características de la Faja Volcánica Mexicana y sus recursos geotérmicos: Geofísica Internacional, v. 24, p. 47–81.

Verma, S.P., and Carrasco-Núñez, G., 2003, Reappraisal of the geology and geochemistry of Volcán Zamorano, Central Mexico: Implications for the discrimination of the Sierra Madre Occidental and Mexican Volcanic Belt province: International Geology Review, v. 45, p. 724–752.

Watanabe, T., Koyaguchi, T., and Seno, T., 1999, Tectonic stress control on ascent and emplacement of magmas: Journal of Volcanology and Geothermal Research, v. 91, p. 65–78, doi: 10.1016/S0377-0273(99)00054-2.

Xu, S.S., Nieto-Samaniego, A.F., and Alaniz-Álvarez, S.A., 2005, Power-law distribution of normal fault size and estimation of extensional strain due to normal faults: The case of the Sierra de San Miguelito, Mexico: Acta Geologica Sinica, v. 79, p. 36–42.

Manuscript Accepted by the Society 29 August 2006

Printed in the USA

Geological Society of America
Special Paper 422
2007

The Caltepec fault zone: Exposed roots of a long-lived tectonic boundary between two continental terranes of southern México

Mariano Elías-Herrera*
Fernando Ortega-Gutiérrez
José Luís Sánchez-Zavala
Consuelo Macías-Romo
Instituto de Geología, Universidad Nacional Autónoma de México (UNAM), Ciudad Universitaria, Deleg. Coyoacán, México, D.F., 04510, México

Amabel Ortega-Rivera[†]
Alexander Iriondo
Centro de Geociencias, Universidad Nacional Autónoma de México (UNAM), Campus Juriquilla, Querétaro, Qto., 76230, México

ABSTRACT

The Caltepec shear zone is a dextral transpressional tectonic boundary between the Oaxacan and Acatlán Complexes, which are crystalline basements of the Zapoteco and Mixteco terranes in southern México, respectively. The terrane boundary (2–6 km wide) reveals protracted and polyphase tectonic activity from at least Early Permian to the present. The major tectonothermal event in the Caltepec fault zone was related to the oblique collision of the metamorphic complexes during the amalgamation of Pangea. An anatectic leucosome and the resulting syntectonic granite (Cozahuico Granite) in the fault zone yielded U-Pb zircon ages of 275.6 ± 1.0 Ma and 270.4 ± 2.6 Ma, respectively. The initial $^{87}Sr/^{86}Sr$ ratios (0.70435–0.70686) and Sm-Nd model ages (T_{DM}) (1.0–1.6 Ga) for the Cozahuico Granite and leucosome indicate a magmatic mixture that originated from melted Proterozoic crust and a component of depleted mantle. The Leonardian age of the cover (Matzitzi Formation) and a $^{40}Ar/^{39}Ar$ cooling age (muscovite) of 268.59 ± 1.27 Ma for mylonitic mica schist at the base of the thrust imply high cooling rates (~180 °C/Ma) and uplift during the Permian. The adjacent sedimentological record indicates intense tectonic reactivation during Early Cretaceous, Paleogene, and Neogene along the long-lived Caltepec fault zone, alternating with periods of relative tectonic quiescence during Triassic, Jurassic, and Mid-Cretaceous times. The trend of the Caltepec fault zone parallel to the

*elias@servidor.unam.mx
[†]Current address: Instituto de Geología, Universidad Nacional Autónoma de México (UNAM), Estación Regional del Noroeste, Hermosillo, Son., 83000, México.

Elías-Herrera, M., Ortega-Gutiérrez, F., Sánchez-Zavala, J.L., Macías-Romo, C., Ortega-Rivera, A., and Iriondo, A., 2007, The Caltepec fault zone: Exposed roots of a long-lived tectonic boundary between two continental terranes of southern México, *in* Alaniz-Álvarez, S.A., and Nieto-Samaniego, Á.F., eds., Geology of México: Celebrating the Centenary of the Geological Society of México: Geological Society of America Special Paper 422, p. 317–342, doi: 10.1130/2007.2422(11).

Oaxaca fault, 50 km to the east, is interpreted as part of a synchronous and dynamically coupled tectonic system that has been releasing tectonic stresses associated with the rupture of Pangea and the evolution of the Pacific margin of southern México from Jurassic to Holocene times.

Keywords: major fault, terrane boundary, Acatlán Complex, Oaxacan Complex, southern México.

INTRODUCTION

Southern México is one of the regions that display more unsolved geological problems than others. Its tectonic evolution is far from being completely understood within a modern plate tectonics framework. The rocks in this region, whose paleogeographic setting has been uncertain for more than 40 yr (e.g., Bullard et al., 1965), are grouped in the Zapoteco, Mixteco, Cuicateco, Chatino, and Guerrero terranes (Fig. 1). All of these rocks record a geological history spanning >1.0 Ga, linked essentially to the Grenville, Appalachian-Alleghenian, and North American Cordilleran orogenic systems, and also to geological events related to the rupture and dispersal of Pangea and to the Cenozoic tectonic evolution of the Pacific basin, which substantially modified the previous scenarios. Southern México is, therefore, an area where crustal deformation, metamorphic and magmatic phenomena, and reactivation of preexisting structures (e.g., Holdsworth et al., 1997, 2001a, 2001b) occurred in great scale and in different crustal levels during the burial and exhumation processes that the region has undergone throughout its history. One of the best examples of major structures with multiple reactivations in this region is the Caltepec fault zone. This fault zone is a fundamental piece in the structural configuration of southern México and constitutes the main subject of this paper. The Caltepec fault zone is a long-lived shear zone with a complex history of both ductile and brittle deformations, which reveals the tectonic boundary between the Paleozoic Acatlán and Mesoproterozoic Oaxacan crystalline basements and delimits the Mixteco and Zapoteco terranes (Fig. 1).

Previous Studies

The geological link between the metamorphic rocks of the Acatlán and Oaxacan regions has been the subject of speculation since the early twentieth century (Ordóñez, 1905, 1906). More than 60 yr passed until the issue of the geology of southern México was taken up again and emphasized as a major problem that needed to be solved (de Cserna, 1970). Nevertheless, because the contact zone between both metamorphic complexes is only exposed at a poorly accessible locality south of Tehuacán, Puebla (Figs. 1 and 2), some details of the nature of this contact were unknown until the mid-1970s. In the first report related to this problem, the contact between the Acatlán and Oaxacan Complexes was described as a north-south–trending zone, ~300 m wide, of vertically foliated cataclasite/mylonite and a tilted, but undeformed, narrow conglomerate wedge with boulders of gneiss, schist, and fossiliferous Cretaceous limestone, confined by vertical normal faults just on the western border of the cataclastic/mylonitic portion (Ortega-Gutiérrez, 1975). The contact was thought to be an important tectonic boundary between crustal blocks with at least two stages of movement (Ortega-Gutiérrez, 1975). The first stage, with undefined pre-Cretaceous age and an uncertain sense of movement, was related to the cataclastic/mylonitic zone and to the intense retrogression of the Oaxacan Complex granulitic rocks. A strong horizontal shortening component with insignificant lateral movement was inferred for this stage (Ortega-Gutiérrez, 1975). The second stage was clearly deduced from the conglomerate tectonic wedge and was essentially characterized by differential post-Cretaceous displacements along normal faults. The contact between the Acatlán and Oaxacan Complexes was also interpreted as a large normal fault involving various periods of activity and tectonic rotation (Ortega-Gutiérrez, 1978).

The tectonic boundary was later described in the area of Los Reyes Metzontla, 5 km north of Caltepec, as a vertical ductile shear zone with north-south trend, for which a southward extension of 200 km was inferred for the region of Juchatengo (Ortega-Gutiérrez, 1981) (Fig. 1). The tectonic boundary was then interpreted as a Paleozoic suture of collided paleocontinents (Ortega-Gutiérrez, 1981). Previously, Hernández-Estévez (1980) had considered the Acatlán Complex as the only metamorphic basement unconformably overlain by the upper Paleozoic Matzitzi Formation in the Los Reyes Metzontla area. In this area, however, the Matzitzi Formation overlies schists belonging to the Acatlán Complex as well as gneisses of the Oaxacan Complex. This important stratigraphic relationship implies that the tectonic juxtaposition of both crystalline complexes occurred before the deposition of the Matzitzi Formation. The presence of the Matzitzi Formation covering both complexes in the area has been previously documented by other authors (e.g., Hernández-Láscares, 2000; García-Duarte, 1999).

As part of the celebration of the 80th anniversary of *Sociedad Geológica Mexicana* in 1984, several papers related to the tectonics of the Mixteca region, in southern México, were published. An important topic discussed during that symposium was the tectonic boundary between the Mixteco and Oaxaca (Zapoteco) terranes, on which articles with alternative interpretations were published (Ramírez-Espinosa, 1984; González-Hervert et al., 1984; Torres-Torres et al., 1984). Ramírez-Espinosa (1984) considered that the Mixteco and Oaxaca (Zapoteco; Sedlock et al., 1993) terranes were accreted during the Early Cretaceous. According to Ramírez-Espinosa (1984), the Middle Jurassic continental and marine rocks (Tecocoyunca Group) covering the entire Mixteco terrane

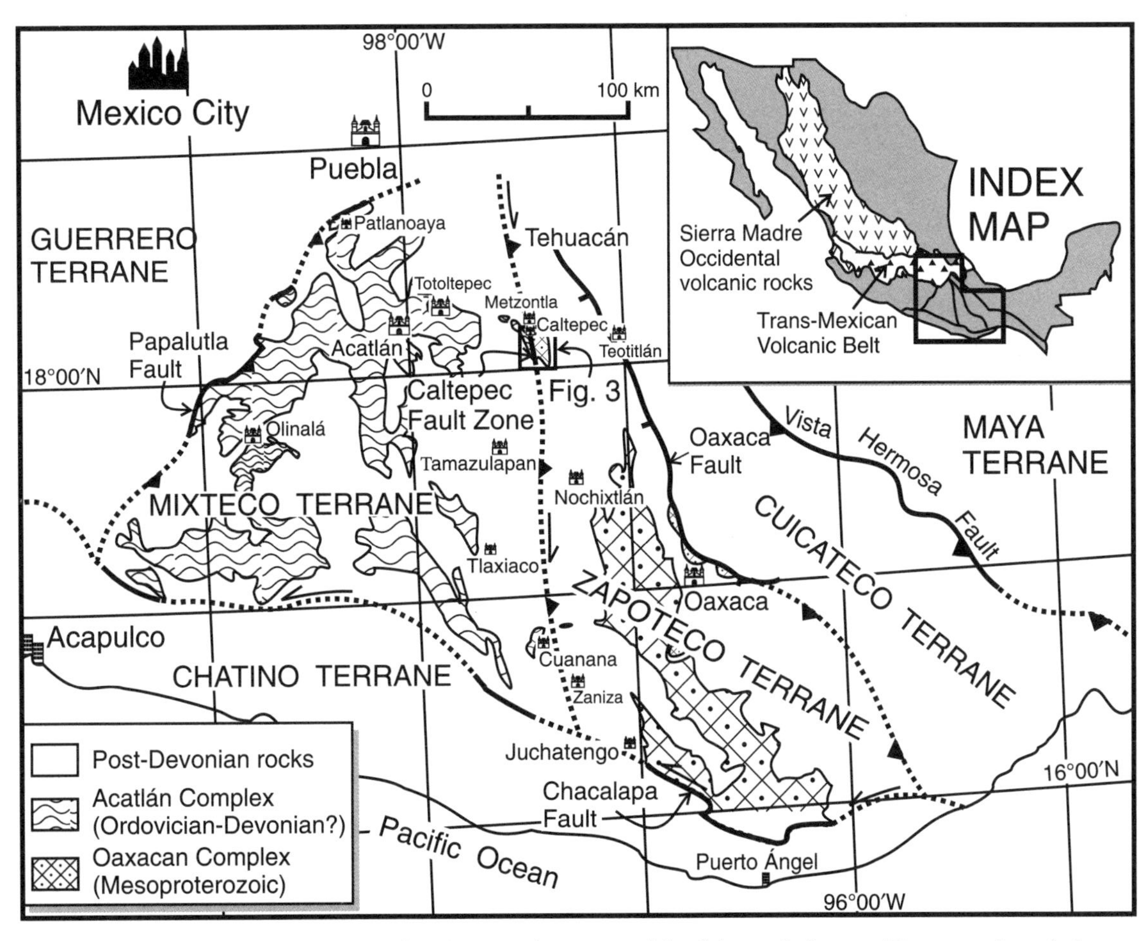

Figure 1. Map of southern México showing the tectonic context of the Caltepec fault zone. The terrane boundaries are after Sedlock et al. (1993). The localities in this map are mentioned in the text.

and their apparent absence on the Zapoteco terrane, and the existence of red beds with transitional variations to Lower Cretaceous carbonate rocks covering both terranes, imply separate geological histories for each terrane until their collision during the Early Cretaceous. González-Hervert et al. (1984) arrived at the same conclusion in their stratigraphic study of the Caltepec-Metzontla area, in which the apparent absence of the Matzitzi Formation overlying the Acatlán Complex was emphasized, and the Lower Cretaceous red beds were described as the unit overlapping the Acatlán and Oaxacan Complexes. González-Hervert et al. (1984) mapped the tectonic contact between these complexes as a very narrow zone with a winding trend and a general north-south orientation, consisting of microfolds and fractures in the rocks of the Acatlán Complex and mylonites in the Precambrian rocks. The nature of the displacement along the contact was not determined.

On the other hand, according to exploration work done by PEMEX in the Tlaxiaco region, the Oaxacan and Acatlán Complexes are juxtaposed by the Tamazulapan fault (López-Ticha, 1985). The Tamazulapan fault (Figs. 1 and 12) was described as a major structure linked to the "consolidation" of ancient terranes, and was considered to be a tectonic boundary between two geological environments, each with its own basement and distinct pre-Cretaceous cover (López-Ticha, 1985). Even though a large portion of the Tamazulapan fault trace is inferred, it has been described in the Tlaxiaco region as a north-south–trending left-lateral fault. Its sinistral nature, a result of Laramide reactivation, was inferred by the structural curvature and the related thrusts seen in the Sierra del Tentzo in southern Puebla by considering the northward extension of the fault to reach this locality (López-Ticha, 1985). A post-Miocene reactivation of the Tamazulapan fault was documented (Torres-Torres et al., 1984). This reactivation was described as strike-slip faulting with a dextral transpressive component affecting volcanic rocks in the Tamazulapan area. Meneses-Rocha et al. (1994) concluded that the Tamazulapan fault was reactivated during the Laramide orogeny as a high-angle thrust fault. More recently, this fault has been mapped (González-Ramos et al., 2000) as a left-lateral fault and as a tectonic boundary between the Mixteco and Oaxaca terranes, the trace of which, though mostly inferred, is north-south–trending. It is worthwhile to mention that the Tamazulapan fault matches the southern extension of the Caltepec fault in the San Juan Teita area, southeast of Tlaxiaco (Figs. 1 and 12).

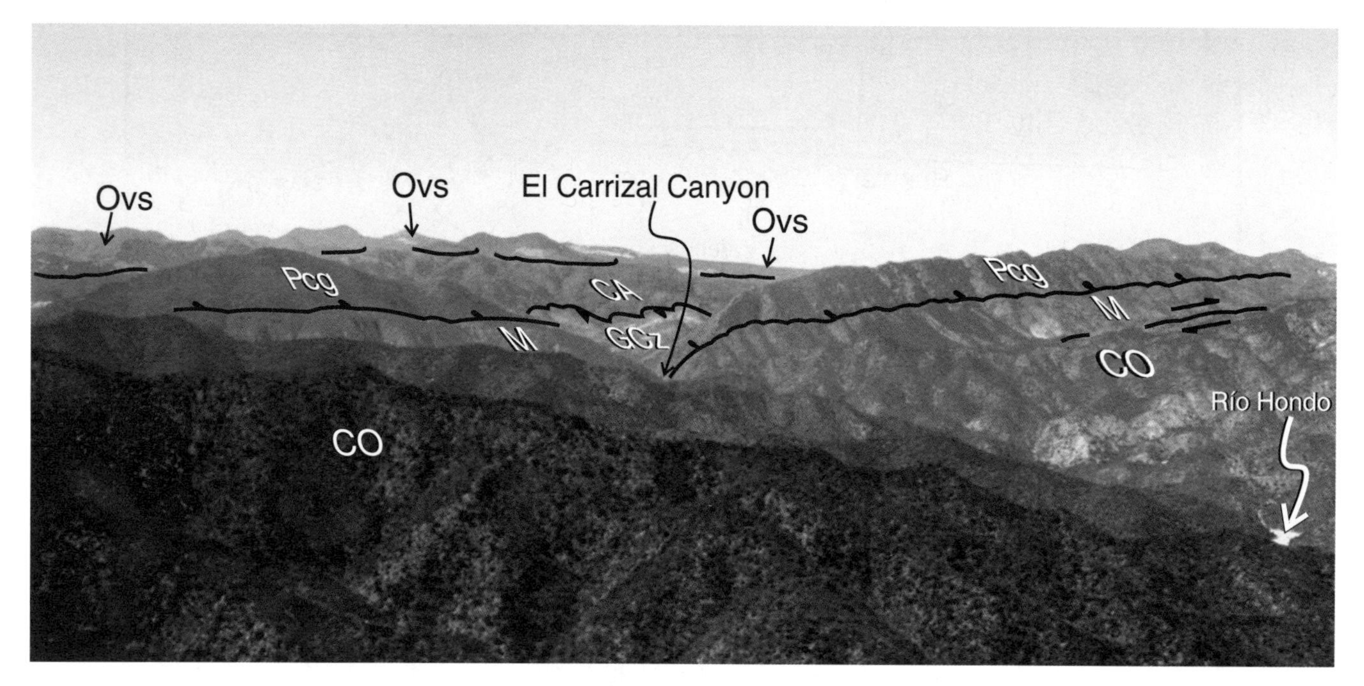

Figure 2. Westward panoramic view of the Caltepec fault zone at the border area between the Puebla and Oaxaca States, southern México. The lithological units in the geomorphologic relief are: CO—Oaxacan Complex; CA—Acatlán Complex; M—mylonite (completely mylonitized granulitic gneisses); GCz—Cozahuico Granite (syntectonic stitching pluton); Pcg—Paleogene conglomerate with fossiliferous Cretaceous limestone, gneiss, and schist boulders; Ovs—Oligocene volcanosedimentary rocks.

In the early 1990s, Yáñez et al. (1991) concluded that the juxtaposition of the Acatlán and Oaxacan Complexes has been the result of the Laurentia-Gondwana collision during the Acadian orogeny in the Middle Devonian. This interpretation was based on geochronological and isotopic data of various units of the Acatlán Complex, and on the assumption that both complexes were intruded and linked by the Esperanza Granitoids, a synorogenic unit for which a U-Pb (zircon) lower intercept age of 371 ± 34 Ma was obtained (Yáñez et al., 1991). Later, in accord with this assumption, the tectonic boundary between the Mixteco and Zapoteco terranes was interpreted to be a Devonian suture, resulting from the collision of the Acatlán and Oaxacan Complexes (Sedlock et al., 1993; Ortega-Gutiérrez, 1993). According to these authors, the collision caused the intense ductile deformation, metamorphism, and syntectonic magmatism in the Acatlán Complex. Sedlock et al. (1993) inferred that the collision probably occurred in a continental margin with a convergent component in an oceanic setting south of the eastern edge of North America (Laurentia) in an uncertain Paleozoic plate tectonics scenario. Ortega-Gutiérrez (1993) interpreted the contact between the complexes as a composite polyorogenic suture for which their original juxtaposition occurred during the Devonian along a major subhorizontal structure in which the Oaxacan Complex, acting as continental crust, was thrust over the Acatlán Complex. The suture involved a nappe, consisting of eclogitized ophiolitic igneous and sedimentary rocks and the Esperanza Granitoids of continental affinity, whose syntectonic emplacement sealed and dated the nappe as an Acadian element (Ortega-Gutiérrez, 1993).

Later, the contact between the Acatlán and Oaxacan crystalline basements in the Caltepec region was recognized as a more intricate tectonic boundary on the basis of geological mapping, structural and microtectonic analyses, and geochronological and petrographical studies (Elías-Herrera and Ortega-Gutiérrez, 1998, 2002). The age of displacement at the contact zone and its extent turned out to be rather different from what was previously considered, and the new data substantially modify the prevailing view on this geological boundary. For the purpose of this paper, the geological map of the Caltepec fault zone is shown in Figure 3, and structural details (modified from Elías-Herrera and Ortega-Gutiérrez, 2002) are included in Figures 4 and 5.

CALTEPEC FAULT ZONE

The tectonic contact between the Acatlán and Oaxacan Complexes is only exposed in the Caltepec area, south of Tehuacán, Puebla (Fig. 1). In this area the contact was reevaluated and redefined as a reworked major fault zone, showing superimposed ductile and brittle deformations with north-south and north-northwest trends. The tectonic contact is 2–6 km thick and has a half-flower structure in cross section (Elías-Herrera and Ortega-Gutiérrez, 1998, 2002). The fault zone (Figs. 2 and 3) from east to west consists of 100–500 m of Proterozoic granulitic gneisses with intense retrogression and is dislocated by subvertical faults; 200–300 m of vertically foliated mylonite of gneissic protolith and subsequent cataclasis; and a 50–800-m-wide tectonic wedge, consisting of Paleogene conglomerate with boulders of schist, deformed

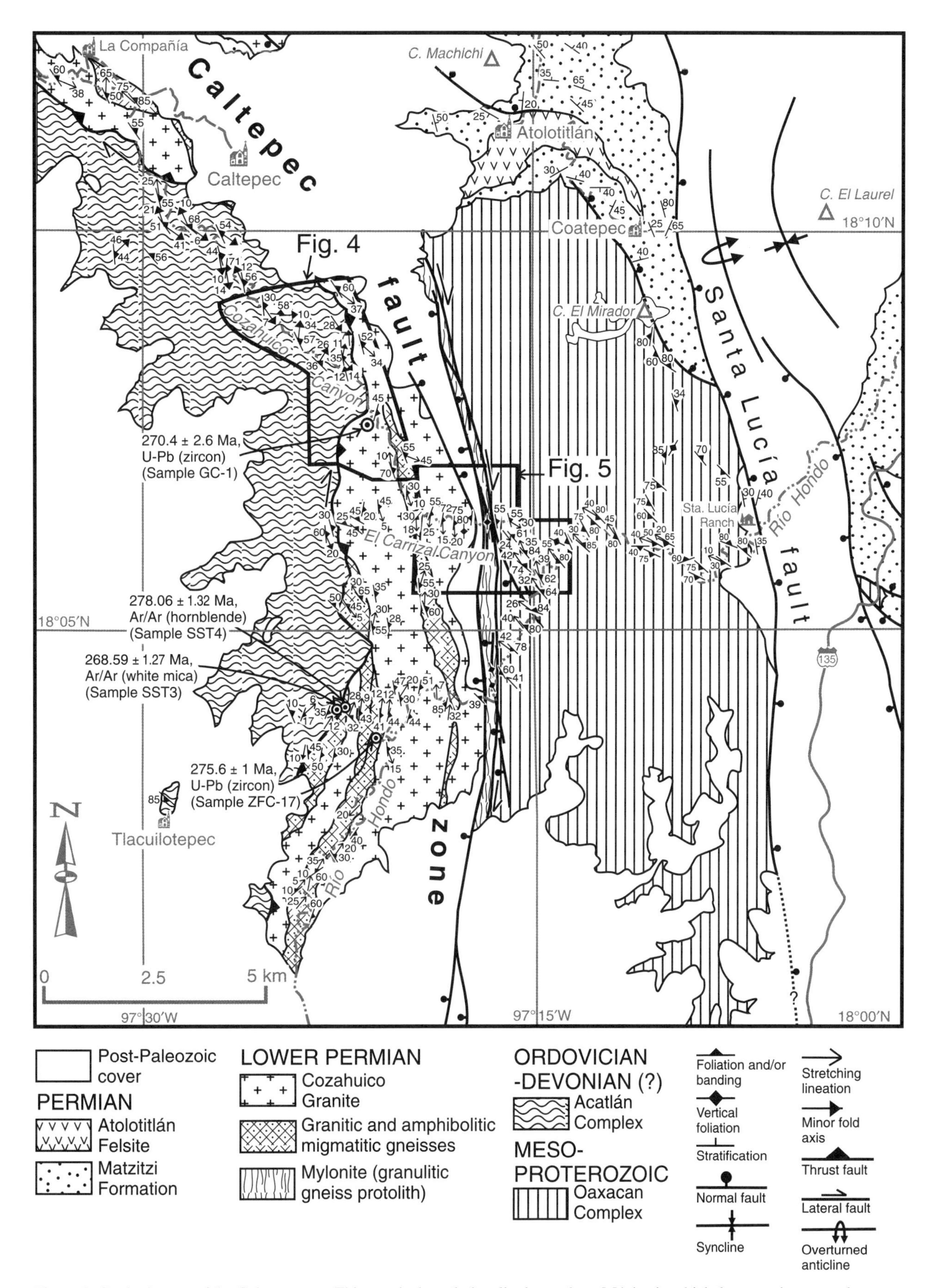

Figure 3. Geologic map of the Caltepec area. This area is the only locality in southern México in which the tectonic contact between the Acatlán and Oaxacan Complexes is exposed.

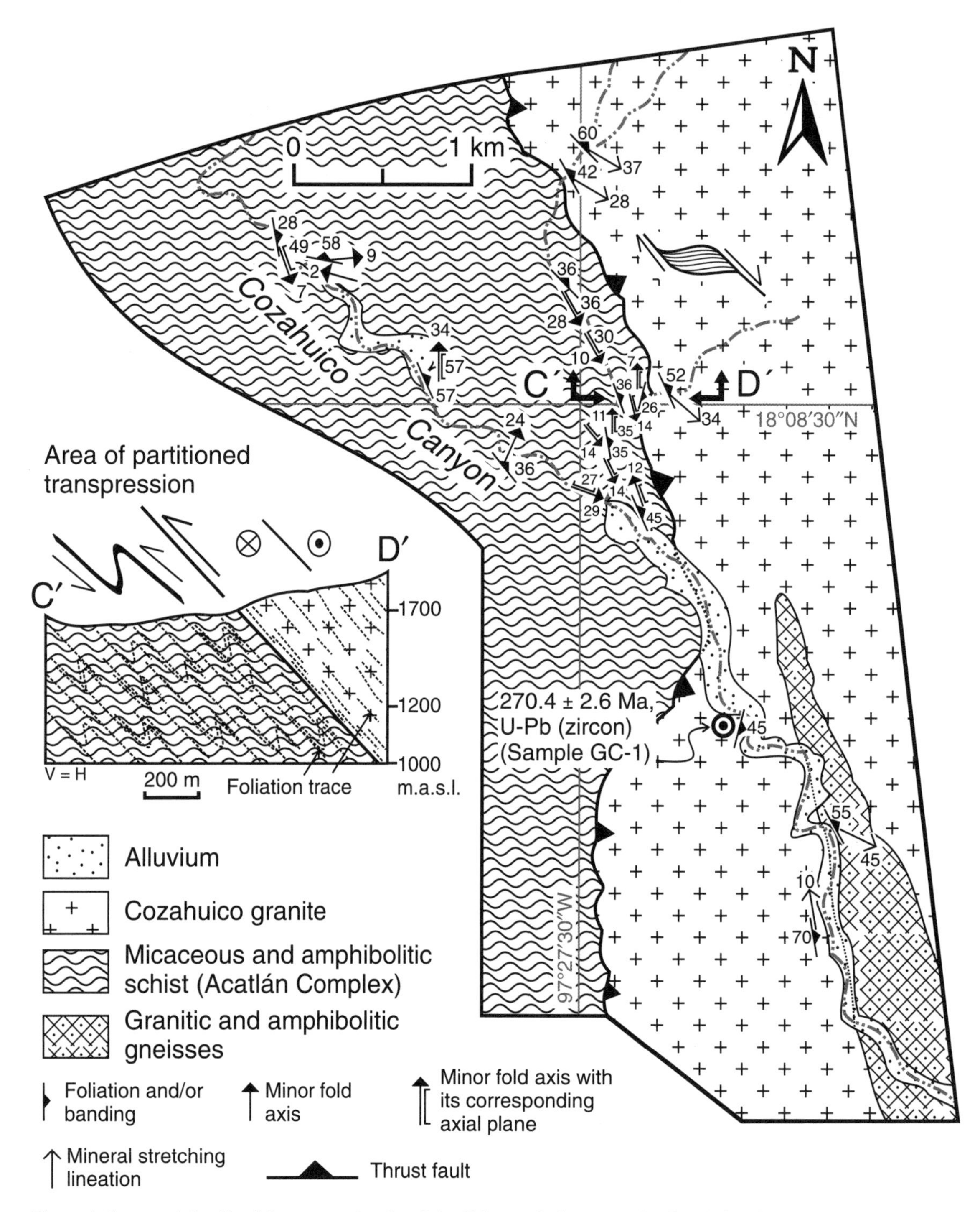

Figure 4. Structural details of the western border of the Caltepec fault zone at the Cozahuico Canyon. The dextral sense of shear in the Cozahuico Granite is illustrated with a schematic mica fish-type feature. For location see Figure 3.

granite, fossiliferous Cretaceous limestone, chert, quartzite, and scarce gneiss fragments. This tectonic wedge is bordered and tilted by vertical faults. In the fault zone, there is a 2–4 km mylonitized granite (Cozahuico Granite) whose foliation varies from vertical to slightly inclined toward the east-northeast, showing later cataclasis, and which to the west is thrust over the Acatlán Complex (Figs. 4 and 5). The last component of this fault zone is a poorly defined mica-garnet schist sector of the Acatlán Complex with a strongly planar structure, dislocated by later faulting, which underlies the Cozahuico Granite. The sense of shear in the ductile deformation (mylonitization) of the Proterozoic gneisses and the Cozahuico Granite is dextral, and was determined by delta and

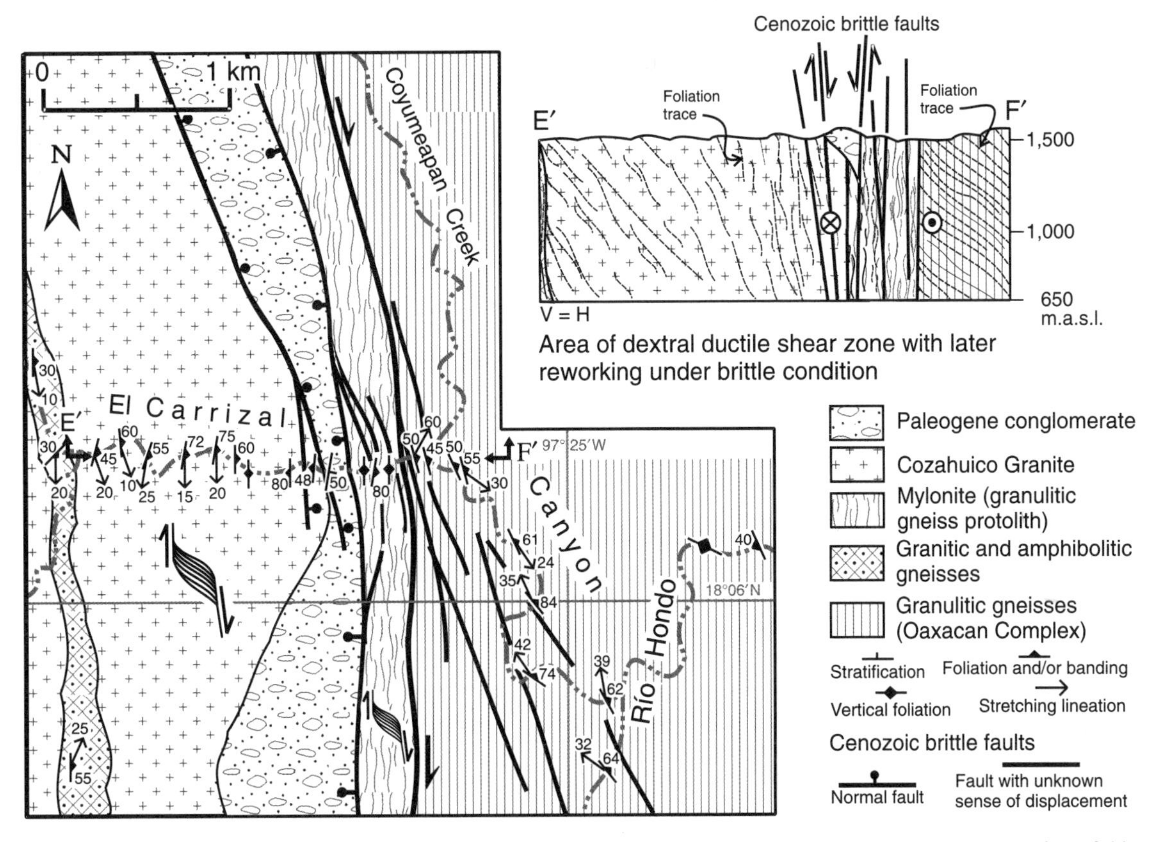

Figure 5. Structural details of the eastern border of the Caltepec fault zone at the El Carrizal Canyon–type section of this fault zone. The dextral movement of the granite and the mylonite is illustrated with a schematic mica fish-type feature. For location see Figure 3.

sigma-type porphyroclasts, crystal grain geometries, S-C microstructures, and antithetic sliding of mineral grains in meso- and microscopic scales. The contractive component defined by the thrusting of the Cozahuico Granite over the Acatlán Complex in a ductile regime, in addition to the dextral shear, indicates that the Caltepec fault zone was a major meso-crustal transpressive structure during at least one portion of its ductile deformation history. The contractive component linked to the thrust was deduced mainly from meso- and microstructures found in mylonitic schists along the tectonic contact with the granite. This structural relation is analyzed in more detail later. Other features such as petrography, textural relations, and implications of the different lithological sectors in the Caltepec fault zone are documented in Elías-Herrera and Ortega-Gutiérrez (2002).

OAXACAN COMPLEX

To the east, outside the fault zone, the Oaxacan Complex is composed of alternating bands of gabbroic-dioritic, tonalitic, and granitic granulitic gneisses (Fig. 6A), as well as pegmatitic facies in some areas. The granulitic mineral assemblage found in the gabbroic gneisses consists of plagioclase + clinopyroxene + orthopyroxene + garnet + titaniferous amphibole + titaniferous biotite ± quartz ± ilmenite. Orthopyroxene has been replaced by bastite. The mineralogy of the granitic gneisses is quartz + perthitic K-feldspar + plagioclase + garnet + titaniferous biotite ± ilmenite ± allanite. The gneisses show banding that mainly strikes northwest-southeast and dips subvertically to the southwest and northeast, and a mineral elongation lineation plunging 20°–50° to the northwest. The gneisses are also dislocated and brecciated locally due to high-angle north-south brittle faulting (Fig. 6B). Near the Caltepec fault zone, the gneissic banding is gradually transposed by subvertical mylonitic foliation with north-northwest trend and by anastomosed ductile shear zones up to one meter thick (Fig. 6C). The mafic granulitic gneisses found in the ductile shear zones are transformed into amphibolitic mylonites. A planar fabric formed by the reduction of grain size due to recrystallization and the lineation of quartz-feldspathic grains and green acicular amphibole crystals overprints the gneissic fabric. Textural relationships of amphibole and plagioclase in the mylonites indicate a dextral movement for the shear (Fig. 6D).

To the east of the Caltepec fault zone the Oaxacan Complex is unconformably overlain by the Matzitzi Formation, Lower Cretaceous red beds, Cenozoic conglomerates, volcanics, and lacustrine sedimentary rocks. All of these rocks also unconformably cover the Acatlán Complex toward the west of the fault zone.

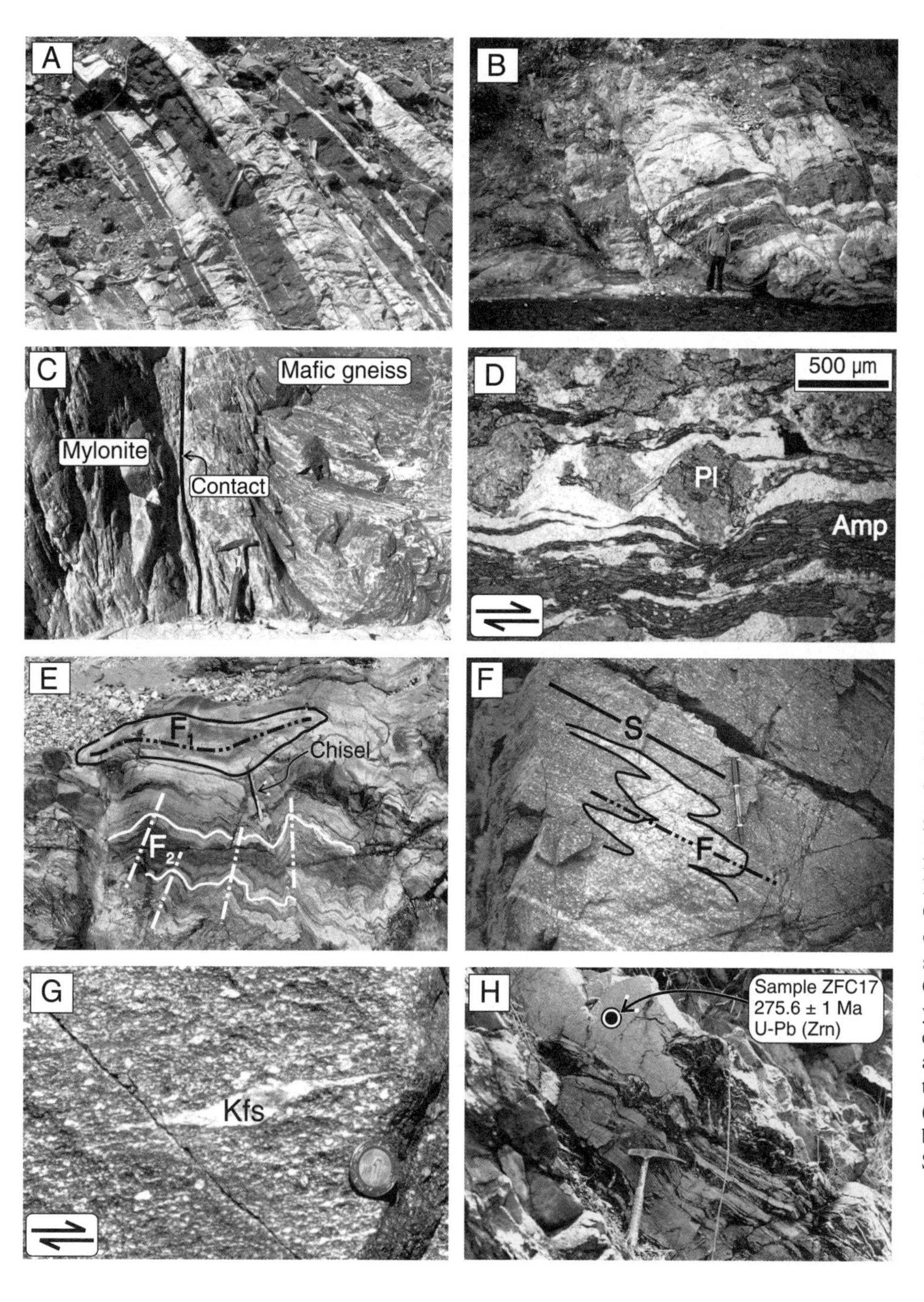

Figure 6. (A) Alternating bands of gabbroic-dioritic gneisses and granites of the Oaxacan Complex in the Río Hondo, east of the Caltepec fault zone. (B) Granitic and mafic gneisses of the Oaxacan Complex dislocated by Cenozoic normal faulting in the El Carrizal Canyon near the Caltepec fault zone. (C) Gabbroic-dioritic granulitic gneiss transformed by ductile shearing into an amphibolitic mylonite in which vertical foliation overprinted gneissic banding; eastern border of the Caltepec fault zone at the El Carrizal Canyon. (D) Amphibolitic mylonite from the eastern sector of the Caltepec fault zone at microscopic scale; the textural relations of the amphibole (Amp) and the plagioclase (Pl) in the mylonite indicate dextral sense of movement. (E) Metasediments of the Acatlán Complex showing at least two superimposed folding phases: sheath fold F_1 affected by later folding F_2 (chisel is 15 cm long); this locality is west of the Caltepec fault zone at the Cozahuico Canyon. (F) Cozahuico Granite with a folded more leucocratic granitic phase whose axial planes (labeled F) are parallel to the mylonitic foliation (S) (marker pen is 12 cm long); Caltepec fault zone at Río Hondo Canyon. (G) Cozahuico Granite with a σ-type K-feldspar porphyroclast (Kfs) showing right-handed shear movement; Caltepec fault zone at El Carrizal Canyon. (H) Banded migmatitic gneisses enclosed within the Cozahuico Granite; amphibole-rich melanosome bands alternate with granitic leucosomatic bands (U-Pb zircon age is indicated for one band); Caltepec fault zone at the Agua Salada ravine, Río Hondo area.

The Matzitzi Formation, as the oldest overlapping sedimentary unit, is a crucial stratigraphic constrain for the tectonic juxtaposition of the metamorphic complexes. This formation is essentially a continental sequence composed of conglomerate, sandstone, and siltstone with abundant fossil flora. On the basis of the plant fossils, a Pennsylvanian age (Silva-Pineda, 1970) or an ambiguous Pennsylvanian-Permian age range (Weber et al., 1987) was previously assigned to the Matzitzi Formation. However, the age of the Matzitzi Formation was defined later as Leonardian, as a result of the discovery of fossil plant, *Lonesomia mexicana* Weber *sp.nov.*, belonging to the gigantopterids (Weber, 1997). This age is strongly supported by the similarity of the flora of the Matzitzi Formation to that of other Leonardian rocks in Texas (Weber, 1997). The tectonic juxtaposition of the Acatlán and Oaxacan Complexes is, therefore, pre-Leonardian and seems to have occurred closer to the Laurentian edge than the Gondwanan, as suggested by the Leonardian flora of the overlapping unit. The Leonardian age corresponds to the Early Permian Artinskian and Kungurian ages of Gradstein et al. (2004).

In the vicinity of Santa Lucía Ranch, located 6 km south-southeast from Coatepec (Fig. 3), the Oaxacan Complex is unconformably covered by Lower Cretaceous red beds. The red beds consist of conglomerates and cross-stratified coarse-grained sandstone, with fragments of gneiss, granite, recycled sandstone, and volcanic rocks. The clastic rocks are gradually interbedded with marls and fossiliferous siltstones of the Barremian Zapotitlán Formation (Calderón-García, 1956). In the Caltepec–La Compañía area, red beds also unconformably cover banded gneisses. These gneisses are indeed xenolithic blocks in the Cozahuico Granite within the Caltepec fault zone (Fig. 3), and they do not correspond to the in situ granulitic Oaxacan Complex, as discussed later.

ACATLÁN COMPLEX

To the west of the Caltepec fault zone, the Acatlán Complex consists of mica schist, amphibolite, quartzo-feldspathic schist, greenschist, quartzite, and some lenses of marble and serpentine bodies. This metamorphic complex is uncomformably covered by Oligocene volcanic rocks and lacustrine sediments, and locally it is covered by the Matzitzi Formation in the Metzontla area and by Lower Cretaceous red beds in the Caltepec and San Simón Tlacuilotepec areas. The mineral assemblage in the micaceous schist is white mica + biotite + garnet with widely distributed chlorite and epidote as retrogressive phases. In the amphibolite, the hornblende ± plagioclase ± epidote ± quartz assemblage suggests that the regional metamorphism reached at least the lower part of the amphibolite facies. Moreover, in some amphibolites the mineral assemblage of sodic-calcic amphibole (bluish-green pleochroism) + epidote (± zoisite) + rutile suggests a high-pressure epidote-amphibolite facies metamorphic condition. In the mica schist, a lot of porphyroblasts of garnet display complex textural relationships that suggest pre- and syntectonic crystallization in relation to the predominant foliation. Intensely folded microscopic metamorphic foliation is commonly observed not only in the micaceous schist, but also in amphibolites. At outcrop scale the metamorphic rocks display evidence of polydeformation with superimposed fold patterns that prove at least two folding events (Fig. 6E) for which orientation and vergence have not been determined.

However, the rocks of the Acatlán Complex near and along the tectonic contact with the Cozahuico Granite show an entirely planar fabric whose foliation completely transposed the previous ones. For instance, in the Cozahuico Canyon, the transposed foliation in the micaceous schist is defined by the reorientation and recrystallization of white mica, biotite, garnet, and quartz, showing a north-northwest strike and a moderate plunge to the northeast (Fig. 4). This foliation is parallel to the axial planes of the tight, asymmetric, minor west-verging folds and to the mylonitic foliation of the underlying Cozahuico Granite. The axes of the minor folds in the schist are parallel to the mineral stretching lineation in the granite. In the Río Hondo area, a quartz–white mica–garnet mylonitized schist is structurally overlain by the Cozahuico Granite and amphibolitic gneisses enclosed in the granite. In this case, the mineral lineation in the schist is closely parallel to the stretching lineation in both gneiss and granite. Mica-fish and S-C microstructures observed in the mylonitic schist indicate a top-to-the-west sense of shear, which is consistent with the westward tectonic transport indicated by the vergence of minor folds in the schist observed in the Cozahuico Canyon. Based on these field data, the western border of the Caltepec fault zone is considered to be a thrust fault with associated ductile deformation. An intrusive relation between the Cozahuico Granite and the metamorphic rocks of the Acatlán Complex in the Caltepec area has not been observed.

COZAHUICO GRANITE

The Cozahuico Granite (Figs. 6F and 6G) is well exposed in the canyon from which it takes its name. It is an extremely elongated medium- to coarse-grained leucocratic pluton, whose contacts are defined by faults and which extends along the Caltepec fault zone (Fig. 3). Because of its field relations, the granite is considered to be an essential unit for understanding the ductile deformation of the fault zone. The granite is delimited by a Paleogene conglomerate and a vertical mylonite zone by means of a normal fault (brittle structure of Cenozoic reactivation) in its eastern border, and by a ductile thrust fault within its western border. The elongated extension of the granite suggests that its emplacement took place along a weakness zone. Because of its internal deformation, the pluton displays an augen gneiss structure, and was originally considered part of the Acatlán Complex and correlated with the Esperanza Granitoids (Ortega-Gutiérrez, 1975), in which porphyroblastic structure is an important feature. Nonetheless, the Cozahuico Granite differs from the Esperanza Granitoids in petrographical, geochemical, isotopic, and geochronological signatures. The Cozahuico Granite has a complex structure formed by a mixture of igneous, metamorphic, and late-alteration minerals associated with later cataclasis. Primary minerals are K-feldspar, quartz, plagioclase, biotite, hornblende, ilmenite, titanite, allanite/epidote, apatite, and zircon. Even though metamorphic and hydrothermal epidote is abundant in the granite, some diagnostic textural features also indicate magmatic epidote in the pluton. Euhedral to subhedral crystals with oscillatory zoning growth around allanite cores, and isolated euhedral crystals within unaltered orthoclase phenocrysts in nondeformed sectors of the granite, are the occurrences in which epidote is considered to have a magmatic origin (e.g., Zen and Hammarstrom, 1984; Keane and Morrison, 1997). The orientation of large (0.3–3-cm-long) euhedral tabular crystals of K-feldspar and plagioclase with growth twins parallel to crystal faces, and medium-sized biotite, ilmenite, and allanite/epidote crystals define a magmatic foliation. The metamorphic mineral assemblage that defines a tectonic or solid-state foliation consists of finely recrystallized quartz, biotite, microcline, muscovite, albite, epidote, clinozoisite, and titanite, with abundant myrmekites surrounding phenocrysts of feldspar along the foliation. In microtectonic terms, the solid-state foliation is parallel to and superimposed on the magmatic foliation, which suggests a syntectonic emplacement (e.g., Paterson et al., 1989; Miller and Paterson, 1994).

Gneisses and Migmatites in the Cozahuico Granite

The Cozahuico Granite contains xenolithic blocks of amphibolite, granite, and migmatitic banded gneisses, whose thickness varies from a few meters to 500 m and up to 7 km in length. The blocks are parallel to the general north-south trend of the granite and fault zone. They show boudinage structures at mappable and outcrop scales. The fabric of the xenolithic gneisses is generally parallel to the magmatic/tectonic foliation and mineral lineation of the host granitic rock, and they also show textural relations indicating dextral kinematics. The gneisses are similar to those of the Oaxacan Complex located away from the fault zone, with alternating mafic and felsic bands and small lens-shaped bodies of impure marble, which display intrusive relations at the same scale as the surrounding banding. These gneisses, however, show no evidence of granulite facies mineral assemblages. The mafic bands consist of hornblende + plagioclase + K-feldspar ± epidote ± biotite, whereas the felsic bands contain monocrystalline quartz (locally blue colored) + polycrystalline feldspar ribbons (mainly microcline) + biotite. The intrusive relation between the Cozahuico Granite and the xenolithic gneisses is clear. Along the contact with the pluton, the gneisses display local hybridization zones suggesting their partial assimilation by the granite; and late deformed and non-deformed granite dikes cut the gneisses.

The migmatitic gneisses consist of alternating bands (<1 cm up to 50 cm in thickness) of melanosome rich in hornblende, biotite, and epidote and granitic leucosome forming the neosome (Fig. 6H). Granitic, granodioritic, and hornblende-rich tonalitic gneisses compose the paleosome. Schlieren, folded, and phlebitic structures are common in the migmatitic gneisses. The granitic neosome in the migmatites is epidote-rich granite with granular hypidiomorphic fabric in which microcline and myrmekite textures are abundant. The epidote in the anatectic granite occurs in anhedral small crystals (0.1–0.6 mm long) associated with K-feldspar, plagioclase, and quartz; in vermicular and myrmekitic intergrowths with quartz; and in euhedral and subhedral crystals with allanite cores included in feldspar and associated with biotite. The epidote also shows textural stability in the melanosome and in hornblende-rich banded granodioritic gneisses, suggesting thermal equilibrium between the granitic leucosome and the mafic migmatitic units. It is worth noting that the mineral assemblage of the neosome is similar to the primary mineralogy of the Cozahuico Granite—they both include magmatic epidote—and that folded bands of Cozahuico Granite and migmatitic units are closely alternated. These mineralogical and structural features strongly suggest a petrogenetic relation of migmatites–Cozahuico Granite.

Based on structural and petrographical data, it is considered that the gneissic blocks correspond to Mesoproterozoic granulitic megaxenoliths, which were strongly deformed and thermally re-equilibrated by metamorphism under amphibolite facies condition and partial melting during the emplacement of the Cozahuico Granite. The ductile deformation, metamorphism, migmatization, and magmatism with primary epidote in the Caltepec fault zone have been, therefore, interpreted as concurrent events, which occurred in the deep roots of this tectonic boundary (Elías-Herrera and Ortega-Gutiérrez, 2000, 2002). In this case, the penetrative deformation, recrystallization, and partial melting were probably triggered simultaneously by the channeling and mobilization of fluids along layers or bands, a common process in intensely foliated rocks and banded migmatites (Mawer et al., 1988). The anatectic magma in the fault was probably generated at minimum pressures around 4.5 kbar (16–17 km in depth) with high H_2O pressures, as suggested by magmatic epidote (e.g., Schmidt and Thompson, 1996). The fluids and heat flow needed to homogenize and melt granulitic gneisses probably derived from the dehydration of the underthrust Acatlán Complex at deeper crustal levels, and/or by regional processes associated with an assumed oblique convergent zone (e.g., Saint Blanquat et al., 1998) in the west during pre-Mesozoic times (Elías-Herrera and Ortega-Gutiérrez, 2002). The tectonothermal inversion observed in the juxtaposition of the migmatitic gneisses (amphibolite facies metamorphism) and the Cozahuico Granite over the schists of the Acatlán Complex (mainly greenschist facies metamorphism) may indicate a large vertical displacement and an important crustal overlap in the western boundary of the Caltepec fault zone. At present exposure levels of the fault zone, thrust faulting along its western border probably represents a late stage of the tectonomagmatic event in which the Cozahuico Granite was emplaced during its cooling period, as suggested by the ductile deformation and the greenschist-lower amphibolite facies metamorphism in the rocks of the Acatlán Complex along the fault zone.

Syntectonic Nature of the Cozahuico Granite

The syntectonic emplacement of plutons is a controversial topic about which important doubts have arisen regarding the significance of the close spatial relation among plutons and faults (Paterson and Schmidt, 1999). Criteria for determining the syntectonic nature of igneous intrusions are commonly ambiguous, and multiple lines of evidence are always required (Paterson and Tobisch, 1988; Paterson et al., 1989; Miller and Paterson, 1994). Notwithstanding these difficulties, the emplacement and ductile deformation of the Cozahuico Granite along the Caltepec fault zone as broadly coeval events are supported by the following arguments: (1) The remarkable elongate shape (2–6 km wide x 25 km long) of the deformed pluton, which has a long axis parallel to the main direction of extension and has borders defined by faults along a terrane boundary, suggests that its emplacement and lateral extension were probably controlled by the trend of the ductile faulting (e.g., Vigneresse, 1995; Román-Bardiel et al., 1997). (2) The foliation in the Cozahuico Granite is concordant with the composite foliation in the footwall schistose rocks in the western border of the fault zone, and hinges of the tight west-southwest–verging minor folds in the underlying schist are parallel to the mineral stretching lineation found in the pluton. Although these relations may be ambiguous, the structural parallelism in this case is interpreted as the result of a simultaneous ductile deformation of the schist and the granite during its cooling stage. The composite foliation in the footwall schistose rocks is a planar structure that was clearly superimposed over earlier foliations. (3) The intrusive relation of the Cozahuico Granite against elongated gneiss blocks with boudinage structures, folded and boudinaged granite veins in the xenolithic gneisses, heterogeneous ductile deformation in granitic veins with strongly and slightly deformed sectors, and folding in xenolithic gneisses with axial planes parallel to the foliation of the surrounding granite are solid evidence of coeval magmatism and deformation. (4) The Cozahuico Granite is cut by nonfoliated pegmatites, mildly foliated medium- to fine-grained granitic veinlets, and aplitic dikes. All of these bodies display tight to close folds with axial planes that are parallel to the mylonitic foliation on the host pluton (Fig. 6F), indicating that late magmatic pulses occurred during cooling and deformation of the Cozahuico Granite. (5) The presence of pervasive microscopic shear zones (S-C structures) throughout the Cozahuico Granite, coupled with the lack of local and narrow shear anastomosing around low-strain domains, also support a syntectonic emplacement rather than pretectonic nature, as suggested by experimental models in deformed granites (e.g., Gapais, 1989). (6) Parallel submagmatic and solid-state foliation, diagnostic feature of syntectonic emplacement (Paterson et al., 1989; Miller and Paterson, 1994), is present in the Cozahuico Granite. Furthermore, the thermo-mechanic aspect involving ductile shearing and syntectonic magmatism is an important and complex issue (e.g., Brown and Solar, 1998; Harrison et al., 1998; Nabelek and Liu, 1999; Leloup et al., 1999) that, for the case of the Caltepec fault zone, has not been addressed until now.

GEOCHRONOLOGY: AGE OF THE COZAHUICO GRANITE, MIGMATITES, AND CALTEPEC FAULT ZONE

Because the syntectonic nature of the Cozahuico Granite is clear, and its emplacement between different metamorphic basements as a stitching pluton probably occurred almost simultaneously with the migmatization in the Caltepec fault zone, the ages of the granite and the neosomes of the migmatites are truly significant. Even though part of the geochronological and isotopic data related to the Cozahuico Granite was previously documented (Ruiz-Castellanos, 1979; Torres-Vargas et al., 1986, 1999), the significance of the pluton as the key element of the Caltepec fault zone and the juxtaposition of the most important crystalline basements in southern México were not fully understood. Ruiz-Castellanos (1979), for example, obtained a Rb-Sr isochron (7 whole rock points) of 269 ± 21 Ma (middle Permian) with an initial $^{87}Sr/^{86}Sr$ ratio of 0.7056 for the deformed granite in the locality of La Compañía in the Caltepec area. This deformed granite clearly corresponds to the northern extension of the Cozahuico Granite. The Permian Rb-Sr age was interpreted ambiguously, either as the emplacement age of the granite, or of its cataclastic deformation. The emplacement of the pluton was, however, considered to be almost simultaneous to its deformation according to the isotopic data (Ruiz-Castellanos, 1979).

It is interesting that this first inference about the syntectonic relation of the granite based on its preliminary Rb-Sr data was found to be consistent with the results obtained from petrographical, geochronological, and structural studies two decades later. Ruiz-Castellanos (1979) also concluded that the "La Compañía granite" is isotopically quite different from the augen-schists of the "Esperanza formation" (Esperanza Granitoids) outcropping elsewhere in the Acatlán Complex. Moreover, the "La Compañía granite" yielded a K-Ar (biotite) age of 167 ± 8 Ma (Torres-Vargas et al., 1986), which may be related to the reworking of the fault zone. Later, a Devonian age for the Cozahuico Granite was implied by assuming that the Esperanza Granitoids were dated by a U-Pb (zircon) lower intersect age of 371 ± 34 Ma in the Acatlán area (Yáñez et al., 1991), and by considering that the granitoid was directly involved in the juxtaposition of the Acatlán and Oaxacan Complexes in the Caltepec fault zone (Yáñez et al., 1991; Sedlock et al., 1993; Ortega-Gutiérrez, 1993). Nonetheless, the age of the Esperanza Granitoids in their type locality turned out to be Late Ordovician–Early Silurian (440 ± 14 Ma; Ortega-Gutiérrez, et al., 1999), and its correlation with the Cozahuico Granite has been invalidated by new geological data.

Because of the lack of adequate isotopic studies clearly establishing the age and ductile deformation of the Cozahuico Granite and the Caltepec fault zone, as previously discussed, we performed geochronological studies by U-Pb (zircon) for this pluton (sample GC1) and for the granitic neosome (sample ZFC17) of the migmatites in the Caltepec fault zone. The results and analytical procedures are shown in Tables 1 and 2, and they were plotted in the traditional U-Pb concordia diagrams (Fig. 7). Six zircon fractions with discordant ages were obtained from the Cozahuico Granite (Table 1 and Fig. 7A). Poorly defined lower and upper intercept ages of 326 ± 118 Ma and 1137 ± 257 Ma, respectively, were obtained with all the discordant fractions. Better constrained lower and upper intercept ages with values of 373 ± 43 Ma and 1301 ± 122 Ma were acquired, discarding the fractions with the highest analytical errors. However, by excluding the most discordant fractions, the ages obtained are 258 ± 11 Ma and 987 ± 11 Ma (Fig. 7A). In any case, the upper intercept Mesoproterozoic data may be interpreted as the ages of the inherited radiogenic component source, whereas the lower intercept ages may be considered to correspond to the time of crystallization of the granite. Thus, the lower intercept age of 373 ± 43 Ma was initially interpreted as the age of emplacement of the Cozahuico Granite and of the transpressive interaction of the Acatlán and Oaxaqueño blocks related to the Acadian orogeny (Elías-Herrera and Ortega-Gutiérrez, 1998), as was previously suggested (Yáñez et al., 1991). Nevertheless, the 258 ± 11 Ma age is closer to the crystallization age, as discussed later, and the corresponding upper

TABLE 1. U-Pb (ZIRCON) GEOCHRONOLOGICAL DATA FOR A MIGMATITIC GNEISS LEUCOGRANITE (NEOSOME) AND THE COZAHUICO GRANITE IN THE CALTEPEC FAULT ZONE*

Fraction†	Weight (mg)	U (ppm)	Pb (ppm)	Observed ratios§			Atomic ratios**			Ages Ma††		
				$^{206}Pb/^{204}Pb$	$^{207}Pb/^{206}Pb$	$^{208}Pb/^{206}Pb$	$^{206}Pb^{\#}/^{238}U$	$^{207}Pb^{\#}/^{235}U$	$^{207}Pb^{\#}/^{206}Pb^{\#}$	$^{206}Pb^{\#}/^{238}U$	$^{207}Pb^{\#}/^{235}U$	$^{207}Pb^{\#}/^{206}Pb^{\#}$ §§
Leucogranite (neosome) in migmatitic gneisses; Río Hondo Canyon, sample ZFC17												
A,W27	0.010	300	12.6	1628	0.06069	0.07396	0.04421 (0.3)	0.31986 (0.6)	0.05247 (0.6)	279.0	282.0	306.0 ± 10
B,W13	0.021	505	26.5	5306	0.06134	0.08129	0.05377 (0.8)	0.43638 (0.8)	0.05886 (0.1)	338.0	368.0	562.0 ± 2
C,W27	0.015	909	39.7	926	0.06730	0.08845	0.04371 (0.3)	0.31179 (0.4)	0.05179 (0.3)	275.5	275.6	276.4 ± 6
D,W35	0.005	879	78.0	3452	0.07207	0.10168	0.08845 (0.4)	0.83295 (0.4)	0.06830 (0.1)	546.3	615.2	877.8 ± 2
F,W15	0.012	779	125.0	1575	0.08455	0.16380	0.14872 (0.4)	1.55274 (0.4)	0.07576 (0.1)	893.5	951.6	1088.8 ± 2
G,W23	0.010	257	40.2	329	0.11474	0.28128	0.12569 (0.6)	1.24615 (0.7)	0.07210 (0.3)	761.7	822.0	988.9 ± 6
H,W13	0.009	495	82.8	363	0.11421	0.18344	0.14390 (0.4)	1.49635 (0.5)	0.07562 (0.3)	865.0	929.2	1084.9 ± 6
E,W7	0.011	244	29.4	2674	0.07248	0.18491	0.11205 (0.4)	1.04414 (0.4)	0.06759 (0.2)	684.6	726.0	856.1 ± 4
Cozahuico Granite; Cozahuico Canyon, sample GC-1 (Coza2)												
1a, abr, elg, dm (8 grn)	0.103	123	9.5	55748	0.0651	0.0614	0.08002 (0.4)	0.71495 (0.4)	0.06480 (0.5)	496.2	547.7	768.0 ± 4
3a, abr, elg, dm (7 grn)	0.023	155	10.4	3835	0.0642	0.0708	0.06893 (0.4)	0.5746 (0.4)	0.06046 (1.3)	430.0	461.0	620.0 ± 8
3b, abr, stb, dm (24 grn)	0.024	259	26.3	75135	0.0721	0.0997	0.10072 (0.4)	0.99834 (0.4)	0.07189 (0.3)	618.6	703.0	982.8 ± 3
pm1, stb, abr (26 grn)	0.021	626	50.0	24144	0.0671	0.0615	0.08178 (0.4)	0.7498 (0.4)	0.06650 (0.6)	507.0	568.0	822.0 ± 5
1b, abr, dm (8 grn)	0.035	184	12.6	3359	0.0930	0.1574	0.116 (2)	1.116 (2)	0.0692 (7)	713.1	761.3	905.6 ± 23
2, dm (1grn)	0.007	107	22.0	212	0.1490	0.2849	0.0895 (9)	0.818 (1)	0.0663 (6)	552.8	607.3	816.4 ± 22

*The ID-TIMS analyses were made by Robert López at the University of California, Santa Cruz.

†Fraction properties: grn—grains; abr—polished; dm—diamagnetics; elg—elongated and thin; stb—short and thick.

§The observed isotopic ratios were corrected for a 1‰ mass fractioning. The fractions were highlighted with the mixed $^{235}U/^{205}Pb$ tracer and were corrected for the effects of the tracer and blank. The 2σ errors in the $^{207}Pb/^{206}Pb$ and $^{208}Pb/^{206}Pb$ ratios are <0.8%, generally better than 0.1%; the errors in the $^{206}Pb/^{204}Pb$ ratio range from 0.3% to 1.9%.

#Radiogenic Pb.

**The decay constants are $^{238}U = 1.55125 \times 10^{-10}$; $^{235}U = 9.48485 \times 10^{-10}$; $^{238}U/^{235}U = 137.88$. The estimated errors for the U/Pb atomic ratios are ±0.4%, based on replica analysis of a single zircon crystal fraction. The errors (2σ, in %) are shown in parentheses.

††The zircon dissolution and the ionic interchange chemistry were modified according to Krogh (1973) and Mattinson (1987) in microcapsules as used by Parrish (1987). The total processed amount from the Pb blank varied between 2 pg and 30 pg, generally averaging <10 pg. The initial Pb compositions correspond to isotopic analysis of selected feldspars with 208:207:206:204 = 18.34:15.57:37.66:1 ratios. The isotopic data were measured using a Daly VG54-30 multicollector mass spectrometer with a pulse counter detector at the University of California, Santa Cruz.

§§The errors in the $^{207}Pb^{\#}/^{206}Pb^{\#}$ ages are 2σ according to PBDAT isotopic data processing software by Ludwig (1991).

TABLE 2. GEOCHRONOLOGIC U-Th-Pb DATA OF POINTED ANALYSIS USING SENSITIVE HIGH-RESOLUTION ION MICROPROBE (SHRIMP) IN ZIRCONS FROM THE COZAHUICO GRANITE (SAMPLE GC1) IN THE CALTEPEC FAULT ZONE*

Grains/ points	Cores/ rims	U†	Th	Th/U	^{206}Pb†	Atomic ratios§			Ages Ma#	Discordance**
		(ppm)	(ppm)		(%)	^{206}Pb/^{238}U	^{207}Pb/^{235}U	^{207}Pb/^{206}Pb	^{206}Pb/^{238}U	(%)
GC1-1	Rim	127	7	0.06	0.612	0.04107 (2.97)	0.32816 (33.44)	0.05795 (33.31)	257 ± 4	
GC1-18	Rim	114	18	0.15	0.160	0.04153 (1.63)	0.29261 (4.18)	0.05110 (3.84)	262 ± 4	
GC1-8	Rim	70	4	0.06	0.433	0.04201 (2.05)	0.27891 (8.13)	0.04815 (7.87)	266 ± 5	
GC1-4	Rim	121	4	0.03	0.478	0.04252 (1.75)	0.33691 (4.06)	0.05746 (3.67)	267 ± 5	
GC1-3	Rim	268	1	0.00	0.109	0.04240 (1.51)	0.31214 (2.97)	0.05339 (2.56)	267 ± 4	
GC1-20	Rim	111	4	0.04	0.090	0.04266 (1.66)	0.31774 (3.78)	0.05402 (3.40)	268 ± 4	
GC1-17	Rim	231	48	0.21	0.246	0.04297 (1.46)	0.32189 (2.64)	0.05434 (2.20)	270 ± 4	
GC1-11	Rim	293	3	0.01	0.118	0.04305 (1.49)	0.31715 (2.83)	0.05343 (2.40)	271 ± 4	
GC1-7	Rim	106	4	0.04	−0.031	0.04388 (1.93)	0.36428 (9.56)	0.06021 (9.36)	274 ± 5	
GC1-5	Rim	104	4	0.04	−0.136	0.04322 (1.80)	0.28412 (6.27)	0.04768 (6.00)	274 ± 5	
GC1-19	Rim	271	3	0.01	0.051	0.04374 (1.45)	0.32654 (3.43)	0.05415 (3.10)	275 ± 4	
GC1-21	Rim	272	5	0.02	−0.174	0.04381 (1.44)	0.30815 (2.56)	0.05101 (2.12)	277 ± 4	
GC1-6	Rim	612	62	0.10	−0.208	0.04520 (1.38)	0.30605 (2.29)	0.04911 (1.82)	286 ± 4	
GC1-2	Core	136	14	0.10	0.847	0.11111 (1.56)	1.04526 (2.63)	0.06823 (2.12)	674 ± 10	29
GC1-15	Core	178	87	0.49	0.723	0.11618 (1.41)	1.11912 (2.16)	0.06986 (1.64)	703 ± 10	30
GC1-9	Core	1145	103	0.09	0.361	0.14529 (1.32)	1.41515 (1.49)	0.07064 (0.68)	872 ± 11	8
GC1-12	Core	698	81	0.12	0.436	0.14690 (1.48)	1.46022 (1.65)	0.07209 (0.73)	880 ± 13	12
GC1-10	Core	689	95	0.14	0.142	0.15642 (1.34)	1.55321 (1.54)	0.07202 (0.75)	935 ± 12	5
GC1-16	Core	407	21	0.05	0.175	0.16282 (1.70)	1.62842 (1.87)	0.07253 (0.77)	971 ± 16	3
GC1-14	Core	1339	161	0.12	−0.106	0.20940 (1.41)	2.31863 (1.56)	0.08031 (0.66)	1227 ± 17	−2

*The analyses were made by Alexander Iriondo using the SHRIMP-RG at Stanford University, California, in co-ownership with U.S. Geological Survey. The followed SHRIMP process was similar to that reported by Williams (1998), and the isotopic data were graphed with IsoplotEx and Squid software (Ludwig, 2001a, 2001b). The primary oxygen ionic ray operated at 2–4 nA and excavated an area 25–30 μm in diameter at a depth of ~1 μm with a sensibility range of 5 to 30 cps/ppm of Pb.

†Common Pb. The analysis of the samples and of the standard zircon used, R33 of 419 Ma, were alternated for a closer control of the U/Pb ratio. The U and Pb concentrations have an uncertainty of 10–20%.

§Corrected atomic ratios for initial Pb using the amount of ^{204}Pb and the corresponding average value for the Earth from Stacey and Kramers (1975). The ^{206}Pb/^{238}U ratios were normalized to the R33 zircon standard. The 2σ errors are presented in parentheses and are expressed in %.

#Absolute errors are shown in the 1σ level in Ma.

**The degree of discordance is the percentage of the distance from the point of analysis to the concordia intercept (equal to its ^{207}Pb/^{206}Pb age) along a straight line extrapolated to the point of origin at 0 Ma.

intercept age of 987 ± 11 Ma is more consistent with the Zapoteco granulitic tectonothermal event of the Oaxacan Complex (Solari et al., 2003), which is probably the source of the inherited radiogenic components. Because of the ambiguity of the data and the results depending on a combination of discordant fractions, we considered that the age of the Cozahuico Granite was not clearly established with the six fractions analyzed.

Eight zircon fractions from the granitic neosome of the migmatitic gneisses were also analyzed (Table 1). Some of the fractions were found to be less discordant. One of them (C, W27) yielded a concordant age of 275.6 ± 1 Ma (Fig. 7B). The linear arrangement of the discordant fractions defines an upper intercept age of 1105 ± 80 Ma. The Early Permian concordant age is considered to be the migmatization age in the Caltepec fault zone. The Mesoproterozoic upper intercept age, similar to that of the Cozahuico Granite, also suggests that the gneisses of the Oaxacan Complex may be the protoliths and/or the sources of the inherited components. Since geological relations suggest that migmatization, partial melting of Mesoproterozoic gneisses, and the generation and emplacement of the Cozahuico Granite were broadly coeval events associated with the oblique collision of the Acatlán and Oaxacan Complexes, the age of the pluton, the ductile deformation along the Caltepec fault, and the collision event were considered to be Early Permian in age (Elías-Herrera and Ortega-Gutiérrez, 2000, 2002). More recent geochronological data, which are discussed in the following section, reinforce this interpretation.

Because the age of the Cozahuico Granite was not clearly determined by means of the conventional U-Pb geochronological method, the same sample (GC1) from the pluton was analyzed by SHRIMP. With this method the age of the pluton was clearly determined to be 270.4 ± 2.6 Ma, and inherited Mesoproterozoic components in the pluton are confirmed. The analyzed zircon crystals were handpicked from the nonmagnetic heavy mineral concentrates, and later mounted on epoxy to cut down half their thicknesses and then be polished. Analyses of 20 points were made in zircons whose ^{206}Pb/^{238}U ages are shown in Table 2 and their corresponding concordia diagrams in Figures 7C and 7D. Of these ages, 13 are concordant or closely concordant and correspond to spot analyses made on the outer rim of zoned zircon crystals (Fig. 8). The remaining seven ages are slightly discordant and represent spot analyses in the older cores of the zoned grains (Fig. 8). The concordant ages vary from 257 ± 4 Ma to 286 ± 4 Ma, and the average concordant age is 270.4 ± 2.6 Ma (Figs. 7C and 7D), which clearly corresponds to the time of crystallization of the Cozahuico Granite. It is important to note that,

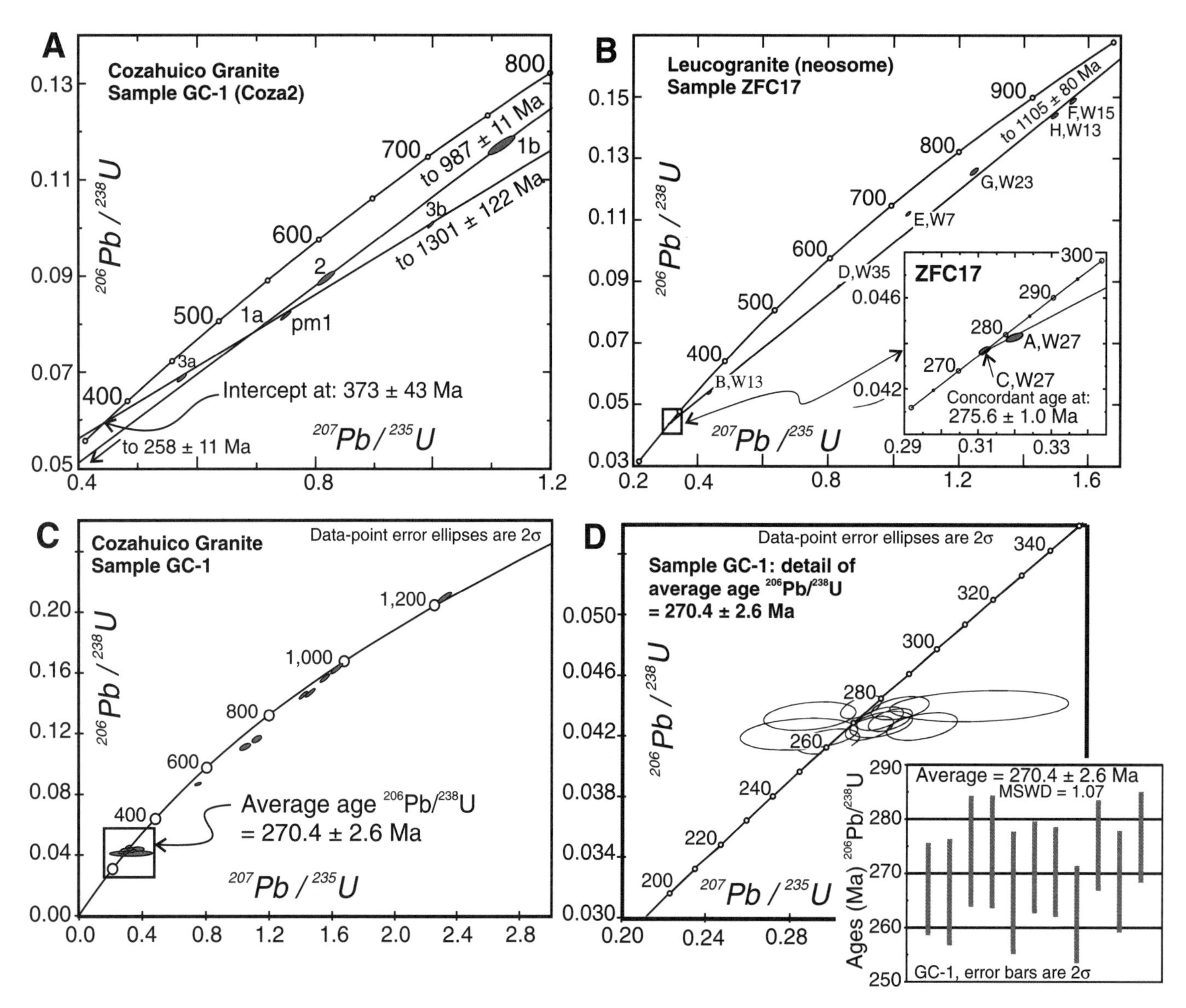

Figure 7. U-Pb concordia diagrams for (A) different discordant zircon fractions from the Cozahuico Granite (sample GC-1); (B) different discordant zircon fractions from granitic leucosome (sample ZFC17) in migmatitic gneisses; in this case, the concordant fraction C, W27 at 275.6 ± 1 Ma defines the age of migmatization in the Caltepec fault zone; (C) SHRIMP analyses in single zircon grains from the Cozahuico Granite (sample GC-1) with detail on the concordant data-point analyses which support an average $^{206}Pb/^{238}U$ age of 270.4 ± 2.6 Ma (D). The data for diagrams A and B, and C and D are shown in Tables 1 and 2, respectively. MSWD—mean square of weighted deviation.

among the discordant ages, the less discordant ones, 971 ± 16 Ma (3% discordant) and 1227 ± 17 Ma (–2% discordant), correspond to zircon cores or inherited zircon grains and are comparable to the granulitic and metaigneous rocks found in the northern portion of the Oaxacan Complex (Solari et al., 2003; Keppie et al., 2003). Moreover, most of the discordant ages have a linear arrangement whose upper intercept defines a ca. 1000 Ma age (Fig. 7C), which coincides with that of the granulitic metamorphism of the Oaxacan Complex.

The crystallization age of 270.4 ± 2.6 Ma for the Cozahuico Granite is slightly younger than the 275.6 ± 1 Ma of the granitic leucosome of the migmatites even within the error intervals. However, aside from the reasonable variation resulting from discrepancies in geochronological zircon methods in different laboratories, both Permian ages are considered to be related to the same tectonothermal event in the Caltepec fault zone as a result of the transpressive collision of the Acatlán and Oaxacan Complexes. The Cozahuico Granite, at least at its northern portion where it has been dated, may thus represent a late phase of this tectonomagmatic event. Taking into account the Early Permian ages and both the migmatites and the Cozahuico Granite in the fault zone in relation to the Leonardian age (Weber, 1997) of the

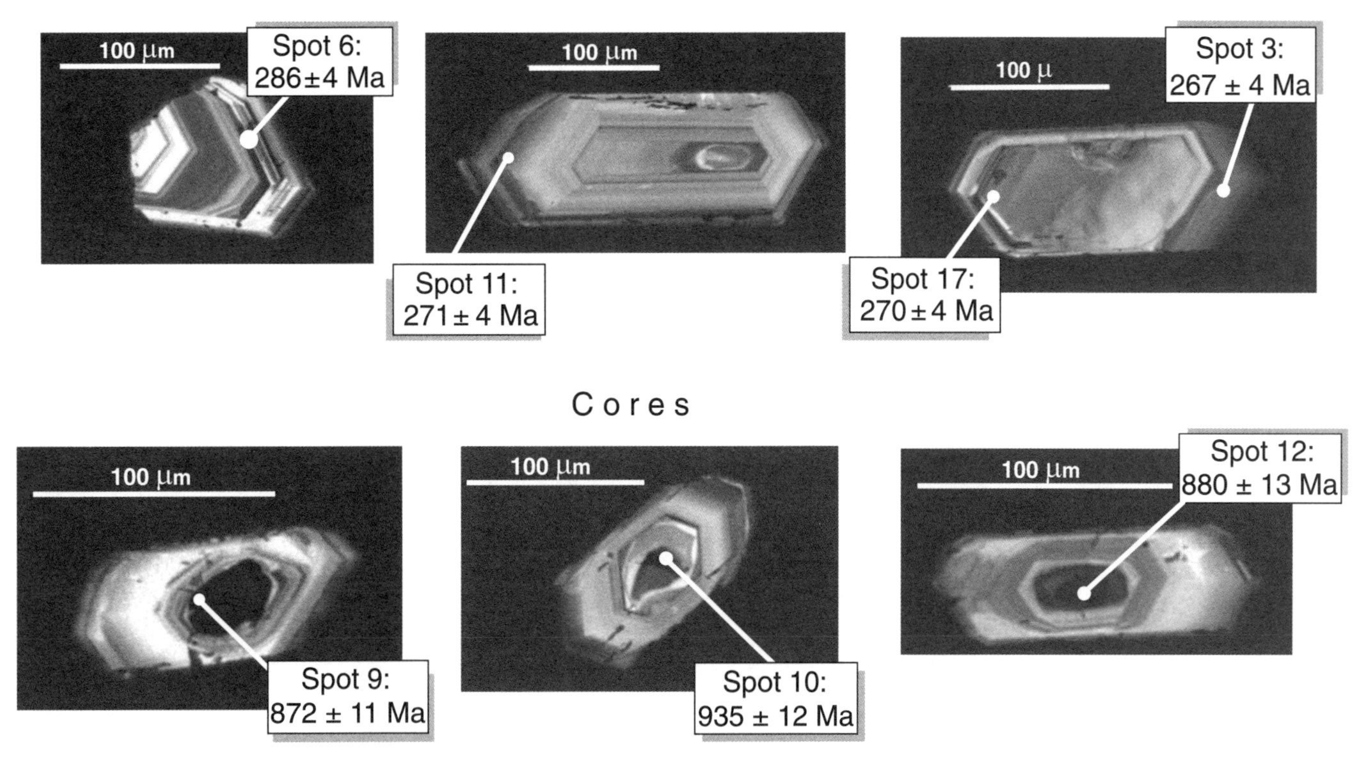

Figure 8. Zircon crystals from the Cozahuico Granite (sample GC-1). The photomicrographs were taken with cathodoluminescence for highlighting and illustrating the internal zoning in the single zircon grains and some of the spots of geochronological analyses by SHRIMP. All data for these analyses are shown in Table 2.

Matzitzi Formation, which is the oldest overlapping nonmetamorphic unit in the Metzontla area, a rapid exhumation for the ductile shear zone is implied. High rates of exhumation for the Caltepec fault zone are in agreement with the coexistence of magmatic and solid-state foliations (e.g., Miller and Paterson, 1994) in the deformed Cozahuico Granite and with the preservation of magmatic epidote as a metastable phase (e.g., Drummond et al., 1997) in this granite and the granitic neosomes. Thermochronological $^{40}Ar/^{39}Ar$ data also support a fast-exhumation model.

$^{40}Ar/^{39}Ar$ Data

The $^{40}Ar/^{39}Ar$ data and details of the analytical process, as well as the corresponding apparent age spectra, are shown in Table 3 and Figure 9, respectively. For the purposes of this study, a plateau age is obtained when the apparent ages of at least three segments or consecutive steps include a minimum of 55% of the released $^{39}Ar_k$ and are consistent, within a 2σ error, with the integrated age of the segments of the plateau. The error values presented in Table 3, the age spectra in Figure 9, and in the isotopic correlation diagrams (not shown) represent the analytical precision in ± 2σ. The entire $^{40}Ar/^{39}Ar$ data are available from M. Elías-Herrera, the first author of this paper. Sample SST4 (hornblende) corresponds to an amphibolitic melanosome band, alternating with granitic leucosome bands, in the migmatitic gneisses within the Cozahuico Granite. The migmatitic gneisses are in the hanging wall of a reverse fault thrust over mylonitized mica schist of the Acatlán Complex in the Agua Salada Canyon. Sample SST3 (muscovite) exactly corresponds to the mica schist located in the footwall of the fault. The location of both samples is shown in Figure 3.

The 278.06 ± 1.32 Ma plateau age for the hornblende (sample SST4) in the amphibolitic gneisses could, initially, be interpreted as a cooling age for the migmatites in the fault zone, and would indicate the time at which the rock crossed the 530 ± 40 °C isotherm (closing temperature for the retention of radiogenic Ar in hornblende; Harrison, 1982). However, the leucosomatic granitic portion yielded a concordant U-Pb (zircon) age of 275.6 ± 1 Ma (Figs. 6H and 7B) that dates the time of anatexis. Thus, the $^{40}Ar/^{39}Ar$ age for the hornblende, slightly higher than the concordant U-Pb (zircon) age for the adjacent leucosome, may rather reflect the excess of ^{40}Ar homogeneously distributed in the structure of the amphibole. The hornblende in the amphibolitic melanosome, as a restitic portion of the migmatites, probably absorbed radiogenic ^{40}Ar released by diffusion from other

TABLE 3. $^{40}Ar/^{39}Ar$ DATA FOR TWO SAMPLES FROM THE CALTEPEC FAULT ZONE†

Segment/ step	Laser power (watts)	Ca/K	^{40}Ar atm (%)	^{39}Ar (%)	$^{40}Ar^*/^{39}Ar_K$	Age (Ma)
Sample SST3 (muscovite, 15 mg), mica schist. Integrated age: 268.31 ± 1.34 Ma; correlation age: 268.53 ± 1.19 Ma (MSWD = 0.892); **plateau age: 268.59 ± 1.27 Ma** (mod. err. 0.92; 95.6% of ^{39}Ar released in six steps).						
1	0.75	0.038	35.15	0.27	12.404 ± 8.962	154.95 ± 107.27
2	1.50	0.025	7.78	1.26	20.042 ± 1.894	244.13 ± 21.57
3	2.00	0.000	0.00	2.92	23.201 ± 0.869	279.77 ± 9.71
4	2.50	0.002	4.16	9.63	22.304 ± 0.281	269.72 ± 3.16
5	3.00	0.000	1.75	17.55	22.062 ± 0.175	267.00 ± 1.96
6	4.00	0.000	0.38	25.37	22.250 ± 0.133	269.11 ± 1.50
7	5.00	0.000	0.54	14.00	22.197 ± 0.216	268.51 ± 2.43
8	6.00	0.001	0.07	13.25	22.290 ± 0.213	269.56 ± 2.39
9	7.00	0.006	0.30	15.75	22.156 ± 0.211	268.06 ± 2.37
Sample SST4 (hornblende, 22 mg), amphibolite. Integrated age: 276.35 ± 1.50 Ma; correlation age: 278.73 ± 2.06 Ma (MSWD = 2.418); **plateau age: 278.06 ± 1.32 Ma** (mod. err. 1.61; 94.2% of ^{39}Ar released in five steps).						
1	2.50	3.725	70.50	2.55	21.687 ± 2.434	263.08 ± 27.48
2	4.00	8.969	37.11	3.28	19.385 ± 1.033	236.91 ± 11.83
3	5.00	11.296	25.29	1.96	22.057 ± 1.359	267.25 ± 15.30
4	6.00	2.435	8.00	12.33	23.136 ± 0.296	279.36 ± 3.32
5	6.50	12.854	2.25	41.45	22.898 ± 0.144	276.70 ± 1.61
6	7.00	12.865	0.96	15.16	23.181 ± 0.247	279.86 ± 2.76
7	7.50	12.707	1.60	23.27	23.152 ± 0.158	279.54 ± 1.77

Note: MSWD—mean square of weighted deviates.

*The $^{40}Ar/^{39}Ar$ analyses were made by Amabel Ortega in the geochronology laboratory at Queen's University, Kingston, Ontario, Canada, using the laser stepwise heating procedure described by Clark et al. (1998). The data were corrected for blanks, mass discrimination, and induced neutron interference.

†The analyzed minerals were concentrated and cleaned by standard grinding, cleaning and selection, and manual separation using binocular microscope from 40–60 mesh fractions in the mineral separation laboratory at Instituto de Geología, UNAM. The separated minerals were charged in aluminum foil packets and radiated in the McMaster nuclear reactor (Hamilton, Ontario).

mineral phases, as the migmatization and the emplacement of the Cozahuico Granite likely reflect a syntectonic high-temperature (~700 °C) isothermal evolution, spanning ca. 6 Ma, linked to the transpressive collision event.

Regarding the 268.59 ± 1.27 Ma $^{40}Ar/^{39}Ar$ plateau age (muscovite) of the mylonitized mica schist (sample SST3) (Fig. 9), a metamorphic cooling age at 300–400 °C (closing temperature range for muscovite; Wijbrans and McDougall, 1986; Hames and Bowring, 1994) may be considered. This $^{40}Ar/^{39}Ar$ age is slightly younger than the 270.4 ± 2.6 Ma U-Pb (zircon) age of the Cozahuico Granite implying a very fast cooling of ~180 °C/Ma (Fig. 10), if the crystallization of mica in the schist was coeval with the syntectonic emplacement of the granite. Because the mylonitized mica schist precisely represents the footwall in the western boundary of the Caltepec fault zone, the age of the mica also dates a cooling stage of the ductile thrusting of the Cozahuico Granite and migmatitic gneisses on the Acatlán Complex. The $^{40}Ar/^{39}Ar$ age of the hornblende (probably with ^{40}Ar excess) in the migmatitic gneisses in the hanging wall from the same locality is inconsistent with a simple cooling history if the $^{40}Ar/^{39}Ar$ age of the mica is considered (Fig. 10).

Thus, the U-Pb and $^{40}Ar/^{39}Ar$ data suggest fast cooling consistent with a high exhumation rate. The fast uplifting of the Oaxacan Complex associated with the transpressive event was stratigraphically recorded in the Matzitzi Formation. The lower portion of this formation in the Coatepec area, 5 km to the east of the Caltepec fault zone, is characterized by a basal conglomerate containing rounded boulders of Proterozoic gneisses of several meters in diameter, indicating high-energy catastrophic deposits related to a very abrupt and unstable topographic relief.

Rb-Sr AND Sm-Nd ISOTOPE ANALYSES FOR MIGMATITES AND THE COZAHUICO GRANITE

The Rb-Sr and Sm-Nd isotope data for the banded migmatitic gneisses and for the Cozahuico Granite are shown in Table 4. Summarized details of the analytical procedure are included in the same table. The analyzed migmatitic gneiss samples belong to the alternating amphibolitic melanosome and granitic leucosome bands, and were collected at the same site of the sample (ZFC17) dated by U-Pb (zircon) at 275.6 ± 1 Ma (Figs. 3 and 6H). Since this age is considered the time of the migmatization, the initial isotopic relations were calculated for 275 Ma. The Sm-Nd model ages (T_{DM}) correspond to the depleted mantle evolutionary model of DePaolo (1981). The melanosome and leucosome bands have initial $^{87}Sr/^{86}Sr$ ratios of 0.70616–0.70686 and negative initial $\varepsilon Nd_{(i)}$ values which vary from −2.7 to −5.8, with T_{DM} ages from 1.2 to 1.6 Ga. The Cozahuico Granite is char-

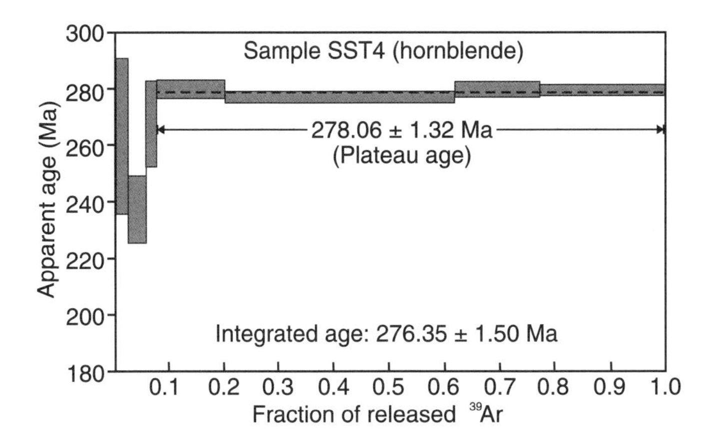

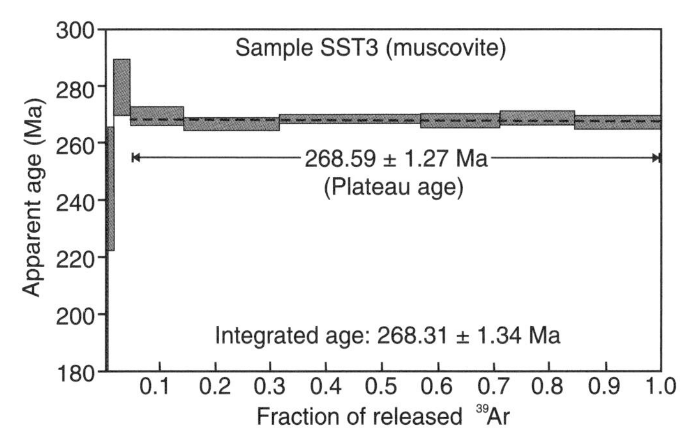

Figure 9. $^{40}Ar/^{39}Ar$ apparent age spectra (Table 3) by stepwise heating of hornblende (sample SST4) and muscovite (sample SST3) of amphibolite and mica schist from the Caltepec fault zone.

acterized by a lower initial $^{87}Sr/^{86}Sr$ ratio of 0.70435–0.70560, a $\varepsilon Nd_{(i)}$ between –3.2 and –3.6, and less varying T_{DM} ages of 1.9–1.2 Ga. Possibly due to chemical and mineralogical fractionation during migmatization (e.g., Zeng et al., 2005), the $^{147}Sm/^{144}Nd$ ratios of the melanosome samples (CO98/58C2 and CO98/58B) and one of the leucosome samples (CO98/58C1) are relatively high (>0.16), and abnormal model ages (>2.0 Ga) are yielded. In these instances the model ages were corrected for fractionation (Milisenda et al., 1994) at 275 Ma.

The initial $^{87}Sr/^{86}Sr \leq 0.70686$ ratios from the migmatitic gneisses and the Cozahuico Granite suggest that during the partial melting event and syntectonic magmatism within the Caltepec fault zone, the involved material corresponded to a poorly differentiated source material from the lower crust or it was simply an immature material. The Rb-Sr data are, however, inconsistent with the Sm-Nd data that indicate evolved crustal material. The Rb/Sr ratio, on which the initial $^{87}Sr/^{86}Sr$ ratios greatly depend, is very low in the gneisses and in the granite. For instance, the granitic leucosome and the Cozahuico Granite, with negative $\varepsilon Nd_{(i)}$ values ranging from –3.16 to –4.96, have Rb/Sr ratios which vary from 0.01 to 0.28. In comparison, Alleghenian granites (ca. 300 Ma) from the southern Appalachians, with similar negative $\varepsilon Nd_{(300\ Ma)}$ values, have 0.12–1.24 Rb/Sr ratios with initial $^{87}Sr/^{86}Sr$ ratios of 0.70642–0.70915 (Samson et al., 1995). Thus, it is probable that some Rb remobilization occurred in the rocks of the Caltepec fault zone, and that the isotopic Rb-Sr system did not remain completely closed during the reworking history of the fault zone. The Rb-Sr and Sm-Nd data, nevertheless, clearly reflect significant involvement of Proterozoic crust in the generation of the Cozahuico Granite.

According to local geological correlations, we interpret that the banded gneisses emplaced within the Cozahuico Granite are indeed Proterozoic gneisses from the Oaxacan Complex. They were completely rehomogenized by migmatization and partial melting during the Early Permian, and they were also involved in the formation of the Cozahuico Granite. The 1.36, 1.48, and 1.62 Ma T_{DM} ages for the granitic leucosome are in agreement with the model ages for the Oaxacan Complex (1.47–1.60 Ga; Ruiz et al., 1988a) and support this interpretation. However, the $\varepsilon Nd_{(i)}$ values for the migmatitic gneisses, as well as for the Cozahuico Granite (Table 4), are incongruent with the isotopic characteristics of the Oaxacan Complex during the Early Permian (Fig. 11). According to the Sm-Nd data of Ruiz et al. (1988a), the estimated $\varepsilon Nd_{(275\ Ma)}$ values for the Oaxacan Complex range from –6.5 to –10.2. Although corrections for Sm/Nd fractioning during the migmatization–partial melting event were made, the isotopic characteristics of the migmatitic gneisses and the Cozahuico Granite, with T_{DM} ages varying from ca. 1.0 to 1.6 Ga, cannot be explained solely by means of partial melting of the gneisses from the Oaxacan Complex.

On the basis of the available isotopic data, the Cozahuico Granite may be considered a result of the partial melting of Proterozoic crust with components of depleted mantle or poorly differentiated lower crust. The zircon cores of the granite with Proterozoic U-Pb ages constitute direct evidence for the involvement of the Oaxacan Complex in the generation of the pluton, which is consistent with the negative $\varepsilon Nd_{(i)}$ values indicating old crustal material as a source. On the other hand, the low initial $^{87}Sr/^{86}Sr$ ratio of 0.70435–0.70560 for the granite, in spite of the probable loss of Rb in the pluton, may reflect the poorly evolved material component. The Oaxacan Complex, as the involved crustal component, is characterized by relatively high Rb/Sr ratios (Ruiz et al., 1988b). The model ages of the granite, ranging around 1.0 Ga, may therefore have no geological meaning, and represent an intermediate value between the Proterozoic age of the continental crust and the involved Paleozoic material from the mantle. If this interpretation is correct, the mantle surely played an important role—not only as an active element in the petrogenesis of the Cozahuico Granite, but also as a heat-flow source in the tectonothermal event linked to the juxtaposition of the crystalline complexes. Mafic rock xenoliths (at centimeter scale) found in the Cozahuico Granite in some localities weakly suggest the presence of mantle magma during the process. In the southern Appalachians, the Alleghenian granites in the Inner Piedmont belt and in the Carolina terrane may represent similar cases to that of the Cozahuico Granite. Many of these plutons are syntectonic with

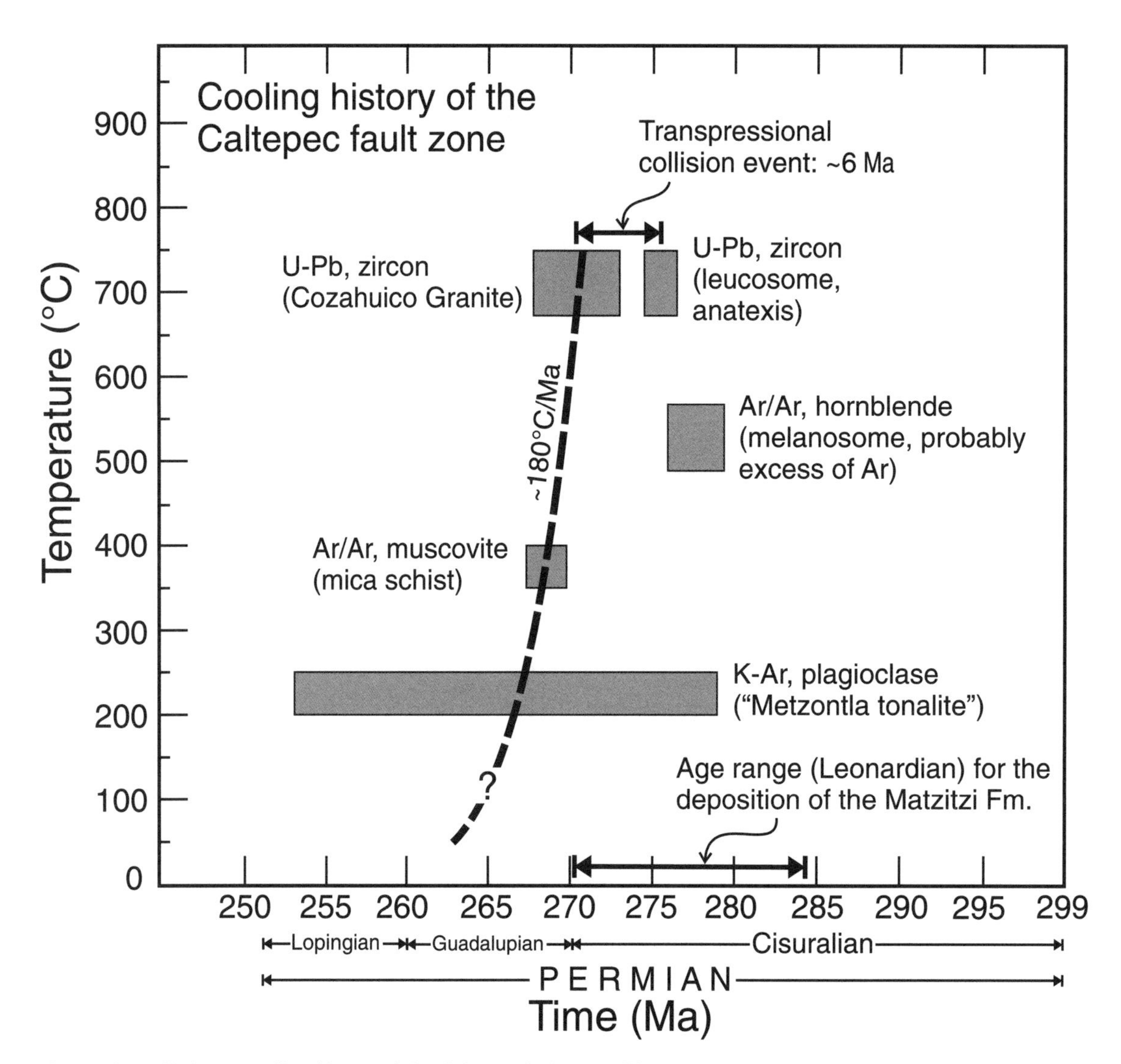

Figure 10. Preliminary cooling history of the Caltepec fault zone. The gray areas represent the error ranges in the geochronological data and in the temperatures for partial melting and magmatism in the fault zone, and closing temperatures for the retention of radiogenic ^{40}Ar (530 ± 40 °C for hornblende; Harrison, 1982; 350–400 °C for muscovite; Wijbrans and McDougall, 1986; Hames and Bowring, 1994; 200–250 °C for plagioclase; McDougall and Harrison, 1988). The K-Ar data for the "Metzontla tonalite" from the Los Reyes Metzontla area are from Torres-Vargas et al. (1986). The subdivisions and epochs in the Permian are according to Gradstein et al. (2004), and the so-called Leonardian age corresponds to the Artinskian and Kungurian ages of these authors. See text for discussion.

and linked to important thrust faults. Their isotopic Sm-Nd ratios, which are comparable to those of the Cozahuico Granite, mainly indicate that their formation was by anatexis of Grenvillian-type evolved continental crust, and they are considered to be collision plutons, not magmatic arc plutons (Samson et al., 1995).

In relation to the migmatitic gneisses, although the U-Pb (zircon) data of the granitic neosome clearly suggest inherited components from Grenvillian rocks, they do not seem to correspond to the isotopic evolution of the Oaxacan Complex (Fig. 11). Nonetheless, this may simply be the result of the lack of Sm-Nd isotopic data of the Oaxacan Complex because of its extent and lithologic diversity. The isotopic evolution of the complex, as seen in Figure 11, corresponds to only eight samples collected in the Oaxaca area. The migmatitic gneisses in Caltepec show good agreement with the Sm-Nd isotopic evolution of the Guichicovi Complex (Weber and Kohler, 1999), as well as with the Huiznopala and Novillo gneisses (Ruiz et al., 1988a) in the Hidalgo and Tamaulipas regions, units that, together with the Oaxacan Complex, shaped the Proterozoic microcontinent Oaxaquia (Ortega-Gutiérrez et al., 1995). The migmatitic gneisses, therefore, correspond to blocks belonging to this microcontinent and may represent fragments of the Oaxacan Complex that were rehomogenized into amphibolite facies and transported to middle crust levels by the Cozahuico Granite along the Caltepec fault zone.

TABLE 4. ISOTOPIC Rb-Sr AND Sm-Nd DATA FROM BANDED MIGMATITIC GNEISSES AND THE COAZHUICO GRANITE IN THE CALTEPEC FAULT ZONE*

Sample	Lithology	Rb (ppm)	Sr (ppm)	$^{87}Rb/^{86}Sr$	$^{87}Sr/^{86}Sr$ ± 1σ	$(^{87}Sr/^{86}Sr)i$**	Sm (ppm)	Nd (ppm)	$^{147}Sm/^{144}Nd$	$^{143}Nd/^{144}Nd$ ± 1σ	$\varepsilon Nd_{(0)}$	$\varepsilon Nd_{(i)}$**	T_{DM}[††] (Ga)
CO98/58A	Granitic leucosome	66.33	317.64	0.604	0.708522 ± 28	0.706159	0.18	0.79	0.135	0.512273 ± 16	–7.12	–4.96	1.48
CO98/58B	Amphibolitic melanosome	126.47	239.04	1.532	0.712356 ± 34	0.706362	6.29	22.85	0.166	0.512403 ± 22	–4.58	–3.53	1.28[§§]
CO98/58C1	Leucosome band	5.60	779.76	0.021	0.706945 ± 32	0.706863	3.01	10.97	0.166	0.512351 ± 19	–5.60	–4.52	1.36[§§]
CO98/58C2	Melanosome band	23.55	446.63	0.153	0.707394 ± 52	0.706796	12.96	43.95	0.178	0.512465 ± 18	–3.37	–2.73	1.22[§§]
CO98/58D	Granitic leucosome	63.33	338.15	0.542	0.708594 ± 44	0.706474	0.25	1.09	0.139	0.512239 ± 18	–7.78	–5.78	1.62
GC-1	Cozahuico Granite	33.50	903.61	0.107	0.704768 ± 37	0.704350	1.07	5.70	0.113	0.512315 ± 20	–6.30	–3.40	1.10
GC-1[†]	Cozahuico Granite						0.99	5.15	0.116	0.512318 ± ?	–6.24	–3.56	1.14
RT-16[§]	"La Compañía granite" (Cozahuico Granite)						2.74	16.75	0.099	0.512301 ± ?	–6.57	–3.16	1.01
RT-20[§]	Cozahuico Granite						0.82	4.08	0.122	0.512331 ± ?	–5.99	–3.60	1.19
634W-669W[#]	"Caltepec granite" (Cozahuico Granite)	97.56	350.33	0.668	0.7080 ± ?	0.7056 ± 4							

*The Sr, Sm, and Nd analyses were made using a FINNIGAN MAT 262 mass spectrometer with an ionic heat source in the Laboratorio Universitario de Geoquímica Isotópica (LUGIS) at the Instituto de Geofísica, Universidad Nacional Autónoma de México. The spectrometer has eight adjustable Faraday collectors and all measurements were made in static mode. The Rb analyses were made with an NBS mass spectrometer at the Instituto de Geología, Universidad Nacional Autónoma de México. All the Sr and Nd isotopic ratios were corrected for mass fractioning by normalizing to $^{86}Sr/^{88}Sr = 0.1194$ and $^{146}Nd/^{144}Nd = 0.7219$, respectively. The LUGIS values for the NBS 987 (Sr) and La Jolla (Nd) standards are $^{87}Sr/^{86}Sr = 0.710233 \pm 16$ ($\pm 1\sigma_{abs}$, n = 179), and $^{143}Nd/^{144}Nd = 0.511882 \pm 22$ ($\pm 1\sigma_{abs}$, n = 81), respectively. The relative uncertainties are $^{87}Rb/^{86}Sr = \pm 2\%$, and $^{147}Sm/^{144}Nd = \pm 1.5\%$ (1σ). The relative reproducibility (1σ) of the Rb, Sr, Sm, and Nd concentrations is of ±4.5%, ±1.8%, ±3.2%, ±2.7%, respectively. The averages of the analytical blanks obtained in the time of the sample analyses are: 0.26 ng for Rb, 4.3 ng for Sr, 1.8 ng for Sm, and 12.7 ng for Nd.

[†]Data from Robert López (personal commun., 15 October 1998) from the University of California at Santa Cruz, USA.

[§]Data reported by Torres-Vargas et al. (1999). The "La Compañía granite" corresponds to the Cozahuico Granite in its northern portion.

[#]Data reported by Ruiz-Castellanos (1979). Average values from six samples. The "Caltepec granite" corresponds to the Cozahuico Granite in its northern portion.

**The initial isotopic $ratios_{(i)}$ were calculated for 275 Ma.

[††]The model ages correspond to DePaolo's (1981) depleted mantle evolution.

[§§]Model ages corrected for fractioning using the equation given by Milisenda et al. (1994).

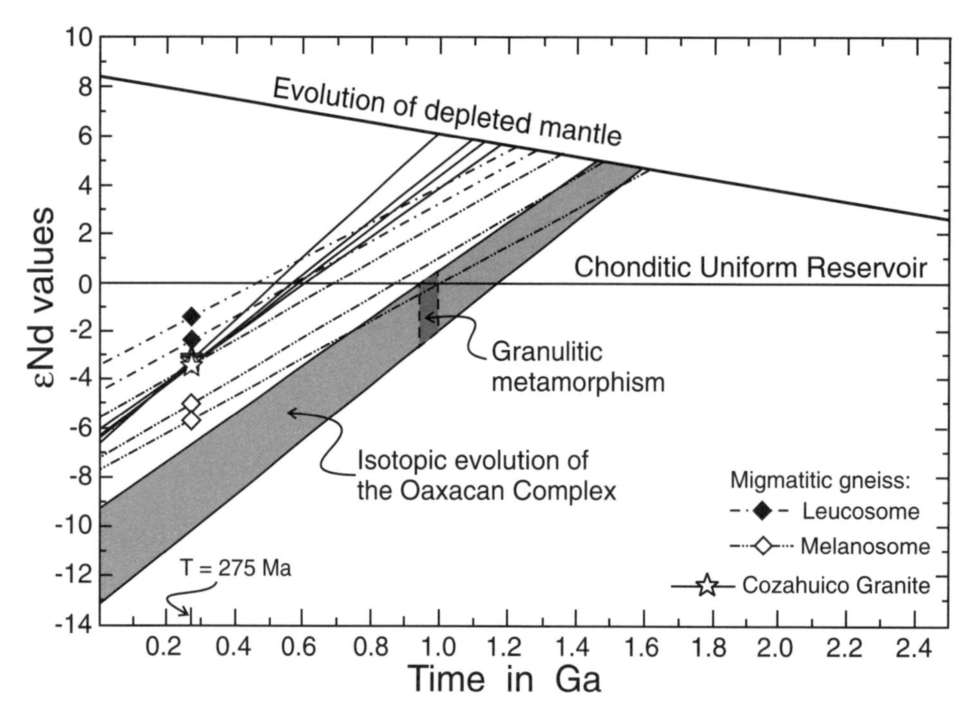

Figure 11. Initial $\varepsilon Nd_{(275\ Ma)}$ values for the Cozahuico Granite and migmatitic gneisses in the Caltepec fault zone in relation to the Sm-Nd isotopic evolution of the Oaxacan Complex (Ruiz et al., 1988a). The negative $\varepsilon Nd_{(275\ Ma)}$ values for the Cozahuico Granite suggest that it was formed by the anatexis of the Oaxacan Complex and a melting component of depleted mantle with $\varepsilon Nd \approx +7.5$.

REGIONAL IMPLICATIONS

From a regional view, the Caltepec fault zone, undoubtedly the tectonic boundary between the two most important crystalline complexes in southern México involving ductile deformation, metamorphism, and syntectonic magmatism, reflects the convergent movements that ultimately configured Pangea in its central-western section. The Early Permian tectonomagmatic event in the Caltepec fault zone coincided with the extinction of the Mississippian-Permian Las Delicias volcanic arc, due to the collision between the Gondwanan Coahuila terrane, and the southern margin of Laurentia, as well as with the final compressive pulses of the Marathon-Ouachita orogeny (Ross, 1986; López, 1997; Carpenter, 1997).

The Caltepec fault zone, despite its limited outcrops encompassing ~20 km in length, must have a regional continuity to the north as far as the Trans-Mexican Volcanic Belt, and to the south up to its abrupt truncation at the Xolapa Complex, the crystalline basement of the Chatino terrane. The fault zone constitutes a terrane boundary for which a high degree of allochthoneity is suggested by the remarkably unrelated nature of the involved metamorphic complexes. Fundamental questions to address in any regional analysis of this tectonic boundary are, Why is the suture that joins both continental crystalline blocks direct even though there are no apparent oceanic elements between the blocks? Is it the result of an unidentified suture in which these elements were eliminated by erosion? How much displacement does the fault zone represent in its ductile deformation? According to our database, these issues may only be partially approached, and future studies are needed to offer more conclusive answers.

The southward prolongation of the Caltepec fault is clearly confirmed in the Tlapiltepec–Tejuapan area, at least in its Neogene reactivation, as discussed below. On the southern extension of the fault, in the Cuanana area (Fig. 1), 150 km south of Caltepec, a gabbroic-dioritic pluton is unconformably overlain by Middle Jurassic sandstone and shale containing fossil plants. The Cuanana pluton shows north-trending mylonitized zones with dextral shear movement (Vega-Carrillo et al., 1998) consistent with the ductile deformation in the Caltepec fault zone. The Cuanana pluton yielded a K-Ar (hornblende) age of 260 ± 20 Ma (Grajales-Nishimura et al., 1986). We recently obtained a U-Pb SHRIMP (zircon) age of 306.7 ± 2.3 Ma (Middle Pennsylvanian) for the Cuanana pluton; thus this pluton may be correlated with the enigmatic Juchatengo Complex (Grajales-Nishimura, 1988; Grajales-Nishimura et al., 1999), located between the Acatlán and the Oaxacan Complexes in southernmost México. The Juchatengo Complex has been interpreted as an upper Paleozoic continental rift adjacent to the Oaxacan Complex (Grajales-Nishimura et al., 1999). However, owing to its probable Pennsylvanian age and its location between two metamorphic complexes, as well as to its oceanic lithotectonic characteristics with poorly evolved igneous rocks, the Juchatengo Complex may correspond to a remnant of oceanic lithosphere that initially separated the crystalline continental blocks now juxtaposed in the Caltepec area (Elías-Herrera and Ortega-Gutiérrez, 2002). If this interpretation is correct, the deformation and greenschist facies regional metamorphism in the Juchatengo Complex must be Early Permian in age and related to the juxtaposition of the Acatlán and Oaxaqueño Complexes.

Assuming that the Acatlán Complex was separated by oceanic crust from the Oaxacan Complex—a southern portion of the microcontinent Oaxaquia—before the Permian, and that they later collided by dextral transpressional faulting (Elías-Herrera and Ortega-Gutiérrez, 2002), the magnitude of the paleogeographic (precollision) displacement and the one recorded in the Caltepec fault zone during the collision cannot be accurately determined. In the first case, the lack of recent paleomagnetic data or of faunal correlations is a poorly addressed problem, and in the second case, the syntectonic magmatic event probably prevented the complete recording of the movements in the fault zone. The paleogeographic displacement between the Acatlán and Oaxaquia blocks may speculatively be estimated at ~600 km, if the relative position of the Acatlán Complex during the Late Mississippian–Early Pennsylvanian until their collision in the Early Permian proposed by Elías-Herrera and Ortega-Gutiérrez (2002) is reasonably valid.

REACTIVATIONS OF THE CALTEPEC FAULT ZONE

The Caltepec fault zone, according to the reactivation criteria established by Holdsworth et al. (1997), is clearly a long-lived polydeformed structure with superimposed ductile and brittle deformation, which was repeatedly affected by subsequent tectonic movements after its consolidation during the Early Permian. The later deformation in the fault zone is essentially brittle, and it is characterized by high-angle faults, numerous minor faults, intense cataclasis, and brecciation, implying changes in the distribution and nature of the involved stresses. The brittle deformation produced fault planes and slickensides with different strikes and dips, and with crystalline developments of fibrous quartz, chlorite, serpentine, and epidote. The minerals related to the cataclasis are tremolite, sericite, albite, epidote, quartz, and carbonates in veinlets, chlorite, Fe-Ti oxides, pyrite, and hematite. The Cozahuico Granite, in some cataclastic sectors, displays pseudotachylites in veinlets and micro-veinlets. The nature, duration, and magnitude of post-Permian activity of the fault zone are unknown, although it seems that Mesozoic and Cenozoic reactivations had a significant influence on the depositional patterns of sedimentary sequences in the region (e.g., López-Ticha, 1985; Meneses-Rocha et al., 1994). The tectonic reactivations of the fault zone throughout its history, as it is known today, are evidenced by stratigraphic, structural, geochronological, and neotectonic criteria.

Mesozoic Reactivations

In the Oaxaca fault, 50 km east of the Caltepec fault zone, the Middle Jurassic dextral movement of the Yucatán block related to the opening of the Gulf of Mexico was clearly recorded (Alaniz-Álvarez et al., 1996). In this event, the Caltepec fault zone was probably reactivated as a transtensive or distensive structure as suggested by clastic deposits in the region. The Middle Jurassic red beds overlying the Acatlán Complex in the nearby area of Chazumba, Oaxaca, indeed imply extensional tectonic activity during this time. However, it is difficult to determine provenance of this activity because the detrital material mainly consists of quartz and quartzite. The apparent absence of Jurassic red beds over the Oaxacan Complex in the area of the Caltepec fault zone suggests moderate tectonic activity; however, the distribution of the Jurassic clastic rocks seems to be controlled by the Caltepec fault in its southern extension (López-Ticha, 1985). The Jurassic reactivation of the Caltepec fault zone may be evidenced by some isotopic data. In its northern extension at the Caltepec-Metzontla area, the K-Ar ages of 167 ± 8 Ma (biotite) and 163 ± 8 Ma (plagioclase) of the Cozahuico Granite and the "Los Reyes dike," respectively (Torres-Vargas et al., 1986), both with intense cataclasis, may correspond to isotopic rehomogenization related to a Middle Jurassic tectonic activity.

On the eastern and western blocks of the Caltepec fault zone, the Upper Jurassic rocks represent low-energy marine deposits, after which continental high-energy conditions were renewed during the Early Cretaceous with the deposit of a thick sequence of conglomerates with large fragments of gneiss and schist on both blocks in the Caltepec-Coatepec area. It is important to point out that in the eastern part of the Zapoteco terrane adjoining the Oaxaca fault, a Lower Cretaceous pelagic sedimentary sequence is found (Alzaga-Ruiz, 1991). This sequence indicates a period of quiescence of the Oaxaca fault for that time, contrasting with the tectonic activity and deposition of clastic continental sediments, which were occurring in the Caltepec region, although in the Tehuacán-Teotitlán area, bordering the northern portion of the Oaxaca fault (Fig. 1), an Early Cretaceous migmatization event affecting an Upper Jurassic–Lower Cretaceous volcanosedimentary sequence (Ángeles-Moreno et al., 2004) suggests an intricate tectonic reactivation of the Oaxaca fault in this region. Thus it seems that during the Mesozoic history of the Caltepec fault zone, its reactivation movements were coupled with those of the Oaxaca fault, accommodating important crustal deformation in southern México.

Cenozoic Reactivations

The Cenozoic tectonics associated with the Caltepec fault zone reflects a process in which the Mixteco terrane at the west of the fault seems to have been the downthrown block during most of the Cenozoic. The Cenozoic reactivation of the Caltepec fault zone is clearly evidenced by the tectonic wedge of Paleogene conglomerate with boulders of Cretaceous fossiliferous limestone (Fig. 5). The conglomerate is bounded by normal faults which affect the original unconformity and have unknown differential displacements that tilted the thick conglomerate strata 50° toward the east, although in the west side of the tectonic wedge a thin crust of the conglomerate was preserved as an unconformable contact on the Cozahuico Granite. The development of these brittle high-angle faults and their extensive lateral propagation were mainly triggered by the preexisting vertical mylonitic foliation in that sector of the fault zone, which show evidence of changes in the kinematic history of older structures. The normal faults that disrupted the Oaxacan Complex gneisses (Fig. 6B) are also part of this reactivation; however, these faults do not display much lateral continuity, contrasting with those formed in the mylonitic zone. The timing of the Cenozoic reactivations in the region is unknown, although it is probable that they were recurrent. An example of this recurrence is the Santa Lucía fault, a structure parallel to the Caltepec fault zone (Figs. 3 and 12); the Santa Lucía fault and the Caltepec fault zone may have been simultaneously active during the Paleocene-Eocene as suggested by the field relations. Nonetheless, the Santa Lucía fault apparently did not affect the Oligocene volcanosedimentary rocks in the area, whereas brittle faults in the Caltepec zone did (Fig. 12).

The volcanosedimentary sequence that covers the tectonic contact between the Acatlán and Oaxacan Complexes was described in detail in the Tamazulapan area, as the Tamazulapan Conglomerate, Yanhuitlán Formation, Llano de Lobos Tuff, Cerro Verde Tuff, Yucudaac Andesite, San Marcos Andesite, and Chilapa Formation (Ferrusquía-Villafranca et al., 1974; Ferrusquía-Villafranca, 1976). The upper part of the Llano de Lobos Tuff and the base of the Yucudaac Andesite yielded K-Ar (biotite and whole rock) ages of 28.9 ± 0.6 Ma and 26.2 ± 0.5 Ma, respectively (Ferrusquía-Villafranca et al., 1974). This sequence becomes as thick as 1000 m and shows several unconformities that may be recording part of the Cenozoic tectonic history of the Caltepec fault zone. For instance, the brittle reactivation of the Caltepec fault in its south extension in the Tlapiltepec-Tejuapan area (Fig. 12) cuts the Oligocene pyroclastic, volcaniclastic, and volcanic rocks of the Llano de Lobos Tuff and the Yucudaac Andesite, indicating Neogene tectonic activity. This activity is defined by a fault with vertical movements and an important lateral component related to the local folding of the volcaniclastic cover in this area. Martiny et al. (2000) proposed that the regional distribution of the Oligocene volcanic rocks in southern México is controlled by strike-slip and oblique faults that as a whole define a stepped system with sinistral displacements.

Recent Tectonic Activity

Recent tectonic activity may be occurring along the Caltepec fault zone, as suggested by the available seismologic data. According to data recorded by the national seismologic network, in just the past 14 yr nearly 100 earthquakes with magnitudes ranging from 3.4 to 7.0, and foci located at depths between 6 km and 178 km, have occurred in the Tehuacán-Nochixtlán region (Servicio Sismológico Nacional, personal commun., 28 April 2004). Given the epicenter locations of these earthquakes, only 21 may be somewhat related to the Caltepec fault system (Fig. 12). However, according to the geometry of the Rivera and Cocos plates subducting under the North American plate in southern México (Pardo and Suárez, 1995), the depth of the Wadati-Benioff zone, which varies from ~80 km to ~50 km in the northern and southern limits, respectively, of the Tehuacán-Nochixtlán region, and taking into account that the crustal thickness is 40–45 km in this portion of southern México (Urrutia-Fucugauchi and Flores-Ruiz, 1996; Campos-Enríquez and Sánchez-Zamora, 2000), the number of earthquakes that may have a direct relation to the Caltepec fault system is significantly diminished.

According to this crustal structure and the depth of the foci (Fig. 12), several of the earthquakes located along the trace of the Caltepec fault consequently occurred on the subducted oceanic plate or in the Wadati-Benioff zone, others took place in the subcontinental lithospheric mantle wedge, and only three occurred in the mid- to lower crust. The earthquake with the greatest magnitude (7) and whose focus was near Caltepec at 69 km depth (Fig. 12) is probably related to the Cocos plate sliding in the Wadati-Benioff zone, whereas the earthquakes in the mantle and crust, with 3.5–4.3 magnitudes, suggest a brittle behavior that may be linked to the extension at depth of the Caltepec fault zone as a lithospheric structure. In areas with low heat flow, as may be the case in the Caltepec fault zone region, seismic rupture has been reported not only in the lower crust, but also in the upper mantle (e.g., Chen and Molnar, 1983). This is congruent with the fact that the northern prolongation of the fault zone coincides with that part of the Trans-Mexican Volcanic Belt that is the narrowest and most distant from the Mesoamerican Trench. At present, however, it is clear that the seismotectonics that may be related to the Caltepec fault zone is an issue that should be approached with further studies.

CONCLUSIONS

According to the current state of knowledge concerning the Caltepec fault zone as discussed here, and in order to clarify what has been achieved up till now, we are listing the main conclusions, while taking into consideration that a full understanding of this structure throughout its entire tectonic history is essential for the complete comprehension of the geological evolution of southern México.

1. The Caltepec fault zone is a long-lived geological boundary or crustal structure that clearly represents the tectonic boundary between the Acatlán and Oaxacan Complexes, crystalline basements of the two most important terranes of southern México, the Mixteco and Zapoteco terranes, respectively.
2. Because of the regional significance of the fault zone and the quality of its outcrops at the Caltepec area, the geological record of its prolonged tectonic activity includes meso- and epicrustal structural features that may be studied in exceptional detail from one block to the other. The Caltepec fault zone is thus an ideal natural laboratory for studying ductile deformation and syntectonic magmatism operating at mid-crustal levels, as well as brittle reactivation mechanisms of large-scale faults.
3. The oldest and deepest activity documented in the fault zone is a tectonomagmatic-metamorphic event characterized by migmatization, syntectonic magmatism, and dextral transpressive shearing, which was dated between

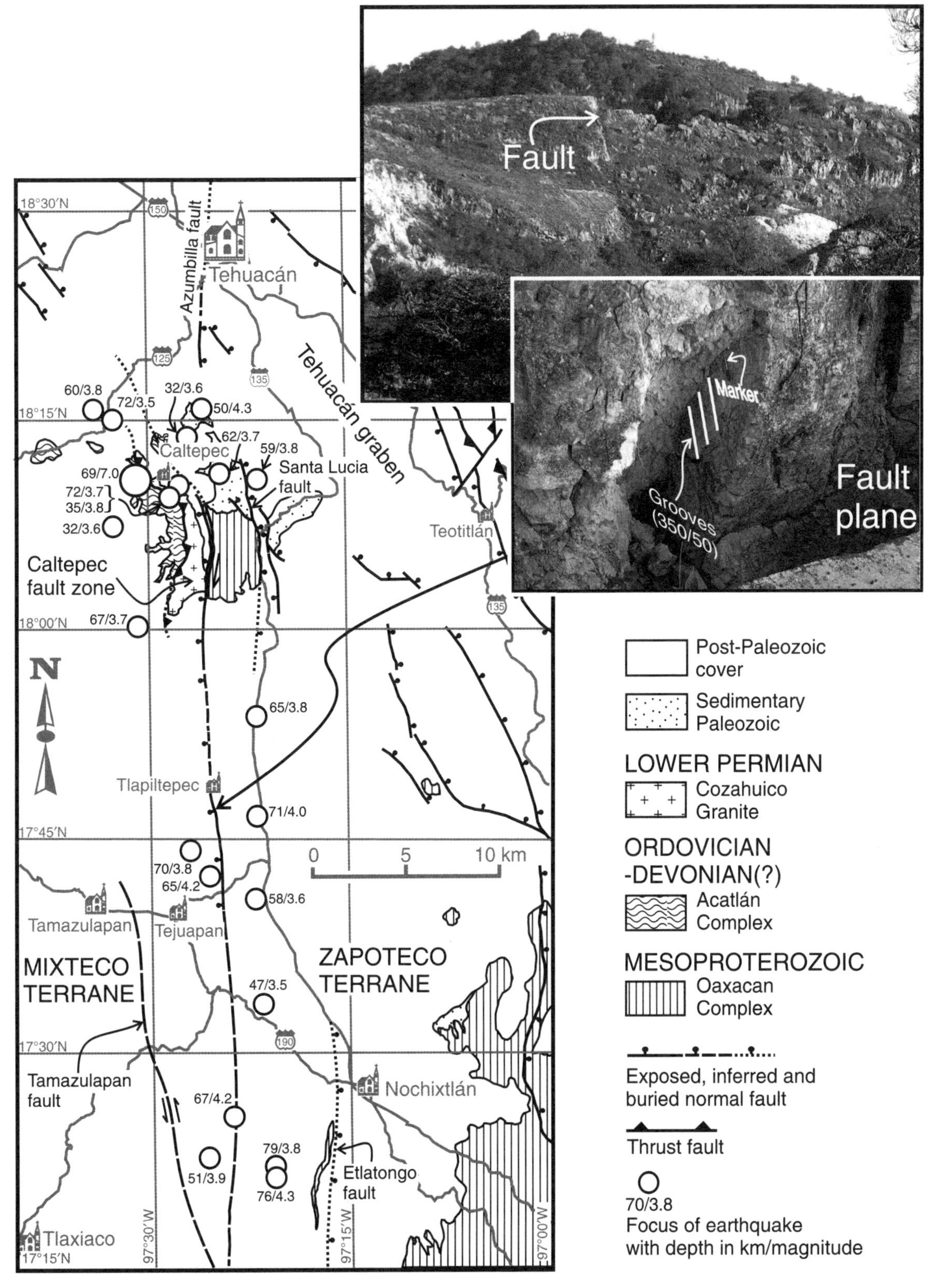

Figure 12. Brittle reactivation of the Caltepec fault zone along its southern extension, in north-central Oaxaca. In the Tlapiltepec-Tejuapan area, northeast of Tamazulapan, the fault cuts Oligocene pyroclastic, volcaniclastic, and volcanic rocks. Structural features suggest a normal movement with hanging wall to the west with a minor lateral component. The trace of Tamazulapan fault with sinistral lateral displacement is after López-Ticha (1985). Earthquakes that occurred between 1990 and 2003 that may be related to the recent tectonic activity of the Caltepec fault zone are included. The depth of the focus and magnitude of most of these earthquakes range from 32 km to 79 km and 3.5–4.3, respectively, and only one had a 7.0 magnitude, with its epicenter near Caltepec village (Servicio Sismológico Nacional, personal commun., 28 April 2004). See text for discussion.

275.6 ± 1.0 Ma (anatectic leucosome) and 270 ± 2.6 Ma (Cozahuico Granite). This event is related to an oblique collision between the southern Oaxaquia (Oaxacan Complex) and Acatlán continental blocks. During the Early Permian, the collision event formed part of a paleogeographically complex collisional front between Laurentia and Gondwana that was related to the final consolidation of Pangea.

4. Because of its petrographical and structural features, the Cozahuico Granite, emplaced within the fault zone, is considered a syntectonic intrusive, resulting from the local melting of Proterozoic crust during the activity of the fault, but with a significant contribution of subcrustal magmas. Rb-Sr and Sm-Nd isotopic data support this conclusion.
5. A high rate of exhumation for the fault zone is suggested by magmatic epidote in the Cozahuico Granite and in the anatectic leucosome, a $^{40}Ar/^{39}Ar$ cooling age of 268.6 ± 1.27 Ma of white mica in a mylonitized mica schist, quite similar to the magmatic and anatectic zircon ages, and by the Leonardian age of the oldest unconformable overlapping Matzitzi Formation.
6. The Caltepec fault zone, after its exposure to the surface ca. 265 Ma, underwent intermittent tectonic activity during the Mesozoic and Cenozoic as indicated by the stratigraphic record in the region. This activity could be linked to the tectonic movements associated with the breaking up of Pangea and the evolution of the southern Pacific margin of México. Neogene reactivations are indicated by crosscutting relations with Oligocene rocks, and neotectonic activity in the fault zone is suggested by the deep foci seismological data, all of which support the theory that the Caltepec fault zone is a long-lived crustal discontinuity.

ACKNOWLEDGMENTS

The CONACYT 36821–T project funded this work. Bodo Weber and Susana Alaniz critically reviewed the manuscript, and their suggestions and comments were very useful in focusing our ideas and improving this paper. We thank Edgar Angeles Moreno for his help in some stages of the fieldwork. Diego Aparicio made several dozen thin sections for petrographical purposes and Sonia Angeles was actively involved in the mineral separation process for dating. A. Iriondo is grateful to Dr. Joe Wooden of the USGS for his support during the SHRIMP zircon dating process. The Servicio Sismológico Nacional in the Instituto de Geofísica, UNAM, kindly provided the earthquake data from the Tehuacán, Puebla, and central Oaxaca regions. Andrew F. Boni and Male Alcayde gently collaborated in the translation of this paper into English.

REFERENCES CITED

Alaniz-Álvarez, S.A., van der Heyden, P., Nieto-Samaniego, A.F., and Ortega-Gutiérrez, F., 1996, Radiometric and kinematic evidence for Middle Jurassic strike-slip faulting in southern Mexico related to the opening of the Gulf of Mexico: Geology, v. 24, p. 443–446, doi: 10.1130/0091-7613(1996)024<0443:RAKEFM>2.3.CO;2.

Alzaga-Ruiz, H., 1991, Estratigrafía y consideraciones paleogeográficas de las rocas del Jurásico Tardío-Cretácico Temprano, en el área de Tomellin-Santiago Nacaltepec, Oaxaca, México: Revista del Instituto Mexicano del Petróleo, v. 23, no. 2, p. 17–27.

Ángeles-Moreno, E., Elías-Herrera, M., Sánchez-Zavala, J.L., Macías-Romo, C., Ortega-Rivera, A., and Iriondo, A., 2004, Terreno Cuicateco, sur de México: evolución tectónica de una cuenca pull-apart para 145–132 Ma: IV Reunión Nacional de Ciencias de la Tierra, Libro de Resúmenes, Centenario de la Sociedad Geológica Mexicana, p. 66–67.

Brown, M., and Solar, G.S., 1998, Shear-zone systems and melts: Feedback relations and self-organization in orogenic belts: Journal of Structural Geology, v. 20, p. 211–227, doi: 10.1016/S0191-8141(97)00068-0.

Bullard, E.C., Everett, J.E., and Smith, A.G., 1965, The fit of continents around the Atlantic, *in* Blackett, P.M.S., Bullard, E.C., and Runcorn, S.K., eds., A Symposium on Continental Drift: Royal Society of London Philosophical Transactions, Series A, Mathematical and Physical Sciences 258, p. 41–51.

Calderón-García, A., 1956, Bosquejo geológico de la región de San Juan Raya, Pue., *in* XX Congreso Geológico Internacional, Estratigrafía del Mesozóico y tectónica del sur de Puebla; Presa de Valsequillo, Sifón de Huexotitlanapa y problemas hidrológicos de Puebla, Libro guía de la excursión A-11: México, El Congreso, p. 9–27.

Campos-Enríquez, J.O., and Sánchez-Zamora, O., 2000, Crustal structure across southern Mexico inferred from gravity data: Journal of South American Earth Sciences, v. 13, p. 479–489, doi: 10.1016/S0895-9811(00)00045-6.

Carpenter, D.L., 1997, Tectonic history of the metamorphic basement rocks of the Sierra del Carmen, Coahuila, Mexico: Geological Society of America Bulletin, v. 109, p. 1321–1332, doi: 10.1130/0016-7606(1997)109<1321:THOTMB>2.3.CO;2.

Chen, W.P., and Molnar, P., 1983, Focal depths of intracontinental and intraplate earthquakes and their implications for the thermal and mechanical properties of the lithosphere: Journal of Geophysical Research, v. 88, no. B5, p. 4183–4214.

Clark, A.H., Archibald, D.A., Lee, A.W., Farrar, E., and Hodgson, C.J., 1998, Laser probe $^{40}Ar/^{39}Ar$ ages of early- and late-stage alteration assemblages, Rosario porphyry copper-molybdenum deposit, Collahuasi District, I Region, Chile: Economic Geology and the Bulletin of the Society of Economic Geologists, v. 93, p. 326–337.

de Cserna, Z., 1970, Reflexiones sobre algunos de los problemas de la geología de la parte centromeridional de México, *in* Segura, R.L., and Torres-Rodríguez, R., eds., Libro-guía de la excursión México-Oaxaca: Sociedad Geológica Mexicana, p. 37–50.

DePaolo, D.J., 1981, Neodymium isotopes in the Colorado Front Range and crust-mantle evolution in the Proterozoic: Nature, v. 291, p. 193–196, doi: 10.1038/291193a0.

Drummond, M.S., Neilson, M.J., Allison, D.T., and Tull, J.F., 1997, Igneous petrogenesis and tectonic setting of granitic rocks from the eastern Blue Ridge and Inner Piedmont, Alabama Appalachians, *in* Sinha, A.K., Whalen, J.B., and Hogan, J.P., eds., The Nature of Magmatism in the Appalachian Orogen: Boulder, Colorado, Geological Society of America Memoir 191, p. 147–164.

Elías-Herrera, M., and Ortega-Gutiérrez, F., 1998, The Caltepec fault zone, southern México: Devonian dextral shear interaction between the Precambrian Oaxacan and Paleozoic Acatlán basement Complexes: International Geological Correlation Program, Project No. 376, Laurentia-Gondwana connections before Pangea: Oaxaca, Universidad Nacional Autónoma de México, Instituto de Geología, Program and Abstracts, p. 15.

Elías-Herrera, M., and Ortega-Gutiérrez, F., 2000, Roots of the Caltepec Fault Zone, southern México: Early Permian epidote-bearing anatexitic granitoids: 2a Reunión Nacional de Ciencias de la Tierra: Resúmenes y Programas, Geos, v. 20, no. 3, p. 323.

Elías-Herrera, M., and Ortega-Gutiérrez, F., 2002, The Caltepec fault zone: An Early Permian dextral transpressional boundary between the Proterozoic Oaxacan and Paleozoic Acatlán Complexes, southern Mexico and regional tectonic implications: Tectonics, v. 21, p. 1–19, doi: 10.1029/2000TC001278.

Ferrusquía-Villafranca, I., 1976, Estudios geológico-paleontológicos en la región Mixteca, Parte 1, Geología del área Tamazulapan, Teposcolula-Yanhuitlán, Mixteca Alta, Estado de Oaxaca: Universidad Nacional Autónoma de México, Instituto de Geología, Boletín, v. 97, 160 p.

Ferrusquía-Villafranca, I., Wilson, J.A., Denison, R.E., Mc Dowell, F.W., and Solorio-Munguia, J., 1974, Tres edades radiométricas oligocénicas y miocénicas de rocas volcánicas de las regiones de la Mixteca Alta y Valle de Oaxaca, Estado de Oaxaca: Boletín de la Asociación Mexicana de Geólogos Petroleros, v. 26, p. 249–262.

Gapais, D., 1989, Shear structure within deformed granites: Mechanical and thermal indicators: Geology, v. 17, p. 1144–1147, doi: 10.1130/0091-7613(1989)017<1144:SSWDGM>2.3.CO;2.

García-Duarte, R., 1999, Evidencias de la naturaleza estructural y relaciones estratigráficas de la Formación Matzitzi en el sur de Puebla, México [bachelor's thesis]: Hermosillo, Sonora, Centro de Estudios Superiores del Estado de Sonora, 90 p.

González-Hervert, M.G., Martínez-González, P.R., Martínez-Graza, J.A., and Rosas-Rojas, R., 1984, Características estratigráficas y estructurales del límite de los terrenos Mixteco y Oaxaca, en la región de Los Reyes Metzontla, Puebla: Boletín de la Sociedad Geológica Mexicana, v. 45, p. 21–32.

González-Ramos, A., Sánchez-Rojas, L.E., Mota-Mota, S., Arceo y Cabrilla, F.A., Soto-Araiza, R., Onofre-Espinosa, L., and Zárate-López, J., 2000, Carta Geológico-Minera Oaxaca: Secretaría de Comercio y Fomento Industrial, Consejo de Recursos Minerales, Pachuca, Hidalgo, carta E14–9, escala 1:250 000, 1 mapa.

Gradstein, F.M., Ogg, J.G., Smith, A.G., Bleeker, W., and Lourens, L.J., 2004, A new Geologic Time Scale with special reference to Precambrian and Neogene: Episodes, v. 27, no. 2, p. 83–100.

Grajales-Nishimura, J.M., 1988, Geology, geochronology, geochemistry, and tectonic implications of the Juchatengo green rocks sequence, state of Oaxaca, southern Mexico [master's thesis]: Tucson, University of Arizona, 145 p.

Grajales-Nishimura, J.M., Torres-Vargas, R., and Murillo-Muñetón, G., 1986, Datos isotópicos K-Ar para rocas ígneas y metamórficas en el estado de Oaxaca, *in* VIII Convención Geológica Nacional, Libro de Resúmenes: México, Sociedad Geológica Mexicana, p. 150–151.

Grajales-Nishimura, J.M., Centeno-García, E., Keppie, J.D., and Dostal, J., 1999, Geochemistry of Paleozoic basalt from the Juchatengo complex of southern Mexico: Tectonic implications: Journal of South American Earth Sciences, v. 12, p. 537–544, doi: 10.1016/S0895-9811(99)00037-1.

Hames, W.E., and Bowring, S.A., 1994, An empirical evaluation of the argon diffusion geometry in muscovite: Earth and Planetary Science Letters, v. 124, p. 161–167, doi: 10.1016/0012-821X(94)00079-4.

Harrison, T.M., 1982, Diffusion of ^{40}Ar in hornblende: Contribution to Mineralogy and Petrology, v. 78, p. 324–331, doi: 10.1007/BF00398927.

Harrison, T.M., Grove, M., Lovera, O.M., and Catlos, E.J., 1998, A model for the origin of Himalayan anatexis and inverted metamorphism: Journal of Geophysical Research, v. 103, no. B11, p. 27,017–27,032, doi: 10.1029/98JB02468.

Hernández-Estévez, S., 1980, Libro-guía de la excursión geológica al borde noroeste de la paleopeninsula de Oaxaca (sureste del Estado de Puebla): México, Sociedad Geológica Mexicana, 11 p.

Hernández-Láscares, D., 2000, Contribución al conocimiento de la estratigrafía de la Formación Matzitzi, área Los Reyes Metzontla-Santiago Coatepec, extremo suroriental del estado de Puebla, México [master's thesis]: Universidad Nacional Autónoma de México, Colegio de Ciencias y Humanidades, Unidad Académica de los Ciclos Profesionales y De Posgrado, 117 p.

Holdsworth, R.E., Butler, C.A., and Roberts, A.M., 1997, The recognition of reactivation during continental deformation: Journal of the Geological Society, London, v. 154, p. 73–78.

Holdsworth, R.E., Hand, M., Miller, J.A., and Buick, I.S., 2001a, Continental reactivation and reworking: an introduction, *in* Miller, J.A., Holdsworth, R.E., Buick, I.S., and Hand, M., eds., Continental Reactivation and Reworking: Geological Society [London], Special Publication 184, p. 1–12.

Holdsworth, R.E., Stewart, M., Imber, J., and Strachan, R.A., 2001b, The structure and rheological evolution of reactivated continental fault zones: A review and case study, *in* Miller, J.A., Holdsworth, R.E., Buick, I.S., and Hand, M., eds., Continental Reactivation and Reworking: Geological Society [London] Special Publication 184, p. 115–137.

Keane, S.D., and Morrison, J., 1997, Distinguishing magmatic from subsolidus epidote: Laser probe oxygen isotope compositions: Contribution to Mineralogy and Petrology, v. 126, p. 265–274, doi: 10.1007/s004100050249.

Keppie, J.D., Dostal, J., Cameron, K.L., Solari, L.A., Ortega-Gutiérrez, F., and López, R., 2003, Geochronology and geochemistry of Grenvillian igneous suites in the northern Oaxacan Complex, southern Mexico—Tectonic implications: Precambrian Research, v. 120, p. 365–389, doi: 10.1016/S0301-9268(02)00166-3.

Krogh, T.E., 1973, A low contamination method for hydrothermal decomposition of zircon and extraction of U and Pb for isotopic age determination: Geochimica et Cosmochimica Acta, v. 37, p. 485–494, doi: 10.1016/0016-7037(73)90213-5.

Leloup, P.H., Ricard, Y., Battaglia, J., and Lacassin, R., 1999, Shear heating in continental strike-slip shear zones: Model and field examples: Geophysical Journal International, v. 136, p. 19–40, doi: 10.1046/j.1365-246X.1999.00683.x.

López, R., 1997, High-Mg andesites from the Gila Band Mountains, southwestern Arizona: Evidence for hydrous melting of lithosphere during Miocene extension; the pre-Jurassic geotectonic evolution of the Coahuila terrane, northwestern Mexico: Grenville basement, a late Paleozoic arc, Triassic plutonism, and the events south of the Ouachita suture [Ph.D. thesis]: Santa Cruz, University of California, 147 p.

López-Ticha, D., 1985, Revisión de la estratigrafía y potencial petrolero de la cuenca de Tlaxiaco: Boletín de la Asociación Mexicana de Geólogos Petroleros, v. 37, no. 1, p. 49–92.

Ludwig, K.R., 1991, PBDAT; forms-DOS, a computer programs for IBM-PC compatibles for processing raw Pb-U-Th isotope data, version 1.20: Washington, D.C., U.S. Geological Survey Open-File Report 88–542, 37 p.

Ludwig, K.R., 2001a, ISOPLOT/Ex, version 2.49, A geochronological toolkit for Microsoft Excel: Berkeley, California, Berkeley Geochronology Center, Special Publication 1, 56 p.

Ludwig, K.R., 2001b, SQUID, version 1.00, A user's manual: Berkeley, California, Berkeley Geochronology Center, Special Publication 2, 17 p.

Martiny, B., Martínez-Serrano, R.G., Morán-Zenteno, D.J., Macías-Romo, C., and Ayuso, R.A., 2000, Stratigraphy, geochemistry and tectonic significance of the Oligocene magmatic rocks of western Oaxaca, southern México: Tectonophysics, v. 318, p. 71–98, doi: 10.1016/S0040-1951(99)00307-8.

Mattinson, J.M., 1987, U-Pb ages of zircons: A basic examination of error propagation: Chemical Geology, v. 66, p. 151–162.

Mawer, C.K., Rubie, D.C., and Bearley, A.J., 1988, A model for rapid melting in crustal shear zones: EOS, Transactions, American Geophysical Union, v. 69, p. 1411.

McDougall, I., and Harrison, T.M., 1988, Geochronology and Thermochronology by the 40Ar/39Ar Method: Oxford, Oxford University Press, Oxford Monographs on Geology and Geophysics 9, 212 p.

Meneses-Rocha, J.J., Monroy-Audelo, M.E., and Gómez-Chavarría, J.C., 1994, Bosquejo paleogeográfico y tectónico del sur de México durante el Mesozoico: Boletín de la Asociación Mexicana de Geólogos Petroleros, v. 44, p. 18–45.

Milisenda, C.C., Liew, T.C., Hofmann, A.W., and Kohler, H., 1994, Nd isotopic mapping of the Sri Lanka basement: Update, and additional constraints from Sr isotopes: Precambrian Research, v. 66, p. 95–110, doi: 10.1016/0301-9268(94)90046-9.

Miller, R.B., and Paterson, S.R., 1994, The transition from magmatic to high-temperature solid-state deformation: implications from the Mount Stuart batholith, Washington: Journal of Structural Geology, v. 16, p. 853–865, doi: 10.1016/0191-8141(94)90150-3.

Nabelek, P.I., and Liu, M., 1999, Leucogranites in the Black Hills of South Dakota: The consequence of shear heating during continental collision: Geology, v. 27, p. 523–526, doi: 10.1130/0091-7613(1999)027<0523:LITBHO>2.3.CO;2.

Ordóñez, E., 1905, Las rocas arcaicas de México: Memoria y Revista de la Sociedad Científica "Antonio Alzate" (México), v. 22, p. 315–331.

Ordóñez, E., 1906, L'Archaique du Cañon de Tomellin, *in* X Congress Geologique Internationale, Guide d'excursion: México, El Congreso 5, 30 p.

Ortega-Gutiérrez, F., 1975, The pre-Mesozoic geology of the Acatlán area, south México [Ph.D. thesis]: Leeds, University of Leeds, 166 p.

Ortega-Gutiérrez, F., 1978, Geología del contacto entre la formación Acatlán paleozoica y el complejo Oaxaqueño precámbrico, al oriente de Acatlán, estado de Puebla: Boletín de la Sociedad Geológica Mexicana, v. 39, p. 27–28.

Ortega-Gutiérrez, F., 1981, Metamorphic belts of southern Mexico and their tectonic significance: Geofísica Internacional, v. 20, p. 177–202.

Ortega-Gutiérrez, F., 1993, Tectonostratigraphic análisis and significance of the Paleozoic Acatlán Complex of southern Mexico, *in* Ortega-Gutiérrez, F., Centeno-García, E., Morán-Zenteno, D.J., and Gómez-Caballero, A., eds., First Circum-Pacific and Circum-Atlantic Terrane Conference: Terrane Geology of Southern Mexico: Guidebook of Field Trip B: Guanajuato, Universidad Nacional Autónoma de México, Instituto de Geología, p. 54–60.

Ortega-Gutiérrez, F., Ruiz, J., and Centeno-García, E., 1995, Oaxaquia, a Proterozoic microcontinent accreted to North America during the late Paleozoic: Geology, v. 23, p. 1127–1130, doi: 10.1130/0091-7613(1995)023<1127:OAPMAT>2.3.CO;2.

Ortega-Gutiérrez, F., Elías-Herrera, M., Reyes-Salas, M., Macías-Romo, C., and López, R., 1999, Late Ordovician-Early Silurian continental collisional orogeny in southern Mexico and its bearing on Gondwana-Laurentia connections: Geology, v. 27, p. 719–722, doi: 10.1130/0091-7613(1999)027<0719:LOESCC>2.3.CO;2.

Pardo, M., and Suárez, G., 1995, Shape of the subducted Rivera and Cocos plates in southern Mexico: Seismic and tectonic implications: Journal of Geophysical Research, v. 100, no. B7, p. 12,357–12,373, doi: 10.1029/95JB00919.

Parrish, R.R., 1987, An improved micro-capsule for zircon dissolution in U-Pb geochronology: Chemical Geology, v. 66, p. 99–102.

Paterson, S.R., and Schmidt, K.L., 1999, Is there a close relationship between faults and plutons?: Journal of Structural Geology, v. 21, p. 1131–1142, doi: 10.1016/S0191-8141(99)00024-3.

Paterson, S.R., and Tobisch, O.T., 1988, Using pluton ages to date regional deformation: Problems with commonly used criteria: Geology, v. 16, p. 1108–1111, doi: 10.1130/0091-7613(1988)016<1108:UPATDR>2.3.CO;2.

Paterson, S.R., Vernon, R.H., and Tobish, O.T., 1989, A review of criteria for the identification of magmatic and tectonic foliations in granitoids: Journal of Structural Geology, v. 11, p. 349–363, doi: 10.1016/0191-8141(89)90074-6.

Ramírez-Espinosa, J., 1984, La acreción de los terrenos Mixteco y Oaxaca durante el Cretácico Inferior, Sierra Madre del Sur de México: Boletín de la Sociedad Geológica Mexicana, v. 45, p. 7–19.

Román-Bardiel, T., Gapais, D., and Brun, J.P., 1997, Granite intrusion along strike-slip zones in experiment and nature: American Journal of Science, v. 297, p. 651–678.

Ross, C.A., 1986, Paleozoic evolution of southern margin of Permian basin: Geological Society of America Bulletin, v. 97, p. 536–554, doi: 10.1130/0016-7606(1986)97<536:PEOSMO>2.0.CO;2.

Ruiz, J., Patchett, P.J., and Ortega-Gutierrez, F., 1988a, Proterozoic and Phanerozoic basement terranes of México from Nd isotopic studies: Geological Society of America Bulletin, v. 100, p. 274–281, doi: 10.1130/0016-7606(1988)100<0274:PAPBTO>2.3.CO;2.

Ruiz, J., Patchett, P.J., and Arculus, R.J., 1988b, Nd-Sr isotope composition of lower crustal xenoliths—Evidence for the origin of mid-tertiary felsic volcanics in Mexico: Contribution to Mineralogy and Petrology, v. 99, p. 36–43, doi: 10.1007/BF00399363.

Ruiz-Castellanos, M., 1979, Rubidium-strontium geochronology of the Oaxaca and Acatlán metamorphic areas of southern Mexico [Ph.D. thesis]: Dallas, University of Texas at Dallas, 188 p.

Saint Blanquat, M., Tikoff, B., Teyssier, C., and Vigneresse, J.L., 1998, Transpressional kinematics and magmatic arcs, *in* Holdsworth, R.E., Strachan, R.A., and Dewey, J.E., eds., Continental Transpressional and Transtensional Tectonics: Geological Society [London] Special Publication 135, p. 327–340.

Samson, S.D., Coler, D.G., and Speer, J.A., 1995, Geochemical and Nd-Sr-Pb isotopic composition of Alleghanian granites of the southern Appalachians: Origin, tectonic setting, and source characterization: Earth and Planetary Science Letters, v. 134, p. 359–376, doi: 10.1016/0012-821X(95)00124-U.

Schmidt, M.W., and Thompson, A.B., 1996, Epidote in calc-alkaline magmas: An experimental study of stability, phase relationships, and the role of epidote in magmatic evolution: American Mineralogist, v. 81, p. 462–474.

Sedlock, R.L., Ortega-Gutiérrez, F., and Speed, R.C., 1993, Tectonostratigraphic terranes and tectonic evolution of Mexico: Geological Society of America Special Paper 278, 153 p.

Silva-Pineda, A., 1970, Plantas del Pensilvánico de la región de Tehuacán, Puebla: Universidad Nacional Autónoma. de México, Instituto de Geología, Paleontología Mexicana, v. 29, 47 p.

Solari, L.A., Keppie, J.D., Ortega-Gutiérrez, F., Cameron, K.L., Lopez, R., and Hames, W.E., 2003, 990 Ma and 1,100 Ma grenvillian tectonothermal events in the northern Oaxacan Complex, southern México: Tectonophysics, v. 365, p. 257–282, doi: 10.1016/S0040-1951(03)00025-8.

Stacey, J.S., and Kramers, J.D., 1975, Approximation of terrestrial lead isotope evolution by a two-stage model: Earth and Planetary Science Letters, v. 26, p. 207–221, doi: 10.1016/0012-821X(75)90088-6.

Torres-Torres, G., Ortega-González, J.V., Gutiérrez-Galicia, L., and Garduño-Monroy, V.H., 1984, Estudio microestructural del límite oriental del terreno Mixteco entre Huajuapan de León y Teposcolula, Oaxaca: Boletín de la Sociedad Geológica Mexicana, v. 45, p. 39–41.

Torres-Vargas, R., Murillo-Muñeton, G., and Grajales-Nishimura, M., 1986, Estudio petrográfico y radiométrico de la porción límite entre los complejos Acatlán y Oaxaqueño, *in* VIII Convención Geológica Nacional, Resúmenes: México, Sociedad Geológica Mexicana, p. 148–149.

Torres-Vargas, R., Ruiz, J., Patchett, P.J., and Grajales, J.M., 1999, Permo-Triassic continental arc in eastern Mexico: Tectonic implications for reconstructions of southern North America, *in* Bartolini, C., Wilson, J.L., and Lawton T.F., eds., Mesozoic Sedimentary and Tectonic History of North-Central Mexico: Geological Society of America Special Paper 340, p. 191–196.

Urrutia-Fucugauchi, J., and Flores-Ruiz, J.H., 1996, Bouguer gravity anomalies and regional crustal structure in central Mexico: International Geology Review, v. 38, p. 176–194.

Vega-Carrillo, J.J., Elías-Herrera, M., and Ortega-Gutiérrez, F., 1998, Complejo plutónico de Cuanana: basamento prejurásico en el borde meridional del terreno Mixteco e interpretación litotectónica, *in* Alaniz-Álvarez, S.A., Ferrari, L., Nieto-Samaniego, Á.F., and Ortega-Rivera, M.A., eds., Libro de resúmenes: México, Sociedad Geológica Mexicana; Instituto Nacional de Geoquímica; Sociedad Mexicana de Geomorfología; Sociedad Mexicana de Mineralogía; Asociación Mexicana de Geólogos Petroleros, Primera Reunión Nacional de Ciencias de la Tierra, p. 145.

Vigneresse, J.L., 1995, Control of granite emplacement by regional deformation: Tectonophysics, v. 249, p. 173–186, doi: 10.1016/0040-1951(95)00004-7.

Weber, B., and Kohler, H., 1999, Sm-Nd, Rb-Sr and U-Pb geochronology of a Grenville terrane in southern Mexico: Origin and geological history of the Guichicovi Complex: Precambrian Research, v. 96, p. 245–262, doi: 10.1016/S0301-9268(99)00012-1.

Weber, R., 1997, How old is the Triassic flora of Sonora and Tamaulipas and news on Leonardian flora in Puebla and Hidalgo, Mexico: Revista Mexicana de Ciencias Geológicas, v. 14, p. 225–243.

Weber, R., Centeno-García, E., and Magallón-Puebla, S.A., 1987, La Formación Matzitzi, estado de Puebla, tiene una edad permocarbonífera, *in* Segundo Simposio sobre la Geología Regional de México, Programa y Resúmenes: México, Universidad Nacional Autónoma de México, Instituto de Geología, p. 57–59.

Wijbrans, J.R., and McDougall, I., 1986, 40Ar/39Ar dating of white micas from an Alpine high-pressure metamorphic belt on Naxos (Greece): The resetting of the argon isotopic system: Contributions to Mineralogy and Petrology, v. 93, p. 187–194, doi: 10.1007/BF00371320.

Williams, I.S., 1998, U-Th-Pb geochronology by ion microprobe, *in* McKibben, M.A., Shanks, W.C., and Ridley, W.I., eds., Applications of Microanalytical Techniques to Understanding Mineralization Processes, Reviews in Economic Geology, v. 7, p. 1–35.

Yáñez, P., Ruiz, J., Patchett, P.J., Ortega-Gutiérrez, F., and Gehrels, G.E., 1991, Isotopic studies of the Acatlán complex, southern Mexico: Implications for Paleozoic North American tectonics: Geological Society of America Bulletin, v. 103, p. 817–828, doi: 10.1130/0016-7606(1991)103<0817:ISOTAC>2.3.CO;2.

Zen, E., and Hammarstrom, J.M., 1984, Magmatic epidote and its petrologic significance: Geology, v. 12, p. 515–518, doi: 10.1130/0091-7613(1984)12<515:MEAIPS>2.0.CO;2.

Zeng, L., Saleeby, J.B., and Asimov, P., 2005, Nd isotope disequilibrium during crustal anatexis: A record from the Goat Ranch migmatite complex, southern Sierra Nevada batholith, California: Geology, v. 33, p. 53–56, doi: 10.1130/G20831.1.

Manuscript Accepted by the Society 29 August 2006

Printed in the USA

Geological Society of America
Special Paper 422
2007

The Chacalapa fault, southern Oaxaca, México

Gustavo Tolson*
Departamento de Geología Regional, Instituto de Geología, Universidad Nacional Autónoma de México, Circuito de la Investigación Científica s/n, Cd. Universitaria, México, D.F. 04510, México

ABSTRACT

Along the coast of Oaxaca between Puerto Angel and Santiago Astata, a suite of rocks is exposed corresponding to the Xolapa metamorphic complex, which is intruded by granitoid igneous rocks of Paleogene to Miocene age. Both of these rock units are in fault-contact along coast-parallel shear zones with granulite facies metamorphic rocks of the Proterozoic age Oaxaca Complex and with cover rocks that unconformably overlie this crystalline basement. The largest of these shear zones is the Chacalapa shear zone exposed to the north of the town of Pochutla and formed by ultramylonites, mylonites, protomylonites, pseudotachylytes, phyllonites, and cataclasites. The kinematics of this shear zone are dominantly left-lateral.

The mylonitic rocks of the Chacalapa shear zone indicate temperatures ~500 °C (as evidenced by crystal-plastic deformation of plagioclase feldspar) that decrease systematically to a strictly brittle deformation regime. Quartz crystallographic textures associated with crystal-plastic deformation processes are locally asymmetric, indicating left-lateral shear. Brittle structures also exhibit left-lateral shear-sense indicators, as do active deformation structures.

The age of the shear zone is constrained between 29 ± 0.2 Ma and 23.7 ± 1.2 Ma. The upper limit is the age of the Huatulco granodiorite, whose northern contact is sheared, and the lower bound is fixed by the K-Ar age of hornblende separated from granodioritic dikes that cut the mylonitic lineation.

Keywords: fault, México, Oaxaca, mylonite, pseudotachylyte, Xolapa.

RESUMEN

A lo largo de la costa de Oaxaca, entre Puerto Ángel y Santiago Astata, afloran rocas metamórficas del Complejo Xolapa intrusionadas por rocas ígneas de edad Paleógeno a Mioceno sin metamorfismo regional. Ambas unidades se encuentran en contacto tectónico a lo largo de zonas de cizalla con rocas Proterozoicas del Complejo Oaxaqueño al norte de Pochutla y con rocas sedimentarias Mesozoicas discordantes que sobreyacen los gneises. La principal de estas zonas de cizalla en el área de estu-

*tolson@servidor.unam.mx

Tolson, G., 2007, The Chacalapa fault, southern Oaxaca, México, *in* Alaniz-Álvarez, S.A., and Nieto-Samaniego, Á.F., eds., Geology of México: Celebrating the Centenary of the Geological Society of México: Geological Society of America Special Paper 422, p. 343–357, doi: 10.1130/2007.2422(12). For permission to copy, contact editing@geosociety.org.

dio es la falla Chacalapa, expuesta al norte de Pochutla, que está constituida por ultramilonitas, milonitas, protomilonitas, pseudotaquilitas, filonitas y cataclasitas en orden cronológico de desarrollo. La cinemática de esta zona de cizalla vertical es predominantemente lateral-izquierda.

Las rocas miloníticas de la falla Chacalapa registran temperaturas de recristalización dinámica de ~500 °C (deformación cristal-plástica de feldespatos) que disminuyen sistemáticamente hasta el régimen netamente quebradizo. Asociadas a la recristalización dinámica se desarrollaron texturas cristalográficas de cuarzo en las rocas miloníticas que, cuando son asimétricas, exhiben un sentido de cizalla izquierdo. Las rocas del régimen quebradizo también registran una cinemática izquierda, al igual que las fallas activas.

La edad de las rocas miloníticas de la falla Chacalapa se ubica entre los 29 ± 0.2 Ma y los 23.7 ± 1.2 Ma. El límite de edad superior lo constituye la edad del intrusivo Huatulco (U-Pb de circones, que se observa milonitizado en su margen septentrional y el límite de edad inferior corresponde a la edad K-Ar de hornblenda procedente de diques granodioríticos porfídicos que truncan la milonita).

Palabras clave: falla, México, Oaxaca, milonita, pseudotaquilita, Xolapa.

INTRODUCTION

Some 10 km to the north of Puerto Angel in the state of Oaxaca is exposed a well-defined E-W–trending belt of mylonitic rocks. This fault zone, known as the Chacalapa Fault, separates the amphibolite facies rocks of the Xolapa Complex to the south from granulite facies rocks of the Oaxaca Complex to the north. This structure was first described in an abstract by Ortega-Gutiérrez and Corona-Esquivel (1986), who reported its existence close to the village of San José Chacalapa on the Puerto Angel–Miahuatlán highway.

The Chacalapa Fault extends some 10 km to the west of San José Chacalapa where it bifurcates and is later truncated by intrusive rocks. Eastward it extends ~40 km, acquiring a brittle character and intersecting the Pacific Coast. The Chacalapa Fault was the central theme of the doctoral thesis of Tolson (1998), who carried out the first systematic mapping of the fault, as well as detailed analysis which will be examined below.

GEOLOGIC SETTING

One of the most striking aspects of the geology of México is the presence of an extensive belt of plutonic rocks exposed along the coast of Guerrero and Oaxaca states in SW México (Fig. 1). These amphibolite facies rocks consist of ortho- and paragneisses of uncertain age cut by batholiths of tonalitic, dioritic, and granitic composition Eocene to Miocene in age (Morán Zenteno, 1992). The metamorphic rocks exposed between Chilpancingo and Acapulco were described in detail and stratigraphically defined by De Cserna (1965) as the Paleozoic Xolapa Complex. The same author also described regionally undeformed granitic rocks cutting the gneiss and observed the truncated nature of regional structures with respect to the coastline.

Sánchez Rubio (1975), Guerrero García et al. (1978), and later Ortega Gutiérrez (1981) documented the importance of migmatitic and orthogneissic rocks within the Xolapa Complex. Ortega Gutiérrez (1981) indicated that the basal metamorphic complex was intruded by plutons exhibiting neither metamorphism nor regional deformation, and also offered a tectonic interpretation of the Xolapa Complex, indicating its high-temperature–low-pressure metamorphic characteristics, and suggesting that it was the mid-crustal section of a magmatic arc.

Morán Zenteno's (1992) doctoral dissertation reported isotopic and geochronological details of Xolapa Complex rocks and intruding rocks to the east of Acapulco, attributing a Late Jurassic–Early Cretaceous age to the orthogneissic rocks and assigning the postmetamorphic plutonism to the Eocene-Oligocene. Subsequently, Corona Chávez (1997) carried out a detailed petrological analysis of the Xolapa Complex rocks, indicating that mineral equilibrium was attained at temperatures >650 °C and at pressures >7 kbar during the metamorphic peak. The history of the truncation and exhumation of the Pacific margin of México is succinctly described in Schaaf et al. (1995) and in Morán Zenteno et al. (1996). The northern limit of the Xolapa Complex occurs along fault zones such as those that may be observed to the north of Acapulco (Riller et al., 1992) and within the present study area, or else along intrusions as in the vicinity of Pinotepa Nacional (Hernández Bernal, 1995; Hernández Bernal and Morán Zenteno, 1996).

Toward the northern part of the study area are exposed rocks of the southern portions of the Oaxaca Complex. These predominantly mafic rocks contain alternating layers of amphibolite, metagabbro, and pyroxenite. The dominant foliation strikes E-W and dips toward the south, although its orientation varies locally, exhibiting folds on a variable (micro-, meso- and macroscopic) scale with N-trending asymmetry.

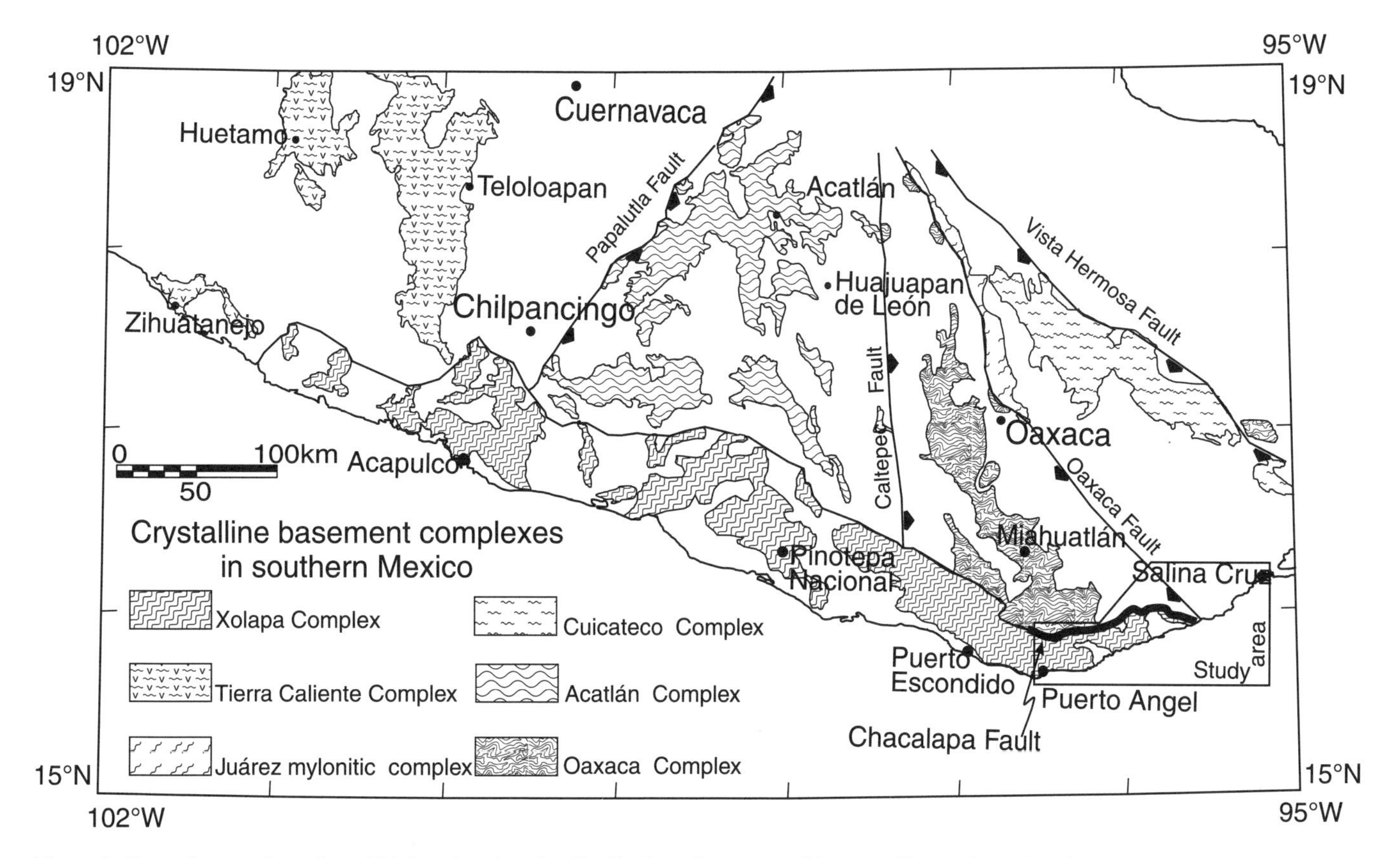

Figure 1. General map of southern México showing the distribution of metamorphic crystalline rocks and regional faults previously reported (modified from Sánchez-Zavala, 2006).

The felsic rocks of the metamorphic basement of this region have previously been considered an alkaline anorthosite (Herz, 1978) and show evidence of synkinematic emplacement. They are constituted by a plagioclase mosaic of intermediate (andesine) composition and lenses of deformed megacrysts with an antiperthitic texture. The quartz content is highly variable and is found in proportions varying from mere traces to up to 30%. Scattered deposits of rutile, ilmenite, and apatite are also observed. In some zones, the titanium concentrations reach an economically significant level, forming deposits such as that of Pluma Hidalgo to the north of Pochutla (Paulson, 1969). In general, there is a variation in the color index of the rocks determined by the content of subhedral to euhedral clinopyroxene crystals largely replaced by chlorite or tremolite.

Alternating with the rock volumes mentioned above are belts of amphibolite, locally pyroxene- and garnet-bearing, augite, and diopside metagabbro, with bands of up to 10 cm of partly chloritized clinopyroxinite. The metamorphic sequence is cross-cut by a generation of mafic to intermediate aphanitic or porphyritic dikes, which are undeformed in the crystal-plastic regime.

Within the area of study, a rock package of basaltic andesite covered by a sequence of continuous conglomerate and sandstone with intercalations of conglomerate overlies the Oaxaca Complex in angular unconformity. Characteristic of this rock package is a greenish color in newly exposed rocks that weathers to a purplish maroon. The sandstone is conformably overlain by calcareous breccia, which is in turn overlain by a massive limestone that forms the region's topographic highs. The limestone contains no index fossils, and during this study only remains of pelecipods and gastropods were identified. This stratigraphic sequence is very similar to that described in a PEMEX (1986) internal report, in which the presence of carbonate rocks was correlated with the Teposcolula Formation of the Albian-Cenomanian. The rocks of this sequence are also affected by the Chacalapa fault system in the study zone, exhibiting greenschist facies metamorphism.

Structural Geology

The structural geology of the area is complex, given that it includes a regional-scale fault that brings into contact two metamorphic rock packages in turn characterized by a history of multiple deformation events. The distinct deformation episodes are manifest in structures of the crystal-plastic regime, structures transitional between crystal-plastic and brittle, and finally strictly brittle features. The relative timing of structures is given by crosscutting relationships.

Regional Structures

The most significant regional structure in the study area is the Chacalapa Fault, orientated in a general east-west direction, which juxtaposes the rocks of the Oaxaca Complex (and its cover sequence) with those of the Xolapa Complex. This tectonic discontinuity extends >50 km along strike.

It possesses characteristics of deformation in the crystal-plastic regime, exhibiting the development of mylonites with recrystallization textures in the solid state, and of deformation in the brittle regime with the development of pseudotachylytes, cataclasites, and gouges. The shear zone is continuous along strike from west of the village of Chacalapa (Lazos Ramírez and Rodríguez Rivera, 1995) to east of Sta. Maria Xadani, but in the vicinity of the village of Xuchil it takes on a braided character, branching around non-mylonitized blocks of the Oaxaca and Xolapa Complexes (Figs. 2, 3, and 4). The thickness of the mylonitic zones varies from some hundreds of meters in the braided region to ~2 km in the vicinity of the town of Sta. Maria Huatulco. This mylonitic zone postdates a series of shear zones found in the rocks of the Xolapa Complex with thicknesses of some tens of meters and along-strike lengths between 1 and 2 km, and crystal-plastic deformation textures; the outcrop pattern of these structures is more complex than that of the Chacalapa Fault, since their inclinations are moderate to low.

The latest structures consist of a series of faults of oblique left-lateral–normal displacement, evident in satellite images, aerial photographs, and in the field. Some of these faults remain active, as the displacement of man-made structures suggests (e.g., Delgado Argote and Carballido Sánchez, 1990).

Mesoscopic Structures

The area studied can be divided into four structural domains: the domain of the Oaxaca Complex and its covering rocks to the north, that of the Xolapa Complex toward the south, the Chacalapa Fault formed by the contact between them, and, finally, the low-grade metamorphic rocks of what Carfantan (1983) has termed the "Chontal Arc" in the northeastern part of the study zone (Fig. 4). The Oaxaca and Xolapa domains are principally characterized by planar structures, in particular the presence of penetrative gneissic foliation at the centimeter scale indicated

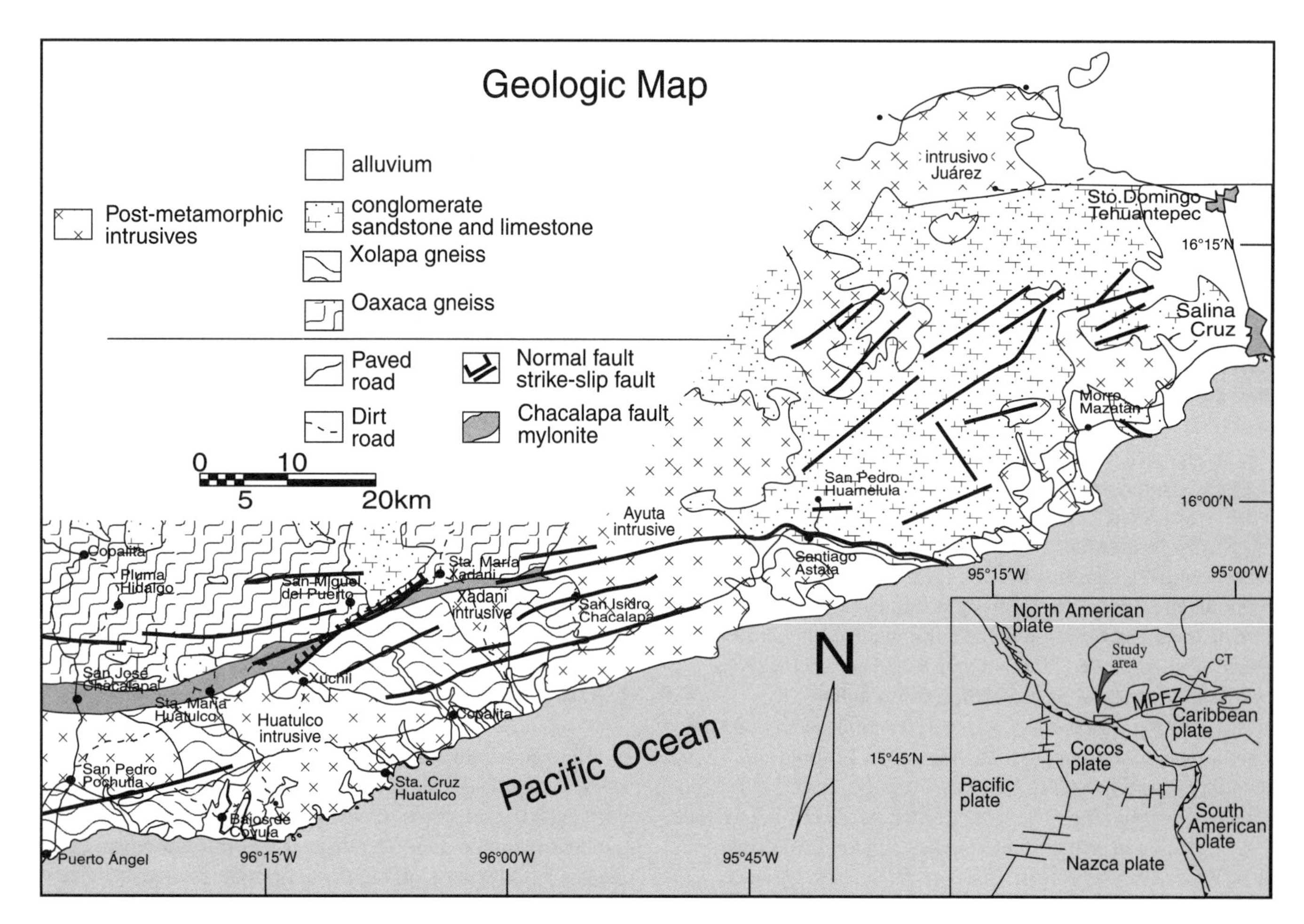

Figure 2. Geologic map of the study area. The northeastern portion is modified from Carfantan (1983). The inset shows the tectonic setting (modified from Morán-Zenteno et al., 1996). Inset shows study area. CT—Cayman Trough; MPFZ—Motagua-Polochic fault zone.

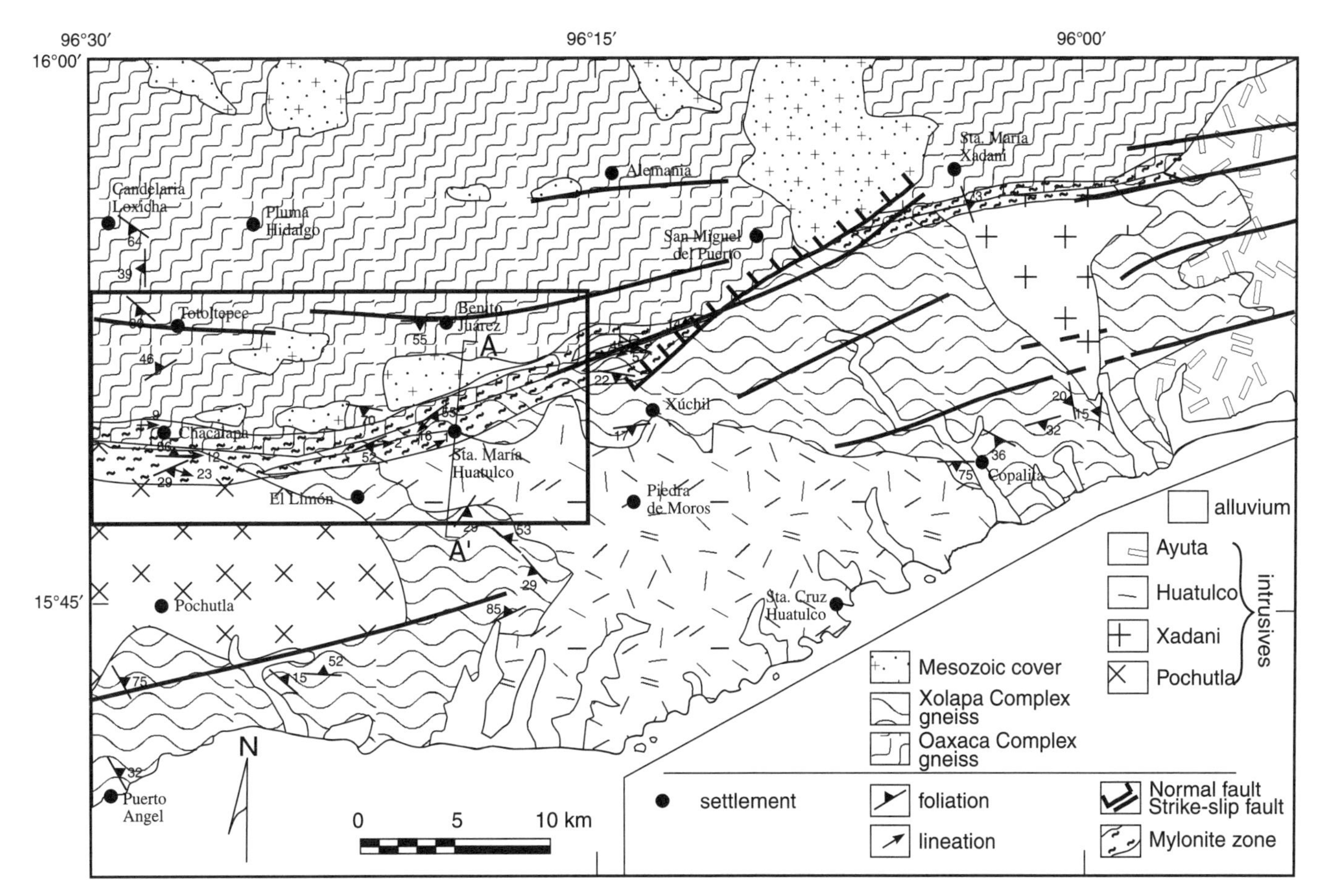

Figure 3. Detailed geologic map of the study area, mapped at a scale of 1:50 000. The area in the rectangle is shown in greater detail in Figure 7.

by an alternation of light and dark minerals. In the rocks of the Xolapa Complex, this foliation has been affected by a late-migmatitic deformation that manifests itself in mylonitic zones of decametric thickness clearly truncated by the Chacalapa Fault. (These mylonitic zones, however, are not the central theme of the present work and will be mentioned only in the context of the evolution of the Xolapa Complex.) In contrast, the mylonitic domain of the Chacalapa Complex has developed a strong stretching lineation (L), while its foliation may be weak or entirely absent.

The eastern segment of the mylonitic zone has previously been described by Ortega-Gutiérrez and Corona Esquivel (1986), and the sector between Puerto Escondido and Puerto Angel has been mapped by Ortega Gutiérrez and co-workers (1990). Regionally, the trace of the fault presents an approximately coast-parallel curvature (Fig. 1), which varies in direction from WNW to ENE. A bachelor's thesis by Lazos Ramírez and Rodríguez Rivera (1995) describes in considerable cartographic and microscopic detail the mylonitic zone from San Jose Chacalapa westward, and documents the braided character of the deformation zone.

The equal area projection of poles to foliation of the Xolapa domain shows a great circle girdle oriented vertically north-south, indicating that the foliation has been folded around a horizontal east-west axis (Fig. 4). A slight asymmetry in the distribution of the planes suggests a northward vergence. This vergence is consistent with the observed presence of tabular shear zones of limited extension—hundreds of meters along strike—oriented more or less E-W (075°–095°), with moderate inclinations (25°–75°) to the south and kinematic indicators of thrusting toward the north.

The projection of poles to foliation for the Oaxaca domain shows the foliation to have a preferential E-W–oriented strike and a moderate to strong inclination to the south (Fig. 4). The poles to bedding planes of the cover rocks of the Oaxaca domain are subparallel to those of the gneissic foliation.

The equal area projection of the mylonitic foliation poles and stretching lineations show that the mylonitic foliations are generally subparallel to those of the Oaxaca and Xolapa domains and that the lineations, subhorizontal and E-W–trending, are contained within the plane of foliation (Fig. 4). The best-exposed section of mylonite is that exposed along the bed of the Huatulco River between Sta. María Huatulco and Benito Juárez. In this section, the transition from non-deformed granitic rocks of the Huatulco intrusion to ultra-mylonite rocks with penetrative mineral stretching lineations at the millimetric scale is clearly observed. This transition is both gradual and complex: the fresh granite is affected by

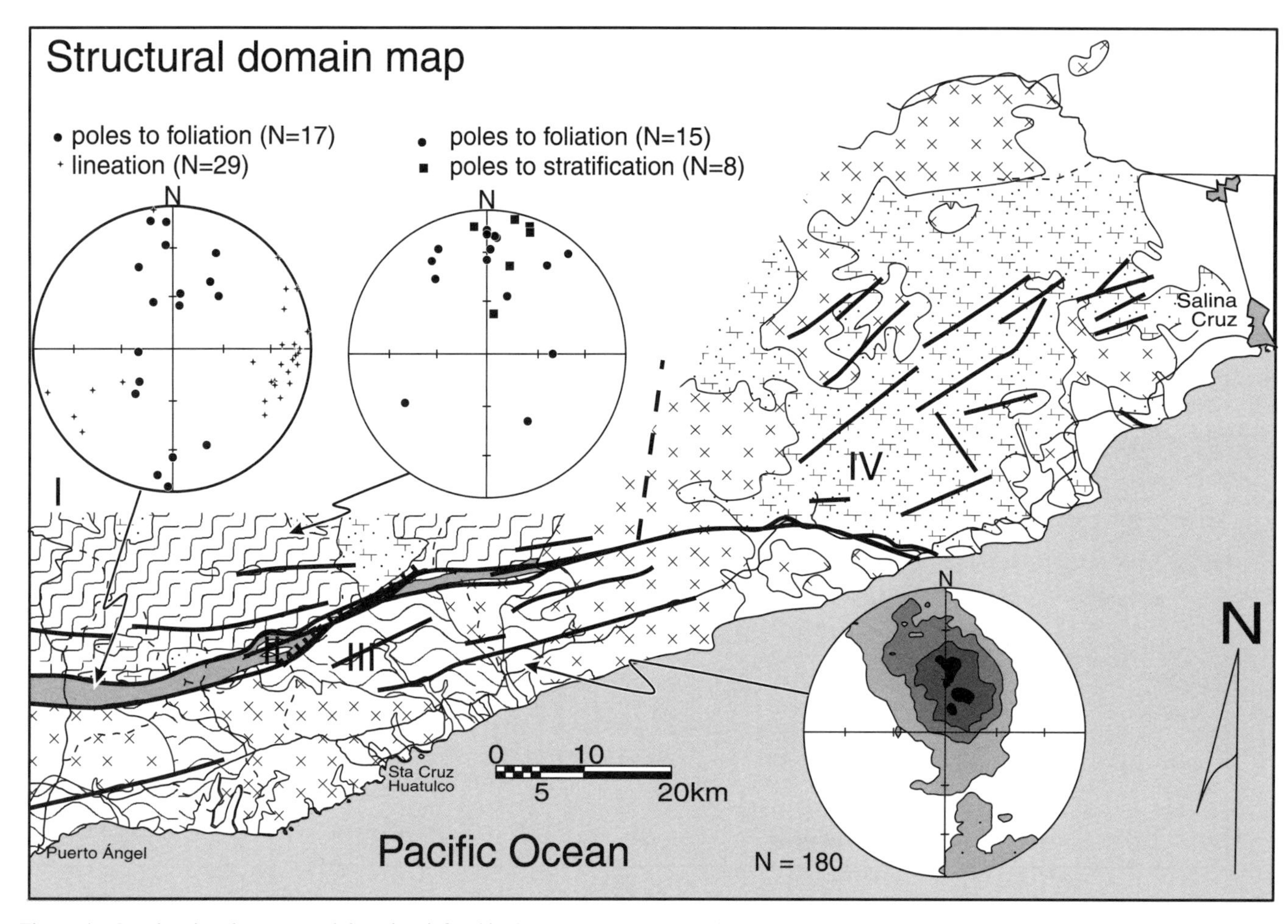

Figure 4. Map showing the structural domains defined in the present paper. Domain I consists mainly of granulitic gneisses of the Oaxaca Complex and its cover rocks. Domain II is defined by L-type mylonitic rocks of the Chacalapa shear zone. Domain III corresponds to locally migmatitic gneisses of the Xolapa Complex and Oligocene age intrusive rocks. Domain IV is a folded volcanic and sedimentary sequence. N = number.

pairs of conjugate faults separated by distances of ~20 m and oriented NNW-SSE and NE-SW. Both the density and thickness of the faults increase gradually as a zone of gouge and breccia with a distinctive blue-gray coloration develops. The fault planes exhibit no lineations that might indicate a relative movement between the blocks, but the cross-cutting relationships between faults suggest that they are dextral-sinistral pairs. The separation between faults continues to diminish until it is reduced to a matter of centimeters. From this point, the rock acquires a highly fractured appearance, and its color darkens as a result of the intense comminution in certain sectors. In a distance on the order of meters, the presence of an extension lineation becomes evident and faults become scarcer. This constitutes the final transition and, from this point onward, the rock is transformed into a true mylonite. In total, the transition sequence extends 300 m along the riverbed. The sequence indicates that the same shear zone trajectory has been reactivated with faulting in the brittle regime.

Metasediment xenoliths of the Xolapa Complex, stretched parallel to the mineral lineation, can be seen within the mylonitized granite. Advancing toward the north, the number of decimetric- to metric-sized xenoliths increases, and finally granite is found only in apophysis within the mylonitized metasediments. This suggests that the mylonite zone in this locality is subparallel to the plane of contact of the granite and the host rock. The development of the mylonitic structures is accompanied by an increase in the proportion of muscovite in the granite that may be due to one of two causes: (1) the reaction and/or dissolution, in the presence of water, of potassium feldspar into quartz and muscovite that is frequently observed in mylonite zones and that corresponds to H_2O pressures >3.5 kbar (Simpson and De Paor, 1993); or (2) the assimilation of metasediments of the Xolapa Complex on the periphery of the Huatulco intrusion, which would impart a more aluminous composition manifest in the growth of muscovite crystals.

In another section, to the south of San Miguel del Puerto, faulted protomylonitic rocks of the Oaxaca Complex intruded by pseudotachylyte are exposed. Farther to the east of this location, on the right bank of the River Copalita, seams of pseudotachylyte up to 30 cm thick are exposed. Along the fault with which these pseudotachylytes are associated, the Copalita River exhibits an apparent displacement of 3.0 km.

In order to estimate strain, the Newfry (Tolson, 1995) and digital image retrodeformation (Weger, 1996) methods are employed. The Newfry method is utilized in the case of rocks of a low degree of deformation, corresponding to the Mesozoic cover, and the image retrodeformation method is applied to plagioclase domains of the mylonitic rocks. The results of strain estimation are presented in a Flinn logarithmic diagram in Figure 5. The orientations and forms of the ellipses XY and YZ are also represented in Figures 6A and 6B, respectively. The Flinn diagram shows that the form of the determined ellipses falls within the domain of prolate ellipses, a result that is corroborated in the field by the characteristics of the L tectonite. In a general shear zone system (with combinations of pure and simple shear), prolate ellipsoids correspond to a transtensive system, as even at high internal distortions the lineation continues horizontally. This does not occur in a transpressive system, because the direction of maximum extension changes from horizontal to vertical with an increase in strain, and the resultant ellipsoids are oblate (Fossen and Tikoff, 1993).

Microscopic Structures

From a microscopic perspective, the rocks studied contain a variety of structures, including schistosity, gneissosity, preferential orientation of granular aggregates, preferential orientation of crystallographic axes, δ and σ porphyroclasts, mica fish, and S-C structures (Figs. 7A and 7B).

The schistosity is defined by flakes of biotite and muscovite in the nonmigmatized metasedimentary rocks of the Xolapa Complex and is, in general, parallel to the gneissosity defined by alternating quartz-feldspathic and mafic domains in the migmatitic and orthogneissic rocks of the complex. Apart from small locally observed crenulations of the micas in both gneisses and schists, neither the gneissosity nor the schistosity provide any information that has not already been discussed in the previous section.

The preferred orientation of granular forms and crystallographic axes, on the other hand, has important implications concerning the thermobaric conditions during deformation. The rocks that exhibit a preferred orientation of granular forms and crystallographic axes are the granitic mylonites of the Huatulco intrusive. Microscopically, these rocks reveal the same gradations

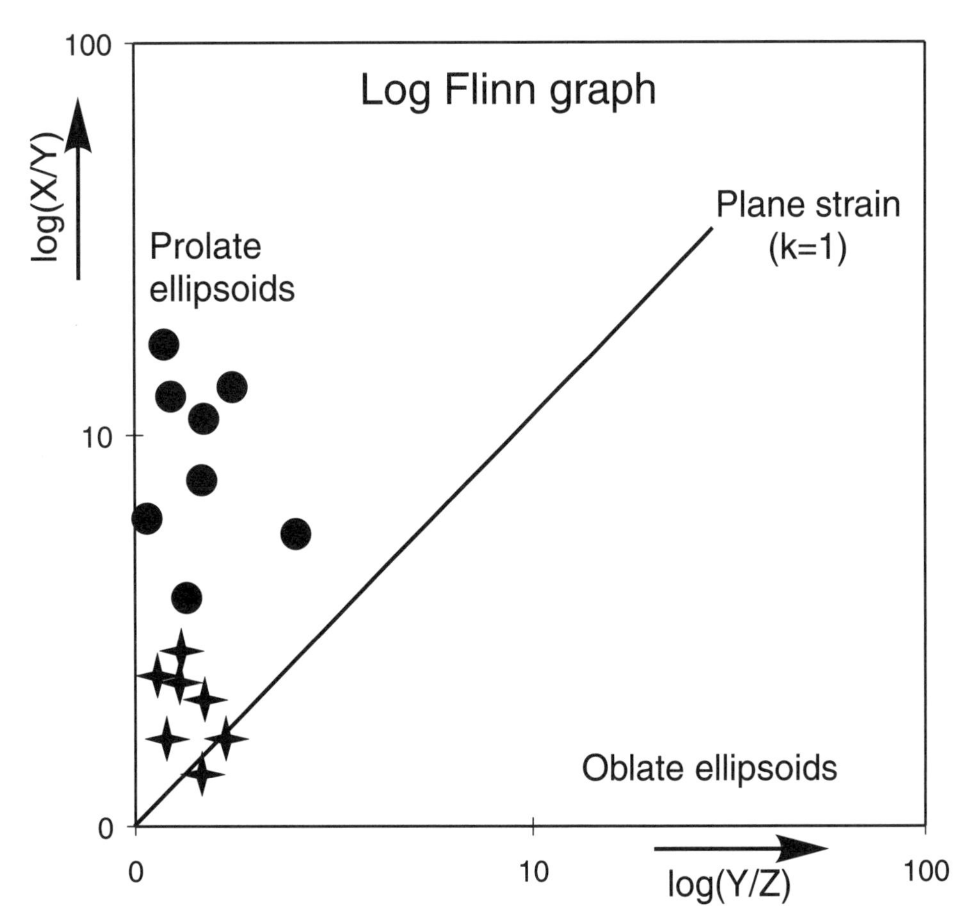

Figure 5. Logarithmic Flinn diagram showing the shape of the strain ellipse within the Chacalapa shear zone determined using the Weger (1996) method (circles) and NewFry (Tolson, 1995). Note that the samples are strongly prolate.

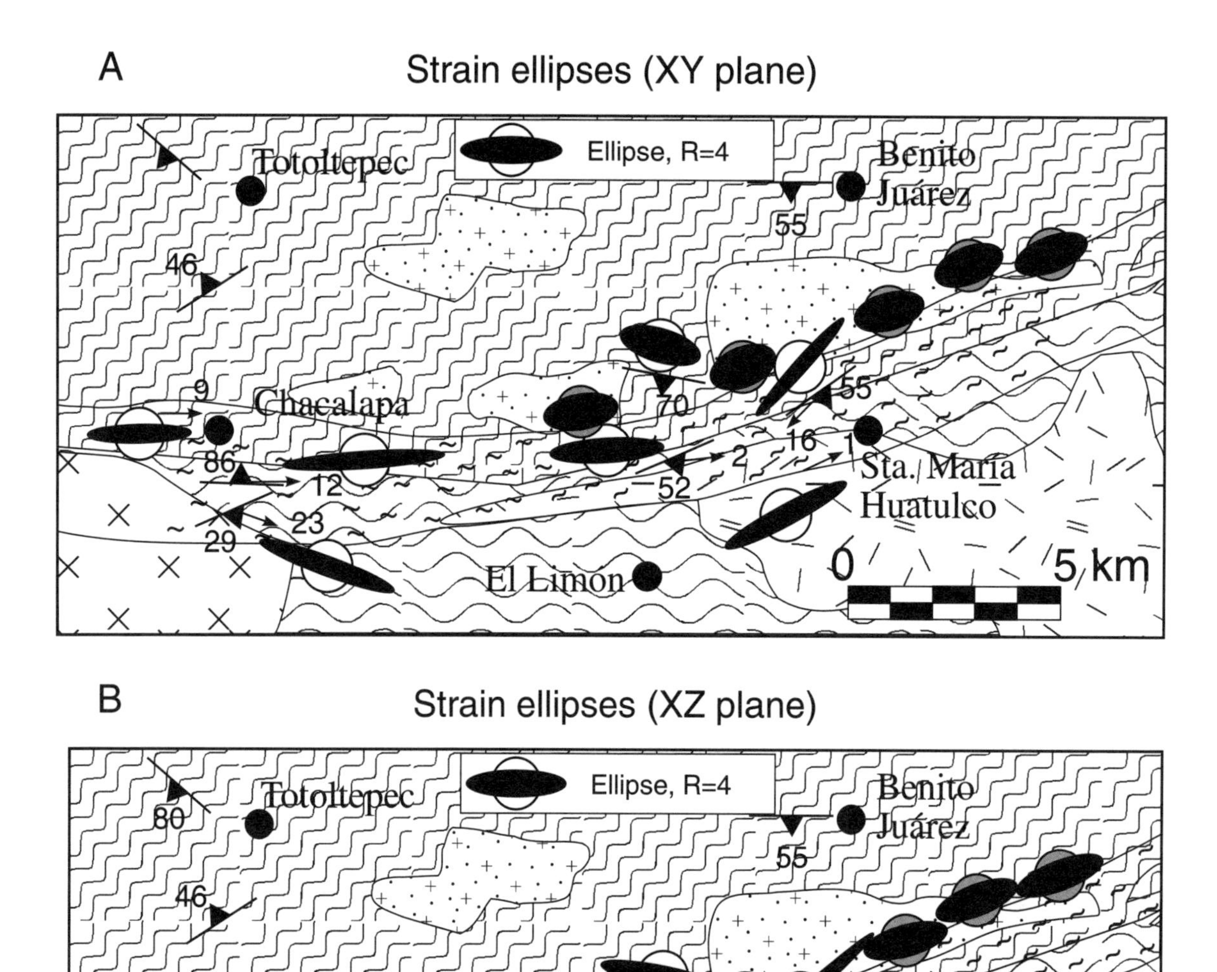

Figure 6. (A) Map showing location of strain determinations and the shape of XY strain ellipses. (B) Map showing location of strain determinations and the shape of XZ strain ellipses. On both maps strain ellipses with blank reference unit circles were determined using NewFry (Tolson, 1995), while those with gray reference unit circles were determined using Weger's (1996) method. R—ellipse axial ratio. See map legend in Figure 4.

in deformation that are observed in outcrops close to the shear zone. Samples from outcrops more distant from the center of the shear zone show evidence of brittle deformation, with fractured plagioclase, comminution zones, and veins filled with secondary material, such as calcite and epidote. The samples of granite from the middle zone of the mylonitic belt, however, contain crystals of plagioclase with folded twin lamellae, and completely recrystallized quartz crystals, with 120° triple unions, without undulose extinction, and with strong lattice preferred orientation.

Following the criteria established by Tullis (1977) and Hirth and Tullis (1992), these microscopic textures suggest a deformation regime with temperatures above ~500 °C. The c-axis crystallographic textures of the quartz exhibit minor circles around the maximum-shortening axis as well as linear maxima oblique to the foliation (Tolson et al., 1993) (Fig. 8). Minor circle textures are common in rocks with oblate deformation ellipsoids (Lister and Dornsiepen, 1982) and the oblique maxima might be interpreted as the result of basal glide in the <1000> direction.

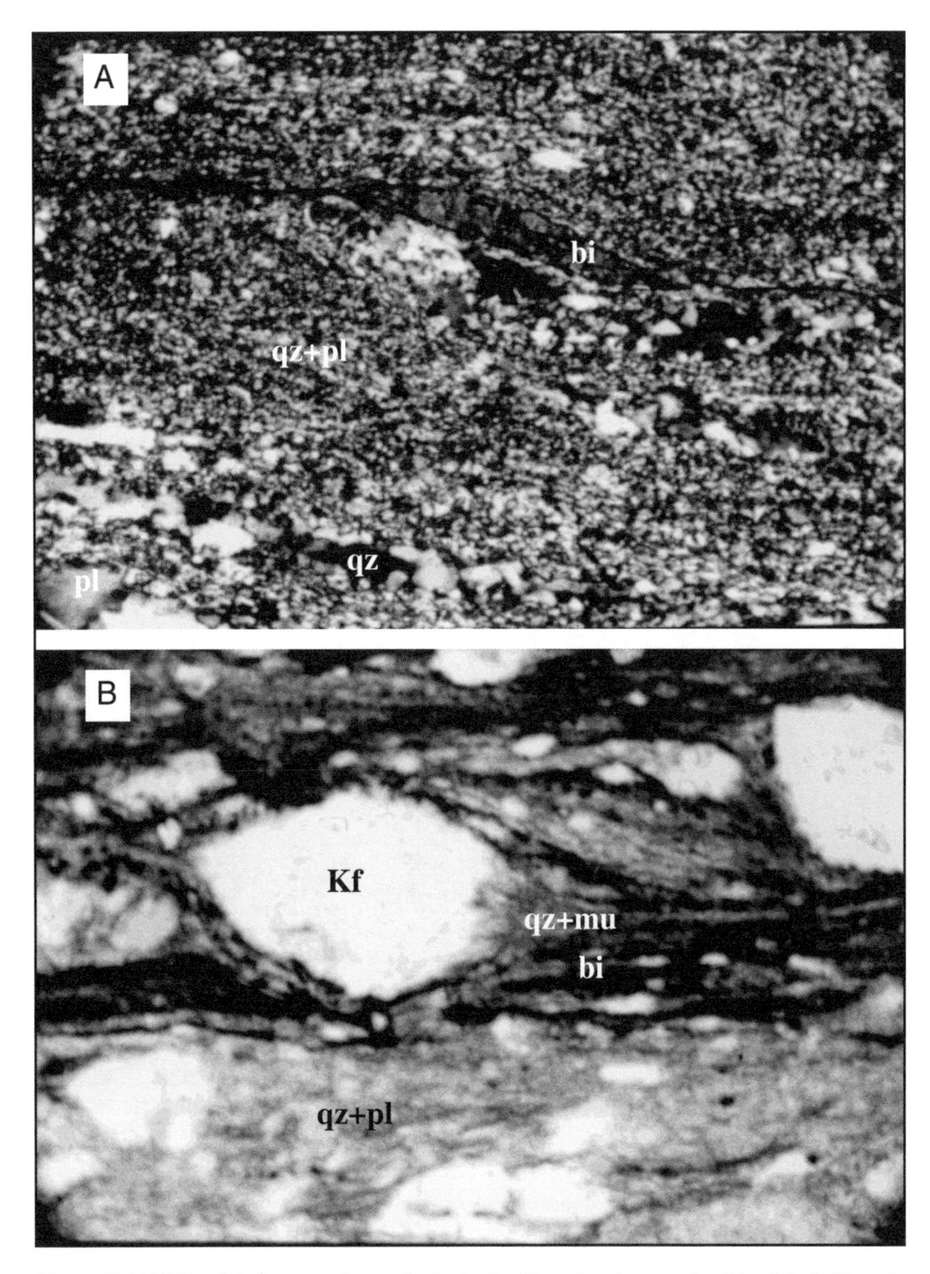

Figure 7. (A) Mica fish from an ultramylonite in the Huatulco river north of Sta. María Huatulco. (B) σ porphyroclasts in the Xolapa gneisses in the bed of the Xuchil river north of the town of Xuchil. In both cases, sinistral kinematics are clear. Abbreviations: bi—biotite; Kf—potassium feldspar; mu—muscovite; pl—plagioclase; qz—quartz.

The two textures have been reported in terranes where the deformation occurred at temperatures >500 °C. The dynamic recrystallization of plagioclase and its crystal-plastic deformation are consistent with temperatures in this range (Passchier and Trouw, 1996).

Also observed in association with the Chacalapa Fault are mylonites with evidence of greenschist facies *P-T* conditions. In these rocks, the assemblage albite + epidote ± chlorite is characteristic, and is observed to be directly associated with the process of deformation (Fig. 9).

Structural Development

Locally, the migmatites of the Xolapa Complex include paleosomatic remnants of pelitic to amphibolitic composition embedded in a granitic neosome. These remains exhibit evidence of a penetrative deformation at the centimetric scale prior to the partial fusion that generated the migmatites of the Xolapa Complex (Fig. 10). This event, here denominated D_n, constitutes the oldest deformation registered by the rocks of the Xolapa Complex in the study area. The second deformation event (D_{n+1})

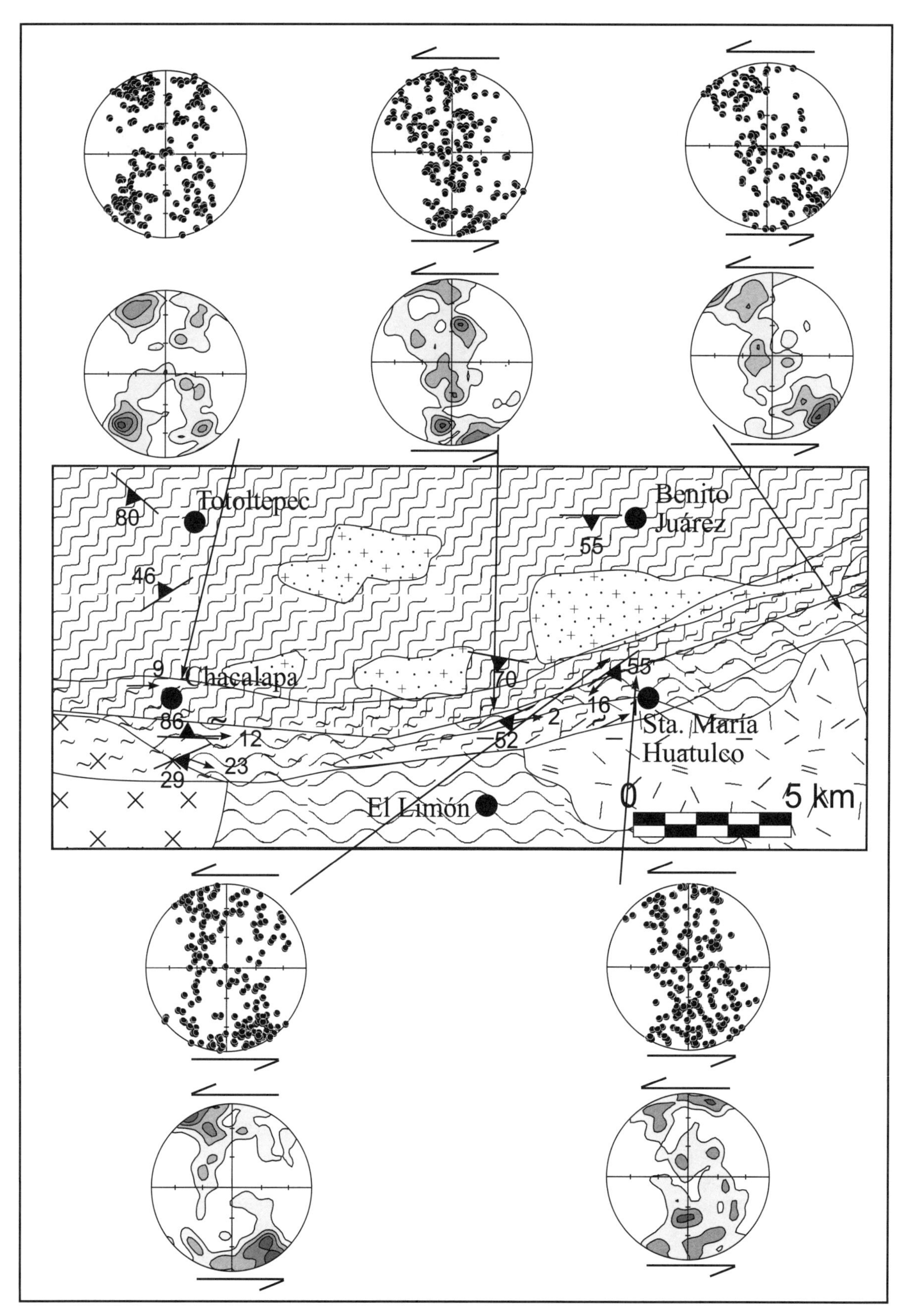

Figure 8. Crystallographic preferred orientation of quartz grains of the Chacalapa shear zone. See map legend in Figure 3.

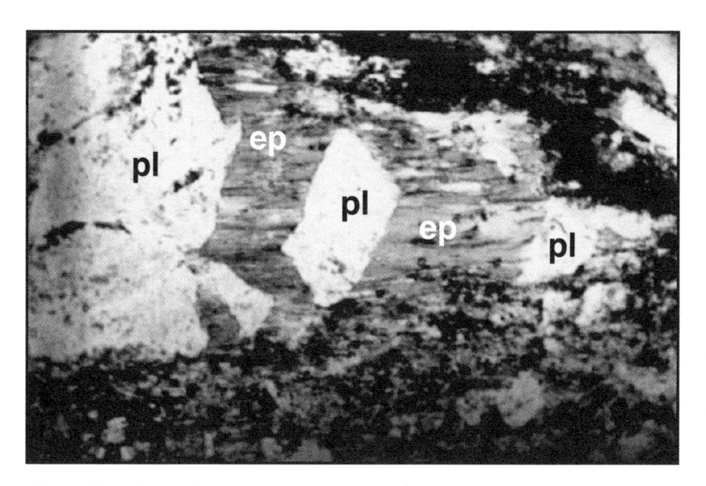

Figure 9. Photomicrograph of protomylonite from the Chacalapa shear zone showing the stretching of a plagioclase (pl) crystal with the growth of epidote fibers (ep) between the fragments.

Figure 10. Outcrop photograph of migmatitic gneiss of the Xolapa Complex. Circled is an amphibolite relic with a preexisting, isoclinally folded foliation. Note the different generations of neosome.

is contemporaneous with the intrusion of synkinematic orthogneisses documented by Morán Zenteno (1992) and is responsible for the development of the foliations in the surrounding/host rocks parallel to the magmatic foliations observed in the orthogneisses. The following deformation event (D_{n+2}) was penecontemporaneous with the migmatitic event, and there is evidence of its continuation immediately following migmatization. The evidence is constituted, for example, by the mylonitic seams of curved trajectories with strikes more or less E-W and low to moderate inclinations to the south and kinematic indicators that suggest north-directed thrusting. These shear zones have the same vergence as the fold asymmetry that affects the gneissic foliation associated with the migmatites, and they are contemporary with it. Also associated with the folds are structures of local intrusion of mobile material that were subsequently involved in the folding. The fourth firmly documented phase of deformation (D_{n+3}) is that associated with the transtensional tectonic regime and with the Chacalapa Fault. This phase of deformation progressively affected the rocks of the Xolapa terrane in both ductile and brittle regimes at the outcrop scale, such that we observe the development of mylonitic rocks (crystal-plastic deformation) in amphibolite and greenschist facies and the formation of pseudotachylytes and cataclasites.

The absolute ages of these deformation events are still unknown but are sufficiently constrained with respect to certain reliably dated magmatic events. The premigmatitic event D_n must be pre–Early Cretaceous if we consider the age of the orthogneissic rocks of the Xolapa Complex obtained by Morán Zenteno (1992) to be reliable. On the other hand, if the inconsistent ages of U/Pb reported by Herrmann et al. (1994) are correct, then D_n might be located toward the terminal Cretaceous. Similarly, the event D_{n+2} could be located in the Early Cretaceous or the middle Paleocene-Eocene, depending upon the data one considers correct. In contrast, transtensional mylonitic event D_{n+3} is known to have occurred between 29 and 23.7 Ma (Tolson, 1998), because the Huatulco stock is affected by shearing, and the mylonitic foliation is truncated by dikes of dacitic composition, and hornblende separated from these latter formations yield the lower age (Fig. 11). The deformation observed in the study area continues to this day along active left-lateral faults.

Kinematics

In the course of the present study, two phases of the development of mylonitic zones have been documented. The first event occurred in the late stages of the migmatitic event of the Xolapa Complex and shows the same vergence as the asymmetry in the folds of migmatitic foliation. The shear zones associated with this event are irregular in layout and have steep to moderate inclinations toward the south. These shear zones best exhibit the development of S-C structures, perhaps characterizing the more transpressive nature of this event in contrast to the rocks of transtensive deformation of the Chacalapa Fault.

The δ and σ porphyroclasts associated with the Chacalapa Fault are notable in the metasedimentary rocks of the Xolapa Complex and in some gneisses of the Oaxaca Complex. Mica fish are common in the majority of mylonitic rocks, while S-C textures become scarce in general, given that tectonite is predominantly of type L (Turner and Weiss, 1963). All such kinematic indicators of the ductile regime imply left-lateral shear. At the same time, the crystallographic textures measured in some samples from the mylonitic zone reinforce this kinematic interpretation, with girdles oblique to the lineation with an asymmetry also indicative of left-lateral movement, although it should be pointed out that there are textures that indicate pure shear (Fig. 8). Faults of the brittle regime also exhibit dominant left-lateral kinematics (Fig. 12), although normal faults with significant displacements have been observed.

Figure 11. Hornblende-bearing quartz diorite porphyritic dike (left) intruding mylonitic rocks of the Chacalapa shear zone (right, with subvertical lineation in photograph), in the bed of the Huatulco River, 1 km north of the town of Sta. María Huatulco.

DISCUSSION AND CONCLUSIONS

Although, as has been demonstrated, values for the age, geometry, and kinematics of the Chacalapa Fault are well constrained, its interpretation in tectonic terms remains uncertain. What is clear is that the Chacalapa Fault separates granulitic rocks of the Oaxaca Complex and its sedimentary cover on the one hand and separates granulitic rocks from the mid-crustal rocks of the Xolapa Complex on the other. For this reason, initial hypotheses suggested that this tectonic limit was similar to detachment faults associated with metamorphic core complexes of the American Cordillera (Robinson et al., 1989). Such interpretations were based on rather superficial and qualitative aspects of the geology of the northern limits of the Xolapa Complex, particularly the juxtaposition of rocks exhibiting crystal-plastic deformation with those deformed in the brittle regime.

However, this model became more sophisticated over time as it was included in tectonic models that put the Chortís block off the coast of Guerrero and its subsequent displacement toward its present position (Meschede, 1994; Morán Zenteno et al., 1996). In its modified form, this model is lent support by the geochronological data relating to crystalline rocks along the Mexican Pacific Coast from Puerto Vallarta, Jalisco, to Salina Cruz, Oaxaca (Morán Zenteno et al., 2000). The data show a general tendency toward a decrease in age toward the southeast, and two more or less linear tendencies can be recognized, which has been cited in support of the gradual extinction of magmatism as the tectonic regime evolved from a lateral distensive tectonic limit to a regime of oblique subduction (Fig. 13).

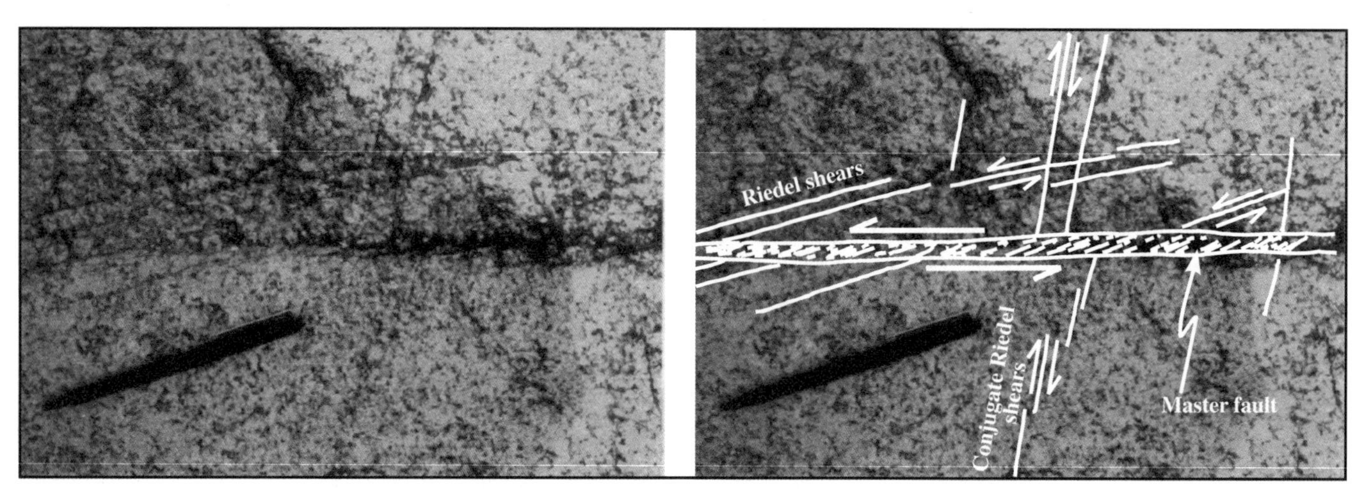

Figure 12. Kinematic indicators of brittle structures within the Ayuta granite.

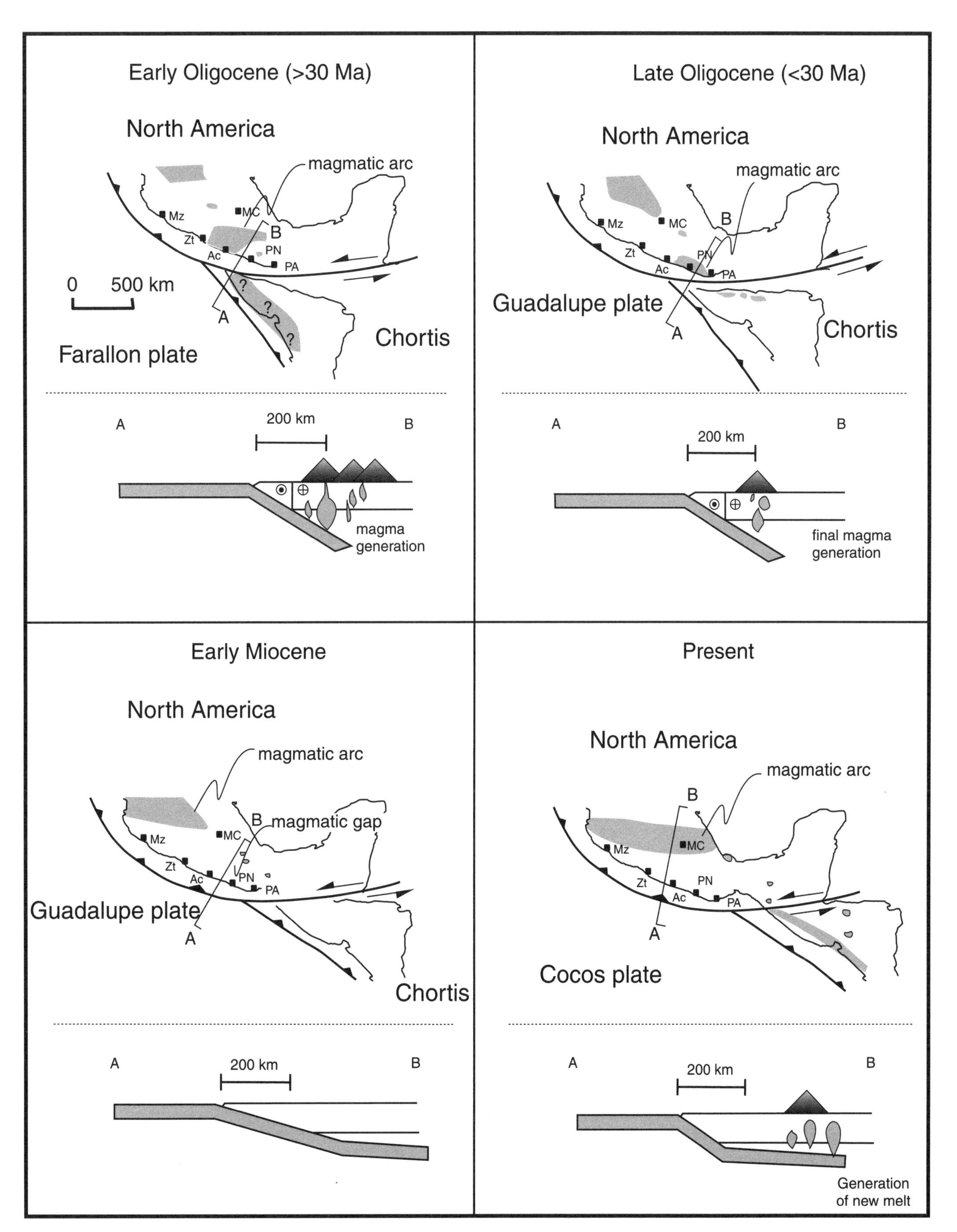

Figure 13. Tectonic model of southern México since the Oligocene showing the pattern of evolution of igneous activity. Ac—Acapulco; MC—México City; Mz—Manzanillo; PA—Puerto Angel; PN—Pinotepa Nacional; ZT—Zihuatanejo.

More recently, the work of Ducea et al. (2004, 2005) has questioned and supplemented the geochronological database relating to the region, to the extent of calling into question the above interpretation. These authors propose that the magmatism along the Mexican Pacific Coast exhibits an episodic case history, characterized by distinct pulses of intense magmatism, reflected in both intrusive and volcanic rocks. Based on the absence of the piercing points common to México and Central America, as well as the pole of rotation of the Caribbean plate, Keppie and Morán-Zenteno (2005) suggest an accretion of the Chortís block by subduction erosion of the Mexican Pacific coastal plain since the Eocene.

Quite independently of the diversity of interpretations concerning the tectonic significance attributed to it, activity along the Chacalapa fault zone has been documented from the Oligocene up to the present day. It presents a suite of progressively younger structures, from mid-crustal to shallow depths, which provide a clear example of flow and strain in a fault zone.

ACKNOWLEDGMENTS

I wish to thank Luigi Solari and Ángel Nieto-Samaniego for constructive reviews that improved the paper. Financial support from the Universidad Nacional Autónoma de México (UNAM) Programa de Apoyo a Proyectos de Investigación e Innovación Tecnológica (PAPIIT), project number IN102602, is gratefully acknowledged.

REFERENCES CITED

Carfantan, J.Ch., 1983, Les ensembles geologiques du Mexique meridional; Evolution geodynamique durant le Mesozoique et le Cenozoique: Geofísica Internacional, v. 22, p. 9–37.

Corona Chávez, P., 1997, Defomazione, Metamorfismo e Meccanismi di Seggregazione Migmatitica nel Complesso Plutonico-Metamorfico del terreno Xolapa, Messico [Ph.D. thesis]: Milano, Italia, Universitá Degli Studi di Milano, 78 p.

De Cserna, Z., 1965, Reconocimiento geológico en la Sierra Madre del Sur de México, entre Chilpancingo y Acapulco, Estado de Guerrero: Universidad Nacional Autónoma de México, Instituto de Geología, Boletín 62, 77 p.

Delgado-Argote, L., and Carballido-Sánchez, E., 1990, Análisis tectónico del sistema transpresivo neogénico entre Macuspana, Tabasco, y Puerto Angel, Oaxaca: Universidad Nacional Autónoma de México, Instituto de Geología, Revista, v. 9, p. 21–32.

Ducea, M.N., Gehrels, G.E., Shoemaker, S., Ruiz, J., and Valencia, V.A., 2004, Geological evolution of the Xolapa Complex, Southern México; Evidence from U-Pb zircon geochronology: Geological Society of America Bulletin, v. 116, p. 1016–1025, doi: 10.1130/B25467.1.

Ducea, M.N., Valencia, V.A., Shoemaker, S., Reiners, P.W., DeCelles, P.G., Fernanda Campa, M., Moran-Zenteno, D., and Ruiz, J., 2005, Rates Of sediment recycling beneath the Acapulco Trench: Constraints from (U-Th)/He thermochronology: Journal of Geophysical Research, v. 109, p. 5665–5672.

Fossen, H., and Tikoff, B., 1993, The deformation matrix for simultaneous simple shearing, pure shearing and volume change, and its applications to transpression-transtension tectonics: Journal of Structural Geology, v. 15, p. 413–422, doi: 10.1016/0191-8141(93)90137-Y.

Guerrero García, J.C., Silver, L.T., and Anderson, T.H., 1978, Estudios geocronológicos en el complejo Xolapa (abs.): Boletín de la Sociedad Geológica Mexicana, v. 39, p. 30–31.

Hernández Bernal, M.S., 1995. Geoquímica y Origen del Batolito de Río Verde, Oaxaca, Terreno Xolapa [M.S. thesis]: México City, Universidad Nacional Autónoma de México, 83 p.

Hernández Bernal, M.S., and Morán Zenteno, D.J., 1996, Origin of the Río Verde Batholith, southern Mexico, as inferred from its geochemical characteristics: International Geology Review, v. 38, p. 361–373.

Herrmann, U.R., Nelson, B.K., and Ratschbacher, L., 1994, The origin of a terrane: U/Pb zircon geochronology and tectonic evolution of the Xolapa complex (southern Mexico): Tectonics, v. 13, p. 455–474, doi: 10.1029/93TC02465.

Herz, N., 1978, Titanium deposits in anorthosite massifs: U.S. Geological Survey Professional Paper 959-D, p. D1–D6.

Hirth, G., and Tullis, J., 1992, Dislocation creep regimes in quartz aggregates: Journal of Structural Geology, v. 14, p. 145–181, doi: 10.1016/0191-8141(92)90053-Y.

Keppie, J.D., and Morán-Zenteno, D.J., 2005, Tectonic Implications of Alternative Cenozoic Reconstructions for Southern Mexico and the Chortis Block: International Geology Review, v. 47, p. 473–491.

Lazos Ramírez, Z.G., and Rodríguez Rivera, R.D., 1995, Estudio Petrológico y Estructural de las Rocas Cristalinas del Área Pochutla-Santo Domingo, Oaxaca [B.S. thesis]: México City, Universidad Nacional Autónoma de México, 90 p.

Lister, G.S., and Dornsiepen, U.F., 1982, Fabric transitions in the Saxony granulite terrain: Journal of Structural Geology, v. 4, p. 81–92, doi: 10.1016/0191-8141(82)90009-8.

Meschede, M., 1994, Tectonic evolution of the northwestern margin of the Caribbean plate in the light of the 'Terrane Concept': Structural and geochemical studies in southern Mexico and Costa Rica: Tübingen Geowissenschaftliche Arbeiten, v. A-22, 113 p.

Morán-Zenteno, D.J., 1992, Investigaciones isotópicas de Rb-Sr y Sm-Nd en rocas cristalinas de la región de Tierra Colorada–Acapulco–Cruz Grande, Estado de Guerrero [Ph.D. thesis]: México City, Universidad Nacional Autónoma de México, 186 p.

Morán-Zenteno, D., Corona Chávez, P., and Tolson, G., 1996, Uplift and subduction erosion in southwestern Mexico since the Oligocene: Pluton geobarometry constraints: Earth and Planetary Science Letters, v. 141, p. 51–66, doi: 10.1016/0012-821X(96)00067-2.

Morán-Zenteno, D.J., Martiny, B., Tolson, G., Solís-Pichardo, G., Alba-Aldave, L., Hernández-Bernal, M.S., Macías-Romo, C., Martínez-Serrano, R.G., Schaaf, P., and Silva Romo, G., 2000, Geocronología y características geoquímicas de las rocas magmáticas terciarias de la Sierra Madre del Sur, *in* Alaniz-Álvarez, S.A., and Ferrari, L., eds., Avances de la Geología Mexicana en la Última Década: Boletín de la Sociedad Geológica Mexicana, v. LIII, núm. 1, p. 27–58.

Ortega-Gutiérrez, F., 1981, Metamorphic belts of southern México and their tectonic significance: Geofísica Internacional, v. 20, no. 3, p. 177–202.

Ortega-Gutiérrez, F., and Corona-Esquivel, R., 1986, La Falla de Chacalapa: Sutura críptica entre los terrenos Zapoteco y Chatino: GEOS, núm. extraordinario, Resúmenes de la Reunión Anual 1986 de la Unión Geofísica Mexicana.

Ortega-Gutiérrez, F., Mitre, L.M., Roldán-Quintana, J., Sanchez-Rubio, G., and De la Fuente, M., 1990, H-3, Acapulco Trench to the Gulf of Mexico across southern Mexico: Boulder, Colorado, Geological Society of America, Decade of North American Geology Continent-Ocean Transect no. 13, one sheet with text, scale 1:500,000.

Passchier, C.W., and Trouw, R.A.J., 1996, Microtectonics: Heidelberg, Springer, 289 p.

Paulson, E.G., 1969, Mineralogy and origin of the titaniferous deposit at Pluma Hidalgo, Oaxaca, Mexico: Economic Geology and the Bulletin of the Society of Economic Geologists, v. 59, p. 753–767.

PEMEX, 1986, Informe Final del Proyecto Mixtepec: Open File Report, 52 p.

Riller, U., Ratschbacher, L., and Frisch, W., 1992, Left-lateral transtension along the Tierra Colorada deformation zone, northern margin of the Xolapa magmatic arc of southern Mexico: Journal of South American Earth Sciences, v. 5, p. 237–249, doi: 10.1016/0895-9811(92)90023-R.

Robinson, K.L., Gastil, R.G., and Campa, M.F., 1989, Early Tertiary extension in southwestern Mexico and exhumation of the Xolapa metamorphic core complex: Geological Society of America Abstracts with Programs, v. 21, no. 6, p. A-92.

Sánchez Rubio, G., 1975, Las migmatitas de Puerto Escondido, Oaxaca [Tesis de Ingeniero Geólogo]: México City, Universidad Nacional Autónoma de México (UNAM), 47 p.

Sánchez Zavala, J.L., 2006. Estratigrafía, sedimentología y análisis de procedencia de la Formación Tecomate y su papel en la evolución del Complejo Acatlán, sur de México [Ph.D. thesis]: México City, Universidad Nacional Autónoma de México (UNAM), 226 p.

Schaaf, P., Morán-Zenteno, D.J., Hernández-Bernal, M.S., Solís Pichardo, G., Tolson, G., and Köhler, H., 1995, Paleogene continental margin truncation in southwestern Mexico: Geochronological evidence: Tectonics, v. 14, p. 1339–1350, doi: 10.1029/95TC01928.

Simpson, C., and De Paor, D.G., 1993, Deformation and kinematics of high strain zones: Journal of Structural Geology, v. 15, p. 1–20, doi: 10.1016/0191-8141(93)90075-L.

Tolson, G., 1995, Using a weighted density function to fit an ellipse to the Fry plot: Geological Society of America Abstracts with Programs, v. 27, no. 6, p. A-71.

Tolson, G., 1998, Deformación, Exhumación y Neotectónica de la Margen Continental de Oaxaca: Datos Estructurales, Petrológicos y Geotermobarométricos [Ph.D. thesis]: México City, Universidad Nacional Autónoma de México (UNAM), 96 p.

Tolson, G., Solís Pichardo, G.N., Morán Zenteno, D., Victoria Morales, A., and Hernández Treviño, J.T., 1993, Naturaleza petrográfica y estructural de las rocas cristalinas en la zona de contacto entre los terrenos Xolapa y Oaxaca, región de Santa María Huatulco, Oaxaca, *in* Delgado Argote, L.A., and Martín Barajas, A., eds., Contribuciones a la Tectónica del Occidente de México: Unión Geofísica Mexicana, Monografía 1, p. 327–349.

Tullis, J., 1977, Preferred orientation of quartz produced by slip during plane strain: Tectonophysics, v. 39, p. 87–102, doi: 10.1016/0040-1951(77)90090-7.

Turner, F.J., and Weiss, L.E., 1963, The Structural analysis of metamorphic tectonites: New York, Wiley, 563 p.

Weger, M., 1996, Duktile Kinematik kontinentaler Kruste am Beispiel del Zentralgneise de westlichen Tauernfensters (Ostalpen, Österreich und Italien) Strainmethodik, Strainverteilung und Geodynamik [Ph.D. thesis]: Munich, Ludwig-Maximilians Universität, 186 p.

Manuscript Accepted by the Society 29 August 2006

Printed in the USA

Geological Society of America
Special Paper 422
2007

Mineralizing processes at shallow submarine hydrothermal vents: Examples from México

Carles Canet*
Rosa María Prol-Ledesma
Departamento de Recursos Naturales, Instituto de Geofísica, Universidad Nacional Autónoma de México, Ciudad Universitaria, Delegación Coyoacán, México D.F. 04510, México

ABSTRACT

Recent studies on shallow submarine hydrothermal vents (at water depths <200 m below sea level [mbsl]) suggest that their activity could have been responsible for the formation of oxide, sulfide, and precious metal–bearing ores.

The boundary between shallow and deep hydrothermal vents has been established at a depth of 200 mbsl, which represents an abrupt change in the environmental parameters and in the structure of the biotic communities. Shallow submarine vents support complex biotic communities, characterized by the coexistence and competition of chemosynthetic and photosynthetic organisms. Some biogeochemical and biomineralizing processes related to chemosynthesis are similar to those described in deep hydrothermal vents and in cold seeps.

Frequently, hydrothermal shallow vent water has lower salinity than seawater. This fact, together with the isotopic compositions, is evidence of a meteoric component in vent water. Venting of exsolved gas, evidenced by continuous bubbling, is a striking feature of shallow submarine hydrothermal systems. In most cases, vent gas is rich in CO_2, but occasionally it can be rich in N_2, CH_4, and H_2S.

In México, shallow submarine hydrothermal venting has been studied in Punta Banda and Bahía Concepción, Baja California Peninsula, and in Punta Mita, Nayarit. The tectonic setting of those hydrothermal systems corresponds to continental margins affected by extension, with high geothermal gradients. These vents do not show obvious links with volcanic activity. Their study has contributed to the understanding of mineralogical and geochemical processes in shallow submarine hydrothermal vents. Those systems could be a potential source of geothermal energy.

Keywords: hydrothermal vents, neritic zone, chemosynthesis, ore deposits.

INTRODUCTION

The study of hydrothermal vents, mainly performed during the last three decades, has contributed to the understanding of the nature of ore-forming processes in seafloor environments and in volcanogenic massive sulfide (VMS) deposits (e.g., Graham et al., 1988; Rona, 1988; Herzig and Hannington, 1995; Humphris et al., 1995a; Parson et al., 1995; Scott, 1997). In addition, their study has elucidated key questions related to the cycle of some metals in oceans (e.g., Gamo et al., 2001), and to the metabolism of extremophile and chemosynthetic biotic communities (e.g., Karl et al., 1980; Jannasch, 1984). Thus, the

*ccanet@geofisica.unam.mx

Canet, C., and Prol-Ledesma, R.M., 2007, Mineralizing processes at shallow submarine hydrothermal vents: Examples from México, *in* Alaniz-Álvarez, S.A., and Nieto-Samaniego, Á.F., eds., Geology of México: Celebrating the Centenary of the Geological Society of México: Geological Society of America Special Paper 422, p. 359–376, doi: 10.1130/2007.2422(13). For permission to copy, contact editing@geosociety.org.

study of hydrothermal vents provided clues to the possible origin of life and the understanding of primitive biosphere (e.g., Corliss et al., 1981; Russell, 1996).

Research on submarine hydrothermal venting dates back to 1965, when hydrothermal brine pools and related metalliferous sediments were discovered in the axial oceanic rift of the Red Sea (Degens and Ross, 1969). These discoveries corresponded to the "Atlantis II Deep" vent site, which is the largest identified submarine metalliferous deposit, with estimated reserves of 94 Mt and grades of 0.45% Cu, 2.07% Zn, 39 ppm Ag, and 0.5 ppm Au (Missack et al., 1989; Herzig, 1999).

In 1976, a low-temperature vent system with associated deposits of nontronite mud and manganese oxides was discovered in the Galapagos Ridge. Later, in 1979, the first black smokers were found at the East Pacific Rise, near latitude 21°N, with vent water rich in dissolved metals at temperatures up to 350 °C (Macdonald et al., 1980). Black smokers are sulfide-rich meter-sized chimney structures that act as discharge conduits for fluids circulating in hydrothermal convection systems within the oceanic crust, and are build up over mounds of sulfates (barite and anhydrite) and base metal sulfides with minor precious metals (Scott, 1997). So far, more than 100 deep-sea hydrothermal vent systems have been discovered in the ocean floor, with fluid temperatures up to 405 °C, mainly in sea-floor spreading settings and, seldom, in seamounts. Most of those sites occur in the Pacific and Atlantic oceans, even though they have been found also in the Indic Ocean and in the Mediterranean Sea (Scott, 1997).

The internal anatomy of deep-sea vents was verified by drilling at the Trans-Atlantic geotraverse (TAG) hydrothermal mound on the Mid-Atlantic Ridge (Humphris et al., 1995b). The structure and morphology, together with the mineralogical and chemical composition observed on deep-sea hydrothermal vent mineralization, suggest their connection to some economic ore deposits that have been important sources of copper, zinc, lead, and gold and also of manganese and iron (e.g., Jorge et al., 1997). Thus, deep-sea hydrothermal vents can be regarded as modern analogues of volcanogenic massive sulfide (VMS) deposits hosted in ophiolites, or Cyprus type VMS deposits (e.g., Sawkins, 1990).

On the other hand, only a few shallow submarine hydrothermal vents (SSHV) have been studied in detail, in spite of being much more accessible and easy to sample. In fact, initially, in comparison with deep-sea vents, research on SSHV did not become known, but now the studies on this subject are growing significantly.

The first publications on SSHV were essentially focused to the specialized biotic communities that colonize these environments (e.g., Tarasov et al., 1985, 1990, 1991, 1993). Results of both these and subsequent studies (e.g., Kamenev et al., 1993, 2004; Hoaki et al., 1995; Tarasov, 2002; Cardigos et al., 2005; Rusch et al., 2005; Tarasov et al., 2005) suggest that specialized prokaryotes are highly diverse and similar to those described in deep-sea vents. In addition, SSHV can be considered natural laboratories to study the interactions between hydrothermal fluids (water and gas), sediments and seabed rocks, and seawater (e.g., Fitzsimons et al., 1997; Botz et al., 1999; Pichler et al., 1999a; Prol-Ledesma et al., 2004; Chen et al., 2005; Forrest et al., 2005; Villanueva-Estrada et al., 2005; Villanueva et al., 2006).

Hydrothermal fluids discharged by shallow submarine vents have lower temperatures than those discharged by black smokers; nevertheless, they can transport dissolved metals in amounts enough high to produce deposits of oxides, sulfides, and precious metals (Martínez-Frías, 1998; Stoffers et al., 1999; Hein et al., 2000; Prol-Ledesma et al., 2002a; Canet et al., 2005b; Jach and Dudek, 2005).

Several published studies have been focused on processes of mineral precipitation in SSHV (Pichler et al., 1999b; Stoffers et al., 1999; Prol-Ledesma et al., 2002a; Canet et al., 2003, 2005a, 2005b; and Alfonso et al., 2005). Those of Canet et al. (2003, 2005a, 2005b) and Alfonso et al. (2005) note the key role of biogeochemical processes in the formation hydrothermal precipitates around SSHV.

Conversely, the consequences of shallow submarine hydrothermal venting on the mass balance in coastal environments, and its supply of hazardous elements to environment, such as arsenic, are not sufficiently known. Likewise, the potential of SSHV as sources of geothermal power has not been evaluated in detail.

DISTRIBUTION OF SHALLOW SUBMARINE HYDROTHERMAL VENTS

SSHV are found in various tectonic settings, mostly in relation to plate margins. Thus, they are located along island arcs and in shallow segments of mid-ocean ridges under the influence of hot spots (Fig. 1) (Fricke et al., 1989; Dando and Leahy, 1993; Hoaki et al., 1995; Fitzsimons et al., 1997; Scott, 1997; Savelli et al., 1999; Stoffers et al., 1999; Geptner et al., 2002). In addition, it is likely that they occur in relation to oceanic intraplate volcanism, where deep-sea vents have been reported, for example, in Loihi seamount, Hawaii (Moyer et al., 1998).

In island arcs, SSHV have been described in Kraternaya Bight, in the Kurile Arc, Russia (Tarasov et al., 1985); in the Bay of Plenty, New Zealand (Stoffers et al., 1999); in the New Britain Islands (Tarasov et al., 1999) and in Ambitle and Lihir (Pichler et al., 1999a, 1998b) in the Bismarck Archipelago, Papua New Guinea; in the Aegean Sea (Sedwick and Stuben, 1996; Dando et al., 1999; Dando et al., 2000) and in the Aeolian Islands, in eastern and central Mediterranean, respectively (Rusch et al., 2005); in Dominica, in the Antilles (Bright, 2004); and along the western margins of the Pacific Ocean (Ferguson and Lambert, 1972; Tarasov et al., 1985, 1990, 1999, 2005; Sarano et al., 1989; Hashimoto et al., 1993; Kamenev et al., 1993; Chen et al., 2005).

In areas with mid-oceanic ridge–hot-spot interaction, SSHV have been reported in Kolbeinsey Ridge, Iceland (Benjamínsson, 1988; Botz et al., 1999), and in João de Castro Seamount, in the Azores Islands (Cardigos et al., 2005).

The geological settings of the above-mentioned sites agree with the close link between most SSHV and recent volcanism.

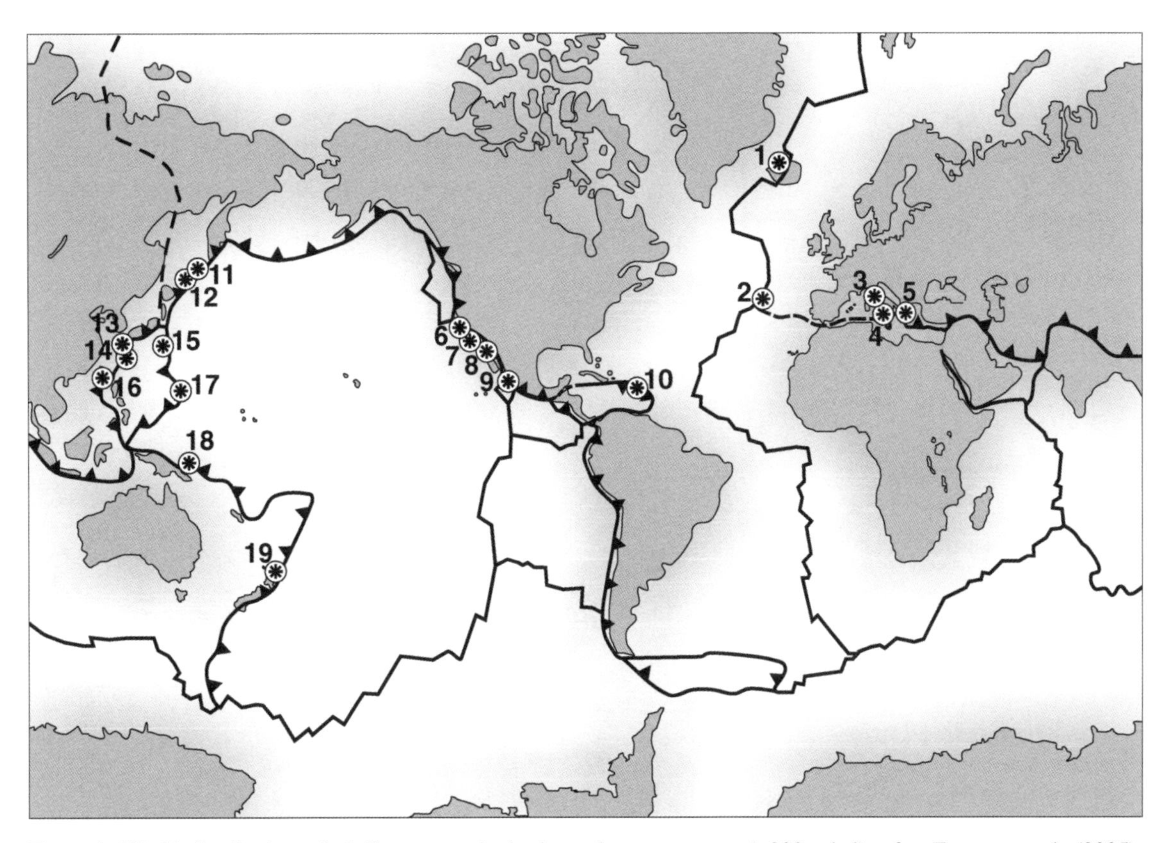

Figure 1. World distribution of shallow-water hydrothermal vent systems (<200 mbsl), after Tarasov et al. (2005). (1) Kolbeinsey, Island; (2) Azores, Portugal; (3) Palinuro and Messino Capes, Tyrrhenian Sea, Italia; (4) Vulcano, Aeolian Islands, Italy; (5) Santorini and Milos, Aegean Sea, Greece; (6) White Point, California, USA; (7) Punta Banda, Baja California; (8) Bahía Concepción, Baja California Sur; (9) Punta Mita, Nayarit, México; (10) Dominica, Antilles; (11) Kraternaya Bight; (12) Kunashir Island, Kuriles, Russia; (13) Kagoshima Bay; (14) Tokora and Iwo Islands; (15) Nishino Island, Japan; (16) Kueishantao Island, Taiwan; (17) Esmeralda Bank Volcano, Marianas; (18) Matupi Harbour and Ambitle and Lihir Islands, Papua New Guinea; and (19) Bay of Plenty, New Zealand.

For instance, there are several SSHV along the Kermadec Arc, north of New Zealand, in the active submarine volcanoes of Rumble III, Rumble V, Macauley Cone, Giggenbach, Ngatoroirangi, Monowai, and Vulkanolog, at depths greater than 130 mbsl (de Ronde et al., 2001; C.E.J. de Ronde, 2004, personal commun.). However, there are also SSHV in continental margins actively affected by extension tectonic processes; for example, in Bahía Concepción (Prol-Ledesma et al., 2004) and Punta Banda (Vidal et al., 1978), both in the Baja California Peninsula; in Punta Mita in Nayarit State, México (Figs. 2 and 3) (Prol-Ledesma et al., 2002a, 2002b); and in White Point in California, USA (Stein, 1984). Likewise, some lakes, placed in intracontinental rift basins, contain systems of underwater hot springs; for example, lakes Baringo in Kenya (Renaut et al., 2002) and Tanganyika in Tanzania (Barrat et al., 2000), and Lake Baikal in Russia (Crane et al., 1991). Similar hydrothermal features have been reported in crater lakes in Taupo, New Zealand (de Ronde et al., 2002), and in Crater Lake in Oregon, USA (Dymond et al., 1989). Even though SSVH can show intermediate geochemical and mineralogical characteristics between deep-sea vents and sublacustrine hot springs (e.g., Schwarz-Schampera et al., 2001; Canet et al., 2003), the latter are clearly closer to continental geothermal systems (Prol-Ledesma et al., 2005).

It has been considered that the maximum depth for SSHV would be ~200 m (Prol-Ledesma et al., 2005). At this depth, which also corresponds to the lower limit of the neritic zone, sharp changes in the environmental parameters and in the structure of the biotic communities occur. Thus, biotic communities inhabiting SSHV clearly differ from those of deep-sea vents by the presence of diatoms, bacterial and algal mats, and phytoplankton.

On the other hand, in comparison to SSHV, deep-sea vents host a vast diversity of highly specialized and symbiotrophic species, and their communities show a more complex structure, with vertical and lateral zoning and a greater biomass (Tarasov et al., 2005). Moreover, formation of large sulfide-rich structures, such as mounds and chimneys, is a distinguishing feature of deep-sea vents.

In addition, the 200 mbsl depth coincides with a sharp change in the slope of the boiling temperature curve for seawater with respect to pressure, which is ~20 bars at this depth (Bischoff and Rosenbauer, 1984; Butterfield et al., 1990).

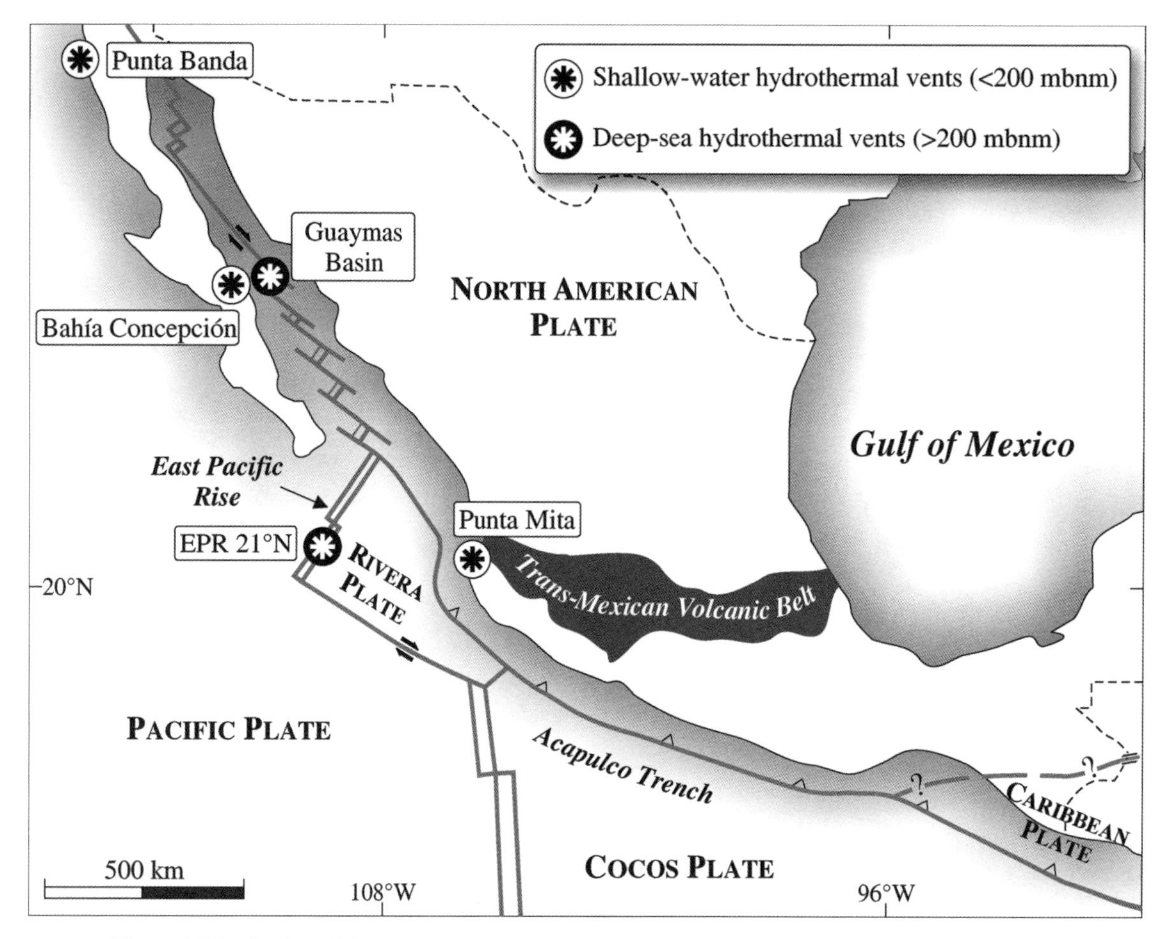

Figure 2. Distribution of deep-sea and shallow submarine hydrothermal vents off the coast of México.

FOSSIL ANALOGUES

The genesis of several ore deposits has been related to SSHV activity. The main examples reported in bibliography are listed here.

In Milos, Greece, there are manganese deposits whose origin has been linked to SSHV, even if they show some features typical of epithermal veins (Hein et al., 2000; Liakopoulos et al., 2001; Naden et al., 2005). They consist in veins of manganese oxides with barite, hosted in volcaniclastic rocks of Pliocene to Pleistocene age, and are located in the coastal area around Cape Vani. In addition, ore mineralization occurs as crusts, cementing sandstone, and replacing limestone. The manganese minerals of the deposits are pyrolusite MnO_2, ramsdellite MnO_2, cryptomelane $K(Mn^{4+},Mn^{2+})_8O_{16}$, hollandite $(Ba,K)Mn_8O_{16}$, coronadite $PbMn_8O_{16}$, and hydrohetaerolite $Zn_2Mn_4O_8{\cdot}(H_2O)$ (Liakopoulos et al., 2001). The ores, with manganese grades up to 60 wt% MnO, have high contents in lead (up to 3.4 wt% Pb), barium (up to 3.1 wt% BaO), zinc (up to 0.8 wt% Zn), arsenic (up to 0.3 wt% As), antimony (up to 0.2 wt% Sb), and silver (up to 10 ppm). These manganese deposits are in the vicinity of Profitis Ilias, a low sulfidation epithermal deposit rich in precious and base metals. The resemblance in style of mineralization, host rock, and geologic setting between both deposits suggests that they would be genetically related (Hein et al., 2000).

Another example of an ore deposit believed to be a fossil SSHV is located in Wafangzi, China, and is the largest manganese deposit known in this country. According to Fan et al. (1999), microbial activity mediated the mineral precipitation processes that led to ore formation.

Jach and Dudek (2005) described a stratabound manganese deposit hosted in encrinites deposited as tempestites during the Early Jurassic, in the Tatra Mountains, Poland. Its mineralogical and geochemical features, together with the presence of specific microbial structures and the inferred sedimentary environment, suggest that this ore deposit formed as a result of venting of hydrothermal fluids in the neritic zone, through synsedimentary faults. The essential minerals of the ores are braunite Mn_7SiO_{12}, caryopilite $(Mn,Mg,Zn,Fe^{2+})_3(Si,As)_2O_5(OH,Cl)_4$, manganiferous calcite, and rhodochrosite $MnCO_3$. The ore body is rich in manganese (up to 62.8 wt% MnO) and barium (up to 4500 ppm), and has low contents of transition metals (Co+Ni+Cu < 0.01 wt%).

The stratabound Ba-Sb-Ag-Fe-Hg ores of Las Herrerías and Valle del Azogue occur in the Neogene-Quaternary volcanic prov-

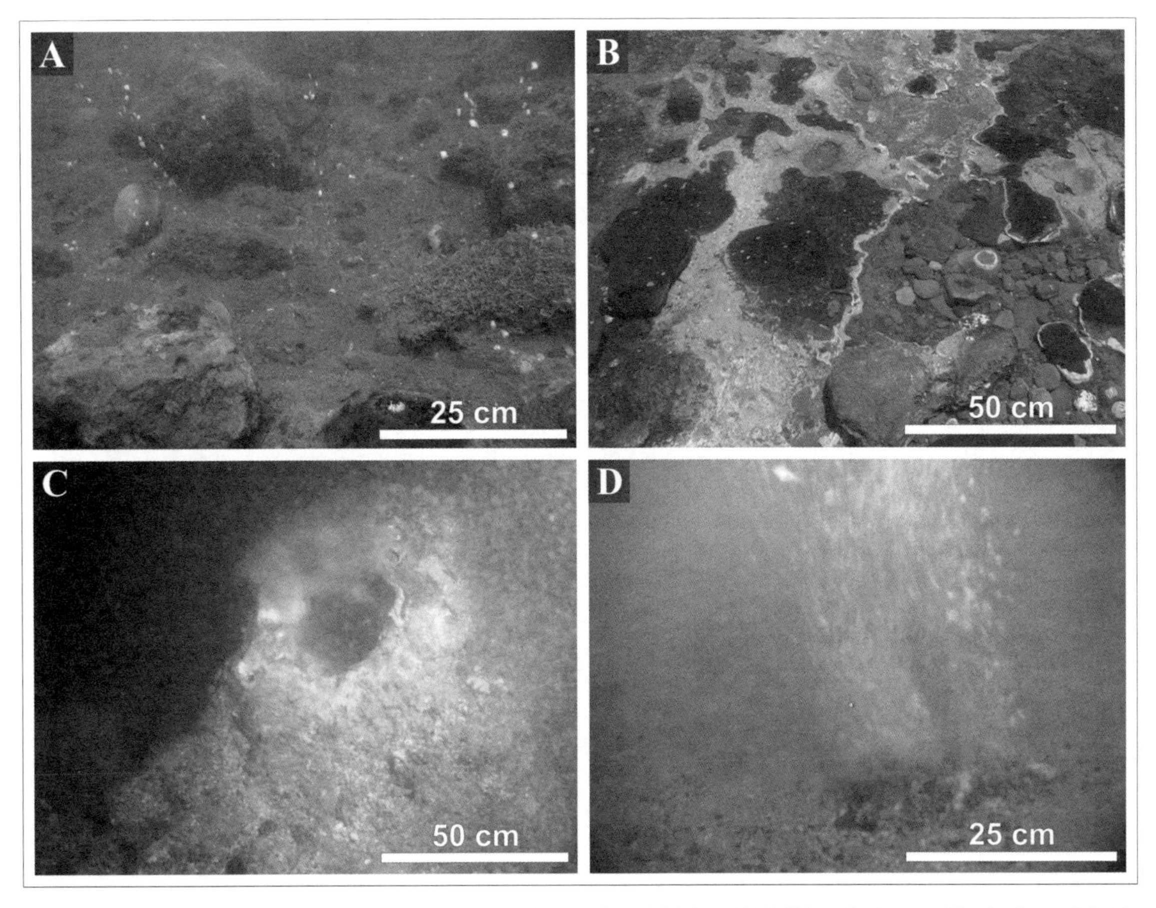

Figure 3. Examples of shallow submarine hydrothermal vents from México. (A) Diffuse discharge of hydrothermal fluids (gas and water) at a depth of 5 mbsl., and (B) silica-carbonate sinter deposit in intertidal hot-springs, Bahía Concepción, Baja California Sur. (C) Calcite chimney, and (D) gas discharge, ~10 mbnm, Punta Mita, Nayarit.

ince of SE Spain (Martínez-Frías, 1998). Orebodies attain up to 10 m in thickness and contain some fossil structures that have been interpreted as vent chimneys (Martínez-Frías, 1998). They are made up of barite and jasper with iron and manganese oxides and hydroxides, with disseminated pyrite, sphalerite, galena, cinnabar, and native silver. The structure of the deposits, the occurrence of chimney-like structures, and their chemical and mineralogical compositions suggest that they formed due to exhalative submarine activity at depths near 200 mbsl (Martínez-Frías, 1998). Those deposits are spatially and temporally related to the epithermal Au, Ag, Fe-Mn, Hg-Sb, and base metal veins of Sierra Almagrera.

In México, there are several manganese deposits close to Bahía Concepción (Baja California Sur) that occur as veins, breccias, and stockworks. They are composed of manganese oxides (pyrolusite, coronadite, and romanechite $(Ba,H_2O,O)_2Mn_5O_{10}$), with dolomite, quartz, and barite, and are hosted principally in Tertiary andesites and, locally, in detrital and carbonate rocks of Pliocene age (Rodríguez-Díaz, 2004). The mineralization arrangement is controlled by Miocene normal faults of NW-SE trend. In the vicinity of these deposits there are numerous manifestations of a SSHV system, at depths from the sea surface to 15 mbsl, with vent temperatures up to 87 °C that are hosted in the same andesitic rocks and related to the same NW-SE faults (Prol-Ledesma et al., 2004).

As a result of the modern hydrothermal venting, there are patches of mineral precipitates, made up of poorly crystallized phases chemically analogous to todorokite $(Mn,Ca,Mg)Mn_3O_7 \cdot H_2O$ and romanechite, with barite, calcite, and opal-A (Canet et al., 2005b). Taking into account the similarities between both fossil and modern mineralization, and the importance of their structural NW-SE fault control, it can be concluded that the metallogenesis of these Late Miocene to Pliocene manganese deposits is analogous to the modern SSHV systems. Likewise, in the same area there is a chert bed of Late Pliocene age, whose origin has been attributed to hydrothermal venting of silica-rich fluids in a mangrove marsh (Ledesma-Vázquez et al., 1997).

ORIGIN AND CIRCULATION OF HYDROTHERMAL FLUIDS

Fluids vented from SSHV show transitional chemical and isotopic compositions between deep-sea vents and continental geothermal systems, both subaerial and sublacustrine (Dymond et al., 1989; Barrat et al., 2000; Schwarz-Schampera et al., 2001, de Ronde et al., 2002; Prol-Ledesma et al., 2002a, 2002b, 2004, 2005; Renaut et al., 2002), and are strongly constrained by the geological and tectonic setting (Prol-Ledesma et al., 2005).

Sampling of vent water in SSHV unavoidably implies mixing with seawater; thus, it is suitable to assume a simple mixing model for estimating the composition of vent water. In this way, a thermal end-member is calculated, which represents vent water previous to the seawater addition through sampling (Table 1) (Prol-Ledesma et al., 2002a, 2002b). The principal hypothesis in the mixing model is that the concentration of magnesium in the thermal end-member is near zero, in agreement with the results of heating seawater experiments carried out by Bischoff and Seyfried (1978). Their results point to an almost complete loss of magnesium in solution due to the precipitation of minerals that scavenge and immobilize it. Depending on the sampling method, it is possible to obtain samples of vent water with <10% of seawater (Villanueva et al., 2006).

Generally, vent water salinities and densities are lower in SSHV than in deep-sea vents (Table 1). This fact, together with the isotopic compositions, point to a meteoric component in SSHV fluids (Prol-Ledesma et al., 2003, 2004, 2005; Villanueva-Estrada et al., 2005). Numerical and conceptual models obtained from several SSHV corroborate that vent water contains meteoric water that interacted in variable degree with the underlying rocks (Fig. 4) (Vidal et al., 1981; Prol-Ledesma et al., 2002b, 2004; Villanueva-Estrada et al., 2005). Nevertheless, SSHV water shows chemical similarities with that from deep-sea vents, both for its concentration in some major ions and for its contents in rare earth elements (Prol-Ledesma, 2003). So, in both cases, vent water is enriched in Si, Ba, Mn, B, and Fe with respect to seawater, it shows enrichment in Eu, and does not exhibit a negative Ce anomaly, since they are reducing fluids (Michard, 1989). The rare earth element (REE) patterns in SSHV can be inherited by the mineral precipitates formed around the vents (Canet et al., 2005b), and they clearly differ

TABLE 1. WATER COMPOSITION OF SOME SHALLOW-WATER SUBMARINE HYDROTHERMAL VENTS

Area	mmol/kg				µmol/kg							mmol/kg			
	Ca	Mg	K	Na	I	Hg	Mn	Cs	Ba	As	Br	Cl	SO_4	HCO_3	Si
Bahía Concepción* (México)	44.5	0.0*	13.0	243.9	5.48	ND	165.07	12.21	8.14	26.91	1550	380.3	4.1	10.3	7.8
Punta Mita* (México)	51.27	0.0*	3.17	110.36	17.35	ND	1.87	6.55	5.84	<1ppm	0.38	283.91	2.81	0.0	2.23
Punta Banda (México)	37.92	3.29	10.0	226.2	ND	ND	15.47	ND	6.84	ND	ND	304.62	3.65	5.69	2.6
Rivari, Milos (Greece)	11.28	54.88	11.02	504.79	ND	ND	4.17	ND	ND	ND	ND	ND	ND	ND	ND
Kueishantao (Taiwan)	9.43	48.7	9.16	433	ND	ND	1.31	ND	ND	ND	ND	509	26.8	ND	0.269

Note: ND—not determined. References: Bahía Concepción, Prol-Ledesma et al. (2004); Punta Mita, Prol-Ledesma et al. (2002a); Punta Banda, Vidal et al. (1978); Rivari, Cronan and Varnavas (1993); Kueishantao, Chen et al. (2005).
*Thermal end-member; Mg concentration is assumed as 0 for the binary mixing model.

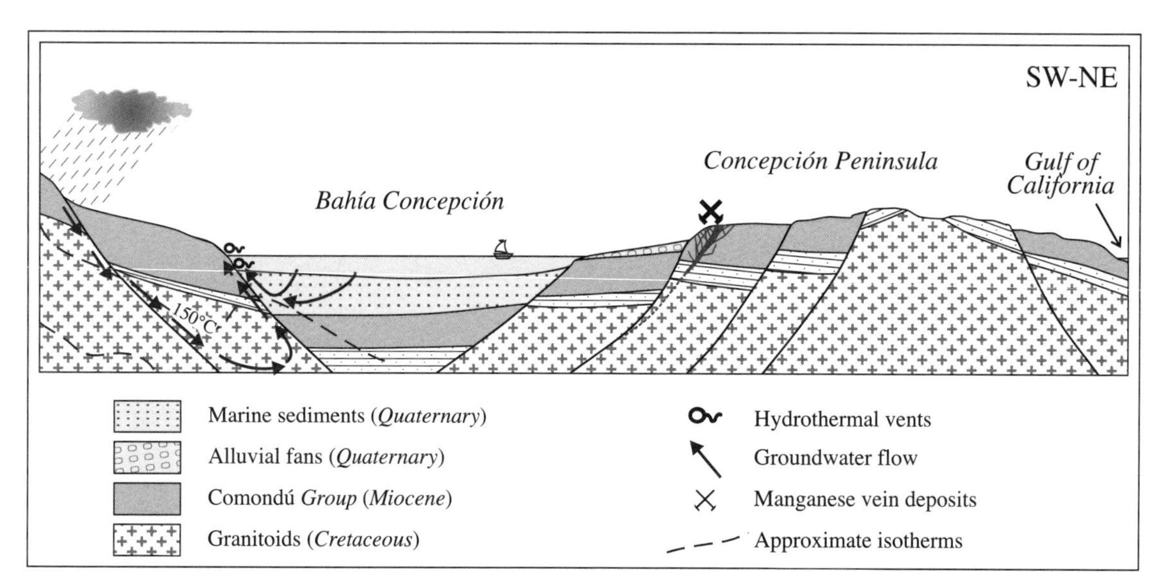

Figure 4. Schematic geological section of Bahía Concepción showing a conceptual hydrologic model for the shallow submarine hydrothermal vents (SSHV) system. The length of the section is ~30 km. After Prol-Ledesma et al. (2004).

from those of hydrothermal fluids in orogenic and continental settings (Michard and Albarède, 1986).

Hydrothermal fluids from deep-sea vents placed on areas of oceanic crust overlaid by thick sedimentary covers; for example, in the Guaymas Basin, Gulf of California, fluids are enriched in Ca and I with respect to seawater (Von Damm et al., 1985). Similarly, in some cases, SSHV fluids can be enriched in these elements (Prol-Ledesma, 2003).

A striking feature of SSHV is the presence of exsolved gas, producing a conspicuous bubbling (Fig. 3; Table 2). Accordingly, some authors (e.g., Tarasov et al., 1990) name the SSHV *gaso-hydrothermal vents*. The composition of the gas phase in SSHV is variable, although it is roughly similar to that of the dissolved gases in deep-sea hydrothermal fluids. Thus, the gas is mainly CO_2 with minor H_2S and CH_4 (Dando et al., 2000; Botz et al., 2002; Amend et al., 2003). Conversely, for vent gas in SSHV systems whose fluids interacted with sedimentary sequences, as in Punta Mita (México), the main exsolved gases are N_2 and CH_4 (Prol-Ledesma et al., 2005), and in SSHV systems directly related to volcanic activity the predominant gas is H_2S (de Ronde et al., 2001).

MINERALOGY AND GEOCHEMISTRY

Potentially, SSHV systems can be expected to generate ore deposits with oxides, sulfides, and precious metals (Figs. 5 and 6) (Martínez-Frías, 1998; Hein et al., 1999; Stoffers et al., 1999; Prol-Ledesma et al., 2002a, 2002b). In SSHV, pressure from the water column (up to 20 bars) is much lower than in deep-sea vents. This fact could trigger the deposition of some metals below the seafloor as a result of deep boiling of hydrothermal fluids.

Mineral assemblages produced by venting of hydrothermal fluids in shallow submarine environments are comparable to those of low to medium temperature deep-sea vents (Canet et al., 2005b), given that they used to be rich in barite, manganese oxides, and iron oxyhydroxides. In addition, they show some resemblances to some low-sulfidation epithermal deposits. Thus, many SSHV deposits are enriched in S, Hg, As, and Sb, a suite of elements that can be found in epithermal deposits (e.g., Bornhorst et al., 1995), although it is also characteristic of mineralization that formed from low-temperature hydrothermal fluids unrelated to volcanism and rich in organic matter (Tritlla and Cardellach, 1997).

In the shallowest manifestations of the SSHV system of Bahía Concepción (Baja California Sur, México) there are sinter-like silicic deposits and banded veins of chalcedony, calcite, and barite (Canet et al., 2005a, 2005b), mineralogically and texturally similar to the sinters and veins that are usually found in association with low sulfidation epithermal deposits (e.g., Hedenquist et al., 1996; Camprubí and Albinson, 2006). Consequently, many SSHV can be considered as a transition between deep-sea vents and epithermal deposits (Schwarz-Schampera et al., 2001).

Poorly crystalline and amorphous phases are prevalent in SSHV deposits. Iron oxyhydroxides, both amorphous and with low crystallinity (protoferrihydrite and ferrihydrite, respectively), have been described in Bahía Concepción, Baja California Sur, México (Canet et al., 2005b), and in Ambitle, Papua New Guinea (Pichler and Veizer, 1999). Iron oxyhydroxides are common in other hydrothermal environments; for example, in deep-sea hydrothermal vents, forming replacive surface crusts over sulfide mounds (e.g., Haymon and Kastner, 1981; Marchig et al., 1999; Hannington et al., 2001; Lüders et al., 2001), in metalliferous sediments, accompanied by manganese oxides (e.g., Daesslé et al., 2000), and in low-temperature hydrothermal vents (Bogdanov et al., 1997).

Precipitation of iron oxyhydroxides depends on pH, Eh, temperature, iron concentration, and microbial activity (Puteanus et al., 1991; Savelli et al., 1999; Mills et al., 2001), whereas the low crystallinity is due to rapid precipitation kinetics (Chao and Theobald, 1976).

Manganese oxides also have been described in SSHV (Canet et al., 2005b), and more often, in metalliferous sediments (Daesslé et al., 2000) and in deep-sea vents (Glasby et al., 1997), where they are usually associated with iron oxyhydroxides and precipitate from low-temperature (up to 25 °C) hydrothermal fluids (Burgath and von Stackelberg, 1995). On the other hand, manganese oxides form hydrogenous deep-sea deposits of nodules and crusts (Nicholson, 1992). In SSHV deposits, poorly crystalline phases chemically equivalent to romanechite and todorokite have been described (Canet et al., 2005b), whereas there are crystalline

TABLE 2. GAS COMPOSITION OF SOME SHALLOW-WATER SUBMARINE HYDROTHERMAL VENTS, IN mmol/mol

Area	CO_2	H_2S	CH_4	N_2	Ar	O_2	He	H_2	References
Bahía Concepción (México)	428.8	ND	21.3	541.2	6.84	1.21	0.439	0.045	Forrest et al. (2005)
Punta Mita (México)	1.9	<0.001	115	880	4.3	<0.001	0.43	0.08	Prol-Ledesma et al. (2002a)
Punta Banda (México)	20	ND	514	443	10	2.9	0.17	0.03	Vidal et al. (1978)
Louise Harbor (Papua New Guinea)	962	8.0	4.6	19.9	ND	4.7	0.021	1.1	Pichler et al. (1999a, 1999b)
Tutum Bay (Papua New Guinea)	949.4	<0.3	14.8	36.8	ND	5.8	0.015	<.01	Pichler et al. (1999a, 1999b)
Kueishantao (Taiwan)	978	20.4	ND	0.871	0.009	0.195	0.28	0.114	Chen et al. (2005)

Note: ND—not detected.

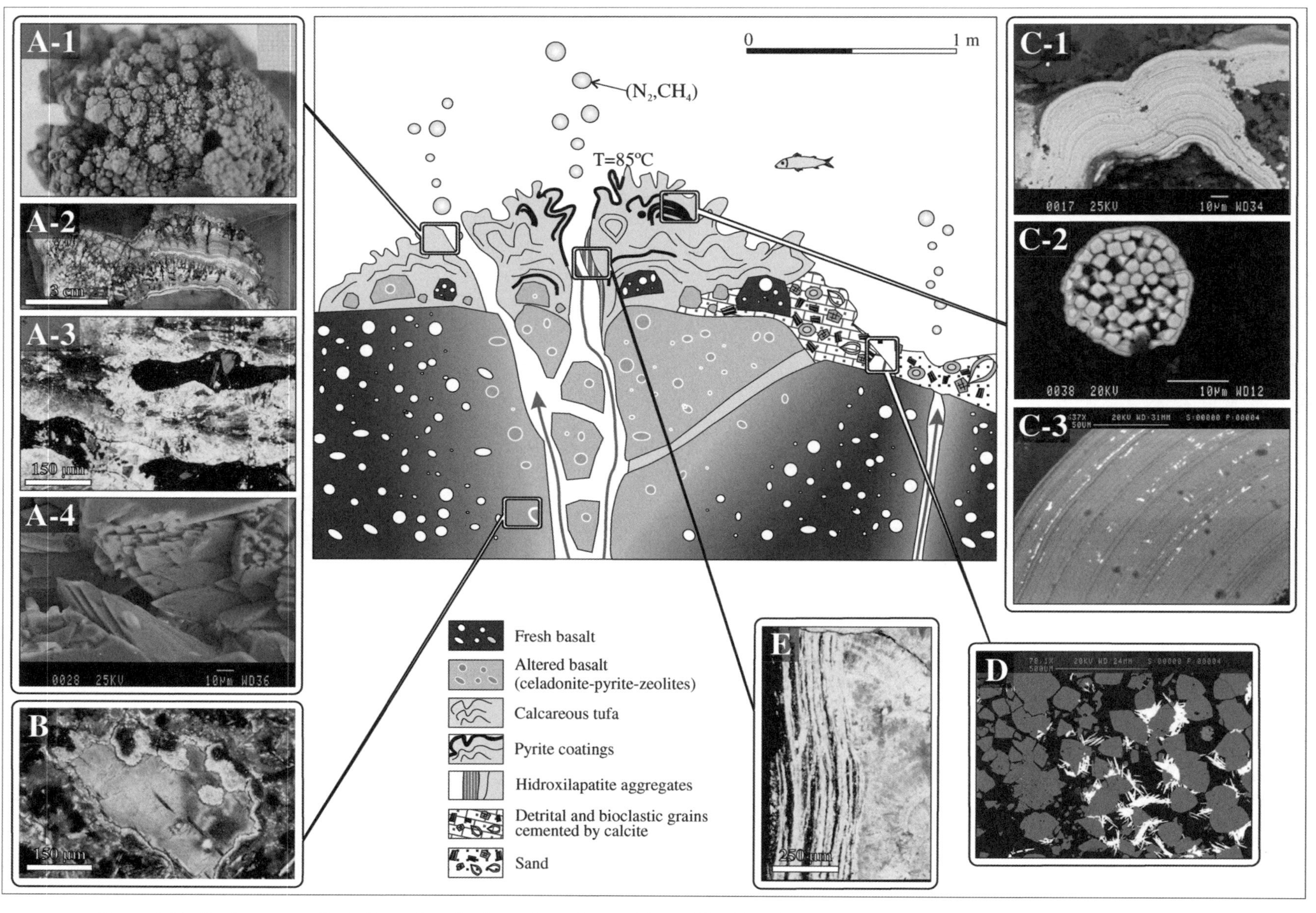

Figure 5. Idealized structure of a shallow submarine hydrothermal vent of Punta Mita (Nayarit, México), showing the associated mineralization. Images show in detail the mineralization. (A) Calcareous tufa: (A-1) hand specimen with calcite arborescent aggregates; (A-2) calcite-pyrite laminated aggregate (section); (A-3) porous aggregates calcite, transmitted light, crossed nichols; (A-4) calcite crystal morphology, scanning electron microscope (SEM) image. (B) Altered basaltic rock with vacuoles filled by heulandite, transmitted light, crossed nichols. (C) Pyrite aggregates: (C-1) pyrite coatings, SEM-backscattered electron (BSE) image; (C-2) framboidal pyrite, SEM-BSE image; (C-3) pyrite (gray) with cinnabar grains (white), SEM-BSE image. (D) Barite tabular crystals (white) associated with fine-grained calcite (gray), SEM-BSE image. (E) Hydroxylapatite-calcite layered aggregates, transmitted light, crossed nichols.

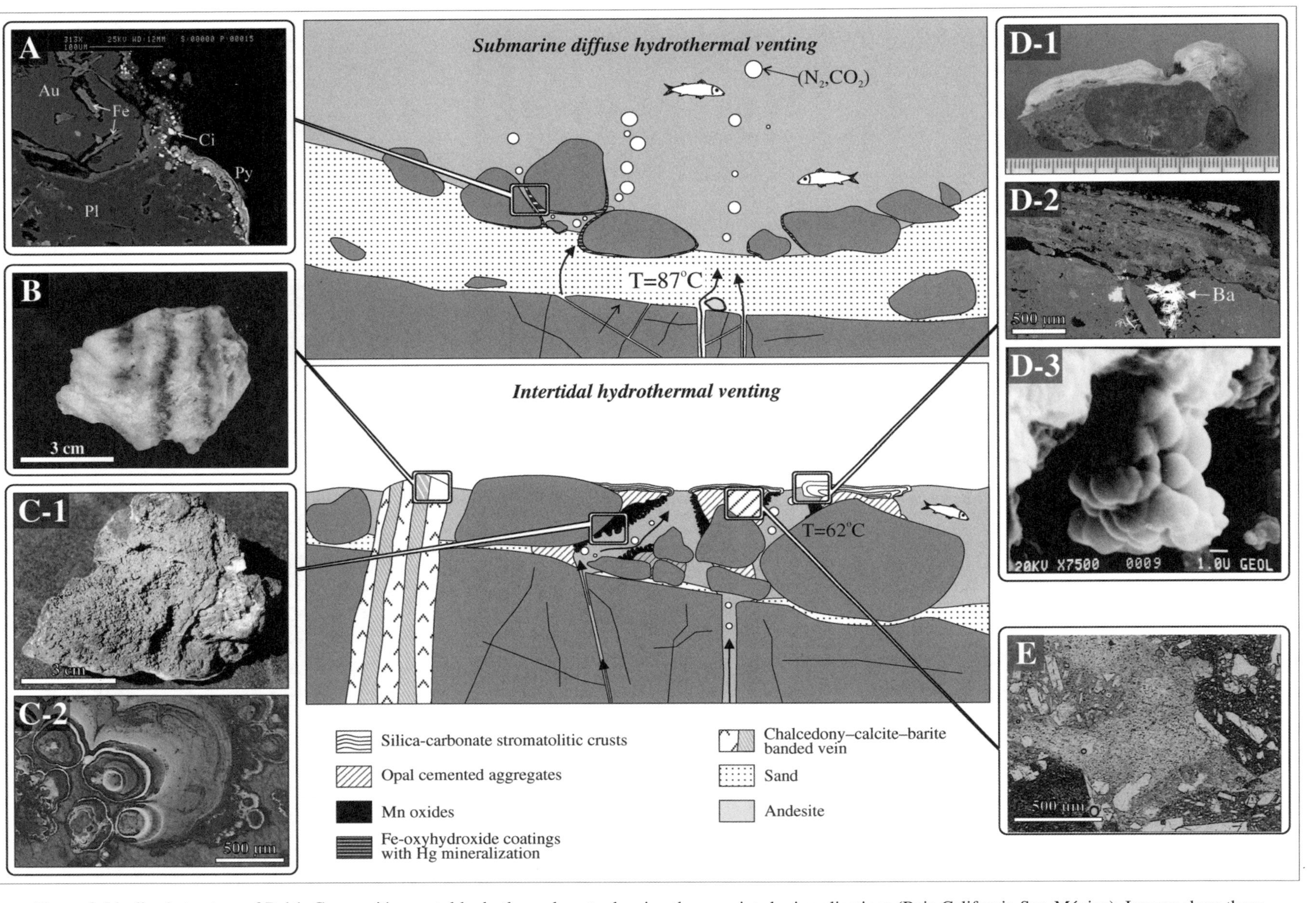

Figure 6. Idealized structure of Bahía Concepción coastal hydrothermal vents showing the associated mineralizations (Baja California Sur, México). Images show these mineralizations in detail. (A) Assemblage of laminated pyrite (Py) with cinnabar grains (Ci) coating a detrital volcanic fragment with augite (Au), plagioclase (Pl), and iron oxide alterations (Fe); SEM-BSE image. (B) Hand specimen of a chalcedony-calcite-barite banded vein. (C) Manganese oxides: (C-1) hand sample showing moss-like texture; (C-2) botryoidal aggregates of a near amorphous phase chemically equivalent to todorokite, reflected light. (D) Silica-carbonate sinter: (D-1) hand specimen (section) showing laminations and a fragment of volcanic rock that act as substrate; (D-2) sinter (laminated, in the upper part) over a volcanic rock (below), with barite (Ba) crystals (white); SEM-BSE image; (D-3) opal-A with microsphere arrangement; SEM image. (E) Fragments of volcanic rocks cemented by opal-A, transmitted light.

species as birnessite $Na_4Mn_{14}O_{27}·9H_2O$ and todorokite in relation to deep-sea hydrothermal vents (Marchig et al., 1999).

In some near shore SSHV, finely laminated sinter-like deposits occur with variable composition: silicic and silica-carbonated (Canet et al., 2005a) and calcareous (Canet et al., 2003). The chemical character of these deposits depends on fluid composition, pH, and temperature, and their precipitation is usually mediated to some degree by biological activity. Opal is the main component of the silicic deposits formed by exhalation of hydrothermal fluids. It has been reported in relation to deep-sea venting of silica-rich hydrothermal fluids; for example, in the Aeolian Arc and in the Tyrrhenian Sea (Savelli et al., 1999) and near the Rodrigues triple point in the Indic Ocean (Halbach et al., 2002). However, silicic deposits are much more abundant in lacustrine hydrothermal vents (e.g., Eugster, 1969; Renaut et al., 2002) and, especially, in subaerial hot springs, where they build up the archetypical sinters; for instance, in the Taupo volcanic zone, New Zealand (Mountain et al., 2003; Rodgers et al., 2004), and in Krisuvik, Iceland (Konhauser et al., 2001). These deposits develop from sodium-chloride thermal waters of neutral, alkaline, or, rarely, acidic pH (e.g., Fournier and Rowe, 1966; Rodgers et al., 2004), with reservoir temperatures over 175 °C (Fournier and Rowe, 1966).

In contrast, silica-carbonated deposits are much more scarce, as they have been found only in the continental geothermal fields of Waikite (Jones et al., 2000) and Ngatamariki (Campbell et al., 2002) in New Zealand, and in the intertidal hot springs of the SSHV system of Bahía Concepción, in Baja California Sur, México (Canet et al., 2005a). In this site, supersaturation of the fluid with respect to silica is attained by cooling and causes the precipitation of opal-A, whereas CO_2 loss triggers supersaturation in calcium carbonate (Canet et al., 2005a).

Calcareous sinters, also called calcareous tufa or hydrothermal travertines, have been reported mainly in subaerial and lacustrine hot springs; for example, in Pyramid Lake, Nevada, USA (Arp et al., 1999), and Yellowstone, Wyoming, USA (Fouke et al., 2000). In these settings, calcite precipitates from alkaline to neutral waters supersaturated in calcium carbonate (Jones et al., 2000). An analogous example from SSHV is found in Punta Mita, in Nayarit, México, that consists in fine-layered calcite aggregates forming at ~10 mbsl (Canet et al., 2003).

Sulfides occur in accessory amounts in SSHV deposits, whereas in deep-sea vents they build up chimney structures and accumulate in massive mounds. The reported sulfides in modern SSHV are pyrite, marcasite, cinnabar, carlinite, realgar, and orpiment (Stoffers et al., 1999; Canet et al., 2005a, 2005b). The occurrence of cinnabar has been confirmed in several SSHV deposits (Halbach et al., 1993; Stoffers et al., 1999; Prol-Ledesma et al., 2002a; Canet et al., 2005b).

In addition, SSHV deposits can contain sulfates (anhydrite, gypsum, and barite), phosphates (hydroxylapatite), and native sulfur and mercury (Stoffers et al., 1999; Canet et al., 2005a, 2005b). Barite precipitation is caused by mixing barium-rich reducing thermal water and SO_4-rich cold seawater (Canet et al., 2005a).

BIOTIC COMMUNITIES AND BIOMINERALIZATION PROCESSES

Generally, in both subaerial and submarine hydrothermal exhalations, mineral deposits form from solutions at temperatures below the upper limit for life development (Jones and Renaut, 1996; Konhauser et al., 2001). The close relationship between hydrothermal venting mineralization and microbial activity has been confirmed in numerous cases (Konhauser et al., 2001). Thermophilic life forms, similar to those that develop in hot spring environments, have been proposed as the common ancestor of life on Earth (Stetter et al., 1990). Therefore, the study of biomineralization processes and their resulting deposits in hydrothermal vents can yield information about life in extreme conditions (Walter and Des Marais, 1993) and about early life forms. In this way, the study of deep-sea black smoker fields and, more recently, of cold seeps have shown complex ecosystems based in chemosynthesis that do not directly depend on the sun's energy supply.

Biogeochemical processes in SSHV sites are significantly different from those of deep-sea hydrothermal vents, since SSHV environments fulfill conditions suitable for both photosynthetic and chemosynthetic metabolisms (Tarasov et al., 1993; Tarasov, 2002). For that reason, SSHV yield an opportunity for studying the relationships and competition interactions between chemosynthetic and photosynthetic organisms.

Prokaryote communities in SSHV are as diverse as in deep-sea vents (Prol-Ledesma et al., 2005), although in SSHV, they comprise few endemic species (Tarasov et al., 2005).

In SSHV environments, biotic communities are specifically adapted to survive in chemically modified habitats, tolerating high sulfur and heavy metal concentrations and high temperatures (Tarasov, 1991). These communities contain photosynthetic species (Dando et al., 1999) and, in some cases, they are hotspots of eukaryote diversity (Morri et al., 1999).

The effects of SSHV activity on planktonic and benthonic communities are not known yet, and biotic communities of SSHV are not as well characterized as those of deep-sea hydrothermal vents.

Some textural, mineralogical, and geochemical characteristics of mineral assemblages formed in SSHV suggest that many mineralizing processes are more or less influenced by microbial activity (Canet et al., 2003, 2005a, 2005b; Alfonso et al., 2005). Thus, microbial activity can promote the precipitation of iron oxyhydroxides (Juniper and Tebo, 1995). In addition, Mita et al. (1994) demonstrated that microbes decisively mediate the precipitation of manganese oxides in subaerial hot springs in Japan. In some cases, microbial activity can cause silica precipitation in sinters (Hinman and Lindstrom, 1996), and it often influences the fabric and textures of these deposits (Canet et al., 2005a). SSHV deposits formed under these influences, for example, the stromatolitic intertidal sinters of Bahía Concepción, in Baja California Sur (México), are similar to those described in lacustrine environments (Walter et al., 1976; Renaut et al., 2002).

In the case of the carbonated deposits of Punta Mita, Nayarit (México), which are related to methane-rich fluids venting, calcite precipitation takes place as a consequence of microbial oxidation of methane (Canet et al., 2003). Besides calcite, these deposits contain pyrite, whose precipitation is caused by a microbial reaction of seawater sulfate reduction coupled to methane oxidation (Alfonso et al., 2005).

SHALLOW SUBMARINE HYDROTHERMAL VENTS IN MÉXICO

Off the Pacific coast of México, the SSHV systems have been reported off Punta Banda (Vidal et al., 1978) and Bahía Concepción (Prol-Ledesma et al., 2004) in Baja California Peninsula, and Punta Mita, in Nayarit (Prol-Ledesma et al., 2002a, 2002b) (Fig. 2).

Unlike SSHV systems off the Western Pacific Margin, which are distributed along island arcs, Mexican examples do not show clear links with modern volcanism. The three described SSHV systems are located in continental margins presently affected by intense tectonic extension, with anomalously high geothermal gradients.

Punta Mita

About 500 m off Punta Mita, in the northern part of Bahía de Banderas (Nayarit), there is a SSHV system. The area lies near the NW limit of the Jalisco Block and in the western end of the Mexican Volcanic Belt. It is affected by the tectonic extension that results from the Rivera Plate displacement (Allan et al., 1991; Ferrari et al., 1994; Kostoglodov and Bandy, 1995). In this area, anomalously high heat flow values have been calculated (Prol-Ledesma and Juárez, 1986).

Hydrothermal activity consists of fluid discharge (gas and water) at 85 °C through a NE-SW–trending fissure that is hosted in basaltic rocks and is partially covered by unconsolidated platform sediments. This activity affects an area of ~1 km^2 at a depth of 10 mbsl (Prol-Ledesma et al., 2002a, 2002b; Canet et al., 2003).

Thermal exsolved gas is composed mainly of N_2 (88%) and CH_4 (12%); it also contains H_2S, CO_2, H_2, Ar, and He in trace amounts (Table 2) (Prol-Ledesma et al., 2002a). Stable isotopic analysis of carbon in methane yields a $\delta^{13}C$ value of −42.8‰ (Vienna Peedee belemnite [VPDB]) (Prol-Ledesma et al., 2002a), suggesting that this gas is not biogenic in origin, but thermogenic (Rooney et al., 1995).

Vent water is less saline than seawater (the calculated salinity of the thermal end-member is 1.51 wt%). It is significantly enriched in Si, Ca, Li, B, Ba, Rb, Fe, Mn, and As with respect to seawater and has lower concentrations of Na, K, Cl, HCO_3^-, SO_4^{2-}, and Br (Table 3). These depleted elements are mainly contributed to the vent discharge by seawater (the calculated SO_4^{2-} concentration in the thermal end-member is 2.81 mmol/kg; Prol-Ledesma et al., 2002a). δD ranges between −11.2 and −16.2‰ (Vienna standard mean ocean water [VSMOW]), and $\delta^{18}O$ between −1.9 and −3.5‰ (VSMOW) (Prol-Ledesma et al., 2002b). These isotopic values of the thermal water agree with a model of seawater mixing with deeply circulating meteoric water whose recharge zone is the adjacent mountain range in the peninsula of Punta Mita.

Water circulation takes place through fault systems related to modern extensive tectonics that affect the entire region. During this process, water interacts with Si- and Ba-rich granitic rocks and with sedimentary layers with organic matter, which act as the source of some elements as Ca, I, and Hg (Prol-Ledesma et al., 2002a) and where methane is generated, similar to deep-sea hydrothermal vents in sediment-covered ocean crust areas (Scott, 1997). Thus, SSHV sites in Punta Mita show temperatures and water compositions similar to those of deep-sea vents and, at the same time, they vent a methane-rich gas, as happens in cold seeps.

Discharging of hydrothermal fluids takes place through discrete seafloor springs separated by some meters and roughly located along a NE-SW–trending fissure. As a consequence,

TABLE 3. CHEMICAL COMPOSITION OF WATER SAMPLES FROM THE PUNTA MITA HYDROTHERMAL VENTS (LOCATION IN FIGS. 1 AND 2)

	mmol/kg				µmol/kg						mmol/kg				
Sample	Ca	Mg	K	Na	I	Hg	Mn	Cs	Ba	Mo	Cl	Br	SO_4	HCO_3	Si
PMM-1	8.23	45.26	10.49	493.73	<0.79	<0.001	<0.02	<0.007	<0.14	0.16	535.93	0.68	26.02	2.29	<0.03
PMM-2	9.73	45.26	9.97	493.73	<0.79	<0.001	0.06	<0.007	<0.14	0.16	535.93	0.74	26.02	2.29	<0.03
PM-1	32.44	22.22	5.88	218.06	8.67	<0.001	1.82	3.54	3.20	0.05	394.89	0.54	13.53	0.88	1.25
PM-2	42.42	10.70	5.37	193.38	11.03	<0.001	1.42	5.42	4.73	0.02	338.48	0.44	7.60	0.46	1.74
PM-3	29.94	24.27	7.16	263.32	8.67	<0.001	1.11	3.16	2.91	0.07	423.10	0.59	15.61	1.23	1.07
PM-4	34.93	18.51	6.14	267.43	14.97	<0.001	1.02	3.91	3.57	0.06	394.89	0.51	12.49	0.95	1.35
PM-5	44.91	4.11	3.58	197.49	14.97	<0.001	1.38	5.49	4.95	0.06	310.27	0.43	5.52	0.23	1.92
Thermal end-member	51.27	0.0*	3.17	110.36	17.35	ND	1.87	6.55	5.84	0.01	283.91	0.38	2.81	0.0	2.23

Note: PMM-1 and *PMM-2* correspond to superficial seawater, sampled far away and above the venting area, respectively. Samples *PM-1* to *PM-5* correspond to vent water. ND—not determined (Prol-Ledesma et al., 2002a, 2002b).

*Mg concentration is assumed as 0 for the binary mixing model.

several mounds of calcareous tufa with Ba, Hg, and Tl mineralization occur (Fig. 5). These mounds are made up of fine-layered calcite aggregates, with arborescent textures developed in the outer area (Fig. 5).

Farther from main active discharge orifices, mineralized structures consist of accumulations of detrital and bioclastic grains cemented by fine-grained calcite. The degree of development of mineralized mounds depends on the venting flow and its age. The largest recognized mound is ~2.5 m in diameter and 0.75 m in height. Currents and waves that at these depths disturb the seafloor constrain the morphology and vertical growth of the mineralized mounds (Canet et al., 2003).

Calcite crystals that compose the mounds are acicular, up to 250 μm in length, and they are low-magnesium calcite ($MgCO_3$ ~4% molar). In addition, there is a later, fine-grained calcite generation, which consists of small rhombohedral crystals of a few microns in length. It cements both detrital grains (basically plagioclase, quartz, and magnetite) and bioclasts (fragments of bivalves, gastropods, foraminifers, echinoderms, and red algae) and, locally, replaces the acicular calcite crystals.

Stable isotopic analyses in calcite aggregates yield $\delta^{13}C$ values between −39.2 and +0.9‰ (VPDB) and $\delta^{18}O$ values between −12.8 and −1.5‰ (VPDB) (Canet et al., 2003). This exceptional depletion of ^{13}C suggests that calcite precipitation is a consequence of microbial oxidation of methane (Canet et al., 2003). On the other hand, $\delta^{18}O$ values are in agreement with a precipitation from thermal water (with $\delta^{18}O$ of −1.9 to −3.5‰ VSMOW; Prol-Ledesma et al., 2002b) at temperatures ranging between 70 °C and 81 °C.

After calcite, the most abundant mineral in the mounds is pyrite, which forms thin layered coatings, up to 500 μm in thickness, covering calcite aggregates, and seldom bivalve fragments, in the inner part of the discharge conduits. In addition, pyrite forms framboidal and spherulitic aggregates up to 100 μm in diameter. It is also present in the suspended particulate material around the discharge areas (Ortega-Osorio et al., 2001).

Sulfur isotopic composition ($\delta^{34}S$) of pyrite grains ranges between −13.3 and −4.9‰ (Alfonso et al., 2005). Both pyrite textures and isotopic composition are consistent with a biogenic precipitation. The isotopic fractionation sulfate-sulfide that results from the analyzed $\delta^{34}S$ values is consistent with a biogenic reduction of seawater sulfate (Ohmoto and Rye, 1979). Therefore, the combination of two coupled microbially mediated reactions, summarized as (1) oxidation of thermal methane, and (2) reduction of seawater sulfate, is the cause of the precipitation of calcite and pyrite, respectively, in the same way as in cold seeps (Kohn et al., 1998; Canet et al., 2006) but at much higher temperatures.

There are disseminated grains of cinnabar, up to 10 μm in size, and carlinite Tl_2S, up to 5 μm in size, associated to layered pyrite aggregates. According to Prol-Ledesma et al. (2002a) thallium forming carlinite is probably scavenged from seawater. As well, there are barite, hydroxylapatite, and baritocalcite $BaCa(CO_3)_2$ in accessory amounts.

Near the sites of hydrothermal discharge, basaltic rocks of the seabed show hydrothermal alteration. Their plagioclase phenocrysts are replaced by zeolites (heulandite and analcime), whereas the augite grains and the groundmass are replaced by a cryptocrystalline assemblage of celadonite and pyrite.

Bahía Concepción

The Bahía Concepción, nearly 40 km in length, is a fault-controlled bay placed in the eastern coast of the Baja California Peninsula (Fig. 2) that hosts a SSHV system that is revealed by several submarine vents (at depths down to 15 mbsl) and intertidal and subaerial hot springs. It is a semi-closed bay, whose NW-SE semigraben configuration results from the extensive tectonics that affected the Gulf of California Province during the Late Miocene (Ledesma-Vázquez and Johnson, 2001).

The area with submarine and intertidal hydrothermal activity extends 700 m along a stretch of rocky shoreline in the western coast of the bay. The cliffs are configured by a system of normal faults that allows hydrothermal fluids to rise to the surface (Forrest et al., 2003).

Two types of the SSHV system surface manifestations are found in this area: (1) a zone of diffuse hydrothermal fluids submarine venting (gas and water) through the sediments of the seabed, at depths between 5 and 15 mbsl; and (2) a cluster of hydrothermal springs and bubbling vents in the intertidal zone (Canet et al., 2005a, 2005b). These manifestations are all NW-SE aligned.

The temperature of vent discharge is up to 87 °C in the submarine diffuse area and 62 °C in the intertidal hot springs, and pH is 5.9 and 6.7, respectively.

Water and gas chemical compositions are shown in Tables 4 and 2, respectively. The exsolved gas is composed largely by CO_2 (44%) and N_2 (54%), although it contains CH_4, Ar, He, H_2, and O_2 in minor amounts (Table 2) (Forrest et al., 2003). $\delta^{13}C$ in CO_2 is −6‰ (Forrest and Melwani, 2003) and−34.3‰ in CH_4. This value is in agreement with a thermogenic origin (Forrest et al., 2005). On the other hand, isotopic composition of N_2, with an average $\delta^{15}N$ value of 1.7‰, suggests that this gas is thermally produced from organic matter in immature sediments, whereas helium isotopes (R/Ra = 1.32) are compatible with a mantle-derived component (Forrest et al., 2005).

Thermal water is a sodium-chloride type and shows high concentrations of Ca, Mn, Si, Ba, B, As, Hg, I, Fe, Li, HCO_3^-, and Sr with respect to seawater (Table 4) (Prol-Ledesma et al., 2004). Chemical geothermometers (Na/Li, Na-K-Ca and Si) yield reservoir temperatures of ~200 °C (Prol-Ledesma et al., 2004). Calculations of the saturation state indicate that the fluid vented by the intertidal hot springs is supersaturated in barite and silica, and that the mixture of this fluid with seawater is subsaturated in calcite and silica and supersaturated in barite (Canet et al., 2005a). These calculations validate the observed mineralogical composition of the vent precipitates in the area. Thus, amorphous silica is the main precipitate in the intertidal hot springs, whereas it lacks

TABLE 4. MAJOR ION COMPOSITION (mmolal) OF WATER SAMPLES FROM THE BAHÍA CONCEPCIÓN HYDROTHERMAL VENTS (LOCATION IN FIGS. 1 AND 2)

Sample	pH	Total alkalinity	Ca	Mg	K	Na	Cl	SO_4	HCO_3	Si
BC-1	5.95	243.3	23.3	35.78	12.7	394.5	458.4	17.0	4.9	3.1
BC-4	6.02	214.1	19.4	41.91	12.5	414.7	500.7	21.2	4.3	2.1
BC-6	5.97	223.8	20.6	40.20	12.5	408.9	493.6	20.6	4.5	2.4
BC-9	7.75	150.8	9.8	58.33	12.48	485.9	527.5	26.6	1.6	0.0
BC-10	6.68	97.3	28.9	25.00	12.2	334.0	409.0	12.4	1.9	4.5
Thermal end-member	ND	ND	44.5	0.0*	13.0	243.9	380.3	4.1	10.3	7.8

Note: Samples: *BC-1*, *BC-4*, and *BC-6*, submarine vent water; *BC-9*, seawater collected as reference far away from the venting area; and *BC-10*, intertidal vent water. ND—not determined (Prol-Ledesma et al., 2004).
*Mg concentration is assumed as 0 for the binary mixing model.

in the submarine vents, where mixing between hydrothermal fluid and seawater is more extensive. On the other hand, supersaturation of the fluid with respect to calcite, which is associated to amorphous silica in the intertidal hot springs, is caused by a process of CO_2 loss (Canet et al., 2005a).

$\delta^{18}O$ and δD (VSMOW) values of thermal water range between −0.3‰ and −3.1‰ and between −0.3‰ and −25.5‰, respectively (Prol-Ledesma et al., 2004). The range of isotopic compositions of the fluid suggests that it has a large meteoric water component, even though mixing with seawater has a decisive role in fluid evolution. From its content in dissolved magnesium, it can be deduced that thermal water discharged in the submarine diffuse vents undergoes mixing with seawater near the ocean bottom in the unconsolidated sediments (Prol-Ledesma et al., 2004).

Neither seafloor mounds nor chimney-like mineralized structures form as a consequence of the submarine discharge of hydrothermal fluids. Around the more active vent areas, millimetric crusts of iron oxyhydroxides form over the detrital blocks of basaltic andesites accumulated in the seabed. Additionally, the sandy sediments are locally covered by yellowish microbial mats. Crusts of iron oxyhydroxide are composed of poorly crystallized ferrihydrite with minor amounts of pyrite and cinnabar. Pyrite forms fine-layered coatings, up to 20 µm thick, and cinnabar occurs as small grains, up to 10 µm, in close association to pyrite (Canet et al., 2005b).

In contrast, around the intertidal hot springs there are conspicuous accumulations of hydrothermal precipitates, forming irregular pavements up to 10 m^2, that remain partially exposed during low tides (Figs. 3A and 3B). These deposits consist of (1) massive aggregates of detrital and bioclastic fragments cemented by opal-A with minor amounts of calcite and barite, (2) manganese oxides, and (3) stromatolitic crusts of silica-carbonate sinter (Fig. 6).

Manganese oxides form moss-like porous aggregates and are composed of poorly crystalline phases chemically equivalent to barium-rich todorokite (>2 wt% BaO) and to romanechite $(Ba,K,Na,Ca,H_2O)_2(Mn^{4+},Mn^{3+},Fe^{3+},Mg,Al,Si)_5O_{10}$ with ~10 wt% BaO (Canet et al., 2005b).

Silica-carbonate sinter crusts overlie the above mentioned precipitates and deposits directly over volcanic cobbles and blocks. They consist of fine-layered bulbous to undulant aggregates, composed essentially of opal-A, calcite, and barite. At the microscale, opal-A forms porous aggregates built up by attached microspheres up to 300 nm in diameter (Canet et al., 2005a).

Isotope analyses of calcite crystals from silica-carbonate sinter show enrichment in ^{13}C with respect to marine carbonates, with $\delta^{13}C$ values up to +9.3‰ (VPDB) and $\delta^{18}O$ values between −2.6‰ and −10.0‰ (VPDB) (Canet et al., 2005a). The ^{13}C enrichment is caused by CO_2 loss by thermal water, the same process that triggers calcite precipitation. $\delta^{18}O$ values are consistent with precipitation of calcite from a mixture of thermal and marine water.

Punta Banda

At Punta Banda, near Ensenada, in the NW coast of Baja California Peninsula (Fig. 2), there is a SSHV coastal system that presents manifestations in two zones of shallow submarine hydrothermal venting at 40 mbsl and in subaerial hot springs.

The discharge of hydrothermal fluids in the submarine vents takes place at a temperature of 102 °C (Vidal et al., 1978). Chemical compositions of vent water and gas are shown in Tables 1 and 2, respectively (Vidal et al., 1978, 1981).

Thermal gas is chemically similar to that of the Punta Mita SSHV, with N_2 and N_4 as principal components. In Punta Banda, however, methane is more abundant, reaching up to 51.4% of the exsolved gas (Vidal et al., 1981).

Thermal water is enriched in SiO_2, HCO_3^-, Ca, K, Li, B, Ba, Rb, Fe, Mn, As, and Zn with respect to seawater (Vidal et al., 1981). Stable isotopes ($\delta^{18}O$ and δD) indicate that thermal water is a mixture of meteoric and marine water. Estimated reservoir temperatures range between 190 °C and 213 °C (Vidal et al., 1981).

According to the geological setting and geochemical characteristics of the hydrothermal fluids, Vidal et al. (1981) reject the influence of magmatic sources and suggest that thermal water is a mixture of ~1:1 local meteoric water and "fossil" seawater.

The most abundant minerals in the submarine hydrothermal deposits are pyrite and gypsum, with high contents of As, Hg, Sb, and Tl in whole rock analysis (Vidal et al., 1978).

CONCLUSIONS AND RECOMMENDATIONS

Most SSHV systems are distributed along volcanic island arcs and show a close relationship to volcanic activity. However, there are also some SSHV systems in continental margins actively affected by tectonic extension and not linked to volcanism. This is the setting of the three main examples of SSHV in México: Punta Banda and Bahía Concepción, in Baja California Peninsula, and Punta Mita, in Nayarit.

The maximum depth in which a submarine hydrothermal system can be considered as a SSHV is 200 mbsl. This depth limit implies sharp changes in the ecological and environmental parameters and corresponds to an increase in the slope of the boiling curve for seawater.

SSHV systems can originate ore deposits rich in Mn, Ba, Pb, Zn, As, Sb, Ag, Hg, and Tl, with oxides, sulfates, sulfides, and native elements.

Thermal water in SSHV can show chemical and isotopic characteristics between deep-sea vents and continental geothermal fields. Generally, it involves a significant component of meteoric water, although the process of mixing with seawater constrains its chemical and isotopic composition.

Venting of exsolved gas, evidenced by continuous bubbling, is a distinctive feature of SSHV systems. Usually, exsolved gas is CO_2-rich, although in particular cases it can be H_2S- or CH_4-rich.

The most common SSHV systems, those directly connected to active volcanism, can be considered transitional systems between epithermal deposits and deep-sea hydrothermal vents.

Mineral assemblages forming in SSHV sites are similar to those from low-temperature deep-sea hydrothermal systems and to those from low-sulfidation epithermal deposits. Amorphous and poorly crystalline phases, including iron oxyhydroxides, manganese oxides, and opal, are widespread in SSHV deposits. In addition, they may contain sulfides (pyrite, cinnabar, carlinite, realgar, and orpiment), sulfates (anhydrite, gypsum, and barite), carbonates (calcite and baritocalcite), phosphates (hydroxylapatite), and native elements (sulfur and mercury).

SSHV environments fulfill appropriate conditions for both photosynthetic and chemosynthetic metabolism. Biogeochemical processes in SSHV unquestionably influence the formation of mineral deposits.

A detailed study of SSHV systems will contribute to understanding processes pertaining to ore formation, to microbially mediated mineral precipitation, to the geochemical cycle of some toxic metals in the oceans (e.g., mercury and arsenic), and to water-rock interaction. Furthermore, the potential of SSHV as sources of geothermal energy should be evaluated.

ACKNOWLEDGMENTS

Funding for the studies that resulted in this article came from the Mexican projects Programa de Apoyo a Proyectos de Investigación e Innovación Tecnológica [PAPIIT] IN-122604 and IN-107003, Consejo Nacional de Ciencia y Tecnología [CONACyT] J-51127-I, 32510, and SEP-2004-C01-46172. Antoni Camprubí Cano encouraged us to write this article, and his comments and exhaustive revision significantly contributed to the improvement of the manuscript. Also, comments from Jordi Tritlla Cambra were very helpful. We thank A. Camprubí, D. Blanco Florido, M. Dando, P. Dando, M.J. Forrest, J. Ledesma Vázquez, A. López Sánchez, A. Melwani, A.A. Rodríguez Díaz, M.A. Torres Vera, and R.E. Villanueva Estrada for helping in different field campaigns. S.I. Franco Sánchez is thanked for her observations on the manuscript. Scanning electron microscope images were obtained from the Serveis Científico-Tècnics of the Universitat de Barcelona (Spain) and from the Instituto de Geofísica and Instituto de Geología of the Universidad Nacional Autónoma de México, with the assistance of R. Fontarnau, C. Linares López, and M. Reyes Salas, respectively.

REFERENCES CITED

Alfonso, P., Prol-Ledesma, R.M., Canet, C., Melgarejo, J.C., and Fallick, A.E., 2005, Isotopic evidence for biogenic precipitation as a principal mineralization process in coastal gasohydrothermal vents, Punta Mita, Mexico: Chemical Geology, v. 224, p. 113–121, doi: 10.1016/j.chemgeo.2005.07.016.

Allan, J.F., Nelson, S.A., Luhr, J.F., Carmichael, J., Wopat, M., and Wallace, P.J., 1991, Pliocene–Recent rifting in SW Mexico and associated volcanism: An exotic terrane in the making, *in* Dauphin, J.P., and Simoneit, R.R.T. eds., The Gulf and Peninsular Provinces of the Californias: American Association of Petroleum Geologists Memoir 47, p. 425–445.

Amend, J.P., Rogers, K.L., Shock, E.L., Gurrieri, S., and Inguaggiato, S., 2003, Energetics of chemolithoautotrophy in the hydrothermal system of Vulcano Island, southern Italy: Geobiology, v. 1, p. 37–58, doi: 10.1046/j.1472-4669.2003.00006.x.

Arp, G., Thiel, V., Reimer, A., Michaelis, W., and Reitner, J., 1999, Biofilm exopolymers control microbialite formation at thermal springs discharging into alkaline Pyramid Lake, Nevada, USA: Sedimentary Geology, v. 126, p. 159–176, doi: 10.1016/S0037-0738(99)00038-X.

Barrat, J.A., Boulegue, J., Tiercelin, J.J., and Lesourd, M., 2000, Strontium isotopes and rare-earth element geochemistry of hydrothermal carbonate deposits from Lake Tanganyika: East Africa: Geochimica et Cosmochimica Acta, v. 64, p. 287–298, doi: 10.1016/S0016-7037(99)00294-X.

Benjamínsson, J., 1988, Jardhiti í sjó og flaedamáli vid Ísland: Natturufraedingurinn, v. 58, p. 153–169.

Bischoff, J.L., and Rosenbauer, R.J., 1984, The critical point and phase boundary of seawater 200–500 °C: Earth and Planetary Science Letters, v. 68, p. 172–180, doi: 10.1016/0012-821X(84)90149-3.

Bischoff, L.B., and Seyfried, W.E., 1978, Hydrothermal chemistry of seawater from 25° to 350°C: American Journal of Science, v. 278, p. 838–860.

Bogdanov, Yu.A., Lisitzin, A.P., Binns, R.A., Gorshkov, A.I., Gurvich, E.G., Dritz, V.A., Dubinina, G.A., Bogdanova, O.Yu., Sivkov, A.V., and Kuptsov, V.M., 1997, Low-temperature hydrothermal deposits of Franklin Seamount, Woodlark Basin, Papua New Guinea: Marine Geology, v. 142, p. 99–117, doi: 10.1016/S0025-3227(97)00043-1.

Bornhorst, T.J., Nurmi, P.A., Rasilainen, K., and Kontas, E., 1995, Trace element characteristics of selected epithermal gold deposits of North America: Special Paper of the Geological Survey of Finland, v. 20, p. 47–52.

Botz, R., Winckler, G., Bayer, R., Schmitt, M., Schmidt, M., Garbe-Schonberg, D., Stoffers, P., and Kristjansson, J.K., 1999, Origin of trace gases in submarine hydrothermal vents of the Kolbeinsey Ridge, north Iceland: Earth and Planetary Science Letters, v. 171, p. 83–93, doi: 10.1016/S0012-821X(99)00128-4.

Botz, R., Wehner, H., Schmitt, M., Worthington, T.J., Schmidt, M., Stoffers, P., and Kristjansson, J.K., 2002, Thermogenic hydrocarbons from the offshore Calypso hydrothermal field, Bay of Plenty, New Zealand: Chemical Geology, v. 186, p. 235–248, doi: 10.1016/S0009-2541(01)00418-1.

Bright, M., 2004, Hydrothermal vent research: http://www.univie.ac.at/marine-biology/hydrothermal/ (Accessed 8 December 2006).
Burgath, K.-P., and von Stackelberg, U., 1995, Sulfide-impregnated volcanics and ferro-manganese incrustations from the southern Lau Basin (southwest Pacific): Marine Georesources and Geotechnology, v. 13, p. 263–308.
Butterfield, A., Massoth, G.J., McDuff, R.E., Lupton, J.E., and Lilley, M.D., 1990, Geochemistry of hydrothermal fluids from Axial Seamount hydrothermal emissions study vent field, Juan de Fuca Ridge: Subseafloor boiling and subsequent fluid-rock interaction: Journal of Geophysical Research, v. 95, p. 12,895–12,921.
Campbell, K.A., Rodgers, K.A., Brotheridge, J.M.A., and Browne, P.R.L., 2002, An unusual modern silica-carbonate sinter from Pavlova spring, Ngatamariki, New Zealand: Sedimentology, v. 49, p. 835–854, doi: 10.1046/j.1365-3091.2002.00473.x.
Camprubí, A., and Albinson, T., 2006, Depósitos epitermales en México: actualización de su conocimiento y reclasificación empírica: Boletín de la Sociedad Geológica Mexicana, Volumen Conmemorativo del Centenario, Revisión de Algunas Tipologías de Depósitos Minerales de México, v. 58, p. 27–81.
Canet, C., Prol-Ledesma, R.M., Melgarejo, J.-C., and Reyes, A., 2003, Methane-related carbonates formed at submarine hydrothermal springs: a new setting for microbially-derived carbonates?: Marine Geology, v. 199, p. 245–261, doi: 10.1016/S0025-3227(03)00193-2.
Canet, C., Prol-Ledesma, R.M., Torres-Alvarado, I., Gilg, H.A., Villanueva, R.E., and Lozano-Santa Cruz, R., 2005a, Silica-carbonate stromatolites related to coastal hydrothermal venting in Bahía Concepción, Baja California Sur, Mexico: Sedimentary Geology, v. 174, p. 97–113, doi: 10.1016/j.sedgeo.2004.12.001.
Canet, C., Prol-Ledesma, R.M., Proenza, J., Rubio-Ramos, M.A., Forrest, M., Torres-Vera, M.A., and Rodríguez-Díaz, A.A., 2005b, Mn-Ba-Hg Mineralization at shallow submarine hydrothermal vents in Bahía Concepción, Baja California Sur, Mexico: Chemical Geology, v. 224, p. 96–112, doi: 10.1016/j.chemgeo.2005.07.023.
Canet, C., Prol-Ledesma, R.M., Escobar-Briones, E., Mortera-Gutiérrez, C., Lozano-Santa Cruz, R., Linares, C., Cienfuegos, E., and Morales-Puente, P., 2006, Mineralogical and geochemical characterization of hydrocarbon seep sediments from the Gulf of Mexico: Marine and Petroleum Geology, v. 23, p. 605–619.
Cardigos, F., Colaço, A., Dando, P.R., Ávila, S.P., Sarradin, P.-M., Tempera, F., Conceição, P., Pascoal, A., and Serrão Santos, R., 2005, Shallow water hydrothermal vent field fluids and communities of the D. João de Castro Seamount (Azores): Chemical Geology, v. 224, p. 153–168, doi: 10.1016/j.chemgeo.2005.07.019.
Chao, T.T., and Theobald, J.P.K., 1976, The significance of secondary iron and manganese oxides in geochemical exploration: Economic Geology and the Bulletin of the Society of Economic Geologists, v. 71, p. 1560–1569.
Chen, A.C., Zhigang, Z., Fu-Wen, K., Tsanyao, F.Y., Bing-Jye, W., and Yueh-Yuan, T., 2005, Tide-influenced Acidic Hydrothermal System off Taiwan: Chemical Geology, v. 224, p. 69–81, doi: 10.1016/j.chemgeo.2005.07.022.
Corliss, J.B., Baross, J.A., and Hoffman, S.E., 1981, An hypothesis concerning the relationship between submarine hot springs and the origin of life on Earth: Oceanologica Acta, v. 4, p. 59–69.
Crane, K., Hecker, B., and Golubev, V., 1991, Hydrothermal vents in Lake Baikal: Nature, v. 350, p. 281, doi: 10.1038/350281a0.
Cronan, D.S., and Varnavas, S.P., 1993, Submarine fumarolic and hydrothermal activity off Milos, Hellenic Volcanic Arc: Terra Abstracts Terra Nova, v. 5, p. 569.
Daesslé, L.W., Cronan, D.S., Marchig, V., and Wiedicke, M., 2000, Hydrothermal sedimentation adjacent to the propagating Valu Fa Ridge, Lau Basin, SW Pacific: Marine Geology, v. 162, p. 479–500, doi: 10.1016/S0025-3227(99)00065-1.
Dando, P.R., and Leahy, Y., 1993, Hydrothermal activity off Milos, Hellenic Volcanic Arc: BRIDGE Newsletter, v. 5, p. 20–21.
Dando, P.R., Stüben, D., and Varnavas, S.P., 1999, Hydrothermalism in the Mediterranean Sea: Progress in Oceanography, v. 44, p. 333–367, doi: 10.1016/S0079-6611(99)00032-4.
Dando, P.R., Aliani, S., Arab, H., Bianchi, C.N., Brehmer, M., Cocito, S., Fowler, S.W., Gundersen, J., Hooper, L.E., Kölbl, R., Kuever, J., Linke, P., Makropoulos, K.C., Meloni, R., Miquel, J.-C., Morri, C., Müller, S., Robinson, C., Schlesner, H., Sievert, S., Stöhr, R., Stüben, D., Thomm, M., Varnavas, S.P., and Ziebis, W., 2000, Hydrothermal studies in the Aegean Sea: Physics and Chemistry of the Earth (B), v. 25, p. 1–8.
de Ronde, C.E.J., Baker, E.T., Massoth, G.J., Lupton, J.E., Wright, I.C., Feely, R.A., and Greene, R.G., 2001, Intra-oceanic subduction-related hydrothermal venting, Kermadec volcanic arc, New Zealand: Earth and Planetary Science Letters, v. 193, p. 359–369, doi: 10.1016/S0012-821X(01)00534-9.
de Ronde, C.E.J., Stoffers, P., Garbe-Schönbergb, D., Christenson, B.W., Jones, B., Manconi, R., Browne, P.R.L., Hissmann, K., Botz, R., Davy, B.W., Schmitt, M., and Battershill, C.N., 2002, Discovery of active hydrothermal venting in Lake Taupo, New Zealand: Journal of Volcanology and Geothermal Research, v. 115, p. 257–275, doi: 10.1016/S0377-0273(01)00332-8.
Degens, E., and Ross, D.A., 1969, Hot brines and recent heavy metal deposits in the Red Sea: New York, Springer-Verlag, 600 p.
Dymond, J., Collier, R.W., and Watwood, M.E., 1989, Bacterial mats from Crater Lake, Oregon and their relationship to possible deep lake hydrothermal venting: Nature, v. 342, p. 673–675, doi: 10.1038/342673a0.
Eugster, H.P., 1969, Inorganic bedded cherts from the Magadi area, Kenya: Contributions to Mineralogy and Petrology, v. 22, p. 1–31, doi: 10.1007/BF00388011.
Fan, D., Ye, J., and Li, J., 1999, Geology, mineralogy, and geochemistry of the Middle Proterozoic Wafangzi ferromanganese deposit, Liaoning Province, China: Ore Geology Reviews, v. 15, p. 31–53, doi: 10.1016/S0169-1368(99)00013-X.
Ferguson, J., and Lambert, I.B., 1972, Volcanic exhalations and metal enrichments at Matupi Harbour, New Britain, T.P.N.G: Economic Geology and the Bulletin of the Society of Economic Geologists, v. 67, p. 25–37.
Ferrari, L., Pasquarè, G., Venegas, S., Castillo, D., and Romero, F., 1994, Regional tectonics of western Mexico and its implications for the northern boundary of the Jalisco block: Geofísica Internacional, v. 33, p. 139–151.
Fitzsimons, M.F., Dando, P.R., Hughes, J.A., Thiermann, F., Akoumianaki, I., and Pratt, S.M., 1997, Submarine hydrothermal brine seeps off Milos, Greece: Observations and geochemistry: Marine Chemistry, v. 57, p. 325–340, doi: 10.1016/S0304-4203(97)00021-2.
Forrest, M.J., and Melwani, A., 2003, Ecological consequences of shallow-water hydrothermal venting along the El Requesón Fault Zone, Bahía Concepción, BCS, México: Geological Society of America Abstracts with Programs, v. 35, no. 6, p. 577–578.
Forrest, M.J., Greene, H.G., Ledesma-Vázquez, J., and Prol-Ledesma, R.M., 2003, Present-day shallow-water hydrothermal venting along the El Requesón fault zone provides possible analog for formation of Pliocene-age chert deposits in Bahía Concepción, BCS, México: Geological Society of America Cordilleran Section Abstracts with Programs, Puerto Vallarta, Jalisco, México, p. 35.
Forrest, M.J., Ledesma-Vázquez, J., Ussler, W., III, Kulongoski, J.T., Hilton, D.R., and Greene, H.G., 2005, Gas geochemistry of a shallow submarine hydrothermal vent associated with El Requesón fault zone in Bahía Concepción, Baja California Sur, México: Chemical Geology, v. 224, p. 82–95, doi: 10.1016/j.chemgeo.2005.07.015.
Fouke, B.W., Farmer, J.D., Des Marais, D.J., Pratt, L., Sturchio, N.C., Burns, P.C., and Discipulo, M.K., 2000, Depositional facies and aqueous-solid geochemistry of travertine-depositing hot springs (Angel Terrace, Mammoth Hot Springs, Yellowstone National Park, U.S.A.): Journal of Sedimentary Research, v. 70, p. 565–585.
Fournier, R.O., and Rowe, J.J., 1966, Estimation of underground temperatures from the silica content of water from hot springs and wet-steam wells: American Journal of Science, v. 264, p. 685–697.
Fricke, H., Giere, O., Stetter, K., Alfredsson, G.A., Kristjansson, J.K., Stoffers, P., and Svavarsson, J., 1989, Hydrothermal vent communities at the shallow subpolar Mid-Atlantic Ridge: Marine Biology, v. 102, p. 425–429, doi: 10.1007/BF00428495.
Gamo, T., Chiba, H., Yamanaka, T., Okudaira, T., Hashimoto, J., Tsuchida, S., Ishibashi, J., Tsunogai, U., Okamura, K., Sano, Y., and Shinjo, R., 2001, Chemical characteristics of newly discovered black-smoker fluids and associated hydrothermal plumes at the Rodriguez Triple Junction, Central Indian Ridge: Earth and Planetary Science Letters, v. 193, p. 371–379, doi: 10.1016/S0012-821X(01)00511-8.
Geptner, A., Kristmannsdottir, H., Kristiansson, J., and Marteinsson, V., 2002, Biogenic saponite from an active submarine hot spring, Iceland: Clay and Clay Minerals, v. 50, p. 174–185, doi: 10.1346/000986002760832775.

Glasby, G.P., Stuben, D., Jeschke, G., Stoffers, P., and Garbe-Schonberg, C.-D., 1997, A model for the formation of hydrothermal manganese crusts from the Pitcairn Island hotspot: Geochimica et Cosmochimica Acta, v. 61, p. 4583–4597, doi: 10.1016/S0016-7037(97)00262-7.

Graham, U.M., Bluth, G.J., and Ohmoto, H., 1988, Sulfide-sulfate chimneys on the East Pacific Rise, 11° and 13°N latitudes. Part I: Mineralogy and paregenesis: Canadian Mineralogist, v. 26, p. 487–504.

Halbach, P., Pracejus, B., and Märten, A., 1993, Geology and mineralogy of massive sulfide ores from the Central Okinawa Trough, Japan: Economic Geology and the Bulletin of the Society of Economic Geologists, v. 88, p. 2210–2225.

Halbach, M., Halbach, P., and Lüders, V., 2002, Sulfide-impregnated and pure silica precipitates of hydrothermal origin from Central Indian Ocean: Chemical Geology, v. 182, p. 357–375, doi: 10.1016/S0009-2541(01)00323-0.

Hannington, M., Herzig, P., Stoffers, P., Scholten, J., Botz, R., Garbe-Schonberg, D., Jonasson, I.R., Woest, W., and the Shipboard Scientific Party, 2001, First observations of high-temperature submarine hydrothermal vents and massive anhydrite deposits off the north coast of Iceland: Marine Geology, v. 177, p. 199–220.

Hashimoto, J., Miura, T., Fujikura, K., and Ossaka, J., 1993, Discovery of vestimentiferan tube-worms in the euphotic zone: Zoological Science, v. 10, p. 1063–1067.

Haymon, R.M., and Kastner, M., 1981, Hot spring deposits on the East Pacific Rise at 21°N: preliminary description of mineralogy and genesis: Earth and Planetary Science Letters, v. 53, p. 363–381, doi: 10.1016/0012-821X(81)90041-8.

Hedenquist, J.W., Izawa, E., Arribas, A., and White, N.C., 1996, Epithermal gold deposits-Styles, characteristics, and exploration: Japan, Resource Geology Special Publication Number 1, Society of Resource Geology, 17 p.

Hein, J.R., Stamatakis, M.G., and Dowling, J.S., 1999, Hydrothermal Mn-oxide deposit rich in Ba, Zn, As, Pb, and Sb, Milos Island, Greece, *in* Stanley, C.J., et al., eds., Mineral Deposits: Processes to Processing: Rotterdam, The Netherlands, Balkema, p. 519–522.

Hein, J.R., Stamatakis, M.G., and Dowling, J.S., 2000, Trace metal-rich Quaternary hydrothermal manganese oxide and barite deposits, Milos Island, Greece: Transactions of the Institute of Mining and the Metallurgy, Section B: Applied Earth Sciences, v. 109, p. B67–B76.

Herzig, P.M., 1999, Economic potential of sea-floor massive sulphide deposits: ancient and modern: Philosophical Transactions of the Royal Society of London series A, v. 357, p. 861–875.

Herzig, P.M., and Hannington, M.D., 1995, Polymetallic massive sulphides at the modern seafloor—A review: Ore Geology Reviews, v. 10, p. 95–115, doi: 10.1016/0169-1368(95)00009-7.

Hinman, N.W., and Lindstrom, R.F., 1996, Seasonal changes in silica deposition in hot spring systems: Chemical Geology, v. 132, p. 237–246, doi: 10.1016/S0009-2541(96)00060-5.

Hoaki, T., Nishijima, M., Miyashita, H., and Maruyama, T., 1995, Dense community of hyperthermophilic sulfur dependent heterotrophs in geothermally heated shallow submarine biotope at Kodakara-Jima island, Kagoshima, Japan: Applied and Environmental Microbiology, v. 61, p. 1931–1937.

Humphris, S.E., Zierenberg, R.A., Mullineaux, L.S., and Thomson, R.E., 1995a, Seafloor Hydrothermal Systems: Physical, Chemical, Biological, and Geological Interactions: Washington D.C., American Geophysical Union Geophysical Monograph 91, 466 p.

Humphris, S.E., Herzig, P.M., Miller, D.J., Alt, J.C., Becker, K., Brown, D., Brugmann, G., Chiba, H., Fouquet, Y., Gemmel, J.B., Guerin, G., Hannington, M.D., Holm, N.G., Honnorez, J.J., Iturrino, G.J., Knott, R., Ludwig, R., Nakamura, K., Petersen, S., Reysenbach, A.L., Rona, P.A., Smith, S., Sturz, A.A., Tivey, M.K., and Zhao, X., 1995b, The internal structure of an active seafloor massive sulphide deposit: Nature, v. 377, p. 713–716, doi: 10.1038/377713a0.

Jach, R., and Dudek, T., 2005, Evidence for hydrothermal origin of Toarcian manganese deposits from Krížna unit, Tatra Mountains, Poland: Chemical Geology, v. 224, p. 136–152, doi: 10.1016/j.chemgeo.2005.07.018.

Jannasch, H.W., 1984, Microbial processes at deep-sea hydrothermal vents, *in* Rona, P.A., Böstrom, K., Laubier, L., and Smith, K.L., eds., Hydrothermal processes at seafloor spreading centers: New York, Plenum Publishing, p. 677–709.

Jones, B., and Renaut, R.W., 1996, Influence of thermophilic bacteria on calcite and silica precipitation in hot springs with water temperatures above 90 °C: Evidence from Kenya and New Zealand: Canadian Journal of Earth Sciences, v. 33, p. 72–83.

Jones, B., Renaut, R.W., and Rosen, M.R., 2000, Trigonal dendritic calcite crystals forming from hot spring waters at Waikite, North Island, New Zealand: Journal of Sedimentary Research, v. 70, p. 586–603.

Jorge, S., Melgarejo, J.-C., and Alfonso, P., 1997, Asociaciones minerales en sedimentos exhalativos y sus derivados metamórficos, *in* Melgarejo, J.-C., ed., Atlas de Asociaciones Minerales en Lámina Delgada: Barcelona, Spain, Edicions de la Universitat de Barcelona, p. 287–308.

Juniper, S.K., and Tebo, B.M., 1995, Microbe-metal interactions and mineral deposition at hydrothermal vents, *in* Karl, D.M., ed., The Microbiology of Deep-Sea Hydrothermal Vents: Boca Raton, Florida, CRC Press, p. 219–253.

Kamenev, G.M., Fadeev, V.I., Selin, N.I., and Tarasov, V.G., 1993, Composition and distribution of macro- and meiobenthos around hydrothermal vents in the Bay of Plenty, New Zealand: New Zealand Journal of Marine and Freshwater Research, v. 27, p. 407–418.

Kamenev, G.M., Kavun, V.Ya., Tarasov, V.G., and Fadeev, V.I., 2004, Distribution of bivalve mollusks Macoma golikovi (Scarlato and Kafanov, 1988) and Macoma calcarea (Gmelin, 1791) in the shallow-water hydrothermal ecosystem of Kraternaya Bight (Yankich Island, Kuril Islands): Connection with feeding type and hydrothermal activity of Ushishir Volcano: Continental Shelf Research, v. 24, p. 75–95, doi: 10.1016/j.csr.2003.09.006.

Karl, D., Wirsen, C., and Jannasch, H., 1980, Deep-sea primary production at the Galapagos hydrothermal vents: Science, v. 207, p. 1345–1347.

Kohn, M.J., Riciputi, L.R., Stakes, D., and Orange, D.L., 1998, Sulfur isotope variability in biogenic pyrite: Reflections of heterogeneous bacterial colonization?: American Mineralogist, v. 83, p. 1454–1468.

Konhauser, K.O., Phoenix, V.R., Bottrell, S.H., Adams, D.G., and Head, I.M., 2001, Microbial-silica interactions in Icelandic hot spring sinter: possible analogues for some Precambrian siliceous stromatolites: Sedimentology, v. 48, p. 415–433, doi: 10.1046/j.1365-3091.2001.00372.x.

Kostoglodov, V., and Bandy, W.L., 1995, Seismotectonic constraints on the convergence rate between the Rivera and North American plates: Journal of Geophysical Research, v. 100, p. 17,977–17,989, doi: 10.1029/95JB01484.

Ledesma-Vázquez, J., Berry, R.W., Johnson, M.E., and Gutiérrez-Sánchez, S., 1997, El Mono chert: A shallow-water chert from the Pliocene Infierno Formation, Baja California Sur, Mexico, *in* Johnson, M.E., and Ledesma-Vázquez, J., Pliocene Carbonates and Related Facies Flanking the Gulf of California, Baja California, Mexico: Geological Society of America Special Paper 318, p. 73–81.

Ledesma-Vázquez, J., and Johnson, M.E., 2001, Miocene–Pleistocene tectonosedimentary evolution of Bahía Concepción region, Baja California Sur (México): Sedimentary Geology, v. 144, p. 83–96, doi: 10.1016/S0037-0738(01)00136-1.

Liakopoulos, A., Glasby, G.P., Papavassiliou, C.T., and Boulegue, J., 2001, Nature and origin of the Vani manganese deposit, Milos, Greece: An overview: Ore Geology Reviews, v. 18, p. 181–209, doi: 10.1016/S0169-1368(01)00029-4.

Lüders, V., Pracejus, B., and Halbach, P., 2001, Fluid inclusions and sulfur isotope studies in probable modern analogue Kuroko-type ores from the Jade hydrothermal field (Central Okinawa Trough, Japan): Chemical Geology, v. 173, p. 45–58, doi: 10.1016/S0009-2541(00)00267-9.

Macdonald, K.C., Becker, K., Spiess, F.N., and Ballard, R.D., 1980, Hydrothermal heat flux of the "black smoker" vents on the East Pacific Rise: Earth and Planetary Science Letters, v. 48, p. 1–7, doi: 10.1016/0012-821X(80)90163-6.

Marchig, V., Stackelberg, U., Wiedicke, M., Durn, G., and Milovanovic, D., 1999, Hydrothermal activity associated with off-axis volcanism in the Peru Basin: Marine Geology, v. 159, p. 179–203, doi: 10.1016/S0025-3227(98)00193-5.

Martínez-Frías, J., 1998, An ancient Ba-Sb-Ag-Fe-Hg–bearing hydrothermal system in SE Spain: Episodes, v. 21, p. 248–251.

Michard, A., 1989, Rare earth element systematics in hydrothermal fluids: Geochimica et Cosmochimica Acta, v. 53, p. 745–750, doi: 10.1016/0016-7037(89)90017-3.

Michard, A., and Albarède, F., 1986, The REE content of some hydrothermal fluids: Chemical Geology, v. 55, p. 51–60, doi: 10.1016/0009-2541(86)90127-0.

Mills, R.A., Wells, D., and Roberts, S., 2001, Genesis of ferromanganese crusts from the TAG hydrothermal field: Chemical Geology, v. 176, p. 283–293, doi: 10.1016/S0009-2541(00)00404-6.

Missack, E., Stoffers, P., and El Goresy, A., 1989, Mineralogy, parageneses, and phase relations of copper-iron sulfides in the Atlantis II Deep: Red Sea: Mineralium Deposita, v. 24, p. 82–91.

Mita, N., Maruyama, A., Usui, A., Higashihara, T., and Hariya, Y., 1994, A growing deposit of hydrous manganese oxide produced by microbial mediation at a hot spring, Japan: Geochemical Journal, v. 28, p. 71–80.

Morri, C., Bianchi, C.N., Cocito, S., Peirano, A., De Biasi, A.M., Aliani, S., Pansini, M., Boyer, M., Ferdeghini, F., Pestarino, M., and Dando, P., 1999, Biodiversity of marine sessile epifauna at an Aegean island subject to hydrothermal activity: Milos, Eastern Mediterranean Sea: Marine Biology, v. 135, p. 729–739, doi: 10.1007/s002270050674.

Mountain, B.W., Benning, L.G., and Boerema, J.A., 2003, Experimental studies on New Zealand hot spring sinters: rates of growth and textural development: Canadian Journal of Earth Sciences, v. 40, p. 1643–1667, doi: 10.1139/e03-068.

Moyer, C.L., Tiedje, J.M., Dobbs, F.C., and Kart, D.M., 1998, Diversity of deep-sea hydrothermal vent Archaea from Loihi Seamount, Hawaii: Deep-Sea Research. Part II, Topical Studies in Oceanography, v. 45, p. 303–317, doi: 10.1016/S0967-0645(97)00081-7.

Naden, J., Kilias, S.P., and Darbyshire, D.P.F., 2005, Active geothermal systems with entrained seawater as modern analogs for transitional volcanic-hosted massive sulfide and continental magmato-hydrothermal mineralization: The example of Milos Island, Greece: Geology, v. 33, p. 541–544, doi: 10.1130/G21307.1.

Nicholson, K., 1992, Contrasting mineralogical-geochemical signatures of manganese oxides: guides to metallogenesis: Economic Geology and the Bulletin of the Society of Economic Geologists, v. 87, p. 1253–1264.

Ohmoto, H., and Rye, R.O., 1979, Isotopes of sulfur and carbon, *in* Barnes, H.L., ed., Geochemistry of Hydrothermal Ore Deposits, 2nd ed.: New York, Wiley & Sons, p. 509–567.

Ortega-Osorio, A., Prol-Ledesma, R.M., Melgarejo, J.-C., Reyes, A., Rubio-Ramos, M.A., and Torres-Vera, M.A., 2001, Study of Hydrothermal Particulate Matter from a Shallow Venting System, offshore Nayarit, Mexico: American Geophysical Union Fall Meeting, San Francisco, California, Abstracts with Programs, p. F1338.

Parson, L.M., Walker, C.L., and Dixon, D.R., 1995, Hydrothermal vents and processes: Geological Society of America Special Publication 87, 411 p.

Pichler, T., and Veizer, J., 1999, Precipitation of Fe(III) oxyhydroxide deposits from shallow-water hydrothermal fluids in Tutum Bay, Ambitle Island, Papua New Guinea: Chemical Geology, v. 162, p. 15–31, doi: 10.1016/S0009-2541(99)00068-6.

Pichler, T., Veizer, J., and Hall, G.E.M., 1999a, The chemical composition of shallow-water hydrothermal fluids in Tutum Bay, Ambitle Island, Papua New Guinea and their effect on ambient seawater: Marine Chemistry, v. 64, p. 229–252, doi: 10.1016/S0304-4203(98)00076-0.

Pichler, T., Giggenbach, W.F., McInnes, B.I.A., Buhl, D., and Duck, B., 1999b, Fe sulfide formation due to seawater-gas-sediment interaction in a shallow water hydrothermal system at Lihir Island, Papua New Guinea: Economic Geology and the Bulletin of the Society of Economic Geologists, v. 94, p. 281–287.

Prol-Ledesma, R.M., 2003, Similarities in the chemistry of shallow submarine hydrothermal vents: Geothermics, v. 32, p. 639–644, doi: 10.1016/j.geothermics.2003.06.003.

Prol-Ledesma, R.M., and Juárez, G., 1986, Geothermal map of Mexico: Journal of Volcanology and Geothermal Research, v. 28, p. 351–362, doi: 10.1016/0377-0273(86)90030-2.

Prol-Ledesma, R.M., Canet, C., Melgarejo, J.-C., Tolson, G., Rubio-Ramos, M.A., Cruz-Ocampo, J.C., Ortega-Osorio, A., Torres-Vera, M.A., and Reyes, A., 2002a, Cinnabar deposition in submarine coastal hydrothermal vents, Pacific Margin of central Mexico: Economic Geology and the Bulletin of the Society of Economic Geologists, v. 97, p. 1331–1340.

Prol-Ledesma, R.M., Canet, C., Armienta, M.A., and Solis, G., 2002b, Vent fluid in the Punta Mita coastal submarine hydrothermal system, Mexico: Geological Society of America Abstracts with Programs, v. 34, no. 6, p. 341.

Prol-Ledesma, R.M., Canet, C., Tolson, G., García-Palomo, A., Miller, R., Rubio-Ramos, M.A., Torres-de León, R., and Huicochea-Alejo, J.S., 2003, Basaltic volcanism and submarine hydrothermal activity in Punta Mita, Nayarit, Mexico, *in* Geologic transects across Cordilleran Mexico, Guidebook for the field trips of the 99th Geological Society of America Cordilleran Section Annual Meeting, Field Trip 7, Puerto Vallarta, Jalisco, Mexico, p. 169–182.

Prol-Ledesma, R.M., Canet, C., Torres-Vera, M.A., Forrest, M.J., and Armienta, M.A., 2004, Vent fluid chemistry in Bahía Concepción coastal submarine hydrothermal system, Baja California Sur, Mexico: Journal of Volcanology and Geothermal Research, v. 137, p. 311–328, doi: 10.1016/j.jvolgeores.2004.06.003.

Prol-Ledesma, R.M., Dando, P.R., and de Ronde, C.E.J., 2005, Preface Special Issue on "Shallow-water Hydrothermal Venting": Chemical Geology, v. 224, p. 1–4, doi: 10.1016/j.chemgeo.2005.07.012.

Puteanus, D., Glasby, G.P., Stoffers, P., and Kunzendorf, H., 1991, Hydrothermal iron-rich deposits from the Teahitia-Mehetia and Macdonald Hot Spot areas, S.W. Pacific: Marine Geology, v. 98, p. 389–409, doi: 10.1016/0025-3227(91)90112-H.

Renaut, R.W., Jones, B., Tiercelin, J.J., and Tarits, C., 2002, Sublacustrine precipitation of hydrothermal silica in rift lakes: evidence from Baringo, central Kenya Rift Valley: Sedimentary Geology, v. 148, p. 235–257, doi: 10.1016/S0037-0738(01)00220-2.

Rodgers, K.A., Browne, P.R.L., Buddle, T.F., Cook, K.L., Greatrex, R.A., Hampton, W.A., Herdianita, N.R., Holland, G.R., Lynne, B.Y., Martin, R., Newton, Z., Pastars, D., Sannazarro, K.L., and Teece, C.I.A., 2004, Silica phases in sinters and residues from geothermal fields of New Zealand: Earth-Science Reviews, v. 66, p. 1–61, doi: 10.1016/j.earscirev.2003.10.001.

Rodríguez-Díaz, A.A., 2004, Caracterización geológica y geoquímica del área mineralizada de manganeso en Bahía Concepción, Baja California [B.Sc. thesis]: México, Universidad Nacional Autónoma de México (UNAM), 82 p.

Rona, P.A., 1988, Hydrothermal mineralization at oceanic ridges: Canadian Mineralogist, v. 26, p. 431–465.

Rooney, M.A., Claypool, G.E., and Chung, H.M., 1995, Modeling thermogenic gas generation using carbon isotope ratios of natural gas hydrocarbons: Chemical Geology, v. 126, p. 219–232, doi: 10.1016/0009-2541(95)00119-0.

Rusch, A., Walpersdorf, E., deBeer, D., Gurrieri, S., and Amend, J.P., 2005, Microbial communities near the oxic/anoxic interface in the hydrothermal system of Vulcano Island, Italy: Chemical Geology, v. 224, p. 169–182, doi: 10.1016/j.chemgeo.2005.07.026.

Russell, M.J., 1996, The generation at hot springs of sedimentary ore deposits, microbialites and life: Ore Geology Reviews, v. 10, p. 199–214, doi: 10.1016/0169-1368(95)00023-2.

Sarano, P., Murphy, R.C., Houghton, B.F., and Hedenquist, J.W., 1989, Preliminary observations of submarine geothermal activity in the vicinity of White Island volcano, Taupo Volcanic zone: New Zealand: Journal of the Royal Society of New Zealand, v. 19, p. 449–459.

Savelli, C., Marani, M., and Gamberi, F., 1999, Geochemistry of metalliferous, hydrothermal deposits in the Aeolian arc Tyrrhenian Sea: Journal of Volcanology and Geothermal Research, v. 88, p. 305–323, doi: 10.1016/S0377-0273(99)00007-4.

Sawkins, F.J., 1990, Metal deposits in relation to plate tectonics: Berlin, Springer Verlag, 461 p.

Schwarz-Schampera, U., Herzig, P.M., Hannington, M.D., and Stoffers, P., 2001, Shallow submarine epithermal-style As-Sb-Hg-Au mineralisation in the active Kermadec Arc, New Zealand, *in* Pietrzyński, A., et al., eds., Mineral deposits at the beginning of the 21st century: Rotterdam, The Netherlands, Balkema, p. 333–335.

Scott, S.D., 1997, Submarine hydrothermal systems and deposits, *in* Barnes, H.L., ed., Geochemistry of hydrothermal ore deposits: New York, John Wiley & Sons, p. 797–935.

Sedwick, P., and Stuben, D., 1996, Chemistry of shallow submarine warm springs in an arc-volcanic setting: Vulcano Island, Aeolian Archipelago, Italy: Marine Chemistry, v. 53, p. 147–161, doi: 10.1016/0304-4203(96)00020-5.

Stein, J.L., 1984, Subtidal gastropods consume sulphur-oxidizing bacteria: Evidence from coastal hydrothermal vents: Science, v. 223, p. 696–698.

Stetter, K.O., Fiala, G., Huber, G., Huber, R., and Segerer, A., 1990, Hyperthermophilic microorganisms: FEMS Microbiology Reviews, v. 75, p. 117–124, doi: 10.1111/j.1574-6968.1990.tb04089.x.

Stoffers, P., Hannington, M., Wright, I., Herzig, P., De Ronde, C., and the Shipboard Scientific Party, 1999, Elemental mercury at submarine hydrothermal vents in the Bay of Plenty, Taupo volcanic zone, New Zealand: Geology, v. 27, p. 931–934, doi: 10.1130/0091-7613(1999)027<0931:EMASHV>2.3.CO;2.

Tarasov, V.G., 1991, Shallow-water vents and ecosystem of the Kraternaya Bight (Ushishir Volcano, Kuriles), v. 1, Functional Parameters, Part 2: Vladivostok, Russia, DVO RAN Press, 196 p. (in Russian)

Tarasov, V.G., 2002, Environment and biota of shallow-water hydrothermal vents of the west Pacific, *in* Gebruk, A.V., ed., Biology of hydrothermal systems: Moscow, KMK Press, p. 264–319 (in Russian).

Tarasov, V.G., Propp, M.V., Propp, L.N., Kamenev, G.M., and Blinov, S.V., 1985, Hydrothermal venting and specific water ecosystem in Kraternaya Caldera (Kuriles): Vladivostok, Russia, DVNTc AN USSR Press, 30 p. (in Russian).

Tarasov, V.G., Propp, M.V., Propp, L.N., Zhirmunsky, A.V., Namsaraev, B.B., Gorlenko, V.M., and Starynin, D.A., 1990, Shallow-water gasohydrothermal vents of Ushishir Volcano and the ecosystem of Kraternaya Bight (The Kurile Islands): Marine Ecology, v. 11, p. 1–23.

Tarasov, V.G., Kondrashev, S.V., and Lastivka, T.V., 1991, Oxygen metabolism of the diatom and bacterial mats of Kraternaya Bight, *in* Tarasov, V.G., ed., Shallow-water Vents and Ecosystem of the Kraternaya Bight (Ushishir Volcano, Kuriles), v. 1, Functional Parameters, Part 2: Vladivostok, Russia, DVO RAN Press, p. 4–19 (in Russian).

Tarasov, V.G., Sorokin, Yu.I., Propp, M.V., Shulkin, V.M., Namsaraev, B.B., Starynin, D.A., Kamenev, G.M., Fadeev, V.I., Malakhov, V.V., and Kosmynin, V.N., 1993, Specifics of structural and functional characteristics of marine ecosystem in zones of shallow-water venting in the West Pacific: Izvestiya RAN, Seriya biologicheskaya (Biology Series), v. 6, p. 914–926 (in Russian).

Tarasov, V.G., Gebruk, A.V., Shulkin, V.M., Kamenev, G.M., Fadeev, V.I., Kosmynin, V.N., Malakhov, V.V., Starynin, D.A., and Obzhirov, A.I., 1999, Effect of shallow-water hydrothermal venting on the biota of Matupi Harbor (Rabaul Caldera, New Britain Island, Papua-New Guinea): Continental Shelf Research, v. 19, p. 79–116, doi: 10.1016/S0278-4343(98)00073-9.

Tarasov, V.G., Gebruk, A.V., Mironov, A.N., and Moskalev, L.I., 2005, Deep-sea and shallow-water hydrothermal vent communities: two different phenomena?: Chemical Geology, v. 224, p. 5–39, doi: 10.1016/j.chemgeo.2005.07.021.

Tritlla, J., and Cardellach, E., 1997, Fluid inclusions in pre-ore minerals from the carbonate-hosted mercury deposits in the Espadán Ranges (eastern Spain): Chemical Geology, v. 137, p. 91–106, doi: 10.1016/S0009-2541(96)00158-1.

Vidal, V.M.V., Vidal, F.V., and Isaacs, J.D., 1978, Coastal submarine hydrothermal activity off northern Baja California: Journal of Geophysical Research, v. 83-B, p. 1757–1774.

Vidal, V.M.V., Vidal, F.V., and Isaacs, J.D., 1981, Coastal submarine hydrothermal activity off northern Baja California 2. Evolutionary history and isotope chemistry: Journal of Geophysical Research, v. 86-B, p. 9451–9468.

Villanueva-Estrada, R.E., Prol-Ledesma, R.M., Torres-Alvarado, I., and Canet, C., 2005, Geochemical modeling of a shallow submarine hydrothermal system at Bahía Concepción, Baja California Sur, México: Proceedings of the World Geothermal Congress, Antalya, Turkey, paper 0892, 5 p.

Villanueva, R.E., Prol-Ledesma, R.M., Torres-Vera, M.A., Canet, C., Armienta, M.A., and de Ronde, C.E.J., 2006, Comparative study of sampling methods and in situ and laboratory analysis for shallow-water submarine hydrothermal systems: Journal of Geochemical Exploration, v. 89, p. 414–419, doi: 10.1016/j.gexplo.2005.11.020.

Von Damm, K.L., Edmond, J.M., Measures, C.I., and Grant, B., 1985, Chemistry of submarine hydrothermal solutions at Guaymas Basin, Gulf of California: Geochimica et Cosmochimica Acta, v. 49, p. 2221–2237, doi: 10.1016/0016-7037(85)90223-6.

Walter, M.R., Bauld, J., and Brock, T.D., 1976, Microbiology and morphogenesis of columnar stromatolites (Conophyton, Vacerrilla) from hot springs in Yellowstone National Park, *in* Walter, M.R., ed., Stromatolites: Elsevier, New York, p. 273–310.

Walter, M.R., and Des Marais, D.J., 1993, Preservation of biological information in thermal spring deposits: Developing a strategy for the search for fossil life on Mars: Icarus, v. 101, p. 129–143, doi: 10.1006/icar.1993.1011.

Manuscript Accepted by the Society 29 August 2006

Geological Society of America
Special Paper 422
2007

Epithermal deposits in México—Update of current knowledge, and an empirical reclassification

Antoni Camprubí*
Centro de Geociencias, Universidad Nacional Autónoma de México, Campus Juriquilla, Carretera 57 km. 15.5, 76023 Santiago de Querétaro, Qro., México

Tawn Albinson*
Exploraciones del Altiplano S.A. de C.V., Sinaloa 106–despacho 302, Colonia Roma, 06700 México D.F., México

ABSTRACT

Epithermal ore deposits have traditionally been the most economically important in México, with renowned world-class deposits like those in the Pachuca–Real del Monte, Guanajuato, Fresnillo, Taxco, Tayoltita, and Zacatecas districts. Whereas in certain areas (like the Great Basin in Nevada) intermediate and low sulfidation deposits have been found to be mutually exclusive in time and space; in the case of epithermal deposits in México, the intermediate and low sulfidation types do not appear to be mutually exclusive and, to the contrary, they coexist in the same regions, formed during the same time spans, and even occur together within a single deposit. These deposits are all Tertiary in age, ranging from middle Eocene to early Miocene, with the possible sole exception of a Paleocene deposit. Their space and time distribution follows the evolution of the continental arc volcanism of the Sierra Madre Occidental and Sierra Madre del Sur. The vast majority of epithermal deposits in México belong to the intermediate (IS) or low (LS) sulfidation types; only a few high sulfidation (HS) deposits have been described in the NW part of the country (e.g., El Sauzal, Mulatos, Santo Niño, La Caridad Antigua, all of them in Sonora and Chihuahua). Because most epithermal deposits in México exhibit composite characteristics of both IS and LS mineralization styles (as well as scarce characteristics of HS), they cannot be simply characterized as IS (polymetallic deposits associated with the most saline brines) or LS deposits (mainly Ag and Au deposits associated with lower salinity brines). Thus, in this paper we propose to use an empirical classification for IS + LS deposits (that is, alkaline/neutral epithermal deposits) into three types of mineralization; namely, A, B, and C. Type A (or IS type) comprises those deposits that generally formed at greater depths from highly saline but unsaturated brines and contain exclusively from top to bottom IS styles of mineralization with a consistent polymetallic character. Type B (or LS-IS type) comprises those deposits that exhibit dominant LS characteristics but have polymetallic IS roots (Zn-Pb-Cu); this is the most widespread type of epithermal mineralization in México. Types A and B generally exhibit

*E-mails: camprubi@geociencias.unam.mx; albinson@prodigy.net.mx

Camprubí, A., and Albinson, T., 2007, Epithermal deposits in México—Update of current knowledge, and an empirical reclassification, *in* Alaniz-Álvarez, S.A., and Nieto-Samaniego, Á.F., eds., Geology of México: Celebrating the Centenary of the Geological Society of México: Geological Society of America Special Paper 422, p. 377–415, doi: 10.1130/2007.2422(14). For permission to copy, contact editing@geosociety.org.

mineralogic and/or fluid inclusion evidence for boiling. Type C (or LS type) comprises those deposits that exhibit only LS styles of mineralization, formed generally by shallow boiling of low salinity fluids, and have relatively high precious metal and low base metal contents. In this paper, we also review other known or attributable aspects of Mexican epithermal deposits, including ore and gangue mineralogy and their evolution in time and space, structure, geothermometry, stable isotopic composition of mineralizing fluids and other components of the deposits, chemistry and sources for mineralizing fluids, and the plausible mechanisms for the mobilization of deep fluid reservoirs and for mineral deposition in the epithermal environment.

Keywords: epithermal, epithermal deposits, México, continental arc volcanism, Tertiary, intermediate sulfidation, low sulfidation, high sulfidation, polymetallic, base metals, precious metals, silver, gold.

INTRODUCTION

In this review, we follow the classification of epithermal deposits proposed by Hedenquist et al. (2000), which uses the "sulfidation state" of the fluids responsible for mineralization. The term "sulfidation state" is used in the sense of Barton (1970), as analogous to the oxidation state, and uses the correlation between temperature and fugacity of S_2 gas to mark the physicochemical boundaries between the types of epithermal deposits, following the stability fields of determined sulfide species. See White and Hedenquist (1990) and Einaudi et al. (2003) for complete discussions about the convenience of such terminology, its problems, and the history of concepts related to the sulfidation state itself.

Thus, we are going to use the terms *high sulfidation* (HS), *intermediate sulfidation* (IS), and *low sulfidation* (LS), to refer to the generic types of epithermal deposits. Aiming to help the reader to follow the present classification of epithermal deposits, we use the following scheme:

1. Acid epithermal deposits
   ⇨ 1.1. High sulfidation type (HS)
2. Alkaline/neutral epithermal deposits
   ⇨ 2.1. Intermediate sulfidation type (IS)
   ⇨ 2.2. Low sulfidation type (LS)
      ⇨ 2.2.1. Related to subalkaline magmas
      ⇨ 2.2.2. Related to alkaline magmas

Some examples of the above types and subtypes of epithermal deposits in México are:

1.1. Mulatos in Sonora (Staude, 2001) and El Sauzal in Chihuahua (Gray, 2001).

2.1. Pachuca–Real del Monte in Hidalgo (Geyne et al., 1963; Dreier, 2005), Fresnillo in Zacatecas (Gemmell et al., 1988; Ruvalcaba-Ruiz and Thompson, 1988; Simmons et al., 1988; Simmons, 1991), Tayoltita in Durango (Clarke and Titley, 1988; Enríquez and Rivera, 2001a, 2001b), and Temascaltepec in the State of México (Camprubí et al., 2001a, 2001b).

2.2.1. San Martín in Querétaro, Pinos in Zacatecas, and Guadalupe de los Reyes in Sinaloa (Albinson et al., 2001).

2.2.2. Ixhuatán in Chiapas (Miranda-Gasca et al., 2005).

The vast majority of the epithermal deposits in México formed under alkaline/neutral chemical regimes, either belonging to the low-sulfidation (LS) or to the intermediate-sulfidation (IS) types. The only epithermal deposits in México known to have formed under an acid chemical regime, or high-sulfidation (HS) deposits, are La Caridad Antigua and Mulatos in Sonora (Staude, 2001; Valencia et al., 2003; Valencia, 2005) and El Sauzal and Santo Niño in Chihuahua (Charest and Castañeda, 1997; Sellepack, 1997; Gray, 2001). The recently described HS epithermal deposit that formed close to the world-class porphyry copper deposit at La Caridad in Sonora (Valencia et al., 2003; Valencia, 2005) occurs in a downthrown fault block, yielded a similar age to the porphyry copper deposit, and could reconstruct to a pre-faulting geologic position close to or directly above the porphyry deposit. Additional HS-type occurrences are also reported in other areas in northwestern México, in particular in the neighboring areas of the Cananea porphyry copper district (L. Ochoa-Landín, 2004, personal commun.), as well as in the Mulatos region (see information about the La India project in www.grayd.com).

SPACE AND AGE DISTRIBUTION—THE LINK WITH VOLCANISM

The epithermal deposits known to date in México are Tertiary in age, generally ranging from Lutetian to Aquitanian-Burdigalian (Camprubí et al., 2003a), or middle Eocene to early Miocene, and their space distribution matches the distribution of the volcanism associated with the evolution of the Sierra Madre Occidental and the Sierra Madre del Sur (see Damon et al., 1981; Clark et al., 1981, 1982; Camprubí et al. 1982, 2003a). See the most complete compilation to date of the stratigraphy and tectonics of the Sierra Madre Occidental in Ferrari et al. (this volume). Magmas associated with the epithermal deposits have been described as calc-alkaline in composition (Damon et al., 1981). Using the distribution of Tertiary volcanism as a guide, the distribution of epithermal deposits in México (Fig. 1) may be outlined following three main age spans (Camprubí et al., 2003a):

1. Deposits older than ca. 40 Ma: examples of deposits that belong to this range of ages are Batopilas in Chihuahua, Topia in Durango, and La Caridad Antigua in Sonora, though the oldest radiometric age obtained to date in any epithermal deposit in México is 58 Ma (Paleocene) at El Barqueño in Jalisco (Camprubí et al., 2006a). This age span broadly coincides with the final stages of the Laramide orogeny in northern México.
2. Deposits with ages between ca. 40 and ca. 27 Ma: the vast majority of epithermal deposits in México formed during this range of ages and is distributed along a NW-SE oriented belt, which extends from Chihuahua to the state of México at a maximum distance east of the Pacific coast of ~600 km. This age range corresponds to the first bimodal andesitic-rhyolitic volcanic episode of the Lower Volcanic Series (LVS) of the Sierra Madre Occidental (Fredrickson, 1974; McDowell and Keizer, 1977). The LVS is largely formed by andesites and, locally, rhyolitic volcanic centers, which are generally associated with important epithermal deposits, as at the San Dimas district (Tayoltita) in Durango and the Pánuco district in Sinaloa. Albinson (1988) identified a preferential range of ages between 35 and 30 Ma for the formation of epithermal deposits in the central part of the Mexican Altiplano, a period that corresponds with the climax of volcanic activity with dominant intermediate compositions. Staude and Barton (2001) suggested that similar deposits may be buried under the Upper Volcanic Series of the Sierra Madre Occidental.
3. Deposits younger than ca. 23 Ma: these deposits are found in the southern part of the Sierra Madre Occidental, north of the Trans-Mexican Volcanic Belt. These deposits as a group show a WNW-ESE arrangement that corresponds with the last volcanic episode in the Sierra Madre Occidental that is clearly ignimbritic, known as Upper Volcanic Series (Fredrickson, 1974; McDowell and Keizer, 1977; Ferrari et al., 2002). The above trend coincides with a change in orientation of the volcanic arc, following the opening of a new trench with a WNW orientation in southern México (Ferrari et al., 1999). Data density to date appears to indicate a sort of "bipolar" distribution of these deposits, with a preferential concentration of deposits in the western portion of the Trans-Mexican Volcanic Belt, like Bolaños and San Martín de Bolaños in Jalisco, El Indio–Huajicori and La Yesca in Nayarit, Mezquital del Oro in Zacatecas (Camprubí et al., 2003a), and a second group of deposits located in the eastern portion of the Trans-Mexican Volcanic Belt at Pachuca–Real del Monte in Hidalgo and Ixtacamaxtitlán in Puebla (McKee et al., 1992; Camprubí et al., 2001c; Tritlla et al., 2004).

The general space distribution of epithermal deposits in México (Fig. 2) defined by Camprubí et al. (1999) includes both the Pb-Zn-Ag and Au-(Ag) metallogenic provinces previously defined by Clark et al. (1981, 1982), which show a partly coincidental distribution. Although in general the deposits with the highest base metal grades are more abundant toward the inner part of the continental margin, it is also evident that the Mexican metallogenic provinces of the Sierra Madre Occidental, the Mesa Central, and the Sierra Madre del Sur host both abundant base metal–rich (comparable to IS) and precious metal–rich (comparable to LS) epithermal deposits.

REGIONAL STRUCTURE

The regional distribution of epithermal deposits in México, as well as deposits belonging to other deposit types (Sn veins, skarns, etc.), indicates they are closely associated in space with regional faults. That is the case of the Mesa Central or Mexican Altiplano, a geological province that is bounded by several major N-S and WNW-ESE fault zones (namely, Bajío fault, Aguascalientes graben, and the large Taxco–San Miguel de Allende, and San Luis–Tepehuanes fault zones) from the surrounding geological provinces of the Sierra Madre Oriental, Sierra Madre Occidental, and the Trans-Mexican Volcanic Belt. Most of the mineral deposits in the Altiplano, including the epithermal deposits, are found close to these limits (Nieto-Samaniego et al., this volume). In addition, several ENE-trending faults at the western margin of the Sierra Madre Occidental may have accommodated transverse displacements during Tertiary extension (Henry and Aranda-Gómez, 2000), and they may represent inherited Laramide faults that reactivated during subsequent extensional tectonics (Horner and Steyrer, 2005). Such long-lived, reactivated fault zones determine the location of the deposits, as they are guides for the emplacement of magmas, the necessary heat and fluid sources that trigger the subsequent hydrothermal activity that leads to mineral deposition. The activity of the large Taxco–San Miguel de Allende and San Luis–Tepehuanes fault zones has been documented, by means of several radiometric determinations (from numerous authors cited in Nieto-Samaniego et al., this volume), to occur during the Oligocene, the Miocene, and the Quaternary, and the location of faulting was inherited from even older structures (Nieto-Samaniego et al., this volume). In many cases, these radiometric ages were obtained from volcanic or subvolcanic rocks post-dating or predating episodes of faulting. In the Mesa Central and its vicinities, the geographic distribution of volcanic centers is broader than the distribution of magmatic-related ore deposits (epithermal, porphyry, skarn, Sn veins, and IOCG [Iron Oxide–Copper–Gold deposits] altogether), but these are preferentially located close to contemporaneously active major fault zones. Furthermore, the ages of ore deposits are roughly coincidental with the ages of major reactivation periods of faulting (Nieto-Samaniego et al., this volume). Such space and time links between faulting, magmatic, and metallogenic activities are more than apparent. The structurally weakest areas in the upper crust, represented by ancient or "dormant" fault zones, may potentially and preferentially allow the emplacement of magmas and the circulation of aqueous fluids, thus favoring the emplacement of ore deposits. Some of them may even host veins, like the renowned Veta Madre in Guanajuato. Large-scale fracture zones, particularly those that traverse magmatic arcs and intersect arc-parallel structural features, are known to be favorable sites for ore-forming processes (Petersen and Vidal, 1996; Richards, 2000;

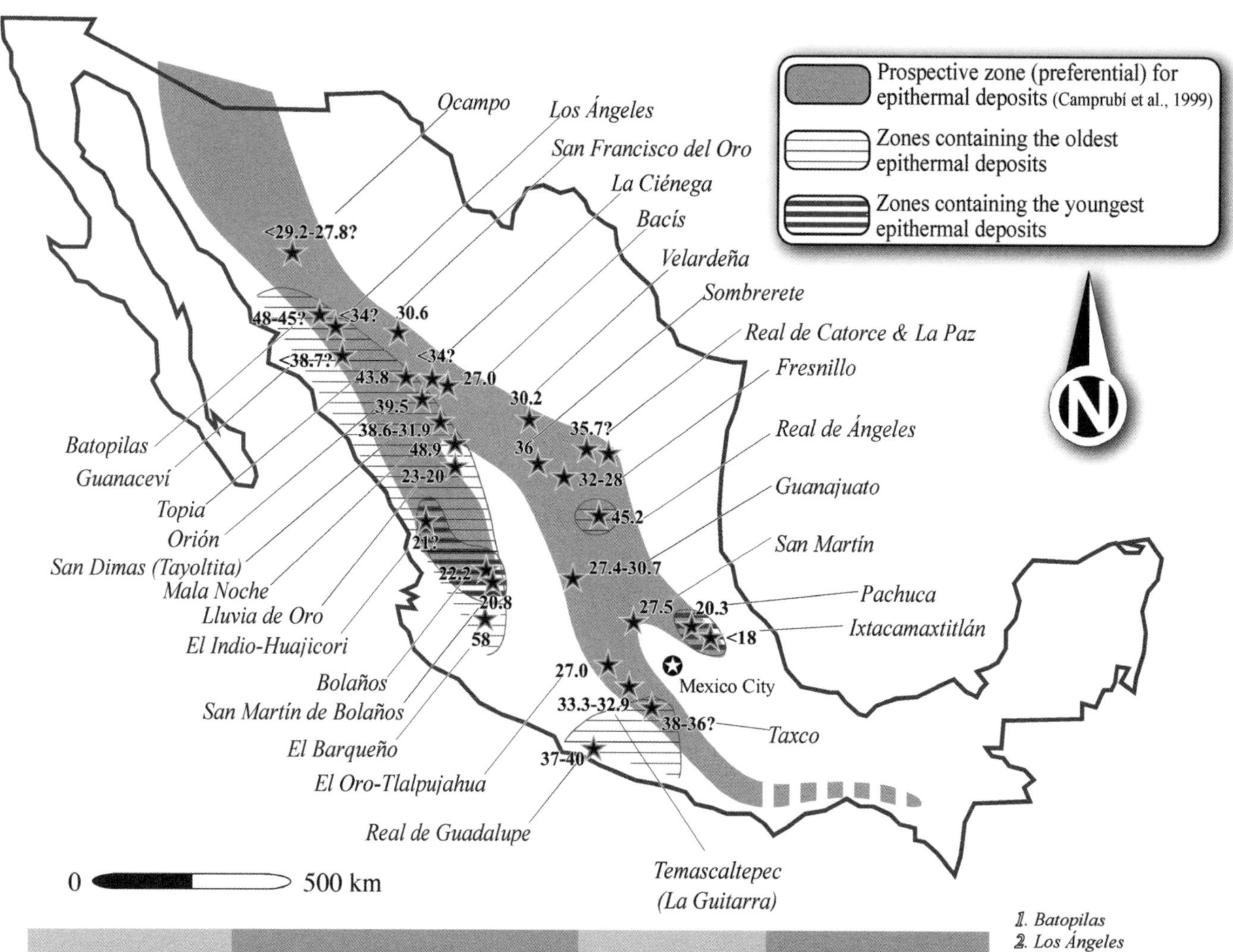

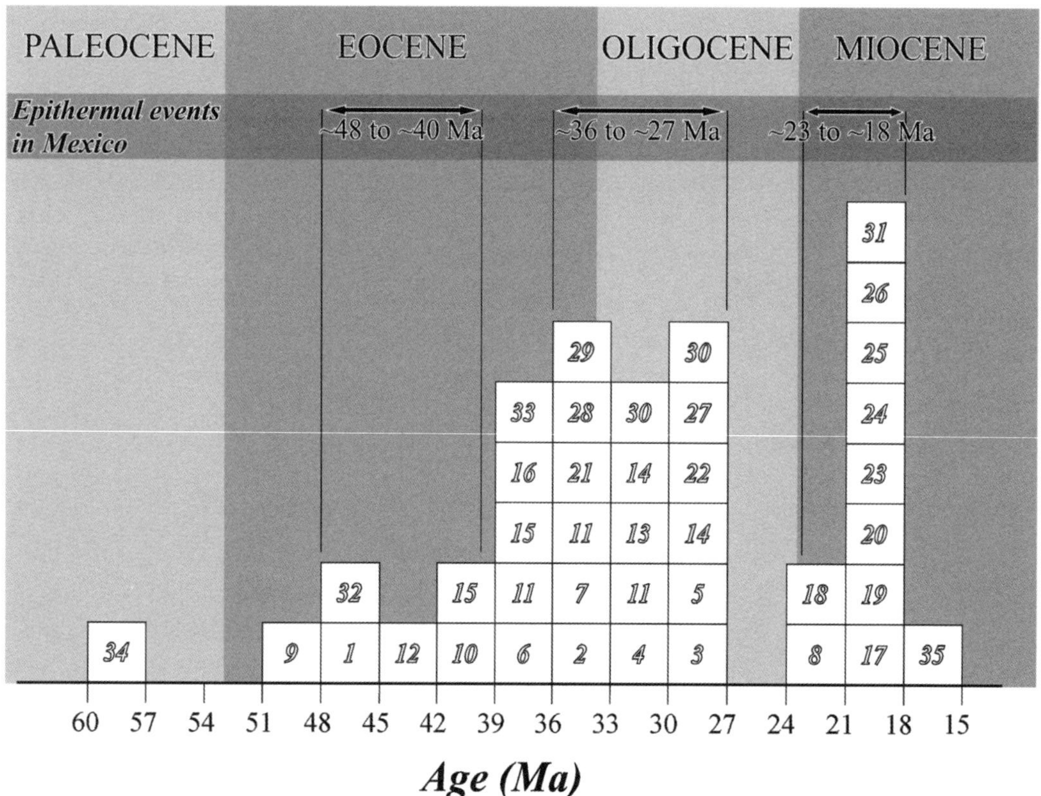

1. Batopilas
2. Los Ángeles
3. Ocampo
4. San Francisco del Oro
5. Bacís
6. Guanaceví
7. La Ciénega
8. Lluvia de Oro
9. Mala Noche
10. Orión
11. San Dimas
12. Topia
13. Velardeña
14. Guanajuato
15. Real de Guadalupe
16. Taxco
17. Pachuca/Real del Monte
18. Bolaños
19. Cinco Minas
20. San Martín de Bolaños
21. Temascaltepec
22. El Oro/Tlalpujahua
23. El Indio/Huajicori
24. El Zopilote
25. La Yesca
26. Santa María del Oro
27. San Martín
28. Real de Catorce
29. Santa María de la Paz
30. Fresnillo
31. Mezquital del Oro
32. Real de Ángeles
33. Sombrerete
34. El Barqueño
35. Ixtacamaxtitlán

Figure 1. Top: Distribution of the dated Mexican low-sulfidation epithermal deposits (modified from Camprubí et al., 2003a). Other deposits not shown are Cinco Minas in Jalisco, Mezquital del Oro in Zacatecas, and La Yesca, El Zopilote, and Santa María del Oro in Nayarit (close to El Indio and Bolaños), whose ages have been attributed to be ca. 21 Ma by Camprubí et al. (2003a). Bottom: Histogram of mineralization ages of epithermal deposits in México, indicating the three preferential age ranges for epithermal mineralization by Camprubí et al. (2003a). Modified from Camprubí et al. (2006a).

Horner and Steyrer, 2005). In México, E-W to ENE-trending fracture zones associated with the Laramide orogeny (ca. 80–40 Ma; Sedlock et al., 1993) have been periodically reactivated during the evolution of the Tertiary magmatic arcs (which progressed eastward until the Eocene and retreated westward until the Miocene; e.g., Ferrari et al., 1999; Camprubí et al., 2003a) and may have played an important role in focusing magmatism and hydrothermalism (Horner and Steyrer, 2005). On satellite images of the Sierra Madre Occidental it is possible to identify easterly to northeasterly structural lineaments with closely associated epithermal deposits and prospects (i.e., the Cuitaboca–San José de Gracia–Guadalupe y Calvo and the San José de Gracia–Baborigame trends between Sinaloa and Chihuahua [Fig. 3] and the Morís–Ocampo–Concheño trend in Chihuahua). Other examples of epithermal deposits whose arrangement may reflect Laramide ENE to NE trends are San Dimas (Tayoltita) and Avino in Durango (Horner and Enríquez, 1999; Horner and Steyrer, 2005).

The origin of the regionally persistent NW orientation of most epithermal veins in México can be associated with three major events involving the rearrangement of crustal plates, as suggested by Starling et al. (1997) and Starling (1999). The oldest and more speculative event proposes the existence of sinistral NW- and WNW-trending strike-slip megashears during the Middle Jurassic (Sedlock et al., 1993; Molina-Garza and Iriondo, this volume), such as the Sonora-Mojave Megashear, whose exis-

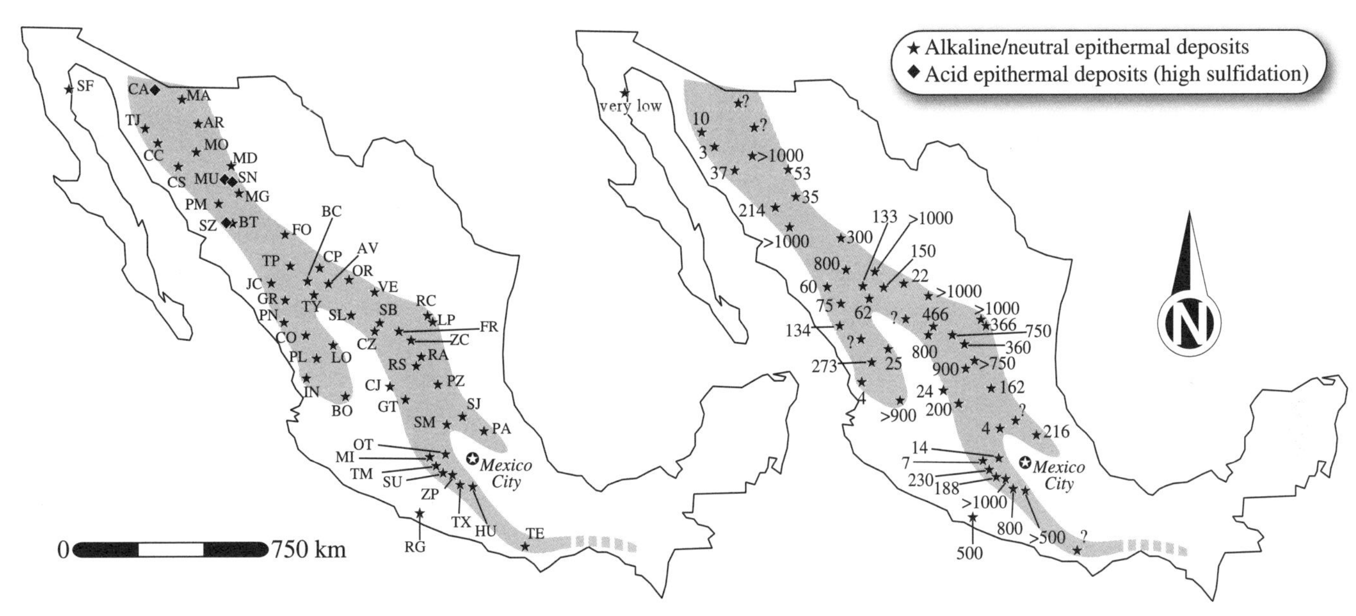

Figure 2. Geographical distribution of significant acid (high sulfidation) and alkaline/neutral epithermal deposits (intermediate and low sulfidation) in México. The map of México at the right exhibits the composite Ag/Au ratios for each deposit (references in Camprubí, 1999; Camprubí et al., 1999). The shaded area contains the prospective zones for epithermal deposits in México, following the Pb-Zn-Ag, Au-Ag, and Hg-Sb belts by Clark et al. (1981 and 1982) and the distribution of known epithermal deposits (Camprubí et al., 1999). See references for most of the deposits in Simmons (1995), White et al. (1995), Camprubí (1999), Camprubí et al. (1999), Hedenquist et al. (2000), and Albinson et al. (2001). AR—Arizpe, Sonora; AV—Avino, Durango; BC—Bacís, Durango; BO—Bolaños and San Martín de Bolaños, Jalisco; BT—Batopilas, Chihuahua; CC—Cerro Colorado, Sonora; CJ—Comanja de Corona, Jalisco; CO—Copala, Sinaloa; CP—Cerro Prieto and Mantos, Durango; CS—Colorada, Sonora; CZ—Colorada, Zacatecas; FO—San Francisco del Oro and Santa Bárbara, Chihuahua; FR—Fresnillo, Zacatecas; GR—Guadalupe de los Reyes, Sinaloa; GT—Guanajuato, Guanajuato; HU—Huautla, Morelos; IN—El Indio, Nayarit; JC—San José del Cobre, Durango/Sinaloa; LO—Lluvia de Oro, Durango; LP—Santa María de la Paz, San Luis Potosí; MA—Magallanes, Sonora; MD—Mineral de Dolores, Chihuahua; MG—Maguaríchic, Chihuahua; MI—Miahuatlán and Ixtapan del Oro, State of México; MO—Moctezuma, Sonora; MU—Mulatos, Sonora; OR—Orito, Durango; OT—El Oro–Tlalpujahua, State of México/Michoacán; PA—Pachuca–Real del Monte, Hidalgo; PL—Plomosas, Sinaloa; PM—Palmarejo, Chihuahua; PN—Pánuco, Sinaloa; PZ—Pozos, Guanajuato; RA—Real de Ángeles, Aguascalientes; RC—Real de Catorce, San Luis Potosí; RG—Real de Guadalupe, Guerrero; RS—Real de Asientos, Aguascalientes; SB—Sombrerete, Zacatecas; SF—San Felipe, Baja California; SJ—San Joaquín, Querétaro; SL—Saladillo, Durango; SM—San Martín, Querétaro; SN—Santo Niño, Chihuahua; SU—Sultepec and Amatepec, State of México; SZ—El Sauzal, Chihuahua; TE—Tejomulco, Oaxaca; TJ—Tajitos, Sonora; TM—Temascaltepec, State of México; TP—Topia, Durango; TX—Taxco, Guerrero; TY—Tayoltita, Durango; VE—Velardeña, Durango; ZC—Zacatecas, Zacatecas; ZP—Zacualpan, State of México.

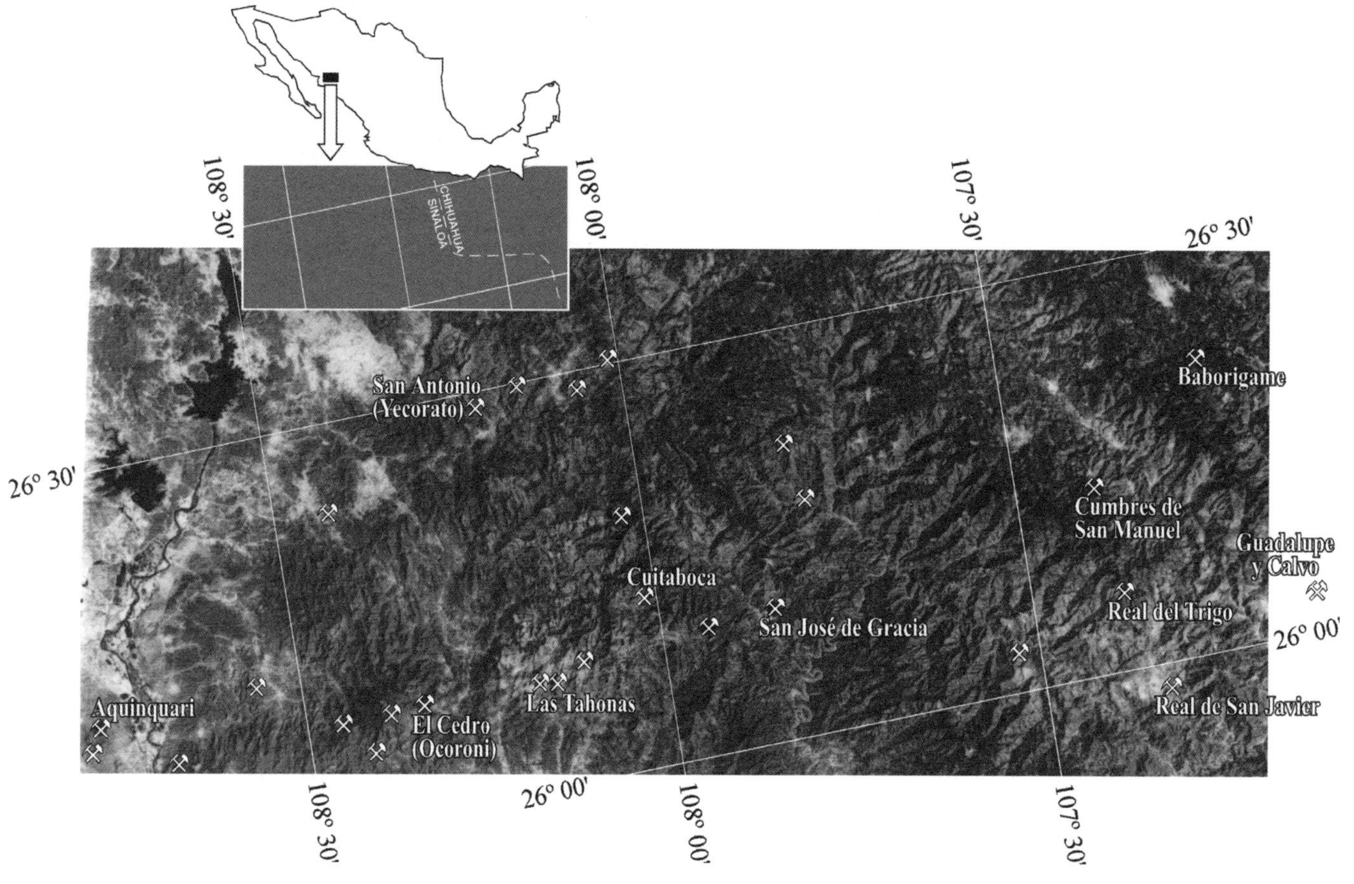

Figure 3. Landsat image of the region around the southern part of the border between Chihuahua and Sinaloa, with the location of epithermal deposits compiled after Consejo de Recursos Minerales (1991). These are both mining districts and prospects, regardless of their size. Notice that the majority of them are found close to E-W or ENE-WSW major lineations.

tence is still under discussion (Molina-Garza and Iriondo, this volume), or any other structures that worked as "bridges" between the basement cratons of North and South America. Identification of such structural features is difficult due to extensive cover by post-Middle Jurassic sedimentary and volcanosedimentary and Tertiary volcanic rocks. Although that tectonic event is not related in age with the formation of epithermal deposits, it could have left major tectonic discontinuities in the basement prone to reactivation during later tectonic events. A more tangible explanation for the NW-striking structural trends in México is the tectonic imprint of the backarc middle Tertiary extension that affected the Mesa Central (Mexican Altiplano) due to the collision and subduction of the Farallon plate (Atwater, 1970) and to which long-lived and dominantly NE-trending extension is related. Such a backarc extensional environment is favorable for the formation of epithermal deposits and is similar to that presently existing in the Taupo Volcanic Zone of New Zealand (Albinson et al., 1995). The third major tectonic event is related to NNW- to N-trending dextral strike-slip tectonics and extension related to the opening of the Gulf of California. Although no documented examples of epithermal deposits of importance are linked to this event, San Felipe in Baja California could represent a possible example.

RANGES OF METAL CONTENTS AND DEPOSIT SIZES

México hosts a high proportion of globally important IS and LS epithermal deposits, particularly those rich in silver (Fig. 2). There are (1) some world-class Ag-Au deposits, such as Pachuca–Real del Monte in Hidalgo (80 Mt); Guanajuato (40 Mt); Fresnillo in Zacatecas (now >70 Mt, with ore reserves included); and Tayoltita in Durango (>20 Mt), all of them grading ≥500 ppm Ag; (2) high-tonnage deposits (between 20 and 85 Mt), but grading <500 ppm Ag, such as Taxco in Guerrero, Real de Ángeles in Zacatecas, or San Francisco del Oro in Chihuahua; and (3) low-tonnage (between 5 and 7 Mt), high-grade deposits, between 800 and 1500 ppm Ag, such as Sombrerete and Colorada in Zacatecas, Real de Catorce in San Luis Potosí, and Bolaños and San Martín de Bolaños in Jalisco. There are also some deposits that have exceptionally high Au contents compared to the rest of epithermal deposits in México, like El Oro–Tlalpujahua in the State of México/Michoacán (over 25 Mt, 7 to ~20 ppm Au, only for the El Oro deposits). See a full review of grades, tonnages, plus other characteristics, for Mexican epithermal deposits in Albinson et al. (2001). In general, large, high-grade silver or high-grade gold deposits are base metal-poor (<5% combined base metals), whereas large low-grade silver deposits are polymetallic and base metal-rich (5%–10% combined base metals).

STRUCTURE OF INTERMEDIATE SULFIDATION AND LOW SULFIDATION EPITHERMAL DEPOSITS

Morphology

In the majority of IS and LS epithermal deposits in México, the ores are found in veins and/or brecciated bodies up to several kilometers long, although precious metal-bearing orebodies are found within vertical intervals generally of 300–600 m (e.g., Geyne et al., 1963; Albinson et al., 2001; Dreier, 2005). The vertical distribution of ores may vary considerably depending on the history of mineral precipitation and the nature of the processes that led to ore deposition. For instance, the Sombrerete deposit in Zacatecas formed at significant depth (>1000 m under the paleosurface) over a vertical interval of >800 m from non-boiling fluids (Albinson, 1988; Albinson et al., 2001), in contrast to precious-metal bonanza deposits, which are related to vertically more restricted shallow ore zones more clearly associated to boiling processes. IS and LS epithermal deposits in México, like elsewhere, usually contain individual veins exhibiting not only a preferential depth range for the occurrence of metallic mineral associations (i.e., Geyne et al., 1963; Dreier, 2005), but also show substantial variation within individual veins, depending on the relative position of proposed feeder channels for mineralizing fluids (Camprubí et al., 2001b). The occurrence of precious metal bonanza grades normally occupies a more restricted central position within the mineralized zone (Fig. 4), both in terms of grade and width, as in Tayoltita (Clarke and Titley, 1988), Topia (Loucks and Petersen, 1988; Loucks et al., 1988), and Bacís in Durango (Albinson et al., 2001); Pachuca–Real del Monte in Hidalgo (Geyne et al., 1963; Dreier, 2005); and Fresnillo in Zacatecas (Gemmell et al., 1988; Simmons, 1991). Most noticeably, it is a common feature that the deposits become richer in base metals with depth. Veins may locally coalesce at depth to form thicker veins. In some cases, the zones where veins coalesce contain the bonanzas with higher grades and/or greater volumes of economic mineralization.

Veins, plus other types of epithermal mineralized bodies, rarely formed under a single hydrothermal episode. The majority of IS and LS epithermal deposits in México have a polyphase and multiepisodic character (e.g., Gemmell et al., 1988; Albinson and Rubio, 2001; Albinson et al., 2001; Camprubí et al., 2001a) and are the product of several phases or stages of vein formation, not all of which are ore-bearing, whose products are megascopically observable, and frequently have distinctive distribution, as well as particular structural, mineralogical, and geochemical characteristics. These stages of mineralization are in turn usually composed via different individual episodes that correspond to discrete pulses of distinct mineralizing fluids that confer to the deposits their characteristic internal banded structure. However, in many cases the apparent complexity of internal banding of gangue and ore minerals appears to be the product of an episodic process of cracking and sealing of veins, in which case the banded texture is not necessarily related to the occurrence of discrete pulses of mineralizing fluids with different origins, as is described in more detail in the Silica Petrography section of this paper.

Sinters

In México, despite the abundant occurrences of IS and LS epithermal deposits that could be susceptible to having associated sinters, very few case examples of surface silica laminated deposits have been reported, either true sinters in the formational sense of the word or non-generic jasperoids whose origin was interpreted to be associated with hydrothermalism that could represent the "tops" of epithermal deposits. According to the available information, some silica stratiform deposits with a hydrothermal origin were described as sinters at Ixtacamaxtitlán in Puebla (Fig. 5) (Camprubí et al., 2001c; Morales-Ramírez et al., 2003); Santa Gertrudis in Sonora (Murray, 1997; Murray and Atkinson, 1997); and Dos Hermanas, Sauz de Caleras, and Cruzalinas in San Luis Potosí (Mason, 1995). At the Sombrerete, Colorada, and Fresnillo districts in Zacatecas, Albinson (1988) described silica deposits with evidence that indicated that these could be the basal parts of sinters, based on the occurrence of diagnostic Hg contents, stratiform jasperoid deposits, and characteristic hydrothermal alteration assemblages, among other aspects. It is not uncommon to find subhorizontal silicified chaotic and blocky silica deposits with Hg mineralization, which could have formed at the position of the paleo-water table, as in presently active systems like Waiotapu in New Zealand (Simmons and Browne, 2000). Other sinters have been reported to occur at El Malacate in Sonora, Lobos in Sinaloa, Ludavina in Baja California, Los Crestones in Durango, and El Salitre (unknown state) in recent mining reports available online (www.miningrecord.com, www.imdex.com/Companies/i/z_IntlNorthair.htm, www.imdex.com/Companies/c/z_Cardero.htm, www.imdex.com/Companies/qr/z_Radius.htm, and www.minefinders.com/Projects/malacate.html).

It is also worth noting the occurrence of calcite-quartz-barite veins and submarine vents with cinnabar and opal-carbonate sinters at Bahía de Concepción in Baja California Sur (Prol-Ledesma et al., 2004; Canet et al., 2005a, 2005b), which are part of a hydrothermal system that could be genetically analogous to epithermal deposits; these manifestations are found in an extensional tectonic environment that is considered one of the most favorable for the formation of LS deposits (Sillitoe and Hedenquist, 2003). Analogous submarine manifestations have been documented worldwide, such as the deposits in the coast of Lihir Island at Papua New Guinea, adjacent to the famous Ladolam LS epithermal deposit; the Kermadec arc in New Zealand; or Milos Island in Greece (Herzig and Hannington, 1995; Sillitoe et al., 1996; Herzig et al., 1999; Schwarz-Schampera et al., 2001; Petersen et al., 2002; Naden et al., 2005).

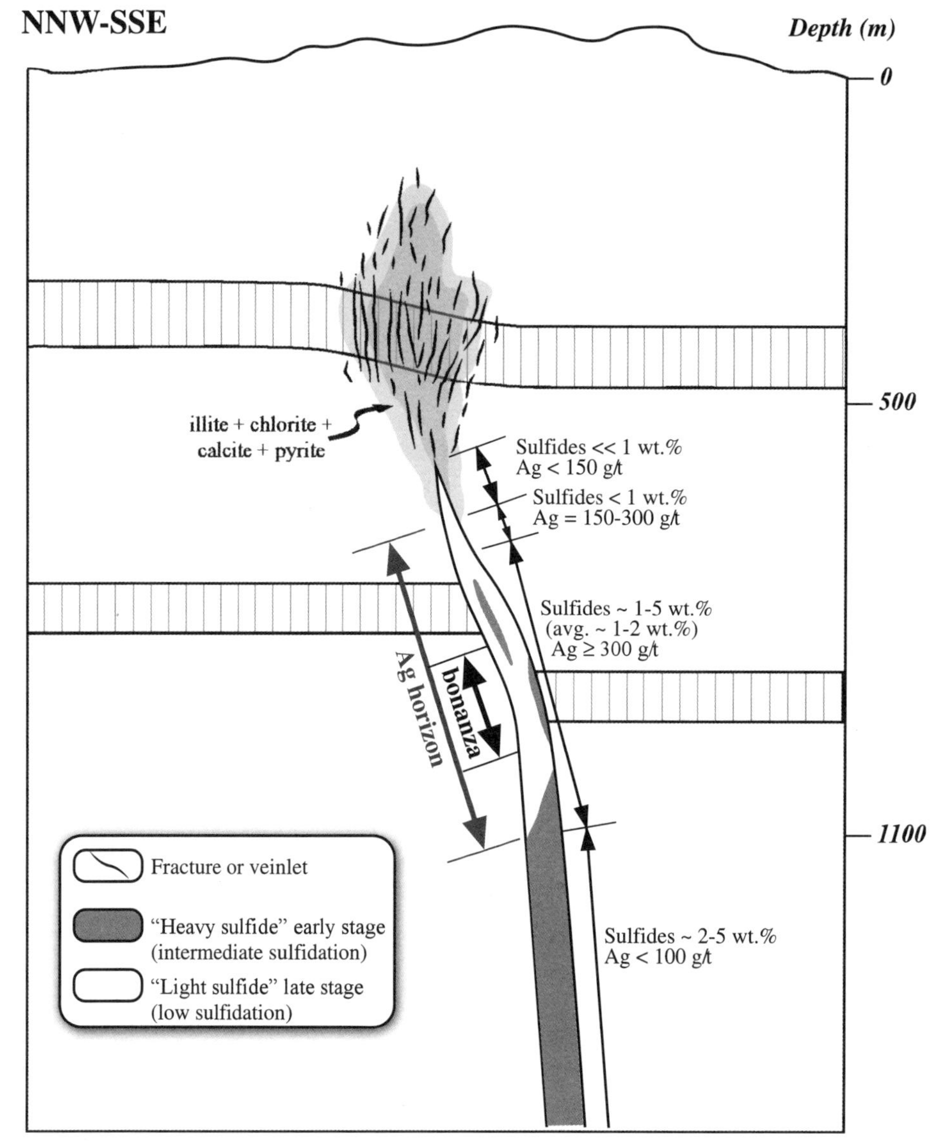

NNW-SSE
Depth (m)
0
500
1100
illite + chlorite + calcite + pyrite
Sulfides << 1 wt.% Ag < 150 g/t
Sulfides < 1 wt.% Ag = 150-300 g/t
Sulfides ~ 1-5 wt.% (avg. ~ 1-2 wt.%) Ag ≥ 300 g/t
bonanza
Ag horizon
Sulfides ~ 2-5 wt.% Ag < 100 g/t
Fracture or veinlet
"Heavy sulfide" early stage (intermediate sulfidation)
"Light sulfide" late stage (low sulfidation)

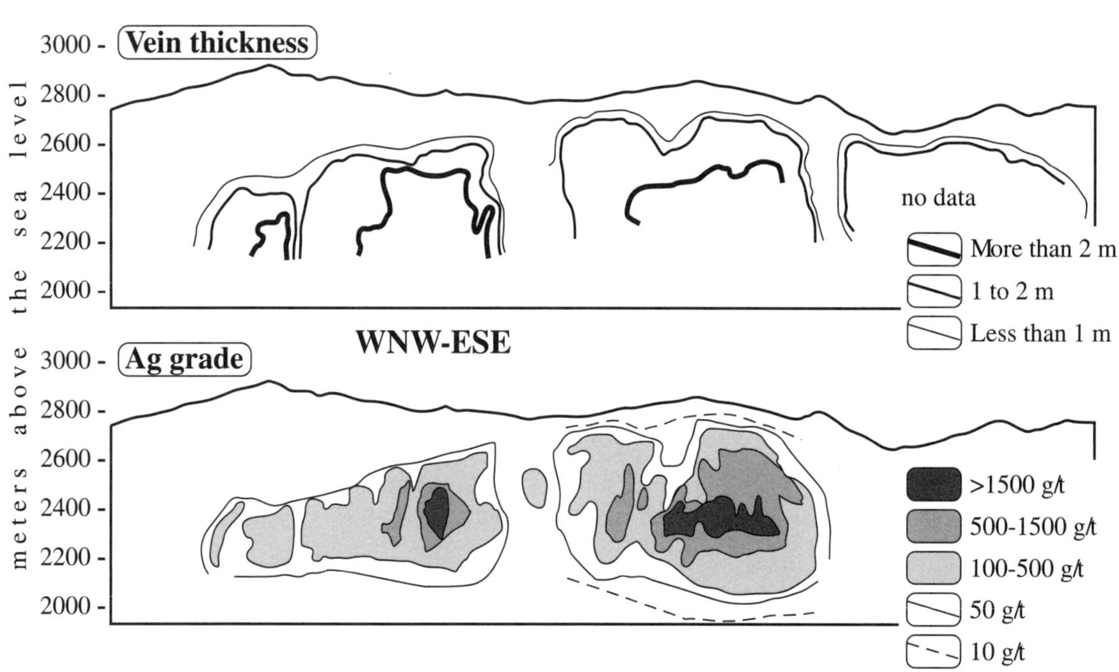

Vein thickness
meters above the sea level
3000
2800
2600
2400
2200
2000
no data
More than 2 m
1 to 2 m
Less than 1 m
WNW-ESE
Ag grade
>1500 g/t
500-1500 g/t
100-500 g/t
50 g/t
10 g/t

Figure 4. Cross section through the Colón–Purísima vein system at the Pachuca–Real del Monte district in Hidalgo, showing the dispersion of the vein at its top through thinner veins and veinlets in a broom-like pattern, and longitudinal sections of the same vein system displaying contours of vein thickness and Ag grade ranges. The veins include both "heavy sulfide" (IS) and "light sulfide" (LS) mineralization stages, with deep IS stages and fragments of these within late LS stages (J. Dreier, 2006, personal commun.) at higher elevations in the deposit. Modified from Dreier (2005). The horizontal scale is the same as the vertical.

"Tops" and "Bottoms" of Ore Zones

The upper and lower limits of orebodies, commonly called "tops" and "bottoms," respectively, are noteworthy features to be addressed in exploration for epithermal deposits in México and elsewhere. The tops of orebodies may be marked by (1) fault closure, (2) lateral dispersion of neutral–chloride fluids, (3) lack of permeability due to the presence of an argillic alteration cap, or (4) an upward drop in metal values due to the presence of a "depleted," deeply boiled neutral pH chloride fluid (Albinson et al., 2001). A fault closure may be due to (a) the lack of movement, or (b) the disappearance of the fault within a mechanically incompetent rock, like the El Cobre vein in the Taxco district (Camprubí et al., 2006c). Fault closures have been documented at the Purísima–Colón vein system in the Pachuca–Real del Monte district (Fig. 4) and in the southeastern part of the Fresnillo district where tension gashes essentially occur as weak fractures 200–300 m above the tops of veins (Simmons, 1991). Numerous other cases, however, show that the faults and/or low-grade veins continue into higher geologic levels above the tops of productive veins, such as the Alacrán vein in the Bolaños district (Lyons, 1988); the Zuloaga vein in the San Martín de Bolaños district (Albinson and Rubio, 2001), both in Jalisco; the Vizcaína fault in the Pachuca–Real del Monte district in Hidalgo (Geyne et al., 1963); the Herrero fault in the Bacís district; and the Centenario fault in the La Ciénega district in Durango (De la Garza et al., 2001). In these districts, the structures persist hundreds of meters above the tops of orebodies, where they exhibit variable amounts of silicification and argillization generally related to the presence of steam-heated waters, as exemplified in the San Martin de Bolaños district (Albinson and Rubio, 2001). The bottoms of orebodies are in turn marked by the lack of ore grade mineralization caused by a drop out of precious metal values in IS and LS deposits. In IS deposits, the thickness of the veins can persist to greater depth where they are typically base-metal rich (Fig. 4).

Vein Stratigraphy and Paragenesis

IS and LS veins, as already noted, formed as a result of several mineralization stages that may be different from each other in internal structure, texture, mineralogy, metal contents and ratios, fluid chemistry, sources for mineralizing fluids, and environment and mechanisms of deposition. Pre-ore stages can be common, as in the Tayoltita, Pachuca–Real del Monte, and Guanajuato districts, and are generally silica-rich, low grade, and volumetrically

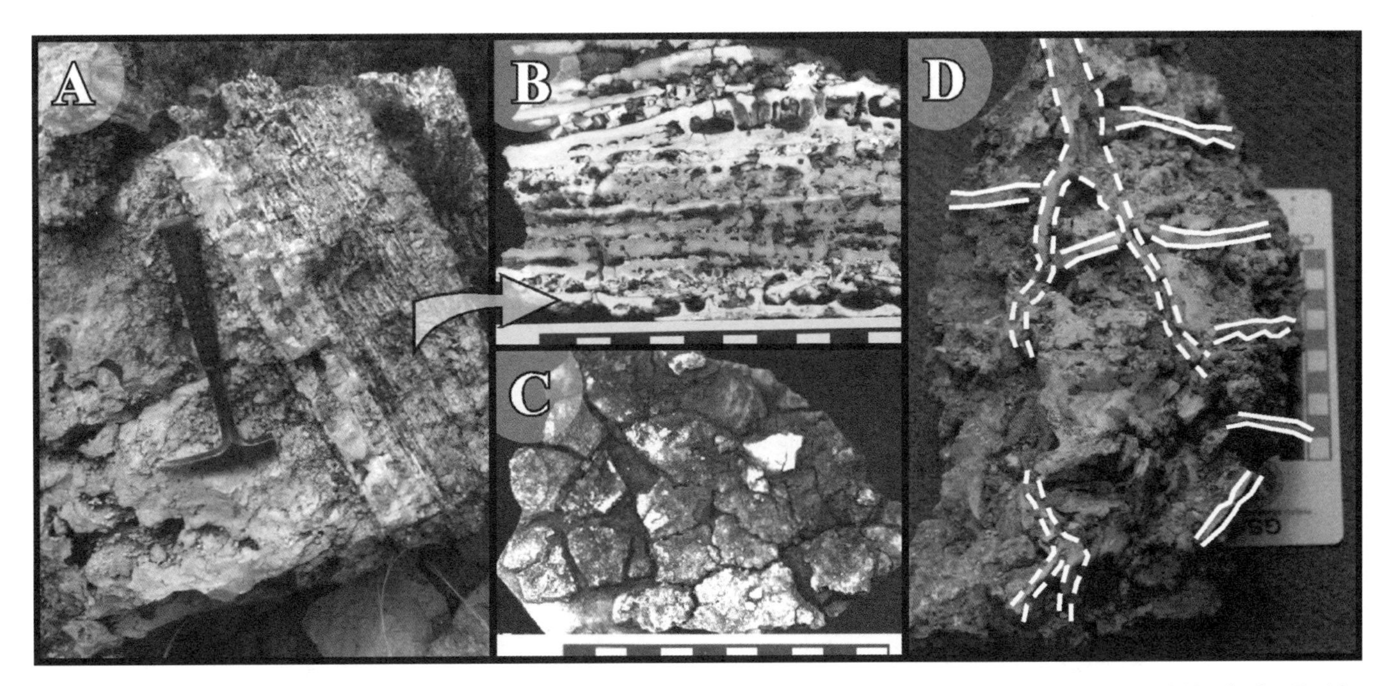

Figure 5. Photographs of sedimentary and vein structures in a hydrothermal sinter deposit at the Cerro el Uno, Ixtacamaxtitlán district, Puebla. (A) Massive and laminated opal. Courtesy of Jordi Tritlla. (B) Detail of opal laminations. (C) Desiccation cracks in an opal layer, evidence for precipitation on a paleosurface and in subaerial conditions or under an intermittent shallow pond. Scale in centimeters. (D) Opal layers, brecciated and recemented by later veins (dashed lines). Scales are in centimeters.

unimportant. Productive stages can be the first, as in the Sombrerete, Temascaltepec, and San Martín de Bolaños districts, and can be sulfide-rich. At the Pachuca–Real del Monte district, however, the second productive stage is sulfide-rich (Dreier, 2005). Although a single important stage of ore formation can lead to important deposits (i.e., Sombrerete, Pachuca–Real del Monte), multistage ore-forming events lead to higher grade deposits such as Fresnillo, where at least three distinct productive stages have been identified (Gemmell et al., 1988). In IS deposits, the productive stages tend to be sulfide rich (>5%) and quartz poor, whereas in LS deposits, the productive stages tend to be sulfide poor (<1%) and rich in quartz and adularia. Although bladed calcite is commonly an important component of LS deposits, it is usually not intimately associated with ore minerals.

The main ore stages are commonly polyphase multibanded and brecciated in both IS and LS ores (Fig. 6). Individual bands

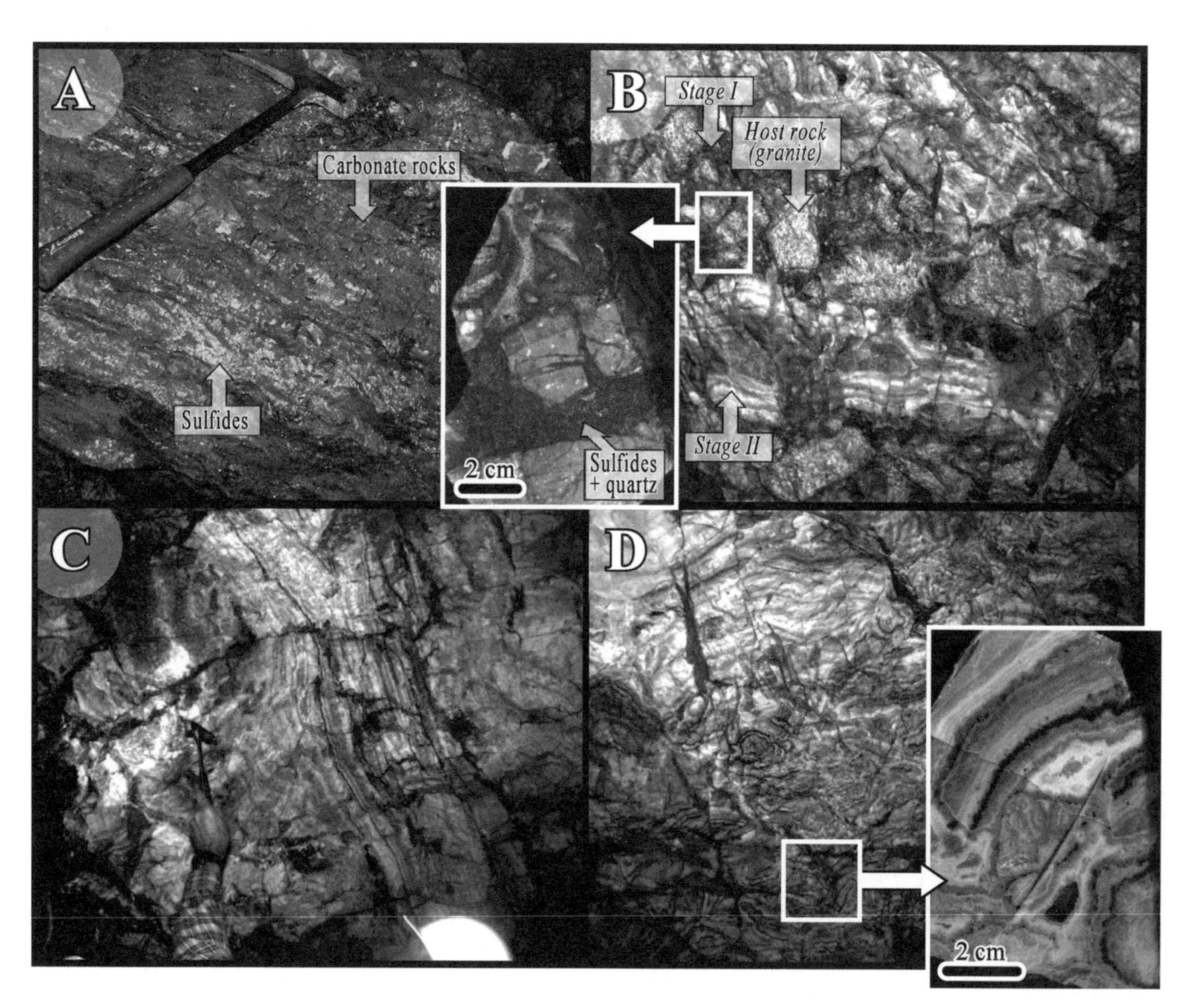

Figure 6. (A) Base metal–rich intermediate sulfidation mineralization hosted by carbonate rocks. El Cobre–Babilonia vein system, Taxco district, Guerrero. (B) Early stage of epithermal mineralization (Stage I), formed by a breccia with propylitized fragments of host granitic rocks, with a matrix of quartz and base metal sulfides, cut by a later stage of mineralization (Stage II), formed by symmetrical bands of crustiform crystalline and microcrystalline quartz. La Guitarra mine, Temascaltepec district, State of México. Courtesy of Noel White. The hand sample (detail) contains fragments of propylitized host granitic rocks and a matrix made up of quartz, base metal sulfides and fluorite. (C) Quartz crustiform bands, corresponding to Stage II of the previous picture. La Guitarra mine, Temascaltepec district, State of México. Courtesy of Noel White. (D) Quartz crustiform brecciated bands with Ag-Au–rich mineralization (black bands). La Guitarra mine, Temascaltepec district, State of México. Courtesy of Noel White. Detail: mineralization in crustiform and brecciated bands and a late band grown on them. Its mineralogy consists basically of cycles of recrystallized opal, chalcedony, jigsaw and crystalline quartz bands, with polymetallic mineralization in the black bands. Notice the occurrence of ore bands both in clasts and in the banded matrix. Scale in centimeters. Stage of mineralization IIB (see Camprubí et al., 2001a).

may contain opal, chalcedony, microcrystalline anhedral to subhedral quartz, and coarse-grained euhedral quartz (including amethyst). Mineralization is generally intimately associated with bands of microcrystalline anhedral quartz (jigsaw quartz), which is interpreted to represent recrystallized originally amorphous silica (Figs. 7E, 7F, 8D, and 8E). Brecciation can be either pre-ore or post-ore, and may be due to structural reactivation or related to the presence of hydrothermal explosion breccias in the shallow epithermal environment (Hedenquist and Henley, 1985). In the existing literature, mineralization is commonly said to be associated with "chalcedony," actually meaning microcrystalline quartz. The term "chalcedony" is a misnomer because true chalcedony is a specific type of a microcrystalline variety of quartz, with a very distinctive extinction pattern under polarized light (Figs. 7A, 7B and 7E). Ore minerals (for reasons unknown) are rarely found in intimate association with primary chalcedony in Mexican epithermal deposits. Ore bands in LS deposits are typically black discrete bands named *ginguro* (an old name used by Japanese miners, the use of which is now borderless) or black ore that consists of fine-grained mixed base-metal sulfides, silver minerals (argentite, aguilarite, naumannite, and sulfosalts) and gold or electrum that are intimately associated with jigsaw quartz and/or adularia (Figs. 7F, 8D, and 8E). Ore associations may be found either slightly after the deposition of minerals that are the products of boiling (Fig. 9C), as in the El Barqueño deposit (Camprubí et al., 2006a), or closely linked to such evidence, as in the El Compás vein in the El Orito vein system in Zacatecas where adularia-rich bands occur before or after the ginguro, which also contain adularia (Figs. 9E and 9F), providing a link between ore formation (the products) and boiling (the process). It is an important feature of LS deposits that the discrete ginguro bands concentrate greater than 90% of the precious metal values of the deposits, with grades running in the hundreds and thousands of ppm Au, in sharp contrast to the greater portions of the veins, which consist of abundant multibanded platy calcite, silica phases, and adularia, generally carrying only background gold values (<1–3 ppm Au). Thus, it is the presence of ginguro bands that actually renders these deposits of economic or bonanza character.

Late stages can typically be represented by (1) crystalline milky quartz to amethyst, and/or by (2) blocky calcite. Stages containing crystalline milky quartz to amethyst are usually barren and can be relatively hot, with fluid inclusion data typically displaying higher homogenization temperatures and lower salinities than earlier productive stages, as shown in Tayoltita (Smith et al., 1982), Sombrerete (Albinson, 1988), the San Nicolás vein in Guanajuato (Abeyta, 2003), and stage IIC at the La Guitarra deposit in Temascaltepec (Fig. 6C; Camprubí et al., 2001a). The origin for such a late barren quartz stage is uncertain, but may be related to upwelling of mostly meteoric fluids that are devoid of metals in solution at depth and may reflect the emplacement of late barren magmas. This stage may be the only phase present in the higher levels (above orebodies) of epithermal deposits, as in the surface expression of the Cedral vein (San Dimas District, Durango) and may be regarded as characteristic of the "tops" of orebodies in applicable districts. Thus, regardless of origin, the presence of this late relatively "hot" stage of barren coarse crystalline quartz can represent an important guide to explorationists in that in some epithermal deposits it can overlie earlier productive vein stages at depth. Final barren calcite "invasion" is generally represented by calcite with both blocky and scarce bladed textures, in crosscutting veins and veinlets, or in vugs with dog-tooth crystal terminations. The late calcite invasion has been interpreted as a product of the collapse of shallow steam-heated carbonate-rich waters (Simpson et al., 2001), either during the final collapse of the hydrothermal system, as in Sombrerete (Albinson, 1988; Rentería, 1992) and the SE portion of the Temascaltepec district (Camprubí, 2003), or during periods of interstage upwelling hydrothermal quiescence, as in Fresnillo (Simmons et al., 1988). This calcite stage is significant in that it reflects the presence of high-level carbonate-rich steam-heated waters and consequently the existence of upwelling boiling neutral-pH chloride fluids at depth. In other cases, lack of late barren stages does not imply absence of a productive ore deposit. For example, in the La Guitarra deposit in the Temascaltepec district, which lacks the late barren blocky calcite "invasion," there is a weak presence of late barren coarse-grained amethyst, and the last stage is represented by silver-bearing sulfosalts in veinlet swarms or stockworks (stage III, Fig. 8C; Camprubí et al., 2001a).

PETROGRAPHY AND MINERAL ASSOCIATIONS

Metallic Minerals

The IS and LS epithermal deposits in México that display more complex ore mineralogy and mineral associations are polymetallic deposits. A characteristic sequence of mineral precipitation as observed in the La Guitarra deposit of the Temascaltepec district in the State of México (Camprubí et al., 2001a) indicates the following metallic mineral associations (Figs. 10A, 10B, and 10D): (1) base metal sulfides, starting with pyrite (and/or arsenopyrite in other deposits), followed by sphalerite (generally rich in FeS), galena, and chalcopyrite, which may contain late argentite-acanthite; (2) Cu ± Ag sulfosalts (generally Ag-tetrahedrite-tennantite minerals), Ag sulfosalts (proustite-pyrargyrite, pearceite-polybasite, stefanite-arsenostefanite, miargyrite, billingsleyite); and (3) Ag-Pb sulfosalts (diaphorite, andorite, freieslebenite, ramdohrite). Portions that match the above evolution have been described in other Mexican deposits, such as Pachuca–Real del Monte (Geyne et al., 1963; Dreier, 2005), Guanajuato (Petruk and Owens, 1974; Vassallo, 1988; Randall et al., 1994), Zacatecas (Ponce and Clark, 1988), La Colorada (Chutas and Sack, 2004), and Fresnillo in Zacatecas (Gemmell et al., 1988; Ruvalcaba-Ruiz and Thompson, 1988), Topia (Loucks et al., 1988) and Velardeña in Durango (Gilmer et al., 1988). Such mineral associations may occur either as passive or as reactive successions. Intimate associations between electrum, native gold or silver, and Ag-tetrahedrite-tennantite or pyrite are commonly observed. In IS and LS epithermal deposits in México, the occur-

Figure 7. Photomicrographs of various silica textures from Mexican epithermal deposits. (A) Plumose primary textures of chalcedony in banded silica from the La Guitarra deposit (Temascaltepec district, State of México), crossed nicols. (B) Primary chalcedony in banded silica from the Altamira vein (La Ciénega district, Durango), crosscut by fine-grained jigsaw quartz through late veinlets, crossed nicols. (C) Bladed calcite crystals replaced by quartz (along basal pinacoids) and interstitial chalcedony from the Altamira vein (La Ciénega district, Durango), plane-polarized light. (D) Replacement textures of bladed calcite by quartz, with scarce calcite relicts, from the La Guitarra deposit (Temascaltepec district, State of México), crossed nicols. (E) Medium-grained jigsaw quartz due to the recrystallization of amorphous silica, crosscut by a chalcedony veinlet, from the El Compás vein, El Orito vein system (Zacatecas district, Zacatecas), crossed nicols. (F) Fine-grained jigsaw quartz due to the recrystallization of amorphous silica in gold-rich bands, from the El Compás vein, El Orito vein system (Zacatecas district, Zacatecas), crossed nicols.

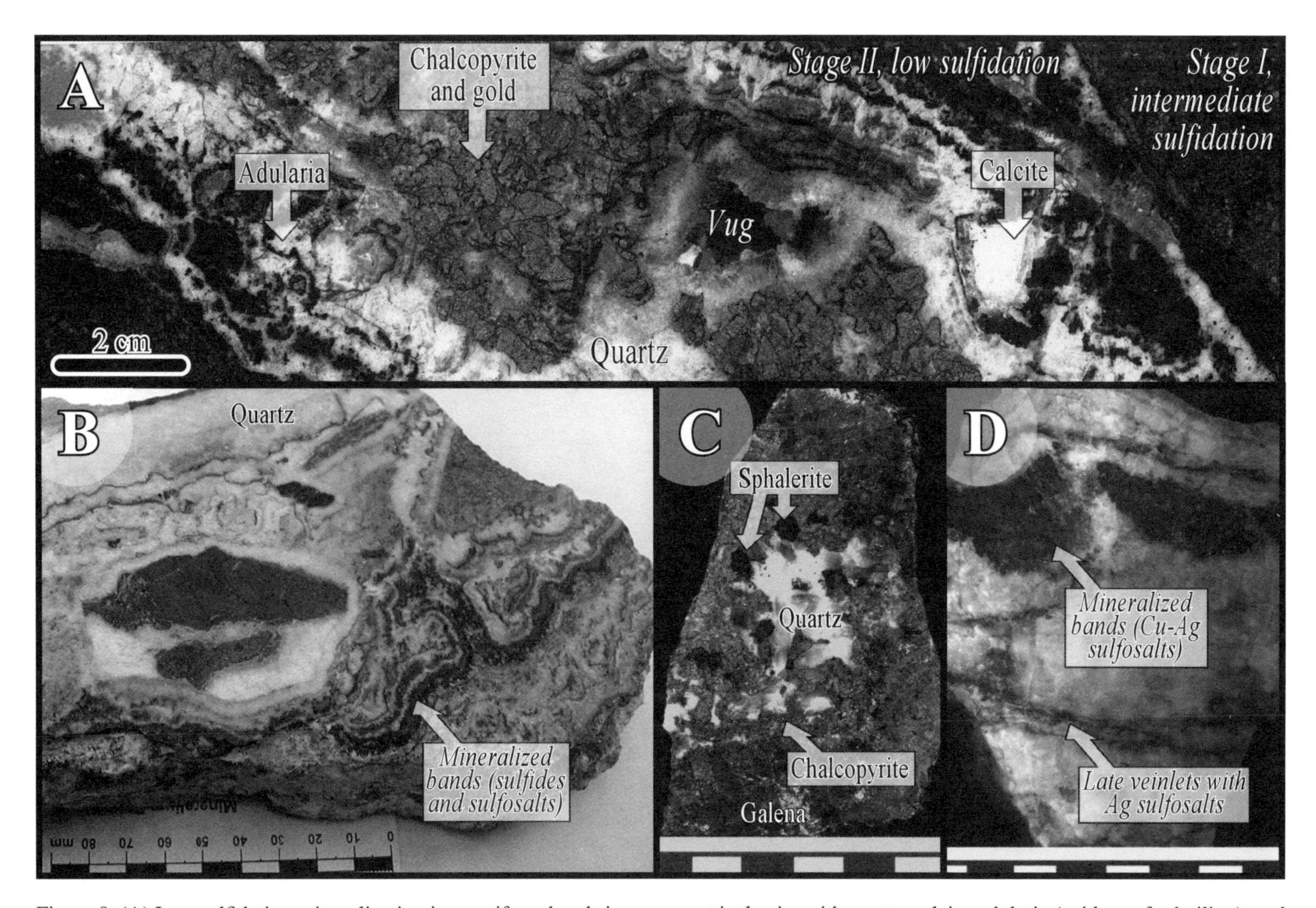

Figure 8. (A) Low sulfidation mineralization in crustiform bands in a symmetrical vein, with quartz, calcite, adularia (evidence for boiling), and an association of chacopyrite and gold. Azteca II vein, El Barqueño district, Jalisco. Courtesy of Eduardo González-Partida. (B) Banded intermediate sulfidation mineralization. Scale in centimeters. Santo Niño vein, Fresnillo district, Durango. (C) Mineralization in base metal sulfides with replacement mantos in carbonate rocks from epithermal veins. Scale in centimeters. Hueyapa vein, Taxco district, Guerrero. (D) Polymetallic mineralization in crustiform bands, and late veinlets. Scale in centimeters. Stages of mineralization IIB and III (see Camprubí et al., 2001a) at the La Guitarra mine, Temascaltepec district, State of México.

rence of enargite, a characteristic mineral of the HS environment, has only been reported in San Felipe in Baja California (Ibarra-Serrano, 1997) and Temascaltepec in the State of México (Camprubí et al., 2001a), as a late mineral to Ag sulfosalt associations. In the case of Se-rich mineral associations, with the occurrence of aguilarite or naumannite as in Guanajuato (Petruk and Owens, 1974; Vassallo, 1988) and San Martín in Querétaro, these minerals usually occur associated with electrum, argentite, and silver sulfosalts (like pearceite-polybasite) after the associations of base metal sulfides with or without argentite. In the Compás vein of the El Orito district in Zacatecas, ore minerals are aguilarite, naumannite, electrum, and silver and mercury halides (iodargyrite, capgaronnite) of suspected primary origin (Albinson, 2006).

In the few HS epithermal deposits described to date in México, the reported ore mineral associations are quite typical for this type of deposit elsewhere, with gold or Ag-bearing gold (namely electrum), pyrite, enargite, tetrahedrite-tennantite, sphalerite, famatinite, luzonite, bismutinite, bornite, digenite, and covellite, among other minerals (Fig. 10C; Sellepack, 1997; Staude, 2001; Valencia et al., 2003; Valencia, 2005). In these deposits, the ore-bearing associations are hosted in cores of vuggy quartz bodies in high-grade veins and veinlets mainly formed by pyrite, enargite, and gold, or barite, gold, and Bi- and Te-minerals, among other minor associations, or are also associated with breccias and advanced argillic alteration fronts, as disseminations (Gray, 2001), where the occurrence of veins is usually representative of the last pulse of mineralization (Charest and Castañeda, 1997; Staude, 2001).

Silica Petrography

The main and most conspicuous gangue mineral in epithermal ore deposits is quartz, and the petrography of silica minerals (i.e., the variety of silica mineral phases in the ore deposits) may be qualitatively diagnostic of formation temperatures, the geological level where deposition was produced, as well as of several

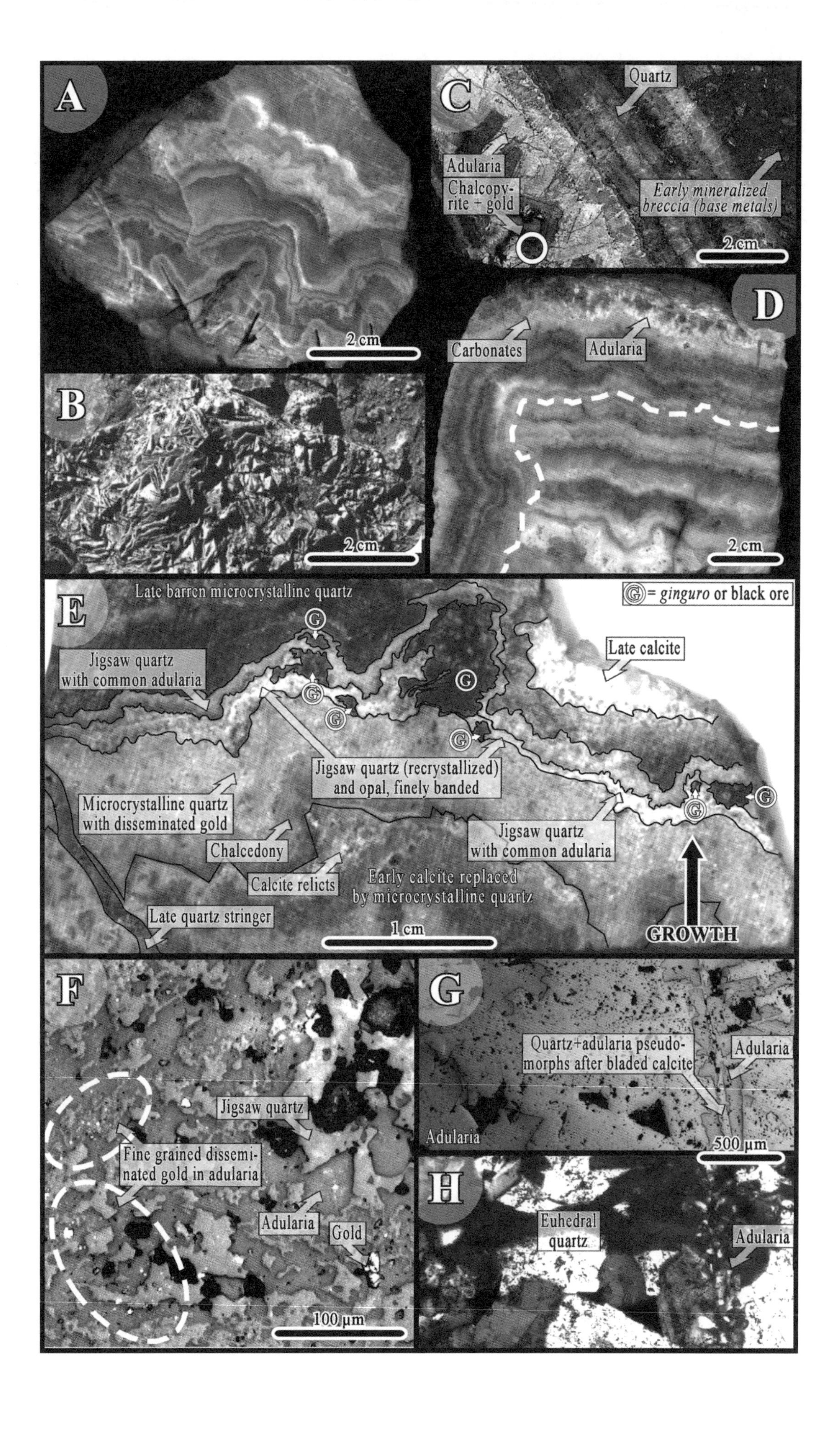

A
2 cm
B
2 cm
C
Quartz
Adularia
Chalcopy-
rite + gold
Early mineralized
breccia (base metals)
2 cm
D
Carbonates
Adularia
2 cm
E
Late barren microcrystalline quartz
= ginguro or black ore
Late calcite
Jigsaw quartz
with common adularia
Jigsaw quartz (recrystallized)
and opal, finely banded
Microcrystalline quartz
with disseminated gold
Chalcedony
Jigsaw quartz
with common adularia
Calcite relicts
Early calcite replaced
by microcrystalline quartz
Late quartz stringer
1 cm
GROWTH
F
Jigsaw quartz
Fine grained dissemi-
nated gold in adularia
Adularia
Gold
100 µm
G
Quartz+adularia pseudo-
morphs after bladed calcite
Adularia
Adularia
500 µm
H
Euhedral
quartz
Adularia

geological processes associated with hydrothermalism. Luckily, the identification of silica mineral phases in epithermal deposits is petrographically simple under conventional transmitted polarized-light optical means, but a correct petrographical observation of these minerals is crucial. Chalcedony exhibits fibrous extinction perpendicular to growth bands (Fig. 7A). Feathery or plumose quartz exhibits an optical extinction in the shape of feathers. The appearance of mosaic quartz, also known as "jigsaw" quartz (Figs. 7E and 7F), either fine- or coarse-grained, is precisely a mosaic or a jigsaw of anhedral to subhedral quartz grains. Crystalline quartz typically occurs as euhedral prismatic crystals, with hexagonal sections. In order to gain familiarity with the different terms and the detailed descriptions of textures and mineralogical species of silica, particularly focused on epithermal deposits, we refer to specific publications in the genre. In the most general paper of the subject, the authors reviewed the textures and types of quartz in several Australian epithermal deposits (Dong et al., 1995). The example of Sleeper in Nevada was also important to the description and interpretation of textures derived from the recrystallization of amorphous silica in LS epithermal deposits (Saunders, 1996). The detailed petrographic description of a sample from the Veta Grande vein in the Zacatecas district proposed that the apparent complexity of veins formed by multiple silica banding may be explained by means of a relatively simple mechanism of cracking and sealing during the formation of veins (Albinson, 1995). The distribution of different silica phases within a vertical profile in the Broadlands geothermal system in New Zealand was described by Simmons and Browne (2000), and in epithermal deposits in México by Albinson et al. (2001).

The following describes the most common silica textures and how they are interpreted. Mosaic or jigsaw quartz originally precipitated as amorphous silica, typically colloform, that recrystallized into a texture with a pattern similar to a jigsaw, since it is a common feature derived from recrystallization to produce anhedral interpenetrative crystals (Fig. 7E). It is also common that amorphous silica exhibits dendritic textures, reflecting the supersaturation of a fluid in silica and its sudden precipitation (rapid growth) from a boiling fluid (Saunders, 1996). This type of quartz is dominantly found in the shallow parts of the epithermal environment (<500 m to the paleosurface) and is the most abundant type in LS epithermal deposits. Crystalline quartz is also present in the shallow geological environment, though it precipitated typically from low-temperature hydrothermal fluids, and contains fluid inclusion associations (FIA's) that are texturally "inmature" (Bodnar et al., 1985). Conversely, Cordilleran hydrothermal deposits formed at greater depths (porphyry copper deposits typically form at depths >2 km, and the deepest portions of IS epithermal deposits form at ~1 km deep) and contain preferentially coarse-grained crystalline quartz, with abundant "mature" fluid inclusions that were trapped under higher temperatures and reflect more rapid kynetics for the fluid inclusion healing process (Bodnar et al., 1985). The epithermal environment at intermediate depths (500–1000 m below the paleosurface) is characterized by the occurrence of a mixed-bag of petrographical silica mineral styles, including jigsaw quartz, chalcedony, feathery quartz, and different styles of crystalline quartz. The multiple repetition of ordered sequences of such petrographical styles of silica, observed in many LS and IS epithermal deposits in México (see Camprubí et al., 2001a, 2001b) and elsewhere, likely reflects successive sealing and opening of the various faults that control vein formation (Albinson, 1995; Sibson, 2001). When open spaces are close to sealed spaces, mineralizing fluids flow slowly and allow crystalline quartz to precipitate. In contrast, when the host structures are reactivated and reopened, presumably by means of seismic activation, phase separation or sudden boiling is favored in upwelling hydrothermal fluids, producing a supersaturation of silica in the fluids and the stabilization of amorphous silica varieties (Simmons and Browne, 2000). Subsequently, as the sealing process advances, chalcedony and plumose quartz precipitate as intermediate silica phases between the initial amorphous silica varieties and the latter crystalline prismatic quartz formed during the final stage of restricted circulation of the fluid. When the process of reopening and sealing of

Figure 9. (A) Barren mineralization with quartz crustiform bands, with phantoms of bladed calcite at its base. Scale in centimeters. Stage of mineralization IIA (see Camprubí et al., 2001a) at the La Guitarra mine, Temascaltepec district, State of México. (B) Bladed calcite phantoms. Peña de Oro vein, El Barqueño district, Jalisco. Courtesy of Eduardo González-Partida. (C) Mineralization in crustiform bands, with quartz, illite-smectite and chlorite, adularia, and an association of chalcopyrite and gold formed immediately on top of adularia crystals. The occurrence of adularia is evidence for boiling mineralizing fluids, and its close association with metallic minerals supports the idea that such mechanism could be responsible for ore precipitation. Azteca II vein, El Barqueño district, Jalisco. (D) Barren stage of mineralization with crustiform and brecciated silica bands cemented by new silica bands. The contour of the clast is marked with a white dashed line. Its mineralogy consists basically in cycles of recrystallized opal (jigsaw quartz), chalcedony, and quartz bands, with carbonates (siderite and ankerite) and adularia. Scale in centimeters. Stage of mineralization IIA (see Camprubí et al., 2001a) at the La Guitarra mine, Temascaltepec district, State of México. (E) Detail of quartz, adularia, and ore bands in a low sulfidation epithermal deposit where the ore is found as patches containing gold and naumannite within adularia-bearing bands. Notice that adularia occurs either in pre-ore bands, post-ore bands, or in ore-bearing bands. Also notice the intimate interbanding between the black ore patches (or ginguro [G]) and the adularia-rich and opal-rich bands. The gray spots within the ginguro are gold grains. The ore patches possibly deposited as ore-bearing silica gel drops, as interpreted from their mammillary appearance. El Compás vein, El Orito vein system, Zacatecas district, Zacatecas. (F) Part of the ginguro in (E) showing the intimate petrographic relationship between adularia and gold, providing evidence that both precipitated from boiling fluids. Notice the distinctive pseudorhombohedral sections of adularia crystals. Plain reflected light. El Compás vein, El Orito vein system, Zacatecas district, Zacatecas. (G) Part of the ore-poor adularia-rich bands formed before the ginguro in (E), illustrating the formation of bladed calcite prior to adularia. Plain reflected light. Notice the distinctive pseudorhombohedral sections of adularia crystals. El Compás vein, El Orito vein system, Zacatecas district, Zacatecas. (H) Same as previous, under transmitted polarized light and crossed nicols.

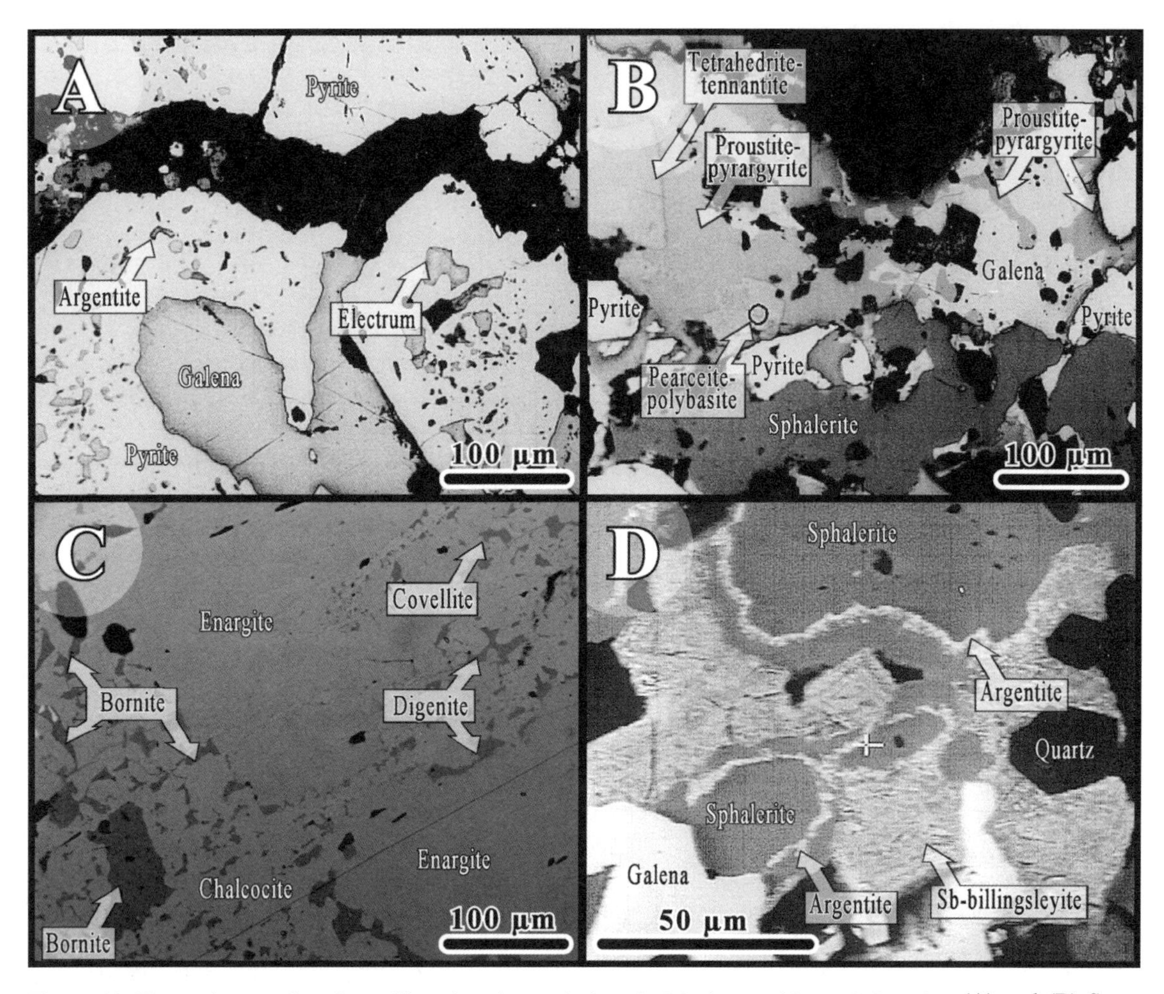

Figure 10. Photomicrographs of metallic mineral associations in Mexican epithermal deposits. (A) and (B) Stage IIB ore bands at the La Guitarra deposit (Temascaltepec district, State of México). Plain reflected light. (C) Copper mineral association at the La Caridad Antigua (Sonora) high sulfidation deposit. Plain polarized light; picture courtesy of Víctor Valencia. (D) Backscattered electron image of early vug filling in the silver mineral assemblage of Stage I at the La Guitarra deposit, Temascaltepec district, State of México, at sample current of 18.3 nA and accelerating voltage of 20 keV.

structures records evidence for the complete sequence, the silica phases show the following order of deposition: (1) amorphous silica (typically dendritic), (2) chalcedony, (3) plumose quartz, and (4) coarse crystalline quartz crystals, with euhedral crystal terminations in the remnant open spaces or vugs (comb quartz) that, subsequently, may be overgrown by amorphous silica corresponding to the next cycle of opening and sealing by silica precipitation. The identification of mosaic quartz in the deepest portions of some epithermal deposits indicates that the precipitation of amorphous silica, though in subordinate amounts, also occurs at significant depths (~1000 m) in the crust.

In conclusion, the association of silica phases found in epithermal deposits can provide diagnostic petrographic evidence indicative of the relative geologic level at which the deposits are exposed (i.e., Albinson et al., 2001). Typically characteristic of silica deposition in the root zones or deep portions of epithermal deposits are (1) the dominant occurrence of coarse crystalline quartz containing fluid inclusions that are petrographically "mature" (trapped at relatively high temperatures, >250 °C), and (2) the occurrence of very subordinate amounts of cryptocrystalline quartz to amorphous silica. In contrast, typically characteristic of silica deposition in the shallow portions of epithermal deposits are (1) the dominant occurrence of criptocrystalline quartz to amorphous silica, and (2) the occurrence of subordinate amounts of coarse crystalline quartz containing fluid inclusions that are petrographically "inmature" (trapped at relatively low temperatures, <200 °C) (Albinson et al., 2001). Thus, fluid inclusion-related studies (microthermometry, stable isotope, He isotope, volatile, and crush-leach studies) can benefit from detailed descriptions of the petrography of associated silica minerals, since both tools are complementary and help to reconstruct the crustal level at which the epithermal deposits formed.

Clay Mineralogy

Clay alteration in IS epithermal deposits shows evidence for the occurrence of HS-type fluids and further contributes to their separation from the LS deposits as an independent type of epithermal deposit. Analyses of clay minerals in some Mexican IS deposits indicate the occurrence of hypogene kaolinite in the deepest portions of these deposits, in contrast to the high-level epithermal environment, where the circulation of acid fluids generated in a steam-heated environment can also form kaolinite. A deep zone with kaolinite and dickite in the Zuloaga vein of the San Martín de Bolaños district in Jalisco (Fig. 11; Albinson and Rubio, 2001) presumably reflects the incursion of relatively low-pH hypogene upwelling fluids. The La Guitarra deposit of the Temascaltepec district in the State of México sporadically exhibits hypogene kaolinite at deep portions of the veins. The El Herrero vein of the Bacís district in Durango exhibits an early stage of hydrothermal activity that formed a hypogene alteration mineral assemblage of quartz-pyrite-kaolinite. This mineral association is restricted in space to the deepest part of the mine (level 18), coincident with the root zones of the precious metal-bearing ore horizon and represents a stage of acid hypogene alteration that predates the formation of precious metal–bearing mineral associations. This constitutes further evidence for the occurrence of early, relatively acid, fluids in the deep portions of IS epithermal deposits. Furthermore, in the Pachuca–Real del Monte district, the Purísima–Colón vein system exhibits kaolinite intimately associated with the ore stage event that formed from a distinctly more acid fluid (pH between ~3.2 and 4) in contrast to pre-ore and post-ore vein forming events, which reflect the presence of near neutral (pH between ~5.5 and 6.1) fluids during the formation of barren vein stages (Dreier, 2005). The early incursion of fluids that generated advanced argillic alteration assemblages in IS epithermal deposits suggests an evolution similar to HS epithermal deposits, which exhibit deep early acid fluids responsible for the development of advanced argillic alteration and a vuggy or massive silica core, and later evolved neutral pH fluids, essentially more IS-like, which are responsible for metallic mineralization.

GEOTHERMOMETRY OF METALLIC MINERAL ASSEMBLAGES

Metallic mineral assemblages can provide important information about the physicochemical conditions (such as temperature) for ore deposition as well. Temperature determinations are very useful in that they are obtained in mineral associations where it is hard to find any fluid inclusions for microthermometry unless an infrared source is used, because all these minerals are opaque. In Mexican deposits, only a few fluid inclusions in proustite-pyrargyrite have been analyzed under plane-polarized light at the Santo Niño vein in Fresnillo (Simmons, 1991). Temperature estimations were obtained from some associations of metallic minerals at the La Guitarra deposit (Camprubí et al., 2001a) that were congruent with fluid inclusion microthermometry and were used to obtain a more complete thermal history of the deposit (Camprubí et al., 2001b).

According to the miargyrite-smithite/trechmannite phase diagram of Ghosal and Sack (1995), the As/(As + Sb) ratio in miargyrites can be used to estimate the minimum formation temperature. These temperature estimations are obtained plotting the highest As/(As + Sb) value of each sample (obtained by electron probe microanalysis) onto the α-miargyrite-trechmannite curve. Thus, miargyrite compositions at the La

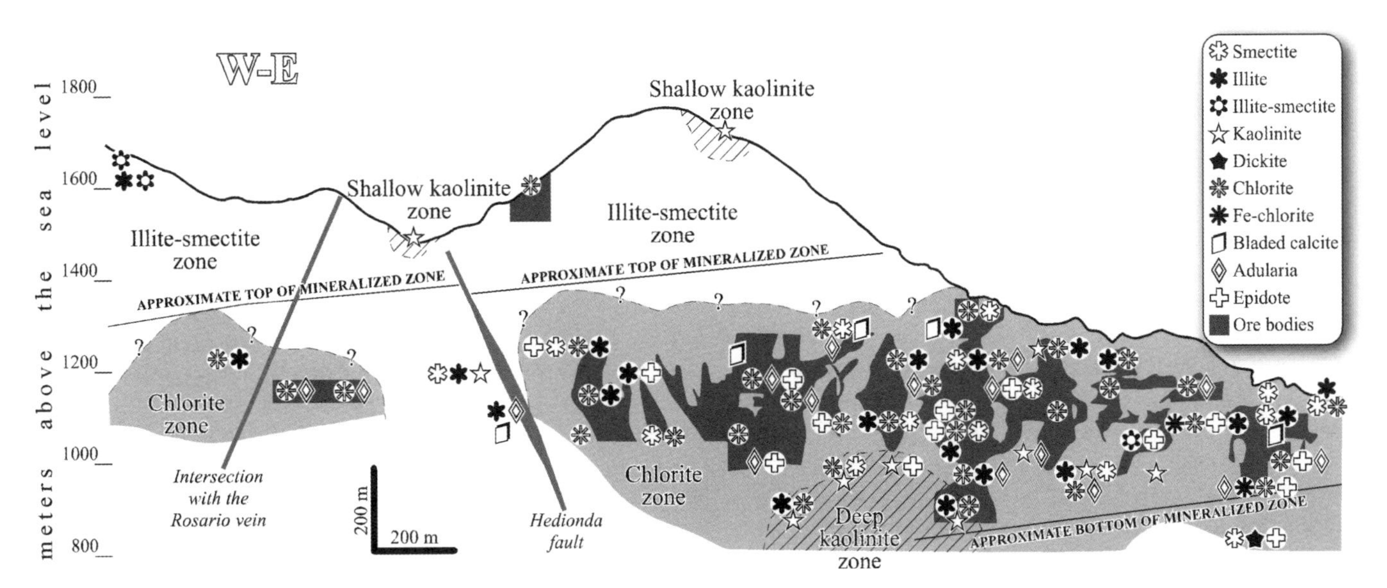

Figure 11. Longitudinal section of the Zuloaga vein of the San Martín de Bolaños district, Jalisco, showing the occurrence of alteration and key gangue minerals using X-Ray Diffraction and Short-Wave Infra-Red Spectroscopy. Modified from Albinson and Rubio (2001). Notice the occurrence of a deep kaolinite zone without connection with shallow kaolinite zones.

Guitarra deposit (Camprubí et al., 2001a) indicate minimum equilibration temperatures of ~170 °C in stage I, 120 °C in stage II, and 180 °C to 240 °C in stage III.

Based on the complex silver mineral parageneses at the La Guitarra deposit, and considering the lack of evidence of disequilibrium or recrystallization, three mineral geothermometers are used: (1) pyrargyrite-pyrostilpnite dimorphic change at 192 ± 5 °C (Keighin and Honea, 1969); (2) proustite-xanthoconite dimorphic change at 192 ± 10 °C (Hall, 1966 and 1967); and (3) stephanite and argentite plus pyrargyrite, or stephanite and argentite plus Sb-billingsleyite equilibria, both at 197 ± 5 °C (Keighin and Honea, 1969). Experimental studies (Keighin and Honea, 1969) have determined that stephanite breaks down when heated at 197 ± 5 °C. The "high-temperature" products of the stephanite breakdown may be argentite plus pyrargyrite, or pyrargyrite plus Sb-billingsleyite in the presence of excess sulfur. Because the mineral assemblage found in stage I of the La Guitarra deposit (Camprubí et al., 2001a) is argentite + Sb-billingsleyite → proustite-pyrargyrite → xanthoconite-pyrostilpnite, no sulfur activity constraints can be drawn from this association. However, this shows a decreasing trend in temperature during mineral deposition.

At the Colorada deposit in Zacatecas (Chutas and Sack, 2004), compositions of the assemblages tetrahedrite-tennantite + proustite-pyrargyrite + sphalerite and tetrahedrite-tennantite + pearceite-polybasite solid solution + $(Ag,Cu)_2S$ solid solution + sphalerite encapsulated in quartz and sphalerite indicate a primary depositional temperature of ~325 °C for metallic mineral associations, using the compositions of tetrahedrite-tennantite minerals and the isotherms calculated by Sack (2000 and 2005) for the Zn/(Zn + Fe)-Ag/(Ag + Cu) compositional field of these minerals. This temperature is congruent with fluid-inclusion data at higher elevations in the mine, where slightly lower temperatures were documented (Albinson, 1988).

FLUID INCLUSION MICROTHERMOMETRY

Albinson et al. (2001) categorized alkaline/neutral (LS + IS) epithermal deposits in México according to the metal contents of the deposits and the ratios between them, by means of which we may distinguish among Zn-Pb-Ag, Ag-Zn-Pb, Ag + Au + Pb + Zn + Cu, Ag-Au, and Au-Ag deposits. These distinctive metal associations correlate broadly with the salinities obtained from fluid inclusion studies carried out in 50 different deposits throughout the country. The precious metal–rich deposits contain fluid inclusions that display much lower salinities than the deposits that are richer in base metals or polymetallic deposits. Salinities in inclusion fluids range between 0 and ~6 wt% NaCl equiv., and between ~2 and ~20 wt% NaCl equiv., in precious metal–rich and polymetallic deposits, respectively (ranges determined after data collected from many sources in Albinson et al., 2001).

The pulsating character of epithermal deposits is reflected not only macroscopically by their complex internal structure (Figs. 6 and 10B–10D) but is also displayed microscopically, even down to the level of mineral associations and to the level of single crystals. This is demonstrated by salinity changes in fluid inclusions that can be correlated with compositional changes in the zonation of a crystal. For instance, in sphalerite crystals from Temascaltepec in the State of México (Camprubí et al., 2001b), whose composition evolved progressively toward higher FeS contents, the change in composition is accompanied by a decrease in salinity of fluid inclusions, consistently zone by zone, that presumably corresponds to discrete brine pulses. Generally speaking, the fluid inclusions analyzed in metallic minerals (sphalerite, pyrargyrite) show higher salinities than fluid inclusions analyzed in the gangue minerals associated with them (Simmons, 1991; Albinson et al., 2001; Camprubí et al., 2001b). Furthermore, the inclusion fluids from mineralization stages with dominant Zn-Pb-Cu-(Ag) mineral associations generally have higher salinities than the inclusion fluids from mineralization stages with dominant Au-Ag mineral associations, and also higher salinities than inclusion fluids in barren stages of the same deposit. This was observed, in some cases, even independently of the host mineral chosen for fluid inclusion microthermometry (Simmons et al., 1988; Simmons, 1991; Camprubí et al., 2001b). Thus, the positive correlation between salinity and the relative content in base metals (with more or less silver) operates at different levels: (1) between different deposits, (2) between different stages of mineralization of the same deposit, and (3) even between different mineral associations of the same stage of mineralization.

The systematic study of fluid inclusions in an epithermal deposit provides estimations of the depth of formation of the many stages of mineralization or mineral associations, depending on the variation with depth of the boiling point of a liquid with a determined salinity (Haas, 1971). These estimations are more precise when it is possible to find fluid inclusion associations (FIAs) that display the convincing coexistence of liquid-rich inclusions and vapor-rich inclusions not associated to necking phenomena (i.e., Simmons, 1991). Although this condition is not commonly found, it represents fluid-inclusion petrographic evidence for the occurrence of boiling at discrete episodes during the lifetime of a paleohydrothermal system. These estimations have been carried out in a few IS and LS epithermal deposits in México; for example, at Real de Catorce in San Luis Potosí (Albinson, 1985), Fresnillo (Simmons et al., 1988), San Martín de Bolaños in Jalisco (Albinson and Rubio, 2001), San Martín in Querétaro (Albinson et al., 2001), Temascaltepec in the State of México (Camprubí et al., 2001b), El Barqueño in Jalisco (Camprubí et al., 2006a), and Pachuca–Real del Monte in Hidalgo (Dreier, 2005). These estimations can be validated with the occurrence of certain geologic features that, when preserved, are known to have formed at the surface or close to it, such as silica or carbonate sinters, alteration zones formed by acid leaching by steam-heated waters, etc. Even when the above shallow features are absent, characteristic mineralogical and petrographic evidence for boiling can be used to anchor a given mineral association on the boiling curve corresponding to the obtained apparent salinity (e.g., Albinson et al., 2001; Camprubí et al., 2001b, 2006a), and thus to obtain an estimation of the depth of formation of that association. Mineralogic and petrographic evidence for boiling are the occurrence of bladed calcite or its pseudomorphs (Figs. 6A, 6B, 7C, and 7D), the occurrence of pseudorhombohedral or pseudoacicular adularia (Figs. 9C and 9D), and the occurrence of truscottite (Browne and Ellis, 1970; Browne, 1978; Henley, 1985; Hedenquist, 1986, 1991; Cathles, 1991; Simmons and Christenson, 1994; Dong and Morrison, 1995; Izawa and Yamashita, 1995; Hedenquist et al., 2000; Simmons and Browne, 2000). Utilizing this evidence, it is possible to estimate at least the position of the paleowater table, if not the position of the paleosurface, during the formation of different mineral associations of an epithermal deposit. Most of the ore zones related to IS and LS deposits in México have been found to reconstruct at depths between 400 and 1000 m below the paleosurfaces (Albinson et al., 2001).

The correlation diagrams of homogenization and freezing point depression temperatures (or salinity) can describe trends of geologic processes, such as precipitation mechanisms. In epithermal deposits, the positive correlation between homogenization temperatures and salinities starting from the hotter and more saline fluid is commonly interpreted as evidence for fluid mixing. Otherwise, inclusion fluids that display cooling along with a slight increase in salinity are interpreted as brines left aside by boiling (commonly named "false brines"), whereas inclusion fluids that display cooling with no apparent change in salinity are interpreted to reflect conductive heat loss.

The distribution of salinities and homogenization temperatures along longitudinal sections of epithermal veins for specific stages can assist the identification of preferential upflow and feeder zones (i.e., Zuloaga vein in San Martín de Bolaños, Jalisco; Albinson and Rubio, 2001). Careful paragenetic control is important for interpreting fluid inclusion data on longitudinal sections since feeder zones change in position over time. The position of preferential feeder zones can also be delineated using the contours of Ag/Pb ratios in whole vein samples or Sb/(Sb + As) in tetrahedrite-tennantite (i.e., Camprubí et al., 2001a), preferentially coupled with fluid inclusion data. The use of fluid inclusion data for this purpose is a method not affected by oxidation processes and allows establishment of a penetrative sampling grid with a better control on time slices and relatively quick obtention of data that can provide vectors into feeder zones, compared with other techniques.

ISOTOPIC COMPOSITION AND ORIGIN OF SULFUR

The host rocks for epithermal deposits in México commonly have different origins, even at the scale of a district, namely magmatic (volcanic, subvolcanic, or plutonic), sedimentary, or metasedimentary rocks. Thus, the isotopic composition of S in the sulfides partly reflects a heterogeneous host rock lithology with its variable isotopic heritance. $\delta^{34}S$ of sulfides in IS and LS epithermal deposits in México are very similar, between 0 and −10‰, in deposits like Guanajuato (Gross, 1975), Fresnillo in Zacatecas (Ruvalcaba-Ruiz and Thompson, 1988), Sultepec, Miahuatlán (González-Partida, 1981), Temascaltepec in the State of México (González-Partida, 1981; Camprubí et al., 2001b), and Real de Ángeles in Aguascalientes (Pearson et al., 1988). The above deposits are hosted or partly hosted by sedimentary or metasedimentary rocks but, in the deposits where host rocks are mostly magmatic, like Velardeña in Durango (Gilmer et al., 1988), $\delta^{34}S$ values reflect magmatic sulfur compositions ~0‰. This is the case with the few HS epithermal deposits in México, with $\delta^{34}S$ values between −7 and 9‰ at El Sauzal in Chihuahua (Sellepack, 1997) and between −3 and 4 ‰ in sulfides and up to 18‰ in sulfates at La Caridad Antigua in Sonora (Valencia, 2005). Additionally, in Guanajuato and Temascaltepec, very or extremely negative isotopic compositions were obtained and thus, in both cases it was likely that the occurrence of sulfur was derived from bacterial sulfate reduction, along with the sources already noted. However, even though in epithermal deposits the sources for sulfur may be diverse, the general conclusion on the subject is that the contribution of magmatic sulfur is essential. That contribution can be produced directly through magmatic fluids released to the brittle crust by a cooling and crystallizing magma or, alternatively, by leaching of preexisting magmatic rocks. Thus, the fundamental contribution of sulfur to assist in the transport of metals to the epithermal environment during the formation of IS and LS deposits is akin to that corresponding to HS epithermal deposits, whose formation is unquestionably controlled by magmatic fluids and other components of the same origin, including sulfur.

ORIGIN OF FLUIDS

The common knowledge of transport and deposition of metals indicates that these are better transported toward their depositional environment by fluids rich in ligands that form chloride or bisulfide complexes in solution. Base metals and silver in the epithermal environment are preferentially transported as chloride complexes (i.e., Ruaya and Seward, 1986), and thus more metals are in solution with increased salinity. Evidence of higher salinities of hydrothermal fluids linked to mineralization is inclusion fluids in sphalerite or pyrargyrite that have consistently higher salinities than inclusion fluids in gangue minerals (Gemmell et al., 1988; Simmons, 1991; Albinson et al., 2001; Camprubí et al., 2001b). In each deposit it is important to evaluate the possible occurrence of evaporites or connate brines on the pathway of the upwelling hydrothermal fluids because, in the absence of these, high salinity fluids that do not correspond to brines left aside by boiling have been described as permissive evidence of magmatic brines (i.e., Gemmell et al., 1988; Albinson et al., 2001; Camprubí et al., 2001b).

Occasionally, chloride brines with anomalously high salinities have been described in IS and LS deposits, compared with other fluid inclusion data obtained in samples from the same deposit and even from the same mineral association (e.g., Camprubí et al., 2001b), regardless of whether the saline inclusion–bearing mineral associations are rich in base metals, gold, silver, or if they are barren. A possible explanation for the occurrence of such brines is that they correspond to residual liquids left aside by boiling, or "false brines," as documented both in fossil epithermal deposits (Sherlock et al., 1995; Scott and Watanabe, 1998) and in analogous geothermal systems where the process has been coined as a local effect of "boiling towards dryness" (Simmons and Browne, 1997).

Based on microthermometry, stable and He isotope, and volatile geochemistry studies in fluid inclusions (Simmons et al., 1988; Benton, 1991; Norman et al., 1991, 1997b; Simmons, 1991, 1995; Albinson et al., 2001; Camprubí et al., 2001b, 2003b, 2006b; Abeyta, 2003; Camprubí, 2003), the analyzed fluids in samples from IS and LS epithermal deposits in México represent a mixture of fluids with diverse origins, following the classification by Norman and Musgrave (1994) and Norman et al. (1997a): (1) shallow meteoric waters; (2) crustal, deeply circulated, evolved meteoric waters; (3) crustal organic waters; (4) magmatic fluids; and (5) evolved magmatic fluids, with higher or lower contents of one or another (Fig. 12). These diverse sources for mineralizing fluids can be identified either in metallic mineral–bearing associations, regardless of their contents in base or precious metals, or in barren portions of the deposits. Nonetheless, it is not less manifest that magmatic fluids are identifiable as major contributors of metallic mineral-bearing associations, either base metal– or precious metal–rich (Simmons et al., 1988; Albinson et al., 2001; Camprubí et al., 2003b, 2006b). Volatile species analyzed in fluid inclusions suggest that the crucial difference between barren and ore bands is the occurrence of ore-bearing fluids that chemically correspond to magmatic fluids, by the use of N_2/Ar ratios, the correlation between N_2/Ar and CO_2/CH_4 ratios, and the position within N_2-He-Ar and N_2-CH_4-Ar compositional fields (Albinson et al., 2001; Camprubí et al., 2003b, 2006b).

The geochemical analysis of volatiles contained in fluid inclusions has allowed a positive correlation of gas species of magmatic affiliation (such as N_2) with the H_2S contents of the fluid inclusions, suggesting that relatively high H_2S contents correspond with increases in the magmatic character of mineralizing fluids (Albinson et al., 2001; Camprubí et al., 2003b), both in individual deposits (i.e., the La Guitarra vein at Temascaltepec) and at a regional scale (Fig. 13). That said, considering that the most likely transport mechanism for gold in solution into the epithermal environment is by means of bisulfide complexes (e.g., Shenberger and Barnes, 1989; Hayashi and Ohmoto, 1991; Gammons and Williams-Jones, 1997) it is plausible or, at least, it cannot be discarded, that the metals deposited in the epithermal deposits may be provided by cooling magmas.

If we refer to Figure 12, it is noticeable that the San Nicolás vein in the Guanajuato district (Abeyta, 2003) exhibits mostly shallow meteoric compositions of volatiles in fluid inclusions. In contrast, volatiles in fluid inclusions from the veins at the Temascaltepec district (Camprubí, 2003; Camprubí et al., 2003b, 2006b) show a wide variety of sources for mineralizing fluids, and shallow meteoric fluids are comparatively underrepresented. At the La Guitarra deposit, the most significant magmatic contributions (represented by the highest N_2/Ar values in Fig. 12) were found in some fluids trapped in sphalerites from stage I (IS stage) and, most noticeably, in ore-bearing quartz bands of stage IIB and in the sulfosalt assemblages from stage III (LS stages). Fluids in sphalerites from stage I also show helium isotope values (R/Ra) consistent with magmatic fluids that were "contaminated" with crustal helium (Camprubí et al., 2006b). This interpretation is in accordance with the wider distribution of volatile data from the IS metallic association, scattered between the compositional ranges of magmatic and crustal fluid in the N_2/Ar–CO_2/CH_4 correlation (Fig. 12), than the distribution of volatile data from LS metallic associations (stages IIB and III), practically confined to the compositional range of magmatic fluids. In addition, the isotopic composition of sulfur (Table 4 and Figure 14 *in* Camprubí et al., 2001b) in sulfides from stage I (IS) and stage IIB (LS) is compatible with mixed crustal (sedimentary or metasedimentary) and magmatic sources (and also bacteriogenic in stage IID), whereas the isotopic composition of sulfur in sulfides from stage III (LS) is compatible with magmatic sources (Camprubí et al., 2001b). Volatile analyses obtained from barren quartz bands from LS stages display the highest component of crustal fluids in the La Guitarra deposit (Fig. 12). This supports the proposition that LS fluids associated with the formation of this deposit correspond to fluids that reflect a relatively long-lasting interaction with crustal rocks and fluids. The exception would be episodic injections of magmatic fluids that led to the formation of metallic mineral associations in the epithermal environment and that conceal the compositional signature of crustal fluids. The close association of magmatic fluids with metallic associations in Mexican epithermal deposits was already noted by Albinson et al. (2001). It may look like a paradox that IS metallic associations have inclusion fluids that underwent a higher degree of

Figure 12. Correlation diagrams between H_2S/SO_2 and CO_2/CH_4 ratios, and between N_2/Ar and CO_2/CH_4 ratios for the samples analyzed from all stages of mineralization at the La Guitarra deposit of the Temascaltepec district (Camprubí et al., 2006b), the SE part of the Temascaltepec district (Camprubí, 2003), and the San Nicolás vein of the Guanajuato district (Abeyta, 2003). Stage I of the La Guitarra deposit formed as a result of an intermediate sulfidation hydrothermal event, whereas all the other mineral associations considered here represent low sulfidation hydrothermal events (see mineralogy in Camprubí et al., 2001a, and fluid inclusions in Camprubí et al., 2001b). The positions of the compositional boundaries were taken from Norman et al. (1997b). N_2/Ar ratios lower than 15 indicate loss of N_2 or addition of Ar, labeled as "excess Ar." The terms "early" and "late" refer to pre- and post-brecciation bands, respectively. In these correlation diagrams N_2/Ar ratios = 0 were skipped, since the absence of N_2 that the corresponding analytical runs recorded do not correspond to an actual lack but to analytical problems. The data from the San Nicolás vein are included as data density contours; the thickness of lines in the pattern corresponds to the highest data density that decreases outward, indicated by thinner lines. Key: a.s.w.—air-saturated water; BMSA—base metal sulfide association; SMA—silver mineral association.

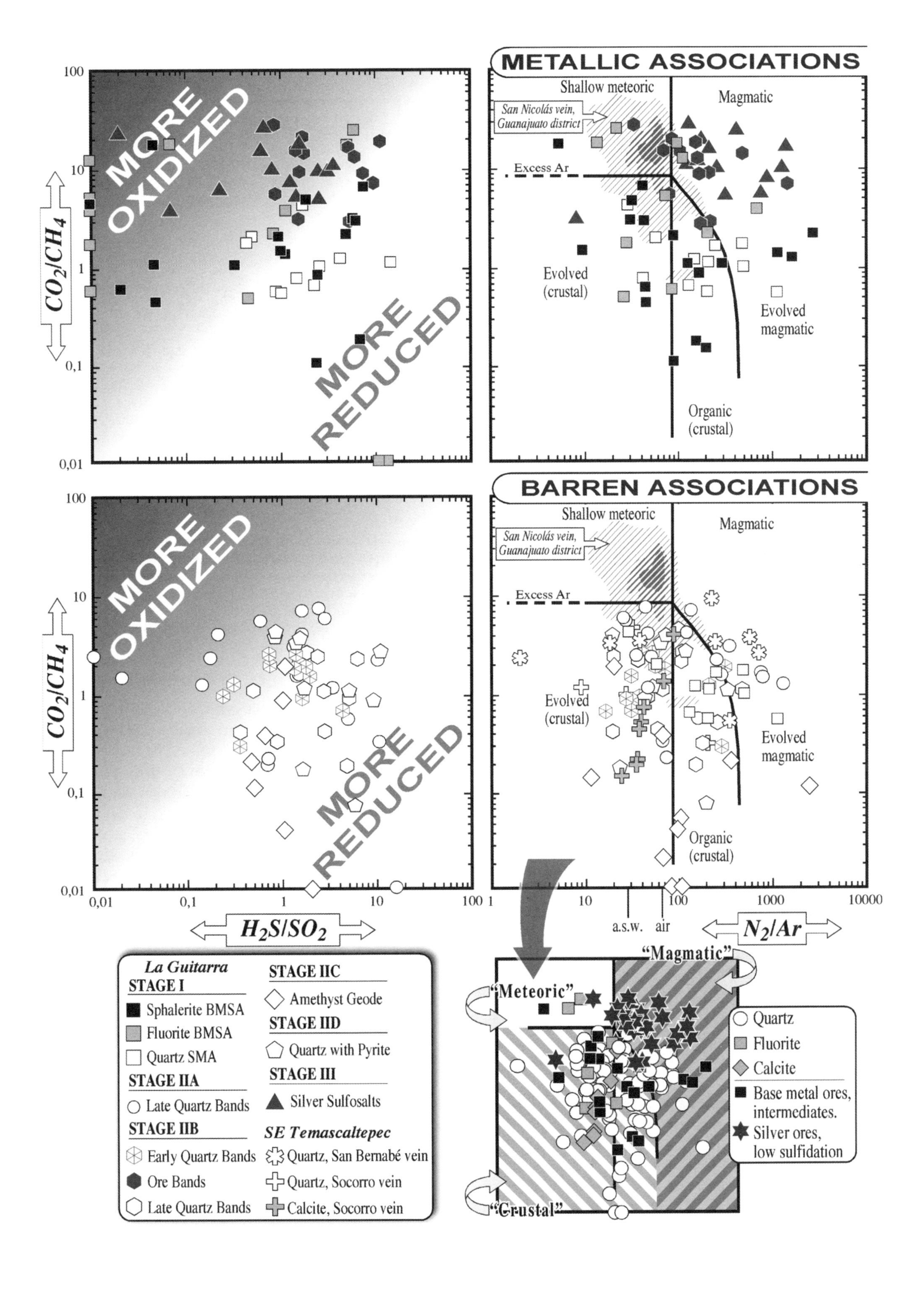

METALLIC ASSOCIATIONS
BARREN ASSOCIATIONS
MORE OXIDIZED
MORE REDUCED
CO2/CH4
H2S/SO2
N2/Ar
Shallow meteoric
Magmatic
San Nicolás vein, Guanajuato district
Excess Ar
Evolved (crustal)
Evolved magmatic
Organic (crustal)
a.s.w.
air
"Magmatic"
"Meteoric"
"Crustal"
Quartz
Fluorite
Calcite
Base metal ores, intermediates.
Silver ores, low sulfidation
La Guitarra
STAGE I
Sphalerite BMSA
Fluorite BMSA
Quartz SMA
STAGE IIA
Late Quartz Bands
STAGE IIB
Early Quartz Bands
Ore Bands
Late Quartz Bands
STAGE IIC
Amethyst Geode
STAGE IID
Quartz with Pyrite
STAGE III
Silver Sulfosalts
SE Temascaltepec
Quartz, San Bernabé vein
Quartz, Socorro vein
Calcite, Socorro vein

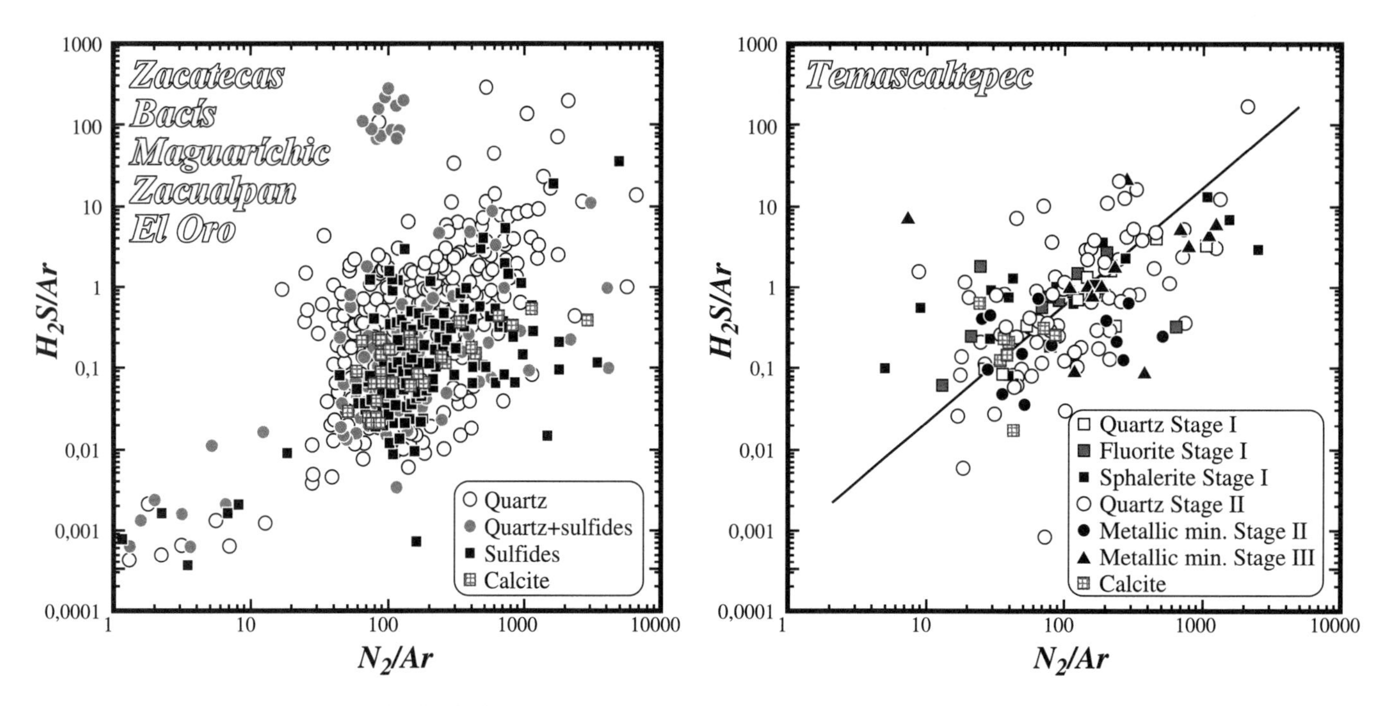

Figure 13. N_2/Ar-H_2S correlation plot for QMS [Quadruple Mass Spectrometry] analyses from fluid inclusions of some Mexican alkaline/neutral epithermal deposits. Left: Data from Bacís in Durango, Maguaríchic in Chihuahua, El Oro and Zacualpan in the State of México, and Zacatecas in Zacatecas state, redrawn and simplified from Albinson et al. (2001). Right: Data from Temascaltepec in the State of México, comprising data from the La Guitarra vein system and from veins of the SE portion of the district, redrawn from Camprubí (2003) and Camprubí et al. (2003b). The quartz dots comprise both analyses from Stage II of the La Guitarra vein system and from barren quartz stages of the San Bernabé and Socorro veins, SE of the district. The calcite samples came from the Socorro vein. The black line in the plot represents the approximate correlation trend of the analyses.

interaction with crustal fluids (and rocks) than fluids associated with LS metallic associations. If the occurrence of brines that carried metals in solution is intimately linked to episodic injections of magmatic fluids, why do LS metallic associations contain mostly magmatic fluids and IS metallic associations do not? The answer to that may be related to different residence periods of those brines at depth. Early IS fluids would have been confined at depth for longer periods than later LS fluids, until tectonic mechanisms triggered the upwelling of IS fluids that were already compositionally equilibrated with crustal rocks and fluids. Subsequently, metal-bearing LS fluids would have resided at depth for shorter periods since they exsolved from cooling magmas, as they would have benefited from faults that already existed (see discussion in the "High sulfidation and intermediate sulfidation epithermal deposits and relation with other deposit types" section). Unfortunately, no salinity data derived from fluid inclusion microthermometry is available from ore-bearing LS associations that could provide additional clues about the IS or LS character of the associated ore-bearing fluids. The mixed magmatic and crustal composition of the fluids associated with the early IS stage at La Guitarra reflect an initially higher interaction with crustal rocks and fluids. The volatile compositions of the San Nicolás vein at Guanajuato (Fig. 12; Abeyta, 2003), in contrast, suggest that this deposit could have formed from dominantly shallow meteoric hydrothermal fluids.

In the La Guitarra deposit, volatile chemistry and salinities in fluid inclusions (Fig. 12) can be related to the FeS content in sphalerites from both IS and LS stages. The highest salinities were recorded in inclusion fluids from the early base metal IS association (stage I) that contain the sphalerites with lowest FeS contents. As already noted, sphalerite grains, in this association with brine pulses, become less saline with time, as these grains become progressively richer in FeS. In contrast, silver mineral associations contain sphalerites with the highest FeS contents, in both IS and LS stages. All the metallic mineral associations (both IS and LS, both base metal sulfide and silver mineral associations) are associated with relatively more oxidized fluids, but the IS base metal sulfide association exhibits the highest crustal influence on mineralizing fluids, while both IS and LS silver mineral associations exhibit the highest magmatic influence on mineralizing fluids. Consequently, in the La Guitarra deposit (1) metallic mineral associations formed after oxidized fluids, (2) base metal sulfide associations (with no silver) formed after saline crustal brines that formed FeS-poor sphalerites, and (3) silver mineral associations (both IS and LS) formed after less saline magmatic brines that formed FeS-rich sphalerites. The last two consequences above reproduce both at small and large scale; that is, from crystal zoning to different mineral associations and to different stages of mineralization. Thus, at La Guitarra, FeS-rich sphalerites or FeS-rich discrete growth zones in sphalerite formed from both IS or LS magmatic fluids, whereas FeS-poor sphalerites or FeS-poor discrete growth zones in sphalerite are more likely to have formed from IS crustal fluids.

Additionally, evidence derived from the analysis of isotopic compositions of oxygen and hydrogen in several IS and LS epithermal deposits in México indicates the occurrence of a substantial contribution of fluids with consistently high oxygen isotopic composition (Albinson et al., 2001; Camprubí et al., 2001b). That contribution cannot be explained by boiling processes or water-rock interaction alone, and the occurrence of significant contributions of magmatic fluids provides the best explanation for the strong shift in composition of the oxygen isotopes toward the magmatic water box. There are also mineralizing fluids whose isotopic compositions can be congruently derived from direct mixing between magmatic fluids and shallow meteoric waters (1) in different stages of mineralization of the same deposit, and (2) in different veins (Camprubí et al., 2001b; Camprubí, 2003). However, more work in this matter is necessary regarding IS and LS epithermal deposits in México, since the only deposit where studies on the O and H isotopic composition of mineralizing fluids have been systematically carried out, based on a detailed and categorical characterization in stages, sub-stages, and particular mineral associations, is the La Guitarra deposit from the Temascaltepec district in the State of México (Camprubí et al., 2001a, 2001b).

MECHANISMS OF PRECIPITATION

Although fluid mixing as a mechanism for precipitation cannot be discarded in most cases, it is generally concluded in the study of most deposits that boiling (after petrographic criteria; i.e., Figs. 9E and 9F) is the most important and most effective mechanism for the precipitation of metallic mineral–bearing associations in IS and LS epithermal deposits in México (Simmons et al., 1988; Simmons, 1991; Albinson and Rubio, 2001; Albinson et al., 2001; Camprubí et al., 2001a, 2001b, 2006a), although the occurrence of boiling does not necessarily lead to the occurrence of metallic mineral–bearing associations (Camprubí et al., 2001a) if the boiling fluids are not endowed with dissolved metals.

Fluid inclusion associations that display the convincing coexistence of liquid-rich and vapor-rich inclusions not associated with necking are evidence for their heterogeneous trapping by the growing host mineral (e.g., van den Kerkhof and Hein, 2001) under boiling conditions. As noted, the strongest mineralogic and petrographic evidence for boiling are the occurrence of bladed calcite or its pseudomorphs (Figs. 7C, 7D, 9A, and 9B), the occurrence of pseudorhombohedral or pseudoacicular adularia (Figs. 9C and 9D), and the occurrence of truscottite. Actualistic models through the evidence displayed by active geothermal systems (i.e., Simmons and Browne, 1997, 2000) were crucial to establish a correspondence between boiling and the above petrographic and mineralogical occurrences, as well as the precipitation of metallic associations, either base metal– or precious metal–rich. In most epithermal deposits in México, boiling has been documented to be the most important mechanism that led to ore deposition because there is a close link in time and space between evidence for boiling and the occurrence of metallic associations, such as the identification of metallic minerals formed directly on pseudorhombohedral crystals of adularia and ore associations within adularia-rich bands (Figs. 9E and 9F; Camprubí et al., 2006a).

Fluid mixing as a mechanism for mineral precipitation in the epithermal environment is commonly the subject of debate and, though sometimes evidence for this process is poor, it is frequently invoked. The environments where fluid mixing may be best invoked as a mechanism for mineral precipitation in epithermal deposits in México are steam-heated horizons, marked by the occurrence of shallow epigenetic argillic alteration blankets or horizons (i.e., Morales-Ramírez et al., 2003). These form where meteoric fluids condense and assimilate water vapor and volatile phases (Sb, As, Hg, CO_2, and H_2S) derived from deeper boiling fluids. The mixing of upwelling neutral chloride-rich ore-bearing fluids with descending oxidized waters or low pH fluids may effectively occur and lead to mineral precipitation due to the desolubilization of bisulfide complexes (Corbett and Leach, 1998). The evidence for such a mechanism would be the presence of kaolinite in the ore stages. In the previously described cases of the Zuloaga vein in the San Martín de Bolaños district and the Colón–Purísima vein system in the Pachuca–Real del Monte district (see the "clay minerals" section) where kaolinite was found as a hypogene alteration mineral or within ore associations, the occurrence of this mineral does not validate mixing as a feasible deposition mechanism, in the sense of Corbett and Leach (1998), because the occurrence of kaolinite is restricted to deep portions of the veins that are detached in space by hundreds of meters with the shallow steam-heated portions of the veins as exemplified in the Zuloaga vein in the San Martín de Bolaños district (Fig. 11; Albinson and Rubio, 2001) and in the El Herrero vein in the Bacís district. In most cases, nonetheless, the occurrence of fluid mixing trends are based on progressive dilution trends in time or space in fluid inclusions (i.e., Camprubí et al., 2001b), which typically invoke the presence of meteoric waters as the low temperature and low salinity end member fluid. By means of the volatile contents in fluid inclusions, the presence of shallow meteoric waters can be inferred at depths of many hundred meters below the paleosurface (Albinson et al., 2001; Camprubí et al., 2003a).

In other cases, especially in mineral associations that are relatively confined at depth (i.e., without any apparent exit to the paleosurface during their formation; e.g., stage III at Temascaltepec; Camprubí et al., 2001a, 2001b), and lacking evidence for boiling, mineral precipitation may be more feasibly explained by means of conductive cooling or, alternatively, as a product of chemical reduction of mineralizing fluids as they react with the country rocks or preexisting vein material. On the basis of textural evidence for replacement of wall rocks and early stage vein silicates and carbonates, Dreier (2005) proposes that the main stage of mineralization at Pachuca–Real del Monte was precipitated as a result of fluid-rock chemical reactions.

HIGH SULFIDATION AND INTERMEDIATE SULFIDATION EPITHERMAL DEPOSITS AND RELATION WITH OTHER DEPOSIT TYPES

A close temporal and spatial relationship has been documented for metalliferous porphyry and epithermal deposits in the Circum-Pacific rim, such as Lepanto (acid epithermal)–Far Southeast (porphyry copper)–Victoria (alkaline/neutral epithermal) in the Philippines (Arribas et al., 1995; Hedenquist et al., 1996, 1998; Sillitoe, 1999), Nevados del Famatina–La Mejicana in Argentina (Losada-Calderón and McPhail, 1996), or Maricunga in Chile (Muntean and Einaudi, 2001). In México, despite the occurrence of more than 30 metalliferous porphyry deposits in the northwestern part of the country (Sillitoe, 1977), in a favorable belt that extends as south as Chiapas (Valencia-Moreno et al., this volume), very little research has focused on the possible links between these deposits and the epithermal deposits in neighboring areas, especially in the southern part of the belt. A good example in México is found at the La Caridad district in Sonora between the conspicuous porphyry copper deposit and the HS epithermal deposit at the La Caridad Antigua (Valencia et al., 2003; Valencia, 2005).

In México, there are also IS vein deposits in the peripheral zones to mineralization centers that are clearly magmatic. Genetic relationships have been defined between porphyry, skarn, and epithermal deposits at the Moctezuma district in Sonora (Deen and Atkinson, 1988; Atkinson, 1990, 1996), as well as between the skarn deposits at San Martín in Zacatecas (Rubin and Kyle, 1988) and several IS veins in the same district (i.e., the Noria de Pantaleón vein), whose formation is possibly linked to the evolution of the skarn deposit (Rubin and Kyle, 1988; Starling et al., 1997; Albinson et al., 2001). Such links in space between deposit types are not uncommon, as there is evidence for analogous links also at Santa María de la Paz in San Luis Potosí (McGibbon, 1979; Gunnesch et al., 1994), Velardeña in Durango (Gilmer et al., 1988), Real de Asientos in Aguascalientes (Rivera, 1993), San Antón de las Minas in Guanajuato, and the Batopilas IS silver-rich veins that occur adjacent to the porphyry deposit of Tahonas in Chihuahua (Wilkerson et al., 1988; Goodell, 1995; Valencia-Moreno et al., this volume). Although the HS El Sauzal deposit (Charest and Castañeda, 1997; Sellepack, 1997; Gray, 2001) is located only ~20 km from the Tahonas porphyry and the Batopilas epithermal deposits, it is unlikely that there is a genetic link between them and El Sauzal. It is more likely that El Sauzal is "stacked" above an independent porphyry of possible similar age to that of Tahonas and that both districts are part of a common metallogenic episode. Vapor-rich fluid inclusions and multiphase fluid inclusions with daughter crystals (with dissolution of daughter halite indicating salinities ranging 30–35 wt% NaCl equiv.) were analyzed in quartz associated with alunite, enargite, and pyrite, from a relatively deep diamond drill (145 m) under the El Sauzal deposit (DDH SZ-96-03). These types of fluid inclusion assemblages are suggestive of the environment that can occur above a high level pluton (Sillitoe and Hedenquist, 2003) and argues for a possible case at El Sauzal of "stacked" porphyry–HS epithermal mineralization types.

A NEW EMPIRICAL CLASSIFICATION FOR INTERMEDIATE AND LOW SULFIDATION EPITHERMAL DEPOSITS IN MÉXICO USING SULFIDATION STATES

Preliminary Considerations

The relative contents in metals (Ag/Au and precious/base metals) and the apparent salinity of mineralizing fluids from fluid inclusion microthermometry allows discrimination among Zn-Pb-Ag, Ag-Zn-Pb, Ag + Au + Pb + Zn + Cu, Ag-Au, and Au-Ag epithermal deposits in México (Albinson et al., 2001) or, more generically, discrimination between polymetallic and precious metal deposits. At a first glance, the first four categories and the last category correspond broadly to the IS and LS types, respectively, according to Hedenquist et al. (2000), Einaudi et al. (2003), and Sillitoe and Hedenquist (2003). Looking in detail, the relationship between Mexican IS and LS epithermal deposits appears tighter than previously suspected and, as will be subsequently argued, several epithermal deposits in México display coexistence between IS and LS mineralization styles. Although Simmons (1991) used the terms *heavy sulfide* and *light sulfide* for the Fresnillo deposits in a different sense (both as polymetallic IS styles), we will use this terminology to indicate relative sulfide contents in the ore assemblages, high and low, as equivalent to IS and LS mineralization styles, respectively, as these terms are very straightforward and descriptive.

Although most epithermal deposits in México still need detailed studies concerning both their ore and gangue mineralogy, and their fluid geochemistry, it appears that the majority of the important alkaline/neutral epithermal deposits belong to the IS type; examples include the world-class deposits at Fresnillo in Zacatecas, Tayoltita in Durango, or Pachuca–Real del Monte in Hidalgo. However, it is commonly not so straightforward to adscribe these deposits to the IS or the LS type, and this section attempts to shed some light on the present confusion concerning the alleged mutual exclusivity of IS and LS epithermal deposits, applied to the case of Mexican deposits. In the metallogenic province of the Great Basin in Nevada, a case has been made for mutually exclusive IS and LS epithermal deposits both in time and space (John, 2001; Sillitoe and Hedenquist, 2003). The IS deposits are dominantly found in the calcalkaline andesitic arc in the westernmost portion of the province and are ca. 38 Ma. In contrast, the LS deposits are spatially restricted to the easternmost portion of the province and are linked to bimodal reduced magmatism (rhyolitic/basaltic) in an extensional rift environment (Northern Nevada Rift) that led to the formation of LS epithermal deposits between 16 and 14 Ma (Sillitoe and Hedenquist, 2003). In the case of the Northern Nevada Rift, the LS deposits are linked in terms of ore-fluid chemistry to degassing of relatively gold-rich reduced basaltic magmas that would source sulfur primarily as H_2S, which can then complex gold and produce a low-sulfidation ore fluid once mixed with abundant meteoric water (John, 2001). In the main metallogenic provinces of México (Sierra Madre Occidental and Mexican Altiplano or Mesa Central) there is no clear time or space divide analogous to the Great Basin setting, and both subtypes of epithermal deposits seem to coexist within the same space and the same time range. This condition is found, not only at the scale of a metallogenic province (Albinson et al., 2001), but also at the scale of mining district, and even to the scale of a single deposit.

The implementation of the concept of "sulfidation state," which is associated with the oxidation state and valence of sulfur and the temperature of magmatic-hydrothermal fluids, demonstrates the link between these parameters and the stability of the associated sulfide minerals (Einaudi et al., 2003). Thus, the epithermal deposits that formed under a sulfidation state of the IS type are characterized by (1) mineralizations with abundant sulfides (the *heavy sulfide* type, with total sulfide mineral contents >5%); (2) the occurrence of sphalerite with compositions dominantly low in FeS; (3) variable amounts of galena, pyrite, chalcopyrite, tetrahedrite-tennantite, and silver sulfosalts; (4) mainly accompanied with crystalline varieties of quartz, Mn carbonates and silicates, fluorite, and relatively scarce adularia as non-sulfide minerals; and (5) dominant alteration styles in which illite changes to sericite with depth, and to propylitic alteration laterally (Einaudi et al., 2003; Sillitoe and Hedenquist, 2003).

The epithermal deposits that formed under a sulfidation state of the LS type, in turn, are characterized by (1) mineralizations poor in sulfides (the *light sulfide* type, with total sulfide mineral contents <1%); (2) with dominant pyrite, arsenopyrite, silver sulfosalts, acanthite, naumannite, and electrum; (3) associated with a gangue, which includes varieties of amorphous silica, multibanded cryptocrystalline, and crystalline quartz, adularia, and bladed calcite; (4) the occurrence of sphalerite with compositions high in FeS; and (5) weakly developed alteration styles, characterized by the occurrence of illite or chlorite, depending on the composition of host rocks (Sillitoe and Hedenquist, 2003).

The fluids responsible for the formation of IS mineralizations in Mexican deposits generally display high temperatures, between 230 °C and 300 °C, with maximum salinities ranging from >7.5 to <23 wt% NaCl equiv., and these deposits formed as deep as 1 km or more. In contrast, the fluids responsible for the formation of LS mineralizations are characterized by temperatures <240 °C, salinities commonly of <3.5 wt% NaCl equiv., although maximum salinities can be as high as 7.5 wt% NaCl equiv., and these deposits generally formed at depths of <300 m (Albinson et al., 2001). Many epithermal deposits in México show coexistence of both mineralization types. Although it is not an example of a large deposit, the epithermal deposit in México studied most extensively to date is the La Guitarra deposit in the Temascaltepec district in the State of México (Camprubí et al., 2001a, 2001b, 2003a, 2003b). Some examples of mineralization styles belonging to the IS type from that deposit (Stage I) are shown in Figures 6A, 6B, 7C, and 7D, and examples of mineralization styles belonging to the LS type (Stage II) are shown in Figures 6C, 6D, 8A, 9C, and 9D.

Low sulfidation epithermal deposits have been classified in the Southwest Pacific rim based on empirical observation and descriptive terminology according to their mineralogy and possible relationship to magmatic source rocks at different crustal levels (Corbett, 2002). Under this classification, arc-related low sulfidation epithermal mineralization ranges from deep intrusive-proximal quartz-sulfide Au ± Cu deposits to carbonate and base metal Au deposits (both styles being transitional to porphyry copper deposits), to distal higher level epithermal quartz Au-Ag deposits. The two deeper styles of mineralization are related to fluids of intermediate sulfidation affinity, whereas the distal shallow style is related to fluids that are low-sulfidation in character. An additional rift-related low sulfidation chalcedony-ginguro style of mineralization is recognized as forming in rift settings where the intrusive connection is deeply seated and uncertain (Corbett, 2002), and represents a type of mineralization that best fits the setting of Mexican low and intermediate sulfidation deposits, especially those in the Mexican Altiplano metallogenic province. As previously stated, the term chalcedony-ginguro, however, is considered a misnomer, since the silica phase most commonly identified as intimately associated with ore-bearing minerals in Mexican deposits is jigsaw quartz (microcrystalline aggregates of anhedral to subhedral quartz due to recrystallization of amorphous silica).

A New Empirical Classification

It is well-known worldwide that a close space and time association can exist between IS epithermal deposits and metalliferous porphyry deposits, skarns, or HS epithermal deposits. The differences between the above types of deposits can be attributed to an evolutionary path related to the variability of oxidation and sulfidation states in mineralizing fluids from a proximal thermal source, namely a porphyry or skarn (Einaudi et al., 2003), toward distal areas. Figure 14A shows the location of different types of Mexican deposits (porphyries, HS, IS) associated with high-level intrusive bodies and is similar to the model presented by Sillitoe and Hedenquist (2003). As already mentioned, there are actually some IS epithermal deposits in México whose environment of formation may be identified in such a scheme, like Santa María de la Paz in San Luis Potosí, Batopilas in Chihuahua, or the Noria de Pantaleón vein at the San Martín "skarn district" in Zacatecas (see lower part of Fig. 14A). Strictly speaking, these vein deposits are not considered epithermal since they form in a deeper environment (lateral to a porphyry intrusive), although in common with IS epithermal veins, they also form from IS-type fluids and therefore can share similar mineralogy and alteration patterns. For example, the calcite-native silver Batopilas veins are characterized by exhibiting relatively low temperature (average Th values 190 °C) and moderate salinity (average 14 wt% NaCl equiv.) fluids with no fluid inclusion evidence of boiling (Wilkerson et al.,

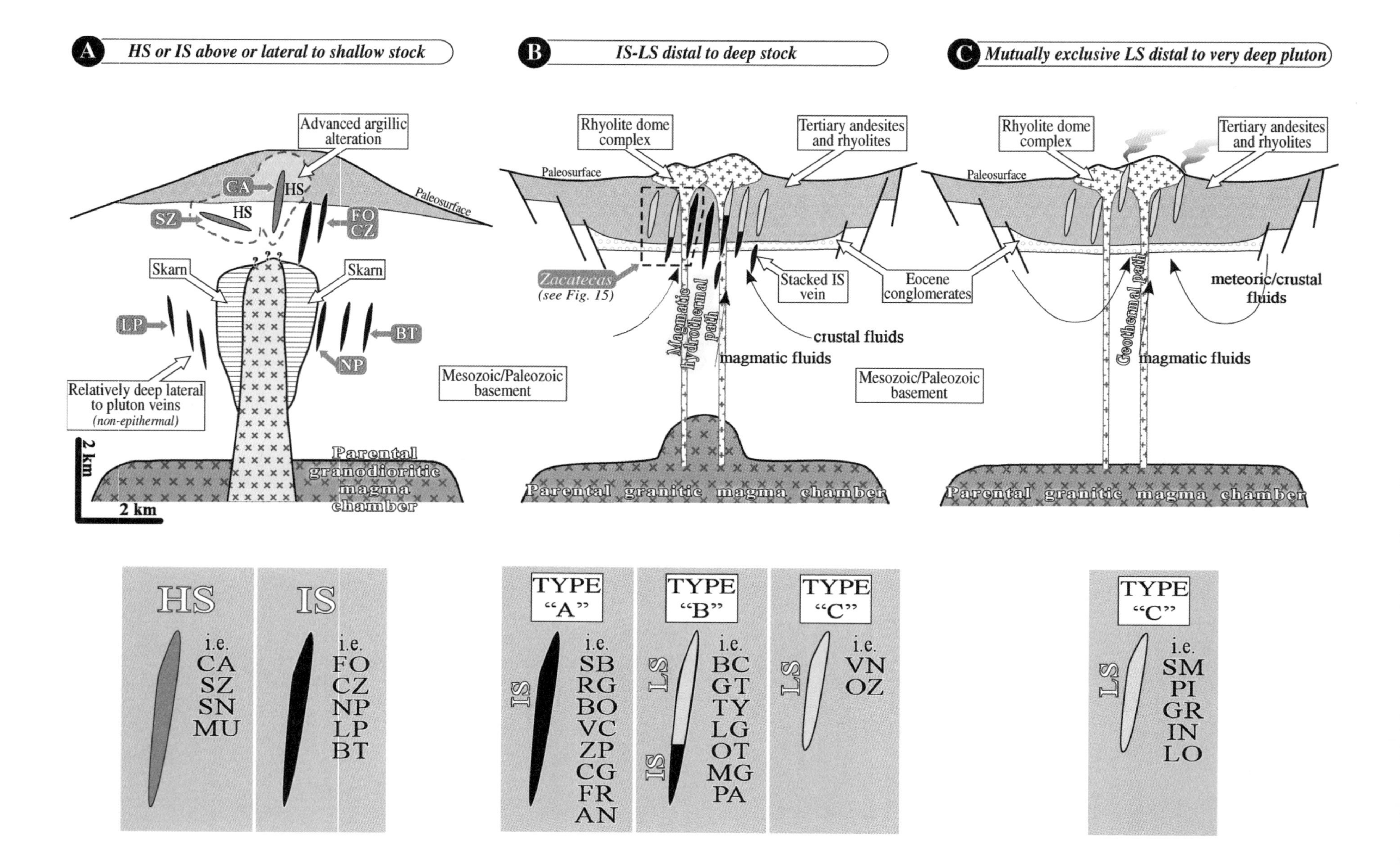
A
HS or IS above or lateral to shallow stock
B
IS-LS distal to deep stock
C
Mutually exclusive LS distal to very deep pluton
Advanced argillic alteration
CA
HS
SZ
HS
FO
CZ
Paleosurface
Skarn
Skarn
LP
BT
NP
Relatively deep lateral to pluton veins
(non-epithermal)
2 km
2 km
Parental granodioritic magma chamber
Mesozoic/Paleozoic basement
Rhyolite dome complex
Tertiary andesites and rhyolites
Paleosurface
Zacatecas
(see Fig. 15)
Stacked IS vein
Eocene conglomerates
Magmatic hydrothermal path
crustal fluids
magmatic fluids
Mesozoic/Paleozoic basement
Parental granitic magma chamber
Rhyolite dome complex
Tertiary andesites and rhyolites
Paleosurface
meteoric/crustal fluids
Geothermal path
magmatic fluids
Parental granitic magma chamber
HS
i.e.
CA
SZ
SN
MU
IS
i.e.
FO
CZ
NP
LP
BT
TYPE "A"
IS
i.e.
SB
RG
BO
VC
ZP
CG
FR
AN
TYPE "B"
LS
IS
i.e.
BC
GT
TY
LG
OT
MG
PA
TYPE "C"
LS
i.e.
VN
OZ
TYPE "C"
LS
i.e.
SM
PI
GR
IN
LO

1988). Figure 14A also shows IS vein deposits located above a high level pluton, that is, in a shallow epithermal environment but genetically related to the porphyry intrusive. Such is the case of the Victoria epithermal IS veins and the Lepanto epithermal HS deposit, which are related to the Far Southeast porphyry copper deposit in the Philippines (Hedenquist et al., 1998, 2001). Analogies of this brand of epithermal IS veins have not yet been well documented in México, mainly due to lack of exposure (tilting and deep erosion of the deposits), or due to lack of exploration into the deeper portion of the deposits. Two possible candidates include the San Francisco del Oro vein system in Chihuahua, which is hosted within an extensive envelope of hornfelsed Cretaceous clastic rocks, indicating the presence of a pluton at depth, and the Colorada veins in Zacatecas, where small bodies of skarn mineralization associated with sills have been reported in deeper explored portions of the vein system (Moller et al., 2001). Both in the Colorada and San Francisco del Oro districts, fluid inclusion evidence for boiling is present (see Appendix A *in* Albinson et al., 2001; Albinson, 1988) and places the depth of formation within the epithermal environment. HS deposits that fit this model include the Caridad Antigua deposit in Sonora, which sits to one side of the La Caridad porphyry copper deposit in a downthrown fault block (Valencia et al., 2003), and the El Sauzal deposit in Chihuahua, which exhibits hypersaline fluids in a quartz-enargite assemblage and may well overlie a high-level pluton (Fig. 14A). To date, no pluton-related IS deposits have been documented as coexisting with HS deposits in México.

It should be emphasized, however, that the majority of IS and LS epithermal deposits in México may be located over deep intrusive bodies, emplaced at depths most likely >5 km (Albinson et al., 2001). Consequently, the genetic models that attempt to explain the full variety of HS, IS, and LS epithermal deposits found in the metallogenic provinces of México must first consider that, as shown in Figure 14B, the starting point for a genetic association between the different types of deposits may be the significant depth of emplacement of the intrusive bodies that may play a parental role to all of them. From this standpoint, and taking into consideration the distribution of the different components of mineralization, either of the IS type or the LS type, it is possible to define three main styles of alkaline/neutral epithermal deposits, which are described in the following paragraphs as types A, B, and C.

"Type A" is characterized by exhibiting exclusively polymetallic IS mineralization (*heavy sulfide*) along the full vertical extent of the deposit (Fig. 14B). Examples of Type A are deposits like Real de Guadalupe and the El Cobre vein at the Taxco district, both in Guerrero, Bolaños and San Martín de Bolaños in Jalisco, Zacualpan in the State of México, and Fresnillo and Sombrerete in Zacatecas. Epithermal IS veins above high-level plutons (Fig. 14A), such as San Francisco del Oro in Chihuahua and Colorada in Zacatecas, can be considered Type A deposits as well. Distinguishing between the two environments (IS epithermal above a high level pluton vs. IS epithermal above a deep pluton) may not be straightforward and may require examination in detail of all aspects of the district geology.

"Type B" is the most widespread in México and exhibits characteristically extensive precious metal LS-type mineralization, with roots of the IS type within a relatively confined vertical distribution (Fig. 14B). In some cases, only specific mineralized bodies in a given deposit or district may display this vertical distribution, such as the Rayas oreshoot of the Veta Madre vein in the Guanajuato district, the Herrero vein in the Bacís district, and

Figure 14. Schematic sections of the formation environments of epithermal deposits in Mexican calcalkaline volcanic arcs. (A) High sulfidation (HS) and/or intermediate sulfidation (IS) epithermal deposits lateral or above shallow parental intrusives. Drawn using the scheme by Sillitoe and Hedenquist (2003) as a base, it exhibits the environment with a relatively high-level intrusive body and the formation of porphyry, skarn, and HS epithermal deposits. The deep veins, lateral to pluton, are not epithermal, although they formed from IS fluids. (B) IS and low sulfidation (LS) epithermal deposits distal to parental intrusives. This is the most common setting in México, where IS and LS epithermal deposits coexist, with or without skarns, associated with deep intrusive bodies and the resulting three types of mineralization proposed in this study (A, B, and C). (C) LS epithermal deposits that are mutually exclusive from IS deposits, distal to very deep pluton. This setting is rarely associated with major volcanic centers and is more frequently associated with domes and extensional settings. The tectonic environment shown in both (B) and (C) corresponds to the backarc setting of the Mexican Altiplano. AN—Angangueo, Michoacán; BC—Bacís, Durango (Smith, 1995; Albinson et al., 2001); BO—Bolaños and San Martín de Bolaños, Jalisco (Lyons, 1988; Scheubel et al., 1988; Albinson and Rubio, 2001); BT—Batopilas and Tahonas, Chihuahua, alkaline epithermal and porphyry copper deposits, respectively (Wilkerson et al., 1988); CA—La Caridad and La Caridad Antigua, Sonora, porphyry and HS epithermal deposit, respectively (Valencia et al., 2003, and Valencia, 2005); CG—La Ciénega, Durango (De la Garza et al., 2001); CZ—Candelaria vein, Colorada, Zacatecas (Albinson, 1985 and 1988); FO—San Francisco del Oro, Chihuahua (Grant and Ruiz, 1988); FR—Fresnillo, Zacatecas (Gemmell et al., 1988; Simmons et al., 1988; Simmons, 1991); GR—Guadalupe de los Reyes, Sinaloa (Allen et al., 2001); GT—Rayas bonanza, Guanajuato district, Guanajuato (Randall et al., 1994); IN—El Indio–Huajicori, Nayarit; LG—La Guitarra at the Temascaltepec district, State of México (Camprubí et al., 2001a, 2001b, 2003b); LO—Lluvia de Oro, Durango (Albinson et al., 2001); LP—Santa María de la Paz, San Luis Potosí (McGibbon, 1979; Gunnesch et al., 1994); MG—Maguaríchic, Chihuahua (Albinson et al., 2001); MU—Mulatos, Sonora (Staude, 2001); NP—Noria de Pantaleón vein in the San Martín skarn district, Zacatecas (Rubin and Kyle, 1988); OT—El Oro–Tlalpujahua, State of México/Michoacán (Flores, 1920; Gómez-Caballero et al., 1977); OZ—El Orito, Zacatecas (Ponce and Clark, 1988); PA—Pachuca–Real del Monte, Hidalgo (Geyne et al., 1963); PI—Pinos, Zacatecas; RG—Real de Guadalupe, Guerrero (Albinson and Parrilla, 1988); SB—Sombrerete, Zacatecas (Albinson, 1988; Albinson et al., 2001); SM—San Martín, Querétaro (Muñoz, 1993); SN—Santo Niño, Chihuahua (Gray, 2001); SZ—El Sauzal, Chihuahua (Charest and Castañeda, 1997; Sellepack, 1997; Gray, 2001); TY—Cinco Señores vein at the Tayoltita district, Durango (Clarke and Titley, 1988); VN—San Nicolás vein at the Guanajuato district, Guanajuato (Abeyta, 2003); VC—El Cobre vein at the Taxco district, Guerrero (Clark, 1990; Camprubí et al., 2006c); ZP—Zacualpan, State of México (Noguez et al., 1988). See additional references in Camprubí (1999) and Camprubí et al. (1999, 2003a), Albinson et al. (2001), and previous sections in this paper.

possibly the San Rafael vein in the El Oro district in the State of México. In other deposits, most of the veins and mineralized bodies display polymetallic IS roots, such the veins of the Temascaltepec district in the State of México, the Pachuca–Real del Monte vein systems in Hidalgo, and the veins of the Zacatecas district. The Colón–Purísima veins in Pachuca–Real del Monte (Fig. 4) and the La Guitarra vein in Temascaltepec (Fig. 15) also show IS mineralization stages either in vein margins or as fragments that were engulfed by late LS stages. The polymetallic IS roots of these deposits may be either rich or poor in silver and gold. It is possible that "Type B" veins in other districts may also have polymetallic IS mineralization at greater depths, but such mineralization remains concealed and unknown due to lack of deeper exploration efforts. A second variation of Type B may also occur and is characterized by exhibiting dominantly IS mineralization, with subordinate LS mineralization that is generally restricted to the uppermost parts of the deposit (Fig. 14B). Examples are the Purísima–Colón vein system in the Pachuca–Real del Monte district in Hidalgo (Fig. 4; Dreier, 2005), Maguaríchic in Chihuahua, the Veta Grande vein in the Zacatecas district, and most of the veins at the southeastern part of the Temascaltepec district in the State of México. In the two types of IS epithermal deposits in México (A and B), the deep polymetallic mineralizations are characterized by containing dominantly Zn-Pb instead of Cu (i.e., Real de Catorce, Fresnillo, Pachuca–Real del Monte, Veta Grande in Zacatecas, San Martín de Bolaños in Jalisco). How-

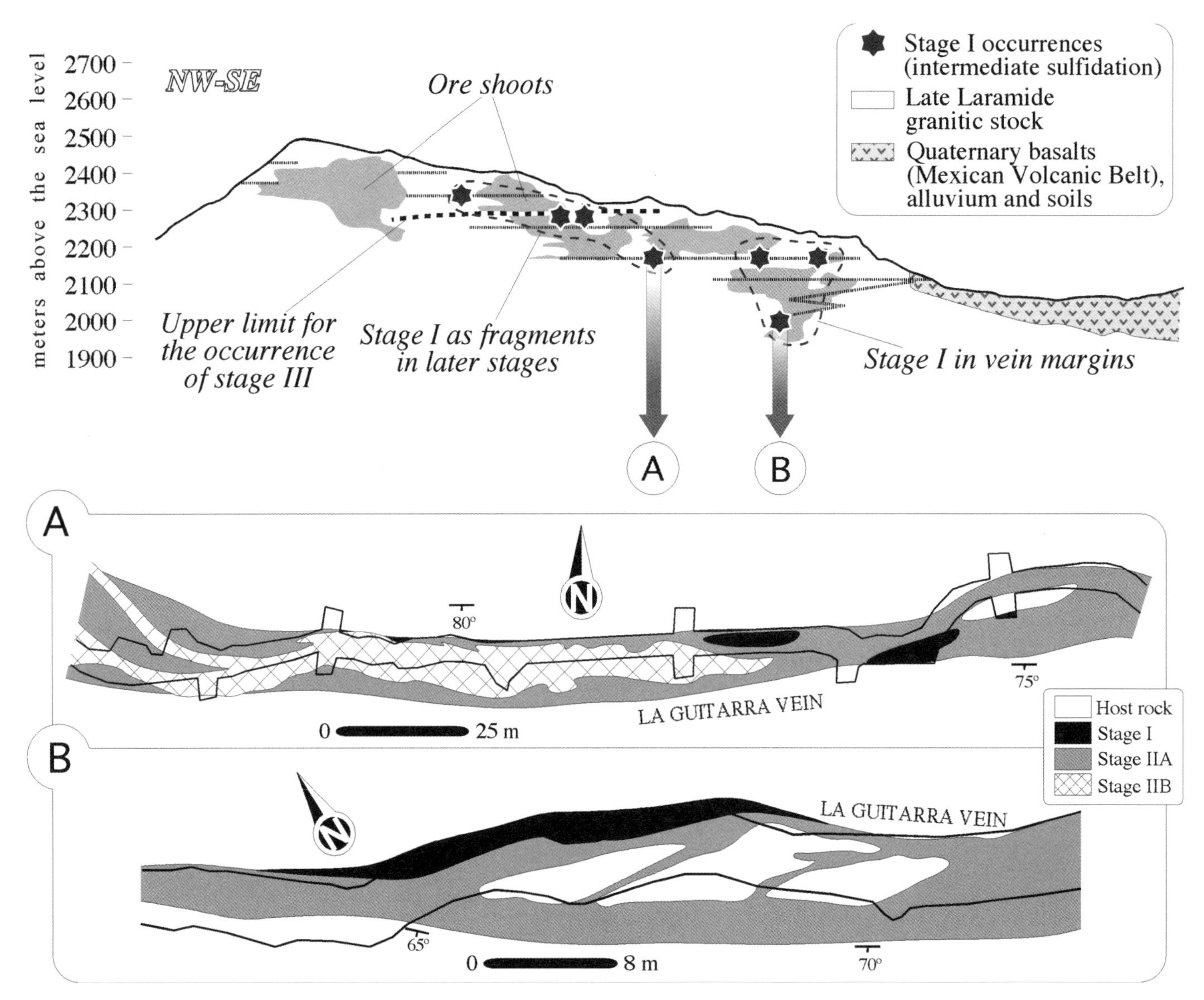

Figure 15. Top: Longitudinal section of the La Guitarra vein of the Temascaltepec district, State of México, with the position of the main occurrences of Stage I (intermediate sulfidation). Bottom (A and B): Mapped mine galleries on the La Guitarra vein where the early intermediate sulfidation stage occurs either as fragments within later low sulfidation stages (A) or in the vein margins (B). The position of ore shoots was obtained from the information available at www.gencoresources.com/geology.html. The location of Stage I exposures and map A are adapted from Camprubí et al. (2001a, 2001b).

ever, the polymetallic roots of some deposits exhibit significant Cu contents, though generally <1%, as is the case of Avino in Durango, the Alacrán vein in the Bolaños district in Jalisco, and the El Cobre vein in the Taxco district in Guerrero. Some of these cases, nonetheless, show higher Cu contents, between 1% and 2.5%, like the Mala Noche vein in the core of the Zacatecas district and the veins in the Angangueo district in Michoacán. Nevertheless, it can be stated that the metallic inheritance of the IS epithermal deposits in México is inclined to be relatively Cu-poor and that this metal occurs in higher contents in deposits formed at higher temperatures and depths, such as skarn deposits.

Types A and B can coexist as individual deposits within a district (Fig. 16), either in veins with the same orientation (i.e., Temascaltepec district, State of México), or may also be individually distributed in neighboring veins with different orientations (i.e., the Zacatecas district or the Taxco district in Guerrero). In the Temascaltepec district, the La Guitarra vein system comprises the NW-SE–striking La Guitarra, Doncellas, and El Salvador veins, and is continuous toward the NW (El Coloso zone) with the Nazareno de Anecas vein. This vein system corresponds to Type B. The veins in the SE portion of the district, on the other hand, namely the Marmajas (El Rincón mine), Candelaria, Santa Ana, Socorro, San Bernabé, Mina de Agua, and Iberia veins, plus several others, have the same orientation, occur apparently in lateral continuity with the La Guitarra vein system, but are suspected to correspond to either Types A or B (Camprubí et al., 2001a; Camprubí, 2003). A similar setting is found in the Bacís district in Durango, where the La Bufa vein (Au-Ag-Cu) corresponds to Type A and the El Herrero vein (Ag-Au) is more akin to Type B, both veins striking NNW.

"Type C" consists solely of LS type mineralization (*light sulfide* style; Fig. 14B) and is a common type of epithermal deposit in México that exhibits a tendency to be relatively small, with maximum tonnages between 1.0 and 3.5 Mt, like Pozos in Guanajuato, Concheño in Chihuahua, Guadalupe de los Reyes in Sinaloa, San Martín in Querétaro, and Pinos in Zacatecas, and commonly exhibits production figures lower than 100,000 tons, like Tambor in Sinaloa, El Indio–Huajicori in Nayarit, Lluvia de Oro in Durango, Benito Juárez and Santa Gertrudis in San Luis Potosí, and Ixtapan del Oro in the State of México (see a full review of productions, metal contents, etc., of several epithermal deposits in México in Table 1 *in* Albinson et al., 2001). There is practically no evidence of association between bimodal basaltic-rhyolitic magmatism and Type C (LS) epithermal deposits in México, unlike the LS epithermal deposits of the Northern Nevada Rift. Given the lack of evidence for reduced magmas at depth, the absence of a distinct parental magma does not support the contention that IS and LS deposits are mutually exclusive in México. On the other hand, LS epithermal deposits in México are commonly genetically linked to rhyolitic hypabyssal igneous activities (i.e., Guadalupe de los Reyes in Sinaloa, San Martín in Querétaro) hosted within calcalkaline or silicic volcanic piles. Additionally, LS mineralizations may be found in peripheral and distal positions to the main IS mineralization, the El Orito veins in the southern part of the Zacatecas district (Fig. 16) being possible examples (Ponce and Clark, 1988; Albinson, 1995), and where K-Ar age data indicate that LS veins of the El Orito vein system represent the final stages (32.7–30.8 Ma; Kapusta, 2006) of protracted hydrothermal activity in a district with predominantly IS vein mineralization (35.5–31.1 Ma; Krueger, 2000) and telescoped in time the Mala Noche vein system (see Fig. 16).

Possible Genesis and Evolution of Epithermal Ore-Bearing Fluids

The different types of IS and LS epithermal deposits in México can be correlated with the classification of epithermal deposits according to their depth of formation (Albinson et al., 2001), in the sense that Type C deposits (LS) are equivalent to shallow boiling deposits (generally named "Hot Spring Type"; i.e., San Martín in Querétaro), whereas Type A deposits (IS) have a tendency to form in deeper environments (i.e., San Martín de Bolaños in Jalisco, Sombrerete in Zacatecas, and Fresnillo—the Proaño veins—in Zacatecas). On the other hand, Type B deposits, which exhibit mineralization styles characteristic of both IS and LS deposits and are the most important types in the Mexican metallogenic provinces, seem to have dominantly formed at intermediate depths that correspond to the deep veins associated with boiling noted in Albinson et al. (2001). Thus, both classifications are similar, though the mineralization type is based on straight observations and integration, and the depth of formation is based on a set of analytical data, interpretations, and analogies.

In order to explain the various apparently complex relationships between contrasting IS and LS mineralization styles, a review will be made of several applicable and possibly genetically related fundamental processes. The first process calls for the presence of "doubly stratified brines," such as those known to occur in the Salton Sea geothermal system (McKibben et al., 1988) and those invoked to occur during the formation of the epithermal deposits in Fresnillo (Simmons, 1991). Under this scenario, the deepest and more saline brine would be injected into shallow crustal levels during periods of faulting or dike intrusions (Simmons, 1991; Sillitoe and Hedenquist, 2003). Unless the multiple dikes were emplaced at deeper non-observable depths, the actual effectiveness of such mechanism is questionable because, in México, although IS and LS epithermal deposits are indeed commonly associated with quartz-rich porphyry dikes, these observable dikes generally predate epithermal mineralizations (i.e., Bacís and La Ciénega in Durango, Zacatecas in Zacatecas, Guadalupe de los Reyes in Sinaloa, and La Guitarra in the State of México), commonly by a few million years and in some cases by tens of millions of years. Thus, although the occurrence of dikes might reflect the last hypabyssal igneous event before hydrothermal activity, the common emplacement of many epithermal veins in close association with pre-mineral dikes may be more reasonably explained due to the fact that the contact between dikes and their host rocks are physical discontinuities that can reactivate as faults and fractures more easily than country rocks without verti-

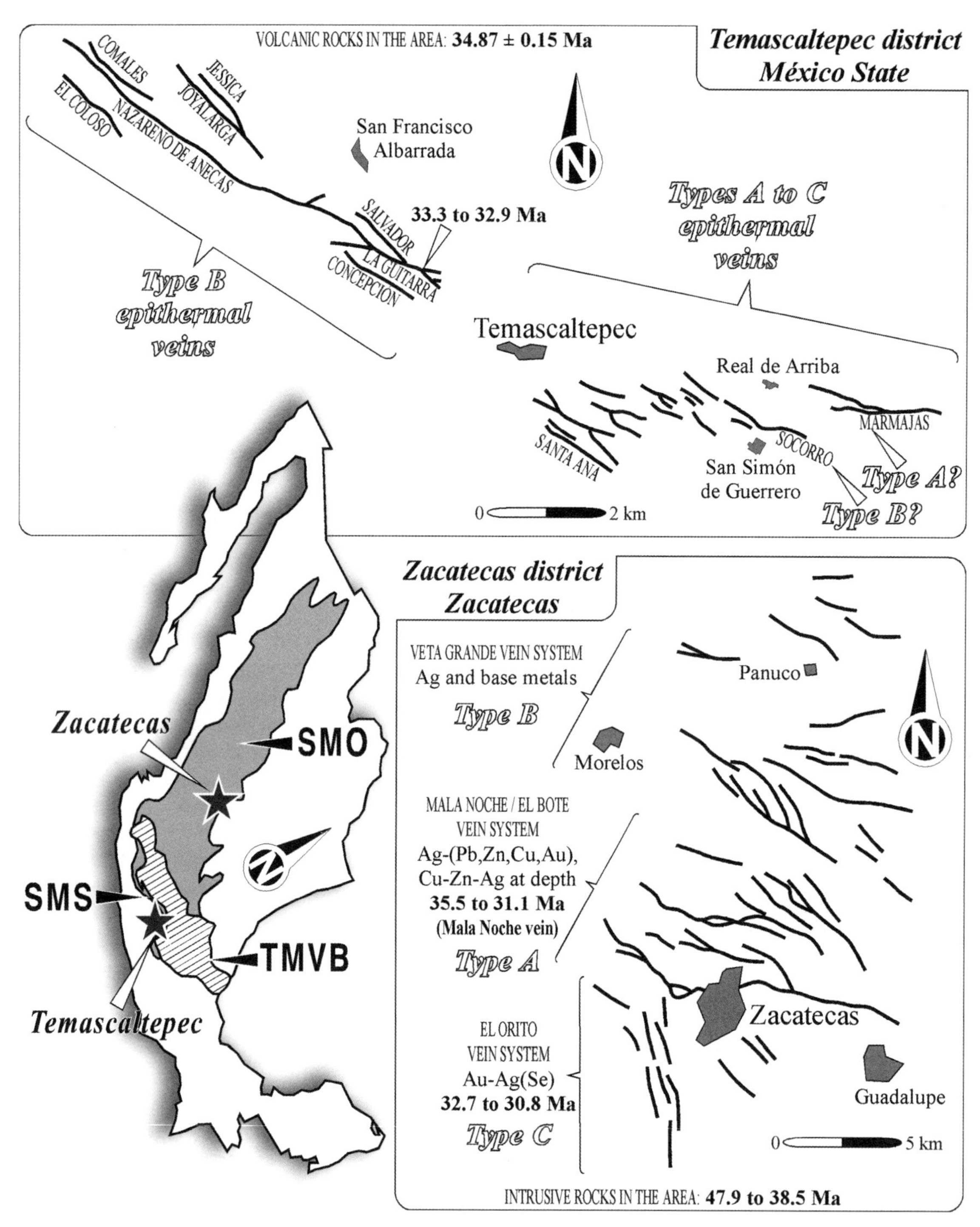

Figure 16. Diagrams showing the main mineralized veins at the Temascaltepec and Zacatecas districts, with the available radiometric ages. The Temascaltepec map was modified from a map on www.gencoresources.com/geology.html, and the Zacatecas map was simplified from Ponce and Clark (1988). The ages from Temascaltepec were obtained using the $^{40}Ar/^{39}Ar$ method by Blatter et al. (2001) for the volcanic rocks and by Camprubí et al. (2003a), and the ages from Zacatecas were obtained using the K-Ar method by Krueger (2000) for the hypabyssal intrusive rocks and the Mala Noche–El Bote vein system and by Kapusta (2006) for the El Orito vein system. TMVB—Trans-Mexican Volcanic Belt; SMO—Sierra Madre Occidental; SMS—Sierra Madre del Sur.

cal discontinuities. An alternative to a dike-driven mechanism is a tectonic mechanism that drives brines into shallow crustal levels as a result of a "seismic pumping" or "suction pump" mechanism (a term coined by Sibson et al., 1975) during periods of structural reactivation (Sibson, 1998, 2001; Sibson and Scott, 1998). If, on the other hand, the brines necessary to carry base metals remained stratified at depth and did not reach shallow levels in the crust, it is possible that some deposits could contain significant IS mineralization at greater depths and would occur vertically separated in space from LS mineralization, or "stacked" in the sense of Barton et al. (1995).

However, it is unlikely that the common coexistence and coincidence of IS and LS styles of mineralization in México is explained solely by the existence of stratified brine reservoirs, especially considering that the detailed study on some deposits indicates that the transitions from IS mineralization to LS mineralization actually reflect a fundamental evolution in time and space of hydrothermal fluids. At the La Guitarra deposit from the Temascaltepec district, the following mineralogic and fluid evolutionary path was established (Camprubí et al., 2001a): (1) Stage I mineralization (Fig. 6B), represented by early deep base metal–rich mineralization with FeS contents in sphalerite up to 0.08 and late silver-rich mineralization with sphalerite with FeS contents up to 0.25 molar, generally low Ag contents (with respect to Stages II and III), virtually no Au (there is no electrum in the paragenesis), and salinities for mineralizing fluids up to 14.4 wt% NaCl equiv.; and (2) Stages II and III (Figs. 6C, 6D, and 7E), represented by multibanded silica (Stage II) with progressively higher contents in precious metals (including Au in electrum in the paragenesis), FeS contents in sphalerite up to 0.16 molar (Stage II) and 0.12 molar (Stage III), and salinities for mineralizing fluids generally between 4 and 5 wt% NaCl equiv. Furthermore, within all stages, similar evolutionary paths are described, as all of them start with base metal minerals and end up with the precipitation of precious metal minerals. For example, sphalerite compositions of stage I evolve from FeS contents of 0.04 molar in the early base metal–rich mineralization to FeS contents up to 0.25 molar associated with the later silver mineral–bearing association. On the other hand, the maximum FeS contents in sphalerite appear to decrease in time from stage I (0.25 molar) to stage II (0.16 molar) to stage III (0.12 molar). Additionally, stage II sphalerites exhibit both maximum and minimum FeS contents in sphalerite increasing with elevation in the mine (Fig. 10 *in* Camprubí et al., 2001a). In sphalerite from the base-metal sulfide association from stage I (the most likely IS mineral association) the brine pulses are more diluted (less saline) with time, and such dilution is coupled with progressively higher FeS contents (i.e., progressive neutralization) in sphalerite (Fig. 6 *in* Camprubí et al., 2001b). These observations indicate that the coincidental occurrence of IS and LS styles of epithermal mineralization is not fortuitous but responds to mineralizing fluids whose initial sulfidation state was within the IS stability field and evolved toward progressively lower sulfidation states as a result of the progressive neutralization of mineralizing fluids by reaction with country rocks (buffering) and earlier vein material that gained a stronger control on the chemistry of fluids as hydrothermal pulses waned. This evolution is similar to that proposed for Lepanto in the Philippines and Julcani in Peru (Einaudi et al., 2003; Sillitoe and Hedenquist, 2003), except that the evolution portrayed for mineralizing fluids in those cases involves HS fluids evolving into IS fluids. In the scheme presented here, we may say that most epithermal deposits in México are hybrids of both IS and LS styles of mineralization, "are born" as products of IS fluids, and "die" as products of LS fluids. They also exhibit some evidence for a less conspicuous component of HS fluids involved in the early stages of some deposits. These HS fluids are manifested by the occurrence of early advanced argillic alteration, in the form of kaolinitization, typically in the deep portions of deposits like San Martín de Bolaños in Jalisco and Bacís in Durango. To date, the presence of hypogene alunite has not been documented in the root zones of IS deposits in México.

Examples of studies that show progressive decrease in sulfidation state as a result of progressive neutralization (or progressive cooling coupled with sulfur consumption) as the hydrothermal system evolves include the Santo Niño vein at the Fresnillo district in Zacatecas (Gemmell et al., 1988; Simmons et al., 1988; Simmons, 1991), which displays similar mineralogical trends. Additionally, the El Herrero vein at the Bacís district in Durango exhibits (1) an early stage of mineralization that is rich in base metal sulfides (with low contents in precious metals) and that is restricted to the deepest levels of the deposit, with salinities for mineralizing fluids between 3.3 and 10.6 wt% NaCl equiv.; and (2) a late stage of precious metal mineralization related to mineralizing fluids with salinities between 0.0 and 4.9 wt% NaCl equiv. (see Appendix *in* Albinson et al., 2001). It is necessary here to point out the occurrence of enargite in the La Guitarra deposit since it is the last mineral to precipitate within a mineral association that started with base metal sulfides followed by silver sulfides, gold and argentian tetrahedrite-tennantite, and silver sulfosalts (Camprubí et al., 2001a). An evolution like this describes a progressive decrease in sulfidation state of sulfide mineralization with a sudden increase in sulfidation state at the end (Fig. 17). This association was found only locally, and it may feasibly respond to the occurrence of a confined hydrothermal event that is clearly extemporaneous and contrary to the general trend. It nevertheless requires explanation. A sudden increase in sulfidation state could in fact occur in response to a final drop in temperature or by excess sulfur in solution following the consumption of most metals in solution within a confined microenvironment.

Two possible pathways of fluid evolution leading to low sulfidation state fluids in Mexican epithermal deposits can be proposed on $f(S_2)$/T diagrams (Fig. 17). The first pathway proposes that IS-type fluids loop downward into LS-type fluids though a process of progressive neutralization and/or sulfur consumption leading to LS styles of mineralization as a natural evolutionary consequence of neutralizing IS-type fluids. In this regard, deposits of Types A and B would result from magmatic hydrothermal

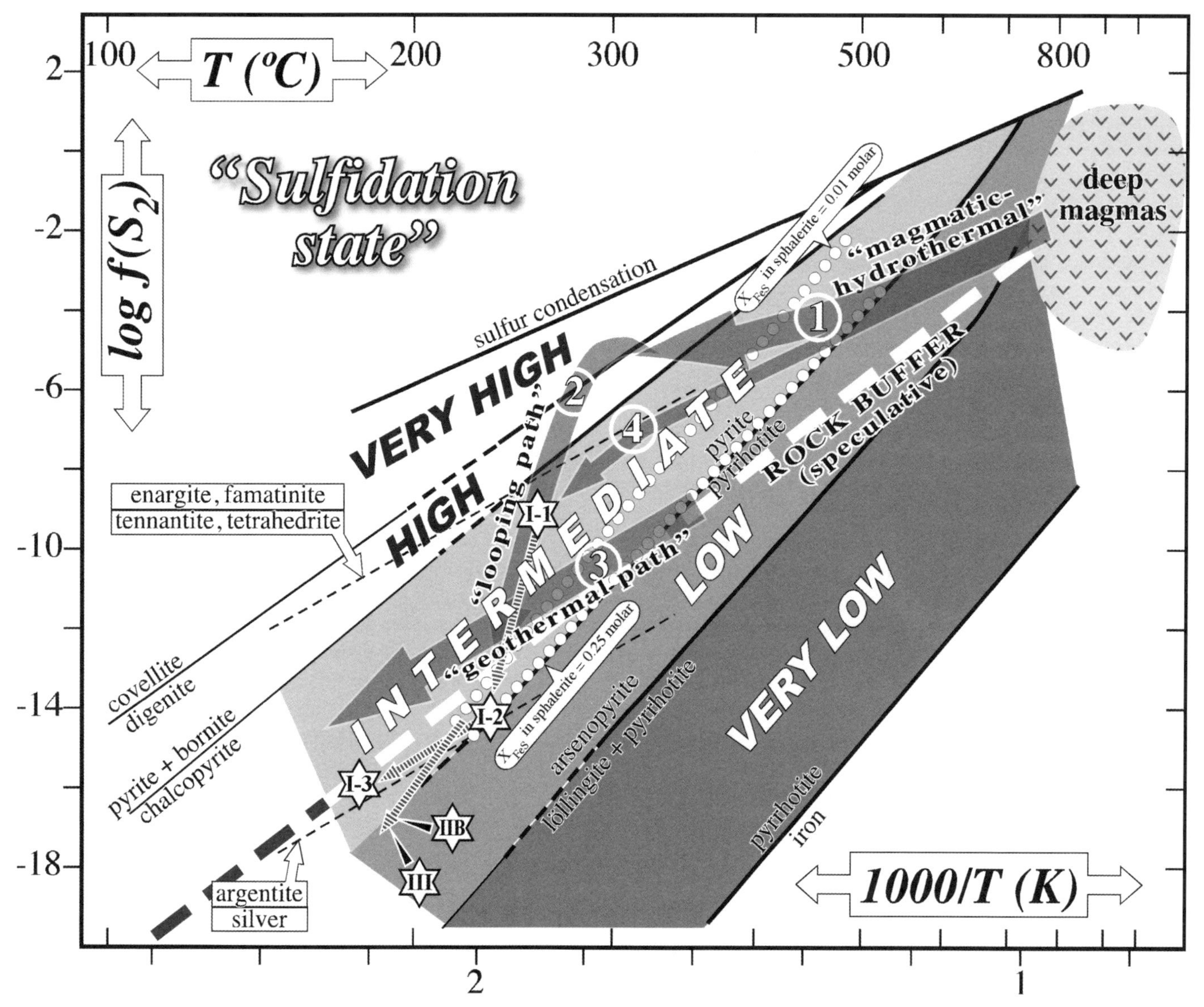

Figure 17. Diagram correlating temperature and sulfur fugacity for the sulfidation states of hydrothermal fluids in the environments of formation of porphyry and epithermal deposits, defined according to the stability fields of key minerals, with the general sulfidation paths determined by the metallic mineral associations at the La Guitarra epithermal deposit, Temascaltepec district, State of México (adapted from Camprubí et al., 2001a, 2001b). The three stages of metallic mineralization of the deposit are noted in Roman numerals. Dotted white lines represent the sulfidation curves for the approximate extreme contents of FeS in sphalerite found in the La Guitarra deposit, deduced from Figure 5 *in* Einaudi et al. (2003). The sulfidation trajectory determined for Stage I starts with the base metal sulfide association (I-1), positioned within the stability field of chalcopyrite and according to the highest Th recorded in this stage; the point where this trajectory changes its slope (I-2) is positioned according to the approximate composition of the sphalerite that is richest in FeS of Stage I, corresponding to the late silver mineral association; the trajectory ends with the lowest Th recorded in this stage (I-3), which remains within the intermediate sulfidation field due to the abundance of argentite and the lack of electrum in the paragenesis. The sulfidation trajectories for Stages IIB and III start slightly above the path between points I-2 and I-3 (due to the low X_{FeS} in early sphalerite) and fall below the stability field of silver, as both Stages IIB and III contain little argentite and have abundant Ag-bearing gold and Au-bearing silver (Camprubí et al., 2001a), which are commonly named "electrum." Notice that the slope of the parallel trajectories for the three stages is lower than the slope of the sulfidation curves of FeS contents in sphalerite, in order to explain the slight decrease in the FeS content of some zoned sphalerite crystals. The trajectory for Stage I also explains the compositional zoning of sphalerites, starting with FeS-poor zones in sphalerites that become progressively richer in FeS, and finally may record a new slight decrease in FeS in the late silver-rich mineral association of Stage I. Trajectories 1, 2, and 3 (white circles) are those described for porphyry copper deposits, base-metal veins associated to porphyries (or "looping path" of magmatic fluids), and active geothermal fluids (or "geothermal path"), respectively, according to Einaudi et al. (2003). Trajectory 4 (in white circle) is a speculative "flattened looping path" that could explain the sulfidation trajectories deduced for the La Guitarra deposit and may also explain the evolution of sulfidation state in other similar epithermal deposits in México. Adapted after Einaudi et al. (2003) and Sillitoe and Hedenquist (2003).

fluids that followed a more flattened "looping path" evolutionary trend (within the intermediate sulfidation field) than the looping path of hydrothermal fluids derived from a high level porphyry intrusive, which can reach higher sulfidation states during their evolution (Einaudi et al., 2003; Sillitoe and Hedenquist, 2003). Some Type C deposits may also represent an endmember product of this trend. On the other hand, Type C deposits may also be a product of hydrothermal fluids that originated from a very deeply seated magma and that repeatedly followed the "geothermal" path on their way to the surface (Fig. 14C). These would represent mature geothermal fluids that followed the rock buffer zone in the $f(S_2)$/T diagram from their exsolution of the magma to the LS epithermal environment. In this regard, the composition of the deeply seated magmas may be less important than the depth of their emplacement. That is, LS deposits formed from fluids that followed the "geothermal path" could be derived from basaltic or calcalkaline magmas alike. From this dual perspective for LS deposits, it can be stated that an important question to address consists of determining which LS deposits in México (or elsewhere) have better chances to host polymetallic IS roots, especially if these are accompanied by significant precious metal–bearing mineralization, which would render them more economically relevant.

Given that there are two distinctive and permissive evolutionary paths of fluids that lead to the formation of precious metal LS deposits, it would appear that LS deposits that were derived from magmatic-hydrothermal fluids that followed the "looping path" would represent the best candidates for underlying IS deposits, whereas LS deposits derived from fluids that followed the "geothermal path" would have questionable, if any, potential to host polymetallic mineralization at depth, and could in fact represent a truly mutually exclusive style of mineralization, in which case exploration at depth would be hopeless. Having said this, can any specific criteria be identified that could be useful at the level of LS deposits to discriminate between both alternative evolutionary pathways? At first glance, the maximum salinity for mineralizing fluids obtained in a LS deposit may represent an important clue to start making predictions about the possible occurrence or lack of associated deep IS deposits. The more common the occurrence of fluid inclusion assemblages with relatively high salinity brines (>7.5 wt% NaCl equiv.) that are not attributable to processes of extreme boiling (Simmons and Browne, 1997) and that truly represent brines derived from deep magmatic sources (that is, not derived from sedimentary brines or evaporites), it is estimated that the higher will be the potential for a deeper occurrence of significant IS mineralization. In the case of IS + LS mineralization styles, which occur overlapped and continuous in space, it would be relatively easier to carry out deeper exploration efforts. On the other hand, exploration for deep IS mineralization, which could occur vertically separated or "stacked" with respect to LS mineralization, represents an important challenge that should be considered in future exploration efforts for epithermal deposits in México and will require careful evaluation if it would be practical to attempt to locate such deposits.

RECOMMENDATIONS FOR FUTURE STUDIES AND EXPLORATION

Ages of Epithermal Deposits and Related Magmatism

A possible time and space link between deep magmatism and IS and LS epithermal deposits (McKee et al., 1992; Conrad et al., 1993; Simmons, 1995; Albinson et al., 2001; Enríquez and Rivera, 2001a; Camprubí et al., 2003a), beyond the evident connection between porphyry, skarn, and epithermal deposit types, is an important topic in the research on ore deposits. Radiometric age dating, particularly $^{40}Ar/^{39}Ar$ ages, allows the determination of high-resolution time relationships between different geological events. Age dating represents an essential tool to trace back the possible igneous sources for epithermal deposits and to determine the time ranges of different mineralization styles or, in other words, the degree of synchronicity between different mineralization styles within a single deposit.

The age and space distribution of the epithermal deposits in México is closely linked to Tertiary volcanic and subvolcanic activity, but the majority of these deposits and the youngest host volcanic or hypabyssal rocks still lack absolute dating and thus the evolution displayed by Camprubí et al. (2003a) must be understood as the starting point of a long-term program of dating of both deposits and host rocks. The present picture may be correct with respect to volcanic stratigraphy (the Lower Volcanic Series and the Upper Volcanic Series of the Sierra Madre Occidental) and their relation with epithermal deposits, but it is insufficient to describe the volcanic and metallogenic evolution within each volcanic series, and particularly absent with respect to the hydrothermal history of individual districts.

Deep HS Episodes in IS + LS Epithermal Deposits

In some deposits, as at San Martín de Bolaños in Jalisco, there is evidence for the incursion of low pH fluids in the deeper parts of the deposit through systematic X-Ray Diffraction (XRD) and Short-Wave Infra-Red Spectroscopy (SWIR) studies. We estimate that semi-systematic analysis of clay minerals in IS + LS epithermal deposits, either by XRD or SWIR, will continue to document that hypogene argillic or advanced argillic alteration assemblages in the deep portions or within IS deposits occur more commonly than suspected to date.

Petrography of Silica for Fluid Inclusions, Not Only Fluid Inclusion Petrography

Fluid inclusion microthermometry in epithermal deposits is an essential tool for obtaining estimations of deposition temperature, the chemistry of mineralizing fluids, and depth from the paleosurface or the paleowater table, and is the starting point to further undertake isotopic or volatile chemistry studies. However, fluid inclusion studies are useless or misleading unless these are based on detailed petrographic studies, not only in those studies leading to a complete characterization of the paragenetic sequence, but also studies that lead to the determination of the petrographic styles of silica minerals. There are textures that are diagnostic for recrystallization of silica, and these must be accounted before starting any fluid inclusion study as, besides criteria from fluid inclusion petrography, the silica petrography sets the most restrictive criteria to fluid inclusion microthermometry.

In addition, more volatile and noble gas isotope analyses in fluid inclusions are required, in order to complete the picture of the origin and evolution of mineralizing fluids in epithermal deposits, especially if the obtained data are linked to specific stages and substages of ore-forming events.

Mineral Geothermometers: The Neglected Ones

In light of recent work about the possible use of the composition of certain metallic minerals in certain mineral associations (i.e., Camprubí et al., 2001a; Chutas and Sack, 2004), what is required and has been lacking are semi-systematic to detailed electron microprobe (EPMA) studies of metallic mineral associations of epithermal deposits. The stability of silver sulfides and sulfosalts (Hall, 1966, 1967; Keighin and Honea, 1969), the As/(As + Sb) ratio in miargyrites (Ghosal and Sack, 1995) and the fahlore geothermometer (Sack, 2000, 2005) are important tools for obtaining temperature estimations for mineral associations with opaque minerals where fluid inclusion microthermometry is possible only in minerals like FeS-poor sphalerite and sometimes pyrargyrite under plane-polarized light, and in some more opaque minerals under infrared light. Furthermore, the composition of fahlore can be used to locate the position of feeder channels during the formation of single stages of mineralization (i.e., Loucks and Petersen, 1988; Camprubí et al., 2001a).

More studies are necessary in FeS content of sphalerite in all types of IS and LS epithermal deposits because the systematic analysis of sphalerite composition is commonly neglected although it may provide valuable information about the chemical evolution of mineralizing fluids.

Regional Metallogenesis

The three styles of IS and LS deposits in México, named types A, B, and C in this paper, may essentially be used for prospecting purposes within a known deposit, but also have a genetic significance, because IS-dominated types commonly record the most likely and highest magmatic influence, and LS-dominated types may not. In this paper, we suggest that Types A and B (perhaps some Type C deposits as well) derived from fluids that followed the "looping path" described by the sulfidation state of magmatic-hydrothermal fluids. In addition, we propose that fluids associated with the "geothermal path" would lead to the formation at high crustal levels of Type C deposits only, and that path could more likely correspond to fluids exsolved from deeper seated magmas than the fluids that followed the "looping path," which most likely exsolved from comparatively shallower magmas. Thus, the petrologic and geochemical characterization of parental magmas and their depth of formation are key problems to address on a regional scale. On a regional scale, it would be helpful to map the location of epithermal deposits in México, indicating, where possible, (1) parental magmas; (2) the type of the deposit, either A, B, C, or high sulfidation; (3) the age of the deposit, the age of the youngest host volcanic or hypabyssal rocks and, if exposed, the age of the parental magma; (4) the regional geochemical characterization of host volcanic and hypabyssal rocks; (5) the physical conditions of emplacement (temperature, pressure, depth) of regional hypabyssal rocks; and (6) the occurrence of different categories of fractures and faults that were active during the formation of epithermal deposits.

One of the distinctive characteristics of epithermal deposits as a deposit type in the Mexican metallogenic provinces is that HS deposits are very scarce compared to other areas with abundant epithermal deposits, such as the Great Basin in the United States, where both IS and HS deposits are well represented in the western calc-alkaline arc (see Figure 17 *in* Simmons et al., 2005). Instead, IS and LS deposits in México are comparatively overrepresented and are hosted within a volcanoplutonic arc that comprises one of the thickest accumulations of andesites, dacites, and rhyolites (Lower Volcanic Series) and of silicic ignimbrites (Upper Volcanic Series) on the globe (Ferrari et al., this volume) in the Sierra Madre Occidental. It is speculated that the ultimate cause for the prolific IS and LS populations of epithermal deposits in México is that the upper part of "parental" stocks are commonly emplaced at greater depths than in most areas of the crust that host epithermal deposits.

ACKNOWLEDGMENTS

The present paper was extracted as a block, with many new additions and the necessary rearrangements, from another paper titled "Depósitos epitermales en México: actualización de su conocimiento y reclasificación empírica," published in the 2006 Boletín de la Sociedad Geológica Mexicana, v. 58, no. 1, p. 27–81, as part of the celebrations of the centennial of the Mexican Geological Society (1904–2004). All the published papers for that occasion were conceived to have the same contents in Spanish and in English. However, in our original paper, we had the intention to offer the first comprehensive review of epithermal deposits in Spanish, in order to facilitate the access to knowledge to non-English or so-so-English readers, especially to students, but who needs another review on epithermal deposits in English? Thus, for our English version, we thought that it was better to take that out of the original paper in Spanish and leave the part that contained something new, that is, the review of Mexican epithermal deposits alone.

The original paper had a long history that began with the Ph.D. thesis of the first author by 1995, as a trial to put some order to the ideas of someone who by then ignored almost everything about epithermal deposits. That trial turned into the making of a review about epithermal deposits that tried to integrate all the available information about them. Hence, the paper came together, though intermittently, over ten years, and it has been the seed for other papers already published (Camprubí et al., 2003a), for fruitful collaborations, and hopefully for future research. Part of this paper comes from the final version of the Ph.D. thesis (Camprubí, 1999), which was financed by the European Union Fund for Research through project CI1*-CT94-0075 (HSMU 12), and was advised by Àngels Canals and Esteve Cardellach, whose support A. Camprubí wholeheartedly acknowledges. One of the previous versions of this paper was used by the first author to obtain tenure at the Universidad Nacional Autónoma de México (UNAM) in 2003 and, for that occasion, it received a general update and amendment.

This work has benefited from additional funding through research projects J32506-T and 46473-F granted by the Consejo Nacional de Ciencia y Tecnología (CONACyT), IN115999 and IN122604 granted by Programa de Apoyo a Proyectos de Investigación e Innovación Tecnológica–Dirección General de Asuntos del Personal Académico (PAPIIT-DGAPA), and annual budgets from the Centro de Geociencias and the Instituto de Geología at UNAM, basically to gather new information since 2000. We are very grateful to Minera Hochschild México S.A. de C.V. and Minas de Bacís S.A. de C.V. for allowing publication of K-Ar ages of epithermal veins in the Zacatecas district. For their critically constructive or motivational comments, suggestions, help in some versions of the text (or parts of it), and, definitely, for their contribution to improve and enrich this paper, special thanks are extended to Víctor A. Valencia, Carles Canet (who also formally refereed final versions of the manuscript), Stuart Simmons, Johannes Horner, Martín Valencia-Moreno, Ángel F. Nieto-Samaniego, Jeff Hedenquist, Lucas Ochoa-Landín, and Rosa María Prol-Ledesma. We are particularly indebted to T.J. Reynolds for his observations and constructive discussions at the eleventh hour of the manuscript. Some of the pictures used in this paper were kindly granted by Noel C. White, Eduardo González-Partida, and Víctor A. Valencia. Last, but not least, we want to congratulate the Sociedad Geológica Mexicana for its first century of existence.

REFERENCES CITED

Abeyta, R.L., 2003, Epithermal gold mineralization of the San Nicolás vein, El Cubo mine, Guanajuato, Mexico: trace element distribution, fluid inclusion microthermometry, and gas chemistry [M.Sc. thesis]: Socorro, New Mexico Institute of Mining and Technology, 130 p.

Albinson, T., 1985, Zoneamientos térmicos y su relación a la distribución de mineral en algunos yacimientos epitermales en México, *in* Memorias Técnicas XXVI Convención Nacional de la AIMMGM, Mazatlán, Sinaloa: México D.F., Asociación de Ingenieros de Minas, Metalurgistas y Geólogos de México (AIMMGM), 17 p.

Albinson, T., 1988, Geologic reconstruction of paleosurfaces in the Sombrerete, Colorada, and Fresnillo district, Zacatecas state, Mexico: Economic Geology and the Bulletin of the Society of Economic Geologists, v. 83, p. 1647–1667.

Albinson, T., 1995, Bosquejo de la evolución estructural e hidrotermal del distrito de Zacatecas, *in* Trabajos Técnicos XXI Convención Nacional Asociación de Ingenieros de Minas, Metalurgistas y Geólogos de México, Acapulco, Guerrero: México D.F., Asociación de Ingenieros de Minas, Metalurgistas y Geólogos de México (AIMMGM), p. 143–170.

Albinson, T., 2006, Petrographic and fluid inclusions of the Compás vein, El Orito, Zacatecas: México D.F., Exploraciones del Altiplano, S.A. de C.V., Private report for Minera Hochschild México, S.A. de C.V., 90 p.

Albinson, T., and Parrilla, L.V., 1988, Geologic, mineralogic and fluid inclusion characteristics of polymetallic veins, Real de Guadalupe mining district, Guerrero, Mexico: Economic Geology and the Bulletin of the Society of Economic Geologists, v. 83, p. 1975–1984.

Albinson, T., and Rubio, M.A., 2001, Mineralogic and thermal structure of the Zuloaga vein, San Martín de Bolaños District, Jalisco, Mexico: Society of Economic Geologists Special Publication Series, v. 8, p. 115–132.

Albinson, T., Simmons, S.F., and Smith, L., 1995, Geotermalismo de Nueva Zelanda y epitermalismo del Altiplano de México: ¿sistemas análogos o existe un eslabón perdido? *in* Trabajos Técnicos XXI Convención Nacional de la AIMMGM, Acapulco, Guerrero: México D.F., Asociación de Ingenieros de Minas, Metalurgistas y Geólogos de México (AIMMGM), p. 102–137.

Albinson, T., Norman, D.I., Cole, D., and Chomiak, B.A., 2001, Controls on formation of low-sulfidation epithermal deposits in Mexico: constraints from fluid inclusion and stable isotope data: Society of Economic Geologists Special Publication Series, v. 8, p. 1–32.

Allen, G., Thurston, B., and Wayne, R., 2001, Geology and gold-silver mineralization in the Guadalupe de los Reyes district, Sinaloa, Mexico: Society of Economic Geologists, Special Publication Series, v. 8, p. 59–70.

Arribas, A., Jr., Hedenquist, J.W., Itaya, T., Okada, T., Concepción, R.A., and Garcia, J.S., Jr., 1995, Contemporaneous formation of adjacent porphyry and epithermal Cu-Au deposits over 300 ka in northern Luzon, Philippines: Geology, v. 23, p. 337–340, doi: 10.1130/0091-7613(1995)023<0337:CFOAPA>2.3.CO;2.

Atkinson, W.W., Jr., 1990, A variety of types of epithermal mineral deposits at Moctezuma: Geological Society of America Abstracts with Programs, v. 22, no. 7, p. 41.

Atkinson, W.W., Jr., 1996, Evidence of a porphyry copper deposit below epithermal veins near Moctezuma, Sonora, Mexico: Geological Society of America Abstracts with Programs, v. 28, no. 7, p. 335.

Atwater, T., 1970, Implications of plate tectonics for the Cenozoic tectonic evolution of western North America: Geological Society of America Bulletin, v. 81, p. 3513–3536.

Barton, M.D., Staude, J.M., Zürcher, L., and Megaw, P.K.M., 1995, Porphyry copper and other intrusion-related mineralization in Mexico, *in* Pierce, F.W., and Bolm, J.G., eds., Porphyry copper deposits of the American Cordillera: Arizona Geological Society Digest, v. 20, p. 487–524.

Barton, P.B., Jr., 1970, Sulfide petrology: Mineralogical Society of America, Special Paper 3, p. 187–198.

Benton, L.D., 1991, Composition and source of the hydrothermal fluids of the Santo Niño vein, Fresnillo, Mexico, as determined from $^{87}Sr/^{86}Sr$, stable isotope, and gas analyses [M.Sc. thesis]: Socorro, New Mexico Institute of Mining and Technology, 55 p.

Blatter, D.L., Carmichael, I.S.E., Deino, A.L., and Renne, P.R., 2001, Neogene volcanism at the front of the central Mexican volcanic belt: basaltic andesites to dacites, with contemporaneous shoshonites and high-TiO_2 lava: Geologic Society of America Bulletin, v. 113, p. 1324–1342, doi: 10.1130/0016-7606(2001)113<1324:NVATFO>2.0.CO;2.

Bodnar, R.J., Reynolds, T.J., and Kuehn, C.A., 1985, Fluid inclusion systematics in epithermal systems: Reviews in Economic Geology, v. 2, p. 73–96.

Browne, P.R.L., 1978, Hydrothermal alteration in active geothermal fields: Annual Review of Earth and Planetary Sciences, v. 6, p. 229–250, doi: 10.1146/annurev.ea.06.050178.001305.

Browne, P.R.L., and Ellis, A.J., 1970, The Ohaaki-Broadlands hydrothermal area, New Zealand: Mineralogy and related geochemistry: American Journal of Science, v. 269, p. 97–131.

Camprubí, A., 1999, Los depósitos epitermales Ag-Au de Temascaltepec (Estado de México), México [Ph.D. thesis]: Barcelona, Universitat de Barcelona, Col·lecció de Tesis Doctorals Microfitxades, v. 3528, 252 p.

Camprubí, A., 2003, Geoquímica de fluidos de los depósitos epitermales del sureste del Distrito de Temascaltepec, Estado de México: Revista Mexicana de Ciencias Geológicas, v. 20, p. 107–123.

Camprubí, A., Prol-Ledesma, R.M., and Tritlla, J., 1999, Comments on "Metallogenic evolution of convergent margins: selected ore deposit models" by S.E. Kesler: Ore Geology Reviews, v. 14, p. 71–76.

Camprubí, A., Canals, À., Cardellach, E., Prol-Ledesma, R.M., and Rivera, R., 2001a, The La Guitarra Ag-Au low sulfidation epithermal system, Temascaltepec district, Mexico: Vein structure, mineralogy, and sulfide-sulfosalt chemistry: Society of Economic Geologists Special Publication Series, v. 8, p. 133–158.

Camprubí, A., Cardellach, E., Canals, À., and Lucchini, R., 2001b, The La Guitarra Ag-Au low sulfidation epithermal system, Temascaltepec district, Mexico: fluid inclusion and stable isotope data: Society of Economic Geologists Special Publication Series, v. 8, p. 159–185.

Camprubí, A., Tritlla, J., Corona-Esquivel, R., Centeno, E., and Terrazas, A., 2001c, The hydrothermal sinter and kaolinite-Au-Ag deposits of Ixtacamaxtitlán (Puebla, México): Preliminary research, *in* Pietrzynski, A., et al., eds., Mineral deposits at the beginning of the 21st century: Lisse, Netherlands, Swets & Zeitlinger, p. 711–714.

Camprubí, A., Ferrari, L., Cosca, M.A., Cardellach, E., and Canals, À., 2003a, Ages of epithermal deposits in Mexico: regional significance and links with the evolution of Tertiary volcanism: Economic Geology and the Bulletin of the Society of Economic Geologists, v. 98, p. 1029–1037.

Camprubí, A., Norman, D.I., and Chomiak, B.A., 2003b, Evidence for fluid sources by quadrupole mass spectrometry in the La Guitarra Ag-Au epithermal deposit, Temascaltepec district, Mexico: Journal of Geochemical Exploration, v. 78–79, p. 593–599, doi: 10.1016/S0375-6742(03)00067-0.

Camprubí, A., González-Partida, E., Iriondo, A., and Levresse, G., 2006a, Mineralogy, fluid characteristics and depositional environment of the Paleocene low-sulfidation epithermal Au-Ag deposits of the El Barqueño district, Jalisco, México: Economic Geology and the Bulletin of the Society of Economic Geologists, v. 101, p. 235–247.

Camprubí, A., Chomiak, B.A., Villanueva-Estrada, R.E., Canals, À., Norman, D.I., Cardellach, E., and Stute, M., 2006b, Fluid sources for the La Guitarra epithermal deposit (Temascaltepec district, México): Volatile and helium isotope analyses in fluid inclusions: Chemical Geology, v. 231, no. 3, p. 252–284.

Camprubí, A., González-Partida, E., and Torres-Tafolla, E., 2006c, Fluid inclusion and stable isotope study of the Cobre–Babilonia polymetallic epithermal vein system, Taxco district, Guerrero, México: Journal of Geochemical Exploration, v. 89, p. 33–38, doi: 10.1016/j.gexplo.2005.11.011.

Canet, C., Prol-Ledesma, R.M., Proenza, J., Rubio-Ramos, M.A., Forrest, M.J., Torres-Vera, M.A., and Rodríguez-Díaz, A.A., 2005a, Mn-Ba-Hg mineralization at shallow submarine hydrothermal vents in Bahía Concepción, Baja California Sur, México: Chemical Geology, v. 224, p. 96–112, doi: 10.1016/j.chemgeo.2005.07.023.

Canet, C., Prol-Ledesma, R.M., Torres-Alvarado, I., Gilg, H.A., Villanueva, R.E., and Lozano-Santa Cruz, R., 2005b, Silica-carbonate stromatolites related to coastal hydrothermal venting in Bahía Concepción, Baja California Sur, México: Sedimentary Geology, v. 174, p. 97–113, doi: 10.1016/j.sedgeo.2004.12.001.

Cathles, L.M., 1991, The importance of vein selvaging in controlling the intensity and character of subsurface alteration in hydrothermal systems: Economic Geology and the Bulletin of the Society of Economic Geologists, v. 86, p. 466–471.

Charest, A., and Castañeda, J., 1997, Geología y modelo yacimiento de oro El Sauzal, Chihuahua, Mex., *in* Trabajos Técnicos XXII Convención Nacional AIMMGM, Acapulco, Guerrero, no. 1: México D.F., Asociación de Ingenieros de Minas, Metalurgistas y Geólogos de México (AIMMGM), v. 137–148.

Chutas, N.I., and Sack, R.O., 2004, Ore genesis at La Colorada Ag-Zn-Pb deposit in Zacatecas, Mexico: Mineralogical Magazine, v. 68, p. 923–937, doi: 10.1180/0026461046860231.

Clark, K.F., 1990, Geology and mineral deposits of the Taxco mining district: Society of Economic Geologists Guidebook Series, v. 6, p. 281–291.

Clark, K.F., Damon, P.E., Shafiqullah, M., Ponce, B.F., and Cárdenas, D., 1981, Sección geológica-estructural a través de la parte sur de la Sierra Madre Occidental, entre Fresnillo y la costa de Nayarit, *in* Memorias Técnicas XIV Convención Nacional AIMMGM, Acapulco, Guerrero: México D.F., Asociación de Ingenieros de Minas, Metalurgistas y Geólogos de México (AIMMGM), p. 69–99.

Clark, K.F., Foster, C.T., and Damon, P.E., 1982, Cenozoic mineral deposits and subduction-related magmatic arcs in Mexico: Geological Society of America Bulletin, v. 93, p. 533–544, doi: 10.1130/0016-7606(1982)93<533:CMDASM>2.0.CO;2.

Clarke, M., and Titley, S.R., 1988, Hydrothermal evolution of silver-gold veins in the Tayoltita mine, San Dimas district, Mexico: Economic Geology and the Bulletin of the Society of Economic Geologists, v. 83, p. 1830–1840.

Conrad, J.E., McKee, E.H., Rytuba, J.J., Nash, J.T., and Utterback, W.C., 1993, Geochronology of the Sleeper deposit, Humboldt County, Nevada: epithermal gold-silver mineralization following emplacement of silicic flow-dome complex: Economic Geology and the Bulletin of the Society of Economic Geologists, v. 88, p. 317–327.

Consejo de Recursos Minerales, 1991, Geological-mining monograph of the state of Sinaloa: México D.F., Consejo de Recursos Minerales, Geological-Mining Monograph Series, Publication M-1e, 159 p.

Corbett, G.J., 2002, Epithermal gold for explorationists: AIG Journal, Applied Geoscientific Practice and Research in Australia, paper 2002-01, p. 1–26.

Corbett, G.J., and Leach, T.M., 1998, Southwest Pacific Rim gold-copper systems: structure, alteration, and mineralization: Society of Economic Geologists Special Publication 6, 238 p.

Damon, P.E., Shafiqullah, M., and Clark, K.F., 1981, Evolución de los arcos magmáticos en México y su relación con la metalogénesis: Revista del Instituto de Geología Universidad Nacional Autónoma de México, v. 5, p. 131–139.

Deen, J.A., and Atkinson, W.W., Jr., 1988, Volcanic stratigraphy and ore deposits of the Moctezuma district, Sonora, Mexico: Economic Geology and the Bulletin of the Society of Economic Geologists, v. 83, p. 1841–1855.

De la Garza, V., Olavide, S., and Villasuso, R., 2001, Geology and ore deposits of the La Ciénega gold district, Durango, Mexico: Society of Economic Geologists Special Publication Series, v. 8, p. 87–93.

Dong, G., and Morrison, G.W., 1995, Adularia in epithermal veins, Queensland: morphology, structural state and origin: Mineralium Deposita, v. 30, p. 11–19, doi: 10.1007/BF00208872.

Dong, G., Morrison, G., and Jaireth, S., 1995, Quartz textures in epithermal veins, Queensland—classification, origin, and implication: Economic Geology and the Bulletin of the Society of Economic Geologists, v. 90, p. 1841–1856.

Dreier, J., 2005, The environment of vein formation and ore position in the Purisima-Colon vein system, Pachuca Real del Monte district, Hidalgo, Mexico: Economic Geology and the Bulletin of the Society of Economic Geologists, v. 100, p. 1325–1348.

Einaudi, M.T., Hedenquist, J.W., and Inan, E.E., 2003, Sulfidation state of fluids in active and extinct hydrothermal systems: transitions from porphyry to epithermal environments: Society of Economic Geologists Special Publication Series, v. 10, p. 285–313.

Enríquez, E., and Rivera, R., 2001a, Timing of magmatic and hydrothermal activity in the San Dimas district, Durango, Mexico: Society of Economic Geologists Special Publication Series, v. 8, p. 33–38.

Enríquez, E., and Rivera, R., 2001b, Geology of the Santa Rita Ag-Au deposit, San Dimas district, Durango, Mexico: Society of Economic Geologists Special Publication Series, v. 8, p. 39–58.

Ferrari, L., López-Martínez, M., Aguirre-Díaz, G., and Carrasco-Núñez, G., 1999, Space-time patterns of Cenozoic arc volcanism in central Mexico: from the Sierra Madre Occidental to the Mexican Volcanic Belt: Geology, v. 27, p. 303–307, doi: 10.1130/0091-7613(1999)027<0303:STPOCA>2.3.CO;2.

Ferrari, L., López-Martínez, M., and Rosas-Elguera, J., 2002, Ignimbrite flare up and deformation in the southern Sierra Madre Occidental, western Mexico: implications for the late subduction history of the Farallon plate: Tectonics, v. 21, p. 17–1/24.

Ferrari, L., Valencia-Moreno, M., and Bryan, S., 2007, this volume, Magmatism and tectonics of the Sierra Madre Occidental and its relation with the evolution of the western margin of North America, *in* Alaniz-Álvarez, S.A., and Nieto-Samaniego, Á.F., eds., Geology of México: Celebrating the Centenary of the Geological Society of México: Geological Society of America Special Paper 422, doi: 10.1130/2007.2422(01).

Flores, T., 1920, Estudio geológico-minero de los distritos de El Oro y Tlalpujahua: Instituto Geológico de México Boletín 37.

Fredrickson, G., 1974, Geology of the Mazatlan area, Sinaloa [Ph.D. thesis]: Austin, University of Texas at Austin, 209 p.

Gammons, C.H., and Williams-Jones, A.E., 1997, Chemical mobility of gold in the porphyry-epithermal environment: Economic Geology and the Bulletin of the Society of Economic Geologists, v. 92, p. 45–59.

Gemmell, J.B., Simmons, S.F., and Zantop, H., 1988, The Santo Niño silver-lead-zinc vein, Fresnillo District, Zacatecas, Mexico: Part I. Structure, vein stratigraphy, and mineralogy: Economic Geology and the Bulletin of the Society of Economic Geologists, v. 83, p. 1597–1618.

Geyne, A.R., Fries, C., Jr., Segerstrom, K., Black, R.F., and Wilson, I.F., 1963, Geology and mineral deposits of the Pachuca-Real del Monte District, State of Hidalgo, México: México D.F., Consejo de Recursos Naturales No Renovables, Publication 5E, 203 p.

Ghosal, S., and Sack, R.O., 1995, As-Sb energetics in argentian sulfosalts: Geochimica et Cosmochimica Acta, v. 59, p. 3573–3579, doi: 10.1016/0016-7037(95)00223-M.

Gilmer, A.L., Clark, K.F., Conde, C., Hernandez, I., Figueroa, J.I., and Porter, E.W., 1988, Sierra de Santa Maria, Velardeña mining district, Durango, Mexico: Economic Geology and the Bulletin of the Society of Economic Geologists, v. 83, p. 1802–1829.

Gómez-Caballero, A., Ponce-Sibaja, B.F., and Miranda-Gasca, M.A., 1977, Informe de evaluación del distrito minero El Oro, Mex.–Tlalpujahua, Mich.: México D.F., Consejo de Recursos Minerales, Gerencia de Estudios Especiales, Unpublished technical report, 80 p.

González-Partida, E., 1981, La province filonienne Au-Ag de Taxco–Guanajuato (Mexique) [Ph.D. thesis]: Nancy, Institut National Polytechnique de Lorraine, 234 p.

Goodell, P., 1995, Porphyry copper deposits along the Batopilas lineament, Chihuahua, Mexico, *in* Pierce, F.W., and Bolm, J.G., eds., Porphyry copper deposits of the American Cordillera: Arizona Geological Society Digest, v. 20, p. 544.

Grant, G.J., and Ruiz, J., 1988, The Pb-Zn-Cu-Ag deposits of the Granadeña mine, San Francisco del Oro-Santa Bárbara district, Chihuahua, Mexico: Economic Geology and the Bulletin of the Society of Economic Geologists, v. 83, p. 1683–1702.

Gray, M.D., 2001, Exploration criteria for high sulfidation gold deposits in México, *in* Corona-Esquivel, R., and Gómez-Godoy, J., eds., Acta de Sesiones de la XXIV Convención Nacional de la AIMMGM: México D.F., Asociación de Ingenieros de Minas, Metalurgistas y Geólogos de México (AIMMGM), p. 68–71.

Gross, W.H., 1975, New ore discovery and source of silver-gold veins, Guanajuato, Mexico: Economic Geology and the Bulletin of the Society of Economic Geologists, v. 70, p. 1175–1189.

Gunnesch, K.A., Torres del Angel, C., Cuba, C.C., and Saez, J., 1994, The Cu-(Au) skarn and Ag-Pb-Zn vein deposits of La Paz, northeastern Mexico: mineralogic, paragenetic, and fluid inclusion characteristics: Economic Geology and the Bulletin of the Society of Economic Geologists, v. 89, p. 1640–1649.

Haas, J.L., Jr., 1971, The effect of salinity on the maximum thermal gradient of a hydrothermal system at hydrostatic pressure: Economic Geology and the Bulletin of the Society of Economic Geologists, v. 66, p. 940–946.

Hall, H.T., 1966, The systems Ag-As-S, Ag-Sb-S, and Ag-Bi-S: Phase relations and mineralogical significance [Ph.D. thesis]: Providence, Rhode Island, USA, Brown University, 172 p.

Hall, H.T., 1967, The pearceite and polybasite series: American Mineralogist, v. 52, p. 1311–1321.

Hayashi, K., and Ohmoto, H., 1991, Solubility of gold in NaCl- and H_2S-bearing aqueous solutions at 250–350 °C: Geochimica et Cosmochimica Acta, v. 55, p. 2111–2126, doi: 10.1016/0016-7037(91)90091-I.

Hedenquist, J.W., 1986, Geothermal systems in the Taupo volcanic zone: their characteristics and relation to volcanism and mineralisation, *in* Smith, I.E.M., ed., Late Cenozoic volcanism in New Zealand: Royal Society of New Zealand Bulletin, v. 23, p. 134–168.

Hedenquist, J.W., 1991, Boiling and dilution in the shallow portion of the Waiotapu geothermal system, New Zealand: Geochimica et Cosmochimica Acta, v. 55, p. 2753–2765, doi: 10.1016/0016-7037(91)90442-8.

Hedenquist, J.W., and Henley, R.W., 1985, Hydrothermal eruptions in the Waiotapu geothermal system, New Zealand: their origin, associated breccias, and relation to precious metal mineralization: Economic Geology and the Bulletin of the Society of Economic Geologists, v. 80, p. 1640–1668.

Hedenquist, J.W., Izawa, E., Arribas, A., Jr., and White, N.C., 1996, Epithermal gold deposits: styles, characteristics, and exploration: Resource Geology Special Publication, v. 1, 18 p.

Hedenquist, J.W., Arribas, A., Jr., and Reynolds, T.J., 1998, Evolution of an intrusion-centered hydrothermal system: Far Southeast-Lepanto porphyry and epithermal Cu-Au deposits, Philippines: Economic Geology and the Bulletin of the Society of Economic Geologists, v. 93, p. 373–404.

Hedenquist, J.W., Arribas, A., Jr., and Urien-Gonzalez, E., 2000, Exploration for epithermal gold deposits: Reviews in Economic Geology, v. 13, p. 245–277.

Hedenquist, J.W., Claveria, R.J.R., and Villafuerte, G.P., 2001, Types of sulfide-rich epithermal deposits, and their affiliation to porphyry systems: Lepanto–Victoria–Far Southeast deposits, Philippines, as examples, *in* ProExplo Congreso, Lima, Perú, April 24–28, 29 p.

Henley, R.W., 1985, The geothermal framework of epithermal deposits: Reviews in Economic Geology, v. 2, p. 1–24.

Henry, C.D., and Aranda-Gómez, J.J., 2000, Plate interactions control middle-late Miocene, proto-Gulf and Basin and Range extension in the southern Basin and Range: Tectonophysics, v. 318, p. 1–26, doi: 10.1016/S0040-1951(99)00304-2.

Herzig, P.M., and Hannington, M.D., 1995, Hydrothermal activity, vent fauna, and submarine gold mineralization at alkaline fore-arc seamounts near Lihir Island, Papua New Guinea, *in* Proceedings Pacific Rim Congress 1995: Australasian Institute of Mining and Metallurgy, p. 279–284.

Herzig, P.M., Petersen, S., and Hannington, M.D., 1999, Epithermal-type gold mineralization at Conical Seamount: a shallow submarine volcano south of Lihir Island, Papua New Guinea, *in* Stanley, C.J., et al., eds., Mineral Deposits: Processes to Processing: Rotterdam, A.A. Balkema, p. 527–530.

Horner, J.T., and Enríquez, E., 1999, Epithermal precious metal mineralization in a strike-slip corridor: The San Dimas district, Durango, Mexico: Economic Geology and the Bulletin of the Society of Economic Geologists, v. 94, p. 1375–1380.

Horner, J.T., and Steyrer, H.-P., 2005, An analogue model of a crustal-scale fracture zone in West-Central Mexico: evidence for a possible control of ore-forming processes: Neues Jahrbuch fur Geologie und Palaontologie, Abhandlungen, v. 236, p. 185–206.

Ibarra-Serrano, A., 1997, Compañía San Felipe S.A. de C.V., Unidad San Felipe-Mexicali, B.C. Geología y tipos de mineralización del yacimiento San Felipe, en Memorias Técnicas XXII Convención Nacional AIMMGM, vol. 1, Acapulco, Guerrero, México: México, D.F., Asociación de Ingenieros de Minas, Metalurgistas y Geólogos de México, p. 219–228.

Izawa, E., and Yamashita, M., 1995, Truscottite from the Hishikari mine, Kagoshima prefecture: Resource Geology, v. 45, no. 252, p. 251.

John, D.A., 2001, Miocene and early Pliocene epithermal gold-silver deposits in the northern Great Basin, western USA: Characteristics, distribution, and relationship to magmatism: Economic Geology and the Bulletin of the Society of Economic Geologists, v. 96, p. 1827–1853.

Kapusta, Y., 2006, K/Ar age dating of whole rocks from El Compas samples, Zacatecas District, Mexico: Geochronology and Isotopic Geotchemistry: Ancaster, Activation Labs, Private Report for Minera Hochschild Mexico, S.A. de C.V., 2 p.

Keighin, C.W., and Honea, R.M., 1969, The system Ag-Sb-S from 600 °C to 200 °C: Mineralium Deposita, v. 4, p. 153–171, doi: 10.1007/BF00208050.

Krueger, D., 2000, K/Ar age dating of whole rocks from the Zacatecas district, Mexico: Geochron Laboratories, Private Report for Minas de Bacis, S.A. de C.V., 9 p.

Losada-Calderón, A.J., and McPhail, D.C., 1996, Porphyry and high-sulfidation epithermal mineralization in the Nevados del Famatina mining district, Argentina: Society of Economic Geologists Special Publication Series, v. 5, p. 91–118.

Loucks, R.R., and Petersen, U., 1988, Polymetallic fissure vein mineralization, Topia, Durango, Mexico: Part II. Silver mineral chemistry and high resolution patterns of chemical zoning in veins: Economic Geology and the Bulletin of the Society of Economic Geologists, v. 83, p. 1529–1559.

Loucks, R.R., Lemish, J., and Damon, P.E., 1988, Polymetallic fissure vein mineralization, Topia, Durango, Mexico: Part I. District geology, geochronology, hydrothermal alteration, and vein mineralogy: Economic Geology and the Bulletin of the Society of Economic Geologists, v. 83, p. 1499–1528.

Lyons, J.I., 1988, Geology and ore deposits of the Bolaños silver district, Jalisco, Mexico: Economic Geology and the Bulletin of the Society of Economic Geologists, v. 83, p. 1560–1582.

Mason, B.E., 1995, A comparative evaluation of three jasperoid occurrences in the Mesa Central province of Mexico, and their possible relation to sediment-hosted precious metals mineralization [M.Sc. thesis]: Fort Collins, Colorado State University, 111 p.

McDowell, F.W., and Keizer, R.P., 1977, Timing of mid-Tertiary volcanism in the Sierra Madre Occidental between Durango city and Mazatlan, Mexico: Geological Society of America Bulletin, v. 88, p. 1479–1487, doi: 10.1130/0016-7606(1977)88<1479:TOMVIT>2.0.CO;2.

McGibbon, D.H., 1979, Origin and paragenesis of ore and gangue minerals, La Paz mining district, San Luis Potosí, Mexico [M.Sc. thesis]: Arlington, University of Texas at Arlington, 86 p.

McKee, E.H., Dreier, J.E., and Noble, D.C., 1992, Early Miocene hydrothermal activity at Pachuca-Real del Monte, Mexico: an example of space-time association of volcanism and epithermal Ag-Au mineralization: Economic Geology and the Bulletin of the Society of Economic Geologists, v. 87, p. 1635–1637.

McKibben, M.A., Andes, J.P., Jr., and Williams, A.E., 1988, Active ore formation at a brine interface in metamorphosed delataic lacustrine sediments: the Salton Sea geothermal system, California: Economic Geology and the Bulletin of the Society of Economic Geologists, v. 83, p. 511–523.

Miranda-Gasca, M.A., Pyle, P., Roldán, J., and Ochoa-Camarillo, H.R., 2005, Gold-silver and copper-gold deposits of Ixhuatán, Chiapas: a new alkalic-rock related metallogenic province of southeastern México, *in* Corona-Esquivel, R., and Gómez-Caballero, A., eds., Acta de Sesiones XXVI Convención Internacional de Minería, Veracruz, México: México D.F., Asociación de Ingenieros de Minas, Metalurgistas y Geólogos de México (AIMMGM), p. 69–70.

Molina-Garza, R.S., and Iriondo, A., 2007, this volume, The Mojave-Sonora megashear: The hypothesis, the controversy, and the current state of knowledge, *in* Alaniz-Álvarez, S.A., and Nieto-Samaniego, Á.F., eds., Geology of México: Celebrating the Centenary of the Geological Society of México: Geological Society of America Special Paper 422, doi: 10.1130/2007.2422(07).

Moller, S.A., Islas, J.E., and Davila, R.T., 2001, New discoveries in the La Colorada district, Zacatecas State, Mexico: Society of Economic Geologists Special Publication Series, v. 8, p. 95–104.

Morales-Ramírez, J.M., Tritlla, J., Camprubí, A., and Corona-Esquivel, R., 2003, Fluid origin of the Ixtacamaxtitlán hydrothermal deposits, Puebla State, Mexico: Journal of Geochemical Exploration, v. 78–79, p. 653–657, doi: 10.1016/S0375-6742(03)00139-0.

Muñoz, F., 1993, Modelo genético de los depósitos de oro proyecto San Martín, Querétaro, *in* Memorias Técnicas XX Convención Nacional AIMMGM, Acapulco, Guerrero: México D.F., Asociación de Minas, Metalurgistas y Geólogos de México (AIMMGM), p. 439–475.

Muntean, J.L., and Einaudi, M.T., 2001, Porphyry-epithermal transition: Maricunga belt, northern Chile: Economic Geology and the Bulletin of the Society of Economic Geologists, v. 96, p. 743–772.

Murray, M., 1997, Structural analysis of disseminated gold deposits in the Santa Gertrudis mining district, Sonora, Mexico [M.Sc. thesis]: Boulder, University of Colorado, 138 p.

Murray, M., and Atkinson, W.W., Jr., 1997, Structural analysis of disseminated gold deposits in the Santa Gertrudis mining district, Sonora, Mexico: Geological Society of America Abstracts with Programs, v. 29, p. 207.

Naden, J., Kilias, S.P., and Darbyshire, D.P.F., 2005, Active geothermal systems with entrained seawater as modern analogs for transitional volcanic-hosted massive sulfide and continental magmato-hydrothermal mineralization: The example of Milos Island, Greece: Geology, v. 33, p. 541–544, doi: 10.1130/G21307.1.

Nieto-Samaniego, Á.F., Alaniz-Álvarez, S.A., and Camprubí, A., 2007, this volume, Mesa Central of México: Stratigraphy, structure, and Cenozoic tectonic evolution, *in* Alaniz-Álvarez, S.A., and Nieto-Samaniego, Á.F., eds., Geology of México: Celebrating the Centenary of the Geological Society of México: Geological Society of America Special Paper 422, doi: 10.1130/2007.2422(02).

Noguez, B.A., Flores, M.J., and Toscano, A.F., 1988, El distrito minero de Zacualpan, Estado de México, *in* Salas, G.P., ed., Geología económica de México: México D.F., Fondo de Cultura Económica, 467–473.

Norman, D.I., and Musgrave, J.A., 1994, N_2-Ar-He compositions in fluid inclusions: Indicators of fluid source: Geochimica et Cosmochimica Acta, v. 58, p. 1119–1131, doi: 10.1016/0016-7037(94)90576-2.

Norman, D.I., Benton, L.D., and Albinson, T., 1991, Calculation of $f(O_2)$ and $f(S_2)$ of ore fluids, and pressure of mineralization from fluid inclusion gas analysis for the Fresnillo, Colorada, and Sombrerete Pb-Zn-Ag deposits, Mexico, *in* Leroy, J.L., and Pagel, M., eds., Source, transport and deposition of metals: Rotterdam, A.A. Balkema, p. 209–212.

Norman, D.I., Moore, J.N., and Musgrave, J.A., 1997a, Gaseous species as tracers in geothermal systems, *in* Proceedings, 22nd Workshop on Geothermal Reservoir Engineering, Stanford University, California, January 27–29.

Norman, D.I., Chomiak, B., Albinson, T., and Moore, J.N., 1997b, Volatiles in epithermal systems: The big picture: Geological Society of America Abstracts with Programs, v. 29, no. 6, p. A-206.

Pearson, M.F., Clark, K.F., and Porter, E.W., 1988, Mineralogy, fluid characteristics, and silver distribution at Real de Ángeles, Zacatecas, Mexico: Economic Geology and the Bulletin of the Society of Economic Geologists, v. 83, p. 1737–1759.

Petersen, S., Herzig, P.M., Hannington, M.D., Jonasson, I.R., and Arribas, A., Jr., 2002, Submarine gold mineralization near Lihir Island, New Ireland fore-arc, Papua New Guinea: Economic Geology and the Bulletin of the Society of Economic Geologists, v. 97, p. 1795–1814.

Petersen, U., and Vidal, C.E., 1996, Magmatic and tectonic controls of the nature and distribution of copper deposits in Peru: Society of Economic Geologists Special Publication Series, v. 5, p. 1–18.

Petruk, W., and Owens, D., 1974, Some mineralogical characteristics of the silver deposits in the Guanajuato mining district, Mexico: Economic Geology and the Bulletin of the Society of Economic Geologists, v. 69, p. 1078–1085.

Ponce, B.F., and Clark, K.F., 1988, The Zacatecas mining district: a Tertiary caldera complex associated with precious and base metal mineralization: Economic Geology and the Bulletin of the Society of Economic Geologists, v. 83, p. 1668–1682.

Prol-Ledesma, R.M., Canet, C., Torres-Vera, M.A., Forrest, M.J., and Armienta, M.A., 2004, Vent fluid chemistry in Bahía Concepción coastal submarine hydrothermal system, Baja California Sur, Mexico: Journal of Volcanology and Geothermal Research, v. 137, p. 311–328, doi: 10.1016/j.jvolgeores.2004.06.003.

Randall, J.A., Saldaña, E., and Clark, K.F., 1994, Exploration in a volcano-plutonic center at Guanajuato, Mexico: Economic Geology and the Bulletin of the Society of Economic Geologists, v. 89, p. 1722–1751.

Rentería, T.D., 1992, Estudio metalogenético de la mina Sombrerete en Zacatecas, México [M.Sc. thesis]: México D.F., Universidad Nacional Autónoma de México, 108 p.

Richards, J.P., 2000, Lineaments revisited: Society of Economic Geologists Newsletter, v. 42, 8 p.

Rivera, R., 1993, Cocientes metálicos e inclusiones fluidas del distrito minero de Real de Asientos, Aguascalientes, *in* Memorias Técnicas de la XX Convención Nacional AIMMGM, Acapulco, Guerrero: México D.F., Asociación de Ingenieros de Minas, Metalurgistas y Geólogos de México (AIMMGM), p. 310–325.

Ruaya, J.R., and Seward, T.M., 1986, The stability of chlorozinc (II) complexes in hydrothermal solutions up to 350 °C: Geochimica et Cosmochimica Acta, v. 50, p. 651–661, doi: 10.1016/0016-7037(86)90343-1.

Rubin, J.N., and Kyle, J.R., 1988, Mineralogy and geochemistry of the San Martín skarn deposit, Zacatecas, Mexico: Economic Geology and the Bulletin of the Society of Economic Geologists, v. 83, p. 1782–1801.

Ruvalcaba-Ruiz, D.C., and Thompson, T.B., 1988, Ore deposits at the Fresnillo mine, Zacatecas, Mexico: Economic Geology and the Bulletin of the Society of Economic Geologists, v. 83, p. 1583–1596.

Sack, R.O., 2000, Internally consistent database for sulfides and sulfosalts in the system Ag_2S-Cu_2S-ZnS-FeS-Sb_2S_3-As_2S_3: Geochimica et Cosmochimica Acta, v. 64, p. 3803–3812, doi: 10.1016/S0016-7037(00)00468-3.

Sack, R.O., 2005, Internally consistent database for sulfides and sulfosalts in the system Ag_2S-Cu_2S-ZnS-FeS-Sb_2S_3-As_2S_3: Update: Geochimica et Cosmochimica Acta, v. 69, p. 1157–1164, doi: 10.1016/j.gca.2004.08.017.

Saunders, J.A., 1996, Retardation of boiling and the genesis of shallow bonanza epithermal gold deposits: evidence from the Sleeper deposit, Nevada: Geological Society of America, Abstracts with Programs, v. 28, no. 7, p. A-94.

Scheubel, F.R., Clark, K.F., and Porter, E.W., 1988, Geology, tectonic environment, and structural controls in the San Martín de Bolaños district, Jalisco, Mexico: Economic Geology and the Bulletin of the Society of Economic Geologists, v. 83, p. 1703–1720.

Schwarz-Schampera, U., Herzig, P.M., and Hannington, M.D., 2001, Shallow submarine epithermal-style As-Sb-Hg-Au mineralization in the active Kermadec arc, New Zealand, *in* Pietrzynski, A., et al., eds., Mineral deposits at the beginning of the 21st century: Rotterdam, A.A. Balkema, p. 333–335.

Scott, A.M., and Watanabe, Y., 1998, "Extreme" boiling model for variable salinity of the Hokko low sulfidation epithermal Au deposit, southwestern Hokkaido, Japan: Mineralium Deposita, v. 33, p. 568–578, doi: 10.1007/s001260050173.

Sedlock, R.L., Ortega-Gutiérrez, F., and Speed, R., 1993, Tectonostratigraphic terranes and tectonic evolution of Mexico: Geological Society of America Special Paper 278, 153 p.

Sellepack, S.M., 1997, The geology and geochemistry of the El Sauzal gold prospect, southwest Chihuahua, Mexico [M.Sc. thesis]: El Paso, University of Texas at El Paso, 89 p.

Shenberger, D.M., and Barnes, H.L., 1989, Solubility of gold in aqueous sulfide solutions from 150 to 350 °C: Geochimica et Cosmochimica Acta, v. 53, p. 269–278, doi: 10.1016/0016-7037(89)90379-7.

Sherlock, R.L., Tosdal, R.M., Lehrman, N.J., Graney, J.R., Losh, S., Jowett, E.C., and Kesler, S.E., 1995, Origin of the McLaughlin mine sheeted vein complex: metal zoning, fluid inclusion, and isotopic evidence: Economic Geology and the Bulletin of the Society of Economic Geologists, v. 90, p. 2156–2181.

Sibson, R.H., 1998, Brittle failure mode plots for compressional and extensional tectonic regimes: Journal of Structural Geology, v. 20, p. 655–660, doi: 10.1016/S0191-8141(98)00116-3.

Sibson, R.H., 2001, Seismogenic framework for hydrothermal transport and ore deposition: Reviews in Economic Geology, v. 14, p. 25–50.

Sibson, R.H., and Scott, J., 1998, Stress/fault controls on the containment and release of overpressured fluids: examples from gold-quartz vein systems in Juneau, Alaska, Victoria, Australia, and Otago, New Zealand: Ore Geology Reviews, v. 13, p. 293–306, doi: 10.1016/S0169-1368(97)00023-1.

Sibson, R.H., Moore, J.McM., and Rankin, A.H., 1975, Seismic pumping—a hydrothermal fluid transport mechanism: Journal of the Geological Society of London, v. 131, p. 653–659.

Sillitoe, R.H., 1977, Metallic mineralization affiliated to subaerial volcanism: A review, *in* Anonymous, Volcanic processes in ore genesis: London, Institution of Mining and Metallurgy–Geological Society of London, p. 99–116.

Sillitoe, R.H., 1999, Styles of high sulfidation gold, silver and copper mineralisation in porphyry and epithermal environments, *in* PACRIM'99, Proc. 7th IEEE Pacific Rim Conference on Communications, Computers and Signal Processing (PACRIM '99), Victoria, B.C., Canada, August 23–25, 1999, p. 29–44.

Sillitoe, R.H., and Hedenquist, J.W., 2003, Linkages between volcanotectonic settings, ore-fluid compositions, and epithermal precious metal deposits: Society of Economic Geologists, Special Publication Series, v. 10, p. 314–343.

Sillitoe, R.H., Hannington, M.D., and Thompson, J.F.H., 1996, High-sulfidation deposits in the volcanogenic massive sulfide environment: Economic Geology and the Bulletin of the Society of Economic Geologists, v. 91, p. 204–212.

Simmons, S.F., 1991, Hydrologic implications of alteration and fluid inclusion studies in the Fresnillo District, Mexico. Evidence for a brine reservoir and a descending water table during the formation of hydrothermal Ag-Pb-Zn orebodies: Economic Geology and the Bulletin of the Society of Economic Geologists, v. 86, p. 1579–1601.

Simmons, S.F., 1995, Magmatic contributions to low-sulfidation epithermal deposits, *in* Thompson, J.F.H., ed., Magmas, fluids and ore deposits: Mineralogical Association of Canada Short Course Series, v. 23, p. 455–477.

Simmons, S.F., and Browne, P.R.L., 1997, Saline fluid inclusions in sphalerite from the Broadlands-Ohaaki geothermal system: a coincidental trapping of fluids boiled towards dryness: Economic Geology and the Bulletin of the Society of Economic Geologists, v. 92, p. 485–489.

Simmons, S.F., and Browne, P.R.L., 2000, Hydrothermal minerals and precious metals in the Broadlands-Ohaaki geothermal system: implications for understanding low-sulfidation epithermal environments: Economic Geology and the Bulletin of the Society of Economic Geologists, v. 95, p. 971–999.

Simmons, S.F., and Christenson, B.W., 1994, Origins of calcite in a boiling geothermal system: American Journal of Science, v. 294, p. 361–400.

Simmons, S.F., Gemmell, J.B., and Sawkins, F.J., 1988, The Santo Niño silver-lead-zinc vein, Fresnillo District, Zacatecas, Mexico: Part II. Physical and chemical nature of ore-forming solutions: Economic Geology and the Bulletin of the Society of Economic Geologists, v. 83, p. 1619–1641.

Simmons, S.F., White, N.C., and John, D.A., 2005, Geological characteristics of epithermal precious and base metal deposits: Economic Geology, 100th Anniversary Volume, p. 485–522.

Simpson, M.P., Simmons, S.F., and Mauk, J.L., 2001, Hydrothermal alteration and hydrologic evolution of the Golden Cross epithermal Au-Ag deposit, New Zealand: Economic Geology and the Bulletin of the Society of Economic Geologists, v. 96, p. 773–796.

Smith, D.M., Jr., Albinson, T., and Sawkins, F.J., 1982, Geologic and fluid-inclusion studies of the Tayoltita silver-gold vein deposit, Durango, Mexico: Economic Geology and the Bulletin of the Society of Economic Geologists, v. 77, p. 1120–1145.

Smith, L., 1995, Evolución dinámica y ocurrencia de mineralización, veta El Herrero, *in* Memorias Técnicas XXI Convención Nacional AIMMGM, Acapulco, Guerrero: México D.F., Asociación de Ingenieros de Minas, Metalurgistas y Geólogos de México (AIMMGM), 13 p.

Starling, T., 1999, Structural and remote sensing analysis of the San Luis del Cordero district, Durango: México D.F., Exploraciones del Altiplano S.A. de C.V., Unpublished report, 18 p.

Starling, T., Uribe, J., and Maldonado, D., 1997, Nuevos conceptos que definen los controles estructurales y la evolución del yacimiento de San Martín, Zacatecas, México, *in* Trabajos técnicos XXII Convención Nacional AIMMGM, Acapulco, Guerrero, no. 1: México D.F., Asociación de Ingenieros de Minas, Metalurgistas y Geólogos de México (AIMMGM), p. 327–338.

Staude, J.-M., 2001, Geology, geochemistry, and formation of Au-(Cu) mineralization and advanced argillic alteration in the Mulatos district, Sonora, Mexico: Society of Economic Geologists Special Publication Series, v. 8, p. 199–216.

Staude, J.-M., and Barton, M.D., 2001, Jurassic to Holocene tectonics, magmatism, and metallogeny of northwestern Mexico: Geological Society of America Bulletin, v. 113, p. 1357–1374, doi: 10.1130/0016-7606(2001)113<1357:JTHTMA>2.0.CO;2.

Tritlla, J., Camprubí, A., Morales-Ramírez, J.M., Iriondo, A., Corona-Esquivel, R., González-Partida, E., Levresse, G., and Carrillo-Chávez, A., 2004, The Ixtacamaxtitlán kaolinite deposit and sinter (Puebla state, Mexico): a magmatic-hydrothermal system telescoped by a shallow paleoaquifer: Geofluids, v. 4, p. 329–340, doi: 10.1111/j.1468-8123.2004.00095.x.

Valencia, V.A., 2005, Evolution of La Caridad porphyry copper deposit, Sonora, and geochronology of porphyry copper deposits in northwest Mexico [Ph.D. thesis]: Tucson, University of Arizona, 192 p.

Valencia, V.A., Ruiz, J., Barra, F., Ochoa-Landín, L., Pérez-Segura, E., and Espinoza, E., 2003, La Caridad porphyry Cu-Mo deposit: a porphyry-epithermal transition in the southwest North America Porphyry Copper Province, *in* Actas 10° Congreso Geológico Chileno: Concepción, Universidad de Concepción, 10 p.

Valencia-Moreno, M., Ochoa-Landín, L., Noguez-Alcántara, B., Ruiz, J., and Pérez-Segura, E., 2007, this volume, Geological and metallogenic characteristics of the porphyry copper deposits in México and their position in the world context, *in* Alaniz-Álvarez, S.A., and Nieto-Samaniego, Á.F., eds., Geology of México: Celebrating the Centenary of the Geological Society of México: Geological Society of America Special Paper 422, doi: 10.1130/2007.2422(16).

van den Kerkhof, A.M., and Hein, U.F., 2001, Fluid inclusion petrography: Lithos, v. 55, p. 27–47, doi: 10.1016/S0024-4937(00)00037-2.

Vassallo, L.F., 1988, Características de la composición mineralógica de las menas de la Veta Madre de Guanajuato: Revista Instituto de Geologia Universidad Nacional Autónoma de México, v. 7, p. 232–243.

White, N.C., and Hedenquist, J.W., 1990, Epithermal environments and styles of mineralization: variations and their causes, and guidelines for exploration: Journal of Geochemical Exploration, v. 36, p. 445–474, doi: 10.1016/0375-6742(90)90063-G.

White, N.C., Leake, M.J., McCaughey, S.N., and Parris, B.W., 1995, Epithermal gold deposits of the southwest Pacific: Journal of Geochemical Exploration, v. 54, p. 87–136, doi: 10.1016/0375-6742(95)00027-6.

Wilkerson, G., Deng, Q., Llavona, R., and Goodell, P., 1988, Batopilas mining district, Chihuahua, Mexico: Economic Geology and the Bulletin of the Society of Economic Geologists, v. 83, p. 1721–1736.

Manuscript Accepted by the Society 29 August 2006

Printed in the USA

Geological Society of America
Special Paper 422
2007

Epigenetic, low-temperature, carbonate-hosted Pb-Zn-Cu-Ba-F-Sr deposits in México: A Mississippi Valley–type classification

Jordi Tritlla*
Gilles Levresse*
Programa de Geofluidos, Centro de Geociencias, Campus U.N.A.M.–Juriquilla, Universidad Nacional Autónoma de México (UNAM), Carr. Qro-SLP km 15,5, Santiago de Querétaro 76230, México

Rodolfo Corona-Esquivel*
Museo de Geología–Instituto de Geología, Universidad Nacional Autónoma de México (UNAM), J. Torres Bodet No. 176, Col. Sta Maria de La Ribera, México D.F. 06400, México

David A. Banks*
School of Earth and Environment, The University of Leeds, Leeds LS2 9JT, UK

Hector Lamadrid*
Posgrado en Ciencias de la Tierra–Programa de Geofluidos, Centro de Geociencias, Campus U.N.A.M.–Juriquilla, Universidad Nacional Autónoma de México (UNAM), Carr. Qro-SLP km 15,5, Santiago de Querétaro 76230, México

Julien Bourdet*
UMR 7566 G2R-CREGU, Université H. Poincaré, B.P. 239 Vandoeuvre-lès-Nancy 54506, France

Porfirio Julio Pinto-Linares*
Instituto Potosino de Ciencia y Tecnología (IPICYT), Depto. Geología Económica, Camino a la Presa San José 2055, Col. Lomas 4 sección CP. 78216, San Luis Potosí, S.L.P. México

ABSTRACT

The low-temperature epigenetic and stratabound Pb-Zn-Cu-Ba-F-Sr–bearing ore deposits enclosed within sedimentary columns historically have been major sources of metals. Exploration companies still find these deposits to be a profitable exploration target due to their simple mineralogy as well as the large tonnage that can present, always considering the mineral districts as a whole.

In northeastern México, several nonmagmatic, low-temperature Pb-Zn-F-Ba deposits have been systematically considered as magmatic-related (skarns, high-temperature replacement deposits, epithermal deposits, etc.). Recently, these deposits

*E-mails: Tritlla: jordit@geociencias.unam.mx; Levresse: glevresse@geociencias.unam.mx; Corona-Esquivel: rcorona@servidor.unam.mx; Banks: eardab@earth.leeds.ac.uk; Lamadrid: lamadrid@geociencias.unam.mx; Bourdet: julien.bourdet@g2r.uhp-nancy.fr; Pinto-Linares: pjpinto@ipicyt.edu.mx.

Tritlla, J., Levresse, G., Corona-Esquivel, R., Banks, D.A., Lamadrid, H., Bourdet, J., and Pinto-Linares, P.J., 2007, Epigenetic, low-temperature, carbonate-hosted Pb-Zn-Cu-Ba-F-Sr deposits in México: A Mississippi Valley–type classification, in Alaniz-Álvarez, S.A., and Nieto-Samaniego, Á.F., eds., Geology of México: Celebrating the Centenary of the Geological Society of México: Geological Society of America Special Paper 422, p. 417–432, doi: 10.1130/2007.2422(15). For permission to copy, contact editing@geosociety.org.

have been restudied and placed within a scenario of deep fluid circulation of basinal brines through the Mesozoic sedimentary series, enriched in Ba, F, and metals during fluid flow and water-rock interactions. These fluids gave rise to a series of stratabound epigenetic ore deposits scattered throughout the whole Mesozoic carbonate platform and can be shown to be unrelated to any period of magmatism. There is no intense alteration to the host rocks. Commonly there is a close association with organic matter, either liquid hydrocarbons or bitumen; they have a very simple mineralogy of barite, celestine, fluorite, sphalerite, galena, and have low formation temperatures (90–105 °C) combined with variable salinities. These characteristics make these deposits similar to the Mississippi Valley–type deposits, possibly most similar to the Alpine-Appalachian subtype.

Keywords: low temperature, carbonate-hosted, Mississippi Valley–type (MVT) deposits; geochemistry; México.

INTRODUCTION

The existence of Mississippi Valley–type (MVT) deposits in México has been largely unrecognized by Mexican ore geologists and mining companies until quite recently, even though these deposits first attracted attention at the end of the nineteenth century. This is partly because these deposits contain small amounts of base metals (e.g., El Diente, Nuevo León; Sierra Mojada, Coahuila), barite (Múzquiz, Coahuila), celestine (distrito de Cuatrociénegas, Coahuila) and/or fluorite (La Azul, Taxco, Guerrero: Tritlla et al., 2001; La Encantada-Buenavista, Coahuila: González-Partida et al., 2003; Tritlla et al., 2004a, 2004b), compared to precious metal deposits (skarn, epithermal) or the larger tonnage base metal deposits (sedimentary exhalative mineral deposits [sedex], volcanogenic massive sulfide [VMS]).

De Cserna (1989) was the first to suggest the presence of MVT deposits in his review of the geology of México, but this was mostly ignored by both the scientific and mining communities in México. Due to the omnipresence of Tertiary intrusive bodies that crosscut the whole of the Mesozoic series in the center and NE of México, the genesis of the MVT ore deposits has been confused with deposits linked with magmatism. The recognition of MVT deposits is important and requires a new assessment of the metallogenic processes that affected the Mesozoic carbonate platform series of central and NE México.

LOW-TEMPERATURE CARBONATE-HOSTED DEPOSITS IN MÉXICO

The vast majority of MVT deposits in México are located within the Mesozoic carbonate sediments outcropping in the states of Guerrero, San Luis Potosí, Coahuila, and Chihuahua (Fig. 1). Due to the lack of research on this type of ore deposition in México, some ore deposits previously classified as magmatic-related can, after reevaluation, probably be reassigned as MVT deposits, broadening their distribution in Mexican territory.

The MVT ore deposits in México are mainly stratabound and represented by barite-bearing, celestine-bearing, fluorite-bearing, and base-metal-bearing ore bodies, the latter always associated with a deep supergenic alteration that, eventually, transformed the whole of the primary sulfides into a secondary nonsulfide Zn-Pb deposit.

The better known deposits are located in the state of Coahuila. Their location is mostly controlled by the position of the paleogeographic structures—mainly basement highs, associated basins, the presence of evaporitic horizons and reefal formations, as well as regional faults (Puente-Solís et al., 2005). Based upon their main substance, these deposits present a marked basin-scale vertical zoning and distribution, probably reflecting the different mobility of the cations throughout the sedimentary pile as well as slightly different genetic mechanisms. The barite-bearing deposits are located in the lower part of the Mesozoic limestone series, enclosed mainly within the Kimmeridgian Olvido Formation and are occasionally reported within the Barremian limestones (Cupido Formation; Puente-Solís et al., 2005) (Fig. 2); Pb-Zn-bearing deposits are found scattered throughout the entire Mesozoic sequence (Fig. 2), even though the main deposits are found within the carbonates of the Cupido Formation, of Hauterivian to Aptian age. Similarly, celestine deposits are mainly located within platform limestones (Aurora and Acatita Formations) of Aptian to Albian age, and mixed fluorite-celestine and fluorite-dominated deposits are located near the top of the Mesozoic sedimentary succession, within the last limestone horizons (Georgetown and Del Rio Formations) of Upper Cretaceous age (Cenomanian; Fig. 2).

Stratiform Barite-Dominated Deposits

North of the state of Coahuila (NE México), a series of stratiform barite deposits appear enclosed within the Olvido Formation (Kimmeridgian). These barite "mantos" (flats) present a significant lateral extension (several hundred meters to some km) and a very constant thickness of between 1 m and 3 m. They are made up exclusively of high-purity microcrystalline barite, with minor amounts of calcite and, locally, barite-celestine (Barosa, 2004, personal commun.).

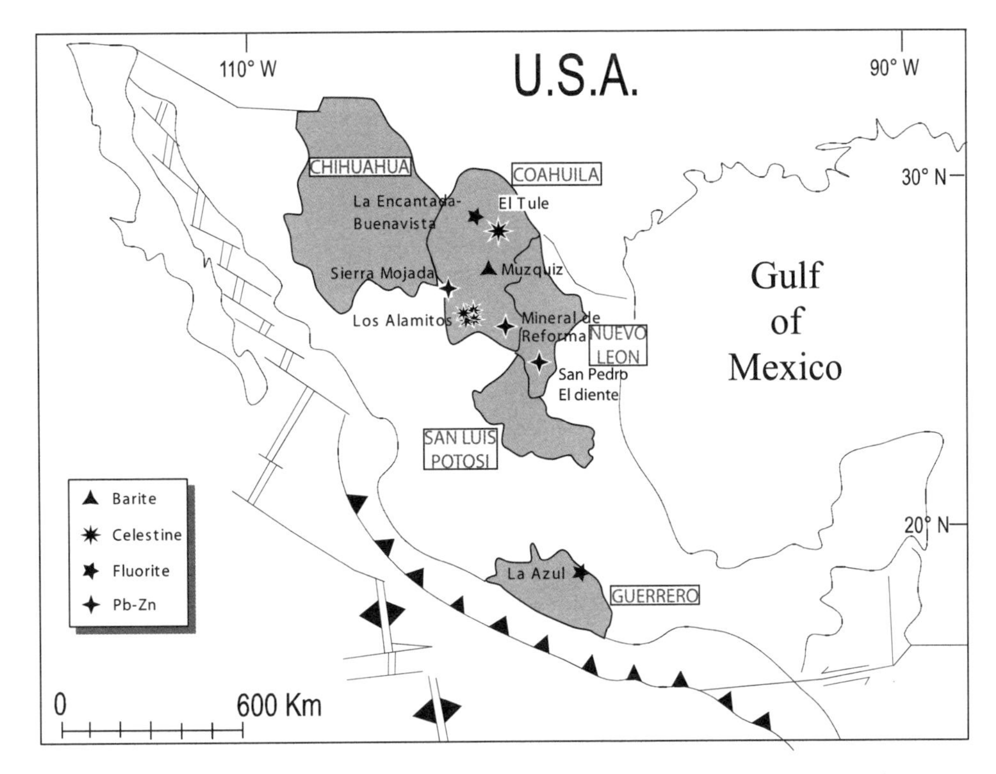

Figure 1. Location of the Mississippi Valley–type celestine, fluorite, and base metal deposits recognized in México.

The main stratiform bodies are mined near the town of Múzquiz (Coahuila). They consist of two stratiform to strata-bound barite mantos (Cocina and Potrero mines), vertically separated from each other by ~50 m, enclosed within the limestones of the Olvido Formation. Both the limestones and the barite horizons show concurrent folds produced during the Laramide Orogeny. This folding produces the partial obliteration of the barite horizons at the fold's hinge, as well as the generation of late fractures filled by coarse blocky calcite. It is noteworthy that the contact between the barite body and the enclosing limestone is always sharp (Figs. 3A and 3B) and marked by a pervasive discoloration zone a few decimeters thick, affecting the limestone. This color alteration grades from whitish at the contact to the original (non-altered) dark gray limestone (Fig. 3C), mainly reflecting the oxidation of the organic matter contained within the limestone due to the excess of sulfate during the barite formation. It has to be pointed out that both the barite and the fresh limestone are fetid (presence of H_2S).

In general, both barite horizons are composed of an isotropic mesh of fine grained, tabular crystals of barite. A close inspection reveals that the occurrence of these barite crystals is closely related to the pseudomorphosis of relic textures: (1) pseudo-stratification, marked by the disposition of residual minerals, mainly clays (Figs. 3A and 3B); (2) convolute surfaces, often boudinaged; (3) changes in the grain size or in the disposition of the pseudo-stratification from bottom to top of the mineralized body (Fig. 3D); (4) banded structures texturally similar to rythmites, made up by the disposition of white and dark barite bands that, occasionally, comprise the whole manto (Fig. 3E); and (5) the most revealing texture of all, globular accumulations of barite, with radial internal textures, with morphologies similar to the "chicken-wire" texture of anhydrite (Fig. 3F), typically from evaporitic series of gypsum that have undergone diagenetic dehydration and compaction. Also, the complete absence of cavities within the barite mass is notable. Locally, some thick calcite veins appear to be arranged close and parallel to the limestone contact within the barite horizon. These calcite veins predate the Laramide orogeny and seem to represent discrete zones of maximum fluid flow during the late stages of barite formation. This entire assembly is post-dated by some calcite veins that are closely related to the late folding and fracturing of the Mesozoic sediments.

Fluid inclusions in barite are scarce, small (<10µm) and are affected by post-trapping changes due to both the perfect cleavage of the barite and the deformation events. Homogenization temperatures (without pressure correction) for the Cocina mine are between 60 and 150 °C, suggesting that these fluid inclusions are affected by leakage. Homogenization temperatures in the Protrero mine have a smaller range of between 60 and 100 °C. In both cases, hydrohalite and ice melting temperatures indicate an electrolyte composition dominated by $CaCl_2$; the calculated total salinity is ~20–23 wt% with 18–22 wt% corresponding to $CaCl_2$ and 1–2 wt% to NaCl (E. González-Partida, 2004, personal commun.; Tritlla et al., 2005).

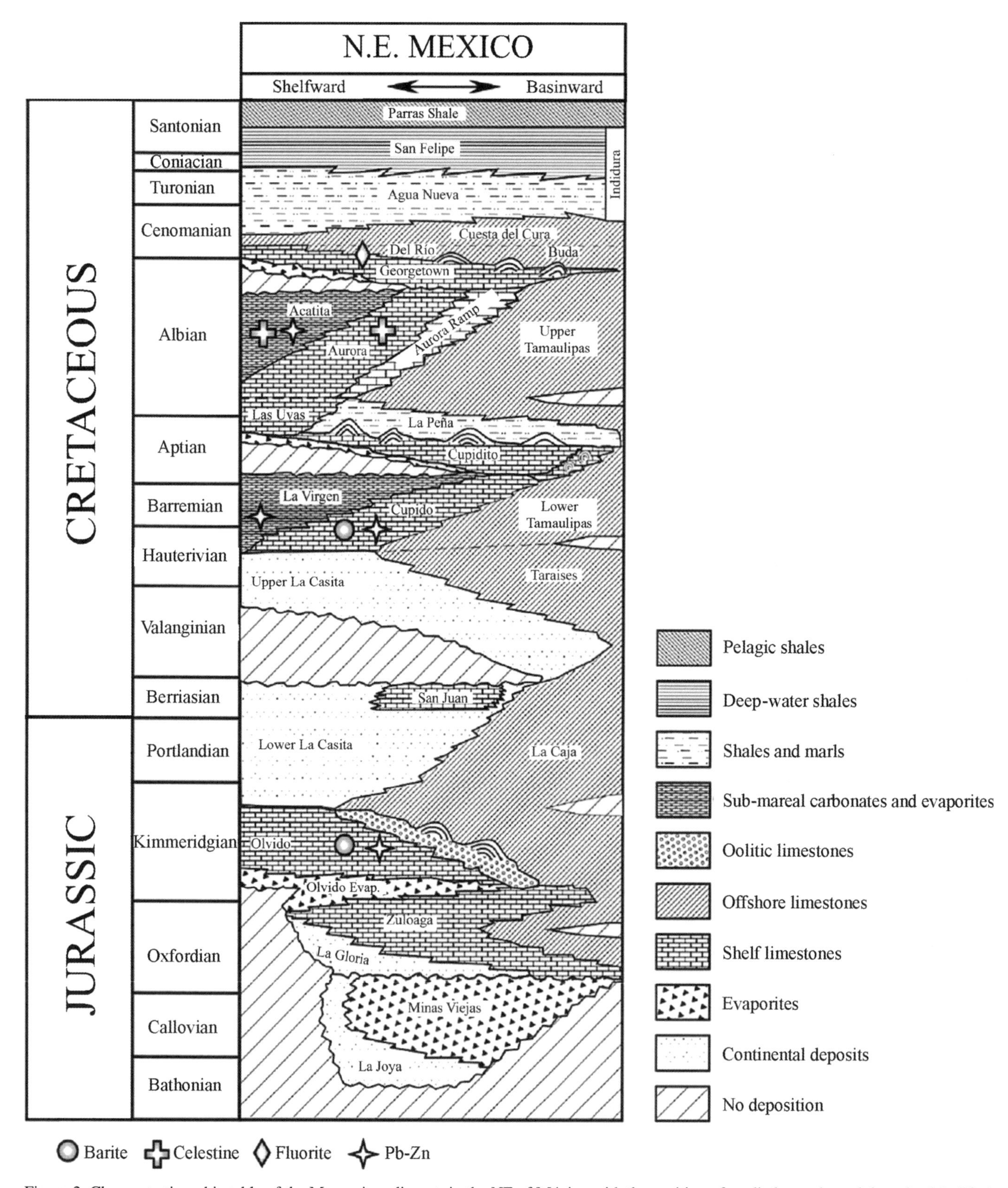

Figure 2. Chronostratigraphic table of the Mesozoic sediments in the NE of México with the position of studied stratabound deposits. Modified after Goldhammer (1999) and Puente-Solís et al. (2005).

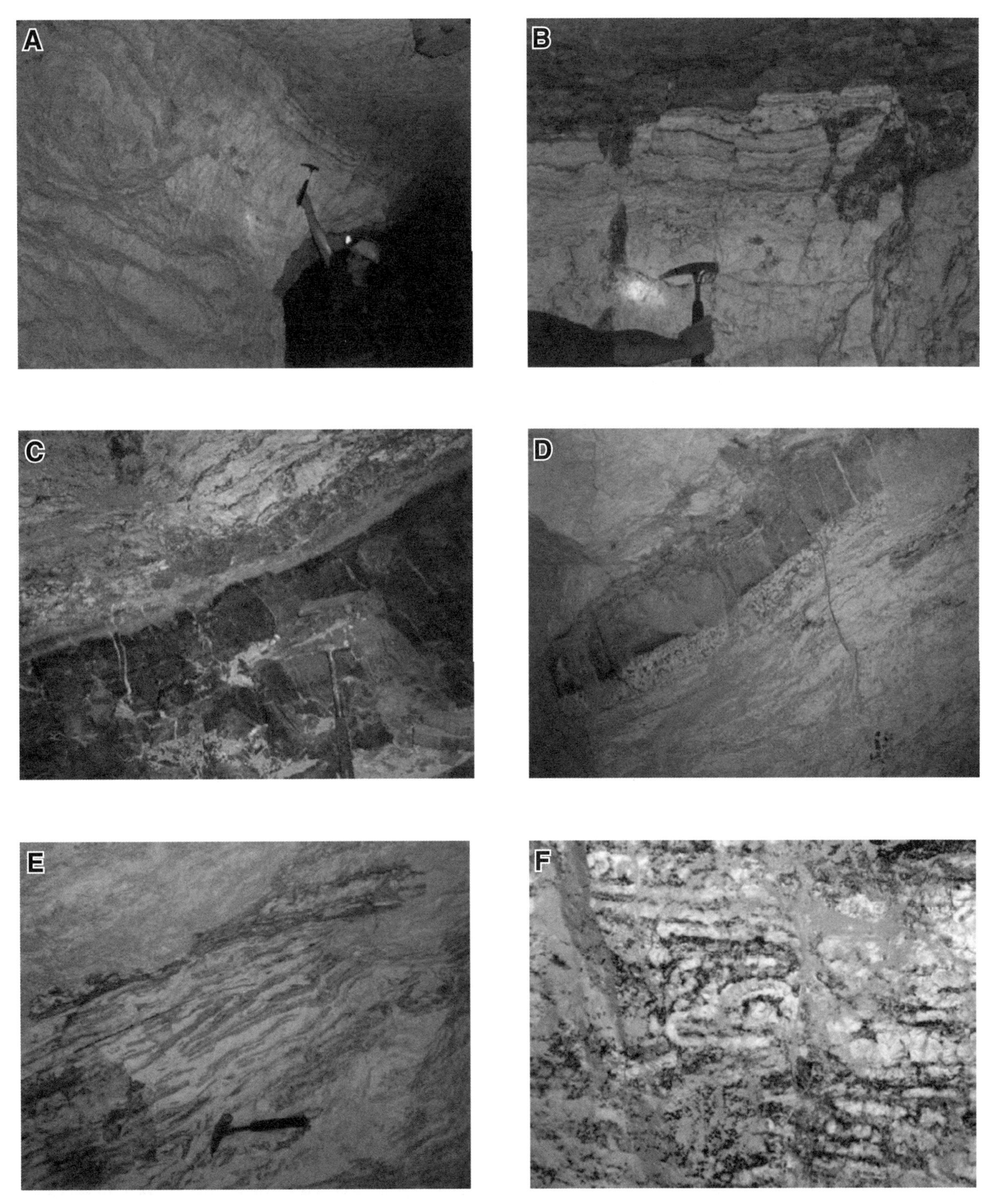

Figure 3. Barite deposits from Múzquiz (Coahuila). (A) General view of the stratiform barite manto, with the presence of rhythmites located both at the top and bottom of the body. (B) Detail of the barite rhythmites close to the contact with the enclosing limestone; this contact is disrupted by later fracturing. (C) Detail of the barite manto lower contact with the limestone; note the intense color alteration, several cm thick, affecting the limestone. (D) Limestone boulder within the barite manto; note the barite facies changing underneath the carbonate block with enterolithic textures, probably reflecting some primary diagenetic textures. (E) Detail of the "zebra" rhythmites developed in barite. (F) Remains of a clearly recognizable "chicken-wire" texture in barite, after anhydrite.

The stratiform character of the barite bodies, their diagenetic-like textures, and the omnipresence of organic matter or its degradation products, both within the barite and limestone, suggest that these barite deposits formed due to the substitution of former anhydritized gypsum horizons, subsequently transformed to barite by highly modified basinal brines that were mobilized during the first stages of the Laramide compression. The origin of Ba is uncertain, although formation waters can scavenge Ba released after feldspar breakdown during diagenesis and transport it in low-sulfate reducing fluids. In our case, the evaporitic horizons acted as a sulfate-rich, oxidant trap for a hot, saline, and Ba-rich solution. Their interaction produced pseudomorphs of the former anhydrite body by barite, partially preserving the original diagenetic textures. Also, the presence of a mobile, hot, sulfate-rich residual fluid, after barite formation, flowing through the evaporitic horizon, could account for oxidation of the organic matter contained within the adjacent limestone, resulting in the discoloration observed. The calcite-filled fractures could also act as channelways for the expulsion of the residual Ca-rich fluids.

Stratabound Celestine-Dominant Deposits

The Mesozoic platform sediments found in the states of Coahuila, San Luis Potosí, and Chihuahua contain one of the biggest accumulations of celestine deposits in the world; yet, this district has received little attention from the scientific and mining communities mainly due to the small and dispersed nature of the single ore bodies.

These celestine-dominant deposits in NE México are mainly found enclosed within the Acatita and Aurora Formations (Albian) in the central part of the Coahuila Platform (Alamitos, Australia, and La Paila Ranges); some other small and isolated celestine deposits are also found north of the San Marcos fault and in the SE margin of the Parras Basin (Puente-Solís et al., 2005).

Previous studies of the celestine deposits in México are scarce. Salas (1973) made the first thorough description of the Sr deposits, based on the mineralized lenses of the La Paila Ranges (Coahuila). Following Salas (1973), these celestine deposits appear as mantos made up of medium-sized, white celestine crystals that contain variable amounts of remnants of the enclosing limestone. When the celestine lenses are pure, it is usual to find pockets and cavities filled with idiomorphic crystals of celestine up to 10 cm in length, with minor quantities of native sulfur, fluorite, and gypsum. Salas (1973) and Rickman (1977) suggested that the La Paila celestine "mantos" are epigenetic and formed after the replacement of the enclosing limestones. Later, Kesler and Jones (1981), based on the S and Sr isotopic composition of gypsum, barite, and celestine (13 samples), concluded that the celestine deposits, formed from Sr derived exclusively from the limestone series, were probably expelled during diagenesis. More recently, Ramos-Rosique (2004), Ramos-Rosique et al. (2005), and Tritlla et al. (2004a, 2005) studied some of the celestine lenses from the Los Alamitos Ranges (El Venado, El Volcán, La Tinaja, La Víbora, El Diablo mines), presenting the first microthermometric data on these deposits and preliminary results on the brine halogen composition.

One of the main features of these deposits is their small to medium size. This not only prevents them being mined on an industrial scale, but also leads to the impossibility of knowing the total number of celestine bodies enclosed within the Mesozoic basins in Coahuila, San Luis Potosí, and Chihuahua. Moreover, it is notable that a huge amount of small mineralized lenses can appear in a single mining area and it is very common to see more than three different celestine-bearing lenses in the same sedimentary column, but in slightly different stratigraphic positions.

The celestine typically appear as stratiform to stratabound bodies within the Cupido (Aptian) and Acatita (Albian) limestone formations, with a general lense-shaped morphology, thicknesses up to 5 m, and total lengths exceeding 500 m (Fig. 4A). They are composed of euhedral to subhedral, prismatic to tabular, blue to black celestine crystals, up to 20 cm in length and 5 cm in width (Fig. 4B), with minor, late selenitic gypsum filling the remaining cavities, and subordinate calcite and traces of native sulfur. Occasionally, a very late, non-tabular celestine generation is found co-precipitating with the late selenitic gypsum. The ubiquitous presence of selenite as a late phase indicates either an excess of sulfate or the cessation of the Sr-bearing brine flow. The crystal size increases from the contact with the limestones (sucrose celestine) to the center of the bodies, forming large, euhedral celestine crystals with abundant intercrystalline voids. As a general pattern, crystals are arranged as rhythmites defined by the alternation of organic matter–rich (fetid, black) and organic matter–poor (blue) celestine (Figs. 4C and 4D). Both black and blue bands are fetid due to the presence of H_2S, probably trapped within fluid inclusions. It is significant that no remnants of precursor gypsum bodies or lenses have been found within the sedimentary series containing the celestine deposits.

A fluid inclusion study was undertaken to reveal the nature of the fluids involved in the genesis of these deposits (Ramos-Rosique, 2004; Ramos-Rosique et al., 2005; Tritlla et al., 2004a, 2005). Fluid inclusions in celestine are abundant but have frequently been affected by post-trapping changes, mainly necking-down and leakage due to the perfect cleavage of the host mineral. Fluid inclusions are mainly biphasic, with a visually estimated degree of filling of 0.9–0.95. Homogenization temperatures are constant, ranging from 80 to 130 °C, with variable calculated salinities ranging between 4 and 13 wt% NaCl eq.

The brine halogen composition, on a Cl/Br versus Na/Br molar ratio diagram, plots beneath but parallel to the trend defined by the evaporation of seawater (Tritlla et al., 2004a) (Fig. 5). This suggests that the electrolytes came from seawater modified by evaporation. These brines probably entered into the sedimentary pile as formation waters. Their salinity is less than would be expected compared with the salinity expected for the degree of seawater evaporation indicated by their halogen ratios. Thus, the fluid has been modified by dilution.

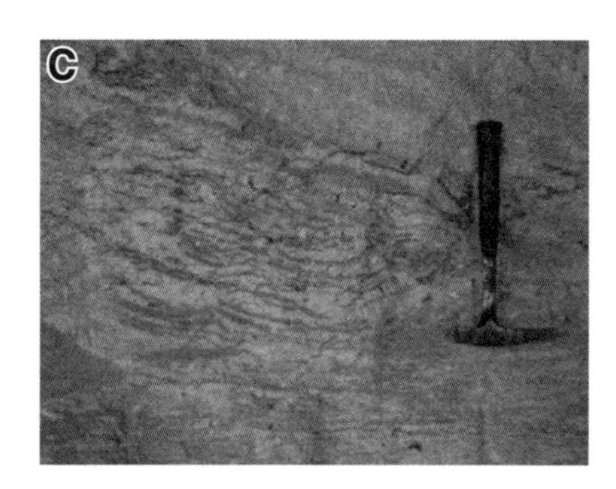

Figure 4. Celestine deposits. (A) General view of the stratabound, lenticular celestine body of La Víbora mine, Los Alamitos Ranges, Coahuila. (B) Celestine rhythmite located in the central part of the ore body, marked by the alternation of blue (O.M.-free) and black (O.M.-bearing) zones mostly within the same decimetric crystals; El Volcán mine, Los Alamitos Ranges, Coahuila. (C) Massive rhythmite zone, La Víbora mine, Los Alamitos Ranges, Coahuila. (D) Detail of the rhythmites located at the edge of the ore body, passing to the recrystallized host rock; La Víbora mine, Los Alamitos Ranges, Coahuila.

An unusual celestine deposit is located at El Tule, north of Muzquiz (Coahuila; Kesler, 1977) enclosed within the limestones of the Buda Formation (Washita Group). It comprises a single stratabound mineralized body whose disposition is controlled by a subhorizontal stratification joint with clear evidences of layer-parallel slip, acquiring a "pinch and swell" overall shape, with local mineralized zones up to 2 m in thickness. The deposit is celestine-dominated and has similar textures to the celestine-bearing deposits discussed above (rhythmites, tabular centimetric to decimetric crystals, fetid, etc.). The main difference is the presence of an early celestine generation with evidence of deformation during crystal growth (crystals bends, undulose extinction), while the latest, dominant celestine generation grew in a deformation-free environment. After celestine precipitation ceased, minor quantities of fluorite formed as a late phase, partially filling the remnant cavities and vugs in passive succession. This fluorite always appear as bluish to colorless, zoned, idiomorphic cubic crystals growing on top of the celestine crystals. No final selenitic gypsum has been recognized in this deposit.

Celestine contains abundant aqueous, two-phase fluid inclusions with evidences of post-trapping changes (necking-down). Homogenization temperatures are between 80 and 120 °C with very variable salinities between 5 and 11 wt% eq. of NaCl (Lamadrid, unpublished personal data). Raman analyses indicate that the gas phase is mainly composed of water vapor; no traces of other gases were found. Data plotted in a Th versus salinity plot suggest fluid mixing as the main mechanism for celestine precipitation, despite the somewhat scattered nature of the data.

Fluorite contains two distinct fluid inclusion types. The brine-bearing fluid inclusions are biphasic (L + V) to polyphasic (L + V + $S_{trapped}$). The trapped solids can be tiny quartz crystals or very small, high-birefringence minerals, thought to be calcite crystals. Raman analyses indicate the presence of variable amounts of CH_4, H_2S, and CO_2 within the gas phase. The hydrocarbon-bearing fluid inclusions are dark brown (heavy oils) and usually polyphasic (L + V + B) due to the presence of variable amounts of solid bitumen. Homogenization temperatures and salinities for the aqueous fluid inclusions are between 120 and 150 °C, and salinities are between 11.7 and 16 wt% eq. of NaCl, respectively (Lamadrid, unpublished personal data), showing much less variation than in the celestine inclusions. In a Th versus salinity plot, the data suggest that fluorite precipitated mainly by cooling after

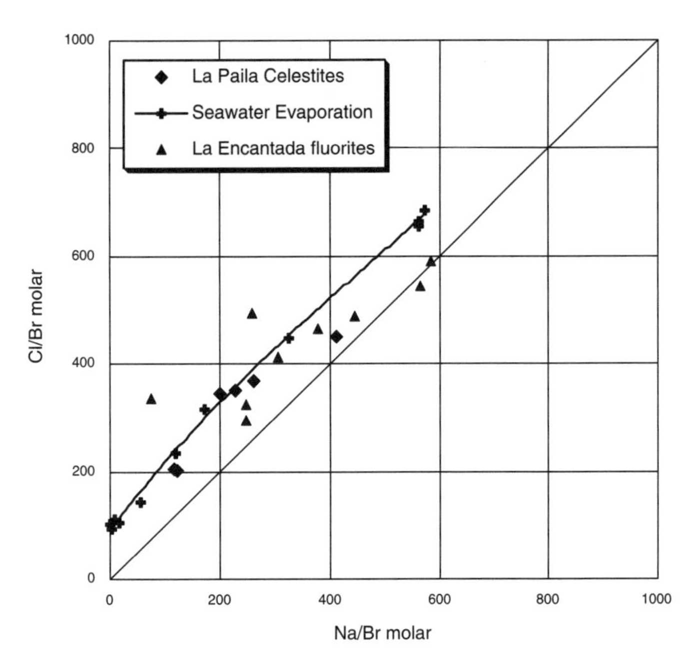

Figure 5. Na/Br versus Cl/Br compositions for the fluorite and celestine fluids. See text for explanation.

mixing of two different fluids. The petrographic analyses give clear evidence of coeval trapping of hydrocarbon-bearing and brine-bearing fluid inclusions within the same growth zone.

Thus, the mineralogical change from celestine to fluorite precipitation is also reflected in the change of the fluid composition and regime. Celestine precipitated within open sedimentary joints during and after the Laramide deformation, partially substituting the enclosing limestone. Fluorite precipitated after a dramatic change in the fluid composition, which probably inhibited the formation of celestine. During the mixing of the brine, most likely enriched in Ca^{2+} remaining after celestine precipitation, with an external emulsion of brine and hydrocarbons, partial degradation of the organic matter by means of thermochemical sulfate reduction (TSR) reactions occurred and generated the CH_4, H_2S, and CO_2 found in the gas phase of the brine-bearing fluid inclusions. An in situ origin for the small amounts of hydrocarbons found is unlikely, as the local source-rock for the organic matter was almost certainly depleted by the excess of sulfate during celestine formation.

Therefore, the El Tule deposit represents a rare example of a deposit that is transitional between the celestine lenses and mantos, well represented in the south of the state of Coahuila (La Paila and Alamitos Ranges), and the fluoritic mantos that are characteristic of the upper Cretaceous sediments in the north of the state of Coahuila (i.e., La Encantada-Buenavista).

Stratabound Fluorite-Dominant Deposits

A series of stratabound fluorite deposits and other occurrences are found in Cretaceous platform carbonates scattered all over México. These deposits are classified as: (1) MVT deposits, of syn- to post-orogenic Laramide age, related to basinal brines and organic matter (oil and gas), including the La Encantada–Buenavista district and the El Tule deposit in the state of Coahuila (Gonzalez-Partida et al., 2003, Tritlla et al., 2004b, 2005) and the La Azul deposit in the state of Guerrero (Tritlla and Levresse, 2006); (2) post-orogenic, nonmagmatic, very low-temperature fluorite bodies, including the Las Cuevas world-class deposit and several minor occurrences in the State of San Luis Potosí (Levresse et al., 2003); and (3) extremely unusual and very small fluorite-bearing skarns developed around F-rich rhyolitic necks in northern Coahuila (Levresse et al., 2006). Due to the ubiquity of magmatic rocks spatially close to these deposits, they were formerly classified as belonging to a single, magmatic-related category. Here, we only discuss the MVT-related typologies.

Fluorite Deposits in the State of Coahuila

La Encantada–Buenavista is the most representative fluorite-dominant MVT district deposit in México. It is located 180 km north of Muzquiz in the state of Coahuila (northern México). The district consists of several fluorite bodies and other occurrences scattered among the different limestone horizons that outcrop in the Sierra de la Encantada; the economic deposits appear mainly within the limestones of the Aurora Formation, of early Cretaceous age (Gonzalez-Partida et al., 2003). This formation is overlain by the Washita Group (Albian-Cenomanian), which is ~130 m thick and split into three formations, from bottom to top: (1) the Georgetown Formation (mudstone to wackestone limestones, 40 m thick); (2) the Del Rio Formation (thin-bedded shale, 10 m thick); and (3) the Buda Formation (impure wackestone-packstone, 80 m thick). The Laramide Orogeny affected the entire sedimentary pile, displaying two major folding styles: (1) asymmetrical and elongated anticline folds, and (2) domal anticlines with steep dips. During the middle Oligocene (Levresse et al., 2006), rhyolitic bodies intruded the sediment, affecting the previously formed fluorite bodies and causing a local enrichment of silica that makes these deposits unsuitable for industrial exploitation.

Because the upper limit for fluorite mineralization is always marked by the occurrence of the first strata of the Washita Group, these sediments are thought to act as a seal or flow barrier for the ore fluids, thus providing an explanation for the anomalously high occurrence of mineralized bodies near the Aurora-Georgetown Formations transition.

Fluorite is found at the intersection of low permeability barriers with low angle faults (layer-parallel slip) developed where stratification joints contain abundant clays (S. Baca, 2004, personal commun.; Tritlla et al., 2004b). The precipitation of fluorite is preceded by a substantial increase in the permeability of the host rock due to hydraulic fracturing (Fig. 6A). The fragments of the enclosing rock, within the brecciated structure, are extensively recrystallized, with some evidence of corrosion rims in the limestone boulders. Fluorite probably precipitated very rapidly, precluding the ongoing carbonate replacement. A thin calcite blanket, a few centimeters to a few decimeters thick, is usually found at the boundary of the fluorite body, smoothly transitioning into the slightly recrys-

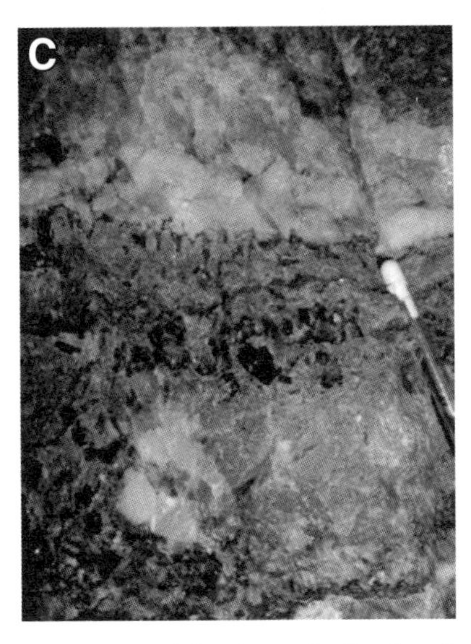

Figure 6. Stratabound fluorite ore bodies from La Encantada–Buenavista (Coahuila). (A) Hydraulic breccia within limestones at the front of a fluorite ore body. (B) Main stratabound fluorite manto exclusively made up of rhythmites of alternating white to colorless and purple fluorite. (C) Detail of the contact between the fluorite body (lower part) and the enclosing limestone (upper part); note the high degree of recrystallization of the limestone, with the development of a pure white calcite "blanket" at the front of the ore body. (D) Late cavity lined by idiomorphic, centimeter-sized purple fluorite crystals.

tallized enclosing limestone (Fig. 6A). The fluorite always appears as white to colorless, corrosion-free subhedral crystals indicative of passive mineral precipitation, usually arranged as rhythmites (Fig. 6B). Later cavities are filled by idiomorphic, cubic, heavily zoned fluorite crystals with a deep-purple color in their outer zones (Figs. 6C and 6D), minor idiomorphic calcite scalenohedrons, and extremely rare barite crystals (Tritlla et al., 2004b).

Aqueous and hydrocarbon-bearing fluid inclusions have been characterized by microthermometry (Gonzalez-Partida et al., 2003), crush-leach, Raman spectroscopy, Fourier Transform Infra-Red (FTIR) spectrometry, and confocal scanning laser microscopy (CSLM), and subsequently modeled using the Petroleum Inclusion Thermodynamics (PIT software; Thiéry et al., 2000).

Fluorite contains abundant primary fluid inclusions up to 50 μm in size, often distributed along growth planes (Fig. 7A). Three fluid inclusion types have been recognized (Figs. 7A and 7B): (1) biphase aqueous fluid inclusions (L + V); (2) brown to dark brown hydrocarbon-bearing fluid inclusions (L + V) that, occasionally, contain solid bitumen (L + V ± S); and (3) polyphase fluid inclusions comprising a saline fluid, liquid hydrocarbon, a gas bubble, variable amounts of solid bitumen and, rarely, minute calcite crystals. These three inclusion types are commonly found to be located in the same growth bands, indicating the coeval trapping of a brine-oil emulsion. Microthermometric data on this deposit (Gonzalez-Partida et al., 2003) indicate homogenization temperatures between 75 and 120 °C, with corresponding salinities between 10.5 and 14.9 wt% eq. NaCl. In some places, fluid inclusions are decrepitated due to the overheating caused by the intrusion of late subvolcanic rhyolite bodies, post-dating the mineralization events.

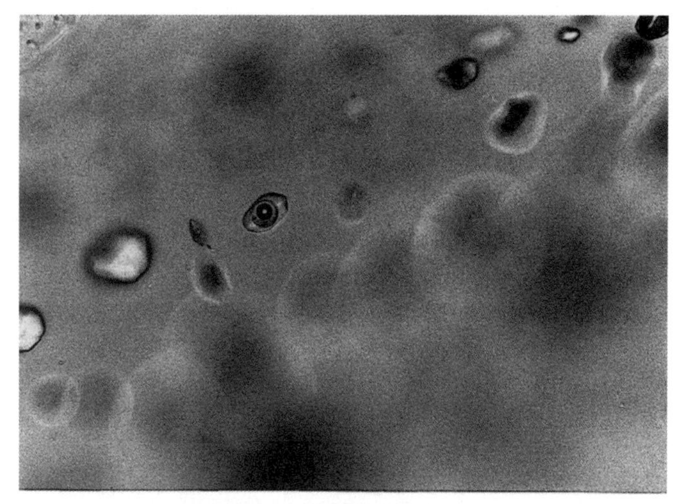

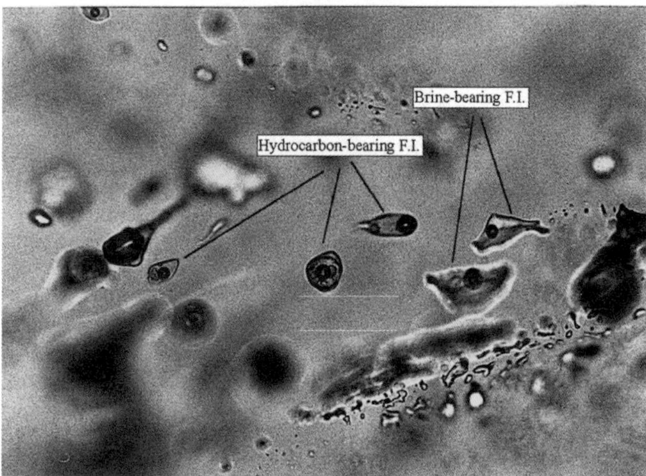

Figure 7. Fluid inclusions in fluorite, La Encantada–Buenavista district (Coahuila). Top: primary polyphase (brine + liquid hydrocarbon + gas + bitumen) fluid inclusions trapped within growth zones in fluorite. Bottom: coeval brine-bearing and hydrocarbon-bearing fluid inclusions trapped within the same growth bands in fluorite. Pictures taken by A. Ramos.

Both microthermometric and halogen analyses of brine-bearing fluid inclusions show that the fluids involved in the genesis of the La Encantada fluorite deposit are of meteoric origin (Tritlla et al., 2004a, 2004b, 2005), probably seawater that has undergone some degree of evaporation. Mixing of fluids is clearly supported by both microthermometric and halogen data, where the salinities are too low for the halogen ratios found, indicating mixing with a low-salinity end-member.

Hydrocarbon-bearing fluid inclusions are composed of yellow to brown "black oils" with low methane (30 mol%) and high CO_2 (5–30 mol%) contents and commonly contain solid bitumen, indicative of CO_2-asphaltene interaction or thermal degradation. Deformation of inclusions is probably due to post-trapping events consistent with pressure variations from 180 to 320 bar. The solubility of fluorite in NaCl solutions increases with temperature at all ionic strengths up to ~100 °C (Richardson and Holland, 1979) and at higher temperature increases with increasing $CaCl_2$ and $MgCl_2$ concentrations in NaCl-dominant solutions. Transport of fluoride is then favored and maximized by (basinal) brines enriched in Ca^2. Hydrocarbon-bearing fluid inclusions repeatedly show the presence of solid bitumen trapped along with heavy oils (Fig. 7), indicative of thermal degradation (Gonzalez-Partida et al., 2003). Both mixing and thermal degradation of hydrocarbon-rich fluids along with hydraulic fracturing of the host rock points to an in situ maturation of organic matter (Tritlla et al., 2004b) by means of TSR, due to the mixing of a saline, oxidized, sulfate-rich fluid, and $CaCl_2$-rich bittern. The latter fluid probably transported fluoride together with an organic matter–rich fluid present in the Cretaceous carbonates.

The products of TSR include H_2S (HS^-) and HCO_3^- (CO_2) (Machel, 2001). The production of volatiles would have resulted in a large volumetric expansion of the fluids (Dubessy and Ramboz, 1986), increasing the fluid pressure and, consequently, favoring the hydraulic fracturing of the enclosing rock. The presence of alkali earth metals and a reduction in the pH of the remaining brine often result in the formation of carbonates, particularly calcite and/or dolomite. In the La Encantada mine, calcite is present as minute solid inclusions, trapped contemporaneously with brine and hydrocarbon-bearing fluid inclusions in fluorite or as centimeter-sized scalenohedrons filling voids within the late-fluorite cavities. If F^- was transported as alkali complexes (CaF^+, for instance) in the bittern, then TSR could have produced sufficient CO_2 after maturation of organic matter to account for the destabilization of F-complexes and fluorite precipitation (Tritlla et al., 2004b).

It is noteworthy that the halogen ratios of the brines present in the celestine-dominated and fluorite-dominated deposits are comparable and overlap in a Cl/Br versus Na/Br plot. This indicates that both types of deposits were formed by the flow of the same highly modified evaporated seawater through the sedimentary pile. As suggested earlier (see the El Tule discussion), the predominance of celestine or fluorite can be controlled by a change in the fluid regime. This was initially controlled by formation waters (Sr-enriched) expelled from and through the sedimentary column during and immediately after the peak of the Laramide deformation. Later, the fluid regime changed to one dominated by gravity-driven flow in an extensive tectonic regime, where F could be scavenged from arkoses or the granitic basement.

Fluorite Deposits in the State of Guerrero

The fluorite deposits in the state of Guerrero are thought to have originated via the same processes discussed in the previous paragraph and are located along a broad NW-SE lineament between Acamixtla, Huajojutla, San Miguel Acuitlapán, and San Francisco de Acuitlapán. All the deposits and occurrences of fluorite are hosted by the limestones of the Morelos Formation (Cretaceous). Several small mines and workings are scattered throughout this zone, but only the La Azul mine open pit is of economic importance.

Research on the La Azul deposit has been carried out by several authors (Skewes-Saunders, 1938; Fowler et al., 1948; Gonzalez-Reyna, 1956; Osborne, 1956, Fernández, 1956; Florenzani, 1974; De Cserna and Fries, 1981; Clark, 1990; Pi et al., 2005; Tritlla et al., 1999, 2001; Tritlla and Levresse, 2006). The main reason for this research is La Azul's proximity (~30 km) to the important Taxco Ag-Zn-Pb-Cu district. The deposit consists of a single stratabound fluorite body that replaces the fossil-rich limestone of the Morelos Formation (Fig. 8A). The upper limit of the mineralization is marked by an irregular reaction front between the fluorite and the limestone. The fluorite mass passes into the fresh limestone by means of a thin recrystallization front (Fig. 8B), a few centimeters thick, with minute fluorite crystals located at calcite grain boundaries and whose abundance rapidly decrease toward the un-recrystallized limestones (Tritlla and Levresse, 2006; Fig. 8C). The lower contact does not outcrop. No fracture-related mineralization is associated with this deposit.

The fluorite contains alternations (rhythmites) of deep purple, black, and white bands (Figs. 8B and 8D) with rare quartz-rich bands and minor quantities of accessory minerals (barite, uraninite; Fig. 8E). Most of the fluorite found at La Azul is characteristically fetid due to the presence of H_2S. The gas is largely contained within the abundant primary biphase fluid inclusions and is interpreted to be a byproduct of organic-matter destruction, as has been recorded in similar fluorite deposits within comparable lithologies throughout México (El Tule deposit and the La Encantada–Buenavista district, Coahuila State; Gonzalez-Partida et al., 2003; Tritlla et al., 2004b). The "black & white" rhythmites contain abundant dark inclusions, probably degraded organic matter. The entire deposit is affected by a high number of hydrothermal fracturing episodes subsequently followed by the precipitation of additional fluorite layers. The substitution of the limestone by fluorite is not complete, and remnant blocks of limestone are found, partially corroded, within the fluorite mass. Carbonate replacement also increases the porosity of the limestone, creating abundant cavities within the fluorite mass, which are occasionally covered by minute cubes of a late yellow fluorite. The rhythmic textures characteristic of this deposit are similar to those found in carbonate-hosted, fluorite-rich deposits throughout México (Gonzalez-Partida et al., 2003; Levresse et al., 2003; Tritlla et al., 2001, 2004b) and worldwide (i.e., MVT deposits).

This deposit, as in the La Encantada–Buenavista fluorite district (Coahuila) discussed above, is intruded by a tertiary rhyolitic dyke, a few meters wide and heavily altered to a mixture of clays, quartz fragments, and some recognizable ferromagnesian minerals. This rhyolite clearly crosscuts and postdates the fluorite manto, including some brecciated fluorite blocks (Fig. 8D). These fluorite blocks as well as the fluorite adjacent to this dyke are heavily recrystallized and have acquired a characteristic grayish to reddish color, in contrast with the original bluish or whitish color found in the undisturbed fluorite. In addition, this recrystallized fluorite does not present the characteristic fetid odor found throughout the rest of the deposit. Some rare small veinlets of remobilized fluorite can also be found crosscutting the grayish recrystallized fluorite blocks. These textures were mistakenly identified as a dissolution breccia by several authors (Fowler et al., 1948; Fernández, 1956; Osborne, 1956; Florenzani, 1974). However, Skewes-Saunders (1938) and Tritlla and Levresse (1999, 2001, 2006) consider this structure to be a piecemeal stoping developed after the intrusion of a felsic subvolcanic body into the rigid fluorite mass.

Fluorite and quartz from La Azul contain abundant, negative crystal-shaped, primary biphase fluid inclusions. Microthermometric studies indicate that the fluids are $NaCl-CaCl_2$–rich fluids with a medium salinity (10–13wt% eq. NaCl) and temperatures (without pressure correction) between 120 and 150 °C (Tritlla and Levresse, 2006). In general, the homogenization temperatures from La Azul have a large range, mainly between 90 and 200 °C as a consequence of the rhyolite dyke intrusion.

Recently, Pi et al. (2005), disregarding the intrusive crosscutting relationships of the limestone, the fluorite body, and the rhyolitic dyke, reported a supposed deposit age of 32 ± 2 Ma (1σ) using the (U-Th)/He method. Tritlla and Levresse (2006) demonstrated that this age does not represent the deposit formation age but a resetting age due to emplacement of the rhyolite.

Stratabound Base-Metal Deposits

The stratabound, low-temperature base metal deposits have been largely ignored until very recently. Information on these deposits is very scarce and old, as they were mined during the second half of the nineteenth and the first decades of the twentieth centuries. Moreover, most of the low-temperature, base-metal carbonate-hosted deposits in México are located in a desert-like environment and have a deep, supergenic alteration, making it difficult to study the primary mineralization features. The primary sulfides are partially or completely replaced by "calamines," a mixture of anhydrous (smithsonite) and hydrated zinc carbonates (hydrozincite). This style of mineralization is typical of the deposits in Sierra Mojada, Sierra de la Purísima, and San Marcos deposits in Coahuila and Minas de San Pedro (also known as "El Diente") in Nuevo León. The base metal deposits (Pb-Zn) are roughly located around the Coahuila paleohighs and do not have the same strong stratigraphical control shown by the nonmetallic deposit. They are mainly hosted by the Cupido Formation but can also be found within the limestones of the Aurora, Acatita, and La Virgen Formations (Upper Jurassic to Cretaceous; Puente-Solís et al., 2005).

The Sierra Mojada was probably the biggest and best known mining district of this kind. It is located in the west central part of the state of Coahuila, near the Coahuila-Chihuahua state border in northern México. The district has a long history of mining activity since its discovery in 1879 and was historically considered a high-grade silver district. The initial discovery was the "Lead Manto" composed of silver-bearing cerussite. Later, in 1906, a Cu-Ag mineralization was discovered, and subsequently, in the 1920s, an oxide-zinc mineralization ("Red Manto") was found. The ores of the district have been selectively mined for bodies of sufficiently high grade to be shipped directly to the smelters (Metalline Mining Company, 2004, www.metalin.com).

Figure 8. La Azul fluorite stratabound body (Guerrero state). (A) Detail of the contact between fluorite mineralization (left) and the enclosing Morelos Formation limestone, heavily recrystallized. (B) Detail of the fluorite replacement front; this is marked by the presence of a thin, white recrystallized calcite rim. (C) Detail under the petrographic microscope of the recrystallization front; the calcite crystals developed well-connected grain borders and triple points, where tiny crystals of fluorite formed. (D) Fluorite rhythmites brecciated by hydraulic fracturing and subsequently recemented by late generations of fluorite. (E) Idiomorphic crystal of fluorite containing several anhedral crystals of uraninite; around these uraninite crystals a metamictic aureole developed due to fluorite lattice destruction. (F) Detail of the rhyolitic dyke, heavily altered and containing fragments of the highly recrystallized fluorite.

The ore deposits are located on the southern margin of the Sabinas Basin and comprise two stratabound mineral systems separated by the east-west–trending Sierra Mojada Fault (Fig. 9). North of the fault, the deposits contain disseminated to massive Cu-Ag-sulfide–bearing mineralization (tetrahedrite, sphalerite, and galena) with celestine-rich zones mostly altered to powdery strontianite. South of the fault, the mineralization is mainly composed of supergene bodies of secondary zinc (hemimorphite, smithsonite, and sauconite) and lead (cerussite) minerals. North of the fault, mineralization is hosted by the Upper Jurassic and lower Cretaceous sediments; in the south, it is hosted by the limestones of the Aurora and La Peña Formations (Middle Cretaceous) (Metalline Mining Company, 2004, www.metalin.com). The deposition of the ores is not associated with any hydrothermal alteration of the host rocks, except for some dolomitization affecting the host rocks of the Red Ore Manto. The absence of a pervasive hydrothermal alteration, the composition of the primary mineralization, and the morphology of the ore bodies suggest that these deposits can be classified as MVT.

The Sierra Mojada is the only MVT district presently undergoing economic reevaluation in México by the Metalline Mining Company. With a total of 7108 ha and its potential for low-grade, high tonnage Zn deposits, mainly calamines, as well as minor Zn-Pb-Cu-Ag primary ores, it may prove to be exploitable.

The Mineral de Reforma District is located in the La Purísima and San Marcos Ranges and includes the Reforma, Ojo de Agua, and Juárez mines. This district acquired some importance between the end of the nineteenth and the beginning of the twentieth centuries. The ore bodies that were exploited are stratabound and enclosed within the oolitic and reef facies of the Cupido Formation (González-Ramos, 1984). The emplacement of the ore deposits is marked by a strong alteration of the host rock by a pervasive dolomitization event, clearly seen as an abrupt change in the color of the rock (Figs. 10A and 10C). This dolostone is fetid (presence of H_2S) and surrounds the base-metal accumulations. The primary mineralization is dominated by galena, sphalerite, barite, and siderite, with traces of chalcopyrite and pyrite. This is almost completely altered to an assemblage of supergene minerals, smithsonite, hydrozincite, limonite, cerussite, etc. (Fig. 10B), and only some rare remnants of primary sulfides are found. Scarce fluid inclusions found within primary minerals indicate formation temperatures of between 100 and 150 °C with salinities between 7.5 and 20 wt% eq. NaCl (González-Ramos, 1984).

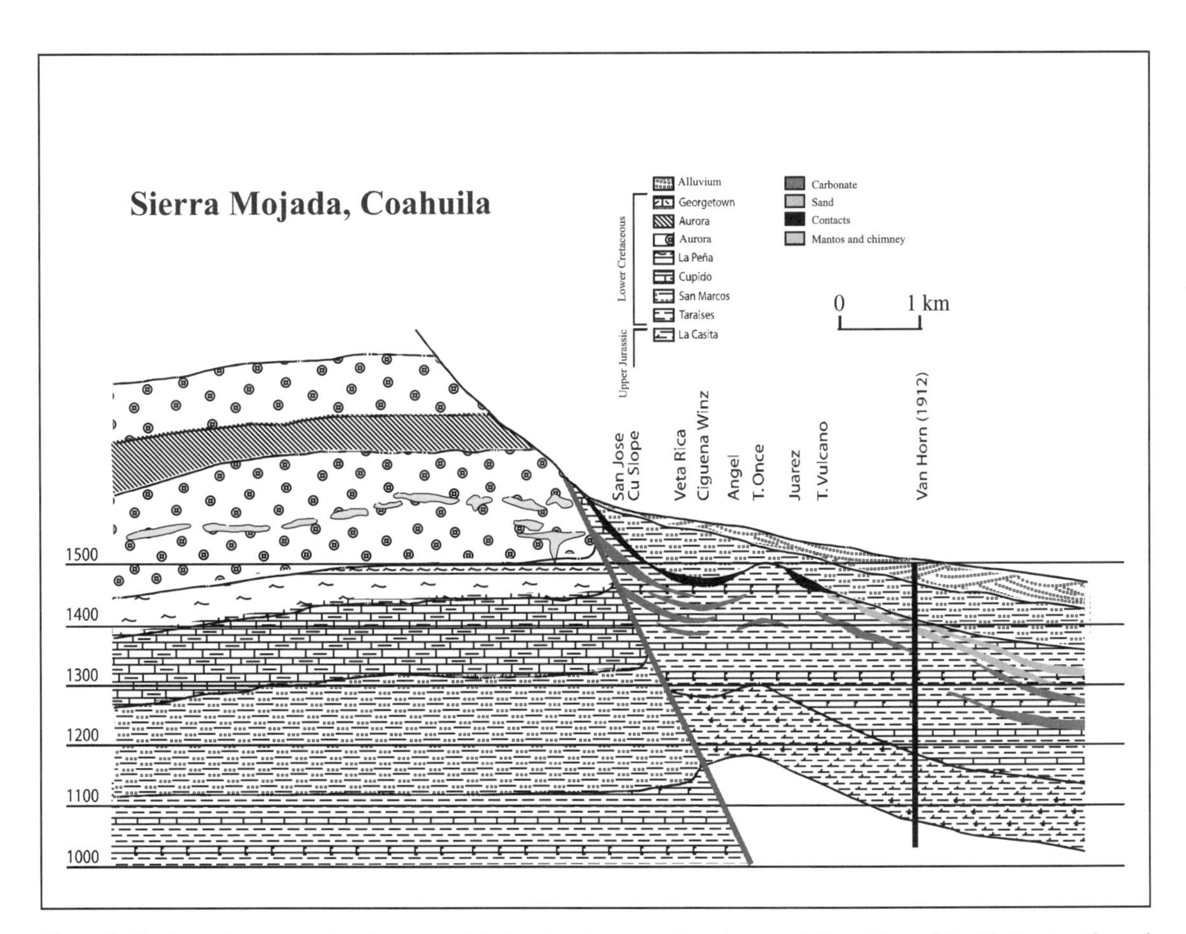

Figure 9. Geological cross section (not to scale) showing the general geology and disposition of the Pb-Zn stratabound deposits at Sierra Mojada (state of Coahuila). Modified after Metalline Mining Company (2004, www.metalin.com).

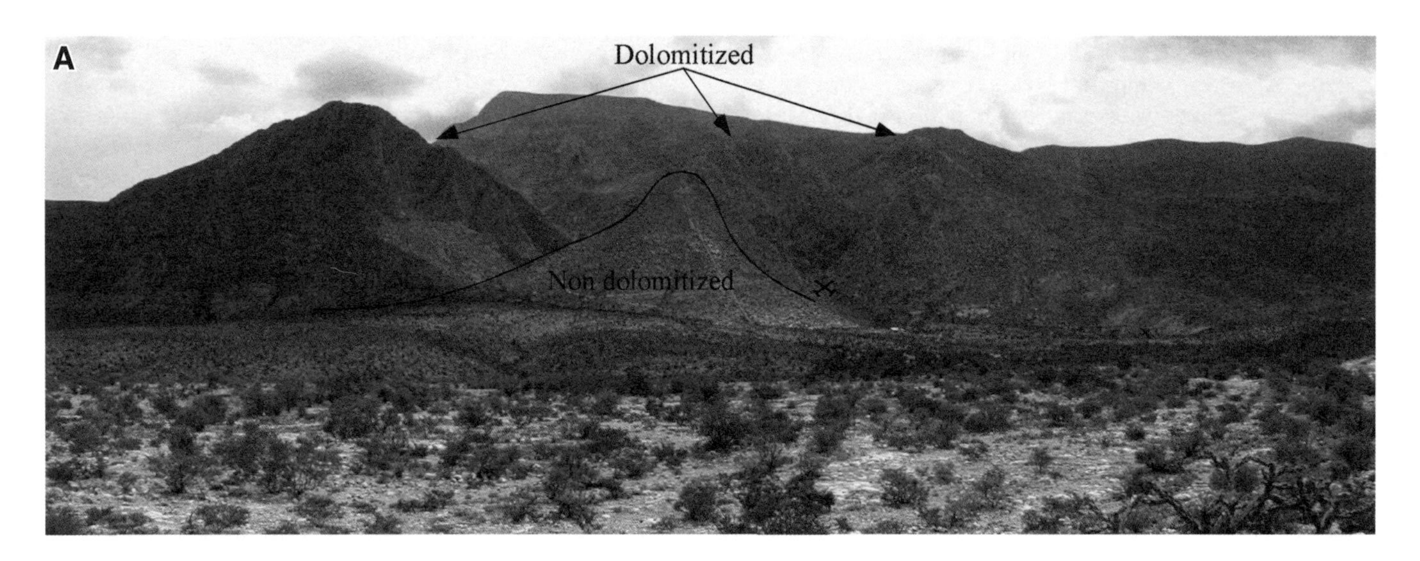

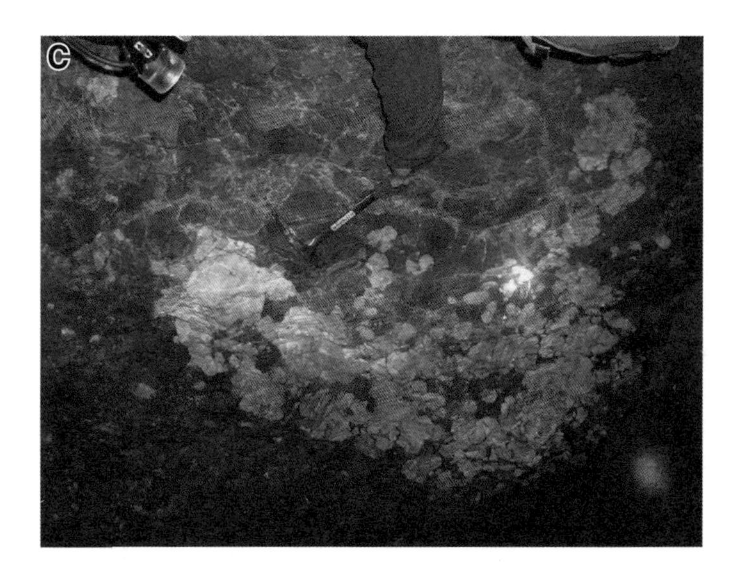

Figure 10. Mineral de Reforma Pb-Zn deposit, Purísima and San Marcos Ranges (Coahuila state). (A) View of the enclosing Mesozoic series; the darkening in color near the mineralization is due to the dolomitization of the limestone. Crossed pickaxes denote the location of the mineshaft. (B) Detail of the heavily oxidized mineralization (calamines). (C) Contact zone between the ore deposit and the enclosing dolostones.

CONCLUSIONS

This review of the geological and geochemical characteristics of the low-temperature, stratabound deposits in México allows us to propose the occurrence of, at least, four groups of deposits: (1) stratiform barite bodies of great purity, located at the lower parts of the sedimentary pile, that replace former anhydrite horizons, preserving some of the evaporite diagenetic textures; (2) stratabound, lens-shaped celestine-dominated bodies enclosed within the limestones of the Cupido Formation or their analogs; (3) stratabound fluorite-dominated mantos and bodies, mainly controlled by low-angle faults and associated hydraulic breccias, within the reefal limestones of the Aurora and Morelos Formations; and (4) Zn-Pb-(Ag-Cu-Ba-Sr) stratabound ore bodies enclosed within limestones that may be dolomitized, of Upper Jurassic and Cretaceous age, often deeply altered to supergene "calamines" and iron oxides.

All of these deposits are closely related to compressional (Laramide) or extensional post-tectonic events that instigated the flow of highly modified basinal brines that interacted with the sedimentary pile at a variety of levels. Locally, the major controls on the ore distribution seem to be (1) presence of reefal limestones rich in organic matter; (2) availability of sulfate, either in basinal brines or, in the case of barite deposits, former evaporitic (anhydrite) horizons; (3) dilation along sedimentary joints due to an increase in pressure from the fluids and crystallization of the minerals; (4) hydraulic fracturing related to fluid overpressure. Halogen systematics, when applied, indicate that these brines are bitterns originating after evaporation of seawater passed halite saturation. These entered the sediments, probably as formation waters that eventually interacted with the sediments along the flow path and during expulsion, drastically changing their cation composition. The involvement of organic matter during the genesis of these deposit is ubiquitous and it is often detected by the presence of liquid hydrocarbons (heavy oils) trapped as hydrocarbon-bearing fluid inclusions, very frequently with bitumen and traces of H_2S, whose abundance is indicative of the thermal degradation of the organic matter.

In all the studied cases, these deposits are clearly genetically unrelated to any magmatic episode whatsoever, even though they can be physically close and intruded by Late Tertiary rhyolitic dykes and domes (La Encantada–Buenavista, Coahuila state; La Azul, Guerrero state) that induce physical and chemical changes in the deposits (silicification, fluid inclusion decrepitation, age resetting, etc.).

The occurrence of mineralized bodies with features similar to those of the classical MVT is much more common than previously thought. As shown in this paper, these deposits have been clearly identified within the undisturbed Mesozoic carbonate sedimentary platforms outcropping in the states of Guerrero, Chihuahua, Coahuila, and Nuevo León. Probably, some other deposits, outcropping in areas where the Mesozoic sediments have been affected by the widespread Tertiary volcanism will be also proven to belong to this type once the disturbing influence of later magmatic events is recognized. Therefore, a thorough revision of the geological and geochemical controls of the carbonate-hosted deposits in México is warranted, especially deposits that do not have any clear and close relationships with magmatism.

ACKNOWLEDGMENTS

We would like to acknowledge Fluorita de México S.A. de C.V., especially Ingenieros (Ing.) Agustín Rodriguez and Samuel Baca for their permission, help, and geological discussions on visiting La Encantada–Buenavista and El Tule districts. We would also like to thank Barosa S.A. de C.V. for the use of their facilities when visiting the Muzquiz barite mines. Special thanks to Ing. C. Martínez-Ramos and A. González-Ramos for their invaluable help when visiting the La Paila and Mineral de Reforma deposits. This work has been financed by the Universidad Nacional Autónoma de México Programa de Apoyo a Proyectos de Investigación e Innovación Tecnológica (UNAM-PAPIIT) projects number IN114002 and IN114106-3.

REFERENCES CITED

Clark, K., 1990, Geology and mineral deposits of the Taxco mining district, *in* Clark, F., ed., Mexican silver deposits: Society of Economic Geology, Guidebook Series 6, p. 281–291.

De Cserna, Z., 1989, An outline of the geology of Mexico, *in* Bally, A.W., and Palmer, A.R., eds., The Geology of North America—An Overview: Boulder, Colorado, Geological Society of America, The Geology of North America, v. A, p. 233–264.

De Cserna, Z., and Fries, C., 1981, Hoja Taxco 14Q-h (7), Geología de los Estados de Guerrero, México y Morelos: México, D.F., Universidad Nacional Autónoma de México, Instituto de Geología, Cartas Geológicas de México serie 1:100,000.

Dubessy, J., and Ramboz, C., 1986, The history of organic nitrogen from early diagenesis to amphibolite facies: Mineralogical, chemical, mechanical and isotopic implications, *in* 5th International Symposium Water-Rock Interaction: Reykjavik, Iceland, Extended Abstracts, p. 170–174.

Fernández, G., 1956, Nota sobre la mina "La Azul," *in* Maldonado-Koerdell, ed., Geología a lo largo de la Carretera entre México D.F., Taxco, Gro. Distrito Minero de Taxco; Visita a un yacimiento de Fluorita en Rocas del Terciario Inferior: México, 20th Congreso Geológico Internacional, Excursiones A-4, C-2, Vigésima Sesión, México, D.F., p. 91–93.

Florenzani, G., 1974, Estudio geológico minero del distrito de Taxco, Edo. de Guerrero, México [B.Sc. Thesis]: Instituto Politécnico Nacional, México, D.F., México, 93 p.

Fowler, G.M., Hernon, R.M., and Stone, E.A., 1948, The Taxco mining district, Guerrero, Mexico, *in* Dunham, K.C., ed., Symposium on the geology, paragenesis and reserves of the ored of lead and zinc: London, UK, 18th International Geological Congress, v. 39, no. 1, p. 107–116.

Goldhammer, R.K., 1999, Mesozoic sequence stratigraphy and paleogeographic evolution of Northeast Mexico, *in* Bartolini, C., et al., eds., Mesozoic sedimentary and tectonic history of north-central Mexico: Geological Society of America Special Paper 340, p. 1–58.

Gonzalez-Partida, E., Carillo-Chavez, A., Grimmer, J.O.W., Pironon, J., Mutterer, J., and Levresse, G., 2003, Fluorite deposits at Encantada-Buenavista, Mexico: Products of Mississippi Valley type processes: Ore Geology Reviews, v. 23, no. 3–4, p. 107–124, doi: 10.1016/S0169-1368(03)00018-0.

González-Ramos, A., 1984, Estudio geológico-geoquímico regional de la Sierra La Purísima, San Marcos, en el Mpio. de Cuatro Cienegas, Coahuila [M.Sc. thesis]: México D.F., México, Facultad de Ingeniería, Universidad Nacional Autónoma de México (UNAM), 128 p.

Gonzalez-Reyna, J., 1956, Riqueza minera y yacimientos minerales de México: México D.F., Departamento de Investigaciones Industriales, Banco de México, 497 p.

Kesler, S.E., 1977, Geochemistry of manto fluorite deposits, northern Coahuila, México: Economic Geology and the Bulletin of the Society of Economic Geologists, v. 72, p. 204–218.

Kesler, S.E., and Jones, L.M., 1981, Sulfur and strontium-isotopic geochemistry of celestine, barite and gypsum from the Mesozoic basins of North Eastern Mexico: Chemical Geology, v. 31, p. 211–224, doi: 10.1016/0009-2541(80)90087-X.

Levresse, G., Gonzalez-Partida, E., Tritlla, J., Camprubí, A., Cienfuegos-Alvarado, E., and Morales-Puente, P., 2003, Fluid origin of the World-class, carbonate-hosted Las Cuevas fluorite deposit (San Luis Potosí, México): Journal of Geochemical Exploration, v. 78–79, p. 537–543, doi: 10.1016/S0375-6742(03)00145-6.

Levresse, G., Tritlla, J., Villareal, J., and Gonzalez-Partida, E., 2006, The El Pilote fluorite skarn: A crucial deposit to understand the origin and mobilization of F from northern Mexico deposits: Journal of Geochemical Exploration, v. 89, p. 205–209, doi: 10.1016/j.gexplo.2005.11.042.

Machel, H.G., 2001, Bacterial and thermochemical sulfate reduction in diagenetic settings—old and new insights: Sedimentary Geology, v. 140, p. 143–175, doi: 10.1016/S0037-0738(00)00176-7.

Metalline Mining Company, 2004, Sierra Mojada, Coahuila, México: http://www.metalin.com/ (last accessed Sept. 2006).

Osborne, T.C., 1956, Geología y depositos minerales del Distrito Minero de Taxco, *in* Maldonado-Koerdell, ed., Geología a lo largo de la Carretera entre México D.F., Taxco, Gro. Distrito Minero de Taxco; Visita a un yacimiento de Fluorita en Rocas del Terciario Inferior: México, 20th Congreso Geológico Internacional, Excursiones A-4, C-2, México, D.F., p. 75–89.

Pi, T., Sole, J., and Taran, Y., 2005, (U-Th)/He dating of fluorite: application to the La Azul fluorspar deposit in the Taxco mining district, México: Mineralium Deposita, v. 39, p. 976–982, doi: 10.1007/s00126-004-0443-y.

Puente-Solís, R., González-Partida, E., Tritlla, J., and Levresse, G., 2005, Distribución de los depósitos estratoligados de barita, celestina, fluorita, plomo-zinc en el Noreste de México, *in* Corona-Esquivel, R., and Gómez-Caballero, J.A., eds., Acta de Sesiones de la XXVI Convención Nacional de Minería, Veracruz, México: Asociación de Ingenieros de Minas, Metalurgistas y Geólogos de México, p. 95–98, ISBN 968-7726-02-4.

Ramos-Rosique, A., 2004, Comportamiento de los fluidos en la génesis de los mantos de celestina en la Sierra de Los Alamitos, Coahuila [B.Sc. Thesis]: Querétaro, México, Facultad de Ingeniería, Universidad Nacional Autónoma de México (UNAM), 54 p.

Ramos-Rosique, A., Villareal-Fuentes, J., González-Partida, E., Tritlla, J., and Levresse, G., 2005, Los yacimientos estratoligados de celestina de El Venado, El Volcán en la Sierra Los Alamitos, Coahuila, México, *in* Corona-Esquivel, R., and Gómez-Caballero, J.A., eds, Acta de Sesiones de la XXVI Convención Nacional de Minería, Veracruz, México: Asociación de Ingenieros de Minas, Metalurgistas y Geólogos de México, p. 99–104, ISBN 968-7726-02-4.

Richardson, C.K., and Holland, H.D., 1979, Fluorite solubility in hydrothermal solutions, an experimental study: Geochimica et Cosmochimica Acta, v. 43, p. 1313–1325, doi: 10.1016/0016-7037(79)90121-2.

Rickman, D.L., 1977, Origin of celestite (strontium sulfate) ores in the Southwestern United States and northern México [M.Sc. Thesis]: Socorro, New Mexico, USA, New Mexico Institute of Mining and Technology, 79 p.

Salas, G., 1973, Geología de los depósitos de celestita de la Sierra de la Paila, Coahuila, *in* Acta de Sesiones de la X Convención Nacional de Minería: Asociación de Ingenieros de Minas, Metalurgistas y Geólogos de México, p. 287–294.

Skewes Saunders, T., 1938, Report upon a deposit of fluorspar near Taxco, Gro: Consejo de Recursos Minerales, México, Archive 120017-08.

Thiéry, R., Pironon, J., Walgenwitz, F., and Montel, F., 2000, PIT (Petroleum Inclusion Thermodynamic): A new modeling tool for the characterization of hydrocarbon fluid inclusions from volumetric and microthermometric measurements: Journal of Geochemical Exploration, v. 69–70, p. 701–704, doi: 10.1016/S0375-6742(00)00085-6.

Tritlla, J., and Levresse, G., 2006, Comments to "(U-Th)/He dating of fluorite: application to the La Azul fluospar deposit in the Taxco mining district, Mexico" (Min. Dep., v. 39, p. 976–982) by Pi et al.: Mineralium Deposita, v. 41 (in press).

Tritlla, J., Camprubí, A., Pi, T., and Corona-Esquivel, R., 1999, Los depósitos de fluorita del distrito de Taxco (Guerrero): Primeros datos, *in* Proceedings, Trabajos Técnicos de la XXIII Convención Nacional, Acapulco, México: Asociación de Ingenieros de Minas, Metalurgistas y Geólogos de México, p. 66.

Tritlla, J., Camprubí, A., and Corona-Esquivel, R., 2001, The Taxco fluorite deposit (Mexico): A new pseudo-chromatographic mechanism for rhythmite formation, *in* Piestrzynski, A., et al., eds., Mineral deposits at the beginning of the 21st century: Rotterdam, Netherlands, Balkema, p. 979–982.

Tritlla, J., Gonzalez-Partida, E., Banks, D., Levresse, G., Baca-Gasca, S., and Rodríguez-Santos, A., 2004a, Fluid characteristics of the stratabound celestite deposits of La Paila Ranges: A modified seawater circulation episode in the Sabinas Basin, Mexico: Eos (Transactions, American Geophysical Union) Fall Meet. Supplement, v. 85, no. 47, Abstract V31A-1422.

Tritlla, J., Gonzalez-Partida, E., Levresse, G., Banks, D., and Pironon, J., 2004b, Fluorite deposits at Encantada-Buenavista, Mexico: Products of Mississippi Valley type processes [Ore Geol. Rev. 23 (2003), 107–124]—A reply: Ore Geology Reviews, v. 25, p. 329–332, doi: 10.1016/j.oregeorev.2004.04.005.

Tritlla, J., Levresse, G., Gónzalez-Partida, E., Corona-Esquivel, R., and Martínez-Ramos, C., 2005, Metalogénia y geoquímica de los fluidos asociados a los depósitos de tipo MVT (Mississippi Valley-type deposit) en el Centro y Norte de México, *in* Corona-Esquivel, R., and Gómez-Caballero, J.A., eds., Acta de Sesiones de la XXVI Convención Internacional de Minería, Veracruz, México: Asociación de Ingenieros de Minas, Metalurgistas y Geólogos de México, p. 113–118.

Van Horn, F.R., 1912, The occurrence of silver, copper, and lead ores at the Veta Rica Mine, Sierra Mojada, Coahuila, Mexico. Transactions of the Society of Mining Engineers of American Institute of Mining, Metallurgical and Petroleum Engineers, Incorporated (AIME), p. 867–881.

Manuscript Accepted by the Society 29 August 2006

Printed in the USA

Geological Society of America
Special Paper 422
2007

Geological and metallogenetic characteristics of the porphyry copper deposits of México and their situation in the world context

Martín Valencia-Moreno*
Estación Regional del Noroeste, Instituto de Geología, Universidad Nacional Autónoma de México (UNAM), Apartado Postal 1039, Hermosillo, Sonora 83000, México

Lucas Ochoa-Landín
Departamento de Geología, Universidad de Sonora, Hermosillo, Sonora 83000, México

Benito Noguez-Alcántara
Servicios Industriales Peñoles S.A. de C.V., Blvd. Navarrete 277, Hermosillo, Sonora 83170, México

Joaquin Ruiz
Department of Geosciences, University of Arizona, Tucson, Arizona 85721, USA

Efrén Pérez-Segura
Departamento de Geología, Universidad de Sonora, Hermosillo, Sonora 83000, México

ABSTRACT

The sustained magmatic activity along the North American Cordillera during late Mesozoic and Paleogene times produced the emplacement of numerous porphyry copper deposits. This activity extended by most of western México, particularly along the northwestern part of the country. This region, along with Arizona and New Mexico in the United States, contains one of the most important centers of copper mineralization on Earth. Most of the Mexican deposits lie in the eastern part of the Laramide magmatic belt (90–40 Ma) and were formed predominantly between 75 and 50 Ma. The largest deposits occur in northeastern Sonora and are represented by Cananea (~30 Mt Cu) and La Caridad (~8 Mt Cu). The copper ores are locally accompanied by molybdenum, tungsten, gold, and other metals. However, the metal distribution is apparently coupled with major changes in the basement of emplacement, which can be roughly separated into three domains: a northern domain characterized by Proterozoic crystalline rocks of North American affinity; a central domain composed of Paleozoic deep-marine basin rocks underlain by the Proterozoic North American rocks; and a southern domain, represented by Mesozoic island-arc–related sequences of the Guerrero terrane. Sr and Nd isotopic data from Laramide plutons along these domains suggest that the basement

*valencia@geologia.unam.mx

Valencia-Moreno, M., Ochoa-Landín, L., Noguez-Alcántara, B., Ruiz, J., and Pérez-Segura, E., 2007, Geological and metallogenetic characteristics of the porphyry copper deposits of México and their situation in the world context, *in* Alaniz-Álvarez, S.A., and Nieto-Samaniego, Á.F., eds., Geology of México: Celebrating the Centenary of the Geological Society of México: Geological Society of America Special Paper 422, p. 433–458, doi: 10.1130/2007.2422(16). For permission to copy, contact editing@geosociety.org.

modified the final composition of the Laramide magmas. Also, the basement seems to have partly controlled the metal commodities along the porphyry copper belt, with relatively larger deposits characterized by Cu-Mo-W mineralization in the northern and central domains, and smaller and more Cu-Au dominated systems in the southern (more oceanic) domain.

Keywords: porphyry copper deposits, arc-related magmatism, granitic plutonism, metallogenesis, basement, México.

INTRODUCTION

Porphyry copper deposits represent the main source of copper and molybdenum in the world, with various notable cases containing more than 1000 Mt of ore mineral grading above 0.5% Cu (Richards, 2003). This has promoted a maintained interest by mining companies and academic institutions to understand the origin and the evolution of these deposits, a fact that is reflected in a continuous evaluation of the state of the art (e.g., Titley and Hicks, 1966; Titley, 1982a; Friedrich et al., 1986; Pierce and Bolm, 1995). From the numerous occurrences of mineralization of this type around the world (Fig. 1), it is clear that the porphyry copper deposits are largely related to tectonic regions characterized by abundant calk-alkaline magmatism associated with Andean-type subduction margins (Fig. 2). Also, there is a narrow relationship between mineralization and the presence of late plutonic stages, which were emplaced at subvolcanic levels in the crust during the cooling of major—and clearly cogenetic—granitic bodies (Fig. 2). The origin of the associated metals is still uncertain, although it is considered that most of the copper was contributed by the asthenospheric mantle trapped above the subduction zone (Fig. 2) (e.g., Sillitoe, 1972; Ruiz and Mathur, 1999). The porphyry copper deposits display a variety of metallic characteristics and lithological associations, but each case exhibits its own complexity. Nevertheless, for the purpose of idealizing its anatomy, Lowell and Guilbert (1970, p. 374) proposed a relatively simple model characterized by lateral and vertical zoning of the ore and alteration mineralogy, which is centered in a relatively small intrusive characterized by a porphyritic texture (Fig. 3). According to these authors, the porphyry copper deposits can be described as "copper and molybdenum sulfide deposits consisting of disseminated and stockwork veinlet sulfide mineralization emplaced in various host rocks that have been altered by hydrothermal solutions into roughly concentric zonal patterns." The conception of this model by Lowell and Guilbert was based on a study carried out in the deposits of San Manuel-Kalamazoo, in Arizona, which was tested and refined with information compiled from a significant number of copper and molybdenum deposits in North and South America. Sillitoe (1973) argued that the mineralizing porphyries are calc-alkaline stocks that are emplaced at depths of 1.5–3 km

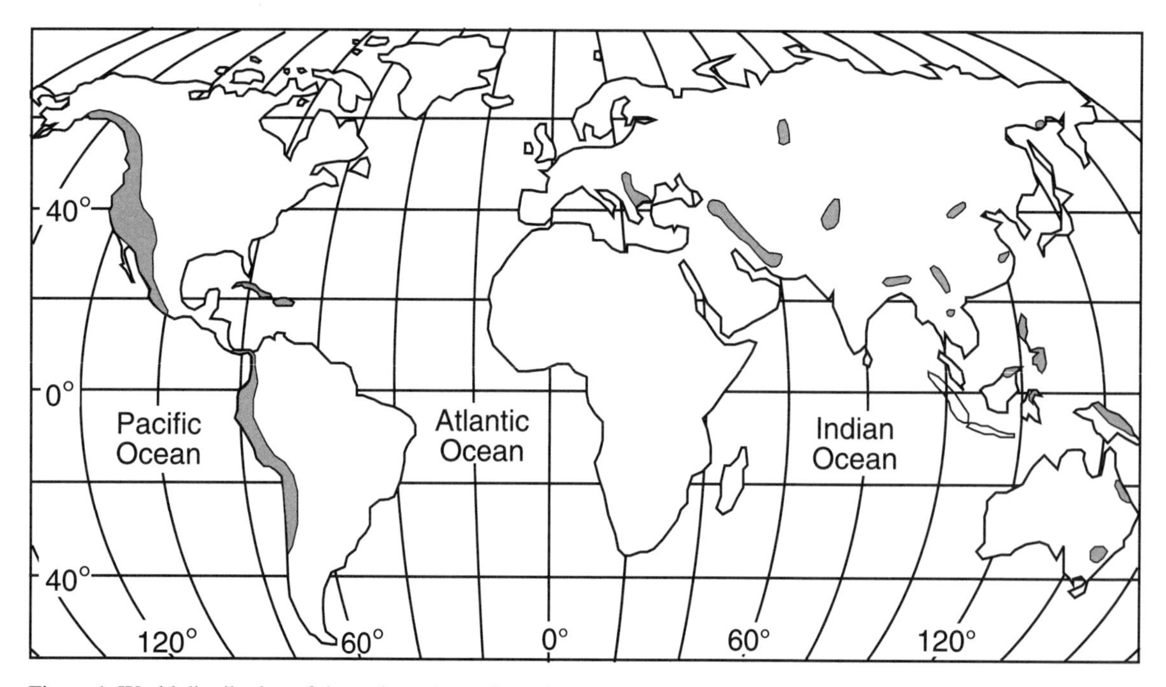

Figure 1. World distribution of the main regions of porphyry copper mineralization (adapted from Sillitoe, 1972; Singer et al., 2005).

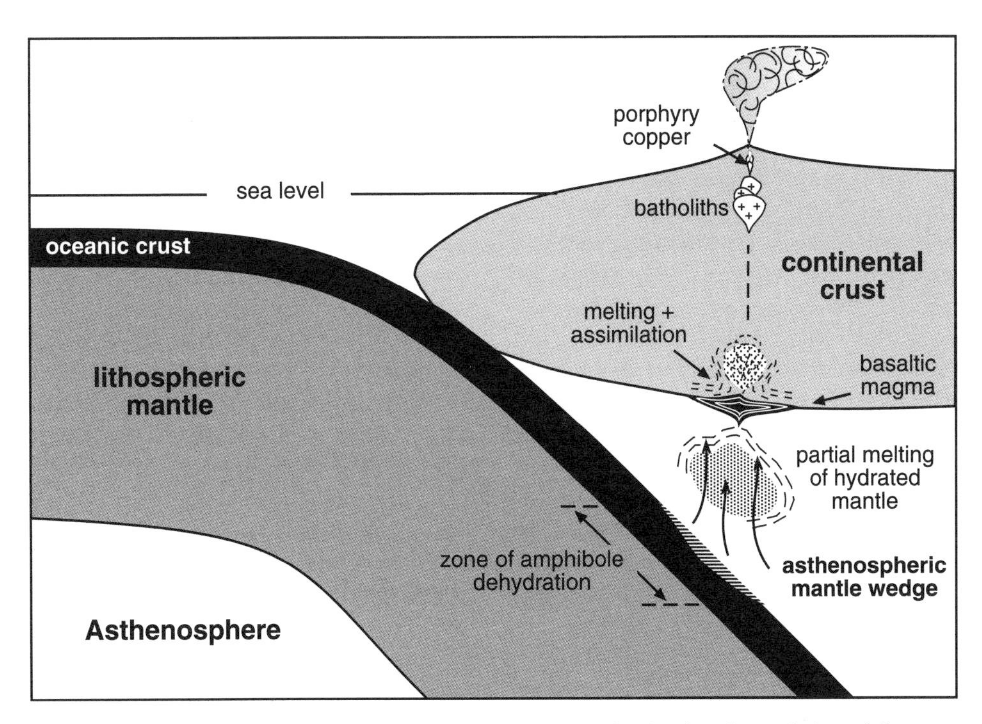

Figure 2. Tectonic model of an Andean-type subduction margin showing the evolution of the magmatic process from its initial origin in the asthenospheric mantle wedge to the subvolcanic and volcanic environments involved in the generation and emplacement of the porphyry copper deposits.

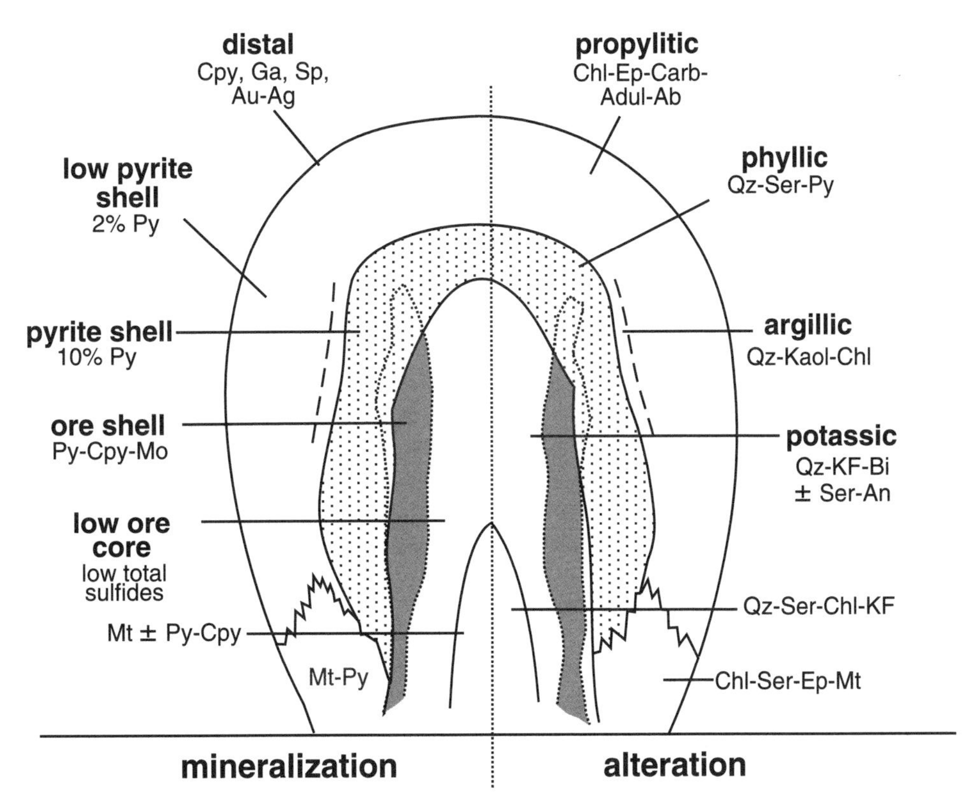

Figure 3. Schematic model for the San Manuel–Kalamazoo orebody in southeastern Arizona, showing the typical vertical and lateral zoning of alteration and mineralization displayed by the porphyry copper systems (according to Lowell and Guilbert, 1970).

in the crust which grade downward to stockwork mineralization and potassic alteration zones in a larger phaneritic intrusive. At shallower depths, the upper part of the system reaches the column of comagmatic volcanic rocks, completing a vertical column of ~8 km for the entire system. Locally, the upper part of the system may grade to epithermal ore deposits, whose connection with the porphyry copper deposits is largely supported by geochemical and geological arguments (e.g., Sillitoe, 1993; Hedenquist et al., 1998; Heinrich et al., 1999; Sillitoe and Hedenquist, 2003; Camprubí and Albinson, this volume). Since porphyry copper deposits are orthomagmatic systems (Hedenquist and Lowenstern, 1994), it is predictable that the age of the mineralization should not differ much from that of the productive plutons. In fact, it is known that the difference between the crystallization ages of the main pluton and those of the porphyry stocks associated with the mineralization is generally smaller that 5.0 m.y. and that the difference between the porphyry stock and the mineralization itself could be smaller than 1.0 m.y. (Ruiz and Mathur, 1999; Zürcher, 2002). In general, the longevity and dynamism of the hydrothermal activity, as well as the presence of favorable physical-chemical conditions in the fluid-rock relation, are important factors for defining the economic characteristics of the deposits (e.g., Clark, 1993). Along with these factors, the repetition and superimposition of the mineralizing events in a system produce a progressive enrichment of the deposit, which is particularly important for the concentrations of copper (Gustafson et. al., 2001). As can be observed in Figure 1, the porphyry copper deposits are not exclusive to a particular region on Earth, although it is clear that the larger associated metallic accumulations are closely related to tectonic regions that involve an important crustal thickness in the magmatic process (Fig. 2). It seems that the conditions for generating this type of mineralization were ideal in the western cordilleras of North and South America (Fig. 4), which constitute the richest and most important copper zones in the world. The more productive metallogenetic epochs in these regions were during the last part of the Mesozoic and the Cenozoic, as shown in Figure 5. In the western cordillera of North America, the porphyry copper deposits have ages mostly in the range of ca. 88–25 Ma, with the greater copper accumulations occurring ca. 56 Ma. In contrast, the South American deposits are relatively younger, exhibiting two important pulses between ca. 64 and 31 Ma and between 20 and 5 Ma, with the most spectacular examples emplaced at the end of both pulses.

Tectonic Framework of the Porphyry Copper Deposits of Southwestern North America

Traditionally, porphyry copper deposits have had a prominent position in the mining activity of southwestern North America, because they usually exhibit large volumes of metal, which can be extracted at relatively low operation costs (Titley, 1982b). A large part of the porphyry copper deposits known in the world is clearly related to the presence of extensive magmatic belts, formed in the upper part of the crust above the subduction zones associated with plate convergent margins of Andean type. This is the case for the western cordillera of North America, where porphyry copper deposits occur from Alaska to southern México (Fig. 4). In this region, the magmatism related to the copper mineralization was active during an extensive time interval. It began ca. 200 Ma in the region of British Columbia, Canada, and gradually migrated south, arriving in northeastern Sonora between 64 and 55 Ma. Subsequently, the migration reached the central and southern region of western México between 45 and 31 Ma and the southernmost tip of the country, in the region of Chiapas, ca. 6 Ma (Damon et. al., 1983a). Although during this time the plutonic activity was very abundant along all the cordillera, the most relevant porphyry copper deposits and the larger number of occurrences are located in southeastern Arizona, USA, and the adjacent regions of New Mexico, USA, and Sonora, México (Figs. 4 and 6). The majority of the deposits have ages between 74 and 55 Ma (Titley, 1990), although older deposits are known in Bisbee, Arizona, and El Arco, Baja California (Fig. 6); nevertheless, is clear that the main metallogenetic pulse occurred during the Laramide event (90–40 Ma; Damon et al., 1983a, 1983b). According to Damon et al. (1983a, 1983b), the deformation and magmatism associated with the Laramide event are products of geodynamic effects in a plate convergent regime, initiated by a marked change in the angle of subduction of the oceanic Farallon plate under North America. During most of the Cretaceous, a relatively high angle of subduction was coupled with a nearly static magmatic axis, located in a zone adjacent to the paleotrench in the Pacific (Coney and Reynolds, 1977; Silver and Chappell, 1988). At the end of the Cretaceous (ca. 70 Ma), the angle of the subducted plate was considerably reduced due to a higher rate of plate convergence. This caused the asthenospheric mantle wedge to retreat in relation to the trench position, which is explained schematically in Figure 7. The removal of the asthenosphere caused a rapid eastern migration of the locus of magmatism, from the coastal regions toward the interior of the continent, reaching its maximum advance between 55 and 40 Ma (Damon et al., 1983a; Humphreys et. al., 2003). During this time span, the coupling zone between the Farallon and North America plates increased from ~200 km to ~1000 km (English et. al., 2003). This subhorizontal geometry of the subduction caused marked compressional efforts in the coupling zone, favoring a notable crustal thickening (Dickinson, 1989; English et. al., 2003). The late stage of this tectonic event (ca. 65 Ma) coincides with intense plutonic activity, as well as with the most productive phase in the generation of porphyry copper deposits of southwestern North America. Despite the large advances achieved at the moment in the knowledge of the origin and evolution of these ore deposits, Titley (1993) considers that there are three fundamental questions regarding the great copper province of Sonora–Arizona–New Mexico that should be attended to: (1) what promoted such a spectacular concentration of copper in this region, particularly during the time interval between 75 and 50 Ma?; (2) which factors controlled the specific location of the deposits in these mining districts?; and

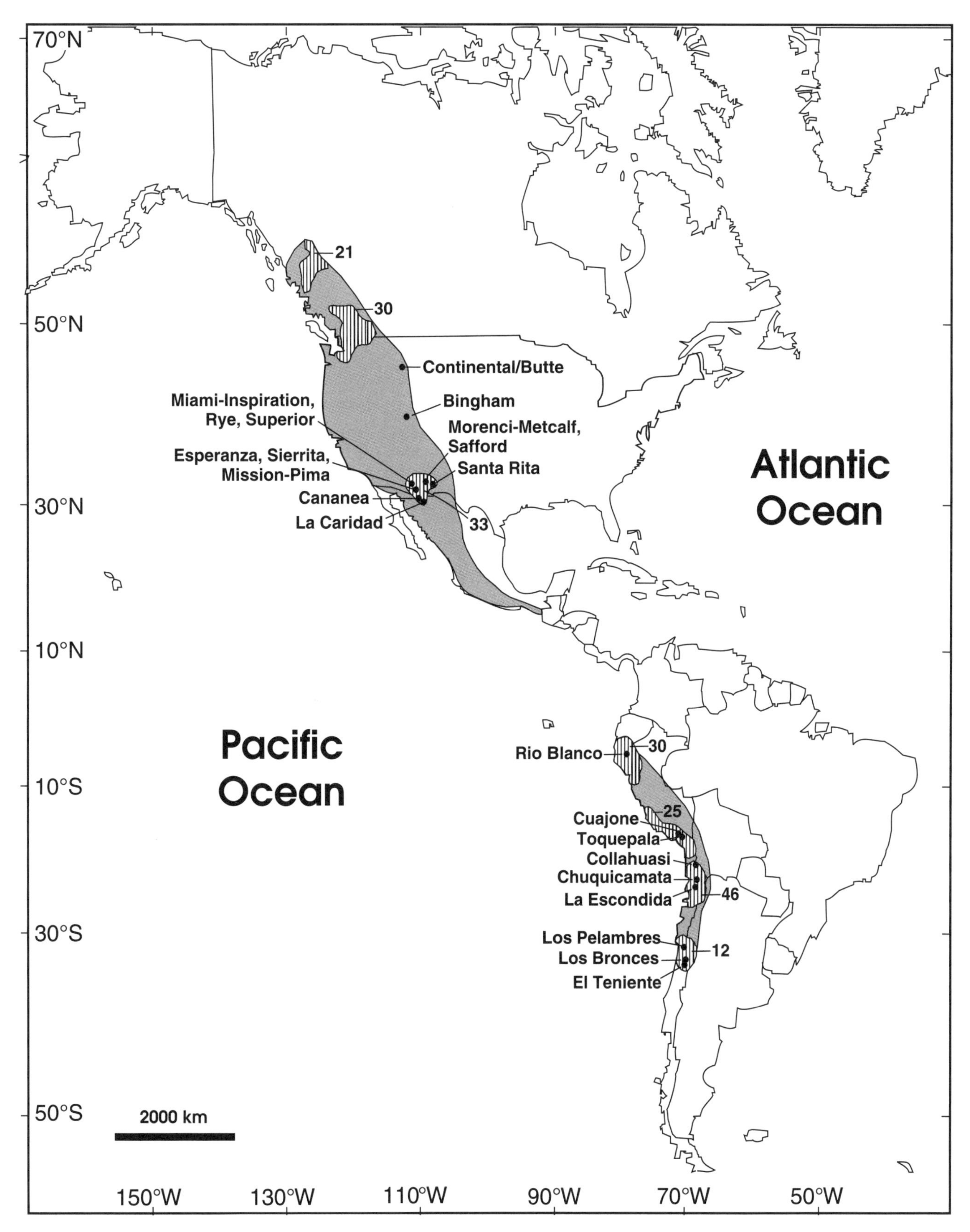

Figure 4. Map of the porphyry copper belts of the western cordilleras of North and South America, showing location of the main deposits. The vertical line pattern represents preferential regions or "clusters" whose number of mineralized centers is indicated.

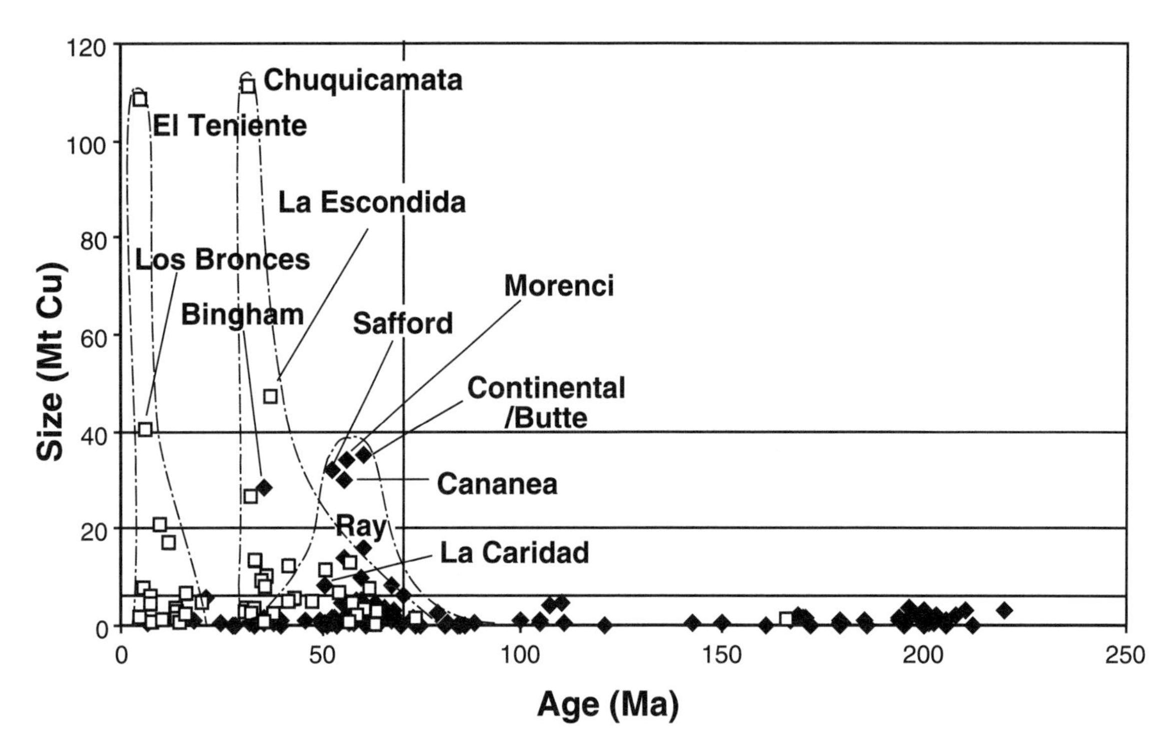

Figure 5. Plot of total copper content against the age of mineralization for porphyry copper deposits of the western cordilleras of North America (solid diamonds) and South America (open squares), according to data in Singer et al. (2005).

(3) which was the source of the metals, particularly for Cu and Mo? Recent studies have provided some clues toward possible answers for these questions. The main features that could have determined favorable metallogenetic conditions in this province appear to include the following:

1. Location near the edge of a thickened continental crust, in continental margins under fast plate convergence and prolonged subduction. An exceptional example is the belt of the central Andes of Peru and Chile, where subduction has been active for >200 m.y. (Clark, 1993; Camus, 2003).
2. Presence of abundant magmatism associated with subduction zones in convergent margins of Andean type. In the case of the metallogenetic province of southwestern North America, this is characterized by a calk-alkaline and metaluminous magmatism, typical of this tectonic environment. Similarly, the presence of magnetite and sphene suggests that the magmas were oxidized and relatively poor in Fe and Ti, which would also favor the presence of elements such as Cu, Pb, Zn, Be, Fe, Mn, U, La, Ce (Keith and Swan, 1996). In contrast, other metals, including Au, Sn, In, Sb, Ti, Yb and the platinum group, seem to be more associated with processes of crystallization in reduced environments (Sillitoe, 1996; Tetsuichi and Katsushiro, 1997).
3. The presence of a thick basement characterized by Proterozoic crystalline rocks and Late Proterozoic and Paleozoic sedimentary platform rocks (Titley, 1982b; Farmer and DePaolo, 1984; Lang and Titley, 1998).
4. The presence of polyphasic intrusive centers, particularly with compositions of diorite-monzonite in the initial phases, granodiorite–quartz monzonite in the intermediate phases (more favorable for the mineralization), and more granitic compositions in the late phases. Such intrusive centers were accompanied by subvolcanic stocks of porphyritic texture, of similar composition in each case (Damon, 1986; Tosdal and Richards, 2001).
5. A subvolcanic environment favorable for the emplacement of the porphyry intrusives, which generally coincides with a stratovolcanic center of andesitic type (Fig. 5.2 *in* Titley 1982b). The associated copper ores are an integral part of the magmatic evolution, and it seems that they are controlled by this subvolcanic environment, as well as by the physical-chemical conditions of the hydrothermal fluids during the emplacement and cooling of the porphyry intrusions (e.g., Titley, 1993; Sillitoe, 1996).

THE PORPHYRY COPPER DEPOSITS OF MÉXICO

The porphyry copper systems of México, including some associated skarn and hydrothermal breccia pipe deposits, occur along a NE-SW–oriented belt exposed along most of the western side of the country. About 60 deposits have been recognized in this belt (Table 1), ~70% of which are located in northwestern México, particularly in the states of Sonora and Sinaloa (Fig. 6). The most relevant localities, considered of great importance at a

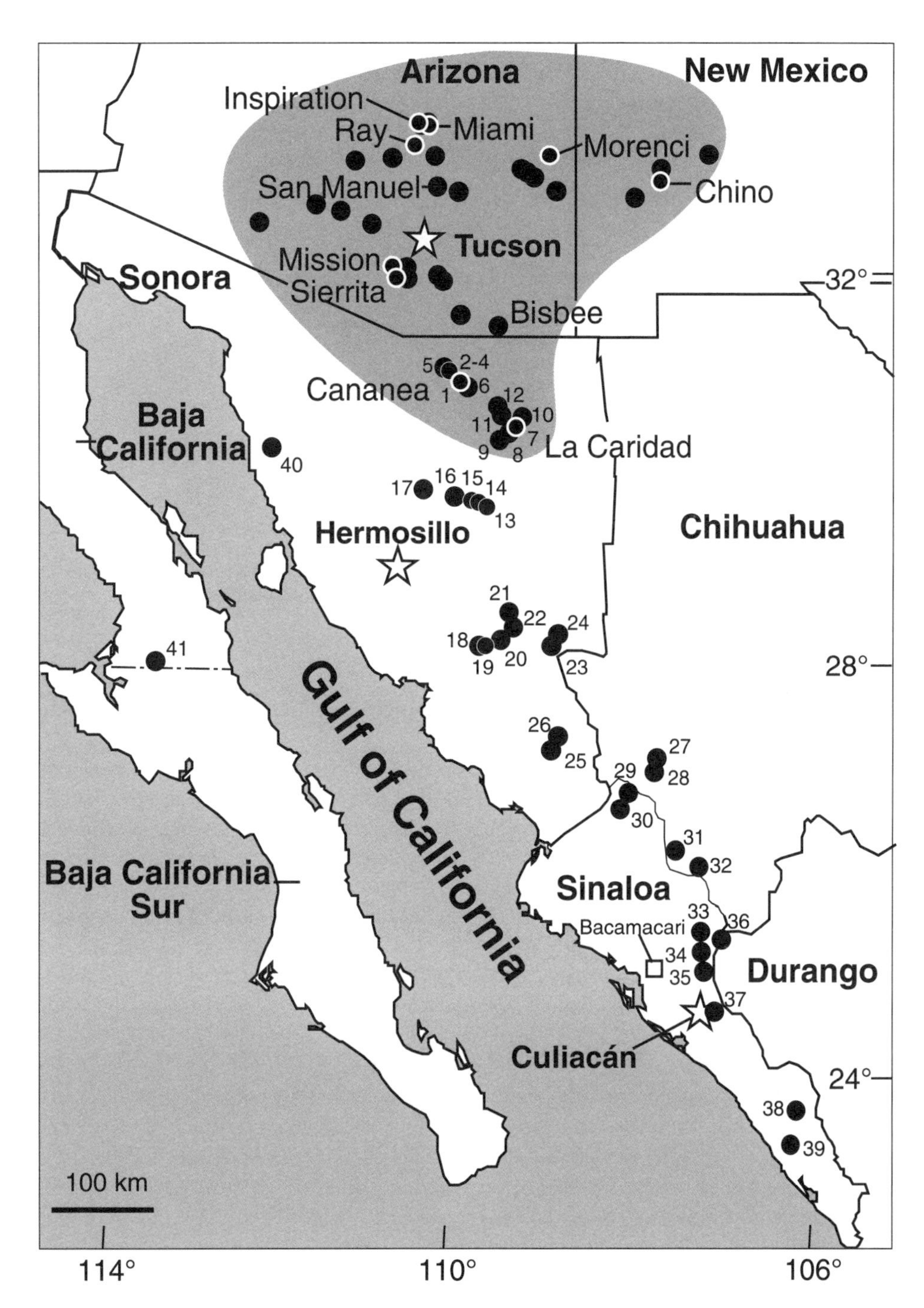

Figure 6. Map of the Laramide porphyry copper belt of northwestern México, showing location of individual deposits (details in Table 1). The shaded zone represents the so-called "great cluster" of the porphyry copper deposits of the western cordillera of North America, indicating the most significant deposits.

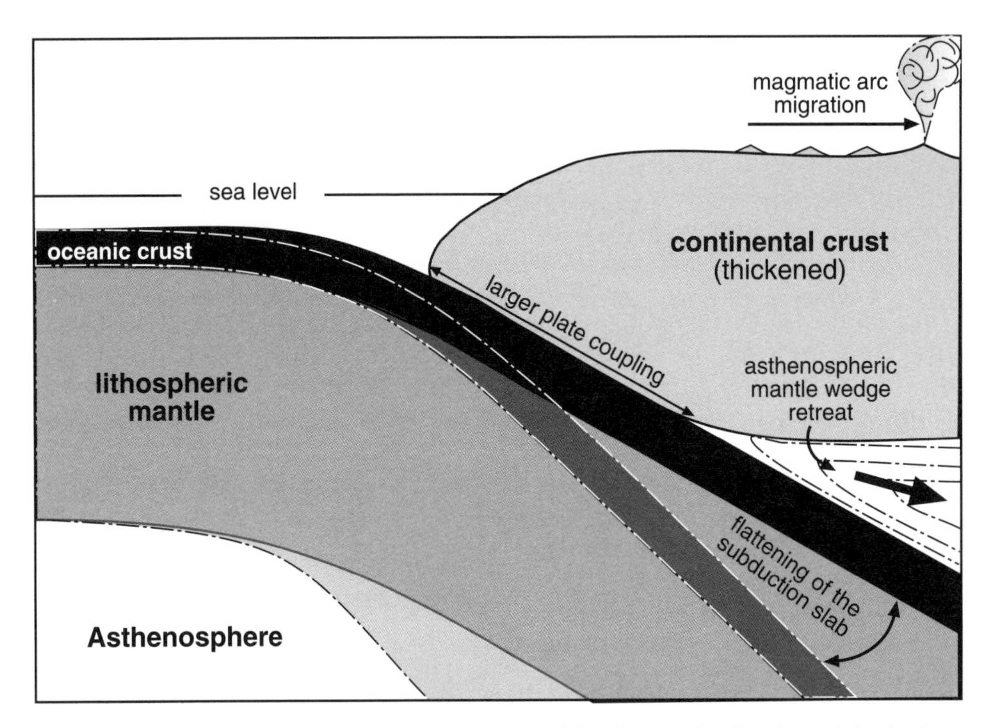

Figure 7. Schematic tectonic model of an Andean-type subduction margin, showing an inland migration of the magmatic arc due to a progressive flattening of the subducted oceanic slab.

global scale, are Cananea and La Caridad, located in northeastern Sonora (Fig. 6). Along with the large deposits of southern Arizona and western New Mexico, in the southwestern United States, the porphyry copper deposits of northwestern México constitute one of the most prominent copper provinces on Earth (Titley, 1993). For that reason, the present work emphasizes the characteristics of the deposits in this region. This province, commonly recognized as "the great cluster" of the porphyry copper deposits of Arizona, Sonora, and New Mexico (Keith and Swan, 1996) (Figs. 2 and 4), includes about fifty localities of demonstrated economic interest, 12 of which have been classified as of giant size (>2.5 Mt Cu; Laznicka, 1999; Clark, 1993). To define the importance of these deposits in México, it suffices to say that the operation of mines associated with these ore systems produces ~95% of the total copper production of the country. Moreover, this represents nearly 3.3% of the copper of this type at a global level (Fig. 8A) and ~18.1% of the potential estimated for the "great cluster" (Fig. 8B). Despite this, the geological and metallogenetic characteristics of the Mexican deposits in the "great cluster" are still little documented. Farther to the south of this rich province, the majority of occurrences are of marginal importance, and the information regarding them is yet more scarce and dispersed. For that reason, this work includes a geological-economic inventory of the most important characteristics, compiled from diverse sources. However, we note that in some cases the deposits are not truly porphyry copper deposits sensu stricto, but they are intrusive-centered systems whose metallic and alteration associations suggest an affinity with this type of metallogenetic process.

Geological Framework of the Mineralization

Although the source of the metals associated with the porphyry copper systems in México is largely unknown, there is an apparent geographical control of the mineralization that suggests the possible participation of the intruded crust (Valencia-Moreno, 1998). Efforts to understand the geotectonic characteristics of the different basement blocks in which the mineralization was emplaced are therefore very important. According to Valencia-Moreno et al. (2001), the Proterozoic crust of North America extends from the southwestern United States across northwestern México, reaching up to southern Sonora near the border with Sinaloa. This basement is characterized by a series of exposures of plutonic rocks and metamorphic rocks of amphibolitic facies, which outcrop mainly in northwestern Sonora, particularly to the south of Caborca (Fig. 9). On the other hand, a deformed sequence of Proterozoic sedimentary and volcanic rocks displaying green schist facies metamorphism is recognized in north and northeastern Sonora. According to Anderson and Silver (1979), both regions constitute two temporal provinces with crystallization ages of 1700–1800 and 1600–1700 Ma, respectively. This time difference served as the main argument to the authors to postulate the existence of a major Jurassic fault with a left lateral displacement, which disrupted part of the North American basement, displacing it to the southeast by ~800 km, which they named the Mojave-Sonora megashear (Fig. 9). Although this model has been traditionally accepted in most paleogeographic configurations of México (e.g., Campa and Coney, 1983; Stewart, 1988; Sedlock et al., 1993), the existence of

TABLE 1. CHARACTERISTICS OF THE PORPHYRY COPPER DEPOSITS OF NORTHWESTERN MEXICO

#	Name	Metals	Style	Intrusive rocks		Age (Ma)	Method	Mineralogy	Ton (x 10^6)	Metal contents	Refs.
				Pre-min.	Porphyry						
1	Cananea	Cu-Mo-Zn	sw, b, sk	gd, mz-di	qz-feld	59.9 ± 2.0	K-Ar (phl)	py, cpy, mo, cc, co, en	7,140	0.42% Cu, 0.008% Mo, 0.58 gr/ton Ag, 0.012 gr/ton Au	1, 2, 3
2	La Mariquita	Cu-Mo	sw, b	gd, mz-di	qz-feld	~63	Re-Os (mo)	py, cpy, cc	100	0.48% Cu	4
3	María	Cu-Mo	sw, b	gd	qz-feld	57.4 ± 1.6	Re-Os (mo)	py, cpy, mo	8.6	1.7% Cu, 0.1% Mo	1, 5
4	Lucy	Mo-Cu	sw	gd	gd	~63	Re-Os (mo)	mo, cpy	-	-	6
5	Milpillas	Cu	sw	gd	qz-feld	63.0–63.1	Re-Os (mo)	cpy, oxides	230	0.85% Cu	
6	El Alacrán	Cu-Mo	sw, b	gd	qz-lat	60.9 ± 0.2	Re-Os (mo)	py, cpy, cc, mo	2.4	0.35% Cu	1, 4, 7
7	La Caridad	Cu-Mo	sw, b	gd, di	qz-mz	53.8–53.6	Re-Os (mo)	py, cpy, mo, cc, sph, gn, co, th	1800	0.452% Cu, 0.0247% Mo	3, 8
8	Pilares	Cu-Mo-W	b	gd, lat	qz-mz	~53	-	py, cpy, mo, sph, sch, sp	147	1.04% Cu	3
9	Bella Esperanza	Cu-Mo	sw, b	gd	qz-mz	55.9 ± 1.2	K-Ar (bi)	py	-	-	9
10	Los Alisos	Cu	sw, b	gd	dac?	-	-	py, cpy, mo, cc, co	-	0.13% Cu	5, 9
11	Florida-Barrigón	Cu-Mo	sw	gd	gr, micro-di	52.4 ± 1.1	K-Ar (ser)	py, cpy, mo	85	0.32% Cu, 0.022% Mo, minor W	5, 9
12	El Batamote	Cu-Mo	sw, b	gd	qz-mz	56.8 ± 1.2	K-Ar (bi)	py, cpy, mo, cc	4.4	0.36% Cu	4, 5, 9
13	San Judas	Mo-(Cu-W)	b, sw	mz	-	40.0 ± 0.9	K-Ar (bi)	mo, py, cpy, te, sph, gn	2	0.25% Cu, 0.2% Mo	2, 9, 10
14	Transvaal	Mo-Cu	b	mz	-	-	-	py, cpy, mo	-	0.4% Cu, 0.2% Mo	4, 5
15	Cobre Rico	Cu	b	gd	-	-	-	-	-	2% Cu	4, 5
16	Washington	Cu-Mo-W	b	gd	-	45.7 ± 1	K-Ar (ser)	py, cpy, sch, mo	1.2	1.8% Cu, 0.106% Mo, 0.14% W, 0.17gr/t Au, 15.8 gr/t Ag	4, 9
17	El Crestón	Mo	sw, b	gd	qz-feld	53.5 ± 1.1	K-Ar (ser)	mo, cpy	100	0.16 Mo, 0.15% Cu	4, 5
18	Lucía	Cu-Mo	sw, d	gd	tn	56.9 ± 1.2	K-Ar (ser)	cpy, py, mo	-	0.5–2% total sulfides	4, 9
19	Suaqui Verde	Cu-Mo	sw, b	gd	qz-di	57.0 ± 0.3	Re-Os (mo)	py, cpy, mo, cc	-	0.1–0.15% Cu	2, 7, 9
20	Cuatro Hermanos	Cu-Mo	sw, b	-	gd, gr	55.7 ± 0.3	Re-Os (mo)	py, cpy, mo, cc	233	0.431% Cu, 0.035% Mo	2, 3, 7
21	San Antonio de La Huerta	Cu-Mo	b	-	dac, micro-di	57.4 ± 1.4	K-Ar (gm)	mo, cc, py, cpy	14.5	0.73% Cu, 0.42 g/t Au	2, 5, 9

(*continued*)

TABLE 1. CHARACTERISTICS OF THE PORPHYRY COPPER DEPOSITS OF NORTHWESTERN MEXICO (*continued*)

#	Name	Metals	Style	Intrusive rocks		Age (Ma)	Method	Mineralogy	Ton (x 10^6)	Metal contents	Refs.
				Pre-min.	Porphyry						
22	Aurora	Cu-Mo	sw, b	cz-mz	qz-mz	55.8 ± 1.8	K-Ar (bi)	py, cpy, mo	-	0.1% Cu, 0.015% Mo	4, 9, 11
23	Los Verdes (San Nicolás)	W-(Mo)	sk	gd	gr, peg	49.6 ± 1.2	K-Ar (bi)	py, cpy, sch, wf	10	1.3% W, minor Mo	2
24	Tres Piedras	Mo-W-Cu	b	gd	gr, peg	55.7 ± 0.8	Ar/Ar (Mu)	py, cpy, bn, mo, sch, wf		-	12
25	Piedras Verdes	Cu-Mo	sw	gd	gd	~60	Re-Os (mo)	py, cpy, mo, cc	105	0.1–0.15% Cu	13, 14
26	Sara Alicia	Au-Co-Cu	sk	gr	-	-	-	sa, er, Au	-	-	4
27	Batopilas: Cerro Colorado	Ag-Au-(Cu-Zn-Pb)	d	gd, cz-di	gd	-	-	cpy, Au, <mo	3	0.3% Cu, 0.4gr/t Au	2
28	Batopilas: Satevo	Au-Ag-Cu	sw	gd, cz-di	micro-qz-di	51.6 ± 1.1	K-Ar (hb)	Ag, th, cc, gn, sph	4	0.4%Cu, 2–3gr/t Au	2, 15
29	Santo Tomás	Cu	d	gd	qz-mz	57.2 ± 1.2	K-Ar (bi)	py, cpy, bn, cc	250	0.45%Cu, 0.05 gr/t Au	2, 9
30	La Reforma	Cu-Zn-Pb-Ag	lr	gd	gr	59.2 ± 1.3	K-Ar (bi)	-	-	-	9
31	La Guadalupana	W-Cu-(Mo)	sw	-	gd	51.5 ± 1.0	K-Ar (ser)	-	-	-	9
32	Cerro Colorado	Cu-Mo	b, d	-	gd	46.3 ± 1.0	K-Ar (ser)	mo, py	-	-	9, 16
33	Tameapa	Cu-Mo	sw, d	gd	qz-mz	57–50	Re-Os (mo)	py, cpy, mo	27	0.37%Cu	2, 9, 17, 18
34	Los Chicharrones	Mo	sw, b	-	qz-mz	56.2 ± 1.2	K-Ar (bi)	mo	-	-	9
35	Las Higueras	Cu-Mo	d	gd	gd	49.0 ± 1.0	K-Ar (ser)	cpy, mo	-	-	9
36	San José del Desierto	Cu-Mo-W	d, b	gd	qz-mz	59.1 ± 1.2	K-Ar (bi)	-	-	-	9
37	Cosalá	Cu-Pb-Zn-Ag-Au	sk	-	gd	58.5 ± 1.2	K-Ar (hb)	py, sph, cpy, arg, gn, th	22	3.4% Cu, 18.6% Zn, 12% Pb, 1415 gr/t Ag, 2.7 gr/t Au	2, 9, 17
38	La Azulita	Cu-Mo	d, b	-	gd	59.5 ± 1.2	K-Ar (bi)	-	0.5	1.2% Cu, 0.01% Mo	9, 17, 19
39	Malpica	Cu-(Mo)	b	gd	gd	54.1 ± 0.3	Re-Os (mo)	cpy, py, sch, mo	14	0.8% Cu	9, 7, 18
40	La Fortuna de Cobre	Cu-(Mo)	sw	gd	qz-mz, qz-feld	Late K (?)	-	cpy, py, mo, cc	-	0.50% Cu, up to 0.026% Mo	20
41	El Arco	Cu-Au	sw	-	di, gd	164.1 ± 0.4	Re-Os (mo)	cc, cpy	600	0.6% Cu, 0.2 g/t Au	21, 22
42	Los Reyes	Cu-W	sk	gd	gd	36.6 ± 0.8	K-Ar (ser)	py, cpy, bn, cc, sch	1	3% Cu, 7 g/t Ag	2, 9
43	San Martín	Cu, Zn, Au, Ag	sk, sw	gd	-	46.2 ± 1.0	K-Ar (bi)	gn, py, cpy, bn, sph	30.5	1.0% Cu, 5.0% Zn, 150 g/t Ag, 0.3–0.7 g/t Au	2, 9, 23

(*continued*)

TABLE 1. CHARACTERISTICS OF THE PORPHYRY COPPER DEPOSITS OF NORTHWESTERN MEXICO (*continued*)

#	Name	Metal	Style	Intrusive rocks		Age (Ma)	Method	Mineralogy	Ton (x 10^6)	Metal contents	Refs.
				Pre-min	Porphyry						
44	La Colorada (Chalchihuites)	Cu, Au, Ag, Pb, Zn	b, m	gd	dac	53.6 ± 1.1	K-Ar (KF)	Carb. Cu, Au, arg, sph, gn	0.4	0.33 g/t Au, 398 g/t Ag, 3.14% Pb, 1.43% Zn, Cu (?)	9, 23
45	Concepcion del Oro	Cu-Au	sk	gd, tn	gr	40.0 ± 1.2	K-Ar (bi)	py, po, cpy, th, sph	2	2% Cu, 1.6 g/t Au	2, 23
46	La Sorpresa	Cu	b	qz-mz	-	54.0	?	py-cpy	<1.0	1.2% Cu	24
47	Tandiguán	Cu	sw, b	-	-	-	-	py, cpy, oxides	-	-	24
48	Tepalcuatita	Cu-Au	sw, b	gd, qz-di	felsic (?)	-	-	py, cpy, Au	27	0.32% Cu, 0.64 ppm Au	24
49	La Verde	Cu-(C0)	b	gd	qz-di	33.4 ± 0.7	K-Ar (hb)	cpy, py, sch	110	0.7% Cu, (Co)	2, 9, 24, 25
50	San Isidro	Cu	b, sw	gd	rdc	32.5 ± 0.7	K-Ar (hb)	py, cpy, gn, sph	-	0.4% Cu	2, 9, 24, 25
51	La Manga de Cuimbo	Cu	sw, b	gr, gd	-	-	-	py, cpy, bn	-	-	24
52	Inguarán	Cu-W	b, sw	qz-mz	-	35.6 ± 0.8	K-Ar (bi)	py, cpy, mo	7	1.0–1.4% Cu, 0.02–0.04% WO_3	2, 25
53	Los Cimientos	Cu-Au	sw, b	gr, gd	qz-mz	62.8	?	py, cc, Au	-	-	24
54	Tiámaro	Cu	sw, b	gd, qz-di	rhy	-	-	py, cpy, bn, cc		0.1% Cu	24, 25
55	Tumbiscatío	Cu-Au (?)	sw	di	-	-	-	cpy, apy	-	1.06% Cu, 28 g/t Ag, 1.25 g/t Au (assay)	2
56	Quiriricuaro	Cu	sk	gr, gd	qz-mz	-	-	py, cpy, oxides	-	-	24
57	El Veladero	Cu	b	gr, gd	qz-mz	-	-	cpy, cc, oxides	-	-	24
58	Las Salinas (Copper King)	Cu	d, sw	-	qz-mz	62.8 ± 1.4	K-Ar (ser)	py, cpy, oxides	-	1.0–1.6% Cu	2, 9, 26
59	Santa Fe	Cu-Au	sk, sw, d	gd	di	2.29 ± 0.1 2.24 ± 0.08	K-Ar (bi), K-Ar (ser)	cpy, bn, arg	0.4	0.6% Cu, 2.6 g/t Au, 120 g/t Ag, 1.30% Pb	2, 9, 27
60	Tolimán	Cu	sw, d	-	qz-mz	5.75 ± 0.1	K-Ar (bi)	py, cpy, bn, cc, <mo	-	<0.8% Cu	2, 9, 27
61	Ixtacamaxtitlán	Cu	sw, sk	tn, di	gd	17.83 ± 0.06	Ar/Ar (bi)	cpy, py, sph	-	-	28

Note: Mineralization style: sw—stockwork and veins; sk—skarn; b—breccia; d—dissemination; lr—limestone replacement; m—manto. Intrusive rocks: qz-feld—quartz-feldespatic porphyry; lat—latite; di—diorite; mz—monzonite; gr—granite; gd—granodiorite; peg—pegmatite; rdc—rhyodacite; rhy—rhyolite; tn—tonalite. Dated materials: bi—biotite; gm—groundmass; KF—potassium feldspar; hb—hornblende; mo—molybdenite; mu—muscovite; phl—phlogopite; ser—sericite. Metallic mineralogy: arg—argentite; apy—arsenopyrite; bn—bornite; cc—chalcocite; co—covellite; cpy—chalcopyrite; en—enargite, er—eritrita; gn—galena; mo—molybdenite; py—pyrite; po—powellita; sa—safflorita; sch—scheelite; sp—specularite; sph—sphalerite; th—tetrahedrite; wf—wolframite. References: 1—Wodzicki, 2001; 2—Barton et al., 1995; 3—Singer et al., 2005; 4—Pérez-Segura, 1985; 5—CRM, 1992; 6—Del Río (personal commun., 2005); 7—Barra et al., 2005; 8—Valencia et al., 2006; 9—Damon et al., 1983a; 10—Scherkenbach et al., 1985; 11—Solano-Rico, 1975; 12—Mead et al., 1988; 13—Dreier and Braun, 1995; 14—Espinosa-Perea, 1999; 15—Wilkerson et al., 1988; 16—Shafiqullah et al., 1983; 17—Clark et al., 1988; 18—Barra et al., 2003; 19—Bustamante-Yáñez, 1986; 20—Salvatierra-Domínguez, 2000; 21—Coolbaugh et al., 1995; 22—Valencia et al., 2004; 23—CRM, 1991; 24—Solano-Rico, 1995; 25—CRM, 1995; 26—CRM, 1999b; 27—CRM, 1999a; 28—Tritlla et al., 2004.

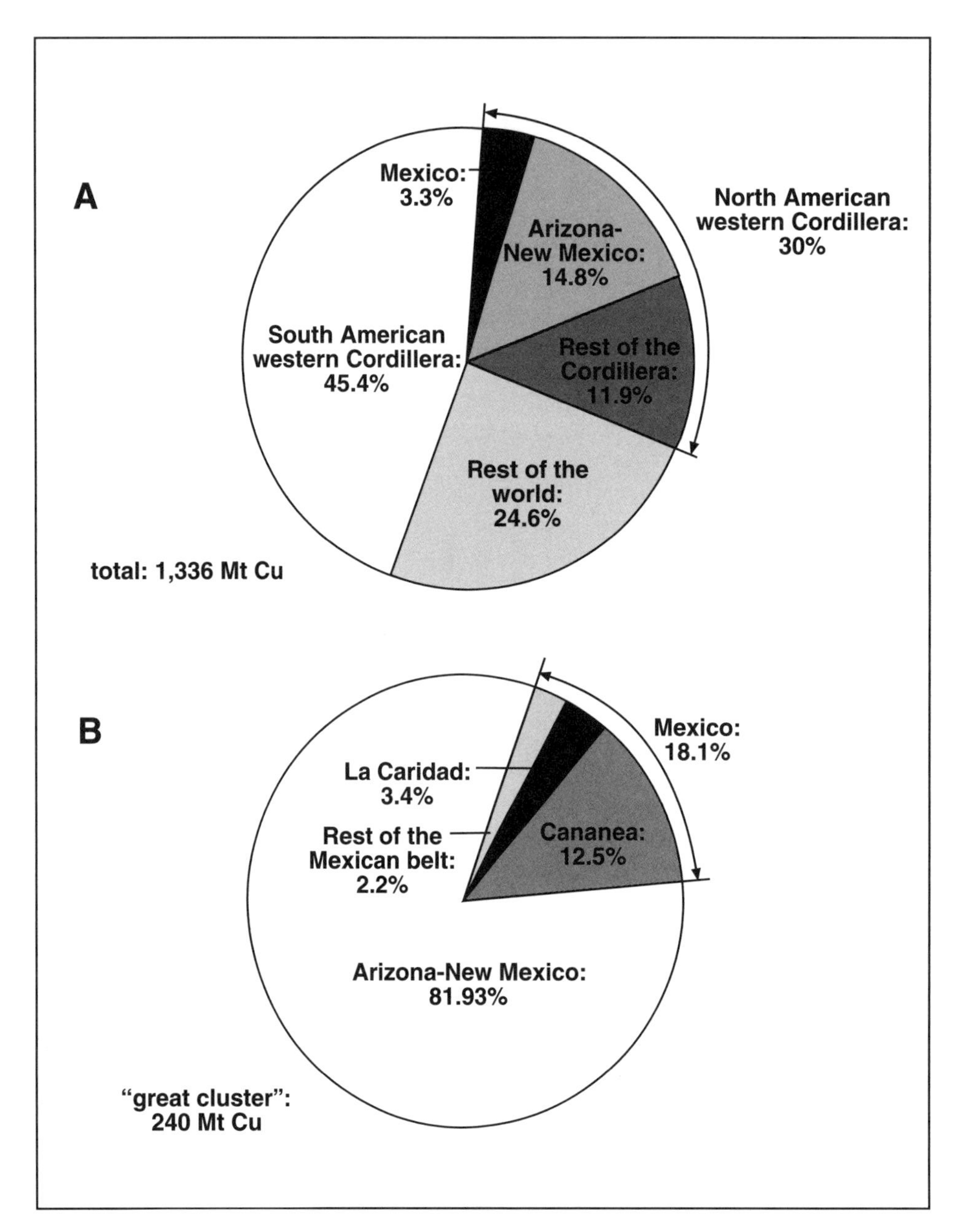

Figure 8. Comparative plots of the porphyry copper resources of northwestern México: (A) at a world-scale, and (B) in the context of the "great cluster" of Arizona–New Mexico–Sonora (estimated from data in Singer et al., 2005).

this megashear, at least in the original terms proposed by Anderson and Silver, has been questioned (e.g., Dickinson and Lawton, 2001; Iriondo et al., 2004). More to the south in central Sonora, the Proterozoic crystalline rocks are very scarce, although they are recognized to the north and east of Hermosillo in the Cerro de Oro region (González-León and Jacques-Ayala, 1988) and in the Sierra de Mazatán (Vega-Granillo and Calmus, 2003) (Fig. 9). In the east-central part of Sonora, particularly to the east of Mazatán (Fig. 9), these rocks are covered by a thick carbonated sequence of Late Proterozoic and the Paleozoic age, which was thrust on the south part by a series of tectonic sheets of Paleozoic deep-marine basin sediments, compressionally transported to the N-NW between the Late Permian and Middle Triassic (Poole et al., 1991; Stewart and Roldán-Quintana, 1991; Valencia-Moreno et al., 1999). Despite their abundance, the exposures of Paleozoic rocks end very abruptly to the south of this tectonic limit. In their place, a sequence of clastic continental and minor marine sediments of Late Triassic age is exposed. This sequence, locally recognized as the Barranca Group, was deposited in a series of E-W rift-related basins (Stewart and Roldán-Quintana, 1991), which were

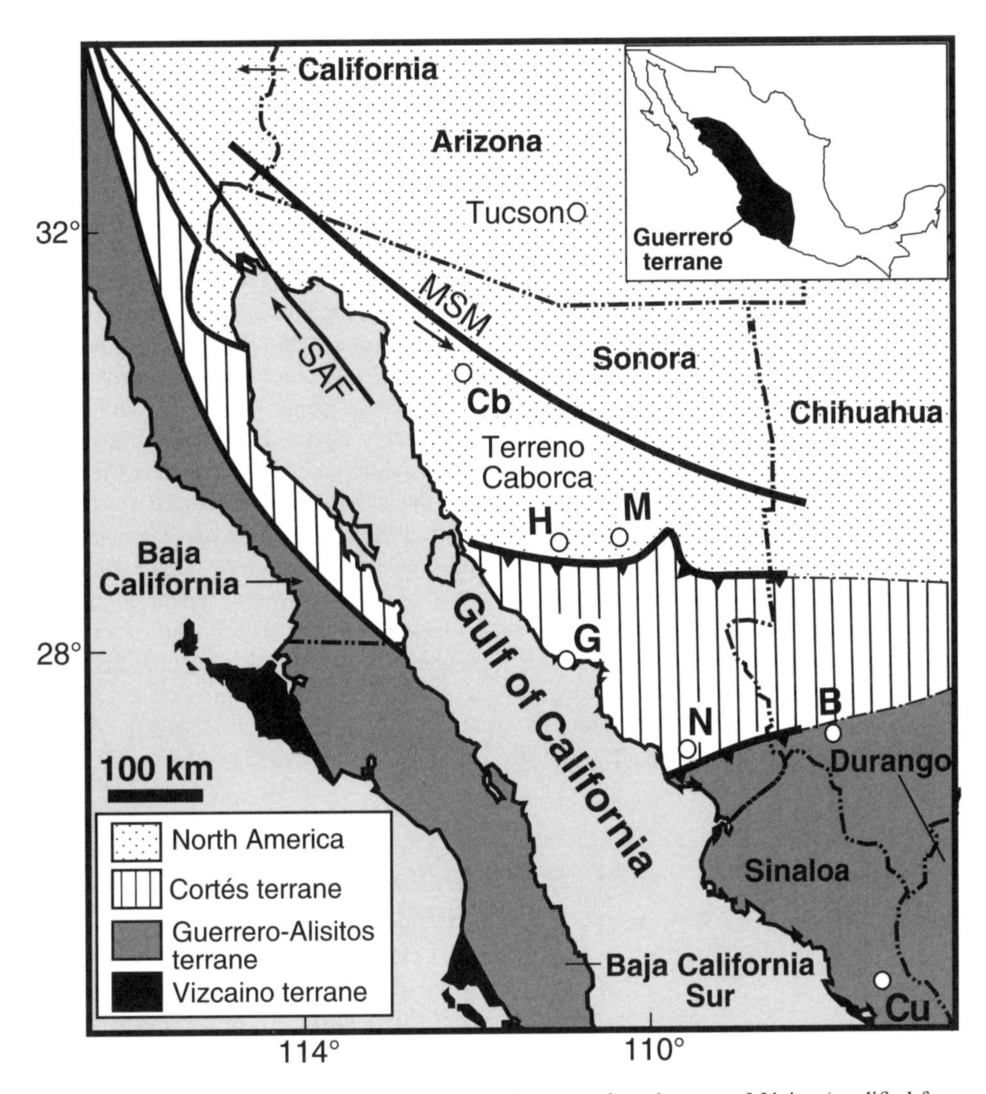

Figure 9. Geological map of the major basement features of northwestern México (modified from Valencia-Moreno et al., 2001). MSM—Mojave-Sonora megashear; SAF—San Andreas fault; Cb—Caborca; H—Hermosillo; M—Mazatán; G—Guaymas; N—Navojoa; B—Batopilas; Cu—Culiacán.

developed after the compressional deformation. Nevertheless, deep marine basin rocks of Paleozoic age outcrop again in southern Sonora and northern Sinaloa (Gastil et al., 1991; Dreier and Braun, 1995), although their geological relationship with the rocks exposed to the north is uncertain. More to the south, the basement of most of western México is dominated by the rocks of the Guerrero terrane (Fig. 9) (Campa and Coney, 1983). This terrane is the largest in the North American Cordillera (Centeno-García et al., 1993) and is composed of various subterranes whose relationships are quite complex (e.g., Coney and Campa, 1987; Sedlock et al., 1993). In general, the Guerrero terrane is characterized by volcanic and volcaniclastic sequences of oceanic affinity, associated with island arcs of Middle Jurassic and Early Cretaceous age, which were accreted to North America in the Late Cretaceous (Campa and Coney, 1983; Centeno-García et al., 1993). The northernmost exposures occur in northern Sinaloa, where meta-andesites, tuffs, pelagic sediments, pillow basalts, and ultramafic rocks of Early Cretaceous age are recognized (Ortega-Gutiérrez, et al., 1979; Servais, et al., 1986), but its northern limit may extend up to southern Sonora (Valencia-Moreno et al., 2001).

Geochemical and Isotopic Characterization of the Host Basement

According to Campa and Coney (1983), the different tectonic terranes recognized in México may have acted as important controls for the accumulation of various metals in the ore deposits known throughout the country. The metals were mostly concentrated and distributed during the activity of various major magmatic events. Particularly, it is considered that an important fraction of the mineralization is clearly associated with the presence of granitic plutons of Laramide age (90–40 Ma) (e.g., Damon et al., 1983a; Clark et al., 1988; Mead et al., 1988; Barton et al., 1995; Staude and Barton, 2001). Without doubt, the most prominent case occurs in the northern part of the porphyry copper belt, which was emplaced in various basement types. It has been noted that the interaction of such basements with the Laramide magmas largely modified the final chemical and isotopic composition of the associated granitic plutons (Valencia-Moreno et al., 2001). Because the porphyry stocks that centered the copper mineralization are clearly late stages of the main cooling granitic plutons (e.g., Damon, 1986), it is quite reasonable to assume that the intruded basement played a very important role in the metallogenesis of the porphyry copper systems. According to the geochemical and isotopic evidence, the composition of the granitic plutons along the Laramide belt of northwestern México was largely modified by contamination with materials derived from the different crustal blocks (Valencia-Moreno et al., 2001, 2003). The results, which are summarized in Figure 10, indicate initial $^{87}Sr/^{86}Sr$ ratios between 0.7070 and 0.7089 in the northern part of the belt underlain by the rocks of the Proterozoic crystalline basement of North America; between 0.7064 and 0.7073 in the central region dominated by Paleozoic deep marine basin rocks and the Triassic sediments of the Barranca Group; and between 0.7026 and 0.7062 in the southern part of the belt dominated by the rocks of the Guerrero terrane. The initial $^{143}Nd/^{144}Nd$ ratios, expressed in units of epsilon neodymium (εNd), revealed a similar variation, with signatures between –5.4 and –4.2 in the northern portion, between –5.1 and –3.4 in the central part, and between –0.9 and +4.2 in the southern part of the granitic belt. As it can be noted, the data reveal a north-south progression in the initial isotope composition, with higher Sr ratios and more negative εNd values in the plutons of the northern part and inversely in the granitoids of the southern part of the belt (Fig. 10A), which suggests that the final composition of the Laramide granitic magmas was considerably modified by processes of crustal assimilation. Along with the radiogenic isotopic evidence, it is interesting to note that the granite rocks of the northern part of the belt showed considerably higher concentrations of rare earth elements (REE) and chondrite-normalized patterns enriched in light REE, displaying a characteristic well-developed negative europium anomaly. On the contrary, the plutons that were emplaced in the southern region underlain by the Guerrero terrane showed much smaller REE enrichments, with flatter normalized slopes and little-developed europium anomalies. The granitic rocks of the central region have geochemical and isotopic signatures that overlap, particularly with the granitoids of the northern part, but in general show intermediate values between these rocks and those of the southern part of the belt (Fig. 10B). Assuming that initial Sr and εNd values of 0.7060

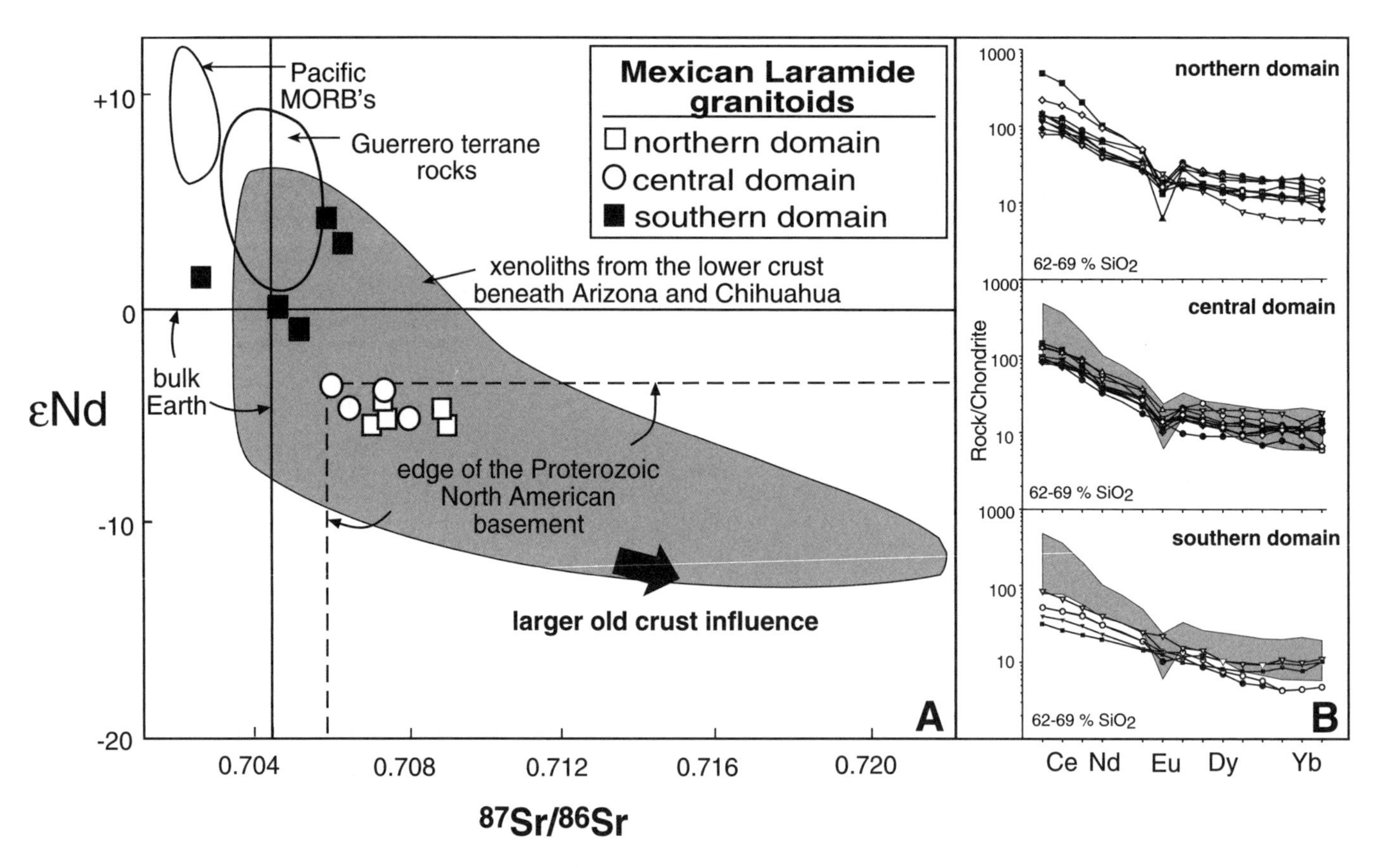

Figure 10. Compositional diagrams for Laramide granitoids of northwestern México: (A) Sm-Nd isotopic correlation diagram, and (B) chondrite-normalized rare earth element plots according to the basement of emplacement. The gray spectra in (B) represent the compositional field of the granitoids emplaced in the northern domain. Modified from Valencia-Moreno et al. (2001, 2003).

and −3.4, respectively, characterize the edge of the North American Proterozoic basement, Valencia-Moreno et al. (2001) proposed that this crust extends beneath central and southern Sonora and disappears close to the boundary with Sinaloa. More to the south, the isotopic values are clearly more primitive, suggesting the existence of a more juvenile crust, which may be represented by the Guerrero terrane and its foundations (Fig. 10A).

SYNOPSIS OF THE PORPHYRY COPPER DEPOSITS OF MÉXICO

The Mexican territory has been subject to a continuous succession of tectonomagmatic events, particularly since the middle of the Mesozoic until recent times. These events have left distinct evidence of their activity, including the generation of numerous mineralized centers (e.g., Staude and Barton, 2001), among which the porphyry copper deposits represent an outstanding example. However, in spite of its relative importance, these deposits have been very little studied. The known porphyry copper occurrences mostly lie on a NW-SE–oriented belt that extends from southwestern United States into northwestern México and continues in that direction through most of western México. The Mexican part of the belt is notably wider in its north portion, particularly in Sonora, mainly due to a larger effect of the Basin and Range extensional tectonics that affected most of southwestern North America during the late Cenozoic (Damon et al. 1983a). Also, along with the adjacent rich copper regions of Arizona and western New Mexico, Sonora contains the most attractive deposits known in the belt. More to the south, the porphyry copper mineralization is still quite continuous, but the metal resources recognized so far are considered of marginal importance. In most cases, the deposits in this region are characterized by extensive zones of potassic, phyllic, propylitic, and argillic hydrothermal alteration, associated with subvolcanic stocks of monzonitic to quartz-dioritic compositions. The mineralization mainly occurs as stockwork zones or is disseminated, especially when hosted in Laramide volcanic rocks of intermediate composition, as well as in the subvolcanic plutons themselves. Also, important ore concentrations have been observed to be associated with breccia-pipe structures and skarn zones, with Mo, Ag, Au, W, Pb and Zn. In northwestern México, the porphyry-related mineralization of Cu-Mo-W is in general more abundant and significant, with major examples represented by the large mines of Cananea and La Caridad in northeastern Sonora. A second type of Cu-Au–dominated porphyries appears to characterize the southern part of the Mexican belt, with more significant examples occuring in Sinaloa, Baja California (once it is restored to its pre-opening of the Gulf of California position; Fig. 11), Michoacán, Guerrero, and Chiapas.

Description of the Deposits

According to previous considerations regarding the nature of the basement in México and the possible metalogenetic implications, the porphyry copper belt has been divided in three domains:

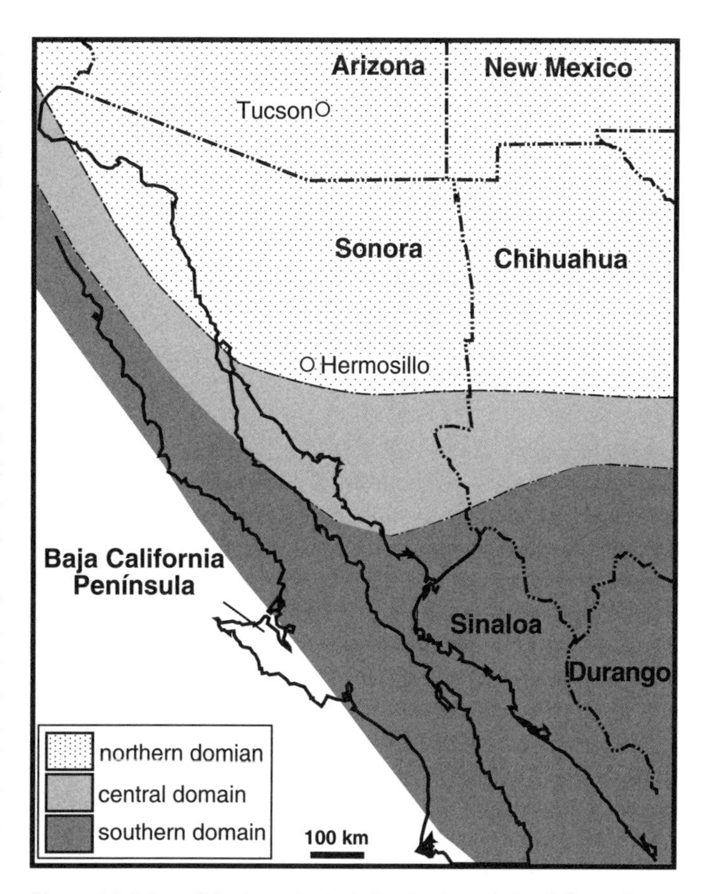

Figure 11. Map of the hypothetical distribution of the different basement domains where the porphyry copper belt of northwestern México was emplaced, showing the of Baja California peninsula restored to its position prior to the opening of the Gulf of California (according to Gastil et al., 1991).

(1) a northern domain, characterized by rocks of the Proterozoic basement of North America and its sedimentary Neoproterozoic and Paleozoic cover; (2) a central domain, composed of Paleozoic rocks of deep marine basin origin underlain by rocks of the North American Proterozoic basement; and (3) a southern domain, whose basement is characterized by the rocks of the Guerrero terrane and its foundations and by an apparent lack of rocks of North American affinity (Fig. 11). The geographical distribution of these domains and their associated deposits is presented in the Figures 6 and 12, and the corresponding labels are recalled in the following description of the different occurrences. A summary of the main geological and economic aspects of the individual deposits is also offered in Table 1.

Northern Domain

District of Cananea. Cananea constitutes the most important mining district of México, and is recognized as one of the principal porphyry copper districts in the world (Bushnell, 1988). It is located in northern Sonora, ~250 km to the northeast of Hermosillo and ~160 km south of Tucson, Arizona (Fig. 6). The mineralized zone extends along a NW-SE–oriented belt that includes

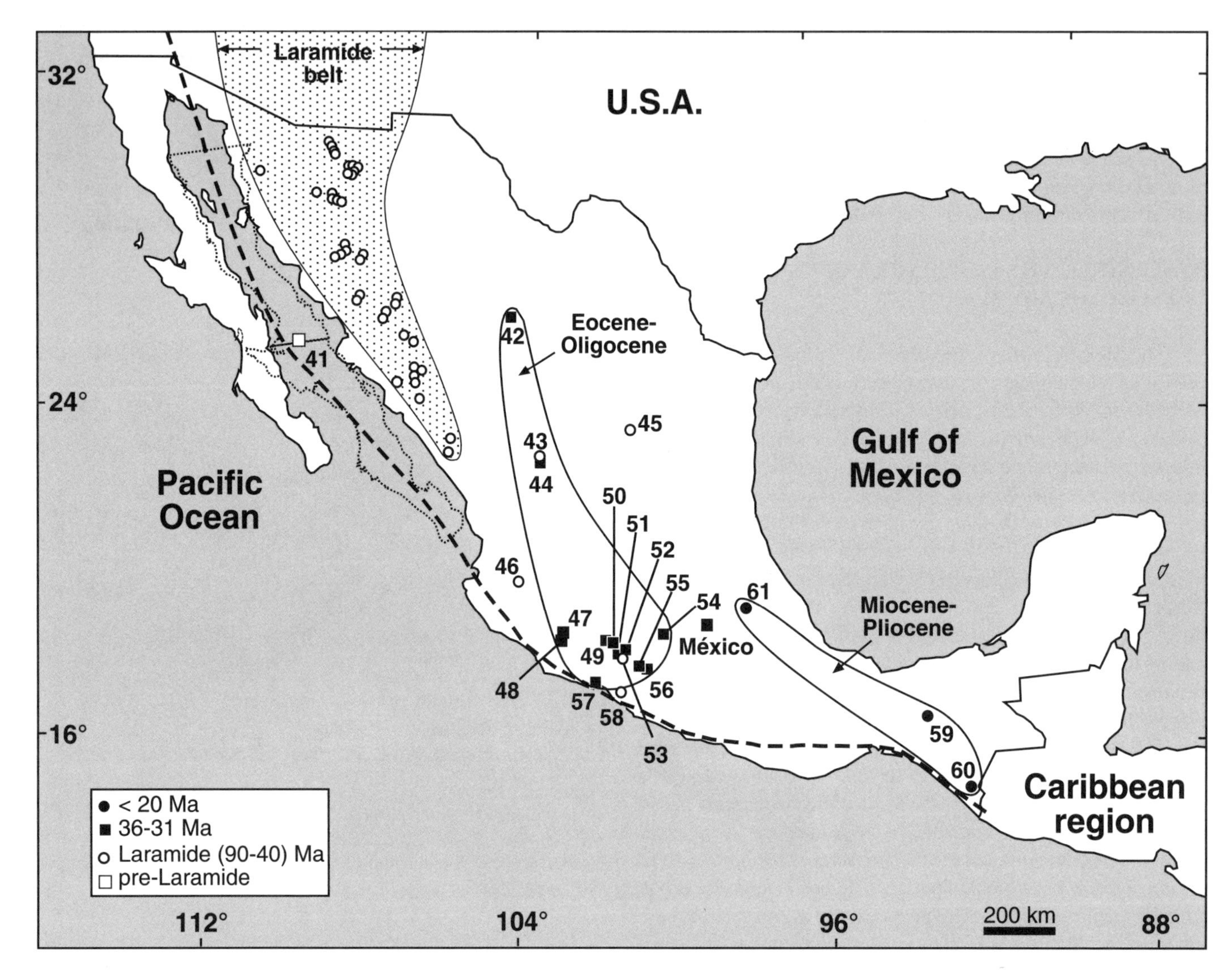

Figure 12. Map of the belt of porphyry copper and associated deposits of México showing the individual sites discussed within the text. The region identified as the "Laramide belt" represents the relatively older and potentially more important zone of the composite belt. The dotted lines represent the limits of the belt, and the regions delimited in the central and southern parts of the country indicate possible temporal subbelts.

the deposits of Cananea (1), Mariquita (2), María (3), Lucy (4), Milpillas (5) and the El Alacrán (6), among the most important. The geology of Cananea exhibits a basement of ~1000 m in thickness, composed of Cambric to Carboniferous quartzite and carbonated rocks, comprising the Bolsa, Abrigo, Escabrosa, and Martin Formations, which are underlain by a Proterozoic granite with a zircon U-Pb age of 1440 ± 15 Ma (Anderson and Silver, 1977). All these rocks unconformably underlie a thick sequence of tuffs and flows of intermediate to felsic composition of possible Triassic (?) and Jurassic age, recognized as the Elenita and Henrietta Formations (Valentine, 1936). These volcanic units are in turn overlain by a 1500-m-thick sequence of andesite and dacite flows of Laramite age belonging to the Mariquita and Mesa Formations (Meinert, 1982; Bushnell, 1988). The Laramide volcanism was accompanied by a series of contemporary intrusive pulses that include the Tinaja diorite, the Cuitaca granodiorite, and the Chivato monzodiorite. Zircon separates from the last two intrusive phases yielded ages at 64 ± 3 and 69 ± 1 Ma, respectively (Anderson and Silver, 1977). Quartz-feldspar and rhyolitic porphyries, some of which are responsible for the mineralization in the Cananea district, were dated between 59.9 ± 2.0 Ma and 57.4 ± 1.6 Ma (Wodzicki, 2001). Most of the copper production in the Cananea mine comes from open pit mining of a >500-m-thick blanket of chalcocite (Meinert, 1982; Wodzicki, 2001). Nevertheless, the highest ore grades are associated with breccia pipes that were mined in the past and continue to be of great economic importance (Bushnell, 1988). Among them, the La Colorada breccia pipe formed an outstanding mineralized structure, which yielded 7 Mt of ore at 6% Cu and 0.4% Mo and interesting values of Au and Ag (Ramón Ayala, 2005, personal commun.). Also, some skarn-type deposits dominated by bornite and chalcopyrite were mined in Puertecitos, northwest of the Cananea mine

(Meinert, 1982; Enaudi, 1982). The total resources of Cananea includes 7140 Mt of ore at 0.42% Cu, 0.008% Mo, and 0.012 gr/ton Au (Singer et al., 2005), which approximately represents a total of 30 Mt of metallic Cu. The mineralization in Cananea took place between 58.5 ± 2.1 Ma and 52.8 ± 2.3 Ma, according to biotite and sericite K-Ar dates, respectively (Barton et al., 1995), although a K-Ar age in phlogopite from the La Colorada breccia yielded an age of 59.9 ± 2.1 Ma (Damon et al., 1983a). Due to its relative importance, the zone of Cananea is the best-studied region within the belt (e.g., Perry, 1933; Valentine, 1936; Meinert, 1982; Bushnell, 1988; Wodzicki, 2001). Nevertheless, the geological characteristics and aspects related to the origin of the mineralization are still poorly known, particularly in terms of an evaluation based on modern geochemical and isotopic techniques. The deposits of Milpillas and Mariquita are located ~14 km to the northwest of the Cananea mine and are separated from each other by a distance of 5 km. Although both deposits are currently being mined, they have been little-documented in the literature. The Mariquita deposit contains 35 Mt of ore at 0.4%–0.6% Cu, while Milpillas has resources of 30 Mt at 2.5% Cu. Both deposits present a geological framework similar to that described for the Cananea mining district, but the oldest rocks in this zone correspond to volcanic flows of the Henrrieta Formation of probable Jurassic age, which are intruded by a quartz-feldspar porphyries associated with the hypogene mineralization. Both deposits exhibit important accumulations of Tertiary gravels that partially cover the mineralized zone in Mariquita and almost totally cover the mineralized zone in Milpillas. An important point to note here is that these continental clastic deposits could have played an important role in the formation of associated exotic copper deposits. Recent studies of Re-Os in molybdenites from Milpillas yielded an approximate age for the mineralization of 63 Ma (Valencia et al., 2006). There are two other smaller but high-grade deposits located near the Cananea mine, known as Maria and Lucy. The former is located ~10 km northwest of Cananea and it is presently exhausted, while the latter is still under mining operation, but their main characteristics have not been documented. Lucy is a relatively simple and very small deposit, with molybdenite and less chalcopyrite hosted in a granodioritic stock, where a Re-Os molybdenite geochronology yielded an age similar to that obtained for Milpillas (Rafael Del Rio, 2005, personal commun.). In the case of María, a fairly detailed description was presented by Wodzicki (2001). The mineralization in this deposit was related to a quartz-feldspar stock and included stockwork with chalcopyrite and molybdenite with ~7 Mt at 0.8% Cu and 0.04% Mo, and a pegmatite breccia body with massive sulfides of ~1.6 Mt at 6% Cu, 0.36% Mo, and 31 gr/ton Ag (Barton et al., 1995). The production concluded in 1996, with a total of 0.65 Mt at 9.45% Cu and 0.232% Mo from the massive sulfide zone, and 0.4 Mt at 1.30% Cu and 0.53% Mo from the pegmatite breccia (Wodzicki, 2001). The El Alacrán (6), located ~17 km southeast of the Cananea mine, is a small deposit of 0.7 Mt with 0.5% Cu (Consejo de Recursos Minerales, 1992), hosted in a 700-m-thick volcanic sequence, composed of flows, tuffs, and agglomerates of intermediate composition. This sequence is in turn intruded by dikes and quartz-feldspar porphyry stocks scattered in a NE-SW direction, which are considered to be responsible for the mineralization. K-Ar ages in biotite from potassic alteration yielded an age of 56.7 ± 1.2 Ma for the hydrothermal alteration (Damon et al., 1983a), which is characterized as a potassic zone overprinted by a penetrative quartz-sericite halo exposed close to the porphyry stocks, which is followed by a peripheral propylitic alteration zone. The mineralization is especially distributed in the potassic and quartz-sericite alteration shells and consists of disseminated and stockwork chalcopyrite and molybdenite bodies. Recent studies of Re-Os in molybdenite set the age for the mineralization at El Alacrán at 60.9 ± 0.2 and 60.8 ± 0.2 Ma (Barra et al., 2005).

District of Nacozari. Nacozari represents the second largest district of the porphyry copper belt in northwestern México and therefore is also a world-scale copper producer. The Nacozari district is located ~185 km to the N-NE of Hermosillo in northeastern Sonora, and its main mining site is located ~1.5 km east of the town of Nacozari de García. The district includes the La Caridad (7), Pilares (8), Bella Esperanza (9), Los Alisos (10), Florida-Barrigón (11) and El Batamote (12) deposits. The geology of the Nacozari district is composed of Early Cretaceous sedimentary rocks of the Bisbee Group, which outcrop mostly in the northwestern part of the district in the Florida-Barrigón area (Theodore and Priego de Wit, 1978). Most of the area is covered by Laramide flows and tuffs of intermediate composition, which are intruded by a series of subvolcanic stocks. La Caridad mine is by far the most important and best-documented deposit within the district (e.g., Echávarri-Pérez, 1971; Livingston, 1973, 1974; Seagart et al., 1974; Berchenbritter, 1976); however, relatively newer geological information is scarce. La Caridad is a significant mineralized zone with an important accumulation of metals, reaching >1800 Mt of ore at 0.452% Cu and 0.0247% Mo (Singer et al., 2005), representing a total volume of 8.14 Mt of metallic Cu. The mineralization is associated with a quartz-monzonite intrusive dated between 54 and 55 Ma (Damon et al., 1983a). This age is slightly older than molybdenite Re-Os dates of 53.8 and 53.6 Ma (Valencia et al., 2004), which suggests that the magmatic-hydrothermal event in La Caridad was relatively short-lived. Presently, exploration in some deposits in the Nacozari district is being carried out, particularly in the Los Alisos and Florida-Barrigón areas, located east and northwest of La Caridad, respectively. Another deposit in the district is the Pilares mine, which was one of the first places in Sonora to develop mining activity. This deposit consists of a volcanic breccia of latitic composition whose fragments are cemented by quartz and specularite ore. In contrast, the Bella Esperanza deposit consists of a stockwork zone and breccias associated with a quartz-monzonite stock dated at 55.9 ± 1.2 Ma (K-Ar in biotite), hosted in Laramide volcanic rocks (Damon et al., 1983a). The deposit is characterized by a remarkable oxidation zone; however, no economic mineralization has been detected yet.

District of Cumobabi. Although presently inactive, some years ago the Cumobabi district was the first producer of molybdenum in México, with a production of 2.9 Mt of ore at 0.245% Mo and 0.165% Cu (Consejo de Recursos Minerales, 1992). This district is located in north-central Sonora, ~150 km northeast of Hermosillo. Many important breccias zones have been recognized in this district, which include San Judas (13), Transvaal (14), Cobre Rico (15), and Washington (16), among others, but the more interesting of them has been the San Judas breccia. The geology of the Cumobabi district is rather simple, in the sense that no exposures of the pre-Laramide basement are known. The older rocks correspond to andesites and dacites intruded by a monzonite pluton, which yielded a biotite K-Ar age of 63.1 ± 1.7 Ma (Scherkenbach et al., 1985). On the other hand, a granodioritic pluton exposed in the area of the Washington mine, located ~25 km west of San Judas, yielded a biotite K-Ar age of 56.4 ± 1.2 Ma (Damon et al., 1983a), which approximately constrains the pre-mineral intrusive event in the Cumobabi district. Biotite and sericite associated with the hydrothermal event yielded K-Ar ages between 56 and 40 Ma (Scherkenbach et al., 1985; Damon et al., 1983a), which suggests a relatively long-lasting period of hydrothermal activity in this district. Nevertheless, two recently obtained Re-Os dates in molybdenites indicate a concordant age of 58.7 ± 0.2 Ma (Barra et al., 2005). According to Scherkenbach et al. (1985), the mineralization in Cumobabi occurred in two main pulses. The first one was associated with potassic alteration and consists of molybdenite, pyrite, quartz, chalcopyrite, anhydrite, and apatite. The second was associated with sericitic alteration and consists of chalcopyrite, ilmenite, anhydrite, tetrahedrite, sphalerite, galena, siderite, and tourmaline. The breccias of the San Judas–Transvaal zone normally contain >10% Mo, while the rocks that host these structures contain <0.01% Mo. In the Washington mine, the breccia structures are dominated by pyrite, chalcopyrite, molybdenite, and scheelite associated with potassium alteration (Simmons and Sawkins, 1983). The Cobre Rico deposit is relatively less documented in literature, but it shows evidence of past mining activity (Scherkenbach et al., 1985). It corresponds to a collapse breccia hosted in volcanic rocks and developed in a zone of sericitic and propylitic alteration (Echávarri-Pérez, 1978; Consejo de Recursos Minerales, 1992).

El Crestón. The El Crestón (17) is an isolated deposit located to the west of the Cumobabi district and ~100 km north of Hermosillo. It is characterized by a metaplutonic complex of Paleoproterozoic age, characterized by gneisses and schists intruded by the Creston granite, which yielded a zircon U-Pb sensitive high-resolution ion microprobe (SHRIMP) age of 1730 Ma (Valenzuela-Navarro et al., 2003, 2005). These old rocks are exposed as a roof pendant on a Laramide granodiorite pluton, which is in turn cut by a porphyry quartz-feldspar stock. This roof pendant structure hosts the main part of the economic mineralization, which has been considered to be a typical molybdenum porphyry deposit (Valenzuela-Navarro et al., 2005). The mineralization essentially consists of molybdenite and chalcopyrite associated with zones of phyllic and potassic alteration (Leon and Miller, 1981). The known resources are of ~100 Mt at 0.16% Mo and less important Cu values (Pérez-Segura, 1985). Sericite separated from fractures in the Creston granite yielded a K-Ar age of 53.5 ± 1.1 Ma (Damon et al., 1983a), which represents the approximate age of the mineralization. More recently, a similar age of 53.6 ± 0.2 Ma was also obtained by Re-Os in molybdenites (Barra et al., 2005).

Central Domain

District of Suaqui Grande. The Suaqui Grande district is located in east-central Sonora, ~150 km E-SE of Hermosillo and includes the deposits of Lucía (18), Suaqui Verde (19), Cuatro Hermanos (20), San Antonio de la Huerta (21), and Aurora (22) as the main mineralized centers. The geology of Suaqui Grande differs from the porphyry copper districts located more to the north in the belt, basically because there are no known exposures of the Proterozoic crystalline basement and its sedimentary cover. The older rocks consist of a strongly deformed sequence of Paleozoic sedimentary rocks deposited in a deep marine basin environment, which are unconformably covered by the Late Triassic clastic-continental and minor marine sedimentary rocks of the Barranca Group. All these rocks are in turn covered by a sequence of Laramide flows and tuffs of intermediate composition, which contain an upper more felsic member, with horizons of locally fosiliferous lacustrine sediments (McDowell et al., 1994, 2001). The Laramide volcanic rocks are particularly abundant in this region of Sonora compared to localities to the north of this zone, where they could have been deeply removed by tectonic uplifting and erosion (Valencia-Moreno et al., 2001). Intrusive rocks displaying compositions between quartz-diorite and granite yielded K-Ar ages between 63.3 ± 3.3 Ma and 49.6 ± 1.2 Ma for this part of the belt (Damon et al., 1983a, 1983b). These plutonic rocks are accompanied by late porphyry stocks of quartz-diorite and quartz-monzonite composition, some of which center the copper mineralization in this district. K-Ar ages obtained in alteration minerals suggest that the main hydrothermal pulse took place between 59 and 53 Ma (Damon et al., 1983a). Although none of the mineralized sites in the Suaqui Grande district are presently considered of economic interest, exhaustive exploration and drilling programs have been carried out in the San Antonio de la Huerta and Cuatro Hermanos deposits. The San Antonio de la Huerta deposit, also known as Luz del Cobre, is included in a sequence of marine basin environment sedimentary rocks of Paleozoic age, which are unconformably covered by the Late Triassic rocks of the Barranca Group. The mineralization in this zone is essentially associated with breccia bodies with molybdenite, chalcopyrite, chalcocite, covellite, and digenite, also including ~207.4 tons of U_3O_8 (Barton et al., 1995; Pérez-Segura, 1985). Moreover, there are also reported Au and Ag values associated with quartz veins and replacement zones exposed close to the breccias (Barton et al., 1995). The Aurora deposit, located slightly south of San Antonio de la Huerta, is one of the lesser known deposits in the district, with maximum estimated grades of 0.1% Cu and 0.015% Mo (Solano-Rico, 1975). In the Cuatro Hermanos deposit, the mineralization is associated with granodiorite and granite porphyry stocks (Zürcher, 2002) and consists of mineralized breccias with pyrite, chalcopyrite, and molybdenite, with a small zone of supergene enrichment. It contains estimated resources of ~212 Mt of ore with 0.43% Cu and 0.022% Mo (Pérez-Segura, 1985; Barton et al., 1995). Recent geochronological studies of Re-Os in molybdenite yielded an age of 55.7 ± 0.3 Ma for this deposit (Barra et al., 2005). In the Suaqui Verde prospect, the mineralization is characterized by pyrite, chalcopyrite, and molybdenite associated with a potassic alteration zone, which is in turn associated with the presence of a quartz-diorite intrusive (Flores-Vázquez et al., 2004). It also shows a small zone of supergene enrichment with 0.1%–0.15% Cu mainly as chalcocite (Pérez-Segura, 1985; Barton et al., 1995), as well as an interesting zone with copper oxides cementing clasts in a conglomerate that covers the orebody, suggesting development of exotic copper deposits. Two recent Re-Os dates in molybdenites separated from this deposit constrain the age of the mineralizing event to 56.8 ± 0.2 and 57.0 ± 0.3 Ma (Barra et al., 2005). Lucía is a small, poorly known deposit located west of Suaqui Verde. It consists of a mineralized zone with Cu-Mo in a stockwork and breccia zones closely related to a tonalite pluton (Pérez-Segura, 1985). Other localities within this district have been mentioned as representing possible porphyry copper deposits, which include San Martín, La Cardeleña, San Ignacio, and Mónica (Consejo de Recursos Minerales, 1992), but their geological and economic characteristics have not been yet documented.

District of San Nicolás. The San Nicolás district is located in eastern Sonora, near the boundary with Chihuahua and ~200 km to the E-SE of the city of Hermosillo. The geology of the district is dominated by a large granodiorite intrusion, which has yielded a hornblende $^{40}Ar/^{39}Ar$ date of 56.7 ± 0.40 Ma (Gans, 1997), and by a coeval volcanic sequence of intermediate composition, covered by thick ignimbrite flows of Oligocene-Miocene age. It is interesting to note that the Suaqui Grande and San Nicolás districts lie in the region dominated by the rift-related Late Triassic Barranca Group basin (Stewart and Roldán-Quintana, 1991). Thus, the apparent E-W orientation and distribution displayed by the different deposits in this zone suggest that the deep-seated structures that bound this basin could have served as an important control for the circulation of the mineralizing fluids. The San Nicolás district includes the deposits of La Verde (23) and Tres Piedras (24), both hosted by the granodiorite pluton. In La Verde, the mineralization consists of a garnet-epidote-quartz-calcite skarn with fluorite, pyrite, chalcopyrite, and powellite. It also presents a series of pegmatite dikes with wolframite, molybdenite, pyrite, chalcopyrite, chalcocite, and scheelite and a greissen-type alteration zone with scheelite and chalcopyrite (Barton et al., 1995). The resources at La Verde are estimated to be 10 Mt with 0.25% Mo, 0.2% Cu, and 0.2% WO_3 (Pérez-Segura, 1985). Although this deposit is presently inactive, during the years between 1916 and 1945 it reached a production of ~90,000 tons of ore with 1.3% WO_3 (Barton et al., 1995). In the Tres Piedras deposit, the mineralization is associated with a breccia body with pyrite, chalcopyrite, bornite, molybdenite, wolframite, and scheelite (Mead et al., 1988).

Southern Domain

Piedras Verdes. Piedras Verdes (25) is an isolated deposit located in southern Sonora, ~35 km to the east of the city of Navojoa (Fig. 6). According to Dreier and Braun (1995), the mineralization of Piedras Verdes is hosted in a sequence of deformed deep marine sediments, which may correlate with a similar sequence exposed in northern Sinaloa (Mullan, 1978; Gastil et al., 1991). These rocks were intruded by a Laramide granodiorite and a family of porphyries associated with the mineralization, which was dated by Re-Os in molybdenite to be ca. 60 Ma (Espinosa-Perea, 1999). According to Dreier and Braun (1995), the Piedras Verdes deposit consists of a stockwork of pyrite, chalcopyrite, and molybdenite with 0.15%–0.3% Cu and values of Mo relatively greater inside of the system, decreasing to <0.1% of Cu and almost no Mo outward. This deposit includes a zone of secondary enrichment from a primary low-grade ore composed of pyrite and chalcopyrite. Estimated resources for this deposit include ~290 Mt of ore, averaging 0.37% Cu (Espinosa-Perea, 1999). To the northeast of Navojoa (Fig. 6) and relatively close to Piedras Verdes, there is a skarn deposit named Sara Alicia (26), which developed at the contact between a sequence of undifferentiated Mesozoic sediments and a Laramide intrusive. This skarn deposit is very poorly documented; nevertheless, its main interest rests on the presence of a rare polymetallic association, characterized by Au-Cu-Co (Pérez-Segura, 1985; Pérez-Segura et al., 1995). This metal association is considered atypical for this region and may possibly indicate significant differences in the nature of the source of the metals in comparison with the porphyry copper districts located more to the north.

District of Batopilas. The district of Batopilas is located in southwestern Chihuahua near the Sonora-Sinaloa-Chihuahua state boundaries, in the region of the town of Batopilas (Fig. 6). The district includes some famous silver mines (Goodell, 1995), which probably are associated with porphyry copper systems (Wilkerson et al., 1988). The oldest rocks in Batopilas consist of a deeply altered andesite and coeval plutonic rocks of granodiorite and quartz-diorite composition (Bagby et al., 1981). These rocks are unconformably covered by a sequence of volcanic breccias, conglomerates, and rhyolitic tuffs of Oligocene age (Wilkerson et al., 1988). Many mineral deposits are known in this district, but only two of them are considered to display typical porphyry copper characteristics: Satevo-Tahonas (27) and Cerro Colorado (28). The former contains Au associated with an oxidation surface locally named the Corralitos gossan, probably related to a micro–quartz-diorite with a biotite K-Ar age of 51.6 ± 1.1 Ma (Wilkerson et al., 1988). The resources are estimated to be 4 Mt of ore with 4% Cu and 2 Mt of ore with 2–3 gr/ton Au and variable Cu concentrations in the upper parts of the oxidation surface (Barton et al., 1995). In the Cerro Colorado deposit, the mineralization is associated with a granodiorite intrusive of domic shape and is mainly found as a stockwork with pyrite and quartz with Au, whose estimated resources include ~3 Mt of ore with 0.3% Cu and 0.4 gr/ton Au (Barton et al., 1995).

District of Choix. The Choix district is located in northwestern Sinaloa, ~25 km N-NW of the town of Choix. It includes the deposits of Santo Tomás (29) and La Reforma (30), both emplaced in the Choix batholith. The Santo Tomás deposit consists of a copper disseminated zone centered by a quartz-monzonite porphyry hosted in Cretaceous limestone and metamorphosed andesite (Bustamante-Yáñez, 1986; Clark et al., 1988). A sericite K-Ar date yielded an age of 57.2 ± 1.2 Ma, which constrains the timing of the mineralization (Damon et al., 1983a). The resources of the deposit are estimated to be 250 Mt with 0.45–0.52% Cu, including a secondary enriched blanket of ~14 Mt with 0.74% Cu and 0.05 gr/ton Au (Barton et al., 1995). There are other mineralized zones around these deposits, including El Magistral, located a few kilometers southeast of Santo Tomás, where there are several breccia bodies with free gold in oxidation zones, particularly in El Orito and El Plátano, where informal mining activity is reported (Bustamante-Yáñez, 1986). La Barranca is another mineralized breccia with Au and Cu values associated with a secondary enrichment surface with evidence of old mining work (Bustamante-Yáñez, 1986). The deposit of La Reforma, located ~7.5 km north of Santo Tomás, contains Zn-Pb-Cu-Ag mineralization in replacement zones in Cretaceous limestones, intruded by a granodiorite and a granite porphyry with biotite K-Ar ages of 59.9 ± 1.3 and 59.2 ± 1.3 Ma, respectively (Damon et al., 1983a; Clark et al., 1988).

La Guadalupana and Cerro Colorado. Very little it is known in the literature about the economic and geologic characteristics of the La Guadalupana (31) and Cerro Colorado (32) prospects. La Guadalupana consists of a small mineralized zone located in southwestern Chihuahua near the border with Sinaloa and consists of Cu, Mo, and W mineralized veins emplaced in a granodiorite intrusive with a biotite K-Ar age of 59.4 ± 1.2 Ma (Damon et al., 1983a). The Cerro Colorado deposit, located ~50 km southeast of La Guadalupana, consists of a breccia-filling body with pyrite and molybdenite, developed in a granodiorite pluton (Bustamante-Yáñez, 1986) whose age obtained by biotite K-Ar dating yielded 48 ± 1.2 Ma (Shafiqullah et al., 1983), which is quite close to the sericite K-Ar age of 46.3 ± 1.2 Ma that characterizes the timing of the mineralizing event (Damon et al., 1983a), suggesting a short time span between magmatic and hydrothermal processes.

District of Tameapa. The Tameapa district is located in the eastern part of Sinaloa, ~120 km to the north of Culicán (Fig. 6), and includes the deposits of Tameapa (33), San José del Desierto (34), and Los Chicharrones (35). The estimated resources in Tameapa are 27 Mt with ~0.37% Cu (Barra et al., 2003). The deposit consists of a stockwork and breccia zones associated with a quartz-monzonite pluton that yielded a biotite K-Ar age of 54.1 ± 1.1 Ma (Damon et al., 1983a). Other ages obtained recently by Re-Os geochronology in molybdenite separates suggest the mineralization occurred in two main pulses at 57 and 53 Ma (Barra et al., 2003). The total sulfide content in the stockwork zone is estimated to be 2% to 7%, while the breccias contain 0.2% Cu and ~0.006% Mo (Barton et al., 1995). The San José del Desierto deposit, located in northwestern Durango, close to the state border between Chihuahua and Sinaloa, consists of a series of breccias and disseminated ore zones with Cu, Mo, and W associated with quartz-diorite and quartz-monzonite stocks (Clark et al., 1988; Bustamante-Yáñez, 1986). Sericite and biotite separates yielded K-Ar ages at 63.3 ± 1.3 and 59.1 ± 1.2 Ma, respectively, which approximately constrain the timing of the mineralization event in San José del Desierto (Damon et al., 1983a). The deposit of Los Chicharrones, located ~15 km south of Tameapa, consists of a stockwork and disseminated ore zones with molybdenite (Bustamante-Yáñez, 1986). The mineralization is associated with a quartz-monzonite stock that yielded a biotite K-Ar age of 56.2 ± 1.2 Ma (Damon et al., 1983a), and which is hosted in a Cretaceous sequence of sediments and interbedded andesite flows. According to the Bustamante-Yáñez report, there are other two nearby orebodies named Virginia-Washington and Pirindongos that present good evidence for Cu and Au mineralization in oxidation zones and that have been sporadically exploited by local gold diggers.

Las Higueras. Las Higueras (36) is an isolated deposit located in the east-central part of Sinaloa, ~30 km N-NE of the city of Culiacán. Its geological and economic characteristics are little known in literature, but the deposit is reported to be a mineralized zone with molybdenite and chalcopyrite closely associated with the presence of quartz veins, but also disseminated in the host rock (Bustamante-Yáñez, 1986). Biotite and sericite K-Ar isotopic analyses yielded dates of 54.9 ± 1.2 and 49.0 ± 1.0 Ma, which are considered to represent the age of the granodiorite pluton associated with the mineralization and the hydrothermal event, respectively (Damon et al., 1983a).

District of Cosalá. The Cosalá district (37) is located in the southeastern portion of Sinaloa, near the border with Durango, ~90 km to the southeast of Culiacán. The Cosalá district includes some mineralized sites, among the most important of which are Nuestra Señora, Promontorio, Mamut, Verde, Provedora, Bolaños, and Cerro San Rafael. The mineralization is mainly hosted in Cretaceous limestone and is closely associated with a granodiorite intrusive that yielded a biotite K-Ar date of 59.0 ± 1.2 Ma (Damon et al., 1983a). The deposit consists of skarn zones, replacement zones, and veins (Barton et al., 1995) with Cu-Pb-Zn-Ag-Au mineralization. It is estimated that the Cosalá district has produced some 15 Mt of ore with 4.1% Cu, 25.5% Zn, 15.2% Pb, 1400 gr/ton Ag, and 3.4 gr/ton Au from the skarn zones; ~5 Mt with 2.5% Cu, 4.8% Zn, 5.7% Pb, 720 gr/ton Ag, and 1 gr/ton Au from the replacement zones; and ~2 Mt of ore with 0.9% Cu, 1.9% Zn, 3.6% Pb, 3270 gr/ton Ag, and 1.2 gr/ton Au from the veins (Barton et al., 1995). Table 1 shows an average of the metal grades from this production.

Malpica. The prospect of Malpica (39), located to the east of Mazatlán in southern Sinaloa, consists of two breccia structures cemented by tourmaline (Bustamante-Yáñez, 1986). These breccias were emplaced in a granodiorite stock cut by a later porphyry phase of the same composition, which yielded biotite and hornblende K-Ar ages of 57.3 ± 0.6 Ma and

54.2 ± 1.2 Ma, respectively (Damon et al., 1983a). The mineralization occurs as breccia filling and stockwork and also as a small zone of supergene enrichment, whose resources are estimated to be 2 Mt of ore with 0.89% Cu oxides (Barton et al., 1995). The age of the mineralization based on Re-Os geochronology in molybdenites is 54.1 ± 0.3 Ma (Barra et al., 2005). The Bustamante-Yañez (1986) report also points out the presence of other two prospects north of Malpica, known as Los Naranjos and La Azulita (38), where interesting breccia structures with Cu mineralization have been observed.

El Arco. The El Arco (41) is a Cu-Au porphyry deposit located in the central part of the Baja California peninsula, near the border between Baja California and Baja California Sur (Fig. 6). The deposit is quite isolated from the rest of the porphyry copper belt of northwestern México, although if the Baja California peninsula is restored to its position prior to the opening of the Gulf of California, a better idea of the real position of El Arco in relation to the main mineralized belt could be obtained (Fig. 12). Recently, Re-Os sulfide dating yielded an age of 164.1 ± 0.4 Ma (Valencia et al., 2004), which is older than the $^{40}Ar/^{39}Ar$ age of 137.3 ± 1.4 Ma reported by Weber and López-Martínez (2002) and even older than the previous hornblende K-Ar age of 107.3 ± 2.4 Ma obtained by Damon et al. (1983a) for a diorite porphyry. This suggests that the mineralization associated with the porphyry copper systems in northwestern México may have had a much wider time-space domain, but the tectonic uplifting and erosion that occurred in this region has left very little evidence of it. The El Arco deposit is hosted by metavolcanic and sedimentary rocks of the Cretaceous-Jurassic Alisitos arc and is considered to be the most typical example of the Cu-Au porphyry deposits in the Mexican belt. The deposit is characterized by a chalcopyrite core surrounded by a pyrite halo developed in zones of potassic and propylitic alteration in andesitic rocks (Weber et al., 2001). There is no published data about the historic production at El Arco, although it is known that it produced gold during the years between 1935 and 1940 and that there has been relatively minor mining activity in other nearby places such as Otilia, El Tigre, and El Águila (Coolbaugh et al., 1995).

Porphyry copper deposits of central and southern México. The continuation of the porphyry copper belt through central and southern México is still poorly documented. Nevertheless, 19 mineralized sites with ore and alteration styles typical of these deposits have been reported (Fig. 12). These sites also correspond to the "southern domain," but due to the lack of more detailed information, this section will give only a brief review of the most important geological and economic aspects, which are summarized in Table 1. Recently, a new deposit named Ixtacamaxtilán (61) has been reported in the state of Puebla, with characteristics suggestive of a porphyry copper environment. This deposit consists of a subvolcanic body crowned by a horizon of kaolinite and cut by quartz veins and a stockwork with pyrite and chalcopyrite, which is associated with a granodiorite porphyry stock with a $^{40}Ar/^{39}Ar$ biotite date of 17.83 ± 0.06 Ma (Tritlla et al., 2004). In general, the porphyry copper occurrences known in the southern domain of the belt represent relatively younger mineralized systems (Fig. 12) whose ages range between 36.6 and 2.3 Ma (Damon et al., 1983a). These deposits were mostly emplaced in rocks of the Guerrero terrane (Campa and Coney, 1983), particularly in the southern portion of it, which has subsequently been named by Sedlock et al. (1993) as the Náhuatl terrane. These rocks are characterized by complexes of oceanic character that were accreted to the North American continent in the Late Cretaceous (e.g., Campa and Coney, 1983; Sedlock et al., 1993; Ortega-Gutiérrez et al., 1994). Regularly, the mineralization is associated with intrusive rocks of variable composition between granite and diorite but with a larger relative abundance of quartz monzonites, as in Inguarán (52), Las Salinas (58), San Martín (43), and Tolimán (60). In most cases, the deposits are small and dominated by copper mineralization, although some of them they have significant concentrations of gold (i.e., in Tumbiscatío [55], Concepción del Oro [45], and in Santa Fe [59]) of tungsten (Inguarán) and of Zn-Pb-Ag-Au (San Martín). The greatest volume of minerals reported for these deposits occurs in La Verde (49), with 110 Mt of ore, averaging 0.7% Cu (Barton et al., 1995). Nevertheless, other significant deposits are San Martín, with 30.5 Mt at 1.0% of Cu; Inguarán, with 7 Mt at 1.0%–1.4% of Cu; and Tepalcuatita (48), with 27 Mt at 0.32% of Cu (Barton et al., 1995; Solano-Rico, 1995; Consejo de Recursos Minerales, 1995, 1999a, 1999b). The mineralization in this region occurs as disseminated ore and in stockwork veinlets (Tolimán and Tumbiscuatío), but particularly as breccia structures (Inguarán and Los Verdes) and skarn bodies (San Martín and Concepción del Oro). The content of primary sulfides is generally low, on the order of 2%–3%, which appear mainly as pyrite, chalcopyrite, and lesser amounts of bornite. None of these deposits exhibits a significant zone of secondary enrichment, as it is frequently observed in the deposits of northwestern México, which largely reduces their economic possibilities.

DISCUSSION

The distribution of the porphyry copper systems in México follows the same NW-SE trend of the cordilleran magmatic belts, with an orientation subparallel to the coast. The El Arco deposit, which appears as an isolated occurrence in the westernmost part of the Mexican belt and whose age may be as old as 160 Ma (Valencia et al., 2004), constitutes a notable exception to the time-space characteristics generally observed in the belt. In the coastal region of Sonora and Sinaloa, the porphyry copper deposits are likewise scarce, although at least one locality, named La Fortuna de Cobre (40), is reported in Sonora (Salvatierra-Domínguez, 2000). Also on the northwest coast of Sinaloa, there is a fissure vein deposit named Bacamacari with Cu-W mineralization (Fig. 6) that yielded a K-Ar biotite age of 87.9 ± 1.8 Ma (Damon et al., 1983a). More to the east, the ages lie in the range of 69–50 Ma. During this time interval, the Laramide magmatism extended south, from the southwestern United States, across eastern Sonora, Sinaloa, and western Chihuahua. This time span clearly represented the most productive stage in the copper province of southwestern North America and is characterized by the famous "great cluster." A relatively small number of the deposits yielded younger isotopic ages (<50 Ma), which are mainly located in the eastern part of the mineralized belt between Chihuahua and Sonora, although they overlap sections with older ages. The younger ages occur in the southern part of the belt, particularly in the states of Zacatecas, Michoacán, Guerrero, and Chiapas (Fig. 12). The deposits in this region seem to comprise two different time subbelts that can be roughly distinguished: one with ages between ca. 37 and 31 Ma, which extends from Guerrero and Michoacán to the north up to northwestern Zacatecas; and the other with ages between 6 and 2 Ma, which is more restricted to the state of Chiapas (Fig. 12).

Spatial Distribution of Metals

The porphyry copper deposits of northeastern Sonora represent the larger accumulation of Cu, Mo, and W in México, but they may also locally contain relatively small but important amounts of Zn, Ag, and Au. The major fraction of these metals is concentrated in two exceptional mining districts, Cananea, accounting for ~30 Mt of Cu, and La Caridad, with >8 Mt of Cu. The former belongs to a very select group of "super-giant" ore deposits (>25 Mt Cu; Laznicka, 1999), which include Chuquicamata, El Teniente, and La Escondida in Chile, and Morenci, Bingham, and Continental-Butte in the United States (Figs. 1 and 5). The latter corresponds to the so-called "giants" deposits (>2.5 Mt Cu, Laznicka; 1999), among which are Miami-Inspiration, Ray, Chino, El Salvador, Toquepala, and Cuajone (Figs. 1 and 5). Comparitively, the districts and prospects that occur in the central and southern domains of the porphyry copper belt of México have a contrasting smaller size and economic importance, and there are no known examples of intermediate magnitude. The copper resources of potentially economic interest in these districts, with the exception of El Arco, Cuatro Hermanos, Piedras Verdes, Santo Tomá,s and Tameapa, barely reach 1.0 Mt of Cu. Farther into central and southern México, the accumulation of metals associated with these deposits is still less significant, although interesting occurrences exist in Inguarán and La Verde in Michoacán, as well as the zones of San Martín and Concepción del Oro in Zacatecas (Table 1). It is interesting to note that the latter occurrence is located to the east of the main belt (Fig. 12) and that its affinity with the porphyry copper systems is not yet clear (Barton et al., 1993). It seems that the tremendous accumulation of metals observed in the northern part of the copper belt, especially in the region of Arizona and northeastern Sonora, could have been largely determined by the preservation of a thick pile of Laramide volcanic rocks, which are widely exposed in this region. Petrogenetic, geochemical, and isotopic studies indicate that these volcanic rocks are consanguineous materials of the main granitic plutons and the late productive intrusive phases (Damon et. al., 1983a; Sillitoe, 1996; Wodzicki, 2001). According to the distribution of the occurrences in the northern part of this region, it seems that the Laramide volcano-plutonic centers formed complexes along major cortical structures. It is probable that these structures, contemporary with, or reactivated during the Laramide deformation, controlled the ascent of magma and hydrothermal fluids to the subvolcanic low-pressure orifices (Tosdal and Richards, 2001). The mineralized centers exhibit various degrees of differentiation, generally with more than one intrusion in each center, although not all of them were associated with the mineralization. These characteristics suggest that the later productive and unproductive stocks were fed episodically by a long-term magma source (Damon, 1986; Stern, 2002). In this way, the concentration and transportation of metals must have occurred during the cooling of the main plutonic bodies, under particularly favorable geochemical and thermodynamics conditions (Keith and Swan, 1996; Mungall, 2002).

Influence of the Basement

According to the inventory of the porphyry copper localities discussed in this work, the lithological composition of the main granitic plutons is very diverse and exhibits no systematic regional control. Nevertheless, the geochemical and isotopic signatures have shown that involvement of the emplaced basement clearly influenced the final composition of the granitic magmas (Fig. 10), a fact that was also noted for the late intrusive phases associated with the mineralization (Damon et al., 1983a). It is estimated that the North American Proterozoic crust continues beneath the Phanerozoic rock sequences to southern Sonora, where the isotopic evidence indicates the presence of a more juvenile basement, represented by the Guerrero terrane (Valencia-Moreno et al., 2001, 2003). The available geochemical information is presently not sufficient to evaluate in a more conclusive way the effects that these major basement changes had on the distribution of the metals associated with the Mexican porphyry copper systems; nevertheless, it seems clear that the region underlain by the North American crust typically promoted the formation of Cu-Mo-W deposits, while in the region dominated by the oceanic rocks of the Guerrero terrane, the deposits are more commonly of Cu-Au, with less significant values of Mo and W. It seems that the deposits of greater economic interest are located in the northern part of the copper belt, decreasing in importance in the central and southern parts, where the deposits generally are marginal in size. The copper has an erratic distribution along the belt, with grades generally between 0.15% and 0.48% for the hypogenic copper, without an apparent control derived from the type of intruded basement. Some mineralized sites, like the Pilares and María mines, show spectacular concentrations above 1% of Cu, although these places represent just local accumulations. Molybdenum also occurs along the entire belt, but the main concentrations are located in the northern and central parts of the belt, with grades normally on the order of 0.015% to 0.035%. Similar to copper, local concentrations of molybdenum, with values as high as 0.16%–0.2%, are known in breccia structures located in the district of Cumobabi, in the north-central part of the belt.

In the southern part of the belt, molybdenum is commonly present; nevertheless, the concentrations are not very attractive and are barely reported in the available literature. On the other hand, gold mineralization grading between 0.2 and 3 gr/ton has been reported, especially in oxidation surfaces in deposits hosted in the southern domain. In the northern part of the porphyry copper belt, gold is reported in the Cananea and Washington mines, but grades are lower than 0.2 gr/ton. However, the San Antonio de la Huerta deposit in the central part of the belt averages 0.42 gr/ton Au (Table 1). According to these observations, the copper mineralization may have been accumulated during the initial stages of the magmatic process and subsequently transported to shallower levels in the crust during the ascent and cooling of the associated subvolcanic stocks. Conversely, molybdenum and tungsten initially could have accumulated in the magma in the same way, but subsequently enriched by interaction with the crust. In the case of gold, it is evident that contribution of a crustal segment with a more mantle affinity can be interpreted to be responsible for the relative enrichments observed in the porphyry copper deposits of the southern domain. The present configuration of the mineralized belt (Fig. 12) and the possible causes for the existence (or not) of important blankets of supergenic enrichment through mechanisms of preservation and exposition of the systems has been already discussed (Barton et al., 1995). This would partly explain the marked differences in the size of the deposits. In most cases, the original systems are incomplete, and only a fraction of them can be observed as remnants of the erosion and extension processes, as in the case of El Crestón. In general, it can be noted that the deposits in the northern domain are mostly hosted in thick volcanic piles of intermediate to felsic composition, which are relatively contemporaneous with the mineralization. These volcanic rocks are very little-preserved in the central and southern parts of the belt, which suggests a relatively major effect by the erosive processes. La Fortuna de Cobre and El Arco deposits, in coastal Sonora and the middle of the Baja California peninsula, respectively, are located to the west and considerably far from the rest of the deposits in the main belt. This suggests that in the western portion, the belt was more exposed to the erosion, and in any case, the Fortuna de Cobre and El Arco deposits are evidence that the belt may have been initially more extensive. This observation becomes more coherent if the Peninsula from Baja California is restored to its position previous to the opening of the Gulf of California (Fig. 12). Just south of the large districts of Cananea and Nacozari, the mineralization is characterized by breccia pipes and stockworks with molybdenum and tungsten in the zones of Cumobabi and El Crestón, which are considered to be relatively deep parts of the system, fed by circulation of water of magmatic origin (Pérez-Segura, 1985). This is coupled with the notable absence of exposures of Laramide volcanic rocks in this zone, broadly located to the north of Hermosillo and south of the district of Nacozari, which may suggest a larger tectonic uplifting and erosion of the upper part of the mineralized structures.

CONCLUSIONS

Despite the lack of systematic information regarding the metal grades in the different deposits along the Mexican porphyry copper belt, it can be concluded that the hypogene copper grades do not show evidence of regional control, derived from the type of basement in which the copper ores were emplaced. On the contrary, the larger values of molybdenum and tungsten occur in the region of the belt dominated by the Proterozoic North American rocks. Also, the systems emplaced in the southern domain underlain by the Guerrero terrane rocks are characterized by relatively more important and regular gold mineralization. On this basis, the porphyry copper deposits of México can be classified in two main groups. A first group, which comprises the northern and central domains and was developed under the significant influence of old continental crust, is characterized by Cu-Mo-W deposits and includes the world-class deposits of Cananea and La Caridad. The second group comprises the southern domain of the belt and exhibits clear genetic relations with a relatively young oceanic basement. It is dominated by Cu-Au mineralization and although it includes a relatively important number of deposits, with the exceptions of El Arco in Baja California (~3.6 Mt Cu) and Santo Tomás (~1.1 Mt Cu), most of the occurrences are relatively small for the standard size expected for this type of ore system.

ACKNOWLEDGMENTS

The initial ideas for this contribution were proposed in the doctoral thesis work of the first author (Consejo Nacional de Ciencia y Tecnología [CONACYT] grant 84734), which has been reworked in the doctoral thesis work of Noguez-Alcántara. Part of the geological and analytic data discussed in this study was produced with funds provided by CONACYT research project I29887T and of the initial results of project grant IN106603 of the Universidad Nacional Autónoma de México (UNAM), Dirección General de Asuntos del Personal Académico Programa de Apoyo a Proyectos de Investigación e Innovación Tecnológica (PAPIIT-DGAPA-UNAM), granted to Valencia-Moreno. We are grateful for the valuable support of the mining companies Minera México S.A. de C.V., Servicios Industriales Peñoles S.A. de C.V., and Minera María S.A. de C.V. during the field work in the different mining units and prospects of their property, as well as for allowing access to their internal reports.

REFERENCES CITED

Anderson, T.H., and Silver, L.T., 1977, U-Pb isotope ages of granitic plutons near Cananea, Sonora: Economic Geology and the Bulletin of the Society of Economic Geologists, v. 72, p. 827–836.

Anderson, T.H., and Silver, L.T., 1979, The role of the Mojave Sonora megashear in the tectonic evolution of northern Sonora, *in* Anderson, T.H., and Roldán-Quintana, J., eds., Geology of northern Sonora: University of Pittsburgh Guidebook, Field Trip 27, p. 59–68.

Bagby, W.C., Cameron, K.L., and Cameron, M., 1981, Contrasting evolution of calc-alkalic volcanic and plutonic rocks of western Chihuahua, Mexico: Journal of Geophysical Research, v. 86, p. 10,402–10,410.

Barra, F., Ruiz, J., and Chesley, J., 2003, The longevity of porphyry copper systems: A Re-Os approach: Geological Society of America Abstracts with Programs, v. 35, no. 6, p. 232.

Barra, F., Ruiz, J., Valencia, V.A., Ochoa-Landín, L., Chesley, J.T., and Zürcher, L., 2005, Laramide porphyry Cu-Mo mineralization in northern Mexico: Age constraints from Re-Os geochronology in molybdenites: Economic Geology and the Bulletin of the Society of Economic Geologists, v. 100, p. 1605–1616.

Barton, M.D., Staude, J.M., Zürcher, L., and Megaw, P.K.M., 1995, Porphyry copper and other intrusion-related mineralization in Mexico, *in* Pierce, F.W., and Bolm, J.G., eds., Porphyry copper deposits of the American Cordillera: Arizona Geological Society Digest, v. 20, p. 487–524.

Berchenbritter, D.K., 1976, The Geology of La Caridad Fault, Sonora, Mexico [M.S. thesis]: Iowa City, University of Iowa, 127 p.

Bushnell, S.E., 1988, Mineralization at Cananea, Sonora, Mexico, and paragenesis and zoning of breccia pipes in quartzofeldspathic rock: Economic Geology and the Bulletin of the Society of Economic Geologists, v. 83, p. 1760–1781.

Bustamante-Yáñez, M.A., 1986, Recursos mineros en el Estado de Sinaloa: Geomimet, v. 142, p. 87–102.

Campa, M.F., and Coney, P.J., 1983, Tectono-stratigraphic terranes and mineral resource distribution in Mexico: Canadian Journal of Earth Sciences, v. 20, p. 1040–1051.

Camprubí, A., and Albinson, T., 2007, this volume, Epithermal deposits in México— Update of current knowledge, and an empirical reclassification, *in* Alaniz-Álvarez, S.A., and Nieto-Samaniego, Á.F., Geology of México: Celebrating the Centenary of the Geological Society of México: Geological Society of America Special Paper 422, doi: 10.1130/2007.2422(14).

Camus, F., 2003, Geología de los sistemas porfíricos en los Andes de Chile: Servicio Nacional de Geología y Minería, Santiago, Chile, 267 p.

Centeno-García, E., Ruiz, J., Coney, P.J., Patchett, P.J., and Ortega-Gutiérrez, F., 1993, Guerrero terrane of Mexico: Its role in the Southern Cordillera from new geochemical data: Geology, v. 21, p. 419–422, doi: 10.1130/0091-7613(1993)021<0419:GTOMIR>2.3.CO;2.

Clark, A.H., 1993, Are outsize porphyry copper deposits either anatomically or environmentally distinctive? *in* Whiting, B.H., et al., eds., Giant ore deposits: Society of Economic Geologists Special Publication 2, p. 213–283.

Clark, K.F., Damon, P.E., and Saffiqullah, M., 1988, Metallization epoch in relation to late Mesozoic and Cenozoic igneous activity, Sinaloa, Mexico, *in* Clark, K.F., Goodell, P.C., and Hoffer, J.M., eds., Stratigraphy, tectonics and resources of parts of Sierra Madre Occidental province, Mexico: El Paso Geological Society, Guidebook for the 1988 Field Conference, p. 343–362.

Coney, P.J., and Campa, M.F., 1987, Lithotectonic terrane map of Mexico: U.S. Geological Survey Miscellaneous Field Studies Map MF-1874-D, scale 1:2,500,000.

Coney, P.J., and Reynolds, S.J., 1977, Cordilleran Benioff zones: Nature, v. 270, p. 403–406, doi: 10.1038/270403a0.

Consejo de Recursos Minerales, 1991, Monografía geológico-minera del Estado de Zacatecas: Pachuca, Hidalgo, Mexico, 154 p.

Consejo de Recursos Minerales, 1992, Monografía geológico-minera del Estado de Sonora: Pachuca, Hidalgo, Mexico, 220 p.

Consejo de Recursos Minerales, 1995, Monografía geológico-minera del Estado de Michoacán: Pachuca, Hidalgo, Mexico, 176 p.

Consejo de Recursos Minerales, 1999a, Monografía geológico-minera del Estado de Chiapas: Pachuca, Hidalgo, Mexico, 180 p.

Consejo de Recursos Minerales, 1999b, Monografía geológico-minera del Estado de Guerrero: Pachuca, Hidalgo, Mexico, 262 p.

Coolbaugh, D.F., Osoria-Hernández, A., Echávarri-Pérez, A., and Martínez-Muller, R., 1995, El Arco porphyry copper deposit, Baja California, México, *in* Pierce, F.W., and Bolm, J.G., eds., Porphyry copper deposits of the American cordillera: Arizona Geological Society Digest, v. 20, p. 525–534.

Damon, P.E., 1986, Batholith-volcano coupling in the metallogeny of porphyry copper deposits, *in* Friedrich G.H., et al., eds., Geology and metallogeny of copper deposits: Berlin, Springer-Verlag, p. 216–234.

Damon, P.E., Clark, K.C., and Shafiqullah, M., 1983a, Geochronology of the porphyry copper deposits and related mineralization of Mexico: Canadian Journal of Earth Sciences, v. 20, p. 1052–1071.

Damon, P.E., Shafiqullah, M., Roldán-Quintana, J., and Cochemé, J.J., 1983b, El batolito Laramide (90–40 Ma) de Sonora: Memorias de la XV Convención Nacional de la AIMMGM, Guadalajara, Jalisco: Asociación de Ingenieros de Minas, Metalurgistas y Geólogos de México (AIMMGM), p. 63–95.

Dickinson, W.R., 1989, Tectonic setting of Arizona through geologic time, *in* Jenny, J.P., and Reynolds, S.J., eds., Geologic Evolution of Arizona: Arizona Geological Society Digest, v. 17, p. 1–16.

Dickinson, W.R., and Lawton, T.F., 2001, Carboniferous to Cretaceous assembly and fragmentation of Mexico: Geological Society of America Bulletin, v. 113, p. 1142–1160, doi: 10.1130/0016-7606(2001)113<1142:CTCAAF>2.0.CO;2.

Dreier, J.E., and Braun, E.R., 1995, Piedras Verdes, Sonora, Mexico: a structurally controlled porphyry copper deposit, *in* Pierce, F.W., and Bolm, J.G., eds., Porphyry copper deposits of the American Cordillera: Arizona Geological Society Digest, v. 20, p. 535–543.

Echávarri-Pérez, A., 1971, Petrografía y alteración del depósito La Caridad, Nacozari, Sonora, México: Memorias de la IX Convención Nacional de la AIMMGM, Hermosillo, Sonora, México: Asociación de Ingenieros de Minas, Metalurgistas y Geólogos de México (AIMMGM), p. 1–33.

Echávarri-Pérez, A., 1978, Metallogenetic map of Sonora, Mexico: Arizona Geological Society Digest, v. XI, p. 145–154.

Enaudi, M., 1982, Description of skarns associated with porphyry copper plutons, southwestern North America, *in* Titley, S.R. (ed.), Advances in geology of the porphyry copper deposits: Tucson, The University of Arizona Press, p. 139–135.

English, J., Johnston, S.T., and Wang, K., 2003, Thermal modeling of Laramide magmatism: Testing of the flat-subduction hypothesis: Earth and Planetary Science Letters, v. 214, p. 619–632, doi: 10.1016/S0012-821X(03)00399-6.

Espinosa-Perea, V.J., 1999, Magmatic evolution and geochemistry of the Piedras Verdes Deposit, Sonora, Mexico [M.S. thesis]: The University of Arizona, Tucson, Arizona, 114 p.

Farmer, G.L., and DePaolo, D.J., 1984, Origin of Mesozoic and Tertiary granites in the western United States and implications for pre-Mesozoic crustal structure: 2. Nd and Sr isotopic studies of unmineralized and Cu- and Mo-mineralized granites in the Precambrian craton: Journal of Geophysical Research, v. 89, p. 10,141–10,160.

Friedrich, G.H., Genkin, A.D., Naldrett, A.J., Ridge, J.D., Sillitoe, R.H., and Vokes, F.M., eds., 1986, Geology and metallogeny of copper deposits: Berlin, Springer-Verlag, 592 p.

Flores-Vázquez, I., Ochoa-Landín, L., Valencia-Moreno, M., Valencia, V., and Del Rio-Salas, R., 2004, Emplacement depths of porphyry copper-related plutons in the Suaqui Verde deposit, east-central Sonora, Mexico: IV Reunión Nacional de Ciencias de la Tierra, Sociedad Geológica Mexicana, p. 190.

Gans, P.B., 1997, Large-magnitude Oligo-Miocene extension in southern Sonora: Implications for the tectonic evolution of northwest Mexico: Tectonics, v. 16, p. 388–408, doi: 10.1029/97TC00496.

Gastil, R.G., Miller, R., Anderson, P., Crocker, J., Campbell, M., Buch, P., Lothringer, C., Leier-Engelhardt, P., DeLattre, M., Hobbs, J., and Roldán-Quintana, J., 1991, The relation between the Paleozoic strata on opposite sides of the Gulf of California, *in* Pérez-Segura. E., and Jacques-Ayala, C., eds., Studies of Sonoran geology: Geological Society of America Special Paper 254, p. 7–17.

González-León, C., and Jacques-Ayala, C., 1988, Estratigrafía de las rocas cretácicas del área de Cerro de Oro, Sonora Central: Boletín del Departamento de Geología, Universidad de Sonora, v. 5, p. 1–23.

Goodell, P., 1995, Porphyry copper deposits along the Batopilas Lineament, Chihuahua, Mexico, *in* Pierce, F.W., and Bolm, J.G., eds., Porphyry copper deposits of the American Cordillera: Arizona Geological Society Digest, v. 20, p. 544.

Gustafson, L.B., Orquera, W., McWilliams, M., Castro, M., Olivares, O., Rojas, G., Maluenda, J., and Méndez, M., 2001, Multiple centers of mineralization in El Indio Muerto District, El Salvador, Chile: Economic Geology and the Bulletin of the Society of Economic Geologists, v. 96, p. 325–350.

Hedenquist, J.W., and Lowenstern, J.B., 1994, The role of magmas in the formation of hydrothermal ore deposits: Nature, v. 370, p. 519–527, doi: 10.1038/370519a0.

Hedenquist, J.W., Arribas, A., Jr., and Reynolds, T.J., 1998, Evolution of an intrusion-centered hydrothermal system: Far Southeast-Lepanto porphyry and epithermal Cu-Au deposits, Philippines: Economic Geology and the Bulletin of the Society of Economic Geologists, v. 93, p. 373–404.

Heinrich, C.A., Gunther, D., Audetat, A., Ulrich, T., and Frischknecht, R., 1999, Metal fractionation between magmatic brine and vapor, and the link between porphyry-style and epithermal Cu-Au deposits: Geology, v. 27, p. 755–758, doi: 10.1130/0091-7613(1999)027<0755:MFBMBA>2.3.CO;2.

Humphreys, E., Hessler, E., Dueker, K., Farmer, G.L., Erslev, E., and Atwater, T., 2003, How Laramide-age hydration of North America lithosphere by the Farallon slab controlled subsequent activity in the western United States: International Geology Review, v. 45, p. 575–595.

Iriondo, A., Premo, W.R., Martínez-Torres, L.M., Budanh, J.R., Atkinson, W.W., Siems, D.F., and Guarás-González, B., 2004, Isotopic, geochemical, and temporal characterization of Proterozoic basement rocks in the Quitovac region, northwestern Sonora, Mexico: Implications for the reconstruction of the southwestern margin of Laurentia: Geological Society of America Bulletin, v. 116, p. 154–170, doi: 10.1130/B25138.1.

Keith, S.B., and Swan, M.M., 1996, The great Laramide porphyry copper cluster of Arizona, Sonora, and New Mexico: The tectonic setting, petrology and genesis of the world class metal cluster, *in* Coyner, A.R., and Fahey, P.L., eds., Geology and ore deposits of the American cordillera: Geological Society of Nevada Symposium Proceedings, Reno-Sparks, Nevada, p. 1667–1747.

Lang, J.R., and Titley, S.R., 1998, Isotopic and geochemical characteristics of Laramide magmatic systems in Arizona and implications for the genesis of porphyry copper deposits: Economic Geology and the Bulletin of the Society of Economic Geologists, v. 93, p. 138–170.

Laznicka, P., 1999, Quantitative relationships among giant deposit of metals: Economic Geology and the Bulletin of the Society of Economic Geologists, v. 94, p. 455–473.

Leon, F.L., and Miller, C.P., 1981, Geology of the Creston molybdenum-copper deposit, *in* Ortlieb, L., and Roldán-Quintana, J., eds., Geology of northwestern Mexico and southern Arizona: Geological Society of America Cordilleran Section Meeting, Field Trip 7, p. 223–238.

Livingston, D.E., 1973, Geology, K-Ar ages and Sr isotropy at La Caridad, Nacozari district, Sonora, Mexico [M.S. thesis]: Tucson, The University of Arizona, 31 p.

Livingston, D.E., 1974, K-Ar ages and isotropy of La Caridad, Sonora, compared to other porphyry copper deposits of the southern Basin and Range province: Geological Society of America Abstract with Programs, v. 6, no. 3, p. 208.

Lowell, J.D., and Guilbert, J.M., 1970, Lateral and vertical alteration zoning in porphyry ore deposits: Economic Geology and the Bulletin of the Society of Economic Geologists, v. 65, p. 373–408.

McDowell, F.W., Roldán-Quintana, J., Amaya-Martínez, R., and González-León, C., 1994, The Tarahumara Formation—A neglected component of the Laramide magmatic arc in Sonora [abs.]: Geos, v. 14, p. 76–77.

McDowell, F.W., Roldán-Quintana, J., and Connelly, J.N., 2001, Duration of Late Cretaceous-early Tertiary magmatism in east-central Sonora, Mexico: Geological Society of America Bulletin, v. 113, p. 521–524, doi: 10.1130/0016-7606(2001)113<0521:DOLCET>2.0.CO;2.

Mead, R.D., Kesler, S.E., Foland, K.A., and Jones, L.M., 1988, Relationship of Sonoran tungsten mineralization to the metallogenic evolution of Mexico: Economic Geology and the Bulletin of the Society of Economic Geologists, v. 83, p. 1943–1965.

Meinert, L.D., 1982, Skarn, manto, and breccia pipe formation in sedimentary rocks of the Cananea mining district, Sonora, Mexico: Economic Geology and the Bulletin of the Society of Economic Geologists, v. 77, p. 919–949.

Mullan, H.S., 1978, Evolution of part of the Nevadan orogen in northwestern Mexico: Geological Society of America Bulletin, v. 89, p. 1175–1188, doi: 10.1130/0016-7606(1978)89<1175:EOPOTN>2.0.CO;2.

Mungall, J.E., 2002, Oxidation of the mantle wedge: Goldschmidt Conference Abstracts, p. A535.

Ortega-Gutiérrez, F., Prieto-Vélez, R., Zúñiga, Y., and Flores, S., 1979, Una secuencia volcano-plutónica sedimentaria cretácica en el norte de Sinaloa: ¿un complejo ofiolítico?: Universidad Nacional Autónoma de México, Instituto de Geología, Revista, v. 3, p. 1–8.

Ortega-Gutiérrez, F., Sedlock, R.L., and Speed, R.C., 1994, Phanerozoic tectonic evolution of Mexico, *in* Speed, R.C., ed., Phanerozoic evolution of North American continent-ocean transitions: Geological Society of America, Decade of North American Geology, Continent-Ocean Transect Volume, p. 265–306.

Pérez-Segura, E., 1985, Carta Metalogenética de Sonora 1:250,000—una interpretación de la metalogenia de Sonora: Gobierno del Estado de Sonora Publicación 7, 64 p.

Pérez-Segura, E., Roldán-Quintana, J., and Amaya-Martínez, R., 1995, Los terrenos tectonoestratigráficos en Sonora y Sinaloa y sus mineralizaciones asociadas: Guías para la exploración minera: Memoria de la XXI Convención Nacional de la AIMMGM: Asociación de Ingenieros de Minas, Metalurgistas y Geólogos de México (AIMMGM), Acapulco, Guerrero, Mexico, 25 p.

Perry, V.D., 1933, Applied geology at Cananea, Sonora: American Institute of Mining and Metallurgical Engineers Transactions, v. 106, p. 701–709.

Pierce, F.W., and Bolm, J.G., eds., 1995, Porphyry copper deposits of the American Cordillera: Arizona Geological Society Digest, v. 20, 656 p.

Poole, F.G., Madrid, R.J., and Oliva-Becerril, F., 1991, Geological setting and origin of the stratiform barite in central Sonora, Mexico, *in* Raines, G.L., Lisle, R.E., Scafer, R.W., and Wilkinson, W.H., eds., Geology and ore deposits of the Great Basin, Reno, Nevada: Geological Society of Nevada, v. 1, p. 517–522.

Richards, J.P., 2003, Tectono-Magmatic Precursors for Porphyry Cu (Mo-Au) Deposit Formation: Economic Geology and the Bulletin of the Society of Economic Geologists, v. 98, p. 1515–1533.

Ruiz, J., and Mathur, R., 1999, Metallogenesis in continental margins: Re-Os evidence from porphyry copper deposits in Chile, *in* Lambert, D.C., and Ruiz, J., eds., Application of radiogenic isotopes to ore deposit research and exploration: Reviews in Economic Geology, v. 12, p. 59–72.

Salvatierra-Domínguez, X., 2000, Petrografía y geoquímica de los depósitos de tipo pórfido cuprífero de La Fortuna de Cobre y El Americano en el noroeste de México [Tesis de maestría]: Hermosillo, Sonora, Universidad de Sonora, 79 p.

Scherkenbach, D.A., Sawkins, F.J., and Seyfried, W.E., 1985, Geologic, fluid inclusion, and geochemical studies of the mineralized breccias at Cumobabi, Sonora, Mexico: Economic Geology and the Bulletin of the Society of Economic Geologists, v. 80, p. 1566–1592.

Seagart, W.E., Sell, J.D., and Kilpatrick, B.E., 1974, Geology and mineralization of La Caridad porphyry copper deposit, Sonora, Mexico: Economic Geology and the Bulletin of the Society of Economic Geologists, v. 67, p. 1069–1077.

Sedlock, R.L., Ortega-Gutierrez, F., and Speed, R.C., 1993, Tectonostratigraphic terranes and tectonic evolution of Mexico: Geological Society of America Special Paper 278, 153 p.

Servais, M., Cuevas-Pérez, E., and Monod, O., 1986, Une section de Sinaloa à San Luis Potosi: nouvelle approche de l'évolucion de Mexique nord-occidental: Bulletin de la Société Géologique de France, v. 8, no. 2, p. 1033–1047.

Shafiqullah, M., Damon, P.E., and Clark, K.E., 1983, K-Ar chronology of Mesozoic-Cenozoic continental magmatic arcs and related mineralization in Chihuahua, *in* Clark, K.F. and Goodel, P.C., eds., Geology and mineral resources of north-central Chihuahua: El Paso Geological Society Guidebook, p. 303–315.

Sillitoe, R.H., 1972, A plate tectonic model for the origin of porphyry copper deposits: Economic Geology and the Bulletin of the Society of Economic Geologists, v. 67, p. 184–197.

Sillitoe, R.H., 1973, The tops and bottoms of porphyry copper deposits: Economic Geology and the Bulletin of the Society of Economic Geologists, v. 68, p. 799–815.

Sillitoe, R.H., 1993, Epithermal models: genetic types, geometrical controls and shallow features, *in* Kirkham, R.V., Sinclair, W.D., Thorpe, R.I., and Duke, J.M., eds., Mineral deposit modeling: Geological Association of Canada Special Paper 40, p. 403–417.

Sillitoe, R.H., 1996, Granites and metal deposits: Episodes, v. 19, p. 126–133.

Sillitoe, R.H., and Hedenquist, J.W., 2003, Linkages between volcanotectonic settings, ore-fluid compositions, and epithermal precious metal deposits: Society of Economic Geologists, Special Publication Series, v. 10, p. 314–343.

Silver, L.T., and Chappell, B.W., 1988, The Peninsular Ranges Batholith: an insight into the evolution of the Cordilleran batholiths of southwestern North America: Transactions of the Royal Society of Edinburgh: Earth Sciences, v. 79, p. 105–121.

Simmons, S.F., and Sawkins, F.J., 1983, Mineralogic and fluid inclusion studies of the Washington Cu-Mo-W-bearing breccia pipe, Sonora, Mexico: Economic Geology and the Bulletin of the Society of Economic Geologists, v. 78, p. 521–526.

Singer, D.A., Berger, V.I., and Moring, B.C., 2005, Porphyry copper deposits of the world: database, map, and grade and tonnage models: U.S. Geological Survey Open-File Report, 2005-1060 (http://pubs.usgs.gov/of/2005/1060/ and http://pubs.usgs.gov/of/2005/1060/PorCu.xls).

Solano-Rico, B., 1975, Some geologic and exploration characteristics of porphyry copper in a volcanic environment, Sonora, Mexico [M.S. thesis]: Tucson, The University of Arizona, 86 p.

Solano-Rico, B., 1995, Some geologic and exploration characteristics of porphyry copper deposits in the Sierra Madre del Sur province, southwestern Mexico, *in* Pierce, F.W., and Bolm, J.G., eds., Porphyry copper deposits of the American Cordillera: Arizona Geological Society Digest, v. 20, p. 545–550.

Staude, J.M., and Barton, M.D., 2001, Jurassic to Holocene tectonics, magmatism, and metallogeny of northwestern Mexico: Geological Society of America Bulletin, v. 113, p. 1357–1374, doi: 10.1130/0016-7606(2001)113<1357:JTHTMA>2.0.CO;2.

Stern, R.J., 2002, Subduction zones: Reviews of Geophysics, v. 40, p. 1–38.

Stewart, J.H., 1988, Latest Proterozoic and Paleozoic southern margin of North America and the accretion of Mexico: Geology, v. 16, p. 186–189, doi: 10.1130/0091-7613(1988)016<0186:LPAPSM>2.3.CO;2.

Stewart, J.H., and Roldán-Quintana, J., 1991, Upper Triassic Barranca Group—nonmarine and shallow marine rift basin deposits of northwestern Mexico, *in* Pérez-Segura, E., and Jaques-Ayala, C., eds., Studies of Sonoran geology: Geological Society of America Special Paper 254, p. 19–36.

Tetsuichi, T., and Katsushiro, T., 1997, Genesis of oxidized and reduced-type granites: Economic Geology and the Bulletin of the Society of Economic Geologists, v. 92, p. 81–86.

Theodore, T.G., and Priego De Wit, M., 1978, Porphyry type metallization and alteration at La Florida de Nacozari, Sonora, Mexico: U.S. Geological Survey Journal of Research, v. 6, p. 59–72.

Titley, S.R., editor, 1982a, Advances in geology of the porphyry copper deposits, southwestern North America: Tucson, University of Arizona Press, 560 p.

Titley, S.R., 1982b, Geologic setting of porphyry copper deposits, *in* Titley, S.R., ed., Advances in geology of the porphyry copper deposits: Tucson, The University of Arizona Press, p. 37–53.

Titley, S.R., 1990, Contrasting metallogenesis and regional settings of Circum-pacific Cu-Au porphyry systems: Proceedings, Pacific Rim 90 Congress, v. II, p. 127–133.

Titley, S.R., 1993, Characteristics of porphyry copper occurrence in the American Southwest, *in* Kirkham, R.V., Sinclair, W.D., Thorpe, R.I., and Duke, J.M., eds., Mineral deposit modeling: Geological Association of Canada Special Paper 40, p. 433–464.

Titley, S.R., and Hicks, C.L., eds., 1966, Geology of the porphyry copper deposits, southwestern North America: Tucson, University of Arizona Press, 287 p.

Tosdal, R.M., and Richards, J.P., 2001, Magmatic and structural controls on the development of porphyry Cu+Mo+Au deposits: Reviews in Economic Geology, v. 14, p. 157–181.

Tritlla, J., Camprubí, A., Morales-Ramírez, J.M., Iriondo, A., Corona-Esquivel, R., González-Partida, E., Levresse, G., and Carrillo-Chávez, A., 2004, The Ixtacamaxtitlán kaolinite deposit and sinter (Puebla State, Mexico): A magmatic-hydrothermal system telescoped by a shallow paleoaquifer: Geofluids, v. 4, p. 329–340, doi: 10.1111/j.1468-8123.2004.00095.x.

Valencia, V., Ruiz, J., Eastoe, C., Gehrels, G., and Barra, F., 2004, Evolución del Pórfido cuprífero de La Caridad, Sonora, México: Basado en análisis de inclusiones fluidas, isótopos de S, O, H, U-Pb y Re-Os: IV Reunión Nacional de Ciencias de la Tierra, Sociedad Geológica Mexicana, p. 189.

Valencia, V., Noguez-Alcántara, B., Barra, F., Ruiz, J., Gehrels, G., Quintanar, F., and Valencia-Moreno, M., 2006, Re-Os Molybdenite and LA-ICP-MS U-Pb zircon geochronology from the Milpillas porphyry copper deposit: Insights for mineralization in the Cananea District, Sonora, Mexico: Revista Mexicana de Ciencias Geológicas, v. 23, p. 39–53.

Valencia-Moreno, M., 1998, Geochemistry of Laramide granitoids and associated porphyry copper mineralization in NW Mexico [Ph.D. thesis]: Tucson, The University of Arizona, 164 p.

Valencia-Moreno, M., Ruiz, J., and Roldán-Quintana, J., 1999, Geochemistry of Laramide granitic rocks across the southern margin of the Paleozoic North American continent, Central Sonora, Mexico: International Geology Review, v. 41, p. 845–857.

Valencia-Moreno, M., Ruiz, J., Barton, M.D., Patchett, P.J., Zürcher, L., Hodkinson, D., and Roldán-Quintana, J., 2001, A chemical and isotopic study of the Laramide granitic belt of northwestern Mexico: Identification of the southern edge of the North American Precambrian basement: Geological Society of America Bulletin, v. 113, p. 1409–1422, doi: 10.1130/0016-7606(2001)113<1409:ACAISO>2.0.CO;2.

Valencia-Moreno, M., Ruiz, J., Ochoa-Landín, L., Martínez-Serrano, R., and Vargas-Navarro, P., 2003, Geochemistry of the coastal Sonora batholith, northwestern Mexico: Canadian Journal of Earth Sciences, v. 40, p. 819–831, doi: 10.1139/e03-020.

Valentine, W.G., 1936, Geology of the Cananea Mountains, Sonora, Mexico: Geological Society of America Bulletin, v. 47, p. 53–63.

Valenzuela-Navarro, L.C., Valencia-Moreno, M., Iriondo, A., and Premo, W., 2003, The El Crestón granite: A new confirmed Paleoproterozoic locality in the Opodepe area, North-Central Sonora, Mexico: Geological Society of America Abstracts with Programs, v. 35, no. 4, p. 83.

Valenzuela-Navarro, L.C., Valencia-Moreno, M., Calmus, T., Ochoa-Landín, L., and González-León, C., 2005, Marco geológico del pórfido de molibdeno El Crestón, Sonora central, México: Revista Mexicana de Ciencias Geológicas, v. 22, p. 345–357.

Vega-Granillo, R., and Calmus, T., 2003, Mazatan metamorphic core complex (Sonora, Mexico): Structures along the detachment fault and its exhumation evolution: Journal of South American Earth Sciences, v. 16, p. 193–204, doi: 10.1016/S0895-9811(03)00066-X.

Weber, B., and López-Martínez, M., 2002, Sr, Nd, Pb isotopes and Ar-Ar dating of the "El Arco" porphyry copper deposit, Baja California: evidence for Cu mineralization within an oceanic island arc: Geological Society of America Abstracts with Program, v. 34, no. 6, p. 88.

Weber, B., Forsythe, L., Romero-Espejel, H., and López-Martínez, M., 2001, Resultados geoquímicos e isotópicos preliminares sobre las características y la formación del pórfido de El Arco, Baja California: Resúmenes, Reunión Anual de la Unión Geofísica Mexicana, Puerto Vallarta, Jalisco, p. 235.

Wilkerson, G., Qingping, D., Lavona, R., and Goodell, P., 1988, Batopilas mining district, Chihuahua, Mexico: Economic Geology and the Bulletin of the Society of Economic Geologists, v. 83, p. 1721–1736.

Wodzicki, W.A., 2001, The evolution of magmatism and mineralization in the Cananea district, Sonora, Mexico: Society of Economic Geologists Special Publication 8, p. 243–263.

Zürcher, L., 2002, Regional setting and magmatic evolution of Laramide porphyry copper systems in western Mexico [Ph.D. thesis]: Tucson, The University of Arizona, 427 p.

Manuscript Accepted by the Society 29 August 2006

Printed in the USA

Index

D

E

F

G

H

I

J

K

L